Fully Revised **Edition**

AF814406

Understanding Physics

JEE Main & Advanced

Mechanics

Volume 1

Fully Revised **Edition**

Understanding Physics

JEE Main & Advanced

Mechanics

Volume 1

DC Pandey
[B.Tech, M.Tech, Pantnagar, ID 15722]

arihant
SUCCESS AND NOTHING LESS

arihant
SUCCESS AND NOTHING LESS

ARIHANT PRAKASHAN (Series), MEERUT

卐 **Administrative & Production Offices**

Regd. Office
'Ramchhaya' 4577/15, Agarwal Road, Darya Ganj, New Delhi -110002
Tele: 011- 47630600, 43518550

Head Office
Kalindi, TP Nagar, Meerut (UP) - 250002
Tel: 0121-7156203, 7156204

卐 **Sales & Support Offices**
Agra, Ahmedabad, Bengaluru, Bareilly, Chennai, Delhi, Guwahati, Hyderabad, Jaipur, Jhansi, Kolkata, Lucknow, Nagpur & Pune.

卐 **PRICE** ₹625.00

卐 **PO No :** TXT-XX-XXXXXXX-X-XX

Published by Arihant Publications (India) Ltd.

For further information about the books published by Arihant, log on to www.arihantbooks.com or e-mail at info@arihantbooks.com

Follow us on

Preface

"The overwhelming response to the previous editions of this book gives me an immense feeling of satisfaction, and I take this as an opportunity to introduce you to the new version of the book, "Mechanics-1."

"Mechanics-1" has undergone meticulous revision to ensure accuracy and relevance. We take pride in affirming that this book serves as a standalone resource, encompassing all essential topics without any extraneous content.

This book is designed with a singular focus on providing concise yet complete theory, coupled with an ample number of practice questions covering all pertinent concepts essential for success in JEE Mains & Advanced.

With each chapter, or as we call them, units, subdivided into modular formats, "Mechanics-1" ensures an organized and structured approach to learning. Each module within a unit is self-contained, comprising theory, illustrative examples, and exercises aimed at reinforcing understanding.

At the end of every unit, readers will find unit exercises presented in two levels, offering ample opportunities for practice and self-assessment. Furthermore, to aid in problem-solving, most exercises are accompanied by helpful hints.

Our commitment to excellence extends to the inclusion of logical problems sourced from reputable resources like **Irodov** and **previous years' question papers.** The level of questions spans from foundational to 1.25 times the highest difficulty level of the JEE Advanced examination, ensuring comprehensive preparation.

In conclusion, I am confident that "Mechanics-1" will prove to be an indispensable companion for aspirants aiming for success in JEE. It is my belief that with dedication and diligent practice using this book, students will be well-equipped to excel in both JEE Mains and Advanced examinations.

I take this opportunity to express my gratitude to **Prof. Lalit Kumar,** who gave me valuable advice and was kind enough to find time for fruitful discussions. I would also like to thank my student Dheeraj Yadav for managing this project as per my expectations. I am also thankful to Mr. Anoop Dhyani and Mr. Ashish, due to their tireless efforts this book could be transformed into the present form.

To err is human, hence despite such tireless efforts, there may still be scope for improvement in this book. So, comments and criticism from readers will be highly appreciated and incorporated in subsequent editions." I also request the readers to drop me a mail at **arihantcorrections@gmail.com,** if you find any corrections in the book. Your feedback is greatly appreciated and will help to improve the quality of the book.

Thankyou for your support and contribution in making this book the **Best it can be.**

DC Pandey
[B.Tech, M.Tech, Pantnagar, ID 15722]

Contents

This book

is dedicated to

my Honourable Grandfather

(Late) Sh. Pitamber Pandey

a Kumaoni poet and a resident of Village

Dhaura (Almora), Uttarakhand

UNIT-1

BASIC MATHEMATICS

Learning Modules

1. Algebra & Logarithm
2. Mensuration, Coordinate Geometry, Percentage & Ratio
3. Trigonometry
4. Differentiation
5. Integration
6. Graphs

MODULE 1

Algebra & Logarithm

Algebra

As far as Physics is concerned, in Algebra we need only simple formulae and their applications.

List of Formulae

(i) **Common formulae**

(a) $(a+b)^2 = a^2 + b^2 + 2ab$

(b) $(a-b)^2 = a^2 + b^2 - 2ab$

(c) $(a+b)^2 = (a-b)^2 + 4ab$

(d) $(a-b)^2 = (a+b)^2 - 4ab$

(e) $(a^2 - b^2) = (a+b)(a-b)$

(f) $(a+b)^3 = a^3 + b^3 + 3ab(a+b)$

(g) $(a-b)^3 = a^3 - b^3 - 3ab(a-b)$

(h) $a^3 + b^3 = (a+b)(a^2 - ab + b^2)$

(i) $a^3 - b^3 = (a-b)(a^2 + ab + b^2)$

(ii) **Quadratic equation and its solution**

An algebraic equation of second order (highest power of the variable is equal to 2) is called a quadratic equation. Equation $ax^2 + bx + c = 0$ is the general quadratic equation.

The general solution of the above quadratic equation or value of variable x is

$$x = \frac{-b \pm \sqrt{b^2 - 4ac}}{2a} \Rightarrow x_1 = \frac{-b + \sqrt{b^2 - 4ac}}{2a}$$

and $$x_2 = \frac{-b - \sqrt{b^2 - 4ac}}{2a}$$

Sum of roots $= x_1 + x_2 = -\frac{b}{a}$ and product of roots $= x_1 x_2 = \frac{c}{a}$

For real roots $b^2 - 4ac \geq 0$ and for imaginary roots $b^2 - 4ac < 0$.

(iii) **Binomial approximation**

If x is very small, then $(1+x)^n \approx 1 + nx$

Here $\approx$ is the sign of nearly equal to.

(iv) **Componendo and dividendo rule**

If $\frac{p}{q} = \frac{a}{b}$, then $\frac{p+q}{p-q} = \frac{a+b}{a-b}$

(v) **Arithmetic progression (AP)**

General form : $a, a+d, a+2d, \ldots, a+(n-1)d$

Here, a = first term, d = common difference

and $[a+(n-1)d]$ is nth term or a_n.

Sum of n terms $S_n = \frac{n}{2}[a + a + (n-1)d]$

$= \frac{n}{2}$ [1st term + nth term]

Further, $n = \frac{a_n - a}{d} + 1$

(vi) **Geometric progression (GP)**

General form : $a, ar, ar^2, \ldots, ar^{n-1}$

Sum of n terms $S_n = \frac{a(1-r^n)}{1-r}; r \neq 1, r < 1$

$S_n = \frac{a(r^n - 1)}{r-1}; r \neq 1, r > 1$

and $S_n = na$ if $r = 1$

Here, a = first term, r = common ratio

Sum of ∞ terms $S_\infty = \frac{a}{1-r}$

$(\because r < 1 \therefore r^\infty \to 0)$

(vii) **Sum of first *n* natural numbers**

$$S_n = 1 + 2 + , \ldots, + n = \frac{n}{2}[1+n] = \left[\frac{n(n+1)}{2}\right]$$

(viii) **Sum of first *n* squared natural numbers**

$$S_{n^2} = 1^2 + 2^2 + 3^2 + , \ldots, + n^2 = \left[\frac{n(n+1)(2n+1)}{6}\right]$$

Example 1 Solve the equation $2x^2 + 5x - 12 = 0$.

Sol. By comparison with the standard quadratic equation

$a = 2, b = 5$

and $c = -12$

$$x = \frac{-5 \pm \sqrt{(5)^2 - 4 \times 2 \times (-12)}}{2 \times 2}$$

$$= \frac{-5 \pm \sqrt{121}}{4}$$

$$= \frac{-5 \pm 11}{4}$$

$$= \frac{+6}{4}, \frac{-16}{4}$$

or $x = \frac{3}{2}, -4$

Example 2 Calculate $\sqrt{0.99}$ by using Binomial approximation.

Sol. $\sqrt{0.99} = (1 \quad 0.01)^{1/2}$

$$\approx 1 - \frac{1}{2}(0.01)$$

$$\approx 1 - 0.005$$

$$\approx 0.995$$

Example 3 Find the sum of given arithmetic progression

$$4 + 8 + 12 + \ldots\ldots + 64$$

Sol. Here, $a = 4, d = 4, n = \frac{a_n - a}{d} + 1$

$$= \frac{64 - 4}{4} + 1 = 16$$

So, sum $= \frac{n}{2}$ [first term + last term]

$$= \frac{16}{2}[4 + 64] = 8(68) = 544$$

Example 4 Find the sum of given geometric series $1 + 2 + 4 + 8 + \ldots\ldots + 256$.

Sol. Here, $a = 1, r = 2, n = 9$

$$[\because 256 = 2^8, a_n = (a)(r)^{n-1} = (1)(2)^{n-1}]$$

So, $S_9 = \frac{a(r^n - 1)}{r - 1} = \frac{(1)(2^9 - 1)}{(2 - 1)}$

$$= 2^9 - 1 = 512 - 1 = 511$$

Example 5 Find $1 + \frac{1}{2} + \frac{1}{4} + \frac{1}{8} + \ldots$upto ∞.

Sol. $a = 1, r = \frac{1}{2}$

So, $S_\infty = \frac{a}{1 - r} = \frac{1}{1 - \frac{1}{2}} = 2$

Logarithm

General Points

1. The exponent or power to which a base must be raised to yield a given number is called **logarithm**.

   For example, we know that,

   $$10^2 = 100$$

   $\Rightarrow$ 2 is the logarithm of 100 to the base 10.

   So, we can also write the above equation as,

   $$\log_{10} 100 = 2$$

2. 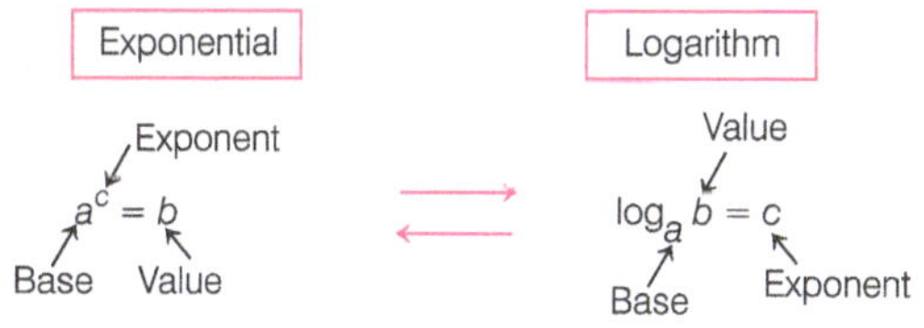

Note *Value is also called* ***argument***.

3. Invented in 17th century to speed up calculations, logarithms vastly reduced the time required for multiplying numbers with many digits. In 19th and 20th century use of calculators and computers made the calculations very fast.
4. **Common logarithm** In common logarithm base is 10. It is simply represented as log. For example,

   $$\log_{10} 100 = \log 100$$

5. **Natural logarithm** In natural logarithm base is '*e*'. Here '*e*' represents the Euler's constant which is approximately equal to 2.7183. Natural logarithm is simply represented as ln.

   For example, $\log_{10} 100 = \ln 100$

6. $x^1 = x \Rightarrow \log_x x = 1$

List of Formulae

(i) $e \approx 2.7183$

(ii) If $e^x = y$, then $x = \log_e y = \ln y$

(iii) If $10^x = y$, then $x = \log_{10} y = \log y$

(iv) $\log_{10} y = 0.4343 \log_e y = 0.4343 \ln y$

(v) $\log(ab) = \log a + \log b$

(vi) $\log\left(\frac{a}{b}\right) = \log a - \log b$

(vii) $\log a^n = n \log a$

(viii) $\log_{10} 2 = 0.3010$

(ix) $\log_{10} 3 = 0.4771$

(x) $\log_e 2 = 0.6931$

(xi) $\log_e 3 = 1.0986$

(xii) $\log_{10} 10 = 1$

(xiii) $\log_e e = 1$

Example 6 Find the values of

(a) $\log_{10} 6$

(b) $\log_{10} 300$

(c) $\log_{10} 10^{-5}$

(d) $\log_{10}(4 \times 10^6)$

Sol. (a) $\log_{10} 6 = \log_{10}(3 \times 2) = \log_{10} 3 + \log_{10} 2$

$= (0.4771) + (0.3010)$

$= 0.7781$ **Ans.**

(b) $\log_{10} 300 = \log_{10}(3 \times 10^2) = \log_{10} 3 + \log_{10} 10^2$

$= \log_{10} 3 + 2\log_{10} 10$

$= (0.4771) + (2)(1)$

$= 2.4771$ **Ans.**

(c) $\log_{10} 10^{-5} = -5\log_{10} 10 = -5(1) = -5$ **Ans.**

(d) $\log_{10}(4 \times 10^6) = \log_{10} 4 + \log_{10} 10^6$

$= \log_{10}(2)^2 + 6\log_{10} 10$

$= 2\log_{10} 2 + 6\log_{10} 10$

$= (2)(0.3010) + (6)(1)$

$= 6.602$ **Ans.**

MODULE 1 Exercise

1. The value of acceleration due to gravity (g) at height h above the surface of earth is given by $g' = \dfrac{gR^2}{(R+h)^2}$. If $h << R$, then prove that $g' = g\left(1 - \dfrac{2h}{R}\right)$.
2. Calculate $(1001)^{1/3}$.
3. Express $5^3 = 125$ in logarithm form.
4. Express $\log_{10} 1 = 0$ in exponential form.
5. Find the log of 32 to the base 4.
6. Find x if $\log_5 (x - 7) = 1$.
7. If $\log_a m = n$, express a^{n-1} in terms of a and m.

Answers

2. 10.003333 **3.** $\log_5 125 = 3$ **4.** $10^0 = 1$ **5.** $\dfrac{5}{2}$ **6.** 12 **7.** $\dfrac{m}{a}$

See hints on page 39.

MODULE 2

Mensuration, Coordinate Geometry, Percentage & Ratio

Mensuration

It is the branch of mathematics that studies the measurement of geometric figures and their parameters like length, volume, shape, surface area, lateral surface area etc.

Mensuration Formulas for 2D Shapes

Shape	Area (Square units)	Perimeter (units)	Figure
Square	a^2	$4a$	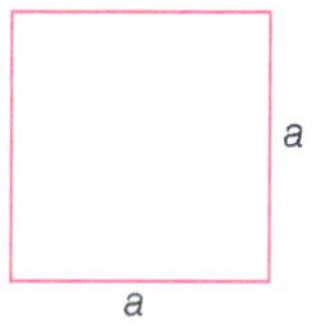
Rectangle	$l \times b$	$2(l+b)$	
Circle	πr^2	$2\pi r$	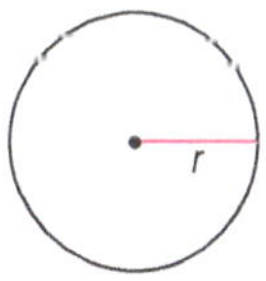
Scalene triangle	$\sqrt{[s(s-a)(s-b)(s-c)]}$ where, $s=(a+b+c)/2$	$a+b+c$	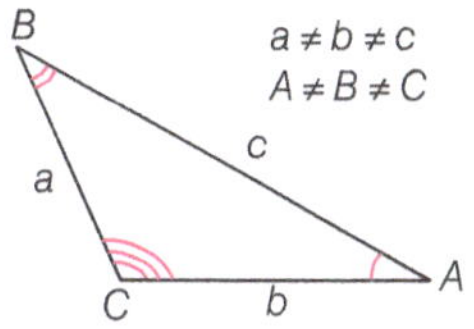
Isosceles triangle	$\frac{1}{2} \times b \times h$	$2a+b$	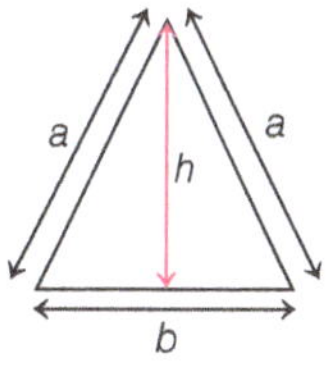

Equilateral triangle	$\left(\frac{\sqrt{3}}{4}\right) \times a^2$	$3a$	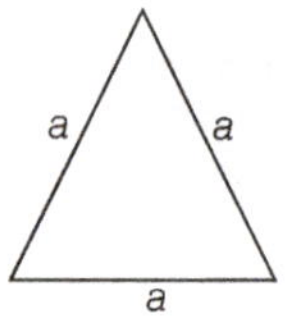
Right angle triangle	$\frac{1}{2} \times b \times h$	$b + \text{hypotenuse} + h$	Hypotenuse $= \sqrt{b^2 + h^2}$ 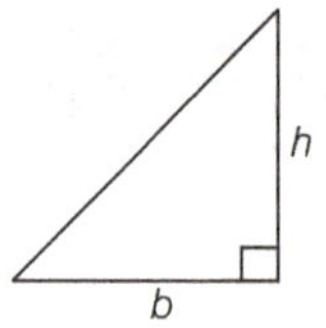
Rhombus	$\frac{1}{2} \times d_1 \times d_2$	$4 \times \text{side}$	Opposite sides are parallel, opposite angles are equals, all sides are equal 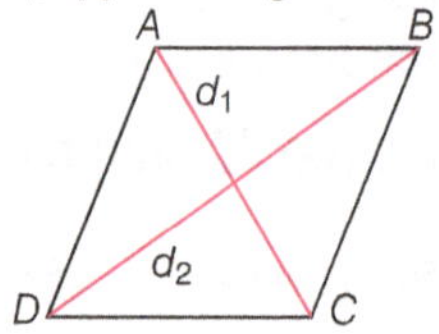
Parallelogram	$b \times h$	$2(l + b)$	Opposite sides are parallel and equal. The sum of adjacent angles is 180° 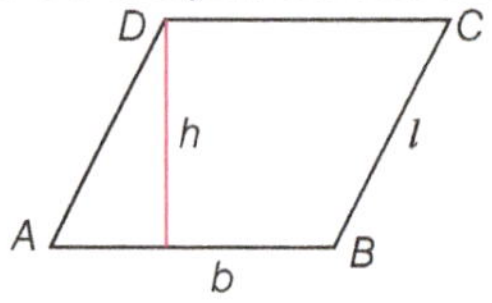
Trapezium	$\frac{1}{2} h(a + c)$	$a + b + c + d$	One pair of opposite sides is parallel to each other. 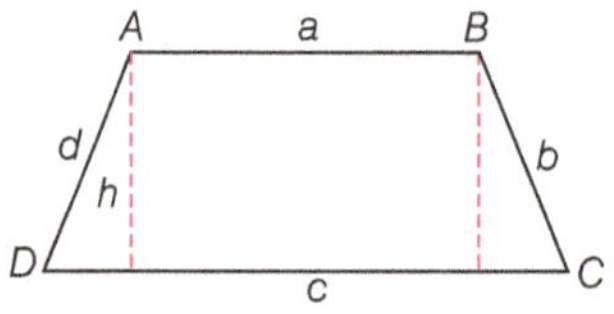
Ellipse	$\pi r_1 r_2$ where, r_1 = semi major axis r_2 = semi minor axis	Very difficult to calculate the perimeter accurately	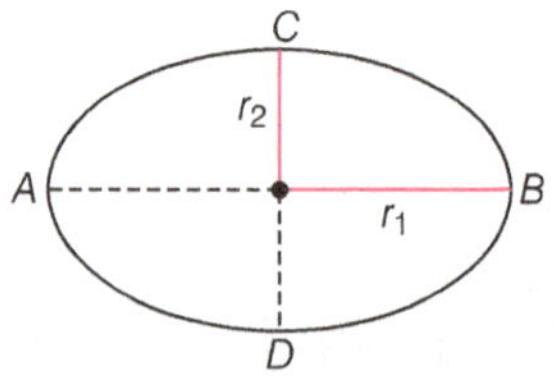 AB = major axis CD = minor axis

Mensuration Formulas for 3D Shapes

Shape	Volume (Cubic units)	Curved surface area or Lateral surface area (Square units)	Total surface area (Square units)	Figure
Cube	a^3	$4a^2$	$6a^2$	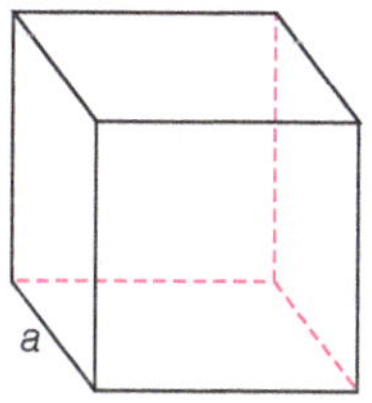
Cuboid	$l \times b \times h$	$2h(l+b)$	$2(lb+bh+hl)$	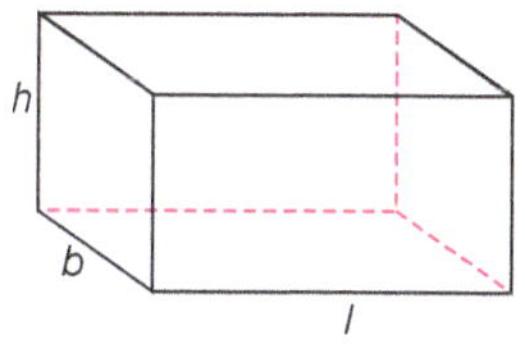
Sphere	$\left(\frac{4}{3}\right)\pi r^3$	$4\pi r^2$	$4\pi r^2$	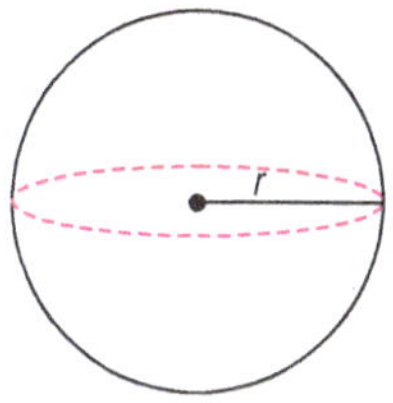
Hemisphere	$\left(\frac{2}{3}\right)\pi r^3$	$2\pi r^2$	$3\pi r^2$	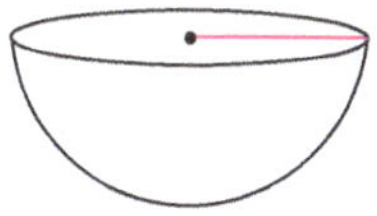
Cylinder	$\pi r^2 h$	$2\pi rh$	$2\pi rh + 2\pi r^2$	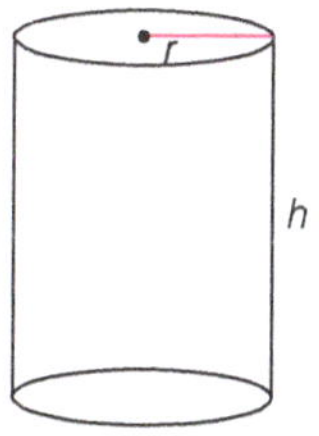
Cone	$\left(\frac{1}{3}\right)\pi r^2 h$	πrl	$\pi r(r+l)$	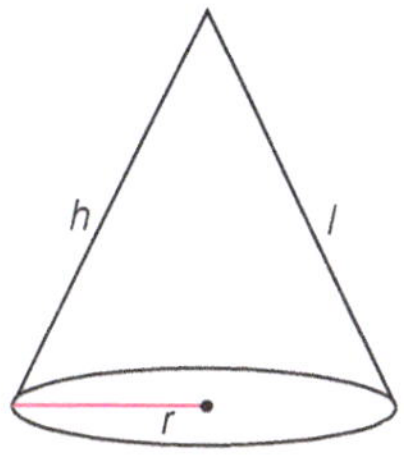

Example 1 What is the circumference of a circle if area of the circle is πcm^2?

Sol. $A = \pi r^2$

$$\Rightarrow \qquad r = \sqrt{\frac{A}{\pi}} = \sqrt{\frac{\pi}{\pi}} = 1\,\text{cm}$$

Now circumference $C = 2\pi r$

$$= (2\pi)(1) = (2\pi)\,\text{cm}$$

Example 2 Find the area of an equilateral triangle whose altitude is given by $\left(\frac{\sqrt{3}}{2}\right)$ cm.

Sol. Given, the altitude of equilateral triangle $= \frac{\sqrt{3}}{2}$ cm

The formula for altitude of equilateral triangle is given by

altitude $= \frac{\sqrt{3}}{2}$(side)

$$\therefore \qquad \frac{\sqrt{3}}{2} = \frac{\sqrt{3}}{2}(\text{side})$$

or, side = 1 unit

So, area of equilateral triangle $= \frac{\sqrt{3}}{4}(\text{side})^2$

$$= \frac{\sqrt{3}}{4}(1)^2 = \frac{\sqrt{3}}{4}\text{cm}^2$$

$\therefore$ Required area of the triangle is $\frac{\sqrt{3}}{4}\text{cm}^2$.

Example 3 The cylindrical milk bottle is of radius 2 cm and height 5 cm. Find the capacity of milk in the given bottle. Also find curved surface area.

Sol. Given, radius $(r) = 2$ cm and height $(h) = 5$ cm

We know that the capacity of the cylindrical bottle is known as the volume of the cylinder.

The volume of the cylinder with radius 'r' and height 'h' is $\pi r^2 h$.

$$\therefore \qquad \text{Volume} = \frac{22}{7} \times 2^2 \times 5$$

$$= \frac{22}{7} \times 20 = \frac{440}{7}$$

$$= 62.86\ \text{cm}^3$$

Hence, the capacity of the given milk bottle is $62.86\ \text{cm}^3$.

Curved surface area

$$= 2\pi rh = (2\pi)(2)(5)$$

$$= (20\pi)\,\text{cm}^2$$

Coordinate Geometry

(i) **Distance formula** The distance between two points (x_1, y_1) and (x_2, y_2) is given by

$$d = \sqrt{(x_2 - x_1)^2 + (y_2 - y_1)^2}$$

Note *In 3-D space,* $d = \sqrt{(x_2 - x_1)^2 + (y_2 - y_1)^2 + (z_2 - z_1)^2}$

(ii) **Slope of a line** The slope of a line joining two points $A(x_1, y_1)$ and $B(x_2, y_2)$ is denoted by m and is given by

$$m = \frac{\Delta y}{\Delta x} = \frac{y_2 - y_1}{x_2 - x_1} = \tan\theta$$

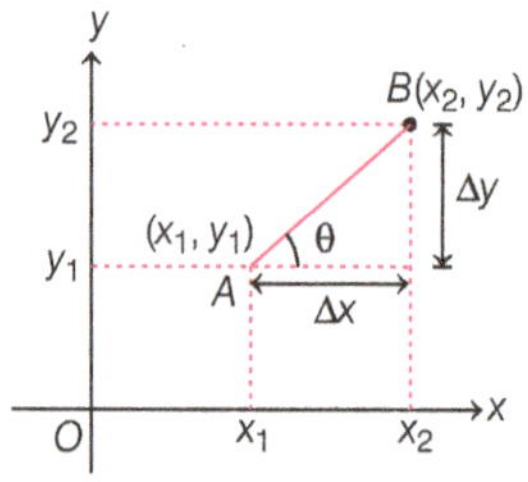

Example 4 Find value of a if distance between the points $(-9\ \text{cm}, a\ \text{cm})$ and $(3\ \text{cm}, 3\ \text{cm})$ is 13 cm.

Sol. By using distance formula,

$$d = \sqrt{(x_2 - x_1)^2 + (y_2 - y_1)^2}$$

$$\text{or,} \qquad d^2 = (x_2 - x_1)^2 + (y_2 - y_1)^2$$

$$\Rightarrow \qquad 13^2 = [3 - (-9)]^2 + (3 - a)^2$$

$$= 12^2 + (3 - a)^2$$

$$\Rightarrow \qquad 13^2 - 12^2 = (3 - a)^2$$

$$\therefore \qquad (3 - a)^2 = 5^2$$

$$\Rightarrow \qquad (3 - a) = \pm 5$$

$$\Rightarrow \qquad a = -2\ \text{cm or } 8\ \text{cm}$$

Example 5 What is the slope of the line joining the points $P(-2, 3)$ and $Q(2, 7)$?

Sol. Let the given points be :

$$P(-2, 3) = (x_1, y_1)$$

$$Q(2, 7) = (x_2, y_2)$$

$$\text{Slope of the line } PQ = \frac{(y_2 - y_1)}{(x_2 - x_1)}$$

$$= \frac{(7 - 3)}{[2 - (-2)]} = \frac{4}{4} = 1$$

Percentage

(i) Suppose original value is x. Then n percentage of x is $\left(\frac{n}{100}\right)(x)$.

For example :

(a) 100% of x is x

(b) 200% of x is $2x$

(c) 500% of x is $5x$

(d) 16% of x is $0.16x$

(e) 3% of x is $0.03x$

Note *In (a) to (e) we have used* $\left(\frac{n}{100}\right)x$.

(ii) Increase has no limits. You can increase my salary 1%, 15%, 200%, 500% etc. depending upon my performance, age and looks ☺. But decrease has limits. Maximum decrease may be 100%. You can decrease my salary. Maximum upto 100% ☹.

(iii) **Percentage decrease** Following table will explain what we want to convey.

S.No.	Percentage decrease	Percentage remaining	New value
1	20%	80%	$0.8x$
2	34%	66%	$0.66x$
3	5%	95%	$0.95x$
4	1%	99%	$0.99x$

(iv) **Percentage increase**

S.No.	Percentage increase	Increased value	New value (= original value + increased value)
1	16%	$0.16x$	$1.16x$
2	125%	$1.25x$	$2.25x$
3	4%	$0.04x$	$1.04x$
4	300%	$3x$	$4x$

Example 6 If original value is 150. Then, find

(a) 10 % (b) 4% (c) 125% (d) 300%

of this value.

Sol. (a) 10% of $150 = \left(\frac{10}{100}\right)(150) = 15$

(b) 4% of $150 = \left(\frac{4}{100}\right)(150) = 6$

(c) 125% of $150 = \left(\frac{125}{100}\right)(150) = 187.5$

(d) 300% of $150 = \left(\frac{300}{100}\right)(150) = 450$

Example 7 Original value is 60. It is

(a) increased by 20%

(b) decreased by 25%

Find new values.

Sol. (a) 20% of $60 = \frac{20}{100}(60) = 12$

So the new value is,

$$60 + 12 = 72$$

(b) After decreasing 25% only 75% is left. So, the new value is,

$$\left(\frac{75}{100}\right)(60) = 45$$

(v) **Percentage change** $= \left(\frac{\text{Change in value}}{\text{Original value}}\right) \times 100$

$$= \left(\frac{\text{Final value} - \text{Initial value}}{\text{Initial value}}\right) \times 100$$

$$= \left[\frac{\text{Final value}}{\text{Initial value}} - 1\right] \times 100$$

Note *Percentage change may be zero, positive or negative.*
If final value > initial value, then percentage change is positive.

Example 8 Find percentage change if value changes from 40 to 50.

Sol. Percentage change $= \left(\frac{\text{Final value}}{\text{Initial value}} - 1\right) \times 100$

$$= \left(\frac{50}{40} - 1\right)(100) = +25\%$$

(vi) If there is a relation between two quantities x and y. Percentage change in x is given and we have to find percentage change in y. Then following two cases are possible :

Case 1. When given percentage change in x is large

In this case first find the new value of y corresponding to changed value of x and then apply the formula,

Percentage change $= \left[\frac{\text{Final value}}{\text{Initial value}} - 1\right] \times 100$

Case 2. When percentage change in x is very small

If we have a formula $y^m = 4x^n$ or $y^m \propto x^n$

Percentage change in x is given and percentage change in y is asked then we can use the approximate result,

$$m\,(\%\text{ change in } y) = n(\%\text{ change in } x)$$

Following example will illustrate this concept.

Example 9 Relation between kinetic energy K and momentum p of a particle is

$$p = \sqrt{2Km}$$

Find percentage change in p, if

(a) K is increased by 1 %

(b) K is decreased by 36 %

Sol. (a) $p \propto \sqrt{K}$ [m = constant] or $p \propto K^{\frac{1}{2}}$

As % change is small,

$$(\%\text{ change in } p) = \frac{1}{2}(\%\text{ change in } K)$$

$$= \frac{1}{2}(+1\%) = +0.5\%$$

Therefore, momentum will increase by 0.5%.

(b) K is decreased by 36%. Therefore, new value of K is 64% of its original value or

$$K' = 0.64K$$

$$\therefore \quad p' = \sqrt{2K'm}$$

$$= \sqrt{2(0.64K)m}$$

$$= 0.8\sqrt{2Km}$$

Now, percentage change in momentum

$$= \left(\frac{\text{Final value}}{\text{Initial value}} - 1 \right) \times 100$$

$$= \left(\frac{0.8\sqrt{2Km}}{\sqrt{2Km}} - 1 \right) \times 100$$

$$= -20\%$$

Therefore, momentum will decrease by 20%.

Ratio Division

Following examples will illustrate the concept what we want to convey.

Example 10 Tom has a height of 4 ft and Jerry 5 ft. We have to divide ₹ 500 between them :

(a) In direct ratio of their heights.

(b) In inverse ratio of their heights.

(c) In direct ratio of square of their heights.

(d) In inverse ratio of square of their heights.

Sol. Let us represent Tom by 1, Jerry by 2, height by h and ruppes by R.

(a) Given $R \propto h$

$\Rightarrow \quad R = kh \qquad$ (k is called proportionality constant)

$$\therefore \quad \frac{R_1}{R_2} = \frac{kh_1}{kh_2} = \frac{h_1}{h_2} = \frac{4}{5}$$

Now total ratio is $(4+5) = 9$

R_1: Out of 9 value of R_1 is 4

$\therefore$ Out of 1 value of R_1 will be $\frac{4}{9}$

$\therefore$ Out of 500 value of R_1 will be

$$\left(\frac{4}{9}\right)(500) = \frac{2000}{9}$$

$$\therefore \quad R_1 = \frac{2000}{9}$$

R_2: Similarly $R_2 = \left(\frac{5}{9}\right)(500) = \frac{2500}{9}$

Note Points

(i) $R \propto h$, means if h is doubled then R is also doubled.

(ii) $\frac{R_1}{R_2} = \frac{h_1}{h_2}$

(iii) $R_1 = \left(\frac{\text{Ratio of } R_1}{\text{Total ratio}}\right)$ *(Amount to be divided)*

Similarly,

$R_2 = \left(\frac{\text{Ratio of } R_2}{\text{Total ratio}}\right)$ *(Amount to be divided)*

(iv) ***Meaning of proportionality constant***

If we have two equations,

$$y = 2x \text{ and } y = 4x$$

then in both equations $y \propto x$.

Therefore in both cases, if x is doubled then y will become two times. But proportionality constant in first case is 2 and in the second case it is 4. Due to the different values of proportionality constants values of y will be different for different values of x and vice-versa.

For example, for $x = 1$,

$y = 2$ in first case and

$y = 4$ in second case.

For $x = 2$,

$y = 4$ in first case and

$y = 8$ in second case

(v) If $R \propto \frac{1}{h}$, then this implies that R will become $\frac{1}{2}$-times if h is doubled and vice-versa. This also implies that $\frac{R_1}{R_2} = \frac{h_2}{h_1}$.

(vi) If $R \propto h^2$, then this implies that R will become 4-times if h is doubled. But $h \propto \sqrt{R}$. So, h will become $\sqrt{2}$-times if R is doubled. So, vice-versa does not apply here.

Further, $R \propto h^2$ means, $\frac{R_1}{R_2} = \left(\frac{h_1}{h_2}\right)^2$

(b) $R \propto \frac{1}{h}$

$$\therefore \quad \frac{R_1}{R_2} = \frac{h_2}{h_1} = \frac{5}{4}$$

$$\therefore \quad R_1 = \left(\frac{5}{9}\right)(500) = \frac{2500}{9}$$

and $$R_2 = \left(\frac{4}{9}\right)(500) = \frac{2000}{9}$$

(c) $R \propto h^2$

$$\Rightarrow \quad \frac{R_1}{R_2} = \left(\frac{h_1}{h_2}\right)^2 = \left(\frac{4}{5}\right)^2 = \frac{16}{25}$$

Total ratio is $16 + 25 = 41$

$$\therefore \quad R_1 = \left(\frac{16}{41}\right)(500) = \frac{8000}{41}$$

and $$R_2 = \left(\frac{25}{41}\right)(500) = \frac{12500}{41}$$

(d) $R \propto \frac{1}{h^2}$

$$\Rightarrow \quad \frac{R_1}{R_2} = \left(\frac{h_2}{h_1}\right)^2 = \left(\frac{5}{4}\right)^2 = \frac{25}{16}$$

Total ratio is $25 + 16 = 41$

$$\therefore \quad R_1 = \left(\frac{25}{41}\right)(500) = \frac{12500}{41}$$

and $$R_2 = \left(\frac{16}{41}\right)(500) = \frac{8000}{41}$$

MODULE 2 Exercise

Mensuration

1. Find the area of a triangle with sides 3 cm, 3 cm and 4 cm.
2. Find the volume of a tent, which is in the form of a right circular cone, with radius and height 24m and 10 m respectively. Also find its curved surface area.
3. Circumference of a circle is 10 cm. Find its area.

Coordinate Geometry

4. (a) If the distance between the points $A(2, -2)$ and $B(-1, x)$ is 5, find the value of x.
   (b) Find slope of a line AB, if $A = (2, 4)$, $B = (6, 10)$.

Percentage Change

5. Momentum of a particle is increased by 2% without change in its mass. Find the percentage change in its kinetic energy.
6. Momentum of a particle is decreased by 20% without change in its mass. Find the percentage change in its kinetic energy.
7. Kinetic energy of a particle is decreased by 3% without change in its mass. Find the percentage change in its momentum.
8. Kinetic energy of a particle is increased by 50% without change in its mass. Find the percentage change in its momentum.

Ratio Division

9. Height of Kartik is 5ft and of Atharva 4ft. Divide ₹1000/- in inverse ratio of square of their heights.
10. Weight of Chintu and Pintu are 60 N and 80 N. Divide ₹ 2000 in direct ratio of square of their weights.

Answers

1. $2\sqrt{5}\,\text{cm}^2$ **2.** $6034.28\,\text{m}^3$, $1961.14\,\text{m}^2$ **3.** $\frac{175}{22}\,\text{cm}^2$ **4.** (a) $x = 2$ or -6 (b) $\frac{3}{2}$

5. $+4\%$ **6.** -36% **7.** -1.5% **8.** $+22\%$ (approximately) **9.** ₹ $\frac{16000}{41}$ and ₹ $\frac{25000}{41}$

10. ₹ 720 and ₹ 1280

See hints on pages 38 and 39.

MODULE 3

Trigonometry

Measurement of an Angle

An angle is a shape in geometry that is formed when two rays are joined together at a single point. The point where two rays meet is called the vertex of the angle and two rays are called the arms of the angle.

Definition of Angle

The angle measurement between the two rays can be denoted using the Greek letter θ, α, β etc.

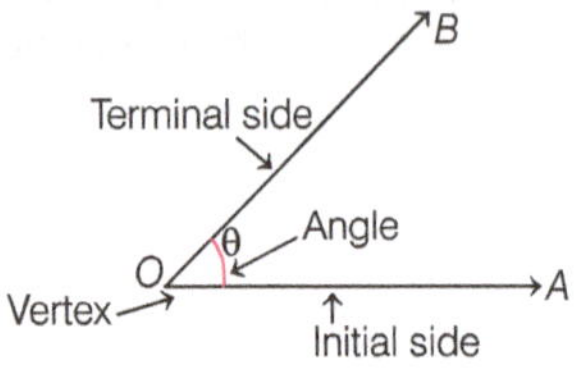

Positive angle If the angle goes in counterclockwise direction, then it is called a positive angle.

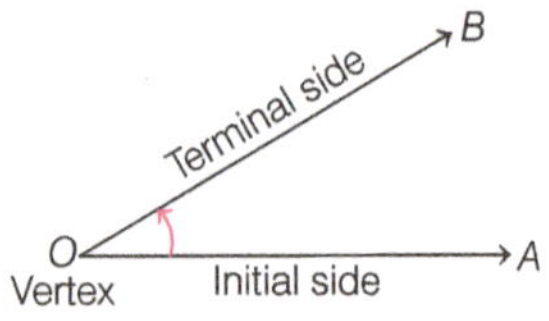

(i) Positive angle

Negative angle If the angle goes in clockwise direction, then it is called a negative angle.

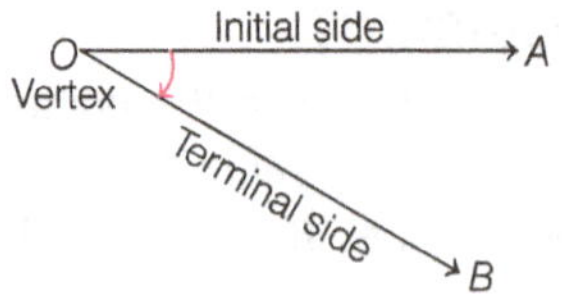

(ii) Negative angle

Units of Angle

A degree usually denoted by °(the degree symbol), is a unit of an angle in which one full rotation equals 360 degrees.

It is not an SI units. The SI unit of an angle is the radian. A full rotation equals 2π radians (or 360°).

Thus,

$$360^\circ = 2\pi \text{ radian}$$

or $$1^\circ = \frac{\pi}{180} \text{ radian}$$

$\therefore$ $$1 \text{ radian} = \left(\frac{180}{\pi}\right)^\circ \approx 57.3^\circ$$

Important POINTS

1. 1 radian is the angle subtended at the centre of a circle by an arc of the circle, whose length is equal to the radius of the circle.

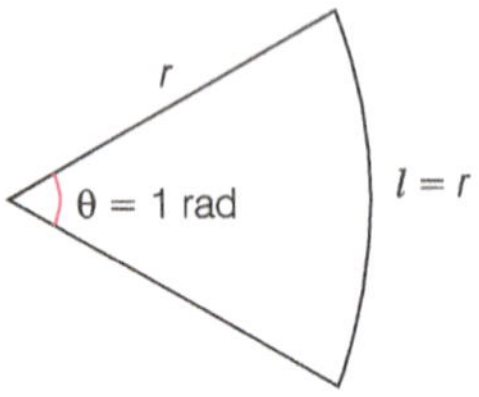

2. 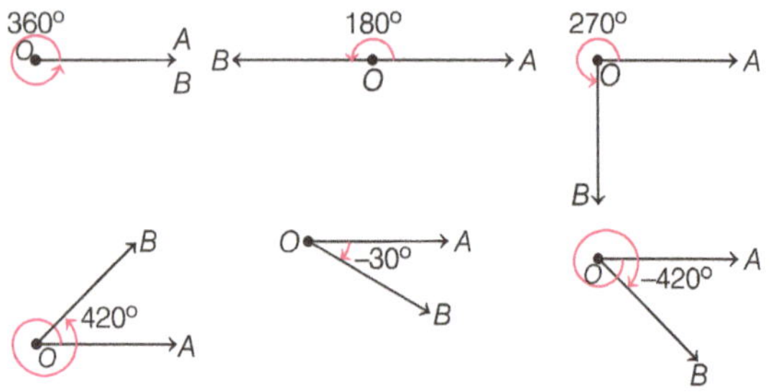

3.

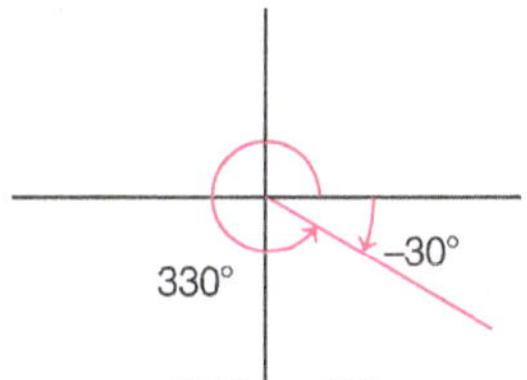

$330° = -30°$

4.

Degree	30°	45°	60°	90°	180°	270°	360°
Radian	$\frac{\pi}{6}$	$\frac{\pi}{4}$	$\frac{\pi}{3}$	$\frac{\pi}{2}$	π	$\frac{3\pi}{2}$	2π

5.

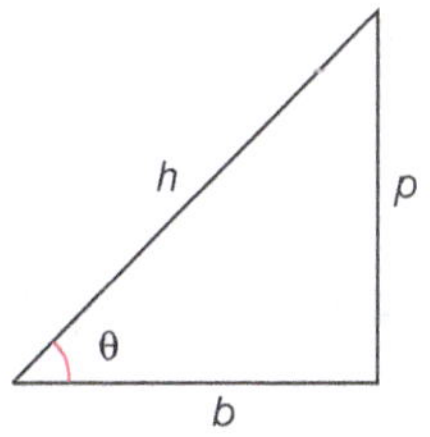

$h^2 = b^2 + p^2 \rightarrow$ Pythagoras theorem

$$\sin\theta = \frac{p}{h},$$

$$\cos\theta = \frac{b}{h},$$

$$\tan\theta = \frac{p}{b}$$

$$\tan\theta = \frac{\sin\theta}{\cos\theta} = \frac{1}{\cot\theta}$$

$$\cot\theta = \frac{\cos\theta}{\sin\theta} = \frac{1}{\tan\theta}$$

$$\sec\ \theta = \frac{1}{\cos\theta} = \frac{h}{b}$$

$$\text{cosec}\ \theta = \frac{1}{\sin\theta} = \frac{h}{p}$$

6. $1' = \left(\frac{1}{60}\right)^{\circ} = 1$ min

$1'' = \left(\frac{1}{60}\right)' = 1$ sec

Values of Different Angles

For 0°

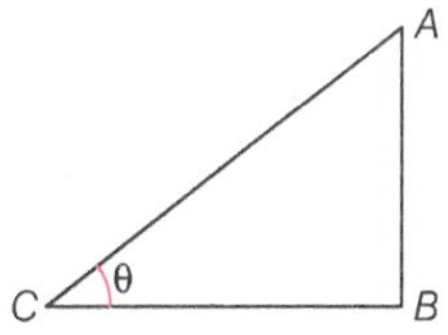

For $\theta = 0°$, we can see that, $AB = 0$ and $BC = AC$

Thus,

$$\sin\theta = \frac{AB}{AC} = \frac{0}{AC} = 0$$

$$\cos\theta = \frac{BC}{AC} = 1 \qquad (\text{as } BC = AC)$$

$$\tan\theta = \frac{AB}{BC} = \frac{0}{BC} = 0$$

Note: $\text{cosec}\ \theta = \frac{1}{\sin\theta} = \frac{1}{0} = \infty.$

Similarly we can find sec θ and cot θ.

For 30° and 60°

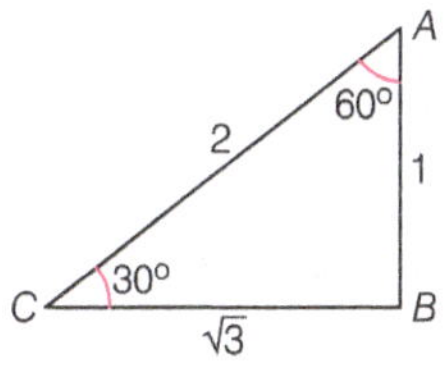

From this triangle we can find any value of 30° or 60°.

For example :

$$\sin 30° = \frac{AB}{AC} = \frac{1}{2}$$

$$\tan 60° = \frac{BC}{AB} = \sqrt{3}$$

For 37° and 53°

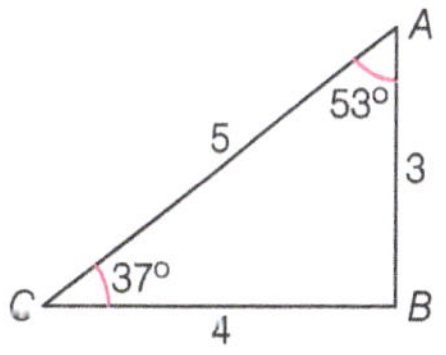

From the shown triangle we can find values of 37° and 53°.

For example :

$$\tan 37° = \frac{AB}{BC} = \frac{3}{4} \quad \text{and} \quad \cos 53° = \frac{AB}{AC} = \frac{3}{5}$$

For 45°

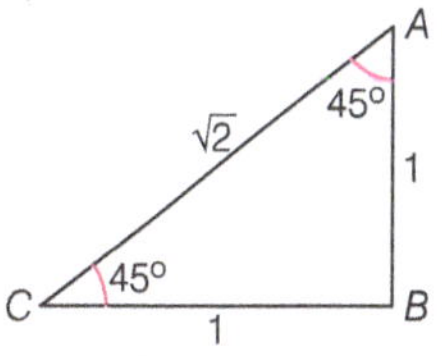

$$\tan 45° = \frac{AB}{BC} = 1$$

$$\sin 45° = \frac{AB}{AC} = \frac{1}{\sqrt{2}} \text{ etc.}$$

For 90°

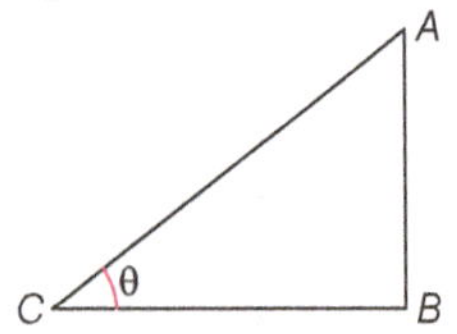

For $\theta = 90°$, we can see that

$$AC = AB \quad \text{and} \quad BC = 0$$

Thus, $\sin 90° = \dfrac{AB}{AC} = 1$ (as $AB = AC$)

and $\tan 90° = \dfrac{AB}{BC} = \dfrac{AB}{0} = \infty$

Note *The values of above angles in tabular form are given below. Value of 180° will be discussed later.*

Table

Angle	$\sin\theta$	$\cos\theta$	$\tan\theta = \dfrac{\sin\theta}{\cos\theta}$
0°	0	1	0
30°	1/2	$\sqrt{3}/2$	$1/\sqrt{3}$
37°	3/5	4/5	3/4
45°	$1/\sqrt{2}$	$1/\sqrt{2}$	1
53°	4/5	3/5	4/3
60°	$\sqrt{3}/2$	1/2	$\sqrt{3}$
90°	1	0	∞
180°	0	−1	0

Four Quadrants and ASTC Rule

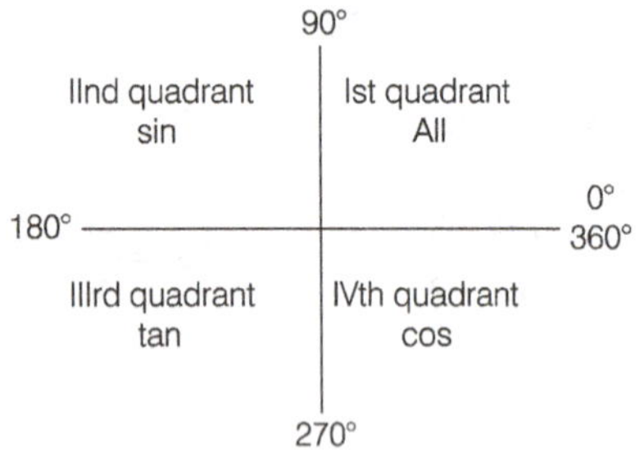

- In first quadrant, all trigonometric ratios are positive.
- In second quadrant, only $\sin\theta$ and $\operatorname{cosec}\theta$ are positive.
- In third quadrant, only $\tan\theta$ and $\cot\theta$ are positive.
- In fourth quadrant, only $\cos\theta$ and $\sec\theta$ are positive.

Remember as **A**ll **S**ilver **T**ea **C**ups.

Note *In Ist quadrant angle θ lies between 0° and 90°. Similarly in IInd quadrant θ lies between 90° and 180° and so on.*

Example 1 Explain whether following values are positive or negative.

(a) cot 20° (b) cos 110° (c) tan 210° (d) sin (–60°)
(e) sin 270°

Sol. (a) 20 ° lies in Ist quadrant. In Ist quadrant all are positive. Hence cot 20° is positive.

(b) 110° lies in 2nd quadrant. Only sine and cosine are positive in 2nd quadrant. Hence, cos 110° is negative.

(c) 210° lies in 3rd quadrant. Only tan and cot are positive in 3rd quadrant. Hence, tan 210° is positive.

(d) –60° or 300° lies in 4th quadrant. Only cos and sec are positive in 4th quadrant. Hence, sin (–60°) is negative.

(e) 270° is the boundary of 3rd and 4th quadrant. Further, value of sin θ in both quadrants is negative. Hence, sin 270° is negative.

Values of Some Other Angles

Following table will help you in finding values of some other angles not given in above table.

$\sin(90° + \theta) = \cos\theta$ $\cos(90° + \theta) = -\sin\theta$ $\tan(90° + \theta) = -\cot\theta$	$\sin(180° - \theta) = \sin\theta$ $\cos(180° - \theta) = -\cos\theta$ $\tan(180° - \theta) = -\tan\theta$	$\sin(-\theta) = -\sin\theta$ $\cos(-\theta) = \cos\theta$ $\tan(-\theta) = -\tan\theta$
$\sin(90° - \theta) = \cos\theta$ $\cos(90° - \theta) = \sin\theta$ $\tan(90° - \theta) = \cot\theta$	$\sin(180° + \theta) = -\sin\theta$ $\cos(180° + \theta) = -\cos\theta$ $\tan(180° + \theta) = \tan\theta$	

Example 2 Find the values of

(a) tan 120° (b) cos 300°
(c) sin 210° (d) cos 180°
(e) sin 270°

Sol. (a) 120° lies in 2nd quadrant. In this quadrant tan is positive. Hence, tan 120° should be positive.

Further,

$$\tan 120° = \tan(180° - 60°) = +\tan 60°$$
$$= +\sqrt{3}$$

Alternate method

$$\tan 120° = \tan(90° + 30°) = +\cot 30°$$
$$= +\sqrt{3}$$

(b) 300° lies in 4th quadrant and cos is positive in this quadrant.

Hence, cos 300° should be positive.

Further,

$$\cos 300° = \cos(-60°) = +\cos 60°$$
$$= +\frac{1}{2}$$

(c) 210° lies in 3rd quadrant and sin is negative in 3rd quadrant. Hence, sin 210° should be negative.

Further,

$$\sin 210° = \sin(180° + 30°) = -\sin 30° = -\frac{1}{2}$$

(d) 180° is the boundary of 2nd and 3rd quadrants and cos is negative in both quadrants.

Hence, cos 180° should be negative.

Further, $\cos 180° = \cos(180° + 0°) = -\cos 0° = -1$

(e) 270° is the boundary of 3rd and 4th quadrants and sin is negative in both quadrants. Hence sin 270° should be negative. Further,

$$\sin 270° = \sin(-90°) = -\sin 90° = -1$$

Alternate method

$\sin 270° = \sin(180° + 90°) = -\sin 90° = -1$

Example 3 Find the values of

(a) cos(–45°) (b) tan 210°

(c) sin 300° (d) cos 120°

Sol. (a) $\cos(-45°) = \cos 45° = \frac{1}{\sqrt{2}}$

(b) $\tan 210° = \tan(180° + 30°) = \tan 30° = \frac{1}{\sqrt{3}}$

(c) $\sin 300° = \sin(-60°) = -\sin 60° = -\frac{\sqrt{3}}{2}$

(d) $\cos 120° = \cos(180° - 60°) = -\cos 60° = -\frac{1}{2}$

Example 4 Find the values of

(a) sin 420° (b) sin 405° (c) tan 390°

Sol. (a) $\sin 420° = \sin(360° + 60°) = \sin 60° = \frac{\sqrt{3}}{2}$

(b) $\sin 405° = \sin(360° + 45°) = \sin 45° = \frac{1}{\sqrt{2}}$

(c) $\tan 390° = \tan(360° + 30°) = \tan 30° = \frac{1}{\sqrt{3}}$

Sine and Cosine Graphs

Sine Graph

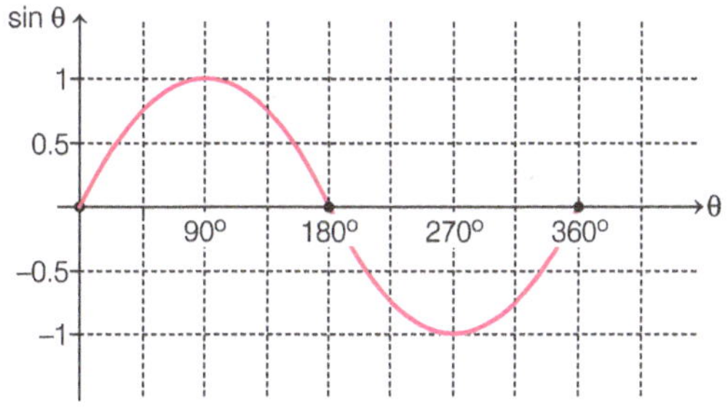

Cosine Graph

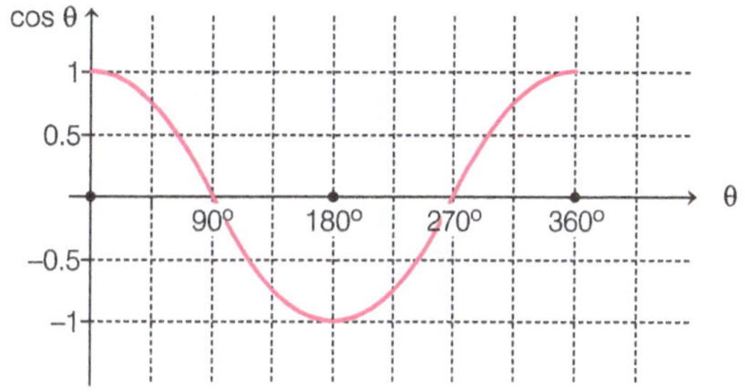

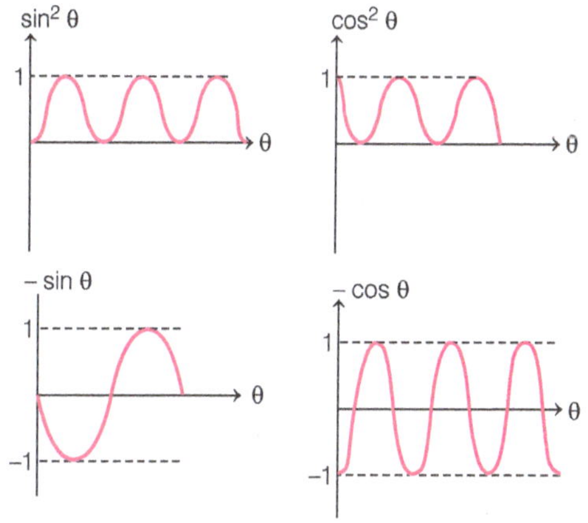

(i) $(\sin\theta)_{max} = (\cos\theta)_{max} = 1$

(ii) $(\sin\theta)_{min} = (\cos\theta)_{min} = -1$

(iii) As θ increases from 0° to 90°, sin θ increases, but cos θ decreases.

(iv) For small values of θ (in radian)

$\sin\theta \approx \theta$, $\tan\theta \approx \theta$ and $\cos\theta \approx 1$

Example 5 Find approximate value of sin 2°.

Sol. For small value of θ (in radian)

$$\sin\theta \approx \theta$$

$$\therefore \quad \sin 2° = \sin\left(\frac{2 \times \pi}{180}\right)\text{radian} = \sin\left(\frac{\pi}{90}\right)\text{rad}$$

$$\approx \frac{\pi}{90} = \frac{22/7}{90}$$

$$\approx 0.035$$

Example 6 Given $y = 4 + 3\sin\theta$

Plot *y versus* θ graph with $(0° \le \theta \le 360°)$.

Sol. $(\sin\theta)_{max} = 1$ at $\theta = 90°$

$\therefore \quad y_{max} = 4 + (3)(1) = 7$

$(\sin\theta)_{min} = -1$ at $\theta = 270°$

$\therefore \quad y_{min} = 4 + 3(-1) = 1$

Therefore *y* varies between 7 and 1 (sinusoidally).

Further, at $\theta = 0°$, $\sin\theta = 0 \Rightarrow y = 4$

at $\theta = 90°$, $\sin\theta = 1 \Rightarrow y = 7$

at $\theta = 180°$, $\sin\theta = 0 \Rightarrow y = 4$
at $\theta = 270°$, $\sin\theta = -1 \Rightarrow y = 1$
at $\theta = 360°$, $\sin\theta = 0 \Rightarrow y = 4$

Therefore y *versus* θ graph is as shown below.

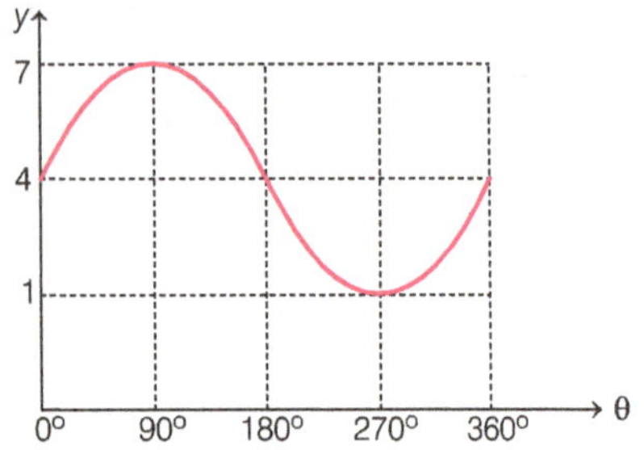

Example 7 Given $y = -2 - 4\cos\theta$

(a) Find minimum and maximum value of y.

(b) Plot y *versus* θ graph $(0° \le \theta \le 360°)$.

Sol. (a) $(\cos\theta)_{max} = 1$ at $\theta = 0°$ and $360°$.

$\therefore \quad y = -2 - 4(1) = -6 = y_{min}$

$(\cos\theta)_{min} = -1$ at $\theta = 180°$

$\therefore \quad y = -2 - 4(-1) = +2 = y_{max}$

$\therefore \quad y_{max} = 2$ and $y_{min} = -6$

(b)

Angle	Value of $\cos\theta$	Value of y where, $y = -2 - 4\cos\theta$
0°	1	−6
90°	0	−2
180°	−1	+2
270°	0	−2
360°	1	−6

Now y *versus* θ graph corresponding to above table is as shown below.

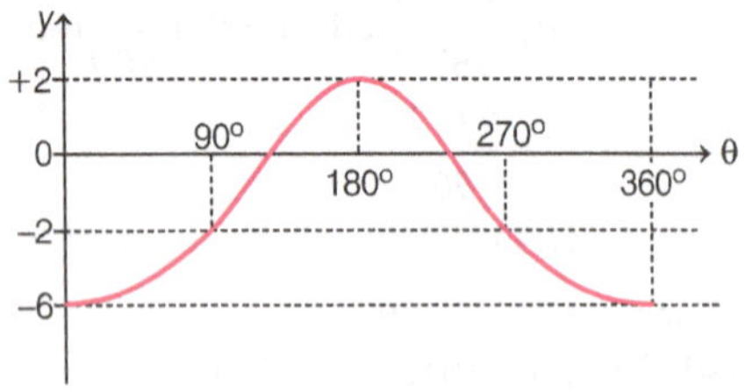

Note *Sinusoidal means sine or cosine curve.*

Some Other Important Formulas of Trigonometry

(i) $\sin^2\theta + \cos^2\theta = 1 \Rightarrow \sin^2\theta = 1 - \cos^2\theta$

and $\cos^2\theta = 1 - \sin^2\theta$

(ii) $\sec^2\theta = 1 + \tan^2\theta \Rightarrow \tan^2\theta = \sec^2\theta - 1$

(iii) $\text{cosec}^2\theta = 1 + \cot^2\theta$

(iv) $\sin(A + B) = \sin A\cos B + \cos A\sin B$

$\sin 2A = 2\sin A\cos A$ [put $B = A$]

$\sin A = 2\sin\frac{A}{2}\cos\frac{A}{2}$

(v) $\sin(A - B) = \sin A\cos B - \cos A\sin B$

$\sin 0° = 0$ [put $B = A$]

(vi) $\cos(A + B) = \cos A\cos B - \sin A\sin B$

$\cos 2A = \cos^2 A - \sin^2 A$ [put $B = A$]

$= 2\cos^2 A - 1$ [$\sin^2 A = 1 - \cos^2 A$]

$= 1 - 2\sin^2 A$

(vii) $\cos A = \cos^2\frac{A}{2} - \sin^2\frac{A}{2} = 2\cos^2\frac{A}{2} - 1 = 1 - 2\sin^2\frac{A}{2}$

(viii) $\cos(A - B) = \cos A\cos B + \sin A\sin B$

$\cos 0° = \cos^2 A + \sin^2 A = 1$ [put $B = A$]

(ix) $\tan(A + B) = \dfrac{\tan A + \tan B}{1 - \tan A\tan B}$

$\tan 2A = \dfrac{2\tan A}{1 - \tan^2 A}$ [put $B = A$]

$\tan A = \dfrac{2\tan\frac{A}{2}}{1 - \tan^2\frac{A}{2}}$

(x) $\tan(A - B) = \dfrac{\tan A - \tan B}{1 + \tan A\tan B}$

$\tan 0° = 0$ [put $B = A$]

(xi) $\sin C + \sin D = 2\sin\frac{C + D}{2}\cos\frac{C - D}{2}$

$\sin C - \sin D = 2\sin\frac{C - D}{2}\cos\frac{C + D}{2}$

$\cos C + \cos D = 2\cos\frac{C + D}{2}\cos\frac{C - D}{2}$

$\cos C - \cos D = 2\sin\frac{C + D}{2}\sin\frac{D - C}{2}$

Example 8 Prove that $1 + \tan^2\theta = \sec^2\theta$.

Sol.

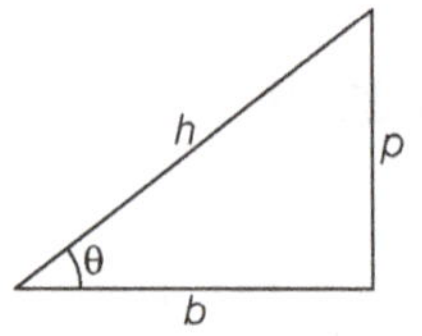

$b^2 + p^2 = h^2$

$\tan\theta = \frac{p}{b}, \sec\theta = \frac{h}{b}$

$1 + \tan^2\theta = 1 + \frac{p^2}{b^2} = \frac{b^2 + p^2}{b^2}$

$= \frac{h^2}{b^2} = \sec^2\theta$ **Hence proved.**

Example 9 Find the value of

(a) $\sin 15^\circ$ (b) $\sin 75^\circ$

Sol. (a) We can write $\sin 15^\circ = \sin(45^\circ - 30^\circ)$

We have $\sin(A - B) = \sin A \cos B - \cos A \sin B$

$$\sin 15^\circ = \sin(45^\circ - 30^\circ)$$
$$= \sin 45^\circ \cdot \cos 30^\circ - \cos 45^\circ \cdot \sin 30^\circ$$
$$= \frac{1}{\sqrt{2}} \cdot \frac{\sqrt{3}}{2} - \frac{1}{\sqrt{2}} \cdot \frac{1}{2} = \frac{\sqrt{3}-1}{2\sqrt{2}}$$
$$\therefore \quad \sin 15^\circ = \frac{\sqrt{3}-1}{2\sqrt{2}}$$

(b) We can write $\sin 75^\circ = \sin(45^\circ + 30^\circ)$

We have $\sin(A + B) = \sin A \cos B + \cos A \sin B$

$$\sin 75^\circ = \sin(45^\circ + 30^\circ)$$
$$= \sin 45^\circ \cdot \cos 30^\circ + \cos 45^\circ \cdot \sin 30^\circ$$
$$= \frac{1}{\sqrt{2}} \cdot \frac{\sqrt{3}}{2} + \frac{1}{\sqrt{2}} \cdot \frac{1}{2} = \frac{\sqrt{3}+1}{2\sqrt{2}}$$
$$\therefore \quad \sin 75^\circ = \frac{\sqrt{3}+1}{2\sqrt{2}}$$

Example 10 Find value of $\cos 120^\circ$, using the formula of $\cos 2A$.

Sol. $2A = 120^\circ$

$$\Rightarrow \quad A = 60^\circ$$

Now, $\cos 2A = 2\cos^2 A - 1$

$$\therefore \quad \cos 120^\circ = 2(\cos 60^\circ)^2 - 1$$
$$= 2\left(\frac{1}{2}\right)^2 - 1$$
$$= \frac{1}{2} - 1 = -\frac{1}{2}$$ **Ans.**

Note: *120° lies in 2nd quadrant. Value of $\cos\theta$ in 2nd quadrant is negative. So, $\cos 120^\circ$ should be negative and we are getting $\cos 120^\circ = -\frac{1}{2}$.*

Important POINTS

(i) $\sin^2\theta \neq \sin\theta^2$ but $\sin^2\theta = (\sin\theta)^2$

(ii) If $\alpha + \beta = 90^\circ$, then

$$\sin\alpha = \cos\beta$$

For example,

$$\sin 2^\circ = \cos 88^\circ$$
$$\cos 15^\circ = \sin 75^\circ$$
$$\sin 10^\circ = \cos 80^\circ$$

(iii) If $\alpha + \beta = 180^\circ$, then

$$\sin\alpha = \sin\beta$$
$$\cos\alpha = -\cos\beta \text{ and}$$
$$\tan\alpha = -\tan\beta$$

For example,

$$\sin 30^\circ = \sin 150^\circ$$
$$\cos 120^\circ = -\cos 60^\circ \text{ and}$$
$$\tan 45^\circ = -\tan 135^\circ$$

(iv) In the below triangle,

$$\frac{a}{\sin A} = \frac{b}{\sin B} = \frac{c}{\sin C}$$

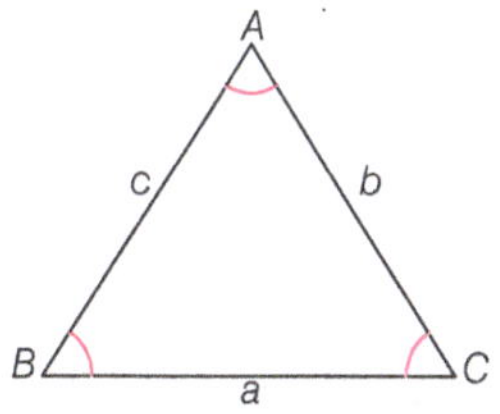

This is called **sine rule.**

(v) If $\tan\theta = 1$, then we can write $\theta = \tan^{-1}(1)$.

Although we know that in this case one answer of θ is 45°.

Similarly if $\sin\theta = \frac{1}{4}$, then $\theta = \sin^{-1}\left(\frac{1}{4}\right)$

If $\cos\theta = -\frac{1}{3}$, then $\theta = \cos^{-1}\left(-\frac{1}{5}\right)$

Miscellaneous Examples

Ex. 1 A circular arc is of length π cm. Find angle subtended by it at the centre in radian and degree.

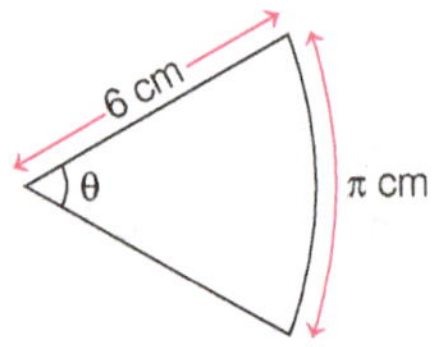

Sol. θ (in radian) $= \dfrac{\text{arc}}{\text{radius}} = \dfrac{\pi \text{ cm}}{6 \text{ cm}} = \dfrac{\pi}{6}$ rad

As $1 \text{ rad} = \dfrac{180°}{\pi}$

So, $\theta = \dfrac{\pi}{6} \times \dfrac{180°}{\pi} = 30°$ **Ans.**

Ex. 2 The moon's distance from the earth is 360000 km and its diameter subtends an angle of 42′ at the eye of the observer. Find the diameter of the moon.

Sol. Here, angle is very small so diameter $\approx$ arc

Here $\theta = 42' = \left(42 \times \dfrac{1}{60}\right)^{\circ}$

$= 42 \times \dfrac{1}{60} \times \dfrac{\pi}{180} = \dfrac{7\pi}{1800}$ rad

So, diameter $= R\theta = 360000 \times \dfrac{7}{1800} \times \dfrac{22}{7}$

$= 4400$ km **Ans.**

Ex. 3 When a clock shows 4 o'clock, how much angle do its minute and hour needles make?

(a) 120° (b) $\dfrac{\pi}{3}$ rad (c) $\dfrac{2\pi}{3}$ rad (d) 160°

Sol.

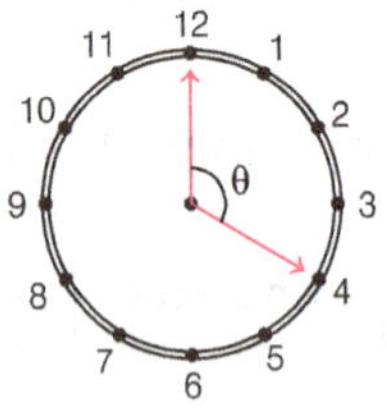

From the above diagram angle

$\theta = 4 \times 30° = 120° = \dfrac{2\pi}{3}$ rad **Ans.**

The correct options are (a) and (c).

Ex. 4 Given that $\sin\theta = \dfrac{1}{4}$. Find the values of $\cos\theta$ and $\tan\theta$.

Sol.

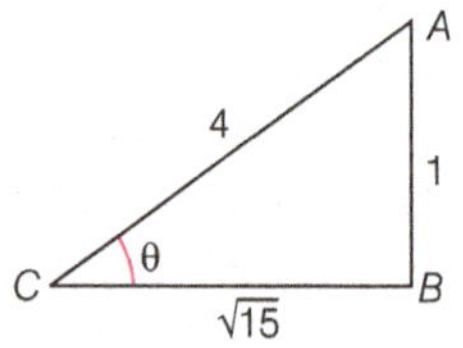

In ΔABC, $\sin\theta = \dfrac{1}{4}$

$\Rightarrow$ $AB = 1$ and $AC = 4$

$\therefore$ $BC = \sqrt{(4)^2 - (1)^2} = \sqrt{15}$

Now, $\cos\theta = \dfrac{BC}{AC} = \dfrac{\sqrt{15}}{4}$ **Ans.**

and $\tan\theta = \dfrac{AB}{BC} = \dfrac{1}{\sqrt{15}}$ **Ans.**

Ex. 5 In right angle

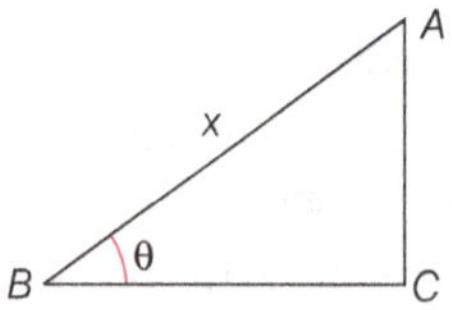

triangle ABC if $AB = x$, then prove that $AC = x\sin\theta$ and $BC = x\cos\theta$.

Sol. (i) $\dfrac{AC}{AB} = \sin\theta$

$\therefore$ $AC = AB\sin\theta = x\sin\theta$

(ii) $\dfrac{BC}{AB} = \cos\theta$

$\therefore$ $BC = AB\cos\theta = x\cos\theta$

MODULE 3 Exercise

1. In a circle of radius 4 m, find the angle of an arc of length 1 m subtended at centre.
2. If $\tan\theta = \frac{2}{5}$, then find the values of $\cos\theta$ and $\sin\theta$.
3. Find approximate value of $\tan 3°$.
4. Which of the following values are positive?
   (a) $\cos 120°$ (b) $\sin 210°$
   (c) $\tan 240°$ (d) $\cos 315°$
5. Find the value of
   (a) $\cos 120°$ (b) $\sin 240°$
   (c) $\tan(-60°)$ (d) $\cot 300°$
   (e) $\tan 330°$ (f) $\cos(-60°)$
   (g) $\sin(-150°)$ (h) $\cos(-120°)$
6. Find the value of $\tan 105°$.
7. Find the value of $\cos 67°$.
8. Find the value of $\cos 105° + \cos 75°$.
9. Find the value of
   (a) $\sec^2\theta - \tan^2\theta$
   (b) $\text{cosec}^2\theta - \cot^2\theta - 1$
   (c) $2\sin 45°\cos 15°$
   (d) $2\sin 15°\cos 45°$
10. Draw the following sinusoidal graphs
    (a) $y = 2 - 3\cos\theta$
    (b) $y = -3 + 4\sin\theta$

Answers

1. (0.25) radian **2.** $\cos\theta = \frac{5}{\sqrt{29}}$, $\sin\theta = \frac{2}{\sqrt{29}}$ **3.** 0.052 **4.** (c) and (d)

5. (a) $-\frac{1}{2}$ (b) $-\frac{\sqrt{3}}{2}$ (c) $-\sqrt{3}$ (d) $-\frac{1}{\sqrt{3}}$ (e) $-\frac{1}{\sqrt{3}}$ (f) $\frac{1}{2}$ (g) $-\frac{1}{2}$ (h) $-\frac{1}{2}$ **6.** $\left(\frac{\sqrt{3}+1}{1-\sqrt{3}}\right)$

7. $\left(\frac{4\sqrt{3}-3}{10}\right)$ **8.** Zero **9.** (a) 1 (b) 0 (c) $\left(\frac{\sqrt{3}+1}{2}\right)$ (d) $\left(\frac{\sqrt{3}-1}{2}\right)$

10.

(a)

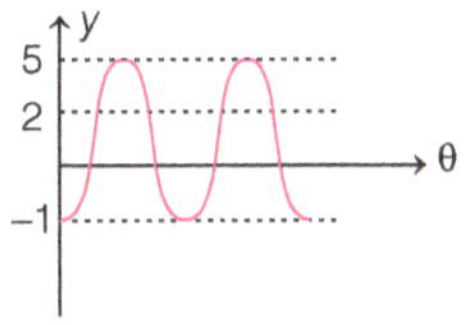

(b)

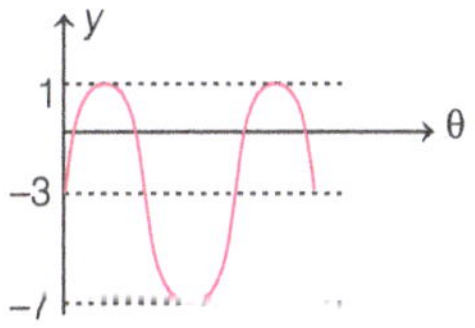

See hints on pages 39 and 40.

MODULE 4

Differentiation

General Points

(i) As far as Physics is concerned we only need formulas of differentiation and their applications. In Mathematics you will read differentiation in detail. So, in Physics you need not to go in detail of differentiation.

(ii) Suppose we have two variables x and y and they are related to each other by some equation. Then we can say that y is a function of x or x is a function of y. They are represented as :

$$y = f(x) \quad \text{or} \quad x = f(y)$$

Let us take the first equation $y = f(x)$ and they are related to each other by the following equation(s) :

$$y = x^2$$

or $$y = \sin x$$

or $$y = x^3 + e^x + 5$$

Now for different values of x we get different values of y.

For example: according to first equation, $y = x^2$,

$y = 0$ if $x = 0$, $y = 1$ if $x = \pm 1$, $y = 4$ if $x = \pm 2$ and so on.

(iii) If y is a function of x and we want to differentiate y with respect to x, then we will write $\frac{dy}{dx}$ or $\frac{d}{dx}(y)$.

Similarly, if y is a function of θ and we want to differentiate y with respect to θ, then we will write $\frac{dy}{d\theta}$ or $\frac{d}{d\theta}(y)$.

(iv) If we want to differentiate y with respect to x two times, then we will write $\frac{d^2y}{dx^2}$.

(v) If $y = f(x)$, then $\frac{dy}{dx}$ is also written as $f'(x)$. Similarly $\frac{d^2y}{dx^2}$ as $f''(x)$.

(vi) *dx* does not mean *d* times x. Similarly $\frac{d}{dx}f(x)$ does not mean $\frac{d}{dx}$ times $f(x)$. "*dx*" (read as *dee-ex*) is an infinitesimally small change in x.

List of Formulas

(i) $\frac{d}{dx}(\text{constant}) = 0$

(ii) $\frac{d}{dx}(x^n) = nx^{n-1}$

(iii) $\frac{d}{dx}(\log_e x)$ or $\frac{d}{dx}(\ln x) = \frac{1}{x}$

(iv) $\frac{d}{dx}(\sin x) = \cos x$

(v) $\frac{d}{dx}(\cos x) = -\sin x$

(vi) $\frac{d}{dx}(\tan x) = \sec^2 x$

(vii) $\frac{d}{dx}(\cot x) = -\operatorname{cosec}^2 x$

(viii) $\frac{d}{dx}(\sec x) = \sec x \tan x$

(ix) $\frac{d}{dx}(\operatorname{cosec} x) = -\operatorname{cosec} x \cot x$

(x) $\frac{d}{dx}(e^x) = e^x$

Note *π or $\frac{22}{7}$ is a constant. Similarly there is an another constant e (≈ 2.7183). So, according to formula (i) of the above list, their differentiation with respect to any variable should be zero. For example,*

$$\frac{d}{dx}(\pi) = 0, \ \frac{d}{dx}(e) = 0 \text{ or } \frac{d}{dy}(e) = 0 \text{ but } \frac{d}{dx}(e^x) = e^x.$$

Example 1 Using formula (ii) from the above table, differentiate $x, 5x, x^2, 6x^2, x^3, 2x^3$ with respect to x.

Sol. (a) $\frac{d}{dx}(x) = \frac{d}{dx}(x^1) = 1x^0 = 1$ or $\frac{d}{dx}(x) = 1$ **Ans.**

(b) $\frac{d}{dx}(5x) = 5 \cdot \frac{d}{dx}(x) = 5(1) = 5$ **Ans.**

Note $\frac{d}{dx}(5) = 0$ *but* $\frac{d}{dx}(5x) \neq 0$

(c) $\frac{d}{dx}(x^2) = 2x^1 = 2x$ or $\frac{d}{dx}(x^2) = 2x$ **Ans.**

(d) $\frac{d}{dx}(6x^2) = 6\frac{d}{dx}(x^2) = 6(2x) = 12x$ **Ans.**

(e) $\frac{d}{dx}(x^3) = 3x^2$ **Ans.**

(f) $\frac{d}{dx}(2x^3) = 2\frac{d}{dx}(x^3) = 2(3x^2) = 6x^2$ **Ans.**

Example 2 Find $\frac{dy}{dx}$, when

(a) $y=\sqrt{x}$

(b) $y=x^5+x^4+7$

(c) $y=x^2+4x^{-1/2}-3x^{-2}$

Sol. (a) $y=\sqrt{x}$

$$\frac{dy}{dx}=\frac{d}{dx}(\sqrt{x})=\frac{d}{dx}(x^{1/2})$$

$$=\frac{1}{2}x^{1/2-1}=\frac{1}{2}x^{-1/2}=\frac{1}{2\sqrt{x}} \quad \textbf{Ans.}$$

(b) $y=x^5+x^4+7$

$$\frac{dy}{dx}=\frac{d}{dx}(x^5+x^4+7)$$

$$=\frac{d}{dx}(x^5)+\frac{d}{dx}(x^4)+\frac{d}{dx}(7)$$

$$=5x^4+4x^3+0=5x^4+4x^3 \quad \textbf{Ans.}$$

(c) $y=x^2+4x^{-1/2}-3x^{-2}$

$$\frac{dy}{dx}=\frac{d}{dx}(x^2+4x^{-1/2}-3x^{-2})$$

$$=\frac{d}{dx}(x^2)+\frac{d}{dx}(4x^{-1/2})-\frac{d}{dx}(3x^{-2})$$

$$=\frac{d}{dx}(x^2)+4\frac{d}{dx}(x^{-1/2})-3\frac{d}{dx}(x^{-2})$$

$$=2x-\frac{1}{2}(4)x^{-3/2}-3(-2)x^{-3}$$

$$=2x-2x^{-3/2}+6x^{-3} \quad \textbf{Ans.}$$

Example 3 Given $y=\frac{5}{\sqrt{x}}+6e^x-2\sin x+\pi$. Find $\frac{dy}{dx}$.

Sol. $$\frac{dy}{dx}=\frac{d}{dx}\left[\frac{5}{\sqrt{x}}+6e^x-2\sin x+\pi\right]$$

$$=5\frac{d}{dx}(x^{-\frac{1}{2}})+6\frac{d}{dx}(e^x)-2\frac{d}{dx}(\sin x)+\frac{d}{dx}(\pi)$$

$$=5\left(-\frac{1}{2}\right)(x^{-\frac{1}{2}-1})+6(e^x)-2(\cos x)+0$$

$$=-\frac{5}{2}x^{-\frac{3}{2}}+6e^x-2\cos x \quad \textbf{Ans.}$$

INTRODUCTORY Exercise 1

1. Find the differentiation of $y=x^3+5x^2+3x+7$ with respect to x.
2. Differentiate $y=x(3x^2-9)$ with respect to x.
3. Given $y=3\sqrt{x}+4\tan x-6\ln x+e$.

   Find $\frac{dy}{dx}$.

Answers

1. $3x^2+10x+3$
2. $9x^2-9$
3. $\frac{3}{2\sqrt{x}}+4\sec^2 x-\frac{6}{x}$

See hints on page 40.

Multiplication and Division Rules

(i) $$\frac{d}{dx}\{f_1(x)\cdot f_2(x)\}=f_1(x)\frac{d}{dx}f_2(x)+f_2(x)\frac{d}{dx}f_1(x)$$

(ii) $$\frac{d}{dx}\frac{f_1(x)}{f_2(x)}=\frac{f_2(x)\frac{d}{dx}f_1(x)-f_1(x)\frac{d}{dx}f_2(x)}{\{f_2(x)\}^2}$$

Example 4 Differentiate the following functions with respect to x

(a) $x\sin x$ (b) $\frac{x}{\sin x}$

Sol. (a) $$\frac{d}{dx}(x\sin x)=x\frac{d}{dx}(\sin x)+\sin x\cdot\frac{d}{dx}(x)$$

$$=x\cos x+\sin x\,(1)$$

$$=x\cos x+\sin x$$

(b) $$\frac{d}{dx}\left(\frac{x}{\sin x}\right)=\frac{\sin x\frac{d}{dx}(x)-x\frac{d}{dx}(\sin x)}{(\sin x)^2}$$

$$=\frac{(\sin x)(1)-x(\cos x)}{\sin^2 x}$$

$$=\frac{\sin x-x\cos x}{\sin^2 x}$$

Example 5 If $y=\tan x$, then using the division rule find $\frac{dy}{dx}$.

Sol. $y=\tan x=\frac{\sin x}{\cos x}$

$$\frac{dy}{dx}=\frac{\cos x\frac{d}{dx}(\sin x)-\sin x\frac{d}{dx}(\cos x)}{(\cos x)^2}$$

$$=\frac{\cos x\cdot\cos x-\sin x(-\sin x)}{\cos^2 x}$$

$$=\frac{\cos^2 x+\sin^2 x}{\cos^2 x}=\frac{1}{\cos^2 x}=\sec^2 x \quad \textbf{Ans.}$$

Example 6 If $y=e^x\ln x$. Find $\frac{dy}{dx}$.

Sol. $$\frac{dy}{dx}=\frac{d}{dx}(e^x\ln x)$$

$$=e^x\frac{d(\ln x)}{dx}+\ln x\cdot\frac{de^x}{dx}$$

$$=e^x\cdot\frac{1}{x}+\ln x\cdot e^x=e^x\left(\frac{1}{x}+\ln x\right)$$

INTRODUCTORY Exercise 2

1. Find the derivatives of $y=(x^2+1)(x^3+3)$.
2. If $y=\sqrt{x}\tan x$, then find $\frac{dy}{dx}$.
3. If $y=(\sin x)\ln x$, then find $\frac{dy}{dx}$.

Answers

1. $5x^4+3x^2+6x$
2. $\sqrt{x}\sec^2 x+\frac{\tan x}{2\sqrt{x}}$
3. $\frac{\sin x}{x}+(\ln x)\cos x$

See hints on page 41.

Three More Concepts of Differentiation

First

Suppose y as a function of x is of following type:

$$y=(6x-4)^3 \Rightarrow y=\sqrt{4x+4}$$

$$y=\frac{1}{\sqrt{3x+2}} \Rightarrow y=\sin(3-6x)$$

$$y=e^{(4x-5)} \text{ etc.}$$

Then multiply with coefficient of x and differentiate by assuming (6x − 4) or (4x + 4) or (3x + 2) or (3 − 6x) or (4x − 5) simply as x.

For example

(i) $\frac{d}{dx}(6x-4)^3=(6)[3(6x-4)^2]=18(6x-4)^2$

(ii) $\frac{d}{dx}(\sqrt{4x+4})=\frac{d}{dx}(4x+4)^{\frac{1}{2}}$

$$=(4)\left[\frac{1}{2}(4x+4)^{\frac{1}{2}-1}\right]$$

$$=2(4x+4)^{-\frac{1}{2}}=\frac{2}{\sqrt{4x+4}}$$

(iii) $\frac{d}{dx}\left[\frac{1}{\sqrt{3x+2}}\right]=\frac{d}{dx}(3x+2)^{-\frac{1}{2}}$

$$=(3)\left(-\frac{1}{2}\right)(3x+2)^{-\frac{1}{2}-1}$$

$$=-\frac{3}{2}(3x+2)^{-\frac{3}{2}}=-\frac{3}{2(3x+2)^{\frac{3}{2}}}$$

(iv) $\frac{d}{dx}\sin(3-6x)=(-6)\cos(3-6x)$

$$=-6\cos(3-6x)$$

(v) $\frac{d}{dx}e^{(4x-5)}=4e^{(4x-5)}$

Second

This is also known as **chain rule.**

Suppose $y=(\cos x^2)$

Here, $y \neq \cos x$

So, first we will differentiate cos function and then x^2 function. Finally we will multiply the results. Thus,

$$\frac{d}{dx}(\cos x^2)=[-\sin x^2][2x]$$

$$=-2x\sin x^2$$

Similarly,

$$\frac{d}{dx}(\sin e^{x^2})=[\cos e^{x^2}][e^{x^2}][2x]$$

$$=2xe^{x^2}\cos e^{x^2}$$

Example 7 If $y=[x^2+3x+2]^3+[\ln x^2]$. Then, find $\frac{dy}{dx}$.

Sol. Using the chain rule, we have

$$\frac{dy}{dx}=3(x^2+3x+2)^{3-1}\cdot(2x+3)+\frac{1}{x^2}\cdot 2x$$

$$=3(2x+3)(x^2+3x+2)^2+\frac{2}{x} \quad \textbf{Ans.}$$

Example 8 If $y=\{(x^2+2x)^{1/2}\}$, then find $\frac{dy}{dx}$.

Sol. Using the chain rule, we have

$$\frac{dy}{dx}=\frac{1}{2}(x^2+2x)^{\frac{1}{2}-1}(2x+2)$$

$$=(x+1)(x^2+2x)^{-1/2} \quad \textbf{Ans.}$$

Third

If $y=f(x)$ and we wish to find $\frac{dy}{dt}$, then

$$\frac{dy}{dt}=\left(\frac{dy}{dx}\right)\left(\frac{dx}{dt}\right)=\left[\frac{d}{dx}f(x)\right]\left(\frac{dx}{dt}\right)$$

Example 9 Given, $y=3x^3$. Find the value of $\frac{dy}{dt}$.

Sol. $\frac{dy}{dt}=\left(\frac{dy}{dx}\right)\left(\frac{dx}{dt}\right)$

$$=\left[\frac{d}{dx}(3x^3)\right]\frac{dx}{dt}=9x^2\left(\frac{dx}{dt}\right)$$

Example 10 If $y=\sin(6x+2)$, then find $\frac{dy}{dz}$.

Sol. $\frac{dy}{dz}=\left(\frac{dy}{dx}\right)\left(\frac{dx}{dz}\right)=\left[\frac{d}{dx}\sin(6x+2)\right]\left(\frac{dx}{dz}\right)$

$$=(6)\cos(6x+2)\left(\frac{dx}{dz}\right)$$

$$=6\cos(6x+2)\frac{dx}{dz}$$

INTRODUCTORY Exercise 3

1. Differentiate the following functions with respect to x.
   (a) $(2x+3)^6$ (b) $\ln(5x+2)$
2. If $y = 4e^{x^2-2x}$, then find $\frac{dy}{dx}$.
3. If $y = (x^2+1)^{1/2}$, then find $\frac{dy}{dx}$.
4. If $y = \cos x^3$, then find $\frac{dy}{dx}$.
5. If $y = x^3$, then find $\frac{dy}{dt}$.

Answers

1. (a) $12(2x+3)^5$ (b) $\frac{5}{5x+2}$ 2. $8(x-1)e^{x^2-2x}$

3. $\frac{x}{(x^2+1)^{1/2}}$ 4. $-3x^2 \sin x^3$ 5. $(3x^2)\frac{dx}{dt}$

See hints on page 41.

Uses of Differentiation

There are mainly two uses of differentiation:

First

Slope of a graph Slope of a graph is just like slope of a hill.

Slope of a straight line Slope of a straight line graph is constant at every point. This constant value may be zero, infinite, positive or negative.

Let us take an example.

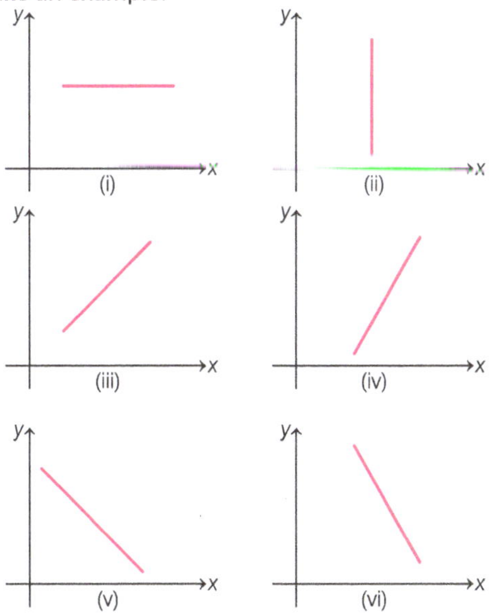

Slope of line in figure (i) is zero. Slope of line in figure (ii) is infinite.

Slopes of lines in figure (iii) and (iv) are of similar nature. They are taken positive slopes. But slope of line in figure (iv) is more than slope of line in figure (iii).

Similarly slopes of lines in figures (v) and (vi) are of similar types. Those are taken negative slopes. But magnitude of slope of line in figure (vi) is more than the magnitude of slope of line in figure (v).

The above nature of slope (zero, infinite, positive, negative, less slope, more slope) is best represented by $\tan\theta$. As we increase θ from 0° to 180°.

$$\tan 0° \text{ or } \tan 180° = 0$$

As θ increases from 0° to 90°, $\tan\theta$ is positive and increases from 0 to ∞.

$$\tan 90° = \infty$$

As θ increases from 90° to 180°, $\tan\theta$ is negative and its magnitude decreases.

Thus the slope of a line is the tangent of the angle made by the line with positive x-axis in the anti-clockwise direction.

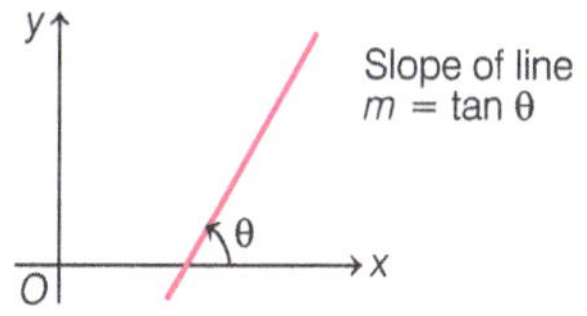

Slope is also represented by m. The following figures will illustrate the theory.

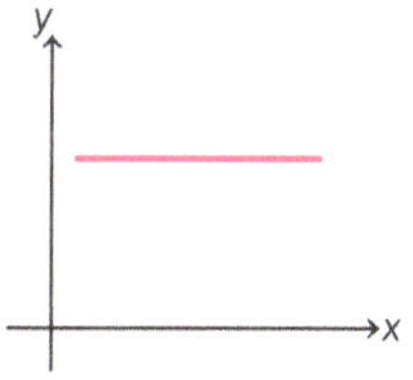

$\theta = 0°$ or 180°

$\therefore$ Slope or $m = \tan 0°$ or $\tan 180°$
$= 0$

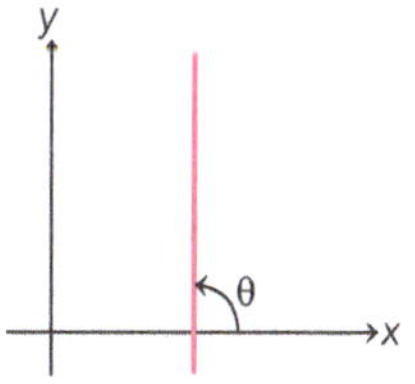

$\theta = 90°$

$\therefore$ Slope or $m = \tan 90° = \infty$

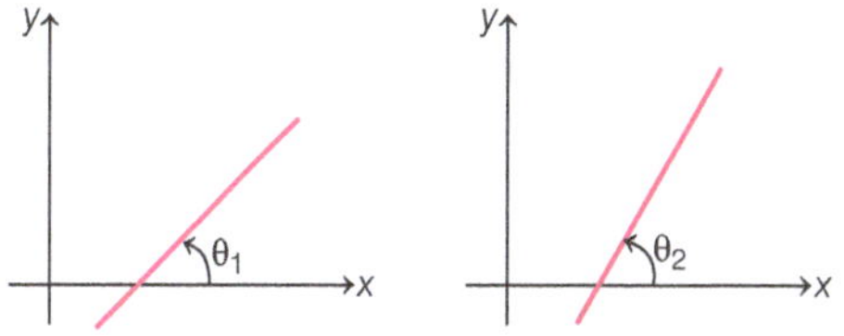

θ_1 and θ_2 both are less than 90°.

So, slope or $\tan\theta$ is positive for both.

But $\theta_2 > \theta_1 \Rightarrow \tan\theta_2$ or $\text{Slope}_2 > \tan\theta_1$ or Slope_1

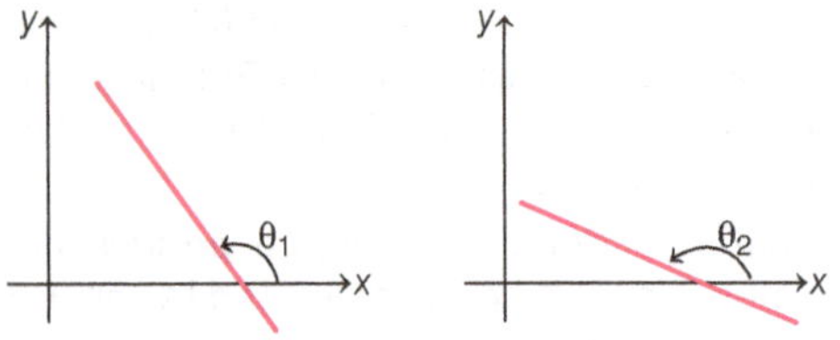

$90° < \theta_1$ and $\theta_2 < 180°$

So, slope or $\tan\theta$ is negative for both. But $\theta_2 > \theta_1$.

$\therefore\ |\tan\theta_2|$ or $|\text{Slope}_2| < |\tan\theta_1|$ or $|\text{Slope}_1|$

Example 11 Find slope of the following lines.

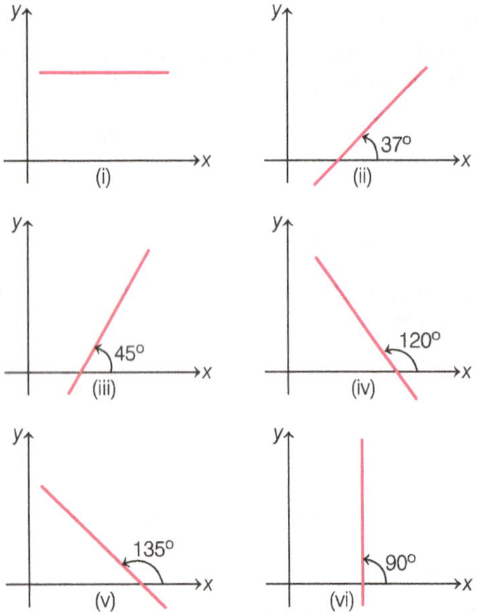

Sol. Slope of line in figure (i) = 0

Slope of line in figure (ii) = $\tan 37° = \frac{3}{4} = 0.75$

Slope of line in figure (iii) $= \tan 45° = 1$

Slope of line in figure (iv) $= \tan 120°$

$= -\tan 60° = -\sqrt{3}$

Slope of line in figure (v) $= \tan 135°$

$= -\tan 45°$

$= -1$

Slope of line in figure (vi) $= \infty$.

Note *Slope can also be calculated as:*

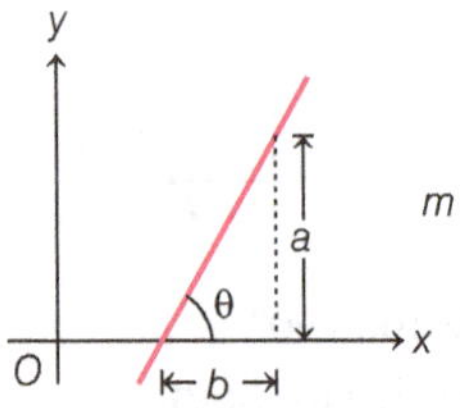

$m = \tan\theta = \frac{a}{b}$

Slope of a General Graph

As we discussed above, slope of a straight line is constant at every point. This constant may be zero, positive, negative or infinite.

Slope of any other graph keeps on changing point to point.

Slope of a curve at some point is the slope of the tangent line at that point.

The following figure will illustrate the concept.

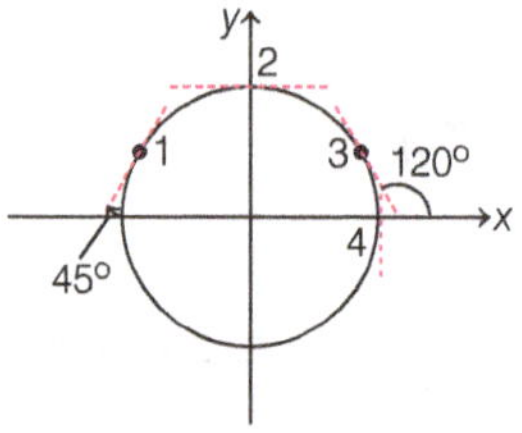

The graph is a circle.

$$(\text{Slope})_{\text{at } 1} = \tan 45° = +1$$
$$(\text{Slope})_{\text{at } 2} = 0$$
$$(\text{Slope})_{\text{at } 3} = \tan 120° = -\tan 60° = -\sqrt{3}$$
$$(\text{Slope})_{\text{at } 4} = \infty$$

Note *If $y = f(x)$, then we can plot a graph between x and y. If we differentiate y with respect to x or if we find $\frac{dy}{dx}$, then the quantity $\frac{dy}{dx}$ is nothing but the slope of the graph. Let us take two examples.*

(i) $y = 4x + 7$

$$\frac{dy}{dx} = 4 = \text{constant}$$

$\therefore$ *Slope = constant*

Since slope is constant. So, graph corresponding to the equation $y = 4x + 7$ is a straight line.

(ii) $y = 2x^2$

$$\frac{dy}{dx} = 4x = f(x)$$

$\therefore$ *Slope = $f(x)$*

At $x = 1$, $\frac{dy}{dx}$ or slope $= 4$

At $x = 2$, $\frac{dy}{dx}$ or slope $= 8$

Second

Maximum or Minimum Value

Suppose y is a function of x. Or $y = f(x)$.

Then we can draw a graph between x and y. Let the graph is as shown in figure.

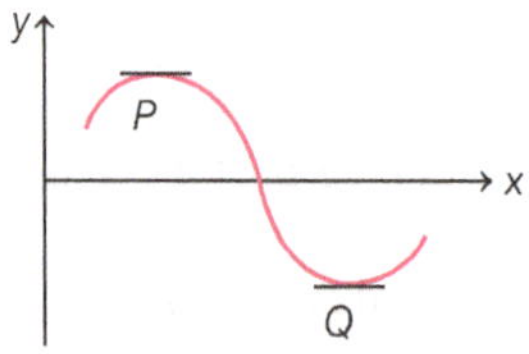

Then from the graph we can see that at maximum or minimum value of y slope $\left(\text{or } \frac{dy}{dx}\right)$ to the graph is zero.

Thus, $\frac{dy}{dx}=0$ at maximum or minimum value of y.

By putting $\frac{dy}{dx}=0$, we will get different values of x. At these values of x, value of y is maximum if $\frac{d^2y}{dx^2}$ (double differentiation of y with respect to x) is negative at this value of x. Similarly, y is minimum if $\frac{d^2y}{dx^2}$ is positive.

Thus, $\frac{d^2y}{dx^2}=-\text{ve}$ for maximum value of y

and $\frac{d^2y}{dx^2}=+\text{ve}$ for minimum value of y

Note *At constant value of y also* $\frac{dy}{dx}=0$ *but in this case* $\frac{d^2y}{dx^2}$ *is zero.*

Example 12 Find maximum or minimum values of y corresponding to following equations.

(a) $y=25x^2+5-10x$

(b) $y=9-(x-3)^2$

Sol. (a) For maximum and minimum value, we can put $\frac{dy}{dx}=0$.

or $\frac{dy}{dx}=50x-10=0$

$\therefore$ $x=\frac{1}{5}$

Further, $\frac{d^2y}{dx^2}=50$

or $\frac{d^2y}{dx^2}$ has positive value at $x=\frac{1}{5}$.

Therefore, y has minimum value at $x=\frac{1}{5}$.

Substituting $x=\frac{1}{5}$ in given equation, we get

$$y_{min}=25\left(\frac{1}{5}\right)^2+5-10\left(\frac{1}{5}\right)=4 \qquad \textbf{Ans.}$$

(b) $y=9-(x-3)^2=9-x^2-9+6x$

or $y=6x-x^2$ $\therefore$ $\frac{dy}{dx}=6-2x$

For minimum or maximum value of y, we will substitute $\frac{dy}{dx}=0$

or $6-2x=0$ or $x=3$

To check whether value of y is maximum or minimum at $x=3$, we will have to check whether $\frac{d^2y}{dx^2}$ is positive or negative.

$$\frac{d^2y}{dx^2}=-2$$

or $\frac{d^2y}{dx^2}$ is negative at $x=3$. Hence, value of y is maximum. This maximum value of y is,

$$y_{max}=9-(3-3)^2=9 \qquad \textbf{Ans.}$$

Note *$(x-3)^2$ is a square quantity. Minimum value of any square quantity is zero.*

$\therefore$ $y_{max}=9-[(x-3)^2]_{min}$

$=9-0=9$

Important POINTS

1. If $y=x^2$, then $\frac{dy}{dx}=2x \Rightarrow dy=(2x)\cdot dx$

   This implies that according to the equation, $y=x^2$, if we change the value of x by a small amount dx, then change in value of y is dy, which is equal to $(2x)\,dx$. Here x is the value at that point.

   For example,

   If $x=2$ and $dx=0.001$, then

   $x'=(x+dx)=2.001$

   Now, $y=x^2=(2)^2=4$

   $y'=x'^2=(2.001)^2=4.004001$

   $\therefore$ $dy=y'-y=0.004001$

   Now, $(2x)\,dx=(2\times 2)(0.001)=0.004$

   So, we can see that, $dy \approx (2x)\,dx$

2. $\frac{dy}{dx}$ means rate of change of y with respect to x. Or, change in the value of y per unit change in x. This is the reason why differentiation of any constant quantity is always zero. Because the constant quantity is not changing.

Miscellaneous Examples

Ex. 1 If $y = \cos^2 x$, then find $\frac{dy}{dx}$.

Sol. We know that $\cos 2x = 2\cos^2 x - 1$

$$\Rightarrow \quad \cos^2 x = \frac{1+\cos 2x}{2}$$

Differentiating both sides w.r.t. x, we get

$$\frac{dy}{dx} = \frac{d}{dx}[\cos^2 x] = \frac{d}{dx}\left[\frac{1+\cos 2x}{2}\right]$$

$$= \frac{d}{dx}\left(\frac{1}{2}\right) + \frac{1}{2}\frac{d}{dx}(\cos 2x)$$

$$= 0 + \frac{1}{2}(-\sin 2x)\,2 = -\sin 2x$$

Alternate method :

$$y = \cos^2 x = (\cos x)^2$$

$$\Rightarrow \quad \frac{dy}{dx} = \frac{d\,[(\cos x)^2]}{dx} = 2(\cos x)^{2-1}(-\sin x)$$

$$= -2\cos x \sin x = -\sin 2x$$

Ex. 2 If $y = \frac{\sin x}{x + \cos x}$, then find $\frac{dy}{dx}$.

Sol.
$$\frac{dy}{dx} = \frac{(x+\cos x)\frac{d(\sin x)}{dx} - \sin x\frac{d(x+\cos x)}{dx}}{(x+\cos x)^2}$$

$$= \frac{(x+\cos x)\cos x - \sin x\,(1-\sin x)}{(x+\cos x)^2}$$

$$= \frac{x\cos x + \cos^2 x - \sin x + \sin^2 x}{(x+\cos x)^2}$$

$$= \frac{x\cos x - \sin x + \sin^2 x + \cos^2 x}{(x+\cos x)^2} = \frac{x\cos x - \sin x + 1}{(x+\cos x)^2}$$

Ex. 3 If $y = \left[\frac{x^2+1}{x+1}\right]$, then find $\frac{dy}{dx}$.

Sol.
$$\frac{dy}{dx} = \frac{(x+1)\frac{d(x^2+1)}{dx} - (x^2+1)\frac{d(x+1)}{dx}}{(x+1)^2}$$

$$= \frac{(x+1)\,2x - (x^2+1)1}{(x+1)^2} = \frac{2x^2+2x-x^2-1}{(x+1)^2}$$

$$= \frac{x^2+2x-1}{(x+1)^2}$$

Ex. 4 If $x = at^4, y = bt^3$, then find $\frac{dy}{dx}$.

Sol. Differentiating x and y with respect to t, we get

$$\frac{dx}{dt} = 4at^3, \frac{dy}{dt} = 3bt^2$$

Dividing $\frac{dy}{dt}$ by $\frac{dx}{dt}$, we get $\frac{dy}{dx} = \frac{3bt^2}{4at^3} = \frac{3b}{4at}$

Ex. 5 If $f(x) = x\cos x$, find $f''(x)$.

Sol. Using the product rule, we have

$$f'(x) = x\frac{d}{dx}(\cos x) + \cos x\frac{d}{dx}(x)$$

$$= -x\sin x + \cos x$$

To find $f''(x)$, we further differentiate $f'(x)$:

$$f''(x) = \frac{d}{dx}(-x\sin x + \cos x)$$

$$= -x\frac{d}{dx}(\sin x) + \sin x\frac{d}{dx}(-x) + \frac{d}{dx}(\cos x)$$

$$= -x\cos x - \sin x - \sin x = -x\cos x - 2\sin x$$

Ex. 6 Find the minimum value of $y = 5x^2 - 2x + 1$.

Sol. For maximum or minimum value, $\frac{dy}{dx} = 0$

$$\Rightarrow \quad 5(2x) - 2(1) + 0 = 0 \Rightarrow x = \frac{1}{5}$$

Now at $x = \frac{1}{5}, \frac{d^2y}{dx^2} = 10$ which is positive so value of y is minimum at $x = \frac{1}{5}$.

Therefore, $y_{min} = 5\left(\frac{1}{5}\right)^2 - 2\left(\frac{1}{5}\right) + 1 = \frac{4}{5}$ **Ans.**

Ex. 7 Divide a number 100 into two parts such that their product is maximum.

Sol. Let the two parts be x and $(100 - x)$

Product, $P = x(100 - x) = 100x - x^2$

For P to be maximum or minimum,

$$\frac{dP}{dx} = 100 - 2x = 0 \Rightarrow x = 50$$

$$\frac{d^2P}{dx^2} = -2$$

i.e. P is maximum at $x = 50$.

Dividing equally the two parts are (50, 50). **Ans.**

Important Result : If we have to divide a number in two parts, so that their product is maximum then divide them in two equal parts. Let us divide a number 10.

(i) In 9 and 1, product is 9.

(ii) In 8 and 2, product is 16

(iii) In 7 and 3, product is 21

(iv) In 6 and 4, product is 24

(v) In 5 and 5, product is 25 and this is maximum.

MODULE 4 Exercise

1. Differentiate the following with respect to x.

$$\sqrt{x} - \frac{1}{\sqrt{x}}$$

2. Differentiate the following with respect to x.

$$(5x^2 + 6)(2x^2 + 4)$$

3. Differentiate the following with respect to x.

$$\sqrt{x}(x^2 + 7)$$

4. Differentiate the following with respect to θ.

$$\frac{\cos\theta}{(1 - \sin\theta)}$$

5. Differentiate the following functions with respect to x.

(a) $x^4 + 3x^2 - 2x$ (b) $x^2 \cos x$

(c) $(6x + 7)^4$ (d) $e^x x^5$

(e) $\frac{(1 + x)}{e^x}$

6. Find the maximum or minimum value of y in the functions given below.

(a) $y = 5 - (x - 1)^2$

(b) $y = (\sin 2x - x)$, where $-\frac{\pi}{2} \le x \le \frac{\pi}{2}$

Answers

1. $\frac{1}{2\sqrt{x}} + \frac{1}{2x^{\frac{3}{2}}} = \frac{1}{2\sqrt{x}}\left(1 + \frac{1}{x}\right)$ 2. $40x^3 + 64x = 8x(5x^2 + 8)$ 3. $\frac{1}{2\sqrt{x}}(7 + 5x^2)$

4. $\frac{1}{1 - \sin\theta}$ 5. (a) $4x^3 + 6x - 2$ (b) $2x\cos x - x^2 \sin x$ (c) $24(6x + 7)^3$ (d) $5e^x x^4 + e^x x^5$ (e) $-xe^{-x}$

6. (a) $y_{max} = 5$ at $x = 1$ (b) $y_{min} = \left(\frac{\pi}{6} - \frac{\sqrt{3}}{2}\right)$ at $x = -\pi/6$ and $y_{max} = \left(\frac{\sqrt{3}}{2} - \frac{\pi}{6}\right)$ at $x = \pi/6$

See hints on pages 41 and 42.

MODULE 5

Integration

General Points

(i) As far as Physics is concerned we only need simple formulae in integration and their applications. In Mathematics you will study integration in detail. So, in Physics you need not to go in detail of **integration**.

(ii) In simple words, integration is the reverse process of differentiation. Differentiation is breaking of something into parts. Integration is summing it up.

(iii) Symbol of integration is $\int f(x)\cdot dx$ or $\int f(t)\, dt$.

(iv) As we said above that integration is reverse process of differentiation. So, let us take an example.

$$\frac{d}{dx}(\sin x) = \cos x \quad \text{...(i)}$$

So, $$\int (\cos x)\cdot dx = \sin x \quad \text{...(ii)}$$

or $$\int (\cos x)\, dx = \sin x + c \quad \text{...(iii)}$$

Here, c is a constant, called constant of integration.

The value of this constant depends upon different conditions given in the problem. For one set of conditions its value may be $\frac{1}{2}$. For another set of conditions it may be $\frac{1}{4}$. So, in general we write it only c.

Now if we differentiate right hand side of equation (ii) or (iii), result is $\cos x$ in both cases.

So, the general solution of $\int (\cos x)\, dx$ is $(\sin x + c)$ not only $\sin x$.

List of Simple Formulas

(i) $\int x^n dx = \frac{x^{n+1}}{n+1} + c \quad (n \neq -1)$

(ii) $\int \frac{dx}{x} = \log_e x + c$ or $\ln x + c$

(iii) $\int \sin x\, dx = -\cos x + c$

(iv) $\int \cos x\, dx = \sin x + c$

(v) $\int e^x dx = e^x + c$

(vi) $\int \sec^2 x\, dx = \tan x + c$

(vii) $\int \text{cosec}^2 x\, dx = -\cot x + c$

(viii) $\int \sec x \tan x\, dx = \sec x + c$

(ix) $\int \text{cosec}\, x \cot x\, dx = -\text{cosec}\, x + c$

Note (i) $\int dx = \int x^0 dx = \frac{x^1}{1} + c = x + c$ or $\int dx = x + c$

(ii) $\int x \cdot dx = \frac{x^{1+1}}{1+1} + c = \frac{x^2}{2} + c$

(iii) $\int x^2 dx = \frac{x^{2+1}}{2+1} + c = \frac{x^3}{3} + c$

Example 1 Integrate the following functions with respect to x.

(a) $\int (5x^2 + 3x - 2)\, dx$ (b) $\int \left(4\sin x - \frac{2}{x}\right) dx$

Sol. (a) $\int (5x^2 + 3x - 2)\, dx = 5\int x^2\, dx + 3\int x dx$

$$-2\int dx = \frac{5x^3}{3} + \frac{3x^2}{2} - 2x + c$$

(b) $\int \left(4\sin x - \frac{2}{x}\right) dx = 4\int \sin x\, dx - 2\int \frac{dx}{x}$

$$= -4\cos x - 2\ln x + c$$

Example 2 Find the value of $\int (2\cos x + 6x^2)\, dx$.

Sol. $\int (2\cos x + 6x^2)\, dx = 2\int (\cos x) dx + 6\int (x^2)\, dx$

$$= 2[\sin x] + 6\left(\frac{x^3}{3}\right) + c = 2\sin x + 2x^3 + c$$

Special Type of Function in Integration

If we have the following types of functions in integration,

(i) $\int (6x-2)^2 dx$

(ii) $\int \sin(2x+4)\cdot dx$

(iii) $\int \left(\frac{1}{-6x-2}\right) dx$

(iv) $\int e^{(5x-2)} dx$

Then we will divide by the coefficient of x and then integrate as usual. In differentiation we have multiplied by the coefficient of x. For example result of the above integrations are as under.

(i) $\int (6x-2)^2\,dx = \left(\frac{1}{6}\right)\frac{(6x-2)^3}{3} + c$

$= \frac{(6x-2)^3}{18} + c$

(ii) $\int \sin(2x+4)\,dx = -\left(\frac{1}{2}\right)\cos(2x+4) + c$

(iii) $\int \left(\frac{1}{-6x-2}\right)dx = \left(\frac{1}{-6}\right)\log_e(-6x-2) + c$

$= \frac{\ln(-6x-2)}{-6} + c$

(iv) $\int e^{(5x-2)}\,dx = \left(\frac{1}{5}\right)(e^{5x-2}) + c$

Definite Integral

When a function is integrated between a lower limit and an upper limit, it is called a definite integral.

If $\frac{d}{dx}f(x) = f'(x)$, then $\int f'(x)dx$ is called indefinite integral and $\int_a^b f'(x)dx$ is called definite integral.

Here, a and b are called lower and upper limits of the variable x.

After carrying out integration, the result is evaluated between upper and lower limits as explained below :

$$\int_a^b f'(x)dx = |f(x)|_a^b = f(b) - f(a)$$

Note *In indefinite integral we get a function with constant of integration. In definite integration we get a value without constant of integration.*

Example 3 Find $I = \int_0^{\pi/4}(\sin x + \cos x)\,dx$.

Sol. $I = \int_0^{\pi/4}(\sin x + \cos x)\,dx = |-\cos x + \sin x|_0^{\pi/4}$

$= \left\{-\cos\left(\frac{\pi}{4}\right) + \sin\left(\frac{\pi}{4}\right)\right\} - \{-\cos(0) + \sin(0)\}$

$= \left\{-\frac{1}{\sqrt{2}} + \frac{1}{\sqrt{2}}\right\} - \{-1+0\} = 1$ **Ans.**

Example 4 The integral $\int_1^5 x^2 dx$ is equal to

(a) $\frac{125}{3}$ (b) $\frac{124}{3}$ (c) $\frac{1}{3}$ (d) 45

Sol. $\int_1^5 x^2\,dx = \left[\frac{x^3}{3}\right]_1^5$

$= \left[\frac{5^3}{3} - \frac{1^3}{3}\right] = \frac{125}{3} - \frac{1}{3}$

$= \frac{124}{3}$

$\therefore$ The correct option is (b).

Area Under a Curve and Definite Integration

Suppose we have a function $y = f(x)$ and the graph between x and y corresponding to the above function is as shown in figure.

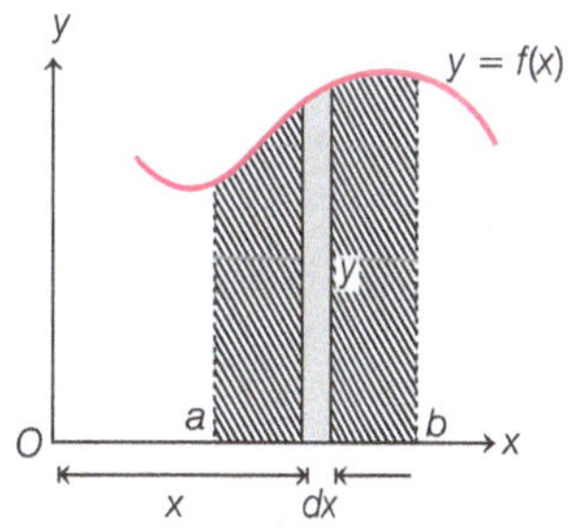

Now, we wish to find area of the graph between $x = a$ and $x = b$, then let us take a small element (or strip) at a general point x of thickness dx. At coordinate x, value of y-coordinate is y which will be given by the function.

Now area of small shown element $= ydx = f(x)dx$

If we sum up all areas between $x = a$ and $x = b$, then $\int_a^b f(x)dx$ = shaded area between curve and x-axis.

Example 5 Given $y = 3x^2$. Find area of the graph between $x = 1$ and $x = 2$.

Sol.

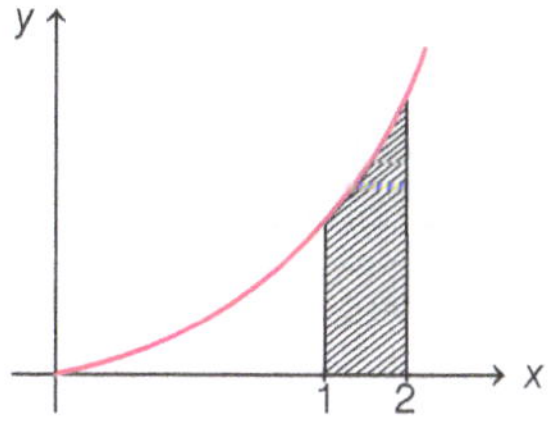

Graph corresponding to the given equation $y = 3x^2$ is as shown in figure. Area of graph between $x = 1$ and $x = 2$ will be :

$A = \int_1^2 f(x)\,dx = \int_1^2 y\,dx = \int_1^2 (3x^2)\,dx$

$= \left[(3)\left(\frac{x^3}{3}\right)\right]_1^2 = [x^3]_1^2$

$= [(2)^3 - (1)^3] = (8-1)$

$= 7$ units

Miscellaneous Examples

Ex. 1 Solve the integral $I = \int_0^{\pi} \sin^2 x dx$.

Sol. $\int_0^{\pi} \sin^2 x \, dx = \int_0^{\pi} \left(\frac{1 - \cos 2x}{2} \right) dx \qquad \left(\because \sin^2 x = \frac{1 - \cos 2x}{2} \right)$

$$= \frac{1}{2}\left[\int_0^{\pi} dx - \int_0^{\pi} (\cos 2x) \, dx \right] = \frac{1}{2}\left[x - \frac{\sin 2x}{2} \right]_0^{\pi}$$

$$= \frac{1}{2}\left[(\pi - 0) - \frac{(\sin 2\pi - \sin 0)}{2} \right] = \frac{\pi}{2}$$

Ex. 2 Solve

(a) $\int_0^3 (ax^2 + bx + c) \, dx$ (b) $\int_{-1}^{1} e^x \, dx$

(c) $\int_{-\pi/2}^{\pi/2} \cos x \, dx$

Sol. (a) $\int_0^3 (ax^2 + bx + c) \, dx = \left(\frac{ax^3}{3} + \frac{bx^2}{2} + cx \right)_0^3$

$$= \frac{a(3)^3}{3} + \frac{b(3)^2}{2} + c(3) - 0 = 9a + \frac{9}{2}b + 3c$$

(b) $\int_{-1}^{1} e^x dx = e^x|_{-1}^{1} = e^1 - e^{-1} = e - \frac{1}{e}$

(c) $\int_{-\pi/2}^{\pi/2} \cos x \, dx = \sin x|_{-\pi/2}^{\pi/2} = \sin\frac{\pi}{2} - \sin\left(\frac{-\pi}{2}\right) = 1 + 1 = 2$

Ex. 3 Integrate the following with respect to x.

(a) x^3 (b) $x - \frac{1}{x}$ (c) $e^{2x} + \frac{1}{x^2}$ (d) $\frac{1}{2x+3}$

(e) $\cos(4x+3)$ (f) $\cos^2 x$

Sol. (a) $\int x^3 \, dx = \frac{x^4}{4} + c \qquad \left[\because \int x^n dx = \frac{x^{n+1}}{n+1} + c \right]$

(b) $\int \left(x - \frac{1}{x} \right) dx \qquad \left[\because \int \frac{1}{x} dx = \ln x + c \right]$

$$= \frac{x^2}{2} - \ln x + c$$

(c) $\int \left(e^{2x} + \frac{1}{x^2} \right) dx$

$$= \frac{e^{2x}}{2} + \frac{x^{-2+1}}{-2+1} + c \qquad \left[\because \int e^{ax} = \frac{e^{ax}}{a} + c \right]$$

$$= \frac{e^{2x}}{2} - \frac{1}{x} + c$$

(d) $\int \frac{dx}{2x+3} = \frac{\ln(2x+3)}{2} + c$

(e) $\int \cos(4x+3) \, dx = \frac{\sin(4x+3)}{4} + c$

(f) $\int \cos^2 x \, dx = \int \left(\frac{1 + \cos 2x}{2} \right) dx$

$$= \int \frac{1}{2} dx + \int \frac{\cos 2x}{2} dx$$

$$= \frac{x}{2} + \frac{\sin 2x}{2 \times 2} + c = \frac{1}{2}\left(x + \frac{\sin 2x}{2} \right) + c$$

MODULE 5 Exercise

1. Integrate the following functions with respect to t

   (a) $\int (3t^2 - 2t) \, dt$

   (b) $\int (4\cos t + t^2) \, dt$

   (c) $\int (2t - 4)^{-4} dt$

   (d) $\int \frac{dt}{(6t - 1)}$

2. Integrate the following functions

   (a) $\int_0^2 2t \, dt$ (b) $\int_{\pi/6}^{\pi/3} \sin x \, dx$

   (c) $\int_4^{10} \frac{dx}{x}$ (d) $\int_0^{\pi} \cos x dx$

   (e) $\int_1^2 (2t - 4) \, dt$

3. Calculate the area enclosed under the curve $f(x) = x^2$ between the limits $x = 2$ and $x = 3$.

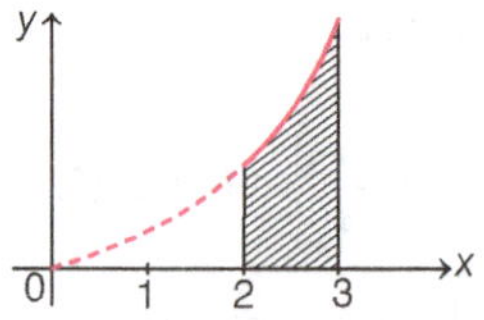

Answers

1. (a) $t^3 - t^2 + c$ (b) $4\sin t + \frac{t^3}{3} + c$ (c) $-\frac{1}{6(2t-4)^3} + c$

   (d) $\frac{\ln(6t-1)}{6} + c$

2. (a) 4 (b) $\frac{\sqrt{3}-1}{2}$ (c) ln(2.5) (d) 0 (e) –1

3. $\frac{19}{3}$

See hints on page 42.

MODULE 6

Graphs

General Points

(i) A graph is always drawn between two quantities, say x and y.

(ii) The graph may be straight line, parabola, ellipse, circle etc.

(iii) The shape of the graph, (whether it is straight line, circle or parabola etc.) depends upon the equation between x and y.

(iv) In differentiation we have already discussed about the slope of a graph (denoted by m).

(v) **x and y-intercepts** The point where the graph cuts the x-axis is called x-intercept C_x and the point where the graph cuts the y-axis is called y-intercept C_y. On x-axis value of y at every point is zero. Therefore, to find x-intercept C_x put $y = 0$ in the given equation (between x and y) and find the value of x. This value is C_x. Similarly to find y-intercept C_y put $x = 0$ in the given equation and find the value of y. This value of y is called the y-intercept C_y. In case of straight line graphs, y-intercept is very important and there it is also represented by C (or c) only.

(iv) In the given equation if $y = 0$ by putting $x = 0$ (or *vice-versa*), then the graph passes through origin.

(vii) In the given equation if y is positive for positive values of x, then graph is in 1st quadrant. Similarly if y is negative for positive values of x, then the graph lies in 4th quadrant.

Example 1 For the given straight line equation, $x - 2y + 4 = 0$, find x and y-intercepts and then draw this straight line graph.

Sol. ***x*-intercept**

Putting $y = 0$ in the given equation, we get $x = -4$.

Therefore x-intercept is, $C_x = -4$

***y*-intercept**

Putting $x = 0$ in the given equation, we get $y = 2$.

Therefore y-intercept is,

$$C_y = 2$$

Graph We have obtained x and y-intercepts. So, the straight line will pass through these two points. The straight line is as shown below.

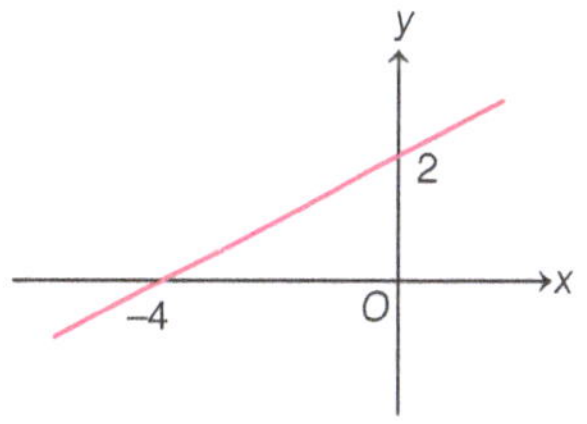

Straight Line Graph

(i) **Linear Equation** A linear equation is an algebraic equation between one or two variables where each variable has a power of 1. Following are some of the examples of linear equations.

$$2y - 3 = 0$$
$$x = 7$$
$$2x - 3y + 8 = 0$$

(ii) Any linear equation always represents a straight line graph.

(iii) A straight line graph may be of following four types.

Type 1.

$$x = \pm c$$

Here c is a constant.

This equation represents a straight line graph parallel to y-axis.

Example 2 Plot a graph corresponding to the equation, $x - 6 = 0$.

Sol. $x - 6 = 0$

$\Rightarrow \quad x = 6$

Therefore value of x at every point on the straight line is 6. The graph is as shown below.

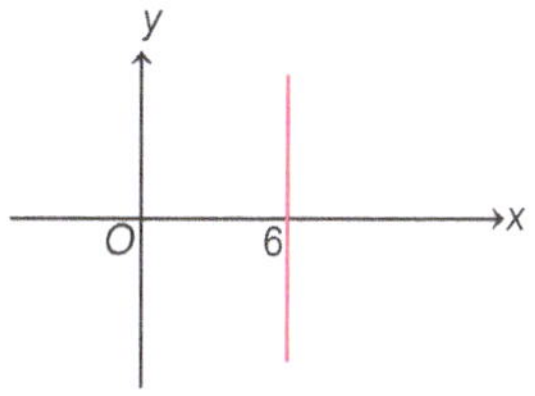

Type 2.

$$\boxed{y = \pm c}$$

Here c is a constant.

This equation represents a straight line graph parallel to x-axis.

Example 3 Plot a graph corresponding to the equation, $y + 2 = 0$.

Sol. $y + 2 = 0 \Rightarrow y = -2$

Therefore value of y at every point on the straight line is -2. The graph is as shown below.

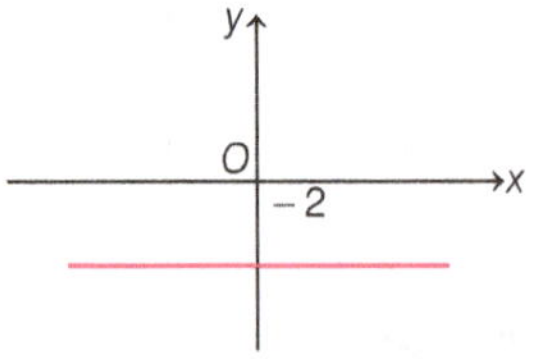

Type 3.

$$\boxed{y = mx}$$

$y = mx$, represents a straight line passing through origin. Here, $m = \tan\theta$, is also called slope of the line, where θ is the angle which the line makes with positive x-axis, when drawn in anti-clockwise direction from the positive x-axis towards the line.

The two possible cases are shown in the figures given below.

In figure (i), $\theta < 90°$. Therefore, $\tan\theta$ or slope of line is positive. In Fig. (ii), $90° < \theta < 180°$. Therefore, $\tan\theta$ or slope of line is negative.

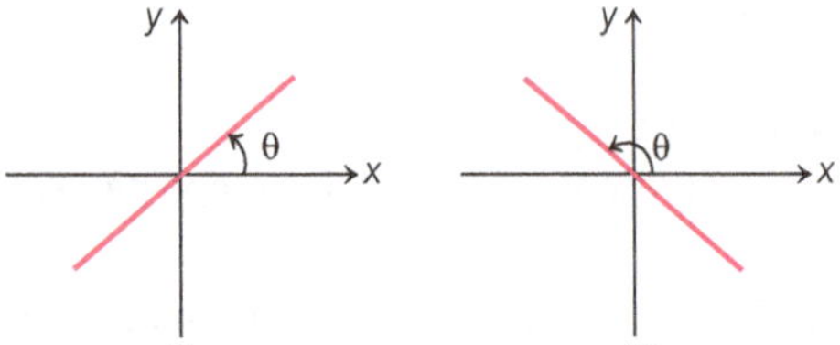

(i) (ii)

Note *That $y = mx$ or $y \propto x$ also means that value of y becomes 2 times, if x is doubled. Or it will ramain $\frac{1}{4}$ times if x becomes $\frac{1}{4}$ times.*

Example 4 Plot straight line graphs corresponding to following equations

(a) $y - 2x = 0$ (b) $y + 4x = 0$

Sol. (a) $y - 2x = 0$

$\Rightarrow \quad y = 2x$

Comparing this with the standard equation, $y = mx$, we get

$$m = \text{slope} = \tan\theta = 2 \text{ or } \theta = \tan^{-1}(2)$$

Slope is positive, so the angle θ is acute ($< 90°$).

The straight line graph is as shown below.

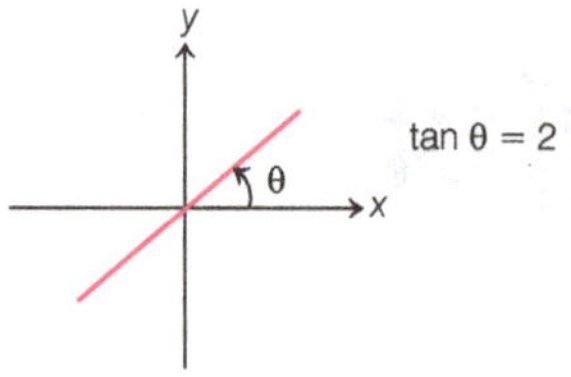

(b) $y + 4x = 0 \Rightarrow y = -4x$

Here, $m = \text{slope} = \tan\theta = -4$

So, $\theta = \tan^{-1}(-4)$

Slope is negative. So the angle θ is obtuse (between $90°$ and $180°$). The straight line graph is as shown below.

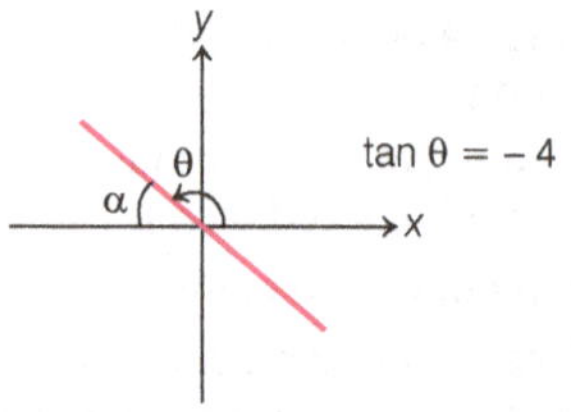

Note

$$\alpha + \theta = 180°$$

$$\Rightarrow \quad \tan\alpha = -\tan\theta = -(-4) = +4$$

Type 4.

$$\boxed{y = mx + c} \quad \text{(General equation of straight line)}$$

$y = mx + c$, represents a straight line not passing through origin. Here, m is the slope of line as discussed above and c the intercept on y-axis (or C_y).

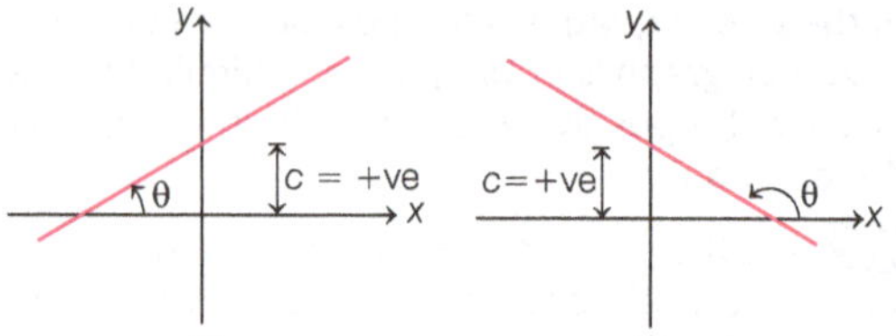

(i) (ii)

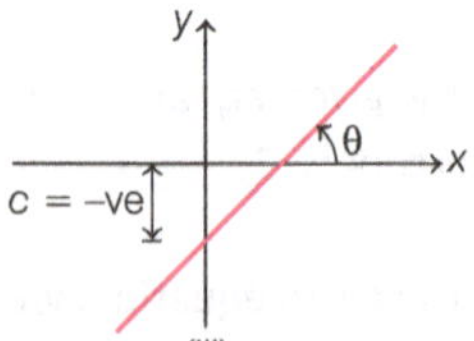

(iii)

In figure (i) : Slope and intercept both are positive.

In figure (ii) : Slope is negative but intercept is positive and

In figure (iii) : Slope is positive but intercept is negative.

Note *(i) That in $y = mx + c$, y does not become two times, if x is doubled.*

(ii) $\frac{x}{a} + \frac{y}{b} = 1$ is equation of a straight line in the intercept form, where a and b are intercepts on x and y axis, respectively.

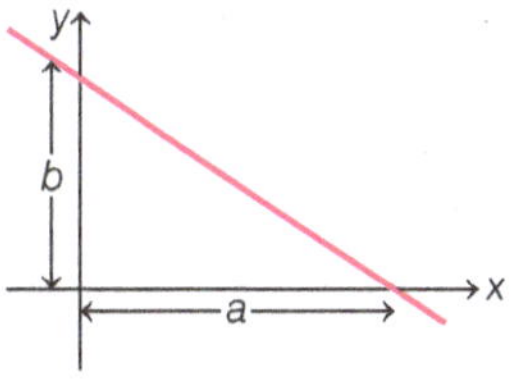

This can also be written as, $\frac{x}{c_x} + \frac{y}{c_y} = 1$

Example 5 Draw straight lines corresponding to following equations.

(a) $y = 4x + 2$

(b) $y = 6x - 4$

Sol. (a) In $y = 4x + 2$, slope is + 4 and y- intercept is 2. The graph is as shown below.

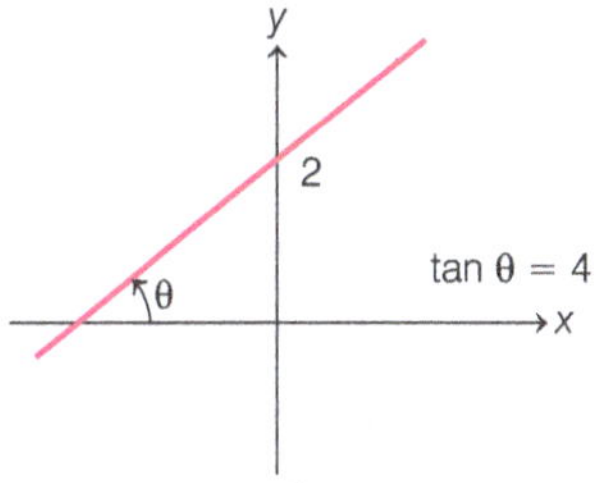

(b) In $y = 6x - 4$, slope is + 6 and y- intercept is - 4. Hence, the graph is as shown below.

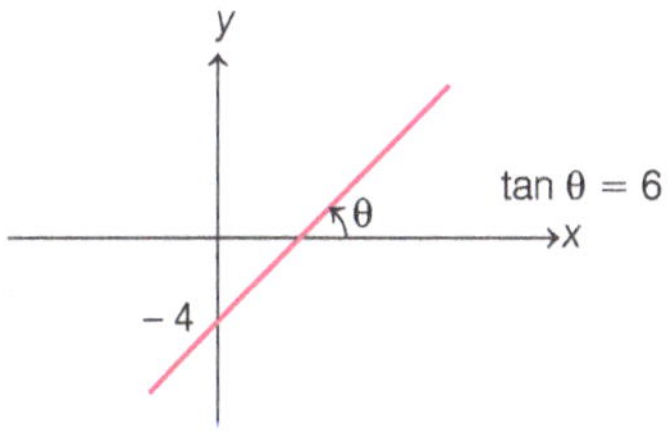

Example 6 Plot the line $2x - 3y = 12$.

Sol. Mothod 1 Given $3y = 2x - 12$

$$\Rightarrow \qquad y = \frac{2}{3}x - 4$$

Comparing $y = mx + c$, we get $m = \tan\theta = \frac{2}{3}$ and $c = -4$.

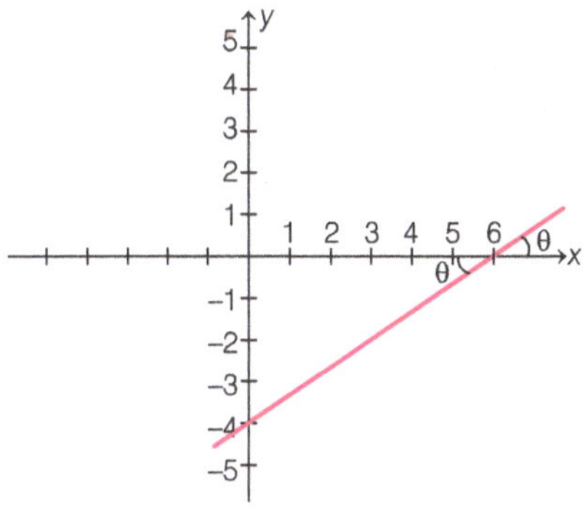

Method 2 Using $\frac{x}{a} + \frac{y}{b} = 1$

$$\Rightarrow \qquad \frac{2x}{12} - \frac{3y}{12} = 1 \Rightarrow \frac{x}{6} + \frac{y}{-4} = 1$$

a or $c_x = 6$ and b or $c_y = -4$

Example 7 Plot the line $-3x - 5y = 15$.

Sol. Method 1 Given $5y = -3x - 15$

When $x = 0$, $y = -3 \Rightarrow c_y = -3$

When $y = 0$, $x = -5 \Rightarrow c_x = -5$

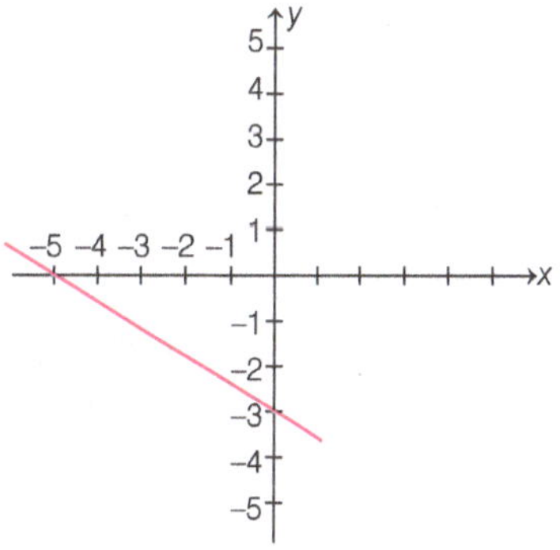

Method 2 Using $\frac{x}{a} + \frac{y}{b} = 1$;

$$\frac{-3x}{15} - \frac{5y}{15} = 1 \Rightarrow \frac{x}{(-5)} + \frac{y}{(-3)} = 1$$

$\therefore a$ or $c_x = -5$ and b or $c_y = -3$

Exponentially Decreasing and Increasing Graphs

(i) $y = Ae^{-Kx}$, represents exponentially decreasing graph. the value of y decreases exponentially from A to 0. The graph is as shown in figure.

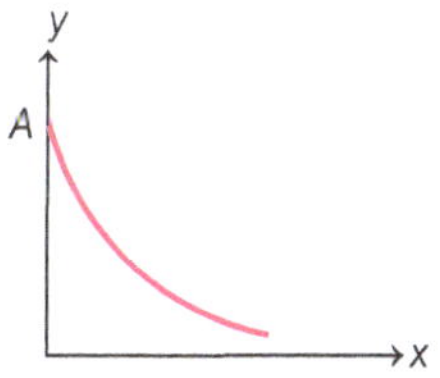

From the graph and the equation, we can see that $y = A$ at $x = 0$ and $y \to 0$ as $x \to \infty$.

(ii) $y = A(1 - e^{-Kx})$, represents an exponentially increasing graph. Value of y increases exponentially from 0 to A. The graph is as shown in figure.

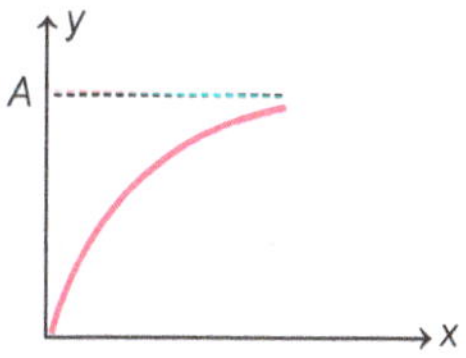

From the graph and the equation, we can see that $y = 0$ at $x = 0$ and $y \to A$ as $x \to \infty$.

Note *In the above equations y and x are variables and A and K are constants. Values of these constants are different in different situations (or problems). Following two examples will illustrate the concept.*

Example 8 Velocity of a particle increases exponentially with time from 2 m/s to 6 m/s. Plot a rough v - t graph and write v - t equation corresponding to it. Check the equation with graph at $t = 0$ and $t = \infty$.

Sol. v-t graph is as shown in figure,

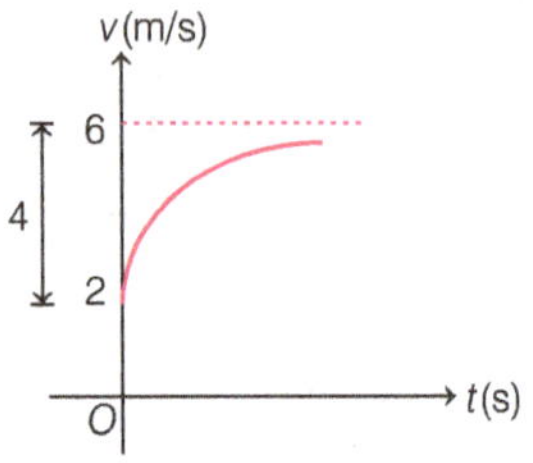

Above 2 m/s, the value 4 m/s increases exponentially. Hence, v-t equation is

$$v = 2 + 4(1 - e^{-kt})$$

Here k is a positive constant, which will depend on factors which are not given in the question.

At $t = 0$, $\quad v = 2 + 4(1 - e^{0})$

$= 2 \text{ m/s} \qquad (\text{as } e^{0} = 1)$

At $t = \infty$, $\quad v = 2 + 4(1 - e^{-\infty})$

$= 6 \text{ m/s} \qquad (\text{as } e^{-\infty} = 0)$

Example 9 Acceleration of a particle decreases exponentially with time from 10 m/s^2 to 6 m/s^2. Plot a rough a-t graph. Write a-t equation and check the equation with graph at $t = 0$ and $t = \infty$.

Sol. a-t graph is as shown in figure.

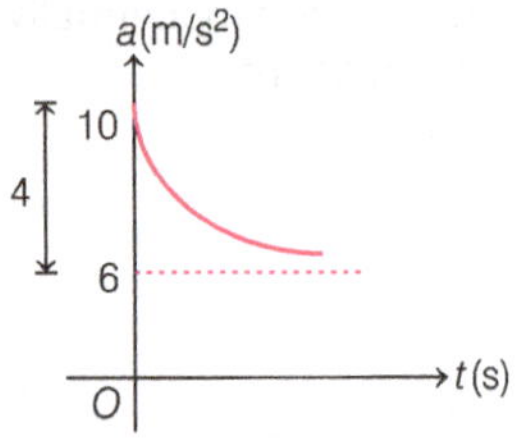

Above 6 m/s^2 the value 4 m/s^2 decreases exponentially. Hence, a - t equation is $a = 6 + 4e^{-kt}$

Here, k is a positive constant.

At $t = 0$, $\quad a = 6 + 4e^{0}$

$= 10 \text{ m/s}^2 \qquad (\text{as } e^{0} = 1)$

At $t = \infty$, $\quad a = 6 + 4e^{-\infty}$

$= 6 \text{ m/s}^2 \qquad (\text{as } e^{-\infty} = 0)$

Other Important Graphs Used in Physics

First

Rectangular Hyperbola

$$xy = \text{constant} \text{ or } y \propto \frac{1}{x} \text{ or } y = \frac{k}{x}$$

Here k is a constant.

For example $y = \frac{2}{x}$ represents a graph as shown in Fig (i).

Graph lies in 1st and 3rd quadrants. The reason is : if x is positive, then according to the equation, $y = \frac{2}{x}$, value of y is also positive. Means 1st quadrant. Similarly if x is negative, then y is also negative. Means 3rd quadrant.

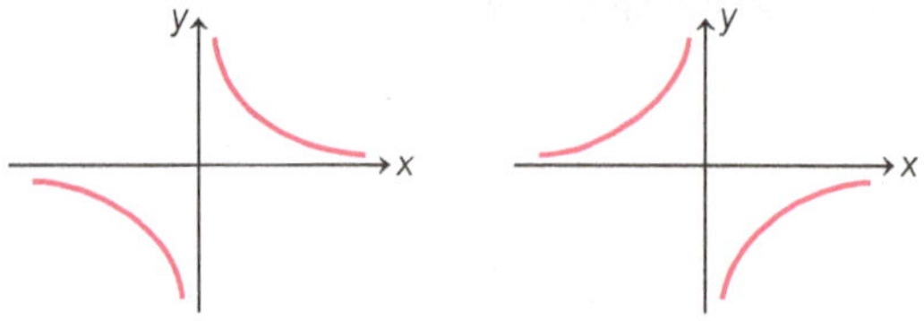

(i) (ii)

From the graph we can see that $y \to 0$ as $x \to \infty$ or $x \to 0$ as $y \to \infty$.

Similarly, $y = -\frac{4}{x}$ represents a similar type of graph in second and fourth quadrants as shown in Fig. (ii).

Note *That in the above case if x is doubled y will become half.*

Second

Parabola All quadratic equations represent parabolic graphs.

(i) $y \propto x^2$ or $y = 2x^2$, etc., represents a parabola passing through origin as shown in Fig. (i).

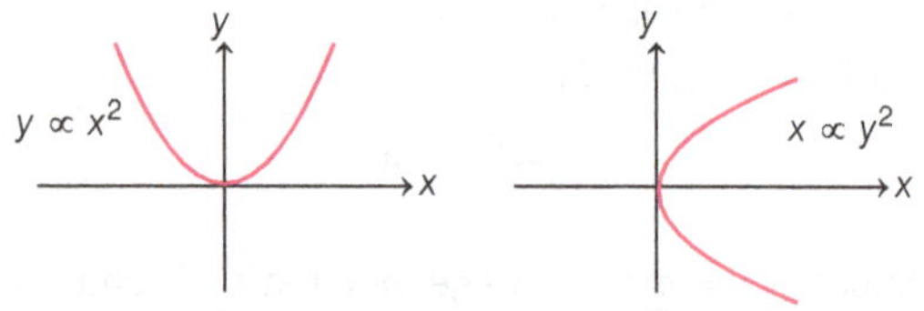

(i) (ii)

Note *That in the parabola $y = 2x^2$ or $y \propto x^2$, if x is doubled, y will become four times.*

Graph $x \propto y^2$ or $x = 4y^2$ is again a parabola passing through origin as shown in figure 1.7 (ii). In this case if y is doubled, x will become four times.

(ii) $y = x^2 + 4$ or $x = y^2 - 6$ will represent a parabola but not passing through origin. In the first equation ($y = x^2 + 4$), if x is doubled, y will not become four times.

Third

Circle

$$\boxed{x^2 + y^2 = a^2} \qquad (a = \text{constant})$$

This equation corresponds to a circle of radius 'a' and centre at origin as shown below.

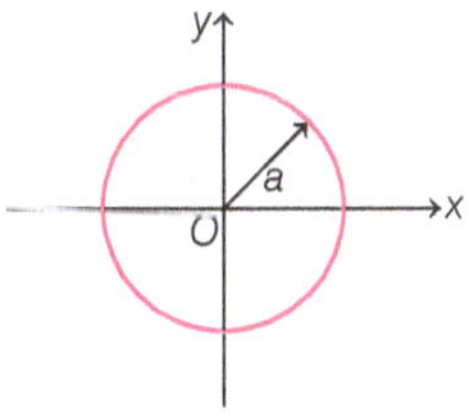

Fourth

Ellipse

$$\boxed{\frac{x^2}{a^2} + \frac{y^2}{b^2} = 1} \qquad (a \text{ and } b \text{ are constants})$$

The above equation represents an ellipse.

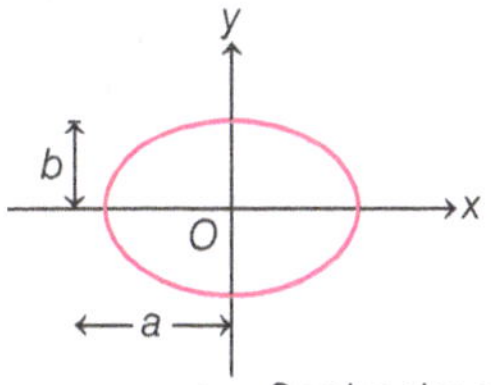

a = Semi major axis
b = Semi minor axis

Note *(i) Circle is a special case of an ellipse, for which semi minor axis = semi major axis*

or $b = a = \text{radius}$

(ii) We have given brief introduction of only those graphs which are used in Physics. In Mathematics you will study all these graphs in detail.

Example 10 Plot the graphs corresponding to the following equations.

(a) $y = 4x^2$

(b) $x = 4y^2$

(c) $y = \dfrac{4}{x}$

(d) $x^2 + y^2 = 16$

(e) $\dfrac{x^2}{16} + \dfrac{y^2}{9} = 1$

Sol. (a) $\mathbf{y = 4x^2}$ Graph is a parabola. In 1st quadrant graph is as shown below.

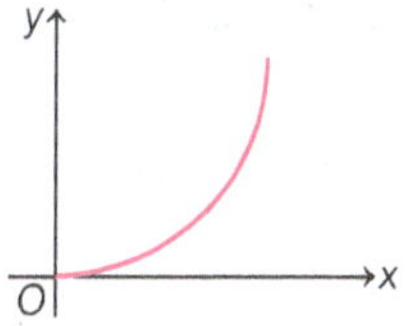

(b) $\mathbf{x = 4y^2}$ Graph is again a parabola. In 1st quadrant graph is as shown below.

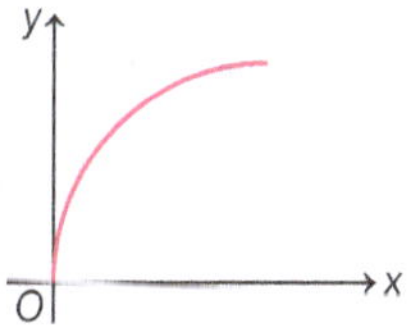

(c) $\mathbf{y = \dfrac{4}{x}}$ Graph is a rectangular hyperbola. In Ist quadrant it is as shown below.

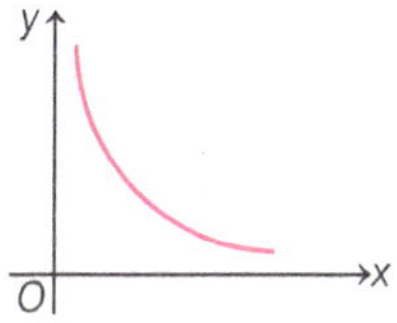

(d) $\mathbf{x^2 + y^2 = 16}$ The equation can be written as

$x^2 + y^2 = (4)^2$

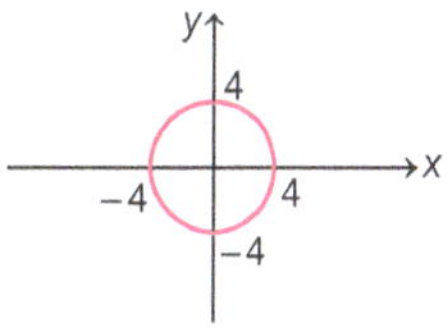

Hence, the graph is a circle with centre at origin and radius equal to 4.

(e) $\mathbf{\dfrac{x^2}{16} + \dfrac{y^2}{9} = 1}$ The given equation can be written as

$$\frac{x^2}{(4)^2} + \frac{y^2}{(3)^2} = 1$$

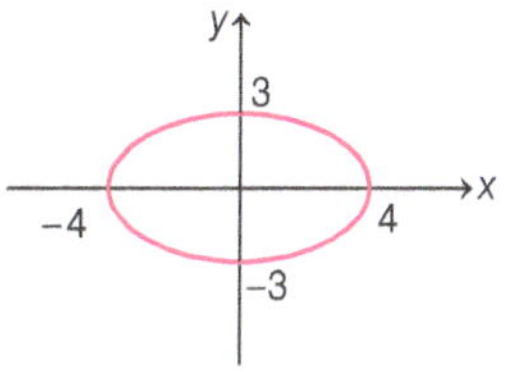

Hence, the graph is an ellipse with semi-major axis = 4, along x-axis and semi-minor axis = 3, along y-axis.

MODULE 6 Exercise

1. For the equations given below, tell the nature of graphs.
   (a) $y = 2x^2$ (b) $y = -4x^2 + 6$
   (c) $y = 6e^{-4x}$ (d) $y = 4(1 - e^{-2x})$
   (e) $y = \frac{4}{x}$ (f) $y = -\frac{2}{x}$
2. Which of the following graphs do not pass through origin?
   (a) $y = 2x + 6$ (b) $y^2 - 2x = 0$
   (c) $y^2 + 3x^2 + 5 = 0$ (d) $y + 6x = 0$
3. Check the quadrants for each part from where the following graphs will pass.
   (a) $y = -4x^2$ (b) $y = \frac{6}{x}$
4. After finding x-intercept C_x and y-intercept C_y, plot the straight line corresponding to the x-y equation given below.
   $$6x - 2y + 5 = 0$$
5. Draw rough graphs corresponding to the equations
   (a) $y = 4x$ (b) $y = -6x$
   (c) $y = x + 4$ (d) $y = -2x + 4$
   (e) $y = 2x - 4$ (f) $y = -4x - 6$
6. For the graphs given below, write down their x-y equations

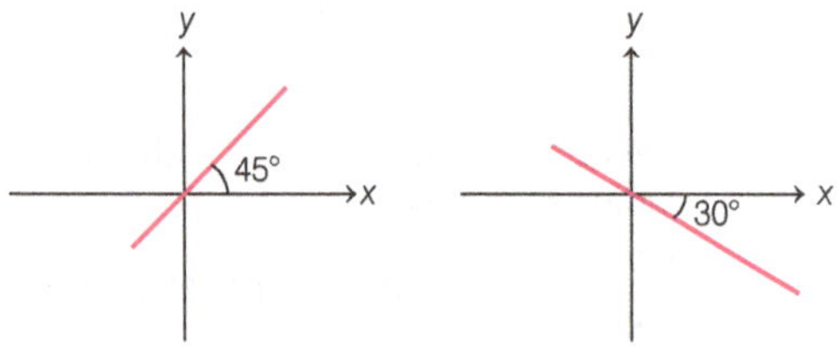

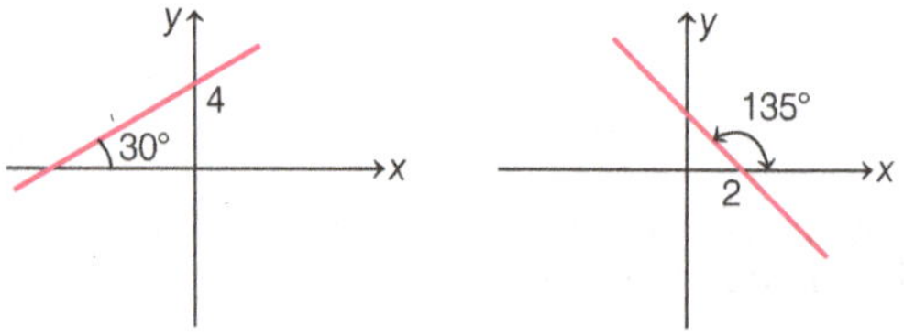

(c) (d)

7. Value of y decreases exponentially from $y = 10$ to $y = 6$. Plot a rough x-y graph.
8. Value of y increases exponentially from $y = -4$ to $y = +4$. Plot a rough x-y graph.
9. The graph shown in figure is exponential. Write down the equation corresponding to the graph.

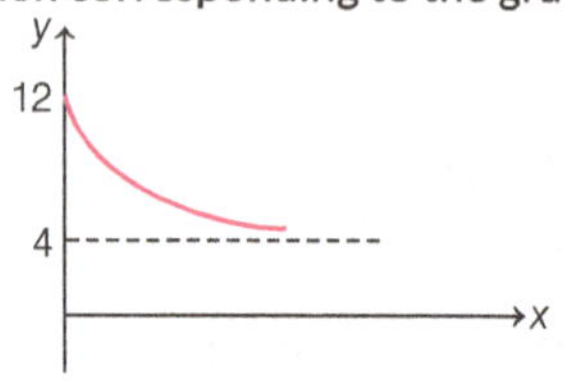

10. The graph shown in figure is exponential. Write down the equation corresponding to the graph.

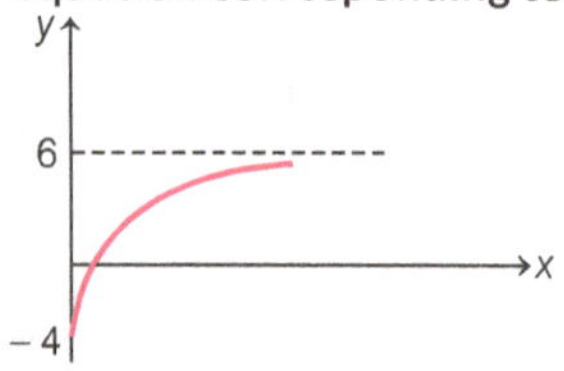

11. In the equation $y = 6 + 4(1 - e^{-kt})$, what is the maximum value of y?
12. In the equation $y = -4 - 6e^{-kt}$, what is the minimum value of y?

Answers

1. (a) Parabola passing through origin (b) Parabola not passing through origin (c) Exponentially decreasing graph
(d) Exponentially increasing graph (e) Rectangular hyperbola in first and third quadrant
(f) Rectangular hyperbola in second and fourth quadrant

2. (a) and (c)

3. (a) III and IV (b) III and I

4.

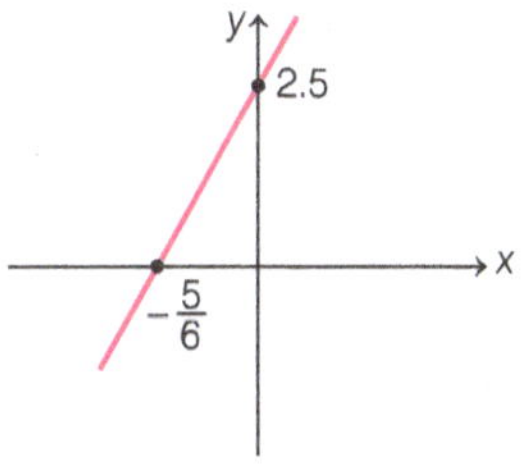

5.

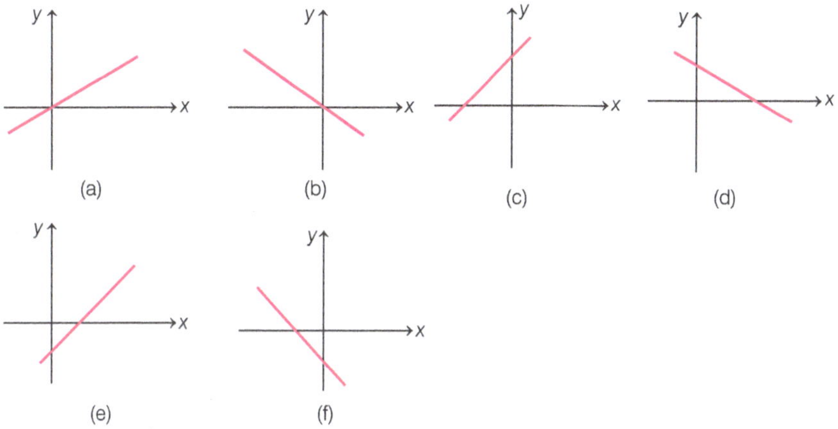

6. (a) $y = x$ (b) $y = -\dfrac{x}{\sqrt{3}}$ (c) $y = \dfrac{x}{\sqrt{3}} + 4$ (d) $y = -x + 2$

7.

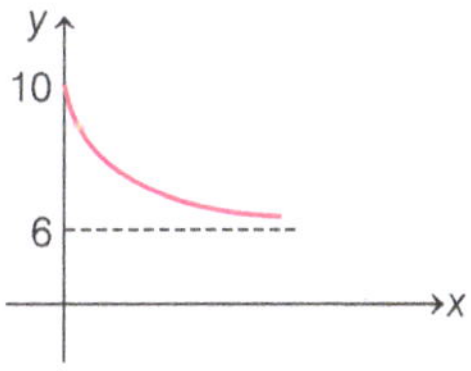

8.

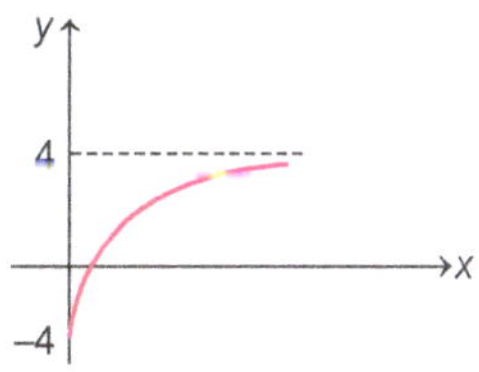

9. $y = 4 + 8e^{-Kx}$. Here, K is a positive constant

10. $y = -4 + 10(1 - e^{-Kx})$. Here, K is positive constant

11. 10

12. –10

See hints on page 42.

Hints & SOLUTIONS

Module 1.

1. We are given

$$g' = g\left(\frac{R}{R+h}\right)^2 = g\left(\frac{1}{1+\frac{h}{R}}\right)^2$$

$$= \frac{g}{\left(1+\frac{h}{R}\right)^2} = g\left(1+\frac{h}{R}\right)^{-2}$$

As we are given $h << R$, hence $\frac{h}{R} << 1$.

Using Binomial approximation,

we can write $g' \approx g\left(1-\frac{2h}{R}\right)$

2. We can write $1001 = 1000\left(1+\frac{1}{1000}\right)$

So that we have

$$(1001)^{1/3} = \left[1000\left(1+\frac{1}{1000}\right)\right]^{1/3} = 10\left[1+\frac{1}{1000}\right]^{1/3}$$

$$= 10(1+0.001)^{1/3} = 10\left(1+\frac{1}{3}\times 0.001\right)$$

$$= 10.003333$$

3. $5^3 = 125$

As we know,

$$a^b = c \Rightarrow \log_a c = b$$

Therefore,

$$\log_5 125 = 3$$

4. Given, $\log_{10} 1 = 0$

By the rule, we know,

$$\log_a c = b \Rightarrow a^b = c$$

Hence, $10^0 = 1$

5. Let, $\log_4 32 = x$

$$4^x = 32$$

$$(2^2)^x = 2\times 2\times 2\times 2\times 2$$

$$2^{2x} = 2^5 \Rightarrow 2x = 5$$

$$x = \frac{5}{2}$$

Therefore, $\log_4 32 = \frac{5}{2}$

6. Given, $\log_5 (x-7) = 1$

Using logarithm rules, we can write

$$5^1 = x-7$$

$$5 = x-7$$

$$x = 5+7$$

$$x = 12$$

7. $\log_a m = n$

$$a^n = m$$

$$\frac{a^n}{a} = \frac{m}{a}$$

$$a^{n-1} = \frac{m}{a}$$

Module 2.

1.

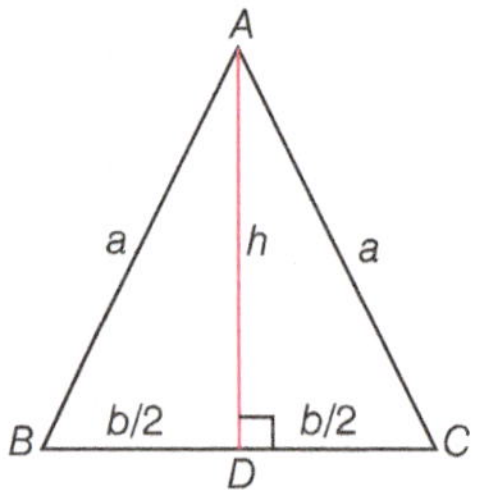

Given, $a = 3$ cm and $b = 4$ cm

The area of the isosceles triangle is given by

$$A = \frac{1}{2}\times b\times h = \frac{1}{2}\times b\times\sqrt{\left(a^2-\frac{b^2}{4}\right)}$$

$$= \frac{1}{2}\times 4\times\sqrt{3^2-\frac{4^2}{4}} = 2\times\sqrt{9-4}$$

$$= 2\sqrt{5}\ \text{cm}^2$$

Hence, the area of the given isosceles triangle is $2\sqrt{5}\ \text{cm}^2$.

2.

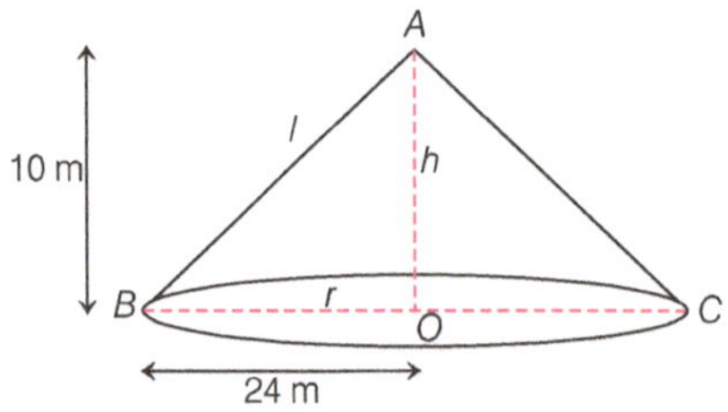

We know that volume of the right circular cone is $\frac{1}{3}\pi r^2 h$.

$\therefore \quad V = \frac{1}{3} \times \frac{22}{7} \times 24^2 \times 10 \Rightarrow V = 6034.28 \text{ m}^3$

Hence, the volume of the tent is 6034.28 m^3.

$$l = \sqrt{h^2 + r^2} = \sqrt{(10)^2 + (24)^2} = 26 \text{ m}$$

Curved surface area $= \pi r l$

$$= \left(\frac{22}{7}\right)(24)(26) = 1961.14 \text{m}^2$$

3. $C = 2\pi r$ (C = Circumference)

$$\Rightarrow \quad r = \frac{C}{2\pi}$$

Now, $A = \pi r^2$

$$= \pi\left(\frac{C}{2\pi}\right)^2 = \frac{C^2}{4\pi} = \frac{(10)^2}{4 \times \frac{22}{7}} = \frac{175}{22} \text{ cm}^2$$

4. (a) Let the given points be :

$A(2, -2) = (x_1, y_1)$

$B(-1, x) = (x_2, y_2)$

Distance between A and $B = \sqrt{[(x_2 - x_1)^2 + (y_2 - y_1)^2]}$

$$\therefore \quad \sqrt{[(-1-2)^2 + (x+2)^2]} = 5$$

Squaring on both sides, we get

$$(3)^2 + (x+2)^2 = 25$$
$$(x+2)^2 = 25 - 9$$
$$(x+2)^2 = 16$$
$$x + 2 = \pm 4$$

When $x + 2 = 4, x = 2$

When $x + 2 = -4, x = -6$

(b) Slope $= \frac{y_2 - y_1}{x_2 - x_1} = \frac{10 - 4}{6 - 2} = \frac{3}{2}$

5. $K = \frac{p^2}{2m}$ or $K \propto p^2$

$\therefore$ % change in K = 2 (% change in p)

$= 2\,(2\%)$

$= 4\%$

6. $K = \frac{p^2}{2m}$

$$p' = 0.8\,p$$

$$\therefore \quad K' = \frac{p'^2}{2m} = \frac{(0.8p)^2}{2m}$$

$$= (0.64)\frac{p^2}{2m}$$

$$= 0.64K$$

Hence, kinetic energy has been decreased by 36%.

7. $p = \sqrt{2Km}$ or $p \propto K^{\frac{1}{2}}$

$\therefore$ (% change in p) $= \frac{1}{2}$ (% change in K)

$= \frac{1}{2}(-3\%)$

$= -1.5\%$

8. $p = \sqrt{2Km}$ ...(i)

$$K' = 1.5\,K$$

$$\therefore \quad p' = \sqrt{2K'\,m}$$

$= \sqrt{2\,(1.5K)\,m}$ ...(ii)

From Eqs. (i) and (ii), we get

$$p' \approx 1.22\,p$$

Therefore percentage change in momentum is approximately 22%.

9. $R \propto \frac{1}{h^2} \Rightarrow \frac{R_1}{R_2} = \left(\frac{h_2}{h_1}\right)^2 = \left(\frac{4}{5}\right)^2 = \frac{16}{25}$

$$\therefore \quad R_1 = \left(\frac{16}{16+25}\right)(1000) = ₹\,\frac{16000}{41}$$

and $R_2 = \left(\frac{25}{16+25}\right)(1000) = ₹\,\frac{25000}{41}$

10. $R \propto w^2$

$$\Rightarrow \quad \frac{R_1}{R_2} = \left(\frac{60}{80}\right)^2 = \frac{9}{16}$$

$$\therefore \quad R_1 = \left(\frac{9}{9+16}\right)(2000) = ₹\,720$$

$$R_2 = \left(\frac{16}{9+16}\right)(2000) = ₹\,1280$$

Module 3.

1. Angle $= \frac{\text{Arc}}{\text{Radius}} = \frac{1}{4} = (0.25)$ radian

2. $x = \sqrt{(5)^2 + (2)^2} = \sqrt{29}$

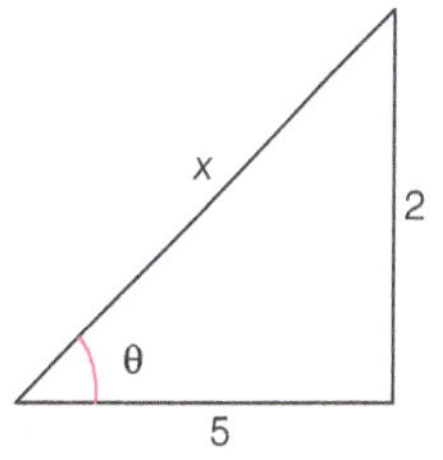

Now $\cos\theta = \frac{5}{x} = \frac{5}{\sqrt{29}}$

and $\sin\theta = \frac{2}{x} = \frac{2}{\sqrt{29}}$

3. $\tan 3° = \tan\left(\frac{3\times\pi}{180}\right)$ rad

$\approx \frac{3\times\pi}{180} = 0.052$

4. 240° lies in 3rd quadrant and tan in 3rd quadrant is positive.

315° lies in 4th quadrant and cos in 4th quadrant is positive.

5. (a) $\cos 120° = -\cos 60° = -\frac{1}{2}$

(b) $\sin 240° = \sin(180° + 60°) = -\sin 60° = -\frac{\sqrt{3}}{2}$

(c) $\tan(-60°) = -\tan 60° = -\sqrt{3}$

(d) $\cot(300°) = \cot(-60°) = -\cot 60° = -\frac{1}{\sqrt{3}}$

(e) $\tan 330° = \tan(-30°) = -\tan 30° = -\frac{1}{\sqrt{3}}$

(f) $\cos(-60°) = \cos 60° = \frac{1}{2}$

(g) $\sin(-150°) = -\sin 150° = -\sin 30° = -\frac{1}{2}$

(h) $\cos(-120°) = \cos 120° = -\cos 60° = -\frac{1}{2}$

6. $\tan 105° = \tan(60° + 45°)$

$= \frac{\tan 60° + \tan 45°}{1-(\tan 60°)(\tan 45°)} = \frac{\sqrt{3}+1}{1-\sqrt{3}}$

7. $\cos 67° = \cos(30° + 37°)$

$= (\cos 30°)(\cos 37°) - (\sin 30°)(\sin 37°)$

$= \left(\frac{\sqrt{3}}{2}\right)\left(\frac{4}{5}\right) - \left(\frac{1}{2}\right)\left(\frac{3}{5}\right)$

$= \left(\frac{4\sqrt{3}-3}{10}\right)$

8. $\cos 105° + \cos 75° = 2\cos\left(\frac{105°+75°}{2}\right)\cos\left(\frac{105°-75°}{2}\right)$

$= 2(\cos 90°)(\cos 35°)$

$= 0$ as $\cos 90° = 0$

9. (a) $\sec^2\theta = 1 + \tan^2\theta$

$\therefore \quad \sec^2\theta - \tan^2\theta = 1$

(b) $\text{cosec}^2\theta = 1 + \cot^2\theta$

$\therefore \quad \text{cosec}^2\theta - \cot^2\theta - 1 = 0$

(c) $2\sin 45° \cos 15°$

$\sin C + \sin D = 2\sin\left(\frac{C+D}{2}\right)\cos\left(\frac{C-D}{2}\right)$

$\therefore \quad \frac{C+D}{2} = 45°$

or $\quad C + D = 90° \quad$ ...(i)

$\frac{C-D}{2} = 15°$

or $\quad C - D = 30° \quad$ ...(ii)

Solving these two equations, we get

$C = 60°, D = 30°$

$\therefore \quad 2\sin 45° \cos 15° = \sin 60° + \sin 30°$

$= \left(\frac{\sqrt{3}+1}{2}\right)$ **Ans.**

(d) Apply $\sin C - \sin D = 2\sin\left(\frac{C-D}{2}\right)\cos\left(\frac{C+D}{2}\right)$

$C = 60°$ and $D = 30°$

$\therefore \quad 2\sin 15° \cos 45° = \sin 60° - \sin 30°$

$= \left(\frac{\sqrt{3}-1}{2}\right)$ **Ans.**

10. (a) $y_{max} = 5$, when $\cos\theta = -1$,

$y_{min} = -1$, when $\cos\theta = +1$

(b) $y_{min} = -7$, when $\sin\theta = -1$,

$y_{max} = +1$, when $\sin\theta = +1$

Module 4.

Introductory Exercise 1

1. Given, $y = x^3 + 5x^2 + 3x + 7$

$\frac{dy}{dx} = \frac{dy}{dx}(x^3 + 5x^2 + 3x + 7)$

$= \frac{d(x^3)}{dx} + \frac{d(5x^2)}{dx} + \frac{d(3x)}{dx} + \frac{d(7)}{dx}$

$= 3x^2 + 5(2x) + 3 + 0$

$= 3x^2 + 10x + 3$

2. Given, $y = x(3x^2 - 9)$

$\Rightarrow \quad y = 3x^3 - 9x$

$\therefore \quad \frac{dy}{dx} = 9x^2 - 9$

3. Given, $y = 3\sqrt{x} + 4\tan x - 6\ln x + e$

$= 3x^{\frac{1}{2}} + 4\tan x - 6\ln x + e$

$\therefore \quad \frac{dy}{dx} = 3\left(\frac{1}{2}\right)x^{\frac{1}{2}-1} + 4\sec^2 x - \frac{6}{x} + 0$

$= \frac{3}{2\sqrt{x}} + 4\sec^2 x - \frac{6}{x}$

Introductory Exercise 2

1. $\frac{d}{dx}[(x^2+1)(x^3+3)] = (x^2+1)\frac{d}{dx}(x^3+3) + (x^3+3)$

$\frac{d}{dx}(x^2+1)$

$= (x^2+1)(3x^2) + (x^3+3)(2x)$

$= 3x^4 + 3x^2 + 2x^4 + 6x$

$= 5x^4 + 3x^2 + 6x$ **Ans.**

2. Using product rule, we get

$$\frac{dy}{dx} = \sqrt{x}\frac{d}{dx}(\tan x) + (\tan x)\frac{d}{dx}(\sqrt{x})$$

$$= \sqrt{x}\sec^2 x + \frac{\tan x}{2\sqrt{x}}$$

3. Using product rule,

$$\frac{dy}{dx} = (\sin x)\frac{d}{dx}(\ln x) + (\ln x)\frac{d}{dx}(\sin x)$$

$$= \frac{\sin x}{x} + (\ln x)\cos x$$

Introductory Exercise 3

1. (a) $\frac{d}{dx}(2x+3)^6$

$= (2)[6(2x+3)^5]$

$= 12(2x+3)^5$

(b) $\frac{d}{dx}\ln(5x+2) = \frac{5}{5x+2}$

2. $\frac{dy}{dx} = 4\frac{d}{dx}e^{(x^2-2x)}$

$= 4e^{(x^2-2x)} \times \frac{d}{dx}(x^2-2x)$

$= 4e^{x^2-2x} \times (2x-2)$

$= 8(x-1)e^{x^2-2x}$ **Ans.**

3. $\frac{dy}{dx} = \frac{d}{dx}(x^2+1)^{1/2}$

$= \frac{1}{2}(x^2+1)^{\left(\frac{1}{2}-1\right)} \times \frac{d}{dx}(x^2+1)$

$= \frac{1}{2}(x^2+1)^{-\frac{1}{2}} \times 2x$

$\therefore \quad \frac{dy}{dx} = \frac{x}{(x^2+1)^{1/2}}$ **Ans.**

4. $\frac{d}{dx}\cos x^3$

$= (-\sin x^3)(3x^2)$

$= -3x^2\sin x^3$ **Ans.**

5. $y = x^3$

$\therefore \quad \frac{dy}{dt} = \frac{dy}{dx}\cdot\frac{dx}{dt} = \left[\frac{d}{dx}(x^3)\right]\left(\frac{dx}{dt}\right)$

$= 3x^2\frac{dx}{dt}$

Exercise

1. $\frac{d}{dx}\left(\sqrt{x} - \frac{1}{\sqrt{x}}\right) = \frac{d}{dx}\left(x^{\frac{1}{2}} - x^{-\frac{1}{2}}\right)$

$= \frac{1}{2}x^{\frac{-1}{2}} + \frac{1}{2}x^{-\frac{3}{2}}$

$= \frac{1}{2\sqrt{x}}\left(1 + \frac{1}{x}\right)$

2. $\frac{d}{dx}[(5x^2+6)(2x^2+4)]$

$= (5x^2+6)\frac{d}{dx}(2x^2+4) + (2x^2+4)\frac{d}{dx}(5x^2+6)$

$= (5x^2+6)(4x) + (2x^2+4)(10x) = 40x^3 + 64x$

3. $\frac{d}{dx}[\sqrt{x}(x^2+7)] = \sqrt{x}\frac{d}{dx}(x^2+7) + (x^2+7)\frac{d}{dx}\sqrt{(x)}$

$= \sqrt{x}(2x) + (x^2+7)\left(\frac{1}{2\sqrt{x}}\right)$

$= \frac{1}{2\sqrt{x}}(7+5x^2)$

4. $\frac{d}{d\theta}\left[\frac{\cos\theta}{1-\sin\theta}\right] = \frac{(1-\sin\theta)\frac{d}{d\theta}(\cos\theta) - \cos\theta\frac{d}{d\theta}(1-\sin\theta)}{(1-\sin\theta)^2}$

$= \frac{(1-\sin\theta)(-\sin\theta) - \cos\theta(-\cos\theta)}{(1-\sin\theta)^2}$

$= \frac{-\sin\theta + \sin^2\theta + \cos^2\theta}{(1-\sin\theta)^2}$

$= \frac{1-\sin\theta}{(1-\sin\theta)^2} = \frac{1}{1-\sin\theta}$

5. (a) $\frac{d}{dx}(x^4+3x^2-2x) = 4x^3+6x-2$

(b) $\frac{d}{dx}(x^2\cos x) = x^2\frac{d}{dx}(\cos x) + (\cos x)\frac{d}{dx}(x^2)$

$= x^2(-\sin x) + (\cos x)(2x)$

$= 2x\cos x - x^2\sin x$

(c) $\frac{d}{dx}(6x+7)^4 = (6)[4(6x+7)^3] = 24(6x+7)^3$

(d) $\frac{d}{dx}e^x x^5 = e^x\frac{d}{dx}(x^5) + x^5\frac{d}{dx}(e^x)$

$= e^x(5x^4) + x^5e^x = 5e^x x^4 + e^x x^5$

(e) $\frac{d}{dx}\left(\frac{1+x}{e^x}\right)=\frac{e^x\frac{d}{dx}(1+x)-(1+x)\frac{d}{dx}(e^x)}{(e^x)^2}$

$$=\frac{1-1-x}{e^x}=-xe^{-x}$$

6. (a) $[(x-1)^2]_{min}=0$ at $x=1$

$\therefore$ $y_{max}=5$ at $x=1$

(b) $y=(\sin 2x-x)$

$$\frac{dy}{dx}=2\cos 2x-1$$

and $\frac{d^2y}{dx^2}=-4\sin 2x$

Putting $\frac{dy}{dx}=0$, we get

$$2\cos 2x-1=0$$

$\therefore$ $\cos 2x=\frac{1}{2}$

or $2x=\pm 60°=\pm\frac{\pi}{3}$

At $2x=+60°$, $\frac{d^2y}{dx^2}$ is -ve, so value of y is maximum.

At $2x=-60°$, $\frac{d^2y}{dx^2}$ is positive. So value of y is minimum.

$\therefore$ $y_{max}=\sin(+60°)-\frac{\pi}{6}$

$$=\left(\frac{\sqrt{3}}{2}-\frac{\pi}{6}\right) \text{ at } 2x=\frac{\pi}{3}$$

or $x=\pi/6$

and $y_{min}=\sin(-60°)+\frac{\pi}{6}$

at $x=-\frac{\pi}{6}$

$y_{min}=\left(\frac{\pi}{6}-\frac{\sqrt{3}}{2}\right)$ **Ans**

Module 5.

1. (a) $\int(3t^2-2t)dt=3\int t^2dt-2\int tdt=\frac{3t^3}{3}-\frac{2t^2}{2}+c$

$$=t^3-t^2+c$$

(b) $\int(4\cos t+t^2)dt=4\int(\cos t)dt+\int t^2dt$

$$=4\sin t+\frac{t^3}{3}+c$$

(c) $\int(2t-4)^{-4}dt=\frac{1}{2}\left[\frac{(2t-4)^{-4+1}}{-4+1}\right]+c$

$$=-\frac{1}{6(2t-4)^3}+c$$

(d) $\int\frac{dt}{6t-1}=\frac{1}{6}\ln(6t-1)+c$

2. (a) $\int_0^2 2tdt=[t^2]_0^2=[(2)^2-(0)^2]=4$

(b) $\int_{\pi/6}^{\pi/3}\sin x dx=[-\cos x]_{\pi/6}^{\pi/3}$

$$=-\left[\cos\frac{\pi}{3}-\cos\frac{\pi}{6}\right]=-\left[\frac{1}{2}-\frac{\sqrt{3}}{2}\right]$$

(c) $\int_4^{10}\frac{dx}{x}=[\ln x]_4^{10}=[\ln 10-\ln 4]=\ln\left(\frac{10}{4}\right)=\ln(2.5)$

(d) $\int_0^{\pi}\cos x dx=[\sin x]_0^{\pi}=[\sin\pi-\sin 0]=[0-0]=0$

(e) $\int_1^2(2t-4)dt=2\int_1^2 tdt-4\int_1^2 dt=[t^2-4t]_1^2$

$$=[(2^2-4\times 2)-(1^2-4\times 1)]$$

$$=[-4+3]=-1$$

3. $\int_2^3 f(x)\,dx=\int_2^3 x^2dx$

$$=\left.\frac{x^3}{3}\right|_2^3=\frac{(3)^3}{3}-\frac{(2)^3}{3}$$

$$=\frac{19}{3}=6.333$$ **Ans.**

Module 6.

2. In (b) and (d) parts we get $y=0$ by putting $x=0$. Hence, graphs (b) and (d) pass through origin.

3. (a) For $x=$ positive, y is negative. Hence IV quadrant.

For $x=$ negative, y is again negative. Hence, III quadrant.

(b) For $x=$ positive, y is positive. Hence, I quadrant.

For $x=$ negative, y is also negative. Hence, III quadrant.

4. For finding C_x, let us put $y=0$, in the given equation. We find,

$$C_x=-\frac{5}{6}$$

For finding C_y, let us put $x=0$, in the given equation. We find,

$$C_y=+2.5$$

UNIT-2

MEASUREMENT & ERRORS

Learning Modules

1. Error Analysis
2. Significant Figures

MODULE 1

Error Analysis

Least Count

The smallest measurement that can be measured accurately by an instrument is called its least count.

Let us take an example.

Example 1 Let us use a centimeter scale (on which only centimeter scales are there) to measure a length AB.

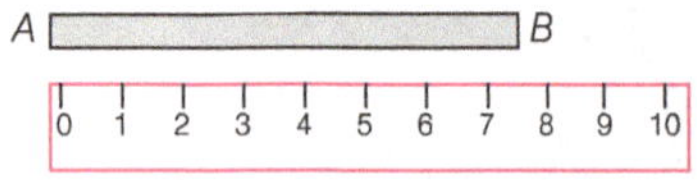

From the figure, we can see that length AB is more than 7 cm and less than 8 cm.

In this case, **Least Count (LC)** of this centimeter scale is 1 cm, as it can measure accurately upto centimeters only. If we note down the length (l) of line AB as $l = 7$ cm then maximum uncertainty or maximum possible error in l can be 1 cm $(= \text{LC})$, because this scale can measure accurately only upto 1 cm.

Instrument	Its least count
mm scale	1 mm
Vernier callipers	0.1 mm
Screw gauge	0.01 mm

Permissible Error due to Least Count

Error in measurement due to the limitation (or least count) of the instrument is called permissible error. Least count of a millimeter scale is 1 mm. Therefore, maximum permissible error in the measurement of a length by a millimeter scale may be 1 mm.

If we measure a length $l = 26$ mm. Then, maximum value of true value may be $(26 + 1)$ mm $= 27$ mm and minimum value of true value may be $(26 - 1)$ mm $= 25$ mm.

Thus, we can write it like,

$$l = (26 \pm 1) \text{ mm}$$

If from any other instrument we measure a length $= 24.6$ mm, then the maximum permissible error (or least count) from this instrument is 0.1 mm. So, we can write the measurement like,

$$l = (24.6 \pm 0.1) \text{ mm}$$

Classification of Errors

Errors can be classified in two ways. First classification is based on the cause of error. Systematic error and random errors fall in this group. Second classification is based on the magnitude of errors. Absolute error, mean absolute error and relative (or fractional) error lie on this group. Now, let us discuss them separately.

Systematic Error

Systematic errors are the errors whose causes are known to us. Such errors can therefore be minimised. Following are few causes of these errors :

(a) Instrumental errors may be due to erroneous instruments. These errors can be reduced by using more accurate instruments and applying zero correction, when required.

(b) Sometimes errors arise on account of ignoring certain facts. For example in measuring time period of simple pendulum error may creep because no consideration is taken of air resistance. These errors can be reduced by applying proper corrections to the formula used.

(c) Change in temperature, pressure, humidity, etc., may also sometimes cause errors in the result. Relevant corrections can be made to minimise their effects.

Random Error

The causes of random errors are not known. Hence, it is not possible to remove them completely. These errors may arise due to a variety of reasons. For example the reading of a sensitive beam balance may change by the vibrations caused in the building due to persons moving in the laboratory or vehicles running nearby. The random errors can be minimized by repeating the observation a large number of times and taking the arithmetic mean of all the observations. The mean value would be very close to the most accurate reading. Thus,

$$a_{\text{mean}} = \frac{a_1 + a_2 + \ldots + a_n}{n} = a_m$$

Absolute Error

The difference between the true value and the measured value of a quantity is called an absolute error. Usually the mean value a_m is taken as the true value.

$$a_m = \frac{a_1 + a_2 + \ldots + a_n}{n}$$

Now let us find the difference of a_m with measured values. Thus,

$$\Delta a_1 = a_m - a_1$$
$$\Delta a_2 = a_m - a_2$$
$$\ldots \quad \ldots \quad \ldots$$
$$\Delta a_n = a_m - a_n$$

Here, $\Delta a_1, \Delta a_2 \dots \Delta a_n$ may be positive or negative but their magnitude $|\Delta a|$ which is always positive is called the absolute error.

Mean Absolute Error

Arithmetic mean of the absolute errors in all the measurements is called the mean absolute error. Thus,

$$\Delta a_{\text{mean}} = \frac{|\Delta a_1| + |\Delta a_2| + \dots + |\Delta a_n|}{n}$$

The final result of measurement can be written as,

$$a = a_m \pm \Delta a_{\text{mean}}$$

This implies that value of a is likely to lie between

$$a_m + \Delta a_{\text{mean}} \text{ and } a_m - \Delta a_{\text{mean}}$$

Relative or Fractional Error

The ratio of mean absolute error to the mean value of the quantity measured is called relative or fractional error. Thus,

$$\text{Relative error} = \frac{\Delta a_{\text{mean}}}{a_m}$$

Relative error expressed in percentage is called as the percentage error, i.e.

$$\text{Percentage error} = \frac{\Delta a_{\text{mean}}}{a_m} \times 100$$

Example 2 The diameter of a wire as measured by screw gauge was found to be 2.620, 2.625, 2.630, 2.628 and 2.626 cm. Calculate

(a) mean value of diameter
(b) absolute error in each measurement
(c) mean absolute error
(d) fractional error
(e) percentage error
(f) Express the result with error limit

Sol. (a) Mean value of diameter with error limit.

$$a_m = \frac{2.620 + 2.625 + 2.630 + 2.628 + 2.626}{5}$$

$= 2.6258$ cm

$= 2.626$ cm (rounding off to three decimal places)

(b) Taking a_m as the true value, the absolute errors in different observations are,

$|\Delta a_1| = |2.626 - 2.620| = + 0.006$ cm
$|\Delta a_2| = |2.626 - 2.625| = + 0.001$ cm
$|\Delta a_3| = |2.626 - 2.630| = + 0.004$ cm
$|\Delta a_4| = |2.626 - 2.628| = + 0.002$ cm
$|\Delta a_5| = |2.626 - 2.626| = 0.000$ cm

(c) Mean absolute error,

$$\Delta a_{\text{mean}} = \frac{|\Delta a_1| + |\Delta a_2| + |\Delta a_3| + |\Delta a_4| + |\Delta a_5|}{5}$$

$$= \frac{0.006 + 0.001 + 0.004 + 0.002 + 0.000}{5}$$

$= 0.0026$ cm $= 0.003$ cm

(rounding off to three decimal places)

(d) Fractional error $= \pm \frac{\Delta a_{\text{mean}}}{a_m} = \frac{\pm 0.003}{2.626} = \pm 0.001$

(e) Percentage error $= \pm 0.001 \times 100 = \pm 0.1\%$

(f) Diameter of wire can be written as,

$d = (2.626 \pm 0.003)$ cm **Ans.**

Combination of Errors

Errors in Sum or Difference

Let $x = a \pm b$

Further, let Δa is the absolute error in the measurement of a, Δb the absolute error in the measurement of b and Δx is the absolute error in the measurement of x.

Then, $x + \Delta x = (a \pm \Delta a) \pm (b \pm \Delta b)$

$= (a \pm b) \pm (\pm \Delta a \pm \Delta b)$

$= x \pm (\pm \Delta a \pm \Delta b)$ or $\Delta x = \pm \Delta a \pm \Delta b$

The four possible values of Δx are $(\Delta a - \Delta b)$, $(\Delta a + \Delta b)$, $(-\Delta a - \Delta b)$ and $(-\Delta a + \Delta b)$.

Therefore, the maximum absolute error in x is,

$$\Delta x = \pm (\Delta a + \Delta b)$$

i.e. the maximum absolute error in sum and difference of two quantities is equal to sum of the absolute errors in the individual quantities.

Note *Here absolute error means its mean value.*

Example 3 The volumes of two bodies are measured to be $V_1 = (10.2 \pm 0.02)$ cm^3 and $V_2 = (6.4 \pm 0.01)$ cm^3. Calculate sum and difference in volumes with error limits.

Sol. $V_1 = (10.2 \pm 0.02)$ cm^3

and $V_2 = (6.4 \pm 0.01)$ cm^3

$\Delta V = \pm (\Delta V_1 + \Delta V_2)$

$= \pm (0.02 + 0.01)$ cm^3

$= \pm 0.03$ cm^3

$V_1 + V_2 = (10.2 + 6.4)$ cm^3 $= 16.6$ cm^3

and $V_1 - V_2 = (10.2 - 6.4)$ cm^3 $= 3.8$ cm^3

Hence, sum of volumes $= (16.6 \pm 0.03)$ cm^3

and difference of volumes $= (3.8 \pm 0.03)$ cm^3 **Ans.**

Errors in a Product

Let $x = ab$

Then, $(x \pm \Delta x) = (a \pm \Delta a)(b \pm \Delta b)$

or $x\left(1 \pm \frac{\Delta x}{x}\right) = ab\left(1 \pm \frac{\Delta a}{a}\right)\left(1 \pm \frac{\Delta b}{b}\right)$

or $1 \pm \frac{\Delta x}{x} = 1 \pm \frac{\Delta b}{b} \pm \frac{\Delta a}{a} \pm \frac{\Delta a}{a} \cdot \frac{\Delta b}{b}$ (as $x = ab$)

or $$\pm\frac{\Delta x}{x}=\pm\frac{\Delta a}{a}\pm\frac{\Delta b}{b}\pm\frac{\Delta a}{a}\cdot\frac{\Delta b}{b}$$

Here, $\frac{\Delta a}{a}\cdot\frac{\Delta b}{b}$ is a small quantity, so can be neglected.

Hence, $\pm\frac{\Delta x}{x}=\pm\frac{\Delta a}{a}\pm\frac{\Delta b}{b}$

Possible values of $\frac{\Delta x}{x}$ are $\left(\frac{\Delta a}{a}+\frac{\Delta b}{b}\right)$, $\left(\frac{\Delta a}{a}-\frac{\Delta b}{b}\right)$,

$\left(-\frac{\Delta a}{a}+\frac{\Delta b}{b}\right)$ and $\left(-\frac{\Delta a}{a}-\frac{\Delta b}{b}\right)$.

Hence, maximum possible value of

$$\frac{\Delta x}{x}=\pm\left(\frac{\Delta a}{a}+\frac{\Delta b}{b}\right)$$

Therefore, maximum fractional error in product of two (or more) quantities is equal to sum of fractional errors in the individual quantities.

Errors in Division

Let $x=\frac{a}{b}$

Then, $x\pm\Delta x=\frac{a\pm\Delta a}{b\pm\Delta b}$ or $x\left(1\pm\frac{\Delta x}{x}\right)=\frac{a\left(1\pm\frac{\Delta a}{a}\right)}{b\left(1\pm\frac{\Delta b}{b}\right)}$

or $$\left(1\pm\frac{\Delta x}{x}\right)=\left(1\pm\frac{\Delta a}{a}\right)\left(1\pm\frac{\Delta b}{b}\right)^{-1} \qquad \left(\text{as } x=\frac{a}{b}\right)$$

As $\frac{\Delta b}{b}<<1$, so expanding binomially, we get

$$\left(1\pm\frac{\Delta x}{x}\right)=\left(1\pm\frac{\Delta a}{a}\right)\left(1\mp\frac{\Delta b}{b}\right)$$

or $$1\pm\frac{\Delta x}{x}=1\pm\frac{\Delta a}{a}\mp\frac{\Delta b}{b}\pm\frac{\Delta a}{a}\cdot\frac{\Delta b}{b}$$

Here, $\frac{\Delta a}{a}\cdot\frac{\Delta b}{b}$ is small quantity, so can be neglected.

Therefore, $$\pm\frac{\Delta x}{x}=\pm\frac{\Delta a}{a}\mp\frac{\Delta b}{b}$$

Possible values of $\frac{\Delta x}{x}$ are $\left(\frac{\Delta a}{a}-\frac{\Delta b}{b}\right)$, $\left(\frac{\Delta a}{a}+\frac{\Delta b}{b}\right)$, $\left(-\frac{\Delta a}{a}-\frac{\Delta b}{b}\right)$ and $\left(-\frac{\Delta a}{a}+\frac{\Delta b}{b}\right)$.

Therefore, the maximum value of

$$\frac{\Delta x}{x}=\pm\left(\frac{\Delta a}{a}+\frac{\Delta b}{b}\right)$$

or the maximum value of fractional error in division of two quantities is equal to the sum of fractional errors in the individual quantities.

Note *From the above expression we can see that,*

$$\Delta x=\pm\left(\frac{\Delta a}{a}+\frac{\Delta b}{b}\right)x$$

Error in Quantity Raised to Some Power

Let, $x=\frac{a^n}{b^m}$. Then, $\ln(x)=n\ln(a)-m\ln(b)$

Differentiating both sides, we get

$$\frac{dx}{x}=n\cdot\frac{da}{a}-m\frac{db}{b}$$

In terms of fractional error we may write,

$$\pm\frac{\Delta x}{x}=\pm n\frac{\Delta a}{a}\mp m\frac{\Delta b}{b}$$

Therefore, maximum value of

$$\frac{\Delta x}{x}=\pm\left(n\frac{\Delta a}{a}+m\frac{\Delta b}{b}\right)$$

Note *Errors in product and division can also be obtained by taking logarithm on both sides* $\left(\text{in } x=ab \text{ or } x=\frac{a}{b}\right)$ *and then differentiating.*

Summary

1. If $x=a\pm b$, then $\Delta x=\pm(\Delta a+\Delta b)$
2. If $x=ab$ or $\frac{a}{b}$, then $\frac{\Delta x}{x}=\pm\left(\frac{\Delta a}{a}+\frac{\Delta b}{b}\right)$

   Or, $\Delta x=\pm\left(\frac{\Delta a}{a}+\frac{\Delta b}{b}\right)x$
3. If $x=\frac{a^n}{b^m}$, then

   $$\frac{\Delta x}{x}=\pm\left(n\frac{\Delta a}{a}+m\frac{\Delta b}{b}\right)$$

   Or, $\Delta x=\pm\left(n\frac{\Delta a}{a}+m\frac{\Delta b}{b}\right)x$
4. In all cases,
   (i) Absolute error $=\Delta x$
   (ii) Fractional error $=\frac{\Delta x}{x}$
   (iii) Percentage error $=\frac{\Delta x}{x}\times 100$
5. All error listed above are maximum possible errors.

Example 4 The mass and density of a solid sphere are measured to be (12.4 ± 0.1) kg and (4.6 ± 0.2) kg/m^3. Calculate the volume of the sphere with error limits.

Sol. Here, $m \pm \Delta m = (12.4 \pm 0.1)$ kg

and $\rho \pm \Delta\rho = (4.6 \pm 0.2)$ kg/m^3

$$\text{Volume } V = \frac{m}{\rho} = \frac{12.4}{4.6}$$

$$= 2.69 \text{ m}^3 = 2.7 \text{ m}^3$$

(rounding off to one decimal place)

As $V = \frac{m}{\rho}$

$$\therefore \quad \frac{\Delta V}{V} = \pm\left(\frac{\Delta m}{m} + \frac{\Delta\rho}{\rho}\right)$$

or
$$\Delta V = \pm\left(\frac{\Delta m}{m} + \frac{\Delta\rho}{\rho}\right) \times V$$

$$= \pm\left(\frac{0.1}{12.4} + \frac{0.2}{4.6}\right) \times 2.7$$

$$= \pm 0.14$$

$\therefore \quad V \pm \Delta V = (2.7 \pm 0.14)$ m^3 **Ans.**

Example 5 The density of a material in the shape of a cube is determined by measuring three sides of the cube and its mass. If the relative errors in measuring the mass and length are respectively 1.5% and 1%, the maximum error in determining the density is: **(2018 Main)**

(a) 6 % (b) 2.5 % (c) 3.5 % (d) 4.5 %

Sol. Given, $\frac{\Delta m}{m} \times 100 = 1.5\%$ and $\frac{\Delta l}{l} \times 100 = 1\%$

$$\rho = \frac{m}{l^3} \Rightarrow \frac{\Delta\rho}{\rho} \times 100 = \frac{\Delta m}{m} \times 100 + \frac{3\Delta l}{l} \times 100$$

$$= 1.5 + 3(1) = 4.5\%$$

$\therefore$ The correct answer is (d).

Method of Solving Complex Problems of Error Analysis

(i) In any problem of error analysis sometimes we have to make a formula according to the given conditions. Otherwise in most of the cases formula will be given in the problem or it may be a standard formula.

(ii) Then we have to find the variables (needed for differentiation) given in the formula. Variables are the ones for which either,

(a) percentage error is given (or asked) Or,

(b) absolute error (Δx or sometimes dx) is given or asked) Or,

(c) the physical quantity is written like $(x \pm \Delta x)$.

(iii) In the final answer either

(a) we have to find the absolute error Δx (or sometimes dx). Or,

(b) we have to find the relative error, $\frac{\Delta x}{x}$ (or sometimes $\frac{dx}{x}$). Or,

(c) we have to find the percentage error $\left(\frac{\Delta x}{x} \times 100\right)$ or sometimes $\left(\frac{dx}{x} \times 100\right)$. Or,

(d) we have to write the quantity in the form $(x \pm \Delta x)$.

(iv) In complex problems which can't be solved by the methods discussed so far, we either

(a) directly differentiate the given equation. Or,

(b) first we take log and then differentiate the equation.

Example of Direct Differentiation

Example 6 Calculate equivalent resistance (with error limits) of two resistors R_1 and R_2 in parallel where, $R_1 = (6 \pm 0.2)$ ohm and $R_2 = (3 \pm 0.1)$ ohm.

Note *Here the variables are R_1 and R_2.*

Sol. In parallel,

$$\frac{1}{R} = \frac{1}{R_1} + \frac{1}{R_2} \quad \text{...(i)}$$

or
$$R = \frac{R_1 R_2}{R_1 + R_2} = \frac{(6)(3)}{6+3} = 2 \text{ ohm}$$

Differentiating Eq. (i), we have

$$-\frac{dR}{R^2} = -\frac{dR_1}{R_1^2} - \frac{dR_2}{R_2^2}$$

Therefore, maximum permissible error in equivalent resistance may be

$$\Delta R = \pm\left(\frac{\Delta R_1}{R_1^2} + \frac{\Delta R_2}{R_2^2}\right)(R^2)$$

Substituting the values we get,

$$\Delta R = \pm\left[\frac{0.2}{(6)^2} + \frac{0.1}{(3)^2}\right](2)^2$$

$$= \pm 0.07 \text{ ohm}$$

$\therefore \quad R = (2 \pm 0.07)$ ohm **Ans.**

Examples of Taking log and then Differentiation

Example 7 Calculate percentage error in determination of time period of a pendulum

$$T = 2\pi\sqrt{\frac{l}{g}}$$

where, l and g are measured with $\pm 1\%$ and $\pm 2\%$.

Note *Here the variables are l, g and T. Formula is already given.*

Sol. $T = 2\pi\sqrt{\frac{l}{g}}$

or $$T = (2\pi)(l)^{+1/2}(g)^{-1/2}$$

Taking logarithm of both sides, we have

$$\ln(T) = \ln(2\pi) + \frac{1}{2}(\ln l) - \left(\frac{1}{2}\right)\ln(g) \quad \text{...(i)}$$

Here, 2π is a constant, therefore $\ln(2\pi)$ is also a constant.

Differentiating Eq. (i), we have

$$\frac{1}{T}\,dT = 0 + \frac{1}{2}\left(\frac{1}{l}\right)(dl) - \frac{1}{2}\left(\frac{1}{g}\right)(dg)$$

or $$\left(\frac{dT}{T}\right)_{max} = \text{Maximum value of}\left(\pm\frac{1}{2}\frac{dl}{l} \mp \frac{1}{2}\frac{dg}{g}\right)$$

$$= \frac{1}{2}\left(\frac{dl}{l}\right) + \frac{1}{2}\left(\frac{dg}{g}\right)$$

This can also be written as

$$\left(\frac{\Delta T}{T}\times 100\right)_{max} = \frac{1}{2}\left[\frac{\Delta l}{l}\times 100\right] + \frac{1}{2}\left[\frac{\Delta g}{g}\times 100\right]$$

or percentage error in time period

$$= \pm\left[\frac{1}{2}(\text{percentage error in } l) + \frac{1}{2}(\text{percentatage error in } g)\right]$$

$$= \pm\left[\frac{1}{2}\times 1 + \frac{1}{2}\times 2\right] = \pm 1.5\%$$ **Ans.**

Note *Since* $T \propto \sqrt{\frac{l}{g}}$ *(Relation between variables only)*

or $T \propto l^{\frac{1}{2}} g^{-\frac{1}{2}}$

So, we can directly write :

(% error in T) $= \frac{1}{2}$ *(% error in l)* $+ \frac{1}{2}$ *(% error in g).*

As we have discussed in the theory, this result comes after taking log and then differentiating.

Example 8 The current voltage relation of diode is given by $I = (e^{1000V/T} - 1)$ mA, where the applied voltage V is in volt and the temperature T is in kelvin. If a student makes an error measuring $\pm$ 0.01 V while measuring the current of 5 mA at 300 K, what will be the error in the value of current in mA? **(2014 Main)**

(a) 0.2 mA (b) 0.02 mA

(c) 0.5 mA (d) 0.05 mA

Sol. Given $I = (e^{1000V/T)} - 1)$ mA, $dV = \pm 0.01$ V

$$T = 300\text{ K}$$

So, $$I = e^{1000V/T} - 1$$

$$I + 1 = e^{1000V/T}$$

Taking $\log_e$ on both sides, we get

$$\ln(I+1) = \frac{1000V}{T}\ln e = \frac{1000V}{T}$$

On differentiating, $$\frac{dI}{I+1} = \frac{1000}{T}dV$$

$$\Rightarrow \quad dI = \left(\frac{1000}{T}\right)(I+1)\,dV = \frac{1000}{300}\times(5+1)\times 0.01$$

$$= 0.2\text{ mA}$$

So, error in the value of current is 0.2 mA.

$\therefore$ The correct option is (a).

Note *Here the variables are V and I. Formula is already given.*

Example 9 The energy of a system as a function of time t is given as $E(t) = A^2\exp(-\alpha t)$, where $\alpha = 0.2\text{ s}^{-1}$. The measurement of A has an error of 1.25%. If the error in the measurement of time is 1.50%, the percentage error in the value of $E(t)$ at $t = 5$ s is **(2015 Adv.)**

Sol. $$E(t) = A^2 e^{-\alpha t} \quad \text{...(i)}$$

Given, $$\alpha = 0.2\text{ s}^{-1}$$

$$\left(\frac{dA}{A}\right)\times 100 = 1.25\%$$

$$\frac{dt}{t}\times 100 = 1.50$$

$$\Rightarrow \quad (dt\times 100) = 1.5\,t$$

$$= 1.5\times 5 = 7.5\text{ sec}$$

Taking $\log_e$ on both sides of Eq. (i), we get

$$\ln E = 2\ln A - \alpha t$$

Now, differentiating this equation, we get

$$\frac{dE}{E} = \pm 2\frac{dA}{A} \pm \alpha dt$$

$$\therefore \quad \left(\frac{dE}{E}\right)\times 100 = \pm\left[2\left(\frac{dA}{A}\right)\times 100 + \alpha(dt\times 100)\right]$$

$$= \pm[2(1.25) + 0.2(7.5)]$$

$$= \pm(2.5 + 1.5] = \pm 4\%$$

Note *Here the variables are E, t and A. Formula is already given.*

Example of First making Formula and then Differentiation

Example 10 A person measures the depth of a well by measuring the time interval between dropping a stone and receiving the sound of impact with the bottom of the well. The error in his measurement of time is $\delta T = 0.01$ s and he measures the depth of the well to be $L = 20$ m. Take the acceleration due to gravity $g = 10\text{ ms}^{-2}$ and the velocity of sound is 300 ms^{-1}. Then the percentage error in the measurement of L, is closest to **(2017 Adv.)**

(a) 1% (b) 5%

(c) 3% (d) 0.2%

Sol. $T = t_1 + t_2$

or $$T = \sqrt{\frac{2L}{g}} + \frac{L}{v_s},\; g = 10\text{ m/s}^2$$

$$T = \sqrt{\frac{L}{5}} + \frac{L}{300}$$

Differentiating this equation, we get

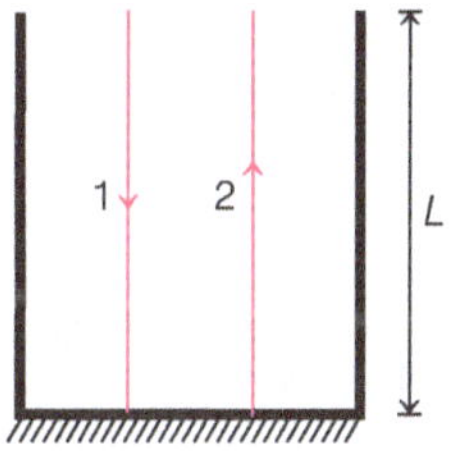

$$dT = \frac{1}{\sqrt{5}}\frac{1}{2}\frac{1}{\sqrt{L}}dL + \left(\frac{1}{300}dL\right)$$

$$dT \text{ or } \delta T = \frac{1}{2\sqrt{5}}\frac{1}{\sqrt{20}}dL + \frac{dL}{300} = 0.01$$

$\Rightarrow$ $$dL\left(\frac{1}{20} + \frac{1}{300}\right) = 0.01$$

or $$dL\left[\frac{16}{300}\right] = 0.01$$

$\therefore$ $$dL \text{ or } \delta L = \frac{3}{16}$$

or $$\frac{\delta L}{L} \times 100 = \frac{3}{16 \times 20} \times 100$$

$$\approx 1\%$$

$\therefore$ The closest answer is (a).

Note *(i) Here the variables are T and L.*

(ii) t_1 *= free fall time* $= \sqrt{\frac{2L}{g}}$

$$t_2 = \frac{\text{distance}}{\text{speed}} = \frac{L}{v_s}$$

Miscellaneous Examples

Ex. 1 The period of oscillation of a simple pendulum is $T = 2\pi\sqrt{\frac{L}{g}}$. Measured value of L is 20.0 cm known to 1 mm accuracy and time for 100 oscillations of the pendulum is found to be 90 s using a wrist watch of 1s resolution. The accuracy in the determination of g is **(2015 Main)**

(a) 3% (b) 2%

(c) 1% (d) 5%

Sol. Time period is given by, $T = \frac{t}{n}$

Further, $$T = 2\pi\sqrt{\frac{L}{g}}$$

$\Rightarrow$ $$g = \frac{(4\pi^2)L}{T^2} = \frac{(4\pi^2)(L)}{\left(\frac{t}{n}\right)^2}$$

$$= (4\pi^2 n^2)\frac{L}{t^2}$$

or $$g \propto \frac{L}{t^2}$$

$L = 20\text{ cm}, \Delta L = 1\text{ mm} = 0.1\text{ cm}$

$t = 90\text{ s}$ and $\Delta t = 1\text{ s}$

Percentage error in the value of 'g' will be

$$\frac{\Delta g}{g} \times 100 = \left(\frac{\Delta L}{L}\right) \times 100 + 2\left(\frac{\Delta t}{t}\right) \times 100$$

$$= \frac{0.1}{20} \times 100 + 2 \times \left(\frac{1}{90}\right) \times 100 = 2.72\%$$

$\therefore$ The nearest answer is 3 %.

$\therefore$ The closest answer is (a).

Note *Here the variables are g, L and t.*

Ex. 2 A student uses a simple pendulum of exactly 1m length to determine g, the acceleration due to gravity. He uses a stop watch with the least count of 1 s for this and records 40 s for 20 oscillations. For this observation, which of the following statement(s) is/are true? **(2010 Adv.)**

(a) Error ΔT in measuring T, the time period, is 0.05 s

(b) Error ΔT in measuring T, the time period, is 1 s

(c) Percentage error in the determination of g is 5%

(d) Percentage error in the determination of g is 2.5%

Sol. $T = \frac{t}{n} = \frac{40\text{ s}}{20} = 2\text{ s}$

Further, $t = nT = 20T$

or $$\Delta t = 20\,\Delta T$$

$\therefore$ $$\frac{\Delta t}{t} = \frac{20\Delta T}{20T} = \frac{\Delta T}{T}$$

or $$\Delta T = \frac{T}{t}\cdot\Delta t = \left(\frac{2}{40}\right)(1)$$

$$= 0.05\text{ s}$$

Further, $$T = 2\pi\sqrt{\frac{l}{g}}$$

or $$g \propto T^{-2}$$

$\therefore$ $$\left[\frac{\Delta g}{g} \times 100\right]_{max} = 2\left[\frac{\Delta T}{T} \times 100\right]$$

$$= \frac{2 \times 0.05}{2} \times 100 = 5\%.$$

$\therefore$The correct options are (a) and (c).

Note *Here the variables are t, T and g. Length of pendulum l is exactly 1m. So, it is not a variable.*

MODULE 1 Exercise

Theory Based, Addition and Subtraction

Single Correct Option

1. The length of a rod is (11.05 ± 0.2) cm. What is the length of the two rods?
   (a) (22.1 ± 0.05) cm
   (b) (22.1 ± 0.1) cm
   (c) (22.10 ± 0.05) cm
   (d) (22.10 ± 0.4) cm

2. If $x = 10.0 \pm 0.1$ and $y = 10.0 \pm 0.1$, then $2x - 2y$ is equal to
   (a) (0.1 ± 0.1)
   (b) zero
   (c) (0.0 ± 0.4)
   (d) (20 ± 0.2)

Multiple Correct Options

3. The speed of two particles is measured to be $v_1 = (10.2 \pm 0.02)\ \text{ms}^{-1}$ and $v_2 = (6.4 \pm 0.01)\ \text{ms}^{-1}$. Then, mark the correct option(s).
   (a) The sum of speeds of particles is $(16.6 \pm 0.03)\ \text{ms}^{-1}$.
   (b) The difference of speeds of particles is $(3.8 \pm 0.03)\ \text{ms}^{-1}$.
   (c) The difference of speeds of particles is $(6.4 \pm 0.01)\ \text{ms}^{-1}$.
   (d) The sum of speeds of particles is $(16.6 \pm 0.01)\ \text{ms}^{-1}$.

Subjective Questions

4. The temperature of two bodies measured by a thermometer are $(20 \pm 0.5)°\text{C}$ and $(50 \pm 0.5)°\text{C}$. Calculate the temperature difference with error limits.

5. The period of oscillation of a simple pendulum in the experiment is recorded as 2.63 s, 2.56 s, 2.42 s, 2.71 s and 2.80 s. Find the mean absolute error.

6. Repeated observations in an experiment gave the values 1.29, 1.33, 1.34, 1.35, 1.32, 1.36, 1.30 and 1.33. Calculate the mean value, mean absolute error, relative error and percentage error.

7. The initial and final temperatures of water as recorded by an observer are $(40.6 \pm 0.2)°$ C and $(78.3 \pm 0.3)°$ C. Calculate the rise in temperature with proper error limits.

Multiplication, Division and Powers

Fill in the Blank

8. For $z = a^2 x^3 y^{1/2}$, where 'a' is a constant. If percentage error in measurement of 'x' and 'y' are 4% and 12%, respectively, then the percentage error for 'z' will be ...... %. **(2022 Main)**

Single Correct Option

9. If error in measurement of radius of a sphere is 1%, what will be the error in measurement of volume?
   (a) 1% (b) $\frac{1}{3}$% (c) 3% (d) None of these

10. The density of a cube is measured by measuring its mass and length of its sides. If the maximum error in the measurement of mass and length are 4% and 3% respectively, the maximum error in the measurement of density will be
   (a) 7% (b) 9% (c) 12% (d) 13%

11. Percentage error in the measurement of mass and speed are 2% and 3%, respectively, the maximum error in the measurement of kinetic energy will be
   (a) 12% (b) 10% (c) 8% (d) 5%

12. A body travels uniformly a distance of (13.8 ± 0.2) m in a time (4.0 ± 0.3) s. The velocity of the body within error limit is
   (a) $(3.45 \pm 0.2)\ \text{ms}^{-1}$ (b) $(3.45 \pm 0.3)\ \text{ms}^{-1}$
   (c) $(3.45 \pm 0.4)\ \text{ms}^{-1}$ (d) $(3.45 \pm 0.5)\ \text{ms}^{-1}$

13. If the error in the measurement of momentum of a particle is (+ 100%), then the error in the measurement of its kinetic energy is
   (a) 100% (b) 200% (c) 300% (d) 400%

14. A wire has a mass (0.3 ± 0.003) g, radius (0.5 ± 0.005) mm and length (6 ± 0.06) cm. The maximum percentage error in the measurement of its density is
   (a) 1 (b) 2 (c) 3 (d) 4

15. In the density measurement of a cube, the mass and edge length are measured as (10.00 ± 0.10) kg and (0.10 ± 0.01) m, respectively. The error in the measurement of density is **(2019 Main)**
   (a) $0.01\ \text{kg/m}^3$ (b) $0.10\ \text{kg/m}^3$
   (c) $0.07\ \text{kg/m}^3$ (d) $0.31\ \text{kg/m}^3$

16. The maximum error in the measurement of resistance, current and time in an electrical circuit are 1%, 2% and 3% respectively. The maximum percentage error in the dissipated heat will be **(2022 Main)**
   (a) 2 (b) 4 (c) 6 (d) 8

Subjective Questions

17. The radius of a sphere is measured to be (1.2 ± 0.2) cm. Calculate its volume with error limits.
18. The radius of a sphere is measured to be (2.1 ± 0.5) cm. Calculate its surface area with error limits.
19. The resistance $R = \frac{V}{I}$, where $V = (100 \pm 5.0)$ V and $I = (10 \pm 0.2)$ A. Find the percentage error in R.
20. Find the percentage error in specific resistance given by $\rho = \frac{\pi r^2 R}{l}$, where r is the radius having value (0.2 ± 0.02) cm, R is the resistance of (60 ± 2) ohm and l is the length of (150 ± 0.1) cm.
21. A physical quantity ρ is related to four variables α, β, γ and η as

$$\rho = \frac{\alpha^3 \beta^2}{\eta \sqrt{\gamma}}$$

The percentage errors of measurements in α, β, γ and η are 1%, 3%, 4% and 2%, respectively. Find the percentage error in ρ.

Miscellaneous Problem

Single Correct Option

22. If error in measuring diameter of a circle is 4%, the error in measuring radius of the circle would be
(a) 2% (b) 8% (c) 4% (d) 1%

Answers

1. (d) **2.** (c) **3.** (a,b) **4.** (30 ± 1)°C **5.** (0.11s) **6.** 1.33, 0.02, $\pm$ 0.02, 1.5% **7.** (37.7 ± 0.5)°C
8. (18%) **9.** (c) **10.** (d) **11.** (c) **12.** (b) **13.** (c) **14.** (d) **15.** (*) **16.** (d)
17. $(7.2 \pm 3.6)\,cm^3$ **18.** $(55.4 \pm 26.4)\,cm^2$ **19.** (7%) **20.** (23.4%) **21.** (13%) **22.** (c)

See hints on page 60.

MODULE 2

Significant Figures

To get some overview of least count and significant figures, let us have some examples.

Example 1 Let us now use a millimeter scale (on which millimeter marks are there). This is also our normal meter scale which we use in our routine life.

From the figure, we can see that length AB is more than 3.3 cm and less than 3.4 cm If we note down the length,

$$l = AB = 3.4 \text{ cm}$$

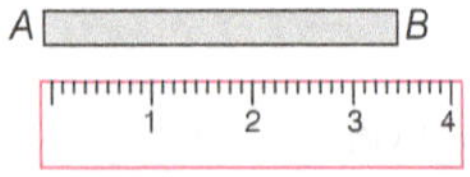

Then, this measurement has two significant figures 3 and 4 in which 3 is absolutely correct and 4 is reasonably correct (doubtful). Least count of this scale is 0.1 cm because this scale can measure accurately only upto 0.1 cm. Further, maximum uncertainty or maximum possible error in l can also be 0.1 cm.

From example 1, we can conclude that:

"In a measured quantity, significant figures are the digits which are absolutely correct plus the first uncertain digit".

INTRODUCTORY Exercise 1

1. If we measure a length $l = 6.24$ cm with the help of a vernier callipers, then
   (a) what is least count of vernier callipers ?
   (b) how many significant figures are there in the measured length ?
   (c) which digits are absolutely correct and which is/are doubtful?
2. If we measure a length $l = 3.267$ cm with the help of a screw gauge, then
   (a) what is maximum uncertainty or maximum possible error in l ?
   (b) how many significant figures are there in the measured length ?
   (c) which digits are absolutely correct and which is/are doubtful ?

Answers

1. (a) 0.01 cm (b) 3
(c) 6 and 2 are absolutely correct and 4 is doubtful.
2. (a) 0.001 cm (b) 4
(c) 3, 2 and 6 are absolutely correct and 7 is doubtful.

See hints on page 54.

Rules for Counting Significant Figures

Rule 1 All non-zero digits are significant. For example, 126.28 has five significant figures.

Rule 2 The zeros appearing between two non-zero digits are significant. For example, 6.025 has four significant figures.

Rule 3 Trailing zeros after decimal places are significant. Measurement $l = 6.400$ cm has four significant figures. Let us take an example in its support.

Table 1

Measurement	Accuracy	l lies between (in cm)	Significant figures	Remarks
$l = 6.4$ cm	0.1 cm	6.3 – 6.5	Two	
$l = 6.40$ cm	0.01 cm	6.39 – 6.41	Three	closer
$l = 6.400$ cm	0.001 cm	6.399 – 6.401	Four	more closer

Thus, the significant figures depend on the accuracy of measurement. More the number of significant figures, more accurate is the measurement.

Rule 4 The powers of ten are not counted as significant figures. For example, 1.4×10^{-7} has only two significant figures 1 and 4.

Rule 5 If a measurement is less than one, then all zeros occurring to the left of last non-zero digit are not significant. For example, 0.0042 has two significant figures 4 and 2.

Rule 6 Change in units of measurement of a quantity does not change the number of significant figures. Suppose a measurement was $l = 72$ mm (two significant figures).

We can write this measurement in other units also (without changing the number of significant figures) :

7.2 cm → Two significant figures.
0.072 m → Two significant figures.
0.000072 km → Two significant figures.
7.2×10^{7} nm → Two significant figures

Rule 7 The terminal or trailing zeros in a number without a decimal point are not significant. This also sometimes arises due to change of unit.

For example, 264 m = 26400 cm = 264000 mm

All have only three significant figures 2, 6 and 4. All trailing zeros are not significant.

Zeroes at the end of a number are significant only if they are behind a decimal point as in Rule-3. Otherwise, it is impossible to tell if they are significant. For example, in the number 8200, it is not clear if the zeros are significant or not.

The number of significant digits in 8200 is at least two, but could be three or four. To avoid uncertainty, use scientific notation to place significant zeros behind a decimal point

8.200×10^3 has four significant digits.

8.20×10^3 has three significant digits.

8.2×10^3 has two significant digits.

Therefore, if it is not expressed in scientific notations, then write least number of significant digits. Hence, in the number 8200, take significant digits as two.

Rule 8 Exact measurements have infinite number of significant figures. For example,

10 bananas in a basket

46 students in a class

speed of light in vacuum

$= 299{,}792{,}458$ m/s (exact)

$\pi = \frac{22}{7}$ (exact)

All these measurements have infinite number of significant figures.

Example 2 Table 2

Measured value	Number of significant figures	Rule number
12376 cm	5	1
6024.7 cm	5	2
0.071 cm	2	5
4100 cm	2	7
2.40 cm	3	3
1.60×10^{14} km	3	4

INTRODUCTORY Exercise 2

1. Count total number of significant figures in the following measurements:
   (a) 4.080 cm (b) 0.079 m
   (c) 950 (d) 10.00 cm
   (e) 4.07080 (f) 7.090×10^5

Answers

1. (a) 4 (b) 2 (c) 2 (d) 4
   (e) 6 (f) 4

See hints on page 54.

Rounding Off a Digit

Following are the rules for rounding off a measurement :

Rule 1 If the number lying to the right of cut off digit is less than 5, then the cut off digit is retained as such. However, if it is more than 5, then the cut off digit is increased by 1.

For example, $x = 6.24$ is rounded off to 6.2 to two significant digits and $x = 5.328$ is rounded off to 5.33 to three significant digits.

Rule 2 If the digit to be dropped is 5 followed by non-zero digits, then the preceding digit is increased by 1.
For example, $x = 14.252$ is rounded off to $x = 14.3$ to three significant digits.

Rule 3 If the digit to be dropped is simply 5 or 5 followed by zeros, then the preceding digit is left unchanged if it is even.
For example, $x = 6.250$ or $x = 6.25$ becomes $x = 6.2$ after rounding off to two significant digits.

Rule 4 If the digit to be dropped is 5 or 5 followed by zeros, then the preceding digit is raised by one if it is odd.
For example, $x = 6.350$ or $x = 6.35$ becomes $x = 6.4$ after rounding off to two significant digits.

Example 3 Table 3

Measured value	After rounding off to three significant digits	Rule
7.364	7.36	1
7.367	7.37	1
8.3251	8.33	2
9.445	9.44	3
9.4450	9.44	3
15.75	15.8	4
15.7500	15.8	4

INTRODUCTORY Exercise 3

1. Round off the following numbers to three significant figures :
   (a) 24572 (b) 24.937
   (c) 36.350 (d) 42.450×10^9
2. Round 742396 to four, three and two significant digits.

Answers

1. (a) 24600 (b) 24.9 (c) 36.4 (d) 42.4×10^9
2. 742400, 742000, 740000

See hints on page 54.

Algebraic Operations with Significant Figures

The final result shall have significant figures corresponding to their number in the least accurate variable involved. To understand this, let us consider a chain of which all links are strong except the one. The chain will obviously break at the weakest link. Thus, the strength of the chain cannot be more than the strength of the weakest link in the chain.

Addition and Subtraction

Suppose, in the measured values to be added or subtracted the least number of digits after the decimal is n. Then, in the sum or difference also, the number of digits after the decimal should be n.

Example 4 $1.2 + 3.45 + 6.789 = 11.439 \approx 11.4$

Here, the least number of significant digits after the decimal is one (in 1.2). Hence, the result will be 11.4 (when rounded off to smallest number of decimal places).

Example 5 $12.63 - 10.2 = 2.43 \approx 2.4$

Multiplication or Division

Suppose in the measured values to be multiplied or divided the least number of significant digits be n. Then in the product or quotient, the number of significant digits should also be n.

Example 6 $1.2 \times 36.72 = 44.064 \approx 44$

The least number of significant digits in the measured values are two (in 1.2). Hence, the result when rounded off to two significant digits become 44. Therefore, the answer is 44.

Example 7 $\dfrac{1101\ \text{ms}^{-1}}{10.2\ \text{ms}^{-1}} = 107.94117647 \approx 108$

Example 8 Find, volume of a cube of side $a = 1.4 \times 10^{-2}$ m.

Sol. Volume $V = a^3$

$$= (1.4 \times 10^{-2}) \times (1.4 \times 10^{-2}) \times (1.4 \times 10^{-2})$$

$$= 2.744 \times 10^{-6}\ \text{m}^3$$

Since, each value of a has two significant figures. Hence, we will round off the result to two significant figures.

$\therefore$ $V = 2.7 \times 10^{-6}\ \text{m}^3$ **Ans.**

Example 9 Radius of a wire is 2.50 mm. The length of the wire is 50.0 cm. If mass of wire was measured as 25 g, then find the density of wire in correct significant figures. [Given, $\pi = 3.14$, exact]

Sol. Given, $r = 2.50$ mm (three significant figures)

$= 0.250$ cm (three significant figures)

Note *Change in the units of measurement of a quantity does not change the number of significant figures.*

Further given that,

$l = 50.0$ cm (three significant figures)

$m = 25$ g (two significant figures)

$\pi = 3.14$ exact (infinite significant figures)

$$\rho = \frac{m}{V} = \frac{m}{\pi r^2 l}$$

$$= \frac{25}{(3.14)(0.250)(0.250)(50.0)}$$

$$= 2.5477\ \text{g/cm}^3$$

But in the measured values, least number of significant figures are two. Hence, we will round off the result to two significant figures.

$\therefore$ $\rho = 2.5\ \text{g/cm}^3$ **Ans.**

INTRODUCTORY Exercise 4

1. Round to the appropriate number of significant digits
   (a) $13.214 + 234.6 + 7.0350 + 6.38$
   (b) $1247 + 134.5 + 450 + 78$
2. Simplify and round to the appropriate number of significant digits
   (a) $16.235 \times 0.217 \times 5$ (b) 0.00435×4.6

Answers

1. (a) 261.2 (b) 1910
2. (a) 20 (b) 0.020

See hints on page 61.

Miscellaneous Examples

Ex. 1 Round off 0.07284 to four, three and two significant digits.

Sol. 0.07284 (four significant digits)

0.0728 (three significant digits)

0.073 (two significant digits)

Ex. 2 Round off 231.45 to four, three and two significant digits.

Sol. 231.4 (four significant digits)

231 (three significant digits)

230 (two significant digits)

Ex. 3 Three measurements are $a = 483, b = 73.67$ and $c = 15.67$.

Find the value $\dfrac{ab}{c}$ to correct significant figures.

Sol. $\dfrac{ab}{c} = \dfrac{483 \times 73.67}{15.67}$

$$= 2270.7472$$

$$= 2.27 \times 10^3$$ **Ans.**

Note *The result is rounded off to least number of significant figures in the given measurement i.e. 3 (in 483).*

Ex. 4 Three measurements are, $a = 25.6, b = 21.1$ and $c = 2.43$. Find the value $a - b - c$ to correct significant figures.

Sol. $a - b - c = 25.6 - 21.1 - 2.43$

$= 2.07 = 2.1$ **Ans.**

Note *In the measurements, least number of significant digits after the decimal is one (in 25.6 and 21.1). Hence, the result will also be rounded off to one decimal place.*

Ex. 5 A thin wire has a length of 21.7 cm and radius 0.46 mm. Calculate the volume of the wire to correct significant figures.

Sol. Given, $l = 21.7$ cm, $r = 0.46$ mm $= 0.046$ cm

Volume of wire $V = \pi r^2 l$

$$= \frac{22}{7}(0.046)^2 (21.7) = 0.1443 \text{ cm}^3$$

$$= 0.14 \text{ cm}^3$$ **Ans.**

Note *The result is rounded off to least number of significant figures in the given measurements i.e. 2 (in 0.46 mm).*

Ex. 6 The area of a square is 5.29 cm^2. The area of 7 such squares taking into account the significant figures is **(2019 Main)**

(a) 37.030cm^2 (b) 37.0cm^2 (c) 37.03cm^2 (d) 37cm^2

Sol. Area of one square $= 5.29 cm^2$

Area of seven such squares

$= 7$ times addition of area of one square

$= 5.29 + 5.29 + 5.29 \ldots 7$ times

$= 37.03 \text{ cm}^2$

As we know that, if in the measured values to be added/subtracted the least number of significant digits after the decimal is *n*.

Then, in the sum or difference also, the number of significant digits after the decimal should be *n*.

Here, number of digits after decimal in 5.29 is 2, so our answer should also contain only two digits after decimal point.

$\therefore$ Area required $= 37.03 cm^2$

$\therefore$ The correct answer is (c).

Note *We can also multiply 5.29 cm^2 by 7 to get total area. In that case we will have to use multiplication rule during rounding off. So, the answer will be different. But addition rule is more general. The reason is, if area were different, then we can't use multiplication formula.*

MODULE 2 Exercise

Least Count and Most Accurate Measurement

Assertion and Reason

Directions : Choose the correct option.

(a) *If both **Assertion** and **Reason** are true and the **Reason** is the correct explanation of the **Assertion**.*

(b) *If both **Assertion** and **Reason** are true but **Reason** is not the correct explanation of **Assertion**.*

(c) *If **Assertion** is true, but the **Reason** is false.*

(d) *If **Assertion** is false but the **Reason** is true.*

(e) *If both **Assertion** and **Reason** are false.*

1. **Assertion :** Out of two measurements $l = 0.7$m and $l = 0.70$ m, the second one is more accurate.

   Reason : In every measurement, mostly the last digit is not accurately known.

2. **Assertion :** The watches having hour hand and minute hand have least count as 1 s.

   Reason : Least count is the smallest measurement that can be measured accurately by an instrument.

Single Correct Option

3. Which of the following measurement is most accurate?

   (a) 40 m (b) 0.04 m
   (c) 4.00×10^{-4} m (d) 4×10^2 m

Significant Number and Round Off

Assertion and Reason

4. **Assertion :** Trailing zeroes in a number with a decimal point is significant.

   Reason : In number 0.800, significant digits are 3.

Single Correct Option

5. The number of significant figures in 3400 is

   (a) 3 (b) 1
   (b) 4 (d) 2

6. The significant figures in the number 6.0023 are

   (a) 2 (b) 5
   (c) 4 (d) 3

7. The respective number of significant figures for the numbers 23.023, 0.0003 and 2.1×10^{-3} are

   (a) 5, 1, 2
   (b) 5, 1, 5
   (c) 5, 5, 2
   (d) 4, 4, 2

8. What is the number of significant figures in 0.310×10^3

   (a) 2 (b) 3
   (c) 4 (d) 6

9. The number of significant figures in 0.06900 is

   (a) 5 (b) 4
   (c) 2 (d) 3

10. The numbers 2.745 and 2.735 on rounding off to 3 significant figures will give
(a) 2.75 and 2.74 (b) 2.74 and 2.73
(c) 2.75 and 2.73 (d) 2.74 and 2.74

Addition and Subtraction

Single Correct Option

11. Find the value of $525.5 + 10.81 - 53.15$
(a) 483.16 (b) 483 (c) 483.2 (d) 483.1

Subjective Questions

12. The mass of a box measured by a grocer's balance is 2.3 kg. Two gold pieces of masses 20.15 g and 20.17 g are added to the box. What is (a) the total mass of the box, (b) the difference in the masses of the pieces to correct significant figures ?

13. Taking into account the significant figures, what is the value of 9.99 m + 0.0099 m?

14. With due regard to significant figures, add the following
(a) 953 and 0.324 (b) 953 and 0.625
(c) 953.0 and 0.324 (d) 953.0 and 0.374

15. With due regard to significant figures, subtract
(a) 0.35 from 7 (b) 0.65 from 7
(c) 0.38 from 7.0 (d) 0.62 from 7.0

16. (a) Add 3.8×10^{-6} to 4.2×10^{-5} with due regard to significant figures.
(b) Subtract 3.2×10^{-6} from 4.7×10^{-4} with due regard to significant figures.
(c) Subtract 1.6×10^{3} from 4.8×10^{4} with due regard to significant figures.

17. A container contains 35 kg of water, 0.2 kg water leaks from the container. Find the amount of water in the container (in kg) with due regard to significant figures.

Multiplication and Division

Single Correct Option

18. Find the value of $\dfrac{3.008 \times 38.8}{2.8768}$
(a) 40.5695 (b) 40.5 (c) 40.57 (d) 40.6

19. The length and breadth of a metal sheet are 3.124 m and 3.002 m, respectively. The area of this sheet upto correct significant figure is
(a) $9.378\,m^2$ (b) $9.37\,m^2$
(c) $9.4\,m^2$ (d) None of these

20. The length, breadth and thickness of a block are given by $l = 12$ cm, $b = 6$ cm and $t = 2.45$ cm.
The volume of the block according to the idea of significant figures should be
(a) $1 \times 10^2\ cm^3$ (b) $2 \times 10^2\ cm^3$
(c) $1.763 \times 10^2\ cm^3$ (d) None of these

21. A body of mass $m = 3.513$ kg is moving along the X-axis with a speed of 5.00 ms^{-1}. The magnitude of its momentum is recorded as
(a) 17.6 kg-ms^{-1} (b) 17.565 kg-ms^{-1}
(c) 17.56 kg-ms^{-1} (d) 17.57 kg-ms^{-1}

22. In Ohm's law experiment, the reading of voltmeter across the resistor is 12.8 V and current through resistor is 0.30 A. Find the resistance of resistor.
(a) $42.67\,\Omega$ (b) $42.667\,\Omega$ (c) $42.3\,\Omega$ (d) $43\,\Omega$

23. A cube has a side of length 1.2×10^{-2}m. What is its volume? **(2003 JEE)**
(a) $1.7 \times 10^{-6}\,m^3$
(b) $1.73 \times 10^{-6}\,m^3$
(c) $1.70 \times 10^{-6}\,m^3$
(d) $1.732 \times 10^{-6}\,m^3$

Subjective Questions

24. A cube has a side of length 2.342 m. Find the volume and surface area of one side in correct significant figures.

25. Find density when a mass of 9.23 kg occupies a volume of $1.1\,m^3$. Take care of significant figures.

Miscellaneous Problems

Single Correct Option

26. The mass of a ball is 1.76 kg. The mass of 25 such balls is
(a) 0.44×10^3 kg (b) 44.0 kg
(c) 44 kg (d) 44.00 kg

27. In the context of accuracy of measurement and significant figures in expressing result of experiment, which of the following is/are correct?
1. Out of two measurements 50.14 cm and 0.00025 ampere, the first one has greater accuracy
2. If one travels 478 km by rail and 397 m by road, the total distance travelled is 478 km.

(a) Only (1) is correct (b) Only (2) is correct
(c) Both are correct (d) None of these

Answers

1. (b) 2. (d) 3. (c) 4. (b) 5. (d)
6. (b) 7. (a) 8. (b) 9. (b) 10. (d)
11. (c) 12. (a) 2.3 kg (b) 0.02 g 13. 10.00 m
14. (a) 953 (b) 954 (c) 953.3 (d) 953.4
15. (a) 7 (b) 6 (c) 6.6 (d) 6.4
16. (a) 4.6×10^{-5} (b) 4.7×10^{-4} (c) 4.6×10^{4} 17. 35 18. (d)
19. (a) 20. (b) 21. (a) 22. (d) 23. (a)
24. $12.85\,m^3$, $5.485\,m^2$ 25. 8.4 kg/m^3 26. (d) 27. (c)

See hints on pages 61-63.

Unit Exercise (Cumulative)

Level 1

Single Correct Option

1. If the random error in the arithmetic mean of 50 observations is α, then the random error in the arithmetic mean of 150 observations would be
(a) α (b) 3α
(c) $\frac{\alpha}{3}$ (d) 2α

2. Let g be the acceleration due to gravity at the earth's surface and K be the rotational kinetic energy of the earth. Suppose the earth's radius decreases by 2%. Keeping all other quantities constant, then (*Attempt this question after completing gravitation and rotation*)
(a) g increases by 2% and K increases by 2%
(b) g increases by 4% and K increases by 4%
(c) g decreases by 4% and K decreases by 2%
(d) g decreases by 2% and K decreases by 4%

3. If the length of the pendulum in pendulum clock increases by 0.1%, then the error in time per day is $\left[T = 2\pi\sqrt{\frac{l}{g}}\right]$ **(2021 JEE)**
(a) 86.4 s (b) 4.32 s
(c) 43.2 s (d) 8.64 s

4. The relative error in the determination of the surface area of a sphere is α. Then the relative error in the determination of its volume is **(2018 JEE)**
(a) $\frac{2}{3}\alpha$ (b) $\frac{2}{3}\alpha$
(c) $\frac{3}{2}\alpha$ (d) α

5. A student measures the time period of 100 oscillations of a simple pendulum four times. The data set is 90 s, 91 s, 95 s and 92 s. If the minimum division in the measuring clock is 1 s, then the reported mean time should be **(2016 JEE)**
(a) (92 ± 1.8) s
(b) (92 ± 3) s
(c) (92 ± 2) s
(d) (92 ± 5.0) s

6. In Ohm's law experiment, the reading of voltmeter across the resistor is 12.8V and current through resistor is 0.30 A. Find the maximum percentage error in measurement of resistance.
(a) 4.1% (b) 4.2%
(c) 5% (d) 10%

7. A student measures the thickness of a human hair by looking at it through a microscope of magnification 100. He makes 20 observations and finds that the average width of the hair is 3.5 mm. What is thickness of the hair?
(a) 0.0035 mm (b) 0.035 mm
(c) 0.01 m (d) 0.7 mm

8. The diameter and height of a cylinder are measured by a meter scale to be 12.6 ± 0.1 cm and 34.2 ± 0.1 cm, respectively. What will be the value of its volume in appropriate significant figures? **(2019 Main)**
(a) $4300 \pm 80\ \text{cm}^3$
(b) $4260 \pm 80\ \text{cm}^3$
(c) $4264.4 \pm 81.0\ \text{cm}^3$
(d) $4264 \pm 81\ \text{cm}^3$

Multiple Correct Options

9. Two resistances are expressed as $R_1 = (3 \pm 0.2)\Omega$ and $R_2 = (6 \pm 0.3)\ \Omega$. Which of the following statement(s) is/are correct?
(a) Equivalent resistance in series combination is $(9 \pm 0.5)\Omega$
(b) Equivalent resistance in series combination is $(9 \pm 0.1)\Omega$
(c) Equivalent resistance in parallel combination is $(2 \pm 0.12)\Omega$
(d) Equivalent resistance in parallel combination is $(2 \pm 0.5)\Omega$

Integer Type Question

10. The focal length of a spherical mirror is given by $\frac{1}{f} = \frac{1}{v} + \frac{1}{u}$, where u is the position of an object from pole of the mirror and v is the position of the image from the pole of the mirror.

If $u = (30 + 0.3)$ cm and $v = (60 + 0.6)$ cm, the maximum percentage error in the measurement of the focal length of the mirror is n %. Find the value of n.

Level 2

Fill in the Blank

11. Three students S_1, S_2 and S_3 perform an experiment for determining the acceleration due to gravity (g) using a simple pendulum. They use different lengths of pendulum and record time for different number of oscillations. The observations are as shown in the table. $\left[T = 2\pi\sqrt{\frac{l}{g}}\right]$

Student No.	Length of pendulum (cm)	No. of oscillations (n)	Total time for n oscillations	Time period (s)
1	64.0	8	128.0	16.0
2	64.0	4	64.0	16.0
3	20.0	4	36.0	9.0

(Least count of length = 0.1 m, least count for time = 0.1 s). If E_1, E_2 and E_3 are the percentage errors in 'g' for students 1, 2 and 3 respectively, then the minimum percentage error is obtained by student no .... **(2021, 2008 JEE)**

Single Correct Option

12. The least count of a stop watch is 0.2 s. The time of 20 oscillations of a pendulum is measured to be 25 s. The percentage error in the time period is
(a) 1.2 % (b) 0.8 %
(c) 1.8 % (d) None of these

13. In a simple pendulum, experiment for determination of acceleration due to gravity (g), time taken for 20 oscillations is measured by using a watch of 1 second least count. The mean value of time taken comes out to be 30 s. The length of pendulum is measured by using a meter scale of least count 1 mm and the value obtained is 55.0 cm. The percentage error in the determination of g is close to **(2019 Main)**
(a) 0.7% (b) 6.8%
(c) 3.5 (d) 0.2%

14. A student performs an experiment to determine the Young's modulus of a wire, exactly 2 m long, by Searle's method. In a particular reading, the student measures the extension in the length of the wire to be 0.8 mm with an uncertainty of $\pm$ 0.05 mm at a load of exactly 1.0 kg. The student also measures the diameter of the wire to be 0.4 mm with an uncertainty of $\pm$ 0.01 mm. Take $g = 9.8\ \text{m/s}^2$ (exact). The Young's modulus obtained from the reading is close to (attempt this question after completing elasticity) **(2007 JEE)**
(a) $(2.0 \pm 0.3)\ 10^{11}\ \text{N/m}^2$
(b) $(2.0 \pm 0.2) \times 10^{11}\ \text{N/m}^2$
(c) $(2.0 \pm 0.1) \times 10^{11}\ \text{N/m}^2$
(d) $(2.0 \pm 0.05) \times 10^{11}\ \text{N/m}^2$

15. A student determined Young's modulus of elasticity using the formula $Y = \frac{MgL^3}{4bd^3\delta}$. The value of g is taken to be $9.8\ \text{m/s}^2$, without any significant error, his observation are as following.

Physical Quantity	Least count of the Equipment used for measurement	Observed value
Mass (M)	1 g	2 kg
Length of bar (L)	1 mm	1 m
Breadth of bar (b)	0.1 mm	4 cm
Thickness of bar (d)	0.01 mm	0.4 cm
Depression (δ)	0.01 m	5 mm

Then the fractional error in the measurement of Y is **(2021 JEE)**
(a) 0.0083 (b) 0.0155
(c) 0.155 (d) 0.083

Multiple Correct Options

16. In an experiment to determine the acceleration due to gravity g, the formula used for the time period of a periodic motion is $T = 2\pi\sqrt{\frac{7(R - r)}{5g}}$. The values of R and r are measured to be (60 ± 1) mm and (10 ± 1) mm, respectively. In five successive measurements, the time period is found to be 0.52 s, 0.56 s, 0.57 s, 0.54 s and 0.59 s. The least count of the watch used for the measurement of time period is 0.01 s. Which of the following statement(s) is (are) true? **(2016 Adv.)**
(a) The error in the measurement of r is 10%
(b) The error in the measurement of T is 3.57%
(c) The error in the measurement of T is 2%
(d) The error in the measurement of g is 11%

17. Given $y = \sin\theta$. The error in measurement of θ is 1°. The fractional error in y
(a) may be zero
(b) may be infinity
(c) will have any value from zero to infinity
(d) must be zero

Passage Based Questions

Passage (Q.18 to Q.19)

If the measurement errors in all the independent quantities are known, then it is possible to determine the error in any dependent quantity. This is done by the use of series expansion and truncating the expansion at the first power of the error. For example, consider the relation $z = x/y$. If the errors in x, y and z are Δx, Δy and Δz respectively, then

$$z \pm \Delta z = \frac{x \pm \Delta x}{y \pm \Delta y} = \frac{x}{y}\left(1 \pm \frac{\Delta x}{x}\right)\left(1 \pm \frac{\Delta y}{y}\right)^{-1}$$

The series expansion for $\left(1 \pm \frac{\Delta y}{y}\right)^{-1}$*, to first power in* $\Delta y/y$*, is* $1 \pm (\Delta y/y)$*. The relative errors in independent variables are always added. So, the error in z will be*

$$\Delta z = z\left(\frac{\Delta x}{x} + \frac{\Delta y}{y}\right)$$

The above derivation makes the assumption that $\Delta x/x \ll 1$, $\Delta y/y \ll 1$*. Therefore, the higher powers of these quantities are neglected.* **(2018 Adv.)**

18. Consider the ratio $r = \frac{(1-a)}{(1+a)}$ to be determined by measuring a dimensionless quantity a. If the error in the measurement of a is Δa $(\Delta a / a \ll 1)$, then what is the error Δr in determining r ?
(a) $\frac{\Delta a}{(1+a)^2}$ (b) $\frac{-2\Delta a}{(1+a)^2}$ (c) $\frac{2\Delta a}{(1-a)^2}$ (d) $\frac{2a\Delta a}{(1-a^2)}$

19. In an experiment, the initial number of radioactive nuclei is 3000. It is found that 1000 ± 40 nuclei decayed in the first 1.0 s. For $|x| \ll 1$, $\ln(1+x) = x$ up to first power in x. The error $\Delta\lambda$, in the determination of the decay constant λ in s^{-1}, is
(a) 0.04 (b) 0.03
(c) 0.02 (d) 0.01

Integer Type Questions

20. In an experiment to find acceleration due to gravity (g) using simple pendulum, time period of 0.5 s is measured from time of 100 oscillation with a watch of 1 s resolution. If measured value of length is 10 cm known to 1 mm accuracy. The accuracy in the determination of g is found to be x %. The value of x is .... **(2022 JEE)**

21. If $y = \ln x$. The percentage error in x is 5%. Find the absolute error in y.

22. If $z = \ln x^2 y$ and percentage errors in x and y are 1% and 2%, respectively. Find the absolute error in z.

Answers

Level 1

1. (c) **2.** (b) **3.** (c) **4.** (c) **5.** (c) **6.** (a) **7.** (b) **8.** (b) **9.** (a,c) **10.** (1)

Level 2

11. (1) **12.** (b) **13.** (b) **14.** (b) **15.** (b) **16.** (a,b,d) **17.** (a,b,c) **18.** (b) **19.** (c) **20.** (5)
21. (0.05) **22.** (0.04)

See hints on pages 63-65.

Hints & SOLUTIONS

Module 1.

1. $l = l_1 + l_2 = 11.05 + 11.05 = 22.10$ cm

$$\Delta l = \Delta l_1 + \Delta l_2 = 0.2 + 0.2 = 0.4 \text{ cm}$$

Hence, $(l \pm \Delta l) = (22.10 \pm 0.4)$ cm

2. Suppose $2x - 2y$ is z. Then,

$$z = 2 \times 10.0 - 2 \times 10.0 = 0.0$$

$$\Delta z = 2\Delta x + 2\Delta y = 2(0.1 + 0.1) = 0.4$$

$\therefore \quad (z \pm \Delta z) = (0.0 \pm 0.4)$

3. (a) Sum $= v_1 + v_2 = (10.2 \pm 0.02) + (6.4 \pm 0.01)$

$$= (16.6 \pm 0.03) \text{ ms}^{-1}$$

(b) Difference $= v_1 - v_2 = (10.2 \pm 0.02) - (6.4 \pm 0.01)$

$$= 3.8 \pm (0.02 + 0.01)$$

$$= (3.8 \pm 0.03) \text{ ms}^{-1}$$

4. $(50°\text{C} \pm 0.5°\text{C}) - (20°\text{C} \pm 0.5°\text{C}) = (30 \pm 1)°\text{C}$

5. Average value $= \dfrac{2.63 + 2.56 + 2.42 + 2.71 + 2.80}{5} = 2.62$ s

Now, $|\Delta T_1| = 2.63 - 2.62 = 0.01$ s

$|\Delta T_2| = 2.62 - 2.56 = 0.06$ s

$|\Delta T_3| = 2.62 - 2.42 = 0.20$ s

$|\Delta T_4| = 2.71 - 2.62 = 0.09$ s

$|\Delta T_5| = 2.80 - 2.62 = 0.18$ s

Mean absolute error

$$\Delta T = \frac{|\Delta T_1| + |\Delta T_2| + |\Delta T_3| + |\Delta T_4| + |\Delta T_5|}{5}$$

$$= \frac{0.54}{5} = 0.108 = 0.11 \text{ s}$$ **Ans.**

6. Here, mean value of quantity measured,

$$x = \frac{1.29 + 1.33 + 1.34 + 1.35 + 1.32 + 1.36 + 1.30 + 1.33}{8}$$

$x = 1.3275 = 1.33$

Absolute errors in measurement are :

$\Delta x_1 = 1.33 - 1.29 = 0.04$; $\quad \Delta x_2 = 1.33 - 1.33 = 0.00$

$\Delta x_3 = 1.33 - 1.34 = -0.01$; $\quad \Delta x_4 = 1.33 - 1.35 = -0.02$

$\Delta x_5 = 1.33 - 1.32 = +0.01$; $\quad \Delta x_6 = 1.33 - 1.36 = -0.03$

$\Delta x_7 = 1.33 - 1.30 = +0.03$; $\quad \Delta x_8 = 1.33 - 1.33 = 0.00$

Mean absolute error, $\overline{\Delta x} = \dfrac{\sum_{i=1}^{i=n} |(\Delta x)_i|}{n}$

$$= \frac{0.04 + 0.00 + 0.01 + 0.02 + 0.01 + 0.03 + 0.03 + 0.00}{8}$$

$$= \frac{0.14}{8} = 0.0175 = 0.02$$

Relative error $= \pm \dfrac{\overline{\Delta x}}{x} = \pm \dfrac{0.02}{1.33} = \pm 0.015 = \pm 0.02$

Percentage error $= \pm 0.015 \times 100 = 1.5\%$ **Ans.**

7. $\theta_1 = (40.6 \pm 0.2)°\text{C}$, $\theta_2 = (78.3 \pm 0.3)°\text{C}$

Rise in temperature, $\theta = \theta_2 - \theta_1 = 78.3 - 40.6 = 37.7°\text{C}$

$$\Delta\theta = \pm(\Delta\theta_1 + \Delta\theta_2) = \pm(0.2 + 0.3) = 0.5°\text{C}$$

Hence, rise in temperature $= (37.7 \pm 0.5)°\text{C}$ **Ans.**

8. $z = a^2 x^3 y^{1/2}$

$$\frac{\Delta z}{z} = \frac{2\Delta a}{a} + \frac{3\Delta x}{x} + \frac{1}{2}\frac{\Delta y}{y}$$

a is constant.

So, $\dfrac{\Delta z}{z} \times 100\% = \left(\dfrac{3\Delta x}{x} + \dfrac{1}{2}\dfrac{\Delta y}{y}\right) \times 100\%$

$$= 3 \times 4 + \frac{1}{2} \times 12$$

$$= 18\%$$

9. $V = \dfrac{4}{3}\pi R^3$

$\therefore \quad$ (% error in V) = 3 (% error in R)

$$= 3(1\%) = 3\%$$

10. $\rho = \dfrac{m}{V} = \dfrac{m}{l^3} = ml^{-3}$

$\therefore$ Maximum % error in ρ = (% error in m) + 3 (% error in l)

11. $K = \dfrac{1}{2}mv^2$

$\therefore \quad$ % error in K = (% error in m) + 2 (% error in v)

12. $v = \dfrac{d}{t} = \dfrac{13.8}{4.0} = 3.45 \text{ ms}^{-1}$

Further, $\dfrac{\Delta v}{v} = \dfrac{\Delta d}{d} + \dfrac{\Delta t}{t}$ or $\Delta v = \left[\dfrac{\Delta d}{d} + \dfrac{\Delta t}{t}\right] v$

$$= \left[\frac{0.2}{13.8} + \frac{0.3}{4}\right](3.45) = 0.3 \text{ m/s}$$

$\therefore \quad (v \pm \Delta v) = (3.45 \pm 0.3) \text{ ms}^{-1}$

13. Since, error in measurement of momentum is + 100%.

$\therefore \quad p_1 = p, p_2 = 2p$

$$K_1 = \frac{p^2}{2m}, K_2 = \frac{(2p)^2}{2m}$$

$$\% \text{ in } K = \left(\frac{K_2 - K_1}{K_1}\right) \times 100$$

$$= \left(\frac{4-1}{1}\right) \times 100 = 300\%$$

14. Density, $\rho = \dfrac{m}{\pi r^2 L}$

$\therefore \quad \dfrac{\Delta\rho}{\rho} \times 100 = \left(\dfrac{\Delta m}{m} + 2\dfrac{\Delta r}{r} + \dfrac{\Delta L}{L}\right) \times 100$

or Maximum percentage error in density

$= \left[\dfrac{0.003}{0.3} \times 100 + \dfrac{2 \times 0.005}{0.5} \times 100 + \dfrac{0.06}{6} \times 100\right] = 4\%$

15. Given, mass $= (10.00 \pm 0.10)$ kg

Edge length $= (0.10 \pm 0.01)$ m

Error in mass,

$$\frac{\Delta M}{M} = \frac{0.1}{10} \quad \text{...(i)}$$

and error in length,

$$\frac{\Delta l}{l} = \frac{0.01}{0.1} \quad \text{...(ii)}$$

Density of the cube is given by

$$\rho = \frac{\text{Mass}}{\text{Volume}} = \frac{M}{l^3}$$

$\therefore$ Permissible error in density is

$$\frac{\Delta\rho}{\rho} = \frac{\Delta M}{M} \pm 3\frac{\Delta l}{l} \quad \text{...(iii)}$$

Substituting the value from Eqs. (i) and (ii) in Eq. (iii),

We get

$$\frac{\Delta\rho}{\rho} = \frac{0.1}{10} + 3 \times \frac{0.01}{0.1} = \frac{1}{100} + \frac{3}{10} = \frac{31}{100}$$

$$\Rightarrow \frac{\Delta\rho}{\rho} = \frac{31}{100} = 0.31$$

Since, $\dfrac{\Delta\rho}{\rho}$ should be unitless quantity.

But, there is no option with unitless error. Hence, no option is correct.

16. Dissipated heat is given as,

$$E_H = I^2 R \times t$$

$$\frac{\Delta E}{E} \times 100 = \frac{2\Delta I}{I} \times 100 + \frac{\Delta R}{R} \times 100 + \frac{\Delta T}{T} \times 100$$

$$= 2 \times 2 + 1 + 3 = 8$$

17. Volume, $V = \dfrac{4}{3}\pi r^3 = \dfrac{4}{3}\left(\dfrac{22}{7}\right)(1.2)^3$

$$= 7.24 \text{ cm}^3 = 7.2 \text{ cm}^3$$

Further, $\dfrac{\Delta V}{V} = 3\left(\dfrac{\Delta r}{r}\right)$

$\therefore \quad \Delta V = 3\left(\dfrac{\Delta r}{r}\right)V = \dfrac{3 \times 0.2 \times 7.2}{1.2}$

$$= 3.6 \text{ cm}^3$$

$\therefore \quad V = (7.2 \pm 3.6) \text{ cm}^3$ **Ans.**

18. $S = 4\pi r^2 = 4\pi(2.1)^2 = 55.4 \text{ cm}^2$

$$\frac{\Delta S}{S} = 2 \times \frac{\Delta r}{r}$$

$\Rightarrow \quad \Delta S = 2 \times \dfrac{\Delta r}{r} \times S$

$$= 2 \times \frac{0.5}{2.1} \times 55.4$$

$$= 26.4 \text{ cm}^2$$

$\therefore \quad (S \pm \Delta S) = (55.4 \pm 26.4) \text{ cm}^2$ **Ans.**

19. The percentage error in V is 5% and in I it is 2%. The total error in R would therefore be $5\% + 2\% = 7\%$.

20. $\dfrac{\Delta\rho}{\rho} \times 100 = \left[2\left(\dfrac{\Delta r}{r}\right) + \left(\dfrac{\Delta R}{R}\right) + \left(\dfrac{\Delta l}{l}\right)\right] \times 100$

$$= \left[\frac{2 \times 0.02}{0.2} + \frac{2}{60} + \frac{0.1}{150}\right] \times 100$$

$= 23.4\%$ **Ans.**

21. % error in $\rho = 3$ (% error in α) $+ 2$ (% error in β)

$$+ \frac{1}{2}(\text{% error in } \gamma) + (\text{% error in } \eta)$$

22. $R = \dfrac{D}{2} \Rightarrow R \propto D$

or (% error in R) = (% error in D) = 4%.

Module 2.

Introductory Exercise 4

1. (a) $13.214 + 234.6 + 7.0350 + 6.38$

$= 261.229$

$= 261.2$

(rounding off to smallest one number of decimal place)

(b) $1247 + 134.5 + 450 + 78$

$= 1909.5 = 1910$

(rounding off to smallest one number of decimal place)

2. (a) $16.235 \times 0.217 \times 5$

$= 17.614975$

$= 20$

(rounding off to minimum one number of significant figure)

(b) 0.00435×4.6

$= 0.02001$

$= 0.020$

(rounding off to minimum two number of significant figures)

Exercise

1. In 0.7 m, significant figure is only one.

In 0.70 m, significant figures are two.

2. Hour hand have least count 1 hour and the minute hand have the least count 1 min.

3. Measurement in (c) option has the maximum significant digits or three.

4. This is a basic rule of a significant number.

7. The reliable digit plus the first uncertain digit is known as significant figures.

For the number 23.023, all the non-zero digits are significant.

For the number 0.0003, the zero(s) on the right of decimal point but to the left of the first non-zero digit are not significant.

For the number 2.1×10^{-3}, significant figures are 2.

8. Number of significant figures are 3, because 10^3 is a decimal multiplier.

9. If the number is less than 1, the zero (s) on the right of decimal point and before the first non-zero digit are not significant.

10. Rounding off 2.745 to 3 significant figures it would be 2.74. Rounding off 2.735 to 3 significant figures it would be again 2.74.

11. $525.5 + 10.81 - 53.15$

$$= 483.16 = 483.2$$

12. (a) Here, mass of the box,

$$m = 2.3 \text{ kg}$$

Mass of first gold piece,

$$m = 20.15 \text{ g} = 0.02015 \text{ kg}$$

Mass of second gold piece,

$$m_2 = 20.17 \text{ g} = 0.02017 \text{ kg}$$

$\therefore$ Total mass = 2.34032 kg

After the decimal point there should be least or one significant figure. Therefore, answer should be 2.3 kg as we have to round off at one digit after decimal point.

(b) $m_1 - m_2 = 20.17 - 20.15 = 0.02 \text{ g}$

In substraction, also the final result should retain as many decimal places as there are in the number with the least decimal places. Therefore, answer should be 0.02g.

13. $9.99 \text{ m} + 0.0099 \text{ m} = 9.999 \text{ m} = 10.00 \text{ m}$

(In proper significant figures)

16. (a) $3.8 \times 10^{-6} + 4.2 \times 10^{-5} = 0.38 \times 10^{-5} + 4.2 \times 10^{-5}$

$$= 4.58 \times 10^{-5} = 4.6 \times 10^{-5}$$

(b) $4.7 \times 10^{-4} - 3.2 \times 10^{-6} = 4.7 \times 10^{-4} - 0.032 \times 10^{-4}$

$$= 4.668 \times 10^{-4} = 4.7 \times 10^{-4} \quad \text{(one decimal place)}$$

(c) $4.8 \times 10^4 - 1.6 \times 10^3 = 4.8 \times 10^4 - 0.16 \times 10^4$

$$= 4.64 \times 10^4 = 4.6 \times 10^4$$

(one decimal place)

17. Mass $= 35 - 0.2 = 34.8 = 35 \text{ kg}$

18. $\dfrac{3.008 \times 38.8}{2.8768} = 40.5695 = 40.6$

(least significant figure is 3)

19. $\therefore \; A = l \times b = 3.124 \times 3.002$

$$= 9.378248 \text{ m}^2 = 9.378 \text{ m}^2$$ **Ans.**

(rounding off to four significant digits)

20. $\therefore \; V = lbt = 12 \times 6 \times 2.45 = 176.4 \text{ cm}^3$

$$= 2 \times 10^2 \text{ cm}^3$$

(rounding off to one significant digit of breadth) **Ans.**

21. Momentum, $p = mv = 3.513 \times 5.00$

$$= 17.565 \text{ kg-ms}^{-1}$$

As the number of significant digits in m is 4 and in v is 3, so p must have 3 (minimum) significant digits. Hence,

$$p = 17.6 \text{ kg - ms}^{-1}$$

22. According to Ohm's law,

$$V = IR \Rightarrow R = \frac{V}{I} = \frac{12.8}{0.30} = 42.667\,\Omega = 43\,\Omega$$

23. $V = l^3 = (1.2 \times 10^{-2} \text{m})^3 = 1.728 \times 10^{-6} \text{ m}^3$

$\because$ Length (l) has two significant figures, the volume (V) should also have two significant figures. Therefore, the correct answer is

$$V = 1.7 \times 10^{-6} \text{ m}^3$$

24. $V = a^3$

$$= (2.342 \text{ m})^3 = 12.84578569 \text{ m}^3$$

$$= 12.85 \text{ m}^3 \quad \text{(rounding off to four significant digits)}$$

$$S = a^2 = (2.342)^2 \text{ m}^2$$

$$= 5.484964 \text{ m}^2$$

$$= 5.485 \text{ m}^2 \quad \text{(rounding off to four significant digits)}$$

Ans.

25. $\rho = \dfrac{m}{V} = \dfrac{9.23}{1.1}$

$$= 9.3909090 \text{ kg/m}^3$$

$$= 8.4 \text{ kg/m}^3$$

(rounding off to two significant digits) **Ans.**

26. If we add 1.76 kg, twenty five times

then, $m_T = 44 \text{kg}$

From the law of significant figures for addition answer should be,

$$m_T = 44.00 \text{kg}$$

Note *In the opinion of author, here the masses are equal. So, we can use multiplication formula. If masses were different, then we will have to add those masses. So, addition formula is more general.*

27. For 50.14 cm, significant number = 4 and for 0.00025, significant numbers = 2

Unit Exercise (Cumulative)

Level 1

1. Random error $= \dfrac{1}{\text{Number of observations}}$

2. $g = \dfrac{GM}{R^2}$ or $g \propto R^{-2}$

$\therefore$ % change in $g = (-2)$ (% change in R)

$= (-2)(-2) = +4\%$

Rotational kinetic energy,

$$K = \frac{L^2}{2I} \quad \text{...(i)}$$

L = angular momentum = constant

But $$I = \frac{2}{5}mR^2 \quad \text{...(ii)}$$

From Eqs. (i) and (ii),

$$K \propto R^{-2}$$

$\therefore$ % change in $K = (-2)$(% change in R)

$= (-2)(-2) = +4\%$ **Ans.**

3. Time period, $T = 2\pi\sqrt{\dfrac{l}{g}}$

$$\therefore \quad \frac{\Delta T}{T} = \frac{1}{2}\frac{\Delta l}{l}$$

Error in time period in one day

$$\Delta t = \left(\frac{\Delta T}{T'}\right)t \approx \left(\frac{\Delta T}{T}\right)t$$

$$\Rightarrow \quad \Delta T = \frac{1}{2} \times \frac{0.1}{100} \times 24 \times 3600$$

$$= \frac{1}{2} \times \frac{0.1}{100} \times 86400$$

$$\Rightarrow \quad \Delta T = 43.2 \text{ s}$$

4. Relative error in surface area, $\dfrac{\Delta S}{S} = 2 \times \dfrac{\Delta r}{r} = \alpha$ and

relative error in volume, $\dfrac{\Delta V}{V} = 3 \times \dfrac{\Delta r}{r}$

$\therefore$Relative error in volume w.r.t. relative error in area

$$\frac{\Delta V}{V} = \frac{3}{2}\alpha$$

5. Sum of all observations = 90 + 91 + 95 + 92 = 368

$$\text{Average} = \frac{368}{4} = 92 \text{ s}$$

$$\Delta T = \frac{|\Delta T_1| + |\Delta T_2| + |\Delta T_3| + |\Delta T_4|}{4}$$

$$= \frac{2 + 1 + 3 + 0}{4} \approx 2$$

As the resolution of measuring clock is 2 therefore the mean time should be (92 ± 2) s.

6. $\therefore \quad R = \dfrac{V}{I}$

$$\therefore \quad \frac{\Delta R}{R} = \frac{\Delta V}{V} + \frac{\Delta I}{I}$$

or $$\frac{\Delta R}{R} = \frac{0.1}{12.8} + \frac{0.01}{0.30} = 0.041$$

Percentage error in resistance is

$$\frac{\Delta R}{R} \times 100 = 4.1\%$$

7. Magnification is 100. Hence, thickness of hair,

$$t = \frac{3.5}{100} = 0.035 \text{ mm}$$

8. Volume of a cylinder of radius 'r' and height h is given by

$$V = \pi r^2 h \text{ or } V = \frac{1}{4}\pi D^2 h$$

where, D is the diameter of circular surface.

Here, $D = 12.6$ cm and $h = 34.2$ cm

$$\Rightarrow \quad V = \frac{\pi}{4} \times (12.6)^2 \times (34.2)$$

$$V = 4262.22 \text{ cm}^3$$

$V = 4260$ (rounded off to three significant numbers)

Now, error calculation can be done as

$$\frac{\Delta V}{V} = 2\left(\frac{\Delta D}{D}\right) + \frac{\Delta h}{h}$$

or $$\Delta V = \left(\frac{2 \times 0.1}{12.6} + \frac{0.1}{34.2}\right)(4262.22)$$

$$\Rightarrow \quad \Delta V = 79.80 \simeq 80 \text{ cm}^3$$

Hence, $$V = 4260 \pm 80 \text{ cm}^3$$

9. (a) In series combination,

$$R = R_1 + R_2 = 3 + 6 = 9\,\Omega$$

$$\Delta R = \Delta R_1 + \Delta R_2 = 0.2 + 0.3 = 0.5$$

$$\therefore \quad R = (9 \pm 0.5)\,\Omega$$

(c) In parallel combination,

$$\frac{1}{R} = \frac{1}{R_1} + \frac{1}{R_2} \text{ or } R = \frac{R_1 R_2}{R_1 + R_2}$$

$$\text{or } R = \frac{3 \times 6}{3 + 6} = 2\,\Omega$$

$$\frac{1}{R} = \frac{1}{R_1} + \frac{1}{R_2}$$

$$\Rightarrow \quad -\frac{1}{R}\cdot\frac{dR}{dx} = \frac{-1}{R_1^2}\cdot\frac{dR_1}{dx} + \frac{-1}{R_2^2}\cdot\frac{dR_2}{dx}$$

$$\Rightarrow \quad \frac{\Delta R}{R^2} = \frac{\Delta R_1}{R_1^2} + \frac{\Delta R_2}{R_2^2}$$

$$\Rightarrow \quad \frac{\Delta R}{4} = \frac{0.2}{9} + \frac{0.3}{36}$$

$\therefore \quad \Delta R = \frac{0.8}{9} + \frac{0.3}{9}$

$\therefore \quad \Delta R = 0.12$

$\therefore \quad R = (2 \pm 0.12)\,\Omega$

10. $\therefore \quad \frac{1}{f} = \frac{1}{v} + \frac{1}{u}$

or $\quad \frac{1}{f} = \frac{1}{30} + \frac{1}{60}$

or $\quad \frac{1}{f} = \frac{3}{60}$ or $f = 20$ cm

Differentiating above equation, we have

$$\frac{\Delta f}{f^2} = \frac{\Delta v}{v^2} + \frac{\Delta u}{u^2}$$

or $\quad \frac{\Delta f}{400} = \frac{0.6}{3600} + \frac{0.3}{900}$

$\Rightarrow \quad \Delta f = 0.2$

Percentage error in focal length

$$= \frac{\Delta f}{f} \times 100 = \frac{0.2}{20} \times 100$$

$$= 1\,\%$$

Level 2

11. Time period, $T = 2\pi\sqrt{\frac{l}{g}}$

$\Rightarrow \quad g = \frac{4\pi^2 l}{T^2} \quad \Rightarrow \quad \frac{\Delta g}{g} = \frac{\Delta l}{l} + \frac{2\Delta T}{T}$

$$\Delta T = \frac{\text{Least count of time } (\Delta T_0)}{\text{Number of oscillations } (n)}$$

$\therefore \quad \frac{\Delta g}{g} = \frac{\Delta l}{l} + \frac{2\Delta T_0}{nT_0}$, T_0 = time for 1 oscillation

$\frac{\Delta g}{g}$ is minimum for highest value of nT_0 and l.

Hence, minimum percentage error in g is obtained by student number 1.

12. $T = \frac{t}{n} \quad \Rightarrow \quad \Delta T = \frac{\Delta t}{n}$

$\therefore \quad \frac{\Delta T}{T} \times 100 = \frac{\Delta t}{t} \times 100$

$$= \frac{0.2}{25} \times 100 = 0.8\,\%$$

13. $T = 2\pi\sqrt{\frac{l}{g}} \quad \Rightarrow \quad g = \frac{4\pi^2 l}{T^2}$

So, fractional error in value of g is

$$\frac{\Delta g}{g} = \frac{\Delta l}{l} + \frac{2\Delta T}{T} \qquad \text{...(i)}$$

Given, $\Delta l = 0.1$ cm, $l = 55$ cm, $\Delta T = 1$ s and T for 20 oscillations $= 30$ s

Substituting above values in Eq. (i), we get

$$\frac{\Delta g}{g} = \frac{0.1}{55} + 2 \times \frac{1}{30}$$

Hence, percentage error in g is

$$= \frac{\Delta g}{g} \times 100 = \frac{10}{55} + \frac{20}{3} = 6.8\,\%$$

14. $Y = \frac{FL}{Al} = \frac{4FL}{\pi d^2 l}$

$$= \frac{(4)(1.0 \times 9.8)(2)}{\pi (0.4 \times 10^{-3})^2 (0.8 \times 10^{-3})}$$

$$= 2.0 \times 10^{11}\ \text{N/m}^2$$

Further $\frac{\Delta Y}{Y} = 2\left(\frac{\Delta d}{d}\right) + \left(\frac{\Delta l}{l}\right)$

$\therefore \quad \Delta Y = \left\{2\left(\frac{\Delta d}{d}\right) + \left(\frac{\Delta l}{l}\right)\right\} Y$

$$= \left\{2 \times \frac{0.01}{0.4} + \frac{0.05}{0.8}\right\} \times 2.0 \times 10^{11}$$

$= 0.225 \times 10^{11}\ \text{N/m}^2$ (By rounding off)

or $\quad (Y + \Delta Y) = (2 \pm 0.2) \times 10^{11}\ \text{N/m}^2$

15. $Y = \frac{MgL^3}{4bd^3\delta}$

$\therefore \quad$ Fraction error

$$\frac{\Delta Y}{Y} = \frac{\Delta M}{M} + \frac{3\Delta L}{L} + \frac{\Delta b}{b} + \frac{3\Delta d}{d} + \frac{\Delta \delta}{\delta}$$

If error is not specified, then least count is only error.

$\Rightarrow \quad \frac{\Delta Y}{Y} = \frac{1}{2000} + \frac{3}{1000} + \frac{1}{400} + \frac{3}{400} + \frac{1}{500}$

$$= \frac{31}{2000} = 0.0155$$

16. Mean time period $= \frac{0.52 + 0.56 + 0.57 + 0.54 + 0.59}{5}$

$= 0.556 \cong 0.56$ sec as per significant figures

Error in reading $= |T_{\text{mean}} - T_1| = 0.04$

$|T_{\text{mean}} - T_2| = 0.00$

$|T_{\text{mean}} - T_3| = 0.01$

$|T_{\text{mean}} - T_4| = 0.02$

$|T_{\text{mean}} - T_5| = 0.03$

Mean error $= 0.1/5 = 0.02$

% error in $T = \frac{\Delta T}{T} \times 100 = \frac{0.02}{0.56} \times 100 = 3.57\%$

% error in $r = \frac{0.001 \times 100}{0.010} = 10\%$

% error in $R = \frac{0.001 \times 100}{0.60} = 1.67\%$

% error in $\frac{\Delta g}{g} \times 100 = \frac{\Delta(R - r)}{R - r} \times 100 + 2 \times \frac{\Delta T}{T} \times 100$

$$= \frac{0.002 \times 100}{0.05} + 2 \times 3.57 = 4\% + 7\% = 11\%$$

17. $y = \sin\theta$ or $\frac{dy}{d\theta} = \cos\theta$

or $dy = \cos\theta d\theta$

or $\Delta y = \cos\theta\,\Delta\theta$

or $\frac{\Delta y}{y} = \frac{\cos\theta\,\Delta\theta}{y} = \frac{\cos\theta\,\Delta\theta}{\sin\theta}$

Fractional error in $y = \frac{\Delta y}{y} = \frac{\Delta\theta}{\tan\theta}$

But $\tan\theta$ has any value from zero to infinity.

Hence, a fractional error will have any value from infinity to zero.

Hence, options (a), (b), (c) are correct.

18. $r = \frac{1-a}{1+a}$

$\ln r = \ln(1-a) - \ln(1+a)$

Differentiating, we get

$$\frac{dr}{r} = -\frac{da}{1-a} - \frac{da}{1+a}$$

or, we can write

$$\frac{\Delta r}{r} = -\left[\frac{\Delta a}{1-a} + \frac{\Delta a}{1+a}\right]$$

$$\frac{\Delta r}{r} = \frac{-2\Delta a}{1-a^2}$$

or $\Delta r = -\left(\frac{2\Delta a}{1-a^2}\right)(r) = \frac{-2\Delta a}{(1+a)^2}$

19. $N = N_0 e^{-\lambda t}$

$\ln N = \ln N_0 - \lambda t$

Differentiating w.r.t. λ, we get

$$\frac{1}{N}\cdot\frac{dN}{d\lambda} = 0 - t$$

$\Rightarrow$ $|d\lambda| = \frac{dN}{Nt} = \frac{40}{2000 \times 1} = 0.02$

20. Time period of simple pendulum is given as

$$T = 2\pi\sqrt{\frac{l}{g}}$$

$\Rightarrow$ $g = \frac{1}{4\pi^2}\frac{T^2}{l}$

$\Rightarrow$ $\frac{\Delta g}{g} = \frac{2\Delta T}{T} + \frac{\Delta l}{l}$

$\Rightarrow$ $\frac{\Delta g}{g} = 2\cdot\frac{1}{100 \times 0.5} + \frac{1\text{ mm}}{10\text{ cm}}$

$\Rightarrow$ $\frac{\Delta g}{g} = \frac{5}{100}$, $\therefore$ $x = 5\%$

21. $y = \ln x$

$$\frac{dy}{dx} = \frac{1}{x} \text{ or } \Delta y = \frac{\Delta x}{x} = 0.05$$

22. $\because$ $z = \ln x^2 y$

$$\Delta z = \left|\frac{\partial}{\partial x}(\ln x^2 y)\,\Delta x\right| + \left|\frac{\partial}{\partial y}(\ln x^2 y)\,\Delta y\right|$$

$$\Delta z = \frac{\partial(\ln x^2 y)}{\partial(x^2 y)}\cdot\frac{\partial(x^2 y)}{\partial x}\Delta x + \frac{\partial(\ln x^2 y)}{\partial(x^2 y)}\cdot\frac{\partial(x^2 y)}{\partial y}\Delta y$$

$$= \frac{1}{x^2 y}2xy\Delta x + \frac{1}{x^2 y}x^2\Delta y$$

$$= 2\frac{\Delta x}{x} + \frac{\Delta y}{y} = 2 \times 0.01 + 0.02$$

$$= 0.04$$

UNIT-3

EXPERIMENTS

Learning Modules

1. Vernier Calipers
2. Screw Gauge

MODULE 1

Vernier Calipers

The device generally used in everyday life for measurement of length is a meter scale (we can also call it mm scale). It can be used for measurement of length with an accuracy of 1 mm. So, the least count of a meter scale is 1 mm. To measure length accurately upto 0.1 mm or 0.01 mm vernier caliper and screw gauge are used.

Vernier caliper has following three parts :

(i) **Main scale** It consists of a steel metallic strip M, graduated in cm and mm at one edge.

It carries two fixed jaws A and C as shown in figure.

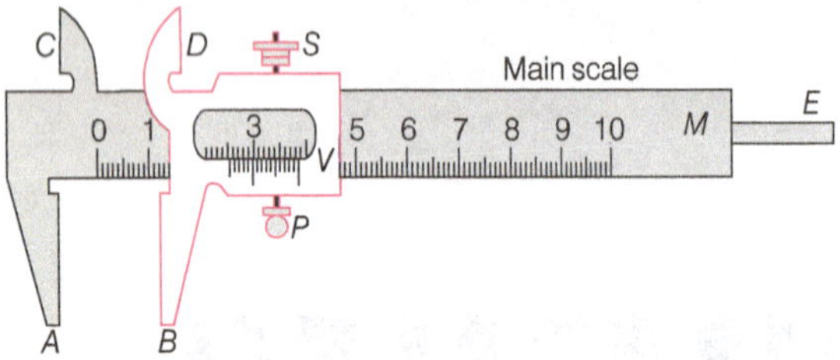

(ii) **Vernier scale** Vernier scale V slides on metallic strip M. It can be fixed in any position by screw S. B and D two movable jaws are fixed with it. When vernier scale is pushed towards A and C, then B touches A and straight side of C will touch straight side of D. In this position, if the instrument is free from error, zeros of vernier scale will coincide with zeros of main scales. To measure the external diameter of an object it is held between the jaws A and B, while the straight edges of C and D are used for measuring the internal diameter of a hollow object.

(iii) **Metallic strip** There is a thin metallic strip E attached to the back side of M and connected with vernier scale. When jaws A and B touch each other, the edge of E touches the edge of M. When the jaws A and B are separated E moves outwards. This strip E is used for measuring the depth of a vessel.

Principle (Theory)

In the common form, the divisions on the vernier scale V are smaller in size than the smallest division on the main scale M, but in some special cases the size of the vernier division may be larger than the main scale division. Let n vernier scale divisions (VSD) coincide with $(n-1)$ main scale divisions (MSD). Then,

$$n \text{ VSD} = (n-1) \text{ MSD}$$

or $$1 \text{ VSD} = \left(\frac{n-1}{n}\right) \text{MSD}$$

$$1 \text{ MSD} - 1 \text{ VSD} = 1 \text{ MSD} - \left(\frac{n-1}{n}\right) \text{MSD} = \frac{1}{n} \text{MSD}$$

The difference between the values of one main scale division and one vernier scale division is known as **Vernier Constant (VC)** or the **Least Count (LC).** This is the smallest distance that can be accurately measured with the vernier scale. Thus,

$$\text{VC} = \text{LC} = 1 \text{ MSD} - 1 \text{ VSD} = \left(\frac{1}{n}\right) \text{MSD}$$

$$= \frac{\text{Smallest division on main scale}}{\text{Number of divisions on vernier scale}}$$

In the ordinary vernier caliper one main scale division be 1 mm and 10 vernier scale divisions coincide with 9 main scale divisions.

$$10 \text{VSD} = 9 \text{MSD}$$

$$\therefore \quad 1 \text{ VSD} = \frac{9}{10} \text{MSD} = 0.9 \text{ mm}$$

$$\text{VC} = 1 \text{ MSD} - 1 \text{ VSD} = 1 \text{ mm} - 0.9 \text{ mm}$$

$$= 0.1 \text{ mm} = 0.01 \text{ cm}$$

Reading a Vernier Caliper

If we have to measure a length AB, the end A is coincided with the zero of main scale, suppose the end B lies between 1.0 cm and 1.1 cm on the main scale. Then,

$$1.0 \text{ cm} < AB < 1.1 \text{ cm}$$

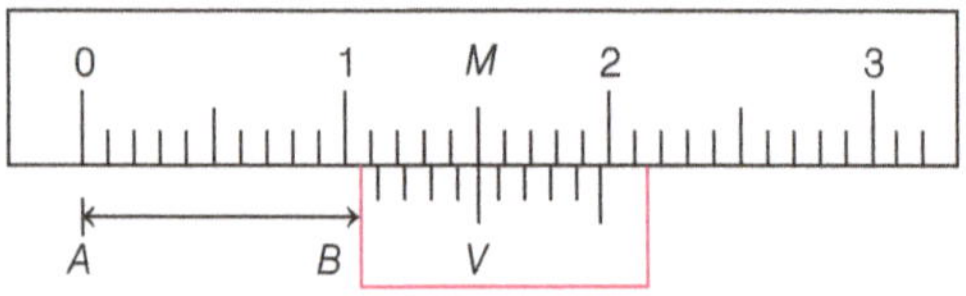

Let 5th division of vernier scale coincides with 1.5 cm of main scale.

Then, $AB = 1.0 + 5 \times \text{VC} = (1.0 + 5 \times 0.01) \text{ cm}$

$= 1.05 \text{ cm}$

Thus, we can make the following formula,

$$\text{Total reading} = N + n \times \text{VC}$$

Here, N = main scale reading before on the left of the zero of the vernier scale.

n = number of vernier division which just coincides with any of the main scale division.

Note *That the main scale reading with which the vernier scale division coincides has no connection with reading.*

Zero Error and Zero Correction

If the zero of the vernier scale does not coincide with the zero of main scale when jaw B touches A and the straight edge of D touches the straight edge of C, then the instrument has an error called **zero error.** Zero error is always algebraically subtracted from measured length.

Zero correction has a magnitude equal to zero error but its sign is opposite to that of the zero error. Zero correction is always algebraically added to measured length.

Zero error $\longrightarrow$ algebraically subtracted

Zero correction $\longrightarrow$ algebraically added

Positive and Negative Zero Errors

If zero of vernier scale lies to the right of the main scale the zero error is positive and if it lies to the left of the main scale the zero error is negative (when jaws A and B are in contact).

$$\text{Positive zero error} = (N + x \times \text{VC})$$

Here, N = main scale reading on the left of zero of vernier scale.

x = vernier scale division which coincides with any main scale division.

When the vernier zero lies before the main scale zero the error is said to be negative zero error. If 3rd vernier scale division coincides with the main scale division, then

$$\text{Negative zero error} = -[0.00\text{ cm} + 3 \times \text{VC}]$$
$$= -[0.00\text{ cm} + 3 \times 0.01\text{ cm}]$$
$$= -0.03\text{ cm}$$

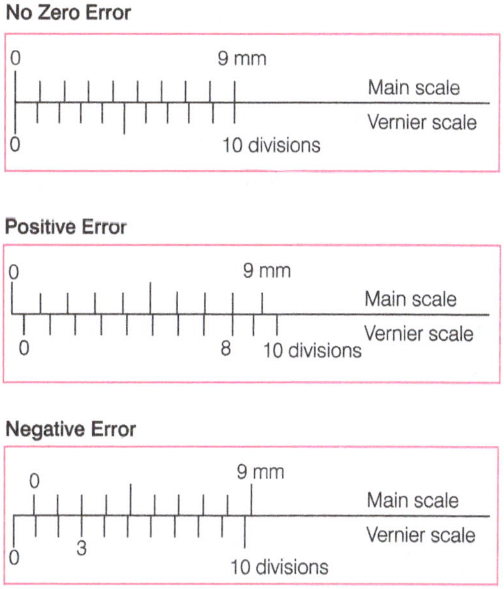

Positive and negative zero error

Summary

1. $\text{VC} = \text{LC} = \dfrac{1}{n}(\text{MSD})$

   $= \dfrac{\text{Smallest division on main scale}}{\text{Number of divisons on vernier scale}}$

   $= 1\text{ MSD} - 1\text{ VSD}$
2. In ordinary vernier caliper, 1 MSD = 1 mm, 1 VSD = 0.9 mm and $n = 10$

   $\therefore \quad \text{VC} \quad \text{or} \quad \text{LC} = \dfrac{1}{10}\text{ mm}$

   or $\quad 1\text{mm} - 0.9\text{mm} = 0.1\text{ mm} = 0.01\text{ cm}$
3. Total reading $= (N + n \times \text{VC})$
4. Zero correction = – zero error
5. Zero error is algebraically subtracted while the zero correction is algebraically added.
6. If zero of vernier scale lies to the right of zero of main scale the error is positive. The actual length in this case is less than observed length.
7. If zero of vernier scale lies to the left of zero of main scale the error is negative and the actual length is more than the observed length.
8. Positive zero error $= (N + x \times \text{VC})$
9. Least count of mm scale is 1mm of vernier caliper is 0.1mm and of screw gauge 0.01mm.

Example 1 N divisions on the main scale of a vernier caliper coincide with $N+1$ divisions on the vernier scale. If each division on the main scale is of a units, determine the least count of the instrument. **(2003 JEE)**

Sol. $(N+1)$ divisions on the vernier scale $= N$ divisions on main scale

$\therefore$ 1 division on vernier scale $= \dfrac{N}{N+1}$ divisions on main scale

Each division on the main scale is of a units.

$\therefore$ 1 division on vernier scale $= \left(\dfrac{N}{N+1}\right) a$ units $= a'$ (say)

Least count = 1 main scale division - 1 vernier scale division

$$= a - a' = a - \left(\frac{N}{N+1}\right) a = \frac{a}{N+1}$$

Example 2 In the diagram shown in figure, find the magnitude and nature of zero error.

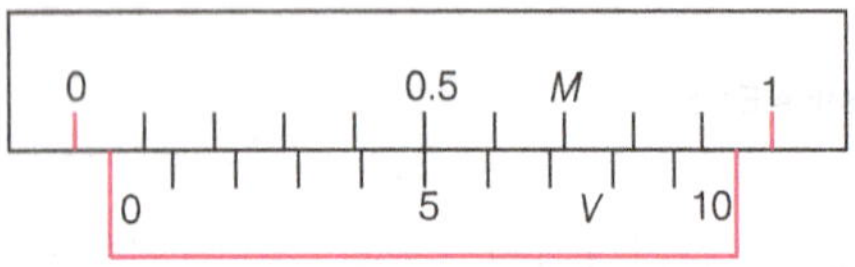

Sol. Here, zero of vernier scale lies to the right of zero of main scale, hence it has positive zero error.

Further, $N = 0,\ x = 5$

LC or VC = 0.01 cm

Hence, Zero error $= N + x \times VC$

$= 0 + 5 \times 0.01 = 0.05$ cm

Zero correction $= -0.05$ cm

$\therefore$ Actual length will be 0.05 cm less than the measured length.

Example 3 The smallest division on main scale of a vernier caliper is 1 mm and 10 vernier divisions coincide with 9 main scale divisions. While measuring the length of a line, the zero mark of the vernier scale lies between 10.2 cm and 10.3 cm and the third division of vernier scale coincides with a main scale division.

(a) Determine the least count of the caliper.

(b) Find the length of the line.

Sol. (a) Least Count (LC)

$$= \frac{\text{Smallest division on main scale}}{\text{Number of divisions on vernier scale}}$$

$$= \frac{1}{10} \text{ mm} = 0.1 \text{ mm}$$

$= 0.01$ cm

(b) $L = N + n\,(LC)$

$= (10.2 + 3 \times 0.01)$ cm

$= 10.23$ cm **Ans.**

INTRODUCTORY Exercise

1. The main scale of a vernier caliper reads 10 mm in 10 divisions. Ten divisions of vernier scale coincide with nine divisions of the main scale. When the two jaws of the caliper touch each other, the fifth division of the vernier coincides with a main scale division and the zero of the vernier is to the right of zero of main scale. When a cylinder is tightly placed between the two jaws, the zero of the vernier scale lies slightly to the right of 3.2 cm and the fourth vernier division coincides with a main scale division. Find diameter of the cylinder.
2. In a vernier caliper, N divisions of the main scale coincide with $N + m$ divisions of the vernier scale. What is the value of m for which the instrument has minimum least count?

Answers

1. 3.19 cm

2. 1

See hints on page 79.

Basic Concept of Vernier Calliper

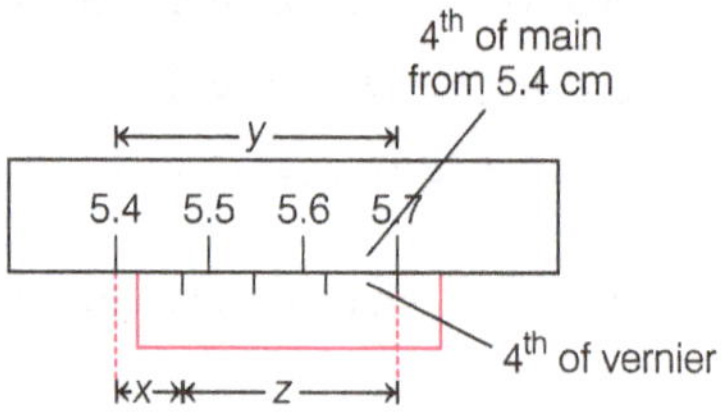

In the above figure,

Reading $= (5.4 \text{ cm}) + x$

$= (5.4 \text{ cm}) + (y - z)$

$= (5.5 \text{ cm}) + (4 \text{ MSD} - 4 \text{ VSD})$

$= (5.4 \text{ cm}) + 4 (\text{MSD} - \text{VSD})$

$= (5.4 \text{ cm}) + 4 (\text{LC})$

Note *In the faulty instrument,*

$$VSD > MSD$$

For example in the above figure 5th of main will coincide with 4th of vernier. So,

Reading $= (5.4 \text{ cm}) + (y - z)$

$= (5.4 \text{ cm}) + 5\text{ MSD} - 4\text{ VSD}$

Following example will illustrate the concept.

Example 4 There are two vernier calipers both of which have 1 cm divided into 10 equal divisions on the main scale. The vernier scale of one of the caliper (C_1) has 10 equal divisions that correspond to 9 main scale divisions. The vernier scale of the other caliper (C_2) has 10 equal divisions that correspond to 11 main scale divisions.

The readings of the two calipers are shown in the figure. The measured values (in cm) by calipers C_1 and C_2 respectively, are **(2016 Adv.)**

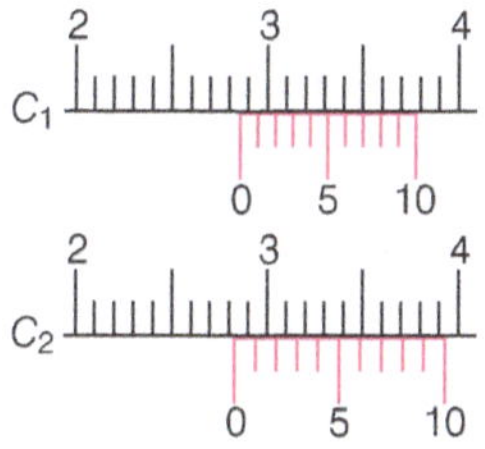

(a) 2.87 and 2.87 (b) 2.87 and 2.83
(c) 2.85 and 2.82 (d) 2.87 and 2.86

Sol. **Caliper C_1** 10 divisions of vernier scale are equal to 9 divisions of main scale. So,

$$\text{VSD} < \text{MSD}$$

So, the measured value can be obtained by normal method. Main scale reading is 2.8 cm. Then 7th of vernier is coinciding with 7th of main (after 2.8 cm). So,

$$\text{Measurement} = (2.8\text{ cm}) + 7\text{ (LC)}$$

$$= (2.8\text{ cm}) + 7\left(\frac{\text{Smallest division on main scale}}{\text{Number of divisions on vernier}}\right)$$

$$= (2.8\text{ cm}) + 7\left(\frac{0.1\text{ cm}}{10}\right) = 2.87\text{ cm}$$

Caliper C_2 10 divisions of vernier scale are equal to 11 divisions of main scale. So,

$$\text{VSD} > \text{MSD}$$

Therefore in this case we will have to find the reading from the basics

$$10\text{ VSD} = 11\text{ MSD} = 11\text{ mm}$$

$$\therefore \quad 1\text{ VSD} = 1.1\text{ mm} = 0.11\text{ cm}$$

$$1\text{MSD} = 1\text{mm} = 0.1\text{cm}$$

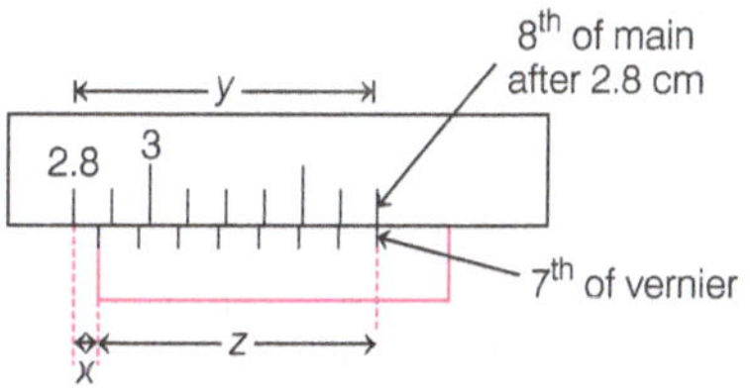

$$\therefore \quad \text{Measurement} = (2.8\text{ cm}) + x$$

$$= (2.8\text{ cm}) + (y - z)$$

$$= (2.8\text{ cm}) + 8\text{ (MSD)} - 7\text{(VSD)}$$

$$= (2.8\text{ cm}) + 8(0.1\text{ cm}) - 7\text{ (0.11 cm)}$$

$$= 2.83\text{ cm}$$

$\therefore$ The correct option is (b).

MODULE 1 Exercise

Least Count, Vernier Constant, VSD, MSD and Zero Error

Single Correct Option

1. The main scale of a vernier caliper has n divisions/cm. n divisions of the vernier scale coincide with $(n-1)$ divisions of main scale. The least count of the vernier caliper is

(a) $\frac{1}{(n+1)(n-1)}$ cm (b) $\frac{1}{n}$ cm
(c) $\frac{1}{n^2}$ cm (d) $\frac{1}{n(n+1)}$ cm

2. For positive error, the correction is

(a) positive (b) negative
(c) nill
(d) may be positive or negative

3. Vernier constant is the

(a) value of one MSD divided by total number of divisions on the main scale
(b) value of one VSD divided by total number of divisions on the vernier scale
(c) total number of divisions on the main scale divided by total number of divisions on the vernier scale
(d) difference between the value of one main scale division and one vernier scale division

4. Select the **incorrect** statement.

(a) If the zero of vernier scale does not coincide with the zero of the main scale, then the vernier caliper is said to be having zero error
(b) Zero correction has a magnitude equal to zero error but sign is opposite to that of zero error
(c) Zero error is positive when the zero of vernier scale lies to the left of the zero of the main scale
(d) Zero error is negative when the zero of vernier scale lies to the left of the zero of the main scale

5. 1 cm of main scale of a vernier caliper is divided into 10 divisions. The least count of the caliper is 0.005 cm, then the vernier scale must have

(a) 10 divisions (b) 20 divisions
(c) 25 divisions (d) 50 divisions

6. The vernier constant of a vernier caliper is 0.001 cm. If 49 main scale divisions coincide with 50 vernier scale divisions, then the value of 1 main scale division is

(a) 0.1 mm
(b) 0.5 mm
(c) 0.4 mm
(d) 1 mm

7. A vernier caliper has 1 mm marks on the main scale. It has 20 equal divisions on the vernier scale which match with 16 main scale divisions. For this vernier caliper, the least count is **(2010 JEE)**
(a) 0.02 mm (b) 0.05 mm (c) 0.1 mm (d) 0.2 mm

8. One main scale division of a vernier calipers is 'a' cm and nth division of the vernier scale coincide with $(n-1)$ th division of the main scale. The least count of the caliper in mm is **(2021 JEE)**
(a) $\frac{10a}{(n-1)}$ (b) $\frac{10na}{(n-1)}$ (c) $\left(\frac{n-1}{10n}\right)a$ (d) $\frac{10a}{n}$

Integer Type Question

9. In a vernier caliper, each cm on the main scale is divided into 20 equal parts. If tenth vernier scale division coincides with ninth main scale division, then the value of vernier constant will be $n \times 10^{-2}$ mm. Find the value of n. **(2022 Main)**

Measurement from Vernier Caliper

Single Correct Option

10. A vernier caliper is having 1 main scale division $= 1$ mm. Ten divisions of vernier scale coincide with nine divisions of main scale. If two jaws of instrument touch each other, the 8th division of vernier scale coincide with a mark of main scale and the zero of the vernier scale lies to the right of the zero of main scale. While measuring the length of cylinder, the zero of the vernier scale lies slightly to the left of 3.2 cm and the fourth vernier division coincides with a main scale division. The length of cylinder is
(a) 3.06 cm (b) 3.15 cm (c) 3.16 cm (d) 3.22 cm

11. In an experiment with vernier calipers of least count 0.1 mm, when two jaws are joined together the zero of vernier scale lies right to the zero of the main scale and 6th division of vernier scale coincides with the main scale division. While measuring the diameter of a spherical bob, the zero of vernier scale lies in between 3.2 cm and 3.3 cm marks and 4th division of vernier scale coincides with the main scale division. The diameter of bob is measured as **(2023 Main)**
(a) 3.18 cm (b) 3.25 cm (c) 3.26 cm (d) 3.22 cm

12. In a vernier caliper, 10 divisions of vernier scale is equal to the 9 divisions of main scale. When both jaws of vernier caliper touch each other, the zero of the vernier scale is shifted to the left of zero of the main scale and 4th vernier scale division exactly coincides with the main scale reading. One main scale division is equal to 1 mm. While measuring diameter of a spherical body, the body is held between two jaws. It is now observed that zero of the vernier scale lies between 30 and 31 divisions of main scale reading and 6th vernier scale division exactly coincides with the main scale reading. The diameter of the spherical body will be **(2022 Main)**
(a) 3.02 cm (b) 3.06 cm
(c) 3.12 cm (d) 3.20 cm

13. The vernier scale used for measurement has a positive zero error of 0.2 mm. If while taking a measurement it was noted that '0' on the vernier scale lies between 8.5 cm and 8.6 cm, vernier coincidence is 6, then the correct value of measurement is.........cm. (least count = 0.01 cm) **(2021 Main)**
(a) 8.54 cm (b) 8.36 cm (c) 8.56 cm (d) 8.58 cm

Integer Type Questions

14. The vernier constant of vernier caliper is 0.1 mm and it has zero error of (– 0.05) cm. While measuring diameter of a sphere, the main scale reading is 1.7 cm and coinciding vernier division is 5. The corrected diameter will be $n \times 10^{-2}$cm. Find n. **(2022 Main)**

15. The diameter of a spherical bob is measured using a vernier caliper. Nine divisions of the main scale, in the vernier calipers, are equal to ten divisions of vernier scale. One main scale division is 1 mm. The main scale reading is 10 mm and 8th division of vernier scale was found to coincide exactly with one of the main scale division. If the given vernier caliper has positive zero error of 0.04 cm, then the radius of the bob is $n \times 10^{-2}$ cm. Find n. **(2021 Main)**

Miscellaneous Problems

Single Correct Option

16. The length of a strip measured with a metre rod is 10.0 cm. Its width measured with a vernier caliper is 1.00 cm. The least count of the metre rod is 0.1 cm and that of vernier caliper 0.01 cm. What will be error in its area?
(a) $\pm 13\%$ (b) $\pm 7\%$ (c) $\pm 4\%$ (d) $\pm 2\%$

17. The length of a cylinder is measured with a metre rod having least count 0.1 cm. Its diameter is measured with vernier caliper having least count 0.01 cm. Given that length is 5.0 cm and radius is 2.0 cm. The maximum percentage error in the calculated value of the volume will be
(a) 1.5% (b) 2.5% (c) 3.5% (d) 4%

18. The internal and external diameters of a hollow cylinder are measured with the help of a vernier caliper. Their values are 4.23 ± 0.01 cm and 3.87 ± 0.01 cm, respectively. The thickness of the wall of the cylinder is
(a) 0.36 ± 0.02 cm (b) 0.18 ± 0.02 cm
(c) 0.36 ± 0.01 cm (d) 0.18 ± 0.01 cm

Answers

1. (c)	2. (b)	3. (b)	4. (d)	5. (b)
6. (b)	7. (d)	8. (d)	9. (5)	10. (a)
11. (a)	12. (c)	13. (a)	14. (180)	15. (52)
16. (d)	17. (b)	18. (d)		

See hints on pages 79 and 80.

MODULE 2

Screw Gauge

Principle of a Micrometer Screw

The least count of vernier caliper ordinarily available in the laboratory is 0.01 cm. When lengths are to be measured with greater accuracy, say upto 0.001 cm, screw gauge and spherometer are used which are based on the principle of **micrometer screw** discussed below.

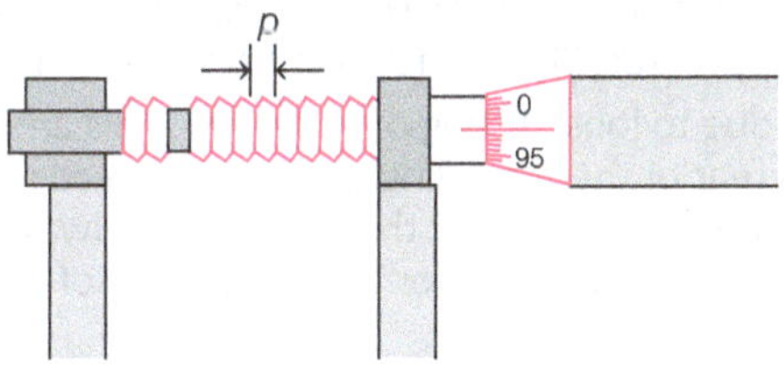

If an accurately cut single threaded screw is rotated in a closely fitted nut, then in addition to the circular motion of the screw there is a linear motion of the screw head in the forward or backward direction, along the axis of the screw. The linear distance moved by the screw, when it is given one complete rotation is called the **pitch** (p) of the screw. This is equal to the distance between two consecutive threads as measured along the axis of the screw.

In most of the cases, it is either 1 mm or 0.5 mm. A circular cap is fixed on one end of the screw and the circumference of the cap is normally divided into 100 or 50 equal parts. If it is divided into 100 equal parts, then the screw moves forward or backward by $\frac{1}{100}\left(\text{or } \frac{1}{50}\right)$ of the pitch, if the circular scale (we will discuss later about circular scale) is rotated through one circular scale division.

It is the minimum distance which can be accurately measured and so called the **Least Count** (LC) of the screw.

Thus, $\text{Least count} = \frac{\text{Pitch}}{\text{Number of divisions on circular scale}}$

If pitch is 1 mm and there are 100 divisions on circular scale then,

$$\text{LC} = \frac{1\text{ mm}}{100} = 0.01\text{ mm}$$
$$= 0.001\text{ cm} = 10\ \mu\text{m}$$

Since, LC is of the order of 10 μm, the screw is called micrometer screw.

Screw Gauge

Screw gauge works on the principle of micrometer screw. It consists of a U-shaped metal frame M. At one end of it is fixed a small metal piece A. It is called stud and it has a plane face. The other end N of M carries a cylindrical hub H. It is graduated in millimeter or half millimeter depending upon the pitch of the screw. This scale is called **linear scale** or **pitch scale.**

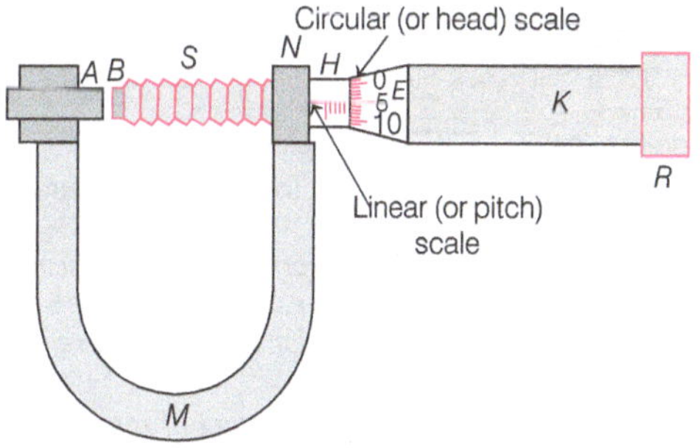

A nut is threaded through the hub and the frame N. Through the nut moves a screw S. The front face B of the screw, facing the plane face A is also plane.

A hollow cylindrical cap K is capable of rotating over the hub when screw is rotated. As the cap is rotated the screw either moves in or out. The surface E of the cap K is divided into 50 or 100 equal parts. It is called the **circular scale** or **head scale.**

In an accurately adjusted instrument when the faces A and B are just touching each other. Zero of circular scale should coincide with zero of linear scale.

To Measure Diameter of a Given Wire Using a Screw Gauge

If, with the wire between plane faces A and B, the edge of the cap lies ahead of N th division of linear scale, and nth division of circular scale lies over reference line.

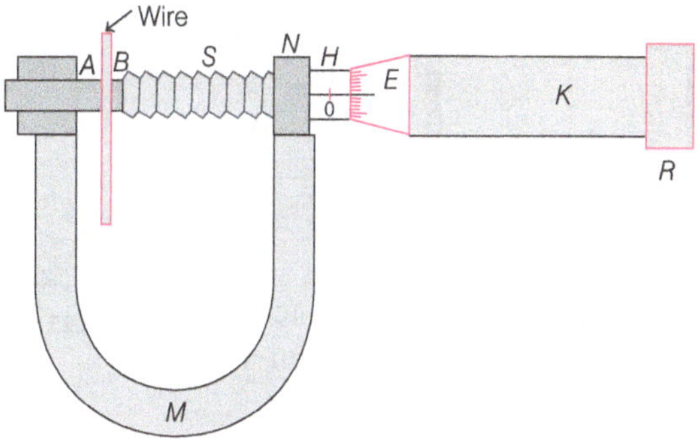

Then, $\text{Total reading} = N + n \times \text{LC}$

Zero Error and Zero Correction

If zero mark of circular scale does not coincide with the zero of the pitch scale when the faces A and B are just touching each other, the instrument is said to possess zero error.

If the zero of the circular scale advances beyond the reference line the zero error is **negative** and zero correction is **positive.** If it is left behind the reference line the zero error is **positive** and zero correction is **negative.**

For example, if zero of circular scale advances beyond the reference line by 5 divisions, zero correction $= +5 \times$ (LC) and if the zero of circular scale is left behind the reference line by 5 divisions, zero correction $= -5 \times$ (LC).

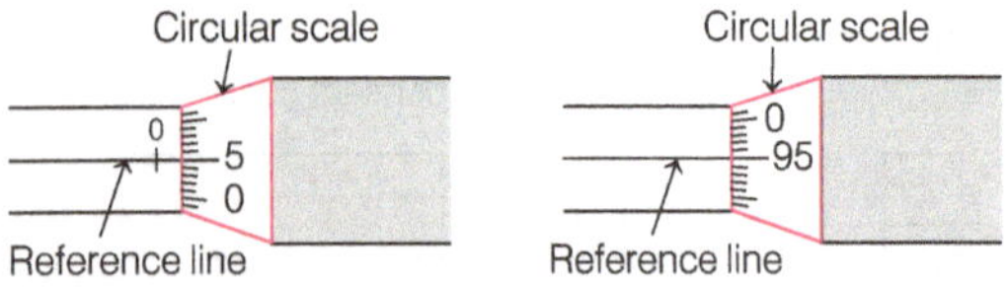

(a) Positive zero error (b) Negative zero error

Note *In negative zero error 95th division of the circular scale is coinciding with the reference line. Hence there are 5 divisions between zero mark on the circular scale and the reference line.*

Following table will explain the concept of zero error more explicity.

Checking for zero error	Observed reading	Actual reading = observed reading − zero error
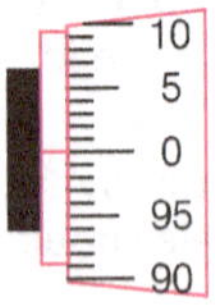 Zero mark on circular scale coincides with reference line on the main scale and reading on main scale is 0 ⇒ No zero error	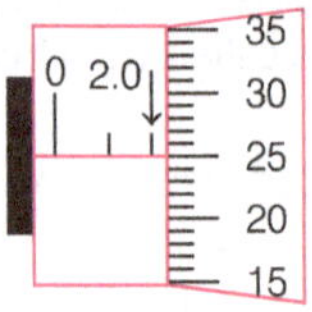 Reading = (2.0 mm) + 25(LC) = (2.0 mm) + 25(0.01 mm) = 2.25 mm	2.25 mm as no zero error correction required.
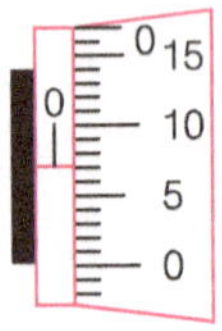 Zero mark on reference line can be seen ⇒ positive zero error. Zero error = 7 (LC) = 7(0.01 mm) = 0.07 mm	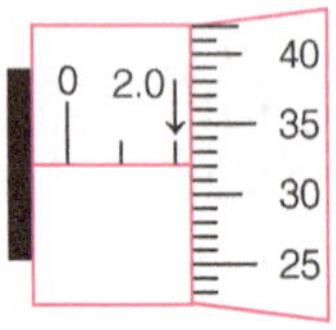 Reading = (2.0 mm) + (32) (LC) = (2.0 mm) + 32(0.01 mm) = 2.32 mm	2.32 − (+0.07) = 2.25 mm
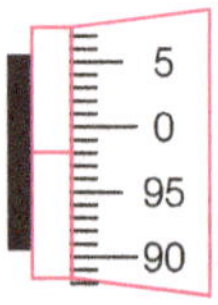 Zero mark on reference line cannot be seen ⇒ negative zero error. Zero error = (−2)(LC) = − 2 (0.01 mm) = − 0.02 mm	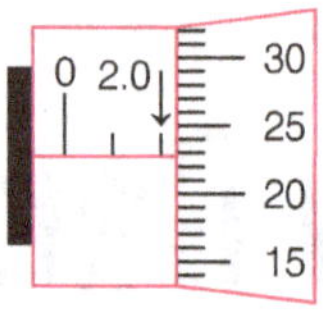 Reading = (2.0 mm) + (23)(LC) = (2.0 mm) + 23 (0.01 mm) = 2.23 mm	2.23 − (− 0.02) = 2.25 mm

Back Lash Error

When the sense of rotation of the screw is suddenly changed. The screw head rotates, but the screw itself does not move forward or backwards. This is known as **back lash error**.

This error is due to loose fitting of the screw. This arises due to wear and tear of the threading due to prolonged use of the screw. To reduce this error the screw must always be rotated in the same direction for a particular set of observations.

Example 1 The pitch of a screw gauge is 1 mm and there are 100 divisions on the circular scale. In measuring the diameter of a sphere there are six divisions on the linear scale and forty divisions on circular scale coincide with the reference line. Find the diameter of the sphere.

Sol. $\text{LC} = \dfrac{1}{100} = 0.01 \text{ mm}$

Linear scale reading $= 6 \text{ (pitch)} = 6 \text{ mm}$

Circular scale reading $= n \text{ (LC)} = 40 \times 0.01 = 0.4 \text{ mm}$

$\therefore$ Total reading $= (6 + 0.4) = 6.4 \text{ mm}$

Example 2 The pitch of a screw gauge is 1 mm and there are 100 divisions on circular scale. When faces A and B are just touching each other without putting anything between the studs, 32nd division of the circular scale coincides with the reference line and zero mark on reference line can be seen. When a glass plate is placed between the studs, the linear scale reads 4 divisions and the circular scale reads 16 divisions. Find the thickness of the glass plate.

Sol. Least count

$$(\text{LC}) = \frac{\text{Pitch}}{\text{Number of divisions on circular scale}} = \frac{1}{100} \text{ mm}$$

$$= 0.01 \text{ mm}$$

As zero is not hidden from circular scale when A and B touches each other. Hence, the screw gauge has positive error.

$$e = +n \text{ (LC)} = 32 \times 0.01 = 0.32 \text{ mm}$$

Linear scale reading $= 4 \times (1 \text{ mm}) = 4 \text{ mm}$

Circular scale reading $= 16 \times (0.01 \text{ mm}) = 0.16 \text{ mm}$

$\therefore$ Measured reading $= (4 + 0.016) \text{ mm} = 4.16 \text{ mm}$

$\therefore$ Absolute reading = Measured reading − e

$= (4.16 - 0.32) \text{ mm} = 3.84 \text{ mm}$

Therefore, thickness of the glass plate is 3.84 mm.

MODULE 2 Exercise

Least Count and Zero Error

Assertion and Reason

Directions : Choose the correct option.

(a) *If both **Assertion** and **Reason** are true and the **Reason** is the correct explanation of the **Assertion**.*

(b) *If both **Assertion** and **Reason** are true but **Reason** is not the correct explanation of **Assertion**.*

(c) *If **Assertion** is true, but the **Reason** is false.*

(d) *If **Assertion** is false but the **Reason** is true.*

(e) *If both **Assertion** and **Reason** are false.*

1. **Assertion** If in five complete rotations of the circular scale, the distance travelled on main scale of the screw gauge is 5 mm and there are 50 total divisions on circular scale, then least count is 0.001 cm.

   Reason Least count

   $$= \frac{\text{Pitch}}{\text{Total divisions on circular scale}}$$

Single Correct Option

2. Screw gauge is said to have a negative error
   (a) when circular scale zero coincides with base line of main scale
   (b) when circular scale zero is above the base line of main scale
   (c) when circular scale zero is below the base line of main scale
   (d) None of the above

3. Least count of screw gauge is defined as
   (a) $\frac{\text{distance moved by circular scale on main scale}}{\text{number of rotations of circular scale}}$
   (b) $\frac{\text{pitch of the screw}}{\text{number of divisions on circular scale}}$
   (c) $\frac{\text{number of rotations of circular scale}}{\text{number of circular scale divisions}}$
   (d) None of the above

4. In four complete rotations, the distance moved by the screw on the linear scale is 2 mm. Its circular scale contains 50 divisions. Find the least count of the screw gauge.
   (a) 0.05 mm (b) 0.01 mm (c) 0.1 mm (d) 0.02 mm

Measurement with No Error

Single Correct Option

5. In a screw gauge, 5 complete rotations of the screw gauge it to move a linear distance of 0.25 cm. There are 100 circular scale divisions. The thickness of a wire measured by this screw gauge gives a reading of 4 main scale divisions and 30 circular scale divisions. Assuming negligible zero error, the thickness of the wire is **(2018 JEE)**
   (a) 0.0430 cm (b) 0.3150 cm (c) 0.4300 cm (d) 0.2150 cm

6. A screw gauge gives the following reading when used to measure the diameter of a wire.
   Main scale reading : 0 mm
   Circular scale reading : 52 divisions
   Given that 1 mm on main scale corresponds to 100 divisions of the circular scale. The diameter of wire from the above data is **(2011 JEE)**
   (a) 0.052 cm (b) 0.026 cm (c) 0.005 cm (d) 0.52 cm

Subjective Question

7. Find the reading of screw gauge as shown in figure. Given that circular scale has 100 divisions and in one complete rotation the screw advances by 1 mm.

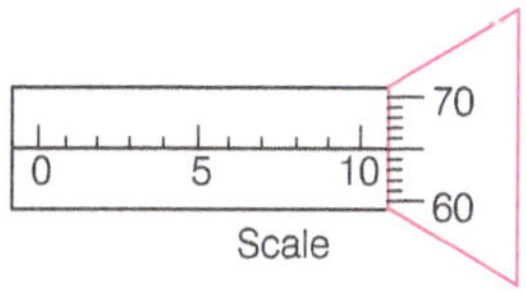

Measurement with Positive Zero Error

Single Correct Option

8. In a screw gauge, fifth division of the circular scale coincides with the reference line when the ratchet is closed. There are 50 divisions on the circular scale, and the main scale moves by 0.5 mm on a complete rotation. For a particular observation the reading on the main scale is 5 mm and the 20th division of the circular scale coincides with reference line. Calculate the true reading. **(2021 JEE)**
   (a) 5.00 mm (b) 5.25 mm (c) 5.15 mm (d) 520 mm

9. The pitch and the number of divisions, on the circular scale for a given screw gauge are 0.5 mm and 100, respectively. When the screw gauge is fully tightened without any object, the zero of its circular scale lies 3 divisions below the reference line.
   The readings of the main scale and the circular scale for a thin sheet are 5.5 mm and 48, respectively the thickness of this sheet is **(2019 Main)**
   (a) 5.950 mm (b) 5.725 mm
   (c) 5.755 mm (d) 5.740 mm

Integer Type Question

10. In a screw gauge, there are 100 divisions on the circular scale and the main scale moves by 0.5 mm on a complete rotation of the circular scale. The zero of circular scale lies 6 divisions below the line of graduation when two studs are brought in contact with each other. When a wire is placed between the studs, 4 linear scale divisions are clearly visible while 46th division the circular scale coincide with the reference line. The diameter of the wire is .... $\times 10^{-2}$mm. **(2023 JEE)**

Measurement with Negative Zero Error

Single Correct Option

11. The distance moved by the screw of a screw gauge is 2 mm in four rotations and there are 50 divisions on its cap. When nothing is put between its jaws, 20th division of circular scale coincides with reference line and zero of linear scale is hidden from circular scale when two jaws touch each other or zero of circular scale is lying above the reference line. When a plate is placed between the jaws, main scale reads 2 divisions and circular scale reads 20 divisions. Thickness of the plate is
(a) 1.1 mm (b) 1.2 mm (c) 1.4 mm (d) 1.5 mm

12. A student measured the diameter of a small steel ball using a screw gauge of least count 0.001 cm. The main scale reading is 5 mm and zero of circular scale division coincides with 25 divisions above the reference level. If screw gauge has a zero error of −0.004 cm, the correct diameter of the ball is
(a) 0.053 cm (b) 0.525 cm (c) 0.521 cm (d) 0.529 cm

13. In an experiment to find out the diameter of wire using screw gauge, the following observations were noted :
(A) Screw moves 0.5 mm on main scale in one complete rotation.
(B) Total divisions on circular scale $=50$
(C) Main scale reading is 2.5 mm
(D) 45th division of circular scale is in the pitch line
(E) Instrument has 0.03 mm negative error
Then the diameter of wire is **(2022 JEE)**
(a) 2.92 mm (b) 2.54 mm
(c) 2.98 mm (d) 3.45 mm

14. A screw gauge with a pitch of 0.5 mm and a circular scale with 50 divisions is used to measure the thickness of a thin sheet of aluminium. Before starting the measurement, it is found that when the two jaws of the screw gauge are brought in contact, the 45th division coincides with the main scale line and that the zero of the main scale is barely visible. What is the thickness of the sheet, if the main scale reading is 0.5 mm and the 25th division coincides with the main scale line? **(2016 Main)**
(a) 0.75 mm (b) 0.80 mm (c) 0.75 mm (d) 0.80 mm

Miscellaneous Problems

Single Correct Option

15. Which of the following is the most precise device for measuring length?
(a) A vernier caliper with 20 divisions on the sliding scale
(b) An optical instrument that can measure length up to a wavelength of light
(c) A screw gauge of pitch 1 mm and 100 divisions on the circular scale
(d) All the above are equally precise

16. A screw gauge has 50 divisions on its circular scale. The circular scale is 4 units ahead of the pitch scale marking, prior to use. Upon one complete rotation of the circular scale, a displacement of 0.5 mm is noticed on the pitch scale. The nature of zero error involved and the least count of the screw gauge, are respectively : **(2020 JEE)**
(a) Negative, 2 μm (b) Positive, 10 μm
(c) Positive, 0.1 mm (d) Positive, 0.1 μm

17. A student measured the length of a rod and wrote it as 3.50 cm. Which instrument did he use to measure it? **(2014 Main)**
(a) A meter scale.
(b) A vernier caliper where the 10 divisions in vernier scale matches with 9 divisions in main scale and main scale has 10 divisions in 1 cm.
(c) A screw gauge having 100 divisions in the circular scale and pitch as 1 mm.
(d) A screw gauge having 50 divisions in the circular scale and pitch as 1 mm.

Integer Type Question

18. Student A and Student B used two screw gauges of equal pitch and 100 equal circular divisions to measure the radius of a given wire. The actual value of the radius of the wire is 0.322 cm. The absolute value of the difference between the final circular scale readings observed by the students A and B is ......
[Figure shows position of reference 'O' when jaws of screw gauge are closed]
Given pitch $=0.1$cm. **(2021 JEE)**

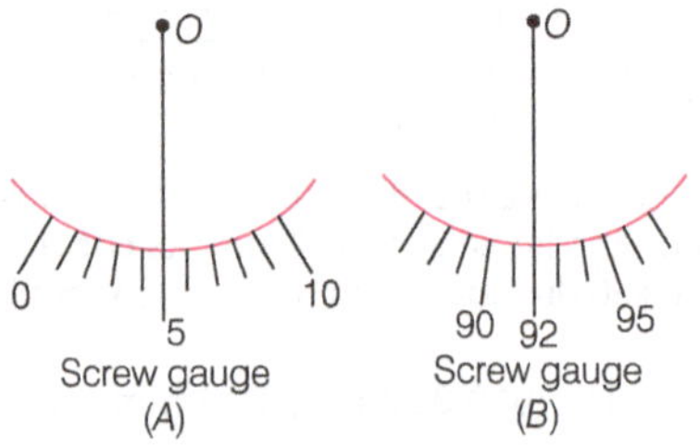

Answers

1. (a)	**2.** (b)	**3.** (b)	**4.** (b)	**5.** (d)	**6.** (a)	**7.** (10.65 mm)	**8.** (c)	**9.** (b)
10. (22)	**11.** (d)	**12.** (d)	**13.** (c)	**14.** (b)	**15.** (b)	**16.** (b)	**17.** (b)	**18.** (13)

See hints on pages 83-85.

Unit Exercise (Cumulative)

Level 1

Single Correct Option

1. A vernier caliper having 1 main scale division = 0.1 cm is designed to have a least count of 0.02 cm. If n be the number of divisions on vernier scale and m be the length of vernier scale, then
(a) $n=10, m=0.5$ cm
(b) $n=9, m=0.4$ cm
(c) $n=10, m=0.8$ cm
(d) $n=10, m=0.2$ cm

2. The diameter of a cylinder is measured using a vernier caliper with no zero error. It is found that the zero of the vernier scale lies between 5.10 cm and 5.15 cm of the main scale. The vernier scale has 50 division equivalent to 2.45 cm. The 24th division of the vernier scale exactly coincides with one of the main scale divisions. The diameter of the cylinder is **(2013 Adv.)**
(a) 5.112 cm (b) 5.124 cm
(c) 5.136 cm (d) 5.148 cm

3. The length of a rectangular plate is measured by a meter scale and is found to be 10.0 cm. Its width is measured by vernier caliper as 1.00 cm. The least count of the meter scale and vernier caliper are 0.1 cm and 0.01 cm, respectively. Maximum permissible error in area measurement is
(a) ± 0.2 cm^2 (b) ± 0.1 cm^2
(c) ± 0.3 cm^2 (d) zero

4. The pitch of screw gauge is 1 mm and there are 100 divisions on circular scale. Screw gauge is zero error free. While measuring diameter of a wire, the main scale reads 1 mm and 48th division on circular scale coincides with the reference line. The length of wire is 5.0 cm. Find the volume of the wire.
(a) 2.6 cm^3 (b) 0.086 cm^3
(c) 0.026 cm^3 (d) 0.036 cm^3

5. In the previous question, minimum possible error in area measurement can be
(a) ± 0.02 cm^2 (b) ± 0.01 cm^2
(c) ± 0.03 cm^2 (d) zero

6. In 10 complete rotations, the distance moved on the linear scale is 1 cm. There are 100 divisions on circular scale. When the gap AB is just closed, the 95th division of circular scale coincides with the reference line. While measuring the diameter of a wire, the linear scale reads 2 mm and 45th division on the circular scale coincides with the reference line. Find cross-sectional area of the wire.
(a) 3.8 cm^2 (b) 0.049 cm^2
(c) 0.098 cm^2 (d) 0.45 cm^2

7. A screw gauge of pitch 0.5 mm is used to measure the diameter of uniform wire of length 6.8 cm, the main scale reading is 1.5 mm and circular scale reading is 7. The calculated curved surface area of wire to appropriate significant figures is : **(2022 JEE)**
(Screw gauge has 50 divisions on its circular scale)
(a) 6.8 cm^2 (b) 3.4 cm^2 (c) 3.9 cm^2 (d) 2.4 cm^2

8. The least count of the main scale of a screw gauge is 1 mm. The minimum number of divisions on its circular scale required to measure 5 μm diameter of a wire is **(2019 Main)**
(a) 50 (b) 200 (c) 500 (d) 100

9. Using a screw gauge of pitch 0.1 cm and 50 divisions on its circular scale, the thickness of an object is measured. It should correctly be recorded as **(2020 Main)**
(a) 2.121 cm (b) 2.124 cm (c) 2.125 cm (d) 2.123 cm

Multiple Correct Options

10. The least count of a meter scale is 1 mm. Which of the following lengths is/are measured by the meter scale?
(a) 2.8 cm (b) 2.80 cm (c) 2.834 cm (d) 2 mm

11. The least count of a wrist watch graduated at every second is
(a) one second, if it has hour hand, minute hand and second hand
(b) one minute, if it has only hour hand and minute hand
(c) one second in any case
(d) one minute in any case

12. In the given figure, main scale is graduated in milimeters. Find the reading, if the least count is 0.1 mm.

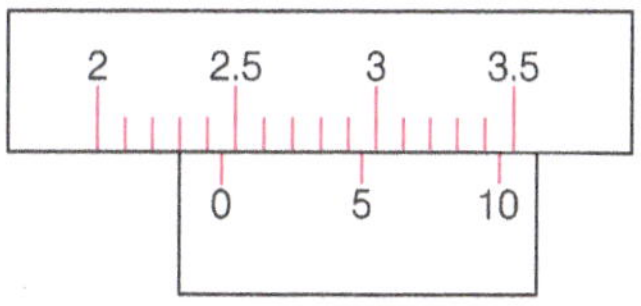

(a) The main scale reading is 2.4 cm
(b) The vernier scale reading is 0.3 mm
(c) The total reading is 2.43 cm
(d) The total reading is 2.4 cm

Subjective Questions

13. The pitch of a screw gauge is 1 mm and there are 100 divisions on the circular scale. While measuring the diameter of a wire, the linear scale reads 1 mm and 47th division on the circular scale coincides with the reference line. The length of the wire is 5.6 cm. Find the curved surface area (in cm^2) of the wire in appropriate number of significant figures. **(2004 JEE)**

14. In a Searle's experiment, the diameter of the wire as measured by a screw gauge of least count 0.001 cm is 0.050 cm. The length, measured by a scale of least count 0.1 cm, is 110.0 cm. When a weight of 50 N is suspended from the wire, the extension is measured to be 0.125 cm by a micrometer of least count 0.001 cm. Find the maximum error in the measurement of Young's modulus of the material of the wire from these data. (*Attempt this question after completing elasticity.*) **(2004 JEE)**

Level 2

Single Correct Option

15. Diameter of a steel ball is measured using a vernier caliper which has divisions of 0.1 cm on its main scale (MS). Ten divisions of its vernier scale (VS) match nine divisions on the main scale. Three such measurements for a ball are given as **(2015 JEE)**

S.No	MS(cm)	VS divisions
1	0.5	8
2	0.5	4
3	0.5	6

If the zero error is – 0.03 cm, then mean corrected diameter is

(a) 0.52 cm (b) 0.59 cm (c) 0.56 cm (d) 0.53 cm

Multiple Correct Options

16. Consider a vernier caliper in which each 1 cm on the main scale is divided into 8 equal divisions and a screw gauge with 100 divisions on its circular scale. In the vernier caliper, 5 divisions of the vernier scale coincide with 4 divisions on the main scale and in the screw gauge, one complete rotation of the circular scale moves it by two divisions on the linear scale. Then **(2015 Adv.)**

(a) if the pitch of the screw gauge is twice the least count of the vernier caliper, the least count of the screw gauge is 0.01 mm

(b) if the pitch of the screw gauge is twice the least count of the vernier caliper, the least count of the screw gauge is 0.005 mm

(c) if the least count of the linear scale of the screw gauge is twice the least count of the vernier caliper, the least count of the screw gauge is 0.01 mm

(d) if the least count of the linear scale of the screw gauge is twice the least count of the vernier caliper, the least count of the screw gauge is 0.005 mm.

Subjective Question

17. The edge of a cube is measured using a vernier caliper. (9 divisions of the main scale is equal to 10 divisions of vernier scale and 1 main scale division is 1 mm). The main scale division reading is 10 and 1st division of vernier scale was found to be coinciding with the main scale. The mass of the cube is 2.736 g. Calculate the density in g/cm^3 upto correct significant figures. **(2005 JEE)**

Integer Type Questions

(Attempt these questions after completing the topic elasticity)

18. During Searle's experiment, zero of the vernier scale lies between 3.20×10^{-2} m and 3.25×10^{-2} m of the main scale. The 20th division of the vernier scale exactly coincides with one of the main scale divisions. When an additional load of 2 kg is applied to the wire, the zero of the vernier scale still lies between 3.20×10^{-2} m and 3.25×10^{-2} m of the main scale but now the 45th division of vernier scale coincides with one of the main scale divisions. The length of the thin metallic wire is 2 m and its cross-sectional area is 8×10^{-7} m^2. The least count of the vernier scale is 1.0×10^{-5} m. The maximum percentage error in the Young's modulus of the wire is n%. Find n. **(2014 Adv.)**

19. A steel wire of diameter 0.5 mm and Young's modulus 2×10^{11} Nm^{-2} carries a load of mass m. The length of the wire with the load is 1.0 m. A vernier scale with 10 divisions is attached to the end of this wire. Next to the steel wire is a reference wire to which a main scale, of least count 1.0 mm, is attached. The 10 divisions of the vernier scale correspond to 9 divisions of the main scale. Initially, the zero of vernier scale coincides with the zero of main scale. If the load on the steel wire is increased by 1.2 kg, the vernier scale division which coincides with a main scale division is n. Find n. (Take, $g = 10$ ms^{-2} and $\pi = 3.2$). **(2018 Adv.)**

Answers

Level 1

1. (c) **2.** (b) **3.** (a) **4.** (b) **5.** (d) **6.** (b) **7.** (b) **8.** (b) **9.** (b) **10.** (a,d)
11. (a,b) **12.** (a,b,c) **13.** (2.6 cm^2) **14.** (1.09×10^{10} N/m^2)

Level 2

15. (b) **16.** (b,c) **17.** (2.66) **18.** (4) **19.** (3)

See hints on pages 82-83.

Hints & SOLUTIONS

Module 1.

Introductory Exercise

1. $LC = \dfrac{\text{Smallest division on main scale}}{\text{Number of divisions on vernier scale}}$

$= \dfrac{1\,\text{mm}}{10} = 0.1\,\text{mm} = 0.01\,\text{cm}$

Positive zero error $= N + x(LC)$

$= 0 + 5 \times 0.01$

$= 0.05\,\text{cm}$

Diameter $= 3.2 + 4 \times 0.01 = 3.24\,\text{cm}$

Actual diameter $= 3.24 - 0.05 = 3.19\,\text{cm}$

2. $(N+M)\text{VSD} = (N)\text{MSD}$

$\Rightarrow \quad 1\,\text{VSD} = \left(\dfrac{N}{N+m}\right)\text{MSD}$

$LC = 1\text{MSD} - 1\text{VSD} = 1\text{MSD} - \left(\dfrac{N}{N+m}\right)\text{MSD}$

$= \dfrac{m}{N+m}\text{MSD} = \left(\dfrac{1}{1+N/m}\right)\text{MSD}$

Now, least count will be minimum for $m = 1$

Exercise

1. As it is given that n divisions of vernier scale coincide with $(n-1)$ divisions of main scale i.e.

$n(\text{VSD}) = (n-1)\text{MSD}$

$\Rightarrow \quad 1\text{VSD} = \dfrac{(n-1)}{n}\text{MSD} \quad \ldots\text{(i)}$

The least count is the difference between one main scale division (MSD) and one vernier scale division (VSD).

$\therefore$ Least Count (LC) $= 1\text{MSD} - 1\text{VSD}$

$= 1\text{MSD} - \dfrac{(n-1)}{n}\text{MSD} \quad$ [From Eq. (i)]

$= \left(1 - \dfrac{(n-1)}{n}\right)\text{MSD} = \dfrac{1}{n}\text{MSD}$

Here, $1\text{MSD} = \dfrac{1}{n}\,\text{cm}$

$\Rightarrow \quad LC = \dfrac{1}{n} \times \dfrac{1}{n}\,\text{cm} = \dfrac{1}{n^2}\,\text{cm}$

5. $LC = \dfrac{1}{n}\text{MSD}$

Here, n = number of vernier scale divisions

$0.005\,\text{cm} = \dfrac{1}{n}\left(\dfrac{1}{10}\,\text{cm}\right)$

$n = \dfrac{0.1}{0.005} = \dfrac{1000}{50}$

$\therefore \quad n = 20$

6. $50\,\text{VSD} = 49\,\text{MSD}$

$1\,\text{VSD} = \dfrac{49}{50}\text{MSD}$

$\therefore \quad VC = 1\,\text{MSD} - 1\,\text{VSD}$

$\therefore \quad VC = 1\,\text{MSD} - \dfrac{49}{50}\text{MSD}$

$\therefore \quad VC = \dfrac{1}{50}\text{MSD}$

$\therefore \quad 1\,\text{MSD} = 50\,(VC)$

$= 50\,(0.001\,\text{cm})$

$\therefore \quad 1\,\text{MSD} = 0.5\,\text{mm}$

7. Least count of vernier caliper

$LC = 1\text{MSD} - 1\,\text{VSD}$

20 divisions of vernier scale = 16 divisions of main scale

$\therefore \quad 1\,\text{VSD} = \dfrac{16}{20}\,\text{mm} = 0.8\,\text{mm}$

$\therefore \quad LC = 1\text{MSD} - 1\text{VSD} = 1\,\text{mm} - 0.8\,\text{mm} = 0.2\,\text{mm}$

8. One main scale division, $1\,\text{MSD} = a\,\text{cm}$

One vernier scale division, $1\,\text{VSD} = \dfrac{(n\times 1)\times(a\,\text{cm})}{n}$

$\therefore \quad$ Least count $= 1\,\text{MSD} - 1\,\text{VSD}$

$= a - \dfrac{(n-1)a}{n} = \dfrac{a(n-n+1)}{n}$

$= \dfrac{a}{n}\,\text{cm} = \dfrac{10a}{n}\,\text{mm} \quad (\because 1\,\text{cm} = 10\,\text{mm})$

9. $1\,\text{MSD} = \dfrac{1}{20}\,\text{cm}$

Now, $10\,\text{VSD} = 9\,\text{MSD}$

$1\,\text{VSD} = \dfrac{9}{10}\text{MSD}$

$= \dfrac{9}{10} \times \dfrac{1}{20}\,\text{cm}$

$1\,\text{VSD} = \dfrac{9}{200}\,\text{cm}$

So, vernier constant $= 1\,\text{MSD} - 1\,\text{VSD}$

$= \dfrac{1}{20}\,\text{cm} - \dfrac{9}{200}\,\text{cm}$

$= \dfrac{1}{200} \times 10\,\text{mm}$

$VC = 5 \times 10^{-2}\,\text{mm}$

10. Least count $= 1\,\text{mm} - 0.9\,\text{mm} = 0.1\,\text{mm}$

Reading = Main scale reading + Vernier scale reading $-|\text{zero error}|$

$= 3.1\,\text{cm} + 4 \times 0.1\,\text{mm} - 0.8\,\text{mm}$

$= (3.1 + 0.04 - 0.08)\,\text{cm}$

$= 3.06\,\text{cm}$

11. Given, least count of vernier caliper, $LC = 0.1$ mm

$\therefore$ Zero error $= 6 \times LC = 0.6$ mm

Positive zero error means error will have to subtracted.

Diameter of spherical bob $= MSR + VSR \times LC - \text{zero error}$

$= [32\text{ mm} + 0.1 \times 4\text{ mm}] - 0.6\text{ mm}$

$= 31.8$ mm

$= 3.18$ cm

12. 1 MSD = 1 mm

9 MSD = 10 VSD

1 VSD = 0.9 MSD = 0.9 mm

LC of vernier caliper $= 1 - 0.9 = 0.1$ mm

$= 0.01$ cm

Zero error $= -(10-4) \times 0.1\text{ mm} = -0.6$ mm

The diameter of the spherical body

$= MSR + VSR - \text{Zero error}$

$= 3\text{ cm} + 6 \times 0.01 - [-0.06]$

$= 3 + 0.06 + 0.06 = 3.12$ cm

13. Correct measured value

$= MSR + (VSD \times LC) - \text{Zero error}$

$= 8.5 + (6 \times 0.01) - 0.2 \times 10^{-1}$

$= 8.54$ cm

14. Measured value of diameter

$= MSR + VSR \times LC - \text{Zero error}$

$= 1.7\text{cm} + 5 \times 0.1\text{mm} - (-0.05\text{cm})$

$= 1.7\text{cm} + 0.05\text{cm} + 0.05\text{cm}$

$= 1.8\text{ cm} = 180 \times 10^{-2}\text{cm}$

15. 9 MSD = 10 VSD $\Rightarrow$ 1 VSD $= \frac{9}{10}$ MSD

LC = 1 MSD − 1 VSD

$= 1\text{ MSD} - \frac{9}{10}\text{ MSD} = \frac{1}{10}\text{ MSD} = \frac{1}{10} \times 1\text{ mm}$

$= 0.1$ mm

Diameter $= MSR + LC \times VSR - \text{Error}$

$= 10 + 0.8 - 0.4 = 10.4$ mm

So, radius $= 5.2\text{ mm} = 5.2 \times 10^{-1}\text{cm}$

$= 52 \times 10^{-2}\text{cm}$

16. $A = I \times \omega$

$$\therefore \frac{\Delta A}{A} \times 100 = \frac{\Delta I}{I} \times 100 + \frac{\Delta\omega}{\omega} \times 100$$

$$= \frac{0.1}{10.0} \times 100 + \frac{0.01}{1.0} \times 100$$

$$= \pm 2\%$$

17. Volume of cylinder

$$V = \pi r^2 L, r = \left(\frac{D}{2}\right)$$

$$\therefore \left(\frac{\Delta V}{V}\right) \times 100 = 2\left(\frac{\Delta D}{D}\right) \times 100 + \left(\frac{\Delta L}{L}\right) \times 100$$

$$= 2\left(\frac{0.01}{4.0}\right) \times 100 + \left(\frac{0.1}{5.0}\right) \times 100$$

$$= 2.5\%$$

18. Thickness = External radius − Internal radius

$\Rightarrow$ $t = R - r$

$\Rightarrow$ $(\Delta t)_{max} = (\Delta R + \Delta r)$

Hence, thickness $= \left(\frac{4.23}{2} - \frac{3.87}{2}\right) \pm \left(\frac{0.01}{2} + \frac{0.01}{2}\right)$

$= 0.18 \pm 0.01$ cm

Module 2.

1. Least count $= \frac{\text{Pitch}}{\text{Total division on circular scale}}$

For 5 revolutions, distance travel 5 mm.

For one revolution, distance travel 1mm.

$\therefore$ $LC = \frac{1\text{ mm}}{50} = 0.02$ mm

Hence, Assertion is incorrect and Reason is correct.

4. Pitch $= \frac{2}{4} = 0.5$ mm

$\therefore$ Least count $= \frac{0.5}{50} = 0.01$ mm

5. Least count $= \frac{\text{Value of 1 part on main scale}}{\text{Number of parts on vernier scale}}$

$= \frac{0.25}{5 \times 100}\text{ cm} = 5 \times 10^{-4}$ cm

Reading $= 4 \times 0.05\text{ cm} + 30 \times 5 \times 10^{-4}\text{cm}$

$= (0.2 + 0.0150)$ cm

$= 0.2150$ cm (Thickness of wire)

6. Least count, $LC = \frac{1}{100}$ mm

Diameter of wire $= MSR + CSR \times LC$

$= 0 + \frac{1}{100} \times 52$

$= 0.52$ mm

$= 0.052$ cm

7. Least count $= \frac{\text{Pitch}}{\text{Number of divisions on circular scale}}$

$= \frac{1\text{ mm}}{100} = 0.01$ mm

Linear scale reading = 10 (pitch) = 10 mm

Circular scale reading $= n(LC) = 65 \times 0.01$

$= 0.65$ mm

$\therefore$ Total reading $= (10 + 0.65)\text{ mm} = 10.65$ mm

8. Least count (LC)

$= \frac{\text{Pitch}}{\text{No. of division on circular scale}} = \frac{0.5}{50} = 0.01$ mm

Zero error $= 5 \times 0.01\,\text{mm}$

True reading $= 5 + 0.01\,\text{mm} \times 20 - 0.01\,\text{mm} \times 5$

$= 5.15\,\text{mm}$

9. For screw gauge, least count is given by the formula,

$$LC = \frac{\text{Pitch}}{\text{Number of divisions}}$$

$\Rightarrow \quad LC = \frac{0.5 \times 10^{-3}}{100}$

$\Rightarrow \quad LC = 0.5 \times 10^{-5}\,\text{m}$

$\Rightarrow \quad LC = 5 \times 10^{-6}\,\text{m}$

$\therefore \quad LC = 5 \times 10^{-3}\,\text{mm}$

The actual value is given by the formula,

$$t = MSR + (CSR \pm \text{error}) \times LC$$

where, MSR = main scale reading = 5.5 mm (Given)

CSR = circular scale reading = 48 (Given)

error = 3 divisions below the mean line

= 3 (Positive error)

For positive error

Formula is :

$$t = MSR + (CSR - \text{error}) \times LC$$

Now, the thickness of the sheet,

$t = 5.5 + (48 - 3) \times (5 \times 10^{-3})$

$\therefore \quad t = 5.725\,\text{mm}$

10. Least count $= \frac{\text{Pitch}}{\text{Number of circular divisions}} = \frac{0.5\,\text{mm}}{100}$

Least count $= 5 \times 10^{-3}\,\text{mm}$

Positive error $= MSR + CSR\,(LC)$

$= 0\,\text{mm} + 6\,(5 \times 10^{-3}\,\text{mm})$

Reading of diameter

$= MSR + CSR(LC) - \text{Positive zero error}$

$= 4 \times 0.5\,\text{mm} + 46\,(5 \times 10^{-3}) - 6\,(5 \times 10^{-3})\,\text{mm}$

$= 2\,\text{mm} + 40 \times 5 \times 10^{-3}\,\text{mm}$

$= 2.2\,\text{mm}$

11. Distance moved in one rotation = 0.5 mm

Least count, LC $= \frac{0.5\,\text{mm}}{50\text{ divisions}} = 0.01\,\text{mm}$

Screw gauge has negative zero error.

This error is (50 – 20) 0.01 mm or (30) (0.01) mm.

Thickness of plate

$= (2 \times 0.5\,\text{mm}) + (30 + 20)\,(0.01\,\text{mm})$

$= 1.5\,\text{mm}$

12. Given, least count of a screw gauge,

LC = 0.001 cm

Main scale reading, MSR = 5 mm = 0.5 cm

Number of coinciding divisions on the circular scale i.e.

Vernier scale reading, VSR = 25

Here, zero error = – 0.004 cm

Final reading obtained from the screw gauge is given as

$= MSR + VSR \times LC - \text{zero error}$

Final reading from the screw gauge

$= 0.5 + 25 \times 0.001 - (-0.004)$

$= 0.5 + 0.025 + 0.004$

$= 0.5 + 0.029 = 0.529\,\text{cm}$

Thus, the diameter of the ball is 0.529 cm.

13. MSR = 2.5 mm

$$CSR = 45 \times \frac{0.5}{50}\,\text{mm} = 0.45\,\text{mm}$$

Diameter reading = Reading of screw gauge

$= MSR + CSR - \text{zero error}$

$= 2.5 + 0.45 - (-0.03)$

$= 2.98\,\text{mm}$

14. Least count $= \frac{\text{Pitch}}{\text{Number of divisions on circular scale}}$

$= \frac{0.5\,\text{mm}}{50}$

$\therefore \quad LC = 0.01\,\text{mm}$

Negative zero error $= (50 - 45)\,LC = 0.05\,\text{mm}$

Measured value = main scale reading + screw gauge reading + Negative zero error

$= 0.5\,\text{mm} + (25 \times 0.01 + 0.05)\,\text{mm}$

$= 0.80\,\text{mm}$

16. Given, number of divisions on circular scale of screw gauge = 50

Pitch = 0.5 mm

Least count of screw gauge

$= \frac{\text{Pitch}}{\text{Number of divisions on circular scale}}$

$= \frac{0.5\,\text{mm}}{50} = 10\,\mu\text{m}$

and nature of zero error is positive.

17. If student measures 3.50 cm, it means that there is an uncertainly of order 0.01 cm.

$$1\text{ MSD} = \frac{1}{10}\,\text{cm} \quad \text{and} \quad 9\text{ MSD} = 10\text{ VSD}$$

LC of vernier caliper = 1 MSD – 1 VSD

$= \frac{1}{10}\left(1 - \frac{9}{10}\right) = \frac{1}{100}\,\text{cm}$

$= 0.01\,\text{cm}$

18. For screw gauge (A) :

Reading = MSR + CSR + Error

or $\quad 0.322 = 0.300 + CSR + 5 \times LC$

$0.322 = 0.300 + CSR + 0.005$

$\therefore \quad CSR = 0.017$

For screw gauge (B) :

Reading = MSR + CSR + Error

$0.322 = 0.200 + CSR + 0.092$

$\therefore \quad CSR = 0.030$

Difference $= 0.030 - 0.017 = 0.013\,\text{cm}$

Absolute value of difference $= \frac{0.013}{0.001} = 13$

Unit Exercise (Cumulative)

Level 1

1. $1\text{ VSD} = \dfrac{0.8\text{ cm}}{10} = 0.08\text{ cm}$

$1\text{ MSD} = 0.1\text{ cm}$

$\therefore \quad LC = 1\text{ MSD} - 1\text{ VSD}$

$= 0.1\text{ cm} - 0.08\text{ cm}$

$= 0.02\text{ cm}$

2. $1\text{ MSD} = 5.15\text{ cm} - 5.10\text{ cm} = 0.05\text{ cm}$

$1\text{ VSD} = \dfrac{2.45\text{ cm}}{50} = 0.049\text{ cm}$

$\therefore \quad LC = 1\text{ MSD} - 1\text{ VSD} = 0.001\text{ cm}$

Hence, diameter of cylinder = (Main scale reading) + (Vernier scale reading) (LC)

$= 5.10 + (24)(0.001) = 5.124\text{ cm}$

3. $A = l \times b = 10 \times 1.0 = 10\text{ cm}^2$

$$\frac{\Delta A}{A} = \frac{\Delta l}{l} + \frac{\Delta b}{b}$$

$$\therefore \quad \Delta A = \pm\left(\frac{\Delta l}{l} + \frac{\Delta b}{b}\right) \times A$$

$$= \pm\left(\frac{0.1}{10.0} - \frac{0.01}{1.00}\right) \times 10$$

$$= \pm 0.2\text{ cm}^2$$

4. $LC = \dfrac{1}{100}\text{ mm} = 0.01\text{ mm}$

Diameter of wire = Reading of main scale + Reading of circular scale

$= 1\text{ mm} + 48 \times LC$

$= 1\text{ mm} + 48 \times 0.01\text{ mm}$

$= 1.48\text{ mm}$

$\therefore$ Volume of wire $= \pi \dfrac{d^2}{4} l$

$$= 3.14 \times \frac{(1.48 \times 10^{-1})^2}{4} \times 5\text{ cm}^3$$

$$= 0.0859732\text{ cm}^3 = 0.086\text{ cm}^3$$

$$= 0.086\text{ cm}^3$$

6. Here, pitch $= \dfrac{1\text{ cm}}{10} = \dfrac{10\text{ mm}}{10} = 1\text{ mm}$

Least count (LC)

$$= \frac{\text{Pitch}}{\text{Number of divisions on circular scale}}$$

$$= \frac{1}{100} = 0.01\text{ mm}$$

Zero error $= -(100 - 95) \times LC$

$= -5 \times 0.01 = -0.05\text{ mm}$

The diameter of wire is

d = Reading of main scale + Reading of circular scale + (Zero error)

$= 2\text{ mm} + (45 \times 0.01 + 0.05)\text{ mm}$

$= 2.5\text{ mm}$

$\therefore \quad$ Cross-sectional area $= \dfrac{\pi d^2}{4}$

$$= \frac{3.14 \times (2.5 \times 10^{-1})^2}{4}\text{ cm}^2$$

$$= 0.049\text{ cm}^2$$

7. $LC = \dfrac{\text{Pitch}}{\text{Number of divisions on circular scale}}$

$$= \frac{0.5\text{ mm}}{50} = 0.01\text{ mm}$$

Length of wire = 6.8 cm

Diameter of wire $= 1.5\text{ mm} + 7 \times LC$

$= 1.5\text{ mm} + 7 \times 0.01$

$= 1.57\text{ mm}$

Curved surface area $= \pi D l$

$= 3.14 \times 6.8 \times 1.57 \times 10^{-1}\text{ cm}^2$

$= 3.4\text{ cm}^2$

8. In a screw gauge,

Least count

$$= \frac{\text{Measure of 1 main scale division (MSD)}}{\text{Number of division on circular scale}}$$

Here, minimum value to be measured/least count is 5 μm.

$= 5 \times 10^{-6}\text{ m}$

$\therefore$ According to the given values,

$$5 \times 10^{-6} = \frac{1 \times 10^{-3}}{N}$$

or $\quad N = \dfrac{10^{-3}}{5 \times 10^{-6}} = \dfrac{1000}{5}$

$= 200$ divisions

9. Thickness = MS Reading + Circular Scale Reading (LC)

Here, $LC = \dfrac{0.1}{50} = 0.002$ cm per division

So, 2.124 cm is the correct answer because any measurement should be integral multiple of LC.

10. Since, least count is 1 mm, so we cannot measure the length less than 1 mm. Hence, option (a) and (d) are correct.

11. (a) The least value of measurement in option (a) is one second. So, least count is one second.

(b) The least value of measurement in option (b) is one minute. So, least count is one minute.

12. (a) The zero mark of the vernier scale is ahead 2.4 cm mark on the main scale. So, the reading of the main scale is 2.4 cm.

(b) The third division of the vernier scale exactly coincides with a mark of the main scale.

(c) Total reading = Main scale reading + Vernier scale reading

$= 2.4\text{ cm} + 0.3\text{ mm}$

$= 2.4\text{ cm} + 0.03\text{ cm}$

$= 2.43\text{ cm}$

13. Least count of screw gauge $= \frac{1\text{mm}}{100} = 0.01\text{ mm}$

Diameter of wire $= (1 + 47 \times 0.01)\text{mm} = 1.47\text{ mm}$

Curved surface area (in cm^2) $= (2\pi)\left(\frac{d}{2}\right)(L)$

or $S = \pi dL$

$= (\pi)(1.47 \times 10^{-1})(5.6)\text{ cm}^2$

$= 2.5848\text{ cm}^2$

Rounding off to two significant digits

$S = 2.6\text{ cm}^2$

14. Young's modulus of elasticity is given by

$$Y = \frac{\text{stress}}{\text{strain}} = \frac{F/A}{l/L} = \frac{FL}{lA} = \frac{FL}{l\left(\frac{\pi d^2}{4}\right)}$$

$$\therefore \quad Y = \frac{50 \times 1.1 \times 4}{(1.25 \times 10^{-3}) \times \pi \times (5.0 \times 10^{-4})^2}$$

$= 2.24 \times 10^{11}\text{ N/m}^2$

Now, $\frac{\Delta Y}{Y} = \frac{\Delta L}{L} + \frac{\Delta l}{l} + 2\frac{\Delta d}{d}$

$= \left(\frac{0.1}{110}\right) + \left(\frac{0.001}{0.125}\right) + 2\left(\frac{0.001}{0.05}\right)$

$= 0.0489$

$\Delta Y = (0.0489)Y$

$= (0.0489) \times (2.24 \times 10^{11})\text{N/m}^2$

$= 1.09 \times 10^{10}\text{ N/m}^2$

Level 2

15. MSD = 0.1 cm

$SD = \frac{9}{10}\text{MSD} = 0.9 \times 0.1 = 0.09\text{ cm}$

Least count = MSD − VSD = 0.1 − 0.09 = 0.01 cm

Total reading

= Main scale reading + Vernier coincidence × LC

Ist reading $= 0.5 + 8 \times 0.01 = 0.58$ cm

2nd reading $= 0.5 + 4 \times 0.01 = 0.54$ cm

3rd reading $= 0.5 + 6 \times 0.01 = 0.56$ cm

Average reading $= \frac{(0.58 + 0.54 + 0.56)}{3} = 0.56$ cm

Corrected reading = Reading − Zero error
$= 0.56 - (-0.03) = 0.59$ cm

16. For vernier caliper

$1\text{ MSD} = \frac{1}{8}\text{ cm}$

$5\text{ VSD} = 4\text{ MSD}$

$\therefore \quad 1\text{ VSD} = \frac{4}{5}\text{ MSD} = \frac{4}{5} \times \frac{1}{8} = \frac{1}{10}\text{ cm}$

Least count of vernier caliper = 1 MSD − 1 VSD

$= \frac{1}{8}\text{ cm} - \frac{1}{10}\text{ cm}$

$= 0.025$ cm

(a) and (b)

Pitch of screw gauge $= 2 \times 0.025 = 0.05$ cm

Least count of screw gauge $= \frac{0.05}{100}\text{ cm} = 0.005$ mm

(c) and (d)

Least count of linear scale of screw gauge = 0.05 cm

Pitch $= 0.05 \times 2 = 0.1$ cm

Least count of screw gauge $= \frac{0.1}{100}\text{ cm} = 0.01$ mm

17. 1 MSD = 1 mm

9 MSD = 10 VSD

$\therefore$ Least count,

$LC = 1\text{ MSD} - 1\text{ VSD} = 1\text{ mm} - \frac{9}{10}\text{ mm} = \frac{1}{10}\text{ mm}$

Measure reading of edge = MSR + VSR (LC)

$= 10 + 1 \times \frac{1}{10} = 10.1$ mm

Volume of cube, $V = (1.01)^3\text{ cm}^3 = 1.030301 = 1.03\text{ cm}^3$

[After rounding off up to 3 significant digits, as edge length is measured up to 3 significant digits]

$\therefore \quad$ Density of cube $= \frac{2.736}{1.03} = 2.6563\text{ g/cm}^3$

$= 2.66\text{ g/cm}^3$

(After rounding off to 3 significant digits)

18. $Y = \frac{FL}{Al}$. Here l is change in length.

$\therefore \quad Y \propto \frac{1}{l}$ or $\frac{\Delta Y}{Y} \times 100$

$= \frac{\Delta l}{l} \times 100 = \frac{(\text{LC})}{(45 - 20)\text{ LC}} \times 100$

$= 4\%$

19. $\Delta l = \frac{Fl}{AY} = \frac{mgl}{\frac{\pi d^2}{4}Y}$

$$= \frac{1.2 \times 10 \times 1}{\frac{\pi}{4} \times (5 \times 10^{-4})^2 \times 2 \times 10^{11}}$$

$= 0.3$ mm

LC of vernier $= \left(1 - \frac{9}{10}\right)\text{mm} = 0.1$ mm

So, 3rd division of vernier scale will coincide with main scale.

UNIT-4

UNITS & DIMENSIONS

Learning Modules

1. Units
2. Dimensions

MODULE 1

Units

To measure a physical quantity we need some standard unit of that quantity. The measurement of the quantity is specified in two parts, the first part gives how many times of the standard unit and the second part gives the name of the unit.

Thus, suppose I say that length of this wire is 5 metre. The numeric part 5 says that it is 5 times of the unit of length and the second part metre says that unit chosen here is metre.

Fundamental and Derived Units

There are a large number of physical quantities and every quantity needs a unit. However, not all the quantities are independent. For example, if a unit of length is defined, a unit of volume is automatically obtained.

Thus, we can define a set of fundamental quantities and all other quantities may be expressed in terms of the fundamental quantities.

Fundamental quantities are only seven in number. Units of all other quantities can be expressed in terms of the units of these seven quantities by multiplication or division.

Many different choices can be made about the fundamental quantities. For example, if we take length and time as the fundamental quantities then speed is a derived quantity and if we take speed and time as fundamental quantities then length is a derived quantity.

Several system of units are in use over the world. The units defined for the fundamental quantities are called fundamental units and those obtained for derived quantities are called the derived units.

SI Units

In 1971, General Conference on Weights and Measures held its meeting and decided a system of units which is known as the International System of Units.

It is abbreviated as SI from the French name Le System International *d'* Unites. This system is widely used throughout the world. Table below gives the seven fundamental quantities and their SI units.

Table 1 Fundamental quantities and their SI units

S.No.	Quantity	SI Unit	Symbol
1.	Length	metre	m
2.	Mass	kilogram	kg
3.	Time	second	s
4.	Electric current	ampere	A
5.	Thermodynamic temperature	kelvin	K
6.	Amount of substance	mole	mol
7.	Luminous intensity	candela	cd

Two **supplementary units**, namely that of plane angle and solid angle are also defined. Their units are radian (rad) and steradian (st) respectively.

(i) **CGS System :** In this system, the units of length, mass and time are centimetre (cm), gram (g) and second (s) respectively. The unit of force is dyne and that of work or energy is erg.

(ii) **FPS System :** In this system, the units of length, mass and time are foot, pound and second. The unit of force in this system is poundal.

Definitions of Fundamental and Supplementary SI Units

(i) **Metre :** $1\ \text{m} = 1{,}650{,}763.73$ wavelengths in vacuum, of radiation corresponding to orange-red light of krypton-86.

(ii) **Second :** $1\ \text{s} = 9{,}192{,}631{,}770$ time periods of a particular radiation from Cesium-133 atom.

(iii) **Kilogram :** 1 kg = mass of 1 litre of water at 4°C.

(iv) **Ampere :** $1\text{A} =$ the current which when flows through two infinitely long straight conductors of negligible cross-section placed at a distance of one metre in vacuum produces a force of 2×10^{-7} N/m between them.

(v) **Kelvin :** $1\ \text{K} = 1/273.16$ part of the thermodynamic temperature of triple point of water.

(vi) **Mole :** $1\ \text{mol} =$ the amount of substance of a system which contains as many elementary particles (atoms, molecules, ions etc.) as there are atoms in 12 g of carbon-12.

Note *JEE aspirants are advised to read both topics of Unit-4 after completing the whole syllabus of 11th and 12th. Because at this stage you do not know all formulae which are required in this unit.*

(vii) **Candela :** 1 cd = the luminous intensity in a perpendicular direction of a surface of $\left(\frac{1}{600000}\right)$ m^2 of a black body at the temperature of freezing platinum under a pressure of 1.013×10^5 N/m^2.

(viii) **Radian :** 1rad = the plane angle between two radii of a circle which cut-off on the circumference, an arc equal in length to the radius.

(ix) **Steradian :** 1st = the solid angle having its vertex at the centre of the sphere, cut-off an area of the surface of sphere equal to that of a square with sides of length equal to the radius of the sphere.

Table 2. SI units of some physical quantities frequently used in physics

S.No.	Physical Quantity	SI Units
1.	Velocity = displacement/time	m/s
2.	Acceleration = velocity/time	m/s^2
3.	Force = mass × acceleration	$kg\text{-}m/s^2$ = newton or N
4.	Work = force × displacement	$kg\text{-}m^2/s^2$ = N-m = joule or J
5.	Energy	J
6.	Torque = force × perpendicular distance	N-m
7.	Power = work/time	J/s or watt
8.	Momentum = mass × velocity	kg-m/s
9.	Impulse = force × time	N-s
10.	Angle = arc/radius	radian or rad
11.	Strain = $\frac{\Delta L}{L}$ or $\frac{\Delta V}{V}$	no units
12.	Stress = force/area	N/m^2
13.	Pressure = force/area	N/m^2
14.	Modulus of elasticity = stress/strain	N/m^2
15.	Frequency = 1/time period	per sec or hertz (Hz)
16.	Angular velocity = angle/time	rad/s
17.	Moment of inertia = (mass) × $(\text{distance})^2$	$kg\text{-}m^2$
18.	Surface tension = force/length	N/m
19.	Gravitational constant = $\frac{\text{force} \times (\text{distance})^2}{(\text{mass})^2}$	$N\text{-}m^2/kg^2$
20.	Angular momentum	$kg\text{-}m^2/s$
21.	Coefficient of viscosity	$N\text{-}s/m^2$
22.	Planck's constant	J-s
23.	Specific heat (*s*)	J/kg-K
24.	Coefficient of thermal conductivity (*K*)	watt/m-K
25.	Gas constant (*R*)	J/mol-K
26.	Boltzmann constant (*k*)	J/K
27.	Wien's constant (*b*)	m-K
28.	Stefan's constant (σ)	$watt/m^2\text{-}K^4$
29.	Electric charge	C
30.	Electric intensity	N/C
31.	Electric potential	volt
32.	Capacitance	farad
33.	Permittivity of free space	$C^2N^{-1}m^{-2}$
34.	Electric dipole moment	C-m
35.	Resistance	ohm
36.	Magnetic field	tesla (T) or weber/m^2 (Wb/m^2)
37.	Coefficient of self induction	henry

Note (i) ***Conversion of units*** *This is based on the fact that the product of the numerical value (n) and its corresponding unit (u) is a constant, i.e.*

$$n[u] = \text{constant}$$

or $$n_1[u_1] = n_2[u_2]$$

(ii) ***Astronomical unit*** *1 AU = mean distance of earth from sun* $\approx 1.5 \times 10^{11}$ *m*

Light year *1 ly = distance travelled by light in vacuum in 1 year* $= 9.46 \times 10^{15}$ *m*

Parsec *1 Parse* $= 3.07 \times 10^{16}$ *m = 3.26 light year*

X-ray unit $1U = 10^{-3}$ *m*

$1 \text{ shake} = 10^{-8}$ s

$1 \text{ Bar} = 10^5 \text{ N/m}^2 = 10^5$ Pa

1 torr = 1 mm of Hg = 133.3 Pa

$1 \text{ barn} = 10^{-28}$ m^2

1 horse power = 746 W

1 pound = 453.6 g = 0.4536 kg

(iii) ***Table 3. Prefixes used for different powers of 10***

Power of 10	Prefix	Symbol	Power of 10	Prefix	Symbol	Power of 10	Prefix	Symbol	Power of 10	Prefix	Symbol
10^{18}	exa	E	10^{-1}	deci	d	10^9	giga	G	10^{-6}	micro	μ
10^{15}	peta	P	10^{-2}	centi	c	10^6	mega	M	10^{-9}	nano	n
10^{12}	tera	T	10^{-3}	milli	m	10^3	kilo	k	10^{-12}	pico	p
10^2	hecto	h	10^{-15}	femto	f	10^1	deca	da	10^{-18}	atto	a

Example 1 Given that $F = 8$ N. Convert it into CGS system.

Sol. $F = 8\dfrac{\text{kg} \times \text{m}}{\text{s}^2}$

$= (8)\dfrac{(10^3 \text{ g})(100 \text{ cm})}{\text{s}^2}$

$= 8 \times 10^5 \dfrac{\text{g - cm}}{\text{s}^2}$ (in CGS system)

This unit $\left(\dfrac{\text{g - cm}}{\text{s}^2}\right)$ is also called dyne.

Thus, $1\text{N} = 10^5$ dyne.

Example 2 A particle is moving with velocity $v = 180$ km/h. Convert the velocity of the particle into m/s.

Sol. $v = 180$ km/h

$= (180)\dfrac{(1000 \text{ m})}{(60 \times 60 \text{ s})}$

$= 180\left(\dfrac{1000}{3600}\right)\dfrac{\text{m}}{\text{s}}$

$= 180 \times \dfrac{5}{18}\dfrac{\text{m}}{\text{s}}$

$\Rightarrow \quad v = 50$ m/s

Note : $\dfrac{1 km}{h} = \dfrac{5}{18}\dfrac{m}{s}$

Example 3 Force acting on a particle is 20 N. If units of length and time are doubled and unit of mass is halved, then find the numerical value of force in the new system of units.

Sol. Force $= 20 \text{ N} = \dfrac{20 \text{ kg - m}}{\text{s}^2}$

If unit of length and time are doubled and the unit of mass is halved.

Then unit of force will be $\left[\dfrac{\frac{1}{2} \times 2}{(2)^2}\right] = \dfrac{1}{4}$ times.

Hence, the numerical value of force will be $\dfrac{1}{4}$ times or answer is 5.

Example 4 Can pressure (p), density (ρ) and velocity (v) be taken as fundamental quantities?

Sol. No, they cannot be taken as fundamental quantities, as they are related to each other by the relation

$$p = \rho v^2$$

Example 5 A new unit of length is chosen such that the speed of light in vacuum is unity. What is the distance between the Sun and the Earth in terms of the new unit if light takes 8 min and 20 s to cover this distance?

Sol. The speed of light in vacuum = 1 unit $= \dfrac{1 \text{ unit of length}}{\text{s}}$

Time taken by the light to cover the distance from the sun to the earth = 8 min and 20 s = 500 s.

Now, distance = speed × time

$= 1\left(\dfrac{\text{unit of length}}{\text{s}}\right)(500 \text{ s})$

$= 500$ length units

MODULE 1 Exercise

Fundamental Units, Derived Units and Theory Based

Assertion and Reason

Directions : Choose the correct option.

(a) *If both **Assertion** and **Reason** are true and the **Reason** is the correct explanation of the **Assertion.***

(b) *If both **Assertion** and **Reason** are true but **Reason** is not the correct explanation of **Assertion.***

(c) *If **Assertion** is true, but the **Reason** is false.*

(d) *If **Assertion** is false but the **Reason** is true.*

(e) *If both **Assertion** and **Reason** are false.*

1. **Assertion :** Velocity, volume and acceleration can be taken as fundamental quantities.

   Reason : All the three are independent from each other.

2. **Assertion :** When we change the unit of measurement of a quantity, its numerical value changes.

   Reason : Smaller the unit of measurement smaller is its numerical value.

Single Correct Option

3. Which of the following sets cannot enter into the list of fundamental quantities in any system of units?
   (a) Length, mass and acceleration
   (b) Length, time and velocity
   (c) Mass, time and velocity
   (d) Length, time and density

4. Temperature can be expressed as a derived quantity in terms of any of the following
   (a) length and mass (b) mass and time
   (c) length, mass and time (d) None of these

Units of Physical Quantity

Single Correct Option

5. Joule $\times$ second is the unit of
   (a) energy (b) momentum
   (c) angular momentum (d) power

6. Which of the following is not equal to watt?
   (a) Joule/second (b) Ampere $\times$ volt
   (c) $(\text{Ampere})^2 \times$ ohm (d) Ampere/volt

7. Which of the following is not the units of surface tension?
   (a) N/m (b) J/m^2
   (c) Kg/s^2 (d) None of these

8. Which of the following quantities has its unit as newton-second?
   (a) Energy (b) Torque
   (c) Momentum (d) Angular momentum

9. A suitable unit for gravitational constant is
   (a) kg-m sec^{-1} (b) Nm^{-1} sec
   (c) $N\text{-}m^2\, kg^{-2}$ (d) kg-m sec^{-1}

10. Which of the following is not a unit of energy?
    (a) W-s (b) kg-m/sec
    (c) N-m (d) Joule

11. Erg-m^{-1} can be the unit of
    (a) force (b) momentum
    (c) power (d) acceleration

12. Ampere-hour is a unit of
    (a) quantity of electricity
    (b) strength of electric current
    (c) power
    (d) energy

13. The unit of impulse is same as the unit of
    (a) momentum
    (b) force
    (c) rate of change of momentum
    (d) torque

Multiple Correct Options

14. The SI unit of the inductance, the henry can be written as **(1998 JEE)**
    (a) weber/ampere (b) volt-second/ampere
    (c) joule/$(\text{ampere})^2$ (d) ohm-second

Match the Columns

15. Match the Column I with Column II and mark the correct option from the codes given below.

	Column I		Column II
(A)	Modulus of rigidity	(p)	Volt-metre per ampere
(B)	Latent heat	(q)	Ohm-metre
(C)	Electric resistivity	(r)	Joule per kg
(D)	Gravitational potential	(s)	Newton per metre^2

Codes

	A	B	C	D		A	B	C	D
(a)	s	r	p,q	r	(b)	p	q	r	s
(c)	q	s	r	p	(d)	p	s	r	p

16. Match Column I with Column II **(2023 JEE)**

	Column I		Column II
A.	Surface tension	(p)	$kg\,m^{-1}s^{-1}$
B.	Pressure	(q)	$kg\,m\,s^{-1}$
C.	Viscosity	(r)	$kg\,m^{-1}s^{-2}$
D.	Impulse	(s)	$kg\,s^{-2}$

Choose the correct answer from the options given below.

	A	B	C	D		A	B	C	D
(a)	s	r	q	p	(b)	s	r	p	q
(c)	r	s	p	q	(d)	q	p	r	s

17. Match Column I with Column II. **(2022 JEE)**

	Column I		Column II
A.	Torque	(p)	Nms^{-1}
B.	Stress	(q)	$J\,kg^{-1}$
C.	Latent heat	(r)	Nm
D.	Power	(s)	$N\,m^{-2}$

Choose the correct answer from the options given below.

	A	B	C	D		A	B	C	D
(a)	r	q	p	s	(b)	r	s	q	p
(c)	s	p	r	q	(d)	q	r	p	s

Subjective Question

18. Give the MKS units for each of the following quantities. **(1980 JEE)**
(a) Young's modulus (b) Magnetic induction
(c) Power of a lens

Conversion of Units (Type-1)

Fill in the Blank

19. Fill in the blanks by suitable conversion of units
(a) $1\ kg\,m^2\,s^{-2} = \ldots\ldots\ g\,cm^2\,s^{-2}$
(b) $3.0\ ms^{-2} = \ldots\ldots kmh^{-2}$
(c) $G = 6.67 \times 10^{-11}\ Nm^2\,(kg)^{-2} = \ldots\ldots (cm)^3\,s^{-2}g^{-1}$

Subjective Questions

20. Young's modulus of steel is $2.0 \times 10^{11}\ N/m^2$. Express it in $dyne/cm^2$.

21. Surface tension of water in the CGS system is 72 dyne/cm. What is its value in SI units?

Conversion of Units (Type-2)

Single Correct Option

22. The units of length, velocity and force are doubled. Which of the following is the correct change in the other units?
(a) Unit of time is doubled
(b) Unit of mass is doubled
(c) Unit of momentum is doubled
(d) Unit of energy is doubled

23. If the unit of length and force be increased four times, then the unit of energy is
(a) increased 4 times (b) increased 8 times
(c) increased 16 times (d) decreased 16 times

Integer Type Question

24. If the acceleration due to gravity is $10\ ms^{-2}$ and the unit of length and time are changed in kilometre and hour respectively. If the numerical value of the acceleration is found to be $x \times 10^2 km\text{-}hr^{-2}$. Find the value of x.

Miscellaneous Problems

Statement Type Question

25. Given below are two statements :
Statement I : Astronomical unit (AU), Parsec (Pc) and Light year (ly) are units for measuring astronomical distances.
Statement II : AU < Parsec (Pc) < ly **(2023 JEE)**
In the light of the above statements choose the most appropriate answer from the options given below :
(a) Both Statement I and Statement II are correct.
(b) Statement I is correct but Statement II is incorrect.
(c) Both Statement I and Statement II are incorrect.
(d) Statement I is incorrect but Statement II is correct.

Single Correct Option

26. Which of the following is smallest unit?
(a) Millimetre (b) Angstrom
(c) Fermi (d) Metre

27. In the equation $\left(p + \frac{a}{V^2}\right)(V - b) = \text{constant}$, the unit of a is
(a) $dyne \times cm^5$ (b) $dyne \times cm^4$
(c) $dyne/cm^3$ (d) $dyne/cm^2$

28. Electric field in a certain region is given by $\mathbf{E} = \left(\frac{A}{x^2}\hat{\mathbf{i}} + \frac{B}{y^3}\hat{\mathbf{j}}\right)$. The SI unit of A and B are : **(2023 JEE)**
(a) Nm^3C^{-1}, $Nm^2\,C^{-1}$ (b) $Nm^2\,C^{-1}$, $Nm^3\,C^{-1}$
(c) $Nm^3\,C$, $Nm^2\,C$ (d) $Nm^2\,C$, $Nm^3\,C$

Answers

1. (e) **2.** (c) **3.** (b) **4.** (d) **5.** (c) **6.** (d) **7.** (d) **8.** (c) **9.** (c) **10.** (b)
11. (a) **12.** (a) **13.** (a) **14.** (a,b,c,d) **15.** (a) **16.** (b) **17.** (b) **18.** (a) N/m^2 (b) Tesla (c) m^{-1}
19. (a) 10^7 (b) 3.9×10^4 (c) 6.67×10^{-8} **20.** $2.0 \times 10^{12}\ dyne/cm^2$ **21.** 0.072 N/m **22.** (c) **23.** (c)
24. (1296) **25.** (b) **25.** (b) **26.** (c) **27.** (b) **28.** (b)

See hints on pages 103-104.

MODULE 2

Dimensions

Dimensions of a physical quantity are the powers to which the fundamental quantities must be raised to represent the given physical quantity.

For example, density $= \dfrac{\text{mass}}{\text{volume}} = \dfrac{\text{mass}}{(\text{length})^3}$

or $\qquad$ density $= (\text{mass})\,(\text{length})^{-3}$ $\qquad$ ...(i)

Thus, the dimensions of density are 1 in mass and −3 in length. The dimensions of all other fundamental quantities are zero.For convenience, the fundamental quantities are represented by one letter symbols. Generally mass is denoted by *M*, length by *L*, time by *T* and electric current by *A*.

The thermodynamic temperature, the amount of substance and the luminous intensity are denoted by the symbols of their units K, mol and cd respectively. The physical quantity that is expressed in terms of the base quantities is enclosed in square brackets.

Thus, Eq. (i) can be written as

$$[\text{density}] = [\text{ML}^{-3}]$$

Such an expression for a physical quantity in terms of the fundamental quantities is called the dimensional formula. Here, it is worthnoting that constants such as 5, π or trigonometrical functions such as sin θ, cos θ, etc., have no units and dimensions.

$$[\sin\theta] = [\cos\theta] = [\tan\theta] = [\log x] = [e^x] = [\text{M}^0\text{L}^0\text{T}^0]$$

Table 4. Dimensional formulae of some physical quantities frequently used in physics

S.No.	Physical Quantity	Dimensional Formula
1.	Velocity = displacement/time	$[\text{M}^0\text{LT}^{-1}]$
2.	Acceleration = velocity/time	$[\text{M}^0\text{LT}^{-2}]$
3.	Force = mass × acceleration	$[\text{MLT}^{-2}]$
4.	Work = force × displacement	$[\text{ML}^2\text{T}^{-2}]$
5.	Energy	$[\text{ML}^2\text{T}^{-2}]$
6.	Torque = force × perpendicular distance	$[\text{ML}^2\text{T}^{-2}]$
7.	Power = work/time	$[\text{ML}^2\text{T}^{-3}]$
8.	Momentum = mass × velocity	$[\text{MLT}^{-1}]$
9.	Impulse = force × time	$[\text{MLT}^{-1}]$
10.	Angle = arc/radius	$[\text{M}^0\text{L}^0\text{T}^0]$
11.	Strain $= \dfrac{\Delta L}{L}$ or $\dfrac{\Delta V}{V}$	$[\text{M}^0\text{L}^0\text{T}^0]$
12.	Stress = force/area	$[\text{ML}^{-1}\text{T}^{-2}]$
13.	Pressure = force/area	$[\text{ML}^{-1}\text{T}^{-2}]$
14.	Modulus of elasticity = stress/strain	$[\text{ML}^{-1}\text{T}^{-2}]$
15.	Frequency = 1/time period	$[\text{M}^0\text{L}^0\text{T}^{-1}]$
16.	Angular velocity = angle/time	$[\text{M}^0\text{L}^0\text{T}^{-1}]$
17.	Moment of inertia = (mass) × (distance)^2	$[\text{ML}^2\text{T}^0]$
18.	Surface tension = force/length	$[\text{ML}^0\text{T}^{-2}]$
19.	Gravitational constant $= \dfrac{\text{force} \times (\text{distance})^2}{(\text{mass})^2}$	$[\text{M}^{-1}\text{L}^3\text{T}^{-2}]$
20.	Angular momentum	$[\text{ML}^2\text{T}^{-1}]$
21.	Coefficient of viscosity	$[\text{ML}^{-1}\text{T}^{-1}]$
22.	Planck's constant	$[\text{ML}^2\text{T}^{-1}]$
23.	Specific heat (*s*)	$[\text{L}^2\text{T}^{-2}\text{K}^{-1}]$
24.	Coefficient of thermal conductivity (*K*)	$[\text{MLT}^{-3}\text{K}^{-1}]$
25.	Gas constant (*R*)	$[\text{ML}^2\text{T}^{-2}\text{K}^{-1}\,\text{mol}^{-1}]$
26.	Boltzmann constant (*k*)	$[\text{ML}^2\text{T}^{-2}\text{K}^{-1}]$
27.	Wien's constant (*b*)	$[\text{LK}]$
28.	Stefan's constant (σ)	$[\text{MT}^{-3}\text{K}^{-4}]$
29.	Electric charge	$[\text{AT}]$
30.	Electric intensity	$[\text{MLT}^{-3}\text{A}^{-1}]$
31.	Electric potential	$[\text{ML}^2\text{T}^{-3}\text{A}^{-1}]$
32.	Capacitance	$[\text{M}^{-1}\text{L}^{-2}\text{T}^4\text{A}^2]$
33.	Permittivity of free space	$[\text{M}^{-1}\text{L}^{-3}\text{T}^4\text{A}^2]$
34.	Electric dipole moment	$[\text{LTA}]$
35.	Resistance	$[\text{ML}^2\text{T}^{-3}\text{A}^{-2}]$
36.	Magnetic field	$[\text{MT}^{-2}\text{A}^{-1}]$
37.	Coefficient of self induction	$[\text{ML}^2\text{T}^{-2}\text{A}^{-2}]$

Example 1 Find the dimensional formula of the following quantities :

(a) Velocity (b) Acceleration
(c) Momentum (d) Force
(e) Work or energy (f) Power
(g) Pressure

Sol. (a) Velocity $[v] = \dfrac{\text{displacement}}{\text{time}}$

$$[v] = \frac{[\text{displacement}]}{[\text{time}]} = \frac{[\text{L}]}{[\text{T}]} = [\text{M}^0\text{LT}^{-1}]$$

(b) Acceleration $[a] = \left[\dfrac{dv}{dt}\right]$

$$[a] = \frac{dv \to \text{kind of velocity}}{dt \to \text{kind of time}}$$

$$= \frac{[\text{LT}^{-1}]}{[\text{T}]} = [\text{LT}^{-2}]$$

(c) Momentum $[P] = [mv]$

$$[P] = [\text{M}]\,[\text{LT}^{-1}]$$
$$= [\text{MLT}^{-1}]$$

(d) Force $[F] = [ma]$

$$[F] = [m]\,[a] = [\text{M}]\,[\text{LT}^{-2}]$$
$$= [\text{MLT}^{-2}]$$

(e) Work or Energy = force × displacement

$$[\text{Work}] = [\text{force}]\,[\text{displacement}]$$
$$= [\text{MLT}^{-2}][\text{L}] = [\text{ML}^2\text{T}^{-2}]$$

(f) Power $= \dfrac{\text{work}}{\text{time}}$

$$\therefore \quad [\text{Power}] = \left[\frac{\text{work}}{\text{time}}\right] = \frac{[\text{ML}^2\text{T}^{-2}]}{[\text{T}]}$$
$$= [\text{ML}^2\text{T}^{-3}]$$

(g) Pressure $= \dfrac{\text{force}}{\text{area}}$

$$\therefore \quad [\text{Pressure}] = \frac{[\text{force}]}{[\text{area}]}$$
$$= \frac{[\text{MLT}^{-2}]}{[\text{L}^2]} = [\text{ML}^{-1}\text{T}^{-2}]$$

Example 2 Find the dimensional formula of the following quantities :

(a) Surface tension, T
(b) Universal constant of gravitation, G
(c) Impulse, J
(d) Torque, τ

The equations involving these equations are :

$$T = F/l,\ F = \frac{Gm_1m_2}{r^2},\ J = F \times t \ \text{ and } \tau = F \times l$$

Sol. (a) $T = \dfrac{F}{l}$

$$\Rightarrow \quad [T] = \frac{[F]}{[l]} = \frac{[\text{MLT}^{-2}]}{[\text{L}]}$$
$$= [\text{MT}^{-2}] \quad \textbf{Ans.}$$

(b) $F = \dfrac{Gm_1m_2}{r^2}$

$$\Rightarrow \quad G = \frac{Fr^2}{m_1m_2}$$

or
$$[G] = \frac{[G][r]^2}{[m]^2} = \frac{[\text{MLT}^{-2}][\text{L}^2]}{[\text{M}^2]}$$
$$= [\text{M}^{-1}\text{L}^3\text{T}^{-2}] \quad \textbf{Ans.}$$

(c) $J = F \times t$

$$\therefore \quad [J] = [F]\,[t]$$
$$= [\text{MLT}^{-2}]\,[\text{T}]$$
$$= [\text{MLT}^{-1}] \quad \textbf{Ans.}$$

(d) $\tau = F \times l$

$$\therefore \quad [\tau] = [F]\,[l]$$
$$= [\text{MLT}^{-2}]\,[\text{L}]$$
$$= [\text{ML}^2\text{T}^{-2}] \quad \textbf{Ans.}$$

Uses of Dimensions

Theory of dimensions have following main uses:

1. **Conversion of units** This is based on the fact that the product of the numerical value (n) and its corresponding unit (u) is a constant, i.e.

$$n\,[u] = \text{constant}$$

or
$$n_1\,[u_1] = n_2\,[u_2]$$

Suppose the dimensions of a physical quantity are a in mass, b in length and c in time. If the fundamental units in one system are M_1, L_1 and T_1 and in the other system are M_2, L_2 and T_2 respectively. Then, we can write

$$n_1\,[\text{M}_1^a\,\text{L}_1^b\,\text{T}_1^c] = n_2\,[\text{M}_2^a\,\text{L}_2^b\,\text{T}_2^c] \quad \text{...(i)}$$

Here, n_1 and n_2 are the numerical values in two systems of units respectively. Using Eq. (i), we can convert the numerical value of a physical quantity from one system of units into the other system.

Example 3 The value of gravitational constant is $G = 6.67 \times 10^{-11}$ N-m^2/kg^2 in SI units. Convert it into CGS system of units.

Sol. The dimensional formula of G is $[\text{M}^{-1}\text{L}^3\text{T}^{-2}]$.

Using Eq. (i), i.e.

$$n_1[\text{M}_1^{-1}\text{L}_1^3\text{T}_1^{-2}] = n_2[\text{M}_2^{-1}\text{L}_2^3\text{T}_2^{-2}]$$

$$n_2 = n_1\left[\frac{\text{M}_1}{\text{M}_2}\right]^{-1}\left[\frac{\text{L}_1}{\text{L}_2}\right]^3\left[\frac{\text{T}_1}{\text{T}_2}\right]^{-2}$$

Here, $n_1 = 6.67 \times 10^{-11}$

$M_1 = 1\,\text{kg}, M_2 = 1\,\text{g} = 10^{-3}\,\text{kg},$

$L_1 = 1\,\text{m}, L_2 = 1\,\text{cm} = 10^{-2}\,\text{m},$

$T_1 = T_2 = 1\,\text{s}$

Substituting in the above equation, we get

$$n_2 = 6.67 \times 10^{-11} \left[\frac{1\,\text{kg}}{10^{-3}\,\text{kg}}\right]^{-1} \left[\frac{1\,\text{m}}{10^{-2}\,\text{m}}\right]^{3} \left[\frac{1\,\text{s}}{1\,\text{s}}\right]^{-2}$$

or $n_2 = 6.67 \times 10^{-8}$

Thus, value of G in CGS system of units is 6.67×10^{-8} dyne cm^2/g^2.

2. **To check the dimensional correctness of a given physical equation.** Every physical equation should be dimensionally balanced. This is called the 'Principle of Homogeneity'. The dimensions of each term on both sides of an equation must be the same. On this basis, we can judge whether a given equation is correct or not. But a dimensionally correct equation may or may not be physically correct.

Example 4 Show that the expression of the time period T of a simple pendulum of length l given by $T = 2\pi\sqrt{\frac{l}{g}}$ is dimensionally correct.

Sol. $T = 2\pi\sqrt{\frac{l}{g}}$

Dimensionally $[\text{T}] = \sqrt{\frac{[\text{L}]}{[\text{LT}^{-2}]}} = [\text{T}]$

As in the above equation, the dimensions of both sides are same. The given formula is dimensionally correct.

Principle of Homogeneity of Dimensions

This principle states that the dimensions of all the terms in a physical expression should be same. For example, in the physical expression $s = ut + \frac{1}{2}at^2$, the dimensions of s, ut and $\frac{1}{2}at^2$ all are same.

Note *The physical quantities separated by the symbols +, −, =, >, < etc., have the same dimensions.*

Example 5 The velocity v of a particle depends upon the time t according to the equation $v = a + bt + \frac{c}{d+t}$. Write the dimensions of a, b, c and d.

Sol. From principle of homogeneity,

$[a] = [v]$

or $[a] = [\text{LT}^{-1}]$

$[bt] = [v]$ or $[b] = \frac{[v]}{[t]} = \frac{[\text{LT}^{-1}]}{[\text{T}]}$

or $[b] = [\text{LT}^{-2}]$

Similarly, $[d] = [t] = [\text{T}]$

Further, $\frac{[c]}{[d+t]} = [v]$

or $[c] = [v][d+t]$

or $[c] = [\text{LT}^{-1}][\text{T}]$

or $[c] = [\text{L}]$

3. **To establish the relation among various physical quantities** If we know the factors on which a given physical quantity may depend, we can find a formula relating the quantity with those factors. Let us take an example.

Example 6 The frequency (f) of a stretched string depends upon the tension F (dimensions of force), length l of the string and the mass per unit length μ of string. Derive the formula for frequency.

Sol. Suppose, that the frequency f depends on the tension raised to the power a, length raised to the power b and mass per unit length raised to the power c. Then,

$$f \propto F^a l^b \mu^c$$

or $f = kF^a l^b \mu^c$ ...(i)

Here, k is a dimensionless constant. Thus,

$$[f] = [F]^a [l]^b [\mu]^c$$

or $[\text{M}^0\text{L}^0\text{T}^{-1}] = [\text{MLT}^{-2}]^a [\text{L}]^b [\text{ML}^{-1}]^c$

or $[\text{M}^0\text{L}^0\text{T}^{-1}] = [\text{M}^{a+c}\text{L}^{a+b-c}\text{T}^{-2a}]$

For dimensional balance, the dimensions on both sides should be same.

Thus, $a + c = 0$ ...(ii)

$a + b - c = 0$ ...(iii)

and $-2a = -1$ ...(iv)

Solving these three equations, we get

$$a = \frac{1}{2}, \quad c = -\frac{1}{2}$$

and $b = -1$

Substituting these values in Eq. (i), we get

$$f = k(F)^{1/2}(l)^{-1}(\mu)^{-1/2}$$

or $f = \frac{k}{l}\sqrt{\frac{F}{\mu}}$

Experimentally, the value of k is found to be $\frac{1}{2}$.

Hence, $f = \frac{1}{2l}\sqrt{\frac{F}{\mu}}$

Limitations of Dimensional Analysis

The method of dimensions has the following limitations :

(i) By this method, the value of dimensionless constant cannot be calculated.

(ii) By this method, the equation containing trigonometrical, exponential and logarithmic terms cannot be analysed.

(iii) This method is useful when a physical quantity depends on other quantities which are multiplied, divided or raised to some powers. It cannot be used if a physical quantity depends on sum or difference of two quantities. For example, we cannot get the relation, $s = ut + \frac{1}{2}at^2$ from dimensional analysis.

Note *There are some physical quantities which have the same dimensions. They are given in tabular form as below :*

S.No.	Physical quantities or combination of physical quantities	Dimensions
1.	Angle, strain, $\sin\theta$, π, e^x	$[M^0L^0T^0]$
2.	Work, Energy, Torque, Rhc	$[ML^2T^{-2}]$
3.	Time, $\frac{L}{R}$, CR, $\sqrt{LC}$	$[M^0L^0T]$
4.	Frequency, ω, $\frac{R}{L}$, $\frac{1}{CR}$, $\frac{1}{\sqrt{LC}}$, velocity gradient, Decay constant. Activity of a radioactive substance	$[M^0L^0T^{-1}]$
5.	Pressure, stress, modulus of elasticity, energy density (energy per unit volume), ε_0E^2, $\frac{B^2}{\mu_0}$	$[ML^{-1}T^{-2}]$
6.	Angular impulse, angular momentum, Planck's constant	$[ML^2T^{-1}]$
7.	Linear momentum, linear impulse	$[MLT^{-1}]$
8.	Wavelength, radius of gyration, Light year	$[M^0LT^0]$
9.	Velocity, $\frac{1}{\sqrt{\varepsilon_0\mu_0}}$, $\sqrt{\frac{GM}{R}}$, $\frac{E}{B}$	$[M^0LT^{-1}]$

Miscellaneous Examples

Ex. 1 Check whether the relation $s = ut + \frac{1}{2}at^2$ is dimensionally correct or not, where symbols have their usual meaning.

Sol. We have $s = ut + \frac{1}{2}at^2$

Checking the dimensions on both sides,

$$\text{LHS} = [M^0L^1T^0]$$

$$\text{RHS} = [LT^{-1}][T] + [LT^{-2}][T^2]$$

$$= [M^0L^1T^0] + [M^0L^1T^0] = [M^0L^1T^0]$$

$\therefore$ LHS = RHS

Hence, the formula is dimensionally correct.

Ex. 2 Find the dimensional formulae of

(a) coefficient of viscosity η

(b) charge q

(c) potential V

(d) capacitance C and

(e) resistance R

Some of the equations containing these quantities are

$$F = -\eta A\left(\frac{\Delta v}{\Delta L}\right), q = It, U = VIt, q = CV \text{ and } V = IR$$

where, A denotes the area, v the velocity, L is the length, I the electric current, t the time and U the energy.

Sol. (a) $\eta = -\frac{F}{A}\frac{\Delta L}{\Delta v}$

$$\therefore \quad [\eta] = \frac{[F][L]}{[A][v]} = \frac{[MLT^{-2}][L]}{[L^2][LT^{-1}]} = [ML^{-1}T^{-1}]$$

(b) $q = It$

$$\therefore \quad [q] = [I][t] = [AT]$$

(c) $U = VIt$

$$\therefore \quad V = \frac{U}{It}$$

or $$[V] = \frac{[U]}{[I][t]} = \frac{[ML^2T^{-2}]}{[A][T]} = [ML^2T^{-3}A^{-1}]$$

(d) $q = CV$

$$\therefore \quad C = \frac{q}{V} \text{ or } [C] = \frac{[q]}{[V]} = \frac{[AT]}{[ML^2T^{-3}A^{-1}]} = [M^{-1}L^{-2}T^4A^2]$$

(e) $V = IR$

$$\therefore \quad R = \frac{V}{I}$$

or $$[R] = \frac{[V]}{[I]} = \frac{[ML^2T^{-3}A^{-1}]}{[A]} = [ML^2T^{-3}A^{-2}]$$

Ex. 3 Write the dimensions of a and b in the relation, $P = \frac{b - x^2}{at}$, where P is power, x is distance and t is time.

Sol. The given equation can be written as,

$$Pat = b - x^2$$

Now, $[Pat] = [b] = [x^2]$

or $[b] = [x^2] = [M^0L^2T^0]$

and $$[a] = \frac{[x^2]}{[Pt]} = \frac{[L^2]}{[ML^2T^{-3}][T]} = [M^{-1}L^0T^2]$$

Ex. 4 The centripetal force F acting on a particle moving uniformly in a circle may depend upon mass (m), velocity (v) and radius (r) of the circle. Derive the formula for F using the method of dimensions.

Sol. Let $F = k\,(m)^x\,(v)^y\,(r)^z$ ...(i)

Here, k is a dimensionless constant of proportionality. Writing the dimensions of RHS and LHS in Eq. (i), we have

$$[MLT^{-2}] = [M]^x\,[LT^{-1}]^y\,[L]^z$$
$$= [M^x L^{y+z} T^{-y}]$$

Equating the powers of M, L and T of both sides, we have,

$$x = 1,\quad y = 2$$

and $y + z = 1$

or $z = 1 - y = -1$

Putting the values in Eq. (i), we get

$$F = kmv^2r^{-1} = k\frac{mv^2}{r}$$

or $F = \dfrac{mv^2}{r}$ (where, $k = 1$)

Ex. 5 If velocity, time and force were chosen as basic quantities, find the dimensions of mass and energy.

Sol. (i) We know that,

$$\text{Force} = \text{mass} \times \text{acceleration}$$
$$= \text{mass} \times \frac{\text{velocity}}{\text{time}}$$

$\Rightarrow$ $\text{mass} = \dfrac{\text{force} \times \text{time}}{\text{velocity}}$

or $[\text{mass}] = \dfrac{[\text{force}] \times [\text{time}]}{[\text{velocity}]} = \dfrac{[F]\,[T]}{[v]}$

$\therefore$ $[\text{mass}] = [FTv^{-1}]$ **Ans.**

(ii) Dimensions of energy are same as the dimensions of kinetic energy

$\therefore$ $[\text{Energy}] = \left[\dfrac{1}{2}mv^2\right] = [m]\,[v]^2$

$$= [FTv^{-1}]\,[v]^2$$
$$= [FTv]$$ **Ans.**

MODULE 2 Exercise

Theory Based Questions

Assertion and Reason

See directions on page 89.

1. **Assertion :** If two physical quantities have same dimensions, then they can be certainly added or subtracted.

   Reason : If the dimensions of both the quantities are same, then both the physical quantities should be similar.

2. **Assertion :** Method of dimensions cannot be used for deriving formulae containing trigonometrical ratios.

   Reason : This is because trigonometrical ratios have no dimensions.

Single Correct Option

3. Choose the **wrong** statements.
   (a) All quantities may be represented dimensionally in terms of the base quantities.
   (b) A base quantity cannot be represented dimensionally in terms of the rest of the base quantities.
   (c) The dimension of a base quantity in other base quantities is always zero.
   (d) The dimension of a derived quantity is never zero in any base quantity

4. Given that T stands for time period and l stands for the length of simple pendulum. If g is the acceleration due to gravity, then which of the following statements about the relation $T^2 = (l/g)$ is correct ?
   (a) It is correct both dimensionally as well as numerically.
   (b) It is neither dimensionally correct nor numerically.
   (c) It is dimensionally correct but not numerically.
   (d) It is numerically correct but not dimensionally.

5. Two quantities A and B have different dimensions. Which mathematical operation given below is physically meaningful?
   (a) A/B (b) $A + B$
   (c) $A - B$ (d) None of these

Dimensions Type-1

Fill in the Blanks

6. Planck's constant has dimensions....... **(1985 JEE)**

7. The dimensions of electrical conductivity is....... **(1997 JEE)**

Single Correct Option

8. Dimension of velocity gradient is
   (a) $[M^0L^0T^{-1}]$ (b) $[ML^{-1}T^{-1}]$
   (c) $[M^0LT^{-1}]$ (d) $[ML^0T^{-1}]$

9. Which of the following is the dimension of the coefficient of friction?
(a) $[M^2L^2T]$ (b) $[M^0L^0T^0]$
(c) $[ML^2T^{-2}]$ (d) $[M^2L^2T^{-2}]$

10. Which of the following sets have different dimensions? **(2005 JEE)**
(a) Pressure, Young's modulus, Stress
(b) Emf, Potential difference, Electric potential
(c) Heat, Work done, Energy
(d) Dipole moment, Electric flux, Electric field

11. The dimensional formula for magnetic flux is
(a) $[ML^2T^{-2}A^{-1}]$ (b) $[ML^3T^{-2}A^{-2}]$
(c) $[M^0L^{-2}T^{-2}A^{-2}]$ (d) $[ML^2T^{-1}A^2]$

12. The dimensional formula for thermal resistance is
(a) $[ML^2T^{-3}K^{-1}]$ (b) $[ML^2T^{-2}K^{-1}]$
(c) $[ML^2T^{-3}K^{-2}]$ (d) $[M^{-1}L^{-2}T^3K]$

13. $[ML^2T^{-3}A^{-1}]$ is the dimensional formula for
(a) capacitance (b) resistance
(c) resistivity (d) potential difference

14. The dimensional representation of specific resistance in terms of charge Q is
(a) $[ML^3T^{-1}Q^{-2}]$ (b) $[ML^2T^{-2}Q^2]$
(c) $[MLT^{-2}Q^{-1}]$ (d) $[ML^2T^{-2}Q^{-1}]$

15. Dimension of $\frac{1}{\mu_0\varepsilon_0}$ should be equal to **(2023 JEE)**
(a) $[L^{-2}T^2]$ (b) $[LT^{-1}]$
(c) $[L^2T^{-2}]$ (d) $[L^{-1}T]$

Multiple Correct Options

16. The dimensions of the quantities in one (or more) of the following pairs are the same. Identify the pair (s). **(1986 JEE)**
(a) Torque and work
(b) Angular momentum and work
(c) Energy and Young's modulus
(d) Light year and wavelength

17. The pairs of physical quantities that have the same dimensions is (are) **(1995 JEE)**
(a) Reynold's number and coefficient of friction
(b) Curie and frequency of a light wave
(c) Latent heat and gravitational potential
(d) Planck's constant and torque

18. Let $[\varepsilon_0]$ denote the dimensional formula of the permittivity of the vacuum and $[\mu_0]$ that of the permeability of the vacuum. If M = mass, L = length, T = time and I = electric current, then **(1998 JEE)**
(a) $[\varepsilon_0] = [M^{-1}L^{-3}T^2I]$
(b) $[\varepsilon_0] = [M^{-1}L^{-3}T^4I^2]$
(c) $[\mu_0] = [MLT^{-2}I^{-2}]$
(d) $[\mu_0] = [ML^2T^{-1}I]$

Match the Columns

19. Match the List I with List II. **(2023 JEE)**

	List I		List II
A.	Angular momentum	I.	$[ML^2T^{-2}]$
B.	Torque	II.	$[ML^{-2}T^{-2}]$
C.	Stress	III.	$[ML^2T^{-1}]$
D.	Pressure gradient	IV.	$[ML^{-1}T^{-2}]$

Choose the correct answer from the options given below:
(a) A-I, B-IV, C-III, D-II (b) A-III, B-I, C-IV, D-II
(c) A-II, B-III, C-IV, D-I (d) A-IV, B-II, C-I, D-III

20. Match List I with List II : **(2023 JEE)**

	List I		List II
A.	Pressure gradient	I.	$[M^0L^2T^{-2}]$
B.	Energy density	II.	$[M^1L^{-1}T^{-2}]$
C.	Electric field	III.	$[M^1L^{-2}T^{-2}]$
D.	Latent heat	IV.	$[M^1L^1T^{-3}A^{-1}]$

Choose the correct answer from the options given below:
(a) A-III, B-II, C-I, D-IV (b) A-II, B-III, C-IV, D-I
(c) A-III, B-II, C-IV, D-I (d) A-II, B-III, C-I, D-IV

21. Match List I with List II : **(2023 JEE)**

	List I		List II
A.	Young's Modulus (Y)	I.	$[ML^{-1}T^{-1}]$
B.	Coefficient of viscosity (η)	II.	$[ML^2T^{-1}]$
C.	Planck's constant (h)	III.	$[ML^{-1}T^{-2}]$
D.	Work function (ϕ)	IV.	$[ML^2T^{-2}]$

Choose the correct answer from the options given below:
(a) A-II, B-III, C-IV, D-I (b) A-III, B-I, C-II, D-IV
(c) A-I, B-III, C-IV, D-II (d) A-I, B-II, C-III, D-IV

22. Match List I with List II : **(2023 JEE)**

	List I		List II
A.	Planck's constant (h)	I.	$[M^1L^2T^{-2}]$
B.	Stopping potential (V_s)	II.	$[M^1L^1T^{-1}]$
C.	Work function (ϕ)	III.	$[M^1L^2T^{-1}]$
D.	Momentum (p)	IV.	$[M^1L^2T^{-3}A^{-1}]$

Choose the correct answer from the options given below:
(a) A-III, B-II, C-II, D-IV (b) A-III, B-IV, C-I, D-II
(c) A-II, B-IV, C-III, D-I (d) A-I, B-III, C-IV, D-II

23. Match the two columns.

	Column I		Column II
(a)	Boltzmann constant	(p)	$[ML^2T^{-1}]$
(b)	Coefficient of viscosity	(q)	$[ML^{-1}T^{-1}]$
(c)	Planck's constant	(r)	$[MLT^{-3}K^{-1}]$
(d)	Thermal conductivity	(s)	$[ML^2T^{-2}K^{-1}]$

24. Match the physical quantities given in Column I with dimensions expressed in terms of mass (M), length (L), time (T) and charge (Q) given in Column II.

	Column I		Column II
(a)	Angular momentum	(p)	$[ML^2T^{-2}]$
(b)	Latent heat	(q)	$[ML^2Q^{-2}]$
(c)	Torque	(r)	$[ML^2T^{-1}]$
(d)	Capacitance	(s)	$[ML^3T^{-1}Q^{-2}]$
(e)	Inductance	(t)	$[M^{-1}L^{-2}T^2Q^2]$
(f)	Resistivity	(u)	$[L^2T^{-2}]$

25. Match List I with List II : **(2023 JEE)**

	List I		List II
A.	Spring constant	I.	$[T^{-1}]$
B.	Angular speed	II.	$[MT^{-2}]$
C.	Angular momentum	III.	$[ML^2]$
D.	Moment of inertia	IV.	$[ML^2T^{-1}]$

Choose the correct answer from the options given below:

(a) A-II, B-I, C-IV, D-III (b) A-IV, B-I, C-III, D-II
(c) A-II, B-III, C-I, D-IV (d) A-I, B-III, C-II, D-IV

26. Match List I with List II : **(2023 JEE)**

	List I		List II
A.	Torque	I.	$[ML^{-2}T^{-2}]$
B.	Stress	II.	$[ML^2T^{-2}]$
C.	Pressure gradient	III.	$[ML^{-1}T^{-1}]$
D.	Coefficient of viscosity	IV.	$[ML^{-1}T^{-2}]$

Choose the correct answer from the options given below:

(a) A-III, B-IV, C-I, D-II (b) A-IV, B-II, C-III, D-I
(c) A-II, B-IV, C-I, D-III (d) A-II, B-I, C-IV, D-III

Subjective Question

27. Write the dimensions of the following in terms of mass, time, length and charge. **(1982 JEE)**

(a) Magnetic flux (b) Rigidity modulus

Dimensions Type-2

Single Correct Option

28. The viscous force F on a sphere of radius a moving in a medium with velocity v is given by $F = 6\pi\eta av$. The dimensions of η are

(a) $[ML^{-3}]$ (b) $[MLT^{-2}]$
(c) $[MT^{-1}]$ (d) $[ML^{-1}T^{-1}]$

29. A force is given by $F = at + bt^2$, where t is the time. The dimensions of a and b are

(a) $[MLT^{-4}]$ and $[MLT]$ (b) $[MLT^{-1}]$ and $[MLT^0]$
(c) $[MLT^{-3}]$ and $[MLT^{-4}]$ (d) $[MLT^{-3}]$ and $[MLT^0]$

30. The ratio of the dimensions of Planck's constant and that of the moment of inertia is the dimension of

(a) frequency (b) velocity
(c) angular momentum (d) time

31. Given that $y = A\sin\left[\left(\frac{2\pi}{\lambda}(ct - x)\right)\right]$, where y and x are measured in metres. Which of the following statements is true ?

(a) The unit of λ is same as that of x and A
(b) The unit of λ is same as that of x but not of A
(c) The unit of c is same as that of $\frac{2\pi}{\lambda}$
(d) The unit of $(ct - x)$ is same as that of $\frac{2\pi}{\lambda}$

32. Suppose refractive index μ is given as $\mu = A + \frac{B}{\lambda^2}$, where A and B are constants and λ is wavelength, then dimensions of B are same as that of

(a) Wavelength (b) Volume
(c) Pressure (d) Area

33. Of the following quantities, which one has the dimensions different from the remaining three? **(1989 JEE)**

(a) Energy density
(b) Force per unit area
(c) Product of charge per unit volume and voltage
(d) Angular momentum per unit mass

34. In the relation $y = r\sin(\omega t - kx)$, the dimensions of ω/k are

(a) $[M^0L^0T^0]$ (b) $[M^0L^1T^{-1}]$
(c) $[M^0L^0T^1]$ (d) $[M^0L^1T^0]$

35. Given that $y = a\sin\omega t + bt + ct^2\cos\omega t$. The unit of abc is same as that of

(a) y (b) y/t
(c) $(y/t)^2$ (d) $(y/t)^3$

36. Write the dimensions of $a \times b$ in the relation $E = \frac{b - x^2}{at}$, where E is the energy, x is the displacement and t is the time.

(a) $[ML^2T]$ (b) $[M^{-1}L^2T^1]$
(c) $[ML^2T^{-2}]$ (d) $[MLT^{-2}]$

37. If E and G respectively denote energy and gravitational constant, then $\frac{E}{G}$ has the dimensions of

(a) $[M^2][L^{-1}][T^0]$ (b) $[M][L^{-1}][T^{-1}]$
(c) $[M][L^0][T^0]$ (d) $[M^2][L^{-2}][T^{-1}]$

38. The dimensions of $\frac{1}{2}\varepsilon_0E^2$ (ε_0 : permittivity of free space; E : electric field) is **(2000 JEE)**

(a) $[MLT^{-1}]$ (b) $[ML^2T^{-2}]$
(c) $[MLT^{-2}]$ (d) $[ML^2T^{-1}]$

Multiple Correct Options

39. L, C and R represent the physical quantities inductance, capacitance and resistance respectively. The combinations which have the dimensions of frequency are

(a) $\frac{1}{RC}$ (b) $\frac{R}{L}$ (c) $\frac{1}{\sqrt{LC}}$ (d) $\frac{C}{L}$

Subjective Question

40. $\frac{\alpha}{t^2} = Fv + \frac{\beta}{x^2}$. Find the dimension formula for $[\alpha]$ and $[\beta]$ (here t = time, F = force, v = velocity, x = distance)

Dimensions Type-3

Single Correct Option

41. If the energy (E), velocity (v) and force (F) be taken as fundamental quantities, then the dimensions of mass will be

(a) $[Fv^{-2}]$ (b) $[Fv^{-1}]$ (c) $[Ev^{-2}]$ (d) $[Ev^{2}]$

42. If frequency f, velocity v and density D are considered fundamental units, the dimensional formula for momentum will be

(a) $[Dvf^2]$ (b) $[Dv^2f^{-1}]$ (c) $[D^2v^2f^2]$ (d) $[Dv^4f^{-3}]$

43. If force F, acceleration a and time T are taken as the fundamental physical quantities, the dimensions of length on this system of units are

(a) $[FAT^2]$
(b) $[FAT]$
(c) $[FT]$
(d) $[AT^2]$

44. If surface tension (S), moment of inertia (I) and Planck's constant (h), were to be taken as the fundamental units, the dimensional formula for linear momentum would be **(2019 Main)**

(a) $[S^{1/2}I^{1/2}h^{-1}]$
(b) $[S^{3/2}I^{1/2}h^{0}]$
(c) $[S^{1/2}I^{1/2}h^{0}]$
(d) $[S^{1/2}I^{3/2}h^{-1}]$

45. If speed (v), acceleration (A) and force (F) are considered as fundamental units, the dimension of Young's modulus will be **(2019 Main)**

(a) $[v^{-4}A^{-2}F]$
(b) $[v^{-2}A^{2}F^{2}]$
(c) $[v^{-2}A^{2}F^{-2}]$
(d) $[v^{-4}A^{2}F]$

Miscellaneous Problems

Single Correct Option

46. If p represents radiation pressure, c represent speed of light and Q represents radiation energy striking a unit area per second, then non-zero integers x, y and z such that $p^xQ^yc^z$ is dimensionless are

(a) $x = 1, y = 1, z = -1$
(b) $x = 1, y = -1, z = 1$
(c) $x = -1, y = 1, z = 1$
(d) $x = 1, y = 1, z = 1$

47. The velocity of a freely falling body changes as g^ph^q where g is acceleration due to gravity and h is the height. The values of p and q are

(a) $1, \frac{1}{2}$
(b) $\frac{1}{2}, \frac{1}{2}$
(c) $\frac{1}{2}, 1$
(d) $1, 1$

48. The speed of a wave produced in water is given by $v = \lambda^a g^b \rho^c$. Where λ, g and ρ are wavelength of wave, acceleration due to gravity and density of water respectively. The values of a, b and c respectively, are **(2023 Main)**

(a) $\frac{1}{2}, \frac{1}{2}, 0$
(b) $1, 1, 0$
(c) $1, -1, 0$
(d) $\frac{1}{2}, 0, \frac{1}{2}$

Subjective Questions

49. A gas bubble, from an explosion under water, oscillates with a period T proportional to $p^a\, d^b\, E^c$, where p is the static pressure, d is the density of water and E is the total energy of the explosion. Find the values of a, b and c. **(1981 JEE)**

50. The energy E of an oscillating body in simple harmonic motion depends on its mass m, frequency n and amplitude a. Using the method of dimensional analysis, find the relation between E, m, n and a.

51. Show dimensionally that an expression, $Y = \frac{MgL}{\pi r^2 l}$ is dimensionally correct, where Y is Young's modulus of the material of wire, L is length of wire, Mg is the weight applied on the wire and l is the increase in the length of the wire.

Answers

1. (e)	2. (a)	3. (d)	4. (c)	5. (a)
6. $[ML^2T^{-1}]$	7. $[M^{-1}L^{-3}T^3A^2]$		8. (a)	9. (b)
10. (d)	11. (a)	12. (d)	13. (d)	14. (a)
15. (c)	16. (a,d)	17. (a,b,c)	18. (b,c)	19. (b)
20. (c)	21. (b)	22. (b)	23. (a-s; b-q; c-p; d-r)	
24. (a- r ; b-u ; c-p ; d-t; e-q ; f-s)			25. (a)	26. (c)
27. (a) $[ML^2T^{-1}Q^{-1}]$ (b) $[ML^{-1}T^{-2}]$			28. (d)	29. (c)
30. (a)	31. (a)	32. (d)	33. (d)	34. (b)
35. (d)	36. (b)	37. (a)	38. (*)	39. (a,b,c)
40. $[ML^2T^{-1}]$, $[ML^4T^{-3}]$		41. (c)	42. (d)	43. (d)
44. (c)	45. (d)	46. (b)	47. (b)	48. (a)
49. $\left(\frac{-5}{6}, \frac{1}{2}, \frac{1}{3}\right)$		50. $E = kmn^2a^2$		

See hints on pages 104-108.

Unit Exercise (Cumulative)

Level 1

Fill in the Blank

1. In the formula $X = 3YZ^2$, X and Z have dimensions of capacitance and magnetic induction, respectively. The dimensions of Y in MKSQ system are .... **(1988 JEE)**

Assertion and Reason

See directions on page 89.

2. **Assertion A :** Product of pressure (p) and time (t) has the same dimension as that of coefficient of viscosity.

   Reason R : Coefficient of viscosity

   $$= \frac{\text{Force}}{\text{Velocity gradient}}$$ **(2022 JEE)**

Single Correct Option

3. The moment of inertia of a body rotating about a given axis is 12.0 kg m^2 in the SI system. What is the value of the moment of inertia in a system of units in which the unit of length is 5 cm and the unit of mass is 10 g?

   (a) 2.4×10^3 (b) 6.0×10^3 (c) 5.4×10^5 (d) 4.8×10^5

4. The density of a material in SI units is 128 kg m^{-3}. In certain units in which the unit of length is 25 cm and the unit of mass is 50 g, the numerical value of density of the material is **(2019 Main)**

   (a) 40 (b) 16 (c) 640 (d) 410

5. Dimensions of 'ohm' are same as

   (a) $\frac{h}{e}$ (b) $\frac{h^2}{e}$ (c) $\frac{h}{e^2}$ (d) $\frac{h^2}{e^2}$

   (where, h is Planck's constant and e is charge)

6. If E = energy, G = gravitational constant, I = impulse and M = mass, then dimensions of $\frac{GIM^2}{E^2}$ are same as that of

   (a) time (b) mass
   (c) length (d) force

7. If m, e, ε_0 and h denote mass of electron, charge of electron, electric permittivity and Planck's constant, respectively. The unit of $\frac{me^4}{\varepsilon_0^2 h^2}$ is same as

   (a) energy (b) power (c) force (d) pressure

8. In the form of G (Universal gravitational constant), h (Planck's constant) and c (speed of light), the time period will be proportional to **(2019 Main)**

   (a) $\sqrt{\frac{Gh}{c^5}}$ (b) $\sqrt{\frac{hc^5}{G}}$ (c) $\sqrt{\frac{c^3}{Gh}}$ (d) $\sqrt{\frac{Gh}{c^3}}$

9. If E and H represents the intensity of electric field and magnetising field respectively, then the unit of E/H will be **(2021 JEE)**

   (a) ohm (b) mho
   (c) joule (d) newton

10. The quantities A and B are related by the relation $A/B = m$, where m is the linear mass density and A is the force, the dimensions of B will be

    (a) same as that of pressure
    (b) same as that of work
    (c) that of momentum
    (d) same as that of latent heat

11. If R, X_L and X_C represent resistance, inductive reactance and capacitive reactance. Then which of the following is dimensionless ? **(2023 JEE)**

    (a) $R\,X_L\,X_C$ (b) $\frac{R}{\sqrt{X_L X_C}}$ (c) $\frac{R}{X_L X_C}$ (d) $R\frac{X_L}{X_C}$

12. Which of the following combinations has the dimension of electrical resistance (ε_0 is the permittivity of vacuum and μ_0 is the permeability of vacuum)? **(2019 Main)**

    (a) $\sqrt{\frac{\mu_0}{\varepsilon_0}}$ (b) $\frac{\mu_0}{\varepsilon_0}$ (c) $\sqrt{\frac{\varepsilon_0}{\mu_0}}$ (d) $\frac{\varepsilon_0}{\mu_0}$

13. Let l, r, c, and v represent inductance, resistance, capacitance and voltage, respectively. The dimension of $\frac{l}{rcv}$ will be **(2019 Main)**

    (a) $[LT^2]$ (b) $[LTA]$
    (c) $[A^{-1}]$ (d) $[LA^{-2}]$

14. In the equation $\left[X + \frac{a}{Y^2}\right][Y - b] = RT$, X is pressure, Y is volume, R is universal gas constant and T is temperature. The physical quantity equivalent to the ratio $\frac{a}{b}$ is : **(2023 Main)**

    (a) Energy
    (b) Impulse
    (c) Pressure gradient
    (d) Coefficient of viscosity

15. Using mass (M), length (L), time (T) and current (A) as fundamental quantities, the dimension of permeability is
(a) $[M^{-1}LT^{-2}A]$ (b) $[ML^{-2}T^{-2}A^{-1}]$
(c) $[MLT^{-2}A^{-2}]$ (d) $[MLT^{-1}A^{-1}]$

16. If the velocity of light c, universal gravitational constant G and Planck's constant h are chosen as fundamental quantities. The dimensions of mass in the new system is **(2023 Main)**
(a) $[h^{1/2}c^{1/2}G^{-1/2}]$ (b) $[h^{-1/2}c^{1/2}G^{1/2}]$
(c) $[h^{1}c^{1}G^{-1}]$ (d) $[h^{1/2}c^{-1/2}G^{1}]$

17. A physical quantity x depends on quantities y and z as follows: $x = Ay + B\tan Cz$, where A, B and C are constants. Which of the following do not have the same dimensions?
(a) x and B (c) C and z^{-1}
(c) y and B/A (d) x and A

Multiple Correct Options

18. Planck's constant h, speed of light c and gravitational constant G are used to form a unit of length L and a unit of mass M. Then, the correct options is/are **(2015 Adv.)**
(a) $M \propto \sqrt{c}$ (b) $M \propto \sqrt{G}$
(c) $L \propto \sqrt{h}$ (d) $L \propto \sqrt{G}$

19. In damped simple harmonic motion, displacement $x = a_0 e^{-\frac{bt}{2m}} \sin(\omega t + \alpha)$, where t is in second. Choose the correct option(s).
(a) Dimensions of $\frac{m}{b}$ are same as t
(b) Dimensions of $a_0\omega$ are $[LT^{-1}]$
(c) b is dimensionless constant
(d) α is dimensionless and unitless

20. A wave is propagating along x-axis in a medium, so that displacement y of particle at any position x, at any time t is given by
$$y = \frac{a}{t^2} e^{-\left[\frac{t}{b} - \frac{x}{c}\right]}$$
where a, b and c are constant. Choose the correct option(s).
(a) dimension of b is [T]
(b) dimension of a is $[LT^2]$
(c) dimension of c is [L]
(d) $e^{-\left(\frac{t}{b} - \frac{x}{c}\right)}$ is dimensionless

Passage Based Questions

Passage (Q.21 to Q.25)

The van der Waals' equation of state for some gases can be expressed as :
$$\left(p + \frac{a}{V^2}\right)(V - b) = RT$$
where p is the pressure, V is the volume, and T is the absolute temperature of the given sample of gas and a, b and R are constants.

21. The dimensions of constant a are
(a) $[ML^5T^{-2}]$ (b) $[ML^{-1}T^{-2}]$
(c) $[L^3]$ (d) $[L^6]$

22. The dimensions of constant b are
(a) $[ML^5T^{-2}]$ (b) $[ML^{-1}T^{-2}]$
(c) $[L^3]$ (d) $[L^6]$

23. Which of the following does not have the same dimensional formula as that for RT?
(a) pV (b) pb
(c) $\frac{a}{V^2}$ (d) $\frac{ab}{V^2}$

24. The dimensional representation of ab/RT is
(a) $[ML^5T^{-2}]$ (b) $[M^0L^3T^0]$
(c) $[ML^{-1}T^{-2}]$ (d) None of these

25. The dimensional formula for RT is same as that of
(a) energy (b) force
(c) specific heat (d) latent heat

Match the Columns

26. Column I gives three physical quantities. Select the appropriate units for the choices given in Column II. Some of the physical quantities may have more than one choice. **(1990 JEE)**

	Column I		Column II
(a)	Capacitance	(p)	ohm-second
(b)	Inductance	(q)	$(\text{coulomb})^2\text{-joule}^{-1}$
(c)	Magnetic induction	(r)	$\text{coulomb (volt)}^{-1}$
		(s)	$\text{newton (ampere-metre)}^{-1}$
		(t)	$\text{volt-second (ampere)}^{-1}$

27. Given β = bulk modulus of elasticity, A = amplitude, K = angular wave number, r = radius, Y = Young's modulus, m = mass, ω = angular frequency, F = force, and S = area. Match Column I with Column II and mark the correct option from the codes given below.

	Column I		Column II
A.	βAk	(p)	Watt/metre^2
B.	$\rho A^2\omega^2C$	(q)	Joule
C.	$\frac{F^2l}{SY}$	(r)	$\text{Newton per metre}^2$
D.	$\sqrt{\frac{r^3}{GM}}$	(s)	Second

Codes :

	A	B	C	D		A	B	C	D
(a)	p	q	r	s	(b)	r	p	q	s
(c)	q	p	s	r	(d)	pq	rs	p	q

28. Some physical quantities are given in Column I and some possible SI units in which these quantities may be expressed are given in Column II. Match the physical quantities in Column I with the units in Column II. **(2007 JEE)**

	Column I		Column II
(a)	GM_eM_s G — Universal gravitational constant, M_e — mass of the earth, M_s — mass of the sun.	(p)	(volt) (coulomb) (metre)
(b)	$\frac{3RT}{M}$ R — Universal gas constant, T — absolute temperature, M — molar mass.	(q)	(kilogram) (metre)^3 (second)$^{-2}$
(c)	$\frac{F^2}{q^2B^2}$ F — Force, q — charge, B — magnetic field.	(r)	(metre)^2 (second)$^{-2}$
(d)	$\frac{GM_e}{R_e}$ G — Universal gravitational constant, M_e — mass of the earth, R_e — radius of the earth.	(s)	(farad) (volt)^2 (kg)$^{-1}$

Subjective Questions

29. Let x and a stand for distance. Is $\int \frac{dx}{\sqrt{a^2 - x^2}} = \frac{1}{a} \sin^{-1} \frac{a}{x}$ dimensionally correct?

30. In the equation $\int \frac{dx}{\sqrt{2ax - x^2}} = a^n \sin^{-1}\left(\frac{x}{a} - 1\right)$. Find the value of n.

Integer Type Questions

31. Speed of light in vacuum is given by 3×10^8 ms^{-1} in SI system. If in a new system, the unit of length and time is increased by a factor of 2 and 4 respectively, then the numerical value of speed of light (In this new system) is found to be $x \times 10^8$m/s. Find the value of x.

32. To find the distance d over which a signal can be seen clearly in foggy conditions, a railway engineer uses dimensional analysis and assumes that the distance depends on the mass density ρ of the fog, intensity (power/area) S of the light from the signal and its frequency f. The engineer finds that d is proportional to $S^{1/n}$. The value of n is **(2014 Adv.)**

Level 2

Single Correct Option

33. A quantity X is given by $\varepsilon_0 L \frac{\Delta V}{\Delta t}$, where ε_0 is the permittivity of free space, L is a length, ΔV is a potential difference and Δt is a time interval. The dimensional formula for X is the same as that of **(2001 JEE)**

(a) resistance (b) charge (c) voltage (d) current

34. Velocity (v) and acceleration (a) in two systems of units 1 and 2 are related as $v_2 = \frac{n}{m^2} v_1$ and $a_2 = \frac{a_1}{mn}$ respectively. Here, m and n are constants. The relations for distance and time in two systems respectively are : **(2022 JEE)**

(a) $\frac{n^3}{m^3} L_1 = L_2$ and $\frac{n^2}{m} T_1 = T_2$

(b) $L_1 = \frac{n^4}{m^2} L_2$ and $T_1 = \frac{n^2}{m} T_2$

(c) $L_1 = \frac{n^2}{m} L_2$ and $T_1 = \frac{n^4}{m^2} T_2$

(d) $\frac{n^2}{m} L_1 = L_2$ and $\frac{n^4}{m^2} T_1 = T_2$

35. In the relation $p = \frac{\alpha}{\beta} e^{-\frac{\alpha Z}{k\theta}}$, p is pressure, Z is distance, k is Boltzmann constant and θ is the temperature. The dimensional formula of β will be **(2004 JEE)**

(a) $[M^0L^2T^0]$ (b) $[ML^2T]$ (c) $[ML^0T^{-1}]$ (d) $[M^0L^2T^{-1}]$

36. If y represents pressure and x represents velocity gradient, then the dimensions of $\frac{d^2y}{dx^2}$ are

(a) $[ML^{-1}T^{-2}]$ (b) $[M^2L^{-2}T^{-2}]$
(c) $[ML^{-1}T^0]$ (d) $[M^2L^{-2}T^{-4}]$

37. If m, e, ε_0, h and c denote mass of electron, charge of electron, Planck's constant and speed of light, respectively. The dimensions of $\frac{me^4}{\varepsilon_0^2 h^3 c}$ are

(a) $[M^0L^0T^{-1}]$ (b) $[M^0L^{-1}T^{-1}]$
(c) $[M^2LT^{-3}]$ (d) $[M^0L^{-1}T^0]$

38. The force of interaction between two atoms is given by $F = \alpha\beta \exp\left(-\frac{x^2}{\alpha kT}\right)$; where x is the distance, k is the Boltzmann constant and T is temperature and α and β are two constants. The dimension of β is **(2019 Main)**

(a) $[MLT^{-2}]$ (b) $[M^0L^2T^{-4}]$
(c) $[M^2LT^{-4}]$ (d) $[M^2L^2T^{-2}]$

39. A physical quantity of the dimensions of length that can be formed out of c, G and $\frac{e^2}{4\pi\varepsilon_0}$ is [c is velocity of light, G is universal constant of gravitation and e is charge]

(a) $\frac{1}{c^2}\left[G\frac{e^2}{4\pi\varepsilon_0}\right]^{1/2}$ (b) $c^2\left[G\frac{e^2}{4\pi\varepsilon_0}\right]^{1/2}$

(c) $\frac{1}{c^2}\left[\frac{e^2}{G\,4\pi\varepsilon_0}\right]^{1/2}$ (d) $\frac{1}{c}G\frac{e^2}{4\pi\varepsilon_0}$

40. Planck's constant (h), speed of light in vacuum (c) and Newton's gravitational constant (G) are three fundamental constants. Which of the following combinations of these has the dimension of length?

(a) $\frac{\sqrt{hG}}{c^{3/2}}$ (b) $\frac{\sqrt{hG}}{c^{5/2}}$ (c) $\sqrt{\frac{hc}{G}}$ (d) $\sqrt{\frac{Gc}{h^{3/2}}}$

Multiple Correct Options

41. If in a new hypothetical system of units, the new unit of mass is 5 kg, new unit of length is 10 metre and new unit of time is 20 seconds, choose the correct statement(s).
(a) 1 SI unit of velocity is equal to 2 new units of velocity.
(b) 1 SI unit of power is equal to 16 new units of power.
(c) 1 SI unit of energy is equal to 5/4 new unit of energy.
(d) 1 SI unit of force is equal to 10 new units of force.

42. If the unit of velocity is 4 ms^{-1}, the unit of acceleration is 24 ms^{-2}, and the unit of force is 6N, then choose the correct statement(s).
(a) Unit of energy is 4J.
(b) Unit of momentum is 1 Ns.
(c) Unit of power is 24 W.
(d) Unit of momentum is 2/3 Ns.

43. In terms of potential difference V, electric current I, permittivity ε_0, permeability μ_0 and speed of light c, the dimensionally correct equations is/are **(2015 Adv.)**

(a) $\mu_0 I^2 = \varepsilon_0 V^2$ (b) $\varepsilon_0 I = \mu_0 V$
(c) $I = \varepsilon_0 cV$ (d) $\mu_0 cI = \varepsilon_0 V$.

Match the Columns

44. Match the two columns.

	Column I		Column II
(a)	$\frac{\text{Magnitude of electric dipole moment}}{\text{Magnitude of magnetic dipole moment}}$	(p)	$[KM^{-1}L^{-3}T^4]$
(b)	(Electric flux) × (Magnetic flux)	(q)	$[K\,mol\,M^{-2}L]$
(c)	$\frac{\text{Gravitational constant}}{\text{Universal gas constant}}$	(r)	$[A^{-2}M^2L^5T^{-5}]$
(d)	$\frac{\text{Inductance} \times \text{Electric permittivity}}{\text{Heat capacity}}$	(s)	$[L^{-1}T]$

45. If G = gravitational constant, M = mass, R = radius, v = speed, λ = linear mass density, μ_0 = magnetic permeability, J = current density, q = charge, B = magnetic field. Match the two columns.

	Column I		Column II
(a)	$\sqrt{\frac{GM}{R}}$	(p)	$[M^0L^0T]$
(b)	λv^2	(q)	$[A^{-1}MT^{-2}]$
(c)	$\mu_0 Jr$	(r)	$[MLT^{-2}]$
(d)	$\frac{m}{qB}$	(s)	$[LT^{-1}]$

Integer Type Questions

46. Imagine a system of units in which the unit of mass is 10 kg, length is 1 km and time is 1 min. If the numerical value of 1 joule in this system is found to be $x \times 10^{-6}$. Find the value of x.

47. In a new system of units mass, acceleration and frequency are taken as fundamental units. If unit of mass is 100 g, unit of acceleration is 2 m/s^2 and unit of frequency is 4 sec^{-1} in the new system of units, then find the value of 0.1 J in this system.

Answers

Level 1

1. $(M^{-3}L^{-2}T^4Q^4)$ 2. (c) 3. (d) 4. (a) 5. (c) 6. (a) 7. (a) 8. (a) 9. (a)
10. (d) 11. (b) 12. (a) 13. (c) 14. (a) 15. (c) 16. (a) 17. (d) 18. (a,c,d) 19. (a,b)
20. (a,b,c,d) 21. (a) 22. (c) 23. (c) 24. (d) 25. (a) 26. (a → q,r ; b → p,t ; c → s) 27. (b)
28. (a → p,q ; b → r,s ; c → r,s ; d → r,s) 29. (No) 30. (Zero) 31. (6) 32. (3)

Level 2

33. (d) 34. (a) 35. (a) 36. (c) 37. (d) 38. (c) 39. (a) 40. (a) 41. (a,b) 42. (a, b, c)
43. (a,c) 44. (a → s ; b → r ; c → q ; d → p) 45. (a → s ; b → r ; c → q ; d → p) 46. (360) 47. (4)

See hints on page 108-113.

Hints & SOLUTIONS

Module 1.

2. 1 cm = 10 mm. Here, mm is smaller unit.

   But its numerical value is more.

3. In option (b), length, time and velocity can be deduced from one another. Therefore, they cannot enter into the list of fundamental quantities in any system.

4. Because temperature is a fundamental quantity.

5. Since, $mvr = n \cdot \frac{h}{2\pi}$ and $E = h\nu$

   So, unit of h = joule second = angular momentum

6. Power, $P = i^2R = Vi = \frac{V^2}{R}$

7. $[N/m] = [J/m^2] = \left[\frac{MLT^{-2}}{L}\right]$

   $= [MT^{-2}] = [kg/s^2]$

8. Momentum = Force × Time = N - s

9. $F = \frac{Gm_1m_2}{d^2} \Rightarrow G = \frac{Fd^2}{m_1m_2} = N\text{-}m^2/kg^2$

10. kg-m/sec is the unit of linear momentum.

11. Energy $(E) = F \times d \Rightarrow F = \frac{E}{d}$, so erg/metre can be the unit of force.

12. Charge = current × time.

13. Momentum = $mv = [MLT^{-1}]$

    Impulse = Force × Time = $[MLT^{-2}] \times [T] = [MLT^{-1}]$

14. (a) $L = \frac{\phi}{i}$ or henry $= \frac{\text{weber}}{\text{ampere}}$

    (b) $e = -L\left(\frac{di}{dt}\right)$

    $\therefore \quad L = -\frac{e}{(di/dt)}$

    or henry $= \frac{\text{volt - second}}{\text{ampere}}$

    (c) $U = \frac{1}{2}Li^2$

    $\therefore \quad L = \frac{2U}{i^2}$

    or henry $= \frac{\text{joule}}{(\text{ampere})^2}$

    (d) $U = \frac{1}{2}Li^2 = i^2Rt$

    $\therefore \quad L = Rt$ or henry = ohm-second

15. (A) Modulus of rigidity $= \frac{\text{Shear stress}}{\text{Shear strain}}$

    Shear strain has no unit,

    But stress $= \frac{F}{A} = Nm^{-2}$

    $\therefore$ The unit of modulus of rigidity is Nm^{-2}

    (B) $\Delta H = mL$

    where ΔH = heat supplied, m = mass, L = latent heat

    $\therefore \quad L = \frac{\Delta H}{m} = Jkg^{-1}$

    (C) According to Ohm's law, $V = IR$

    where V = potential difference. Its unit is volt.

    I = electric current. Its unit is ampere.

    $\therefore$ Electric resistance, $R = \frac{V}{I}$ = Volt per ampere or ohm

    $\therefore \quad R = \rho\frac{l}{A}$, where, l = length of rod and

    A = area of cross-section of rod

    $\therefore$ Electric resistivity, $\rho = \frac{RA}{l} = \frac{\text{ohm - m}^2}{\text{m}}$

    = ohm-metre

    = volt-metre per ampere

    (D) Gravitational potential is gravitational potential energy per unit mass. Thus, its unit is joule per kilogram.

16. (A) Surface tension $= \frac{F}{l} = \frac{MLT^{-2}}{L} = MT^{-2} = kgs^{-2}$

    (B) Pressure $= \frac{F}{A} = \frac{MLT^{-2}}{L^2} = kgm^{-1}s^{-2}$

    (C) Viscosity $= \frac{F}{A\left(\frac{dv}{dz}\right)} = \frac{MLT^{-2}}{L^2\left(\frac{LT^{-1}}{L}\right)}$

    $= ML^{-1}T^{-1} = kg\,m^{-1}s^{-1}$

    (D) Impulse $= \int F dt = MLT^{-2} \times T$

    $= MLT^{-1} = kg\,ms^{-1}$

17. Torque $= F(N) \times r\,(m) \rightarrow Nm$

    Stress $= \frac{\text{Force (N)}}{\text{Area}(m^2)} \rightarrow N/m^2$

Latent heat $=\dfrac{\text{Energy (J)}}{\text{Mass (kg)}} \to \text{J kg}^{-1}$

Power $=\dfrac{\text{Work (Nm)}}{\text{Time (s)}} \to \text{N ms}^{-1}$

18. (a) Young's Modulus $=\dfrac{F/A}{\Delta l/l}$

Hence, the MKS units are N/m^2.

(b) Tesla

(c) Power of a lens (in dioptre) $=\dfrac{1}{f(\text{in metre})}$

19. (a) $1\,\text{kgm}^2\text{s}^{-2} = 1\times10^{-3}\,\text{g}\,(10^2\,\text{cm})^2\,\text{s}^{-2} = 10^7\,\text{g cm}^2\text{s}^{-2}$

(b) $3\,\text{ms}^{-2} = 3\times10^{-3}\text{km}\left(\dfrac{1}{60\times60}\text{h}\right)^{-2}$

(c)
$$\begin{aligned} G &= 6.67\times10^{-11}\,\text{Nm}^2\,\text{kg}^{-2}\\ &= 6.67\times10^{-11}\,(\text{kgms}^{-2})\,\text{m}^2\,\text{kg}^{-2}\\ &= 6.67\times10^{-11}\,\text{m}^3\text{s}^{-2}\text{kg}^{-1}\\ &= 6.67\times10^{-11}(100\,\text{cm})^3\text{s}^{-2}(1000\,\text{g})^{-1}\\ &= 6.67\times10^{-8}\,\text{cm}^3\text{s}^{-2}\text{g}^{-1}\end{aligned}$$

20. $1\text{N} = 10^5$ dyne

$$1\text{m}^2 = 10^4\,\text{cm}^2$$

$$\therefore\quad 2.0\times10^{11}\frac{\text{N}}{\text{m}^2} = \frac{2.0\times10^{11}\times10^5}{10^4}\frac{\text{dyne}}{\text{cm}^2} = 2.0\times10^{12}\,\text{dyne/cm}^2$$

21. $1\text{ dyne} = 10^{-5}\,\text{N}$

$$1\text{ cm} = 10^{-2}\,\text{m}$$

$$\therefore\quad 72\,\text{dyne/cm} = \frac{72\times10^{-5}}{10^{-2}}\frac{\text{N}}{\text{m}} = 0.072\frac{\text{N}}{\text{m}}$$

22. Since, units of length, velocity and force are doubled.

Hence, $[m] = \dfrac{[\text{force}][\text{time}]}{[\text{velocity}]}, [\text{time}] = \dfrac{[\text{length}]}{[\text{velocity}]}$

23. Energy = force × distance, so if both are increased by 4 times, then energy will increase by 16 times.

24.
$$\begin{aligned} n_2 &= n_1\left[\frac{L_1}{L_2}\right]^1\left[\frac{T_1}{T_2}\right]^{-2} = 10\left[\frac{\text{m}}{\text{km}}\right]^1\left[\frac{\text{sec}}{\text{hr}}\right]^{-2}\\ &= 10\left[\frac{\text{m}}{10^3\text{m}}\right]^1\left[\frac{\text{sec}}{3600\,\text{sec}}\right]^{-2}\\ &= 129600 = 1296\times10^2\end{aligned}$$

25. $1\text{ AU} = 1.496\times10^{11}\,\text{m}$

1 parsec $= 3.262$ light year $= 3.08\times10^{16}$ m

1 light year = distance travelled by light in one year

$$= 9.46\times10^{15}\text{m}$$

$\therefore$ AU < ly < parsec

26. 1 fermi $= 10^{-15}$ metre

27. Units of a and pV^2 are same and equal to dyne × cm^4.

28. Electric field in a certain region is given by,

$$\mathbf{E} = \frac{A}{x^2}\hat{\mathbf{i}} + \frac{B}{y^3}\hat{\mathbf{j}}$$

$$\left[\frac{A}{x^2}\right] = \text{NC}^{-1} \Rightarrow [A] = \text{Nm}^2\text{C}^{-1}$$

$$\left[\frac{B}{y^3}\right] = \text{NC}^{-1} \Rightarrow [B] = \text{Nm}^3\text{C}^{-1}$$

Module 2.

4. The correct relation for time period of simple pendulum is $T = 2\pi(l/g)^{1/2}$. So the given relation is numerically incorrect as the factor of 2π is missing. But it is correct dimensionally.

5. Quantities having different dimensions can only be divided or multiplied but they cannot be added or subtracted.

6. $E = h\nu$

$$\therefore\quad h = \frac{E}{\nu}\quad\text{or}\quad [h] = \frac{[E]}{[\nu]} = \frac{[\text{ML}^2\text{T}^{-2}]}{[\text{T}^{-1}]} = [\text{ML}^2\text{T}^{-1}]$$

7. Electrical conductivity, $\sigma = \dfrac{J}{E} = \dfrac{i/A}{F/q}$

$$= \frac{qi}{FA} = \frac{(it)(i)}{FA} = \frac{i^2t}{FA}$$

$$[\sigma] = \frac{[\text{A}^2][\text{T}]}{[\text{MLT}^{-2}][\text{L}^2]} = [\text{M}^{-1}\text{L}^{-3}\text{T}^3\text{A}^2]$$

8. Velocity gradient is change in velocity per unit depth.

9. Coefficient of friction is unitless and dimensionless.

10. Dipole moment = (charge) × (distance)

Electric flux = (electric field) × (area)

Hence, the correct option is (d).

11. $\phi = BA = \dfrac{F}{IL}\cdot A = \left[\dfrac{\text{MLT}^{-2}}{\text{AL}}\right][\text{L}^2] = [\text{ML}^2\text{T}^{-2}\text{A}^{-1}]$

12. Heat current $= \dfrac{dQ}{dt} = \dfrac{\text{TD}}{R}$

Here, TD = temperature difference

and R = thermal resistance $= \dfrac{\text{TD}}{dQ/dt}$

$\therefore \quad [R]=[TD]\left[\frac{dt}{dQ}\right]=[K]\left[\frac{T}{ML^2T^{-2}}\right]$

$=[M^{-1}L^{-2}T^3K]$

13. Power, $P=Vi$

$\therefore \quad [V]=\left[\frac{P}{i}\right]=\left[\frac{ML^2T^{-3}}{A}\right]$

$=[ML^2T^{-3}A^{-1}]$

14. Since, $R=\frac{\rho l}{A}$, where ρ is a specific resistance.

$\therefore \quad [\rho]=\left[\frac{RA}{l}\right], R=\frac{V}{i}, V=\frac{W}{Q}$

$[\rho]=[ML^3T^{-1}Q^{-2}]$

15. Velocity of light in vacuum is given by $c=\frac{1}{\sqrt{\mu_0\varepsilon_0}}$

$\Rightarrow \quad \frac{1}{\mu_0\varepsilon_0}=c^2$

$\Rightarrow \quad \left[\frac{1}{\mu_0\varepsilon_0}\right]=[c^2]=[LT^{-1}]^2=[L^2T^{-2}]$

16. (a) Torque and work both have the dimension $[ML^2T^{-2}]$.

(d) Light year and wavelength both have the dimensions of length, i.e. [L].

17. Reynold's number and coefficient of friction are dimensionless quantities.

Curie is the number of atoms decaying per unit time and frequency is the number of oscillations per unit time.

Latent heat and gravitational potential both have the same dimension corresponding to energy per unit mass.

18. $F=\frac{1}{4\pi\varepsilon_0}\cdot\frac{q_1q_2}{r^2}$

$[\varepsilon_0]=\frac{[q_1][q_2]}{[F][r^2]}=\frac{[IT]^2}{[MLT^{-2}][L^2]}$

$=[M^{-1}L^{-3}T^4I^2]$

Speed of light, $c=\frac{1}{\sqrt{\varepsilon_0\mu_0}}$

$\therefore \quad [\mu_0]=\frac{1}{[\varepsilon_0][c]^2}=\frac{1}{[M^{-1}L^{-3}T^4I^2][LT^{-1}]^2}$

$=[MLT^{-2}I^{-2}]$

19. We have

Angular momentum $=[ML^2T^{-1}]$

Torque $=[ML^2T^{-2}]$

Stress $=[ML^{-1}T^{-2}]$

Pressure gradient $=\frac{dp}{dx}=[ML^{-2}T^{-2}]$

20. Dimension of pressure gradient

$\frac{dp}{dx}=\frac{\left(\frac{F}{A}\right)}{L}=\frac{[ML^{-1}T^{-2}]}{[L]}=[M^1L^{-2}T^{-2}]$

Energy density $(u)=\frac{(E)}{(V)}=\frac{[ML^2T^{-2}]}{[L^3]}$

$=[M^1L^{-1}T^{-2}]$

Electric field $(E)=\frac{(F)}{(q)}=\frac{[MLT^{-2}]}{[AT]}$

$=[M^1L^1T^{-3}A^{-1}]$

Latent heat $(L)=\frac{(Q)}{(m)}=\frac{[ML^2T^{-2}]}{[M]}$

$=[M^0L^2T^{-2}]$

21. $Y=\frac{\text{Stress}}{\text{Strain}}=\frac{Fl}{A\Delta l}$

So, $[Y]=\frac{[F][l]}{[A\Delta l]}=\frac{MLT^{-2}L}{L^2L}=ML^{-1}T^{-2}$

$F=6\pi\eta rv \Rightarrow [\eta]=\frac{[F]}{[6\pi rv]}$

$\Rightarrow \quad [\eta]=\frac{MLT^{-2}}{LLT^{-1}}=ML^{-1}T^{-1}$

$E=h\nu \Rightarrow h=\frac{E}{\nu} \Rightarrow [h]=\frac{[E]}{[\nu]}$

$\Rightarrow \quad [h]=\frac{ML^2T^{-2}}{T^{-1}}=ML^2T^{-1}$

Work function has same dimension as that of energy.

So, $[\phi]=ML^2T^{-2}$

22. $[h]=ML^2T^{-1}$

$[V_s]=\frac{[W]}{[Q]}=\frac{ML^2T^{-2}}{AT}=ML^2T^{-3}A^{-1}$

$[\phi]=ML^2T^{-2}$

$[p]=MLT^{-1}$

23. (a) $U=\frac{1}{2}kT$

$\Rightarrow \quad [ML^2T^{-2}]=[k][K]$

$\Rightarrow \quad [k]=[ML^2T^{-2}K^{-1}]$

(b) $F=\eta A\frac{dv}{dx}$

$\Rightarrow \quad [\eta]=\frac{[MLT^{-2}]}{[L^2LT^{-1}L^{-1}]}$

$=[ML^{-1}T^{-1}]$

(c) $E=h\nu$

$\Rightarrow \quad [ML^2T^2]=[h][T^{-1}]$

$\Rightarrow \quad [h]=[ML^2T^{-1}]$

(d) $\dfrac{dQ}{dt}=\dfrac{kA\Delta\theta}{l}$

$$\Rightarrow [k]=\frac{[ML^2T^{-3}L]}{[L^2K]}=[MLT^{-3}K^{-1}]$$

24. Angular momentum $L=I\omega$

$\therefore \quad [L]=[I\omega]=[ML^2][T^{-1}]=[ML^2T^{-1}]$

Latent heat, $L=\dfrac{Q}{m}$ (as $Q=mL$)

$$\Rightarrow [L]=\left[\frac{Q}{m}\right]=\left[\frac{ML^2T^{-2}}{M}\right]=[L^2T^{-2}]$$

Torque $\tau=F\times r_\perp$

$\therefore \quad [\tau]=[F\times r_\perp]=[MLT^{-2}][L]=[ML^2T^{-2}]$

Capacitance $C=\dfrac{1}{2}\dfrac{q^2}{U}$ $\left(\text{as } U=\dfrac{1}{2}\dfrac{q^2}{C}\right)$

$$\therefore [C]=\left[\frac{q^2}{U}\right]=\left[\frac{Q^2}{ML^2T^{-2}}\right]=[M^{-1}L^{-2}T^2Q^2]$$

Inductance $L=\dfrac{2U}{i^2}$ $\left(\text{as } U=\dfrac{1}{2}Li^2\right)$

$$\therefore [L]=\left[\frac{U}{i^2}\right]=\left[\frac{Ut^2}{Q^2}\right] \quad \left(\text{as } i=\frac{Q}{t}\right)$$

$$=\left[\frac{ML^2T^{-2}T^2}{Q^2}\right]=[ML^2Q^{-2}]$$

Resistivity $\rho=\dfrac{RA}{l}$ $\left(\text{as } R=\rho\dfrac{l}{A}\right)$

$$=\left[\frac{H}{i^2t}\right]\left[\frac{A}{l}\right] \quad (\text{as } H=i^2Rt)$$

$$=\left[\frac{Ht}{Q^2}\right]\left[\frac{A}{l}\right] \quad \left(\text{as } i=\frac{Q}{t}\right)$$

$$=\left[\frac{ML^2T^{-2}TL^2}{Q^2L}\right]=[ML^3T^{-1}Q^{-2}]$$

The correct table is as under

Column I	Column II
Angular momentum	$[ML^2T^{-1}]$
Latent heat	$[L^2T^{-2}]$
Torque	$[ML^2T^{-2}]$
Capacitance	$[M^{-1}L^{-2}T^2Q^2]$
Inductance	$[ML^2Q^{-2}]$
Resistivity	$[ML^3T^{-1}Q^{-2}]$

25. Angular momentum,

$$L=mvr=[M][LT^{-1}][L]=[ML^2T^{-1}]$$

Moment of inertia,

$$I=mr^2=[M][L^2]=[ML^2]$$

Spring constant

$$k=\frac{F}{x}=\frac{[MLT^{-2}]}{[L]}=[MT^{-2}]$$

Angular speed

$$\omega=\frac{\theta}{t}=\frac{1}{T}=[T^{-1}]$$

26. A. Torque $\vec{\tau}=\mathbf{r}\times\mathbf{F}$

$$=[L][MLT^{-2}]=[ML^2T^{-2}]$$

B. Stress $=\dfrac{F}{A}=\dfrac{MLT^{-2}}{L^2}$

$$=[ML^{-1}T^{-2}]$$

C. Pressure gradient $=\dfrac{\Delta p}{\Delta x}=\dfrac{(F/A)}{(L)}=\dfrac{MLT^{-2}}{L^3}$

$$=[ML^{-2}T^{-2}]$$

D. Coefficient of viscosity $F=6\pi\eta rv$

$$MLT^{-2}=\eta[L][LT^{-1}]$$

or $\eta=\dfrac{MLT^{-2}}{L^2T^{-1}}$

or $\eta=[ML^{-1}T^{-1}]$

27. (a) Magnetic flux, $\phi=Bs=\left(\dfrac{F}{qv}\right)(s)$ $[\because F=Bqv]$

$$\therefore [\phi]=\left[\frac{Fs}{qv}\right]=\left[\frac{MLT^{-2}L^2}{QLT^{-1}}\right]$$

$$=[ML^2T^{-1}Q^{-1}]$$

(b) Rigidity modulus $=\left[\dfrac{F}{A}\right]=\left[\dfrac{MLT^{-2}}{L^2}\right]$

$$=[ML^{-1}T^{-2}]$$

28. $[\eta]=\left[\dfrac{F}{av}\right]=\left[\dfrac{MLT^{-2}}{L^2T^{-1}}\right]=[ML^{-1}T^{-1}]$

29. $a=\dfrac{F}{t}, b=\dfrac{F}{t^2}$

30. $L=I\omega=\dfrac{nh}{2\pi}$

$$\therefore \quad \frac{h}{I}=[\omega]$$

31. $\because \dfrac{2\pi x}{\lambda}$ is dimensionless.

$$\therefore \quad [\lambda]=[x]=[L]=[A]$$

32. As $\mu=\dfrac{\text{Velocity of light in vacuum}}{\text{Velocity of light in medium}}$

The refractive index μ is dimensionless, so each term on the RHS of given equation should be dimensionless. Hence, B/λ^2 is dimensionless. This implies that dimensions of B will have dimensions of λ^2, i.e. cm^2, i.e. area.

33. Energy density $=\dfrac{\text{Energy}}{\text{Volume}}=[ML^{-1}T^{-2}]$

Force/Area $=[ML^{-1}T^{-2}]$[Charge / Volume]$\times$[Voltage]

$$=\frac{Q}{\text{Volume}}\times\frac{W}{Q}=\frac{\text{Work}}{\text{Volume}}=[ML^{-1}T^{-2}]$$

The dimensions of (d) are different, i.e.

$$\frac{[ML^2T^{-1}]}{[M]}=[M^0L^2T^{-1}]$$

Hence, correct choice is (d).

34. $y=r\sin(\omega t-kx)$

Here, $\omega t=$ angle

$\Rightarrow \qquad \omega=\dfrac{1}{T}=T^{-1}$

Similarly, $kx=$ angle

$\Rightarrow \qquad k=\dfrac{1}{x}=L^{-1}$

$\therefore \qquad \dfrac{\omega}{k}=\dfrac{T^{-1}}{L^{-1}}=LT^{-1}$

Or, simply ω/k represents wave velocity.

$\dfrac{\omega}{k}=\dfrac{2\pi f}{2\pi/\lambda}=f\lambda=v$, where f is frequency.

35. $y=a\sin\omega t+bt+ct^2\cos\omega t$

Here, $a=y, b=y/t, c=y/t^2$

$\therefore \qquad a\times b\times c=y\times y/t\times y/t^2=(y/t)^3$

36. Here, b and $x^2=L^2$ have same dimensions.

Also, $a=\dfrac{L^2}{E\times t}=\dfrac{L^2}{[ML^2T^{-2}]T}$

$=M^{-1}T^1 \Rightarrow a\times b=[M^{-1}L^2T^1]$

37. Dimensional formula of energy $[E]=[ML^2T^{-2}]$ ......(i)

Dimensional formula of gravitational constant

$[G]=[M^{-1}L^3T^{-2}]$ ......(ii)

On dividing Eq. (i) by Eq. (ii), we get

Hence, $\left[\dfrac{E}{G}\right]=[M^2L^{-1}T^0]$

38. $\dfrac{1}{2}\varepsilon_0E^2$ is the expression of energy density (Energy per unit volume)

$$\left[\frac{1}{2}\varepsilon_0E^2\right]=\left[\frac{ML^2T^{-2}}{L^3}\right]=[ML^{-1}T^{-2}]$$

Note *From this question, students can take a lesson that in IIT-JEE also, questions may be wrong or there may be no correct answer/answers in the given choices. So, don't waste time if you are confident.*

39. CR and $\dfrac{L}{R}$ both are time constants. Their units is second.

$\therefore \quad \dfrac{1}{CR}$ and $\dfrac{R}{L}$ have the SI unit (second)$^{-1}$.

Further, resonance frequency $\omega=\dfrac{1}{\sqrt{LC}}$

40. Since, dimension of Fv

$=[Fv]=[MLT^{-2}][LT^{-1}]=[ML^2T^{-3}]$

So, $\left[\dfrac{\beta}{x^2}\right]$ should also be $[ML^2T^{-3}]$

$$\frac{[\beta]}{[x^2]}=[ML^2T^{-3}]$$

$[\beta]=[ML^4T^{-3}]$ **Ans.**

and $\left[Fv+\dfrac{\beta}{x^2}\right]$ will also have dimension $[ML^2T^{-3}]$, so LHS should also have the same dimension $[ML^2T^{-3}]$.

So, $\dfrac{[\alpha]}{[t^2]}=[ML^2T^{-3}]$

$[\alpha]=[ML^2T^{-1}]$ **Ans.**

41. Let $E^av^bF^c\propto m$

Then, $[ML^2T^{-2}]^a[LT^{-1}]^b[MLT^{-2}]^c=[M]$

Equating the powers, we get

$a=1, b=-2$

and $\qquad c=0$ **Ans.**

42. Momentum, $p=mv=MLT^{-1}=ML^{-3}L^4T^{-4}T^3=[Dv^4f^{-3}]$

43. $AT^2=[LT^{-2}\times T^2]=[M^0LT^0]$

44. Let the dimensional formula of linear momentum is,

$[p]=[h]^a[S]^b[I]^c$ ...(i)

$\Rightarrow \qquad [MLT^{-1}]=[ML^2T^{-1}]^a[MT^{-2}]^b[ML^{+2}]^c$

For dimensional balance, the dimensions on both sides should be same.

Thus, equating dimensions, we have

$a+b+c=1$

$2(a+c)=1 \quad$ or $\quad a+c=\dfrac{1}{2}$

$-a-2b=-1 \quad$ or $\quad a+2b=1$

Solving these three equations, we get

$a=0, b=\dfrac{1}{2}, c=\dfrac{1}{2}$

$\therefore \qquad p=h^0S^{\frac{1}{2}}I^{\frac{1}{2}}$ or $p=S^{\frac{1}{2}}I^{\frac{1}{2}}h^0$

45. Let dimensions of Young's modulus is expressed in terms of speed, acceleration and force as;

$[Y]=[v]^\alpha[A]^\beta[F]^\gamma$ ...(i)

$[ML^{-1}T^{-2}]=[LT^{-1}]^\alpha[LT^{-2}]^\beta[MLT^{-2}]^\gamma$

$=[M^\gamma L^{\alpha+\beta+\gamma}T^{-\alpha-2\beta-2\gamma}]$

Now comparing the powers, we get

$$\gamma = 1$$
$$\alpha + \beta + \gamma = -1$$
and $$-\alpha - 2\beta - 2\gamma = -2$$

Solving these equations, we get

$$\alpha = -4,\ \beta = 2,\ \gamma = 1$$

Substituting and α, β and γ in Eq. (i), we get

$$[Y] = [v^{-4}A^2F]$$

46. Since, $p^xQ^yc^z$ is dimensionless. Therefore,

$$[ML^{-1}T^{-2}]^x[MT^{-3}]^y[LT^{-1}]^z = [M^0L^0T^0]$$

Only option (b) satisfies this expression.

So, $x = 1, y = -1, z = 1$

47. $v \propto g^ph^q$ (given)

By substituting the dimension of each quantity and comparing the powers in both sides, we get

$$[LT^{-1}] = [LT^{-2}]^p[L]^q$$

$\Rightarrow$ $p + q = 1, -2p = -1$

$\therefore$ $p = \frac{1}{2}, q = \frac{1}{2}$

48. $v = \lambda^ag^b\rho^c$

Using dimension formula,

$$[M^0L^1T^{-1}] = [L^1]^a[L^1T^{-2}]^b[M^1L^{-3}]^c$$

$\Rightarrow$ $[M^0L^1T^{-1}] = [M^cL^{a+b-3c}T^{-2b}]$

$\therefore$ $c = 0, a + b - 3c = 1, -2b = -1 \Rightarrow b = \frac{1}{2}$

Now $a + b - 3c = 1 \Rightarrow a + \frac{1}{2} - 0 = 1$

$\Rightarrow$ $a = \frac{1}{2}$

$\therefore$ $a = \frac{1}{2}, b = \frac{1}{2}, c = 0$

49. $[T] = [p^ad^bE^c] = [ML^{-1}T^{-2}]^a[ML^{-3}]^b[ML^2T^{-2}]^c$

Equating the powers of both sides, we have

$$a + b + c = 0 \quad \text{... (i)}$$
$$-a - 3b + 2c = 0 \quad \text{... (ii)}$$
$$-2a - 2c = 1 \quad \text{... (iii)}$$

Solving these three equations, we have

$$a = -\frac{5}{6}, b = \frac{1}{2} \text{ and } c = \frac{1}{3}$$

50. $[E] = k[m]^x[n]^y[a]^z$

where, k is a dimensionless constant.

$\therefore$ $[ML^2T^{-2}] = k[M]^x[T^{-1}]^y[L]^z$

Solving, we get

$$x = 1, y = 2 \text{ and } z = 2$$

$\therefore$ $E = kmn^2a^2$

51. $[Y] = [ML^{-1}T^{-2}]$

$$\left[\frac{MgL}{\pi r^2 l}\right] = \left[\frac{MLT^{-2}L}{L^2L}\right] = [ML^{-1}T^{-2}]$$

Dimensions of RHS and LHS are same. Therefore, the given equation is dimensionally correct.

Unit Exercise (Cumulative)

Level 1

1. $[X] = [C] = [M^{-1}L^{-2}T^2Q^2]$

$[Z] = [B] = [MT^{-1}Q^{-1}]$ $\left(\because \text{Given } Y = \frac{X}{3Z^2}\right)$

$\therefore$ $[Y] = \frac{[M^{-1}L^{-2}T^2Q^2]}{[MT^{-1}Q^{-1}]^2} = [M^{-3}L^{-2}T^4Q^4]$

2. (c) As $F_{drag} = 6\pi\eta rv \Rightarrow \eta = \frac{F}{6\pi rv}$

So, unit of η is N-s/m^2.

Now, unit of $p \times t$ is N-s/m^2.

So, Assertion is correct.

Also, $F = \eta A\frac{dv}{dx} \Rightarrow \eta = \frac{F}{A\frac{dv}{dx}}$.

So Reason is incorrect.

3. $n_2 = n_1\left(\frac{M_1}{M_2}\right)^a\left(\frac{L_1}{L_2}\right)^b\left(\frac{T_1}{T_2}\right)^c$

Dimensional formula of moment of inertia $= [ML^2T^0]$

$\therefore$ $a = 1, b = 2, c = 0$

$n_1 = 12.0, M_1 = 1\text{ kg}, M_2 = 10\text{ g}$

$L_1 = 1\text{ m}, L_2 = 5\text{ cm}, T_1 = 1\text{ s}, T_2 = 1\text{ s}$

$$n_2 = 12.0\left(\frac{1\text{ kg}}{10\text{ g}}\right)^1\left(\frac{1\text{ m}}{5\text{ cm}}\right)^2\left(\frac{1\text{ s}}{1\text{ s}}\right)^0$$

$$= 12 \times \left(\frac{1000\text{ g}}{10\text{ g}}\right)^1\left(\frac{100\text{ cm}}{5\text{ cm}}\right)^2 \times 1$$

$$= 12 \times 100 \times 400 = 4.8 \times 10^5.$$

4. To convert a measured value from one system to another system, we use

$$N_1u_1 = N_2u_2$$

where, N is numeric value and u is unit.

We get

$$128 \cdot \frac{\text{kg}}{\text{m}^3} = N_2\frac{50\text{ g}}{(25\text{ cm})^3} \quad \left[\because \text{density} = \frac{\text{mass}}{\text{volume}}\right]$$

$$\Rightarrow \frac{128 \times 1000\text{ g}}{100 \times 100 \times 100\text{ cm}^3} = \frac{N_2 \times 50\text{ g}}{25 \times 25 \times 25\text{ cm}^3}$$

$\Rightarrow \quad N_2 = \dfrac{128 \times 1000 \times 25 \times 25 \times 25}{50 \times 100 \times 100 \times 100} = 40$

5. Dimension of (ohm) $R = \dfrac{h}{e^2}$; (e = charge = current × time)

$= \dfrac{[Et]}{[it]^2} = \dfrac{P}{i^2} = R$ as $P = \dfrac{E}{t}$

6. $\left[\dfrac{GIM^2}{E^2}\right] = \dfrac{[M^{-1}L^3T^{-2}][MLT^{-1}][M^2]}{[ML^2T^{-2}]^2} = [T]$

7. The energy of electron in first orbit of H-atom is

$E = \dfrac{-me^4}{8\varepsilon_0^2 h^2} \Rightarrow \left[\dfrac{me^4}{\varepsilon_0^2 h^2}\right] = [E] =$ energy

8. Let $t = k(G)^a(h)^b(c)^c$ ...... (i)

Here k is a dimensionless constant.

For calculating the values of a, b and c, compare the dimensional formula for both side.

LHS t = time $= [M^0L^0T^1]$

RHS $G = \dfrac{F \cdot r^2}{m_1 m_2} = \dfrac{\text{kg ms}^{-2} \times \text{m}^2}{(\text{kg})^2} = [M^{-1}L^3T^{-2}]^a$

$h = [M^1L^2T^{-1}]^b$

$c = \dfrac{d}{t} = [M^0L^1T^{-1}]^c$

Compare both side for powers of M, L and T,

$M \Rightarrow 0 = -a + b$ ...(ii)

$L \Rightarrow 0 = 3a + 2b + c$ ...(iii)

$T \Rightarrow 1 = -2a - b - c$ ...(iv)

Solving Eqs. (ii), (iii), (iv), we get

$a = \dfrac{1}{2}, \quad b = \dfrac{1}{2}$ and $c = \dfrac{-5}{2}$

So, put these values in Eq. (i)

$t = kG^{1/2}h^{1/2}c^{-5/2} \Rightarrow t = k\sqrt{\dfrac{Gh}{c^5}}$

So, $\quad t \propto \sqrt{\dfrac{Gh}{c^5}}$

9. Unit of $\dfrac{E}{H}$ is

$\dfrac{\text{volt / metre}}{\text{ampere / metre}} = \dfrac{\text{volt}}{\text{ampere}} = \text{ohm}$

10. $\dfrac{A}{B} = m \Rightarrow B = \dfrac{A}{m} = \dfrac{\text{Force}}{\text{Linear density}} = \dfrac{MLT^{-2}}{ML^{-1}}$

$\therefore \quad B = [M^0L^2T^{-2}]$

Latent heat $= \dfrac{\text{Heat energy}}{\text{Mass}} = \dfrac{ML^2T^{-2}}{M} = [M^0L^2T^{-2}]$

Thus, B has same dimensions as that of latent heat.

11. As resistance, inductive reactance and capacitive reactance have same unit ohm. So, all three have same dimension. Thus, $\dfrac{R}{\sqrt{X_L X_C}}$ is dimensionless.

12. Let $[R] = [\varepsilon_0]^\alpha [\mu_0]^\beta$ ...(i)

$[R] = [M^1L^2T^{-3}A^{-2}]$, $[\varepsilon_0] = [M^{-1}L^{-3}T^4A^2]$,

$[\mu_0] = [M^1L^1T^{-2}A^{-2}]$

Now, from Eq. (i), we get

$[M^2L^2T^{-3}A^{-2}] = [M^{-1}L^{-3}T^4A^2]^\alpha [M^1L^1T^{-2}A^{-2}]^\beta$

$[M^1L^2T^{-3}A^{-2}] = [M^{-\alpha+\beta}L^{-3\alpha+\beta}T^{4\alpha-2\beta}A^{2\alpha-2\beta}]$

Comparing both sides, we get

$-\alpha + \beta = 1$ ...(ii)

$-3\alpha + \beta = 2$ ...(iii)

$4\alpha - 2\beta = -3$ ...(iv)

$2\alpha - 2\beta = -2$ ...(v)

Solving above equations, we get

$\alpha = \dfrac{-1}{2}$ and $\beta = +\dfrac{1}{2}$

$\therefore \quad [R] = [\varepsilon_0]^{-1/2}[\mu_0]^{1/2} = \left[\sqrt{\dfrac{\mu_0}{\varepsilon_0}}\right]$

13. $\left[\dfrac{l}{r}\right] = [t]$

and cv = charge or q

$\therefore \quad \left[\dfrac{l}{rcv}\right] = \left[\dfrac{t}{q}\right]$

$= \left[\dfrac{1}{\text{current}}\right] = [A^{-1}]$

14. From the given equation it is clear,

X and $\dfrac{a}{Y^2}$ have same dimensions

Y and b have same dimensions

$\therefore \quad [a] = [ML^5T^{-2}]$ and $[b] = [L^3]$

$\therefore \quad \dfrac{[a]}{[b]} = [ML^2T^{-2}]$

The dimensions of energy.

15. $B = \dfrac{\mu_0}{2\pi}\dfrac{i}{r}$

But $\quad B = \dfrac{F}{il}$ $\quad (\because F = ilB)$

$\therefore \quad \dfrac{F}{il} = \dfrac{\mu_0}{2\pi}\dfrac{i}{r}$

$\therefore \quad [\mu_0] = \left[\dfrac{F}{i^2}\right] = [MLT^{-2}A^{-2}]$ **Ans.**

16. Let $m = Kh^a c^b G^c$

$\Rightarrow \quad [m] = [h]^a [c]^b [G]^c$

$\Rightarrow \quad M^1 = [ML^2T^{-1}]^a [LT^{-1}]^b [M^{-1}L^3T^{-2}]^c$

$\Rightarrow \quad 1 = a - c, 2a + b + 3c = 0, -a - b - 2c = 0$

On solving above equations, we get

$$a = \frac{1}{2}, b = \frac{1}{2} \text{ and } c = -\frac{1}{2}$$

17. $x = Ay + B \tan Cz$

From the dimensional homogeneity

$$[x] = [Ay] = [B]$$

$$\Rightarrow \quad \left[\frac{x}{A}\right] = [y] = \left[\frac{B}{A}\right]$$

$$[Cz] = [M^0L^0T^0] = \text{Dimensionless}$$

18. $M \propto h^a c^b G^c$

$M^1 \propto [ML^2\,T^{-1}]^a [LT^{-1}]^b [M^{-1}L^3\,T^{-2}]^c$

$\propto M^{a-c} L^{2a+b+3c} T^{-a-b-2c}$

$$a - c = 1 \quad \text{...(i)}$$

$$2a + b + 3c = 0 \quad \text{...(ii)}$$

$$a + b + 2c = 0 \quad \text{...(iii)}$$

On solving Eqs. (i), (ii), (iii), we get

$$a = \frac{1}{2}, b = +\frac{1}{2}, c = -\frac{1}{2}$$

$\therefore M \propto \sqrt{c}$ only $\rightarrow$ (a) is correct.

In the same way we can find that, $L \propto h^{1/2} c^{-3/2} G^{1/2}$

$L \propto \sqrt{h}, L \propto \sqrt{G} \rightarrow$ (c), (d) are also correct.

19. (a) $\dfrac{bt}{2m} = [M^0L^0T^0] \;\Rightarrow\; [t] = \left[\dfrac{m}{b}\right]$

(b) $[b] = [a] = [L] \;\Rightarrow\; [\omega t] = [M^0L^0T^0]$

$\therefore \quad [\omega] = [T^{-1}]$

$\therefore \quad [a_0\,\omega] = [LT^{-1}]$

20. $y = \dfrac{a}{t^2} e^{-\left[\frac{t}{b} - \frac{x}{c}\right]}$

$$\left[\frac{t}{b}\right] = \left[\frac{x}{c}\right] = M^0L^0T^0$$

$\Rightarrow \quad [b] = [T]$ and $[c] = [L]$ and $\left[\dfrac{a}{t^2}\right] = y$

$\Rightarrow \quad [a] = [LT^{-2}]$ and $e^{-\left(\frac{t}{b} - \frac{x}{L}\right)}$ is dimensionless

21. $[a] = [pV^2] = \left[\dfrac{FV^2}{A}\right] = \dfrac{[MLT^{-2}L^6]}{[L^2]} = [ML^5T^{-2}]$

22. $[b] = [V] = [L^3]$

23. According to given equation, $pV + \dfrac{a}{V} - \dfrac{ab}{V^2} = RT$

As dimensions of all the terms on LHS must be equal to dimensions of term RT on RHS, a/V^2 does not have the same dimensional formula as that of RT.

24. $\left[\dfrac{ab}{V^2}\right] = [RT] \;\Rightarrow\; \left[\dfrac{ab}{RT}\right] = [V^2] = [L^6]$

25. $[RT] = [pV] = \left[\dfrac{F}{A} \cdot V\right] = [ML^2T^{-2}]$

So the dimensional formula of RT is same as that of energy.

26. $t \equiv \dfrac{L}{R}$

$\therefore \quad L \equiv tR \equiv \text{ohm - second}$

$$U \equiv \frac{q^2}{2C}$$

$\therefore \quad C \equiv \dfrac{q^2}{U} \equiv \text{coulomb}^2/\text{joule}$

$$q \equiv CV$$

$\therefore \quad C \equiv \dfrac{q}{V} \equiv \text{coulomb /volt}$

$$L \equiv \frac{-e}{di/dt}$$

$\therefore \quad L \equiv \dfrac{e\,(dt)}{(di)} \equiv \text{volt-second/ampere}$

$$F = ilB$$

$\therefore \quad B \equiv \dfrac{F}{il} \equiv \text{newton/ampere-metre}$

Column I	Column II
Capacitance	coulomb/(volt)$^{-1}$(coulomb)^2 joule^{-1}
Inductance	ohm-second, volt-second/ampere^{-1}
Magnetic induction	newton (ampere-metre)$^{-1}$

27. (A) Pressure amplitude in pressure wave $= \beta Ak$

$\therefore \quad \beta Ak = \text{Pressure} = \dfrac{\text{Force}}{\text{Area}} = \text{Nm}^{-2}$

(B) Intensity of wave $= \dfrac{1}{2}\rho A^2 \omega^2 c = \dfrac{\text{Power}}{\text{Area}} = \text{Wm}^{-2}$

(C) $\dfrac{F^2 l}{SY} = \dfrac{\text{N}^2\text{m}}{\text{m}^2\text{N/m}^2} = \text{N/m} = \text{J}$

(D) $\sqrt{\dfrac{r^3}{GM}} = \sqrt{\dfrac{\text{m}^3}{\dfrac{\text{Nm}^2}{\text{kg}^2} \times \text{kg}}}$

$$= \sqrt{\frac{\text{kg m}}{\text{N}}} = \sqrt{\frac{\text{kg m}}{\text{kg m/s}^2}} = \text{second}$$

28. (a) $F = \dfrac{GM_eM_s}{r^2}$

= Gravitational force between sun and earth

$\Rightarrow \quad GM_eM_s = Fr^2$

$\therefore \quad [GM_eM_s] = [Fr^2]$

$= [MLT^{-2}][L^2] = [ML^3T^{-2}]$

(b) $v_{rms} = \sqrt{\dfrac{3RT}{M}}$ = rms speed of gas molecules

$\therefore \quad \dfrac{3RT}{M} = v_{rms}^2$

or $\left[\dfrac{3RT}{M}\right] = [v_{rms}^2] = [LT^{-1}]^2 = [L^2T^{-2}]$

(c) $F = Bqv$ = magnetic force on a charged particle

$\therefore \quad \dfrac{F}{Bq} = v$ or $\left[\dfrac{F^2}{B^2q^2}\right] = [v]^2 = [L^2T^{-2}]$

(d) $v_o = \sqrt{\dfrac{GM_e}{R_e}}$ = orbital velocity of earth's satellite

$\therefore \quad \dfrac{GM_e}{R_e} = v_o^2$ or $\left[\dfrac{GM_e}{R_e}\right] = [v_o^2] = [L^2T^{-2}]$

(p) $W = qV$

$\Rightarrow$ (Coulomb) (Volt) = Joule

or [(Volt) (Coulomb) (Metre)] = [(Joule) (Metre)]

$= [ML^2T^{-2}][L] = [ML^3T^{-2}]$

(q) [(kilogram) (metre)^3 (second)$^{-2}$] $= [ML^3T^{-2}]$

(r) [(metre)^2 (second)$^{-2}$] $= [L^2T^{-2}]$

(s) $U = \dfrac{1}{2}CV^2$

$\Rightarrow$ (farad)(volt)^2 = Joule

or [(farad) (volt)^2 (kg)$^{-1}$] = [(Joule)(kg)$^{-1}$]

$= [ML^2T^{-2}][M^{-1}] = [L^2T^{-2}]$

29. $\left[\dfrac{dx}{\sqrt{a^2-x^2}}\right]$ is dimensionless.

$\left[\dfrac{1}{a}\sin^{-1}\left(\dfrac{a}{x}\right)\right]$ has the dimension of $[L^{-1}]$.

30. $\left[\dfrac{dx}{\sqrt{2ax-x^2}}\right]$ is dimensionless.

Therefore, $\left[a^n \sin^{-1}\left(\dfrac{x}{a}-1\right)\right]$ should also be dimensionless.

Hence, $n = 0$.

31. As $n_1u_1 = n_2u_2$

$\Rightarrow \quad n_2 = \left(\dfrac{u_1}{u_2}\right)n_1 = \left(\dfrac{L_1T_1^{-1}}{L_2T_2^{-1}}\right)(n_1)$

$= \left(\dfrac{1}{2}\times 4\right)(3\times 10^8) = 6\times 10^8$

32. Let $d = k(\rho)^a(S)^b(f)^c$

where, k is a dimensionless constant. Then,

$$[L] = \left[\frac{M}{L^3}\right]^a\left[\frac{ML^2T^{-2}}{L^2T}\right]^b\left[\frac{1}{T}\right]^c$$

Equating the powers of M and L, we have

$0 = a + b$ ...(i)

$1 = -3a$ ...(ii)

Solving these two equations, we get

$b = \dfrac{1}{3} \Rightarrow n = 3$

Level 2

33. $C = \dfrac{\Delta q}{\Delta V} = \dfrac{\varepsilon_0 A}{d}$

or $\quad \varepsilon_0 = \dfrac{A}{L} = \dfrac{\Delta q}{\Delta V}$

or $\quad \varepsilon_0 = \dfrac{(\Delta q)L}{A.(\Delta V)}$

$X = \varepsilon_0 L\dfrac{\Delta V}{\Delta t} = \dfrac{(\Delta q)L}{A(\Delta V)}L\dfrac{\Delta V}{\Delta t}$

But $\quad [A] = [L^2]$

$\therefore \quad X = \dfrac{\Delta q}{\Delta t}$ = current

34. Given, $v_2 = \dfrac{n}{m^2}v_1 \Rightarrow \dfrac{L_2}{T_2} = \dfrac{n}{m^2}\cdot\dfrac{L_1}{T_1}$ ...(i)

and $a_2 = \dfrac{a_1}{mn} \Rightarrow \dfrac{L_2}{T_2^2} = \dfrac{L_1}{T_1^2}\cdot\dfrac{1}{mn}$ ...(ii)

Dividing Eq. (i) by Eq. (ii), we get

$$\frac{L_2/T_2}{L_2/T_2^2} = \frac{n/m^2\cdot L_1/T_1}{L_1/T_1^2\cdot 1/mn}$$

$\Rightarrow \quad T_2 = \dfrac{n^2}{m}\cdot T_1$

As $\quad L_2 = \dfrac{T_2}{T_1}\cdot L_1\dfrac{n}{m^2} = \dfrac{n^2}{m}L_1\dfrac{n}{m^2} = \dfrac{n^3}{m^3}L_1$

35. $\left[\dfrac{\alpha Z}{k\theta}\right] = [M^0 l^0 T^0]$

$[\alpha] = \left[\dfrac{k\theta}{Z}\right]$

Further $\quad [p] = \left[\dfrac{\alpha}{\beta}\right]$

$\therefore \quad [\beta] = \left[\dfrac{\alpha}{p}\right] = \left[\dfrac{k\theta}{Zp}\right]$

Dimensions of $k\theta$ are that to energy. Hence,

$[\beta] = \left[\dfrac{ML^2T^{-2}}{LML^{-1}T^{-2}}\right]$

$= [M^0L^2T^0]$

36. As $\frac{d^2s}{dt^2}$ is acceleration or $\frac{m}{s^2}$ or $[LT^{-2}]$.

In the similar way,

$$\left[\frac{d^2y}{dx^2}\right]=\left[\frac{y}{x^2}\right]=\frac{[ML^{-1}T^{-2}]}{[T^{-1}]^2}=[ML^{-1}T^0]$$

37. In Bohr's model, $\frac{1}{\lambda}=\frac{me^4}{\varepsilon_0^2h^3c}\left(\frac{1}{n_1^2}-\frac{1}{n_2^2}\right)$

where, λ = wavelength, n_1 and n_2 are principal quantum numbers.

$$\therefore \quad \left[\frac{me^4}{\varepsilon_0^2h^3c}\right]=[L^{-1}]=[M^0L^{-1}T^0]$$

38. Exponential terms are dimensionless, so

$$\left(\frac{-x^2}{\alpha kT}\right)=[M^0L^0T^0]$$

or $$\frac{[L^2]}{[\alpha][ML^2T^{-2}]}=[M^0L^0T^0]$$

or $$[\alpha]=[M^{-1}T^2]$$

Now, $$[F]=[\alpha\beta]$$

$$[MLT^{-2}]=[M^{-1}T^2][\beta]$$

or $$[\beta]=[M^2LT^{-4}]$$

39. Dimensions of length is related as,

$$L=[c]^x[G]^y\left[\frac{e^2}{4\pi\varepsilon_0}\right]^z \Rightarrow \left[\frac{e^2}{4\pi\varepsilon_0}\right]=[ML^3T^{-2}]$$

$$\Rightarrow \quad [L]=[LT^{-1}]^x[M^{-1}L^3T^{-2}]^y[ML^3T^{-2}]^z$$

$$\Rightarrow \quad [L]=[L^{x+3y+3z}M^{-y+z}T^{-x-2y-2z}]$$

Comparing both sides, $-y+z=0 \Rightarrow y=z$ ... (i)

$$x+3y+3z=1 \quad \text{... (ii)}$$

$$-x-4z=0 \quad \text{... (iii)}$$

Solving Eqs. (i), (ii) and (iii), we get

$$z=y=\frac{1}{2}, x=-2$$

Hence, $L=c^{-2}\left[G\cdot\frac{e^2}{4\pi\varepsilon_0}\right]^{\frac{1}{2}}$

40. In terms of h, c and G, let length be expressed as $L=(h)^a(c)^b(G)^c$

Writing dimensions on both sides, we get

$$[M^0LT^0]=[ML^2T^{-1}]^a[LT^{-1}]^b[M^{-1}L^3T^{-2}]^c$$

$$=M^{a-c}L^{2a+b+3c}T^{-a-b-2c}$$

On comparing powers of M, L and T on both sides, we get

$$a-c=0,\ 2a+b+3c=1 \text{ and } -a-b-2c=0$$

On solving, we get

$$a=c=\frac{1}{2} \text{ and } b=-\frac{3}{2}$$

$\therefore$ Dimensions of length is

$$L=(h)^{1/2}(c)^{-3/2}(G)^{1/2}=\frac{\sqrt{hG}}{c^{3/2}}$$

41. (a) Let $1\text{ m/s}=x\left(\frac{10\text{ m}}{20\text{ s}}\right) \Rightarrow x=2$

(b) $1\text{ W}=x\text{W}'$

$$x=1\frac{W}{W'}=1\left(\frac{M_1L_1^2T_1^{-3}}{M_2L_2^2T_2^{-3}}\right)$$

$$=1\left(\frac{1\text{ kg}}{5\text{ kg}}\right)\left(\frac{1\text{ m}}{10\text{ m}}\right)^2\left(\frac{1\text{ s}}{20\text{ s}}\right)^2=16$$

(c) $1\text{ J}=x\text{J}'$

$$x=1\frac{J}{J'}=1\left(\frac{M_1L_1^2T_1^{-2}}{M_2L_2^2T_2^{-2}}\right)$$

$$=1\left(\frac{1\text{ kg}}{5\text{ kg}}\right)\left(\frac{1\text{ m}}{10\text{ m}}\right)^2\left(\frac{1\text{ s}}{20\text{ s}}\right)^2=\frac{4}{5}$$

(d) $1\text{ N}=x\text{N}'$

$$x=1\frac{N}{N'}=1\left(\frac{M_1L_1T_1^{-2}}{M_2L_2T_2^{-2}}\right)$$

$$=1\left(\frac{1\text{kg}}{5\text{kg}}\right)\left(\frac{1\text{m}}{10\text{m}}\right)\left(\frac{1\text{s}}{20\text{s}}\right)^{-2}=8$$

42. Unit of mass $=\frac{\text{Force}}{\text{Acceleration}}=\frac{6}{24}=\frac{1}{4}\text{ kg}$

Unit of time $=\frac{\text{Velocity}}{\text{Acceleration}}=\frac{4}{24}=\frac{1}{6}\text{ s}$

Unit of energy $=(\text{Mass})(\text{Velocity})^2=\left(\frac{1}{4}\right)(4)^2=4\text{ J}$

Unit of momentum $=\text{Force}\times\text{Time}=6\times\frac{1}{6}=1\text{ Ns}$

Unit of power $=\frac{\text{Energy}}{\text{Time}}=\frac{4}{1/6}=24\text{ W}$

43. (a) Energy $U=\frac{1}{2}LI^2=\left(\frac{1}{2}\frac{\mu_0N^2A}{l}\right)I^2$

Energy $U=\frac{1}{2}CV^2=\left(\frac{1}{2}\varepsilon_0\frac{A}{d}\right)V^2$

$\Rightarrow \mu_0\frac{A}{l}I^2$ and $\varepsilon_0\frac{A}{d}V^2$ have same dimension.

So, μ_0I^2 and ε_0V^2 have same dimension.

(c) $Q=CV$, $\frac{Q}{t}=I=\frac{CV}{t}=\left(\frac{\varepsilon_0A}{l}\right)\frac{V}{t}$

$$\therefore \quad I=\varepsilon_0\left(\frac{A}{lt}\right)V$$

But, $\frac{A}{lt}$ have units of speed.

So, $I = \varepsilon_0 cV$

44. $\frac{\text{Electric dipole moment}}{\text{Magnetic dipole moment}} = \frac{ql}{ml}$

where, $q =$ charge $= [AT]$,

$m =$ pole strength $= [AL]$,

$l =$ length $= [L]$

$\therefore \frac{\text{Electric dipiole moment}}{\text{Magnetic dipole moment}} = \frac{q}{m}$

$= \frac{[AT]}{[AL]} = [L^{-1}T]$

Thus, (a) → (s)

(b) (Electric flux) × (Magnetic flux) = (EA) (BA)

where, $E =$ electric field, $A =$ area,
$B =$ magnetic field.

∴ Electric flux × Magnetic flux $= EBA^2$

$= [A^{-1}MLT^{-3}][A^{-1}MT^{-2}][L^4]$

$= [A^{-2}M^2L^5T^{-5}]$

Thus, (b) → (r)

(c) $\frac{G}{R} = \frac{[M^{-1}M^3T^{-2}]}{[mol^{-1}M][L^2T^{-2}K^{-1}]} = [K\,mol\,M^{-2}L]$

Thus, (c) → (q)

(d) $\frac{\text{Inductance} \times \text{Electric permittivity}}{\text{Heat capacity}}$

$= [A^{-2}ML^2T^{-2}] \times \frac{[A^2M^{-1}L^{-3}T^4]}{[K^{-1}ML^2T^{-2}]}$

$= [KM^{-1}L^{-3}T^4]$

Thus, (d) → (p)

45. (a) $\because \frac{GMm}{R^2} = F$

$\therefore \frac{GM}{R} = \frac{FR}{m} = \frac{[MLT^{-2}L]}{[M]}$

$= [M^2T^{-2}]$

$\therefore \sqrt{\frac{GM}{R}} = [LT^{-1}]$

Thus, (a) → (s)

(b) $\lambda v^2 = [ML^{-1}][LT^{-1}]^2 = [MLT^{-2}]$

Thus, (b) → (r)

(c) Magnetic field, $B = \mu_0 Jr$

$\therefore [\mu_0 Jr] = \theta[B] = [A^{-1}MT^{-2}]$

Thus, (c) → (q)

(d) $\left[\frac{m}{qB}\right] = \frac{[M]}{[ATA^{-1}MT^{-2}]} = [T] = [M^0L^0T]$

Thus, (d) → (p)

46. $n_1u_1 = n_2u_2 \Rightarrow u_2 = \left(\frac{n_1}{n_2}\right)u_1$

$= \left(\frac{M_1}{M_2}\right)^1\left(\frac{L_1}{L_2}\right)^2\left(\frac{T_1}{T_2}\right)^{-2} u_1$

$= \left(\frac{1\,kg}{10\,kg}\right)\left(\frac{1\,m}{1\,km}\right)^2\left(\frac{1\,s}{1\,min}\right)^{-2} \times 1$

$= \frac{1}{10} \times 10^{-6} \times 60 \times 60$

$= 360 \times 10^{-6}$

47. Energy $\propto m^x a^y f^z$

$[ML^2T^{-2}] = [M]^x[LT^{-2}]^y[T^{-1}]^z$

$x = 1, y = 2$ and $z = -2$

Energy $\propto \frac{ma^2}{f^2}$

$0.1\,J = xJ'$

$x = (0.1)\frac{J}{J'} = (0.1)\left(\frac{M_1}{M_2}\right)^1\left(\frac{a_1}{a_2}\right)^2\left(\frac{f_1}{f_2}\right)^{-2}$

$= (0.1)\left(\frac{1\,kg}{100\,g}\right)\left(\frac{1\,m/s^2}{2\,m/s^2}\right)^2\left(\frac{1\,sec^{-1}}{4\,sec^{-1}}\right)^{-2}$

$= (0.1)(10)\left(\frac{1}{4}\right)(16) = 4$

UNIT-5

VECTORS

Learning Modules

1. Vector and Scalar Quantities
2. Addition and Subtraction of Vectors
3. Multiplication of Vectors
4. Representation of a Vector in Terms of Unit Vectors
5. Left Over Topics of Vectors

MODULE 1

Vector & Scalar Quantities

Any physical quantity is either a scalar or a vector. A scalar quantity can be described completely by its magnitude only. Addition, subtraction, division or multiplication of scalar quantities can be done according to the ordinary rules of algebra. Mass, volume, density, etc., are few examples of scalar quantities.

A vector quantity has both magnitude and direction.

Vector mathematics is different from scalar mathematics. In addition, subtraction or multiplication of two vectors, the angle between two vectors (this is also the angle between directions of those two vectors) plays a very important role.

Let us take an example. Mass is a scalar quantity. Suppose we have two blocks A and B of masses 2 kg and 3kg. Now these two blocks are kept in different positions as shown below:

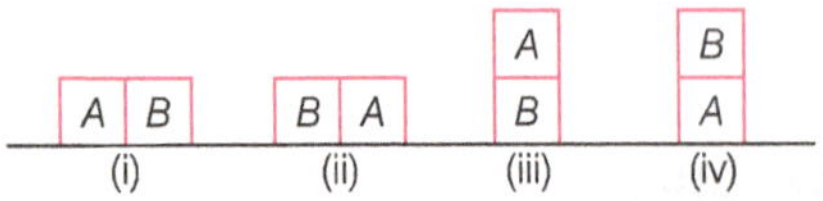

In all four cases total mass will be 5 kg.

Now let us take an example of a force, which is a vector quantity. Suppose we have two forces of magnitude 2N and 3N. These two forces are applied on a block in following two manners.

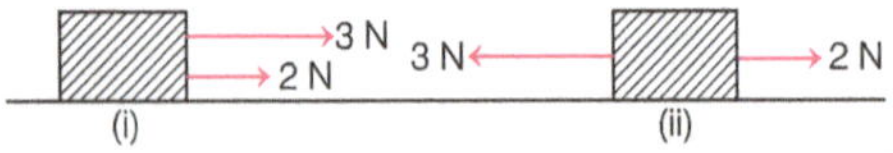

We can easily see that net force on the block in first case is 5N (towards right) and in the second case net force is 1N (towards left). From this example we can see that the magnitude and direction of resultant in the two cases are not same. It is depending upon the directions of two forces. Or, we can say the angle between them.

Thus, a scalar quantity has only magnitude and a vector quantity has both magnitude as well as direction. Secondly, scalar mathematics (addition, subtraction and multiplication) is different from vector mathematics. The result not only depend upon magnitudes of vectors but also upon their directions or the angle between the vectors.

1. **Scalar quantities** Mass, volume, distance, speed, density, work, power, energy, length, gravitation constant (G), specific heat, specific gravity, charge, current, potential, time, electric or magnetic flux, pressure, surface tension, temperature.

2. **Vector quantities** Displacement, velocity, acceleration, force, weight, acceleration due to gravity (g), gravitational field strength, electric field, magnetic field, dipole moment, torque, linear momentum, angular momentum.

General Points Regarding Vectors

Vector Notation

Usually a vector is represented by a bold capital letter or an arrow over it, as **A**, **B**, **C** or $\vec{A}, \vec{B}, \vec{C}$ etc.

Note *We will use the first method to represent a vector.*

The magnitude of a vector **A** is represented by A or $|\mathbf{A}|$ and it is always positive (zero in some special cases).

Graphical Representation of a Vector

Graphically a vector is represented by an arrow (drawn to a chosen scale) parallel to the direction of the vector. The length and the direction of the arrow thus represent the magnitude and the direction of the vector respectively.

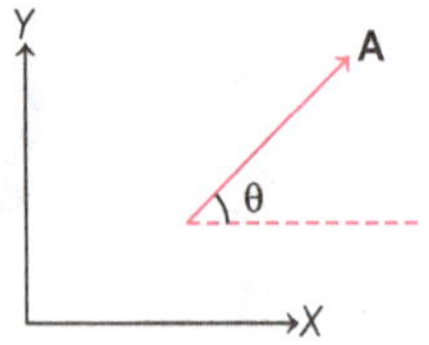

Thus, the arrow shown in the above figure represents a vector **A** in x-y plane making an angle θ with x-axis.

Steps Involved in Representing a Vector

(i) By choosing a proper scale, draw a line whose length is proportional to the magnitude of the vector.

(ii) Indicate the direction of the vector by marking an arrow head at one end of the line.

Example Displacement vector

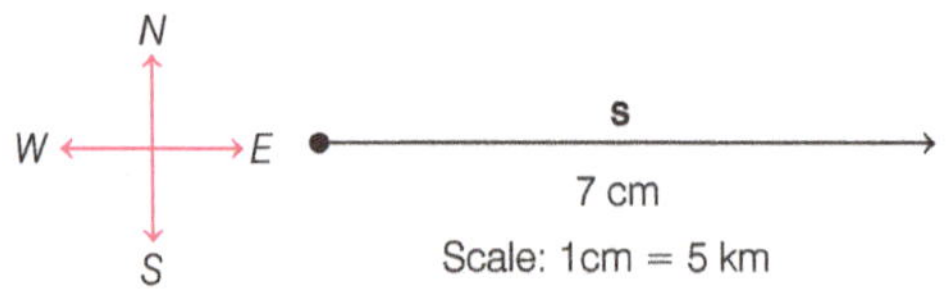

Graphical representation of a vector

In the figure shown above, **s** is a displacement vector of magnitude 35 km towards east.

Note *A vector can be displaced from one position to another. During the displacement if we do not change direction and magnitude then the vector remains unchanged.*

Angle between Two Vectors (θ)

As we discussed above that in vector algebra, angle between two vectors is very important.

To find angle between two vectors both the vectors are drawn from one point in such a manner that arrows of both the vectors are outwards from that point. Now, the smaller angle between them is called the angle between two vectors.

For example, in the figure given below, angle between **A** and **B** is 60° not 120°.Because in Fig.(a), they are wrongly drawn while in Fig. (b) they are drawn as we desire.

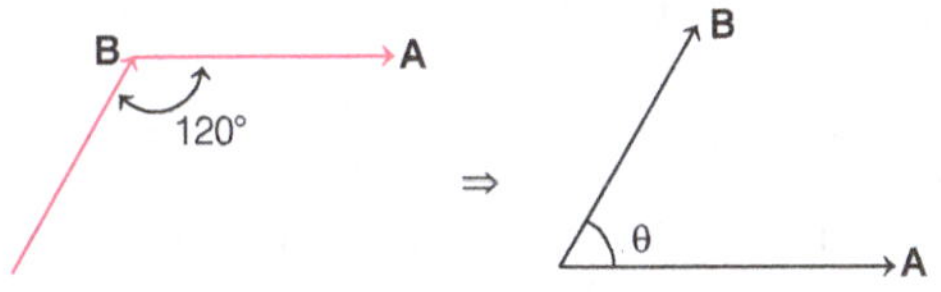

(a) (b)

Note $0° \le \theta \le 180°$

Different Types of Vectors

Unit Vector

A vector of unit magnitude is called a unit vector and the notation for it in the direction of **A** is $\hat{\mathbf{A}}$ read as '*A* cap or *A* caret'.

Thus, $\mathbf{A} = A\hat{\mathbf{A}}$ or $\hat{\mathbf{A}} = \dfrac{\mathbf{A}}{|\mathbf{A}|} = \dfrac{\mathbf{A}}{A}$

A unit vector merely indicates a direction. Unit vector along *x*, *y* and *z*-directions are $\hat{\mathbf{i}}$, $\hat{\mathbf{j}}$ and $\hat{\mathbf{k}}$.

Zero Vector or Null Vector

A vector having zero magnitude is called a null vector or zero vector.

Note *(i) Zero vector has no specific direction.*
(ii) Zero vectors are only of mathematical importance.

Equal Vectors

Vectors are said to be equal if both vectors have same magnitude and direction.

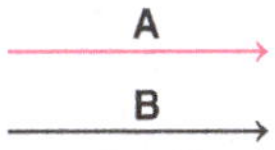

A = B

Parallel Vectors

Vectors are said to be parallel if they have the same direction.

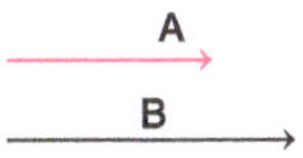

The vectors **A** and **B** shown in above figure represent parallel vectors.

Note *Two equal vectors are always parallel but, two parallel vectors may not be equal vectors.*

Anti-parallel Vectors

Vectors are said to be anti-parallel if they act in opposite directions.

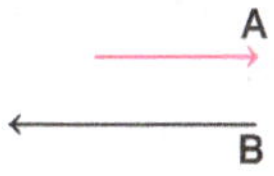

The vectors **A** and **B** shown in above figure are anti-parallel vectors.

Negative Vector

The negative vector of any vector is a vector having equal magnitude but in opposite direction.

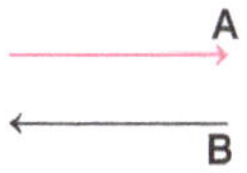

$\mathbf{A} = -\mathbf{B}$

or $\mathbf{B} = -\mathbf{A}$

Concurrent Vectors (Co-initial Vectors)

Vectors having the same initial points are called concurrent vectors or co-initial vectors.

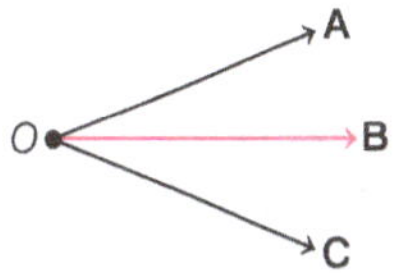

A, **B** and **C** are concurrent at point *O*.

Coplanar Vectors

The vectors lying in the same plane are called coplanar vectors.

(a) (b)

The vectors **A** and **B** as shown in figure (a) are coplanar vectors. The vectors **A** and **B** shown in figure (b) are concurrent coplanar vectors.

Orthogonal Vectors

Two vectors are said to be orthogonal if the angle between them is 90°.

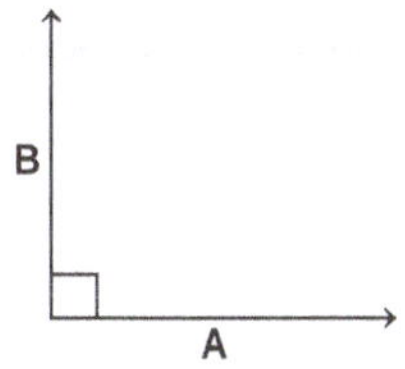

The vectors shown in above figure, **A** and **B** are orthogonal to one another.

Multiplication and Division of Vectors by Scalars

The product of a vector **A** and a scalar m is a vector $m\mathbf{A}$ whose magnitude is m times the magnitude of **A** and which is in the direction or opposite to **A** according as the scalar m is positive or negative. Thus,

$$|m\mathbf{A}| = |m| A$$

Further, if m and n are two scalars, then

$$(m+n)\mathbf{A} = m\mathbf{A} + n\mathbf{A} \quad \text{and} \quad m(n\mathbf{A}) = n(m\mathbf{A}) = (mn)\mathbf{A}$$

The division of vector **A** by a non-zero scalar m is defined as the multiplication of **A** by $\frac{1}{m}$.

Example 1

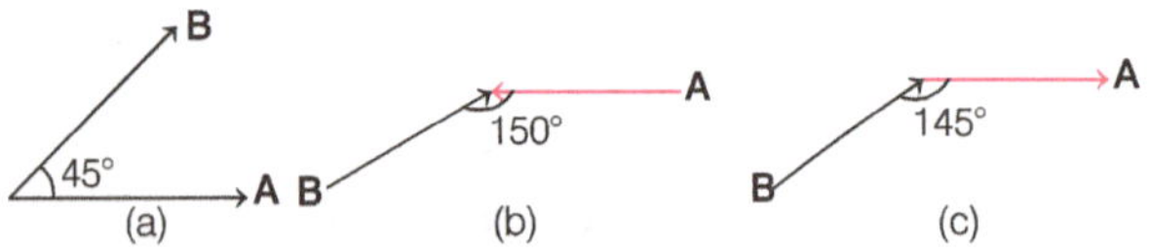

In the shown figures (a), (b) and (c), find the angle between **A** and **B**.

Sol. If we draw both the vectors from one point with their arrows outwards, then they can be shown as below.

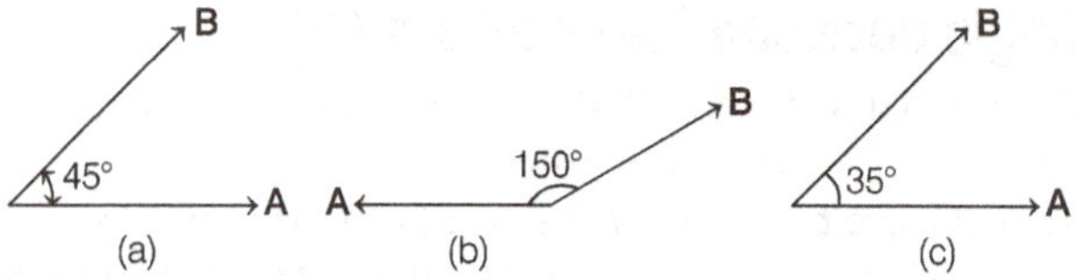

In Fig. (a), $\theta = 45°$

In Fig. (b), $\theta = 150°$ and

In Fig. (c), $\theta = 35°$

Example 2 What is the angle between **a** and $-\frac{3}{2}\mathbf{a}$.

Sol. $-\frac{3}{2}\mathbf{a}$ has a magnitude equal to $\frac{3}{2}$ times the magnitude of **a** and its direction is opposite to **a**.

Therefore, **a** and $-\frac{3}{2}\mathbf{a}$ are antiparallel to each other or angle between them is 180°.

Important POINTS

1. Moment of inertia of a rigid body is neither a vector nor a scalar. It is a matrix. If rigid body undergoes only rotations about all three axes then this is 3×3 matrix. If you are dealing with rotations about 2 axes then it is 2×2 matrix. If you are a student, then generally (99% cases) you will deal with rotations about a single axis then it is 1×1 matrix. You can treat this 1×1 matrix as a scalar.

Note *At this stage you can also call matrix as a tensor. Although there is some difference between the two.*

2. Surface tension is a scalar quantity because it has no specific direction. It is perpendicular to a line (both sides of this line) on the surface of a liquid. So it's the line that determines the direction not the surface tension. Same is the case with pressure. It is also a scalar quantity as it has no specific direction. At any point pressure acts in all directions. For example, in a gas chamber pressure acts on all walls of the container.
3. Stress is neither a scalar nor a vector quantity. It is a tensor.
4. To qualify as a vector, a physical quantity must not only possess magnitude and direction but must also satisfy the parallelogram law of vector addition. The finite rotation of a rigid body about a given axis is not a vector quantity because the two finite rotations of the body do not add up in accordance with the law of vector addition. However, if the rotation be small or infinitesimal, it may be regarded as a vector quantity. Same is the case with current, which is a scalar quantity because two currents are not added by vector method.
5. Area can behave either as a scalar or a vector and how it behaves depends on circumstances.
6. Vectors associated with a linear or directional effect are called polar vectors or simply as vectors and those associated with rotation about an axis are referred to as axial vectors. Thus, force, linear velocity and linear acceleration are polar vectors and angular velocity, angular acceleration are axial vectors.

MODULE 1 Exercise

True False

1. Angle between two vectors is 270°. Is this statement true or false?

Single Correct Option

2. Which one of the following is a scalar quantity?
   (a) Dipole moment (b) Electric field
   (c) Acceleration (d) Work
3. Which one of the following is not a vector quantity?
   (a) Torque (b) Displacement
   (c) Velocity (d) Speed
4. Which one is a vector quantity?
   (a) Time
   (b) Temperature
   (c) Magnetic flux
   (d) Magnetic field intensity
5. Which one of the following statement is **false**?
   (a) A vector cannot be displaced from one point to another point
   (b) Distance is a scalar quantity but displacement is a vector quantity
   (c) Momentum, force and torque are vector quantities
   (d) Mass, speed and energy are scalar quantities
6. A vector is not changed, if
   (a) it is rotated through an angle
   (b) it is multiplied by a scalar
   (c) it is divided by a scalar
   (d) it is displaced parallel to itself
7. The forces, which meet at one point but their lines of action do not lie in one plane, are called
   (a) non-coplanar, non-concurrent forces
   (b) non-coplanar, concurrent forces
   (c) coplanar, concurrent forces
   (d) coplanar, non-concurrent forces

Subjective Questions

8. What is the angle between 2**a** and 4**a**?
9. What is the angle between 3**a** and –5**a**? What is the ratio of magnitudes of two vectors?
10. Find the angle between **A** and **B** in following figures.

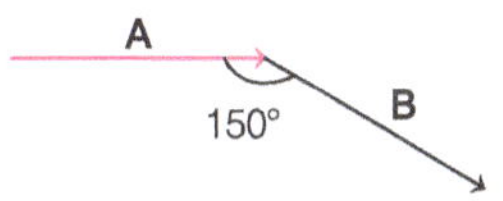

(i)

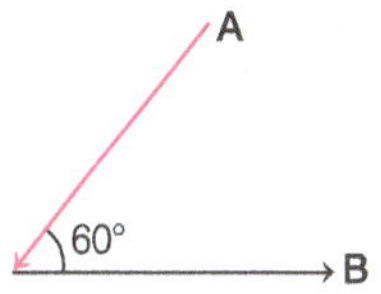

(ii)

Answers

1. (False) **2.** (d) **3.** (d) **4.** (d) **5.** (a) **6.** (d) **7.** (b) **8.** (0°) **9.** (180°, 0.6)
10. (i) 30° (ii) 120°

MODULE 2

Addition and Subtraction of Vectors

Addition of Two Vectors

There are three mothods of vector addition:

(i) The parallelogram law

(ii) The triangle law

(iii) The polygon law

The Parallelogram Law of Vector Addition

Given We are given two vectors **A** and **B**.

Known **A** and **B** are given. It means five quantities are known.

(i) Magnitude of **A** or |**A**| or A

(ii) Magnitude of **B** or |**B**| or B

(iii) Direction of **A**

(iv) Direction of **B**

(v) From directions of **A** and **B** we can find angle between **A** and **B**. Or, the angle θ.

We have to find Resultant of **A** and **B** or **A** + **B** or **R** (= **A** + **B**).

Since we have to find a vector. So, we have to find its magnitude |**R**| or R and its direction.

Method Since we can displace a vector from one point to another point without changing its direction and magnitude. So, draw both vectors from one point so that their arrows are outwards from that point as shown in figure.

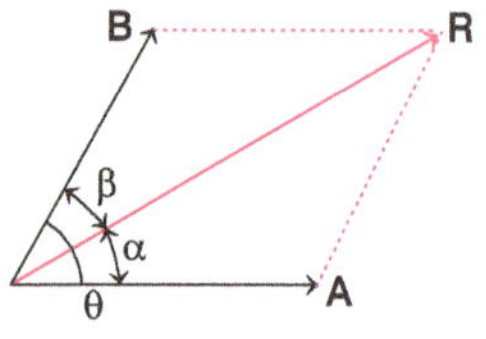

Let **R** be the resultant of two vectors **A** and **B**. According to parallelogram law of vector addition, the resultant **R** is the diagonal of the parallelogram of which **A** and **B** are the adjacent sides as shown in above figure.

Magnitude Magnitude of **R** is given by

$$R = \sqrt{A^2 + B^2 + 2AB\cos\theta} \quad \text{...(i)}$$

Here, θ = angle between **A** and **B**.

Direction Directions of **A** and **B** are known. So, direction of **R** can be given with respect to these known directions.

The direction of **R** can be found by angle α or β of **R** with **A** or **B**.

Here, $\boxed{\tan\alpha = \dfrac{B\sin\theta}{A + B\cos\theta}}$ and $\boxed{\tan\beta = \dfrac{A\sin\theta}{B + A\cos\theta}}$ ...(ii)

Special cases

Case 1. When the given vectors act in the same direction ($\theta = 0°$).

$$R = \sqrt{A^2 + B^2 + 2AB\cos 0°} = \sqrt{A^2 + B^2 + 2AB}$$

$$= \sqrt{(A+B)^2} = A + B = R_{max} \qquad [\because \cos 0° = 1]$$

R is in the direction of **A** or **B**.

Case 2. When the given vectors act in opposite directions ($\theta = 180°$).

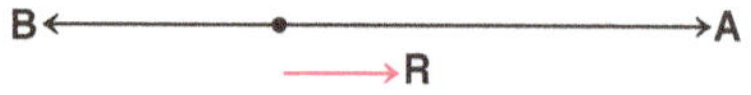

$$R = \sqrt{A^2 + B^2 + 2AB\cos 180°} = \sqrt{A^2 + B^2 - 2AB}$$

$$= \sqrt{(A-B)^2} \qquad [\because \cos 180° = -1]$$

$$= \pm (A - B) = A \sim B = R_{min}$$

R is in the direction of **A** as $A > B$.

Case 3. When the given vectors **A** and **B** act at right angle to each other ($\theta = 90°$).

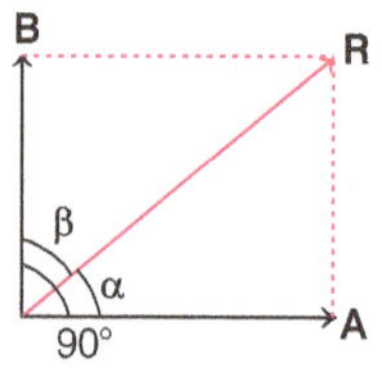

$$\tan\alpha = \frac{B}{A} \quad \text{or} \quad \tan\beta = \frac{A}{B}$$

$$R = \sqrt{A^2 + B^2 + 2AB\cos 90°}$$

$$= \sqrt{A^2 + B^2} \qquad [\because \cos 90° = 0]$$

In all other cases magnitude and direction of **R** can be calculated by using Eqs. (i) and (ii).

Proof of Parallelogram Law

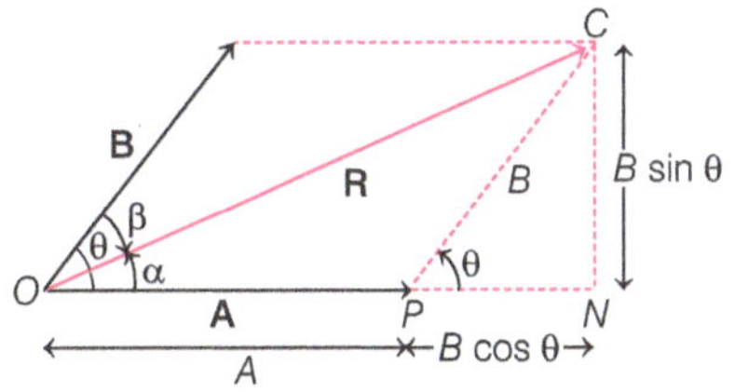

Magnitude

Since, $R^2 = ON^2 + CN^2 = (OP + PN)^2 + CN^2$

$$= (A + B\cos\theta)^2 + (B\sin\theta)^2$$

$$= A^2 + 2AB\cos\theta + B^2\cos^2\theta + B^2\sin^2\theta$$

$$= A^2 + 2AB\cos\theta + B^2(\cos^2\theta + \sin^2\theta)$$

$$= A^2 + B^2 + 2AB\cos\theta$$

$\therefore \quad R = |\mathbf{R}| = |\mathbf{A} + \mathbf{B}| = \sqrt{A^2 + B^2 + 2AB\cos\theta}$

Direction From triangle OCN we can see that,

$$\tan\alpha = \frac{CN}{ON} = \frac{CN}{OP + PN} = \frac{B\sin\theta}{A + B\cos\theta}$$

Similarly, we can find $\tan\beta$.

The Triangle Law of Vector Addition

In the triangle law of vector addition, when the magnitude and direction of two vectors are represented by the two sides of a triangle such that the tail of the second vector is at the head of first vector then the third side will represent the resultant of vector and direction will be from tail of first vector to head of second vector.

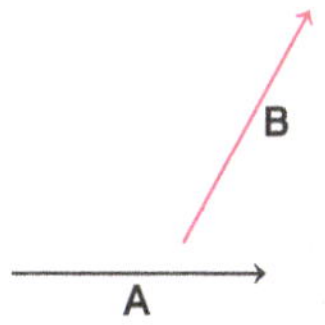

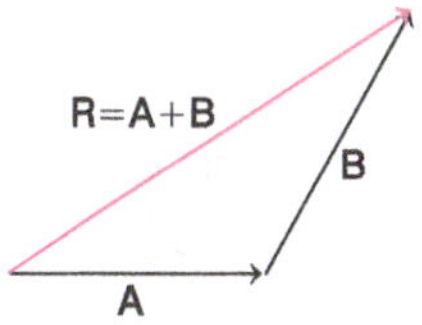

The Polygon Law of Vector Addition

Polygon law of vector addition states that if a number of vectors can be represented in magnitude and direction by the sides of a polygon taken in the same order, then their resultant is represented in magnitude and direction by the closing side of the polygon taken in the opposite order.

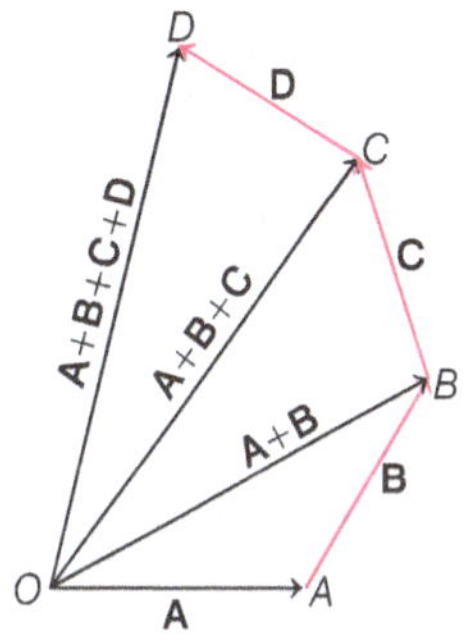

Important POINTS

1. $\mathbf{A} + \mathbf{B} = \mathbf{B} + \mathbf{A}$ (Commutative law)
2. $\mathbf{A} + (\mathbf{B} + \mathbf{C}) = (\mathbf{A} + \mathbf{B}) + \mathbf{C}$ (Associative law)
3. Parallelogram law of vector addition and triangle law of vector addition can be applied for two vectors, whereas polygon law of vector can be applied for any number of vectors.
4. In parallelogram law of vector addition we have direct formula for magnitude and direction of resultant. However, polygon law of vector addition is a graphical method. So, we don't have direct formula.
5. In polygon law, if initial and final points coincide. Or, it becomes a closed polygon then the resultant will be zero. Or, it is a null vector.
6. In polygon law of vector addition we can take any order. Like,

$$\mathbf{A} + \mathbf{B} + \mathbf{C} + \mathbf{D} = \mathbf{C} + \mathbf{A} + \mathbf{D} + \mathbf{B}$$

7. In polygon law, the resultant can intersect any of the vector. Like,

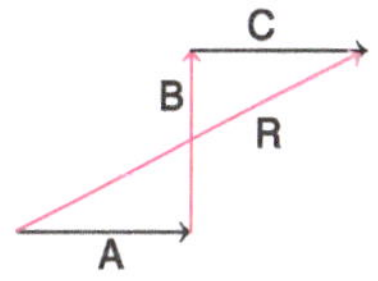

$$\mathbf{R} = \mathbf{A} + \mathbf{B} + \mathbf{C}$$

8. If $|\mathbf{A}| > |\mathbf{B}|$, then $\mathbf{R}$ is more inclined towards $\mathbf{A}$. Or, $\alpha < \beta$, as shown below.

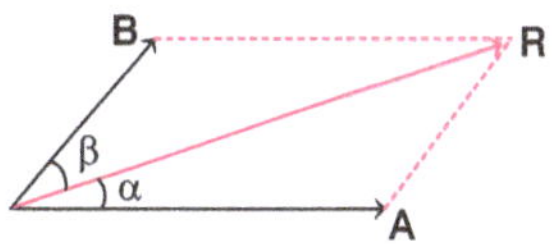

9. If $|\mathbf{A}| = |\mathbf{B}| = A$ (say), then resultant will pass through centre. Or,

$$\alpha = \beta = \frac{\theta}{2} \quad \text{and} \quad \boxed{R = 2A\cos\frac{\theta}{2}}$$

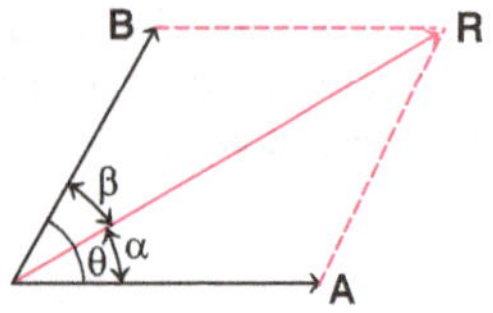

Proof : Substituting $B = A$ in the expression,

$$R = \sqrt{A^2 + B^2 + 2AB\cos\theta}$$

We have,

$$R = \sqrt{A^2 + A^2 + 2A(A)\cos\theta}$$
$$= \sqrt{2A^2 + 2A^2\cos\theta}$$
$$= \sqrt{2A^2(1 + \cos\theta)}$$
$$= \sqrt{2A^2\left[1 + \left(2\cos^2\frac{\theta}{2} - 1\right)\right]}$$
$$= \sqrt{4A^2\cos^2\frac{\theta}{2}} = 2A\cos\frac{\theta}{2}$$

Note For $\theta = 0°$, $R = 2A$
$\theta = 60°$, $R = \sqrt{3}A$
$\theta = 90°$, $R = \sqrt{2}A$
$\theta = 120°$, $R = A$
and $\theta = 180°$, $R = 0$

10. If n-vectors have equal magnitudes and angle between two successive vectors is $\frac{360°}{n}$, then value of their resultant is zero.

For example, if $n = 4$, then $\frac{360°}{4} = 90°$. The result can be proved by following two methods.

Method 1 :

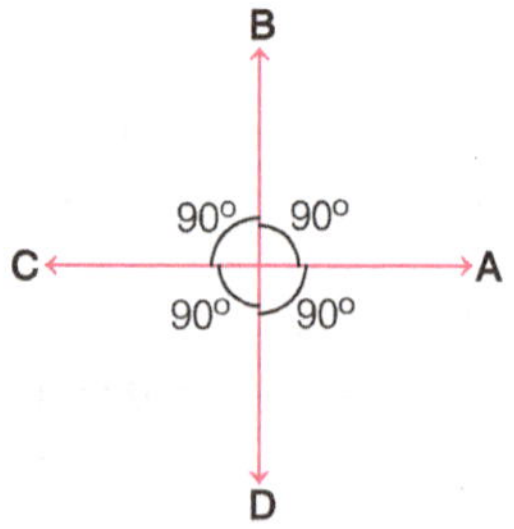

$$|\mathbf{A}| = |\mathbf{B}| = |\mathbf{C}| = |\mathbf{D}|$$

In the above figure **A** cancels **C** and **B** cancels **D**. Therefore resultant will be zero.

Note: *In this method we will get $\frac{n}{2}$ pairs of equal and opposite vectors. Therefore resultant is zero.*

Method 2 : If we add the four vectors according to polygon law of vector addition, then we will get a regular closed polygon of n-sides (a square) as shown below.

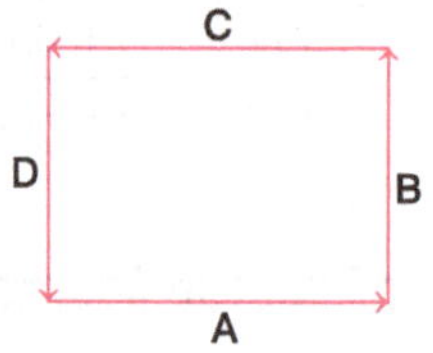

Initial and final points are coinciding. So, the resultant is zero.

Note *If n is even then we can apply both methods but if n is odd then we can prove the above result only by method-2.*

Exercise: Prove the above result for odd number $n = 3$.

Sol. Here, $\frac{360°}{n} = \frac{360°}{3} = 120°$

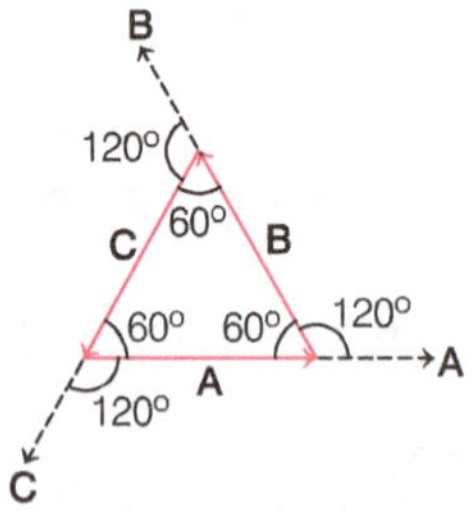

In the above figure we can see that, angle between **A** and **B** or angle between **B** and **C** or angle between **A** and **C** is 120°.

When we add **A**, **B** and **C** as per polygon law we get closed equilateral triangle ($A = B = C$). Initial and final points are coinciding. Hence, the resultant is zero.

Example 1 Two vectors have magnitudes 10 units and 4 units. Find maximum and minimum value of their resultant.

Sol. Given, $A = 10$ and $B = 4$

$$R_{max} = A + B \quad \text{(when } \theta = 0°)$$
$$= 10 + 4 = 14 \text{ units}$$
$$R_{min} = A \sim B \quad \text{(when } \theta = 180°)$$
$$= 10 - 4 = 6 \text{ units}$$

$\therefore$ Magnitude of **R** or $|\mathbf{R}|$ or R always lies between 14 and 6.

Example 2 Two vectors have equal magnitudes $A = 4$ units. Find the magnitude of their resultant if angle between these two vectors θ is :

(a) 0° (b) 60°
(c) 90° (d) 120°
(e) 180°

Sol. When magnitude of two vectors is same (= A), then magnitude of their resultant is given by,

$$R = 2A\cos\frac{\theta}{2}$$

(a) $\theta = 0°, A = 4$ units

$\therefore \quad R = 2 \times 4 \times \cos 0° = 8$ units

(b) $\theta = 60°$, $\therefore$ $\dfrac{\theta}{2} = 30°$

$\therefore$ $R = 2 \times 4 \times \cos 30°$

$= 2 \times 4 \times \dfrac{\sqrt{3}}{2}$

$= 4\sqrt{3}$ units

(c) $\theta = 90°$, $\therefore$ $\dfrac{\theta}{2} = 45°$

$\therefore$ $R = 2 \times 4 \times \cos 45°$

$= 2 \times 4 \times \dfrac{1}{\sqrt{2}} = 4\sqrt{2}$ units

(d) $\theta = 120°$, $\therefore$ $\dfrac{\theta}{2} = 60°$

$\therefore$ $R = 2 \times 4 \times \cos 60°$

$= 2 \times 4 \times \dfrac{1}{2} = 4$ units

(e) $\theta = 180°$, $\therefore$ $\dfrac{\theta}{2} = 90°$

$\therefore$ $R = 2 \times 4 \times \cos 90°$

$= 0$ (as $\cos 90° = 0$)

Example 3 Given, $|\mathbf{A}|$ or $A = 3$ units, $|\mathbf{B}|$ or $B = 4$ units. Find magnitude and direction of $\mathbf{R}$ $(= \mathbf{A} + \mathbf{B})$, if angle θ between $\mathbf{A}$ and $\mathbf{B}$ is,

(a) $\theta = 0°$ (b) $\theta = 90°$

(c) $\theta = 60°$ (d) $\theta = 180°$

Sol. (a) At $\theta = 0°$,

$R = R_{max} = A + B$

$= 4 + 3 = 7$ units

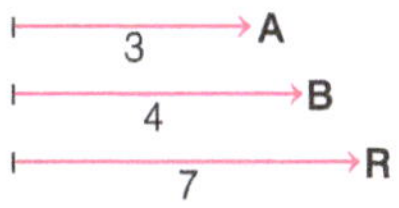

Direction of **R** is along **A** or **B**.

(b) At $\theta = 90°$

$R = \sqrt{A^2 + B^2}$

$= \sqrt{3^2 + 4^2} = \sqrt{9 + 16} = \sqrt{25}$

$= 5$ units

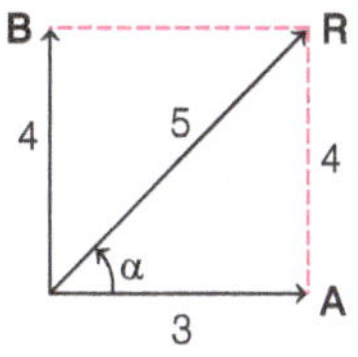

$\tan \alpha = \dfrac{4}{3} \Rightarrow \alpha = 53°$

So, **R** is at 53° from **A** towards **B**.

(c) $\theta = 60°$

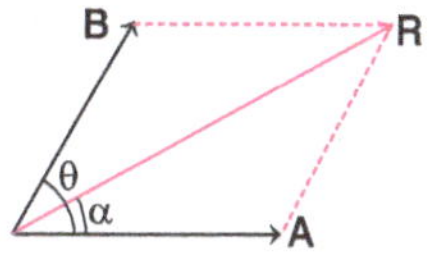

Magnitude $R = \sqrt{A^2 + B^2 + 2AB\cos\theta}$

$= \sqrt{3^2 + 4^2 + 2(3)(4)\cos 60°}$

$= \sqrt{9 + 16 + 24 \times \dfrac{1}{2}} = \sqrt{37}$ units

Direction $\tan \alpha = \dfrac{B \sin \theta}{A + B\cos\theta}$

$= \dfrac{4 \sin 60°}{3 + 4\cos 60°}$

$= \dfrac{4 \times \dfrac{\sqrt{3}}{2}}{3 + 4 \times \dfrac{1}{2}} = \dfrac{2\sqrt{3}}{5}$

$\therefore$ $\alpha = \tan^{-1}\left(\dfrac{2\sqrt{3}}{5}\right)$

So, **R** is at an angle α from **A** towards **B**.

(d) At $\theta = 180°$

$R = A \sim B$

Here, $B > A$

So, $R = B - A$

$= 4 - 3 = 1$ unit

Direction

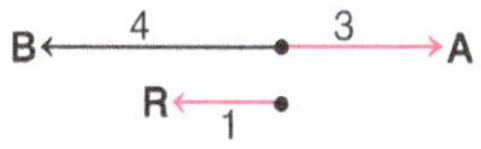

Direction of **R** is towards **B** or towards $-$**A**.

Example 4 Displacement is a vector quantity represented by **s**. So, two or moe than two displacements can be added by polygon law of vector addition.

Anoop first moves 3m towards east then 4m towards north and finally 1m towards west. Find total displacement of Anoop from the starting point.

Sol.

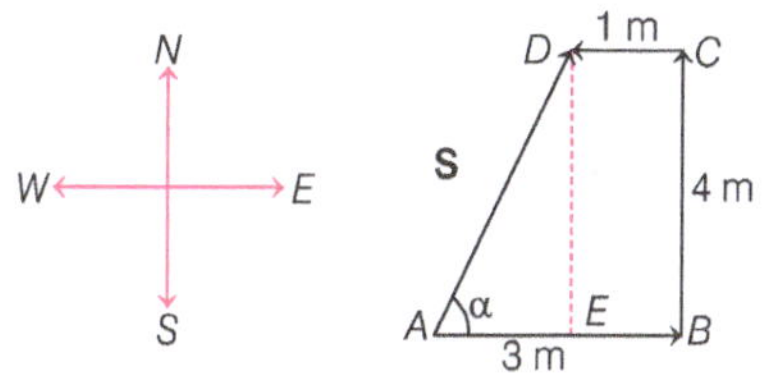

Using polygon law of vector addition, we can see that **s** is along AD.

$|\mathbf{s}|$ or $s = AD = \sqrt{AE^2 + ED^2}$

$= \sqrt{(AB - EB)^2 + BC^2}$

$$= \sqrt{(AB - DC)^2 + BC^2} = \sqrt{(3-1)^2 + 4^2}$$

$$= \sqrt{2^2 + 4^2} = \sqrt{4 + 16} = \sqrt{20} \text{ m}$$

$$\tan \alpha = \frac{DE}{AE} = \frac{4}{2} = 2$$

$$\therefore \quad \alpha = \tan^{-1}(2)$$

So, net displacement is $\sqrt{20}$ m from east towards north at an angle α.

INTRODUCTORY Exercise 1

1. Two vectors have magnitudes 6 units and 8 units, respectively. Find magnitude of resultant of two vectors if angle between two vectors is
   (a) 0° (b) 180° (c) 60° (d) 120°
   (e) 90°
2. Find the resultant of the three vectors shown in figure.

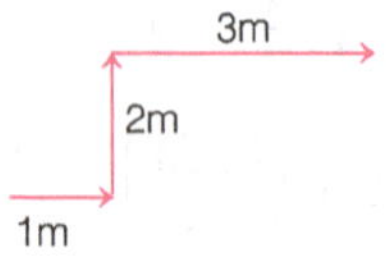

3. Let **A** and **B** be the two vectors of magnitude 10 unit each. If they are inclined to the x-axis at angles 30° and 60° towards y-axis, find the resultant.
4. Four forces of magnitude $P, 2P, 3P$ and $4P$ act along the four sides of a square $ABCD$ in cyclic order. Find the magnitude of resultant force.
5. Five vectors have equal magnitudes of 4 unit each angle between two sucessive vectors is 72°.
   (a) Find resultant of these 5 vectors.
   (b) If one vector is removed, then find resultant of remaining four vectors.

Answers

1. (a) 14 units (b) 2 units (c) $2\sqrt{37}$ units (d) $2\sqrt{13}$ units (e) 10 units
2. $\sqrt{20}$ m
3. $20 \cos 15°$ unit at 45° with x-axis
4. $2\sqrt{2}P$
5. (a) 0° (b) 4 units

See hints on page 150.

Subtraction of Two Vectors

Given we are given two vectors **A** and **B**.

Known **A** and **B** are given. It means five quantities are known.

(i) Magnitude of **A** or $|\mathbf{A}|$ or A
(ii) Magnitude of **B** or $|\mathbf{B}|$ or B
(iii) Direction of **A**
(iv) Direction of **B**
(v) From directions of **A** and **B** we can find angle between **A** and **B**. Or, the angle θ.

We have to find Subtraction of two vectors $\mathbf{S} = \mathbf{A} - \mathbf{B}$.
Since we have to find a vector. So, we have to find its magnitude $|\mathbf{S}|$ or S and its direction.

Method

(i) $|\mathbf{B}| = |-\mathbf{B}| = B$ (say)
(ii) Angle between **A** and **B** is θ. So, angle between **A** and $-$**B** is $(180° - \theta)$.

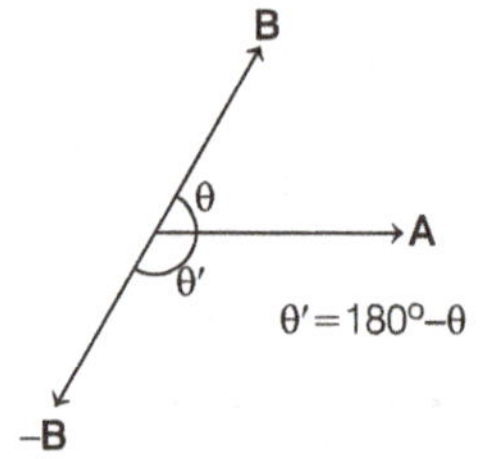

(iii) $\cos \theta' = \cos(180° - \theta) = -\cos \theta$ and
$\sin \theta' = \sin(180° - \theta) = \sin \theta$
(iv) $\mathbf{S} = \mathbf{A} - \mathbf{B} = \mathbf{A} + (-\mathbf{B})$.
So, $\mathbf{A} - \mathbf{B}$ is nothing but addition of **A** and $-$**B**. Addition we have already done. Therefore **S** can be obtained as below :

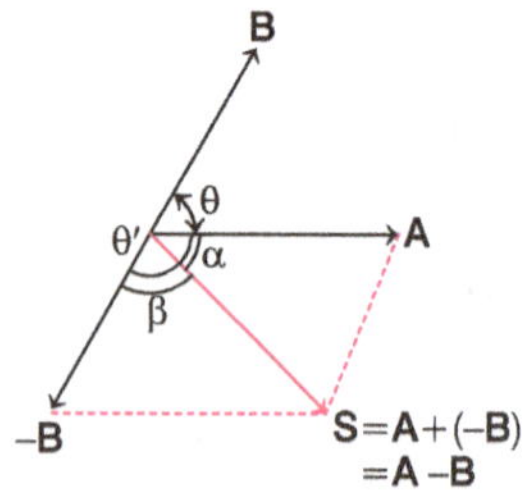

Magnitude $S = \sqrt{|\mathbf{A}|^2 + |-\mathbf{B}|^2 + 2|\mathbf{A}||-\mathbf{B}|\cos\theta'}$

or $$S = \sqrt{A^2 + B^2 - 2AB\cos\theta} \quad \text{...(i)}$$

Direction Directions of **A** and **B** (or $-$**B**) are known. So, we can give direction of **S** from **A** or $-$**B** by using the following formula.

$$\tan \alpha = \frac{B \sin \theta'}{A + B\cos \theta'}$$

or $$\tan \alpha = \frac{B \sin \theta}{A - B\cos \theta} \quad \text{...(ii)}$$

and $$\tan \beta = \frac{A \sin \theta'}{B + A\cos \theta'}$$

or $$\tan \beta = \frac{A \sin \theta}{B - A\cos \theta} \quad \text{...(iii)}$$

Note *In the above expressions θ is the angle between **A** and **B**. It is not the angle between **A** and $-$**B**.*

Important POINTS

1. **A** – **B** or **B** – **A** can also be found by making triangles as shown in figures (a) and (b).

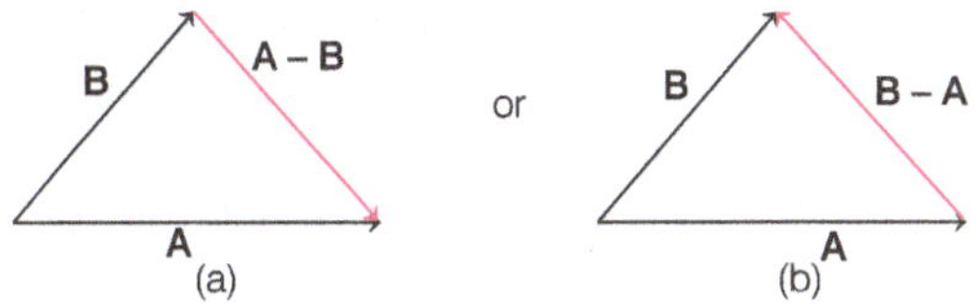

(a) (b)

2. In the figure shown,

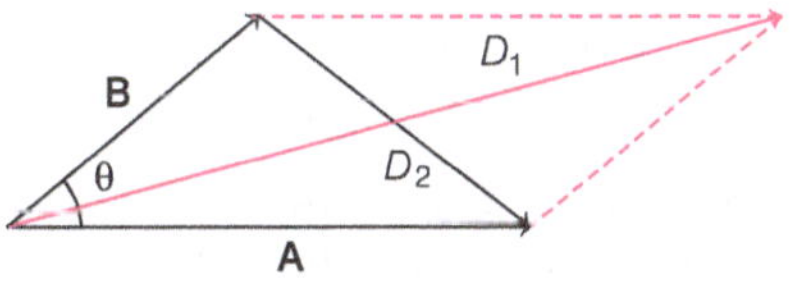

diagonal $D_1 = |\mathbf{A} + \mathbf{B}$ or $\mathbf{R}| = \sqrt{A^2 + B^2 + 2AB\cos\theta}$

diagonal $D_2 = |\mathbf{A} - \mathbf{B}$ or $\mathbf{S}| = \sqrt{A^2 + B^2 - 2AB\cos\theta}$

$$D_1 = D_2 = \sqrt{A^2 + B^2} \text{ if } \theta = 90^\circ$$

Otherwise, $D_1 \neq D_2$

3. If **A** and **B** (or – **B** also) have same magnitudes (say A), then **S** will make equal angles from **A** and – **B**.

or $$\alpha = \beta = \frac{\theta'}{2}$$

and using equation (i) we can find magnitude of **S**.

$\therefore$ $$S = \sqrt{A^2 + B^2 - 2\,AB\cos\theta}$$

$$= \sqrt{A^2 + A^2 - 2AA\cos\theta}$$

(as $B = A$)

$$= \sqrt{2A^2(1 - \cos\theta)}$$

$$= \sqrt{2A^2\left[1 - \left(1 - 2\sin^2\frac{\theta}{2}\right)\right]}$$

$$= \sqrt{4A^2\sin^2\frac{\theta}{2}}$$

$$= 2A\sin\frac{\theta}{2}$$

Thus, $$\boxed{S = 2A\sin\frac{\theta}{2}}$$

For
$\theta = 0^\circ,\ S = 0$
$\theta = 60^\circ,\ S = A$
$\theta = 90^\circ,\ S = \sqrt{2}\,A$
$\theta = 120^\circ, S = \sqrt{3}\,A$
$\theta = 180^\circ, S = 2A$

4. Thus, if $|\mathbf{A}| = |\mathbf{B}| = |-\mathbf{B}| = A$ (say)

then,

$$R = |\mathbf{R}| = |\mathbf{A} + \mathbf{B}| = 2A\cos\frac{\theta}{2} \text{ and}$$

$$S = |\mathbf{S}| = |\mathbf{A} - \mathbf{B}| = |\mathbf{A} + (-\mathbf{B})| = 2A\sin\frac{\theta}{2}$$

θ	$R = 2A\cos\frac{\theta}{2}$	$S = 2A\sin\frac{\theta}{2}$
0°	2A	0
60°	$\sqrt{3}\,A$	A
90°	$\sqrt{2}\,A$	$\sqrt{2}\,A$
120°	A	$\sqrt{3}\,A$
180°	0	2A

Note *As θ increases from 0° to 180°, value of R decreases from 2A to 0 and S increases from 0 to 2A.*

Example 5 Given $|\mathbf{A}|$ or $A = 3$ units, $|\mathbf{B}|$ or $B = 4$ units. Find magnitude and direction of **S** or **A** – **B** or **A** + (–**B**), if angle θ between **A** and **B** is

(a) 0° (b) 60° (c) 90° (d) 180°

Sol. (a) $\theta = 0^\circ$:

Magnitude

$$S = \sqrt{A^2 + B^2 - 2AB\cos\theta} \quad \text{...(i)}$$

$$= \sqrt{3^2 + 4^2 - 2(3)(4)\cos 0^\circ}$$

$$= 1 \text{ unit}$$

Direction

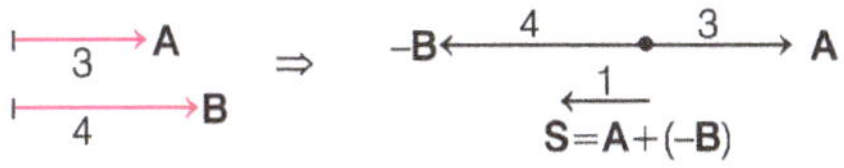

Thus **S** is in the direction of – **B** or – **A**.

(b) $\theta = 60^\circ$:

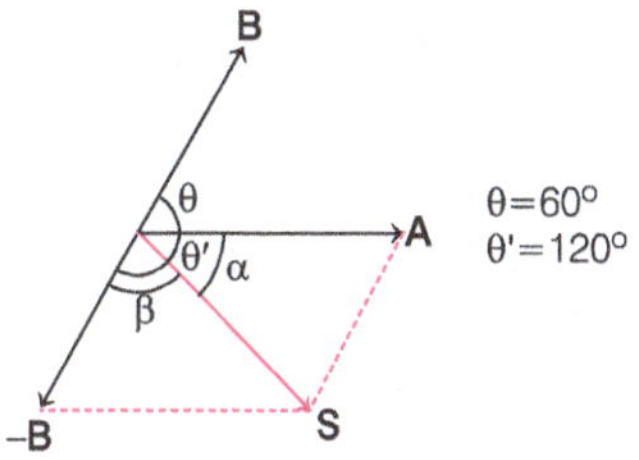

Magnitude: Using equation (i),

$$S = \sqrt{3^2 + 4^2 - 2(3)(4)\cos 60^\circ}$$

$$= \sqrt{9 + 16 - 12}$$

$$= \sqrt{13} \text{ units}$$

Direction: $\tan \alpha = \dfrac{B \sin \theta}{A - B \cos \theta} = \dfrac{4 \sin 60°}{3 - 4 \cos 60°}$

$$= \frac{4 \times \frac{\sqrt{3}}{2}}{3 - 4 \times \frac{1}{2}} = 2\sqrt{3}$$

$\therefore \qquad \alpha = \tan^{-1} (2\sqrt{3})$

So, **S** has magnitude $\sqrt{13}$ units and it is making an angle α from **A** towards $-$**B**.

(c) $\theta = 90°$:

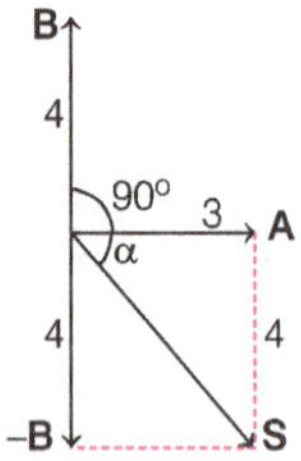

Magnitude : From Pythagoras theorem or equation (i) for $\theta = 90°$.

$$S = \sqrt{A^2 + B^2} \qquad (\because \cos 90° = 0)$$

$$= \sqrt{3^2 + 4^2} = 5 \text{ units}$$

Direction : $\qquad \tan \alpha = \dfrac{4}{3}$

$\therefore \qquad \alpha = 53°$

So, **S** has magnitude of 5 units and it is making an angle of 53° from **A** towards $-$**B**.

(d) $\theta = 180°$:

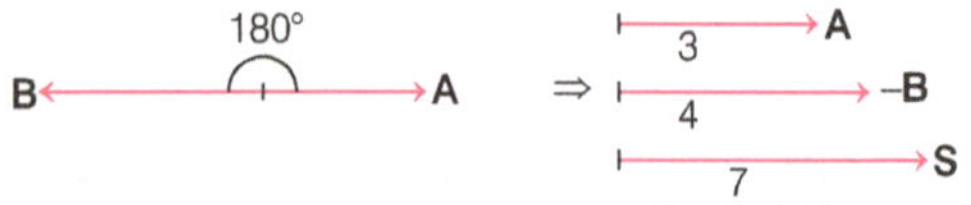

S=A+(–B)

Magnitude : $\qquad S = A + B = 7$ units

Note *Magnitude can also be obtained by putting $\theta = 180°$ in equation (i).*

Direction : From the figure we can see that direction of **S** is towards **A** or $-$**B**.

Example 6 Given $|\mathbf{A}|$ or $A = |\mathbf{B}|$ or $B = 4$ units. Find magnitude of **S** (or **A** $-$ **B**) when angle between **A** and **B** is :

(a) 0° (b) 60° (c) 90° (d) 120°
(e) 180°

Sol. Here magnitudes of **A** and **B** are same. So, we can use the direct result,

$$S = 2A \sin \frac{\theta}{2}$$

(a) For $\theta = 0°$, $\qquad S = 0$

(b) For $\theta = 60°$, $\qquad S = A = 4$ units

(c) For $\theta = 90°$, $\qquad S = \sqrt{2} A = 4\sqrt{2}$ units

(d) For $\theta = 120°$, $\qquad S = \sqrt{3} A = 4\sqrt{3}$ units

(e) For $\theta = 180°$, $\qquad S = 2A = 8$ units

Example 7 A particle is moving in a circle with constant speed of 4 m/s. Find magnitude of change in velocity when particle is rotated by 60°.

Sol. (i) Speed is constant but not the velocity. Because velocity is a vector quantity (having direction also). Since direction is continuously changing, so velocity is continuously changing.

(ii) Change in velocity is represented by $\Delta \mathbf{v}$

$\Delta \mathbf{v}$ = Final velocity $-$ Initial velocity

$$= \mathbf{v}_f - \mathbf{v}_i$$

(iii) $|\mathbf{v}_f| = |\mathbf{v}_i|$ = constant speed

$= 4$ m/s $= v$ (say)

(iv) It is subtraction of two vectors of same magnitude. So, we can use the formula,

$$|\Delta \mathbf{v}| = |\mathbf{v}_f - \mathbf{v}_i| = 2v \sin \frac{\theta}{2}$$

Here, $\theta = 60°$

$\Rightarrow \qquad \dfrac{\theta}{2} = 30°$

$\therefore \qquad |\Delta \mathbf{v}| = 2 \times 4 \times \sin 30°$

$= 4$ m/s **Ans.**

Note

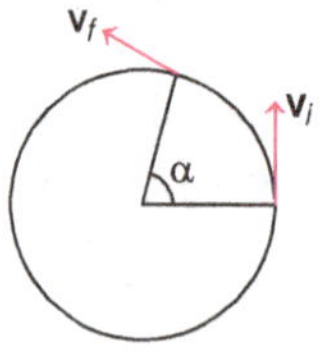

Given $\alpha = 60°$. From geometry we can show that angle between $\mathbf{v}_i$ and $\mathbf{v}_f$ is also $\theta = 60°$.

INTRODUCTORY Exercise 2

1. Two vectors **A** and **B** have magnitudes 6 units and 8 units, respectively. Find $|\mathbf{A} - \mathbf{B}|$, if the angle between two vectors is
   (a) 0° (b) 180°
   (c) 60° (d) 120°
   (e) 90°
2. For what angle between **A** and **B**, $|\mathbf{A} + \mathbf{B}| = |\mathbf{A} - \mathbf{B}|$?
3. $|\mathbf{A} - \mathbf{B}| = |\mathbf{A}| + |\mathbf{B}|$, then what is the angle between **A** and **B**.
4. Given, $|\mathbf{A}| = |\mathbf{B}| = |\mathbf{A} - \mathbf{B}|$, then what is the angle between **A** and **B**.

Answers

1. (a) 2 units (b) 14 units (c) $2\sqrt{13}$ units (d) $2\sqrt{37}$ units
(e) 10 units

2. 90° **3.** 180° **4.** 60°

See hints on page 150.

Miscellaneous Examples

Ex. 1 If **a** and **b** are the vectors **AB** and **BC** determined by the adjacent sides of a regular hexagon. What are the vectors determined by the other sides taken in order ?

Sol. Given **AB** = **a** and **BC** = **b**

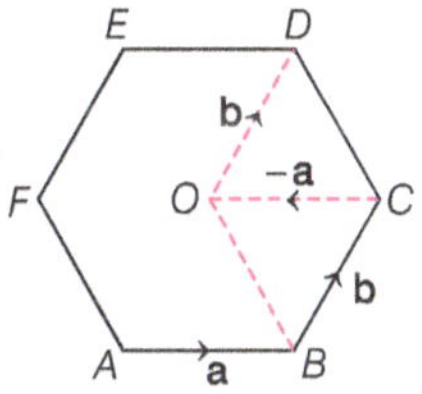

From the method of vector addition (or subtraction) we can show that,

$$\mathbf{CD} = \mathbf{CO} + \mathbf{OD} = -\mathbf{a} + \mathbf{b} = \mathbf{b} - \mathbf{a}$$

Similarly, $\mathbf{DE} = -\mathbf{AB} = -\mathbf{a}$

$\mathbf{EF} = -\mathbf{BC} = -\mathbf{b}$

and $\mathbf{FA} = -\mathbf{CD} = \mathbf{a} - \mathbf{b}$

Ex. 2 The greatest and least resultant of two forces acting at a point is 10 N and 6 N, respectively. If each force is increased by 3 N, find the resultant of new forces when acting at a point at an angle of 90° with each other.

Sol. Let A and B be the magnitudes of two forces.

$$R_{max} = A + B = 10 \quad \text{...(i)}$$

$$R_{min} = A - B = 6 \quad \text{...(ii)}$$

Solving Eqs. (i) and (ii), we get

$$A = 8\text{ N and } B = 2\text{ N}$$

When each force is increased by 3 N then

$$A' = A + 3 = 8 + 3 = 11\text{ N, then}$$

$$B' = B + 3 = 2 + 3 = 5\text{ N}$$

As the new forces are acting at an angle of 90° (i.e. $\theta = 90°$)

So, $R' = \sqrt{A'^2 + B'^2}$

$$= \sqrt{(11)^2 + (5)^2}$$

$$= \sqrt{146}\text{ N}$$

Ex. 3 Two forces whose magnitudes are in the ratio 3 : 5 give a resultant of 28 N. If the angle of their inclination is 60°, find the magnitude of each force.

Sol. Let A and B be the magnitudes of two forces.

Given, $\frac{A}{B} = \frac{3}{5}$, $R = 28$ N and $\theta = 60°$

So, let $A = 3x$ and $B = 5x$

$\because \quad R = \sqrt{A^2 + B^2 + 2AB\cos\theta}$

$\Rightarrow \quad 28 = \sqrt{(3x)^2 + (5x)^2 + 2(3x)(5x)\cos 60°}$

$$= \sqrt{9x^2 + 25x^2 + 15x^2} = 7x$$

$\Rightarrow \quad x = \frac{28}{7} = 4$

Hence, forces are

$$A = 3 \times 4 = 12\text{ N}, B = 5 \times 4 = 20\text{ N}$$

MODULE 2 Exercise

Addition and Subtraction of Vectors

Single Correct Option

1. Minimum number of vectors of unequal magnitudes which can give zero resultant are
(a) two (b) three
(c) four (d) more than four

2. Two vectors of equal magnitudes have a resultant equal to either of them, then the angle between them will be
(a) 30° (b) 120° (c) 60° (d) 150°

3. 12 different coplanar forces (all of equal magnitude) maintain a body in equilibrium, then the angle between any two adjacent forces is
(a) 15° (b) 30° (c) 45° (d) 60°

4. Which of the sets given below may represent the magnitudes of three vectors adding to zero?
(a) 2, 4, 8 (b) 4, 8, 16
(c) 1, 2, 1 (d) 0.5, 1, 2

5. The resultant of **A** and **B** makes an angle α with **A** and β with **B**, then
(a) α is always less than β
(b) $\alpha < \beta$, if $A < B$
(c) $\alpha < \beta$, if $A > B$
(d) $\alpha < \beta$, if $A = B$

6. Let $\mathbf{C} = \mathbf{A} + \mathbf{B}$
(a) $|\mathbf{C}|$ is always greater than $|\mathbf{A}|$
(b) It is possible to have $|\mathbf{C}| < |\mathbf{A}|$ and $|\mathbf{C}| < |\mathbf{B}|$
(c) C is always equal to $A + B$
(d) C is never equal to $A + B$

7. Let the angle between two non-zero vectors **A** and **B** be 120° and its resultant be **C**.
(a) C must be equal to $|A - B|$
(b) C must be less than $|A - B|$
(c) C must be greater than $|A - B|$
(d) C may be equal to $|A - B|$

8. The magnitudes of vectors **A**, **B** and **C** are 3, 4 and 5 units, respectively. If $\mathbf{A} + \mathbf{B} = \mathbf{C}$, the angle between **A** and **B** is
(a) $\frac{\pi}{2}$ (b) $\cos^{-1}(0.6)$
(c) $\tan^{-1}\left(\frac{7}{5}\right)$ (d) $\frac{\pi}{4}$

9. The resultant of **P** and **Q** is perpendicular to **P**. What is the angle between **P** and **Q**?
(a) $\cos^{-1}(P/Q)$ (b) $\cos^{-1}(-P/Q)$
(c) $\sin^{-1}(P/Q)$ (d) $\sin^{-1}(-P/Q)$

10. If the sum of two unit vectors is a unit vector, then magnitude of difference is
(a) $\sqrt{2}$ (b) $\sqrt{3}$
(c) $1/\sqrt{2}$ (d) $\sqrt{5}$

11. Two forces P and Q of magnitudes $2F$ and $3F$, respectively are at an angle θ with each other. If the force Q is doubled, their resultant also gets doubled. Then, the angle θ is **(2019 Main)**
(a) 60° (b) 120°
(c) 30° (d) 90°

12. If **a** and **b** are two units vectors inclined at an angle of 60° to each other, then
(a) $|\mathbf{a}+\mathbf{b}| > 1$ (b) $|\mathbf{a}+\mathbf{b}| < 1$
(c) $|\mathbf{a}-\mathbf{b}| > 1$ (d) $|\mathbf{a}-\mathbf{b}| < 1$

13. If **A** and **B** are two vectors such that $|\mathbf{A} + \mathbf{B}| = 2|\mathbf{A} - \mathbf{B}|$, the angle between vectors **A** and **B** is
(a) 45° (b) 60°
(c) 30° (d) data insufficient

14. Two vectors **A** and **B** have equal magnitudes. If magnitude of **A** + **B** is equal to n times the magnitude of **A** − **B**, then the angle between **A** and **B** is **(2019 Main)**
(a) $\cos^{-1}\left(\frac{n-1}{n+1}\right)$ (b) $\cos^{-1}\left(\frac{n^2-1}{n^2+1}\right)$
(c) $\sin^{-1}\left(\frac{n-1}{n+1}\right)$ (d) $\sin^{-1}\left(\frac{n^2-1}{n^2+1}\right)$

15. A vector **a** is turned without a change in its length through a small angle $\Delta\theta$. The value of $|\Delta \mathbf{a}|$ and Δa are respectively.
(a) $0, a\Delta\theta$ (b) $a\Delta\theta, 0$
(c) $0, 0$ (d) $a\Delta\theta, a\Delta\theta$

16. Two vectors **A** and **B** have equal magnitudes. If magnitude of **A** + **B** is equal to two times the magnitude of **A** − **B**, then the angle between **A** and **B** will be **(2022 Main)**
(a) $\sin^{-1}\left(\frac{3}{5}\right)$ (b) $\sin^{-1}\left(\frac{1}{3}\right)$
(c) $\cos^{-1}\left(\frac{3}{5}\right)$ (d) $\cos^{-1}\left(\frac{1}{3}\right)$

Subjective Question

17. Two vectors have magnitudes 3 unit and 4 unit respectively. What should be the angle between them if the magnitude of the resultant is (a) 1 unit, (b) 5 unit and (c) 7 unit.

Miscellaneous Problems

Single Correct Option

18. Which is correct ?

(a) $|\mathbf{a}-\mathbf{b}|=|\mathbf{a}|\sim|\mathbf{b}|$ (b) $|\mathbf{a}-\mathbf{b}|\le|\mathbf{a}|\sim|\mathbf{b}|$
(c) $|\mathbf{a}-\mathbf{b}|\ge|\mathbf{a}|\sim|\mathbf{b}|$ (d) $|\mathbf{a}-\mathbf{b}|>|\mathbf{a}|\sim|\mathbf{b}|$

19. Given that $\mathbf{P}+\mathbf{Q}+\mathbf{R}=0$. Two out of the three vectors are equal in magnitude. The magnitude of the third vector is $\sqrt{2}$ times that of the other two. Which of the following can be the angles between these vectors?

(a) 90°, 135°, 135° (b) 45°, 45°, 90°
(c) 30°, 60°, 90° (d) 45°, 90°, 135°

20. The resultant of two forces $3P$ and $2P$ is R. If the first force is doubled, then the resultant is also doubled. The angle between the two forces is

(a) 60° (b) 120°
(c) 30° (d) 135°

21. In the vector diagram shown in figure,

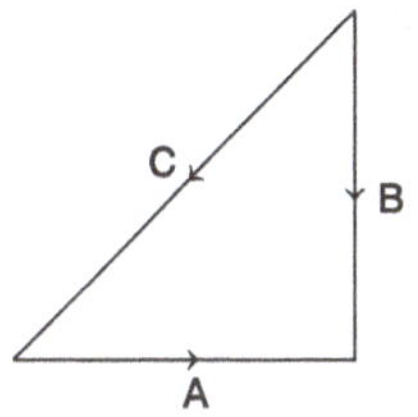

(a) $\mathbf{A}+\mathbf{B}=\mathbf{C}$ (b) $\mathbf{B}+\mathbf{C}=\mathbf{A}$
(c) $\mathbf{C}+\mathbf{A}=\mathbf{B}$ (d) $\mathbf{C}+\mathbf{A}+\mathbf{B}=0$

22. The resultant of $\mathbf{A}$ and $\mathbf{B}$ is $\mathbf{R}_1$. On reversing the vector $\mathbf{B}$, the resultant becomes $\mathbf{R}_2$. What is the value of $R_1^2+R_2^2$?

(a) A^2+B^2 (b) A^2-B^2
(c) $2(A^2+B^2)$ (d) $2(A^2-B^2)$

23. In the diagram shown in figure,

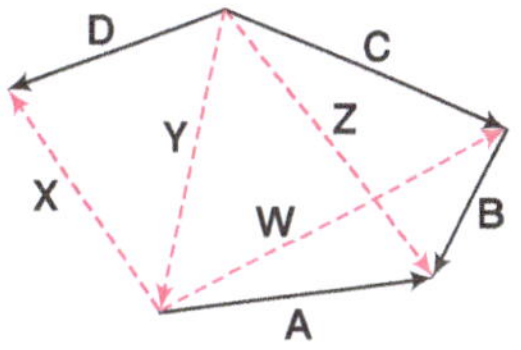

(a) $\mathbf{X}=\mathbf{A}+\mathbf{B}-\mathbf{C}+\mathbf{D}$ (b) $\mathbf{Y}=\mathbf{B}+\mathbf{C}+\mathbf{A}$
(c) $\mathbf{Z}=\mathbf{B}+\mathbf{C}$ (d) $\mathbf{W}=\mathbf{A}+\mathbf{B}$

Multiple Correct Options

24. Two vectors $\mathbf{P}$ and $\mathbf{Q}$ are arranged in such a way that they form adjacent sides of a parallelogram as shown in the figure below.

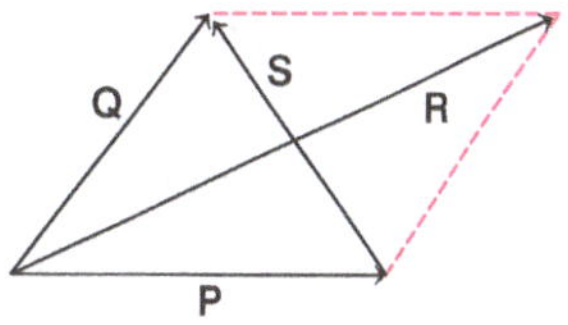

Which of the following options have the correct relationship?

(a) $\mathbf{Q}=\mathbf{R}+\mathbf{S}$ (b) $\mathbf{R}=\mathbf{P}+\mathbf{Q}$ (c) $\mathbf{R}=\mathbf{P}+\mathbf{S}$ (d) $\mathbf{S}=\mathbf{Q}-\mathbf{P}$

Answers

1. (b)	**2.** (b)	**3.** (b)	**4.** (c)	**5.** (c)	**6.** (b)	**7.** (c)	**8.** (a)	**9.** (b)	**10.** (b)
11. (b)	**12.** (a)	**13.** (d)	**14.** (b)	**15.** (b)	**16.** (c)	**17.** (a) 180° (b) 90° (c) 0°			**18.** (c)
19. (a)	**20.** (b)	**21.** (c)	**22.** (c)	**23.** (c)	**24.** (b,d)				

See hints on pages 150-151.

MODULE 3

Multiplication of Vectors

Two Types of Multiplication

There are limited number of physical quantities (like speed, velocity and force etc.). Some of them are scalars and the remaining quantities are vectors.

Now let us focus on vectors. There are limited number of physical quantities which are obtained after multiplying two vector quantities. Let us take three examples.

Example 1 Force and displacement are vector quantities. The product of these two vectors is work and work is a scalar quantity.

Example 2 If we multiply position vector with force vector than we will get torque and torque is a vector quantity.

Example 3 If we multiply force vector with velocity vector, then we will not get any physical quantity.

So, if we multiply two vector quantities then either we will get,

(i) a scalar quantity (example-1), or

(ii) a vector quantity (example-2)

In rest of the cases we will not get any known physical quantity (example-3).

Accordingly, there are two types of multiplications between two vectors. First is dot product or scalar product and the second one is cross product or vector product.

Note *Only four mathematical operations are possible between two vectors. Addition, subtraction, dot product and cross product.*

Dot Product or Scalar Product

Given We are given two vectors **A** and **B**.

Known **A** and **B** are given. It means five quantities are known.

(i) Magnitude of **A** or $|\mathbf{A}|$ or A

(ii) Magnitude of **B** or $|\mathbf{B}|$ or B

(iii) Direction of **A**

(iv) Direction of **B**

(v) From directions of **A** and **B** we can find angle between **A** and **B**. Or, the angle θ.

We have to find Dot product or scalar product of **A** and **B**. This is denoted as $\mathbf{A}\cdot\mathbf{B}$. Since the product is a scalar quantity, so it has only magnitude. This magnitude is $AB\cos\theta$. Thus,

$$\boxed{\mathbf{A}\cdot\mathbf{B} = AB\cos\theta}$$

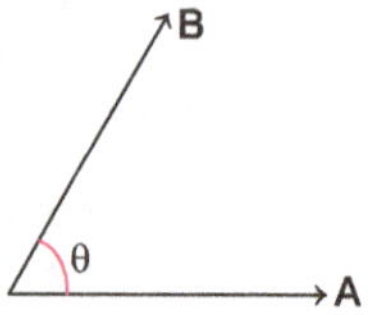

Note *In the above expression A and B are magnitudes of **A** and **B**, which are always positive. So, the sign of dot product depends upon the value of $\cos\theta$. Following three cases are possible.*

(i) If θ is acute ($<90°$), then $\cos\theta$ is positive. So, dot product is positive.

(ii) If θ is obtuse ($>90°$), then $\cos\theta$ is negative. So, dot product is negative.

(iii) If θ is 90°, then $\cos\theta$ is zero. Therefore dot product is zero. Thus, dot product of two perpendicular vectors is always zero.

Special Cases

1. If $\theta = 0°$, $\mathbf{A}\cdot\mathbf{B} = AB$ = maximum value $[\because \cos 0° = 1]$
2. If $\theta = 180°$, $\mathbf{A}\cdot\mathbf{B} = -AB$ = maximum negative value $[\because \cos 180° = -1]$
3. If $\theta = 90°$, $\mathbf{A}\cdot\mathbf{B} = 0$ = minimum value $[\because \cos 90° = 0]$
   So if two vectors are perpendicular, then their dot product is zero.
4. If θ is acute, then $\mathbf{A}\cdot\mathbf{B}$ is positive. $[\because \cos\theta$ is positive when θ is acute$]$
5. If θ is obtuse, then $\mathbf{A}\cdot\mathbf{B}$ is negative. $[\because \cos\theta$ is negative when θ is obtuse$]$

Important Points

(i) $\mathbf{A}\cdot\mathbf{B} = \mathbf{B}\cdot\mathbf{A}$ (ii) $\mathbf{A}\cdot(\mathbf{B}+\mathbf{C}) = \mathbf{A}\cdot\mathbf{B} + \mathbf{A}\cdot\mathbf{C}$

(iii) $\mathbf{A}\cdot\mathbf{A} = A^2$

Example 1 Prove that $\mathbf{A}\cdot\mathbf{A} = A^2$

Sol. We have to find dot product of two same vectors **A** and **A**. Angle between them is 0°.

Thus, $\mathbf{A}\cdot\mathbf{A} = (A)(A)\cos 0° = A^2$ **Hence proved.**

Example 2 Given, $|\mathbf{A}|$ or $A = 3$ units, $|\mathbf{B}|$ or $B = 4$ units. Find $\mathbf{A}\cdot\mathbf{B}$, if angle between **A** and **B** is

(a) $\theta = 0°$ (b) $\theta = 60°$ (c) $\theta = 90°$ (d) $\theta = 120°$

(e) $\theta = 180°$

Sol. $\mathbf{A}\cdot\mathbf{B} = AB\cos\theta$

$= (3)(4)\cos\theta = 12\cos\theta$

(a) $\theta = 0°$, $\cos 0° = 1$

$\therefore \quad \mathbf{A} \cdot \mathbf{B} = 12$ units

(b) $\theta = 60°$, $\cos 60° = \frac{1}{2}$

$\therefore \quad \mathbf{A} \cdot \mathbf{B} = 6$ units

(c) $\theta = 90°$, $\cos 90° = 0$

$\therefore \quad \mathbf{A} \cdot \mathbf{B} = 0$

(d) $\theta = 120°$, $\cos 120° = -\cos 60° = -\frac{1}{2}$

$\therefore \quad \mathbf{A} \cdot \mathbf{B} = -6$ units

(e) $\theta = 180°$, $\cos 180° = -1$

$\therefore \quad \mathbf{A} \cdot \mathbf{B} = -12$ units

Note *As θ increases from 0° to 180°, value of* $\mathbf{A} \cdot \mathbf{B}$ *changes from* $+AB$ *to* $-AB$.

Cross Product or Vector Product

Given We are given two vectors **A** and **B**.

Known **A** and **B** are given. It means five quantities are known.

(i) Magnitude of **A** or $|\mathbf{A}|$ or A

(ii) Magnitude of **B** or $|\mathbf{B}|$ or B

(iii) Direction of **A**

(iv) Direction of **B**

(v) From directions of **A** and **B**, we can find angle between **A** and **B**. Or, the angle θ.

We have to find Cross product or vector product of **A** and **B**, denoted by $\mathbf{A} \times \mathbf{B}$. Since, product is a vector quantity, so it has both magnitude as well as direction.

Magnitude Magnitude of $\mathbf{A} \times \mathbf{B}$ or $\mathbf{B} \times \mathbf{A}$ is given by,

$$|\mathbf{A} \times \mathbf{B}| = |\mathbf{B} \times \mathbf{A}| = AB \sin \theta$$

As discussed in dot product, AB is always positive. Further, angle between two vectors θ always lies between 0° and 180°, means first and second quadrant and $\sin \theta$ is positive in both quadrants.

Therefore the magnitude $AB \sin \theta$ is always positive. Otherwise also, magnitude of a vector quantity is always positive except for $\theta = 0°$ or 180°.

For these two angles $\sin \theta$ is zero. Hence, magnitude of cross product is zero. A vector whose magnitude is zero is called a null vector. Thus, cross product of two parallel or antiparallel vectors is a null vector.

Direction Direction of $\mathbf{A} \times \mathbf{B}$ is perpendicular to both **A** and **B**. Now two vectors are basically two lines and from two lines we can make a plane.

Thus $\mathbf{A} \times \mathbf{B}$ is perpendicular to the plane formed by **A** and **B**, so that it becomes perpendicular to both **A** and **B** separately. Now perpendicular to any plane, further there are two directions.

The actual direction can be obtained by following three methods.

Method 1. As shown in the figure when a right-handed screw is rotated from **A** towards **B** (through the smaller angle between them), the screw moves in the direction of $\mathbf{A} \times \mathbf{B}$.

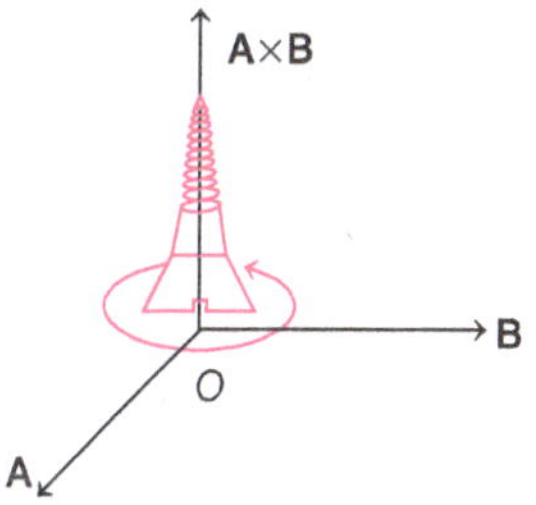

Method 2. With your right-hand, point your index finger along vector **A**, and point your middle finger along vector **B**. The cross product $\mathbf{A} \times \mathbf{B}$ is in the direction of your thumb.

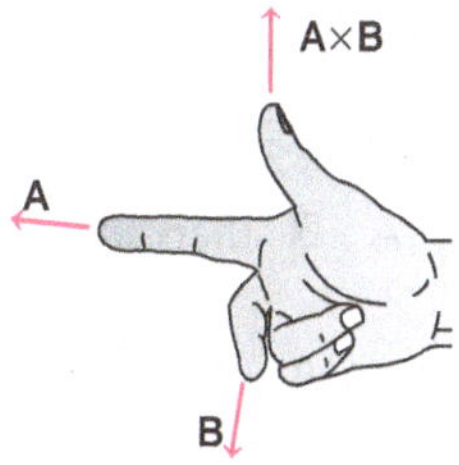

Method 3. If we hold the right-hand out with the fingers pointing in the direction of **A**, then curl the fingers towards vector **B**, the thumb points in the direction of the cross product $\mathbf{A} \times \mathbf{B}$ as shown in figure.

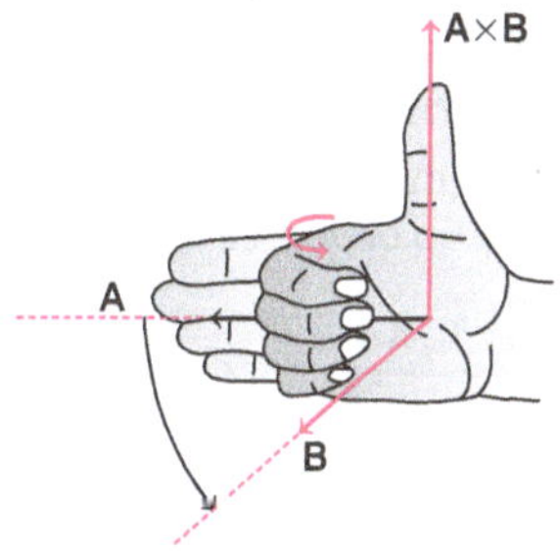

Note *In most of the cases* ***A*** *and* ***B*** *(or any two vectors) are in our plane of paper. So* $\mathbf{A} \times \mathbf{B}$ *or* $\mathbf{B} \times \mathbf{A}$ *are perpendicular to this plane or perpendicular to the plane of paper, either inwards [shown by ⊗] or outwards [shown by ⊙].*

To find whether it is ⊗ or ⊙, draw both vectors from one point so that their arrows are outwards with respect to that point. Now to find the direction of $\mathbf{A} \times \mathbf{B}$ *rotate a screw from* ***A*** *towards* ***B*** *via smaller angle. Clockwise rotation means ⊗ direction and anticlockwise rotation means ⊙ direction. Following figures will illustrate the concept.*

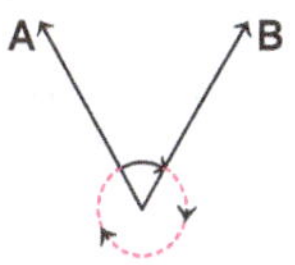

$\mathbf{A} \times \mathbf{B}$ is in ⊗ direction

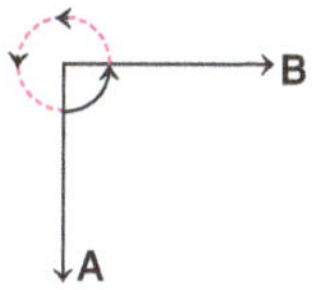

$\mathbf{A} \times \mathbf{B}$ is in ⊙ direction

Important POINTS

(i) $(\mathbf{A} \times \mathbf{B}) = -(\mathbf{B} \times \mathbf{A})$

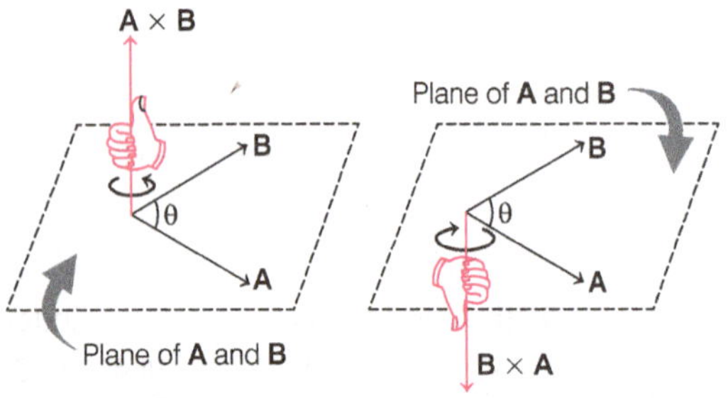

Because $|\mathbf{A} \times \mathbf{B}| = |\mathbf{B} \times \mathbf{A}| = AB \sin\theta$ and directions are opposite.

(ii) $\mathbf{A} \times (\mathbf{B} + \mathbf{C}) = (\mathbf{A} \times \mathbf{B}) + (\mathbf{A} \times \mathbf{C})$

(iii) If $|\mathbf{A} + \mathbf{B}| = |\mathbf{A} - \mathbf{B}|$, then angle between **A** and **B** is 90°.

Proof : Given $|\mathbf{A} + \mathbf{B}| = |\mathbf{A} - \mathbf{B}|$

Squaring both sides, we get $|\mathbf{A} + \mathbf{B}|^2 = |\mathbf{A} - \mathbf{B}|^2$

$$\Rightarrow \quad A^2 + B^2 + 2AB\cos\theta = A^2 + B^2 - 2AB\cos\theta$$

$$\Rightarrow \quad A^2 + B^2 + 2\mathbf{A}\cdot\mathbf{B} = A^2 + B^2 - 2\mathbf{A}\cdot\mathbf{B}$$

$$\Rightarrow \quad 4\mathbf{A}\cdot\mathbf{B} = 0 \Rightarrow \mathbf{A}\cdot\mathbf{B} = 0$$

Hence, **A** is perpendicular to **B**.

(iv) We can find the addition of two vectors using dot product.

If $\mathbf{R} = \mathbf{A} + \mathbf{B}$

Then, $|\mathbf{R}| = |\mathbf{A} + \mathbf{B}|$

or $|\mathbf{R}|^2 = |\mathbf{A} + \mathbf{B}|^2$

or $R^2 = (\mathbf{A} + \mathbf{B})\cdot(\mathbf{A} + \mathbf{B}) = A^2 + B^2 + 2\mathbf{A}\cdot\mathbf{B}$

or $R = \sqrt{A^2 + B^2 + 2\mathbf{A}\cdot\mathbf{B}} = \sqrt{A^2 + B^2 + 2AB\cos\theta}$

Similarly, we can find subtraction of two vectors also.

Miscellaneous Examples

Ex. 1 A is towards north and **B** is towards west. Find the direction of $\mathbf{A} \times \mathbf{B}$.

Sol. $\mathbf{A} \times \mathbf{B}$ is perpendicular to both **A** (means north direction) and **B** (means west direction). So, it should be either vertically up or vertically down.

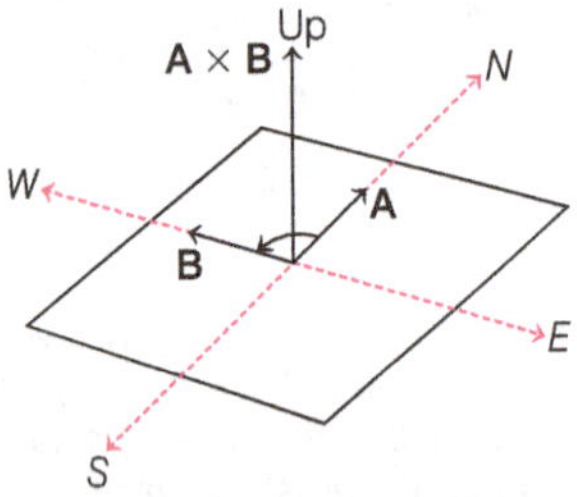

From the figure we can see that if we rotate the screw from **A** towards **B** through smaller angle then it moves in upward direction. So, $\mathbf{A} \times \mathbf{B}$ is vertically upwards.

Ex. 2 **A** is vertically upwards and **B** is towards east. Find the direction of $\mathbf{A} \times \mathbf{B}$.

Sol. From the figure shown below, we can see that if we rotate a right-handed screw from **A** towards **B** (*via* smaller angle), then it advances towards north. So, $\mathbf{A} \times \mathbf{B}$ is towards north.

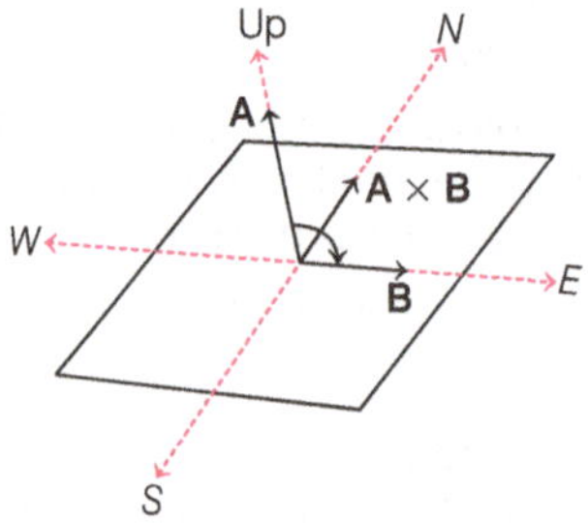

Ex. 3 Given $|\mathbf{A}|$ or $A = 3$ units, $|\mathbf{B}|$ or $B = 4$ units. Find magnitude of $(\mathbf{A} \times \mathbf{B})$ if angle between **A** and **B** is :

(a) $\theta = 0°$ (b) $\theta = 60°$

(c) $\theta = 90°$ (d) $\theta = 120°$

(e) $\theta = 180°$

Sol. $|\mathbf{A} \times \mathbf{B}| = AB\sin\theta$

$$= (3)(4)\sin\theta$$

$$= 12\sin\theta$$

(a) $\theta = 0° \Rightarrow \sin\theta = 0$

$\therefore \quad |\mathbf{A} \times \mathbf{B}| = 0$

(b) $\theta = 60° \Rightarrow \sin\theta = \frac{\sqrt{3}}{2}$

$\therefore \quad |\mathbf{A} \times \mathbf{B}| = 6\sqrt{3}$ units

(c) $\theta = 90°$

$\Rightarrow \quad \sin\theta = 1$

$\therefore \quad |\mathbf{A} \times \mathbf{B}| = 12$ units

(d) $\theta = 120°$

$\Rightarrow \quad \sin 120° = \sin 60° = \dfrac{\sqrt{3}}{2}$

$\therefore \quad |\mathbf{A} \times \mathbf{B}| = 6\sqrt{3}$ units

(e) $\theta = 180°$

$\Rightarrow \quad \sin\theta = 0$

$\therefore \quad |\mathbf{A} \times \mathbf{B}| = 0$

Note *From the above example we can see that as θ increases from 0° to 180°, magnitude of $\mathbf{A} \times \mathbf{B}$ first increases from 0 to 90° and then it decreases from 90° to 180°.*

Ex. 4 If $\mathbf{a} \times \mathbf{b} = \mathbf{b} \times \mathbf{c} \neq 0$ with $\mathbf{a} \neq -\mathbf{c}$ then show that $\mathbf{a} + \mathbf{c} = k\mathbf{b}$, where k is scalar.

Sol. $\mathbf{a} \times \mathbf{b} = \mathbf{b} \times \mathbf{c}$

$\mathbf{a} \times \mathbf{b} = -\mathbf{c} \times \mathbf{b}$

$\therefore \quad \mathbf{a} \times \mathbf{b} + \mathbf{c} \times \mathbf{b} = 0$

$(\mathbf{a} + \mathbf{c}) \times \mathbf{b} = 0$

Hence, $\mathbf{a} + \mathbf{c}$ is parallel or antiparallel to $\mathbf{b}$.

$\therefore \quad \mathbf{a} + \mathbf{c} = k\mathbf{b}$ **Hence proved.**

Ex. 5 Prove that $|\mathbf{a} \times \mathbf{b}|^2 = a^2b^2 - (\mathbf{a} \cdot \mathbf{b})^2$.

Sol. Let $|\mathbf{a}| = a$, $|\mathbf{b}| = b$ and θ be the angle between them.

$$|\mathbf{a} \times \mathbf{b}|^2 = (ab\sin\theta)^2 = a^2b^2\sin^2\theta$$

$$= a^2b^2(1 - \cos^2\theta) = a^2b^2 - (a \cdot b\cos\theta)^2$$

$$= a^2b^2 - (\mathbf{a} \cdot \mathbf{b})^2$$ **Hence proved.**

Ex. 6 Given $|\mathbf{A}|$ or $A = 3$ units and $|\mathbf{B}|$ or $B = 4$ units. Plot a graph between :

(i) $\mathbf{A} \cdot \mathbf{B}$ *versus* θ (ii) $|\mathbf{A} \times \mathbf{B}|$ *versus* θ

Sol. (i) $\mathbf{A} \cdot \mathbf{B} = AB\cos\theta$

$= (3)(4)\cos\theta$

$= 12\cos\theta$

So, $\mathbf{A} \cdot \mathbf{B}$ *versus* θ graph is just like $\cos\theta$ graph with maximum value + 12 and minimum value − 12 with $0° \le \theta \le 180°$

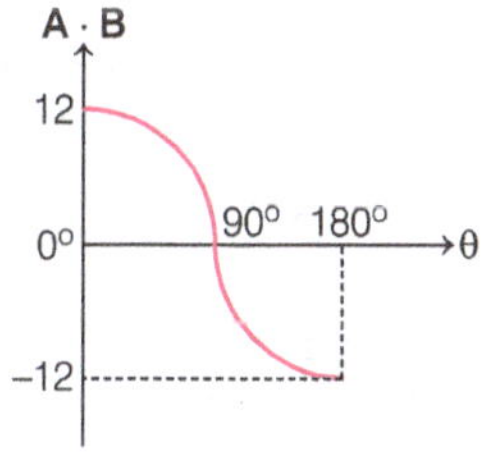

(ii) $|\mathbf{A} \times \mathbf{B}| = AB\sin\theta$

$= 12\sin\theta$

So, it is basically $\sin\theta$ graph as shown below.

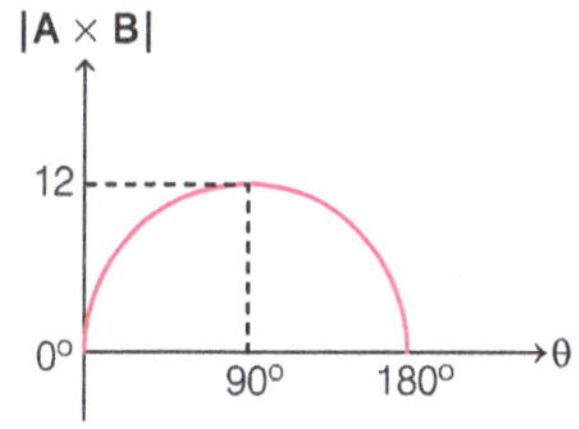

MODULE 3 Exercise

Dot Product

Single Correct Option

1. Three vectors satisfy the relation $\mathbf{A} \cdot \mathbf{B} = 0$ and $\mathbf{A} \cdot \mathbf{C} = 0$, then $\mathbf{A}$ is parallel or antiparallel to
   (a) $\mathbf{C}$ (b) $\mathbf{B}$
   (c) $\mathbf{B} \times \mathbf{C}$ (d) $\mathbf{B} \cdot \mathbf{C}$

2. If $\mathbf{a}_1$ and $\mathbf{a}_2$ are two non-collinear unit vectors and if $|\mathbf{a}_1 + \mathbf{a}_2| = \sqrt{3}$, then the value of $(\mathbf{a}_1 - \mathbf{a}_2) \cdot (2\mathbf{a}_1 + \mathbf{a}_2)$ is
   (a) 2 (b) $\frac{3}{2}$ (c) $\frac{1}{2}$ (d) 1

3. Let $|\mathbf{A}_1| = 3, |\mathbf{A}_2| = 5$ and $|\mathbf{A}_1 + \mathbf{A}_2| = 5$. The value of $(2\mathbf{A}_1 + 3\mathbf{A}_2) \cdot (3\mathbf{A}_1 - 2\mathbf{A}_2)$ is **(2019 Main)**
   (a) -106.5 (b) -112.5
   (c) -99.5 (d) -118.5

Cross Product

Single Correct Option

4. A vector $\mathbf{A}$ points vertically upward and $\mathbf{B}$ points towards north. The vector product $\mathbf{A} \times \mathbf{B}$ is
   (a) along west (b) along east
   (c) zero (d) vertically downward

5. If $\mathbf{a}$ and $\mathbf{b}$ are two vectors, then the value of $(\mathbf{a} + \mathbf{b}) \times (\mathbf{a} - \mathbf{b})$ is
   (a) $2(\mathbf{b} \times \mathbf{a})$ (b) $-2(\mathbf{b} \times \mathbf{a})$
   (c) $\mathbf{b} \times \mathbf{a}$ (d) $\mathbf{a} \times \mathbf{b}$

Multiple Correct Options

6. The magnitude of the vector product of two vectors $|\mathbf{A}|$ and $|\mathbf{B}|$ may be
   (a) greater than AB (b) equal to AB
   (c) less than AB (d) equal to zero

Subjective Question

7. If $\mathbf{A}$, $\mathbf{B}$ and $\mathbf{C}$ are mutually perpendicular, then show that $\mathbf{C} \times (\mathbf{A} \times \mathbf{B}) = 0$.

Miscellaneous Problems

Single Correct Option

8. If $|\mathbf{A} \times \mathbf{B}| = \sqrt{3}\mathbf{A} \cdot \mathbf{B}$, then the value of $|\mathbf{A} + \mathbf{B}|$ is
   (a) $(A^2 + B^2 + AB)^{1/2}$ (b) $\left(A^2 + B^2 + \frac{AB}{\sqrt{3}}\right)^{1/2}$
   (c) $(A + B)$ (d) $(A^2 + B^2 + \sqrt{3}AB)^{1/2}$

9. If the angle between the vectors $\mathbf{A}$ and $\mathbf{B}$ is θ, the value of the product $(\mathbf{B} \times \mathbf{A}) \cdot \mathbf{A}$ is equal to
   (a) $BA^2 \cos\theta$ (b) $BA^2 \sin\theta$
   (c) $BA^2 \sin\theta \cos\theta$ (d) zero

10. What is the value of $(\mathbf{A} + \mathbf{B}) \cdot (\mathbf{A} \times \mathbf{B})$?
    (a) 0 (b) $A^2 - B^2$
    (c) $A^2 + B^2 + 2AB$ (d) None of these

11. If $\mathbf{A}$ is perpendicular to $\mathbf{B}$, then
    (a) $\mathbf{A} \times \mathbf{B} = 0$ (b) $\mathbf{A} \cdot [\mathbf{A} + \mathbf{B}] = A^2$
    (c) $\mathbf{A} \cdot \mathbf{B} = AB$ (d) $\mathbf{A} \cdot [\mathbf{A} + \mathbf{B}] = A^2 + AB$

Match the Columns

12. Column I shows some vector equations. Match Column I with the value of angle between $\mathbf{A}$ and $\mathbf{B}$ given in Column II.

	Column I		Column II
(a)	$\lvert\mathbf{A} \times \mathbf{B}\rvert = \lvert\mathbf{A} \cdot \mathbf{B}\rvert$	(p)	zero
(b)	$\mathbf{A} \times \mathbf{B} = \mathbf{B} \times \mathbf{A}$	(q)	$\frac{\pi}{2}$
(c)	$\lvert\mathbf{A} + \mathbf{B}\rvert = \lvert\mathbf{A} - \mathbf{B}\rvert$	(r)	$\frac{\pi}{4}$
(d)	$\mathbf{A} + \mathbf{B} = \mathbf{C}$ and $A + B = C$	(s)	$\frac{3\pi}{4}$

Subjective Question

13. Prove that $\mathbf{A} \cdot (\mathbf{A} \times \mathbf{B}) = 0$.

Answers

1. (c) **2.** (c) **3.** (d) **4.** (a) **5.** (a) **6.** (b,c,d) **8.** (a) **9.** (d) **10.** (a) **11.** (b)
12. (a $\rightarrow$ r,s; b $\rightarrow$ p; c $\rightarrow$ q; d $\rightarrow$ p)

See hints on pages 151-152.

MODULE 4

Representation of a Vector in Terms of Unit Vectors

Two Methods of Representing any Vector Quantity

Any vector quantity can be represented in two ways:

First

We can give its magnitude and direction separately. In the previous topics we have represented the vector quantities by using this method only.

Second

We can represent the given vector in terms of $\hat{i}$, $\hat{j}$ and $\hat{k}$. Here $\hat{i}$, $\hat{j}$ and $\hat{k}$ are unit vectors along positive x, y and z-directions.

Let us take an example :

If a force of magnitude 5N is acting at 37° from positive x-axis towards positive y-axis then this force vector in terms of $\hat{i}$, $\hat{j}$ and $\hat{k}$ can be written as,

$$\mathbf{F} = (4\hat{i} + 3\hat{j})\,\text{N}$$

From this expression of **F** we can find its magnitude, as well as direction. To understand the method of representing a vector in terms of $\hat{i}$, $\hat{j}$ and $\hat{k}$ we will have to first understand the meaning of component of a vector or resolution of a vector in components.

Resolution of a Vector in Components

Resolving a vector in components means breaking the vector in two or more than two parts (called components), so that net effect of that vector remains unchanged. Let us take one simple example :

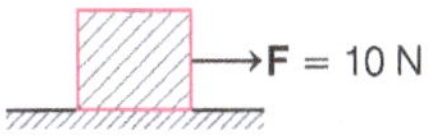

Suppose a force of 10N is applied on a block as shown in figure. Then this force can be resolved (or broken) in a number of ways as shown below :

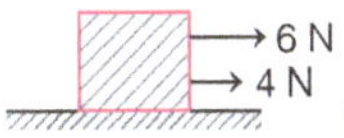

or
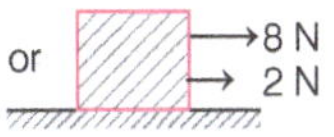

or
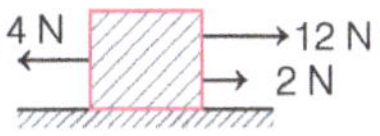

In all three figures shown above, the net force is 10N.

Note *A vector can be resolved in any number of components. But normally it is resolved maximum in three components, along x, y and z-axis. When the components are perpendicular to each other, they are called rectangular components.*

Rectangular Components

Let us find the rectangular components of a vector along x and y-axis.

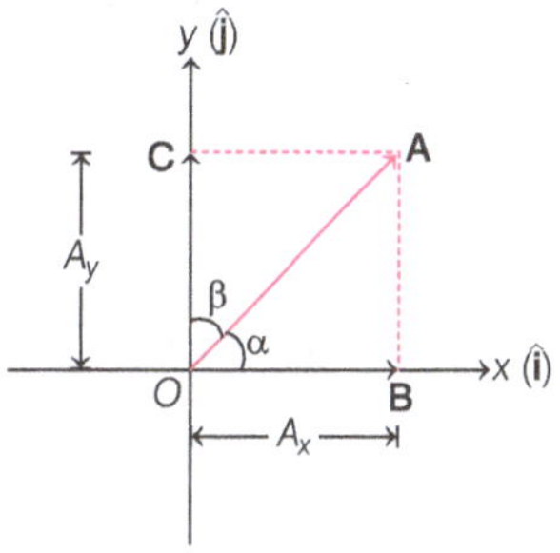

In the figure shown, we can see that

$$\mathbf{A} = \mathbf{B} + \mathbf{C} \quad \text{(From parallelogram law of vector addition)}$$

So, we can say that we have resolved **A** into two rectangular components **B** (along x-axis) and **C** (along y-axis).

Here,

$$|\mathbf{B}| = A_x = x\text{-component of } \mathbf{A} \text{ and}$$

$$|\mathbf{C}| = A_y = y\text{-component of } \mathbf{A}$$

Now, unit vectors along x and y-axis are $\hat{i}$ and $\hat{j}$. So, we can write,

$$\mathbf{B} = A_x\hat{i} \quad \text{and}$$

$$\mathbf{C} = A_y\hat{j}$$

$$\therefore \quad \mathbf{A} = \mathbf{B} + \mathbf{C} = A_x\hat{i} + A_y\hat{j}$$

Thus we have represented **A** in terms of $\hat{i}$ and $\hat{j}$.

Values of A_x and A_y

In the figure shown above,

$$A_x = A\cos\alpha \text{ or } A\sin\beta \quad \text{and}$$

$$A_y = A\cos\beta \text{ or } A\sin\alpha$$

Proof

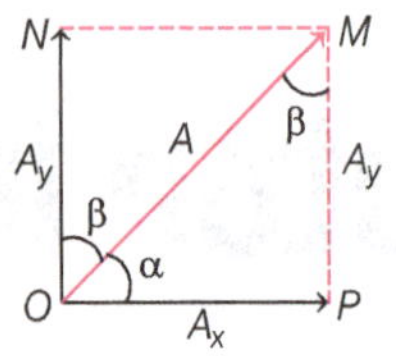

In triangle *OPM*,

$$\cos\alpha = \frac{A_x}{A} \quad \Rightarrow \quad A_x = A\cos\alpha$$

$$\sin\alpha = \frac{A_y}{A} \quad \Rightarrow \quad A_y = A\sin\alpha$$

$$\cos\beta = \frac{A_y}{A} \quad \Rightarrow \quad A_y = A\cos\beta$$

Similarly, $\sin\beta = \frac{A_x}{A} \quad \Rightarrow \quad A_x = A\sin\beta$

Representation of a Vector in Terms of $\hat{i}$, $\hat{j}$ and $\hat{k}$

Basically there are two cases :

First Representing a vector in terms of $\hat{i}$, $\hat{j}$ and $\hat{k}$.

Let us take few examples.

Example 1 Velocity of a particle is 4 m/s along positive *x*-axis. Represent this in terms of $\hat{i}$, $\hat{j}$ and $\hat{k}$.

Sol. $\mathbf{v} = (4\hat{i})$ m/s

Example 2 Acceleration of a particle is $6\ m/s^2$ towards negative *z*-axis. Represent it in terms of $\hat{i}$, $\hat{j}$ and $\hat{k}$.

Sol. $\mathbf{a} = (-6\hat{k})\ m/s^2$

Example 3 A force of 5N is acting at 37° from negative *x*-axis towards positive *y*-axis. Represent it in terms of $\hat{i}$, $\hat{j}$ and $\hat{k}$.

Sol.

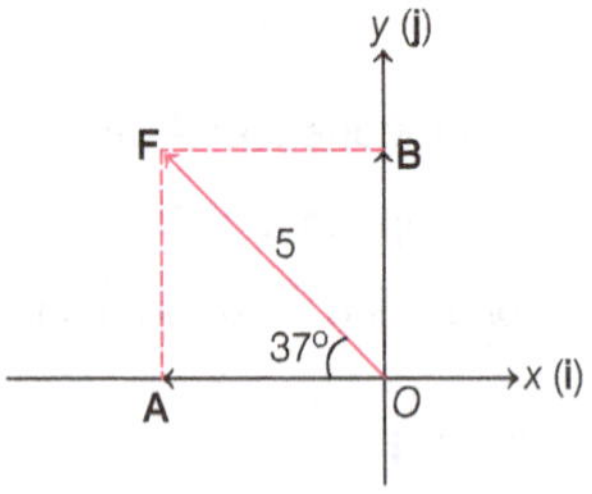

$$OA = 5\cos 37° = 5 \times \frac{4}{5} = 4\text{ N}$$

$$\mathbf{OA} = -4\,\hat{i}$$

$$OB = 5\sin 37° = 5 \times \frac{3}{5} = 3\text{ N}$$

$$\mathbf{OB} = 3\,\hat{j}$$

Now, $\mathbf{F} = \mathbf{OA} + \mathbf{OB}$

$$= (-4\hat{i} + 3\hat{j})\text{ N}$$

Second Finding magnitude and direction of a vector if it is given in terms of $\hat{i}$, $\hat{j}$ and $\hat{k}$.

Let us take some simple examples.

Example 4 Force acting on a particle is, $\mathbf{F} = (-6\,\hat{j})$ N. What is magnitude and direction of this force.

Sol. Magnitude is 6N and direction is along $-\hat{j}$ or along negative *y*-axis.

Example 5 Velocity of a particle is $\mathbf{v} = (3\hat{i} - 4\hat{j})$ m/s. Find its magnitude and direction.

Sol.

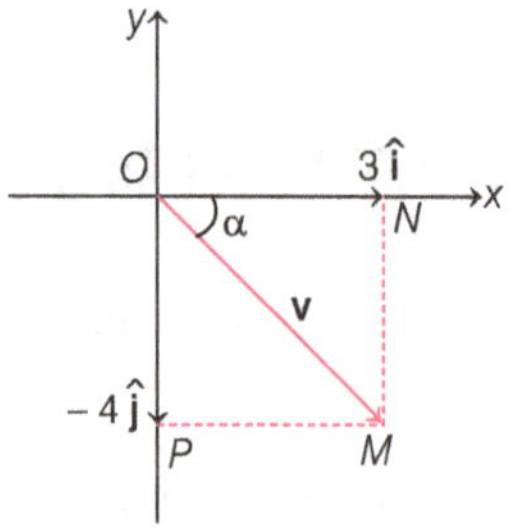

Magnitude : $|\mathbf{v}| = v = OM = \sqrt{OP^2 + ON^2}$

$$= \sqrt{4^2 + 3^2}$$

$$= 5\text{ m/s}$$

Direction : $\tan\alpha = \frac{MN}{ON} = \frac{4}{3} \Rightarrow \alpha = 53°$

Thus, given velocity vector has magnitude of 5 m/s. This is also speed of the particle. The magnitude of velocity vector is at 53° from positive *x*-axis towards negative *y*-axis.

Note *In all above examples, we have either taken one dimensional vectors or two dimensional vectors. The concept can be extended to three dimensional vectors too. You will learn more about them in mathematics. But we will take a brief idea about them in next article.*

Vector Mathematics in Terms of $\hat{i}$, $\hat{j}$ and $\hat{k}$

Addition, subtraction and multiplication (dot product and cross product) of vectors become simple if we represent the vectors in terms of $\hat{i}$, $\hat{j}$ and $\hat{k}$.

Let $\mathbf{A} = a_x\hat{i} + a_y\hat{j} + a_z\hat{k}$ and

$\mathbf{B} = b_x\hat{i} + b_y\hat{j} + b_z\hat{k}$, then :

Magnitude of a Vector

$$|\mathbf{A}| \text{ or } A = \sqrt{a_x^2 + a_y^2 + a_z^2}$$

Similarly we can find $|\mathbf{B}|$ or *B*.

Note *Magnitude of a vector can't be negative.*

Direction of a Vector

In the above article we have discussed about the direction of one dimensional or two dimensional vectors. In three dimensional vector, direction cosines are given for the direction.

For example, suppose α, β and γ are the angles of vector **A** with positive-x, positive-y and positive-z directions, then :

$$\cos\alpha = \frac{a_x}{A}, \cos\beta = \frac{a_y}{A} \text{ and } \cos\gamma = \frac{a_z}{A}$$

Here $\cos\alpha$, $\cos\beta$ and $\cos\gamma$ are the direction cosines of **A**.

Note

(i) In the above expressions,

$$A = |\mathbf{A}| = \sqrt{a_x^2 + a_y^2 + a_z^2}$$

Here, A is always positive. Now suppose, a_x is positive, a_y is zero and a_z is negative. Then, $\cos\alpha$ will be positive, $\cos\beta$ will be zero and $\cos\gamma$ will be negative. So, α is acute, β is 90° and γ is obtuse.

(ii) $\cos^2\alpha + \cos^2\beta + \cos^2\gamma = 1$

Proof

$$\cos^2\alpha + \cos^2\beta + \cos^2\gamma = \frac{a_x^2}{A^2} + \frac{a_y^2}{A^2} + \frac{a_z^2}{A^2}$$

$$= \frac{(a_x^2 + a_y^2 + a_z^2)}{A^2} = \frac{(a_x^2 + a_y^2 + a_z^2)}{(a_x^2 + a_y^2 + a_z^2)}$$

$$= 1$$

(iii) $\sin^2\alpha + \sin^2\beta + \sin^2\gamma = 2$

Proof

$$\cos^2\alpha + \cos^2\beta + \cos^2\gamma = 1$$

$$\Rightarrow \quad (1-\sin^2\alpha) + (1-\sin^2\beta) + (1-\sin^2\gamma) = 1$$

$$\Rightarrow \quad 1+1+1-1 = \sin^2\alpha + \sin^2\beta + \sin^2\gamma$$

or $\quad \sin^2\alpha + \sin^2\beta + \sin^2\gamma = 2$

Addition of Vectors

$$\mathbf{R} = \mathbf{A} + \mathbf{B} = (a_x\hat{i} + a_y\hat{j} + a_z\hat{k}) + (b_x\hat{i} + b_y\hat{j} + b_z\hat{k})$$

$$= (a_x + b_x)\hat{i} + (a_y + b_y)\hat{j} + (a_z + b_z)\hat{k}$$

Note *Resultant is a vector. So, it should have $\hat{i}$, $\hat{j}$ and (or) $\hat{k}$. In **R** we can see that we have $\hat{i}$, $\hat{j}$ and $\hat{k}$.*

Subtraction of Vectors

$$\mathbf{S} = \mathbf{A} - \mathbf{B} = (a_x\hat{i} + a_y\hat{j} + a_z\hat{k}) - (b_x\hat{i} + b_y\hat{j} + b_z\hat{k})$$

$$= (a_x - b_x)\hat{i} + (a_y - b_y)\hat{j} + (a_z - b_z)\hat{k}$$

Note ***S*** *is also a vector quantity. So, it should also have $\hat{i}$, $\hat{j}$ and (or) $\hat{k}$.*

Dot Product

$$\mathbf{A}\cdot\mathbf{B} = (a_x\hat{i} + a_y\hat{j} + a_z\hat{k}) \cdot (b_x\hat{i} + b_y\hat{j} + b_z\hat{k})$$

$$= (a_x b_x + a_y b_y + a_z b_z)$$

Note *Dot product is a scalar quantity. So, it should not have $\hat{i}$, $\hat{j}$ and (or) $\hat{k}$.*

Cross Product

Cross product is a vector quantity. So, again it should have $\hat{i}$, $\hat{j}$ and (or) $\hat{k}$.

We have $\mathbf{A} = a_x\hat{i} + a_y\hat{j} + a_z\hat{k}$

and $\quad \mathbf{B} = b_x\hat{i} + b_y\hat{j} + b_z\hat{k}$

The cross product of **A** and **B** can be written in terms of determinant.

$$\mathbf{A}\times\mathbf{B} = \begin{vmatrix} \hat{i} & \hat{j} & \hat{k} \\ a_x & a_y & a_z \\ b_x & b_y & b_z \end{vmatrix}$$

The above determinant can be solved by the method discussed below.

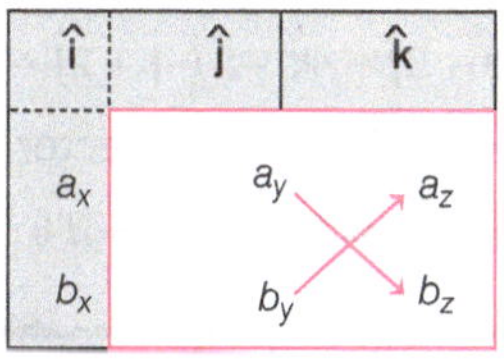

$$= \hat{i}\,(a_y b_z - b_y a_z) \quad \text{...(i)}$$

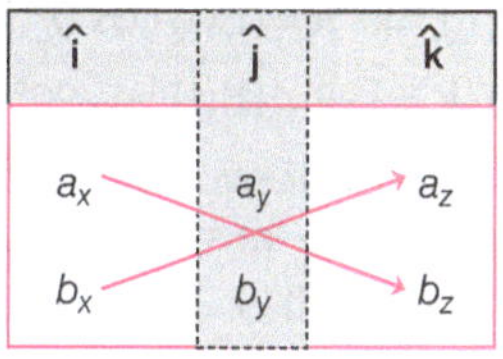

$$= -\hat{j}\,(a_x b_z - b_x a_z) \quad \text{...(ii)}$$

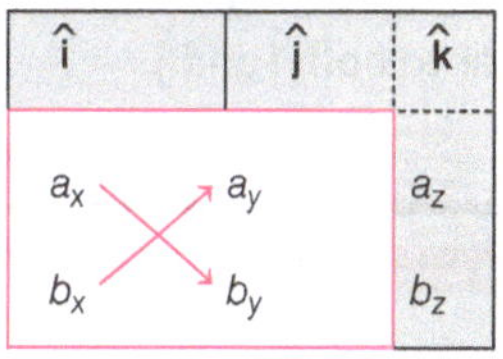

$$= \hat{k}\,(a_x b_y - b_x a_y) \quad \text{...(iii)}$$

Now adding right hand side of Eqs. (i), (ii) and (iii), we get

$$\mathbf{A}\times\mathbf{B} = \begin{vmatrix} \hat{i} & \hat{j} & \hat{k} \\ a_x & a_y & a_z \\ b_x & b_y & b_z \end{vmatrix}$$

$$= \hat{i}(a_y b_z - b_y a_z) - \hat{j}(a_x b_z - b_x a_z) + \hat{k}(a_x b_y - b_x a_y)$$

Note *We can also write coefficient of $\hat{j}$ as, $+\hat{j}\,(b_x a_z - a_x b_z)$*

Example 6 Given, $\mathbf{A} = 2\hat{i} - 3\hat{j} + 4\hat{k}$ and $\mathbf{B} = -\hat{i} + 2\hat{j} + 3\hat{k}$

Find,

(a) $|\mathbf{A}|$ or A
(b) Direction cosines of **A**
(c) $2\mathbf{A} + 3\mathbf{B}$
(d) $\mathbf{A} - 2\mathbf{B}$
(e) $\mathbf{A}\cdot\mathbf{B}$
(f) $\mathbf{A}\times\mathbf{B}$

Sol. (a) $|\mathbf{A}|$ or $A = \sqrt{(2)^2 + (-3)^2 + (4)^2}$

$$= \sqrt{4+9+16} = \sqrt{29}$$

(b) Direction cosines of **A**

$$\cos\alpha = \frac{a_x}{A} = \frac{2}{\sqrt{29}} = +\text{ve} \Rightarrow \alpha \text{ is acute}$$

$$\cos\beta = \frac{a_y}{A} = \frac{-3}{\sqrt{29}} = -\text{ve} \Rightarrow \beta \text{ is obtuse}$$

$$\cos\gamma = \frac{a_z}{A} = \frac{4}{\sqrt{29}} = +\text{ve} \Rightarrow \gamma \text{ is acute}$$

(c) $2\mathbf{A} + 3\mathbf{B} = 2(2\hat{i} - 3\hat{j} + 4\hat{k}) + 3(-\hat{i} + 2\hat{j} + 3\hat{k})$

$= (\hat{i} + 17\hat{k}) \rightarrow$ a vector quantity

(d) $\mathbf{A} - 2\mathbf{B} = (2\hat{i} - 3\hat{j} + 4\hat{k}) - 2(-\hat{i} + 2\hat{j} + 3\hat{k})$

$= (4\hat{i} - 7\hat{j} - 2\hat{k}) \rightarrow$ a vector quantity

(e) $\mathbf{A} \cdot \mathbf{B} = (2\hat{i} - 3\hat{j} + 4\hat{k}) \cdot (-\hat{i} + 2\hat{j} + 3\hat{k})$

$= (2)(-1) + (-3)(2) + (4)(3)$

$= -2 - 6 + 12$

$= 4 \rightarrow$ a scalar quantity

Note *$\mathbf{A} \cdot \mathbf{B}$ is positive. So, angle between **A** and **B** is acute.*

(f) $$\mathbf{A} \times \mathbf{B} = \begin{vmatrix} \hat{i} & \hat{j} & \hat{k} \\ 2 & -3 & 4 \\ -1 & 2 & 3 \end{vmatrix}$$

$= \hat{i}[(-3)(3) - (2)(4)] - \hat{j}[(2)(3) - (-1)(4)] + \hat{k}[(2)(2) - (-1)(-3)]$

$= \hat{i}(-9 - 8) - \hat{j}(6 + 4) + \hat{k}(4 - 3)$

$= (-17\hat{i} - 10\hat{j} + \hat{k}) \rightarrow$ a vector quantity

Important POINTS

(i) $\hat{i} \cdot \hat{i} = \hat{j} \cdot \hat{j} = \hat{k} \cdot \hat{k} = (1)(1)\cos 0° = 1$

(ii) $\hat{i} \cdot \hat{j} = \hat{j} \cdot \hat{k} = \hat{i} \cdot \hat{k} = (1)(1)\cos 90° = 0$

(iii) Cross product of two parallel or antiparallel vectors is a null vector. Thus,

$(\hat{i} \times \hat{i})$ or $\hat{i} \times (-\hat{i})$ or $(\hat{j} \times \hat{j})$ etc. is a null vector.

(iv) $|\mathbf{A} \times \mathbf{B}| = AB \sin\theta$

So, $|\hat{i} \times \hat{j}| = (1)(1)\sin 90° = 1$

Since, $|\hat{i} \times \hat{j}|$ is 1. So, $\hat{i} \times \hat{j}$ is a unit vector perpendicular to both $\hat{i}$ and $\hat{j}$.

This unit vector is actually $\hat{k}$.

Thus, $\hat{i} \times \hat{j} = \hat{k}$

Similarly,

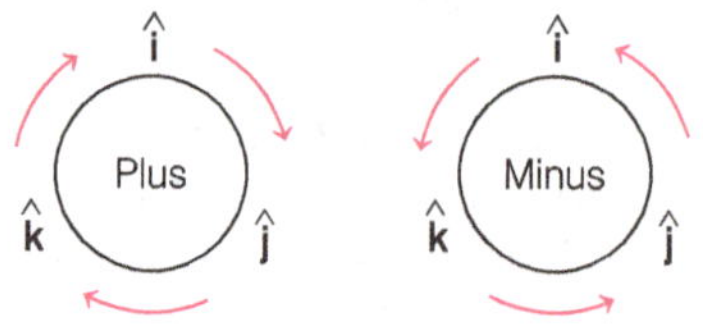

$\hat{i} \times \hat{j} = -\hat{j} \times \hat{i} = \hat{k}$

$\hat{j} \times \hat{k} = -\hat{k} \times \hat{j} = \hat{i}$ and $\hat{k} \times \hat{i} = -\hat{i} \times \hat{k} = \hat{j}$

Miscellaneous Examples

Ex. 1 Given $\mathbf{A} = 2\hat{i} - 3\hat{k}$ and $\mathbf{B} = 4\hat{j} + \hat{k}$. Find:

(a) $\mathbf{A} \cdot \mathbf{B}$ (b) $\mathbf{A} \times \mathbf{B}$

Sol. (a) **A** and **B** can be written as,

$\mathbf{A} = 2\hat{i} + 0\hat{j} - 3\hat{k}$ and

$\mathbf{B} = 0\hat{i} + 4\hat{j} + \hat{k}$

$\therefore \mathbf{A} \cdot \mathbf{B} = (2)(0) + (0)(4) + (-3)(1)$

$= -3$

Note *$\mathbf{A} \cdot \mathbf{B}$ is negative. So, angle between **A** and **B** is obtuse.*

(b) Since **A** and **B** have only two coefficients each. So, we can find their cross product without making determinant as below :

$\mathbf{A} \times \mathbf{B} = (2\hat{i} - 3\hat{k}) \times (4\hat{j} + \hat{k})$

$= (2)(4)(\hat{i} \times \hat{j}) + 2(\hat{i} \times \hat{k}) - (3)(4)(\hat{k} \times \hat{j}) - 3(\hat{k} \times \hat{k})$

$= (8\hat{k} - 2\hat{j} + 12\hat{i})$ **Ans.**

Note *$(\hat{i} \times \hat{j}) = \hat{k}$, $(\hat{i} \times \hat{k}) = -\hat{j}$, $(\hat{k} \times \hat{j}) = -\hat{i}$ and $(\hat{k} \times \hat{k})$ is a null vector*

Ex. 2 Given $\mathbf{a} = \hat{i} + 3\hat{j} - 2\hat{k} \times -\hat{i} + 3\hat{k}$

Find the magnitude of **a**.

Sol. $\mathbf{a} = (\hat{i} + 3\hat{j} - 2\hat{k}) \times (-\hat{i} + 3\hat{k})$

$$\mathbf{a} = \begin{vmatrix} \hat{i} & \hat{j} & \hat{k} \\ 1 & 3 & -2 \\ -1 & 0 & 3 \end{vmatrix}$$

$= \hat{i}(9 - 0) - \hat{j}(3 - 2) + \hat{k}(0 + 3)$

$= 9\hat{i} - \hat{j} + 3\hat{k}$

$|\mathbf{a}| = \sqrt{9^2 + 1^2 + 3^2} = \sqrt{91}$

Ex. 3 Let a force **F** be acting on a body free to rotate about a point *O* and let **r** the position vector of any point *P* on the line of action of the force. Then torque (τ) of this force about point *O* is defined as

$\tau = \mathbf{r} \times \mathbf{F}$

Given, $\mathbf{F} = (2\hat{i} + 3\hat{j} - \hat{k})$ N and $\mathbf{r} = (\hat{i} - \hat{j} + 6\hat{k})$ m

Find the torque of this force.

Sol. $\tau = \mathbf{r} \times \mathbf{F} = \begin{vmatrix} \hat{i} & \hat{j} & \hat{k} \\ 1 & -1 & 6 \\ 2 & 3 & -1 \end{vmatrix}$

$$= \hat{i}(1-18) - \hat{j}(-1-12) + \hat{k}(3+2)$$

or $\tau = (-17\hat{i} + 13\hat{j} + 5\hat{k})$ N-m

Ex. 4 Show that the vectors $\mathbf{a} = 3\hat{i} - 2\hat{j} + \hat{k}$, $\mathbf{b} = \hat{i} - 3\hat{j} + 5\hat{k}$ and $\mathbf{c} = 2\hat{i} + \hat{j} - 4\hat{k}$ form a right angled triangle.

Sol. We have $\mathbf{b} + \mathbf{c} = (\hat{i} - 3\hat{j} + 5\hat{k}) + (2\hat{i} + \hat{j} - 4\hat{k})$

$$= 3\hat{i} - 2\hat{j} + \hat{k} = \mathbf{a}$$

Also, we observe that no two of these vectors are parallel.

Further, $\mathbf{a} \cdot \mathbf{c} = (3\hat{i} - 2\hat{j} + \hat{k}) \cdot (2\hat{i} + \hat{j} - 4\hat{k}) = 6 - 2 - 4 = 0$

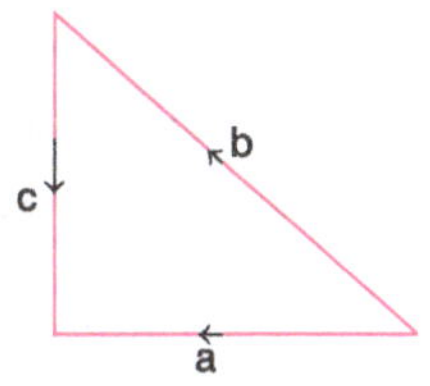

Dot product of two non-zero vectors is zero. Hence, they are perpendicular. So they form a right angled triangle.

$a = |\mathbf{a}| = \sqrt{9+4+1} = \sqrt{14}$,

$b = |\mathbf{b}| = \sqrt{1+9+25} = \sqrt{35}$

and $c = |\mathbf{c}| = \sqrt{4+1+16} = \sqrt{21}$

$\Rightarrow$ $a^2 + c^2 = b^2$ **Hence proved.**

MODULE 4 Exercise

Problems Based on Basic Concepts

Single Correct Option

1. The angles which the vector $\mathbf{A} = 3\hat{i} + 6\hat{j} + 2\hat{k}$ makes with the coordinate axes are
   (a) $\cos^{-1}\frac{3}{7}, \cos^{-1}\frac{6}{7}$ and $\cos^{-1}\frac{2}{7}$
   (b) $\cos^{-1}\frac{4}{7}, \cos^{-1}\frac{5}{7}$ and $\cos^{-1}\frac{3}{7}$
   (c) $\cos^{-1}\frac{3}{7}, \cos^{-1}\frac{4}{7}$ and $\cos^{-1}\frac{1}{7}$
   (d) None of the above

Multiple Correct Options

2. If unit vector $\mathbf{a} = a_1\hat{i} + a_2\hat{j} + a_3\hat{k}$. The possible values of a_1, a_2 and a_3 are
   (a) $a_1 = \frac{2}{3}, a_2 = \frac{2}{3}$ and $a_3 = \frac{1}{3}$
   (b) $a_1 = 0.3, a_2 = 0.4$ and $a_3 = \frac{\sqrt{3}}{2}$
   (c) $a_1 = 1, a_2 = 1$ and $a_3 = 1$
   (d) $a_1 = -\frac{2}{3}, a_2 = \frac{2}{3}$ and $a_3 = -\frac{1}{3}$

3. A force of $\sqrt{3}$ N makes equal angles with x-axis, y-axis and z-axis. The possible value of force are
   (a) $(\hat{i} + \hat{j} + \hat{k})$ N (b) $(-\hat{i} + \hat{j} - \hat{k})$ N
   (c) $(-\hat{i} - \hat{j} - \hat{k})$ N (d) $(-\hat{i} + \hat{j} + \hat{k})$ N

4. Find the component of a 50 N force which makes an angle of 45° with z-axis and whose projection in x-y plane makes 45° with the x-axis.
   (a) $F_x = 25$ N
   (b) $F_y = 25\sqrt{2}$ N
   (c) $F_y = 25$ N
   (d) $F_z = 25\sqrt{2}$ N

Subjective Questions

5. A force **F** has magnitude of 15 N. Direction of **F** is at 37° from negative x-axis towards positive y-axis. Represent **F** in terms of $\hat{\mathbf{i}}$ and $\hat{\mathbf{j}}$.

6. Find magnitude and direction of a vector, $\mathbf{A} = (6\hat{i} - 8\hat{j})$.

7. Resolve a weight of 10 N in two directions which are parallel and perpendicular to a slope inclined at 30° to the horizontal.

8. Resolve horizontally and vertically a force $\mathbf{F} = 8$ N which makes an angle of 45° with the horizontal.

9. Find magnitude and direction cosines of a vector, $\mathbf{A} = (3\hat{i} - 4\hat{j} + 5\hat{k})$

10. Resolve a force $\mathbf{F} = 10$ N along x and y-axes. Where this force vector is making an angle of 60° from negative x-axis towards negative y-axis?

Addition and Subtraction of Vectors

Single Correct Option

11. When vector $\mathbf{A} = 2\hat{i} + 3\hat{j} + 2\hat{k}$ is subtracted from vector **B**, it gives a vector equal to $2\hat{j}$. Then the magnitude of vector **B** will be **(2023 Main)**
(a) $\sqrt{13}$ (b) 3
(c) $\sqrt{6}$ (d) None of these

Subjective Questions

12. Obtain the magnitude of $2\mathbf{A} - 3\mathbf{B}$ if
$\mathbf{A} = \hat{i} + \hat{j} - 2\hat{k}$ and $\mathbf{B} = 2\hat{i} - \hat{j} + \hat{k}$

13. Find magnitude of $\mathbf{A} - 2\mathbf{B} + 3\mathbf{C}$, where $\mathbf{A} = 2\hat{i} + 3\hat{j}, \mathbf{B} = \hat{i} + \hat{j}$ and $\mathbf{C} = \hat{k}$.

Dot Product and Cross Product

Single Correct Option

14. In a clockwise system,
(a) $\hat{j} \times \hat{k} = \hat{i}$ (b) $\hat{k} \cdot \hat{i} = 1$ (c) $\hat{i} \cdot \hat{i} = 0$ (d) $\hat{j} \times \hat{j} = 1$

Match the Columns

15. Match the Column I with Column II and select the option from the given code below.

	Column I		Column II
A	$2\hat{i} \times (\hat{i} + \hat{j})$	(p)	$2\hat{k}$
B	$4\hat{i} \times (2\hat{i} - \hat{j})$	(q)	0
C	$3\hat{i} \times (3\hat{i} - 4\hat{j})$	(r)	$-4\hat{k}$
D	$2(\hat{i} + \hat{j}) \times 2\hat{i}$	(s)	$-12\hat{k}$

	A	B	C	D		A	B	C	D
(a)	p	q	r	s	(b)	q	p	s	r
(c)	p	r	s	r	(d)	s	r	q	p

Subjective Questions

16. Find the values of
(a) $(4\hat{i}) \times (-6\hat{k})$
(b) $(3\hat{j}) \cdot (-4\hat{j})$
(c) $(2\hat{i}) \cdot (-4\hat{k})$

17. Find $(2\mathbf{A}) \times (-3\mathbf{B})$, if $\mathbf{A} = 2\hat{i} - \hat{j}$ and $\mathbf{B} = (\hat{j} + \hat{k})$

18. If $\mathbf{A} = 2\hat{i} - 3\hat{j} + 7\hat{k}, \mathbf{B} = \hat{i} + 2\hat{k}$ and $\mathbf{C} = \hat{j} - \hat{k}$, then find $\mathbf{A} \cdot (\mathbf{B} \times \mathbf{C})$.

Miscellaneous Problems

Single Correct Option

19. If a unit vector is represented by $0.5\hat{i} + 0.8\hat{j} + c\hat{k}$, the value of c is
(a) 1 (b) $\sqrt{0.11}$
(c) $\sqrt{0.01}$ (d) 0.39

20. Given $\mathbf{A} = 3\hat{i} + 4\hat{j}$ and $\mathbf{B} = 6\hat{i} + 8\hat{j}$, which of the following statement is correct?
(a) $\mathbf{A} \times \mathbf{B} = 0$ (b) $\frac{|\mathbf{A}|}{|\mathbf{B}|} = \frac{1}{3}$
(c) $|\mathbf{A}| = 15$ (d) $\mathbf{A} \cdot \mathbf{B} = 48$

21. The vector that must be added to the vector $\hat{i} - 3\hat{j} + 2\hat{k}$ and $3\hat{i} + 6\hat{j} - 7\hat{k}$ so that the resultant vector is a unit vector along the y-axis is
(a) $4\hat{i} + 2\hat{j} + 5\hat{k}$ (b) $-4\hat{i} - 2\hat{j} + 5\hat{k}$
(c) $3\hat{i} + 4\hat{j} + 5\hat{k}$ (d) Null vector

Multiple Correct Options

22. If $\mathbf{a} = 3\hat{i} + 4\hat{j}$ and $\mathbf{b} = 4\hat{i} - 3\hat{j}$. Then
(a) the magnitude of resultant of **a** and **b** is $5\sqrt{2}$ units
(b) the angle between **a** and **b** is 90°
(c) the magnitude of resultant of **a** and **b** is 5 units
(d) the angle between **a** and **b** is zero

Answers

1. (a) **2.** (a, b, d) **3.** (a, c) **4.** (a, c, d) **5.** $(-12\hat{i} + 9\hat{j})$ N **6.** 10 units, 53°
7. $w_{||} = 5$ N, $w_{\perp} = 5\sqrt{3}$ N **8.** $4\sqrt{2}$ N, $4\sqrt{2}$ N **9.** $5\sqrt{2}$ units, $\cos\alpha = \frac{3}{5\sqrt{2}}$, $\cos\beta = \frac{-4}{5\sqrt{2}}$ and $\cos\gamma = \frac{1}{\sqrt{2}}$
10. $F_x = -5$ N, $F_y = -5\sqrt{3}$ N **11.** (d) **12.** $\sqrt{90}$ **13.** $\sqrt{10}$ units **14.** (a) **15.** (c)
16. (a) $24\hat{j}$ (b) -12 (c) Zero **17.** $6\hat{i} + 12\hat{j} - 12\hat{k}$ **18.** Zero **19.** (b) **20.** (a) **21.** (b) **22.** (a, b)

See hints on pages 152-154.

MODULE 5

Left Over Topics of Vectors

Position Vector and Displacement Vector

Position Vector

To locate the position of any point P in a plane or space, generally a fixed point of reference called the origin O is taken. The vector **OP** is called the position vector of P with respect to O as shown in figure. If coordinates of point P are (x, y) then position vector of point P with respect to point O is

$$\mathbf{OP} = \mathbf{r} = \mathbf{OM} + \mathbf{MP} = x\hat{\mathbf{i}} + y\hat{\mathbf{j}}$$

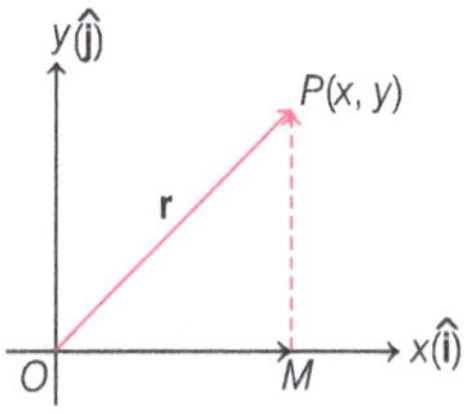

Note *(i) For a point P, there is one and only one position vector with respect to the origin O.*

(ii) Position vector of a point P changes if the position of the origin O is changed.

(iii) If coordinates of point P are (x, y, z), then $\mathbf{r} = x\hat{\mathbf{i}} + y\hat{\mathbf{j}} + z\hat{\mathbf{k}}$

Displacement Vector

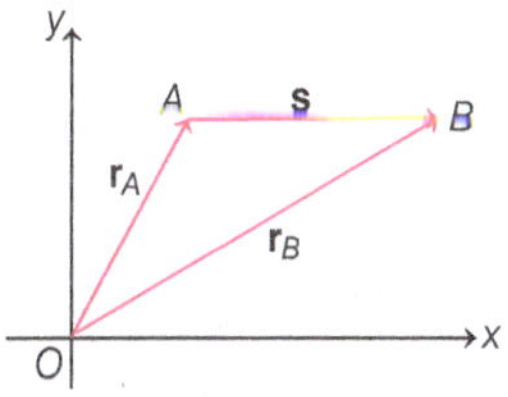

Suppose a body is displaced from A to B and the coordinates of point A are (x_1, y_1, z_1) and B are (x_2, y_2, z_2).

Then, position vector of $A = \mathbf{r}_A = \mathbf{OA} = x_1\hat{\mathbf{i}} + y_1\hat{\mathbf{j}} + z_1\hat{\mathbf{k}}$

Position vector of $B = \mathbf{r}_B = \mathbf{OB} = x_2\hat{\mathbf{i}} + y_2\hat{\mathbf{j}} + z_2\hat{\mathbf{k}}$

and $\mathbf{AB} = \mathbf{OB} - \mathbf{OA} = \mathbf{r}_B - \mathbf{r}_A =$ displacement vector (**s**)

$$= (x_2 - x_1)\hat{\mathbf{i}} + (y_2 - y_1)\hat{\mathbf{j}} + (z_2 - z_1)\hat{\mathbf{k}}$$

Example 1 Work done by a force **F** on a body is $W = \mathbf{F} \cdot \mathbf{s}$, where **s** is the displacement of body. Given that under a force $\mathbf{F} = (2\hat{\mathbf{i}} + 3\hat{\mathbf{j}} + 4\hat{\mathbf{k}})$ N a body is displaced from position vector $\mathbf{r}_1 = (2\hat{\mathbf{i}} + 3\hat{\mathbf{j}} + \hat{\mathbf{k}})$ *m* to the position vector $\mathbf{r}_2 = (\hat{\mathbf{i}} + \hat{\mathbf{j}} + \hat{\mathbf{k}})$ *m*. Find the work done by this force.

Sol. The body is displaced from $\mathbf{r}_1$ to $\mathbf{r}_2$. Therefore, displacement of the body is

$$\mathbf{s} = \mathbf{r}_2 - \mathbf{r}_1$$
$$= (\hat{\mathbf{i}} + \hat{\mathbf{j}} + \hat{\mathbf{k}}) - (2\hat{\mathbf{i}} + 3\hat{\mathbf{j}} + \hat{\mathbf{k}})$$
$$= (-\hat{\mathbf{i}} - 2\hat{\mathbf{j}})\ \text{m}$$

Now, work done by the force is $W = \mathbf{F} \cdot \mathbf{s}$

$$= (2\hat{\mathbf{i}} + 3\hat{\mathbf{j}} + 4\hat{\mathbf{k}}) \cdot (-\hat{\mathbf{i}} - 2\hat{\mathbf{j}})$$
$$= (2)(-1) + (3)(-2) + (4)(0) = -8\ \text{J}$$

Angle between Two Vectors

As we discussed earlier, in vector mathematics, angle between two vectors in very important.

In some of the cases you can directly find angle between two vectors just by observation. Following table will explain this point more explicitly.

S.No	A	B	Angle between A and B	Reason
1	$6\hat{\mathbf{i}}$	$2\hat{\mathbf{i}}$	0°	both are along $\hat{\mathbf{i}}$
2	$4\hat{\mathbf{j}}$	$-6\hat{\mathbf{j}}$	180°	**A** is along $\hat{\mathbf{j}}$ and **B** is along $-\hat{\mathbf{j}}$
3	$2\hat{\mathbf{i}} - 3\hat{\mathbf{j}}$	$6\hat{\mathbf{k}}$	90°	**A** is in x-y plane and **B** is along $\hat{\mathbf{k}}$
4	$6\hat{\mathbf{i}}$	$4\hat{\mathbf{i}} + 4\hat{\mathbf{j}}$	45°	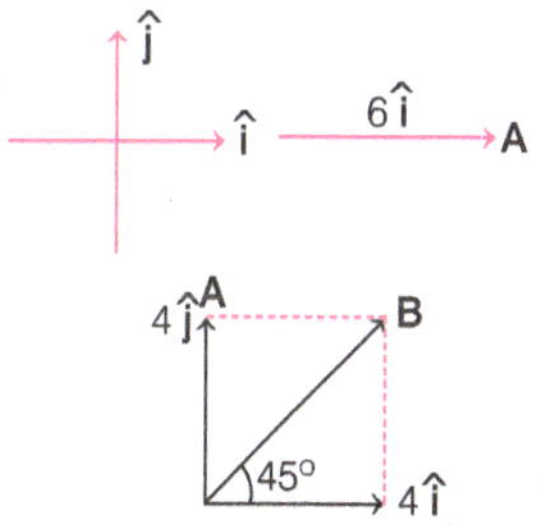
5	$4\hat{\mathbf{j}}$	$2\hat{\mathbf{i}} - 2\hat{\mathbf{j}}$	135°	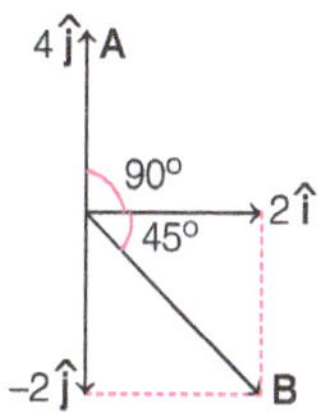

In all other cases, we can find the angle between them by using the following relation.

$$\mathbf{A} \cdot \mathbf{B} = AB \cos \theta$$

$$\therefore \quad \cos\theta = \frac{\mathbf{A}\cdot\mathbf{B}}{AB} \quad \text{...(i)}$$

or

$$\boxed{\theta = \cos^{-1}\left(\frac{\mathbf{A}\cdot\mathbf{B}}{AB}\right)}$$

Note (i) *In equation (i), the denominator AB is positive. The numerator* $\mathbf{A}\cdot\mathbf{B}$ *may be positive, negative or zero. Accordingly* $\cos\theta$ *may be positive, negative or zero. Therefore* θ *may be acute, obtuse or 90°.*

(ii) *Two vectors are perpendicular if their dot product is zero.*

(iii) *Two vectors can be shown parallel (or antiparallel) to one another if :*

(a) *The coefficients of* $\hat{\mathbf{i}}$, $\hat{\mathbf{j}}$ *and* $\hat{\mathbf{k}}$ *of both the vectors bear a constant ratio. For example, a vector* $\mathbf{A} = a_1\hat{\mathbf{i}} + b_1\hat{\mathbf{j}} + c_1\hat{\mathbf{k}}$ *is parallel (or antiparallel) to another vector* $\mathbf{B} = a_2\hat{\mathbf{i}} + b_2\hat{\mathbf{j}} + c_2\hat{\mathbf{k}}$ *if :* $\frac{a_1}{a_2} = \frac{b_1}{b_2} = \frac{c_1}{c_2}$ *= constant. If this constant has positive value, then the vectors are parallel and if the constant has negative value then the vectors are antiparallel.*

OR

(b) *The cross product of both the vectors is a null vector. For instance,* **A** *and* **B** *are parallel (or antiparallel) to each other if*

$$\mathbf{A}\times\mathbf{B} = \begin{vmatrix} \hat{\mathbf{i}} & \hat{\mathbf{j}} & \hat{\mathbf{k}} \\ a_1 & b_1 & c_1 \\ a_2 & b_2 & c_2 \end{vmatrix} = \text{a null vector.}$$

(iv) $|\mathbf{A}\times\mathbf{B}| = AB\sin\theta$

$$\Rightarrow \quad \sin\theta = \frac{|\mathbf{A}\times\mathbf{B}|}{AB}$$

$$\therefore \quad \theta = \sin^{-1}\left[\frac{|\mathbf{A}\times\mathbf{B}|}{AB}\right]$$

We could also find the angle θ *from the above relation, but this will not give the correct answer.*

Reason is simple. $|\mathbf{A}\times\mathbf{B}|$ *and AB both are positive. So* $\sin\theta$ *from the above relation always comes out to be positive. Suppose this is* $\frac{1}{2}$, *then sin 30° and sin 150° both are* $\frac{1}{2}$. *So,* θ *may be either 30° or 150°. But this is not the case with* $\cos\theta$. *If angle is 30°, then we will get* $\cos\theta = \frac{\sqrt{3}}{2}$ *and if angle is 150° then* $\cos\theta$ *will be* $-\frac{\sqrt{3}}{2}$.

Example 2 Find the angle between two vectors $\mathbf{A} = 2\hat{\mathbf{i}} + \hat{\mathbf{j}} - \hat{\mathbf{k}}$ and $\mathbf{B} = \hat{\mathbf{i}} - \hat{\mathbf{k}}$.

Sol.

$$A = |\mathbf{A}| = \sqrt{(2)^2 + (1)^2\,(-1)^2} = \sqrt{6}$$

$$B = |\mathbf{B}| = \sqrt{(1)^2 + (-1)^2} = \sqrt{2}$$

$$\mathbf{A}\cdot\mathbf{B} = (2\hat{\mathbf{i}} + \hat{\mathbf{j}} - \hat{\mathbf{k}})\cdot(\hat{\mathbf{i}} - \hat{\mathbf{k}})$$

$$= (2)(1) + (1)(0) + (-1)(-1) = 3$$

$$\text{Now, } \cos\theta = \frac{\mathbf{A}\cdot\mathbf{B}}{AB} = \frac{3}{\sqrt{6}\cdot\sqrt{2}}$$

$$= \frac{3}{\sqrt{12}} = \frac{\sqrt{3}}{2}$$

$$\therefore \quad \theta = 30°$$

Example 3 Prove that the vectors $\mathbf{A} = 2\hat{\mathbf{i}} - 3\hat{\mathbf{j}} + \hat{\mathbf{k}}$ and $\mathbf{B} = \hat{\mathbf{i}} + \hat{\mathbf{j}} + \hat{\mathbf{k}}$ are mutually perpendicular.

Sol.

$$\mathbf{A}\cdot\mathbf{B} = (2\hat{\mathbf{i}} - 3\hat{\mathbf{j}} + \hat{\mathbf{k}})\cdot(\hat{\mathbf{i}} + \hat{\mathbf{j}} + \hat{\mathbf{k}})$$

$$= (2)(1) + (-3)(1) + (1)(1) = 0 = AB\cos\theta$$

$\therefore \cos\theta = 0$ (as $A \neq 0, B \neq 0$)

or $\theta = 90°$

or the vectors **A** and **B** are mutually perpendicular.

Example 4 Prove that if coefficients of $\hat{\mathbf{i}}$, $\hat{\mathbf{j}}$ and $\hat{\mathbf{k}}$ of two vectors bear a constant ratio, then the vectors are either parallel or antiparallel.

Sol. Let $\mathbf{A} = (\hat{\mathbf{i}} + \hat{\mathbf{j}} + \hat{\mathbf{k}})$

$\mathbf{B} = (2\hat{\mathbf{i}} + 2\hat{\mathbf{j}} + 2\hat{\mathbf{k}})$ and

$\mathbf{C} = (-3\hat{\mathbf{i}} - 3\hat{\mathbf{j}} - 3\hat{\mathbf{k}})$

Then, we can see that $\mathbf{B} = 2\mathbf{A}$. So, **B** is along **A** with magnitude two times. So, **B** is parallel to **A**. Here the ratio of coefficients is $+\frac{1}{2}$.

Similarly we can see that, $\mathbf{C} = -3\mathbf{A}$. So, **C** is antiparallel to **A** with magnitude three times. Now the ratio of coefficients is $-\frac{1}{3}$.

Example 5 Show that the vector $\mathbf{A} = \hat{\mathbf{i}} - \hat{\mathbf{j}} + 2\hat{\mathbf{k}}$ is parallel to a vector $\mathbf{B} = 3\hat{\mathbf{i}} - 3\hat{\mathbf{j}} + 6\hat{\mathbf{k}}$.

Sol. A vector **A** is parallel to an another vector **B** if it can be written as

$$\mathbf{A} = m\mathbf{B}$$

Here,

$$\mathbf{A} = (\hat{\mathbf{i}} - \hat{\mathbf{j}} + 2\hat{\mathbf{k}}) = \frac{1}{3}(3\hat{\mathbf{i}} - 3\hat{\mathbf{j}} + 6\hat{\mathbf{k}})$$

or

$$\mathbf{A} = \frac{1}{3}\mathbf{B}$$

This implies that **A** is parallel to **B** and magnitude of **A** is $\frac{1}{3}$ times the magnitude of **B**.

Component of a Vector along Another Vector

Component of **A** along $\mathbf{B} = A\cos\theta = \frac{\mathbf{A}\cdot\mathbf{B}}{B}$

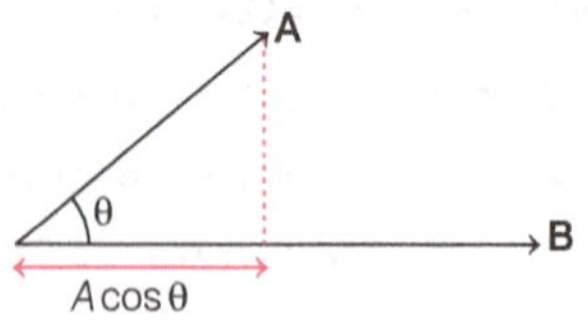

Similarly, component of **B** along **A**

$$= B\cos\theta = \frac{\mathbf{A}\cdot\mathbf{B}}{A}$$

Component of **A** along **B** = component of **B** along **A** if $|\mathbf{A}| = |\mathbf{B}|$ or $A = B$. Otherwise they are not equal.

Note *In the above figure $A \cos\theta$ is the magnitude of component of **A** along **B**. If we wish to find this component in vector form then we will have to multiply this value with a unit vector along **B** which is given by,*

$$\hat{B} = \frac{B}{B}$$

*Thus, component of **A** along **B** in vector form* $= (A\cos\theta)\,\hat{B}$

$$= \left(\frac{A \cdot B}{B}\right)\left(\frac{B}{B}\right) = \left(\frac{A \cdot B}{B^2}\right)(B)$$

Example 6 Given $\mathbf{A} = 2\hat{i} + 3\hat{j}$ and $\mathbf{B} = 3\hat{j} - 4\hat{k}$. Find :

(a) Magnitude of component of **A** along **B**

(b) Component of A along B in vector form

Sol. $\mathbf{A} = 2\hat{i} + 3\hat{j} + 0\hat{k}$

$$\mathbf{B} = 0\hat{i} + 3\hat{j} - 4\hat{k}$$

$$\mathbf{A} \cdot \mathbf{B} = (2)(0) + (3)(3) + (0)(-4) = 9$$

$$B = |\mathbf{B}| = \sqrt{3^2 + (-4)^2} = 5$$

(a) Magnitude of component of **A** along **B** $= A\cos\theta$

$$= \frac{\mathbf{A} \cdot \mathbf{B}}{B} = \frac{9}{5}$$ **Ans.**

(b) Component of **A** along **B** in vector form $= \left(\frac{\mathbf{A} \cdot \mathbf{B}}{B^2}\right)(\mathbf{B})$

$$= \frac{9}{(5)^2}(3\hat{i} - 4\hat{k})$$

$$= \frac{27}{25}\hat{i} - \frac{36}{25}\hat{k}$$ **Ans.**

Unit Vector Perpendicular to A and B

As we know that $\mathbf{A} \times \mathbf{B}$ and $\mathbf{B} \times \mathbf{A}$ are perpendicular to both **A** and **B**. So, if we divide $\mathbf{A} \times \mathbf{B}$ or $\mathbf{B} \times \mathbf{A}$ by their magnitudes, then we will get a unit vector which is perpendicular to both **A** and **B**. Thus, a unit vector (say $\hat{\mathbf{n}}$) perpendicular to both **A** and **B** will be given by,

$$\hat{\mathbf{n}} = \pm \frac{(\mathbf{A} \times \mathbf{B})}{|\mathbf{A} \times \mathbf{B}|}$$

Example 7 Find a unit vector perpendicular to $\mathbf{A} = 2\hat{i} + 3\hat{j} + \hat{k}$ and $\mathbf{B} = \hat{i} - \hat{j} + \hat{k}$ both.

Sol. As we have read, $\mathbf{C} = \mathbf{A} \times \mathbf{B}$ is a vector perpendicular to both **A** and **B**. Hence, a unit vector $\hat{\mathbf{n}}$ perpendicular to **A** and **B** can be written as

$$\hat{\mathbf{n}} = \pm \frac{\mathbf{C}}{C} = \pm \frac{\mathbf{A} \times \mathbf{B}}{|\mathbf{A} \times \mathbf{B}|}$$

Here, $$\mathbf{A} \times \mathbf{B} = \begin{vmatrix} \hat{i} & \hat{j} & \hat{k} \\ 2 & 3 & 1 \\ 1 & -1 & 1 \end{vmatrix}$$

$$= \hat{i}(3+1) + \hat{j}(1-2) + \hat{k}(-2-3) = 4\hat{i} - \hat{j} - 5\hat{k}$$

Further,

$$|\mathbf{A} \times \mathbf{B}| = \sqrt{(4)^2 + (-1)^2 + (-5)^2} = \sqrt{42}$$

$\therefore$ The desired unit vector is

$$\hat{\mathbf{n}} = \pm \frac{\mathbf{A} \times \mathbf{B}}{|\mathbf{A} \times \mathbf{B}|} \quad \text{or} \quad \hat{\mathbf{n}} = \pm \frac{1}{\sqrt{42}}(4\hat{i} - \hat{j} - 5\hat{k})$$

Area of Triangle or Parallelogram

Triangle

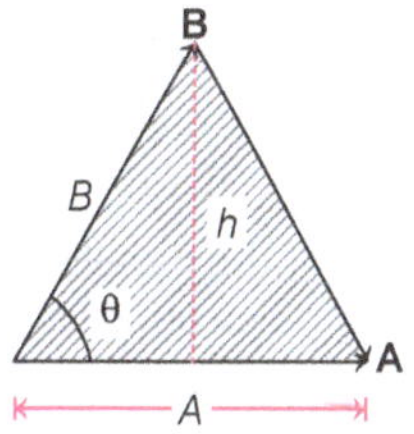

$$\frac{h}{B} = \sin\theta \Rightarrow h = B\sin\theta$$

If two sides of a triangle are represented by two vectors **A** and **B** as shown in figure, then area of this triangle is given by,

$$\text{Area} = \frac{1}{2}|\mathbf{A} \times \mathbf{B}|$$

Proof : $$\text{Area} = \frac{1}{2}\text{(base) (height)}$$

$$= \frac{1}{2}(A)(h)$$

$$= \frac{1}{2}(A)(B\sin\theta)$$

$$= \frac{1}{2}|\mathbf{A} \times \mathbf{B}| \qquad [\because |\mathbf{A} \times \mathbf{B}| = AB\sin\theta]$$

Parallelogram

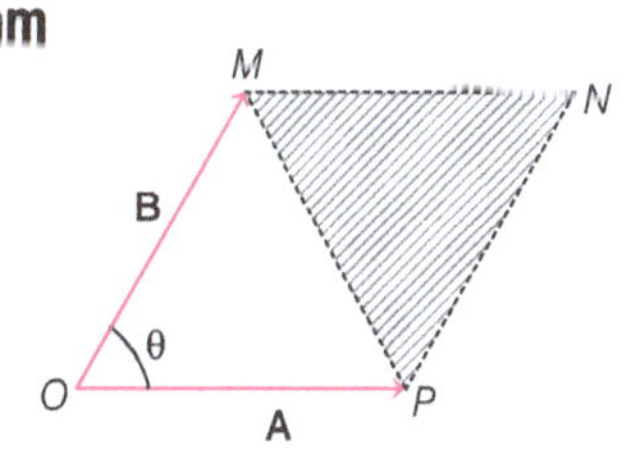

Area of parallelogram *OMNP*

$$= 2\,(\text{Area of triangle } OMP)$$

$$= 2\left[\frac{1}{2}|\mathbf{A} \times \mathbf{B}|\right]$$

$$= |\mathbf{A} \times \mathbf{B}|$$

Example 8 Calculate the area of a triangle if two adjacent sides of the triangle are represented by vectors $\mathbf{A} = 3\hat{i} + 4\hat{j}$ and $\mathbf{B} = -3\hat{i} + 7\hat{j}$.

Sol. We know that the half of magnitude of the cross product of two vectors gives the area of the triangle.

$$\mathbf{A}\times\mathbf{B}=\begin{vmatrix}\hat{i} & \hat{j} & \hat{k}\\ 3 & 4 & 0\\ -3 & 7 & 0\end{vmatrix}$$

$$=\hat{i}(0-0)-\hat{j}(0-0)+\hat{k}(21+12)=33\hat{k}$$

Taking magnitude $|\mathbf{A}\times\mathbf{B}|=\sqrt{33^2}=33$

So, area of triangle $=\frac{1}{2}|\mathbf{A}\times\mathbf{B}|=\frac{33}{2}$ units.

Example 9 Calculate the area of the parallelogram if adjacent sides of the parallelogram are given by the vectors $\mathbf{A}=\hat{i}+2\hat{j}+3\hat{k}$ and $\mathbf{B}=2\hat{i}-3\hat{j}+\hat{k}$.

Sol. We know that area of the parallelogram is equal to magnitude of the cross product of given vectors. Now,

$$\mathbf{A}\times\mathbf{B}=\begin{vmatrix}\hat{i} & \hat{j} & \hat{k}\\ 1 & 2 & 3\\ 2 & -3 & 1\end{vmatrix}$$

$$=\hat{i}(2+9)+\hat{j}(6-1)+\hat{k}(-3-4)=11\hat{i}+5\hat{j}-7\hat{k}$$

So, area of parallelogram

$$|\mathbf{A}\times\mathbf{B}|=\sqrt{11^2+5^2+(-7)^2}$$

$$=\sqrt{195}\text{ units}$$

Example 10 Find area of triangle, if coordinates of three vertices are (0, 0), (2, 4), (4, 2).

Sol.

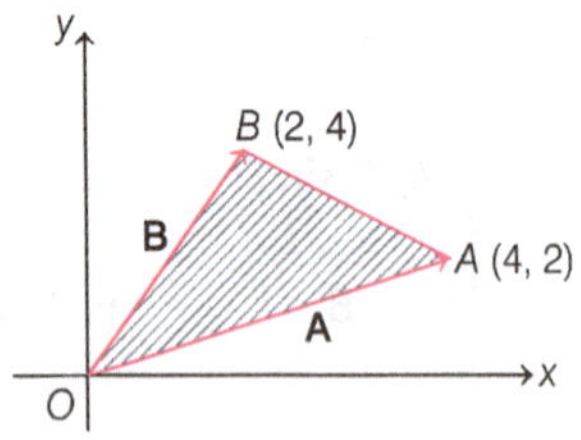

$$\mathbf{A}=\mathbf{r}_A=\mathbf{OA}=(4\hat{i}+2\hat{j})$$

$$\mathbf{B}=\mathbf{r}_B=\mathbf{OB}=(2\hat{i}+4\hat{j})$$

$$\mathbf{A}\times\mathbf{B}=(4\hat{i}+2\hat{j})\times(2\hat{i}+4\hat{j})$$

$$=(4)(2)(\hat{i}\times\hat{i})+(4)(4)(\hat{i}\times\hat{j})+(2)(2)(\hat{j}\times\hat{i})+(2)(4)(\hat{j}\times\hat{j})$$

$$=0+16\hat{k}-4\hat{k}+0$$

$$=12\hat{k}$$

$\therefore$ $|\mathbf{A}\times\mathbf{B}|=12$ units

$\therefore$ Area of triangle $=\frac{1}{2}|\mathbf{A}\times\mathbf{B}|$

$=6$ units **Ans.**

Scalar Triple Product

(i) $\mathbf{A}\cdot(\mathbf{B}\times\mathbf{C})$ is called scalar triple product. It is a scalar quantity. We can show that $\mathbf{A}\cdot(\mathbf{B}\times\mathbf{C})=(\mathbf{A}\times\mathbf{B})\cdot\mathbf{C}=\mathbf{B}\cdot(\mathbf{C}\times\mathbf{A})$.

(ii) The volume of a parallelopiped bounded by vectors **A**, **B** and **C** can be obtained by $(\mathbf{A}\times\mathbf{B})\cdot\mathbf{C}$.

(iii) If three vectors are coplanar then the volume of the parallelopiped bounded by these three vectors should be zero or we can say that their scalar triple product should be zero.

(iv) If $\mathbf{A}=a_1\hat{i}+a_2\hat{j}+a_3\hat{k}$, $\mathbf{B}=b_1\hat{i}+b_2\hat{j}+b_3\hat{k}$ and $\mathbf{C}=c_1\hat{i}+c_2\hat{j}+c_3\hat{k}$ then $\mathbf{A}\cdot(\mathbf{B}\times\mathbf{C})$ is also written as [**ABC**] and it has the following value :

$$[\mathbf{ABC}]=\begin{vmatrix}a_1 & a_2 & a_3\\ b_1 & b_2 & b_3\\ c_1 & c_2 & c_3\end{vmatrix}$$

= Volume of parallelopiped whose adjacent sides are along **A**, **B** and **C**.

Example 11 Find [**A B C**] when $\mathbf{A}=2\hat{i}-3\hat{j}+4\hat{k}$, $\mathbf{B}=\hat{i}+2\hat{j}-\hat{k}$ and $\mathbf{C}=3\hat{i}-\hat{j}+2\hat{k}$.

Sol. Given,

$$[\mathbf{ABC}]=\begin{vmatrix}2 & -3 & 4\\ 1 & 2 & -1\\ 3 & -1 & 2\end{vmatrix}$$

$$=2(4-1)-(-3)(2+3)+4(-1-6)$$

$$=2(3)+3(5)+4(-7)$$

$$=6+15-28$$

$$=-7$$

Example 12 Find the volume of the parallelopiped whose coterminous edges are given by the vectors $\hat{i}-\hat{j}+3\hat{k}$, $\hat{i}+\hat{j}-2\hat{k}$ and $2\hat{i}+\hat{j}+4\hat{k}$.

Sol. Volume of the parallelopiped

$$=|\mathbf{A\ B\ C}|$$

$$[\mathbf{A\ B\ C}]=\begin{vmatrix}1 & -1 & 3\\ 1 & 1 & -2\\ 2 & 1 & 4\end{vmatrix}$$

$$=1(4+2)-1(-4-4)+3(1-2)$$

$$=6+8-3=11$$

Example 13 If $\mathbf{A}=2\hat{i}-3\hat{j}+7\hat{k}$, $\mathbf{B}=\hat{i}+2\hat{j}$ and $\mathbf{C}=\hat{j}-\hat{k}$. Find $\mathbf{A}\cdot(\mathbf{B}\times\mathbf{C})$.

Sol. $\mathbf{A}\cdot(\mathbf{B}\times\mathbf{C})=[\mathbf{ABC}]=$ volume of parallelopiped

$$=\begin{vmatrix}2 & -3 & 7\\ 1 & 2 & 0\\ 0 & 1 & -1\end{vmatrix}=2(-2-0)+3(-1-0)+7(1-0)$$

$$=-4-3+7=0$$

Therefore **A**, **B** and **C** are coplanar vectors.

Note *Three vectors are coplanar if their scalar triple product is zero.*

Important POINTS

(i) $\mathbf{A} = a_x\hat{\mathbf{i}} + a_y\hat{\mathbf{j}} + a_z\hat{\mathbf{k}}$ is a null vector (of zero magnitude) if individually, a_x, a_y and a_z are zero. Thus,

$$a_x = a_y = a_z = 0$$

(ii)

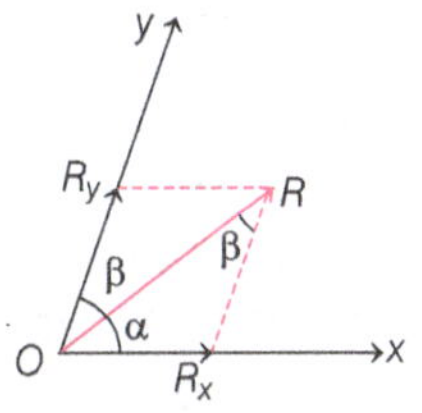

Vector **R** has been resolved in two axes x and y not perpendicular to each other. Applying sine law in the triangle shown , we have

$$\frac{R}{\sin[180-(\alpha+\beta)]} = \frac{R_x}{\sin\beta} = \frac{R_y}{\sin\alpha}$$

or $$R_x = \frac{R\sin\beta}{\sin(\alpha+\beta)}$$

and $$R_y = \frac{R\sin\alpha}{\sin(\alpha+\beta)}$$

If $\alpha+\beta = 90°, R_x = R\sin\beta$ and $R_y = R\sin\alpha$.

(iii) Negative sign in a vector merely represents its direction. It has nothing to do with its magnitude. For example : **A** and −**A** both have same magnitude. Only their directions are opposite.

(iv) If all vectors are one dimensional, then we can write them like scalars by assigning a sign convention. Again positive and negative signs will represent two opposite directions. For example if two or more than two forces are acting only along x-axis.

Then, we can write different forces like,

$$F_1 = 4\text{N} \text{ and } F_2 = -6\text{N}$$

So, F_1 force is of 4N acting towards positive x-direction and F_2 force is of 6N acting towards negative x-direction, if we have taken positive x-direction as positive direction for all force vectors.

Miscellaneous Examples

Ex. 1 Find component of vector **A** + **B** along (i) x-axis, (ii) **C**. Given $\mathbf{A} = \hat{\mathbf{i}} - 2\hat{\mathbf{j}}, \mathbf{B} = 2\hat{\mathbf{i}} + 3\hat{\mathbf{k}}$ and $\mathbf{C} = \hat{\mathbf{i}} + \hat{\mathbf{j}}$.

Sol. $\mathbf{A}+\mathbf{B} = (\hat{\mathbf{i}} - 2\hat{\mathbf{j}}) + (2\hat{\mathbf{i}} + 3\hat{\mathbf{k}}) = 3\hat{\mathbf{i}} - 2\hat{\mathbf{j}} + 3\hat{\mathbf{k}}$

(i) Component of **A** + **B** along x-axis is 3.

(ii) Component of **A** + **B** = **R** (say) along **C** is

$$R\cos\theta = \frac{\mathbf{R}\cdot\mathbf{C}}{C} = \frac{(3\hat{\mathbf{i}} - 2\hat{\mathbf{j}} + 3\hat{\mathbf{k}})\cdot(\hat{\mathbf{i}} + \hat{\mathbf{j}})}{\sqrt{(1)^2+(1)^2}}$$

$$= \frac{3-2}{\sqrt{2}} = \frac{1}{\sqrt{2}}$$

Ex. 2 Find the angle that the vector $\mathbf{A} = 2\hat{\mathbf{i}} + 3\hat{\mathbf{j}} - \hat{\mathbf{k}}$ makes with y-axis.

Sol. $$\cos\theta = \frac{A_y}{A} = \frac{3}{\sqrt{(2)^2+(3)^2+(-1)^2}} = \frac{3}{\sqrt{14}}$$

$$\therefore \quad \theta = \cos^{-1}\left(\frac{3}{\sqrt{14}}\right)$$

Ex. 3 Find the magnitude of resultant of three vectors **OA**, **OB** and **OC** as shown in figure. Radius of circle is 'R'.

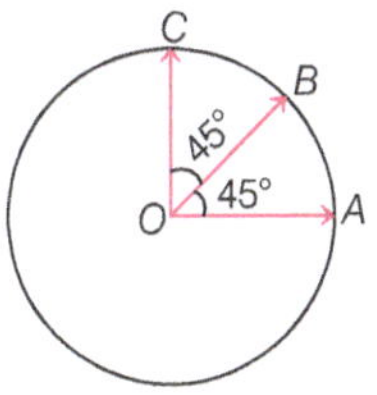

Sol. $OA = OC$

OA + **OC** is along **OB**, (bisector) and its magnitude is $2R\cos 45° = R\sqrt{2}$

(**OA** + **OC**) + **OB** is along **OB** and its magnitude is $R\sqrt{2} + R = R(1+\sqrt{2})$

Ex. 4 Let **A**, **B** and **C** be the unit vectors. Suppose that $\mathbf{A}\cdot\mathbf{B} = \mathbf{A}\cdot\mathbf{C} = 0$ and the angle between **B** and **C** is $\frac{\pi}{6}$ then prove that $\mathbf{A} = \pm 2(\mathbf{B}\times\mathbf{C})$.

Sol. Since, $\mathbf{A}\cdot\mathbf{B} = 0, \mathbf{A}\cdot\mathbf{C} = 0$

Hence, $$(\mathbf{B}+\mathbf{C})\cdot\mathbf{A} = 0$$

So, **A** is perpendicular to (**B** + **C**). Further, **A** is a unit vector perpendicular to the plane of vectors **B** and **C**,

$$\mathbf{A} = \pm\frac{\mathbf{B}\times\mathbf{C}}{|\mathbf{B}\times\mathbf{C}|}$$

$$|\mathbf{B}\times\mathbf{C}| = |\mathbf{B}||\mathbf{C}|\sin\frac{\pi}{6} = 1\times 1\times\frac{1}{2} = \frac{1}{2}$$

$$\mathbf{A} = \pm\frac{\mathbf{B}\times\mathbf{C}}{|\mathbf{B}\times\mathbf{C}|} = \pm 2(\mathbf{B}\times\mathbf{C})$$

Ex. 5 A particle moves on a given line with a constant speed v. At a certain time, it is at a point P on its straight line path. O is a fixed point. Show that (**OP** × **v**) is independent of the position P.

Sol. Let $\mathbf{v} = v\hat{\mathbf{i}}$

$$\mathbf{OP} = x\hat{\mathbf{i}} + y\hat{\mathbf{j}}$$

Take $\mathbf{OP} \times \mathbf{v} = (x\hat{i} + y\hat{j}) \times v\hat{i}$

$= -yv\hat{k}$

$=$ constant (because y is constant)

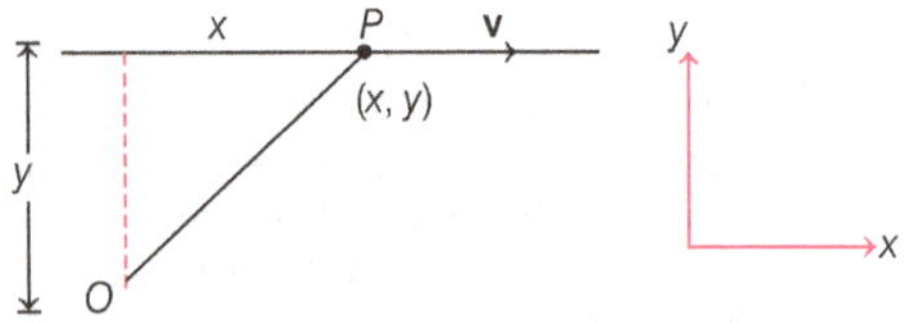

Hence, $\mathbf{OP} \times \mathbf{v}$ is independent of position of P.

Ex. 6 Prove that the mid-point of the hypotenuse of right angled triangle is equidistant from its vertices.

Sol. Here, $\angle CAB = 90°$,

let D be the mid-point of hypotenuse, we have

$$\mathbf{BD} = \mathbf{DC}$$
$$\mathbf{AB} = \mathbf{AD} + \mathbf{DB}$$
$$\mathbf{AC} = \mathbf{AD} + \mathbf{DC} = \mathbf{AD} + \mathbf{BD}$$

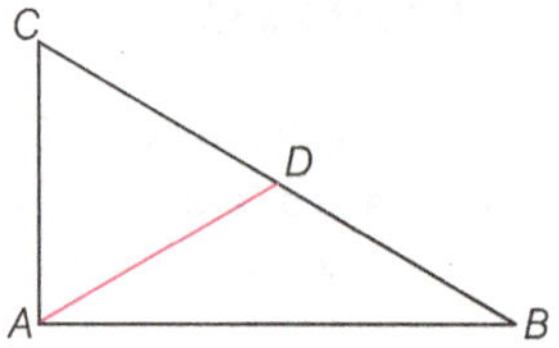

Since, $\angle BAC = 90°$

$\Rightarrow$ $\mathbf{AB} \perp \mathbf{AC}$

$$(\mathbf{AD} + \mathbf{DB}) \cdot (\mathbf{AD} + \mathbf{BD}) = 0$$
$$(\mathbf{AD} - \mathbf{BD}) \cdot (\mathbf{AD} + \mathbf{BD}) = 0$$
$$AD^2 - BD^2 = 0$$

$\therefore$ $AD = BD$ also $BD = DC$

$\because$ D is mid-point of BC.

Thus, $|AD| = |BD| = |DC|$.

So, mid-point of hypotenuse or D is equidistant from A, B and C.

MODULE 5 Exercise

Fill in the Blank

1. If $\mathbf{A} = (2\hat{i} + 3\hat{j} - \hat{k})$ m and $\mathbf{B} = (\hat{i} + 2\hat{j} + 2\hat{k})$ m, then the magnitude of component of vector **A** along vector **B** will be........m. **(2022 Main)**

Single Correct Option

2. If a vector $2\hat{i} + 3\hat{j} + 8\hat{k}$ is perpendicular to the vector $4\hat{i} - 4\hat{j} + \alpha\hat{k}$, then the value of α is
   (a) -1 (b) $\frac{1}{2}$ (c) $-\frac{1}{2}$ (d) 1

3. Unit vector parallel to the resultant of vectors $\mathbf{A} = 4\hat{i} - 3\hat{j}$ and $\mathbf{B} = 8\hat{i} + 8\hat{j}$ will be
   (a) $\frac{24\hat{i} + 5\hat{j}}{13}$ (b) $\frac{12\hat{i} + 5\hat{j}}{13}$
   (c) $\frac{6\hat{i} + 5\hat{j}}{13}$ (d) None of these

4. The sum of two vectors **A** and **B** is at right angles to their difference. Then,
   (a) $A = B$ (b) $A = 2B$
   (c) $B = 2A$
   (d) **A** and **B** have the same direction

5. If the vectors $(\hat{i} + \hat{j} + \hat{k})$ and $3\hat{i}$ form two adjacent sides of a triangle, the area of the triangle is
   (a) $\sqrt{3}$ unit (b) $2\sqrt{3}$ unit
   (c) $\frac{3}{\sqrt{2}}$ unit (d) $3\sqrt{2}$ unit

6. If three vectors along coordinate axis represent the adjacent sides of a cube of length b, then the unit vector along its diagonal passing through the origin will be
   (a) $\frac{\hat{i} + \hat{j} + \hat{k}}{\sqrt{2}}$ (b) $\frac{\hat{i} + \hat{j} + \hat{k}}{\sqrt{36}}$ (c) $\hat{i} + \hat{j} + \hat{k}$ (d) $\frac{\hat{i} + \hat{j} + \hat{k}}{\sqrt{3}}$

7. If $\mathbf{A} = 3\hat{i} + 4\hat{j}$ and $\mathbf{B} = 7\hat{i} + 24\hat{j}$, the vector having the same magnitude as **B** and parallel to **A** is
   (a) $5\hat{i} + 20\hat{j}$ (b) $15\hat{i} + 10\hat{j}$ (c) $20\hat{i} + 15\hat{j}$ (d) $15\hat{i} + 20\hat{j}$

8. The angle between $\mathbf{A} + \mathbf{B}$ and $\mathbf{A} \times \mathbf{B}$ is
   (a) 0 (b) $\frac{\pi}{4}$ (c) $\frac{\pi}{2}$ (d) π

9. The projection of a vector $\mathbf{r} = 3\hat{i} + \hat{j} + 2\hat{k}$ on the x - y plane has magnitude
   (a) 3 (b) 4 (c) $\sqrt{14}$ (d) $\sqrt{10}$

10. Given, $\mathbf{A} = A\cos\theta\hat{i} + A\sin\theta\hat{j}$. A vector **B**, which is perpendicular to **A**, is given by
    (a) $B\cos\theta\hat{i} - B\sin\theta\hat{j}$ (b) $B\sin\theta\hat{i} - B\cos\theta\hat{j}$
    (c) $B\cos\theta\hat{i} + B\sin\theta\hat{j}$ (d) $B\sin\theta\hat{i} + B\cos\theta\hat{j}$

11. The angle which the vector $\mathbf{A} = 2\hat{i} + 3\hat{j}$ makes with the y-axis, where $\hat{i}$ and $\hat{j}$ are unit vectors along x and y-axes respectively, is
    (a) $\cos^{-1}(3/5)$ (b) $\cos^{-1}(2/3)$
    (c) $\tan^{-1}(2/3)$ (d) $\sin^{-1}(2/3)$

12. Given $\mathbf{A} = 4\hat{i} + 6\hat{j}$ and $\mathbf{B} = 2\hat{i} + 3\hat{j}$. Which of the following is correct?

(a) $\mathbf{A} \times \mathbf{B} = 0$ (b) $\mathbf{A} \cdot \mathbf{B} = 24$

(c) $\left|\frac{\mathbf{A}}{\mathbf{B}}\right| = \frac{1}{2}$ (b) $\mathbf{A}$ and $\mathbf{B}$ are antiparallel

13. Given $\mathbf{A} = 2\hat{i} + p\hat{j} + q\hat{k}$ and $\mathbf{B} = 5\hat{i} + 7\hat{j} + 3\hat{k}$. If $\mathbf{A} \parallel \mathbf{B}$, then the values of p and q are respectively

(a) $\frac{14}{5}$ and $\frac{6}{5}$ (b) $\frac{14}{3}$ and $\frac{6}{5}$ (c) $\frac{6}{5}$ and $\frac{1}{3}$ (d) $\frac{3}{4}$ and $\frac{1}{4}$

14. The component of vector $\mathbf{A} = 2\hat{i} + 3\hat{j}$ along the vector $\hat{i} + \hat{j}$ is

(a) $\frac{5}{\sqrt{2}}$ (b) $10\sqrt{2}$ (c) $5\sqrt{2}$ (d) 5

15. In the cube of side 'a' shown in the figure, the vector from the central point of the face $ABOD$ to the central point of the face $BEFO$ will be

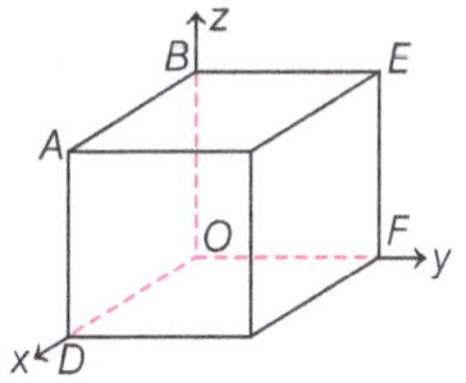

(a) $\frac{1}{2}a(\hat{i} - \hat{k})$

(b) $\frac{1}{2}a(\hat{j} - \hat{i})$

(c) $\frac{1}{2}a(\hat{j} - \hat{k})$

(d) $\frac{1}{2}a(\hat{k} - \hat{i})$ **(2019 Main)**

16. A vector in x - y plane makes an angle of 30° with y-axis. The magnitude of y-component of vector is $2\sqrt{3}$. The magnitude of x-component of the vector will be **(2023 Main)**

(a) $\frac{1}{\sqrt{3}}$ (b) 6 (c) $\sqrt{3}$ (d) 2

Multiple Correct Options

17. If $\mathbf{A} = 2\hat{i} + \hat{j} + \hat{k}$ and $\mathbf{B} = \hat{i} + \hat{j} + \hat{k}$ are two vectors, then the unit vector

(a) perpendicular to $\mathbf{A}$ is $\frac{-\hat{j} + \hat{k}}{\sqrt{2}}$

(b) parallel to $\mathbf{A}$ is $\frac{2\hat{i} + \hat{j} + \hat{k}}{\sqrt{6}}$

(c) perpendicular to $\mathbf{B}$ is $\left(\frac{-\hat{j} + \hat{k}}{\sqrt{2}}\right)$

(d) parallel to $\mathbf{A}$ is $\frac{\hat{i} + \hat{j} + \hat{k}}{\sqrt{3}}$

18. If $\mathbf{A}$ is perpendicular to $\mathbf{r} = x\hat{i} + y\hat{j}$. The possible values of $\mathbf{A}$ are

(a) $x\hat{i} - y\hat{j}$ (b) $y\hat{i} - x\hat{j}$ (c) $+x\hat{j} - y\hat{i}$ (d) $\hat{k}$

19. If the resultant of three forces $\mathbf{F}_1 = p\hat{i} + 3\hat{j} - \hat{k}$, $\mathbf{F}_2 = -5\hat{i} + \hat{j} + 2\hat{k}$ and $\mathbf{F}_3 = 6\hat{i} - \hat{k}$ acting on a particle has magnitude equal to 5 units, then the value(s) of p is (are)

(a) -6 (b) -4 (c) 2 (d) 4

20. The vector $\hat{i} + x\hat{j} + 3\hat{k}$ is rotated through a certain angle and doubled in magnitude, then it becomes $4\hat{i} + (4x - 2)\hat{j} + 2\hat{k}$. The value of x may be

(a) $-\frac{2}{3}$ (b) $\frac{1}{3}$ (c) $\frac{2}{3}$ (d) 2

Subjective Questions

21. The work done by a force $\mathbf{F}$ during a displacement $\mathbf{r}$ is given by $\mathbf{F} \cdot \mathbf{r}$. Suppose a force of 12 N acts on a particle in vertically upward direction and the particle is displaced through 2.0 m in vertically downward direction. Find the work done by the force during this displacement.

22. Obtain the angle between $\mathbf{A} + \mathbf{B}$ and $\mathbf{A} - \mathbf{B}$, if $\mathbf{A} = 2\hat{i} + 3\hat{j}$ and $\mathbf{B} = \hat{i} - 2\hat{j}$.

23. Find the area of the parallelogram whose adjacent sides are represented by $2\hat{i} + 4\hat{j} - 6\hat{k}$ and $\hat{i} + 2\hat{k}$.

24. The x and y-components of vector $\mathbf{A}$ are 4 m and 6 m, respectively. The x and y-components of vector $\mathbf{A} + \mathbf{B}$ are 10 m and 9 m, respectively. Calculate for the vector $\mathbf{B}$ the following :

(a) its x and y-components

(b) its length

(c) the angle it makes with x-axis

25. If two vectors are $\mathbf{A} = 2\hat{i} + \hat{j} - \hat{k}$ and $\mathbf{B} = \hat{j} - 4\hat{k}$. By calculation, prove that $\mathbf{A} \times \mathbf{B}$ is perpendicular to both $\mathbf{A}$ and $\mathbf{B}$.

26. The resultant of two vectors $\mathbf{a}$ and $\mathbf{b}$ has magnitude 2 N. The magnitude of $\mathbf{a}$ is $2\sqrt{3}$ N and makes an angle of 30° with the resultant. Find the magnitude of $\mathbf{b}$ (in newton).

Integer Type Question

27. If the projection of $2\hat{i} + 4\hat{j} - 2\hat{k}$ on $\hat{i} + 2\hat{j} + \alpha\hat{k}$ is zero. Then, the value of α will be **(2022 Main)**

Answers

1. (2)	**2.** (b)	**3.** (b)	**4.** (a)	**5.** (c)
6. (d)	**7.** (d)	**8.** (c)	**9.** (d)	**10.** (b)
11. (c)	**12.** (a)	**13.** (a)	**14.** (a)	**15.** (b)
16. (d)	**17.** (a,b,c)	**18.** (b,c,d)	**19.** (b,c)	**20.** (a,d)
21. (–24 J)	**22.** $\cos^{-1}\left(\frac{\mathbf{R} \cdot \mathbf{S}}{RS}\right)$		**23.** 13.4 units	

24. (a) (6m, 3m) (b) $3\sqrt{5}$ m (c) $\theta = \tan^{-1}\left(\frac{1}{2}\right)$ **26.** 2 N

27. (5)

See hints on pages 154-156.

Unit Exercise (Cumulative)

Level 1

Single Correct Option

1. The sum of magnitudes of two forces is 16 N. If their resultant is normal to the smaller force and has a magnitude of 8 N, then two forces are
(a) 6 N, 10 N (b) 8 N, 8 N
(c) 4 N, 12 N (d) 2 N, 14 N

2. A car travels 6 km towards north at an angle of 45° to the east and then travels a distance of 4 km towards north at an angle of 135° to the east. How far is the point from the starting point? What angle does the straight line joining its initial and final position makes with the east? **(2008 JEE)**
(a) $\sqrt{50}$ km and $\tan^{-1}(5)$ (b) 10 km and $\tan^{-1}(\sqrt{5})$
(c) $\sqrt{52}$ km and $\tan^{-1}(5)$ (d) $\sqrt{52}$ km and $\tan^{-1}(\sqrt{5})$

3.. The resultant of two forces, one double the other in magnitude, is perpendicular to the smaller of the two forces. The angle between the two forces is
(a) 120° (b) 60° (c) 90° (d) 150°

4. Given that $P = Q = R$. If $\mathbf{P} + \mathbf{Q} = \mathbf{R}$, then the angle between **P** and **R** is θ_1. If $\mathbf{P} + \mathbf{Q} + \mathbf{R} = 0$, then the angle between **P** and **R** is θ_2. What is the relation between θ_1 and θ_2?
(a) $\theta_1 = \theta_2$ (b) $\theta_1 = \dfrac{\theta_2}{2}$
(c) $\theta_1 = 2\theta_2$ (d) None of these

5. In the figure shown *ABCDEF* is a regular hexagon. What is the value of $\mathbf{AB} + \mathbf{AC} + \mathbf{AD} + \mathbf{AE} + \mathbf{AF}$?

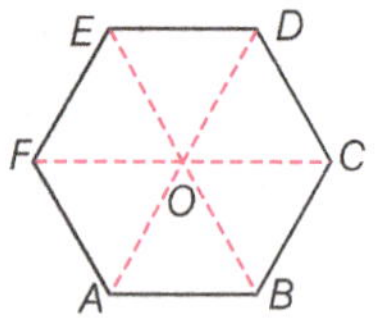

(a) **AO** (b) **2AO**
(c) **4AO** (d) **6AO**

6. Figure shows three vectors **p**, **q** and **r**, where *C* is the mid-point of *AB*. Then, which of the following relation is correct?

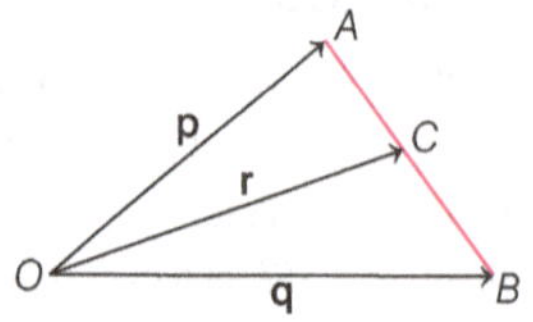

(a) $\mathbf{p} + \mathbf{q} = 2\mathbf{r}$ (b) $\mathbf{p} + \mathbf{q} = \mathbf{r}$
(c) $\mathbf{p} - \mathbf{q} = 2\mathbf{r}$ (d) $\mathbf{p} - \mathbf{q} = \mathbf{r}$

7. Given that $\mathbf{A} + \mathbf{B} = \mathbf{C}$ and that **C** is perpendicular to **A** further if $|\mathbf{A}| = |\mathbf{C}|$, then what is the angle between **A** and **B** ?
(a) $\dfrac{2\pi}{5}$ (b) $\dfrac{5\pi}{6}$ (c) $\dfrac{3\pi}{4}$ (d) $\dfrac{2\pi}{3}$

8. If $\mathbf{a} + \mathbf{b} + \mathbf{c} = 0$. The angle between **a** and **b**; and **b** and **c** are 150° and 120°, respectively. Then, the magnitude of vectors **a**, **b** and **c** are in ratio of
(a) $1 : 2 : 3$ (b) $1 : 2 : \sqrt{3}$
(c) $\sqrt{3} : 2 : 1$ (d) $2 : \sqrt{3} : 1$

9. Three vectors **P**, **Q** and **R** are shown in the figure. Let *S* be any point on the vector **R**. The distance between the points *P* and *S* is $b|\mathbf{R}|$. The general relation among vectors **P**, **Q** and **S** is **(2017 Adv.)**

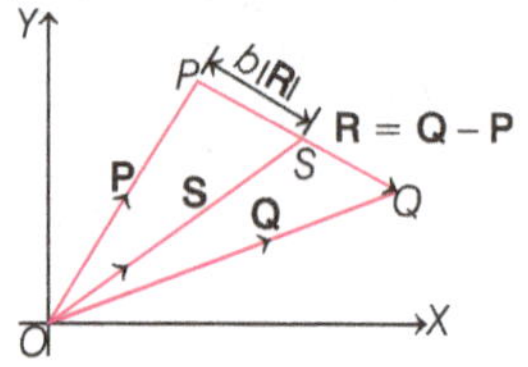

(a) $\mathbf{S} = (1 - b^2)\mathbf{P} + b\mathbf{Q}$ (b) $\mathbf{S} = (b - 1)\mathbf{P} + b\mathbf{Q}$
(c) $\mathbf{S} = (1 - b)\mathbf{P} + b\mathbf{Q}$ (d) $\mathbf{S} = (1 - b)\mathbf{P} + b^2\mathbf{Q}$

10. The value of $\hat{\mathbf{i}} \times (\hat{\mathbf{i}} \times \mathbf{a}) + \hat{\mathbf{j}} \times (\hat{\mathbf{j}} \times \mathbf{a}) + \hat{\mathbf{k}} \times (\hat{\mathbf{k}} \times \mathbf{a})$ is
(a) **a** (b) $\mathbf{a} \times \hat{\mathbf{k}}$ (c) $-2\,\mathbf{a}$ (d) $-\mathbf{a}$

11. The angle between $\mathbf{P} + \mathbf{Q}$ and $\mathbf{P} - \mathbf{Q}$ will be
(a) 90° (b) between 0° and 180°
(c) 180° only (d) None of these

12. The three vectors $\mathbf{A} = 3\hat{\mathbf{i}} - 2\hat{\mathbf{j}} + \hat{\mathbf{k}}$, $\mathbf{B} = \hat{\mathbf{i}} - 3\hat{\mathbf{j}} + 5\hat{\mathbf{k}}$ and $\mathbf{C} = 2\hat{\mathbf{i}} + \hat{\mathbf{j}} - 4\hat{\mathbf{k}}$ form
(a) an equilateral triangle (b) isosceles triangle
(c) a right angled triangle (d) no triangle

Multiple Correct Options

13. Given two vectors $\mathbf{A} = 3\hat{\mathbf{i}} + 4\hat{\mathbf{j}}$ and $\mathbf{B} = \hat{\mathbf{i}} + \hat{\mathbf{j}}$. θ is the angle between **A** and **B**. Which of the following statements is/are correct?
(a) $|\mathbf{A}|\cos\theta\left(\dfrac{\hat{\mathbf{i}} + \hat{\mathbf{j}}}{\sqrt{2}}\right)$ is the component of **A** along **B**.
(b) $|\mathbf{A}|\sin\theta\left(\dfrac{\hat{\mathbf{i}} - \hat{\mathbf{j}}}{\sqrt{2}}\right)$ is the component of **A** perpendicular to **B**.
(c) $|\mathbf{A}|\cos\theta\left(\dfrac{\hat{\mathbf{i}} - \hat{\mathbf{j}}}{\sqrt{2}}\right)$ is the component of **A** along **B**.
(d) $|\mathbf{A}|\sin\theta\left(\dfrac{\hat{\mathbf{i}} + \hat{\mathbf{j}}}{\sqrt{2}}\right)$ is the component of **A** perpendicular to **B**.

14. A ray of light is incident along the vector $\frac{1}{\sqrt{2}}\hat{i} - \frac{1}{\sqrt{2}}\hat{j} + \hat{k}$ on a plane mirror placed in x-y plane.
(a) The normal on incidence point is along z-axis
(b) The angle of incidence is 30°
(c) The angle of reflection is 60°
(d) The angle of incidence is 45°

Match the Columns

15. Match the information given in Column I with Column II. Select the correct option from the codes given below.

	Column I		Column II
A.	The component of vector $6\hat{i} - 3\hat{j} - 6\hat{k}$ parallel to the vector $\hat{i} + \hat{j} + \hat{k}$.	p.	0
B.	The component of vector $6\hat{i} - 3\hat{j} - 6\hat{k}$ perpendicular to the vector $\hat{i} + \hat{j} + \hat{k}$.	q.	$-\sqrt{3}$
C.	The component of vector $6\hat{i} - 3\hat{j} - 6\hat{k}$ perpendicular to the vector $2\hat{i} - \hat{j} - 2\hat{k}$.	r.	$7\hat{i} - 2\hat{j} - 5\hat{k}$
D.	The component of vector $6\hat{i} - 3\hat{j} - 6\hat{k}$ parallel to the vector $2\hat{j} - \hat{k}$.	s.	$-\hat{i} - \hat{j} - \hat{k}$

Codes

	A	B	C	D		A	B	C	D
(a)	s	r	q	p	(b)	s	r	p	p
(c)	p	q	r	s	(d)	q,s	p	r	p

16. If $\mathbf{a} = \hat{i} - 2\hat{j} - 3\hat{k}$, $\mathbf{b} = 2\hat{i} + \hat{j} - \hat{k}$ and $\mathbf{c} = \hat{i} + 3\hat{j} - 2\hat{k}$.
Match the Column I with Column II and select the correct option from the given code below.

	Column I		Column II
A.	$\lvert(\mathbf{a} \times \mathbf{b}) \times \mathbf{c}\rvert$	p.	$35\sqrt{3}$
B.	$\lvert\mathbf{a} \times (\mathbf{b} \times \mathbf{c})\rvert$	q.	20
C.	$\lvert\mathbf{a} \cdot (\mathbf{b} \times \mathbf{c})\rvert$	r.	$3\sqrt{10}$
D.	$\lvert(\mathbf{a} \times \mathbf{b}) (\mathbf{b} \cdot \mathbf{c})\rvert$	s.	$5\sqrt{26}$

Codes

	A	B	C	D		A	B	C	D
(a)	p	q	r	s	(b)	q	p	s	r
(c)	s	r	q	p	(d)	p,q	q,r	r	p,s

Subjective Questions

17. The resultant of vectors **OA** and **OB** is perpendicular to **OA** as shown in figure. Find the angle AOB.

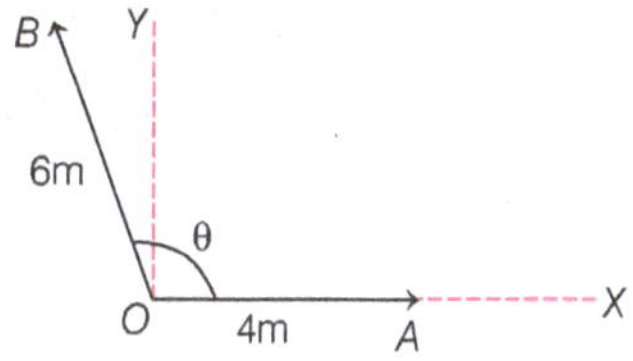

18. When two non-zero vectors **a** and **b** are perpendicular to each other, their magnitude of resultant is R. When they are opposite to each other, their resultant is of magnitude $\frac{R}{\sqrt{2}}$. Find the value of $\left(\frac{a}{b} + \frac{b}{a}\right)$.

19. Prove by the method of vectors that in a triangle $\frac{a}{\sin A} = \frac{b}{\sin B} = \frac{c}{\sin C}$.

Integer Type Qyestion

20. Vectors $a\hat{i} + b\hat{j} + \hat{k}$ and $2\hat{i} - 3\hat{j} + 4\hat{k}$ are perpendicular to each other when $3a + 2b = 7$, the ratio of a to b is $\frac{x}{2}$. The value of x is.......... **(2023 Main)**

Level 2

Statement Type Question

21. **Statement I :** Two forces $(\mathbf{P} + \mathbf{Q})$ and $(\mathbf{P} - \mathbf{Q})$ where $\mathbf{P} \perp \mathbf{Q}$, when act at an angle θ_1 to each other, the magnitude of their resultant is $\sqrt{3(P^2 + Q^2)}$, when they act at an angle θ_2, the magnitude of their resultant becomes $\sqrt{2(P^2 + Q^2)}$. This is possible only when $\theta_1 < \theta_2$.
Statement II : In the situation given above. $\theta_1 = 60°$ and $\theta_2 = 90°$.
In the light of the above statements, choose the most appropriate answer from the options given below: **(2021 Main)**
(a) Statement I is false but Statement II is true
(b) Both Statement I and Statement II are true
(c) Statement I is true but Statement II is false
(d) Both Statement I and Statement II are false

Answers

Level 1

1. (a)	**2.** (c)	**3.** (a)	**4.** (b)	**5.** (d)	**6.** (a)	**7.** (c)	**8.** (c)	**9.** (c)	**10.** (c)
11. (b)	**12.** (c)	**13.** (a,b)	**14.** (a,d)	**15.** (b)	**16.** (c)	**17.** $\cos^{-1}\left(\frac{-2}{3}\right)$		**18.** (4)	**20.** (1)

Level 2

21. (b)

See hints on pages 156-158.

Hints & SOLUTIONS

Module 2.

Introductory Exercise 1

1. Apply $R = \sqrt{A^2 + B^2 + 2AB\cos\theta}$

2. 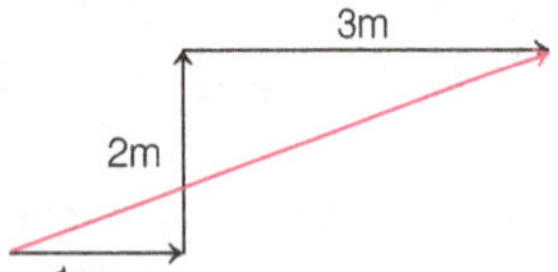

$$|\mathbf{S}| \text{ or } S = \sqrt{(1+3)^2 + (2)^2} = \sqrt{20}\text{ m}$$

3. $R = 2A\cos\dfrac{\theta}{2}$

$$= (2)(10)\cos\frac{30°}{2}$$

$$= 20\cos 15°$$

Resultant passes through the centre.

4. 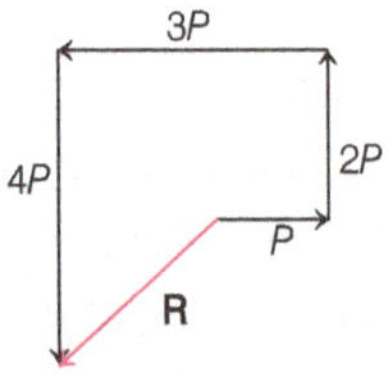

$$|\mathbf{R}| = \sqrt{(2P)^2 + (2P)^2}$$

$$= 2\sqrt{2}P$$

Introductory Exercise 2

1. Apply $S = \sqrt{A^2 + B^2 - 2AB\cos\theta}$

2. $R = S \Rightarrow \sqrt{A^2 + B^2 + 2AB\cos\theta}$

$$= \sqrt{A^2 + B^2 - 2AB\cos\theta}$$

Solving, we get $\cos\theta = 0$ or $\theta = 90°$

Exercise

2. $R = \sqrt{A^2 + A^2 + 2AA\cos\theta}$

$R = A$ at $\theta = 120°$ **Ans.**

3. For given condition, $\theta = \dfrac{360°}{n} = 30°$

4. $\mathbf{A} + \mathbf{B} + \mathbf{C} = 0$

$\Rightarrow \quad \mathbf{C} = -(\mathbf{A} + \mathbf{B})$ or $|\mathbf{C}| = |-(\mathbf{A}+\mathbf{B})|$

or $\quad A \sim B \le C \le A + B$

5. Resultant is always inclined towards a vector of larger magnitude.

6. $A \sim B \le C \le A + B$

7. $A \sim B \le C \le A + B$

8. $\because \mathbf{C} = \mathbf{A} + \mathbf{B}$

Let the angle between **A** and **B** be θ.

This implies that **C** is the resultant of **A** and **B**.

$\therefore \quad C = \sqrt{A^2 + B^2 + 2AB\cos\theta}$

$\Rightarrow \quad 5 = \sqrt{9 + 16 + 2 \times 3 \times 4 \times \cos\theta}$

$\Rightarrow \quad 5 = \sqrt{25 + 24\cos\theta}$

$\Rightarrow \quad \cos\theta = 0$

$\Rightarrow \quad \theta = \dfrac{\pi}{2}$

9. 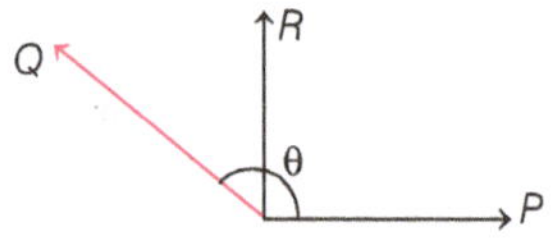

$\Rightarrow \quad \tan 90° = \dfrac{Q\sin\theta}{P + Q\cos\theta} \Rightarrow P + Q\cos\theta = 0$

$\cos\theta = \dfrac{-P}{Q}$

$\therefore \quad \theta = \cos^{-1}\left(\dfrac{-P}{Q}\right)$

10. Use, $R = 2A\cos\dfrac{\theta}{2}$

and $S = 2A\sin\dfrac{\theta}{2}$

Here, $A = 1$

11. **In first case :**

$$R^2 = (2F)^2 + (3F)^2 + 2(2F)(3F)\cos\theta$$

or $\quad R^2 = 13F^2 + 12F^2\cos\theta \quad$ ...(i)

In second case :

$$(2R)^2 = (2F)^2 + (6F)^2 + 2 \times 2 \times 6F^2\cos\theta$$

$\Rightarrow \quad 4R^2 = 40F^2 + 24F^2\cos\theta \quad$ ... (ii)

Solving these two equations, we get

$$\theta = 120°$$

12. $|\mathbf{a} + \mathbf{b}| = 2A\cos\dfrac{\theta}{2} = (2)(1)\cos 30° = \sqrt{3}$

$|\mathbf{a} - \mathbf{b}| = 2A\cos\dfrac{\theta}{2} = (2)(1)\cos 30° = 1$

13. $\sqrt{A^2+B^2+2AB\cos\theta} = 2\sqrt{A^2+B^2-2AB\cos\theta}$

From this equation, we cannot find θ.

14. Let θ be the angle between **A** and **B**.

$$|\mathbf{A}+\mathbf{B}| = n|\mathbf{A}-\mathbf{B}|$$

$$\Rightarrow \sqrt{A^2+B^2+2AB\cos\theta} = n\sqrt{A^2+B^2+2AB\cos(180^\circ-\theta)}$$

$$|\mathbf{A}| = |\mathbf{B}| = A = B = x$$

$$2x^2(1+\cos\theta) = n^2\cdot 2x^2(1-\cos\theta)$$

$$1+\cos\theta = n^2 - n^2\cos\theta$$

$$(1+n^2)\cos\theta = n^2-1$$

or $$\cos\theta = \frac{n^2-1}{n^2+1}$$

or $$\theta = \cos^{-1}\left(\frac{n^2-1}{n^2+1}\right)$$

15. $|\Delta\mathbf{a}| = \sqrt{a^2+a^2-2\cdot a\cdot a\cdot\cos(\Delta\theta)}$

$$= 2a\sin\left(\frac{\Delta\theta}{2}\right)$$

For small angles, $\sin\frac{\Delta\theta}{2} \approx \frac{\Delta\theta}{2}$

$\therefore$ $$|\Delta\mathbf{a}| = 2a\times\frac{\Delta\theta}{2} = a\cdot\Delta\theta$$

16. Given, $|\mathbf{A}+\mathbf{B}| = |\mathbf{A}-\mathbf{B}|\times 2$

$$\Rightarrow |\mathbf{A}+\mathbf{B}|^2 = |\mathbf{A}-\mathbf{B}|^2\times 4$$

$$A^2+A^2+2A^2\cos\theta = 4(A^2+A^2-2A^2\cos\theta)$$

Solving we get

$$\cos\theta = \frac{3}{5}$$

18. $\mathbf{a}-\mathbf{b}$ is nothing but addition of **a** and $-\mathbf{b}$.

So, the magnitude of $\mathbf{a}-\mathbf{b}$ will lie between $|\mathbf{a}|+|\mathbf{b}|$ and $|\mathbf{a}|-|\mathbf{b}|$.

19.

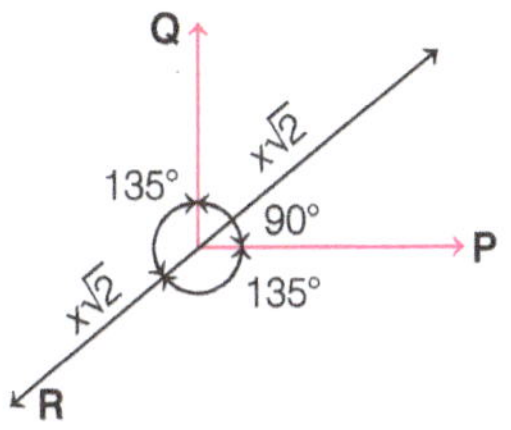

$$|\mathbf{P}| = |\mathbf{Q}| = x$$
$$|\mathbf{R}| = \sqrt{2}\,x$$

20. $$R = \sqrt{(3P)^2+(2P)^2+2(3P)(2P)\cos\theta} \quad \text{...(i)}$$

$$2R = \sqrt{(6P)^2+(2P)^2+2(6P)(2P)\cos\theta} \quad \text{...(ii)}$$

Solving these two equations, we get

$$\cos\theta = -\frac{1}{2}$$

or $$\theta = 120^\circ$$

21. From polygon law of vector addition, $\mathbf{C}+\mathbf{A} = \mathbf{B}$.

22. $$R_1^2 = A^2+B^2+2AB\cos\theta$$

$$R_2^2 = A^2+B^2-2AB\cos\theta$$

$\therefore$ $$R_1^2+R_2^2 = 2(A^2+B^2)$$

23. Apply polygon law.

24. From parallelogram law of vector addition, $\mathbf{P}+\mathbf{Q} = \mathbf{R}$

Also, $\mathbf{P}+\mathbf{S} = \mathbf{Q}$ (triangle rule)

$\therefore$ $$\mathbf{S} = \mathbf{Q}-\mathbf{P}$$

Module 3.

1. $\mathbf{A}\cdot\mathbf{B} = 0$

$\therefore$ $$\mathbf{A}\perp\mathbf{B}$$

$$\mathbf{A}\cdot\mathbf{C} = 0$$

$\therefore$ $$\mathbf{A}\perp\mathbf{C}$$

A is perpendicular to both **B** and **C** and $\mathbf{B}\times\mathbf{C}$ is also perpendicular to both **B** and **C**. Therefore, **A** is parallel to $\mathbf{B}\times\mathbf{C}$.

2. Since, $\mathbf{a}_1$ and $\mathbf{a}_2$ are non-collinear.

$\therefore$ $$a_1 = a_2 = 1$$

and $$|\mathbf{a}_1+\mathbf{a}_2| = \sqrt{3}$$

$$\Rightarrow a_1^2+a_2^2+2a_1a_2\cos\theta = (\sqrt{3})^2$$

$$\Rightarrow 1+1+2\cos\theta = 3$$

$$\Rightarrow \cos\theta = \frac{1}{2}$$

Now, $(\mathbf{a}_1-\mathbf{a}_2)\cdot(2\mathbf{a}_1+\mathbf{a}_2)$

$$= 2a_1^2 - a_2^2 - a_1a_2\cos\theta$$

$$= 2-1-\frac{1}{2} = \frac{1}{2}$$

3. $5 = \sqrt{(3)^2+(5)^2+2\times3\times5\cos\theta}$

$\therefore$ $$\cos\theta = -\frac{3}{10}$$

Now, $(2\mathbf{A}_1+3\mathbf{A}_2)\cdot(3\mathbf{A}_1-2\mathbf{A}_2)$

$$= 6|\mathbf{A}_1|^2 - 6|\mathbf{A}_2|^2 + 5\mathbf{A}_1\cdot\mathbf{A}_2$$

$$= 6|\mathbf{A}_1|^2 - 6|\mathbf{A}_2|^2 + 5|\mathbf{A}_1||\mathbf{A}_2|\cos\theta$$

$$= 6\times9 - 6\times25 + 5\times3\times5\times\left(\frac{-3}{10}\right)$$

$$= -118.5$$

5. $(\mathbf{a}+\mathbf{b})\times(\mathbf{a}-\mathbf{b}) = \mathbf{a}\times\mathbf{a} - \mathbf{a}\times\mathbf{b} + \mathbf{b}\times\mathbf{a} - \mathbf{b}\times\mathbf{b}$

$$= 2(\mathbf{b}\times\mathbf{a})$$

As $\mathbf{a}\times\mathbf{a}$ and $\mathbf{b}\times\mathbf{b}$ are two null vectors and $-\mathbf{a}\times\mathbf{b} = \mathbf{b}\times\mathbf{a}$

6. $|\mathbf{A}\times\mathbf{B}| = AB\sin\theta$

$$0 \le \sin\theta \le 1$$

$$\therefore \quad 0 \le |\mathbf{A}\times\mathbf{B}| \le AB$$

7. $\mathbf{A}\times\mathbf{B}$ is perpendicular to both **A** and **B**. So, this is parallel or antiparallel to **C**. Now, cross product of two parallel or antiparallel vectors is zero. Hence,

$$\mathbf{C}\times(\mathbf{A}\times\mathbf{B}) = 0.$$

8. $|\mathbf{A}\times\mathbf{B}| = \sqrt{3}\,\mathbf{A}\cdot\mathbf{B}$

$$\Rightarrow \quad AB\sin\theta = \sqrt{3}\,AB\cos\theta$$

$$\text{or} \quad \tan\theta = \sqrt{3}$$

$$\Rightarrow \quad \theta = 60^\circ$$

$$|\mathbf{A}+\mathbf{B}| = \sqrt{A^2+B^2+2AB\cos 60^\circ}$$

$$= \sqrt{A^2+B^2+AB}$$

9. $(\mathbf{B}\times\mathbf{A})$ is $\perp$ to both **A** and **B**.

$$\therefore \quad (\mathbf{B}\times\mathbf{A})\cdot\mathbf{A} = 0$$

10. $(\mathbf{A}\times\mathbf{B})$ is perpendicular to both **A** and **B** or $(\mathbf{A}+\mathbf{B})$ and dot product of two perpendicular vectors is zero.

11. If **A** is perpendicular to **B**, then $\mathbf{A}\cdot\mathbf{B} = 0$ and $\mathbf{A}\times\mathbf{B} \ne 0$.

12. (a) $AB\sin\theta = |AB\cos\theta|$

$$\therefore \quad \tan\theta = \pm 1$$

$$\Rightarrow \quad \theta = 45^\circ \text{ or } 135^\circ$$ **Ans.**

(b) If **A** and **B** are parallel or antiparallel to each other, then their cross product is a null vector and they may be said to be equal otherwise for any other angle,

$$\mathbf{A}\times\mathbf{B} = -\mathbf{B}\times\mathbf{A}$$

(c) $\sqrt{A^2+B^2+2AB\cos\theta} = \sqrt{A^2+B^2-2AB\cos\theta}$

$$\therefore \quad \cos\theta = 0$$

$$\text{or} \quad \theta = 90^\circ$$ **Ans.**

(d) $A+B=C$ only at $\theta = 0^\circ$

13. $\mathbf{A}\times\mathbf{B}$ is perpendicular to **A**. Now, dot product of two perpendicular vectors is zero. Hence,

$$\mathbf{A}\cdot(\mathbf{A}\times\mathbf{B}) = 0$$

Module 4.

1. $A = |\mathbf{A}| = \sqrt{(3)^2+(6)^2+(2)^2}$

$$= 7$$

$$\alpha = \cos^{-1}\left(\frac{A_x}{A}\right) = \cos^{-1}\left(\frac{3}{7}\right)$$

= angle of **A** with positive x-axis.

Similarly, β and γ angles.

2. The magnitude of unit vector is 1.

$$\therefore \quad |a| = 1 \text{ or } \sqrt{a_1^2+a_2^2+a_3^2} = 1$$

This equation is satisfied by options (a), (b) and (d).

3. $\because \alpha = \beta = \gamma$

But $\cos^2\alpha + \cos^2\beta + \cos^2\gamma = 1$

or $3\cos^2\alpha = 1$ or $\cos\alpha = \pm\frac{1}{\sqrt{3}}$

$$\therefore \quad F = F\cos\alpha\,\hat{\mathbf{i}} + F\cos\beta\,\hat{\mathbf{j}} + F\cos\gamma\,\hat{\mathbf{k}}$$

$$\text{or} \quad F = \pm(\hat{\mathbf{i}}+\hat{\mathbf{j}}+\hat{\mathbf{k}})\,\text{N}$$

4. $F_z = 50\cos 45^\circ = 25\sqrt{2}$ N

$$F_{xy} = 50\sin 45^\circ = 25\sqrt{2}\,\text{N}$$

$$\therefore \quad F_x = F_{xy}\cos 45^\circ = 25\sqrt{2}\times\frac{1}{\sqrt{2}} = 25\,\text{N}$$

$$\text{and} \quad F_y = F_{xy}\sin 45^\circ = 25\sqrt{2}\times\frac{1}{\sqrt{2}} = 25\,\text{N}$$

5. The given force is as shown in figure.

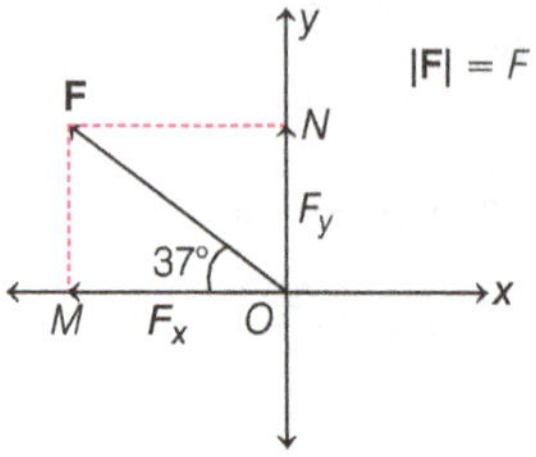

$|\mathbf{F}| = F = 15$ N

Let us find its x and y-components.

$$F_x = F\cos 37^\circ$$

$$= 15\times\frac{4}{5} = 12\,\text{N} \quad \text{(along negative } x\text{-axis)}$$

$$F_y = F\sin 37^\circ = 15\times\frac{3}{5}$$

$$= 9\,\text{N} \quad \text{(along positive } y\text{-axis)}$$

From parallelogram law of vector addition, we can see that

$$\mathbf{F} = \mathbf{OM} + \mathbf{ON}$$

$$= F_x(-\hat{\mathbf{i}}) + F_y(\hat{\mathbf{j}})$$

$$= (-12\hat{\mathbf{i}} + 9\hat{\mathbf{j}})\,\text{N}$$ **Ans.**

6. **Magnitude of A**

$$|\mathbf{A}| \text{ or } A = \sqrt{(6)^2+(-8)^2} = 10 \text{ units}$$ **Ans.**

Direction of A Vector **A** can be shown as

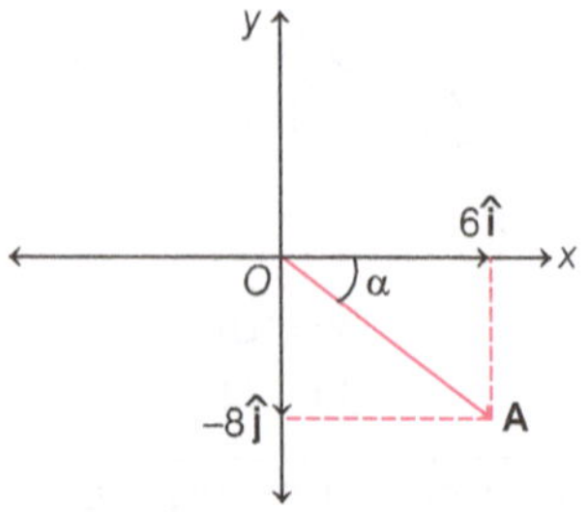

$$\tan\alpha = \frac{8}{6} = \frac{4}{3}$$

$\therefore \quad \alpha = \tan^{-1}\left(\frac{4}{3}\right) = 53°$

Therefore, **A** is making an angle of 53° from positive x-axis towards negative y-axis.

7. Component perpendicular to the plane

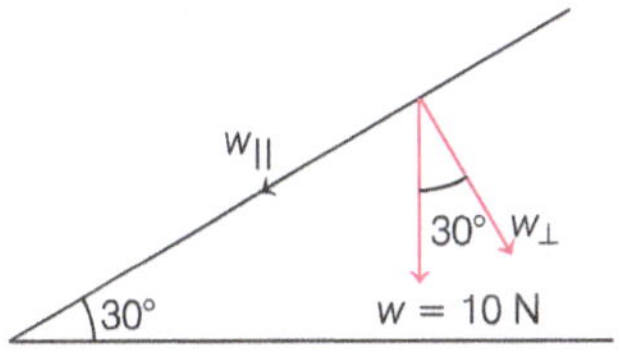

$$w_{\perp} = w\cos 30° = (10)\frac{\sqrt{3}}{2} = 5\sqrt{3}\text{ N}$$

and component parallel to the plane

$$w_{||} = w\sin 30° = (10)\left(\frac{1}{2}\right) = 5\text{ N}$$

Ans.

8. Horizontal component of **F** is

$$F_H = F\cos 45° = (8)\left(\frac{1}{\sqrt{2}}\right) = 4\sqrt{2}\text{ N}$$

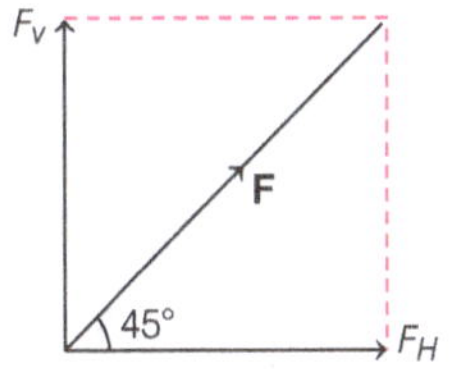

and vertical component of **F** is

$$F_V = F\sin 45° = (8)\left(\frac{1}{\sqrt{2}}\right) = 4\sqrt{2}\text{ N}$$

9. A or $|\mathbf{A}| = \sqrt{(3)^2 + (-4)^2 + (5)^2} = 5\sqrt{2}$ units

Directions of cosines are,

$$\cos\alpha = \frac{A_x}{A} = \frac{3}{5\sqrt{2}}$$

$$\cos\beta = \frac{A_y}{A} = \frac{-4}{5\sqrt{2}}$$

and $$\cos\gamma = \frac{A_z}{A} = \frac{1}{\sqrt{2}}$$

10.

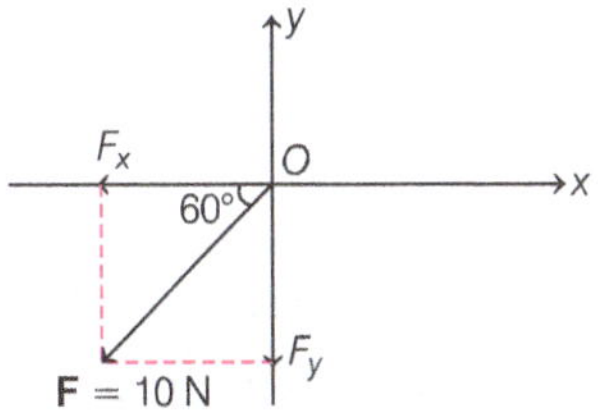

$F_x = 10\cos 60° = 5\text{ N}$ (along negative x-direction)

$F_y = 10\sin 60° = 5\sqrt{3}\text{ N}$ (along negative y-direction)

11. Given,

$$\mathbf{B} - \mathbf{A} = 2\hat{j}$$

$$\Rightarrow \quad \mathbf{B} = \mathbf{A} + 2\hat{j}$$

$$\Rightarrow \quad \mathbf{B} = 2\hat{i} + 3\hat{j} + 2\hat{k} + 2\hat{j}$$

$$\Rightarrow \quad \mathbf{B} = 2\hat{i} + 5\hat{j} + 2\hat{k}$$

$$|\mathbf{B}| = \sqrt{(2)^2 + (5)^2 + (2)^2} = \sqrt{4 + 25 + 4} = \sqrt{33}$$

12. $2\mathbf{A} - 3\mathbf{B} = 2(\hat{i} + \hat{j} - 2\hat{k}) - 3(2\hat{i} - \hat{j} + \hat{k}) = -4\hat{i} + 5\hat{j} - 7\hat{k}$

$\therefore$ Magnitude of $2\mathbf{A} - 3\mathbf{B} = \sqrt{(-4)^2 + (5)^2 + (-7)^2} = \sqrt{16 + 25 + 49} = \sqrt{90}$

13. $\mathbf{A} - 2\mathbf{B} + 3\mathbf{C} = (2\hat{i} + 3\hat{j}) - 2(\hat{i} + \hat{j}) + 3\hat{k} = (\hat{j} + 3\hat{k})$

$\therefore \quad |\mathbf{A} - 2\mathbf{B} + 3\mathbf{C}| = \sqrt{(1)^2 + (3)^2} = \sqrt{10}$ units

14. (a) is correct option according to question,

(b) $\hat{k}\cdot\hat{i} = 0$ as $\theta = 90°$

(c) $\hat{i}\cdot\hat{i} = 1$

(d) $\hat{j}\times\hat{j} = 0$

15. (a) $2\hat{i}\times(\hat{i} + \hat{j}) = 2\hat{i}\times\hat{i} + 2\hat{i}\times\hat{j} = 2\hat{k}$

(b) $4\hat{i}\times(2\hat{i} - \hat{j}) = 4\hat{i}\times 2\hat{i} - 4\hat{i}\times\hat{j} = 0 - 4\hat{k} = -4\hat{k}$

(c) $3\hat{i}\times(3\hat{i} - 4\hat{j}) = 3\hat{i}\times 3\hat{i} - 12(\hat{i}\times\hat{j}) = 0 - 12\hat{k} = -12\hat{k}$

(d) $2(\hat{i} + \hat{j})\times 2\hat{i} = 4(\hat{i}\times\hat{j}) + 4(\hat{j}\times\hat{i}) = 0 + (-4\hat{k}) = -4\hat{k}$

17. $2\mathbf{A} = 4\hat{i} - 2\hat{j}$

$$-3\mathbf{B} = -3\hat{j} - 3\hat{k}$$

$$\Rightarrow \quad (2\mathbf{A})\times(-3\mathbf{B}) = \begin{vmatrix} \hat{i} & \hat{j} & \hat{k} \\ 4 & -2 & 0 \\ 0 & -3 & -3 \end{vmatrix}$$

$$= \hat{i}(6 - 0) + \hat{j}(0 + 12) + \hat{k}(-12 - 0)$$

$$= 6\hat{i} + 12\hat{j} - 12\hat{k}$$

18. $$\mathbf{A}\cdot(\mathbf{B}\times\mathbf{C}) = \begin{vmatrix} 2 & -3 & 7 \\ 1 & 0 & 2 \\ 0 & 1 & -1 \end{vmatrix}$$

$$= 2(0 - 2) - 3(0 + 1) + 7(1 - 0)$$

$$= -4 - 3 + 7 = 0$$

19. $\sqrt{(0.5)^2 + (0.8)^2 + c^2} = 1$ **Ans.**

20. $|\mathbf{A}| = \sqrt{9 + 16} = 5$ and $|\mathbf{B}| = \sqrt{36 + 64} = 10.$

$\mathbf{B} = 2\mathbf{A}$ or **B** is parallel to **A**. Ratio of their coefficients are equal, so they are parallel. Or their cross product is zero.

21. Unit vector along y-axis is $\hat{j}$, so the required vector

$$=\hat{j}-[(\hat{i}-3\hat{j}+2\hat{k})+(3\hat{i}+6\hat{j}-7\hat{k})]$$

$$=-4\hat{i}-2\hat{j}+5\hat{k}$$

22. (a) The resultant $=R=a+b=7\hat{i}+\hat{j}$

$\therefore$ The magnitude of resultant

$$=\sqrt{49+1}=\sqrt{50}$$

$$=5\sqrt{2}\text{ units}$$

(b) $\because \quad a\cdot b=|a||b|\cos\theta$

$$\therefore \quad \cos\theta=\frac{a\cdot b}{|a||b|}$$

$$=\frac{12-12}{\sqrt{9+16}\sqrt{16+9}}$$

$$=0$$

$$\Rightarrow \quad \theta=90^\circ$$

Module 5.

1. Given

$\mathbf{A}=(2\hat{i}+3\hat{j}-\hat{k})\text{ m}$ and $\mathbf{B}=(\hat{i}+2\hat{j}+2\hat{k})\text{ m}$

Component of vector $\mathbf{A}$ along vector $\mathbf{B}=\mathbf{A}\cdot\mathbf{B}$

$$=\frac{\mathbf{A}\cdot\mathbf{B}}{|\mathbf{B}|}=\frac{2+6-2}{\sqrt{1^2+2^2+2^2}} \qquad \left(\because \mathbf{B}=\frac{\mathbf{B}}{|\mathbf{B}|}\right)$$

$$=\frac{6}{3}=2\text{ m}$$

2. $(2\hat{i}+3\hat{j}+8\hat{k})\cdot(4\hat{i}-4\hat{j}+\alpha\hat{k})=0$

$$\therefore \quad 8-12+8\alpha=0$$

$$\therefore \quad \alpha=\frac{1}{2} \qquad \textbf{Ans.}$$

3. $\mathbf{R}=\mathbf{A}+\mathbf{B}=12\hat{i}+5\hat{j}$

$$R=|\mathbf{R}|=\sqrt{(12)^2+(5)^2}$$

$$=13$$

$$\hat{\mathbf{R}}=\frac{\mathbf{R}}{R}=\frac{12\hat{i}+5\hat{j}}{13}$$

4. $(\mathbf{A}+\mathbf{B})\perp(\mathbf{A}-\mathbf{B})$

$$\therefore \quad (\mathbf{A}+\mathbf{B})\cdot(\mathbf{A}-\mathbf{B})=0$$

$$\therefore \quad A^2+BA\cos\theta-AB\cos\theta-B^2=0$$

or $\qquad A=B \qquad$ **Ans.**

5. Area of triangle $=\frac{1}{2}AB\sin\theta=\frac{|\mathbf{A}\times\mathbf{B}|}{2}$

$$\mathbf{A}\times\mathbf{B}=(\hat{i}+\hat{j}+\hat{k})\times 3\hat{i}$$

$$=-3\hat{k}+3\hat{j}$$

$$|\mathbf{A}\times\mathbf{B}|=\sqrt{(-3)^2+(3)^2}$$

$$=3\sqrt{2}$$

$$\therefore \quad \text{Area}=\frac{3\sqrt{2}}{2}=\frac{3}{\sqrt{2}}\text{ unit}$$

6. Diagonal vector, $\mathbf{A}=b\hat{i}+b\hat{j}+b\hat{k}$

or $\qquad A=\sqrt{b^2+b^2+b^2}=\sqrt{3}b$

$$\therefore \quad \hat{\mathbf{A}}=\frac{\mathbf{A}}{A}=\frac{\hat{i}+\hat{j}+\hat{k}}{\sqrt{3}}$$

7. A vector parallel to $\mathbf{A}$ will be $n\mathbf{A}$ or $(3n\hat{i}+4n\hat{j})$

Now, $|n\mathbf{A}|=|\mathbf{B}|$ is given.

Hence, $\qquad n\sqrt{9+16}=\sqrt{49+576}$

or $\qquad n=5$

$$\therefore \quad n\mathbf{A}=15\hat{i}+20\hat{j}$$

8. $\therefore \quad (\mathbf{A}+\mathbf{B})\cdot(\mathbf{A}\times\mathbf{B})=|\mathbf{A}+\mathbf{B}||\mathbf{A}\times\mathbf{B}|\cos\alpha$

$$\mathbf{A}\cdot(\mathbf{A}\times\mathbf{B})+\mathbf{B}\cdot(\mathbf{A}\times\mathbf{B})=|\mathbf{A}+\mathbf{B}||\mathbf{A}\times\mathbf{B}|\cos\alpha$$

$$0+0=|\mathbf{A}+\mathbf{B}||\mathbf{A}\times\mathbf{B}|\cos\alpha$$

$$\Rightarrow \quad \cos\alpha=0 \quad \Rightarrow \alpha=\frac{\pi}{2}\text{ rad}$$

9. Projection on x-y plane $=\sqrt{3^2+1^2}=\sqrt{10}$

10. Clearly, $\mathbf{B}$ should be either in the second quadrant or the fourth quadrant. In none of the given options, we have $-\hat{i}$ term. So the second quadrant is ruled out. Also $\mathbf{B}$ should make an angle of $(90^\circ-\theta)$ with the x- axis (see figure). So, B should be

$$B\cos(90^\circ-\theta)\,\hat{i}-B\sin(90^\circ-\theta)\hat{j}=B\sin\theta\,\hat{i}-B\cos\theta\,\hat{j}$$

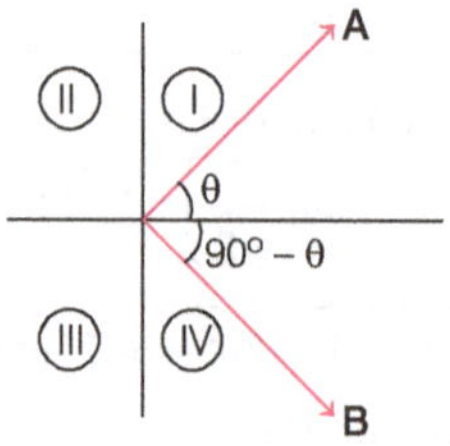

11. See figure,

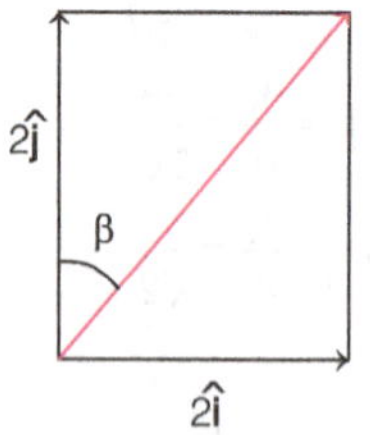

$$\tan\beta=\frac{2}{3} \text{ or } \beta=\tan^{-1}\left(\frac{2}{3}\right)$$

12. $\mathbf{A}\times\mathbf{B}=(4\hat{i}+6\hat{j})\times(2\hat{i}+3\hat{j})=12(\hat{i}\times\hat{j})+12(\hat{j}\times\hat{i})$

$$=12(\hat{i}\times\hat{j})-12(\hat{i}\times\hat{j})=0$$

Again, $\mathbf{A}\cdot\mathbf{B}=(4\hat{i}+6\hat{j})\cdot(2\hat{i}+3\hat{j})=8+18=26$

Again, $\left|\frac{\mathbf{A}}{\mathbf{B}}\right| = \frac{\sqrt{16+36}}{\sqrt{4+9}} \neq \frac{1}{2}$

Also, $\mathbf{B} = \frac{1}{2}\mathbf{A} \Rightarrow$ **A** and **B** are parallel and not antiparallel.

13. $\mathbf{A} \times \mathbf{B} = \begin{vmatrix} \hat{i} & \hat{j} & \hat{k} \\ 2 & p & q \\ 5 & 7 & 3 \end{vmatrix} = 0$

or $\hat{i}(3p-7q) + \hat{j}(5q-6) + \hat{k}(14-5p) = 0$

$3p = 7q, 5q - 6 = 0 \quad \text{or} \quad q = \frac{6}{5}$

14. $\frac{\mathbf{A}\cdot\mathbf{B}}{|\hat{i}+\hat{j}|} = \frac{(2\hat{i}+3\hat{j})(\hat{i}+\hat{j})}{\sqrt{2}} = \frac{2+3}{\sqrt{2}} = \frac{5}{\sqrt{2}}$

15. In the given cube, coordinates of point G (centre point of $ABOD$) are $x_1 = \frac{a}{2}, y_1 = 0, z_1 = \frac{a}{2}$

and coordinates of point H (centre point of $BEFO$) are

$$x_2 = 0,\ y_2 = \frac{a}{2},\ z_2 = \frac{a}{2}$$

So, vector GH is

$$\mathbf{GH} = (x_2 - x_1)\hat{i} + (y_2 - y_1)\hat{j} + (z_2 - z_1)\hat{k}$$

$$= -\frac{a}{2}\hat{i} + \frac{a}{2}\hat{j} = \frac{a}{2}(\hat{j} - \hat{i})$$

16.

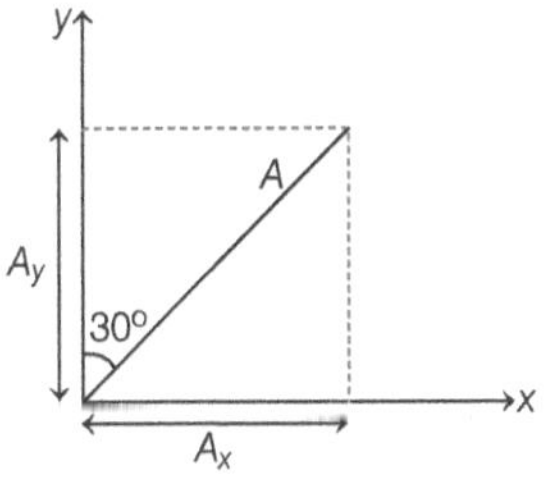

$$A_y = A\cos 30° = 2\sqrt{3}$$

$\Rightarrow \quad A\frac{\sqrt{3}}{2} = 2\sqrt{3} \Rightarrow A = 4$

Now, $\quad A_x = A\sin 30° = 4 \times \frac{1}{2} = 2$

17.

$$\mathbf{A} \times \mathbf{B} = \begin{vmatrix} \hat{i} & \hat{j} & \hat{k} \\ 2 & 1 & 1 \\ 1 & 1 & 1 \end{vmatrix}$$

$$= \hat{i}(1-1) - \hat{j}(2-1) + \hat{k}(2-1)$$

$$= -\hat{j} + \hat{k}$$

The unit vector perpendicular to **A** and **B** is $\left(\frac{-\hat{j}+\hat{k}}{\sqrt{2}}\right)$.

Any vector whose magnitude is K (constant) times $(2\hat{i} + \hat{j} + \hat{k})$ is parallel to **A**.

So, unit vector $\frac{2\hat{i}+\hat{j}+\hat{k}}{\sqrt{6}}$ is parallel to **A**.

18. If **A** and **r** are perpendicular to each other.

Then, $\mathbf{A}\cdot\mathbf{r} = 0$

This condition is satisfied by options (b), (c) and (d).

19. $\mathbf{R} = \mathbf{F}_1 + \mathbf{F}_2 + \mathbf{F}_3 = (p+1)\hat{i} + 4\hat{j} + 0\hat{k}$

$$R = \sqrt{(p+1)^2 + 4^2}$$

$\Rightarrow \quad 5^2 = (p+1)^2 + 4^2$

$\Rightarrow \quad (p+1)^2 = 9$

$\Rightarrow \quad p + 1 = \pm 3$

$\Rightarrow \quad p = 2, -4$

20. Magnitude of the new vector is twice that of the other.

$\therefore \quad \sqrt{4^2 + (4x-2)^2 + 2^2} = 2\sqrt{1^2 + x^2 + 3^2}$

Solving the above equation gives $x = -\frac{2}{3}, 2$.

21. The angle between the force **F** and the displacement **r** is 180°. Thus, the work done is

$$W = \mathbf{F}\cdot\mathbf{r} = Fr\cos\theta$$

$$= (12\text{ N})(2.0\text{ m})(\cos 180°)$$

$$= -24\text{ N-m} = -24\text{ J}$$

22. $\mathbf{A} + \mathbf{B} = \mathbf{R} = 3\hat{i} + \hat{j}$,

$$\mathbf{A} - \mathbf{B} = \mathbf{S} = \hat{i} + 5\hat{j}$$

Now, angle between **R** and **S** is

$$\theta = \cos^{-1}\left(\frac{\mathbf{R}\cdot\mathbf{S}}{RS}\right)$$

23. $\mathbf{A} \times \mathbf{B} = \begin{vmatrix} \hat{i} & \hat{j} & \hat{k} \\ 2 & 4 & -6 \\ 1 & 0 & 2 \end{vmatrix}$

$$= \hat{i}(8-0) + \hat{j}(-6-4) + \hat{k}(0-4)$$

$$= 8\hat{i} - 10\hat{j} - 4\hat{k}$$

Area of parallelogram $= |\mathbf{A} \times \mathbf{B}|$

$$= \sqrt{(8)^2 + (10)^2 + (4)^2}$$

$= 13.4$ units **Ans.**

24. (a) $\mathbf{A} + \mathbf{B} = \mathbf{R}$

$$R_x = A_x + B_x$$

$\therefore \quad B_x = R_x - A_x = 10 - 4 = 6\text{ m}$

Similarly,

$$B_y = R_y - A_y = 9 - 6 = 3\text{ m}$$

(b) $B = \sqrt{B_x^2 + B_y^2} = \sqrt{(6)^2 + (3)^2} = 3\sqrt{5}\text{ m}$

(c) $\quad \tan\theta = \frac{B_y}{B_x} = \frac{3}{6} = \frac{1}{2}$

$\therefore \quad \theta = \tan^{-1}\left(\frac{1}{2}\right)$

25. Find $\mathbf{A} \times \mathbf{B}$ and then prove that $(\mathbf{A} \times \mathbf{B}) \cdot \mathbf{A} = 0$ and $(\mathbf{A} \times \mathbf{B}) \cdot \mathbf{B} = 0$.

It means $(\mathbf{A} \times \mathbf{B}) \perp$ to both **A** and **B**.

26. $\because \quad \mathbf{R} = \mathbf{a} + \mathbf{b}$

$\therefore \quad \mathbf{b} = \mathbf{R} - \mathbf{a}$

$\Rightarrow \quad b^2 = R^2 + a^2 - 2Ra\cos 30°$

$\Rightarrow \quad b^2 = 4 + 12 - 2 \times 2 \times 2\sqrt{3} \times \frac{\sqrt{3}}{2}$

$\Rightarrow \quad b^2 = 16 - 12 = 4$

$\therefore \quad b = 2\,\text{N}$

27. Here, $\mathbf{a} \cdot \mathbf{b} = 0$

$\therefore \quad \mathbf{a} \cdot \mathbf{b} = (2\hat{i} + 4\hat{j} - 2\hat{k}) \cdot (\hat{i} + 2\hat{j} + \alpha\hat{k}) = 0$

$2 \times 1 + 4 \times 2 - 2 \times \alpha = 0$

$\therefore \quad \alpha = 5$

Unit Exercise (Cumulative)

Level 1

1.

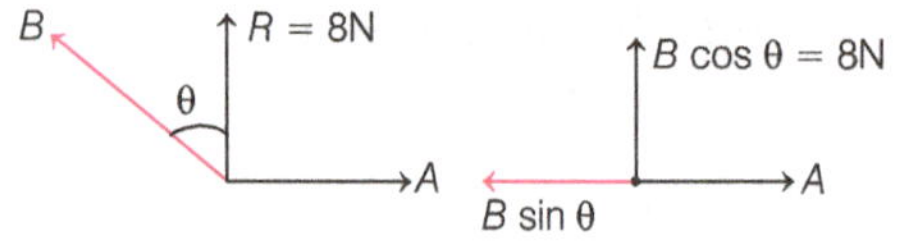

$$A + B = 16 \qquad \text{...(i)}$$

$$B\sin\theta = A$$

and $\quad B\cos\theta = 8$

Squaring and adding these two equations, we get

$$B^2 = A^2 + 64 \qquad \text{...(ii)}$$

Solving Eqs. (i) and (ii), we get

$A = 6\,\text{N}$ and $B = 10\,\text{N}$ **Ans.**

2. Net distance travelled along x-direction,

$$s_x = 6\cos 45° - 4\cos 45° = 2 \times \frac{1}{\sqrt{2}}$$

$$= \sqrt{2}\,\text{km}$$

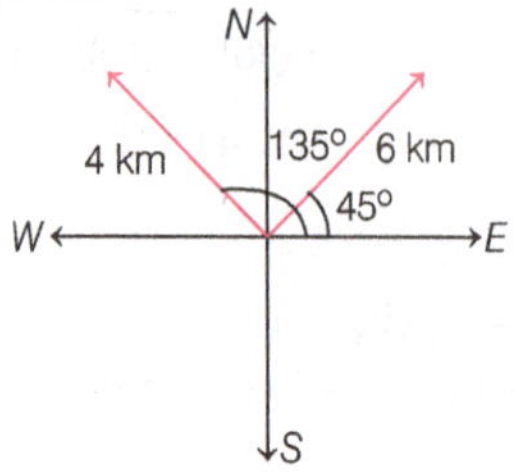

Net distance travelled along y-direction

$$s_y = 6\sin 45° + 4\sin 45°$$

$$= 10 \times \frac{1}{\sqrt{2}} = 5\sqrt{2}\,\text{km}$$

$\therefore$ Net distance travelled from starting point,

$$|\mathbf{s}| = \sqrt{s_x^2 + s_y^2}$$

$$= \sqrt{(\sqrt{2})^2 + (5\sqrt{2})^2} = \sqrt{52}\,\text{km}$$

Angle which the resultant makes with the east direction

$$\tan\theta = \frac{y\text{-component}}{x\text{-component}} = \frac{5\sqrt{2}}{\sqrt{2}}$$

$\therefore \quad \theta = \tan^{-1}(5)$

3.

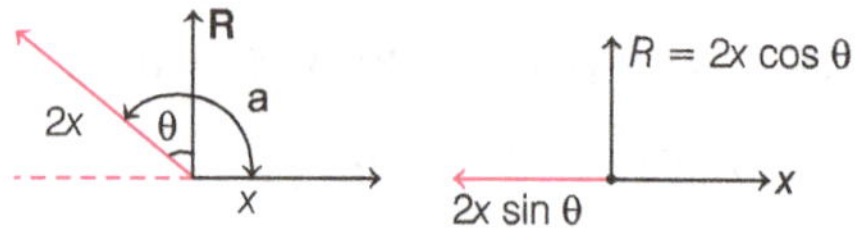

$$2x\sin\theta = x$$

$\therefore \quad \theta = 30°$

or $\quad \alpha = 120°$ **Ans.**

4. $P = Q = R$ and $\mathbf{P} + \mathbf{Q} = \mathbf{R}$

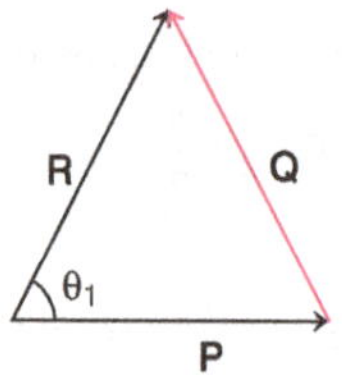

$$\theta_1 = 60°$$

If $\quad P = Q = R$ and $\mathbf{P} + \mathbf{Q} + \mathbf{R} = 0$

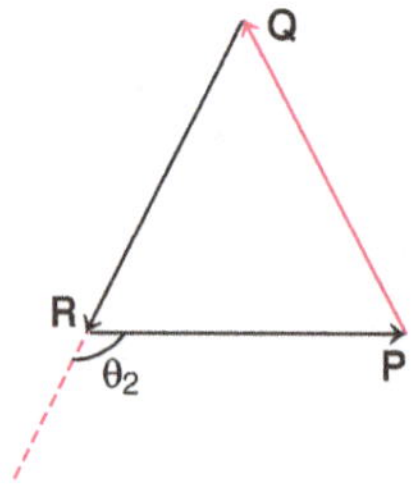

$$\theta_2 = 120°$$

5. $\mathbf{AB} + \mathbf{AC} + \mathbf{AD} + \mathbf{AE} + \mathbf{AF}$

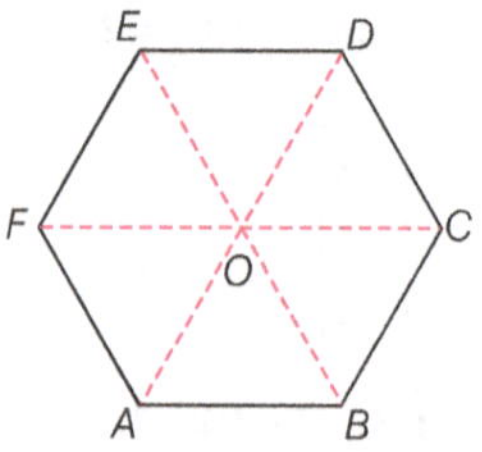

$= \mathbf{AB} + (\mathbf{AB} + \mathbf{BC}) + (\mathbf{AB} + \mathbf{BC} + \mathbf{CD}) + (\mathbf{AF} + \mathbf{FE}) + \mathbf{AF}$

$= (\mathbf{AB} + \mathbf{CD}) + (\mathbf{BC}) + (\mathbf{BC}) + 2(\mathbf{AB} + \mathbf{AF}) + (\mathbf{FE})$

$= \mathbf{AO} + \mathbf{AO} + \mathbf{AO} + 2\,\mathbf{AO} + \mathbf{AO}$

$= 6\,\mathbf{AO}$

6. $\mathbf{OB} + \mathbf{BC} = \mathbf{r}$

or $\mathbf{q} + \mathbf{BC} = \mathbf{r}$ ...(i)

$\mathbf{OA} + \mathbf{AC} = \mathbf{r}$

or $\mathbf{p} + \mathbf{AC} = \mathbf{r}$ ...(ii)

Adding these two equations, we get $\mathbf{p} + \mathbf{q} = 2\mathbf{r}$, as **AC** and **BC** are equal and opposite vectors.

7.

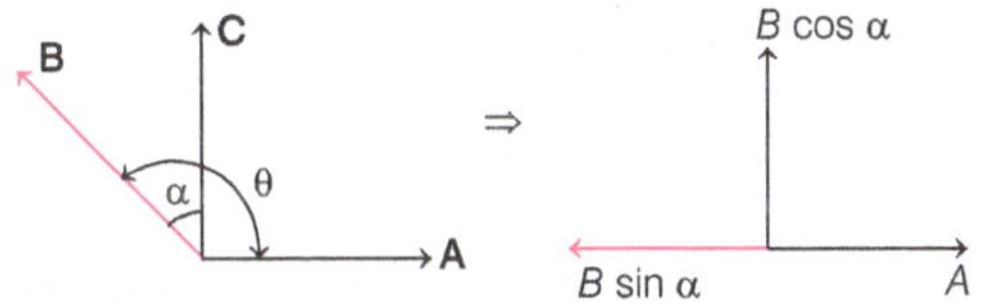

C is perpendicular to **A**. Therefore,

$$B \sin \alpha = A \quad \text{...(i)}$$

Further, $C = B \cos \alpha = A$ ...(ii)

Dividing Eq. (i) by Eq. (ii), we get

$$\tan \alpha = 1 \quad \text{or} \quad \alpha = 45°$$

Therefore, angle between **A** and **B** is

$$\theta = 90° + \alpha = 135° \quad \text{or} \quad \frac{3\pi}{4}$$

8. According to Lami's theorem,

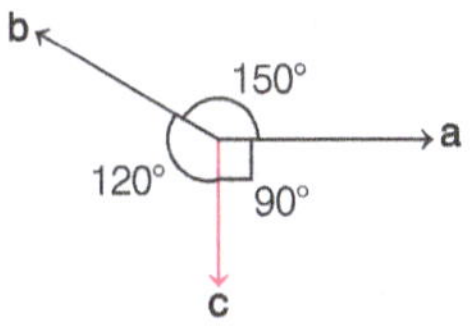

$$\frac{a}{\sin 120°} = \frac{b}{\sin 90°} = \frac{c}{\sin 150°} \quad \text{or} \quad \frac{2a}{\sqrt{3}} = b = 2c$$

$$\therefore \quad a = \frac{\sqrt{3}}{2} b \quad \text{and} \quad c = b/2$$

$$\therefore \quad a : b : c = \frac{\sqrt{3}}{2} b : b : b/2 = \sqrt{3} : 2 : 1$$

9. $\mathbf{S} = \mathbf{P} + b\mathbf{R} = \mathbf{P} + b(\mathbf{Q} - \mathbf{P}) = \mathbf{P}(1 - b) + b\mathbf{Q}$

10. Let $\mathbf{a} = a_1\hat{\mathbf{i}} + a_2\hat{\mathbf{j}} + a_3\hat{\mathbf{k}}$, Then,

$$\hat{\mathbf{i}} \times \mathbf{a} = \hat{\mathbf{i}} \times (a_1\hat{\mathbf{i}} + a_2\hat{\mathbf{j}} + a_3\hat{\mathbf{k}}) = a_2\hat{\mathbf{k}} - a_3\hat{\mathbf{j}}$$

$$\hat{\mathbf{j}} \times \mathbf{a} = \hat{\mathbf{j}} \times (a_1\hat{\mathbf{i}} + a_2\hat{\mathbf{j}} + a_3\hat{\mathbf{k}}) = -a_1\hat{\mathbf{k}} + a_3\hat{\mathbf{i}} = a_3\hat{\mathbf{i}} - a_1\hat{\mathbf{k}}$$

$$\hat{\mathbf{k}} \times \mathbf{a} = \hat{\mathbf{k}} \times (a_1\hat{\mathbf{i}} + a_2\hat{\mathbf{j}} + a_3\hat{\mathbf{k}}) = a_1\hat{\mathbf{j}} - a_2\hat{\mathbf{i}}$$

Now, $\hat{\mathbf{i}} \times (\hat{\mathbf{i}} \times \mathbf{a}) + \hat{\mathbf{j}} \times (\hat{\mathbf{j}} \times \mathbf{a}) + \hat{\mathbf{k}} \times (\hat{\mathbf{k}} \times \mathbf{a})$

$$= \hat{\mathbf{i}} \times (a_2\hat{\mathbf{k}} - a_3\hat{\mathbf{j}}) + \hat{\mathbf{j}} \times (a_3\hat{\mathbf{i}} - a_1\hat{\mathbf{k}}) + \hat{\mathbf{k}} \times (a_1\hat{\mathbf{j}} - a_2\hat{\mathbf{i}})$$

$$= -a_2\hat{\mathbf{j}} - a_3\hat{\mathbf{k}} - a_3\hat{\mathbf{k}} - a_1\hat{\mathbf{i}} - a_1\hat{\mathbf{i}} - a_2\hat{\mathbf{j}}$$

$$= -2(a_1\hat{\mathbf{i}} + a_2\hat{\mathbf{j}} + a_3\hat{\mathbf{k}})$$

$$= -2\mathbf{a}$$

11. $(\mathbf{P} + \mathbf{Q}) \cdot (\mathbf{P} - \mathbf{Q})$

$$= P^2 + PQ \cos \theta - PQ \cos \theta - Q^2 = P^2 - Q^2$$

Since, dot product may be positive (if $P > Q$), negative (if $Q > P$) or zero (if $P = Q$). Therefore, angle between $(\mathbf{P} + \mathbf{Q})$ and $(\mathbf{P} - \mathbf{Q})$ may be acute, obtuse or 90°.

12. $\mathbf{A} = 3\hat{\mathbf{i}} - 2\hat{\mathbf{j}} + \hat{\mathbf{k}}, \mathbf{B} = \hat{\mathbf{i}} - 3\hat{\mathbf{j}} + 5\hat{\mathbf{k}}, \mathbf{C} = 2\hat{\mathbf{i}} + \hat{\mathbf{j}} - 4\hat{\mathbf{k}}$

$$|\mathbf{A}| = \sqrt{3^2 + (-2)^2 + 1^2} = \sqrt{9 + 4 + 1} = \sqrt{14}$$

$$|\mathbf{B}| = \sqrt{1^2 + (-3)^2 + 5^2} = \sqrt{1 + 9 + 25} = \sqrt{35}$$

$$|\mathbf{A}| = \sqrt{2^2 + 1^2 + (-4)^2} = \sqrt{4 + 1 + 16} = \sqrt{21}$$

As $B = \sqrt{A^2 + C^2}$, therefore ABC will be right angled triangle.

13. Component of **A** along **B** is $|\mathbf{A}| \cos \theta \, \hat{\mathbf{B}}$ for θ being the angle between the vectors.

Also $\hat{\mathbf{B}} = \frac{\hat{\mathbf{i}} + \hat{\mathbf{j}}}{\sqrt{2}}$. So choice (a) is correct.

The vector $(\hat{\mathbf{i}} - \hat{\mathbf{j}})$ is perpendicular to the vector $(\hat{\mathbf{i}} + \hat{\mathbf{j}})$.

So, the other resolved component is $|\mathbf{A}| \sin \theta \left(\frac{\hat{\mathbf{i}} - \hat{\mathbf{j}}}{\sqrt{2}} \right)$.

14. The mirror is in x-y plane. So, normal will be perpendicular to x-y plane (i.e. z-axis).

If angle of incidence is θ, then $\cos \theta = (\hat{\mathbf{e}}_1 \cdot \hat{\mathbf{n}})$.

Here, $\mathbf{e}_1$ = unit vector along incidence ray

$$= \frac{\frac{1}{\sqrt{2}}\hat{\mathbf{i}} - \frac{1}{\sqrt{2}}\hat{\mathbf{j}} + \hat{\mathbf{k}}}{\sqrt{\left(\frac{1}{\sqrt{2}}\right)^2 + \left(\frac{-1}{\sqrt{2}}\right)^2 + 1^2}} = \frac{1}{2}\hat{\mathbf{i}} - \frac{1}{2}\hat{\mathbf{j}} + \frac{1}{\sqrt{2}}\hat{\mathbf{k}}$$

$\hat{\mathbf{n}}$ = unit vector along normal on incidence point $= \hat{\mathbf{k}}$

$$\therefore \quad \cos \theta = |\hat{\mathbf{e}}_1 \cdot \hat{\mathbf{n}}_1| = \frac{1}{\sqrt{2}} \Rightarrow \theta = 45°$$

15. A. Let $\mathbf{a} = 6\hat{\mathbf{i}} - 3\hat{\mathbf{j}} - 6\hat{\mathbf{k}}$ and $\mathbf{b} = \hat{\mathbf{i}} + \hat{\mathbf{j}} + \hat{\mathbf{k}}$

$$\therefore \quad a_{||} = \frac{(\mathbf{a} \cdot \mathbf{b})\mathbf{b}}{|b|^2} = \frac{(6 - 3 - 6)(\hat{\mathbf{i}} + \hat{\mathbf{j}} + \hat{\mathbf{k}})}{(\sqrt{1^2 + 1^2 + 1^2})^2} = -\hat{\mathbf{i}} - \hat{\mathbf{j}} - \hat{\mathbf{k}}$$

B. $\mathbf{a} = \mathbf{a}_{||} + \mathbf{a}_{\perp}$

Here, $\mathbf{a}_{||}$ = component of a along b,

$\mathbf{a}_{\perp}$ = component of a perpendicular to b.

$$\therefore \quad \mathbf{a}_{\perp} = \mathbf{a} - \mathbf{a}_{||} = 6\hat{\mathbf{i}} - 3\hat{\mathbf{j}} - 6\hat{\mathbf{k}} - (-\hat{\mathbf{i}} - \hat{\mathbf{j}} - \hat{\mathbf{k}})$$

$$= 7\hat{\mathbf{i}} - 2\hat{\mathbf{j}} - 5\hat{\mathbf{k}}$$

C. Here, $\mathbf{b} = 2\hat{\mathbf{i}} - \hat{\mathbf{j}} - 2\hat{\mathbf{k}}$

$$\therefore \quad \mathbf{a}_{||} = \frac{(\mathbf{a} \cdot \mathbf{b})\mathbf{b}}{|b|^2} = \frac{(12 + 3 + 12)(2\hat{\mathbf{i}} - \hat{\mathbf{j}} - 2\hat{\mathbf{k}})}{(\sqrt{4 + 1 + 4})^2}$$

$$= \frac{27}{9}(2\hat{\mathbf{i}} - \hat{\mathbf{j}} - 2\hat{\mathbf{k}}) = 6\hat{\mathbf{i}} - 3\hat{\mathbf{j}} - 6\hat{\mathbf{k}} = 0$$

$$\therefore \quad \mathbf{a}_{\perp} = \mathbf{a} - \mathbf{a}_{||}$$

$$= (6\hat{\mathbf{i}} - 3\hat{\mathbf{j}} + 6\hat{\mathbf{k}}) - (6\hat{\mathbf{i}} - 3\hat{\mathbf{j}} - 6\hat{\mathbf{k}}) = 0$$

D. Here, $\mathbf{a} = 2\hat{\mathbf{j}} - \hat{\mathbf{k}}$

$$\therefore \quad \mathbf{a}_{\parallel} = \frac{(\mathbf{a}\cdot\mathbf{b})\mathbf{b}}{|b|^2} = \frac{(-6+6)(2\hat{\mathbf{j}}-\hat{\mathbf{k}})}{(\sqrt{4+1})^2} = 0$$

16. A. $$\mathbf{a}\times\mathbf{b} = \begin{vmatrix} \hat{\mathbf{i}} & \hat{\mathbf{j}} & \hat{\mathbf{k}} \\ 1 & -2 & -3 \\ 2 & 1 & -1 \end{vmatrix} = \hat{\mathbf{i}}(2+3) - \hat{\mathbf{j}}(-1+6) + \hat{\mathbf{k}}(1+4)$$

$$= 5\hat{\mathbf{i}} - 5\hat{\mathbf{j}} + 5\hat{\mathbf{k}}$$

$$\therefore \quad (\mathbf{a}\times\mathbf{b})\times\mathbf{c} = \begin{vmatrix} \hat{\mathbf{i}} & \hat{\mathbf{j}} & \hat{\mathbf{k}} \\ 5 & -5 & 5 \\ 1 & 3 & -2 \end{vmatrix}$$

$$= \hat{\mathbf{i}}(10-15) - \hat{\mathbf{j}}(-10-5) + \hat{\mathbf{k}}(15+5)$$

$$= -5\hat{\mathbf{i}} + 15\hat{\mathbf{j}} + 20\hat{\mathbf{k}}$$

$$\therefore \quad |(\mathbf{a}\times\mathbf{b})\times\mathbf{c}| = \sqrt{25+225+400} = \sqrt{650}$$

$$= \sqrt{25+26} = 5\sqrt{26}$$

B. $$\mathbf{b}\times\mathbf{c} = \begin{vmatrix} \hat{\mathbf{i}} & \hat{\mathbf{j}} & \hat{\mathbf{k}} \\ 2 & 1 & -1 \\ 1 & 3 & -2 \end{vmatrix} = \hat{\mathbf{i}}(-2+3) - \hat{\mathbf{j}}(-4+1) + \hat{\mathbf{k}}(6-1)$$

$$= \hat{\mathbf{i}} + 3\hat{\mathbf{j}} + 5\hat{\mathbf{k}}$$

$$\therefore \quad \mathbf{a}\times(\mathbf{b}\times\mathbf{c}) = \begin{vmatrix} \hat{\mathbf{i}} & \hat{\mathbf{j}} & \hat{\mathbf{k}} \\ 1 & -2 & -3 \\ 1 & 3 & 5 \end{vmatrix}$$

$$= \hat{\mathbf{i}}(-10+9) - \hat{\mathbf{j}}(5+3) + \hat{\mathbf{k}}(3+2)$$

$$= -\hat{\mathbf{i}} - 8\hat{\mathbf{j}} + 5\hat{\mathbf{k}}$$

$$\therefore \quad |\mathbf{a}\times(\mathbf{b}\times\mathbf{c})| \times \sqrt{(-1)^2+(-8)^2+5^2}$$

$$= \sqrt{1+64+25} = \sqrt{90} = 3\sqrt{10}$$

C. $$\mathbf{b}\times\mathbf{c} = \begin{vmatrix} \hat{\mathbf{i}} & \hat{\mathbf{j}} & \hat{\mathbf{k}} \\ 2 & 1 & -1 \\ 1 & 3 & -2 \end{vmatrix} = \hat{\mathbf{i}}(-2+3) - \hat{\mathbf{j}}(-4+1) + \hat{\mathbf{k}}(6-1)$$

$$= \hat{\mathbf{i}} + 3\hat{\mathbf{j}} + 5\hat{\mathbf{k}}$$

$$\therefore \quad \mathbf{a}\cdot(\mathbf{b}\times\mathbf{c}) = (\hat{\mathbf{i}} - 2\hat{\mathbf{j}} + 3\hat{\mathbf{k}})\cdot(\hat{\mathbf{i}} + 3\hat{\mathbf{j}} + 5\hat{\mathbf{k}})$$

$$= 1 - 6 - 15 = -20$$

$$\therefore \quad |\mathbf{a}\cdot(\mathbf{b}\times\mathbf{c})| = 20$$

D. From solution of A, $\mathbf{a}\times\mathbf{b} = 5\hat{\mathbf{i}} - 5\hat{\mathbf{j}} + 5\hat{\mathbf{k}}$

$$\mathbf{b}\cdot\mathbf{c} = (2\hat{\mathbf{i}} + \hat{\mathbf{j}} - \hat{\mathbf{k}})\cdot(\hat{\mathbf{i}} + 3\hat{\mathbf{j}} - 2\hat{\mathbf{k}}) = 2+3+2 = 7$$

$$\therefore \quad |(\mathbf{a}\times\mathbf{b})(\mathbf{b}\cdot\mathbf{c})| = 7|a\times b|$$

$$= 7\sqrt{5^2+(-5)^2+5^2} = 35\sqrt{3}$$

17. Their x-components should be equal and opposite.

$$\therefore \quad 6\cos\theta = -4 \quad \text{or} \quad \theta = \cos^{-1}\left(-\frac{2}{3}\right)$$

18. If **a** and **b** are perpendicular to each other.

$$R = \sqrt{a^2+b^2}$$

If a and b are opposite to each other,

$$\frac{R}{\sqrt{2}} = a - b \quad \Rightarrow \quad \frac{R^2}{2} = a^2 + b^2 - 2ab$$

$$\Rightarrow \quad a^2 + b^2 = 2a^2 + 2b^2 - 4ab$$

$$\Rightarrow \quad a^2 + b^2 - 4ab = 0$$

$$\Rightarrow \quad \frac{a^2+b^2}{ab} = 4 \quad \text{or} \quad \frac{a}{b} + \frac{b}{a} = 4$$

19. From polygon law of vector addition we can see that,

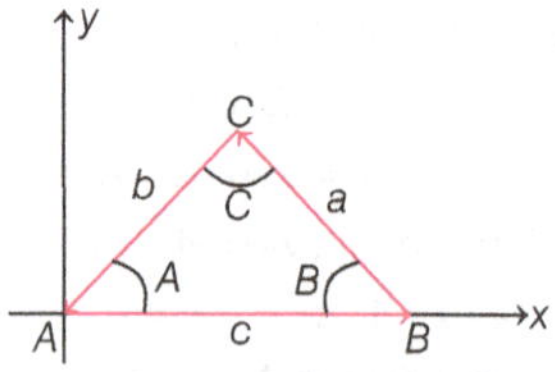

$$\mathbf{AB} + \mathbf{BC} + \mathbf{CA} = 0$$

$$\therefore \quad c\hat{\mathbf{i}} + (a\sin B\hat{\mathbf{j}} - a\cos B\hat{\mathbf{i}}) + (-b\cos A\hat{\mathbf{i}} - b\sin A\hat{\mathbf{j}}) = 0$$

$$\therefore \quad (c - a\cos B - b\cos A)\hat{\mathbf{i}} + (a\sin B - b\sin A)\hat{\mathbf{j}} = 0$$

Putting coefficient of $\hat{\mathbf{j}} = 0$, we find that

$$\frac{a}{\sin A} = \frac{b}{\sin B}$$

Now, taking B as origin and BC as the x-axis, we can prove other relation.

20. Given $\frac{a}{b} = \frac{x}{2}$ and $3a + 2b = 7$

If **a** and **b** are perpendicular

$$\mathbf{a}\cdot\mathbf{b} = 0 \quad \Rightarrow \quad 2a - 3b + 4 = 0$$

$$\Rightarrow \quad 2a - 3b = -4 \qquad \text{...(i)}$$

Also, $3a + 2b = 7$ ...(ii)

Solving Eqs. (i) and (ii), we get

$$a = 1, b = 2$$

So, $\frac{a}{b} = \frac{1}{2} \quad \Rightarrow \quad x = 1$

Level 2

21. $\mathbf{A} = \mathbf{P} + \mathbf{Q}$

$$\mathbf{B} = \mathbf{P} - \mathbf{Q}$$

$$\mathbf{P} \perp \mathbf{Q}$$

$$|\mathbf{A}| = |\mathbf{B}| = \sqrt{P^2 + Q^2}$$

$$|\mathbf{A} + \mathbf{B}| = \sqrt{2(P^2+Q^2)(1+\cos\theta)}$$

For $$|\mathbf{A} + \mathbf{B}| = \sqrt{3(P^2+Q^2)}$$

$$\theta_1 = 60°$$

For $$|\mathbf{A} + \mathbf{B}| = \sqrt{2(P^2+Q^2)}$$

$$\theta_2 = 90° \qquad \text{Ans.}$$

UNIT-6

KINEMATICS

Learning Modules

1. Introduction
2. Basic Definitions
3. Uniform Motion
4. One Dimensional Motion with Uniform Acceleration
5. One Dimensional Motion with Non-Uniform Acceleration
6. Motion in Two and Three Dimensions
7. Graphs
8. Relative Motion

MODULE 1

Introduction

Introduction to Mechanics and Kinematics

Mechanics is the branch of physics which deals with the motion of particles or bodies in space and time. Position and motion of a body can be determined only with respect to other bodies.

Motion of the body involves **position** and **time**. For practical purposes a coordinate system, e.g. the cartesian system is fixed to the reference body and position of the body is determined with respect to this reference body. For calculation of time generally clock is used.

Kinematics is the branch of mechanics which deals with the motion regardless of the causes producing it. The study of causes of motion is called **dynamics.**

General Points of Motion

1. **Velocity** is a **vector** quantity and **speed** is a **scalar** quantity. Magnitude of velocity vector is called its speed at that particular instant.

   Speed can't be negative but velocity can be negative. Positive and negative signs in velocity vector only represent their opposite directions.

   For example if two velocities are $+4$ m/s and -4 m/s, then both have speed of 4 m/s but their directions are opposite.

2. Meaning of **acceleration** in simple words is nothing but change in velocity in unit time (or in one second). For example if velocity is $+6$ m/s at $t = 0$ and accelerations is -2 m/s^2, then it implies that velocity will change by 2 m/s in every second in negative direction.

   So, at 1 s velocity will become $+4$ m/s. Similarly at 2 s it will become $+2$ m/s and so on.

3. If **acceleration** is **negative**, then it does not mean speed is always decreasing or there is retardation.

   It simply represents negative, direction of acceleration. If velocity and acceleration both have same sign (positive or negative), then speed will be increasing. If they have opposite signs then speed will be decreasing.

4. If we have two accelerations, say $a_1 = +2$ m/s^2 and $a_2 = -2$ m/s^2 then both motions are called **accelerated motions.** The only difference is first acceleration is in positive direction and the second acceleration is in negative direction.

Note *In all the points discussed above we have not used vector sign with vector quantities (like velocity and acceleration). The reason is, we are explaining the concepts for one dimensional vectors. This is the reason why we are using plus and minus signs with them.*

5. If a particle is in motion then its coordinates or we can say **position vector** will be changing with time. Or, we can say that they are functions of time. In one dimensional motion, position can be represented by a single coordinate (normally x) and this coordinate will be a function of time if the particle is in motion.

6. **Direction of velocity** is in the **direction of motion.** But direction of acceleration is not necessarily in the direction of motion. Direction of acceleration is in the direction of net force acting on the body. For example, if we say that a body is moving due east, it means velocity of the body is towards east. From the above statement, we cannot find the direction of acceleration.

7. Motion in a straight line is called a **one dimensional motion.**

8. Motion which is not one dimensional is called a **curvilinear motion**. Circular motion and projectile motion are the examples of curvilinear motion.

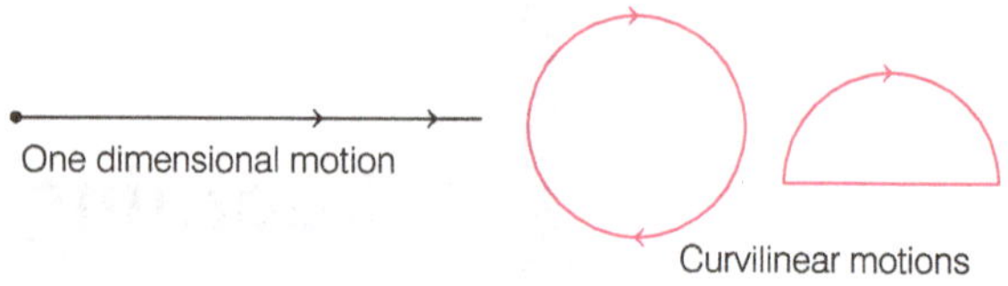

9. **Direction of velocity** at any point on a curvilinear path is tangential to the path. With direction of acceleration there is no such condition. As we have stated earlier also, it is in the direction of net force. For example, in the figure shown below a particle is moving on a curvilinear path.

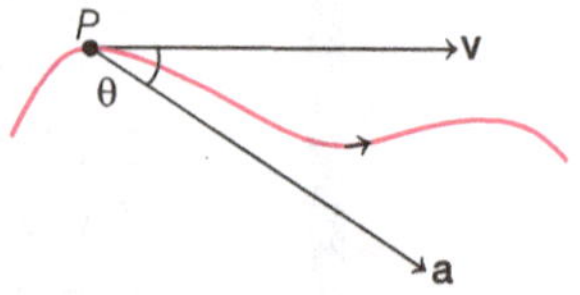

At point P, velocity of the particle is tangential to the path but acceleration is making an angle θ with velocity. If θ is acute $(0° \le \theta < 90°)$, then speed (which is also magnitude of velocity vector) of the particle increases. If θ is obtuse $(90° < \theta \le 180°)$, then speed of the particle decreases.

Note *$\theta = 90°$ is a special case when speed remains constant. This point we shall discuss in our later discussions.*

Example 1 Velocity of a particle at some instant is $\mathbf{v} = (3\hat{\mathbf{i}} + 4\hat{\mathbf{j}} + 5\hat{\mathbf{k}})$ m/s. Find speed of the particle at this instant.

Sol. Magnitude of velocity vector at any instant of time is the speed of particle. Hence,

Speed $= v$

or $|\mathbf{v}| = \sqrt{(3)^2 + (4)^2 + (5)^2}$

$= 5\sqrt{2}$ m/s **Ans.**

Example 2 "A lift is ascending with decreasing speed". What are the directions of velocity and acceleration of the lift at the given instant.

Sol. (i) Direction of motion is the direction of velocity. Lift is ascending (means it is moving upwards). So, direction of velocity is upwards.

(ii) Speed of lift is decreasing. So, direction of acceleration should be in opposite direction or it should be downwards.

Example 3 In the following cases state whether speed of a particle is increasing or decreasing

(a) velocity and acceleration both are positive

(b) velocity and acceleration both are negative

(c) velocity is positive and acceleration is negative

(d) velocity is negative and acceleration is positive

Ans. (a) Increasing (b) Increasing (c) Decreasing (d) Decreasing

Example 4 Find velocity at 1s, 2s, 3s and 4s if

(a) velocity at $t = 0$ is $+4$ m/s and constant acceleration is $-2\,\text{m/s}^2$.

(b) velocity at $t = 0$ is -2 m/s and constant acceleration is $-4\,\text{m/s}^2$.

Ans.

Part	Velocity at $t = 0$ (in m/s)	Acceleration (in m/s^2)	Velocity at (in m/s) 1s	2s	3s	4s
(a)	+4	−2	+2	0	−2	−4
(b)	−2	−4	−6	−10	−14	−18

Classification of Motion

A motion can be classified in following two ways :

First

(i) One dimensional (1-D)
(ii) Two dimensional (2-D)
(iii) Three dimensional (3-D)

In one dimensional motion, particle (or a body) moves in a straight line. In two dimensional motion, it moves in a plane and in three dimensional motion body moves in space.

Second

(i) Uniform
(ii) Uniformly accelerated
(iii) Non-uniformly accelerated

In uniform motion, acceleration is zero and velocity of the particle is constant. In uniformly accelerated motion, acceleration is constant but non-zero; and in non-uniformly accelerated motion acceleration of the particle is not constant.

Equations, $\mathbf{v} = \mathbf{u} + \mathbf{a}t$ etc. can be applied only for uniformly accelerated motion. If the motion is one dimensional then these equations can be written as, $v = u + at$ etc. For solving a problem of non-uniform acceleration, either integration of differentiation is required.

Note *In uniform motion, velocity of particle is constant, therefore acceleration is zero. Velocity is constant, means its magnitude (or speed) is constant and direction of velocity is fixed. So, the particle moves in a straight line. Hence, it is always one dimensional motion.*

Example 5 Give two examples of two dimensional motion.

Sol. Two dimensional motion takes place in a plane. Its two examples are circular motion and projectile motion.

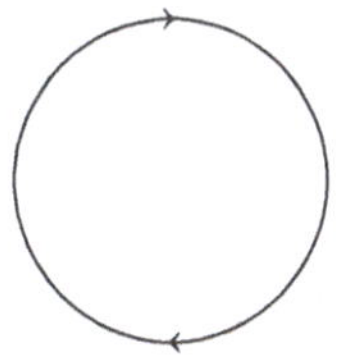

Circular motion

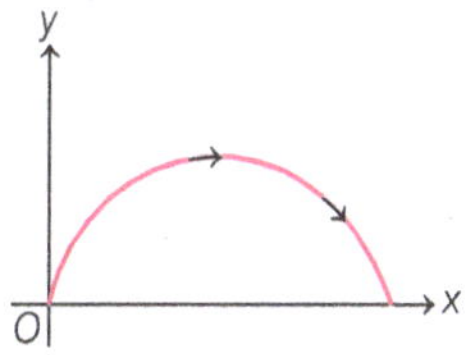

Projectile motion

Normally, the plane of circular motion is either horizontal or vertical and plane of projectile motion is always vertical. Plane of circular motion may be inclined also.

Example 6 Velocity of a particle is $\mathbf{v} = (2\hat{\mathbf{i}} + 3\hat{\mathbf{j}} - 4\hat{\mathbf{k}})$ m/s and its acceleration is zero. State whether it is 1-D, 2-D or 3-D motion?

Sol. Since, acceleration of the particle is zero. Therefore, it is uniform motion or motion in a straight line. So, it is one dimensional motion.

Example 7 Projectile motion is a two dimensional motion with constant acceleration. Is this statement true or false?

Sol. True. Projectile motion takes place in a plane. So, it is two dimensional. For small heights, its acceleration is constant (= acceleration due to gravity). Therefore, it is a two dimensional motion with constant acceleration.

MODULE 1 Exercise

Multiple Correct Options

1. The body will speed up if
   (a) velocity and acceleration are in the same direction
   (b) velocity and acceleration are in opposite directions
   (c) velocity and acceleration are in perpendicular direction
   (d) velocity and acceleration are acting at acute angle w.r.t. each other.

2. For a particle moving in straight line with increasing speed the appropriate sign of acceleration a and velocity v can be
   (a) $a > 0$ and $v > 0$ (b) $a < 0$ and $v < 0$
   (c) $a > 0$ and $v < 0$ (d) $a < 0$ and $v > 0$

Match the Columns

3. Match the following two columns :

	Column I		Column II
(a)	$\mathbf{v} = -2\hat{\mathbf{i}}, \mathbf{a} = -4\hat{\mathbf{j}}$	(p)	speed increasing
(b)	$\mathbf{v} = 2\hat{\mathbf{i}}, \mathbf{a} = 2\hat{\mathbf{i}} + 2\hat{\mathbf{j}}$	(q)	speed decreasing
(c)	$\mathbf{v} = -2\hat{\mathbf{i}}, \mathbf{a} = +2\hat{\mathbf{i}}$	(r)	speed constant
(d)	$\mathbf{v} = 2\hat{\mathbf{i}}, \mathbf{a} = -2\hat{\mathbf{i}} + 2\hat{\mathbf{j}}$	(s)	Nothing can be said

Subjective Questions

4. "A lift is descending with increasing speed". What are the directions of velocity and acceleration in the given statement?

5. Velocity and acceleration of a particle at some instant are
$$\mathbf{v} = (3\hat{\mathbf{i}} - 4\hat{\mathbf{j}} + 2\hat{\mathbf{k}})\ \text{m/s}$$
and $$\mathbf{a} = (2\hat{\mathbf{i}} + \hat{\mathbf{j}} - 2\hat{\mathbf{k}})\ \text{m/s}^2$$
   (a) What is the value of dot product of **v** and **a** at the given instant?
   (b) What is the angle between **v** and **a**, acute, obtuse or 90°?
   (c) At the given instant, whether speed of the particle is increasing, decreasing or constant?

6. Velocity and acceleration of a particle are
$$\mathbf{v} = (2\hat{\mathbf{i}} - 4\hat{\mathbf{j}})\ \text{m/s}$$
and $$\mathbf{a} = (-2\hat{\mathbf{i}} + 4\hat{\mathbf{j}})\ \text{m/s}^2$$
Which type of motion is this ?

7. Velocity and acceleration of a particle are
$$\mathbf{v} = (2\hat{\mathbf{i}})\ \text{m/s}$$
and $$\mathbf{a} = (4t\hat{\mathbf{i}} + t^2\hat{\mathbf{j}})\ \text{m/s}^2$$
where, t is the time. Which type of motion is this ?

8. In the above question, can we use $\mathbf{v} = \mathbf{u} + \mathbf{a}t$ equation directly?

9. If motion is taking place under gravity and a particle is projected downwards from a tower, then acceleration (due to gravity) is negative if upward direction is taken positive. Then does it mean motion is retarded ?

10. Find velocity at 1s, 2s and 3s, if velocity at $t = 0$ is -6 m/s and acceleration is $+3$ m/s^2.

11. In the above example, in which interval speed of the particle will be
   (a) decreasing
   (b) increasing
   Given reasons.

Answers

1. (a, d) **2.** (a, b) **3.** (a-r ; b-p ; c-q ; d-q) **4.** Both downwards **5.** (a) -2 m^2/s^3 (b) Obtuse (c) Decreasing
6. One dimensional with constant acceleration **7.** Two dimensional with non-uniform acceleration **8.** No **9.** No
10. -3 m/s, 0, $+3$ m/s **11.** (a) 0-2s (b) 2s onwards

See hints on page 239.

MODULE 2

Basic Definitions

Position of a Particle

Position of a particle in one dimensional motion can be represented by a single coordinate (say x). In two or three dimensional motion position can be represented by its two (or three) coordinates or the position vector.

Position Vector and Displacement Vector

If coordinates of point A are (x_1, y_1, z_1) and B are (x_2, y_2, z_2). Then, position vector of A

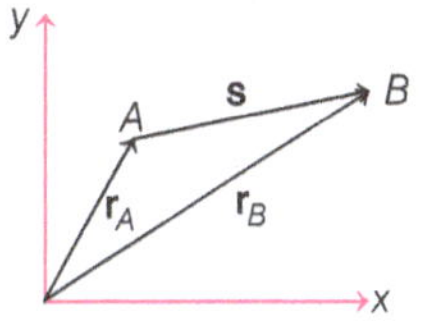

$$\mathbf{r}_A = \mathbf{OA} = x_1\hat{\mathbf{i}} + y_1\hat{\mathbf{j}} + z_1\hat{\mathbf{k}}$$

Position vector of B

$$\mathbf{r}_B = \mathbf{OB} = x_2\hat{\mathbf{i}} + y_2\hat{\mathbf{j}} + z_2\hat{\mathbf{k}}$$

and displacement vector, $\mathbf{s} = \mathbf{AB} = \mathbf{r}_B - \mathbf{r}_A$

$$= (x_2 - x_1)\hat{\mathbf{i}} + (y_2 - y_1)\hat{\mathbf{j}} + (z_2 - z_1)\hat{\mathbf{k}}$$

Note *If a particle is in motion then its coordinates, position vector and displacement vector change with time. Or, we can say that they are functions of time.*

Distance and Displacement

Distance is the actual path length covered by a moving particle or body in a given time interval, while displacement is the change in position vector, i.e. a vector joining initial to final positions. If a particle moves from A to C (see figure) through a path ABC. Then, distance travelled is the actual path length ABC, while the displacement is

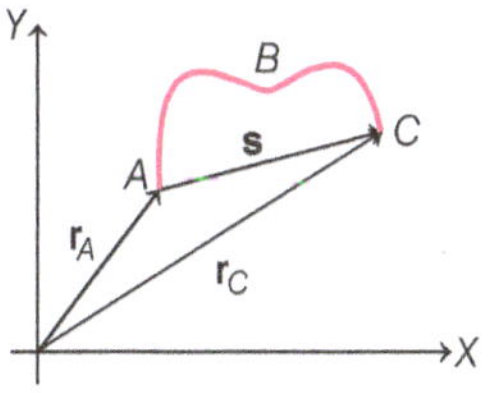

$$\mathbf{s} = \mathbf{AC} = \mathbf{r}_C - \mathbf{r}_A$$

If a particle moves in a straight line without change in direction, the magnitude of displacement is equal to the distance travelled, otherwise, it is always less than it. Thus,

$$|\text{displacement}| \le \text{distance}$$

Average Speed and Average Velocity

The average speed of a particle in a given time interval is defined as the ratio of total distance travelled to the total time take.

The average velocity is defined as the ratio of total displacement to the total time taken.

Thus, $$v_{av} = \text{average speed} = \frac{\text{total distance}}{\text{total time}}$$

and $$\mathbf{v}_{av} = \text{average velocity} = \frac{\text{total displacement}}{\text{total time}}$$

$$= \frac{\mathbf{s}}{\Delta t} \text{ or } \frac{\mathbf{s}}{t} \text{ or } \frac{\Delta \mathbf{r}}{\Delta t} = \frac{\mathbf{r}_f - \mathbf{r}_i}{\Delta t}$$

Here, $\mathbf{r}_f$ = final position vector and $\mathbf{r}_i$ = initial position vector

Instantaneous Velocity and Instantaneous Speed

Instantaneous velocity and instantaneous speed are defined at a particular instant and are given by

$$\mathbf{v}_i \text{ or simply } \mathbf{v} = \lim_{\Delta t \to 0} \frac{\Delta \mathbf{s}}{\Delta t}$$

$$\text{or } \frac{d\mathbf{s}}{dt} \text{ or } \frac{d\mathbf{r}}{dt}$$

Here, $\mathbf{r}$ is position vector of the body (or particle) at a general time t.

Magnitude of instantaneous velocity at any instant is called its instantaneous speed at that instant. Thus,

$$\text{Instantaneous speed} = v = |\mathbf{v}| = \left|\frac{d\mathbf{s}}{dt} \text{ or } \frac{d\mathbf{r}}{dt}\right|$$

Average and Instantaneous Acceleration

Average acceleration is defined as the ratio of change in velocity, i.e. $\Delta \mathbf{v}$ to the time interval Δt in which this change occurs. Hence,

$$\mathbf{a}_{av} = \frac{\Delta \mathbf{v}}{\Delta t} = \frac{\mathbf{v}_f - \mathbf{v}_i}{\Delta t}$$

The instantaneous acceleration is defined at a particular instant and is given by

$$\mathbf{a} = \lim_{\Delta t \to 0} \frac{\Delta \mathbf{v}}{\Delta t} = \frac{d\mathbf{v}}{dt}$$

Here, $\mathbf{v}_f$ = final velocity and $\mathbf{v}_i$ = initial velocity vector

Important POINTS

(i) If motion is one dimensional (let along x-axis) then all vector quantities (displacement, velocity and acceleration) can be treated like scalars by assuming one direction as positive and the other as negative. In this case, all vectors along positive direction are given positive sign and the vectors in negative direction are given negative sign.

(ii) For example, displacement, instantaneous velocity, instantaneous acceleration, average velocity and average acceleration in this case be written as

$$s = \Delta r = r_f - r_i \quad \text{or} \quad x_f - x_i$$

$$v = \frac{ds}{dt} \quad \text{or} \quad \frac{dx}{dt}$$

$$a = \frac{dv}{dt}$$

$$v_{av} = \frac{\Delta s}{\Delta t} = \frac{r_f - r_i}{\Delta t} = \frac{x_f - x_i}{\Delta t}$$

and $$a_{av} = \frac{\Delta v}{\Delta t} = \frac{v_f - v_i}{\Delta t}$$

(iii) If any physical quantity (either scalar or vector) is constant then its average value in any interval of time is equal to that constant quantity.

(iv) In uniform circular motion speed of particle is constant. But velocity keeps on changing.

Time period, $$T = \frac{2\pi R}{v}$$

Here, R is radius and v is constant speed of particle.

Summary

Distance and Displacement

(i) Distance = ABC

(ii) Displacement = $\mathbf{AC} = \mathbf{s}$

$= \mathbf{r}_f - \mathbf{r}_i$

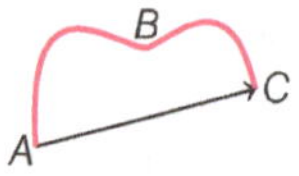

(iii) In 1-D motion,

$$s = x_f - x_i$$

(iv) Distance ≥ |Displacement|

Instantaneous and Average Speed

(i) Instantaneous speed, $v = |\mathbf{v}|$

(ii) Average speed $= \dfrac{\text{total distance}}{\text{total time}}$

Instantaneous and Average Velocity

(i) Instantaneous velocity $\mathbf{v} = \dfrac{d\mathbf{s}}{dt} = \dfrac{d\mathbf{r}}{dt}$

(ii) Average velocity $\mathbf{v}_{av} = \dfrac{\text{total displacement}}{\text{total time}}$

$$= \frac{\mathbf{s}}{t} = \frac{\mathbf{r}_f - \mathbf{r}_i}{t}$$

(iii) In 1-D motion,

instantaneous velocity $= \dfrac{ds}{dt} = \dfrac{dx}{dt}$ and

average velocity $= \dfrac{s}{t} = \dfrac{x_f - x_i}{t}$

Instantaneous and Average Acceleration

(i) Instantaneous acceleration, $\mathbf{a} = \dfrac{d\mathbf{v}}{dt}$

(ii) Average acceleration $\mathbf{a}_{av} = \dfrac{\Delta \mathbf{v}}{\Delta t} = \dfrac{\mathbf{v}_f - \mathbf{v}_i}{t}$

(iii) In 1-D motion,

instantaneous acceleration, $a = \dfrac{dv}{dt}$ and

average acceleration, $a_{av} = \dfrac{\Delta v}{\Delta t} = \dfrac{v_f - v_i}{t}$

Miscellaneous Examples

Ex. 1 A particle is moving Eastwards with a velocity of 5 m/s. In 10 s, the velocity changes to 5 m/s Northwards. The average acceleration in this time is **(1982 JEE)**

(a) zero

(b) $\dfrac{1}{\sqrt{2}}$ m/s^2 towards North-East

(c) $\dfrac{1}{\sqrt{2}}$ m/s^2 towards North-West

(d) $\dfrac{1}{2}$ m/s^2 towards North

Sol.

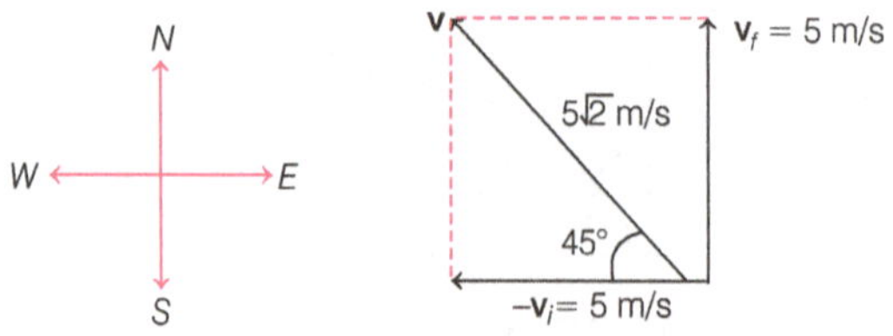

$$\mathbf{a}_{av} = \frac{\Delta \mathbf{v}}{\Delta t} = \frac{\mathbf{v}_f - \mathbf{v}_i}{\Delta t} = \frac{\mathbf{v}_f + (-\mathbf{v}_i)}{\Delta t}$$

$\Delta \mathbf{v} = 5\sqrt{2}$ m/s in North-West direction.

$$\mathbf{a}_{av} = \frac{5\sqrt{2}}{10} = \frac{1}{\sqrt{2}} \text{ m/s}^2 \text{ (in North-West direction)}$$

$\therefore$ The correct option is (c).

Ex. 2 The position vector of a particle changes with time according to the relation $\mathbf{r}(t) = 15t^2\hat{\mathbf{i}} + (4 - 20t^2)\hat{\mathbf{j}}$. What is the magnitude of the acceleration (in ms^{-2}) at $t = 1$ s? **(2019 Main)**

(a) 50 (b) 100 (c) 25 (d) 40

Sol. $\mathbf{r} = 15t^2\hat{\mathbf{i}} + (4 - 20t^2)\hat{\mathbf{j}}$

Velocity of particle is

$$\mathbf{v} = \frac{d\mathbf{r}}{dt} = 30t\,\hat{\mathbf{i}} - 40\,t\,\hat{\mathbf{j}}$$

Acceleration of particle is

$$\mathbf{a} = \frac{d}{dt}(\mathbf{v}) = 30\hat{\mathbf{i}} - 40\hat{\mathbf{j}}$$

So, magnitude of acceleration at $t = 1$ s is :

$$|\mathbf{a}| = \sqrt{30^2 + (-40)^2}$$

$$= 50 \text{ ms}^{-2}$$

$\therefore$ The correct option is (a).

Ex. 3 The position of a particle as a function of time t, is given by $x(t) = at + bt^2 - ct^3$ where a, b and c are constants. When the particle attains zero acceleration, then its velocity will be **(2019 Main)**

(a) $a + \frac{b^2}{2c}$ (b) $a + \frac{b^2}{4c}$ (c) $a + \frac{b^2}{3c}$ (d) $a + \frac{b^2}{c}$

Sol. Position of particle is, $x(t) = at + bt^2 - ct^3$

So, its velocity is, $v = \frac{dx}{dt} = a + 2bt - 3ct^2$

and acceleration is, $a = \frac{dv}{dt} = 2b - 6ct$

When acceleration becomes zero,

$$2b - 6ct = 0 \Rightarrow t = \frac{b}{3c}$$

Substituting this 't' in the expression of velocity, we get

$$v = a + \frac{b^2}{3c}$$

$\therefore$ The correct option is (c).

Ex. 4 In three dimensional system, the coordinates of a particle (in motion) are given below

$$x = a\cos\omega t$$
$$y = a\sin\omega t$$
$$z = a\omega t$$

The speed of the particle will be **(2019 Main)**

(a) $\sqrt{2}\,a\omega$ (b) $2\,a\omega$ (c) $a\omega$ (d) $\sqrt{3}\,a\omega$

Sol. $\mathbf{r} = x\hat{\mathbf{i}} + y\hat{\mathbf{j}} + z\hat{\mathbf{k}}$

$\Rightarrow$ $\mathbf{r} = a\cos\omega t\hat{\mathbf{i}} + a\sin\omega t\hat{\mathbf{j}} + a\,\omega t\hat{\mathbf{k}}$

$\mathbf{r} = a[\cos\omega t\hat{\mathbf{i}} + \sin\omega t\hat{\mathbf{j}} + \omega t\hat{\mathbf{k}}]$

$\therefore$ $\mathbf{v} = \frac{d\mathbf{r}}{dt}$

$\Rightarrow$ $\mathbf{v} = (-a\omega\sin\omega t\hat{\mathbf{i}} + a\omega\cos\omega t\hat{\mathbf{j}} + a\omega\hat{\mathbf{k}})$

The magnitude of velocity is speed.

$\therefore$ Speed $= |\mathbf{v}| = \sqrt{v_x^2 + v_y^2 + v_z^2}$

or $|\mathbf{v}| = \sqrt{(-a\omega\sin\omega t)^2 + (a\omega\cos\omega t)^2 + (a\omega)^2}$

$= \omega a\sqrt{(\sin^2\omega t + \cos^2\omega t) + 1}$

$= \sqrt{2}\,\omega a$

$\therefore$ The correct option is (a).

Ex. 5 A particle is moving with constant speed in a circular path. When the particle turns by an angle 90°, the ratio of instantaneous velocity to its average velocity is $\pi : x\sqrt{2}$. The value of x will be **(2023 Main)**

(a) 2 (b) 5 (c) 1 (d) 7

Sol.

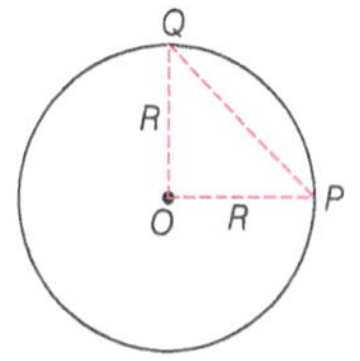

$$PQ = R\sqrt{2}$$

Let v be the instantaneous velocity = constant speed = v (say)

Average velocity

$$= \frac{\text{Displacement}}{\text{Time}} = \frac{R\sqrt{2}}{t}$$

$$= \frac{\sqrt{2}R}{T/4} = \frac{4\sqrt{2}R}{T} = \frac{4\sqrt{2}R}{(2\pi R/v)}$$

$$= \frac{2\sqrt{2}v}{\pi}$$

$\therefore$ The ratio is $\frac{v}{(2\sqrt{2}v/\pi)}$

$$= \frac{\pi}{2\sqrt{2}}$$

$\therefore$ The correct option is (a).

Ex. 6 A body is moving with constant speed, in a circle of radius 10 m. The body completes one revolution in 4s. At the end of 3rd second, the displacement of body (in m) from its starting point is **(2023 Main)**

(a) 30 (b) 15π (c) 5π (d) $10\sqrt{2}$

Sol.

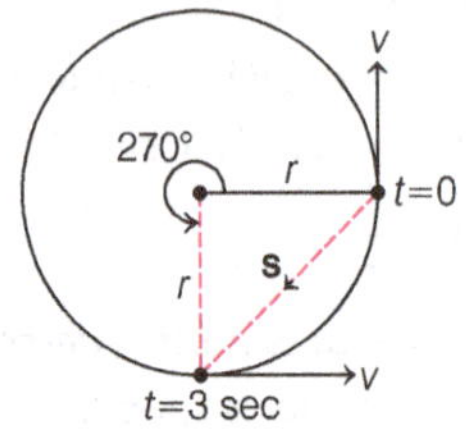

$$t = 3\text{ s} = \frac{3T}{4}$$

So, the particle rotates 270°.

$$|\mathbf{s}| = \sqrt{r^2 + r^2} = \sqrt{2}\, r = \sqrt{2} \times 10$$
$$= 10\sqrt{2}\text{ m}$$

$\therefore$ The correct option is (d).

Ex. 7 A particle is moving in a circle of radius 4 cm with constant speed of 1 cm/s. Find (a) time period of the particle (b) average speed, average velocity and average acceleration in a time interval from $t = 0$ to $t = T/4$. Here, T is the time period of the particle. Give only their magnitudes.

Sol. (a) $T = \dfrac{2\pi R}{v} = \dfrac{2\pi(4)}{1} = 25.13\text{ s}$ **Ans.**

(b) (i) Since speed is constant. Therefore,

average speed = constant speed

= 1.0 cm/s **Ans**

(ii) Average velocity $= \dfrac{s}{t} = \dfrac{\sqrt{2}R}{(T/4)} = \dfrac{4\sqrt{2}R}{T}$

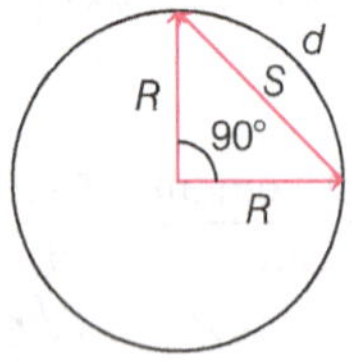

$$= \frac{(4)(\sqrt{2})(4.0)}{25.13} = 0.9\text{ cm/s}$$ **Ans.**

(iii) Velocity vector will rotate 90°

$$\therefore \quad |\Delta \mathbf{v}| = \sqrt{v^2 + v^2 - 2vv\cos 90^\circ}$$
$$= \sqrt{2}\, v$$
$$\therefore \quad |\mathbf{a}_{av}| = \frac{|\Delta \mathbf{v}|}{\Delta t} = \frac{\sqrt{2}\, v}{(T/4)} = \frac{4\sqrt{2}\, v}{T}$$
$$= \frac{(4\sqrt{2})(1.0)}{25.13} = 0.23\text{ cm/s}^2$$ **Ans.**

Ex. 8 Although **s** and **r** are not equal, still we write,

$$\mathbf{v} = \frac{d\mathbf{s}}{dt} = \frac{d\mathbf{r}}{dt}$$

Give the reason why?

Sol.

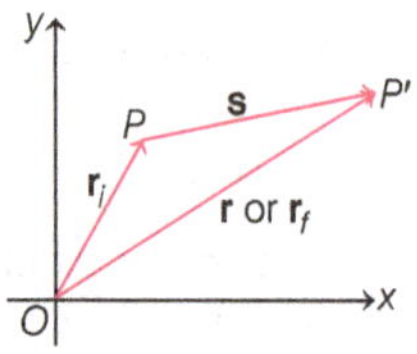

In the figure shown, a particle is in motion. At $t = 0$, particle is at P and at time t it is at P'.

$\mathbf{OP} = \mathbf{r}_i$ = initial position vector = a constant or fixed vector

$\mathbf{OP'} = \mathbf{r}$ or $\mathbf{r}_f$ or $\mathbf{r}(t)$ = position vector at time t

$\mathbf{PP'} = \mathbf{s}$ = displacement vector

$= \mathbf{r}_f - \mathbf{r}_i = \mathbf{r} - \mathbf{r}_i$

Here **s** and **r** are changing with time.

Further, $\mathbf{s} = \mathbf{r} - \mathbf{r}_i$

Differentiating this equation with respect to time, we get

$$\frac{d\mathbf{s}}{dt} = \frac{d\mathbf{r}}{dt} - \frac{d\mathbf{r}_i}{dt}$$
$$= \frac{d\mathbf{r}}{dt} - \frac{d}{dt}\text{(a constant vector)}$$
$$= \frac{d\mathbf{r}}{dt} - 0$$
$$\therefore \quad \frac{d\mathbf{s}}{dt} = \frac{d\mathbf{r}}{dt}$$ **Hence proved.**

Ex. 9 In one second, a particle goes from point A to point B moving in a semicircle. Find the magnitude of the average velocity. **(1999 JEE)**

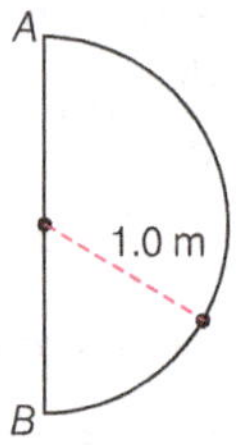

Sol. $|\mathbf{v}_{av}| = \left|\dfrac{\mathbf{s}}{t}\right| = \dfrac{AB}{\Delta t}$ m/s

$$= \frac{2.0}{1.0}\text{ m/s} = 2\text{ m/s}$$

Ex. 10 A particle is moving along x-axis. In the table, speed of particle at different time intervals is shown.

Time interval (in sec)	**Speed of particle** (in m/s)
0 - 2	2
2 - 5	3
5 - 10	4
10 - 15	2

Find total distance travelled by the particle and its average speed.

Sol. Distance = speed × time

$\therefore$ Total distance $= (2\times 2) + (3)(3) + (5)(4) + (5)(2)$ $= 43$ m

Total time taken is 15 s. Hence,

$$\text{Average speed} = \frac{\text{Total distance}}{\text{Total time}}$$

$$= \frac{43}{15} = 2.87 \text{ m/s}$$

Ex. 11 A particle is moving along x-axis. Its x-coordinate varies with time as,

$$x = 2t^2 + 4t - 6$$

Here, x is in metres and t in seconds. Find average velocity between the time interval $t = 0$ to $t = 2$ s.

Sol. In 1-D motion, average velocity can be written as

$$v_{av} = \frac{\Delta s}{\Delta t} = \frac{x_f - x_i}{\Delta t} = \frac{x_{2\text{ sec}} - x_{0\text{ sec}}}{2-0}$$

$$= \frac{[2(2)^2 + 4(2) - 6] - [2(0)^2 + 4(0) - 6]}{2}$$

$$= 8 \text{ m/s} \qquad \textbf{Ans.}$$

Ex. 12 A particle is moving in x-y plane. Its x and y coordinates vary with time as

$$x = 2t^2 \quad \text{and} \quad y = t^3$$

Here, x and y are in metres and t in seconds. Find average acceleration between a time interval from $t = 0$ to $t = 2$ s.

Sol. The position vector of the particle at any time t can be given as

$$\mathbf{r} = x\hat{\mathbf{i}} + y\hat{\mathbf{j}} = 2t^2\hat{\mathbf{i}} + t^3\hat{\mathbf{j}}$$

The instantaneous velocity is $\mathbf{v} = \dfrac{d\mathbf{r}}{dt} = (4t\hat{\mathbf{i}} + 3t^2\hat{\mathbf{j}})$

$$\text{Now, } \mathbf{a}_{av} = \frac{\Delta \mathbf{v}}{\Delta t} = \frac{\mathbf{v}_f - \mathbf{v}_i}{\Delta t} = \frac{\mathbf{v}_{2\text{ sec}} - \mathbf{v}_{0\text{ sec}}}{2-0}$$

$$= \frac{[(4)(2)\hat{\mathbf{i}} + (3)(2)^2\hat{\mathbf{j}}] - [(4)(0)\hat{\mathbf{i}} + (3)(0)^2\hat{\mathbf{j}}]}{2}$$

$$= (4\hat{\mathbf{i}} + 6\hat{\mathbf{j}}) \text{ m/s}^2 \qquad \textbf{Ans.}$$

MODULE 2 Exercise

Fill in the Blank

1. A particle moves in a circle of radius R. In half the period of revolution its displacement is ...... and distance covered is .... **(1983 JEE)**

Single Correct Option

2. A point traversed $\frac{3}{4}$th of the circle of radius R in time t. The magnitude of the average velocity of the particle in this time interval is

(a) $\dfrac{\pi R}{t}$ (b) $\dfrac{3\pi R}{2t}$ (c) $\dfrac{R\sqrt{2}}{t}$ (d) $\dfrac{R}{\sqrt{2}t}$

3. The position vector of a particle is $\mathbf{r} = a\cos\omega t\hat{\mathbf{i}} + a\sin\omega t\hat{\mathbf{j}}$.

The velocity of the particle is

(a) parallel to position vector
(b) perpendicular to position vector
(c) directed towards origin
(d) directed away from the origin

4. A particle moving with uniform speed v, changes its direction by an angle θ in time t. Magnitude of its average acceleration during this time is

(a) zero (b) $\dfrac{2v}{t}\sin\dfrac{\theta}{2}$

(c) $\dfrac{v\sqrt{2}}{t}$ (d) $\dfrac{2v}{t}\cos\dfrac{\theta}{2}$

5. As shown in the figure, a particle is moving with constant speed π m/s. Considering its motion from A to B, the magnitude of the average velocity is **(2023 Main)**

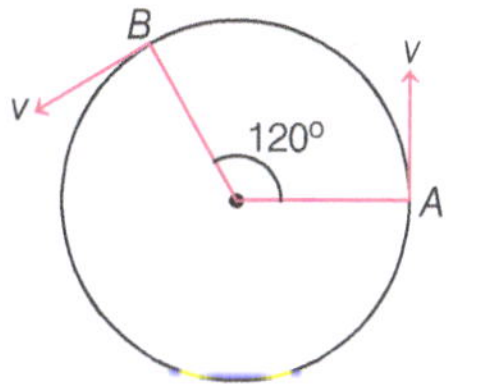

(a) π m/s (b) $\sqrt{3}$ m/s
(c) $2\sqrt{3}$ m/s (d) $1.5\sqrt{3}$ m/s

6. A person moved from A to B on a circular path as shown in figure. If the distance travelled by him is 60 m, then the magnitude of displacement would be (Given $\cos 135° = -0.7$) **(2022 Main)**

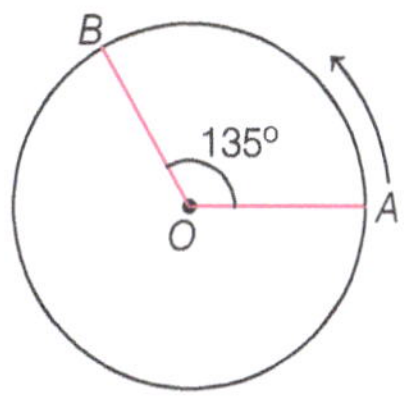

(a) 42 m (b) 47 m
(c) 19 m (d) 40 m

Multiple Correct Options

7. Average acceleration is in the direction of
   (a) initial velocity
   (b) final velocity
   (c) change in velocity
   (d) final velocity if initial velocity is zero

Subjective Questions

8. Average speed is always equal to magnitude of average velocity. Is this statement true or false?
9. When a particle moves with constant velocity its average velocity, its instantaneous velocity and its speed all are equal. Is this statement true or false?
10. (a) Is it possible to be accelerating if you are travelling at constant speed?
    (b) Is it possible to move on a curved path with zero acceleration, constant acceleration, variable acceleration?
11. (a) What does $\left|\frac{dv}{dt}\right|$ and $\frac{d|v|}{dt}$ represent?
    (b) Can these be equal?
12. Say Yes or No.
    (a) Can an object moving towards North have acceleration towards South?
    (b) Can an object reverse the direction of its motion even though it has constant acceleration?
    (c) Can an object reverse the direction of its acceleration even though it continues to move in the same direction?
    (d) Average speed is the magnitude of average velocity.
    (e) At any instant of time, the directions of change in velocity and acceleration are different.

Integer Type Question

13. A person starts his journey from centre '*O*' of the park and comes back to the same position following path *OPQO* as shown in the figure. The radius of path taken by person is 200 m and he takes 3 min 58 sec to complete his journey. The average speed of the person is......... ms^{-1}. (Take $\pi = 3.14$) **(2022 Main)**

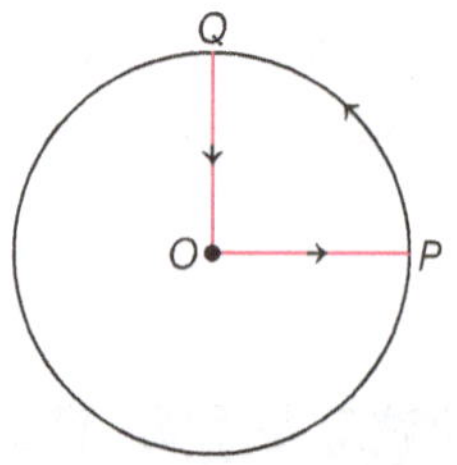

Answers

1. $(2R, \pi R)$ **2.** (c) **3.** (b) **4.** (b) **5.** (d) **6.** (b) **7.** (c, d) **8.** (False) **9.** (True)
10. (a) Yes, in uniform circular motion (b) No, yes (projectile motion), yes
11. (a) Magnitude of total acceleration and tangential acceleration (b) equal in 1-D motion
12. (a) Yes (b) Yes (c) Yes (d) No (e) No **13.** (3)

See hints on pages 239-240.

MODULE 3

Uniform Motion

As we have discussed earlier also, in uniform motion **velocity** of the particle is **constant** and **acceleration** is **zero**. Velocity is constant means its magnitude (called speed) is constant and direction is fixed. Therefore, motion is 1-D in same direction.

If velocity is along positive direction, then displacement is also along positive direction. Therefore, distance travelled (d) is equal to the displacement (s).

If velocity is along negative direction then displacement is also negative and distance travelled in this case is the magnitude of displacement.

Equations involved in this motion are :

(i) Velocity (may be positive or negative) = constant

(ii) Speed, v = constant

(iii) Acceleration = 0

(iv) Displacement (may be positive or negative) = velocity × time

(v) Distance = speed × time or $d = vt$

(vi) Distance and speed are always positive, whereas displacement and velocity may be positive or negative.

(vii) Most of the problems of average speed are based on uniform motion.

Note *In the problems of average speed (of uniform motion), we have three quantities, d, v and t. If two of them are given, we can find the third unknown quantity. Following examples will illustrate the concept.*

Example 1 A particle travels first half of the total distance with constant speed v_1 and second half with constant speed v_2. Find the average speed during the complete journey.

Sol.

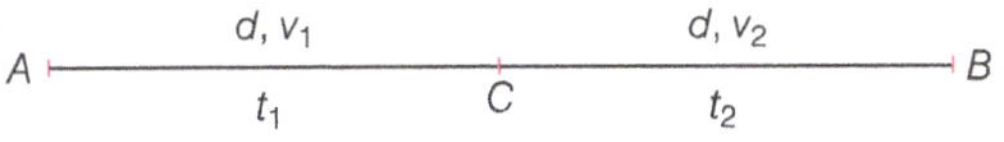

$$t_1 = \frac{d}{v_1}$$

and $$t_2 = \frac{d}{v_2}$$

$$\text{Average speed} = \frac{\text{total distance}}{\text{total time}}$$

$$= \frac{d + d}{t_1 + t_2} = \frac{2d}{(d/v_1) + (d/v_2)}$$

$$= \frac{2v_1 v_2}{v_1 + v_2}$$ **Ans.**

Example 2 A particle travels first half of the total time with speed v_1 and second half time with speed v_2. Find the average speed during the complete journey.

Sol.

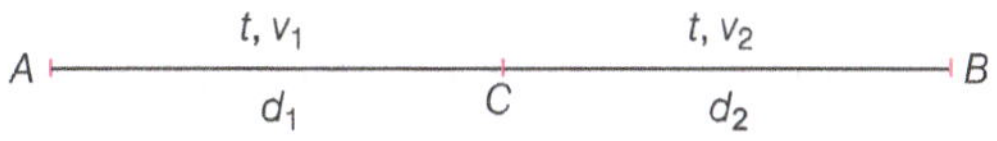

$$d_1 = v_1 t \quad \text{and} \quad d_2 = v_2 t$$

$$\text{Average speed} = \frac{\text{total distance}}{\text{total time}} = \frac{d_1 + d_2}{t + t}$$

$$= \frac{v_1 t + v_2 t}{2t} = \frac{v_1 + v_2}{2}$$ **Ans.**

Example 3 A particle travels first half of the total distance with speed v_1. In second half distance, constant speed in $\frac{1}{3}$rd time is v_2 and in remaining $\frac{2}{3}$rd time constant speed is v_3. Find average speed during the complete journey.

Sol.

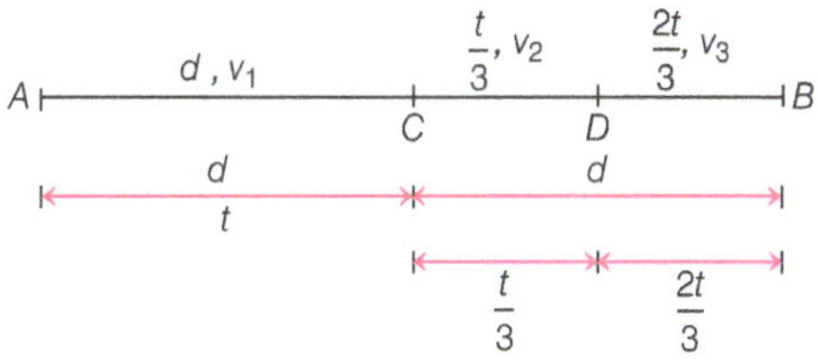

$$CD + DB = d$$

$$\Rightarrow \quad v_2\left(\frac{t}{3}\right) + \left(\frac{2t}{3}\right)(v_3) = d$$

or $$t = \frac{3d}{v_2 + 2v_3} \quad \text{...(i)}$$

Further, $$t_{AC} = \frac{d}{v_1}$$

Now, $$\text{Average speed} = \frac{\text{total distance}}{\text{total time}} = \frac{d + d}{t_{AC} + t_{CD} + t_{DB}}$$

$$= \frac{2d}{\frac{d}{v_1} + \frac{t}{3} + \frac{2t}{3}} = \frac{2d}{\left(\frac{d}{v_1} + t\right)}$$

Substituting value of t from Eq. (i), we have

$$\text{Average speed} = \frac{2d}{(d/v_1) + (3d/v_2 + 2v_3)}$$

$$= \frac{2v_1(v_2 + 2v_3)}{3v_1 + v_2 + 2v_3}$$ **Ans.**

MODULE 3 Exercise

Single Correct Option

1. A vehicle travels 4 km with speed of 3 km/h and another 4 km with speed of 5 km/h, then its average speed is **(2023 Main)**

(a) 425 km/h (b) 3.50 km/h (c) 4.00 km/h (d) 3.75 km/h

2. During the first 18 min of a 60 min trip, a car has an average speed of $11\,\text{ms}^{-1}$. What should be the average speed for remaining 42 min so that car is having an average speed of $21\,\text{ms}^{-1}$ for the entire trip?

(a) $25.3\,\text{ms}^{-1}$ (b) $29.2\,\text{ms}^{-1}$ (c) $31\,\text{ms}^{-1}$ (d) $35.6\,\text{ms}^{-1}$

3. If a particle travels n equal distances with speeds $v_1, v_2, \ldots, v_n$, then the average speed v of the particle will be

(a) $v = \dfrac{v_1 + v_2 + \ldots + v_n}{n}$

(b) $v = \dfrac{n(v_1 v_2 \ldots\ldots v_n)}{v_1 + v_2 + v_3 + \ldots + v_n}$

(c) $\dfrac{1}{v} = \dfrac{1}{n}\left(\dfrac{1}{v_1} + \dfrac{1}{v_2} + \ldots + \dfrac{1}{v_n}\right)$

(d) $v = \sqrt{v_1^2 + v_2^2 + \ldots + v_n^2}$

4. A moving car possesses average velocities of $5\,\text{ms}^{-1}$, $10\,\text{ms}^{-1}$ and $15\,\text{ms}^{-1}$ in the first, second and third seconds, respectively. What is the total distance covered by the car in these 3 s?

(a) 15 m (b) 30 m
(c) 55 m (d) None of these

5. If a car covers $\dfrac{2}{5}$th of the total distance with v_1 speed and $\dfrac{3}{5}$th of the total distance with v_2, then average speed is

(a) $\dfrac{1}{2}\sqrt{v_1 v_2}$ (b) $\dfrac{v_1 + v_2}{2}$ (c) $\dfrac{2v_1 v_2}{v_1 + v_2}$ (d) $\dfrac{5v_1 v_2}{3v_1 + 2v_2}$

6. One car moving on a straight road covers one-third of the distance with 20 km/hr and the rest with 60 km/hr. The average speed is

(a) 40 km/hr (b) 80 km/hr (c) $46\dfrac{2}{3}$ km/hr (d) 36 km/hr

7. A man walks on a straight road from his home to a market 2.5 km away with a speed of 5 km/h. Finding the market closed, he instantly turns and walks back home with a speed of 7.5 km/h. The average speed of the man over the interval of time 0 to 40 min is equal to

(a) 5 km/h (b) $\dfrac{25}{4}$ km/h
(c) $\dfrac{30}{4}$ km/h (d) $\dfrac{45}{8}$ km/h

8. An object moves with speed v_1, v_2 and v_3 along a line segment AB, BC and CD respectively as shown in figure. Where $AB = BC$ and $AD = 3\,AB$. Then average speed of the object will be **(2023 Main)**

(a) $\dfrac{v_1 v_2 v_3}{3(v_1 v_2 + v_2 v_3 + v_3 v_1)}$ (b) $\dfrac{3v_1 v_2 v_3}{(v_1 v_2 + v_2 v_3 + v_3 v_1)}$

(c) $\dfrac{(v_1 + v_2 + v_3)}{3}$ (d) $\dfrac{(v_1 + v_2 + v_3)}{3v_1 v_2 v_3}$

Subjective Questions

9. A particle moves in a straight line with constant speed of 4 m/s for 2s, then with 6 m/s for 3s. Find the average speed of the particle in the given time interval.

10. A particle travels half of the time with constant speed 2 m/s. In remaining half of the time it travels $\dfrac{1}{4}$th distance with constant speed of 4 m/s and $\dfrac{3}{4}$th distance with 6 m/s. Find average speed during the complete journey.

11. Figure shows a particle starting from point A, travelling upto B with a speed v, then upto point C with a speed $2v$, and finally upto A with a speed of $3v$. Determine its average speed.

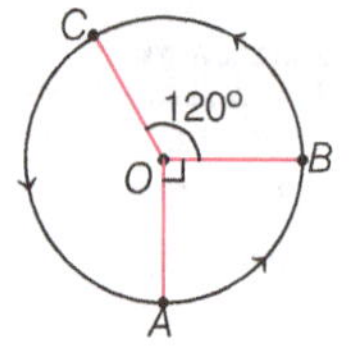

Integer Type Question

12. A horse rider covers half the distance with 5 m/s speed. The remaining part of the distance was travelled with speed 10 m/s for half the time and with speed 15 m/s for other half of the time. The mean speed of the rider averaged over the whole time of motion is $\dfrac{x}{7}$ m/s. The value of x is .... **(2023 Main)**

Answers

1. (d)	**2.** (a)	**3.** (c)	**4.** (b)	**5.** (d)	**6.** (d)	**7.** (d)	**8.** (b)	**9.** 5.2 m/s
10. $\dfrac{11}{3}$ m/s	**11.** 1.8 v	**12.** (50)						

See hints on pages 240-241.

MODULE 4

One Dimensional Motion with Uniform Acceleration

As we have discussed earlier that equations like **v = u + at** etc. can be applied directly with constant (or uniform) acceleration.

Further, in one dimensional motion, all vector quantities (displacement, velocity and acceleration) can be treated like scalars by using sign convention method. In this method, onc direction is taken as positive and the other as the negative and then all vector quantities are written with paper signs.

In most of the cases, we will take following sign convention.

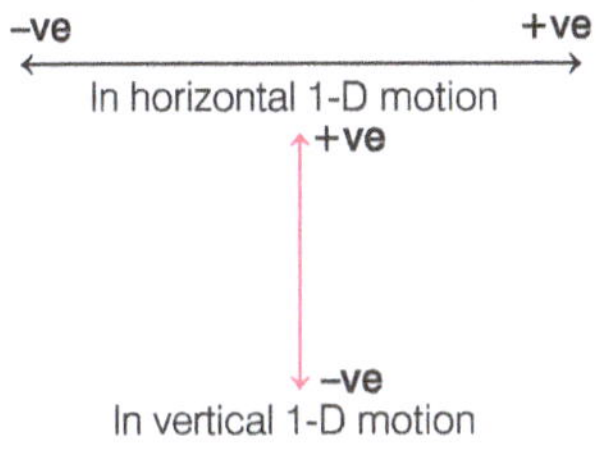

The equations used in 1-D motion with uniform acceleration are:

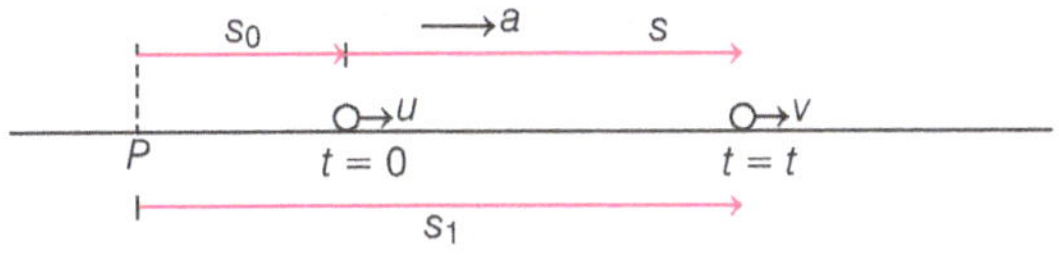

$$v = u + at \quad \text{...(i)}$$

$$v^2 = u^2 + 2as \quad \text{...(ii)}$$

$$s = ut + \frac{1}{2}at^2 \quad \text{...(iii)}$$

$$s_1 = s_0 + ut + \frac{1}{2}at^2 \quad \text{...(iv)}$$

$$s_t = u + at - \frac{1}{2}a \quad \text{...(v)}$$

In the above equations,

u = initial velocity, v = velocity

at time t, a = constant acceleration s = displacement measured from the starting point.

Here, starting point means the point where the particle was at $t = 0$. It is not the point where $u = 0$.

s_1 = displacement measured from any other point, say P, where P is not the starting point.

s_0 = displacement of the starting point from P.

s_t = displacement (not the distance in t^{th} second).

Key Points

(i) In most of the cases, displacement is measured from the starting point. Therefore Eq. (iii) or $s = ut + \frac{1}{2}at^2$ is used.

(ii) For small heights, if the motion is taking place under gravity then acceleration is always constant (= acceleration due to gravity). This is $9.8\ m/s^2 (\approx 10\ m/s^2)$ in downward direction. According to our sign convention downward direction is negative. Therefore,

$$a = g = -9.8\ m/s^2 \approx -10\ m/s^2$$

(iii) One dimensional motion (with constant acceleration) can have following three cases:

Case 1 Initial velocity is zero.

Case 2 Initial velocity is parallel to constant acceleration.

Case 3 Initial velocity is antiparallel to constant acceleration.

In first two cases, motion is only accelerated and direction of motion does not change. In the third case, motion is first retarded (till the velocity becomes zero) and then accelerated in opposite direction.

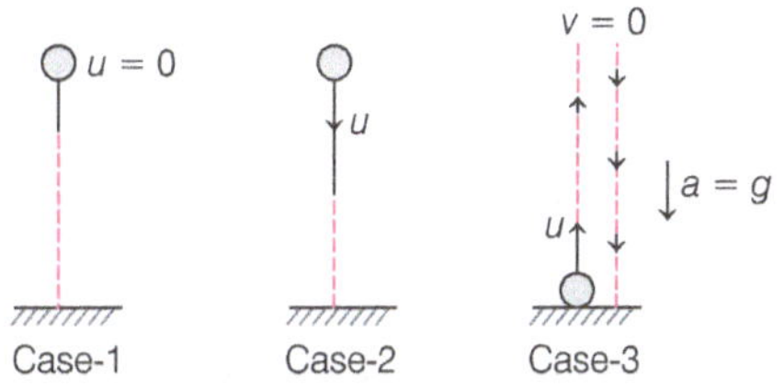

(iv) In most of the problems of time calculations, $s = ut + \frac{1}{2}at^2$ equation is very useful. But s has to be measured from the starting point.

(v) $s = ut + \frac{1}{2}at^2$ is a quadratic equation in t. We will get two values of t. Values may be positive, negative or imaginary. Ignore negative and imaginary values. Negative value means according to equation particle was at that position before $t = 0$ and imaginary value means particle will never reach that position. For example if you throw a particle upwards and I ask you that at what time particle will be at 10 km height. Then $s = ut + \frac{1}{2}at^2$ will give imaginary value of t.

(vi) In case 3, we need not to apply two separate equations, one for retarded motion (when motion is upwards) and the other for accelerated motion (when motion is downwards). Problem can be solved by applying the equation only one time, provided s

(in $s = ut + \frac{1}{2}at^2$) is measured from the starting point and all vector quantities are substituted with proper signs.

Example 1 A ball is thrown upwards from the top of a tower 40 m high with a velocity of 10 m/s. Find the time when it strikes the ground. Take $g = 10\text{ m/s}^2$.

Sol. In the problem, $u = +10$ m/s, $a = -10\text{ m/s}^2$

and $s = -40$ m (at the point where stone strikes the ground)

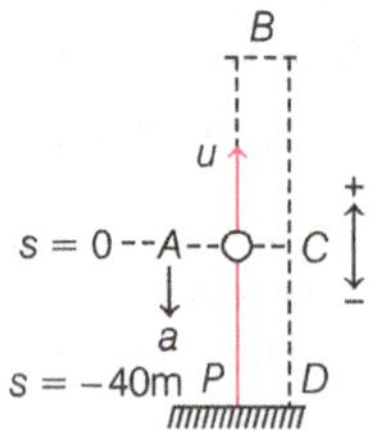

Substituting in $s = ut + \frac{1}{2}at^2$,

we have

$$-40 = 10t - 5t^2$$

or $5t^2 - 10t - 40 = 0$

or $t^2 - 2t - 8 = 0$

Solving this equation, we get

$$t = 4\text{ s} \text{ and } -2\text{ s}$$

Taking the positive value $t = 4$ s **.Ans.**

Note *The significance of $t = -2$ s can be explained as:*

$$t_{AB} = t_{BC} = 1\text{ s}$$
$$t_{PA} = t_{CD} = 2\text{ s}$$

At point	At time
P	−2 s
A	0
B	1 s
C	2 s
D	4 s

Example 2 A ball is thrown upwards from the ground with an initial speed of u. The ball is at a height of 80 m at two times, the time interval being 6 s. Find u. Take $g = 10\text{ m/s}^2$.

Sol. Here, $u = u$ m/s, $a = g = -10\text{ m/s}^2$ and $s = 80$ m.

Substituting the values in $s = ut + \frac{1}{2}at^2$,

we have $80 = ut - 5t^2$

or $5t^2 - ut + 80 = 0$

or $t = \frac{u + \sqrt{u^2 - 1600}}{10}$

and $= \frac{u - \sqrt{u^2 - 1600}}{10}$

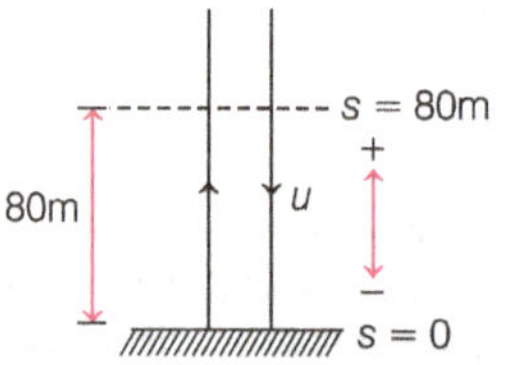

Now, it is given that

$$\frac{u + \sqrt{u^2 - 1600}}{10} - \frac{u - \sqrt{u^2 - 1600}}{10} = 6$$

or $\frac{\sqrt{u^2 - 1600}}{5} = 6$ or $\sqrt{u^2 - 1600} = 30$

or $u^2 - 1600 = 900$

$\therefore$ $u^2 = 2500$ or $u = \pm 50$ m/s

Ignoring the negative sign, we have $u = 50$ m/s

Example 3 Prove the result, $s_t = u + at - \frac{1}{2}a$

Sol. $s_t = s_{0\text{ to }t} - s_{0\text{ to }(t-1)}$

$$s_t = \left[ut + \frac{1}{2}at^2\right] - \left[u(t - t) + \frac{1}{2}a(t - 1)^2\right]$$

Solving we get, $s_t = u + at - \frac{1}{2}a$

Important POINTS

In motion under gravity, we can use the following results directly in MCQs.

(i) If a particle is projected upwards with velocity u, then

(a) maximum height attained by the particle, $h = \frac{u^2}{2g}$

(b) time of ascent = time of descent $= \frac{u}{g}$

$\therefore$ Total time of flight $= \frac{2u}{g}$

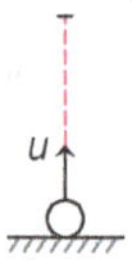

(ii) If a particle is released from rest from a height h (also called free fall), then

(a) velocity of particle at the time of striking with ground, $v = \sqrt{2gh}$

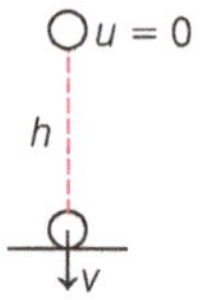

(b) time of descent (also called free fall time) $t = \sqrt{\frac{2h}{g}}$

(iii) $s=\frac{1}{2}gt^2=5t^2$ (if $g=10 m/s^2$)

$$s_{1s}=5(1)^2=5 \text{ m}$$
$$s_{2s}=5(2)^2=20 \text{ m}$$
$$s_{3s}=5(3)^2=45 \text{ m and so on.}$$

Note *In the above results, air resistance has been neglected and we have already substituted the signs of u, g etc. So, you have to substitute only their magnitudes.*

(iv) If $u=0$ and $a=$constant, then

$$s_{0-t}:s_{t-2t}:s_{2t-3t}=1:3:5$$

For example, $s_{0-1s}:s_{1s-2s}:s_{2s-3s}=1:3:5$

or, $s_{0-2s}:s_{2s-4s}:s_{4s-6s}=1:3:5$

Exercise : Prove the above result.

(v) If a particle is just dropped from a moving body then just after dropping, velocity of the particle (not acceleration) is equal to the velocity of the moving body at that instant.

For example, if a stone is dropped from a train moving with velocity 20 m/s, then initial velocity of the stone is 20 m/s in horizontal direction in the direction of motion of train. But, after dropping, it comes under gravity. Therefore, its acceleration is g downwards.

(vi) A particle is thrown upwards with velocity u. Suppose it takes time t to reach its highest point, then distance travelled in last second is independent of u.

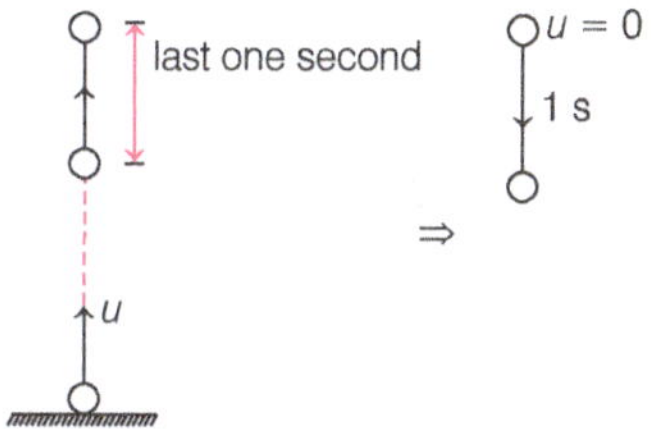

This is because this distance is equal to the distance travelled in first second of a freely falling object. Thus,

$$s=\frac{1}{2}gt^2$$
$$=\frac{1}{2}\times 10\times (1)^2=5 \text{ m}$$

Exercise : A particle is thrown upwards with velocity $u(>20 \text{ m/s})$. Prove that distance travelled in last 2 s is 20 m.

(vii) In a car race initial and final velocities are zero. Car first accelerates with acceleration a and then it retards with retardation na. Then

(a) retardation time $=\dfrac{\text{acceleration time}}{n}$ and

(b) retardation distance $=\dfrac{\text{acceleration distance}}{n}$

Exercise : Prove the above results.

Difference between Distance (d) and Displacement (s)

The s in equations of motion ($s=ut+\frac{1}{2}at^2$ and $v^2=u^2+2as$) is really the displacement not the distance. They have different values only when u and a are of opposite signs or $u\uparrow\downarrow a$.

Let us take the following two cases :

Case 1 When u is either zero or parallel to a, then motion is simply accelerated and in this case distance is equal to displacement. So, we can write

$$\boxed{d=|s|=\left|ut+\frac{1}{2}at^2\right|}$$

Case 2 When u is antiparallel to a, the motion is first retarded then accelerated in opposite direction. So, distance is either greater than or equal to displacement ($d\geq|s|$). In this case, first find the time when velocity becomes zero. Say it is t_0.

$$0=\pm u\mp at_0$$
$$\therefore \quad t_0=\left|\frac{u}{a}\right|$$

Now, if the given time $t\leq t_0$, distance and displacement are equal. So, $d=|s|=\left|ut+\frac{1}{2}at^2\right|$

For $t>t_0$, distance is greater than displacement. $d=d_1+d_2$

Here, $d_1=$ distance travelled before coming to rest $=\left|\frac{u^2}{2a}\right|$

$d_2=$ distance travelled in remaining time $t-t_0$

$$=\frac{1}{2}|a(t-t_0)^2|$$

$$\therefore \quad \boxed{d=\left|\frac{u^2}{2a}\right|+\frac{1}{2}|a(t-t_0)^2|}$$

Note *The displacement is still* $s=ut+\frac{1}{2}at^2$.

Example 4 A particle is projected vertically upwards with velocity 40 m/s. Find the displacement and distance travelled by the particle in (a) 2 s (b) 4 s (c) 6 s Take $g=10 m/s^2$.

Sol. Here, u is positive (upwards) and a is negative (downwards). So, first we will find t_0, the time when velocity becomes zero.

$$t_0=\left|\frac{u}{a}\right|=\frac{40}{10}$$
$$=4 \text{ s}$$

(a) $t<t_0$. Therefore, distance and displacement are equal.

$$d = s = ut + \frac{1}{2}at^2$$

$$= 40 \times 2 - \frac{1}{2} \times 10 \times 4 = 60 \text{ m}$$

(b) $t = t_0$. So, again distance and displacement are equal.

$$d = s = 40 \times 4 - \frac{1}{2} \times 10 \times 16 = 80 \text{ m}$$

(c) $t > t_0$. Hence, $d > |s|$, $s = 40 \times 6 - \frac{1}{2} \times 10 \times 36 = 60 \text{ m}$

While $d = \left|\frac{u^2}{2a}\right| + \frac{1}{2}|a(t - t_0)^2|$

$$= \frac{(40)^2}{2 \times 10} + \frac{1}{2} \times 10 \times (6-4)^2$$

$$= 100 \text{ m}$$

Typed Problems

Type 1. Collision of two particles or overtaking of one particle by the other particle.

Concept

(i) If two particles start from the same point and they collide, then their displacements are same or

$$s_1 = s_2$$

If they start from different points, t.hen

$$s_1 \neq s_2$$

(ii) If they start their journeys simultaneously, then their time of journeys are same or

$$t_1 = t_2 = t \quad \text{(say)}$$

otherwise their time of journeys are different

$$t_1 \neq t_2$$

How to Solve? (In 1-D motion)

- Take one direction as positive and the other as negative.
- Without considering, the given directions of their initial velocities and accelerations assume that both particles are moving along positive direction.

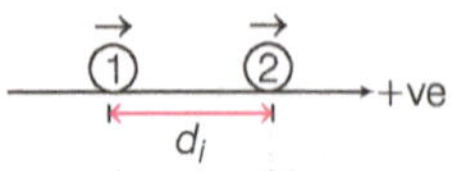

Assume this

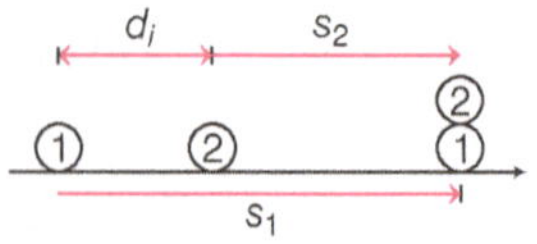

At the time of collision/overtaking

From the second figure, we can see that particle-1 (which is behind the particle-2) will collide (or overtake) particle-2 if it travels an extra distance d_i (= initial distance between them) or

$$s_1 = s_2 + d_i \quad \text{...(i)}$$

If motion is uniformly accelerated, then for s we can write

$$s = ut + \frac{1}{2}at^2$$

Now, u and a are vector quantities so, in Eq. (i) we will substitute them with sign.

By putting proper values in Eq. (i) we can find their time of collision. Same method can be applied in vertical motion also.

Ex. 1 Two particles are moving along x-axis. Particle-1 starts from $x = -10$ m with velocity 4 m/s along negative x-direction and acceleration 2 m/s^2 along positive x-direction. Particle-2 starts from $x = +2$ m with velocity 6 m/s along positive x-direction and acceleration 2 m/s^2 along negative x-direction.

(a) Find the time when they collide.

(b) Find the x-coordinate where they collide. Both start simultaneously.

Sol. (a)

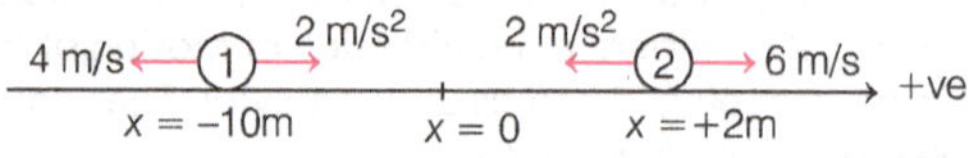

Given values

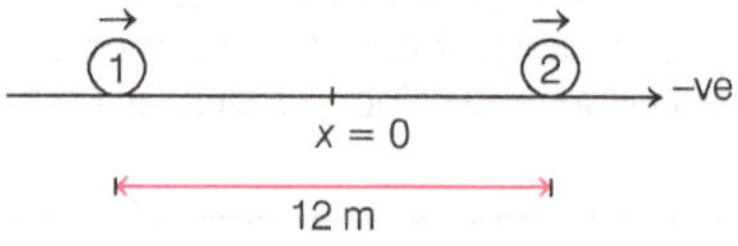

Assume this

Particle-1 is behind the particle-2 at a distance of 12m. So particle-1 will collide particle-2, if

$s_1 = s_2 + 12$

$\therefore \quad u_1t + \frac{1}{2}a_1t^2 = u_2t + \frac{1}{2}a_2t^2 + 12$

But now we will substitute the values of u_1, u_2, a_1 and a_2 with sign

$\therefore \quad (-4)t + \frac{1}{2}(+2)t^2 = (+6)t + \frac{1}{2}(-2)t^2 + 12$

Solving this equation, we get positive value of time,

$$t = 6 \text{ s} \qquad \textbf{Ans.}$$

(b) At the time of collision, $s_1 = u_1t + \frac{1}{2}a_1t^2$

$$= (-4)(6) + \frac{1}{2}(+2)(6)^2 = +12 \text{ m}$$

At the time of collision, x-coordinate of particle - 1 :

$$x_1 = (\text{Initial } x\text{-coordinate of particle-1}) + s_1$$

$$= -10 + 12 = +2 \text{ m}$$

Since, they collide at the same point. Hence,

$$x_2 = x_1 = +2 \text{ m} \qquad \textbf{Ans.}$$

Note *This was also the starting x-coordinate of particle-2.*

Exercise : Find their velocities at the time of collision.

Ans. $v_1 = +8$ m/s, $v_2 = -6$ m/s

Type 2. To find minimum distance between two particles moving in a straight line.

Concept

If two particles are moving along positive directions as shown in figure.

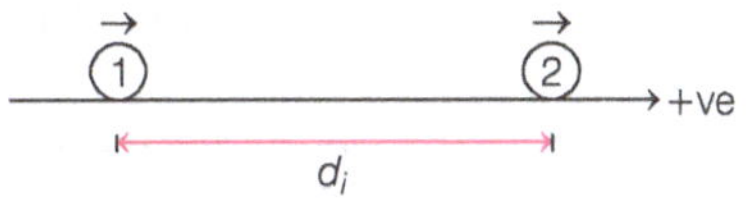

From the general experience, we can understand that distance between them will increase if $v_2 > v_1$ and distance between them will decrease if $v_1 > v_2$.

Therefore, in most of the problems we will get minimum distance, when $v_1 = v_2$

How to Solve?

By putting, $v_1 = v_2$ or, $u_1 + a_1 t = u_2 + a_2 t$ (if a = constant)

find the time when they are closest to each other.

In this time, particle-1 should travel some extra distance and whatever is the extra displacement (of particle-1), that will be subtracted from the initial distance between them to get the minimum distance.

$$\therefore \quad d_{min} = d_i - \Delta s = d_i - (s_1 - s_2)$$

For s, we can use $ut + \frac{1}{2}at^2$ if acceleration is constant.

Ex. 2 Two particles are moving along x-axis. Particle-1 is 40 m behind particle-2. Particle-1 starts with velocity 12 m/s and acceleration 4 m/s^2 both in positive x-direction. Particle-2 starts with velocity 4 m/s and acceleration 12 m/s^2 also in positive x-direction. Find

(a) the time when distance between them is minimum.

(b) the minimum distance between them.

Sol.

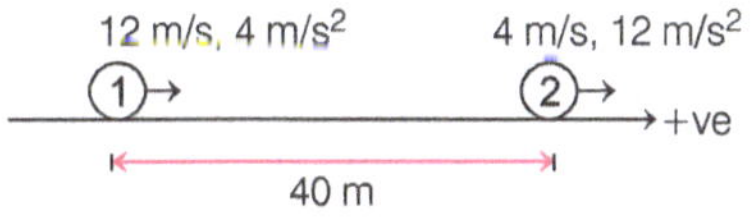

(a) As discussed above, distance between them is minimum, when

$$v_1 = v_2$$

or $$u_1 + a_1 t = u_2 + a_2 t$$

Substituting the values with sign we have,

$$(+12) + (4)t = (+4) + (12)t$$

$$\therefore \quad t = 1 \text{ s} \qquad \textbf{Ans.}$$

(b) **In 1 sec**

$$s_1 = u_1 t + \frac{1}{2}a_1 t^2$$

$$= 12 \times 1 + \frac{1}{2} \times 4 \times (1)^2 = 14\text{m}$$

and $$s_2 = u_2 t + \frac{1}{2}a_2 t^2$$

$$= 4 \times 1 + \frac{1}{2} \times 12 \times (1)^2$$

$$= 10 \text{ m}$$

Extra displacement of particle-1 with respect to 2 is

$$\Delta s = s_1 - s_2 = 14 - 10 = 4 \text{ m}$$

$\therefore$ Minimum distance between them

$$= d_i - \Delta s = 40 - 4$$

$$= 36 \text{ m} \qquad \textbf{Ans.}$$

Type 3. Comparison with standard equations.

Ex. 3 Velocity of a particle moving in a straight line varies with its displacement as $v = (\sqrt{4 + 4s})$ m/s. Displacement of particle at time $t = 0$ is $s = 0$. Find displacement of particle at time $t = 2$ s.

Sol. Squaring the given equation, we get

$$v^2 = 4 + 4s$$

Now, comparing it with $v^2 = u^2 + 2as$, we get

$$u = 2 \text{ m/s}$$

and $$a = 2 \text{ m/s}^2$$

$\therefore$ Displacement at $t = 2$s is

$$s = ut + \frac{1}{2}at^2$$

or $$s = (2)(2) + \frac{1}{2}(2)(2)^2$$

or $$s = 8 \text{ m} \qquad \textbf{Ans.}$$

Type 4. Based on difference between distance and displacement.

Concept

There is no direct formula for calculation of distance. In the formula,

$$s = ut + \frac{1}{2}at^2$$

s = displacement, not the distance

So, you will have to convert the given distance into proper displacement and then apply the above equation.

Following example will illustrate the concept.

Ex. 4 A particle is moving along x-axis. At time $t = 0$, its x-coordinate is $x = -4$ m. Its velocity-time equation is $v = 8 - 2t$ where, v is in m/s and t in seconds.

(a) At how many times, particle is at a distance of 8 m from the origin?

(b) Find those times.

Sol. (a) Comparing the given v-t equation with $v = u + at$, we have, $u = 8$ m/s and $a = -2$ m/s^2 = constant

Now, motion of the particle is as shown below.

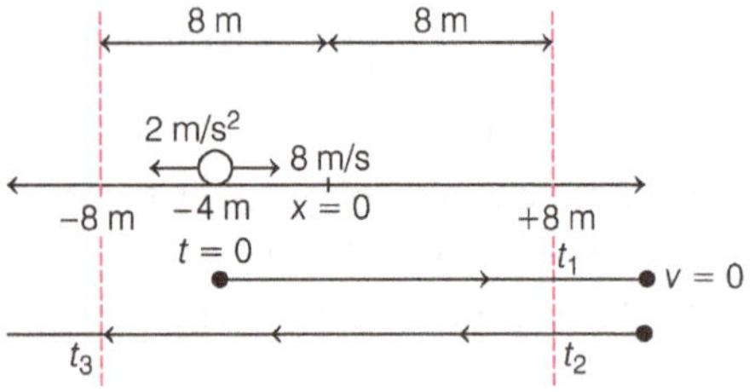

Now, 8 m distance from origin will be at two coordinates $x = 8$ m and $x = -8$ m. From the diagram, we can see that particle will cross these two points three times, t_1, t_2 and t_3.

(b) t_1 and t_2 **:** At $x = 8$ m, displacement from the starting point is

$$s = x_f - x_i = 8 - (-4) = 12 \text{ m}$$

Substituting in $s = ut + \frac{1}{2}at^2$, we have

$$12 = 8t - \frac{1}{2} \times 2 \times t^2$$

Solving this equation, we get

smaller time $t_1 = 2$ s **Ans.**

and larger time $t_2 = 6$ s **Ans.**

t_3 **:** At $x = -8$ m, displacement from the starting point is

$$s = x_f - x_i = -8 - (-4) = -4 \text{ m}$$

Substituting in $s = ut + \frac{1}{2}at^2$, we have

$$-4 = 8t - \frac{1}{2} \times 2 \times t^2$$

Solving this equation, we get the positive time,

$t_3 = 8.47$ s **Ans.**

Note *We can see that,* $t_3 > t_2 > t_1$

Miscellaneous Examples

Ex. 1 An open lift is moving upwards with velocity 10 m/s. It has an upward acceleration of 2 m/s^2. A ball is projected upwards with velocity 20 m/s relative to ground. Find

(a) time when ball again meets the lift

(b) displacement of lift and ball at that instant.

(c) distance travelled by the ball upto that instant.

(Take $g = 10$ m/s^2)

Sol. (a) At the time when ball again meet the lift,

$$s_L = s_B$$

$$\therefore \quad 10t + \frac{1}{2} \times 2 \times t^2 = 20t - \frac{1}{2} \times 10t^2$$

Solving this equation, we get

$$t = 0 \quad \text{and} \quad t = \frac{5}{3} \text{ s}$$

$\therefore$ Ball will again meet the lift after $\frac{5}{3}$ s.

(b) At this instant

$$s_L = s_B = 10 \times \frac{5}{3} + \frac{1}{2} \times 2 \times \left(\frac{5}{3}\right)^2 = \frac{175}{9} \text{ m} = 19.4 \text{ m}$$

(c) For the ball u is antiparallel to a. Therefore, we will first find t_0, the time when its velocity becomes zero.

$$t_0 = \left|\frac{u}{a}\right| = \frac{20}{10} = 2 \text{ s}$$

As $t \left(= \frac{5}{3} \text{ s}\right) < t_0$, distance and displacement are equal

or $\quad d = 19.4$ m

Ex. 2 A particle starts with an initial velocity and passes successively over the two halves of a given distance with constant accelerations a_1 and a_2 respectively. Show that the final velocity is the same as if the whole distance is covered with a uniform acceleration $\frac{(a_1 + a_2)}{2}$.

Sol.

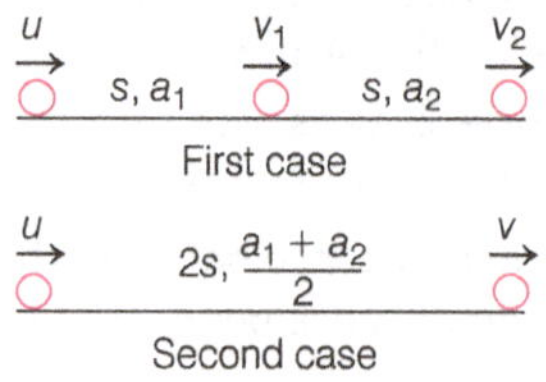

In the first case,

$$v_1^2 = u^2 + 2a_1 s \qquad \text{...(i)}$$

$$v_2^2 = v_1^2 + 2a_2 s \qquad \text{...(ii)}$$

Adding Eqs. (i) and (ii), we have

$$v_2^2 = u^2 + 2\left(\frac{a_1 + a_2}{2}\right)(2s) \qquad \text{...(iii)}$$

In the second case,

$$v^2 = u^2 + 2\left(\frac{a_1 + a_2}{2}\right)(2s) \qquad \text{...(iv)}$$

From Eqs. (iii) and (iv), we can see that

$v_2 = v$ **Hence proved.**

Ex. 3 In a car race, car A takes a time t less than car B at the finish and passes the finishing point with speed v more than that of the car B. Assuming that both the cars start from rest and travel with constant acceleration a_1 and a_2 respectively. Show that $v = \sqrt{a_1 a_2}\, t$. **(2019 Main)**

Sol. Let A takes t_1 second, then according to the given problem B will take $(t_1 + t)$ seconds. Further, let v_1 be the velocity of B at finishing point, then velocity of A will be $(v_1 + v)$. Writing equations of motion for A and B.

$$v_1 + v = a_1 t_1 \quad \ldots\text{(i)}$$

and $$v_1 = a_2(t_1 + t) \quad \ldots\text{(ii)}$$

(i) – (ii) gives,

$$v = (a_1 - a_2) t_1 - a_2 t \quad \ldots\text{(iii)}$$

Total distance travelled by both the cars is equal.

or $$s_A = s_B \quad \text{or} \quad \frac{1}{2} a_1 t_1^2 = \frac{1}{2} a_2 (t_1 + t)^2$$

Solving this equation, we get

$$t_1 = \frac{\sqrt{a_2}\, t}{\sqrt{a_1} - \sqrt{a_2}}$$

Substituting this value of t_1 in Eq. (iii), we get the desired result

$$v = (\sqrt{a_1 a_2})\, t$$

Ex. 4 An open elevator is ascending with constant speed $v = 10$ m/s. A ball is thrown vertically up by a boy on the lift when he is at a height $h = 10$ m from the ground. The velocity of projection is $v = 40$ m/s with respect to ground. Find

(a) the maximum height attained by the ball.
(b) the time taken by the ball to meet the elevator again.
(c) time taken by the ball to reach the ground after crossing the elevator.

Sol. (a) Velocity of ball = 40 m/s (upwards)

$$\therefore \quad h_{max} = h_i + h_f$$

Here, h_i = initial height = 10 m

and h_f = further height attained by ball

$$= \frac{u^2}{2g} = \frac{(40)^2}{2 \times 10} = 80 \text{ m}$$

$$\therefore \quad h_{max} = (10 + 80) \text{ m} = 90 \text{ m} \qquad \textbf{Ans.}$$

(b) The ball will meet the elevator again when displacement of lift = displacement of ball

or $$10 \times t = 40 \times t - \frac{1}{2} \times 10 \times t^2$$

or $$t = 6 \text{ s} \qquad \textbf{Ans.}$$

(c) Let t_0 be the total time taken by the ball to reach the ground. Then,

$$-10 = 40 \times t_0 - \frac{1}{2} \times 10 \times t_0^2$$

Solving this equation we get,

$$t_0 = 8.24 \text{ s}$$

Therefore, time taken by the ball to reach the ground after crossing the elevator,

$$= (t_0 - t) = 2.24 \text{ s}$$

Ex. 5 From an elevated point A, a stone is projected vertically upwards. When the stone reaches a distance h below A, its velocity is double of what it was at a height h above A. Show that the greatest height attained with respect to point A by the stone is $\frac{5}{3}h$.

Sol. Let u be the velocity with which the stone is projected vertically upwards.

Given that, $$v_{-h} = 2v_h$$

or $$(v_{-h})^2 = 4v_h^2$$

$$\therefore \quad u^2 - 2g(-h) = 4(u^2 - 2gh)$$

$$\therefore \quad u^2 = \frac{10gh}{3}$$

Now, $$h_{max} = \frac{u^2}{2g} = \frac{5h}{3} \qquad \textbf{Hence proved.}$$

Ex. 6 A particle is moving in a straight line with constant acceleration. If x, y and z be the distances described by a particle during the pth, qth and rth second respectively, prove that

$$(q - r)x + (r - p)y + (p - q)z = 0$$

Sol. As $$s_t = u + at - \frac{1}{2} a = u + \frac{a}{2}(2t - 1)$$

$$\therefore \quad x = u + \frac{a}{2}(2p - 1) \quad \ldots\text{(i)}$$

$$y = u + \frac{a}{2}(2q - 1) \quad \ldots\text{(ii)}$$

$$z = u + \frac{a}{2}(2r - 1) \quad \ldots\text{(iii)}$$

Subtracting Eq. (iii) from Eq. (ii),

$$y - z = \frac{a}{2}(2q - 2r)$$

or $$q - r = \frac{y - z}{a}$$

or $$(q - r)x = \frac{1}{a}(yx - zx) \quad \ldots\text{(iv)}$$

Similarly, we can show that

$$(r - p)y = \frac{1}{a}(zy - xy) \quad \ldots\text{(v)}$$

and $$(p - q)z = \frac{1}{a}(xz - yz) \quad \ldots\text{(vi)}$$

Adding Eqs. (iv), (v) and (vi), we get

$$(q - r)x + (r - p)y + (p - q)z = 0$$

MODULE 4 Exercise

Fill in the Blanks

1. A ball of mass 0.5 kg is dropped from a height of 10m. The height, at which the magnitude of velocity becomes equal to the magnitude of acceleration due to gravity, is........m. (Use $g = 10\ m/s^2$) **(2022 Main)**

2. A ball is projected vertically upward with an initial velocity of $50\ ms^{-1}$. At $t = 2s$, another ball is projected vertically upward with same velocity. At t =........s, second ball will meet the first ball. ($g = 10\ ms^{-2}$) **(2022 Main)**

3. A ball is dropped from the top of a 100 m high tower on a planet. In the last $\frac{1}{2}$ s before hitting the ground, it covers a distance of 19 m. Acceleration due to gravity (in ms^{-2}) near the surface on that planet is........ **(2020 Main)**

True False

4. Two balls of different masses are thrown vertically upwards with the same speed. They pass through the point of projection in their downward motion with the same speed (Neglect air resistance). **(1983 JEE)**

5. A particle moves in a straight line with uniform acceleration. Its velocity at time $t = 0$ is v_1 and at time $t = t$ is v_2. The average velocity of the particle in this time interval is $\frac{v_1 + v_2}{2}$. Is this statement true or false?

Assertion and Reason

Directions : Choose the correct option.

(a) *If both **Assertion** and **Reason** are true and the **Reason** is the correct explanation of the **Assertion**.*

(b) *If both **Assertion** and **Reason** are true but **Reason** is not the correct explanation of **Assertion**.*

(c) *If **Assertion** is true, but the **Reason** is false.*

(d) *If **Assertion** is false but the **Reason** is true.*

(e) *If both **Assertion** and **Reason** are false.*

6. **Assertion :** Average velocity can't be zero in case of uniform acceleration.

   Reason : For average velocity to be zero, a non zero velocity should not remain constant.

7. **Assertion :** A body is dropped from height h and another body is thrown vertically upwards with a speed $\sqrt{gh}$. They meet at height $h/2$.

   Reason : The time taken by both the blocks in reaching the height $h/2$ is same.

8. **Assertion :** Two bodies of unequal masses m_1 and m_2 are dropped from the same height. If the resistance offered by air to the motion of both bodies is the same, the bodies will reach the earth at the same time.

   Reason : For same air resistance, acceleration of fall of masses m_1 and m_2 will be different.

Single Correct Option

9. A body starts with an initial velocity of $10\ ms^{-1}$ and is moving along a straight line with constant acceleration. When the velocity of the particle is $50\ ms^{-1}$, the acceleration is reversed in direction. The velocity of the particle when it again reaches the starting point is

   (a) $70\ ms^{-1}$ (b) $60\ ms^{-1}$
   (c) $10\ ms^{-1}$ (d) $30\ ms^{-1}$

10. A ball is thrown vertically upwards from the ground. If T_1 and T_2 are the respective time taken in going up and coming down, and the air resistance is not ignored, then

    (a) $T_1 > T_2$ (b) $T_1 = T_2$
    (c) $T_1 < T_2$ (d) nothing can be said

11. When a ball is thrown up vertically with velocity v_0, it reaches a maximum height of h. If one wishes to triple the maximum height then the ball should be thrown with velocity

    (a) $\sqrt{3}\ v_0$ (b) $3\ v_0$ (c) $9\ v_0$ (d) $\frac{3}{2} v_0$

12. Two particles P and Q simultaneously start moving from point A with velocities 15 m/s and 20 m/s, respectively. The two particles move with accelerations equal in magnitude but opposite in direction. At some instant velocity of P is 30 m/s. The velocity of Q at this instant will be

    (a) 30 m/s (b) 5 m/s
    (c) 20 m/s (d) 15 m/s

13. A particle starting from rest with constant acceleration travels a distance x in first 2 s and a distance y in next 2 s, then

    (a) $y = x$ (b) $y = 2x$
    (c) $y = 3x$ (d) $y = 4x$

14. A particle moving in a straight line has velocity-displacement equation as $v = 5\sqrt{1 + s}$. Here v is in ms^{-1} and s in metres. Select the correct alternative.

    (a) Particle is initially at rest.
    (b) Initially velocity of the particle is 5 m/s and the particle has a constant acceleration of $12.5\ ms^{-2}$.
    (c) Particle moves with a uniform velocity
    (d) None of the above

15. Water drops fall at regular intervals from a tap 5 m above the ground. The third drop is leaving the tap at an instant when the first drop touches the ground. How far above the ground is the second drop at that instant? ($g = 10\ ms^{-2}$) **(1995 JEE)**
(a) 1.25 m (b) 2.50 m (c) 3.75 m (d) 4.00 m

16. A stone is dropped from the top of a tower and one second later, a second stone is thrown vertically downward with a velocity $20\ ms^{-1}$. The second stone will overtake the first after travelling a distance of ($g = 10\ ms^{-2}$)
(a) 13 m (b) 15 m (c) 11.25 m (d) 19.5 m

17. A ball is released from the top of a tower of height h metre. It takes T second to reach the ground. What is the position of the ball in $T/3$ second?
(a) $\frac{h}{9}$ metre from the ground
(b) $(7h/9)$ metre from the ground
(c) $(8h/9)$ metre from the ground
(d) $(17h/18)$ metre from the ground

18. A particle is projected vertically upwards and reaches the maximum height H in time T. The height of the particle at any time t $(<T)$ will be
(a) $g(t-T)^2$ (b) $H-g(t-T)^2$
(c) $\frac{1}{2}g(t-T)^2$ (d) $H-\frac{1}{2}g(T-t)^2$

19. A stone is thrown vertically upwards. When stone is at a height half of its maximum height, its speed is $10 ms^{-1}$; then the maximum height attained by the stone is ($g = 10\ ms^{-2}$)
(a) 25 m (b) 10 m (c) 15 m (d) 20 m

20. A stone is allowed to fall freely from rest. The ratio of the times taken to fall through the first metre and the second metre distance is
(a) $\sqrt{2}-1$ (b) $\sqrt{2}+1$
(c) $\sqrt{2}$ (d) None of these

21. A stone thrown upward with a speed u from the top of the tower reaches the ground with a speed $3u$. The height of the tower is
(a) $3u^2/g$ (b) $4u^2/g$
(c) $6u^2/g$ (d) $9u^2/g$

22. A particle is dropped under gravity from rest from a height h ($g = 9.8\ m/s^2$) and it travels a distance $9h/25$ in the last second, the height h is
(a) 100 m (b) 122.5 m (c) 145 m (d) 167.5 m

23. A ball is thrown vertically upward with a speed v from a height h metre above the ground. The time taken for the ball to hit the ground is
(a) $\frac{v}{g}\sqrt{1-\frac{2hg}{v^2}}$ (b) $\frac{v}{g}\sqrt{1+\frac{2hg}{v^2}}$
(c) $\sqrt{1+\frac{2hg}{v^2}}$ (d) $\frac{v}{g}\left[1+\sqrt{1+\frac{2hg}{v^2}}\right]$

24. A body thrown vertically up from the ground passes the height of 10.2 m twice in an interval of 10 s. What was its initial velocity?
(a) 52 m/s (b) 61 m/s (c) 45 m/s (d) 26 m/s

25. A particle is thrown upwards from ground. It experiences a constant resistance force which can produce a retardation of $2\ ms^{-2}$. The ratio of time of ascent to time of descent is ($g = 10\ ms^{-2}$)
(a) 1 : 1 (b) $\sqrt{\frac{2}{3}}$ (c) $\frac{2}{3}$ (d) $\sqrt{\frac{3}{2}}$

26. Two bodies A and B start from rest from the same point with a uniform acceleration of $2\ m/s^2$. If B starts one second later, then the two bodies are separated, at the end of the next second, by
(a) 1 m (b) 2 m (c) 3 m (d) 4 m

27. Two objects are moving along the same straight line. They cross a point A with an acceleration a, $2a$ and velocity $2u$, u at time $t = 0$. The distance moved by the objects when one overtakes the other is
(a) $\frac{6u^2}{a}$ (b) $\frac{2u^2}{a}$ (c) $\frac{4u^2}{a}$ (d) $\frac{8u^2}{a}$

28. A body moving with a uniform acceleration crosses a distance of 15 m in the 3rd second and 23 m in the 5th second. The displacement in 10 s will be
(a) 150 m (b) 200 m (c) 250 m (d) 300 m

29. A body is moving with uniform velocity of $8\ ms^{-1}$. When the body just crossed another body, the second one starts and moves with uniform acceleration of $4\ m/s^2$. The time after which two bodies meet, will be
(a) 2 s (b) 4 s (c) 6 s (d) 8 s

30. A car starts from rest and moves on straight line with constant acceleration a_0. After time t_0, brakes are applied which cause a deceleration of magnitude equal to its initial acceleration. The distance travelled by the car before coming to rest is
(a) $\frac{1}{2}a_0t_0^2$ (b) $2a_0t_0^2$ (c) $a_0t_0^2$ (d) $\frac{2}{3}a_0t_0^2$

31. A train starting from rest accelerates uniformly for 100 s, then comes to a stop with a uniform retardation in the next 200 s. During the motion, it covers a distance of 3 km. Choose the correct option.
(a) Its acceleration is $0.2\ m/s^2$
(b) Its retardation is $0.1\ m/s^2$
(c) The maximum velocity is 20 m/s
(d) All of the above

32. A point moves with uniform acceleration and v_1, v_2 and v_3 denote the average velocities in the three successive intervals of time t_1, t_2 and t_3. Which of the following relation is correct?
(a) $(v_1-v_2):(v_2-v_3)=(t_1-t_2):(t_2+t_3)$
(b) $(v_1-v_2):(v_2-v_3)=(t_1+t_2):(t_2+t_3)$
(c) $(v_1-v_2):(v_2-v_3)=(t_1-t_2):(t_1-t_3)$
(d) $(v_1-v_2):(v_2-v_3)=(t_1-t_2):(t_2-t_3)$

33. Two identical balls are shot upward one after another at an interval of 2s along the same vertical line with same initial velocity of 40 ms^{-1}. The height at which the balls collide is
(a) 50 m (b) 75 m (c) 100 m (d) 125 m

34. A body starts from rest with uniform acceleration a, its velocity after n seconds is v. The displacement of the body in last 3s is (assume total time of journey from 0 to n second)
(a) $\frac{v(6n-9)}{2n}$ (b) $\frac{2v(6n-9)}{n}$
(c) $\frac{2v(2n+1)}{n}$ (d) $\frac{2v(n-1)}{n}$

35. A particle starting from rest and moving with a uniform acceleration along a straight line covers distances a and b in first p and next q seconds. The acceleration of the particle is
(a) $\frac{a+b}{2(p+q)}$ (b) $\frac{2b}{(q+2p)q}$
(c) $\frac{2a}{p(p+2p)}$ (d) $\frac{a+b}{(p+q)^2}$

36. A small block slides without friction down an inclined plane starting from rest. Let s_n be the distance travelled from $t=n-1$ to $t=n$. Then, $\frac{s_n}{s_{n+1}}$ is **(2004 JEE)**
(a) $\frac{2n-1}{2n}$ (b) $\frac{2n+1}{2n-1}$
(c) $\frac{2n-1}{2n+1}$ (d) $\frac{2n}{2n+1}$

37. A particle is moving with speed $v=b\sqrt{x}$ along positive x-axis. Calculate the speed of the particle at time $t=\tau$ (assume that the particle is at origin at $t=0$). **(2019 Main)**
(a) $\frac{b^2\tau}{4}$ (b) $\frac{b^2\tau}{2}$ (c) $b^2\tau$ (d) $\frac{b^2\tau}{\sqrt{2}}$

38. A goods train accelerating uniformly on a straight railway track, approaches an electric pole standing on the side of track. Its engine passes the pole with velocity u and the guard's room passes with velocity v. The middle wagon of the train passes the pole with a velocity **(2021 JEE)**
(a) $\frac{u+v}{2}$ (b) $\frac{1}{2}\sqrt{u^2+v^2}$
(c) $\sqrt{uv}$ (d) $\sqrt{\left(\frac{u^2+v^2}{2}\right)}$

39. A scooter accelerates from rest for time t_1 at constant rate a_1 and then retards at constant rate a_2 for time t_2 and comes to rest. The correct value of $\frac{t_1}{t_2}$ will be **(2021 Main)**
(a) $\frac{a_1+a_2}{a_1}$ (b) $\frac{a_1}{a_2}$ (c) $\frac{a_2}{a_1}$ (d) $\frac{a_1+a_2}{a_2}$

40. A bullet of mass 20 g has an initial speed of 1 ms^{-1}, just before it starts penetrating a mud wall of thickness 20 cm. If the wall offers a mean resistance of 2.5×10^{-2} N, the speed of the bullet after emerging from the other side of the wall is close to **(2019 Main)**
(a) 0.1 ms^{-1} (b) 0.7 ms^{-1}
(c) 0.3 ms^{-1} (d) 0.4 ms^{-1}

41. An automobile travelling at 40 km/h, can be stopped at a distance of 40 m by applying brakes. If the same automobile is travelling at 80 km/h, the minimum stopping distance, in metres, is (assume no skidding) **(2018 Main)**
(a) 75 m (b) 160 m
(c) 100 m (d) 150 m

42. A car starting from rest, accelerates at the rate f through a distance s, then continues at constant speed for time t and then decelerates at the rate $\frac{f}{2}$ to come to rest. If the total distance traversed is $15s$, then **(2005 JEE)**
(a) $s=\frac{1}{6}ft^2$ (b) $s=ft$
(c) $s=\frac{1}{4}ft^2$ (d) $s=\frac{1}{72}ft^2$

43. If a body loses half of its velocity on penetrating 3 cm in a wooden block, then how much will it penetrate more before coming to rest? **(2002 JEE)**
(a) 1 cm (b) 2 cm
(c) 3 cm (d) 4 cm

44. A ball is thrown vertically upward with an initial velocity of 150 m/s. The ratio of velocity after 3s and 5s is $\frac{x+1}{x}$. The value of x is
(Take, $g=10\ ms^{-2}$) **(2023 Main)**
(a) 6 (b) 5
(c) −5 (d) 10

45. A ball is thrown up vertically with a certain velocity so that, it reaches a maximum height h. Find the ratio of the times in which it is at height $\frac{h}{3}$ while going up and coming down respectively. **(2022 Main)**
(a) $\frac{\sqrt{2}-1}{\sqrt{2}+1}$ (b) $\frac{\sqrt{3}+\sqrt{2}}{\sqrt{3}-\sqrt{2}}$
(c) $\frac{\sqrt{3}-1}{\sqrt{3}+1}$ (d) $\frac{1}{3}$

46. A ball is released from a height h. If t_1 and t_2 be the time required to complete first half and second half of the distance respectively. Then, choose the correct relation between t_1 and t_2. **(2022 Main)**
(a) $t_1=(\sqrt{2})t_2$
(b) $t_1=(\sqrt{2}-1)t_2$
(c) $t_2=(\sqrt{2}+1)t_1$
(d) $t_2=(\sqrt{2}-1)t_1$

47. A NCC parade is going at a uniform speed of 9 km/h under a mango tree on which a monkey is sitting at a height of 19.6 m. At any particular instant, the monkey drops a mango. A cadet will receive the mango whose distance from the tree at time of drop is (Given, $g = 9.8 \text{ ms}^{-2}$) **(2022 Main)**

(a) 5 m (b) 10 m
(c) 19.8 m (d) 24.5 m

48. A balloon was moving upwards with a uniform velocity of 10 m/s. An object of finite mass is dropped from the balloon when it was at a height of 75 m from the ground level. The height of the balloon from the ground when object strikes the ground was around (Take, $g = 10 \text{ ms}^{-2}$) **(2021 Main)**

(a) 300 m (b) 200 m
(c) 125 m (d) 250 m

49. A stone is dropped from the top of a building. When it crosses a point 5 m below the top, another stone starts to fall from a point 25 m below the top. Both stones reach the bottom of building simultaneously. The height of the building is **(2021 Main)**

(a) 35 m (b) 45 m
(c) 25 m (d) 50 m

50. A helicopter rises from rest on the ground vertically upwards with a constant acceleration g. A food packet is dropped from the helicopter when it is at a height h. The time taken by the packet to reach the ground is close to [g is the acceleration due to gravity] **(2020 Main)**

(a) $t = \frac{2}{3}\sqrt{\left(\frac{h}{g}\right)}$ (b) $t = 1.8\sqrt{\frac{h}{g}}$

(c) $t = \sqrt{\frac{2h}{3g}}$ (d) $t = 3.4\sqrt{\left(\frac{h}{g}\right)}$

51. A parachutist after bailing out falls 50 m without friction. When parachute opens, it decelerates at 2 m/s^2. He reaches the ground with a speed of 3 m/s. At what height, did he bail out? **(2005 JEE)**

(a) 182 m (b) 91 m
(c) 111 m (d) 293 m

52. From a building two balls A and B are thrown such that A is thrown upwards and B downwards (both vertically) with same speed. If v_A and v_B are their respective velocities on reaching the ground, then **(2002 JEE)**

(a) $v_B > v_A$
(b) $v_A = v_B$
(c) $v_A > v_B$
(d) their velocities depend on their masses

Multiple Correct Options

53. A particle is moving along a straight line. The displacement of the particle becomes zero in a certain time ($t > 0$). The particle does not undergo any collision.

(a) The acceleration of the particle may be always zero
(b) The acceleration of the particle may be uniform
(c) The velocity of the particle must be zero at some instant
(d) The acceleration of the particle must change its direction

54. A rocket is fired vertically up from the ground with a resultant acceleration of 10 m/s^2. The fuel is finished in 1 min and it continues to move up ($g = 10 \text{ m/s}^2$)

(a) the maximum height reached by rocket from ground is 18 km
(b) the maximum height reached by rocket from ground is 36 km
(c) the time from $t = 0$ when rocket is again at ground is $(180 + 30\sqrt{2})$ s
(d) the time from $t = 0$ when rocket is again at ground is $(120 + 60\sqrt{2})$ s

Subjective Questions

55. An automobile and a truck start from rest at the same instant, with the automobile initially at some distance behind the truck. The truck has a constant acceleration of 2.2 m/s^2 and the automobile has an acceleration of 3.5 m/s^2. The automobile overtakes the truck when it (truck) has moved 60 m.

(a) How much time does it take the automobile to overtake the truck ?
(b) How far was the automobile behind the truck initially?
(c) What is the speed of each during overtaking ?

56. Equation $s_t = u + at - \frac{1}{2}a$ does not seem dimensionally correct, why?

57. A particle is projected vertically upwards. What is the value of acceleration

(i) during upward journey,
(ii) during downward journey and
(iii) at highest point?

58. A ball is thrown vertically upwards. Which quantity remains constant among speed, kinetic energy, velocity and acceleration?

59. A particle moves rectilinearly with initial velocity u and constant acceleration a. Find the average velocity of the particle in a time interval from $t = 0$ to $t = t$ second of its motion.

60. Find the average velocity of a particle released from rest from a height of 125 m over a time interval till it strikes the ground. Take $g = 10 \text{ m/s}^2$.

61. A particle starts with an initial velocity 2.5 m/s along the positive x-direction and it accelerates uniformly at the rate 0.50 m/s^2.

(a) Find the distance travelled by it in the first two seconds

(b) How much time does it take to reach the velocity 7.5 m /s?

(c) How much distance will it cover in reaching the velocity 7.5 m/s?

62. A ball is projected vertically upward with a speed of 50 m/s. Find (a) the maximum height, (b) the time to reach the maximum height, (c) the speed at half the maximum height. Take $g = 10 \text{ m/s}^2$.

63. A particle starting from rest has a constant acceleration of 4 m/s^2 for 4 s. It then retards uniformly for next 8 s and comes to rest. Find during the motion of particle

(a) average acceleration

(b) average speed and

(c) average velocity

64. Two particles A and B start moving simultaneously along the line joining them in the same direction with acceleration of 1 m/s^2 and 2 m/s^2 and speeds 3 m/s and 1 m/s respectively. Initially, A is 10 m behind B. What is the minimum distance between them?

65. In one dimensional motion, position of a particle from a fixed point (say P) is $+2m$ at time $t = 0$ and the particle is at P at $t = 10$ s. If velocity of particle is zero at $t = 6$ s, determine the constant acceleration a and velocity at $t = 10$ s.

66. The velocity of a particle moving in the positive direction of the x-axis varies as $v = \alpha\sqrt{x}$, where α is a positive constant. Assuming that at the moment $t = 0$ the particle was located at the point $x = 0$, find :

(a) the time dependence of the velocity and the acceleration of the particle

(b) the mean velocity of the particle averaged over the time that the particle takes to cover the first s metres of the path. **(Irodov 1.22)**

Integer Type Question

67. A ball is thrown vertically upwards with a velocity of 19.6 ms^{-1} from the top of a tower. The ball strikes the ground after 6s. The height from the ground up to which the ball can rise will be $\left(\frac{k}{5}\right)$ m. The value of k is .... (Use $g = 9.8 \text{ m/s}^2$) **(2022 Main)**

Answers

1. (5m)	**2.** (6s)	**3.** (8 m/s^2)	**4.** (True)	**5.** True	**6.** (d)	**7.** (a)	**8.** (d)	**9.** (a)	**10.** (c)
11. (a)	**12.** (b)	**13.** (c)	**14.** (b)	**15.** (c)	**16.** (c)	**17.** (c)	**18.** (d)	**19.** (b)	**20.** (b)
21. (b)	**22.** (b)	**23.** (d)	**24.** (a)	**25.** (b)	**26.** (c)	**27.** (a)	**28.** (c)	**29.** (b)	**30.** (c)
31. (d)	**32.** (b)	**33.** (b)	**34.** (a)	**35.** (b)	**36.** (c)	**37.** (b)	**38.** (d)	**39.** (c)	**40.** (b)
41. (b)	**42.** (d)	**43.** (a)	**44.** (b)	**45.** (b)	**46.** (d)	**47.** (a)	**48.** (c)	**49.** (b)	**50.** (d)
51. (d)	**52.** (b)	**53.** (b,c)	**54.** (b,d)	**55.** (a) 7.39s (b) 35.5 m (c) 16.2 m/s, 25.9 m/s				**56.** (See the hints)	

57. (Always g) **58.** Acceleration **59.** $u + \frac{1}{2}at$ **60.** 25 m/s (downwards)

61. (a) 6.0 m (b) 10 s (c) 50 m **62.** (a) 125 m (b) 5 s (c) Approximately 35 m/s **63.** (a) Zero (b) 8m/s (c) 8 m/s

64. 8 m **65.** 0.2 m/s^2, 0.8 m/s **66.** (a) $v = \frac{\alpha^2}{2}t$, $a = \frac{\alpha^2}{2}$ (b) $v_{av} = \frac{\alpha\sqrt{s}}{2}$ **67.** (392)

See hints on pages 241-249.

MODULE 5

One Dimensional Motion with Non-Uniform Acceleration

When acceleration of a particle is not constant we take the help of differentiation or integration.

Equations of Differentiation

(a) $$v = \frac{ds}{dt} \quad \text{...(I)}$$

If the motion is taking place along x-axis is then this equation can be written as,

$$v = \frac{dx}{dt}$$

Here, v is the instantaneous velocity and x is the x coordinate at a general time t.

(b) $$a = \frac{dv}{dt} \quad \text{...(ii)}$$

Here, a is the instantaneous acceleration of the particle. Further, a can also be written as

$$a = \frac{dv}{dt} = \left(\frac{ds}{dt}\right)\left(\frac{dv}{ds}\right) = v\left(\frac{dv}{ds}\right) \qquad \left[\text{as } \frac{ds}{dt} = v\right]$$

$$\therefore \quad a = v\left(\frac{dv}{ds}\right) \quad \text{...(iii)}$$

Equations of Integration

(c) $$\int ds = \int v dt \quad \text{...(iv)}$$

or $$s = \int v dt \quad \text{...(v)}$$

In the above equations, v should be either constant or function of t.

(d) $$\int dv = \int a dt \quad \text{...(vi)}$$

or $$\Delta v = v_f - v_i = \int a dt \quad \text{...(vii)}$$

In the above equations a should be either constant or function of time t.

(e) $$\int v dv = \int a ds \quad \text{...(viii)}$$

In the above equation a should be either constant or function of s.

Note *(i) To convert s-t equation into v-t equation and then v-t equation into a-t equation differentiation will be done.*

$s\text{-}t \rightarrow v\text{-}t \rightarrow a\text{-}t$ *(differentiation)*

(ii) To convert a-t equation into v-t equation and then v-t equation into s-t equation, integration (with some limits) is done. By limit we mean, the value of physical quantity which we will get after integration should be known at some given time.

For example, after integrating v (w.r.t. time) we will get displacement s. Therefore, to get complete s function value of s should be known at some given time. Otherwise constant of integration remains as an unknown.

Thus, $a\text{-}t \rightarrow v\text{-}t \rightarrow s\text{-}t$ *(integration with limits)*

Derivations of Equations with Constant Acceleration

(i) $\boldsymbol{v = u + at}$ **:**

For one dimensional motion with $a =$ constant.

We can write, $dv = a\, dt$ $\qquad \left(\text{as } a = \frac{dv}{dt}\right)$

Integrating both sides, we have

$$\int dv = a \int dt \qquad (\text{as } a = \text{constant})$$

At $t = 0$, velocity is u and at $t = t$ velocity is v. Hence,

$$\int_u^v dv = a \int_0^t dt$$

$$\therefore \quad [v]_u^v = a\,[t]_0^t \quad \text{or} \quad v - u = at$$

$$\therefore \quad v = u + at$$ **Hence proved.**

Alternate method: $\int dv = a \int dt$. If we integrate both sides we will get, $v = at + c$. Here c is constant of integration. Now, $v = u$ at $t = 0$. Substituting in $v = at + c$, we get $c = u$. Now substituting value of c in the equation $v = at + c$ we get, $v = u + at$.

(ii) $\boldsymbol{s = ut + \frac{1}{2}at^2}$ **:**

We can write $ds = v\, dt$ $\qquad \left(\text{as } v = \frac{ds}{dt}\right)$

$$= (u + at)\, dt \qquad (\text{as } v = u + at)$$

At time $t = 0$ suppose $s = 0$ and at $t = t$, displacement is s, then

$$\int_0^s ds = \int_0^t (u + at)\, dt$$

$$\therefore \quad [s]_0^s = \left[ut + \frac{1}{2}at^2\right]_0^t$$

or $$s = ut + \frac{1}{2}at^2$$ **Hence proved.**

(iii) $v^2 = u^2 + 2as$:

We can also write, $v \cdot dv = a \cdot ds$ $\left(\text{as } a = v \cdot \frac{dv}{ds}\right)$

When $s = 0$, v is u and at $s = s$, velocity is v. Therefore,

$$\int_u^v v \cdot dv = a \int_0^s ds$$

or $$\left[\frac{v^2}{2}\right]_u^v = a[s]_0^s \quad \text{(as } a = \text{constant)}$$

∴ $$\frac{v^2}{2} - \frac{u^2}{2} = as$$

or $$\boxed{v^2 = u^2 + 2as}$$ **Hence proved.**

Example 1 x-coordinate *versus* time equation of a particle moving along x-axis is

$x = 20 + t^3 - 12t$ (SI units)

(a) Find, position and velocity of particle at time $t = 0$.

(b) State whether the motion is uniformly accelerated or not.

(c) Find position of particle when velocity of particle is zero.

Sol. (a) $x = 20 + t^3 - 12t$ ...(i)

At $t = 0$, $x = 20 + 0 - 0 = 20$ m

Velocity of particle at time t can be obtained by differentiating Eq. (i) w.r.t. time i.e.

$$v = \frac{dx}{dt} = 3t^2 - 12 \quad \text{...(ii)}$$

At $t = 0$, $v = 0 - 12 = -12$ m/s

(b) Differentiating Eq. (ii) w.r.t. time t, we get the acceleration $a = \frac{dv}{dt} = 6t$

As acceleration is a function of time, the motion is non-uniformly accelerated.

(c) Substituting $v = 0$ in Eq. (ii), we have $0 = 3t^2 - 12$

Positive value of t comes out to be 2 s from this equation. Substituting $t = 2$ s in Eq. (i), we have $x = 20 + (2)^3 - 12(2)$

or $x = 4$ m

Example 2 Velocity-time equation of a particle moving in a straight line is,

$$v = (10 + 2t + 3t^2) \quad \text{(SI units)}$$

Find

(a) displacement of particle from a fixed position at time $t = 1$ s, if it is given that displacement is 20 m at time $t = 0$.

(b) acceleration-time equation.

Sol. (a) The given equation can be written as,

$$v = \frac{ds}{dt} = (10 + 2t + 3t^2)$$

or $$ds = (10 + 2t + 3t^2)\, dt$$

or $$\int_{20}^{s} ds = \int_0^1 (10 + 2t + 3t^2)\, dt$$

or $$s - 20 = [10t + t^2 + t^3]_0^1$$

$$= (10 + 1 + 1) = 12$$

∴ $$s = 20 + 12 = 32 \text{ m}$$

(b) Acceleration-time equation can be obtained by differentiating the given equation w.r.t. time.

Thus, $$a = \frac{dv}{dt} = \frac{d}{dt}(10 + 2t + 3t^2)$$

or $$a = 2 + 6t$$

Typed Problems

Type 1. If a particle is given some initial velocity and there is only a retardation which is proportional to its instantaneous velocity, then its velocity decreases exponentially.

Example 3 Initial velocity of a particle is v_0. It has a retardation which is directly proportional to its instantaneous velocity v and varies as : $a = -rv$ where r is a positive constant. Then find :

(a) velocity as a function of time t.

(b) total distance travelled before coming to rest.

Sol. (a) Given, $a = -rv$

∴ $$\frac{dv}{dt} = -rv$$

∴ $$\frac{dv}{v} = -r\,dt$$

or $$\int_{v_0}^{v} \frac{dv}{v} = -r \int_0^t dt$$

∴ $$[\ln v]_{v_0}^{v} = -r[t]_0^t$$

⇒ $$[\ln v - \ln v_0] = -r[t - 0]$$

⇒ $$\log_e \frac{v}{v_0} = -rt$$

⇒ $$\frac{v}{v_0} = e^{-rt}$$

or $$v = v_0 e^{-rt}$$

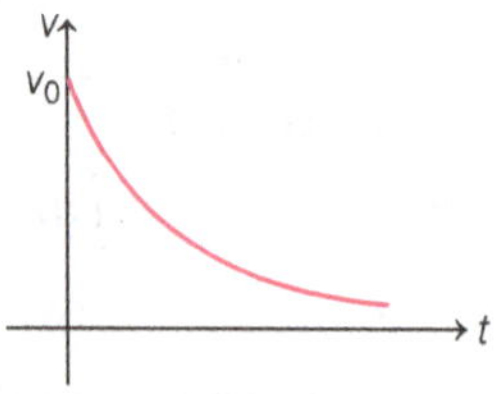

From the above expression we can see that velocity of particle decreases exponentially from v_0 to 0 as shown in above figure.

(b) Given, $a = -rv$

∴ $$v \cdot \frac{dv}{ds} = -rv \quad \text{or} \quad dv = -r\,ds$$

or $$\int_{v_0}^{0} dv = -r \int_0^s ds$$

$$\therefore \quad [v]_{v_0}^{0} = -r\,[s]_0^s$$

$$\Rightarrow \quad [0 - v_0] = -r\,[s - 0]$$

$$\therefore \quad s = \frac{v_0}{r}$$

$\therefore$ Total displacement (or distance) before coming to rest is $\frac{v_0}{r}$.

Type 2. Initial velocity of particle is zero. Acceleration has two parts. One is positive and constant. Other is negative and proportional to instantaneous velocity *v*, then velocity increases exponentially.

Example 4 Initial velocity of a particle is zero and its acceleration is given as,

$$a = \alpha - \beta v$$

Find velocity as a function of time *t*. Here α and β are positive constants.

Note *Here acceleration α is constant and retardation $-\beta v$ is proportional to v.*

Sol. Given,

$$a = \alpha - \beta v$$

$$\therefore \quad \frac{dv}{dt} = \alpha - \beta v$$

$$\text{or} \quad \frac{dv}{\alpha - \beta v} = dt$$

$$\Rightarrow \quad \int_0^v \frac{dv}{\alpha - \beta v} = \int_0^t dt$$

$$\Rightarrow \quad \left[-\frac{1}{\beta}\right][\ln(\alpha - \beta v)]_0^v = [t]_0^t$$

$$\Rightarrow \quad \left[-\frac{1}{\beta}\right][\ln(\alpha - \beta v) - \ln \alpha] = t$$

$$\Rightarrow \quad \log_e\left(\frac{\alpha - \beta v}{\alpha}\right) = -\beta t$$

$$\Rightarrow \quad 1 - \frac{\beta}{\alpha} v = e^{-\beta t}$$

$$\therefore \quad v = \frac{\alpha}{\beta}(1 - e^{-\beta t}) \qquad \textbf{Ans.}$$

From the above expression we can see that velocity of particle increases exponentially as shown below.

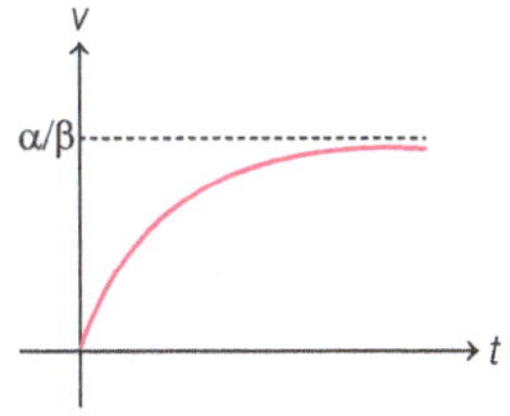

Here, maximum velocity $\frac{\alpha}{\beta}$ is called the terminal velocity.

You will find the reference of this terminal velocity in the topic of viscosity.

Type 3. Acceleration has two parts. One is constant and positive. Other is negative and proportional to distance travelled. Initial velocity is zero. Then velocity first increases to a maximum value and then decreases to zero.

Example 5 A particle free to move along *x*-axis starts from the origin with initial velocity zero. Acceleration of the particle is :

$$a = \alpha - \beta x$$

Here α and β are positive constants.

Find

(a) maximum velocity during the motion of the particle.

(b) maximum *x*-coordinate of the particle.

Note *Here acceleration α is positive and constant and retardation $-\beta x$ is proportional to x.*

Sol. Given, $a = \alpha - \beta x$

$$\Rightarrow \quad v \cdot \frac{dv}{dx} = \alpha - \beta x$$

$$\Rightarrow \quad v\,dv = (\alpha - \beta x)\,dx$$

$$\text{or} \quad \int_0^v v \cdot dv = \int_0^x (\alpha - \beta x)\,dx$$

$$\Rightarrow \quad \left[\frac{v^2}{2}\right]_0^v = \left[\alpha x - \frac{\beta x^2}{2}\right]_0^x$$

$$\Rightarrow \quad \frac{v^2}{2} = \alpha x - \beta \frac{x^2}{2}$$

$$\text{or} \quad v = \sqrt{2\alpha x - \beta x^2} \qquad \text{...(i)}$$

The above expression gives *v versus x* equation.

(a) $v^2 = 2\alpha x - \beta x^2$

v will be maximum when v^2 is maximum.

Or, $\frac{d}{dx}(v^2)$ is zero.

$$\therefore \quad \frac{d}{dx}(v^2) = \frac{d}{dx}(2\alpha x - \beta x^2) = 0$$

$$\text{or} \quad 2\alpha - 2\beta x = 0$$

$$\therefore \quad x = \frac{\alpha}{\beta}$$

At $x = \frac{\alpha}{\beta}$,

$\frac{d^2}{dx^2}(v^2)$ comes out to be negative. So, v^2 or *v* is maximum.

Putting $x = \frac{\alpha}{\beta}$ in Eq. (i), we get

$$v_{max} = \frac{\alpha}{\sqrt{\beta}} \qquad \textbf{Ans.}$$

(b) Initial velocity is zero. Then velocity increases to v_{max} at $x = \frac{\alpha}{\beta}$. Then velocity decreases from v_{max} to 0 at x_{max}. So, putting $v = 0$ in Eq. (i),

we have $x = 0$ and $x = \frac{2\alpha}{\beta}$

Thus, $x_{max} = \frac{2\alpha}{\beta}$ **Ans.**

Note (i) *From $x = 0$ to $x = \frac{\alpha}{\beta}$, net acceleration a is positive because $\alpha > \beta x$. So, velocity is increasing. At $x = \frac{\alpha}{\beta}$, acceleration $a = 0$, because $\alpha = \beta x$. So, velocity is maximum at $x = \frac{\alpha}{\beta}$.*

From $x = \frac{\alpha}{\beta}$ to $x = \frac{2\alpha}{\beta}$, acceleration a is negative. Because, $\beta x > \alpha$. So, velocity is decreasing.

(ii) *The above three types of situations (discussed in three types) you will get at many places in physics. We have taken the nature of acceleration. But in different other situations acceleration may be replaced by the net force acting on the particle. Results will remain same. Because F is simply ma. So, nature of acceleration is same as nature of net force.*

MODULE 5 Exercise

Assertion and Reason

See directions on page 178.

1. **Assertion :** We know the relation $a = v \cdot \frac{dv}{ds}$.
Therefore, if velocity of a particle is zero, then acceleration is also zero.
Reason : In the above equation, *a* is the instantaneous acceleration.

2. **Assertion :** Starting from rest with zero acceleration, if acceleration of particle increases at a constant rate of 2 ms^{-3}, then velocity should increase at constant rate of 1 ms^{-2}.
Reason : For the given condition,
$$\frac{da}{dt} = 2\ ms^{-3}$$
$$\therefore \quad a = 2t$$

Single Correct Option

3. The displacement of a particle moving in a straight line depends on time as $x = \alpha t^3 + \beta t^2 + \gamma t + \delta$. The ratio of initial acceleration to its initial velocity depends
(a) only on α and γ (b) only on β and γ
(c) only on α and β (d) only on α

4. The acceleration of a particle is increasing linearly with time *t* as *bt*. The particle starts from the origin with an initial velocity v_0. The distance travelled by the particle in time *t* will be
(a) $v_0t + \frac{1}{6}bt^3$ (b) $v_0t + \frac{1}{3}bt^3$
(c) $v_0t + \frac{1}{3}bt^2$ (d) $v_0t + \frac{1}{2}bt^2$

5. If the displacement of a particle varies with time as $\sqrt{x} = t + 3$,
(a) velocity of the particle is inversely proportional to *t*
(b) velocity of particle varies linearly with *t*
(c) velocity of particle is proportional to $\sqrt{t}$
(d) initial velocity of the particle is zero

6. A particle moves along a straight line. Its position at any instant is given by $x = 32t - \frac{8t^3}{3}$ where *x* is in metres and *t* in seconds. Find the acceleration of the particle at the instant when particle is at rest.
(a) $-16\ ms^{-2}$ (b) $-32\ ms^{-2}$ (c) 32 ms^{-2} (d) 16 ms^{-2}

7. A body starts moving with a velocity $v_0 = 10\ ms^{-1}$. It experiences a retardation equal to $0.2v^2$. Its velocity after 2s is given by
(a) $+2\ ms^{-1}$ (b) $+4\ ms^{-1}$
(c) $-2\ ms^{-1}$ (d) $+6\ ms^{-1}$

8. A body of mass 10 kg is being acted upon by a force $3t^2$ and an opposing constant force of 32 N. The initial speed is 10 ms^{-1}. The velocity of body after 5 s is
(a) 14.5 ms^{-1} (b) 6.5 ms^{-1}
(c) 3.5 ms^{-1} (d) 4.5 ms^{-1}

9. The position of a particle along *x*-axis at time *t* is given by $x = 2 + t - 3t^2$. The displacement and the distance travelled in the interval $t = 0$ to $t = 1$ are respectively
(a) 2, 2 (b) −2, 2.5
(c) 0, 2 (d) None of these

10. The instantaneous velocity of a particle moving in a straight line is given as $v = \alpha t + \beta t^2$, where α and β are

constants. The distance travelled by the particle between 1s and 2s is **(2021 Main)**

(a) $3\alpha + 7\beta$ (b) $\frac{3}{2}\alpha + \frac{7}{3}\beta$

(c) $\frac{\alpha}{2} + \frac{\beta}{3}$ (d) $\frac{3}{2}\alpha + \frac{7}{2}\beta$

11. The position of a particle as a function of time t, is given by $x(t) = at + bt^2 - ct^3$ where a, b and c are constants. When the particle attains zero acceleration, then its velocity will be **(2019 Main)**

(a) $a + \frac{b^2}{4c}$ (b) $a + \frac{b^2}{3c}$ (c) $a + \frac{b^2}{c}$ (d) $a + \frac{b^2}{2c}$

12. An object moving with a speed of 6.25 m/s, is decelerated at a rate given by $\frac{dv}{dt} = -2.5\sqrt{v}$ where v is the instantaneous speed. The time taken by the object, to come to rest, would be **(2011 JEE)**

(a) 2s (b) 4s (c) 8s (d) 1s

Multiple Correct Options

13. The displacement (x) of a particle depends upon time t as $x = \alpha t^2 - \beta t^3$. Choose the **incorrect** statements from the following.

(a) The particle never returns to its starting point

(b) The particle comes to rest after time $\frac{2\alpha}{3\beta}$

(c) The initial velocity of the particle is zero

(d) The initial acceleration of the particle is zero

Match the Columns

14. A particle is moving along x-axis. Its x-coordinate varies with time as :

$$x = -20 + 5t^2$$

For the given equation match the following two columns :

	Column I		Column II
(a)	Particle will cross the origin at	(p)	zero second
(b)	At what time velocity and acceleration are equal	(q)	1 s
(c)	At what time particle changes its direction of motion	(r)	2 s
(d)	At what time velocity is zero	(s)	None of the above

Subjective Questions

15. Velocity (in m/s) of a particle moving along x-axis varies with time as, $v = (10 + 5t - t^2)$

At time $t = 0, x = 0$. Find

(a) acceleration of particle at $t = 2$ s and

(b) x-coordinate of particle at $t = 3$ s

16. A particle is moving with a velocity of $v = (3 + 6t + 9t^2)$ cm/s. Find out

(a) the acceleration of the particle at $t = 3$ s.

(b) the displacement of the particle in the interval $t = 5$s to $t = 8$ s.

17. The motion of a particle along a straight line is described by the function $x = (2t - 3)^2$, where x is in metres and t is in seconds. Find

(a) the position, velocity and acceleration at $t = 2$ s.

(b) the velocity of the particle at origin.

18. x-coordinate of a particle moving along this axis is $x = (2 + t^2 + 2t^3)$. Here, x is in metres and t in seconds. Find

(a) position of particle from where it started its journey,

(b) initial velocity of particle and

(c) acceleration of particle at $t = 2$ s.

19. The velocity of a particle moving in a straight line is directly proportional to 3/4th power of time elapsed. How does its displacement and acceleration depend on time?

20. Velocity of a particle moving along positive x-direction is $v = (40 - 10t)$ m/s. Here, t is in seconds. At time $t = 0$, the x-coordinate of particle is zero. Find the time when the particle is at a distance of 60 m from origin.

21. A stone is released from an elevator going up with an acceleration of $\frac{g}{2}$. What is the acceleration of stone just after release?

Integer Type Question

22. The distance x covered by a particle in one dimensional motion varies with time t as $x^2 = at^2 + 2bt + c$. If the acceleration of the particle depends on x as x^{-n}, where n is an integer, the value of n is .... **(2020 Main)**

Answers

1. (d) **2.** (d) **3.** (b) **4.** (a) **5.** (b) **6.** (b) **7.** (a) **8.** (b) **9.** (d) **10.** (b)
11. (b) **12.** (a) **13.** (a, d) **14.** (a-r ; b-q ; c-s ; d-p) **15.** (a) $1 m/s^2$ (b) 43.5 m
16. (a) $60 cm/s^2$ (b) 1287 cm **17.** (a) $x = 1.0$ m, $v = 4$ m/s, $a = 8$ m/s^2 (b) zero **18.** (a) $x = 2.0$ cm (b) Zero (c) $26 ms^{-2}$
19. $s \propto t^{7/4}$ and $a \propto t^{-1/4}$ **20.** $t_1 = 2$ s, $t_2 = 6$ s, $t_3 = 2(2 + \sqrt{7})$ s **21.** g (downwards) **22.** (3)

See hints on pages 249-251.

MODULE 6

Motion in Two and Three Dimension

The motion of a particle thrown in a vertical plane at some angle with horizontal ($\neq 90°$) is an example of two dimensional motion.

Similarly, a circular motion is also an example of 2-D motion.

A two dimensional motion takes place in a plane. Motion of an aeroplane or a bird is a three dimensional motion, which takes place in space.

According to nature of acceleration we can classify two or three dimensional motion in following two types.

Uniform Acceleration

Equations of motion for uniformly accelerated motion ($\mathbf{a}$ = constant) are used in vector form as below.

$$\mathbf{v} = \mathbf{u} + \mathbf{a}t, \mathbf{s} = \mathbf{u}t + \frac{1}{2}\mathbf{a}t^2, \mathbf{v}\cdot\mathbf{v} = \mathbf{u}\cdot\mathbf{u} + 2\,\mathbf{a}\cdot\mathbf{s}$$

Here, $\mathbf{u}$ = initial velocity of particle, $\mathbf{v}$ = velocity of particle at time t and

$\mathbf{s}$ = displacement of particle in time t

Note *If initial position vector of a particle is $\mathbf{r}_0$, then position vector at time t can be written as* $\mathbf{r} = \mathbf{r}_0 + \mathbf{s} = \mathbf{r}_0 + \mathbf{u}t + \frac{1}{2}\mathbf{a}t^2$

Non-Uniform Acceleration

When acceleration is not constant then we will have to go for differentiation or integration. The commonly used equations in differentiation are:

(i) $\mathbf{v} = \frac{d\mathbf{s}}{dt}$ or $\frac{d\mathbf{r}}{dt}$

(ii) $\mathbf{a} = \frac{d\mathbf{v}}{dt}$

Here, $\mathbf{v}$ and $\mathbf{a}$ are instantaneous velocity and acceleration vectors. The equations of integration are

(iii) $\int d\mathbf{s} = \int \mathbf{v}\,dt$ and

(iv) $\int d\mathbf{v} = \int \mathbf{a}\,dt$

Note *A two or three dimensional motion can also be solved by component method.*

For example, in two dimensional motion (in x-y plane) the motion can be resolved along x and y directions. Now, along these two directions we can use sign method, as we have used in one-dimensional motion (but separately). By separately we mean, when we are looking the motion along x-axis we need not to bother about the motion along y-axis.

Example 1 A particle of mass 1 kg has a velocity of 2 m/s. A constant force of 2 N acts on the particle for 1 s in a direction perpendicular to its initial velocity. Find the velocity and displacement of the particle at the end of 1 s.

Sol. Force acting on the particle is constant. Hence, acceleration of the particle will also remain constant.

$$a = \frac{F}{m} = \frac{2}{1} = 2\text{ m/s}^2$$

Since, acceleration is constant. We can apply

$$\mathbf{v} = \mathbf{u} + \mathbf{a}t$$

and $$\mathbf{s} = \mathbf{u}t + \frac{1}{2}\mathbf{a}t^2$$

Refer figure (a), $\mathbf{v} = \mathbf{u} + \mathbf{a}t$

Here, $\mathbf{u}$ and $\mathbf{a}t$ are two mutually perpendicular vectors. So,

$$|\mathbf{v}| = \sqrt{(|\mathbf{u}|)^2 + (|\mathbf{a}t|)^2}$$
$$= \sqrt{(2)^2 + (2)^2}$$
$$= 2\sqrt{2}\text{ m/s}$$
$$\alpha = \tan^{-1}\frac{|\mathbf{a}t|}{|\mathbf{u}|} = \tan^{-1}\left(\frac{2}{2}\right)$$
$$= \tan^{-1}(1) = 45°$$

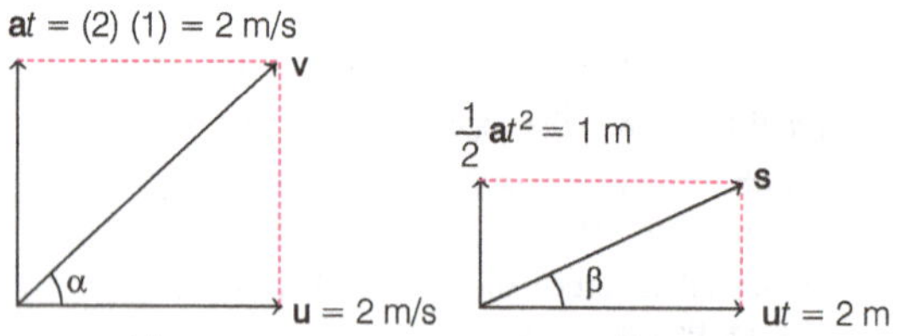

(a) (b)

Thus, velocity of the particle at the end of 1s is $2\sqrt{2}$ m/s at an angle of 45° with its initial velocity.

Refer figure (b), $\mathbf{s} = \mathbf{u}t + \frac{1}{2}\mathbf{a}t^2$

Here, $\mathbf{u}t$ and $\frac{1}{2}\mathbf{a}t^2$ are also two mutually perpendicular vectors. So,

$$|\mathbf{s}| = \sqrt{(|\mathbf{u}t|)^2 + (|\frac{1}{2}\mathbf{a}t^2|)^2}$$
$$= \sqrt{(2)^2 + (1)^2} = \sqrt{5}\text{ m}$$

and $\beta = \tan^{-1}\dfrac{\left|\frac{1}{2}at^2\right|}{|ut|}$

$= \tan^{-1}\left(\dfrac{1}{2}\right)$

Thus, displacement of the particle at the end of 1 s is $\sqrt{5}$ m at an angle of $\tan^{-1}\left(\dfrac{1}{2}\right)$ from its initial velocity.

Example 2 Velocity and acceleration of a particle at time $t = 0$ are $\mathbf{u} = (2\hat{i} + 3\hat{j})$ m/s and $\mathbf{a} = (4\hat{i} + 2\hat{j})$ m/s^2 respectively. Find the velocity and displacement of particle at $t = 2$s.

Sol. Here, acceleration $\mathbf{a} = (4\hat{i} + 2\hat{j})$ m/s^2 is constant.

So, we can apply

$$\mathbf{v} = \mathbf{u} + \mathbf{a}t \quad \text{and} \quad \mathbf{s} = \mathbf{u}t + \frac{1}{2}\mathbf{a}t^2$$

Substituting the proper values, we get

$$\mathbf{v} = (2\hat{i} + 3\hat{j}) + (2)(4\hat{i} + 2\hat{j})$$
$$= (10\hat{i} + 7\hat{j}) \text{ m/s}$$

and $$\mathbf{s} = (2)(2\hat{i} + 3\hat{j}) + \frac{1}{2}(2)^2(4\hat{i} + 2\hat{j})$$
$$= (12\hat{i} + 10\hat{j}) \text{ m}$$

Therefore, velocity and displacement of particle at $t = 2$ s are $(10\hat{i} + 7\hat{j})$ m/s and $(12\hat{i} + 10\hat{j})$ m respectively.

Example 3 Velocity of a particle in x-y plane at any time t is
$\mathbf{v} = (2t\hat{i} + 3t^2\hat{j})$ m/s

At $t = 0$, particle starts from the coordinates (2m, 4m). Find
(a) acceleration of the particle at $t = 1$ s.
(b) position vector and coordinates of the particle at $t = 2$ s.

Sol. (a) $\mathbf{a} = \dfrac{d\mathbf{v}}{dt} = \dfrac{d}{dt}(2t\hat{i} + 3t^2\hat{j})$

$= (2\hat{i} + 6t\hat{j})$ m/s^2

At $t = 1$ s,

$\mathbf{a} = (2\hat{i} - 6\hat{j})$ m/s^2 **Ans.**

(b) $\int d\mathbf{s} = \int \mathbf{v}\, dt$

or $\mathbf{s} = \int \mathbf{v}\, dt = \int (2t\hat{i} + 3t^2\hat{j})\, dt$

$\therefore$ $\mathbf{r}_f - \mathbf{r}_i = \int_{\text{initial}}^{\text{final}} (2t\hat{i} + 3t^2\hat{j})\, dt$

or $\mathbf{r}_{2\text{ sec}} - \mathbf{r}_{0\text{ sec}} = \int_0^2 (2t\hat{i} + 3t^2\hat{j})\, dt$

$\therefore$ $\mathbf{r}_{2\text{ sec}} = \mathbf{r}_{0\text{ sec}} + [t^2\hat{i} + t^3\hat{j}]_0^2$

$= (4\hat{i} + 2\hat{j}) + (8\hat{i} + 4\hat{j})$

$= (6\hat{i} + 12\hat{j})$ m **Ans.**

Therefore, coordinates of the particle at $t = 2$ s are (6 m, 12 m). **Ans.**

Typed Problems

Type 1. To find trajectory of a particle.

Method 1

In this type, a particle will be moving in x-y plane. Its x and y-coordinates as function of time will be given in the question and we have to find trajectory (or x-y relation) of the particle.

How to Solve?

From the given x and y coordinates (as function of time) just eliminate t and find x-y relation. This is a general method which can be applied anywhere in whole physics.

Example 4 A particle is moving in x-y plane with its x and y coordinates varying with time as, $x = 2t$ and $y = 10t - 16t^2$. Find trajectory of the particle.

Sol. Given, $x = 2t$

$\Rightarrow$ $t = \dfrac{x}{2}$

Now, $y = 10t - 16t^2$

Substituting value of t in this equation, we have

$$y = 10\left(\frac{x}{2}\right) - 16\left(\frac{x}{2}\right)^2$$

or $y = 5x - 4x^2$

This is the required equation of trajectory of the particle. This is a quadratic equation. Hence, the path of the particle is a parabola.

Method 2

With the help of given conditions in the question we first make an equation of following type :

$$f_1(x)\, dx = f_2(y)\, dy$$

Then we integrate this equation with proper limits to find x-y relation or trajectory of the particle.

Example 5 A particle moves in the plane x - y with velocity $\mathbf{v} = \alpha\hat{i} + \beta x\hat{j}$, where $\hat{i}$ and $\hat{j}$ are the unit vectors of the x and y-axes, and α and β are constants. At the initial moment of time, the particle was located at the point $x = y = 0$. Find the equation of the particle's trajectory $y(x)$. **(Irodov 1.35)**

Sol. $v_x = \frac{dx}{dt} = \alpha$ ...(i)

$$v_y = \frac{dy}{dt} = \beta x \quad ...(ii)$$

$\frac{(i)}{(ii)}$ gives,

$$\frac{dx}{dy} = \frac{\alpha}{\beta x} \quad \text{or} \quad (\beta x)dx = \alpha dy$$

Integrating, we get

$$\frac{\beta x^2}{2} = \alpha y + c$$

At $x = 0, y = 0$

$\Rightarrow$ $c = 0$

$\therefore$ The equation of trajectory is

$$\frac{\beta x^2}{2} = \alpha y \quad \text{or} \quad y = \frac{\beta x^2}{2\alpha}$$

Example 6 A balloon starts rising from the surface of the Earth. The ascension rate is constant and equal to v_0. Due to the wind the balloon gathers the horizontal velocity component $v_x = \alpha y$, where α is a constant and y is the height of ascent. Find how the following quantities depend on the height of ascent

(a) the horizontal drift of the balloon $x(y)$

(b) the total acceleration of the balloon. **(Irodov 1.34)**

Sol. (a) $v_y = \frac{dy}{dt} = v_0$ ...(i)

$$v_x = \frac{dx}{dt} = \alpha y \quad ...(ii)$$

$\frac{(i)}{(ii)}$ gives,

$$\frac{dy}{dx} = \frac{v_0}{\alpha y} \quad \text{or} \quad \alpha y dy = v_0 dx$$

Integrating this equation, we get

$$\frac{\alpha y^2}{2} = v_0 x + c$$

At $x = 0, y = 0 \Rightarrow c = 0$

Therefore the drift of the balloon is,

$$x = \frac{\alpha y^2}{2v_0}$$

(b) **Total acceleration**

Differentiating Eqs. (i) and (ii) w.r.t. time, we get

$$a_y = \frac{dv_y}{dt} = 0$$

and $$a_x = \frac{dv_x}{dt} = \alpha\left(\frac{dy}{dt}\right) = \alpha v_0$$

$\therefore$ Total acceleration,

$$a = \sqrt{a_x^2 + a_y^2}$$

or $$a = \alpha v_0$$

Type 2. Two dimensional motion by component method.

Concept

We will first find the components of all vector quantities along x, y and z-directions. Then, deal different axes separately as one dimension by assigning proper signs to all vector quantities. While dealing x-direction, we don't have to bother about y and z-directions.

Example 7 A particle is moving in x-y plane. Its initial velocity and acceleration are $\mathbf{u} = (4\hat{i} + 8\hat{j})$ m/s and $\mathbf{a} = (2\hat{i} - 4\hat{j})$ m/s^2. Find

(a) the time when the particle will cross the x-axis.

(b) x-coordinate of particle at this instant.

(c) velocity of the particle at this instant.

Initial coordinates of particle are (4m, 10m).

Sol.

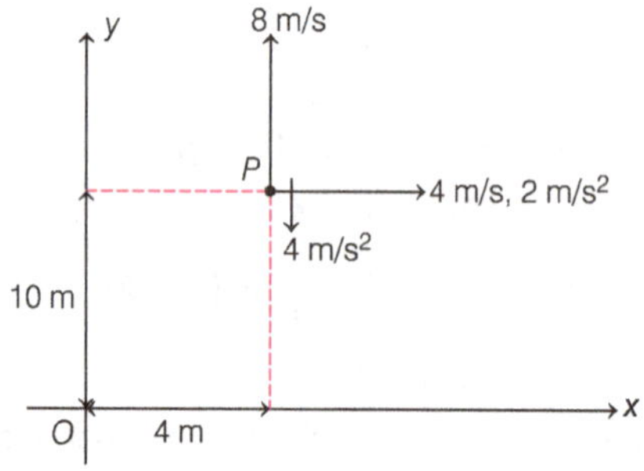

Particle starts from point P. Components of its initial velocity and acceleration are as shown in figure.

(a) At the time of crossing the x-axis, its y-coordinate should be zero or its y-displacement (w.r.t initial point P) is -10 m.

Using the equation,

$$s_y = u_y t + \frac{1}{2} a_y t^2$$

$$-10 = 8t - \frac{1}{2} \times 4 \times t^2$$

Solving this equation, we get positive value of time,

$$t = 5 \text{ s}$$ **Ans.**

(b) **x-coordinate of particle at time t :**

x = initial x-coordinate + displacement along x-axis

or $x = x_i + s_x$ (at time t)

$$= x_i + u_x t + \frac{1}{2} a_x t^2$$

Substituting the proper values, we have

$$x = 4 + (4 \times 5) + \frac{1}{2} \times 2 \times (5)^2 = 49 \text{ m}$$ **Ans.**

(c) Since, given acceleration is constant, so we can use,

$$\mathbf{v} = \mathbf{u} + \mathbf{a}t$$

$\therefore$ $$\mathbf{v} = (4\hat{i} + 8\hat{j}) + (2\hat{i} - 4\hat{j})(5)$$

$$= (14\hat{i} - 12\hat{j}) \text{m/s}$$ **Ans.**

MODULE 6 Exercise

Single Correct Option

1. A particle moves along the curve $y = \frac{x^2}{2}$. Here x varies with time as $x = \frac{t^2}{2}$. Where x and y are measured in metres and t in seconds. At $t = 2$ s, the velocity of the particle (in ms^{-1}) is
(a) $4\hat{i} + 6\hat{j}$ (b) $2\hat{i} + 4\hat{j}$ (c) $4\hat{i} + 2\hat{j}$ (d) $4\hat{i} + 4\hat{j}$

2. The horizontal and vertical displacements of a particle moving along a curved line are given by $x = 5t$ and $y = 2t^2 + t$. Time after which its velocity vector makes an angle of 45° with the horizontal is
(a) 0.5 s (b) 1 s (c) 2 s (d) 1.5 s

3. The initial velocity of a particle of mass 2 kg is $(4\hat{i} + 4\hat{j})$ m/s. A constant force of $-20\,\hat{j}$ N is applied on the particle. Initially, the particle was at (0, 0). Find the x-coordinate of the point where its y-coordinate is again zero.
(a) 3.2 m (b) 6m (c) 4.8 m (d) 1.2 m

4. A particle moves from the point $(2.0\hat{i} + 4.0\hat{j})$ m at $t = 0$ with an initial velocity $(5.0\hat{i} + 4.0\hat{j})\,ms^{-1}$. It is acted upon by a constant force which produces a constant acceleration $(4.0\hat{i} + 4.0\hat{j})\,ms^{-2}$. What is the distance of the particle from the origin at time 2 s? **(2019 Main)**
(a) 5 m (b) $20\sqrt{2}$ m (c) $10\sqrt{2}$ m (d) 15 m

5. A particle has an initial velocity of $3\hat{i} + 4\hat{i}$ and an acceleration of $0.4\,\hat{i} + 0.3\,\hat{j}$. Its speed after 10 s is **(2009 JEE)**
(a) $7\sqrt{2}$ units (b) 7 units
(c) 8.5 units (d) 10 units

6. The coordinates of a moving particle at any time 't' are given by $x = \alpha t^3$ and $y = \beta t^3$. The speed of the particle at time 't' is given by **(2003 JEE)**
(a) $3t\sqrt{\alpha^2 + \beta^2}$ (b) $3t^2\sqrt{a^2 + \beta^2}$
(c) $t^2\sqrt{\alpha^2 + \beta^2}$ (d) $\sqrt{\alpha^2 + \beta^2}$

Multiple Correct Options

7. A particle moves in x-y plane and at time t particle is at a point $(t^2, t^3 - 2t)$, then which of the following is/are correct?
(a) At $t = 0$, particle is moving parallel to y-axis
(b) At $t = 0$, direction of velocity and acceleration are perpendicular
(c) At $t = \sqrt{\frac{2}{3}}$, particle is moving parallel to x-axis
(d) At $t = 0$, particle is at rest

Match the Columns

8. x and y-coordinates of a particle moving in x-y plane are,
$$x = 1 - 2t + t^2 \text{ and } y = 4 - 4t + t^2$$
For the given situation match the following two columns :

	Column I		Column II
(a)	y-component of velocity when it crosses the y-axis	(p)	+ 2 SI unit
(b)	x-component of velocity when it crosses the x-axis	(q)	− 2 SI units
(c)	Initial velocity of particle	(r)	+ 4 SI units
(d)	Initial acceleration of particle	(s)	None of these

Subjective Questions

9. Velocity of a particle at time $t = 0$ is 2 m/s. A constant acceleration of 2 m/s^2 acts on the particle for 2 s at an angle of 60° with its initial velocity. Find the magnitude of velocity and displacement of particle at the end of $t = 2$ s.

10. Velocity of a particle at any time t is $\mathbf{v} = (2\hat{i} + 2t\hat{j})$ m/s. Find acceleration and displacement of particle at $t = 1$ s. Can we apply $\mathbf{v} = \mathbf{u} + \mathbf{a}t$ or not?

11. Acceleration of a particle in x-y plane varies with time as
$$\mathbf{a} = (2t\hat{i} + 3t^2\hat{j})\ m/s^2$$
At time $t = 0$, velocity of particle is 2 m/s along positive x-direction and particle starts from origin. Find velocity and coordinates of particle at $t = 1$ s.

12. A particle is moving in x-y plane. At time $t = 0$, particle is at (1m, 2m) and has velocity $(4\hat{i} + 6\hat{j})$ m/s. At $t = 4$ s, particle reaches at (6m, 4m) and has velocity $(2\hat{i} + 10\hat{j})$ m/s. In the given time interval, find
(a) average velocity,
(b) average acceleration and
(c) from the given data, can you find average speed.

13. At time $t = 0$, a particle is at (2m, 4m). It starts moving towards positive x-axis with constant acceleration 2 m/s^2 (initial velocity = 0). After 2 s, an additional acceleration of 4 m/s^2 starts acting on the particle in negative y-direction also. Find after next 2 s.
(a) velocity and
(b) coordinates of particle.

14. A particle starts from the origin at $t = 0$ with a velocity of $8.0\hat{j}$ m/s and moves in the x-y plane with a constant acceleration of $(4.0\,\hat{i} + 2.0\hat{j})$ m/s^2. At the instant the particle's x-coordinate is 29 m, what are (a) its y-coordinate and (b) its speed ?

15. A particle leaves the origin with an initial velocity $\mathbf{v} = (3.00\,\hat{\mathbf{i}})$ m/s and a constant acceleration $\mathbf{a} = (-1.00\,\hat{\mathbf{i}} - 0.500\,\hat{\mathbf{j}})$ m/s^2. When the particle reaches its maximum x-coordinate, what are (a) its velocity and (b) its position vector?

16. A radius vector of a point A relative to the origin varies with time t as $\mathbf{r} = \alpha t\hat{\mathbf{i}} - \beta t^2\hat{\mathbf{j}}$, where α and β are positive constants, and $\hat{\mathbf{i}}$ and $\hat{\mathbf{j}}$ are the unit vectors of the x and y axes. Find : **(Irodov 1.24)**
(a) the equation of the point's trajectory $y(x)$, plot this function
(b) the time dependence of the velocity **v** and acceleration **a** vectors, as well as of the moduli of these quantities
(c) the time dependence of the angle θ between the vectors **a** and **v**
(d) the mean velocity vector averaged over the first t seconds of motion and the modulus of this vector.

17. A point moves in the plane x - y according to the law $x = \alpha t,\ y = \alpha t(1 - \beta t)$, where α and β are positive constants, and t is time. Find:
(a) the equation of the point's trajectory $y(x)$; plot this function
(b) the velocity v and the acceleration a of the point as functions of time
(c) the moment t_0 at which the velocity vector forms an angle $\pi/4$ with the acceleration vector. **(Irodov 1.25)**

18. A point moves in the plane x - y according to the law $x = \alpha \sin \omega t,\ y = \alpha(1 - \cos \omega t)$, where α and ω are positive constants, Find :
(a) the distance d traversed by the point during the time τ,
(b) the angle between the point's velocity and acceleration vectors. **(Irodov 1.26)**

19. Find the magnitude and direction of the force acting on the particle of mass m during its motion in the plane x - y according to the law $x = a \sin \omega t$, $y = b \cos \omega t$, where a, b and ω are constants. **(Irodov 1.95)**

20. A point moves in the plane x-y according to the law $x = a \sin \omega t,\ y = b \cos \omega t$, where a, b and ω are positive constants. **(Irodov 4.14)**
Find :
(a) the trajectory equation $y(x)$ of the point and the direction of its motion along this trajectory;
(b) the acceleration **a** of the point as a function of its radius vector **r** relative to the origin of coordinates.

21. Find the trajectory equation $y(x)$ of a point if it moves according to the following laws:
(a) $x = a \sin \omega t,\ y = a \sin 2\omega t$
(b) $x = a \sin \omega t,\ y = a \cos 2\omega t$.
Plot these trajectories. **(Irodov 4.15)**

22. A body is projected up such that its position vector varies with time as $\mathbf{r} = \{3t\hat{\mathbf{i}} + (4t - 5t^2)\hat{\mathbf{j}}\}$ m. Here, t is in seconds. Find the time and x-coordinate of particle when its y-coordinate is zero.

Answers

1. (d) **2.** (b) **3.** (a) **4.** (b) **5.** (a)
6. (b) **7.** (a, b, c) **8.** (a-q; b-p; c-s; d-s)
9. $2\sqrt{7}$ m/s, $4\sqrt{3}$ m **10.** $(2\hat{\mathbf{j}})$ m/s^2, $(2\hat{\mathbf{i}} + \hat{\mathbf{j}})$ m, yes
11. $\mathbf{v} = (3\hat{\mathbf{i}} + \hat{\mathbf{j}})$ m/s, coordinates $= \left(\frac{7}{3}\text{m}, \frac{1}{4}\text{m}\right)$
12. (a) $(1.25\hat{\mathbf{i}} + 0.5\hat{\mathbf{j}})$ m/s (b) $(-0.5\hat{\mathbf{i}} + \hat{\mathbf{j}})$ m/s^2 (c) No
13. (a) $(8\hat{\mathbf{i}} - 8\hat{\mathbf{j}})$ m/s (b) (18 m, − 4 m)
14. (a) 45 m (b) 22 m/s **15.** (a) $(-1.5\hat{\mathbf{j}})$ m/s (b) $(4.5\,\hat{\mathbf{i}} - 2.25\hat{\mathbf{j}})$ m
16. (a) $y = -\frac{\beta x^2}{\alpha^2}$ (b) $\mathbf{v} = \alpha\hat{\mathbf{i}} - 2\beta t\hat{\mathbf{j}}$, $\mathbf{a} = -2\beta\hat{\mathbf{j}}$, $|\mathbf{v}| = \sqrt{\alpha^2 + 4\beta^2 t^2}$, $|\mathbf{a}| = 2\beta$ (c) $\tan\theta = \frac{\alpha}{2\beta t}$ (d) $\mathbf{v}_{av} = \alpha\hat{\mathbf{i}} - \beta t\hat{\mathbf{j}}$, $|\mathbf{v}_{av}| = \sqrt{\alpha^2 + \beta^2 t^2}$
17. (a) $y = x - \frac{\beta x^2}{\alpha}$ (b) $|\mathbf{v}| = \alpha\sqrt{1 + (1 - 2\beta t)^2}$, $|\mathbf{a}| = 2\alpha\beta =$ constant (c) $t_0 = \frac{1}{\beta}$
18. (a) $d = \alpha\omega\tau$ (b) $\theta = \frac{\pi}{2}$
19. $\mathbf{F} = -m\omega^2\mathbf{r}$, $|\mathbf{F}| = m\omega^2\sqrt{x^2 + y^2}$
20. (a) $\frac{x^2}{a^2} + \frac{y^2}{b^2} = 1$ clockwise (b) $\mathbf{a} = -\omega^2\mathbf{r}$
21. (a) $y^2 = 4x^2(1 - x^2/a^2)$ (b) $y = a(1 - 2x^2/a^2)$

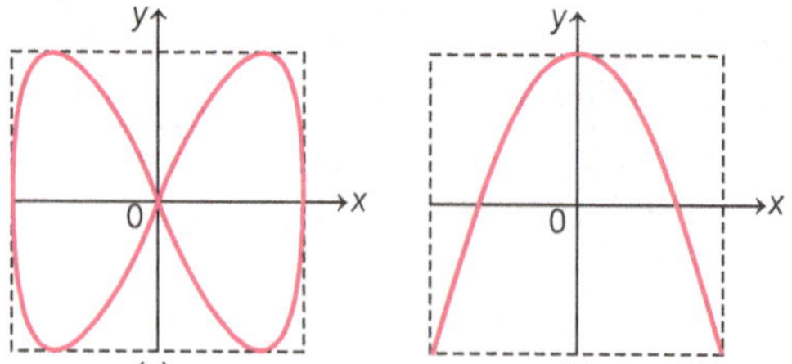

(a) (b)

22. (0.8s, 2.4 m)

See hints on pages 251-255.

MODULE 7

Graphs

Before studying graphs of kinematics, let us first discuss some general points of graphs :

(i) Mostly a graph is drawn between two variable quantities (say x and y). In kinematics, the commonly used graphs are s-t, v-t, a-t, a-s and v-s.

(ii) Equation between x and y will decide the shape of graph whether it is straight line, circle, parabola or rectangular hyperbola etc. If the equation is linear, graph is a straight line. If equation is quadratic then graph is a parabola. In kinematics, most of the graphs are straight line or parabola.

(iii) By putting $x = 0$ in y-x equation if we get $y = 0$, then graph passes through origin, otherwise not.

(iv) If $z = \frac{dy}{dx}$, then the value of z at any point can be obtained by the slope of the graph at that point.

For example,

instantaneous velocity $v = \frac{ds}{dt}$ = slope of s-t graph and,

instantaneous acceleration $a = \frac{dv}{dt}$ = slope of v-t graph

and,

(v) If $dz = y\,dx$.

Then, $\int dz = \int y\,dx$

$\Rightarrow$ z or $z_f - z_i$ or Δz = area under y-x graph, with projection along x-axis.

For example,

$$ds = vdt$$

$\Rightarrow$ Displacement $s = \int vdt$ = area under v-t graph with projection along t-axis.

Further, $dv = adt$

$\Rightarrow$ $v_f - v_i$ or $\Delta v = \int adt$ = area under a-t graph with projection along t-axis.

These results have been summarized in following table :

Table 1.

Name of Graph	Slope	Area
s-t	v	No physical quantity
v-t	a	s
a-t	Rate of change of acceleration	$v_f - v_i$ or Δv

For better understanding of three graphs (s-t, v-t and a-t) of kinematics, we have classified the one dimensional motion in following four types.

Uniform Motion

Equations

The three equations of uniform motion are given below.

$a = 0$, v = constant and $s = vt$ or $s = s_0 + vt$

Key Points

(i) s-t equation is linear. Therefore, s-t graph is straight line.

(ii) In $s = vt$, displacement is measured from the starting point ($t = 0$). Corresponding to this equation s-t graph passes through origin, as $s = 0$ when $t = 0$. In $s = s_0 + vt$, displacement is measured from any other point (say P) and s_0 is the displacement of starting point ($t = 0$) from P.

(iii) Slope of s-t graph = v. As v = constant, therefore slope of s-t graph = constant.

(iv) Slope of v-t graph = a. As $a = 0$, therefore slope of v-t graph = 0.

The corresponding graphs are as shown below.

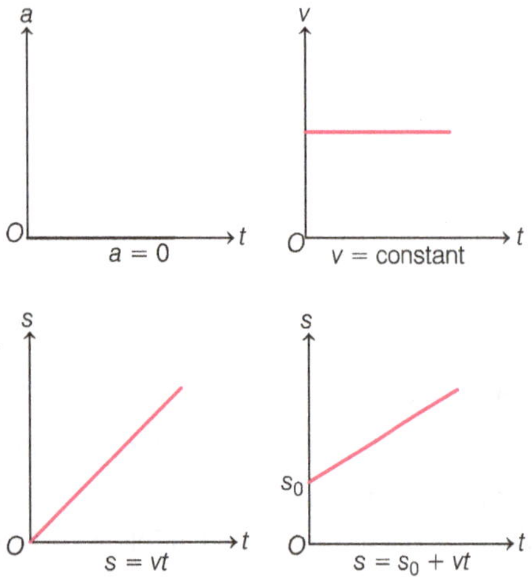

Example 1 s-t graph of a particle in motion is as shown below.

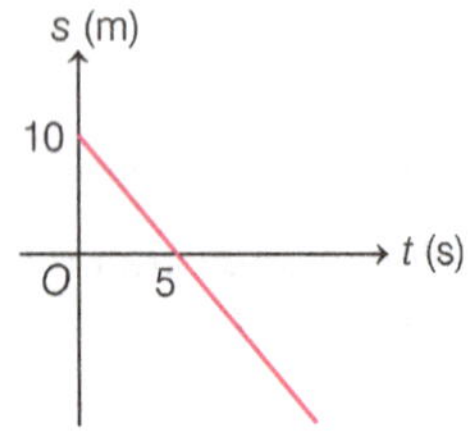

(a) State, whether the given graph represents a uniform motion or not.

(b) Find velocity of the particle.

Sol. (a) $v =$ slope of s-t graph. Since, the given s-t graph is a straight line and slope of a straight line is always constant.

Hence, velocity is constant. Therefore, the given graph represents a uniform motion.

(b) $v =$ slope of s-t graph $= \frac{-10}{5} = -2$ m/s **Ans.**

Example 2 A particle is moving along x-axis. Its x-coordinate versus time graph is as shown below.

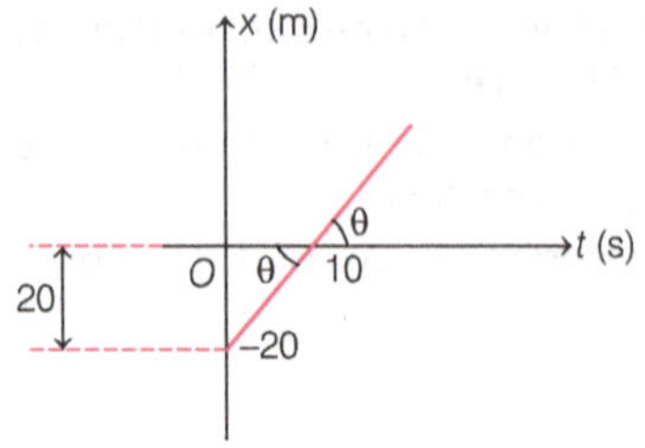

Draw some conclusions from the given graph.

Sol. The conclusions drawn from the graph are as under:

(i) x-t graph is a straight line, slope of which $\left(v = \frac{dx}{dt}\right)$ is positive and constant. Therefore, velocity is positive and constant.

(ii) $v = \frac{dx}{dt} =$ slope of x-t graph $= \tan\theta$

$$= +\frac{20}{10} = +2 \text{ m/s}$$

Therefore, velocity is 2 m/s along positive x-direction.

(iii) At $t = 0$, $x = -20$ m and at $t = 10$ s, $x = 0$

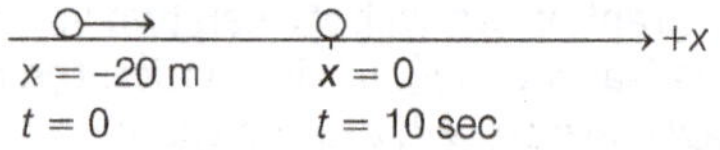

Note *In 10s particle has moved from –20 to 0. Or, it has travelled 20 m along positive x-direction. So, velocity should be 2m/s along positive x-direction.*

Uniformly Accelerated Motion

Equations

$a =$ constant (and positive)

$v = u + at$ or $v = at$, if $u = 0$

$s = ut + \frac{1}{2}at^2$ or $s = \frac{1}{2}at^2$ if $u = 0$

or $s = s_0 + ut + \frac{1}{2}at^2$ if $s_0 \neq 0$

or $s = s_0 + \frac{1}{2}at^2$ if $s_0 \neq 0$ but $u = 0$

Key Points

(i) v-t equation is linear. Therefore, v-t graph is a straight line. Further, $v = at$ is a straight line passing through origin (as $v = 0$ when $t = 0$)

(ii) All s-t equations are quadratic. Therefore, all s-t graphs should be parabolic.

(iii) Slope of s-t graph gives the instantaneous velocity. Therefore, initial slope of s-t graph gives initial velocity u. In this case, we are considering only accelerated motion (in which speed keeps on increasing in positive direction). Therefore, velocity is positive and continuously increasing. Hence, slope of s-t graph should be positive and keep on increasing.

(iv) Slope of v-t graph gives instantaneous acceleration. Now, acceleration is positive and constant. Therefore, slope of v-t graph should be positive and constant.

Graphs

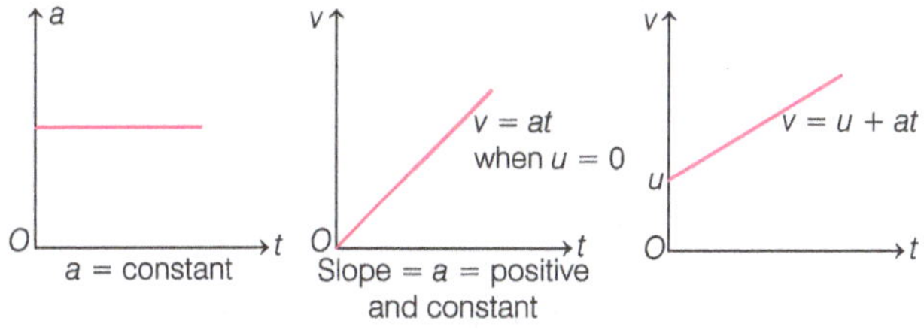

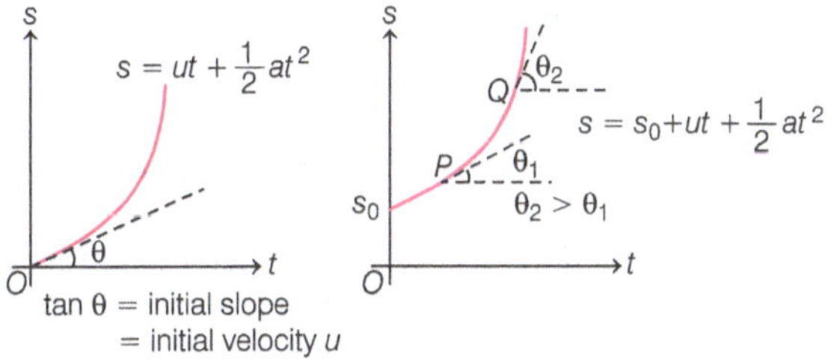

From P to Q slope is increasing (positive at both points). Therefore, velocity is positive and increasing.

Uniformly Retarded Motion

We are considering the case when initial velocity is positive and a constant acceleration acts in negative direction (till the velocity becomes zero).

Equations

$$a = \text{constant (and negative)}$$

$$v = u - at$$

$$s = ut - \frac{1}{2}at^2$$

Key Points

(i) In this case, u cannot be zero. Therefore, straight line of v-t graph cannot pass through origin. Further, initial slope of s-t graph cannot be zero.

(ii) Velocity is positive but keeps on decreasing from u to zero.

(iii) Slope of v-t graph gives instantaneous acceleration. Acceleration is constant and negative. Therefore, slope of v-t graph should be negative and constant.

(iv) Initial slope of s-t graph will give you initial velocity u. Final slope of s-t graph will give you final velocity which is zero. In between these two, velocity is positive and decreasing. Therefore, slope of s-t graph should be positive and decreasing.

Graphs

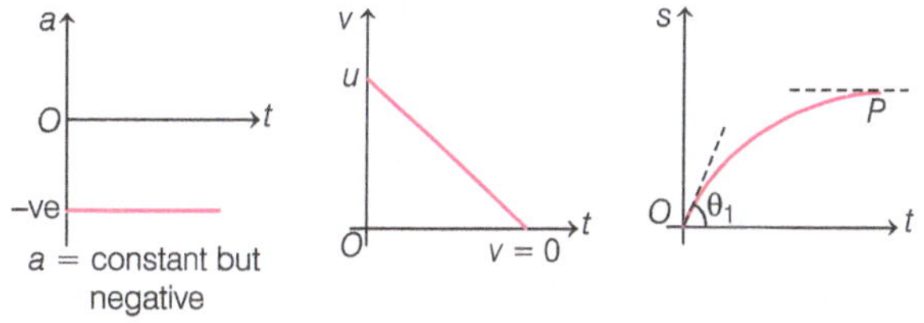

$\tan\theta_1 = \text{initial slope} = u$

From O to P, slope or v is positive but decreasing.

At P, slope $= 0$, therefore $v = 0$

Example 3 A particle is projected upwards with velocity 30 m/s. Taking upward direction as positive and $g = 10\ \text{m/s}^2$, plot a - t, v - t and s - t graphs till velocity becomes zero.

Sol.

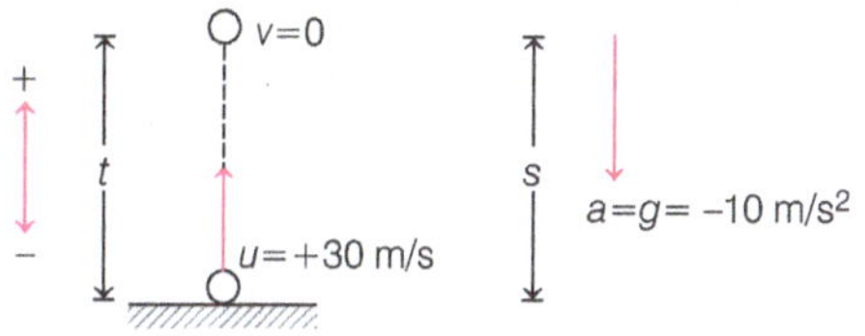

$$t = \frac{u}{g} = \frac{30}{10} = 3\ \text{s}$$

$$h = s = \frac{u^2}{2g} = \frac{(30)^2}{2 \times 10} = 45\ \text{m}$$

a - t, v - t and s - t graphs are as shown below.

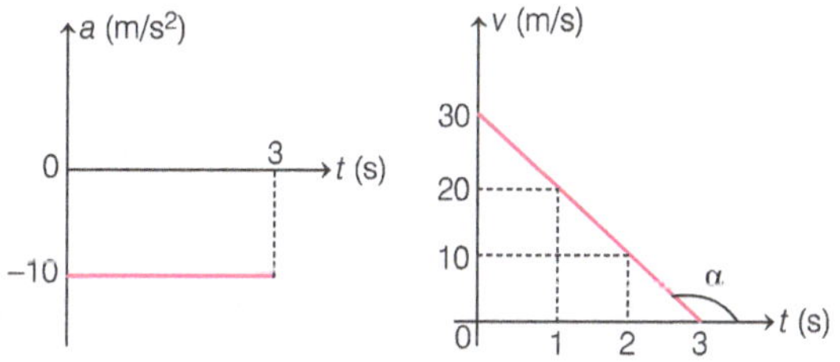

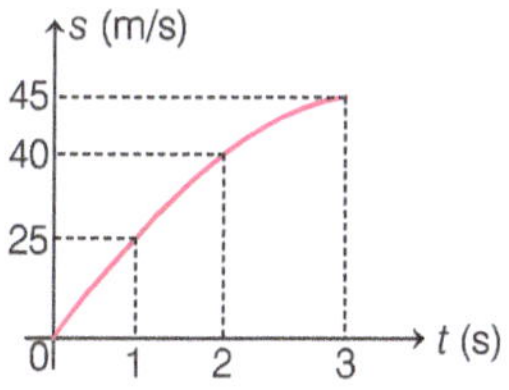

Key Points

(i) $\tan\alpha = \text{Slope of } v\text{-}t \text{ graph} = a = -10\ \text{m/s}^2$

(ii) Slope of s - t graph $= v$

$\therefore \quad \text{Slope}_{t=0} = v_{t=0} = u = +30\ \text{m/s}$

$$\text{Slope}_{1s} = v_{1s} = +20 \text{ m/s}$$
$$\text{Slope}_{2s} = v_{2s} = +10 \text{ m/s}$$
$$\text{Slope}_{3s} = v_{3s} = 0$$

(iii) $s_{0\text{-}1s} = 25 \text{ m}$

$s_{1s\text{-}2s} = (40 - 25) \text{ m} = 15 \text{ m}$

$s_{2s\text{-}3s} = (45 - 40) \text{ m} = 5 \text{ m}$

Note *$s_{2s\text{-}3s}$ = displacement in last second of upward journey = 5 m.*

Uniformly Retarded and then Accelerated Motion in Opposite Direction

If a particle is projected upwards then first it is retarded in upward (or positive) direction. At highest point its velocity becomes zero and finally it is accelerated in downward (or negative) direction.

Throughout the motion, its acceleration is downwards and constant (= acceleration due to gravity). Therefore, it is negative and constant.

If air resistance is neglected, then speed of the particle at the time of projection is equal to speed at the time of striking with the ground.

But velocities are in opposite directions. So, their signs are different. During retardation, velocity is upwards (therefore positive) but decreasing. During acceleration, velocity is downwards (therefore negative) and increasing.

Upward journey time is equal to the downward journey time.

Finally, the particle returns to the ground. Therefore final displacement is zero. In upward journey, displacement increases (parabolically) in positive direction. In downward journey, it decreases. But displacement from the starting point (ground) is still positive. Slope of *s*-*t* graphs gives the instantaneous velocity. In upward journey, velocity is positive and decreasing.

Therefore, slope is positive and decreasing. At highest point velocity is zero. Therefore, slope is zero. In downward journey, velocity is negative and increasing. Therefore, slope is negative and increasing.

Graphs

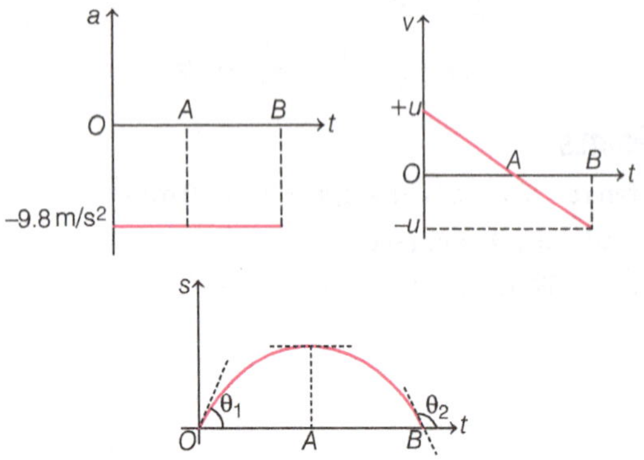

In the above graphs,

(i) $a = -9.8 \text{ m/s}^2$ if the motion is taking place under gravity.

(ii) *O* is the starting point, where

$$v = +u \Rightarrow \text{slope of } s\text{-}t \text{ graph} = \tan \theta_1 = u$$

(iii) *A* is the highest point, where

$$v = 0 \Rightarrow \text{slope of } s\text{-}t \text{ graph} = 0$$

(iv) *B* is the point where particle again strikes the ground. At this point,

$$v = -u \Rightarrow \text{slope of } s\text{-}t \text{ graph} = \tan \theta_2 = -u$$

At this point,

$$s = 0$$

(v) Upwards journey time t_{OA} = downward journey time t_{AB}

(vi) In upward motion (from *O* to *A*), velocity is positive and decreasing. Therefore, slope of *s*-*t* graph is positive and decreasing.

(vii) In downward motion (from *A* to *B*), velocity is negative and increasing. Therefore, slope of *s*-*t* graph is negative and increasing.

Example 4 A particle is projected upwards with velocity 20 m/s. Taking upward direction as the positive direction and $g = 10 \text{ m/s}^2$, plot *a* - *t*, *v* - *t* and *s* - *t* graphs till the particle returns to the ground.

Sol.

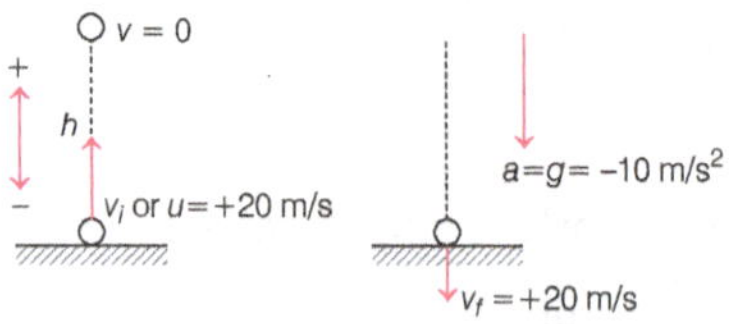

$$t_{up} = t_{down} = \frac{u}{g} = \frac{20}{10} = 2\text{s}$$

$$t_{Total} = t = 4 \text{ s}$$

$$h = \frac{u^2}{2g} = \frac{(20)^2}{2 \times 10} = 20 \text{ m}$$

a - *t*, *v* - *t* and *s* - *t* graphs are as shown below.

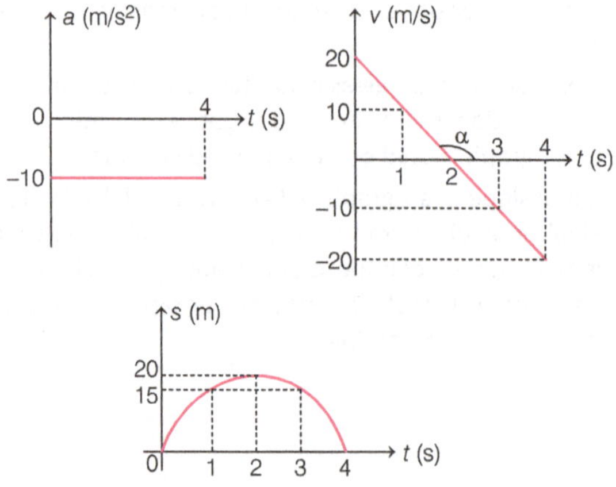

Key Points

(i) Area of a-t graph $= \Delta v = v_f - v_i$

Area $= (-10)(4) = -40$ m/s

$v_f - v_i = -20 - 20 = -40$ m/s

(ii) $\tan \alpha =$ Slope of v-t graph

$= a = -10$ m/s^2

(iii) Slope of s-t graph $= v$

$\therefore \quad \text{Slope}_{t=0} = v_{t=0} = u = +20$ m/s

Similarly, $\text{slope}_{t=4s} = v_{t=4s} = -20$ m/s

(iv) $s_{0\text{-}1s} = 15$ m

$s_{1s\text{-}2s} = (20 - 15) = 5$ m

= displacement in last second of upward journey

Important POINTS

1. Slope of v-t or s-t graph can't be infinite, because infinite slope of v-t graph means infinite acceleration. Similarly, infinite slope of s-t graph means infinite velocity. Hence, the following graphs are not possible.

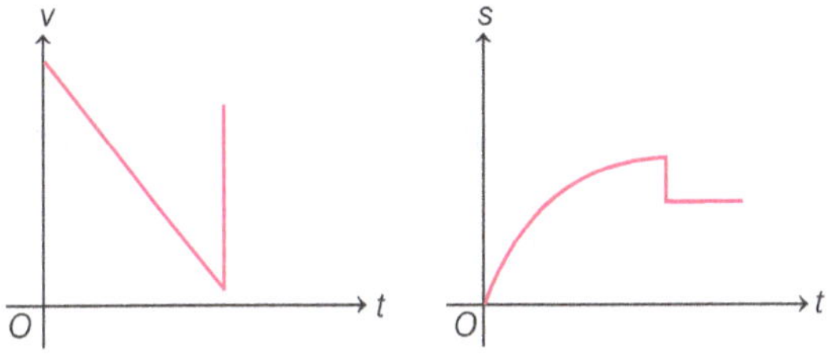

2. At one time (say t_0), two values of velocity or displacement are not possible. Hence, the following graphs are not acceptable.

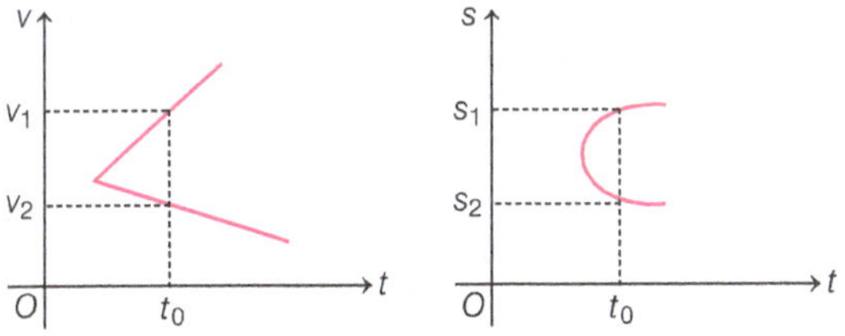

Note *At one position, two values of v or a can exist. So, following graph is acceptable.*

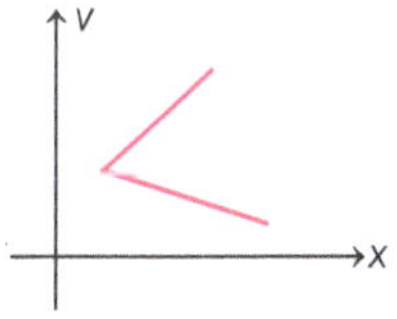

3. Time never returns. Therefore, on time axis we will always move ahead. On this ground, following graph cannot exist in real life .

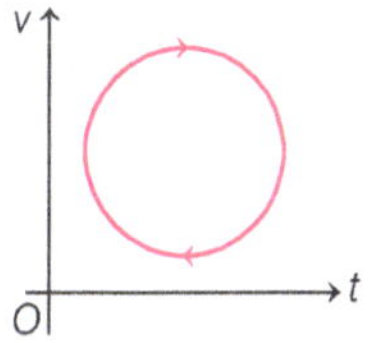

4. Different values of displacements corresponding to given v-t graph can be obtained just by calculating area under v-t graph. There is no need of using equations like $s = ut + \frac{1}{2}at^2$ etc.

Typed Problems

Type 1. From the slope of s-t graph we can find the signs of v and a.

Concept

Slope of s-t graph gives you velocity. Following example will illustrate the concept.

Example 5 From the s-t graph given below find signs of v and a.

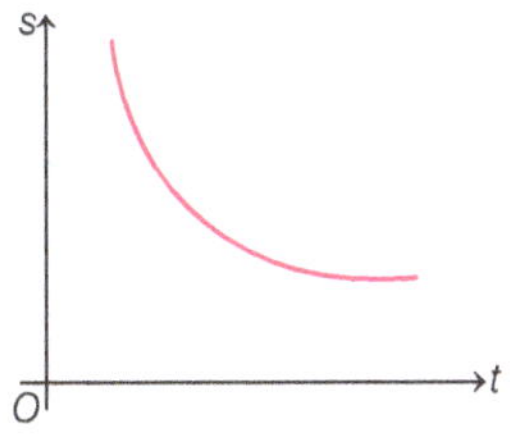

Sol. On time axis we have to move always ahead. So, let us check slope of s-t graph at any two times (say 4 PM and 5 PM)

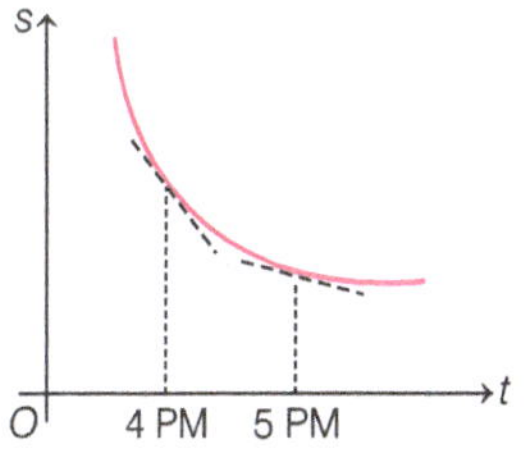

At both points slope is negative. So, velocity is negative. Further, magnitude of slope (means magnitude of velocity or speed) is decreasing. Hence, acceleration is positive.

In type-3, we have called this type of motion as R^{-1}. You can check the s-t graph corresponding to R^{-1} in type-3. It is same as given in this example.

Type 2. To convert s-t (or x-t) graph into v-t graph.

Concept

If we differentiate displacement, we get velocity. Therefore, the method discussed here is a general method, which can be applied in all those problems where we get the result after differentiation.

For example :

$$s\text{-}t \text{ or } x\text{-}t \rightarrow v\text{-}t$$
$$v\text{-}t \rightarrow a\text{-}t$$
$$p\text{-}t \rightarrow F\text{-}t$$
$$V\text{-}x \rightarrow E\text{-}x$$

Here, p = linear momentum and

$F = \text{force} = \dfrac{dp}{dt}$

V = electric potential (SI unit volt)

and E = electric field (SI unit N/C or V/m)

$= -\dfrac{dV}{dx}$

Let us take one example of last formula $E = -\dfrac{dV}{dx} = -$Slope of V - x graph. So, to convert V - x graph into E - x graph we have to see the slope of V - x graph.

Example 6 V - x graph along x -axis is as given below. Convert this graph into E - x graph.

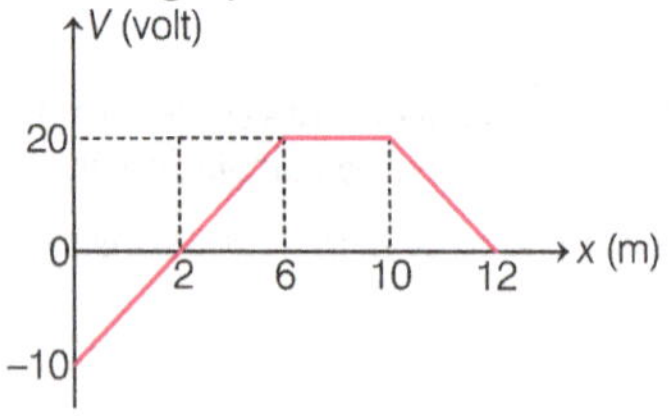

Sol.

Value of x (in m)	Slope of V - x graph (in V/m or N/C)	$E = -$Slope of V - x graph (in V/m or N/C)
0-6	+ 5	− 5
6-10	0	0
10-12	−10	+10

Therefore E - x graph is as shown below :

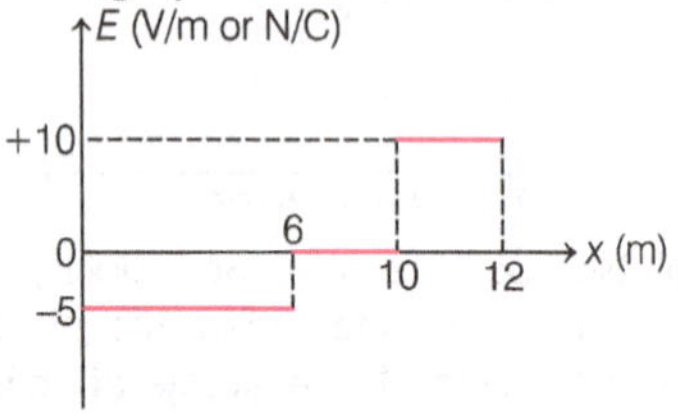

Example 7 A particle is moving along x-axis. Its x-coordinate *versus* time graph is as shown below.

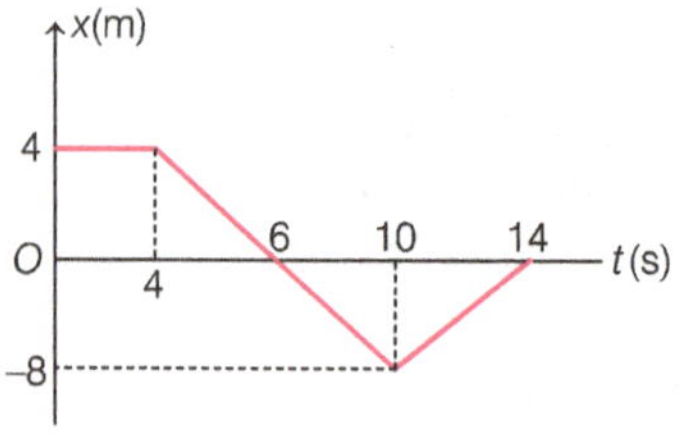

Plot v - t graph corresponding to this.

Sol. $v = \dfrac{dx}{dt}$ = slope of x - t graph. Slope for different time intervals is given in following table.

Time interval	Slope of x-t graph (= v)
0 − 4 s	0
4 s −10 s	− 2 m/ s
10 s −14 s	+ 2 m/s

v - t graph corresponding to this table is as shown below.

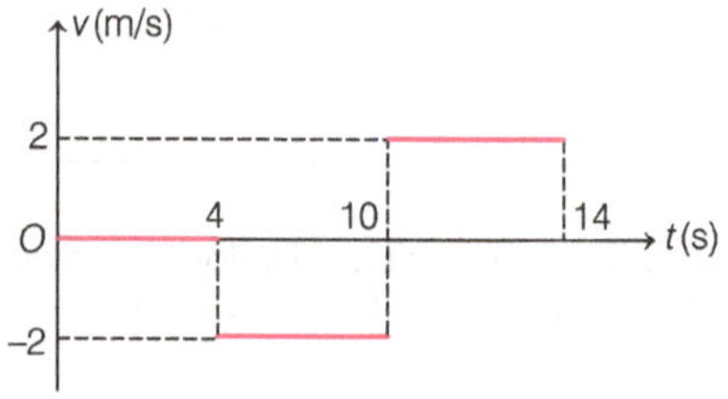

Type 3. To convert given v-t graph into s-t graph (For $a = 0$ or a = constant)

Concept

(i) If we integrate velocity, we get displacement. Therefore, the method discussed in this type is a general method, which can be applied in all those problems where we get the result after integration.

For example

$$v\text{-}t \longrightarrow s\text{-}t$$
$$a\text{-}t \longrightarrow v\text{-}t$$
$$F\text{-}t \longrightarrow p\text{-}t$$

Here, p = linear momentum and F is force ($F = \dfrac{dp}{dt}$ or $dp = Fdt$).

(ii) For zero or constant acceleration, we can classify any one dimensional motion into six types. Corresponding v-t and s-t graphs are as shown below.

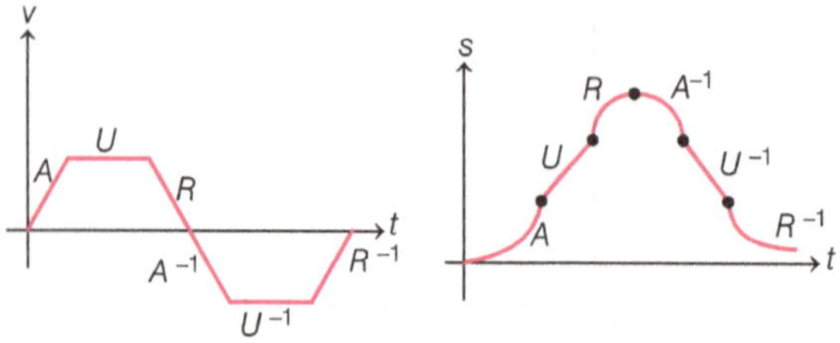

(iii) The explanation of these six motions is as under.

Motion type	About the motion	Velocity or Slope of *s-t* graph ($v = ds/dt$)
A	Accelerated in positive direction	positive and increasing
U	Uniform in positive direction	positive and constant
R	Retarded in positive direction	positive and decreasing
A^{-1}	Accelerated in negative direction	negative and increasing
U^{-1}	Uniform in negative direction	negative and constant
R^{-1}	Retarded in negative direction	negative and decreasing

(iv) In A, U and R motions, velocity is positive (above t-axis). Therefore, body is moving along positive direction. In A^{-1}, U^{-1} and R^{-1} motions, velocity is negative (below t-axis). Therefore, body is moving along negative direction.

Note *In type -1 we have already explained R^{-1} motion.*

Exercise : Using type-1, justify all motions corresponding to A, U, R etc. in s - t graph.

How to Solve?

Mark A, U, R, A^{-1}, U^{-1} or R^{-1} in the given v-t graph for different time intervals.

Calculate area (= displacement) under v-t graph for different time intervals.

Plot s-t graph according to A, U, R etc.

Keep on adding area for further displacements.

Example 8 Velocity-time graph of a particle moving along x-axis is as shown below.

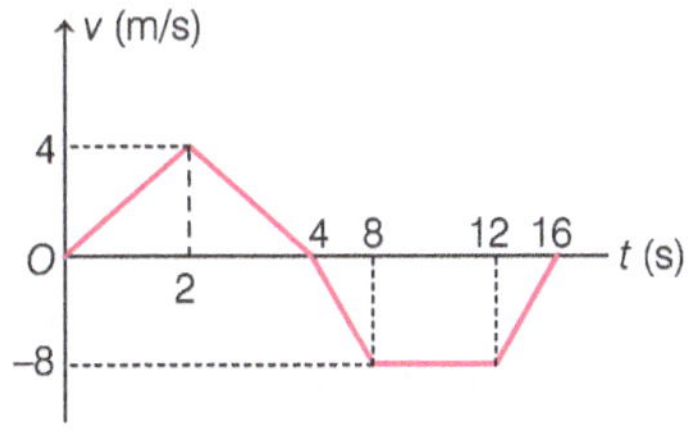

At time $t = 0$, x-coordinate of the particle is $x = 10$ m.

(a) Plot x-coordinate versus time graph.

(b) Find average velocity and average speed of the particle during the complete journey.

(c) Find average acceleration of the particle between the time interval from $t = 2$ s to $t = 8$ s.

Sol. (a) Let us first mark A, U, R etc. in the given v-t diagram and calculate their area (= displacement) in different time intervals.

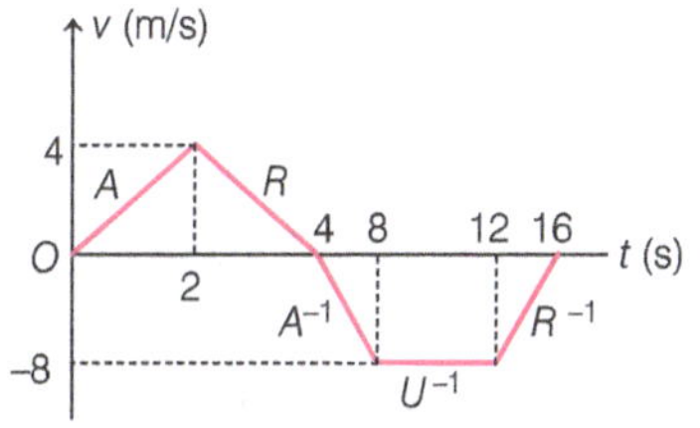

Time interval	Area or displacement	Final x-coordinate at the end of intervals $x = x_i + s$
0 - 2 s	+ 4 m	10 + 4 = 14 m
2 s - 4 s	+ 4 m	14 + 4 = 18 m
4 s - 8 s	– 16 m	18 – 16 = 2 m
8 s - 12 s	– 32 m	2 – 32 = – 30 m
12 s - 16 s	– 16 m	–30 – 16 = – 46 m

Corresponding x-t graph is as shown below.

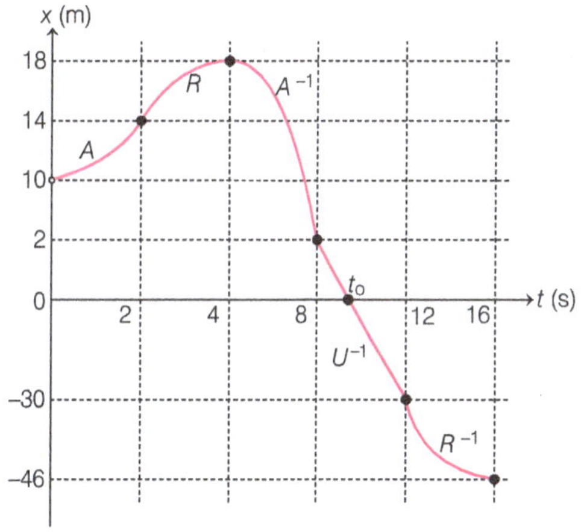

(b) Total displacement $= 4 + 4 - 16 - 32 - 16 = -56$ m

This is also equal to $x_f - x_i = -46 - 10 = -56$ m

Total distance $= 4 + 4 + 16 + 32 + 16 = 72$ m

Total time = 16 s

$$\text{Now, Average velocity} = \frac{\text{Total displacement}}{\text{Total time}}$$

$$= -\frac{56}{16} = -3.5 \text{ m/s} \qquad \textbf{Ans.}$$

$$\text{Average speed} = \frac{\text{Total distance}}{\text{Total time}}$$

$$= \frac{72}{16} = 4.5 \text{ m/s} \qquad \textbf{Ans.}$$

(c) $$\text{Average acceleration} = \frac{\Delta v}{\Delta t}$$

$$= \frac{v_f - v_i}{\Delta t} = \frac{v_{8\,s} - v_{2\,s}}{8 - 2}$$

$$= \frac{-8 - 4}{6} = -2 \text{ m/s}^2 \qquad \textbf{Ans.}$$

Exercise : *Find the time t_0 when x-coordinate of the particle is zero.*

Ans. 8.25 s

Type 4. To convert *v-s* graph into *a-s* graph.

Method

First make v-s equation (if it is a straight line graph) from the given v-s graph. Then, with the help of this v-s equation and standard equation $a = v \cdot \frac{dv}{ds}$, make a-s equation. Now draw a-s graph corresponding to this a-s equation.

Example 9 Corresponding to given v - s graph of a particle moving in a straight line, plot a - s graph.

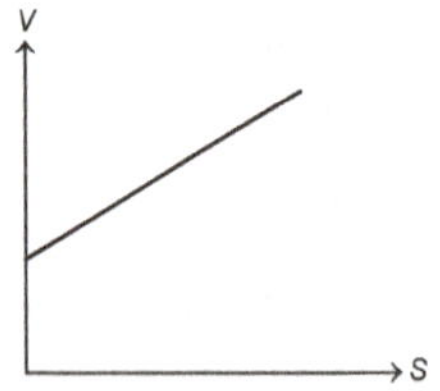

Sol. The given v - s graph is a straight line with positive slope (say m) and positive intercept (say c). Therefore, v-s equation is

$$v = ms + c \Rightarrow \frac{dv}{ds} = m$$

Now, $$a = v \cdot \frac{dv}{ds} = (ms + c)(m)$$

$$a = m^2 s + mc$$

a-s equation is a linear equation. Therefore, a-s graph is also a straight line with positive slope $(= m^2)$ and positive intercept $(= mc)$. Hence, a-s graph is as shown below.

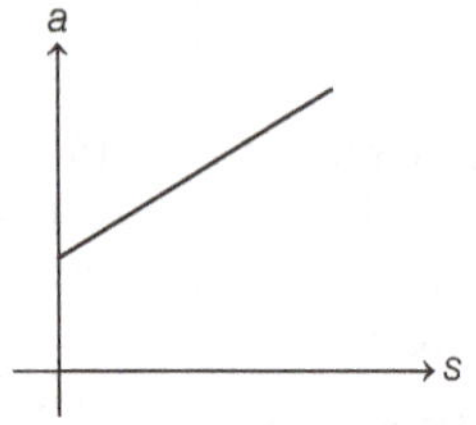

Type 5. Acceleration is constant in different time intervals. But this constant value is different in different time intervals.

Method

Since acceleration is constant. Therefore, we can use the equations like $v = u + at$ etc. But the acceleration is different for different time intervals. So, we will have to change the equations every time because initial velocity u will change every time.

So, the simple way is, **first draw v - t graph** (which is straight line for constant acceleration) and then **area of v - t graph** gives you further **displacement**. You can also convert this v - t graph into s - t (or x - t) graph by the method discussed in type-3.

Example 10 A particle is moving along x -axis. At $t = 0$ velocity is 4 m/s along positive x-axis and x-coordinate of particle is –6m. Constant acceleration in different time intervals is given below.

Time interval (in s)	Acceleration (in m/s^2)
0-2	+ 2
2-6	+ 4
6-10	– 5

Find velocity and x-coordinate at the end of 10s.

Sol. From 0-2s, acceleration is $+2 \text{ m/s}^2$. So, in every second velocity will change by 2 m/s in positive direction. Therefore at the end of 2s, it will change by 4 m/s in positive x-direction. Initial velocity was 4 m/s in positive x-direction. Hence at 2s it will become 8 m/s along positive x-direction. Using the same logic for different time intervals, we can plot v - t graph as shown below.

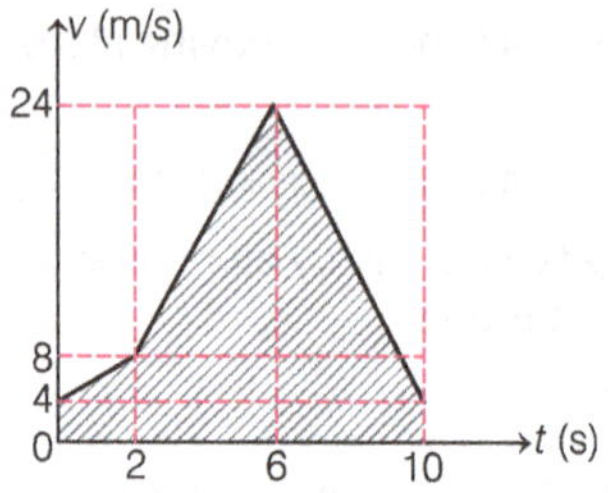

(i) From the graph we can see that,

$$v_{10s} = 4 \text{ m/s}$$ **Ans.**

(ii) Area of v - t graph $= s = x_f - x_i$

$$\therefore x_f = \text{Area} + x_i$$

$$= \left[\frac{1}{2}(4+8)(2) + \frac{1}{2}(8+24)(4) + \frac{1}{2}(24+4)(4)\right] - 6$$

$$= (12 + 64 + 56) - 6$$

$$= 126 \text{ m}$$ **Ans.**

Exercise : Using type-3, convert this v - t graph into x - t graph.

Type 6. Use of graph in a car race situation when initial velocity is zero and final velocity is also zero. Car accelerates for some time and then retards for rest of the time.

Concepts

(i) If retardation is n-times the acceleration then :

$$\text{retardation time} = \frac{\text{acceleration time}}{n}$$

and retardation displacement

$$= \frac{\text{acceleration displacement}}{n}$$

(ii) Area of v - t graph = displacement

Example 11 A car accelerates from rest at a constant rate α for some time, after which it decelerates at a constant rate β, to come to rest. If the total time elapsed is t seconds, then evaluate (a) the maximum velocity reached and (b) the total distance travelled.

Sol. (a) Let the car accelerates for time t_1 and decelerates for time t_2. Then,

$$t = t_1 + t_2 \qquad \text{...... (i)}$$

and corresponding velocity-time graph will be as shown in figure.

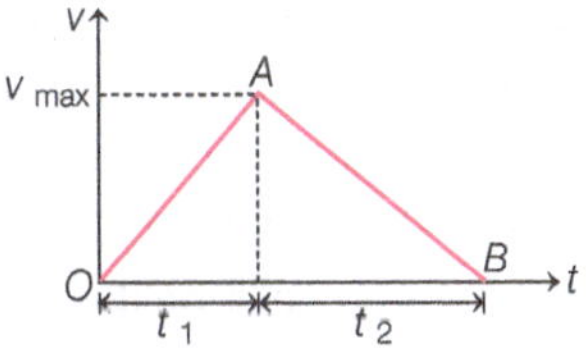

From the graph,

$$\alpha = \text{slope of line } OA = \frac{v_{max}}{t_1}$$

or $$t_1 = \frac{v_{max}}{\alpha} \qquad \text{...(ii)}$$

and $$\beta = -\text{ slope of line } AB = \frac{v_{max}}{t_2}$$

or $$t_2 = \frac{v_{max}}{\beta} \qquad \text{...(iii)}$$

From Eqs. (i), (ii) and (iii), we get

$$\frac{v_{max}}{\alpha} + \frac{v_{max}}{\beta} = t$$

or $$v_{max}\left(\frac{\alpha + \beta}{\alpha\beta}\right) = t$$

or $$v_{max} = \frac{\alpha\beta t}{\alpha + \beta} \qquad \textbf{Ans.}$$

(b) Total distance = total displacement

$$= \text{area under } v\text{-}t \text{ graph}$$

$$= \frac{1}{2} \times t \times v_{max}$$

$$= \frac{1}{2} \times t \times \frac{\alpha\beta t}{\alpha + \beta}$$

or $$\text{Distance} = \frac{1}{2}\left(\frac{\alpha\beta t^2}{\alpha + \beta}\right) \qquad \textbf{Ans.}$$

Note *This problem can also be solved by using equations of motion ($v = u + at$ etc.). Try it yourself.*

Type 7. Based on complex graphs.

Key Points:

(i) Speed, time, distance, height from ground can't be negative. So a graph between any two physical quantities (from this list) will always lie in first quadrant.

(ii) Velocity, acceleration and displacement are vector quantities. So, they will come with sign. According to our sign convention, upward direction is positive and downward direction is negative.

(iii) Displacement may first increase and then decrease. But distance always keeps on increasing (as long as the particle is in motion).

(iv) For constant acceleration, v-t graph is a straight line and s-t graph is a parabola. For any other situation we will have to see the equation between the two physical quantities.

(v) In MCQ type questions initial and final coordinates are also very important for eliminating some of the wrong options. Let us take an example:

Suppose a particle is projected upwards with velocity 20 m/s. Then initial coordinates in v-t graph will be, (+20 m/s, 0).

So, with these important points let us take an example.

Example 12 A particle is projected upwards with velocity u from a tower of height h. Let it strikes the ground with velocity v (in downward direction). Collision time is negligible and in each collision speed remains half. Plot a rough (without calculations) graph of :

(a) velocity *versus* time

(b) speed *versus* time

(c) displacement *versus* time and

(d) distance *versus* time

Sol. (a) **Velocity *versus* time**

(i) a = constant

$\therefore$ v-t graph is straight line

(ii) Initial coordinates are $(+u, 0)$

(iii) Coordinates after long time ($t \to \infty$) are $(0, \infty)$

(iv)

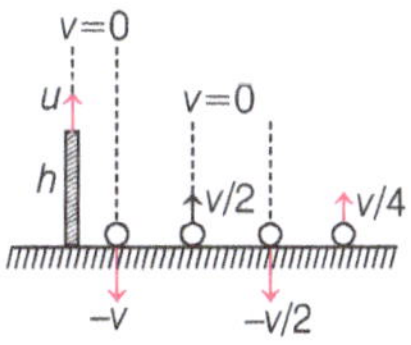

Velocity changes like this:

$u \longrightarrow 0$

$0 \longrightarrow -v$

$+v/2 \longrightarrow 0$

$0 \longrightarrow v/2$ and so on

(v) At highest point (say H), $v = 0$ and at point of collision (say C) velocity suddenly changes its direction and magnitude remains half. So, velocity *versus* time graph is as shown below.

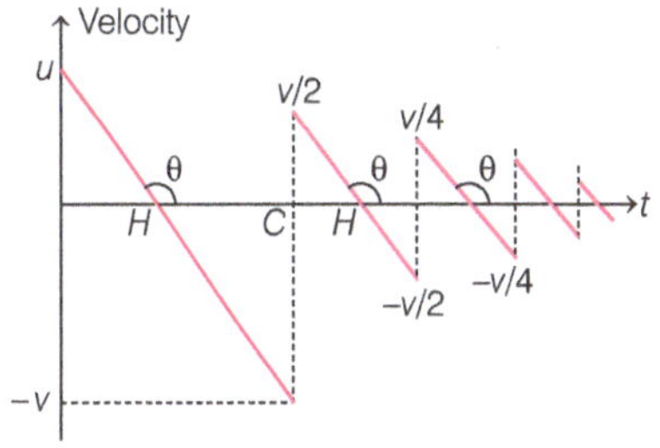

Slope of each line $= \tan\theta$

$$= a = -9.8\ \text{m/s}^2 \ \text{ or } \ 10\ \text{m/s}^2$$

From the graph we can see that, $v \to 0$ as $t \to \infty$.

So, the final coordinates are $(0, \infty)$.

(b) Speed *versus* time

(i) Speed and time both are positive. So, graph remains in first quadrant only.

(ii) Sequence is like this :

$u \to 0$

$0 \to v$

$\frac{v}{2} \to 0$

$0 \to \frac{v}{2}$ and so on.

So, speed-time graph is as shown below.

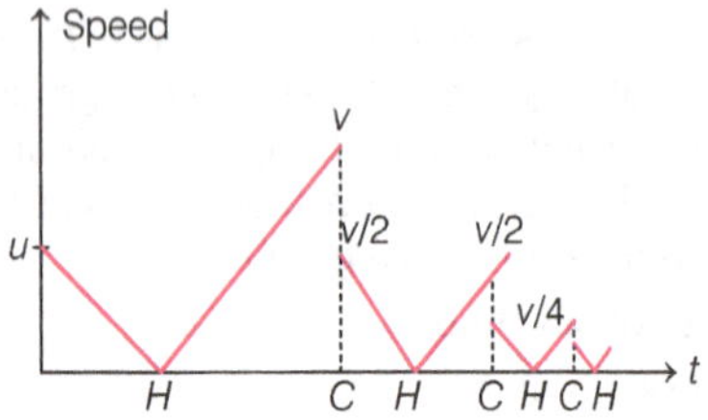

(c) Displacement *versus* time

(i) Displacement s is measured from the point where the particle was at $t = 0$. So, initial coordinates are $s = 0, t = 0$ or $(0, 0)$. Hence, the graph will start from origin.

(ii)

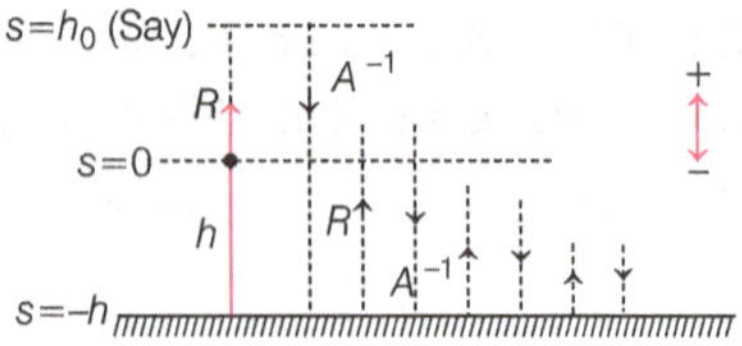

Particle will never cross the line $s = -h$.

After longtime $(t \to \infty)$, $s = -h$. So, the final coordinates are $(-h, \infty)$.

(iii) If we compare type-3 questions, then sequence (in s - t graph) is like this :

$R \to A^{-1} \to R \to A^{-1}$ and so on.

So, s - t graph is as shown below :

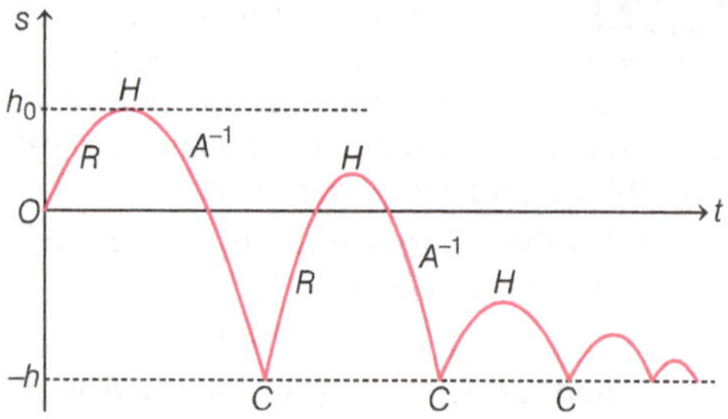

(iv) At highest point, slope of s - t graph or $v = 0$.

(v) At collision points slope of s - t graph changes its sign and magnitude. For example at first collision, slope or velocity just before collision is $-v$ and just after collision it is $+\frac{v}{2}$.

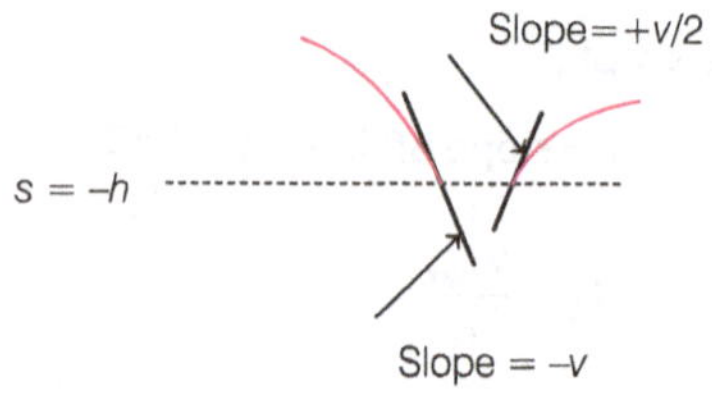

(vi) After long time $(t \to \infty)$ we can see from the graph that $s \to -h$. So, the final coordinates are $(-h, \infty)$.

(d) Distance *versus* time

(i) Distance (d) and time (t) both are positive. So, the graph will remain in first quadrant.

(ii) Distance keeps on increasing. But slowly speed will decrease. So, rate of increase of distance will be less.

(iii) In distance, there is no positive or negative direction. So, the sequence is like :

$R \to A \to R \to A$ and so on.

So, the graph is as shown here.

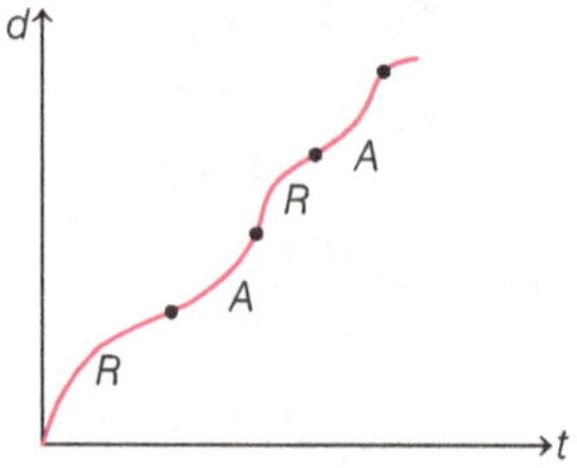

Miscellaneous Examples

Ex. 1 Acceleration-time graph of a particle moving in a straight line is as shown in figure. Velocity of particle at time $t=0$ is 2 m/s. Find the velocity at the end of fourth second.

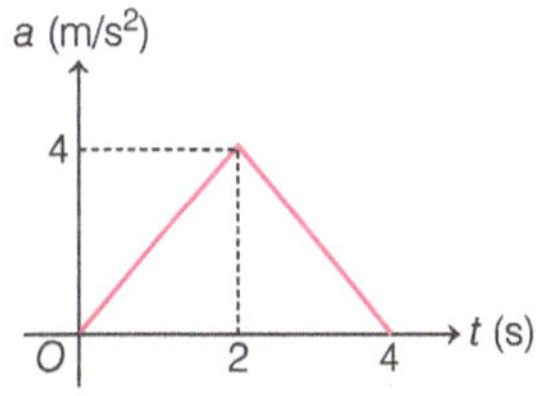

Sol. Change in velocity = area under a-t graph

Hence, $v_f - v_i = \frac{1}{2}(4)(4)$

$= 8$ m/s

$\therefore$ $v_f = v_i + 8 = (2+8)$ m/s

$= 10$ m/s

Ex. 2 The acceleration *versus* time graph of a particle moving along a straight line is shown in the figure. Draw the respective velocity-time graph. Given $v=0$ at $t=0$.

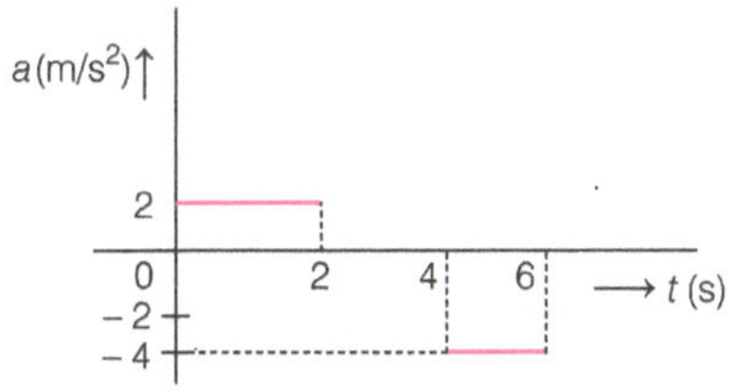

Sol. From $t=0$ to $t=2$ s, $a=+2\text{ m/s}^2$

$\therefore$ $v = at = 2t$

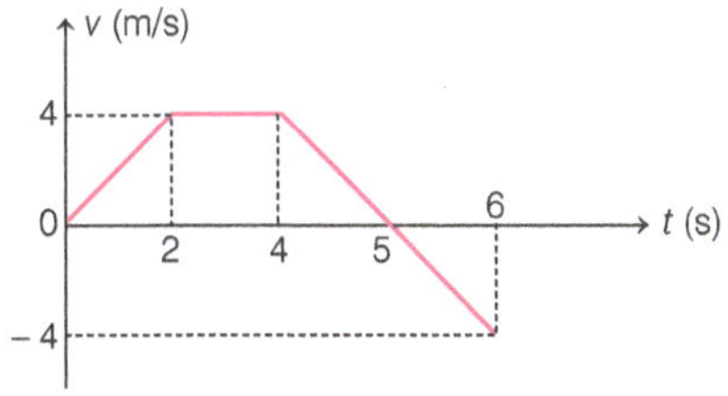

or v-t graph is a straight line passing through origin with slope 2 m/s^2.

At the end of 2 s,

$v = 2 \times 2 = 4$ m/s

From $t = 2$ to 4 s, $a = 0$.

Hence, $v = 4$ m/s will remain constant.

From $t = 4$ to 6 s, $a = -4\text{ m/s}^2$.

Hence, $v = u - at = 4 - 4t$ (with $t=0$ at 4 s)

$v = 0$ at $t = 1$ s or at 5 s from origin.

At the end of 6 s (or $t = 2$ s) $v = -4$ m/s. Corresponding v-t graph is as shown in figure.

Ex. 3 A rocket is fired vertically upwards with a net acceleration of 4 m/s^2 and initial velocity zero. After 5 s its fuel is finished and it decelerates with g. At the highest point its velocity becomes zero. Then, it accelerates downwards with acceleration g and return back to ground. Plot velocity-time and displacement-time graphs for the complete journey. Take $g = 10\text{ m/s}^2$.

Sol. In the graphs,

$v_A = at_{OA} = (4)(5) = 20$ m/s

$v_B = 0 = v_A - gt_{AB}$

$\therefore$ $t_{AB} = \frac{v_A}{g} = \frac{20}{10} = 2$ s

$\therefore$ $t_{OAB} = (5+2)$ s $= 7$ s

Now, s_{OAB} = area under v-t graph between 0 to 7 s

$= \frac{1}{2}(7)(20) = 70$ m

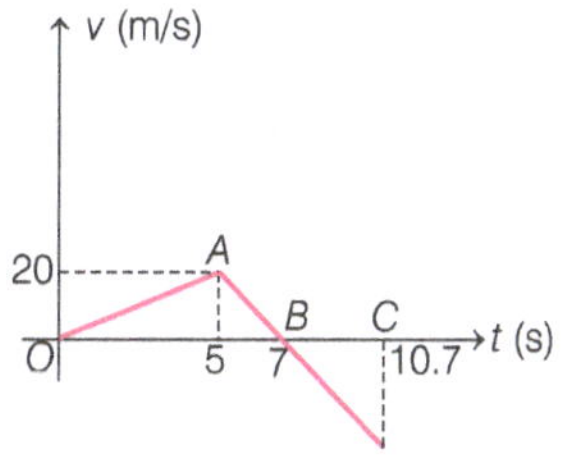

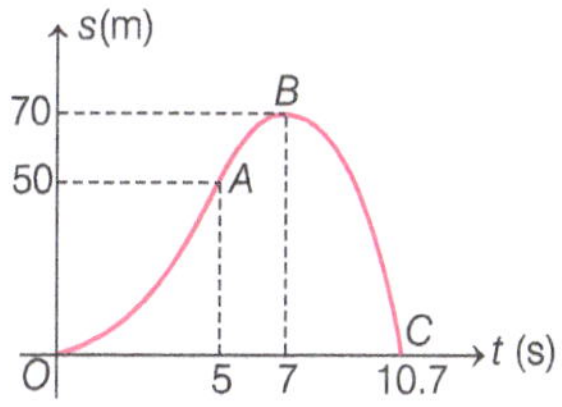

Further, $|s_{OAB}| = |s_{BC}| = \frac{1}{2}gt_{BC}^2$

$\therefore$ $70 = \frac{1}{2}(10)t_{BC}^2$

$\therefore$ $t_{BC} = \sqrt{14} \approx 3.7$ s

$\therefore$ $t_{OABC} = 7 + 3.7 = 10.7$ s

Also, s_{OA} = area under v-t graph between OA

$= \frac{1}{2}(5)(20) = 50$ m

MODULE 7 Exercise

Assertion and Reason

See directions on page 178.

1. **Assertion :** In v-t graph shown in figure, average velocity in time interval from 0 to t_0 depends only on v_0. It is independent of t_0.

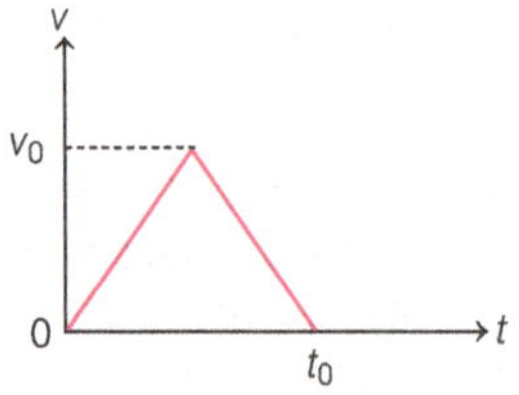

Reason : In the given time interval, average velocity is $\frac{v_0}{2}$.

2. **Assertion :** In the x-t diagram shown in figure, the body starts in positive direction but not from $x = 0$.

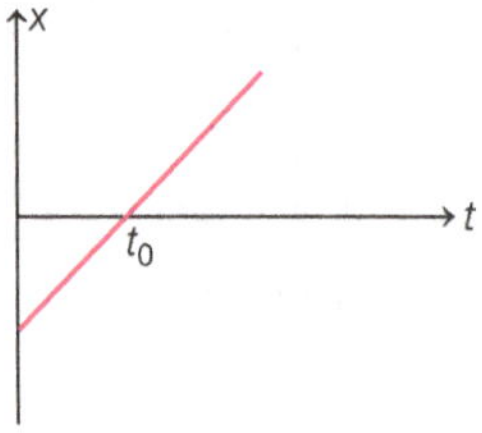

Reason : At $t = t_0$, velocity of body changes its direction of motion.

3. Given below are two statements :

Statement I : Area under velocity-time graph gives the distance travelled by the body in a given time.

Statement II : Area under acceleration-time graph is equal to the change in velocity in the given time.

In the light of given statements, choose the correct answer from the options given below. **(2023 Main)**

(a) Both Statement I and Statement II are true.
(b) Statement I is correct but Statement II is false.
(c) Statement I is incorrect but Statement II is true.
(d) Both Statement I and Statement II are false.

Single Correct Option

4. The acceleration-time (a-t) graph of a particle moving in a straight line is as shown in the figure. The velocity-time graph of the particle would be

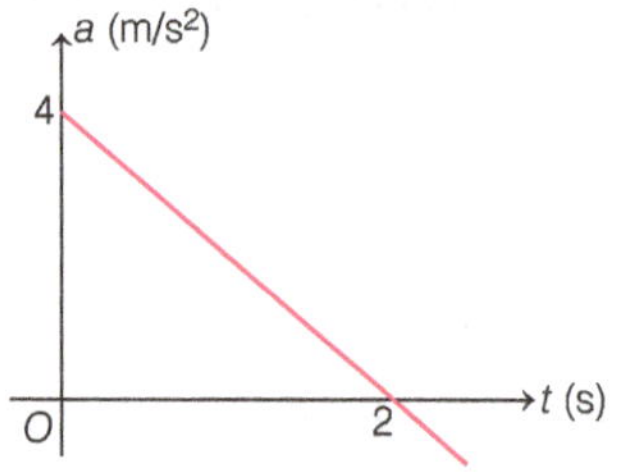

(a) a straight line (b) a parabola
(c) a circle (d) an ellipse

5. Acceleration-time graph for a particle moving in a straight line is as shown in figure. Change in velocity of the particle from $t = 0$ to $t = 6$ s is

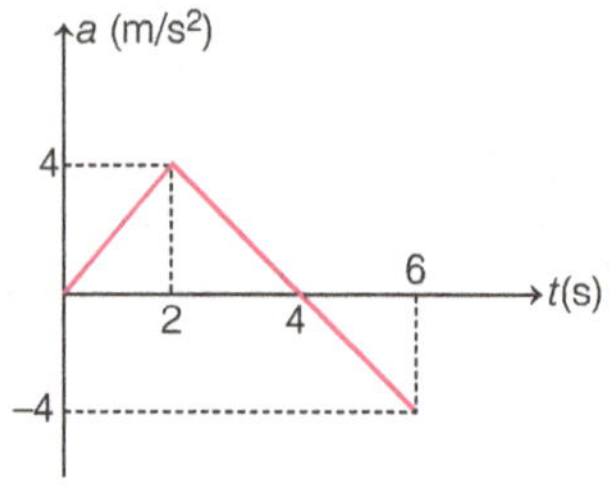

(a) $10\,m/s$ (b) $4\,m/s$ (c) $12\,m/s$ (d) $8\,m/s$

6. The velocity-time graph for a particle moving along x-axis is shown in the figure. The corresponding displacement-time graph is correctly shown by

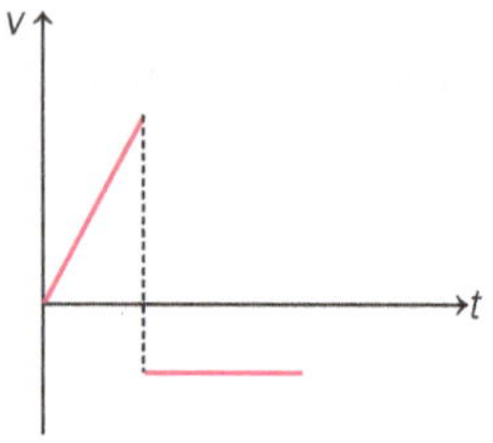

(a)
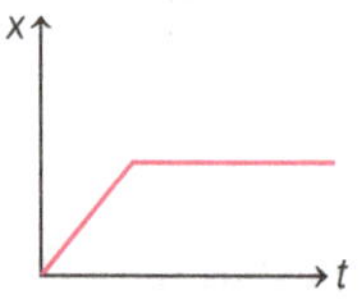

(b)
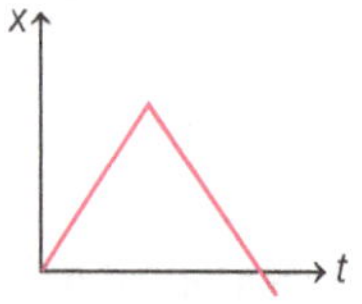

(c)
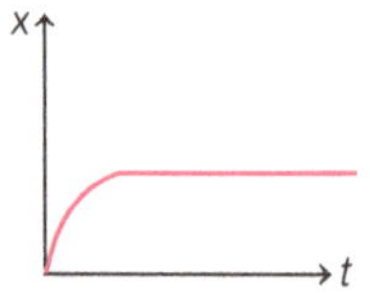

(d)
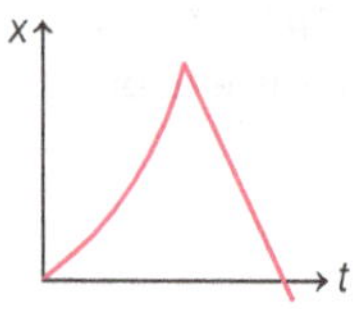

7. Acceleration *versus* time graph of a body starting from rest is shown in the figure. The velocity *versus* time graph of the body is given by

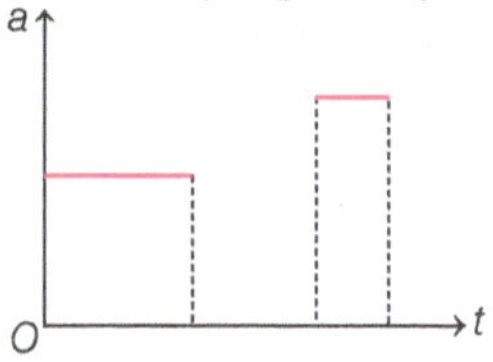

(a)
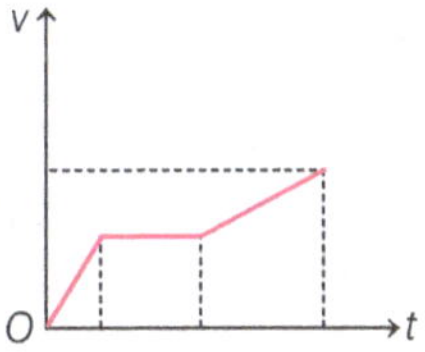

(b)
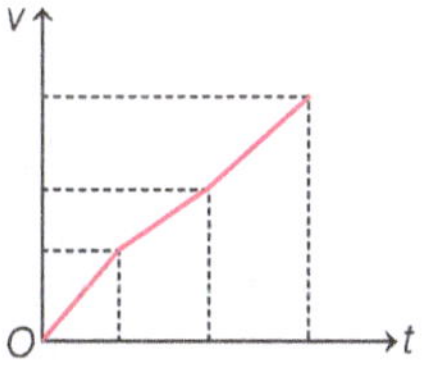

(c)
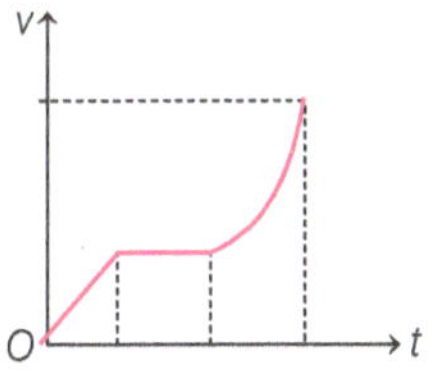

(d)
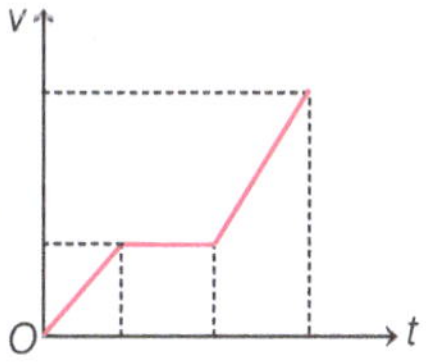

8. If the velocity v of a particle moving along a straight line decreases linearly with its displacement s from 20 ms^{-1} to a value approaching zero at $s = 30$ m, then acceleration of the particle at $s = 15$ m is

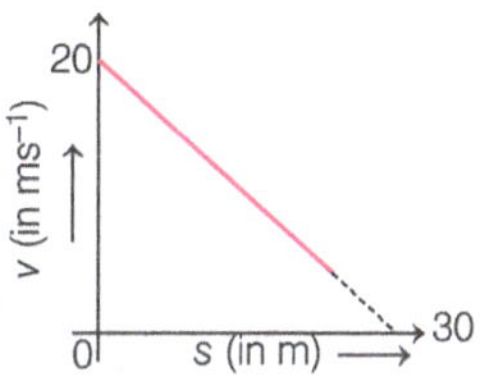

(a) $\frac{2}{3} ms^{-2}$ (b) $-\frac{2}{3} ms^{-2}$ (c) $\frac{20}{3} ms^{-2}$ (d) $-\frac{20}{3} ms^{-2}$

9. v^2 *versus* s graph of a particle moving in a straight line is as shown in figure. From the graph some conclusions are drawn. State which statement is **wrong**.

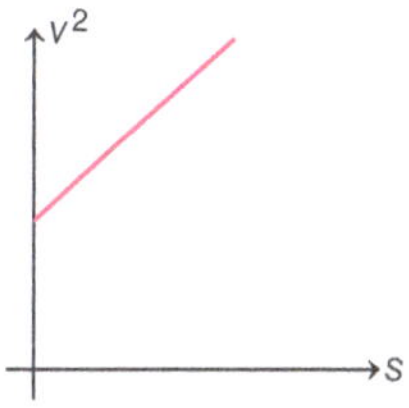

(a) The given graph shows a uniformly accelerated motion
(b) Initial velocity of the particle is zero
(c) Corresponding s-t graph will be a parabola
(d) None of the above

10. Two particles A and B start from rest and move for equal time on a straight line. Particle A has an acceleration of 2 m/s^2 for the first half of the total time and 4 m/s^2 for the second half. The particle B has acceleration 4 m/s^2 for the first half and 2 m/s^2 for the second half. Which particle has covered larger distance?
(a) A
(b) B
(c) Both have covered the same distance
(d) Data insufficient

11. A particle starts from rest and traverses a distance l with uniform acceleration, then moves uniformly over a further distance $2l$ and finally comes to rest after moving a further distance $3l$ under uniform retardation. Assuming entire motion to be rectilinear motion, the ratio of average speed over the journey to the maximum speed on its ways is
(a) $\frac{1}{5}$ (b) $\frac{2}{5}$ (c) $\frac{3}{5}$ (d) $\frac{4}{5}$

12. The acceleration-time graph of a particle moving along a straight line is as shown in figure. At what time the particle acquires its initial velocity?

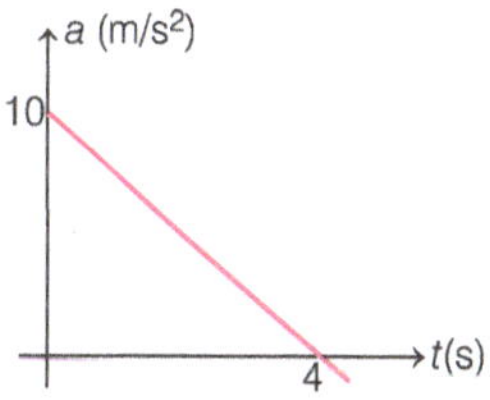

(a) 12 s (b) 5 s
(c) 8 s (d) 16 s

13. A particle is moving along a straight line whose velocity-displacement graph is as shown in the figure. What is the magnitude of acceleration when displacement is 3 m ?

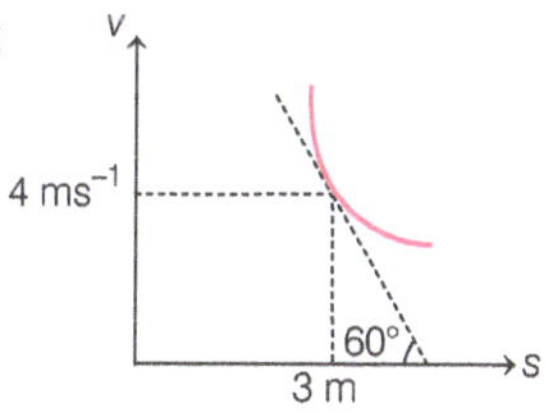

(a) $4\sqrt{3}$ ms^{-2} (b) $3\sqrt{3}$ ms^{-2}
(c) $\sqrt{3}$ ms^{-2} (d) $\frac{4}{\sqrt{3}}$ ms^{-2}

14. The displacement x of a particle in a straight line motion is given by $x = 1 - t - t^2$. The correct representation of the motion is

(a)
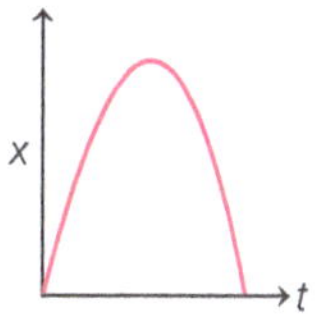

(b)
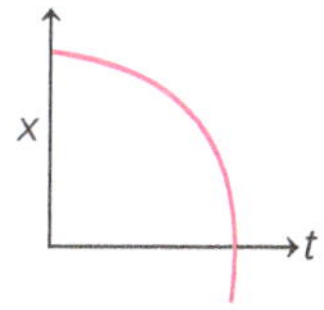

(c)
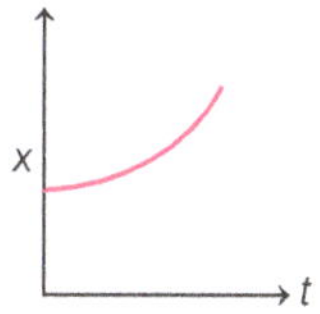

(d)
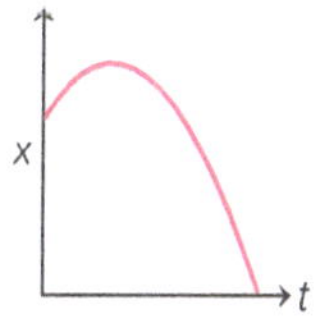

15. A particle starts from rest. Its acceleration (a) *versus* time (t) is as shown in the figure. The maximum speed of the particle will be **(2004 JEE)**

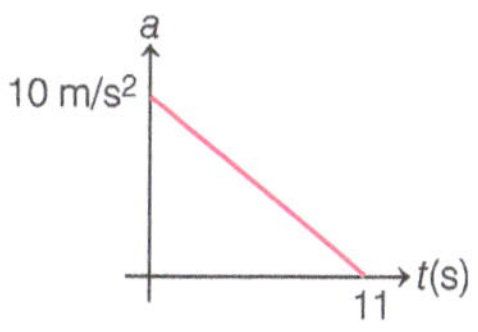

(a) 110 m/s (b) 55 m/s (c) 550 m/s (d) 660 m/s

16. The given graph shows the variation of velocity with displacement. Which one of the graph given below correctly represents the variation of acceleration with displacement? **(2005 JEE)**

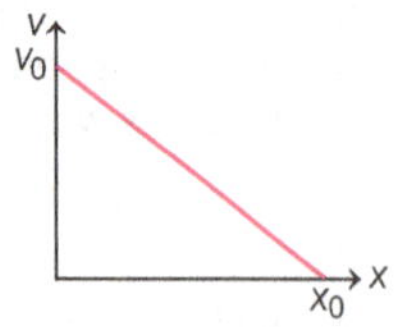

(a)
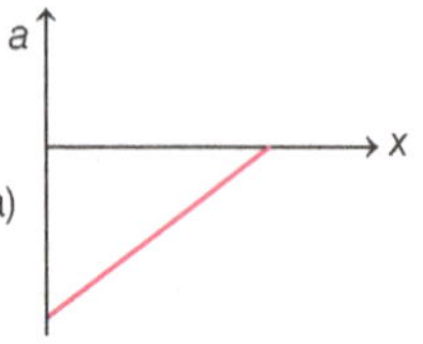

(b)
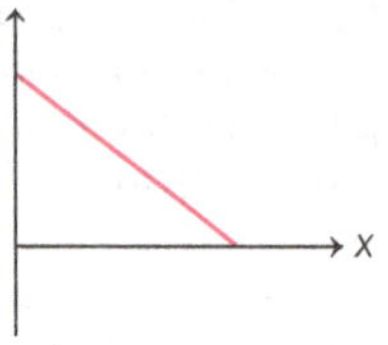

(c)
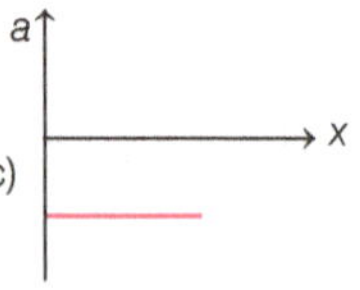

(d)
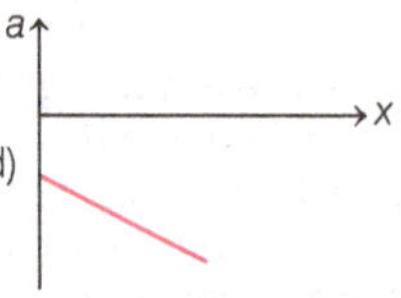

17. A body is thrown vertically upwards. Which one of the following graphs correctly represent the velocity *versus* time? **(2017 Main)**

(a)
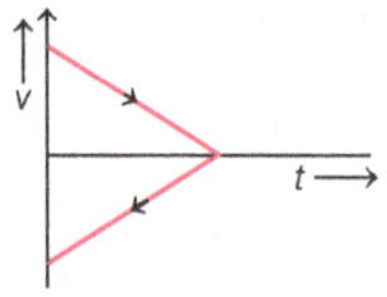

(b)
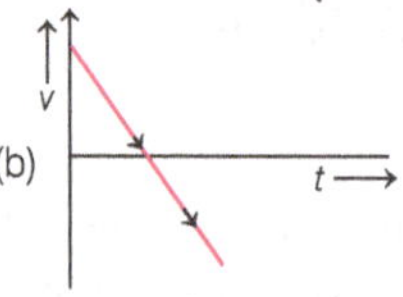

(c)
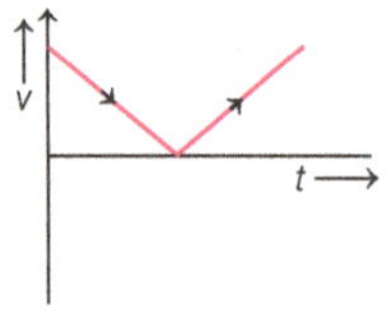

(d)
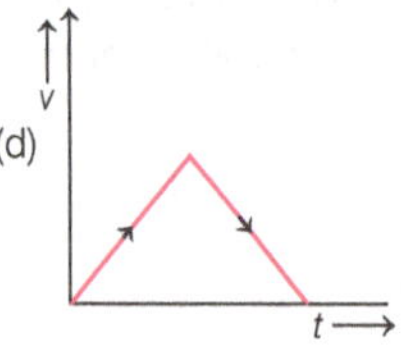

18. A particle starts from origin O from rest and moves with a uniform acceleration along the positive X-axis. Identify all figures that correctly represent the motion qualitatively.
(a = acceleration, v = velocity, x = displacement, t = time) **(2019 Main)**

(A)
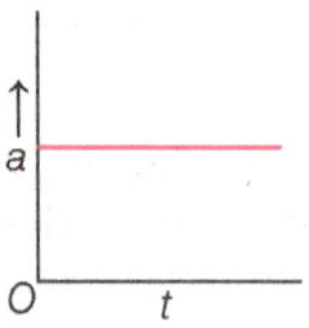

(B)
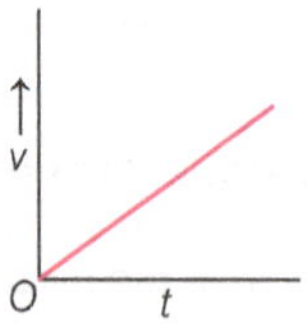

(C)
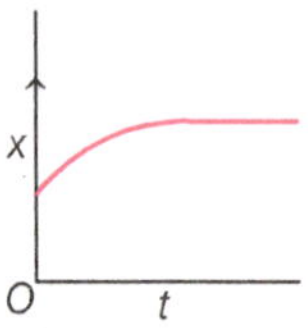

(D)

(a) (A)
(b) (A), (B), (C)
(c) (B), (C)
(d) (A), (B), (D)

19. A particle starts from the origin at time $t = 0$ and moves along the positive x-axis. The graph of velocity with respect to time is shown in figure. What is the position of the particle at time $t = 5$ s? **(2019 Main)**

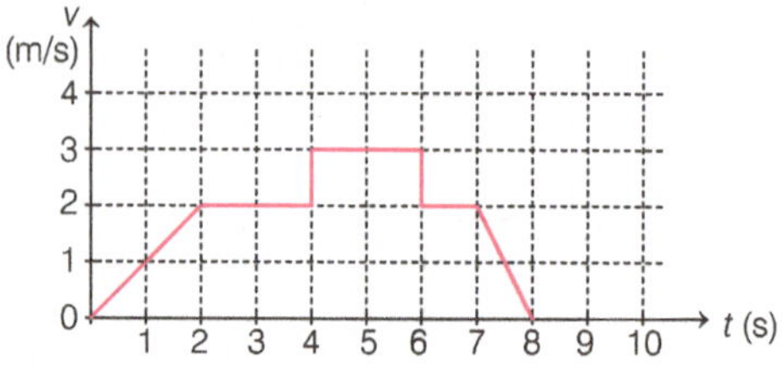

(a) 6 m (b) 3 m (c) 10 m (d) 9 m

20. From the v - t graph shown, the ratio of distance to displacement in 25s of motion **(2023 Main)**

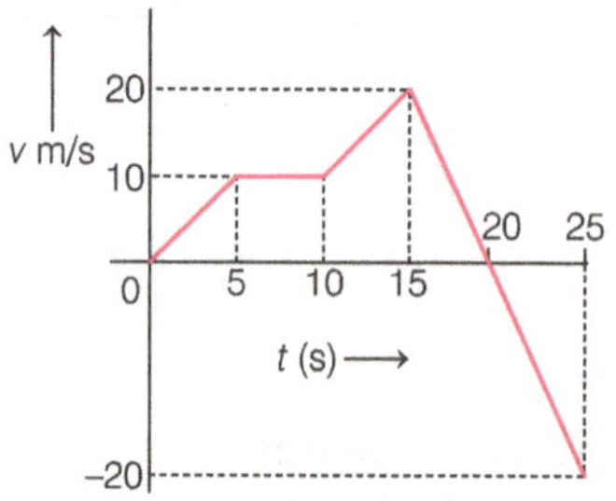

(a) $\frac{3}{5}$ (b) $\frac{1}{2}$ (c) $\frac{5}{3}$ (d) 1

21. The velocity (v) and time (t) graph of a body in a straight line motion is shown in the figure. The point S is at 4.333 s. The total distance covered by the body in 6 s is **(2020 Main)**

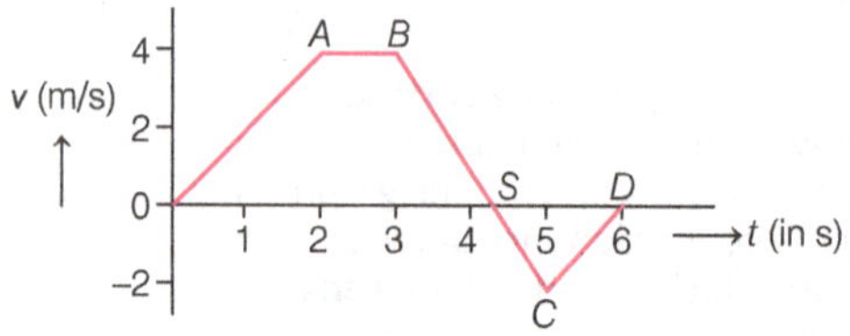

(a) $\frac{37}{3}$ m (b) 12 m (c) 11 m (d) $\frac{49}{4}$ m

22. The distance travelled by a body moving along a line in time t is proportional to t^3. The acceleration-time (a - t) graph for the motion of the body will be **(2012 Main)**

(a)
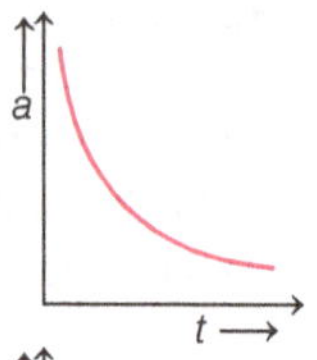

(b)
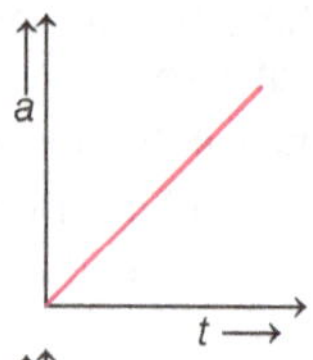

(c)
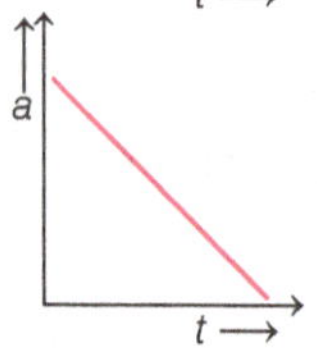

(d)
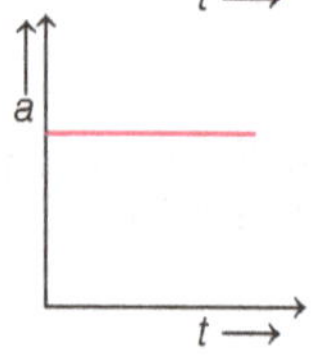

23. A bullet is shot vertically downwards with an initial velocity of 100 m/s from a certain height. Within 10 s, the bullet reaches the ground and instantaneously comes to rest due to the perfectly inelastic collision. The velocity-time curve for total time $t = 20$ s will be (Take $g = 10\text{ m/s}^2$) **(2022 Main)**

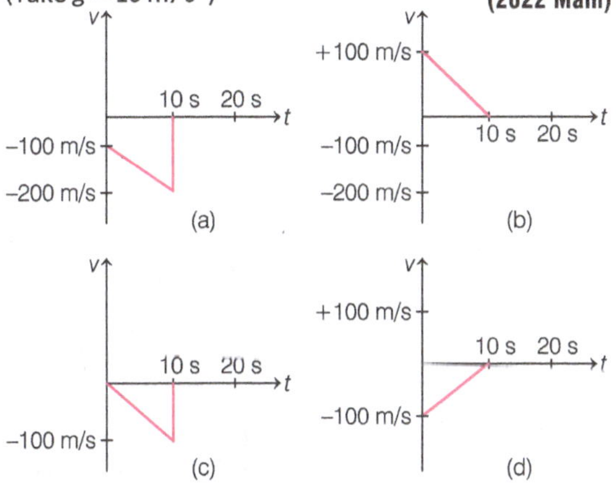

24. The position, velocity and acceleration of a particle moving with a constant acceleration can be represented by **(2021 Main)**

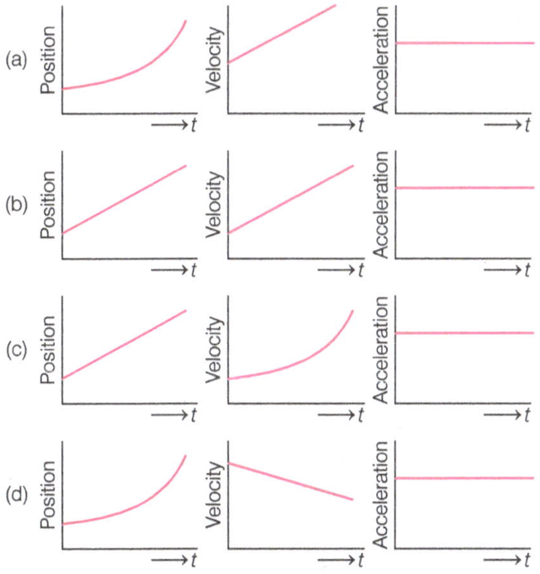

Multiple Correct Options

25. A particle is resting over a smooth horizontal floor. At $t = 0$, a horizontal force starts acting on it. Magnitude of the force increases with time according to the law $F = \alpha t$, where α is a positive constant. According to the shown figure, which of the following statements are correct ?

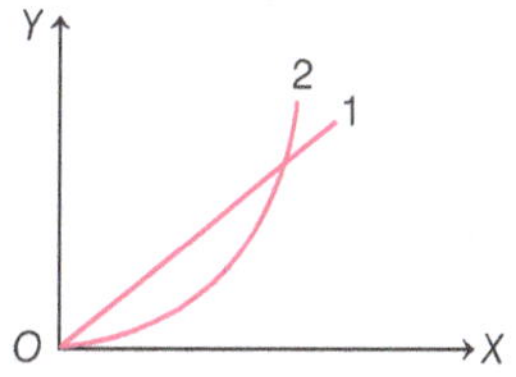

(a) Curve 1 can be the plot of acceleration against time
(b) Curve 2 can be the plot of velocity against time
(c) Curve 2 can be the plot of velocity against acceleration
(d) Curve 1 can be the plot of displacement against time

26. The figure shows the velocity (v) of a particle plotted against time (t).

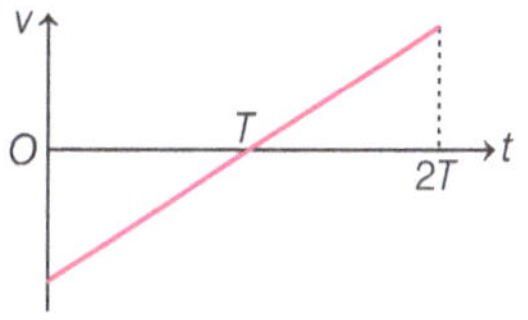

(a) The particle changes its direction of motion at some point
(b) The acceleration of the particle remains constant
(c) The displacement of the particle is zero
(d) The initial and final speeds of the particle are the same

27. Two particles P and Q are moving along x-axis. Their position-time graph is shown in the figure.

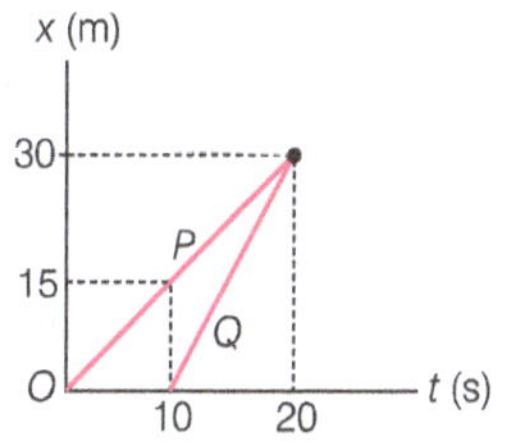

Choose the correct option(s).
(a) Particle P starts from point O at $t = 0$
(b) Particle Q starts from point O at $t = 10$ s
(c) Particle Q overtakes the particle P at $x = 30$ m and $t = 20$ s
(d) Particle Q can never overtake particle P

Match the Columns

28. Match the Column-I with Column-II :

	Column-I		Column-II
A.	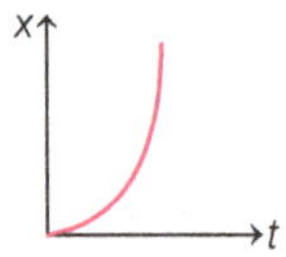	I.	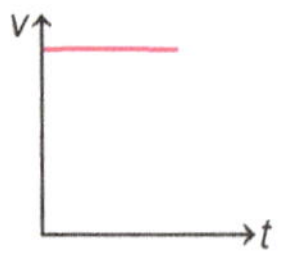
B.	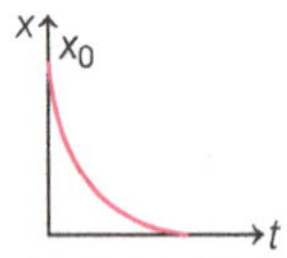	II.	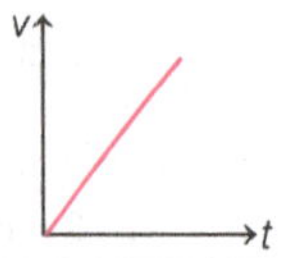
C.	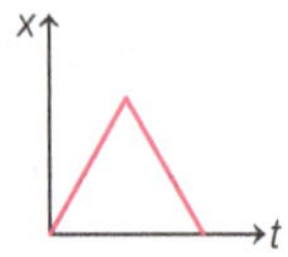	III.	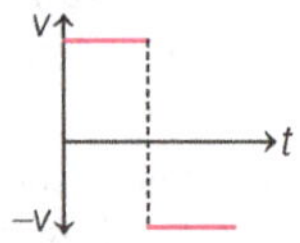
D.	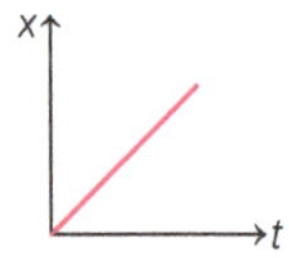	IV.	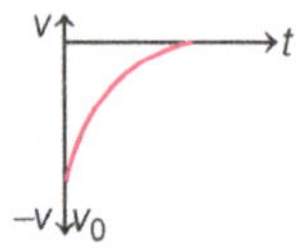

Choose the correct answer from the options given below : **(2023 Main)**

(a) A-II, B-IV, C-III, D-I (b) A-I, B-II, C-III, D-IV
(c) A-II, B-III, C-IV, D-I (d) A-I, B-III, C-IV, D-II

29. Match the following two columns :

	Column I		Column II
(a)	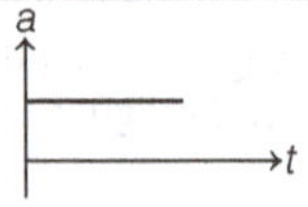	(p)	Speed must increase
(b)	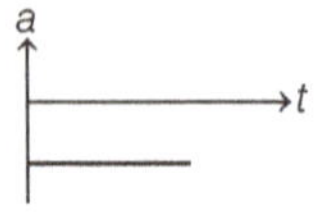	(q)	Speed must decrease
(c)	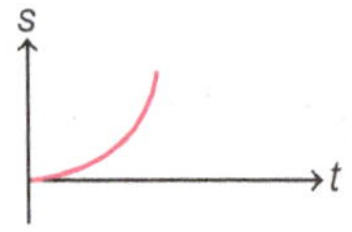	(r)	Speed may be increasing
(d)	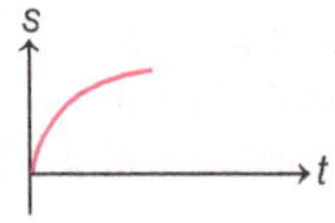	(s)	Speed may be decreasing

30. The velocity-time graph of a particle moving along x-axis is shown in figure. Match the entries of Column I with the entries of Column II.

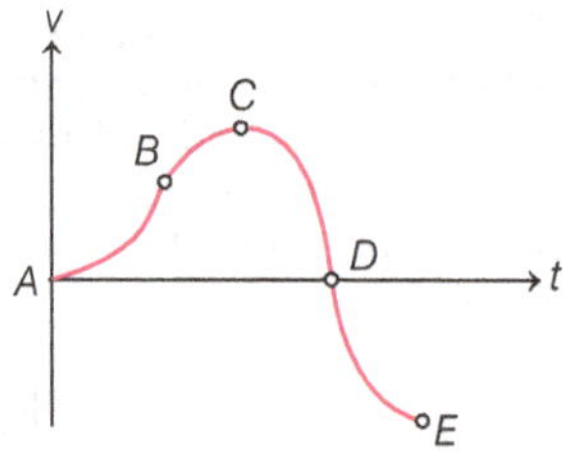

	Column I		Column II
(a)	For AB, particle is	(p)	moving in +ve x-direction with increasing speed
(b)	For BC, particle is	(q)	moving in +ve x-direction with decreasing speed
(c)	For CD, particle is	(r)	moving in –ve x-direction with increasing speed
(d)	For DE, particle is	(s)	moving in –ve x-direction with decreasing speed

31. Corresponding to velocity-time graph in one dimensional motion of a particle as shown in figure, match the following two columns.

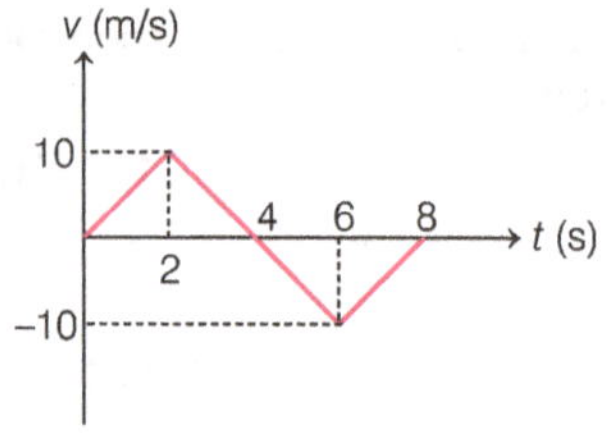

	Column I		Column II
(a)	Average velocity between zero second and 4 s	(p)	10 SI units
(b)	Average acceleration between 1 s and 4 s	(q)	2.5 SI units
(c)	Average speed between zero second and 6 s	(r)	5 SI units
(d)	Rate of change of speed at 4 s	(s)	None of these

32. A car accelerates from rest at a constant rate of $10\ ms^{-2}$ for some time after which it decelerates at constant rate of $5 ms^{-2}$ to come to rest. If total time elapsed is 15 s. Match the following columns.

	Column I		Column II
(a)	Maximum velocity of car (in ms^{-1}).	(p)	10
(b)	Distance (in m) travelled by car, at the instant when its velocity is half of its maximum velocity.	(q)	25
(c)	Average velocity (in ms^{-1}) during total time of motion.	(r)	50
(d)	The time (in s) during which the particle is in the condition of deceleration.	(s)	None of the above

Subjective Questions

33. Two particles A and B are moving along x-axis. Their x-coordinate *versus* time graphs are as shown in figure

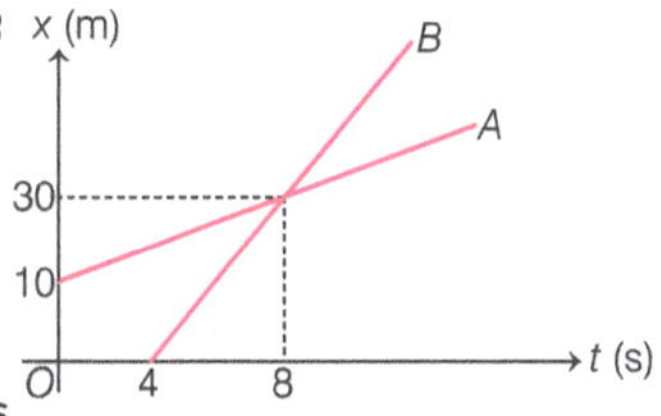

(a) Find the time when the particles start their journey and the x-coordinate at that time.

(b) Find velocities of the two particles.

(c) When and where the particles meet with each other.

34. Figure shows the graph of the x-coordinate of a particle going along the x-axis as a function of time. Find (a) the average velocity during 0 to 10 s, (b) instantaneous velocity at 2, 5, 8 and 12 s.

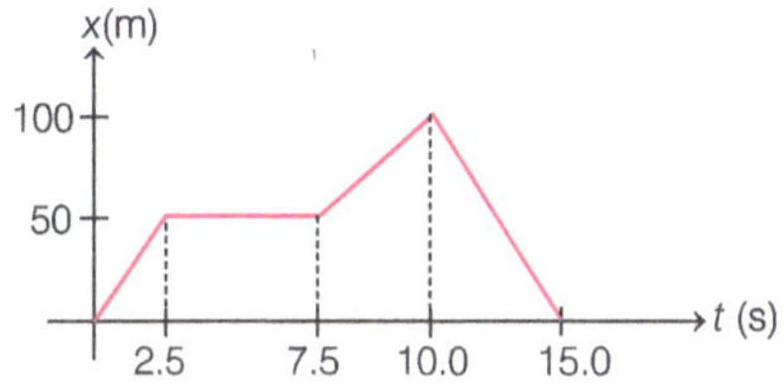

35. Displacement-time graph of a particle moving in a straight line is as shown in figure.

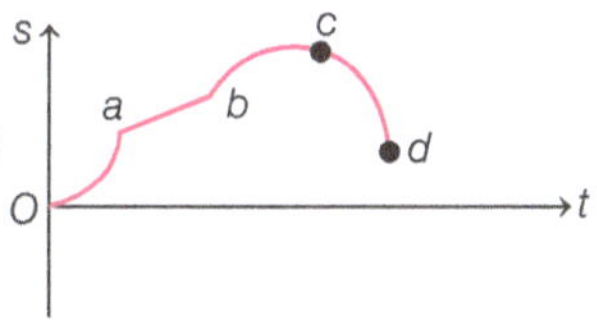

(a) Find the sign of velocity in regions oa, ab, bc and cd.

(b) Find the sign of acceleration in the above region.

36. Velocity-time graph of a particle moving in a straight line is shown in figure. In the time interval from $t = 0$ to $t = 14$ s, find

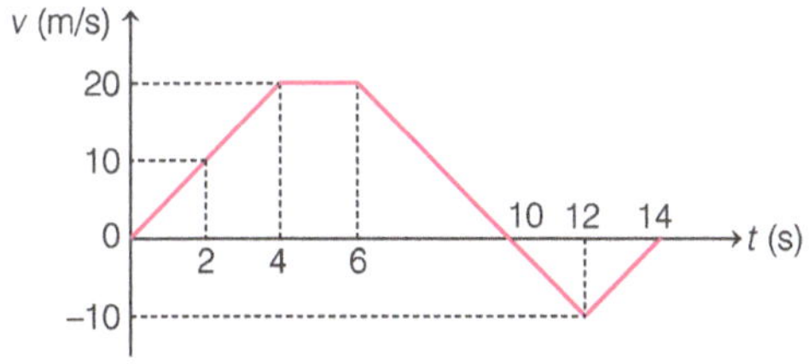

(a) average velocity and

(b) average speed of the particle.

37. Acceleration-time graph of a particle moving in a straight line is as shown in figure. At time $t = 0$, velocity of the particle is zero. Find

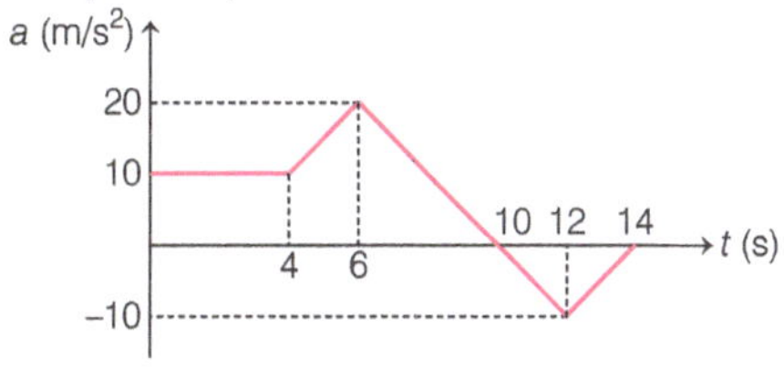

(a) average acceleration in a time interval from $t = 6$ s to $t = 12$ s,

(b) velocity of the particle at $t = 14$ s.

38. A particle of mass m is released from a certain height h with zero initial velocity. It strikes the ground elastically (direction of its velocity is reversed but magnitude remains the same). Plot the graph between its kinetic energy and time till it returns to its initial position.

39. A ball is dropped from a height of 80 m on a floor. In each collision, the ball loses half of its speed. Plot the speed-time graph and velocity-time graph of its motion till two collisions with the floor.

(Take, $g = 10\ \text{ms}^{-2}$)

40. The acceleration of particle varies with time as shown.

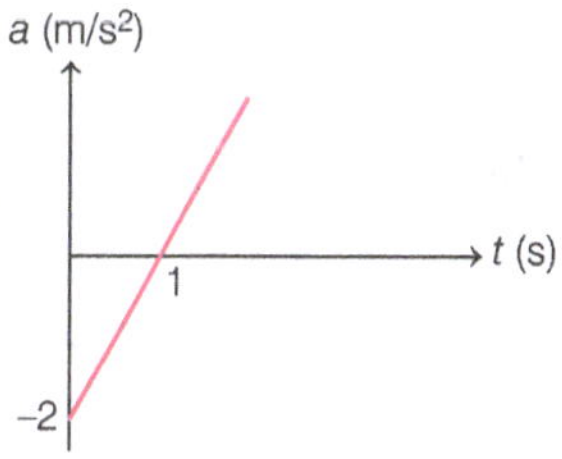

(a) Find an expression for velocity in terms of t.

(b) Calculate the displacement of the particle in the interval from $t = 2$ s to $t = 4$ s.

Assume that $v = 0$ at $t = 0$..

41. A point travels along the x-axis with a velocity whose projection v is presented as a function of time by the plot in figure.

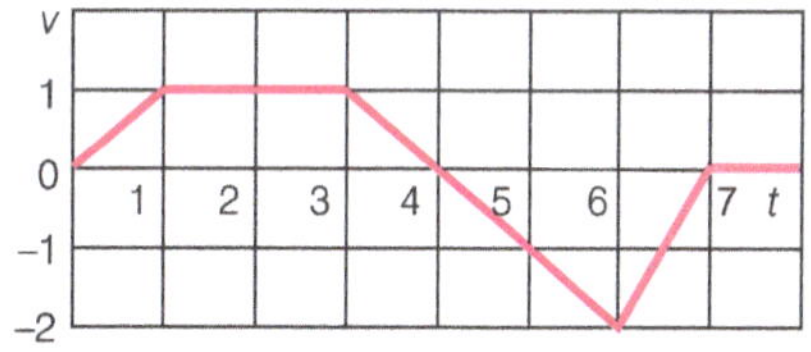

Assuming the coordinate of the point $x = 0$ at the moment $t = 0$, draw the approximate time dependence plots for the acceleration a, the x-coordinate and the distance covered d. **(Irodov 1.18)**

Integer Type Questions

42. If a particle is moving on a straight line, then its velocity-time graph is sinusoidal as shown in the figure.

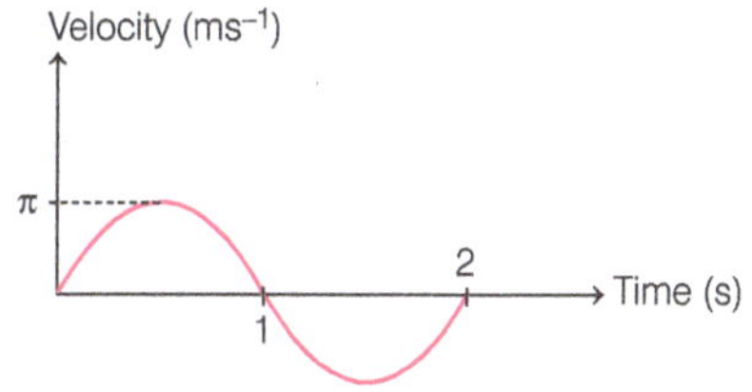

Find distance (in m) travelled by the particle in 2 s

43. A particle is moving with constant acceleration 'a'. Following graph shows v^2 *versus* x (displacement) plot. The acceleration of the particle is……..m/s^2.

(2021 Main)

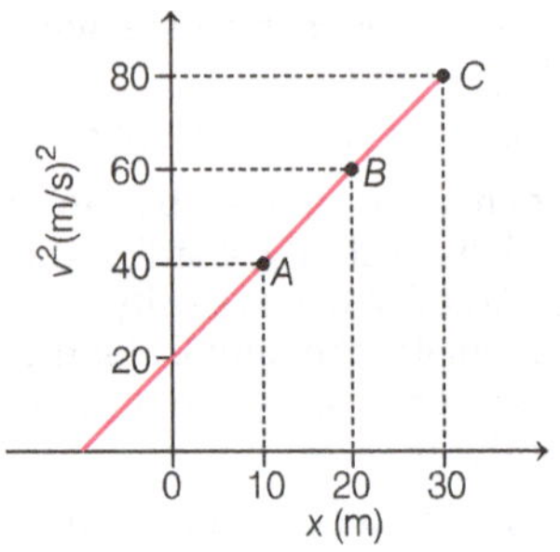

44. The speed *versus* time graph for a particle is shown in the figure. The distance travelled (in m) by the particle during the time interval $t = 0$ to $t = 5$ s will be ....

(2020 Main)

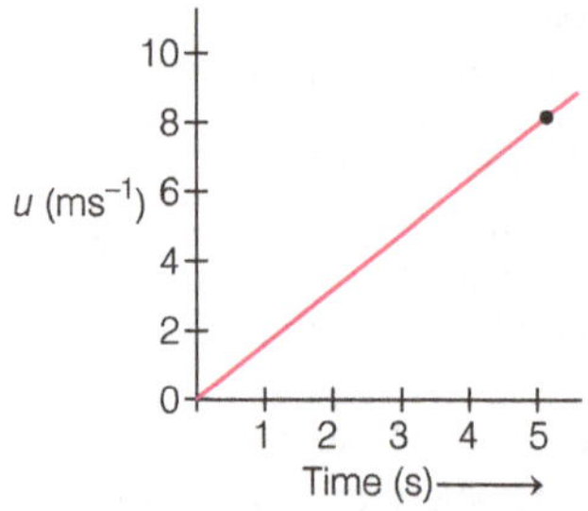

Answers

1. (a) **2.** (c) **3.** (c) **4.** (b) **5.** (b) **6.** (d) **7.** (d) **8.** (d) **9.** (b) **10.** (b)
11. (c) **12.** (c) **13.** (a) **14.** (b) **15.** (b) **16.** (a) **17.** (b) **18.** (d) **19.** (d) **20.** (c)
21. (a) **22.** (b) **23.** (a) **24.** (a) **25.** (a,b,c) **26.** (a, b, c, d) **27.** (a, b, c) **28.** (a)
29. (a- r, s; b-r, s; c-p; d-q) **30.** (a-p; b-p; c-q; d - r) **31.** (a - r; b - q; c - r; d- r) **32.** (a - r; b - s; c - q; d- p)
33. (a) Particle A starts at $t = 0$ from $x = 10$ m. Particle B starts at $t = 4$s from $x = 0$ (b) $v_A = +25$ m/s, $v_B = +7.5$ m/s
(c) They strike at $x = 30$m and $t = 8$ s. **34.** (a) 10 m/s (b) 20 m/s, zero, 20 m/s, –20 m/s
35. (a) positive, positive, positive, negative (b) positive, zero, negative, negative
36. (a) $\frac{50}{7} ms^{-1}$ (b) $10 ms^{-1}$ **37.** (a) $-5 ms^{-2}$ (b) $90 ms^{-1}$

38.

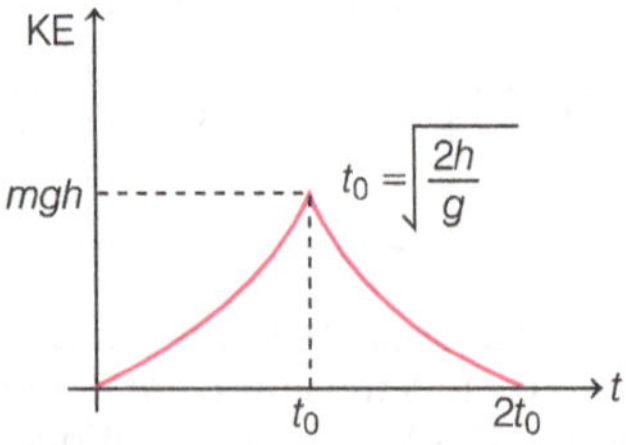

39.

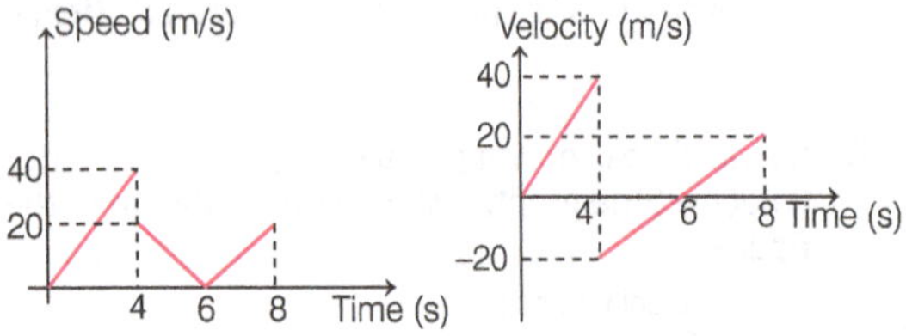

40. (a) $v = t^2 - 2t$ (b) 6.67 m **41.** See the hints **42.** (4) **43.** (1) **44.** (20)

See hints on pages 255-259.

MODULE 8

Relative Motion

The word 'relative' is a very general term, which can be applied to physical, non-physical, scalar or vector quantities. For example, my height is 167 cm while my wife's height is 162 cm. If I ask you what is my height relative to my wife, your answer will be 5 cm.

What did you do? You simply subtracted my wife's height from my height. The same concept is applied everywhere, whether it is a relative velocity, relative acceleration or anything else. So, from the above discussion we may now conclude that relative velocity of A with respect to B (written as $\mathbf{v}_{AB}$) is

$$\mathbf{v}_{AB} = \mathbf{v}_A - \mathbf{v}_B$$

Similarly, relative acceleration of A with respect to B is

$$\mathbf{a}_{AB} = \mathbf{a}_A - \mathbf{a}_B$$

If it is a one dimensional motion we can treat the vectors as scalars just by assigning the positive sign to one direction and negative to the other. So, in case of a one dimensional motion the above equations can be written as

$$v_{AB} = v_A - v_B$$

and

$$a_{AB} = a_A - a_B$$

Further, we can see that

$$\mathbf{v}_{AB} = -\mathbf{v}_{BA} \text{ or } \mathbf{a}_{BA} = -\mathbf{a}_{AB}$$

Note *Relative velocity of A w.r.t. B ($\mathbf{v}_{AB}$) means velocity of B is assumed to be zero. Or, B is assumed to be at rest and A is moving with $\mathbf{v}_{AB}$. But this is not the case with $\mathbf{a}_{AB}$. In this case acceleration of B may be assumed zero. But zero acceleration does not mean B is at rest. So, acceleration of B is assumed zero and acceleration of A is equal to $\mathbf{a}_{AB}$.*

Example 1 Anoop is moving due east with a velocity of 1 m/s and Dhyani is moving due east with a velocity of 2 m/s. What is the velocity of Anoop with respect to Dhyani?

Sol. It is a one dimensional motion. So, let us choose the east direction as positive and the west as negative. Now, given that

v_A = velocity of Anoop = 1 m/s

and v_D = velocity of Dhyani = 2 m/s

Thus, v_{AD} = velocity of Anoop with respect to Dhyani

$= v_A - v_D = 1 - 2 = -1$ m/s

Hence, velocity of Anoop with respect to Dhyani is 1 m/s due west.

Example 2 Ashish is moving due east with a velocity of 1 m/s and Anoop is moving due west with a velocity of 2 m/s. What is the velocity of Ashish with respect to Anoop?

Sol. It is a one dimensional motion. So, let us choose the east direction as positive and the west as negative. Now, given that

v_A = velocity of Ashish = 1 m/s

and v_D = velocity of Anoop = −2 m/s

Thus, v_{AD} = velocity of Ashish with respect to Anoop

$= v_A - v_D = 1 - (-2) = 3$ m/s

Hence, velocity of Ashish with respect to Anoop is 3 m/s due east.

Note *From the above two examples we can see that, if two vectors are in opposite directions then in magnitude of their relative value they are added. If they are in same direction then they are subtracted.*

Example 3 Car A has an acceleration of $2\ \text{m/s}^2$ due east and car B, $4\ \text{m/s}^2$ due north. What is the acceleration of car B with respect to car A?

Sol. It is a two dimensional motion. Therefore,

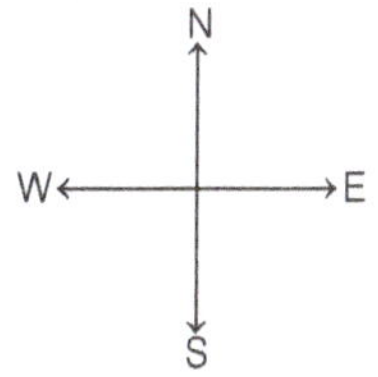

$\mathbf{a}_{BA}$ = acceleration of car B with respect to car A

$= \mathbf{a}_B - \mathbf{a}_A = \mathbf{a}_B + (-\mathbf{a}_A)$

Here, $\mathbf{a}_B$ = acceleration of car B

$= 4\ \text{m/s}^2$ (due north)

and $\mathbf{a}_A$ = acceleration of car A

$= 2\ \text{m/s}^2$ (due east)

$$|\mathbf{a}_{BA}| = \sqrt{(4)^2 + (2)^2} = 2\sqrt{5}\ \text{m/s}^2$$

and $\alpha = \tan^{-1}\left(\dfrac{4}{2}\right) = \tan^{-1}(2)$

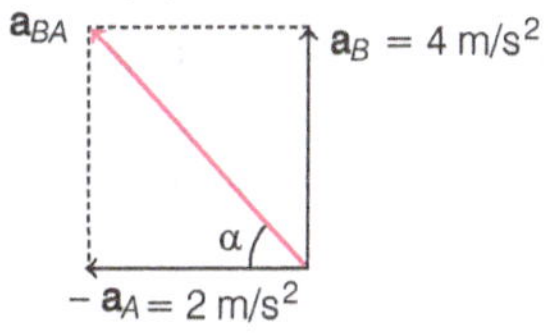

Thus, $\mathbf{a}_{BA}$ is $2\sqrt{5}\ \text{m/s}^2$ at an angle of $\alpha = \tan^{-1}(2)$ from west towards north.

INTRODUCTORY Exercise 1

1. Find v_{12} and v_{21} in following two figures.

(a) 

(b) 

2. Two particles are moving along x-axis. Their x-coordinate *versus* time graph are as shown below.

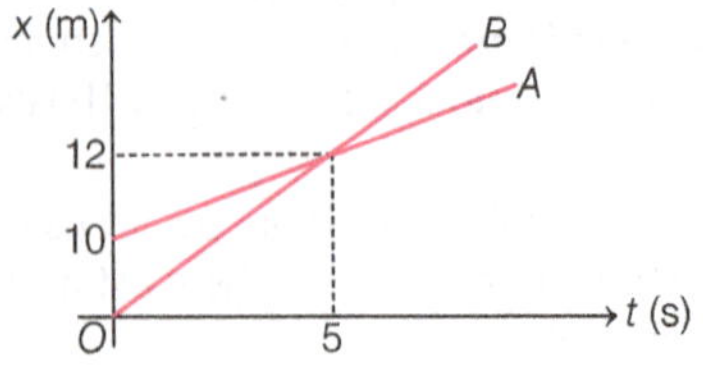

Find velocity of A w.r.t. B.

3. Two balls A and B are projected vertically upwards with different velocities. What is the relative acceleration between them?

4. Four particles A, B, C and D are moving with constant speed v each. At the instant shown, relative velocity of A with respect to B, C and D are in directions.

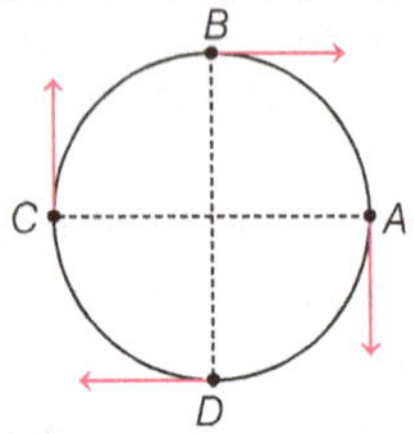

(a) 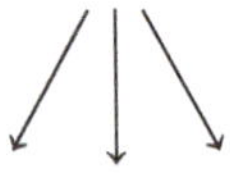 (b) 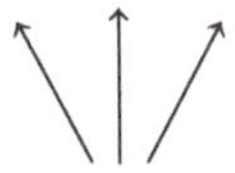

(c) 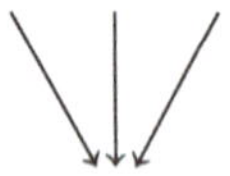 (d) 

Answers

1. (a) $v_{12} = 1$ m/s (rightwards), $v_{21} = 1$ m/s (leftwards)
   (b) $v_{12} = 7$ m/s (rightwards), $v_{21} = 7$ m/s (leftwards)
2. -2 m/s 3. Zero 4. (a)

See hints on page 259.

Applications of Relative Motion

The topic relative motion is very useful in two and three dimensional motion. Some of them are given below:

(i) **Problems of collision in one dimension**

We can assume one particle at rest and find the relative value of initial velocity and acceleration. Then we can apply $s = ut + \frac{1}{2}at^2$.

This value is very useful specially when both particles are moving under gravity. Because in that case relative acceleration is zero. Relative motion is uniform. So, we can directly use,

$$\text{time} = \frac{\text{distance}}{\text{speed}}$$

Here speed = magnitude of relative velocity.

Following example will illustrate the concept.

Example 4 Two particles 1 and 2 are thrown in the directions shown in figure simultaneously with velocities 5 m/s and 20 m/s. Initially, particle 1 is at a height of 20 m from the ground. Taking upwards as the positive direction, find

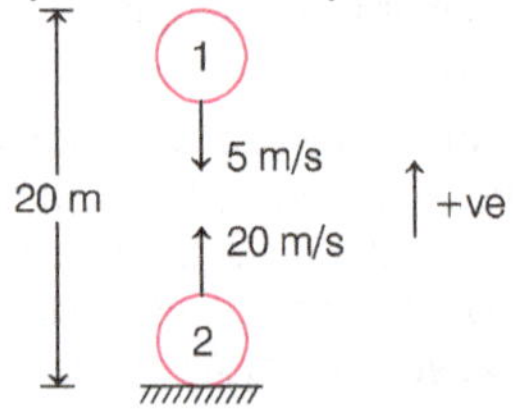

(a) acceleration of 1 with respect to 2
(b) initial velocity of 2 with respect to 1
(c) velocity of 1 with respect to 2 after time $t = \frac{1}{2}$ s
(d) time when the particles collide with each other.

Sol. (a) $a_1 = a_2 = -g$

$\therefore \quad a_{12} = a_1 - a_2 = 0$ **Ans.**

(b) $u_{21} = u_2 - u_1 = 20 - (-5)$

$= +25$ m/s **Ans.**

(c) $u_{12} = -u_{21} = -25$ m/s

Since, $a_{12} = 0$

$\therefore \quad u_{12} = \text{constant}$

$= -25$ m/s **Ans.**

(d) $a_{21} = 0$. Therefore, relative motion between them is uniform with constant velocity 25 m/s.

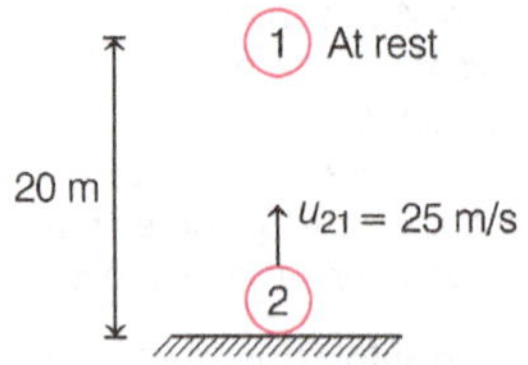

$\therefore \quad t = \frac{d}{u_{21}} = \frac{20}{25} = 0.8$ s **Ans.**

Note *In a general case you will have to find relative acceleration also. Then you have to apply,*

$$s = ut + \frac{1}{2}at^2$$

Here u and a are relative values with proper signs.

Example 5 Car A and car B start moving simultaneously in the same direction along the line joining them. Car A moves with a constant acceleration $a = 4$ m/s^2, while car B moves with a constant velocity $v = 1$ m/s. At time $t = 0$, car A is 10 m behind car B. Find the time when car A overtakes car B.

Sol. Given, $u_A = 0$, $u_B = 1\,\text{m/s}$, $a_A = 4\,\text{m/s}^2$ and $a_B = 0$

Assuming car B to be at rest, we have

$$u_{AB} = u_A - u_B = 0 - 1 = -1\,\text{m/s}$$

$$a_{AB} = a_A - a_B = 4 - 0 = 4\,\text{m/s}^2$$

Now, the relative motion of A with respect to B is as shown below.

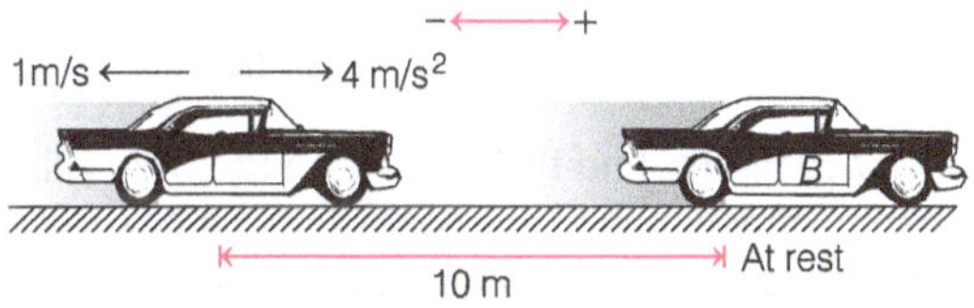

To find the time when A overtakes B, let us substitute the proper values in equation $s = ut + \frac{1}{2}at^2$, we get

$$10 = -t + \frac{1}{2}(4)(t^2)$$

or $\quad 2t^2 - t - 10 = 0$

or
$$t = \frac{1 \pm \sqrt{1+80}}{4} = \frac{1 \pm \sqrt{81}}{4} = \frac{1 \pm 9}{4}$$

or $\quad t = 2.5\,\text{s}$ and $-2\,\text{s}$

Ignoring the negative value, the desired time is 2.5 s.

Example 6 An elevator car whose floor-to-ceiling distance is equal to 2.7 m starts ascending with constant accelerating 1.2 ms^{-2}. After 2.0 s a bolt begins falling from the ceiling of the car. Find :

(a) the bolt's free fall time

(b) the displacement and the distance covered by the bolt during the free fall in the reference frame fixed to the elevator shaft. **(Irodov 1.15)**

Sol.

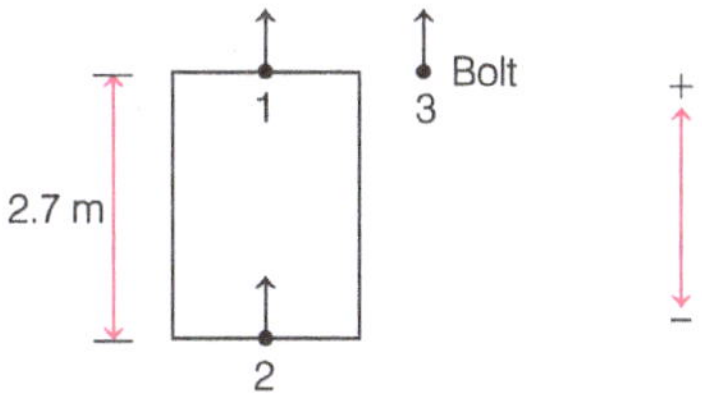

Note *For clear understanding, bolt has been shown outside the elevator.*

(a) At 2 sec, $v_1 = v_2 = v_3 = at = 1.2 \times 2 = 2.4\,\text{m/s}$

Bolt has to collide with the floor. Or, 3 has to collide with 2.

Since $v_2 = v_3$. So relative initial velocity is zero.

Acceleration of 2 is $1.2\,\text{m/s}^2$ upwards and acceleration of 3 is $9.8\,\text{m/s}^2$ downwards. Hence relative acceleration of 3 with respect to 2 is $(9.8 + 1.2)\,\text{m/s}^2$ or $11\,\text{m/s}^2$ downwards. So, for 3 to collide with 2, we can apply

$$s = \frac{1}{2}at^2 \qquad \text{(relative motion)}$$

or
$$2.7 = \frac{1}{2} \times 11 \times t^2$$

Solving this equation, we get

$$t = 0.7\,\text{s}$$

Alternate method (absolute approach)

2 will collide with 3 if displacement of 2 is 2.7 m more than displacement of 3.

$$\therefore \quad s_2 = s_3 + 2.7$$

$$\left(ut + \frac{1}{2}at^2\right)_2 = \left(ut + \frac{1}{2}at^2\right)_3 + 2.7$$

Now substituting the vector quantities u and a with sign,

$$2.4t + \frac{1}{2} \times 1.2 \times t^2 = 2.4t - \frac{1}{2} \times 9.8 \times t^2 + 2.7$$

From here also, we find

$$t = 0.7\,\text{s}$$

(b) Frame fixed to elevator shaft means with respect to ground.

Displacement

$$s = ut + \frac{1}{2}at^2 = 2.4 \times 0.7 - \frac{1}{2} \times 9.8 \times (0.7)^2$$

or $\quad s \approx 0.7\,\text{m}$

Distance

Let velocity of bolt will become zero after time t_0, then

$$v = u + at$$

or $\quad 0 = 2.4 - 9.8t_0$

or $\quad t_0 = 0.24\,\text{s}$

$$\therefore \quad d_{0 \to 0.7\,\text{s}} = d_{0 \to 0.24\,\text{s}} + d_{0.24\,\text{s} \to 0.7\,\text{s}}$$

$$= \frac{v_3^2}{2g} + \frac{1}{2}g(0.7 - 0.24)^2$$

$$= \frac{(2.4)^2}{2 \times 9.8} + \frac{1}{2} \times 9.8 \times (0.46)^2$$

$$= 1.3\,\text{m}$$

$$\therefore \quad d = 1.3\,\text{m}$$

Exercise

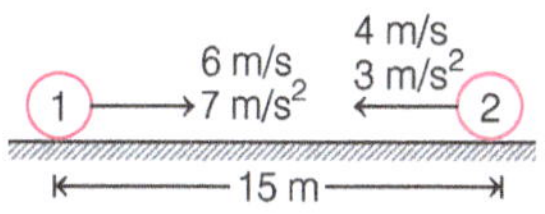

In the above figure, using the concept of relative motion, find the time when they will collide.

Ans. 1s.

(ii) **Problems of collision in two or three dimensions**

Following example will illustrate the concept.

Example 7 Two particles, 1 and 2 move with constant velocities $\mathbf{v}_1$ and $\mathbf{v}_2$. At the initial moment their radius vectors are equal to $\mathbf{r}_1$ and $\mathbf{r}_2$. How must these four vectors be interrelated for the particles to collide? **(Irodov 1.5)**

Sol. Let us assume particle-2 at rest. Position vector of 2 with respect to 1 is $\mathbf{r}_2 - \mathbf{r}_1$ and velocity of 1 with respect to 2 is $\mathbf{v}_1 - \mathbf{v}_2$.

Now, if particle-2 is assumed to be at rest then particle-1 will collide particle-2, if relative velocity of 1 with respect to 2 is along position vector of 2 with respect to 1.

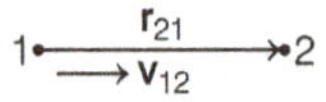

The best way to meet the above condition is, we equate the unit vectors of two.

Unit vector is obtained by dividing a vector by its magnitude. Hence the condition is,

$$\frac{\mathbf{r}_2 - \mathbf{r}_1}{|\mathbf{r}_2 - \mathbf{r}_1|} = \frac{\mathbf{v}_1 - \mathbf{v}_2}{|\mathbf{v}_1 - \mathbf{v}_2|}$$

or

$$\frac{\mathbf{r}_1 - \mathbf{r}_2}{|\mathbf{r}_1 - \mathbf{r}_2|} = \frac{\mathbf{v}_2 - \mathbf{v}_1}{|\mathbf{v}_2 - \mathbf{v}_1|}$$

(iii) **Problems of two trains crossing each other or overtaking each other :**

In this case we can directly apply the formula,

$$\boxed{t = \frac{l_1 + l_2}{v_r}}$$

Here l_1 and l_2 are the lengths of two trains and v_r is the magnitude of relative velocity. In such cases motion is uniform in most of the problems. So, we don't have to find the relative acceleration.

Example 8

In the above figure, find the time when two trains will cross each other.

Sol. Here $l_1 = 100$ m, $l_2 = 150$ m

As velocities are in opposite directions.

$\therefore \quad v_r = (6 + 4)$ m/s $= 10$ m/s

$$\therefore \quad t = \frac{l_1 + l_2}{v_r}$$

$$= \frac{100 + 150}{10} = 25 \text{ s}$$ **Ans.**

Example 9

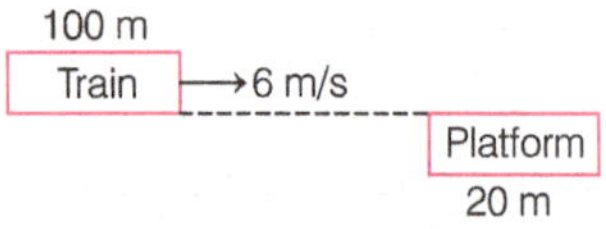

In the above figure, find the time when the train will cross the platform.

Sol. Here $l_1 = 100$ m, $l_2 = 20$ m and $v_r = 6$ m/s because velocity of platform is zero.

$$\therefore \quad t = \frac{l_1 + l_2}{v_r} = \frac{100 + 20}{6} = 20 \text{ s}$$ **Ans.**

Example 10

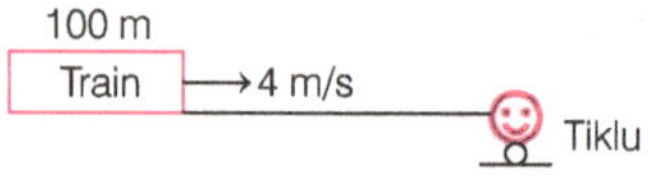

In the above figure, find the time when the train will cross Tiklu.

Sol. Here, $l_1 = 100$ m, $l_2 = 0$, as length of Tiklu is almost zero and $v_r = 4$ m/s, because Tiklu is at rest.

$$\therefore \quad t = \frac{l_1 + l_2}{v_r} = \frac{100 + 0}{4}$$

$$= 25 \text{ s}$$ **Ans.**

INTRODUCTORY Exercise 2

1. A 100 m long train crosses a man travelling at 5 km/h, in opposite direction in 7.2 s then the velocity of train is
   (a) 40 km/h (b) 25 km/h
   (c) 20 km/h (d) 45 km/h
2. What are the speeds of two objects if, when they move uniformly towards each other, they get 4 m closer in 1s and when they move uniformly in the same direction with the original speeds, they get 4.0 m closer in 10 s?
   (a) 2.8 m/s and 1.2 m/s (b) 5.2 m/s and 4.6 m/s
   (c) 3.2 m/s and 2.1 m/s (d) 2.2 m/s and 1.8 m/s

Answers

1. (d) **2.** (d)

See hints on page 259.

(iv) **Escalator-person (in Malls) or travelator-person (in Airports) problems**

In these problems we have mainly three velocities.

(a) v_1 = absolute velocity of escalator

(b) v_2 = velocity of person with respect to escalator

(c) v_{net} = net velocity of person with respect to ground

For finding the net velocity we can come across following four cases.

S.No.	Case	Net Velocity (v_{net})
1.	Only escalator is in motion	v_1
2.	Only person is in motion	v_2
3.	Both escalator and person are moving in same direction	$v_1 + v_2$
4.	Both escalator and person are moving in opposite directions	$v_1 \sim v_2$

After finding v_{net}, we can use the formula,

$$\text{time} = \frac{\text{distance}}{\text{speed}} \quad \text{or} \quad t = \frac{d}{v_{net}}$$

Note *(i) Motion is one dimensional. So, we have not used vector sign in velocity. Only sign method is sufficient.*

(ii) In most of the cases motion is uniform. So, we have not used acceleration in any of the above four cases.

(iii) $v_{PE} = v_P - v_E$

$\therefore$ $$\underset{\uparrow \atop v_{net}}{v_P} = \underset{\uparrow \atop v_1}{v_E} + \underset{\uparrow \atop v_2}{v_{PE}}$$

Here P stands for person and E for escalator.

Now, we can see that,

If $v_2 = 0$, $v_{net} = v_1$

If $v_1 = 0$, $v_{net} = v_2$

If both have same sign (means same direction), $v_{net} = v_1 + v_2$ *and if both have opposite signs (means opposite directions), then* $v_{net} = v_1 \sim v_2$.

Example 11 A person walks up a stationary 15 m long escalator in 90 s. When standing on the same escalator, now moving, the person is carried up in 60 s. How much time would it take that person to walk up the moving escalator? Does the answer depend on the length of the escalator?

Sol. Let l be the length of escalator.

Using $t = \frac{l}{v_{net}}$ three times with the symbols used in the text discussed above.

$$t_1 = \frac{l}{v_2} \quad \Rightarrow \quad v_2 = \frac{l}{t_1}$$

$$t_2 = \frac{l}{v_1} \quad \Rightarrow \quad v_1 = \frac{l}{t_2}$$

$$t_3 = \frac{l}{v_1 + v_2} = \frac{l}{\frac{l}{t_2} + \frac{l}{t_1}} = \frac{t_1 t_2}{t_1 + t_2} = \frac{(90)(60)}{(90 + 60)}$$

$= 36$ s **Ans.**

We can see that $t_3 = \frac{t_1 t_2}{t_1 + t_2}$ is independent of l. So, the answer does not depend upon the length of escalator.

(v) **Minimum distance between two particles in motion**

In these type of problems, one body is assumed to be at rest and the relative motion of the other body is considered. By assuming so, two body problem is converted into one body problem and the solution becomes easy. Following example will illustrate the statement.

Example 12 Two ships A and B are 10 km apart on a line running south to north. Ship A farther north is streaming west at 20 km/h and ship B is streaming north at 20 km/h. What is their distance of closest approach and how long do they take to reach it?

Sol. Ships A and B are moving with same speed 20 km/h in the directions shown in figure. It is a two dimensional, two body problem with zero acceleration.

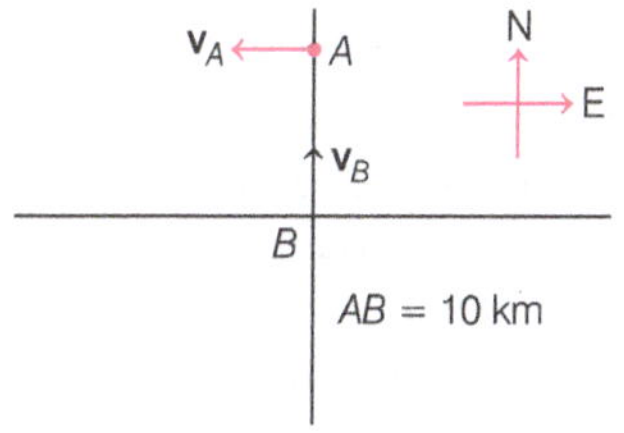

Let us find $\mathbf{v}_{BA}$.

$\therefore$ $$\mathbf{v}_{BA} = \mathbf{v}_B - \mathbf{v}_A = \mathbf{v}_B + (-\mathbf{v}_A)$$

Here, $$|\mathbf{v}_{BA}| = \sqrt{(20)^2 + (20)^2} = 20\sqrt{2} \text{ km/h}$$

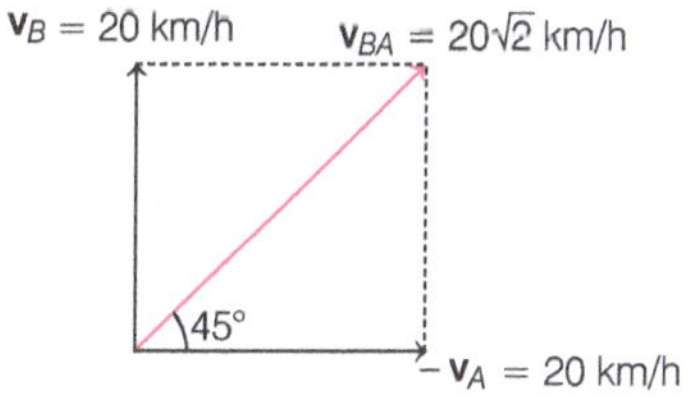

i.e. $\mathbf{v}_{BA}$ is $20\sqrt{2}$ km/h at an angle of 45° from east towards north. Thus, the given problem can be simplified as below. A is assumed to be at rest and B is moving with $\mathbf{v}_{BA}$ in the direction shown in figure below.

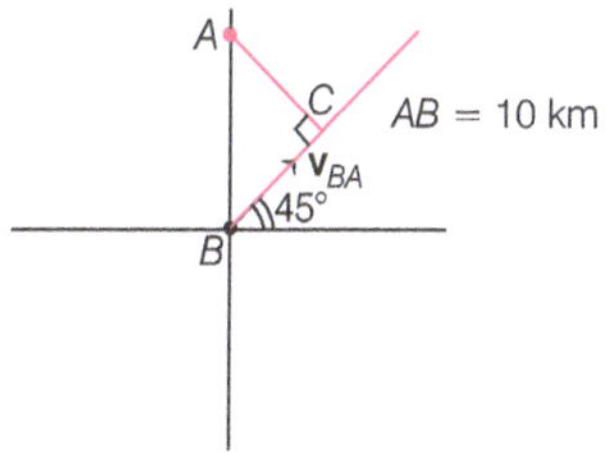

The minimum distance between the two is

$$d_{min} = AC = AB \sin 45°$$

$$= 10\left(\frac{1}{\sqrt{2}}\right) \text{ km} = 5\sqrt{2} \text{ km}$$ **Ans.**

and the desired time is

$$t = \frac{BC}{|\mathbf{v}_{BA}|} = \frac{5\sqrt{2}}{20\sqrt{2}} \quad (BC = AC = 5\sqrt{2} \text{ km})$$

$$= \frac{1}{4} \text{ h} = 15 \text{ min}$$ **Ans.**

(vi) **Rain Umbrella Problems**

In these type of problems, we again come across three terms $\mathbf{v}_r$, $\mathbf{v}_m$ and $\mathbf{v}_{rm}$

Here, $\mathbf{v}_r$ = velocity of rain

$\mathbf{v}_m$ = velocity of man (it may be velocity of cyclist or velocity of motorist also)

and $\mathbf{v}_{rm}$ = velocity of rain with respect to man.

Here, $\mathbf{v}_{rm}$ is the velocity of rain which appears to the man.

So, the man should hold his umbrella in the direction of $\mathbf{v}_{rm}$ or $\mathbf{v}_r - \mathbf{v}_m$, to save him from rain.

Example 13 A man is walking with 3 m/s, due east. Rain is falling vertically downwards with speed 4 m/s. Find the direction in which man should hold his umbrella, so that rain does not wet him.

Sol. As we have discussed above, man should hold his umbrella in the direction of $\mathbf{v}_{rm}$ or $\mathbf{v}_r - \mathbf{v}_m$ or $\mathbf{v}_r + (-\mathbf{v}_m)$.

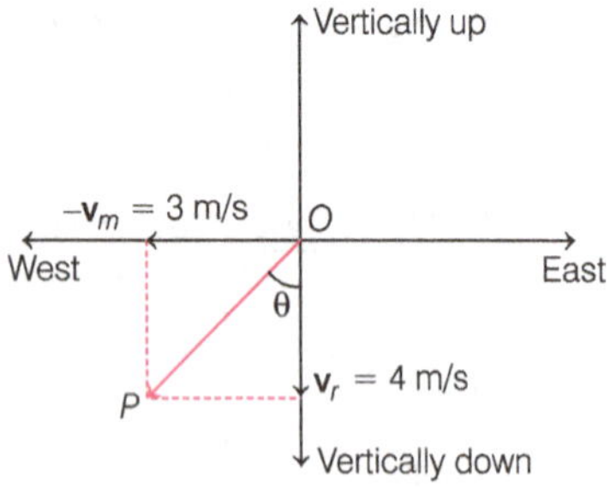

$$OP = \mathbf{v}_r + (-\mathbf{v}_m) = \mathbf{v}_r - \mathbf{v}_m = \mathbf{v}_{rm}$$

$$\Rightarrow \quad \tan\theta = \frac{3}{4} \quad \Rightarrow \quad \theta = \tan^{-1}\left(\frac{3}{4}\right) = 37°$$

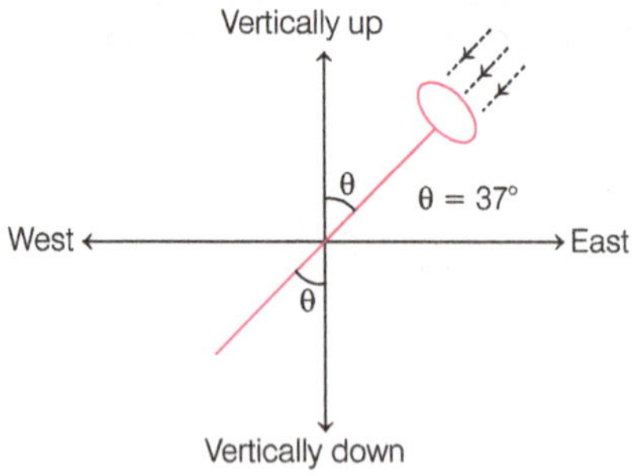

Therefore, man should hold his umbrella at an angle of 37° east of vertical (or 37° from vertical towards east).

Example 14 To a man walking at the rate of 3 km/h the rain appears to fall vertically downwards. When he increases his speed to 6 km/h it appears to meet him at an angle of 45° with vertical. Find the speed of rain.

Sol. Let $\hat{\mathbf{i}}$ and $\hat{\mathbf{j}}$ be the unit vectors in horizontal and vertical directions respectively.

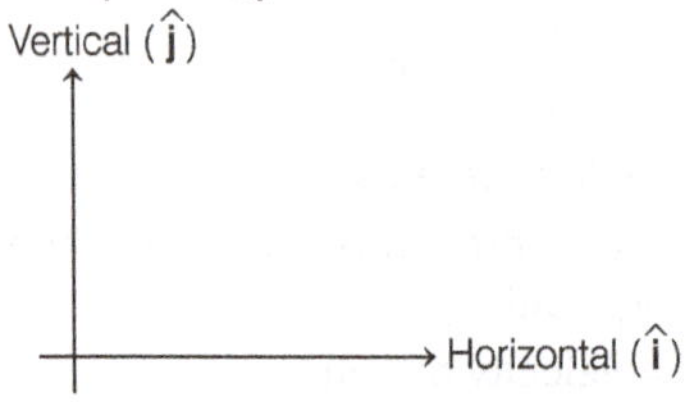

Let velocity of rain $\quad \mathbf{v}_r = a\hat{\mathbf{i}} + b\hat{\mathbf{j}}$ ...(i)

Then, speed of rain will be $|\mathbf{v}_r| = \sqrt{a^2 + b^2}$ ...(ii)

In the first case,

$$\mathbf{v}_m = \text{velocity of man} = 3\hat{\mathbf{i}}$$

$$\therefore \quad \mathbf{v}_{rm} = \mathbf{v}_r - \mathbf{v}_m = (a-3)\,\hat{\mathbf{i}} + b\hat{\mathbf{j}}$$

It seems to be in vertical direction.

Hence, $\quad a - 3 = 0 \quad$ or $\quad a = 3$

In the second case, $\quad \mathbf{v}_m = 6\hat{\mathbf{i}}$

$$\therefore \quad \mathbf{v}_{rm} = (a-6)\hat{\mathbf{i}} + b\hat{\mathbf{j}} = -3\hat{\mathbf{i}} + b\hat{\mathbf{j}}$$

This seems to be at 45° with vertical. Hence, $|b| = 3$

Therefore, from Eq. (ii) speed of rain is

$$|\mathbf{v}_r| = \sqrt{(3)^2 + (3)^2} = 3\sqrt{2} \text{ km/h}$$ **Ans.**

(vii) **River-Boat Problems :** In these type of problems we come across three terms :

(a) $\mathbf{v}_r$ = absolute velocity of river

(b) $\mathbf{v}_{br}$ = velocity of boatman with respect to river. Or, velocity of boatman in still water.

This is also called rowing velocity.

(c) $\mathbf{v}_b$ = absolute velocity of boatman (with respect to ground)

Concept

$$\mathbf{v}_{br} = \mathbf{v}_b - \mathbf{v}_r \quad \Rightarrow \quad \mathbf{v}_b = \mathbf{v}_{br} + \mathbf{v}_r$$

Or, the net velocity is the vector sum of the rowing velocity and river velocity.

1-D Problems

For finding the net velocity in 1-D problems we come across following four cases. Since we are discussing 1-D problems, so there is no need of using vector sign on vector quantities. Only sign method is sufficient.

These four cases are just like escalator-person problems discussed before.

No.	Case	Net velocity of boatman (v_b)
1	Rowing is zero	v_r
2	River velocity is zero. Or, water is still	v_{br}
3	Rowing is in the same direction of river current	$v_{br} + v_r$
4	Rowing is in the opposite direction of river current	$v_{br} \sim v_r$

2-D Problems

In 2-D problems we come across following cases.

(a) General Case

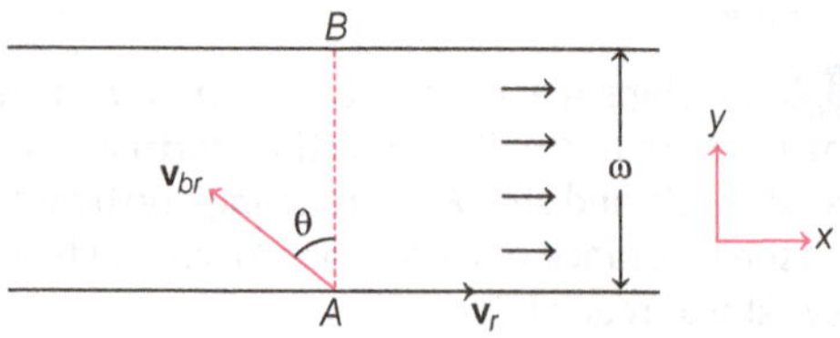

A boatman starts from point A on one bank of a river with velocity $\mathbf{v}_{br}$ in the direction shown in figure. River is flowing along positive x-direction with velocity $\mathbf{v}_r$. Width of the river is ω, then

$$\mathbf{v}_b = \mathbf{v}_r + \mathbf{v}_{br}$$

Therefore, $$v_{bx} = v_{rx} + v_{brx} = v_r - v_{br} \sin\theta$$

and $$v_{by} = v_{ry} + v_{bry}$$
$$= 0 + v_{br} \cos\theta = v_{br} \cos\theta$$

Now, time taken by the boatman to cross the river is

$$t = \frac{\omega}{v_{by}} = \frac{\omega}{v_{br}\cos\theta} \quad \text{or} \quad \boxed{t = \frac{\omega}{v_{br}\cos\theta}} \qquad \text{...(i)}$$

Further, displacement along x-axis when he reaches the other bank (also called drift) is

$$x = v_{bx}\, t = (v_r - v_{br}\sin\theta)\frac{\omega}{v_{br}\cos\theta}$$

or $$\boxed{x = (v_r - v_{br}\sin\theta)\frac{\omega}{v_{br}\cos\theta}} \qquad \text{...(ii)}$$

(b) Shortest Time

First approach

From Eq. (i) we can see that time (t) will be minimum when $\theta = 0°$, i.e. the boatman should steer his boat perpendicular to the river current.

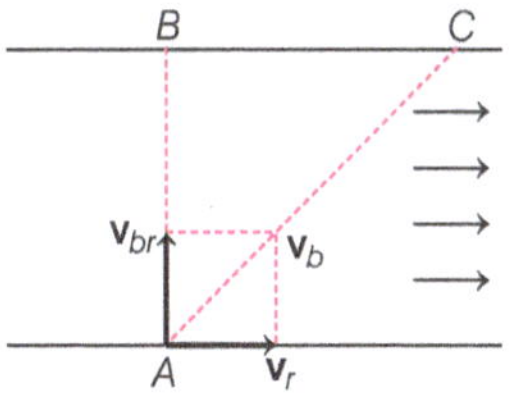

Also, $$t_{\min} = \frac{\omega}{v_{br}}$$

as $$\cos\theta = 1$$

In this case, net velocity,

$$v_b \text{ or } |\mathbf{v}_b| = \sqrt{v_r^2 + v_{br}^2}$$

Drift $= BC = (v_r)(t)$

From the above figure, we can see that expression of time t can be written as,

$$t = \frac{AB}{v_{br}} = \frac{BC}{v_r} = \frac{AC}{v_b}$$

Second approach

From equation (i) we can see that a component of $\mathbf{v}_{br}$ perpendicular to river current which is $v_{br}\cos\theta$ is responsible for crossing the river. So, if we want to cross the river in shortest time, then why to take the help of component ? Whole $\mathbf{v}_{br}$ should be perpendicular to the river current as shown in the figure.

(c) To reach a point *B* which is just opposite of *A*

First approach

In this case, the drift (x) should be zero.

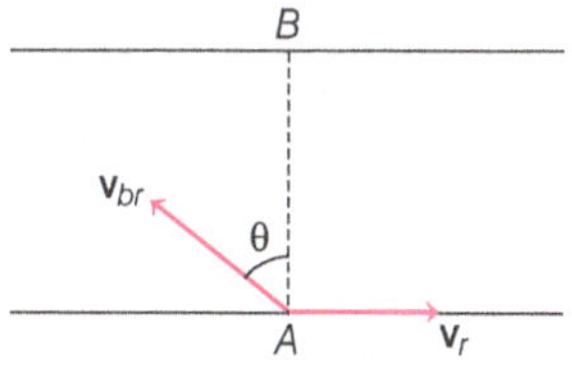

$$\therefore \qquad x = 0 \quad \text{or} \quad (v_r - v_{br}\sin\theta)\frac{\omega}{v_{br}\cos\theta} = 0$$

or $$v_r = v_{br}\sin\theta$$

or $$\boxed{\sin\theta = \frac{v_r}{v_{br}} \quad \text{or} \quad \theta = \sin^{-1}\left(\frac{v_r}{v_{br}}\right)}$$

Hence, to reach point B the boatman should row at an angle $\theta = \sin^{-1}\left(\frac{v_r}{v_{br}}\right)$ upstream from AB.

As, $\sin\theta \not> 1$.

So, if $v_r \ge v_{br}$, the boatman can never reach at point B. Because if $v_r = v_{br}$, $\sin\theta = 1$ or $\theta = 90°$ and it is just impossible to reach at B if $\theta = 90°$. Moreover, it can be seen that $v_b = 0$ if $v_r = v_{br}$ and $\theta = 90°$. Similarly, if $v_r > v_{br}$, $\sin\theta > 1$, i.e. no such angle exists. Practically, it is not possible to reach at point B if river velocity (v_r) is too high or $v_r \ge v_{br}$.

Second approach

To reach at point B, which is just opposite to the starting point A, net velocity of boatman $\mathbf{v}_b$ or the vector sum of $\mathbf{v}_r$ and $\mathbf{v}_{br}$ should be along AB. The velocity diagram is as below:

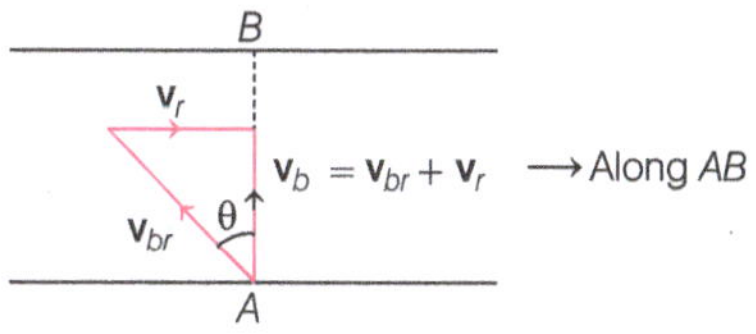

From the diagram we can see that,

$$|\mathbf{v}_b| \text{ or } v_b = \sqrt{v_{br}^2 - v_r^2} \qquad \text{...(i)}$$

Time, $$t = \frac{\omega}{v_b} = \frac{\omega}{\sqrt{v_{br}^2 - v_r^2}}$$

drift $x = 0$ and $\sin\theta = \frac{v_r}{v_{br}}$ or $\theta = \sin^{-1}\left(\frac{v_r}{v_{br}}\right)$

From Eq. (i), we can see that this case is possible if,

$$v_{br} > v_r$$

otherwise, v_b is either zero or imaginary.

(d) To reach a particular point *P*

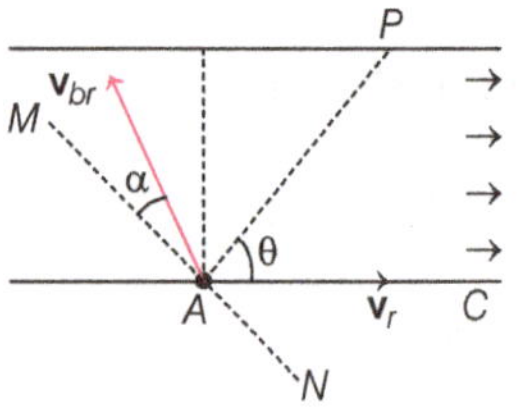

Suppose, starting from *A* we want to reach point *P* which is at angle θ from downstream direction *AC*.

Concept

Net velocity $\mathbf{v}_b$ which is vector sum of $\mathbf{v}_r$ and $\mathbf{v}_{br}$ should be along *AP*. Or, their components perpendicular to *AP* (in the direction of *MN*) should cancel.

Approach

Suppose $\mathbf{v}_{br}$ is kept at an angle α from *MN* as shown.

Then, perpendicular to *AP* (or along *MN*) we have to cancel the components of $\mathbf{v}_r$ and $\mathbf{v}_{br}$. So,

$$v_{br}\cos\alpha = v_r \sin\theta$$

From here we can find the direction of rowing velocity or we can find α.

Now components along *AP* are added and give the net velocity or v_b. Thus,

Net velocity $v_b = v_{br}\sin\alpha + v_r\cos\theta$ and time $t = \dfrac{AP}{v_b}$.

Example 15 Width of a river is 30 m, river velocity is 2 m/s and rowing velocity is 5 m/s at 37° from the direction of river current (a) find the time taken to cross the river, (b) drift of the boatman while reaching the other shore.

Sol.

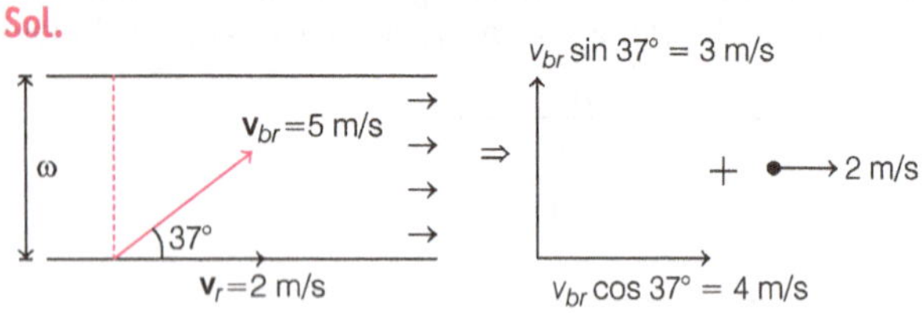

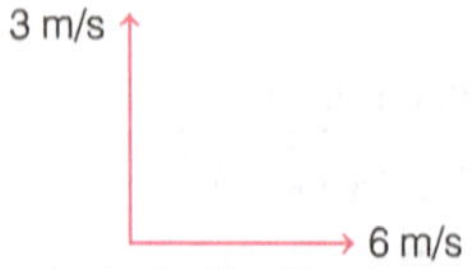

Net velocity of boatman

(a) Time taken to cross the river,

$$t = \frac{\omega}{3} = \frac{30}{3} = 10 \text{ s} \qquad \textbf{Ans.}$$

(b) Drift along the river (downstream)

$$x = (6)(t) = 6 \times 10$$

$$= 60 \text{ m} \qquad \textbf{Ans.}$$

Example 16 Width of a river is 30 m, river velocity is 4 m/s and rowing velocity of boatman is 5 m/s.

(a) Make the velocity diagram for crossing the river in shortest time. Then, find this shortest time, net velocity of boatman and drift along the river.

(b) Can the boatman reach a point just opposite on the other shore? If yes then make the velocity diagram, the direction in which he should row his boat and the time taken to cross the river in this case.

(c) How long will it take him to row 10 m up stream and then back to his starting point?

Sol. (a) Shortest time

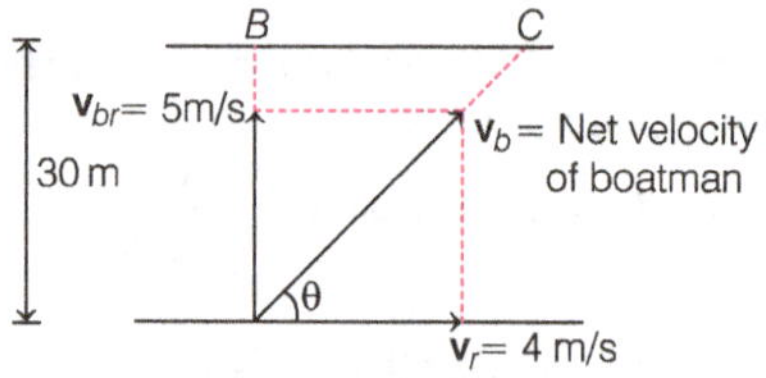

(i) $$t = \frac{30}{5} = 6 \text{ s} = t_{min} \qquad \textbf{Ans.}$$

(ii) $$|\mathbf{v}_b| \text{ or } v_b = \sqrt{(5)^2 + (4)^2} = \sqrt{41} \text{ m/s} \qquad \textbf{Ans.}$$

$$\tan\theta = \frac{5}{4} \Rightarrow \theta = \tan^{-1}\left(\frac{5}{4}\right) \qquad \textbf{Ans.}$$

(iii) $$\text{Drift} = BC = (4)(t) = 4 \times 6 = 24 \text{ m} \qquad \textbf{Ans.}$$

(b) Since, $v_{br} > v_r$, so this case is possible. Velocity diagram is as under.

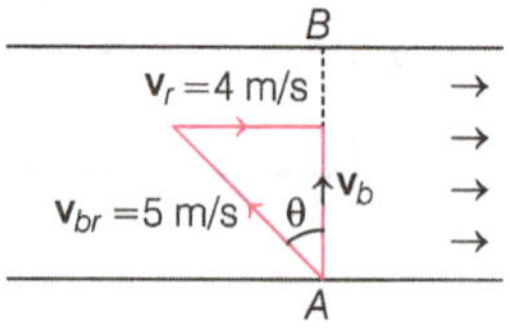

Net velocity $|\mathbf{v}_b|$ or $v_b = \sqrt{(5)^2 - (4)^2} = 3$ m/s along *AB*

$$\sin\theta = \frac{4}{5} \Rightarrow \theta = \sin^{-1}\frac{4}{5} = 53° \qquad \textbf{Ans.}$$

$$t = \frac{AB}{v_b} = \frac{30}{3} = 10 \text{ sec} \qquad \textbf{Ans.}$$

Note *If the boatman wants to return to the same point A, then diagram is as under*

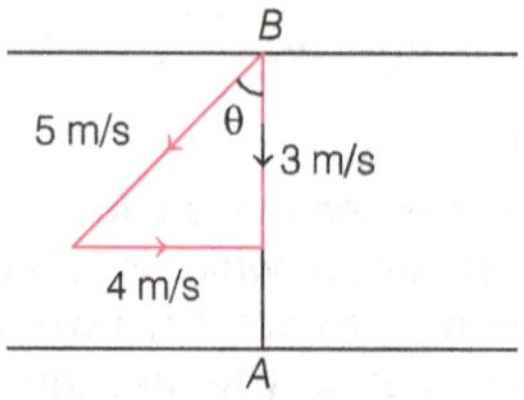

$$t_{BA} = \frac{BA}{3} = \frac{30}{3} = 10 \text{ s}$$

(c)

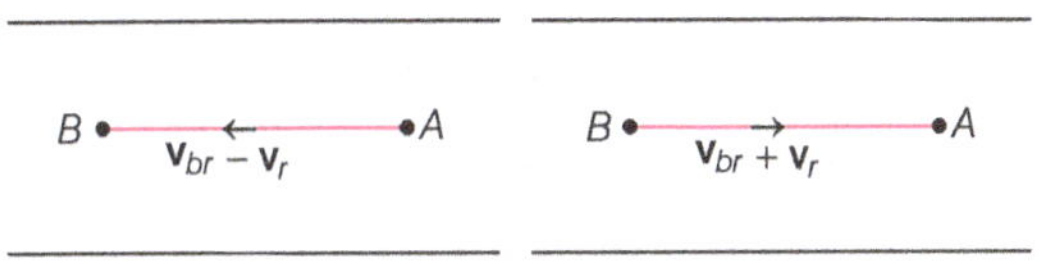

$$t = t_{AB} + t_{BA} = \frac{AB}{v_{br} - v_r} + \frac{BA}{v_{br} + v_r}$$

or
$$t = \frac{10}{5-4} + \frac{10}{5+4}$$

$$= \frac{100}{9}\text{ s} \qquad \textbf{Ans.}$$

(e) Shortest path in river-boat problems

Path length travelled by the boatman when he reaches the opposite shore is

$$s = \sqrt{\omega^2 + x^2}$$

Here, ω = width of river which is constant. So, for s to be minimum modulus of x (drift) should be minimum. Now, two cases are possible.

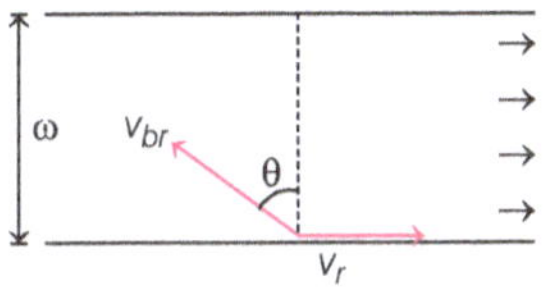

When $v_r < v_{br}$: In this case $x = 0$, when $\theta = \sin^{-1}\left(\frac{v_r}{v_{br}}\right)$

or
$$\boxed{s_{min} = \omega \quad \text{at} \quad \theta = \sin^{-1}\left(\frac{v_r}{v_{br}}\right)}$$

When $v_r > v_{br}$: In this case x is minimum, where $\frac{dx}{d\theta} = 0$

or
$$\frac{d}{d\theta}\left\{\frac{\omega}{v_{br}\cos\theta}(v_r - v_{br}\sin\theta)\right\} = 0$$

or
$$\frac{\omega}{v_{br}}[v_r \sec\theta\tan\theta - v_{br}\sec^2\theta] = 0$$

or
$$v_r \tan\theta - v_{br}\sec\theta = 0$$

or
$$v_r \sin\theta - v_{br} = 0$$

or
$$\theta = \sin^{-1}\left[\frac{v_{br}}{v_r}\right]$$

Now, at this angle we can find x_{min} and then s_{min} which comes out to be

$$\boxed{s_{min} = \omega\left(\frac{v_r}{v_{br}}\right) \quad \text{at} \quad \theta = \sin^{-1}\left(\frac{v_{br}}{v_r}\right)}$$

(viii) **Aircraft wind problems**

This is similar to river boat problems. So, all cases discussed in river - boat problems can be applied here also. The only difference is that $\mathbf{v}_{br}$ is replaced by $\mathbf{v}_{aw}$ (velocity of aircraft with respect to wind or velocity of aircraft in still air which is also known as steering velocity) $\mathbf{v}_r$ is replaced by $\mathbf{v}_w$ (velocity of wind) and $\mathbf{v}_b$ is replaced by $\mathbf{v}_a$ (absolute velocity of aircraft). Further, $\mathbf{v}_a = \mathbf{v}_{aw} + \mathbf{v}_w$.

In this case, some of the problems are slightly different. The given variables are

(i) Complete wind velocity $\mathbf{v}_w$
(ii) Steering speed or $|\mathbf{v}_{aw}|$
(iii) Starting point (say A) and destination point (say B)

We have to find direction of $\mathbf{v}_{aw}$ (or steering velocity) and the time taken in moving from A to B.

Concept

Net velocity of aircraft $\mathbf{v}_a$ or vector sum of $\mathbf{v}_w$ and $\mathbf{v}_{aw}$ should be along AB.

To solve such problems, we can apply the following steps:

(i) Take starting point A as the origin.
(ii) Wind velocity vector is completely given. So, draw $\mathbf{v}_w$ from point A.
(iii) Draw another vector $\mathbf{v}_a$ starting from A in a direction from A to B.
(iv) In above two steps we have already made two sides of a triangle in vector form. Complete the third side. This represents $\mathbf{v}_{aw}$. While completing the triangle for finding direction of $\mathbf{v}_{aw}$, polygon law of vector addition is to be followed, so that

$$\mathbf{v}_w + \mathbf{v}_{aw} = \mathbf{v}_a$$

(v) Applying, sine law in this triangle. We can find direction of $\mathbf{v}_{aw}$ and the net velocity of aircraft $\mathbf{v}_a$. Now,

time taken, $\quad t = \frac{AB}{|\mathbf{v}_a|}$ or $\frac{AB}{v_a}$

The following example will illustrate the above theory.

Example 17 An aircraft flies at 400 km/h in still air. A wind of $200\sqrt{2}$km/h is blowing from the south towards north. The pilot wishes to travel from A to a point B north east of A. Find the direction he must steer and time of his journey if $AB = 1000$ km.

Sol. Given that $\mathbf{v}_w = 200\sqrt{2}$ km/h (towards north)

$\mathbf{v}_{aw} = 400$ km/h. Here $\mathbf{v}_a$ should be along AB or in north-east direction. Thus, the direction of $\mathbf{v}_{aw}$ should be such as the resultant of $\mathbf{v}_w$ and $\mathbf{v}_{aw}$ is along AB or in north-east direction.

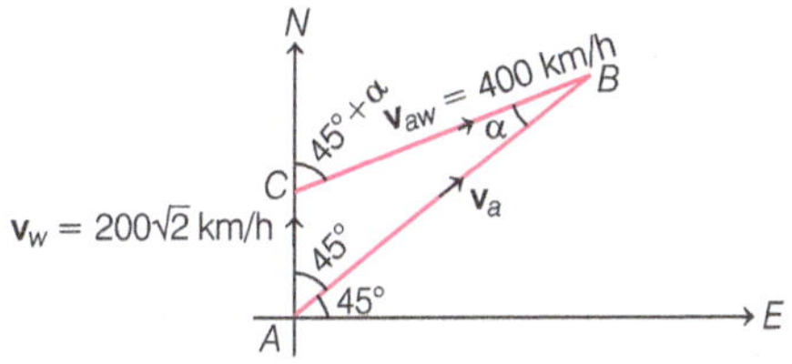

Let $\mathbf{v}_{aw}$ makes an angle α from AB as shown in figure. Applying sine law in triangle ABC, we get

$$\frac{CB}{\sin 45^\circ} = \frac{AC}{\sin \alpha}$$

or $$\sin \alpha = \left(\frac{AC}{CB}\right) \sin 45^\circ$$

$$= \left(\frac{200\sqrt{2}}{400}\right) \frac{1}{\sqrt{2}} = \frac{1}{2}$$

$\therefore$ $$\alpha = 30^\circ$$

Therefore, the pilot should steer in a direction at an angle of $(45^\circ + \alpha)$ or 75° from north towards east. Further,

$$\frac{|\mathbf{v}_a|}{\sin (180^\circ - 75^\circ)} = \frac{400}{\sin 45^\circ}$$

or $$|\mathbf{v}_a| = \frac{\sin 105^\circ}{\sin 45^\circ} \times (400) \text{ km/h}$$

$$= \left(\frac{\sin 75^\circ}{\sin 45^\circ}\right)(400) \text{ km/h}$$

$$= \left(\frac{0.9659}{0.707}\right)(400) \text{ km/h}$$

$$= 546.47 \text{ km/h}$$

$\therefore$ The time of journey from A to B is

$$t = \frac{AB}{|\mathbf{v}_a|} = \frac{1000}{546.47} \text{ h}$$

$$t = 1.83 \text{ h}$$ **Ans.**

Note *(i) In the above problem we have used polygon law of vector addition for the vector addition.*

$$\mathbf{v}_w + \mathbf{v}_{aw} = \mathbf{v}_a$$

We can also use parallelogram law of vector addition as shown in the figure given below.

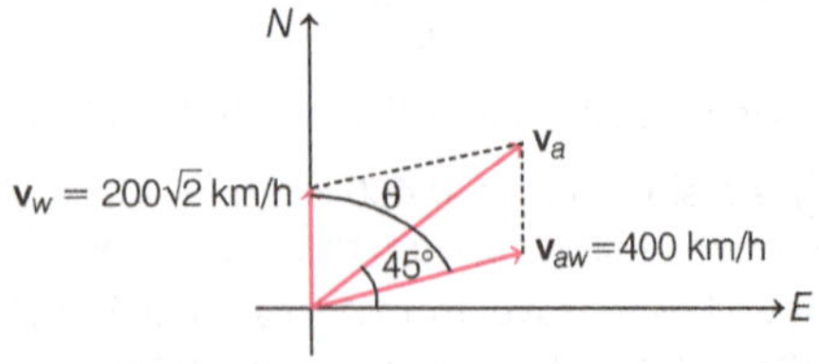

I am leaving the calculation part for the students. After calculations we can see that θ comes out to be 75° and $|\mathbf{v}_a| = 546.47$ km/h.

(ii) We can also use the concept discussed in river-boat problems.

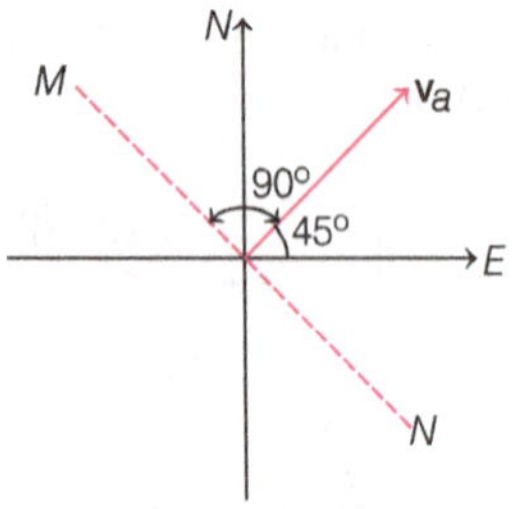

Net velocity is along N-E direction.

So, components of $\mathbf{v}_a$ and $\mathbf{v}_{aw}$ perpendicular to N-E direction (along MN) should cancel each other.

For instance, in above example $\mathbf{v}_a$ has to be along AB and we know that $\mathbf{v}_a = \mathbf{v}_{aw} + \mathbf{v}_w$. Therefore, sum of components of $\mathbf{v}_{aw}$ and $\mathbf{v}_w$ perpendicular to line AB (shown as dotted) should be zero.

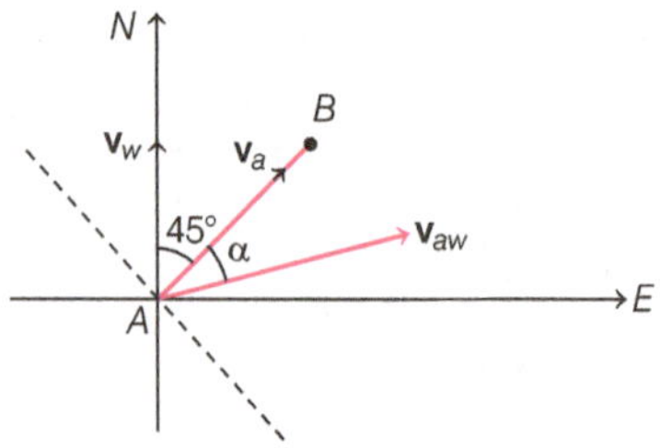

or $|\mathbf{v}_{aw}| \sin \alpha = |\mathbf{v}_w| \sin 45^\circ$

or $$\sin \alpha = \frac{|\mathbf{v}_w|}{|\mathbf{v}_{aw}|} \sin 45^\circ$$

$$= \left(\frac{200\sqrt{2}}{400}\right)\left(\frac{1}{\sqrt{2}}\right) = \frac{1}{2}$$

$\therefore$ $$\alpha = 30^\circ \text{ or } 45^\circ + \alpha = 75^\circ$$

Therefore $\mathbf{v}_{aw}$ is at 75° from north towards east.

Now, $|\mathbf{v}_a| = |\mathbf{v}_{aw}| \cos \alpha + |\mathbf{v}_w| \cos 45^\circ$

$$= (400) \cos 30^\circ + (200\sqrt{2})\left(\frac{1}{\sqrt{2}}\right)$$

$$= (400)\frac{\sqrt{3}}{2} + 200 = 346.47 + 200$$

$$= 546.47 \text{ km/h}$$

$\therefore$ Time of journey from A to B will be

$$t = \frac{AB}{|\mathbf{v}_a|} = \frac{1000}{546.47} = 1.83 \text{ h.}$$

Left over topics from all the units discussed so far

(i) **Difference sign (~) :**

Suppose $A = 6$ and $B = 2$. Then :

$A - B = 4$ and $B - A = -4$

But, $A \sim B = B \sim A = 4$

(ii)

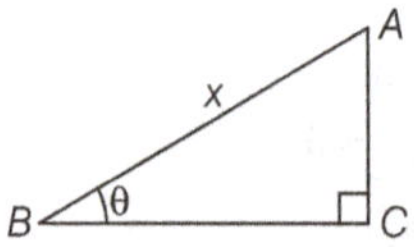

In the right angle triangle as shown in figure,

$$BC = x \cos \theta \quad \text{and}$$

$$AC = x \sin \theta$$

(iii) If y (may be velocity, acceleration etc.) is a function of time or $y = f(t)$ and we want to find the average value of y, between a time interval of t_1 and t_2. Then,

$\langle y \rangle_{t_1 \text{ to } t_2}$ = average value of y between t_1 and t_2

$$= \frac{\int_{t_1}^{t_2} f(t)\, dt}{t_2 - t_1} \quad \text{or} \quad \langle y \rangle_{t_1 \text{ to } t_2} = \frac{\int_{t_1}^{t_2} f(t)\, dt}{t_2 - t_1}$$

If $f(t)$ is a linear function of t, then

$$y_{av} = \frac{y_f + y_i}{2}$$

Here, y_f = final value of y and y_i = initial value of y

At the same time, we should not forget the basic formulas like,

$$v_{av} = \frac{\text{total displacement}}{\text{total time}}$$

and $$a_{av} = \frac{\text{change in velocity}}{\text{total time}}$$

Example 18 In one dimensional uniformly accelerated motion, find average velocity between a time interval from $t = 0$ to $t = t$.

Sol. We can solve this problem by three methods.

Method 1. $v = u + at$ or $v = f(t)$

$$\therefore \quad \langle v \rangle_{0-t} = \frac{\int_0^t (u + at)\, dt}{t - 0}$$

$$= \frac{ut + \frac{1}{2}at^2}{t} = u + \frac{1}{2}at$$

Method 2. Since, v is a linear function of time, we can write

$$v_{av} = \frac{v_f + v_i}{2} = \frac{(u + at) + u}{2} = u + \frac{1}{2}at$$

Method 3. $$v_{av} = \frac{\text{Total displacement}}{\text{Total time}}$$

$$= \frac{ut + \frac{1}{2}at^2}{t} = u + \frac{1}{2}at$$

(iv) Angle between velocity vector **v** and acceleration vector **a** decides whether the speed of particle is increasing, decreasing or constant.

Speed increases, if $0° \le \theta < 90°$

Speed decreases, if $90° < \theta \le 180°$

Speed is constant, if $\theta = 90°$

The angle θ between **v** and **a** can be obtained by the relation,

$$\theta = \cos^{-1}\left(\frac{\mathbf{v} \cdot \mathbf{a}}{va}\right)$$

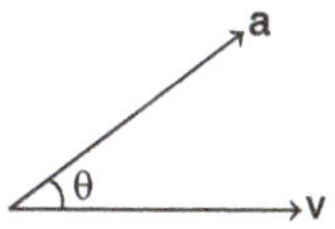

Hence, speed of a particle increases if dot product of **v** and **a** is positive, speed decreases, if the dot product is negative and speed remains constant if dot product is zero.

(v) The magnitude of instantaneous velocity is called the instantaneous speed, i.e.

$$v = |\mathbf{v}| = \left|\frac{d\mathbf{r}}{dt}\right|$$

Speed is not equal to $\frac{dr}{dt}$, i.e. $v \ne \frac{dr}{dt}$

where, r is the modulus of radius vector **r** because in general $|d\mathbf{r}| \ne dr$. For example, when **r** changes only in direction, i.e. if a point moves in a circle, then r = constant, $dr = 0$ but $|d\mathbf{r}| \ne 0$.

Miscellaneous Examples

Ex. 1 Figure shows a rod of length l resting on a wall and the floor. Its lower end A is pulled towards left with a constant velocity v. Find the velocity of the other end B downward when the rod makes an angle θ with the horizontal.

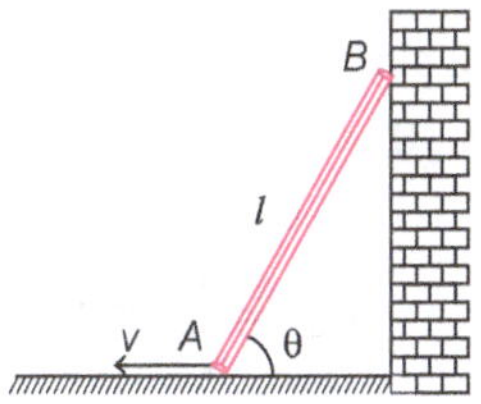

Sol. Let the distance from the corner to the point A is x and that up to B is y. Then,

$$v = \frac{dx}{dt} \quad \text{and} \quad v_B = -\frac{dy}{dt}$$

(– sign denotes that y is decreasing)

Further, $$x^2 + y^2 = l^2$$

Differentiating with respect to time t

$$2x\frac{dx}{dt} + 2y\frac{dy}{dt} = 0$$

$$\therefore \quad x\left(\frac{dx}{dt}\right) = y\left(\frac{-dy}{dt}\right)$$

$$\therefore \quad xv = yv_B$$

$$v_B = \frac{x}{y}v = v \cot\theta \qquad \textbf{Ans.}$$

Ex. 2 Three points are located at the vertices of an equilateral triangle whose side equals a. They all start moving simultaneously with velocity v constant in modulus, with the first point heading continuously for the second, the second for the third, and the third for the first. How soon will the points converge?

(Irodov 1.12)

Sol.

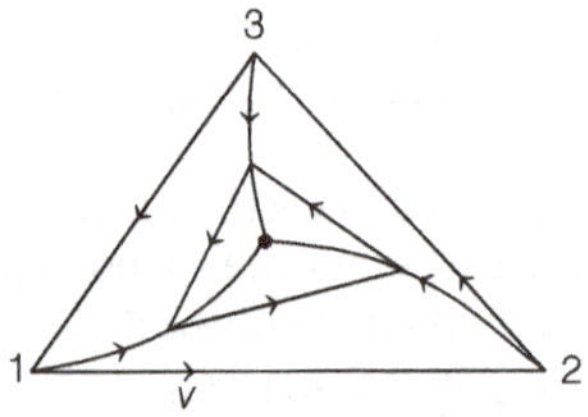

From symmetry, we can say that they will meet at centre of the equilateral triangle. They will always make an equilateral triangle whose size will keep on decreasing.

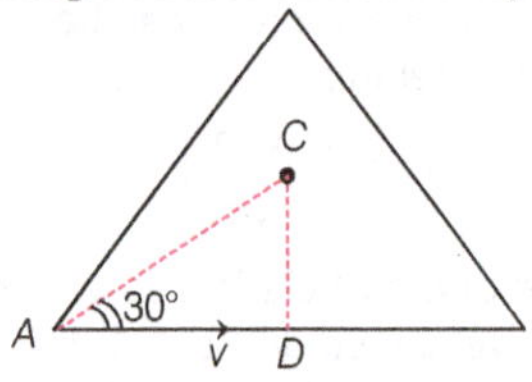

Now let us take a particle at A.

$$AC = AD \sec 30° = \left(\frac{a}{2}\right)\left(\frac{2}{\sqrt{3}}\right) = \frac{a}{\sqrt{3}}$$

Component of v along AC is always $v \cos 30°$ or $\frac{\sqrt{3}v}{2}$.

$$\therefore \quad t = \frac{AC}{v \cos 30°} = \frac{a/\sqrt{3}}{\sqrt{3}\,v/2} = \frac{2a}{3v}$$

$$\therefore \quad t = \frac{2a}{3v}$$

Alternate solution

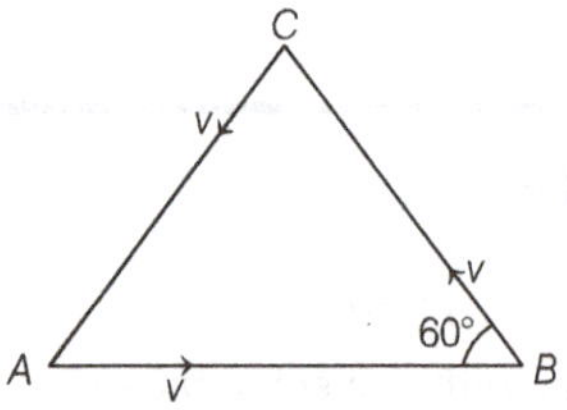

Along AB, magnitude of relative velocity between A and B is,

$$v + v\cos 60° = v + \frac{v}{2} = \frac{3v}{2}$$

Now this is the rate (which is constant) by which distance between A and B decreases from initial value d to final value 0.

$$\therefore \quad t = \frac{d}{3v/2} = \frac{2d}{3v}$$

Ex. 3 A man wants to reach point B on the opposite bank of a river flowing at a speed as shown in figure. What minimum speed relative to water should the man have so that he can reach point B? In which direction should he swim?

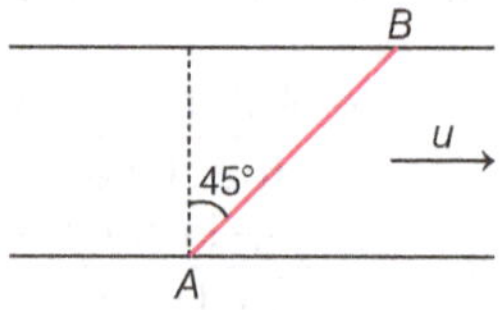

Sol. Let v be the speed of boatman in still water.

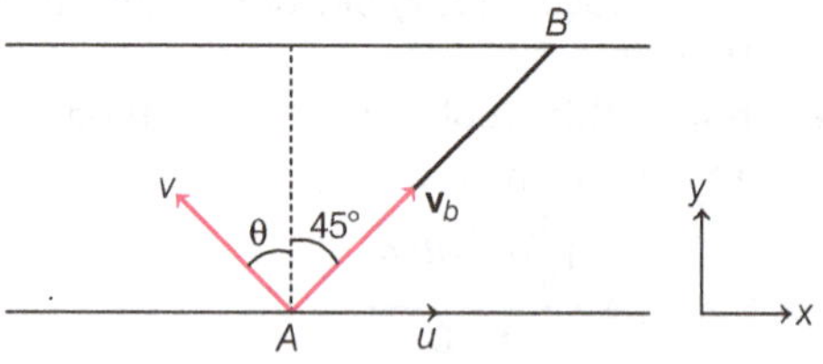

Resultant of v and u should be along AB. Components of $\mathbf{v}_b$ (absolute velocity of boatman) along x and y-directions are,

$$v_x = u - v\sin\theta$$

and $$v_y = v\cos\theta$$

Further, $$\tan 45° = \frac{v_y}{v_x}$$

or $$1 = \frac{v\cos\theta}{u - v\sin\theta}$$

$$\therefore \quad v = \frac{u}{\sin\theta + \cos\theta}$$

$$= \frac{u}{\sqrt{2}\left[\frac{1}{\sqrt{2}}\sin\theta + \frac{1}{\sqrt{2}}\cos\theta\right]}$$

$$= \frac{u}{\sqrt{2}[\cos 45° \sin\theta + \sin 45° \cos\theta]}$$

$$= \frac{u}{\sqrt{2}\sin(\theta + 45°)}$$

v is minimum at, $\theta + 45° = 90°$

or $\theta = 45°$ **Ans.**

and $v_{min} = \frac{u}{\sqrt{2}}$ at $\theta + 45° = 90°$ **Ans.**

MODULE 8 Exercise

Relative Motion

Assertion and Reason

See directions on page 178.

1. **Assertion :** Displacement-time equation of two particles moving in a straight line are, $s_1 = 2t - 4t^2$ and $s_2 = -2t + 4t^2$. Relative velocity between the two will go on increasing.

   Reason : If velocity and acceleration are of same sign, then speed will increase.

2. **Assertion** For an observer looking out through the window of a fast moving train, the nearby objects appear to move in the opposite direction to the train, while the distant objects appear to be stationary.

   Reason If the observer and the object are moving at velocities $\mathbf{v}_1$ and $\mathbf{v}_2$, respectively with reference to a laboratory frame, the velocity of the object with respect to the observer is $\mathbf{v}_2 - \mathbf{v}_1$. **(2008 JEE)**

Single Correct Option

3. Figure shows two ships moving in xy-plane with velocities v_A and v_B. The ships move such that B always remains north of A. The ratio $\frac{v_A}{v_B}$ is equal to

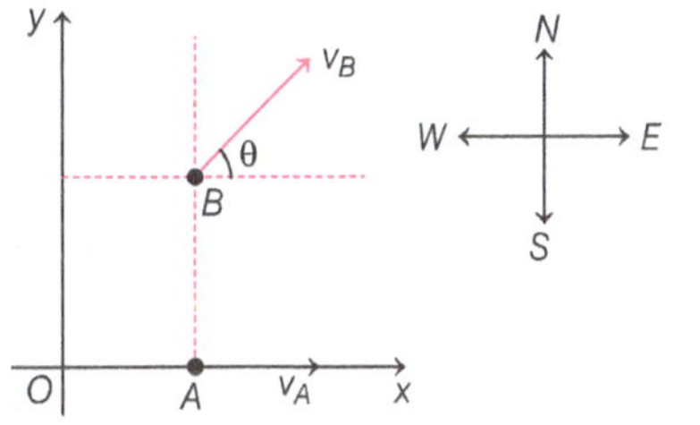

(a) $\cos\theta$ (b) $\sin\theta$
(c) $\sec\theta$ (d) $\text{cosec}\,\theta$

4. A long horizontal belt is moving from left to right with a uniform speed 2 m/s. There are two ink marks A and B on the belt 60 m apart. An insect runs on the belt to and fro between A and B such that its speed relative to belt is constant and equals 4 m/s. When the insect is moving on the belt in the direction of motion of the belt, its speed as observed by a person standing on ground will be

   (a) 6 m/s (b) 2 m/s (c) 1.5 m/s (d) 4 m/s

5. In the above question, if A lies to the left of B, then

   (a) time taken by insect to travel from A to B and time taken by it to travel from B to A are equal
   (b) time taken by insect to travel from A to B is less than the time taken by it to travel from B to A
   (c) time taken by insect to travel from A to B is more than the time taken by it to travel from B to A
   (d) None of the above

6. In Q.4, if A lies to the left of B then time taken by the insect to travel from B to A will be

   (a) 12 s (b) 15 s (c) 18 s (d) 21 s

7. Train A and train B are running on parallel tracks in the opposite directions with speeds of 36 km/h and 75 km/h, respectively. A person is walking in train A in the direction opposite to its motion with a speed of 1.8 km/h with respect to train A. Speed (in ms^{-1}) of this person as observed from train B will be close to : (Take the distance between the tracks as negligible) **(2020 Main)**

   (a) 29.5 ms^{-1} (b) 28.5 ms^{-1}
   (c) 31.5 ms^{-1} (d) 30.5 ms^{-1}

Multiple Correct Options

8. Ship A is located 4 km north and 3 km east of ship B. Ship A has a velocity of 20 kmh^{-1} towards the south and ship B is moving at 40 kmh^{-1} in a direction 37° north of east. x and y-axes are along east and north directions respectively, then :

   (a) Velocity of A relative to B is $(-32\hat{\mathbf{i}} - 44\hat{\mathbf{j}})$ km/h
   (b) Position of A relative to B as a function of time is given by $\mathbf{r}_{AB} = [(3-32t)\hat{\mathbf{i}} + (4-44t)\hat{\mathbf{j}}]$ km
   (c) Velocity of A relative to B is $(32\hat{\mathbf{i}} - 44\hat{\mathbf{j}})$ km/h
   (d) Position of A relative to B as a function of time is given by $(32t\hat{\mathbf{i}} - 44t\hat{\mathbf{j}})$ km

Subjective Question

9. A ship moves along the equator to the east with velocity $v_s = 30\ \text{kmh}^{-1}$. The southeastern wind blows at an angle $\phi = 60°$ to the equator with velocity $v_w = 15\ \text{kmh}^{-1}$. Find the wind velocity v_{ws} relative to the ship and the angle θ between the equator and the wind direction in the reference frame fixed to the ship. **(Irodov 1.6)**

Collision of Two Particles

Single Correct Option

10. A car is standing 200 m behind a bus, which is also at rest. The two start moving at the same instant but with different forward accelerations. The bus has an acceleration of 2 m/s^2 and the car has acceleration 4 m/s^2. The car will catch up with the bus after a time: **(2017 Main)**

    (a) $\sqrt{110}$ s (b) $\sqrt{120}$ s (c) $10\sqrt{2}$ s (d) 15 s

Subjective Question

11. A ball is thrown vertically upward from the 12 m level with an initial velocity of 18 m/s. At the same instant an open platform elevator passes the 5 m level, moving upward with a constant velocity of 2 m/s. Determine ($g = 9.8 \text{ m/s}^2$)
(a) when and where the ball will meet the elevator,
(b) the relative velocity of the ball with respect to the elevator when the ball hits the elevator.

Train Problems

Single Correct Option

12. A passenger train of length 60 m travels at a speed of 80 km/hr. Another freight train of length 120 m travels at a speed of 30 km/hr. The ratio of times taken by the passenger train to completely cross the freight train when : (i) they are moving in the same direction and (ii) in the opposite direction is **(2019 Main)**
(a) $\frac{3}{2}$ (b) $\frac{25}{11}$ (c) $\frac{11}{5}$ (d) $\frac{5}{2}$

13. Two trains 'A' and 'B' of length 'l' and '$4l$' are travelling into a tunnel of length 'L' in parallel tracks from opposite directions with velocities 108 km/h and 72km/h respectively. If train 'A' takes 35 s less time than train 'B' to cross the tunnel, then length 'L' of tunnel is (Given $L = 60l$) **(2023 Main)**
(a) 1200 m (b) 2700 m (c) 1800 m (d) 900 m

14. A passenger sitting in a train A moving at 90 km/h observes another train B moving in the opposite direction for 8s. If the velocity of the train B is 54 km/h, then length of train B is **(2023 Main)**
(a) 80 m (b) 200 m (c) 120 m (d) 320 m

Minimum Distance

Single Correct Option

15. Two particles are moving along two long straight lines, in the same plane with same speed equal to 20 cm/s. The angle between the two lines is 60° and their intersection point is O. At a certain moment, the two particles are located at distances 3m and 4m from O and are moving towards O. Subsequently, the shortest distance between them will be
(a) 50 cm (b) $40\sqrt{2}$ cm (c) $50\sqrt{2}$ cm (d) $50\sqrt{3}$ cm

16. Ship A is sailing towards north-east with velocity $\mathbf{v} = (30\hat{\mathbf{i}} + 50\hat{\mathbf{j}})$ km/h, where $\hat{\mathbf{i}}$ points east and $\hat{\mathbf{j}}$ north. Ship B is at a distance of 80 km east and 150 km north of Ship A and is sailing towards west at 10 km/h. A will be at minimum distance from B in **(2019 Main)**
(a) 4.2 h (b) 2.6 h (c) 3.2 h (d) 2.2 h

Subjective Question

17. Two particles 1 and 2, move with constant velocities v_1 and v_2 along two mutually perpendicular straight lines toward the intersection point O. At the moment $t = 0$ the particles were located at the distances l_1 and l_2 from the point O. How soon will the distance between the particles become the smallest? What is it equal to? **(Irodov 1.16)**

Rain Problems

Single Correct Option

18. A man standing on a road has to hold his umbrella at 30° with the vertical to keep the rain away. He throws the umbrella and starts running at 10 km/h. He finds that raindrops are hitting his head vertically. What is the speed of rain with respect to ground?
(a) $10\sqrt{3}$ km/h (b) 20 km/h
(c) $\frac{20}{\sqrt{3}}$ km/h (d) $\frac{10}{\sqrt{3}}$ km/h

19. A girl standing on road holds her umbrella at 45° with the vertical to keep the rain away. If she starts running without umbrella with a speed of $15\sqrt{2}$ kmh^{-1}, the rain drops hit her head vertically. The speed of rain drops with respect to the moving girl is **(2022 Main)**
(a) 30 kmh^{-1} (b) $\frac{25}{\sqrt{2}}$ kmh^{-1} (c) $\frac{30}{\sqrt{2}}$ kmh^{-1} (d) 25 kmh^{-1}

River Problems

Single Correct Option

20. A swimmer crosses a river of width d flowing at velocity v. While swimming, he heads himself always at an angle of 120° with the river flow and on reaching the other end he finds a drift of $d/2$ in the direction of flow of river. The speed of the swimmer with respect to the river is
(a) $(2-\sqrt{3})v$ (b) $2(2-\sqrt{3})v$
(c) $4(2-\sqrt{3})v$ (d) $(2+\sqrt{3})v$

21. On a calm day a boat can travel a certain distance in a lake and return in time T_0 with a speed V. On a rough day there is a uniform current at speed v to help the onward journey and impede the return journey. If the time taken to go across and return on the rough day be T, then T/T_0 is
(a) $1 - v^2/V^2$ (b) $\frac{1}{1 - v^2/V^2}$
(c) $1 + v^2/V^2$ (d) $\frac{1}{1 + v^2/V^2}$

22. A river is flowing from West to East at a speed of 5 m/min. A man on the South bank of the river capable of swimming at 10 m/min in still water, wants to swim across the river in the shortest time. He should swim in a direction **(1983 JEE)**
(a) due North
(b) 30° East of North
(c) 30° West of North
(d) 60° East of North

23. The stream of a river is flowing with a speed of 2 km/h. A swimmer can swim at a speed of 4 km/h. What should be the direction of the swimmer with respect to the flow of the river to cross the river straight ? **(2019 Main)**

(a) $60°$ (b) $120°$
(c) $90°$ (d) $150°$

Subjective Questions

24. A river 400 m wide is flowing at a rate of 2.0 m/s. A boat is sailing at a velocity of 10.0 m/s with respect to the water in a direction perpendicular to the river.
(a) Find the time taken by the boat to reach the opposite bank.
(b) How far from the point directly opposite to the starting point does the boat reach the opposite bank?

25. A man crosses a river in a boat. If he crosses the river in minimum time he takes 10 min with a drift 120 m. If he crosses the river taking shortest path, he takes 12.5 min, find
(a) width of the river
(b) velocity of the boat with respect to water
(c) speed of the current

26. A river is 20 m wide. River speed is 3 m/s. A boat starts with velocity $2\sqrt{2}$ m/s at angle 45° from the river current (relative to river)
(a) Find the time taken by the boat to reach the opposite bank.
(b) How far from the point directly opposite to the starting point does the boat reach the opposite bank?

27. A man in a boat crosses a river from point A. If he rows perpendicular to the banks he reaches point C ($BC = 120$m) in 10 min. If the man heads at a certain angle α to the straight line AB (AB is perpendicular to the banks) against the current he reaches point B in 12.5 min. Find the width of the river ω, the rowing velocity u, the speed of the river current v and the angle α. Assume the velocity of the boat relative to water to be constant and the same magnitude in both cases.

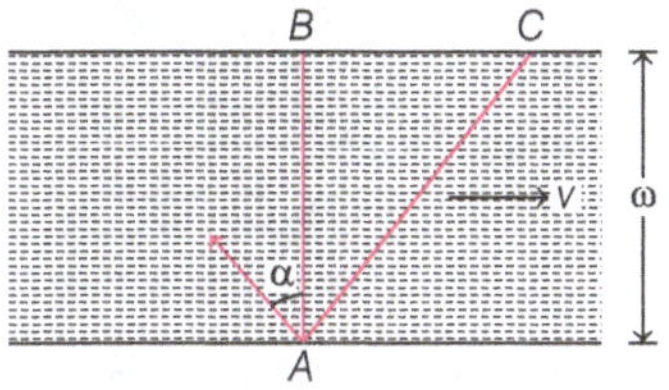

28. Two swimmers leave point A on one bank of the river to reach point B lying right across on the other bank. One of them crosses the river along the straight line AB while the other swims at right angles to the stream and then walks the distance that he has been carried away by the stream to get to point B. What was the velocity u of his walking if both swimmers reached the destination simultaneously ? The stream velocity $v_R = 2.0\ \text{kmh}^{-1}$ and the velocity v_{SR} of each swimmer with respect to water equals $2.5\ \text{kmh}^{-1}$. **(Irodov 1.7)**

29. A boat moves relative to water with a velocity which is $n = 2.0$ times greater than the river flow velocity. At what angle to the stream direction must the boat move to minimize drifting? **(Irodov 1.9)**

Integer Type Questions

30. The speed of a swimmer is $4\ \text{kmh}^{-1}$ in still water. If the swimmer makes his strokes normal to the flow of river of width 1 km, he reaches a point 750 m down the stream on the opposite bank. The speed of the river water is .... kmh^{-1}. **(2023 Main)**

31. A person is swimming with a speed of 10 m/s at an angle of 120° with the flow and reaches to a point directly opposite on the other side of the river. The speed of the flow is 'x' m/s. The value of 'x' to the nearest integer is .... **(2021 Main)**

32. A swimmer can swim with velocity of 12 km/h in still water. Water flowing in a river has velocity 6 km/h. The direction with respect to the direction of flow of river water he should swim in order to reach the point on the other bank just opposite to his starting point is .... degree. (Round off to the nearest integer) **(2021 Main)**

Aircraft Problems

Subjective Questions

33. An aeroplane has to go from a point A to another point B, 500 km away due 30° east of north. Wind is blowing due north at a speed of 20 m/s. The steering-speed of the plane is 150 m/s. (a) Find the direction in which the pilot should head the plane to reach the point B. (b) Find the time taken by the plane to go from A to B.

34. An aeroplane has to go from a point P to another point Q, 1000 km away due north. Wind is blowing due east at a speed of 200 km/h. The air speed of plane is 500 km/h.
(a) Find the direction in which the pilot should head the plane to reach the point Q.
(b) Find the time taken by the plane to go from P to Q.

Miscellaneous Problems

Fill in the Blanks

35. Four persons K, L, M, N are initially at the four corners of a square of side d. Each person now moves with a uniform speed v in such a way that K always moves directly towards L, L directly towards M, M directly towards N and N directly towards K. The four persons will meet at a time .... **(1984 JEE)**

36. A spotlight S rotates in a horizontal plane with constant angular velocity of 0.1 rad/s. The spot of

light P moves along the wall at a distance of 3 m. The velocity of the spot P when $\theta = 45°$ (see figure) is....... m/s. **(1987 JEE)**

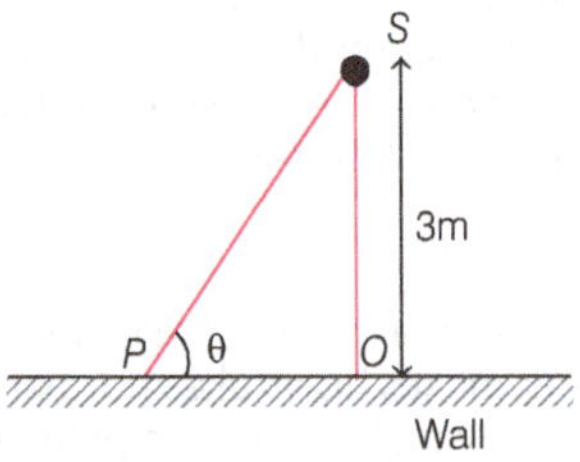

Single Correct Option

37. An ant is at a corner of a cubical room of side a. The ant can move with a constant speed u. The minimum time taken to reach the farthest corner of the cube is

(a) $\frac{3a}{u}$ (b) $\frac{\sqrt{3}a}{u}$

(c) $\frac{\sqrt{5}a}{u}$ (d) $\frac{(\sqrt{2}+1)a}{u}$

38. A cart is moving horizontally along a straight line with constant speed $30\,ms^{-1}$. A particle is to be fired vertically upwards from the moving cart in such a way that it returns to the cart at the same point from where it was projected after the cart has moved 80 m. At what speed (relative to the cart) must the projectile be fired? (Take $g = 10\,ms^{-2}$)

(a) $10\,ms^{-1}$ (b) $10\sqrt{8}\,ms^{-1}$

(c) $\frac{40}{3}\,ms^{-1}$ (d) None of these

39. In the arrangement shown in the figure, the ends P and Q of an unstretchable string move downwards with uniform speed U. Pulleys A and B are fixed. Mass M moves upwards with a speed **(1982 JEE)**

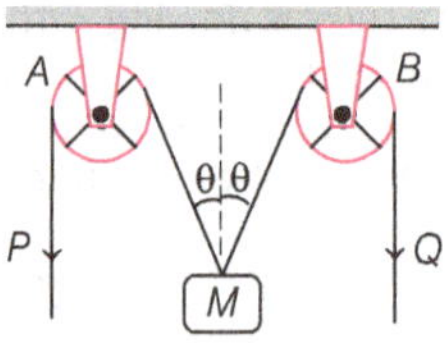

(a) $2U\cos\theta$ (b) $\frac{U}{\cos\theta}$

(c) $\frac{2U}{\cos\theta}$ (d) $U\cos\theta$

Passage Based Questions

Passage (Q.40 to Q.42)

Two buses A and B are running over a two line bridge with the same speed $20\,ms^{-1}$ in opposite directions. Both buses approach a car C moving with a speed of $15\,ms^{-1}$ in the same direction of bus A. To avoid accident, the driver of bus A decides to overtake car C just before the bus B does, at the instant when buses A and B are 500 m apart and provides required acceleration $2\,ms^{-2}$ to the bus.

40. Find the distance of car C from the bus B when driver of bus A starts accelerating.

(a) 100 m (b) 200 m (c) 350 m (d) 400 m

41. Find the distance of car C from the bus A, when driver of bus A starts accelerating.

(a) 100 m (b) 150 m (c) 200 m (d) 350 m

42. After what time, the buses A and B will cross each other?

(a) 10 s (b) 35 s

(c) 20 s (d) 15 s

Answers

1. (d) **2.** (b) **3.** (a) **4.** (a) **5.** (a) **6.** (b) **7.** (a) **8.** (a, b)
9. $|\mathbf{v}_{ws}| \approx 40$ km/h, $\theta \approx 19°$ **10.** (c) **11.** (a) 3.65 s, at 12.30 m level (b) $19.8\,ms^{-1}$ (downwards) **12.** (c)
13. (c) **14.** (d) **15.** (d) **16.** (b) **17.** $d_{min} = \frac{|l_1v_2 - l_2v_1|}{\sqrt{v_1^2+v_2^2}}, t = \frac{l_1v_1 + l_2v_2}{v_1^2+v_2^2}$ **18.** (b) **19.** (c) **20.** (c)
21. (b) **22.** (a) **23.** (b) **24.** (a) 40 s (b) 80 m **25.** (a) 200 m (b) 20 m/min (c) 12 m/min
26. (a) 10 s (b) 50 m **27.** 200 m, 20 m/min, 12m/ min, 36° 50′ **28.** $u = 3$ km/h **29.** $\alpha = 120°$
30. (13) **31.** (5) **32.** (120) **33.** (a) $\sin^{-1}(1/15)$ east of the line AB (b) 50 min
34. (a) At an angle $\theta = \sin^{-1}(0.4)$ west of north (b) $\frac{10}{\sqrt{21}}h$ **35.** (d/v) **36.** (0.6) **37.** (c) **38.** (c) **39.** (b)
40. (c) **41.** (b) **42.** (a)

See hints on page 259-266.

Unit Exercise (Cumulative)

Level 1

Fill in the Blanks

1. A body moves in the southern direction for 10s at the speed of 10 ms^{-1}. It then starts moving in the eastern direction at the speed of 20 ms^{-1} for 10s. The magnitude of the average velocity is......... The average speed is......... . The total displacement will be..........

2. A car covers AB distance with first one-third at velocity $v_1\,ms^{-1}$, second one-third at $v_2\,ms^{-1}$ and last one-third at $v_3\,ms^{-1}$. If $v_3 = 3v_1$, $v_2 = 2v_1$ and $v_1 = 11\,ms^{-1}$ then the average velocity of the car is ....ms^{-1}. **(2022 Main)**

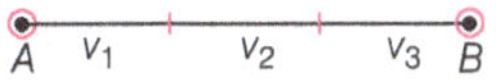

Assertion and Reason

See directions on page 178.

3. **Assertion :** Acceleration of a moving particle can change its direction without any change in direction of velocity.
   Reason : If the direction of change in velocity vector changes, the direction of acceleration vector also changes.

Single Correct Option

4. The length of a seconds hand in watch is 1 cm. The change in velocity of its tip in 15 s is
   (a) zero (b) $\frac{\pi}{30\sqrt{2}}$ cm/s
   (c) $\frac{\pi}{30}$ cm/s (d) $\frac{\pi\sqrt{2}}{30}$ cm/s

5. Taxies leave station X for station Y every 10 min. Simultaneously, a taxi also leaves station Y for station X every 10 min. The taxies move at the same constant speed and go from X and Y or *vice-versa* in 2h. How many taxies coming from the other side will meet each taxi enroute from Y and X ?
   (a) 24 (b) 23
   (c) 12 (d) 11

6. A lift performs the first part of its ascent with uniform acceleration a and the remaining with uniform retardation $2a$. If t is the time of ascent, find the depth of the shaft.
   (a) $\frac{at^2}{4}$ (b) $\frac{at^2}{3}$ (c) $\frac{at^2}{2}$ (d) $\frac{at^2}{8}$

7. A ball is thrown vertically upwards from the ground and a student gazing out of the window sees it moving upward past him at 10 ms^{-1}. The window is at 15 m above the ground level. The velocity of ball 3 s after it was projected from the ground is (Take $g = 10\,ms^{-2}$)
   (a) 10 m/s, up (b) 20 ms^{-1}, up
   (c) 20 ms^{-1}, down (d) 10 ms^{-1}, down

8. A body dropped from the top of a tower covers a distance $7x$ in the last second of its journey where x is the distance covered in first second. How much time does it take to reach the ground?
   (a) 3 s (b) 4 s (c) 5 s (d) 6 s

9. A stone is dropped from a height h. Simultaneously, another stone is thrown up from the ground which reaches a height $4h$. The two stones will cross each other after time
   (a) $\sqrt{\frac{h}{8g}}$ (b) $\sqrt{8gh}$ (c) $\sqrt{2gh}$ (d) $\sqrt{\frac{h}{2g}}$

10. Two balls are projected simultaneously with the same speed from the top of a tower, one vertically upwards and the other vertically downwards. They reach the ground in 9 s and 4 s, respectively. The height of the tower is ($g = 10\,m/s^2$)
    (a) 90 m (b) 180 m (c) 270 m (d) 360 m

11. Two trains are moving with velocities $v_1 = 10\,ms^{-1}$ and $v_2 = 20\,ms^{-1}$ on the same track in opposite directions. After the application of brakes if their retarding rates are $a_1 = 2\,ms^{-2}$ and $a_2 = 1\,ms^{-2}$ respectively, then the minimum distance of separation between the trains to avoid collision is
    (a) 150 m (b) 225 m
    (c) 450 m (d) 300 m

12. A body is thrown vertically upwards from the top A of a tower. It reaches the ground in t_1 s. If it is thrown vertically downwards from A with the same speed it reaches the ground in t_2 s. If it is allowed to fall freely from A, then the time it takes to reach the ground is given by
    (a) $t = \frac{t_1 + t_2}{2}$ (b) $t = \frac{t_1 - t_2}{2}$
    (c) $t = \sqrt{t_1 t_2}$ (d) $t = \sqrt{\frac{t_1}{t_2}}$

13. A ball is dropped on the floor from a height of 80 m. It rebounds to a height of 20 m. If the ball is in contact with the floor for 0.1 s, the average acceleration during contact is

(a) 400 m/s^2 (b) 500 m/s^2
(c) 600 m/s^2 (d) 800 m/s^2

14. Two particles move in the same straight line starting at the same moment from the same point. The first moves with constant velocity u and the second with constant acceleration f. During the time elapses before the second catches the first, greatest distance between the particles is

(a) $\frac{u}{f}$ (b) $\frac{u^2}{2f}$ (c) $\frac{f}{2u^2}$ (d) $\frac{f}{u^2}$

15. A body is projected vertically upwards. If t_1 and t_2 be the times at which it is at height h above the point of projection while ascending and descending respectively, then h is

(a) $\frac{1}{2}g t_1 t_2$ (b) $g t_1 t_2$
(c) $2g t_1 t_2$ (d) $4g t_1 t_2$

16. In the previous problem, the velocity of projection is

(a) $\frac{1}{2}g(t_1 + t_2)$ (b) $g(t_1 + t_2)$
(c) $2g(t_1 + t_2)$ (d) $4g(t_1 + t_2)$

17. A ball is thrown vertically upwards with a speed u. It reaches a point B at a height h (lower than the maximum height) after time t_1. It returns to the ground after time t_2 from the instant it was at B during the upward journey. Then, $t_1 t_2$ is equal to

(a) $2h/g$ (b) h/g (c) $h/2g$ (d) $h/4g$

18. Two bodies move in a straight line towards each other at initial velocities v_1 and v_2 and with constant accelerations a_1 and a_2 directed against their initial velocities. The maximum initial separation d_{max} between the bodies for which they will meet during the motion is

(a) $\frac{v_1^2}{a_1} + \frac{v_2^2}{a_2}$ (b) $\frac{(v_1 + v_2)^2}{2(a_1 + a_2)}$
(c) $\frac{v_1 v_2}{\sqrt{a_1 a_2}}$ (d) $\frac{v_1^2 - v_2^2}{2(a_1 - a_2)}$

19. A thief in a stolen car passes through a police check post at his top speed of 90 kmh^{-1}. A motorcycle cop, reacting after 2 s, accelerates from rest at 5 ms^{-2}. His top speed being 108 kmh^{-1}. Find the maximum separation between policemen and thief.

(a) 112.5 m (b) 115 m
(c) 116.5 m (d) None of these

20. The vertical height of point P above the ground is twice that of Q. A particle is projected downward with a speed of 5 m/s from P and at the same time another particle is projected upward with the same speed from Q. Both particles reach the ground simultaneously, then

(a) $PQ = 30 \text{ m}$
(b) time of flight of stones $= 3 \text{ s}$
(c) Both (a) and (b) are correct
(d) Both (a) and (b) are wrong

21. A target is made of two fixed plates, one of wood and the other of iron. The thickness of the wooden plate is 4 cm and that of iron plate is 2 cm. A bullet fired goes through the wood first and then penetrates 1 cm into iron. A similar bullet fired with the same velocity from opposite direction goes through iron first and then penetrates 2 cm into wood. If a_1 and a_2 be the retardations offered to the bullet by wood and iron plates respectively, then

(a) $a_1 = 2a_2$ (b) $a_2 = 2a_1$
(c) $a_1 = a_2$ (d) Data insufficient

22. A juggler throws balls vertically upwards with same initial velocity in air. When the first ball reaches its highest position, he throws the next ball. Assuming the juggler throws n balls per second, the maximum height the balls can reach is **(2022 Main)**

(a) $g/2n$ (b) g/n
(c) $2gn$ (d) $g/2n^2$

23. Two balls A and B are placed at the top of 180 m tall tower. Ball A is released from the top at $t = 0$ s. Ball B is thrown vertically down with an initial velocity 'u' at $t = 2$ s. After a certain time, both balls meet 100 m above the ground. Find the value of 'u' in ms^{-1}. [Use $g = 10 \text{ ms}^{-2}$] **(2022 Main)**

(a) 10 (b) 15
(c) 20 (d) 30

24. A particle moves on a straight line under deceleration of bv^2, initial velocity is v_0. The distance covered in time t is

(a) $bv_0 t$ (b) $\frac{1}{b}[\ln(1 + bv_0 t)]$
(c) $\frac{\ln bv_0 t}{b}$ (d) $b \ln bv_0 t$

25. Equation of motion of a body is $\frac{dv}{dt} = -4v + 8$, where v is the velocity in ms^{-1} and t is the time in second. Initial velocity of the particle was zero. Then,

(a) the initial rate of change of acceleration of the particle is 8 ms^{-2}
(b) the terminal speed is 2 ms^{-1}
(c) Both (a) and (b) are correct
(d) Both (a) and (b) are wrong

26. A particle of mass m moves on positive x-axis under the influence of force acting towards the origin given by $-kx^2\hat{\mathbf{i}}$. If the particle starts from rest at $x = a$, the speed it will attain when it crosses the origin is

(a) $\sqrt{\frac{k}{ma}}$ (b) $\sqrt{\frac{2k}{ma}}$
(c) $\sqrt{\frac{ma}{2k}}$ (d) None of these

27. A particle starts from rest. Its acceleration at an instant t is $a = \frac{10}{v+1}$ ms^{-2}, where v is instantaneous velocity. The distance travelled after which the particle attain a velocity of 15 ms^{-1} is

(a) 123.75 m
(b) 1237.5 m
(c) 200 m
(d) 375 m

28. A ball is thrown upward with an initial velocity v_0 from the surface of the earth. The motion of the ball is affected by a drag force equal to $m\gamma v^2$ (where, m is mass of the ball, v is its instantaneous velocity and γ is a constant). Time taken by the ball to rise to its maximum height is **(2019 Main)**

(a) $\frac{1}{\sqrt{2\gamma g}} \tan^{-1}\left(\sqrt{\frac{2\gamma}{g}}\, v_0\right)$

(b) $\frac{1}{\sqrt{\gamma g}} \tan^{-1}\left(\sqrt{\frac{\gamma}{g}}\, v_0\right)$

(c) $\frac{1}{\sqrt{\gamma g}} \sin^{-1}\left(\sqrt{\frac{\gamma}{g}}\, v_0\right)$

(d) $\frac{1}{\sqrt{\gamma g}} \ln\left(1+\sqrt{\frac{\gamma}{g}}\, v_0\right)$

29. A particle moves in the x-y plane with velocity $v_x = 8t - 2$ and $v_y = 2$. If it passes through the point $x = 14$ and $y = 4$ at $t = 2$ s, the equation of the path is

(a) $x = y^2 - y + 2$
(b) $x = y^2 - 2$
(c) $x = y^2 + y - 6$
(d) None of these

30. The equation of trajectory of a particle is $x = \sqrt{3}y - 5y^2$. The particle is at origin at $t = 0$. Its acceleration is $\mathbf{a} = -10\,\hat{\mathbf{i}}$. Its initial speed is

(a) 1 ms^{-1}
(b) 2 ms^{-1}
(c) 10 ms^{-1}
(d) cannot be determined

31. Starting from the origin at time $t = 0$, with initial velocity $5\hat{\mathbf{j}}$ ms^{-1}, a particle moves in the x-y plane with a constant acceleration of $(10\hat{\mathbf{i}} + 4\hat{\mathbf{j}})$ ms^{-2}. At time t its coordinates are (20m, y_0m). The values of t and y_0 are respectively **(2020 Main)**

(a) 4s and 52m
(b) 2s and 18m
(c) 2s and 24m
(d) 5s and 25m

32. A particle moves such that its position vector $\mathbf{r}(t) = \cos\omega t\,\hat{\mathbf{i}} + \sin\omega t\,\hat{\mathbf{j}}$ where ω is a constant and t is time. Then which of the following statements is true for the velocity $\mathbf{v}(t)$ and acceleration $\mathbf{a}(t)$ of the particle? **(2020 Main)**

(a) $\mathbf{v}$ is perpendicular to $\mathbf{r}$ and $\mathbf{a}$ is directed away from the origin
(b) $\mathbf{v}$ and $\mathbf{a}$ both are perpendicular to $\mathbf{r}$
(c) $\mathbf{v}$ and $\mathbf{a}$ both are parallel to $\mathbf{r}$
(d) $\mathbf{v}$ is perpendicular to $\mathbf{r}$ and $\mathbf{a}$ is directed towards the origin

33. A lift starts from rest. Its acceleration is plotted against time. When it comes to rest its height above its starting point is

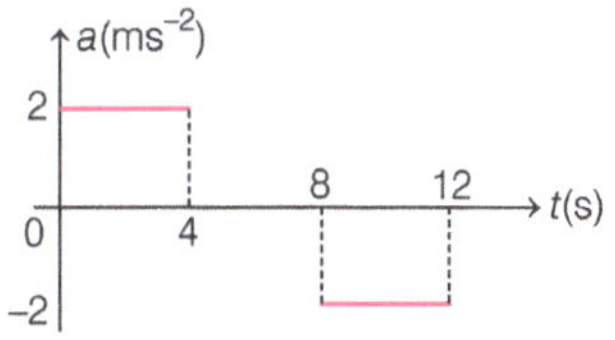

(a) 20 m
(b) 64 m
(c) 32 m
(d) 36 m

34. The figure shows velocity-time graph of a particle moving along a straight line. Identify the correct statement.

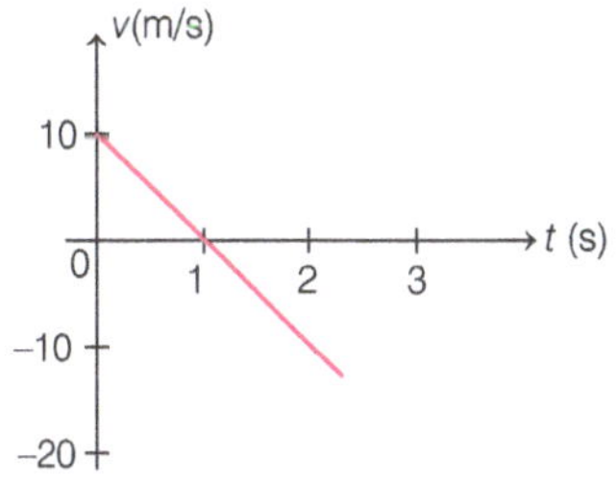

(a) The particle starts from the origin
(b) The particle crosses its initial position at $t = 2$s
(c) The average speed of the particle in the time interval, $0 \le t \le 2$s is zero
(d) All of the above

35. A car is travelling on a straight road. The maximum velocity the car can attain is 24 ms^{-1}. The maximum acceleration and deceleration it can attain are 1 ms^{-2} and 4 ms^{-2} respectively. The shortest time the car takes from rest to rest in a distance of 200 m is,

(a) 22.4 s
(b) 30 s
(c) 11.2 s
(d) 5.6 s

36. Speed-time graph of two cars A and B approaching towards each other is shown in figure. Initial distance between them is 60 m. The two cars will cross each other after time

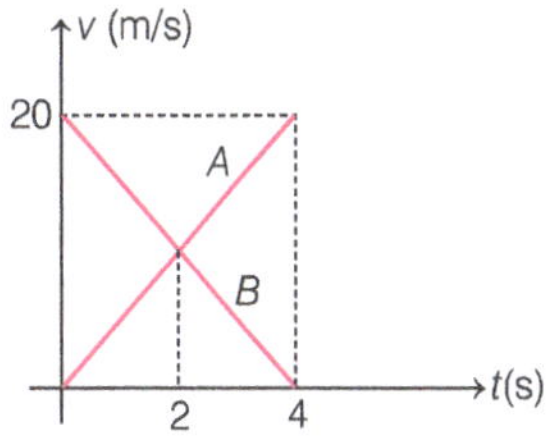

(a) 2 s
(b) 3 s
(c) 1.5 s
(d) $\sqrt{2}$ s

37. A ball is dropped vertically from a height d above the ground. It hits the ground and bounces up vertically to a height $d/2$.

Neglecting subsequent motion and air resistance, its velocity v varies with height h above the ground as **(2000 JEE)**

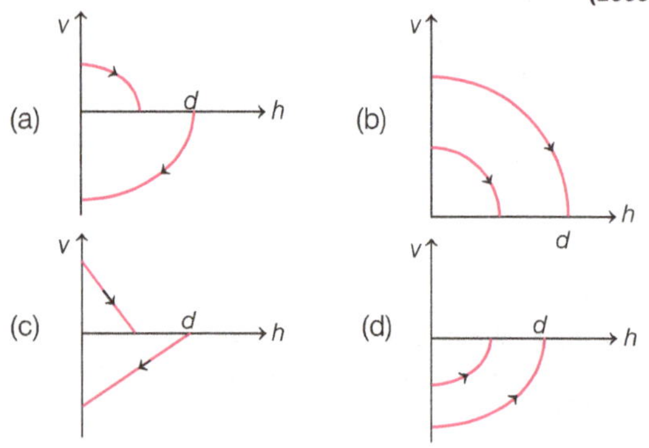

38. All the graphs below are intended to represent the same motion. One of them does it **incorrectly**. Pick it up. **(2018 Main)**

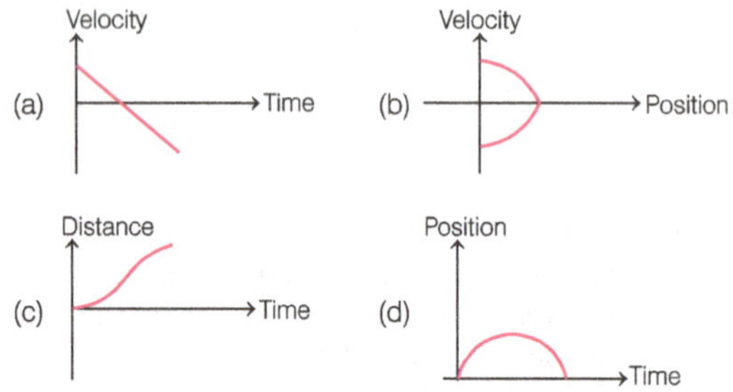

39. The velocity-time graphs of a car and a scooter are shown in the figure. (i) The difference between the distance travelled by the car and the scooter in 15 s and (ii) the time at which the car will catch up with the scooter are, respectively **(2018 Main)**

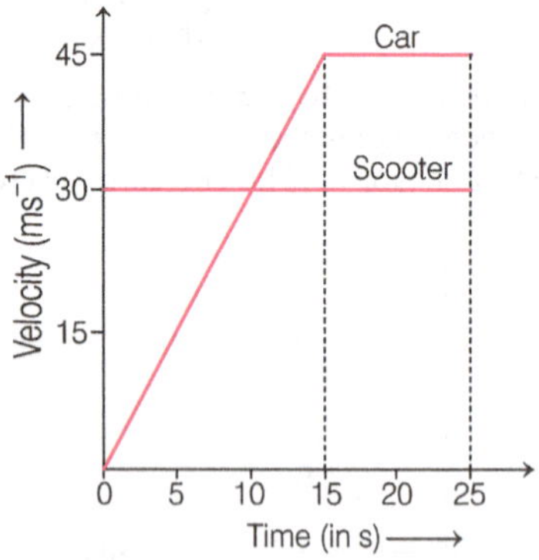

(a) 337.5 m and 25 s
(b) 225.5 m and 10 s
(c) 112.5 m and 22.5 s
(d) 11.25 m and 15 s

40. A body is at rest at $x = 0$. At $t = 0$, it starts moving in the positive x-direction with a constant acceleration. At the same instant, another body passes through $x = 0$ moving in the positive x-direction with a constant speed. The position of the first body is given by $x_1(t)$ after time t and that of the second body by $x_2(t)$ after the same time interval. Which of the following graphs correctly describes $(x_2 - x_1)$ as a function of time?

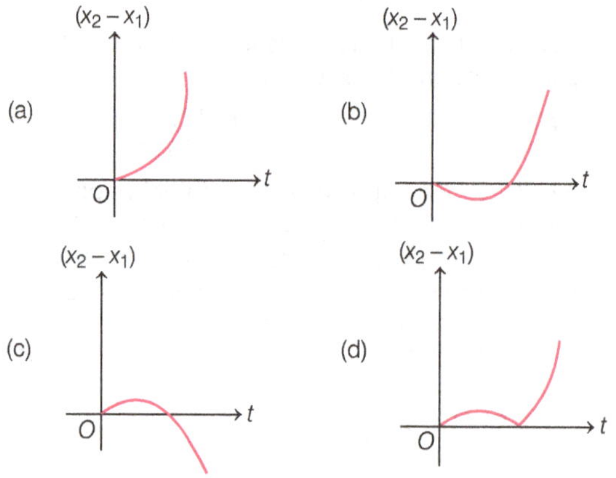

41. Two particles start simultaneously from the same point and move along two straight lines, one with uniform velocity v and other with a uniform acceleration a. If α is the angle between the lines of motion of two particles then the relative velocity will be perpendicular to the velocity of first particle at time t given by

(a) $\frac{v}{a}\sin\alpha$
(b) $\frac{v}{a}\sec\alpha$
(c) $\frac{v}{a}\tan\alpha$
(d) $\frac{v}{a}\cot\alpha$

42. Two stones are thrown up simultaneously from the edge of a cliff 240 m high with initial speed of 10 m/s and 40 m/s, respectively. Which of the following graph best represents the time variation of relative position of the second stone with respect to the first? (Assume stones do not rebound after hitting the ground and neglect air resistance. (Take $g = 10\text{m/s}^2$) **(2015 Main)**

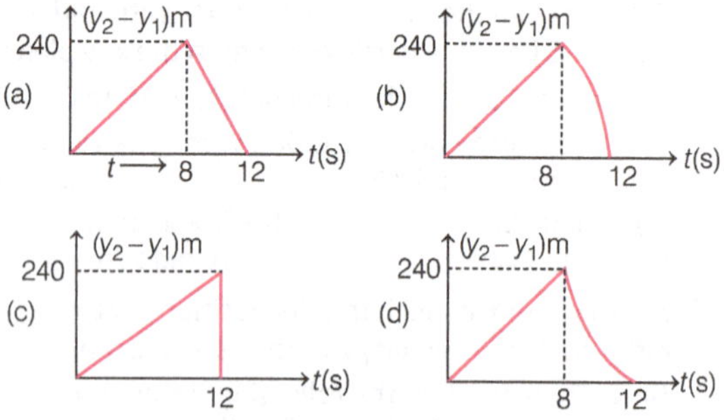

43. Two particles A and B are placed in gravity free space at (0, 0, 0) m and (30, 0, 0) m respectively. Particle A is projected with a velocity $(5\hat{i} + 10\hat{j} + 5\hat{k})\text{ms}^{-1}$, while particle B is projected with a velocity $(10\hat{i} + 10\hat{j} + 5\hat{k})\text{ ms}^{-1}$ simultaneously. Then,

(a) they will collide at (10, 20, 10) m
(b) they will collide at (10, 10, 10) m
(c) they will never collide
(d) they will collide at 2 s

44. Anoop (A) hits a ball along the ground with a speed u in a direction which makes an angle 30° with the line joining him and the fielder Babul (B). Babul runs to intercept the ball with a speed $\frac{2u}{3}$. At what angle θ should he run to intercept the ball ?

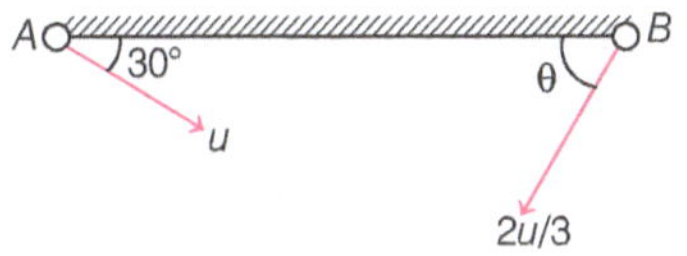

(a) $\sin^{-1}\left[\frac{\sqrt{3}}{2}\right]$ (b) $\sin^{-1}\left[\frac{2}{3}\right]$
(c) $\sin^{-1}\left[\frac{3}{4}\right]$ (d) $\sin^{-1}\left[\frac{4}{5}\right]$

45. When a man moves down an inclined plane with a constant speed $5\ \text{ms}^{-1}$ which makes an angle of 37° with the horizontal, he finds that the rain is falling vertically downward. When he moves up the same inclined plane with the same speed, he finds that the rain makes an angle $\theta = \tan^{-1}\left(\frac{7}{8}\right)$ with the horizontal.

The speed of the rain is
(a) $\sqrt{116}\ \text{ms}^{-1}$ (b) $\sqrt{32}\ \text{ms}^{-1}$
(c) $5\ \text{ms}^{-1}$ (d) $\sqrt{73}\ \text{ms}^{-1}$

46. When a car is at rest, its driver sees raindrops falling on it vertically. When driving the car with speed v, he sees that raindrops are coming at an angle 60° from the horizontal. On further increasing the speed of the car to $(1+\beta)\,v$, this angle changes to 45°. The value of β is close to **(2020 Main)**
(a) 0.73 (b) 0.41
(c) 0.37 (d) 0.50

47. Velocity of the river with respect to ground is v_0. Width of the river is d. A swimmer swims (with respect to water) perpendicular to the current with acceleration $a = 2t$ (where t is time) starting from rest from the origin O at $t = 0$. The equation of trajectory of the path followed by the swimmer is

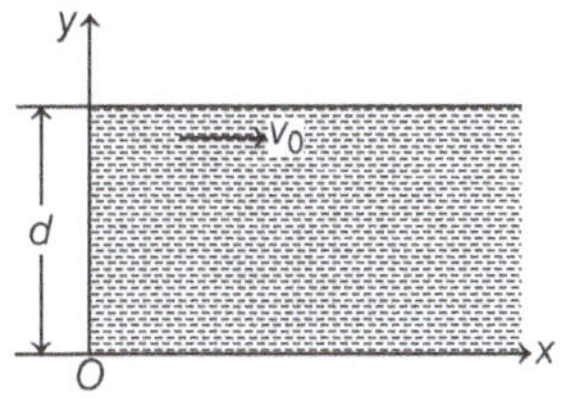

(a) $y = \frac{x^3}{3v_0^3}$ (b) $y = \frac{x^2}{2v_0^2}$
(c) $y = \frac{x}{v_0}$ (d) $y = \sqrt{\frac{x}{v_0}}$

48. Three particles start simultaneously from a point on a horizontal smooth plane. First particle moves with speed v_1 towards east, second particle moves towards north with speed v_2 and third-one moves towards north east. The velocity of the third particle, so that the three always lie on a line, is
(a) $\frac{v_1 + v_2}{\sqrt{2}}$ (b) $\sqrt{v_1 v_2}$
(c) $\frac{v_1 v_2}{v_1 + v_2}$ (d) $\sqrt{2}\left(\frac{v_1 v_2}{v_1 + v_2}\right)$

49. Two towns A and B are connected by a regular bus service with a bus leaving in either direction every T minutes. A man cycling with a speed of $20\ \text{kmh}^{-1}$ in the direction A to B notices that a bus goes past him every 18 min in the direction of his motion, and every 6 min in the opposite direction. What is the period T of the bus service and with what speed (assumed constant) do the buses ply on the road?
(a) $T = 9$ min, speed $= 40\ \text{kmh}^{-1}$
(b) $T = 10$ min, speed $= 40\ \text{kmh}^{-1}$
(c) $T = 9$ min, speed $= 20\ \text{kmh}^{-1}$
(d) $T = 10$ min, speed $= 20\ \text{kmh}^{-1}$

50. A particle P is sliding down a frictionless hemispherical bowl. It passes the point A at $t = 0$. At this instant of time, the horizontal component of its velocity is v. A bead Q of the same mass as P is ejected from A at $t = 0$ along the horizontal string AB, with the speed v. Friction between the bead and the string may be neglected. Let t_P and t_Q be the respective times taken by P and Q to reach the point B. Then **(1993 JEE)**

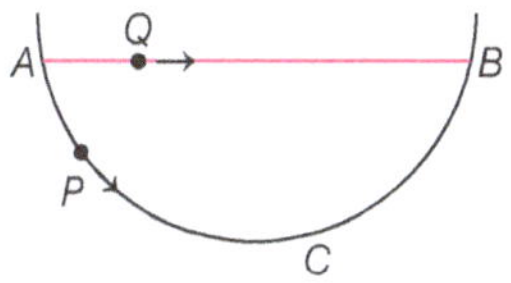

(a) $t_P < t_Q$
(b) $t_P = t_Q$
(c) $t_P > t_Q$
(d) $\frac{t_P}{t_Q} = \frac{\text{length of arc } ACB}{\text{length of chord } AB}$

51. An object moves with constant acceleration $\mathbf{a}$. Which of the following expressions are also constant ?
(a) $\frac{d|\mathbf{v}|}{dt}$ (b) $\left|\frac{d\mathbf{v}}{dt}\right|$
(c) $\frac{d(\mathbf{v}^2)}{dt}$ (d) $\frac{d\left(\frac{\mathbf{v}}{|\mathbf{v}|}\right)}{dt}$

Multiple Correct Options

52. For a moving particle, which of the following options may be correct?
(a) $|\mathbf{v}_{av}| < v_{av}$ (b) $|\mathbf{v}_{av}| > v_{av}$
(c) $\mathbf{v}_{av} = 0$ but $v_{av} \neq 0$ (d) $\mathbf{v}_{av} \neq 0$ but $v_{av} = 0$
Here, $\mathbf{v}_{av}$ is average velocity and v_{av} the average speed.

53. Choose the correct statements in the following :
(a) A body having a constant velocity still can have varying speed.

(b) A body having a constant speed can have varying velocity.
(c) A body having constant speed can have an acceleration.
(d) If velocity and acceleration are in the same direction, then distance is equal to displacement.

54. Which of the following statements is/are correct?
(a) If the velocity of a body changes, it must have some acceleration.
(b) If the speed of a body changes, it must have some acceleration.
(c) If the body has acceleration, its speed must change.
(d) If the body has acceleration, its speed may change.

55. For a particle moving along the x-axis, mark the correct statement(s).
(a) If x is positive and is increasing with the time, then average velocity of the particle is positive.
(b) If x is negative and becoming positive after some time, then the velocity of the particle is always positive.
(c) If x is negative and becoming less negative as time passes, then the average velocity of the particle is positive.
(d) If x is positive and is increasing with time, then the velocity of the particle is always positive.

56. Mark the correct statement(s).
(a) A particle can have zero displacement and non-zero average velocity.
(b) A particle can have zero displacement and non-zero velocity.
(c) A particle can have zero acceleration and non-zero velocity.
(d) A particle can have zero velocity and non-zero acceleration.

57. An object may have
(a) varying speed without having varying velocity.
(b) varying velocity without having varying speed.
(c) non-zero acceleration without having varying velocity.
(d) non-zero acceleration without having varying speed.

58. Which of the following statements are correct?
(a) A body can have zero velocity and still be accelerated
(b) A body can have a constant velocity and still have a varying speed
(c) A body can have a constant speed and still have a varying velocity
(d) The direction of the velocity of a body can change when its acceleration is constant

59. A car is moving with uniform acceleration along a straight line between two stops X and Y. Its speed at X and Y are $2\,ms^{-1}$ and $14\,ms^{-1}$, Then
(a) its speed at mid-point of XY is $10\,ms^{-1}$
(b) its speed at a point A such that $XA:AY=1:3$ is $5\,ms^{-1}$
(c) the time to go from X to the mid-point of XY is double of that to go from mid-point to Y
(d) the distance travelled in first half of the total time is half of the distance travelled in the second half of the time

60. At time $t=0$, a car moving along a straight line has a velocity of $16\,ms^{-1}$. It slows down with an acceleration of $-0.5t\,ms^{-2}$, where t is in second. Mark the correct statement (s).
(a) The direction of velocity changes at $t=8$ s
(b) The distance travelled in 4 s is approximately 58.67 m
(c) The distance travelled by the particle in 10 s is 94 m
(d) The speed of particle at $t=10$ s is $9\,ms^{-1}$

61. A body is projected from origin such that its position vector varies with time as $\mathbf{r}=[3t\hat{\mathbf{i}}+(4t-5t^2)\hat{\mathbf{j}}]$ m and t is time in second. Then,
(a) x-coordinate of particle is 2.4 m when y-coordinate is zero
(b) speed of projection is 5 m/s
(c) angle of projection with x-axis is $\tan^{-1}(4/3)$
(d) time when particle is again at x-axis is 0.8 s

62. The coordinates of a particle moving in a plane are given by $x(t)=a\cos(pt)$ and $y(t)=b\sin(pt)$ where $a,b\,(<a)$ and p are positive constants of appropriate dimensions. Then, **(1999 JEE)**
(a) the path of the particle is an ellipse
(b) the velocity and acceleration of the particle are normal to each other at $t=\pi/2p$
(c) the acceleration of the particle is always directed towards a focus
(d) the distance travelled by the particle in time interval $t=0$ to $t=\pi/2p$ is a

63. A train starts from rest at $s=0$ and is subjected to an acceleration as shown in figure. Then,

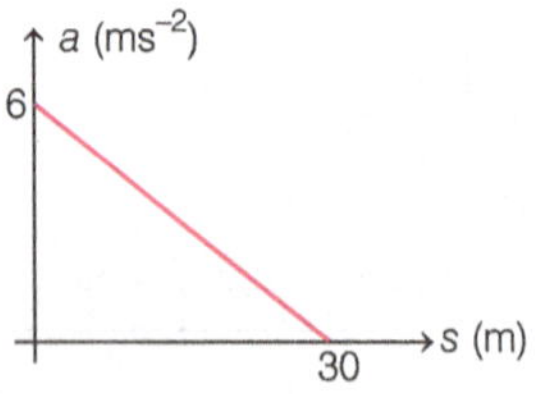

(a) velocity at the end of 10 m displacement is $20\,ms^{-1}$
(b) velocity of the train at $s=10$ m is $10\,ms^{-1}$
(c) the maximum velocity attained by train is $\sqrt{180}\,ms^{-1}$
(d) the maximum velocity attained by the train is $15\,ms^{-1}$

Passage Based Questions

Passage 1 (Q.64 to Q.66)

A situation is shown, in which two objects A and B start their motion from same point in same direction. The graph of their velocities against time is drawn. u_A and u_B are the initial velocities of A and B respectively. T is the time at which their velocities become equal after start of motion. You cannot use the data of one question while solving another question of the same set. So all the questions are independent of each other.

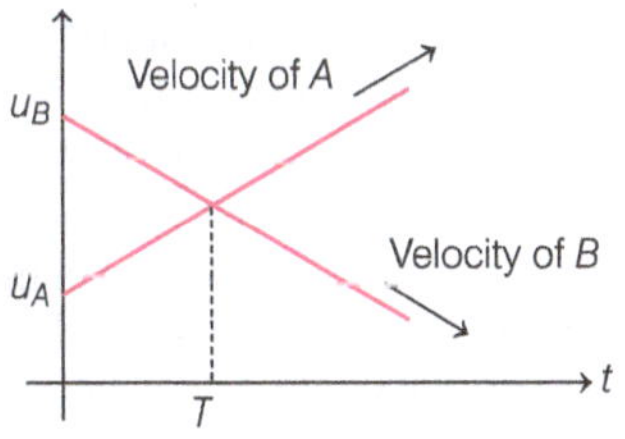

64. If the value of T is 4 s, then the time after which A will meet B is
    (a) 12 s
    (b) 6 s
    (c) 8 s
    (d) data insufficient

65. Let v_A and v_B be the velocities of the particles A and B respectively at the moment A and B meet after start of the motion. If $u_A = 5\,\text{ms}^{-1}$ and $u_B = 15\,\text{ms}^{-1}$, then the magnitude of the difference of velocities v_A and v_B is
    (a) $5\,\text{ms}^{-1}$
    (b) $10\,\text{ms}^{-1}$
    (c) $15\,\text{ms}^{-1}$
    (d) data insufficient

66. After 10 s of the start of motion of both objects A and B, find the value of velocity of A if $u_A = 6\,\text{ms}^{-1}$ and $u_B = 12\,\text{ms}^{-1}$. At time T, velocity of A is $8\,\text{ms}^{-1}$ where $T = 4\,\text{s}$.
    (a) $12\,\text{ms}^{-1}$
    (b) $10\,\text{ms}^{-1}$
    (c) $15\,\text{ms}^{-1}$
    (d) None of these

Passage 2 (Q.67 to Q.68)

If particles A and B are moving with velocities $\mathbf{v}_A$ and $\mathbf{v}_B$ respectively. The relative velocity of A with respect to B is defined as $\mathbf{v}_{AB} = \mathbf{v}_A - \mathbf{v}_B$. The driver of a car travelling Southward at 30 kmh^{-1} observes that wind appears to be coming from the West. The driver of another car travelling Southward at 50 kmh^{-1} observes that wind appears to be coming from the South-West.

67. Mark the correct option.
    (a) Wind is blowing in a direction $\tan^{-1}\left(\frac{3}{2}\right)$ South of East
    (b) Wind is blowing in a direction $\tan^{-1}\left(\frac{3}{2}\right)$ North of West
    (c) Wind is blowing towards West
    (d) Wind is blowing towards North

68. The speed of wind is
    (a) 10 k kmh^{-1}
    (b) $10\sqrt{13}\,\text{kmh}^{-1}$
    (c) $5\sqrt{3}\,\text{kmh}^{-1}$
    (d) 25 kmh^{-1}

Passage 3 (Q. 69 to Q. 72)

An elevator without a ceiling is ascending up with an acceleration of 5 ms^{-2}. A boy on the elevator shoots a ball in vertical upward direction from a height of 2 m above the floor of elevator. At this instant the elevator is moving up with a velocity of 10 ms^{-1} and floor of the elevator is at a height of 50 m from the ground. The initial speed of the ball is 15 ms^{-1} with respect to the elevator. Consider the duration for which the ball strikes the floor of elevator in answering following questions. ($g = 10\,\text{ms}^{-2}$)

69. The time in which the ball strikes the floor of elevator is given by
    (a) 2.13 s
    (b) 2.0 s
    (c) 1.0 s
    (d) 3.12 s

70. The maximum height reached by ball, as measured from the ground would be
    (a) 73.65 m
    (b) 116.25 m
    (c) 82.56 m
    (d) 63.25 m

71. Displacement of ball with respect to ground during its flight would be
    (a) 16.25 m
    (b) 8.76 m
    (c) 20.24 m
    (d) 30.56 m

72. The maximum separation between the floor of elevator and the ball during its flight would be
    (a) 12 m
    (b) 15 m
    (c) 9.5 m
    (d) 7.5 m

Subjective Questions

73. A clock has its second hand 2.0 cm long. Find the average speed and modulus of average velocity of the tip of the second hand in 15 s.

74. Can a body have
    (a) Zero instantaneous velocity and yet be accelerating?
    (b) Zero average speed but not-zero average velocity?
    (c) Negative acceleration and yet be speeding up?
    (d) Magnitude of average velocity be equal to average speed?

75. A particle moves in a straight line with velocity v for time t_0. Then, its velocity becomes $2v$ for next t_0 time. Finally, its velocity becomes $3v$ for time T. If average velocity during the complete journey was 2.5 v, then find T in terms of t_0.

76. A student starts from his house with a speed of 12 kmh^{-1} and reaches the school 3 min late. Next day he increases speed by 1 kmh^{-1} and reaches the school 3 min earlier. Find the distance between his house and school.

77. A point mass starts moving in a straight line with constant acceleration. After time t_0 the acceleration changes its sign, remaining the same in magnitude. Determine the time T from the beginning of motion in which the point mass returns to the initial position.

78. To stop a car, first you require a certain reaction time to begin braking; then the car slows under the constant braking deceleration. Suppose that the total distance moved by your car during these two phases is 56.7 m when its initial speed is 80.5 km/h and 24.4 m when its initial speed is 48.3 km/h. What are
(a) your reaction time and
(b) the magnitude of the deceleration?

79. A particle moves along a horizontal path, such that its velocity is given by $v = (3t^2 - 6t)$ m/s, where t is the time in seconds. If it is initially located at the origin O, determine the distance travelled by the particle in time interval from $t = 0$ to $t = 3.5$ s and the particle's average velocity and average speed during the same time interval.

80. At the moment $t = 0$, a particle leaves the origin and moves in the positive direction of the x-axis. Its velocity varies with time as $\mathbf{v} = \mathbf{v}_0(1 - t/\tau)$, where $\mathbf{v}_0$ is the initial velocity vector whose modulus equals $v_0 = 10.0\ \text{cms}^{-1}$ $\tau = 5.0$ s. Find : **(Irodov 1.21)**
(a) the x-coordinate of the particle at the moments of time 6.0, 10 and 20 s
(b) the moments of time when the particle is at the distance 10.0 cm from the origin
(c) the distance d covered by the particle during the first 4.0 and 8.0 s; draw the approximate plot $d(t)$.

81. A point moves rectilinearly with deceleration whose modulus depends on the velocity v of the particle as $a = \alpha\sqrt{v}$, where α is a positive constant. At the initial moment the velocity of the point is equal to v_0. What distance will it traverse before it stops? What time will it take to cover the distance? **(Irodov 1.23)**

82. A motorboat of mass m moves along a lake with velocity v_0. At the moment $t = 0$, the engine of the boat is shutdown. Assuming the resistance of water to be proportional to the velocity of the boat $\mathbf{F} = -r\mathbf{v}$, find:
(a) how long the motorboat moved with the shutdown engine
(b) the velocity of the motorboat as a function of the distance covered with the shutdown engine, as well as the total distance covered till the complete stop
(c) the mean velocity of the motorboat over the time interval (beginning with the moment $t = 0$), during which its velocity decreases η times. **(Irodov 1.100)**

83. Having gone through a plank of thickness h, a bullet changed its velocity from v_0 to v. Find the time of motion of the bullet in the plank, assuming the resistance force to be proportional to the square of the velocity. **(Irodov 1.101)**

84. A body of mass m is thrown straight up with velocity v_0. Find the velocity v' with which the body comes down if the air drag equals kv^2, where k is a constant and v is the velocity of the body. **(Irodov 1.104)**

85. Velocity-time graph of a particle moving in a straight line is shown in figure. At time $t = 0$, displacement of the particle from mean position is 10 m. Find

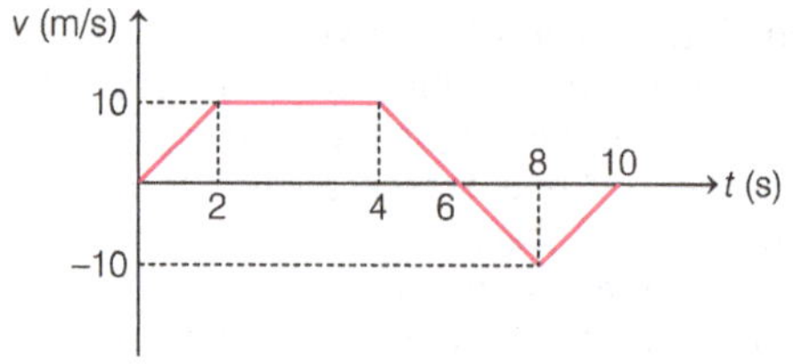

(a) acceleration of particle at $t = 1$ s, 3 s and 9 s.
(b) position of particle from mean position at $t = 10$ s.
(c) write down s-t equation for time interval
(i) $0 \le t \le 2$ s, (ii) $4\text{ s} \le t \le 8$ s

86. The acceleration-displacement graph of a particle moving in a straight line is as shown in figure. Initial velocity of particle is zero. Find the velocity of the particle when displacement of the particle is $s = 12$ m.

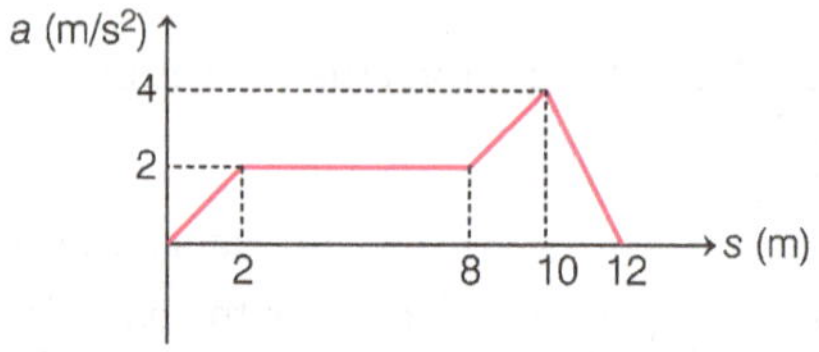

87. A train stopping at two stations 4 km apart takes 4 min on the journey from one station to the other. Assuming that it first accelerates with a uniform acceleration x and then with uniform retardation y. Prove that $\frac{1}{x} + \frac{1}{y} = 2$.

88. A point moves rectilinearly in one direction. Figure shows the distance s traversed by the point as a function of the time t. Using the plot find :

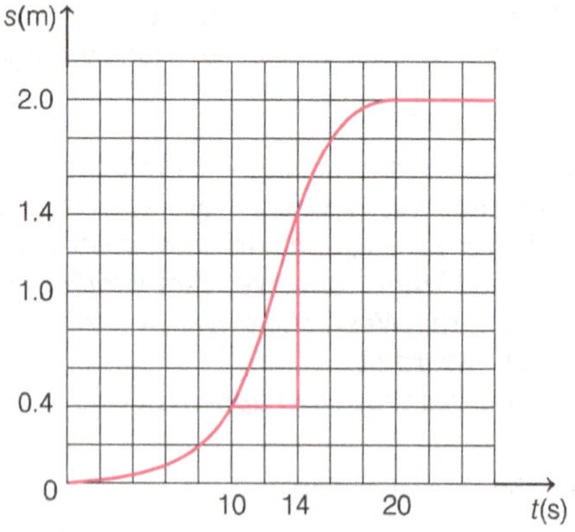

(a) the average velocity of the point during the time of motion
(b) the maximum velocity
(c) the time moment t_0 at which the instantaneous velocity is equal to the mean velocity averaged over the first t_0 seconds. **(Irodov 1.4)**

89. A man wishes to cross a river of width 120 m by a motorboat. His rowing speed in still water is 3 m/s and his maximum walking speed is 1 m/s. The river flows with velocity of 4 m/s.
   (a) Find the path which he should take to get to the point directly opposite to his starting point in the shortest time.
   (b) Also, find the time which he takes to reach his destination.

90. A motorboat going downstream overcame a raft at a point A. After 1 h it turned back and after some time passed the raft at a distance $l = 6.0$ km from the point A. Find the flow velocity assuming the power of the engine to be constant. **(Irodov 1.1)**

91. Two boats A and B move away from a buoy anchored at the middle of a river along the mutually perpendicular straight lines; the boat A along the river, and the boat B across the river. Having moved off an equal distance from the buoy the boats returned. Find the ratio of times of motion of boats $\frac{\tau_A}{\tau_B}$ if the velocity of each boat with respect to water is $\eta = 1.2$ times greater than the stream velocity. **(Irodov 1.8)**

92. From point A located on a highway one has to get by car as soon as possible to point B located in the field at a distance l from the highway. It is known that the car moves in the field η times slower than on the highway. At what distance from point D one must turn off the highway? **(Irodov 1.17)**

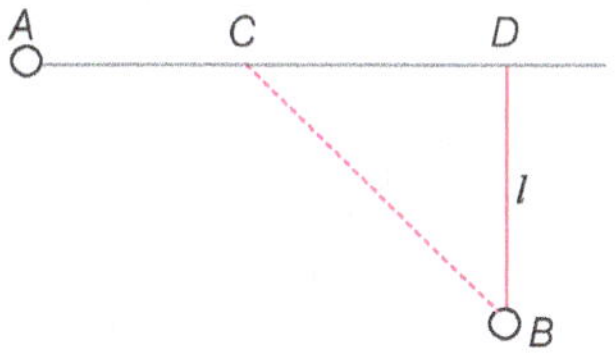

Integer Type Questions

93. A truck is moving at a speed of 72 kmh^{-1} on a straight road. The driver can produce deceleration of 2 ms^{-2} by applying brakes. The stopping distance of truck is $13x$ m, if the reaction time of the driver is 0.2 s. Find the value of x.

94. A train is moving on straight track with velocity $v_0 = 13.5$ ms^{-1}. To stop the train at a particular station, driver apply brakes at $t = 0$, which produces a retardation proportional to velocity of the train. The speed of train reduces 50% in the first $t_0 = 4 \ln 2$ s. Find velocity of train (in ms^{-1}) at $t = 4$ s. (Given, $e = 2.7$)

95. A ball is released from rest from top of a tower. The retardation due to air resistance is bv, where b is 10 per second and velocity v is in ms^{-1}. The velocity of ball at $t = \frac{1}{10}$ s is $\frac{n}{27}$ ms^{-1}. Find the value of n. (Given, $e = 2.7$)

96. A ball is projected upward with initial velocity v_0. The retardation due to air resistance is bv, where b is positive constant. The time taken by the particle to reach the maximum height is t_0.
If $\frac{bv_0}{g} = 2$, then find the value of e^{bt_0}.

97. A swimmer wants to cross a river from point A to point B. Line AB makes an angle of 30° with the flow of river. Magnitude of velocity of the swimmer is same as that of the river. The angle θ with the line AB should be .... degrees, so that swimmer reaches B. **(2021 Main)**

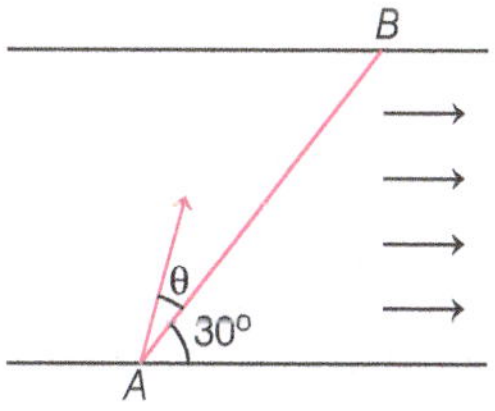

98. A student used to go to his coaching institute every evening and he takes 20 min. Once on his way, he realised that he had forgotten his book at home. He knew that if he continued walking to the coaching at the same speed, he would be there 8 min before the right time, so he went back home for the book and arrived the coaching 10 min late. If he had walked all the way with his usual speed, the fraction of the way to coaching institute had he covered at that moment when he turned back is $\frac{x}{20}$. Find the value of x.

Level 2

Single Correct Option

99. A car is travelling on a road. The maximum velocity the car can attain is 24 ms^{-1} and the maximum deceleration is 4 ms^{-2}. If car starts from rest and comes to rest after travelling 1032 m in the shortest time of 56s, the maximum acceleration that the car can attain is
(a) 6 ms^{-2}
(b) 1.2 ms^{-2}
(c) 12 ms^{-2}
(d) 3.6 ms^{-2}

100. A particle of mass m is initially situated at point P inside a hemispherical surface of radius r as shown in the figure. A horizontal acceleration of magnitude a_0 is suddenly produced on the particle in the horizontal direction. If gravitational acceleration is neglected, then time taken by the particle to touch the sphere again is

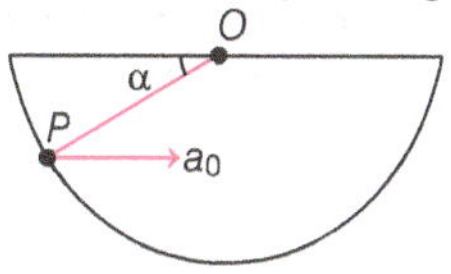

(a) $\sqrt{\frac{4r \sin \alpha}{a_0}}$
(b) $\sqrt{\frac{4r \tan \alpha}{a_0}}$
(c) $\sqrt{\frac{4r \cos \alpha}{a_0}}$
(d) None of these

Multiple Correct Options

101. The speed of a train increases at a constant rate α from zero to v and then remains constant for an interval and finally decreases to zero at a constant rate β. The total distance travelled by the train is l. The time taken to complete the jouney is t. Then,

(a) $t = \frac{l(\alpha+\beta)}{\alpha\beta}$ (b) $t = \frac{l}{v} + \frac{v}{2}\left(\frac{1}{\alpha}+\frac{1}{\beta}\right)$

(c) t is minimum when $v = \sqrt{\frac{2l\alpha\beta}{(\alpha-\beta)}}$

(d) t is minimum when $v = \sqrt{\frac{2l\alpha\beta}{(\alpha+\beta)}}$

102. Starting from rest, a particle is first accelerated for time t_1 with constant acceleration a_1 and then stops in time t_2 with constant retardation a_2. Let v_1 be the average velocity in this case and s_1 the total displacement. In the second case it is accelerating for the same time t_1 with constant acceleration $2a_1$ and comes to rest with constant retardation a_2 in time t_3. If v_2 is the average velocity in this case and s_2 the total displacement, then

(a) $v_2 = 2v_1$ (b) $2v_1 < v_2 < 4v_1$
(c) $s_2 = 2s_1$ (d) $2s_1 < s_2 < 4s_1$

103. A particle of mass m moves on the x-axis as follows : it starts from rest at $t = 0$ from the point $x = 0$ and comes to rest at $t = 1$ at the point $x = 1$. No other information is available about its motion at intermediate times ($0 < t < 1$). If α denotes the instantaneous acceleration of the particle, then **(1993 JEE)**

(a) α cannot remain positive for all t in the interval $0 \le t \le 1$
(b) $|\alpha|$ cannot exceed 2 at any point in its path
(c) $|\alpha|$ must be ≥ 4 at some point or points in its path
(d) α must change sign during the motion, but no other assertion can be made with the information given

Match the Columns

104. When a ball is dropped from a certain height, it experiences a retardation due to air resistance which is equal to $(5\,v)\,\text{ms}^{-2}$, where v is instantaneous velocity. A student drops three balls A, B and C one by one. He observes that ball A moves with constant velocity, but ball B moves with increasing speed for some time and then moves with constant velocity. He also observes that ball C moves with decreasing speed and after some time it moves with constant velocity. In Column I, some results are given related to Column II. Match the following columns.

	Column I		Column II
(a)	Ball A	(p)	Initial velocity is $2\ \text{ms}^{-1}$
(b)	Ball B	(q)	Initial velocity may be zero
(c)	Ball C	(r)	Initial velocity is greater than $2\ \text{ms}^{-1}$
		(s)	Final velocity is $2\ \text{ms}^{-1}$

Subjective Questions

105. A particle travels in a straight line, such that for a short time $2\text{ s} \le t \le 6\text{ s}$, its motion is described by $v = (4/a)\ \text{m/s}$, where a is in m/s^2. If $v = 6\ \text{m/s}$ when $t = 2\text{s}$, determine the particle's acceleration when $t = 3\text{s}$.

106. A particle starts from rest to move along x-axis. The acceleration of the particle is $a = (t - x)\ \text{ms}^{-2}$. During motion, maximum acceleration of the particle is $a_0 = 2\ \text{ms}^{-2}$. Find the velocity (in ms^{-1}) of the particle at $t = \frac{\pi}{3}$ s.

107. A radius vector of a particle varies with time t as $\mathbf{r} = \mathbf{w}t\,(1 - \alpha t)$, where $\mathbf{w}$ is a constant vector and α is a positive factor. Find :
(a) the velocity $\mathbf{v}$ and the acceleration $\mathbf{a}$ of the particle as functions of time;
(b) the time interval Δt taken by the particle to return to the initial points, and the distance d covered during that time. **(Irodov 1.20)**

108. The speed of a particle moving in a plane is equal to the magnitude of its instantaneous velocity, $v = |\mathbf{v}| = \sqrt{v_x^2 + v_y^2}$.
(a) Show that the rate of change of the speed is $\frac{dv}{dt} = (v_x a_x + v_y a_y)/\sqrt{v_x^2 + v_y^2}$.
(b) Show that the rate of change of speed can be expressed as $\frac{dv}{dt} = \frac{\mathbf{v}\cdot\mathbf{a}}{v}$, and use this result to explain why $\frac{dv}{dt}$ is equal to a_t the component of $\mathbf{a}$ that is parallel to $\mathbf{v}$.

109. A particle of mass m moves in a certain plane P due to a force $\mathbf{F}$ whose magnitude is constant and whose vector rotates in that plane with a constant angular velocity ω. Assuming the particle to be stationary at the moment $t = 0$, find:
(a) its velocity as a function of time
(b) the distance covered by the particle between two successive stops, and the mean velocity over this time. **(Irodov 1.105)**

110. A particle moves in a straight line with constant acceleration a. The displacements of particle from origin in times t_1, t_2 and t_3 are s_1, s_2 and s_3 respectively. If times are in AP with common difference d and displacements are in GP, then prove that $a = \frac{(\sqrt{s_1} - \sqrt{s_3})^2}{d^2}$.

111. A car is to be hoisted by elevator to the fourth floor of a parking garage, which is 14 m above the ground. If the elevator can have maximum acceleration of $0.2\ \text{m/s}^2$ and maximum deceleration of $0.1\ \text{m/s}^2$ and can reach a maximum speed of 2.5 m/s, determine the shortest time to make the lift, starting from rest and ending at rest.

112. The v-s graph describing the motion of a motorcycle is shown in figure. Construct the a-s graph of the motion and determine the time needed for the motorcycle to reach the position $s = 120$ m. Given $\ln 5 = 1.6$.

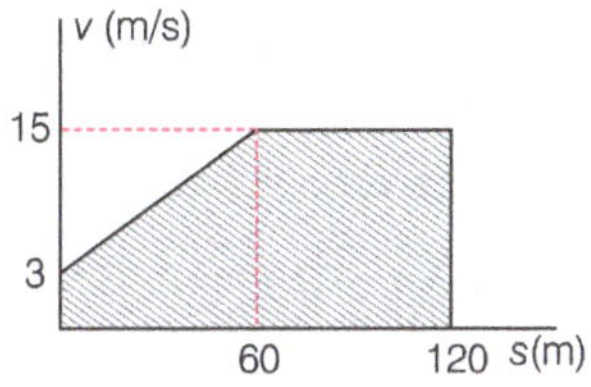

113. A car starts moving rectilinearly, first with acceleration $a = 5.0 \text{ ms}^{-2}$ (the initial velocity is equal to zero), then uniformly, and finally, decelerating at the same rate a and finally comes to a stop. The total time of motion equals $\tau = 25$ s. The average velocity during that time is equal to $\langle v \rangle = 72 \text{ km h}^{-1}$. How long does the car move uniformly? **(Irodov 1.3)**

114. On a frictionless horizontal surface, assumed to be the x-y plane, a small trolley A is moving along a straight line parallel to the y-axis (see figure) with a constant velocity of $(\sqrt{3}-1)$ m/s.

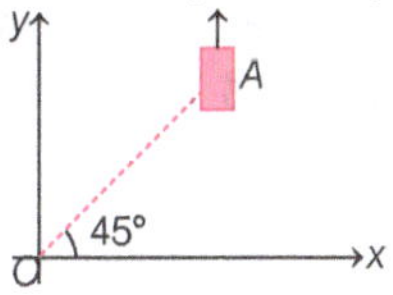

At a particular instant when the line OA makes an angle of 45° with the x-axis, a ball is thrown along the surface from the origin O. Its velocity makes an angle ϕ with the x-axis and it hits the trolley. **(2002 JEE)**

(a) The motion of the ball is observed from the frame of the trolley. Calculate the angle θ made by the velocity vector of the ball with the x-axis in this frame.

(b) Find the speed of the ball with respect to the surface, if $\phi = 4\theta/3$.

115. The current velocity of river grows in proportion to the distance from its bank and reaches the maximum value v_0 in the middle. Near the banks the velocity is zero. A boat is moving along the river in such a manner that the boatman rows his boat always perpendicular to the current. The speed of the boat in still water is u. Find the distance through which the boat crossing the river will be carried away by the current. The width of the river is c. Also determine the trajectory of the boat.

116. A river of width a with straight parallel banks flows due north with speed u. The points O and A are on opposite banks and A is due east of O. Coordinate axes OX and OY are taken in the east and north directions respectively. A boat, whose speed is v relative to water, starts from O and crosses the river. If the boat is steered due east and u varies with x as :

$u = x(a-x)\dfrac{v}{a^2}$. Find

(a) equation of trajectory of the boat,

(b) time taken to cross the river,

(c) absolute velocity of boatman when he reaches the opposite bank,

(d) the displacement of boatman when he reaches the opposite bank from the initial position.

117. A river of width ω is flowing with a uniform velocity v. A boat starts moving from point P also with velocity v relative to the river. The direction of resultant velocity is always perpendicular to the line joining boat and the fixed point R. Point Q is on the opposite side of the river. P, Q and R are in a straight line. If $PQ = QR = \omega$, find (a) the trajectory of the boat, (b) the drifting of the boat and (c) the time taken by the boat to cross the river.

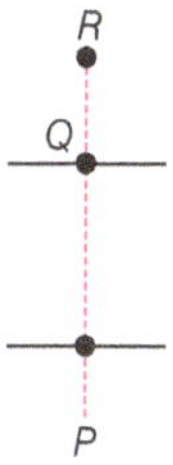

118. A man with some passengers in his boat, starts perpendicular to flow of river 200 m wide and flowing with 2 m/s. Speed of boat in still water is 4 m/s. When he reaches half the width of river the passengers asked him that they want to reach the just opposite end from where they have started.

(a) Find the direction along which he must row to reach the required end.

(b) How many times more time, it would take to that if he would have denied the passengers?

119. A child in danger of drowning in a river is being carried downstream by a current that flows uniformly at a speed of 2.5 km/h. The child is 0.6 km from shore and 0.8 km upstream of a boat landing when a rescue boat sets out. If the boat proceeds at its maximum speed of 20 km/h with respect to the water, what angle does the boat velocity v make with the shore? How long will the boat take to reach the child?

120. A launch plies between two points A and B on the opposite banks of a river always following the line AB. The distance S between points A and B is 1200 m. The velocity of the river current $v = 1.9$ m/s is constant over the entire width of the river. The line AB makes an angle $\alpha = 60°$ with the direction of the current. With what velocity u and at what angle β to the line AB should the launch move to cover the distance AB and back in a time $t = 5$ min? The angle β remains the same during the passage from A to B and from B to A.

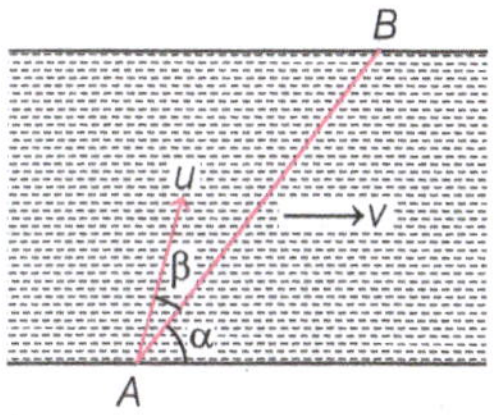

121. Point A moves uniformly with velocity v so that the vector v is continuously "aimed" at point B which in its turn moves rectilinearly and uniformly with velocity $u < v$. At the initial moment of time $v \perp u$ and the points are separated by a distance l. How soon will the points converge? **(Irodov 1.13)**

Integer Type Questions

122. Airplanes A and B are flying with constant velocity in the same vertical plane at angles 30° and 60° with respect to the horizontal respectively as shown in figure. The speed of A is $100\sqrt{3} \text{ ms}^{-1}$. At time $t = 0$ s, an observer in A finds B at a distance of 500 m. This observer sees B moving with a constant velocity perpendicular to the line of motion of A. If at $t = t_0$, A just escapes being hit by B, t_0 in seconds is .... **(2014 JEE**

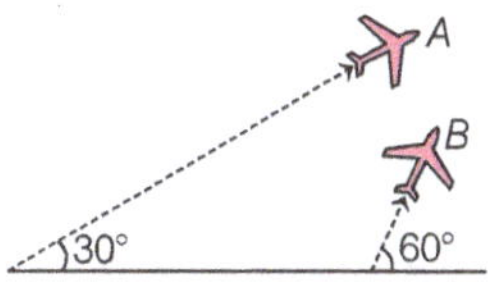

123. A vertical wind screen of a vehicle is made of two parts, the upper one is 25 cm vertically long and covers the top 5 cm of lower fixed part. The upper part is hinged at upper end, so that it can be opened outward. The car is running on the horizontal road at $20\sqrt{3}$ kmh^{-1}. The rain is falling vertically with speed of 20 kmh^{-1}. The maximum angle through which the upper part can be opened outward is θ. If $\sin\left(\frac{\pi}{3}+\theta\right)=\frac{\sqrt{12}}{n}$, then find the value of n.

Answers

Level 1

1. $5\sqrt{5}$ m/s, 15 m/s, $100\sqrt{5}$ m **2.** 18 m/s **3.** (a or b) **4.** (d) **5.** (b) **6.** (b) **7.** (d) **8.** (b)

9. (a) **10.** (b) **11.** (b) **12.** (c) **13.** (c) **14.** (b) **15.** (a) **16.** (a) **17.** (a) **18.** (b)

19. (c) **20.** (c) **21.** (b) **22.** (d) **23.** (d) **24.** (b) **25.** (b) **26.** (d) **27.** (a) **28.** (b)

29. (a) **30.** (b) **31.** (b) **32.** (d) **33.** (b) **34.** (b) **35.** (a) **36.** (b) **37.** (a) **38.** (c)

39. (c) **40.** (c) **41.** (b) **42.** (b) **43.** (c) **44.** (c) **45.** (b) **46.** (a) **47.** (a) **48.** (d)

49. (a) **50.** (a) **51.** (b) **52.** (a,c) **53.** (b,c,d) **54.** (a,b,d) **55.** (a,c,d) **56.** (b,c,d) **57.** (b,d) **58.** (a,c,d)

59. (a,c) **60.** (a,b,c,d) **61.** (a,b,c,d) **62.** (a,b,c) **63.** (b,c) **64.** (c) **65.** (b) **66.** (d) **67.** (a) **68.** (b)

69. (a) **70.** (c) **71.** (d) **72.** (c) **73.** $\frac{\pi}{15}$ cm/s, $\frac{2\sqrt{2}}{15}$ cm/s **74.** (a) Yes (b) No (c) Yes (d) Yes **75.** $4\,t_0$

76. 15.6 km **77.** $(3.414)\,t_0$ **78.** 0.74 s, 6.2 m/s^2 **79.** 14.125 m, 1.75 m/s, 4.03 m/s

80. (a) 0.24, 0 and – 2.0 m (b) 1.1, 9 and 11s (c) 24 cm, 34 cm **81.** $s=\left(\frac{2}{3\alpha}\right)v_0^{3/2},\ t=\frac{2\sqrt{v_0}}{\alpha}$

82. (a) Motorboat will continue to move until $t\to\infty$ (b) $v=v_0-\frac{r}{m}s,\ s_{total}=\frac{mv_0}{r}$ (c) $\langle v\rangle=\frac{v_0}{\ln\eta}\left(1-\frac{1}{\eta}\right)$ **83.** $t=\frac{h(v_0-v)}{v_0 v\ln\left(\frac{v_0}{v}\right)}$

84. $v'=\frac{v_0}{\sqrt{\frac{1+kv_0^2}{mg}}}$ **85.** (a) 5 ms^{-2}, zero, 5 ms^{-2} (b) $s=30$ m (c) (i) $s=10+2.5t^2$ (ii) $s=40+10(t-4)-2.5(t-4)^2$

86. $4\sqrt{3}$ m/s **88.** (a) $v_{av}=10$ cm/s (b) $v_{max}=25$ cm/s (c) $t_0=16$ s

89. (a) $90°+\sin^{-1}(315)$ from river current (b) 2 min 40 s **90.** $v_r=3$ km/h **91.** $\frac{\tau_A}{\tau_B}\approx 1.8$

92. $x=\frac{l}{\sqrt{\eta^2-1}}$ **93.** (8) **94.** (5) **95.** (17) **96.** (3) **97.** (30) **98.** (9)

Level 2

99. (b) **100.** (c) **101.** (b,d) **102.** (a,d) **103.** (a,c) **104.** (a- p,s ; b- q, s; c-r, s) **105.** 0.603 m/s^2

106. (1 m/s) **107.** $\mathbf{v}=\mathbf{w}(1-2\alpha t)$, $\mathbf{a}=-2\alpha\mathbf{w}$ = constant (b) $\Delta t=\frac{1}{\alpha}$, $d=\frac{w}{2\alpha}$ **108.** (See the hints)

109. (a) $|\mathbf{v}|=\frac{2F}{m\omega}\left|\sin\frac{\omega t}{2}\right|$ (b) $d=\frac{8F}{m\omega^2},\ \frac{4F}{\pi m\omega}$ **111.** 20.5 s **112.** 12.0 s, for the graph; see the hints **113.** (15 s)

114. (a) 45° (b) 2 m/s

115. $\frac{cv_0}{2u},\ y^2=\frac{ucx}{v_0}$ **116.** (a) $y=\frac{x^2}{2a}-\frac{x^3}{3a^2}$ (b) $\frac{a}{v}$ (c) v (due east) (d) $a\hat{\mathbf{i}}+\frac{a\hat{\mathbf{j}}}{6}$

117. (a) Circle (b) $\sqrt{3}\omega$ (c) $1.317\,\frac{\omega}{v}$

118. (a) At an angle $(90°+2\theta)$ from river current (upstream). Here, $\theta=\tan^{-1}(1/2)$ (b) 4/3

119. 37°, 3 min **120.** $u=8\ ms^{-1}$, $\beta=12°$ **121.** $t=\frac{vl}{v^2-u^2}$ **122.** (5) **123.** (5)

See hints on pages 266-288.

Hints & SOLUTIONS

Module 1.

1. The body will speed up if the angle between velocity and acceleration is acute.

2. For increasing speed, velocity and acceleration both should be in same direction, either both towards right or both towards left.

3. If $\mathbf{v} \cdot \mathbf{a} = 0$, speed is constant because angle between $\mathbf{v}$ and $\mathbf{a}$ in this case is 90°.

   If $\mathbf{v} \cdot \mathbf{a} =$ positive then speed is increasing because angle between $\mathbf{v}$ and $\mathbf{a}$ in this case is acute.

   If $\mathbf{v} \cdot \mathbf{a} =$ negative then speed is decreasing because angle between $\mathbf{v}$ and $\mathbf{a}$ in this case is obtuse.

5. (a) $\mathbf{v} \cdot \mathbf{a} = 6 - 4 - 4 = -2\,\text{m}^2/\text{s}^3$

   (b) Since dot product is negative. So angle between $\mathbf{v}$ and $\mathbf{a}$ is obtuse.

   (c) As angle between $\mathbf{v}$ and $\mathbf{a}$ at this instant is obtuse, speed is decreasing.

6. $\mathbf{v}$ and $\mathbf{a}$ both are constant vectors. Further, these two vectors are antiparallel. So, it is one dimensional.

7. $\mathbf{a}$ is function of time and $\mathbf{v}$ and $\mathbf{a}$ are neither parallel nor antiparallel. So, motion is two dimensional with non-uniform acceleration.

9. Velocity and acceleration both are negative (because both are downwards). So, speed will be increasing. Otherwise, also motion may be called accelerated if there is acceleration. It hardly matters whether it is positive or negative.

11. From 0-2s, velocity is negative and acceleration is positive. From 2s onwards, velocity and acceleration both are positive.

Module 2.

1. Displacement $= AC = 2R$

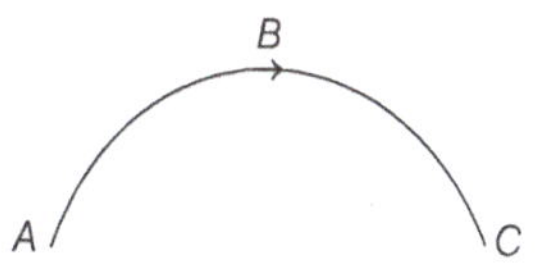

   Distance $= ABC = \pi R$

2. $v_{av} = \dfrac{\text{Displacement}}{\text{Time}} = \dfrac{\sqrt{2}R}{t}$

3. $\mathbf{r} = (a\cos\omega t)\,\hat{\mathbf{i}} + (a\sin\omega t)\,\hat{\mathbf{j}}$

   $$\therefore \quad \mathbf{v} = \frac{d\mathbf{r}}{dt} = (-a\omega\sin\omega t)\,\hat{\mathbf{i}} + (a\omega\cos\omega t)\,\hat{\mathbf{j}}$$

   $$\mathbf{r} \cdot \mathbf{v} = 0$$

   $$\therefore \quad \mathbf{r} \perp \mathbf{v}$$

4. $$a_{av} = \left|\frac{\Delta \mathbf{v}}{\Delta t}\right| = \frac{|\mathbf{v}_f - \mathbf{v}_i|}{t}$$

   $$= \frac{\sqrt{v^2 + v^2 - 2v \cdot v \cdot \cos\theta}}{t}$$

   $$= \frac{2v(\sin\theta/2)}{t}$$

5. $$|\langle \mathbf{v} \rangle| = \frac{|\mathbf{r}_f - \mathbf{r}_i|}{\Delta t}$$

   $$= \frac{2R\cos\left[\dfrac{\pi-\theta}{2}\right]}{\dfrac{2\pi R}{3v}} = \frac{2R\cos\left[\dfrac{180°-120°}{2}\right]}{2\pi R}$$

   $$= 3\cos 30°$$

   $$= 1.5\sqrt{3}\ \text{m/s}$$

6. $d = R\theta$

   $$\Rightarrow \quad 60 = R\left(\frac{3\pi}{4}\right)$$

   $$\Rightarrow \quad R = \frac{60 \times 4}{3\pi} = \frac{80}{\pi}\ \text{m}$$

   $$\therefore \text{Displacement} = \sqrt{R^2 + R^2 - 2R^2\cos 135°}$$

   $$= \sqrt{2R^2 - 2R^2(-0.7)}$$

   $$= \sqrt{3.4R^2} = \sqrt{3.4\left(\frac{80}{\pi}\right)^2} \approx 47\ \text{m}$$

7. Since average acceleration = change in velocity/time, so average acceleration is in the direction of change in velocity. Also if the initial velocity is zero, then the final velocity and change in velocity will be in the same direction.

8. Distance may be greater than or equal to magnitude of displacement.

9. Constant velocity means constant speed in same direction. Further if any physical quantity has a constant value (here, the velocity) then its average value in any interval of time is equal to that constant value.

10. (a) On a curvilinear path (a path which is not a straight line), even if speed is constant, velocity will change due to change in direction.

    $$\therefore \quad \mathbf{a} \neq 0$$

    (b) (i) On a curved path velocity will definitely change (at least due to change in direction).

    $$\therefore \quad \mathbf{a} \neq 0$$

(ii) In projectile motion, path is curved, but acceleration is constant $(=g)$.

(iii) Variable acceleration on curved path is definitely possible.

11. (a) $\left|\frac{dv}{dt}\right|$ is the magnitude of total acceleration.

While $\frac{d|v|}{dt}$ represents the time rate of change of speed (called the tangential acceleration, a component of total acceleration) as $|\mathbf{v}| = v$.

(b) These two are equal in case of one dimensional motion.

12. (a) Yes, if its velocity slows down towards North.

(b) Yes, if velocity and acceleration are in opposite directions.

(c) Yes, if an accelerating object starts decelerating.

(d) No, because average speed can also be greater than average velocity.

(e) No, change in velocity takes place in the direction of acceleration.

13. $v_{av} = \frac{s_{total}}{t_{total}} = \frac{200 + 200 + \frac{2\pi \times 200}{4}}{238}$

$$= \frac{400 + 100\pi}{238} = \frac{714}{238} = 3 \text{ m/s}$$

Module 3.

1. Given, first distance travelled,

$$d_1 = 4 \text{km}$$

Second distance travelled, $d_2 = 4 \text{km}$

$$v_{av} = \frac{\text{Distance}}{\text{Time}} = \frac{d_1 + d_2}{\frac{d_1}{v_1} + \frac{d_2}{v_2}}$$

Here, v_1 and v_2 are the velocities during first and second distance travelled.

$$\Rightarrow \quad v_{av} = \frac{4+4}{\frac{4}{3} + \frac{4}{5}}$$

$$\Rightarrow \quad v_{av} = \frac{15}{4} = 3.75 \text{ km/h}$$

2. $v_{av} = \frac{d}{t} = \frac{d_1 + d_2}{t_1 + t_2}$

$$\Rightarrow \quad 21 = \frac{(18)(11)(60) + (42)(v)(60)}{(60)(60)}$$

$$\Rightarrow \quad v = 25.3 \text{ m/s} \qquad \textbf{Ans.}$$

3. $t_1 = \frac{d}{v_1}, t_2 = \frac{d}{v_2}, \ldots, t_n = \frac{d}{v_n}$

Average speed = (Total distance)/(Total time)

$$\Rightarrow \quad v = \frac{nd}{t_1 + t_2 + \ldots + t_n}$$

$$= \frac{nd}{\frac{d}{v_1} + \frac{d}{v_2} + \ldots + \frac{d}{v_n}} = \frac{n}{\frac{1}{v_1} + \frac{1}{v_2} + \ldots + \frac{1}{v_n}}$$

Taking reciprocal, we get

$$\frac{1}{v} = \frac{1}{n}\left(\frac{1}{v_1} + \frac{1}{v_2} + \ldots + \frac{1}{v_n}\right)$$

4. Distance covered $= d = v_{av} \times$ time

For first second : $d_1 = 5 \times 1 = 5 \text{m}$

For second second : $d_2 = 10 \times 1 = 10 \text{m}$

For third second : $d_3 = 15 \times 1 = 15 \text{m}$

Total distance travelled

$$d = d_1 + d_2 + d_3 = 5 + 10 + 15$$
$$= 30 \text{m}$$

5. Average speed $= \frac{\text{Total distance travelled}}{\text{Total time taken}}$

$$= \frac{d}{\frac{2d/5}{v_1} + \frac{3d/5}{v_2}} = \frac{5v_1v_2}{3v_1 + 2v_2}$$

6. Average speed $= \frac{\text{Total distance}}{\text{Total time}} = \frac{d}{t_1 + t_2}$

$$= \frac{d}{\frac{d/3}{v_1} + \frac{2d/3}{v_2}} = \frac{1}{\frac{1}{3 \times 20} + \frac{2}{3 \times 60}}$$

$$= 36 \text{ km/hr}$$

7. A man walks from his home to market with a speed of 5 km/h.

Distance = 2.5 km and time $= \frac{d}{v} = \frac{2.5}{5} = \frac{1}{2}$ hr

and the distance travelled in 10 min with a speed of 7.5 km/hr during his return journey is :

$$\text{Distance} = 7.5 \times \frac{10}{60} = 1.25 \text{ km}$$

So, average speed $= \frac{\text{total distance}}{\text{total time}}$

$$= \frac{(2.5 + 1.25) \text{ km}}{(40/60) \text{ hr}} = \frac{45}{8} \text{ km/hr}$$

8. Consider,

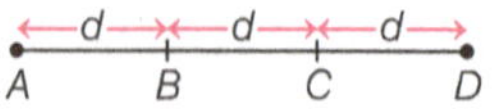

$$AB = d$$
$$BC = d$$
$$2d + CD = 3d$$
$$\Rightarrow \quad CD = 3d - 2d = d$$

Average speed of the object $\langle v \rangle = \frac{\text{Total distance}}{\text{Total time}}$

$$\langle v \rangle = \frac{3d}{\frac{d}{v_1}+\frac{d}{v_2}+\frac{d}{v_3}} = \frac{3v_1v_2v_3}{v_1v_2+v_2v_3+v_3v_1}$$

9. Average speed $= \frac{d}{t} = \frac{d_1+d_2}{t_1+t_2}$

$$= \frac{v_1t_1+v_2t_2}{t_1+t_2}$$

$$= \frac{(4\times 2)+(6\times 3)}{2+3}$$

$$= 5.2 \text{ m/s}$$ **Ans.**

10.

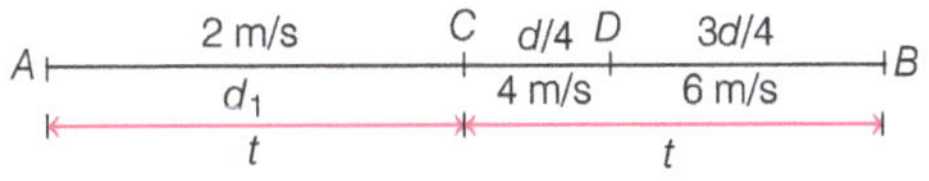

$$2t = d_1 \quad \text{...(i)}$$

$$\frac{(d/4)}{4}+\frac{(3d/4)}{6} = t$$

$$\Rightarrow \quad \frac{3d}{16} = t \quad \text{...(ii)}$$

Average speed $= \frac{\text{Total distance}}{\text{Total time}}$

$$= \frac{d_1+d}{2t} = \frac{2t+\frac{16t}{3}}{2t}$$

$$= \frac{11}{3} \text{ m/s}$$ **Ans.**

11. $t_1 = \frac{AB}{v} = \frac{\frac{\pi}{2}r}{v}, t_2 = \frac{BC}{2v} = \frac{\frac{2\pi}{3}r}{2v}$

$$t_3 = \frac{CA}{3v} = \frac{\frac{5\pi}{6}r}{3v}$$

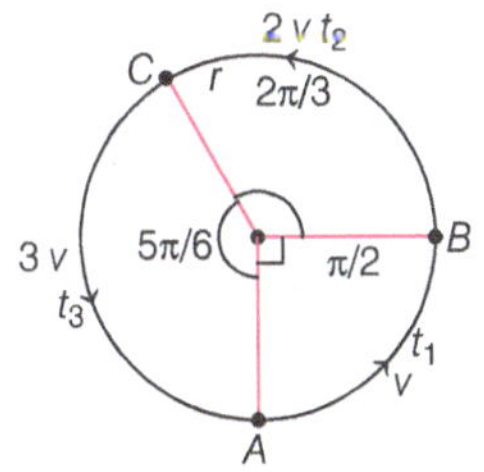

Average speed $= \frac{2\pi r}{t_1+t_2+t_3} = 1.8\,v$

12.

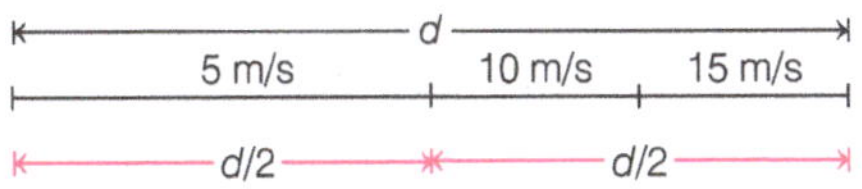

Average velocity $= \frac{d}{\frac{d/2}{5}+(t+t)}$

$$\because \quad \frac{d}{2} = 10t+15t$$

$$\Rightarrow \quad d = 50t$$

Average velocity $= \frac{d}{\frac{d/2}{5}+\frac{d}{25}} = \frac{d}{d}\left(\frac{1}{\frac{1}{10}+\frac{1}{25}}\right) = \frac{50}{7}$ m/s

Module 4.

1. $v^2 = u^2 + 2gh'$

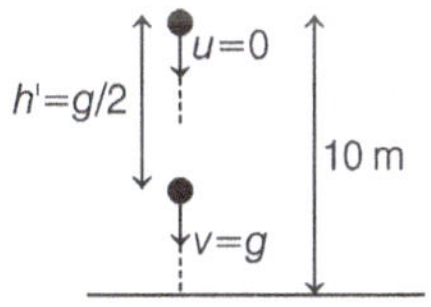

$$g^2 = 0^2 + 2gh'$$

$$h' = \frac{g}{2}$$

So, required height

$$= 10 - h' = 10 - \frac{g}{2}$$

$$= 10 - \frac{10}{2} = 5 \text{ m}$$

2. Let the two balls meet at $t = t$ sec.

Then, first ball get 't' sec and second ball get $(t-2)$ sec and they will meet at same height.

So, $\quad h_1 = h_2$

$$\Rightarrow \quad 50t - \frac{1}{2}gt^2 = 50(t-2) - \frac{1}{2}g(t-2)^2$$

$$\Rightarrow \quad 0 = -100 + \frac{1}{2}g[t^2-(t-2)^2]$$

$$\Rightarrow \quad 20 = 4t - 4$$

$$\Rightarrow \quad 24 = 4t \quad \Rightarrow \quad t = 6 \text{ sec}$$

3. Time to travel 81 m is t sec.

Time to travel 100 m is $\left(t+\frac{1}{2}\right)$ sec.

$$\therefore \quad 81 = \frac{1}{2}\times a \times t^2 \quad \Rightarrow \quad t = 9\sqrt{\frac{2}{a}}$$

and $\quad 100 = \frac{1}{2}\times a \times \left(t+\frac{1}{2}\right)^2$

$$\Rightarrow \quad t+\frac{1}{2} = 10\sqrt{\frac{2}{a}}$$

$$\therefore \quad 9\sqrt{\frac{2}{a}}+\frac{1}{2} = 10\sqrt{\frac{2}{a}}$$

$$\Rightarrow \quad \frac{1}{2} = \sqrt{\frac{2}{a}}$$

$$\Rightarrow \quad a = 8 \text{ m/s}^2$$

4. Acceleration due to gravity is independent of the mass of the body.

5. $v = u + at$

i.e. v-t function is linear. In linear function,

$$\text{average value} = \frac{(\text{final value} + \text{initial value})}{2}$$

$$\therefore \quad v_{av} = \frac{v_f + v_i}{2} = \frac{v_1 + v_2}{2} \quad \textbf{Ans.}$$

6. If a particle is projected upwards, then $s = 0$ when it returns back to its initial position.

$$\therefore \quad \text{Average velocity} = \frac{s}{t} = 0$$

But its acceleration is constant $= g$

7. $t_1 = \sqrt{\frac{2(h/2)}{g}} = \sqrt{\frac{h}{g}}$

Velocity of second particle at height $\frac{h}{2}$ is,

$$v = \sqrt{u^2 - 2g\,(h/2)} = \sqrt{gh - gh} = 0$$

or $\frac{h}{2}$ is its highest point.

$$\therefore \quad t_2 = \frac{u}{g} = \frac{\sqrt{gh}}{g} = \sqrt{\frac{h}{g}}$$

$\Rightarrow \quad t_1 = t_2$

8. $a = \frac{mg - F}{m} = g - \frac{F}{m} =$ depends on m

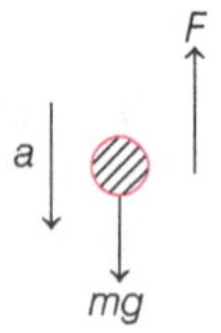

Since $\quad m_1 \neq m_2$

$\therefore \quad a_1 \neq a_2$

or $\quad t_1 \neq t_2$

9. $(50)^2 = (10)^2 + 2as \quad$ ...(i)

$$v^2 = (50)^2 + 2(-a)(-s) = (50)^2 + 2as \quad \text{...(ii)}$$

Solving these two equations, we get

$$v = 70\ \text{m/s}$$

10. Air resistance (let F) is always opposite to motion (or velocity)

Retardation in upward journey

$$a_1 = \frac{F_1 + mg}{m} = g + \frac{F_1}{m}$$

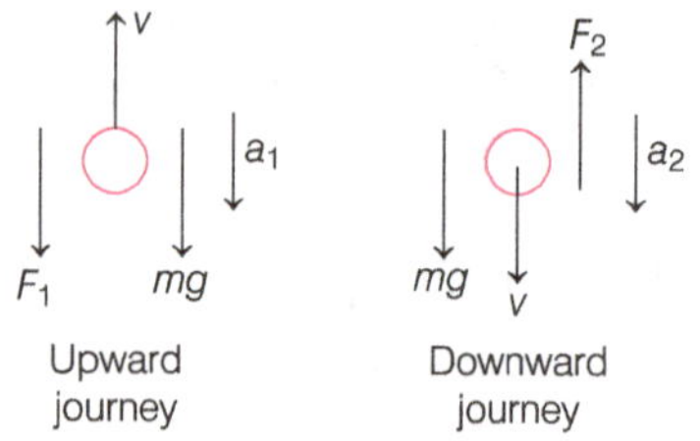

Acceleration in downward journey

$$a_2 = \frac{mg - F_2}{m} = g - \frac{F_2}{m}$$

Since, $\quad a_1 > a_2 \quad \Rightarrow \quad T_1 < T_2$

11. $h = \frac{u^2}{2g}$ or $u \propto \sqrt{h}$

12. For P, $\quad 30 = 15 + at$

or $\quad at = 15\ \text{m/s}$

For Q, $\quad v = 20 - at$

or $\quad v = 20 - 15 = 5\ \text{m/s}$

13. $\frac{1}{2} \times (a)(2)^2 = x \quad$ or $\quad 2a = x$

Now, $\frac{1}{2}(a)(4)^2 - \frac{1}{2}(a)(2)^2 = 6\,a = 3x$

14. $v^2 = 25 + 25\,s$

or $\quad v^2 = (5)^2 + 2\,(12.5)\,s$

Now compare with $v^2 = u^2 + 2as$

15. $t = \sqrt{\frac{2h}{g}} = \sqrt{\frac{2 \times 5}{10}} = 1\ \text{s}$

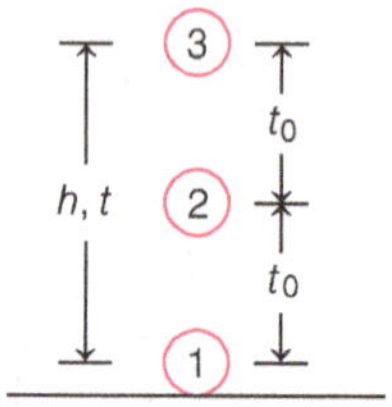

Let t_0 is the interval between two drops. Then

$$2t_0 = t$$

$$\therefore \quad t_0 = 0.5\ \text{s}$$

2nd drop has taken t_0 time to fall. Therefore distance fallen,

$$d = \frac{1}{2} g t_0^2$$

$$= \left(\frac{1}{2}\right)(10)(0.5)^2$$

$$= 1.25\ \text{m}$$

$\therefore \quad$ Height from ground $= h - d$

$$= 5 - 1.25$$

$$= 3.75\ \text{m} \quad \textbf{Ans.}$$

16. $\frac{1}{2}gt^2 = 20(t-1) + \frac{1}{2}g(t-1)^2$

Solving this equation, we get

$\therefore \quad t = 1.5\text{ s}$

Now, $\quad d = 20(t-1) + \frac{1}{2}g(t-1)^2$

$= 11.25\text{ cm}$ **Ans.**

17. $h = \frac{1}{2}gT^2$

At $\frac{T}{3}$ second, distance fallen

$$d = \frac{1}{2}g\left(\frac{T}{3}\right)^2 = \frac{h}{9}$$

$\therefore$ Height from ground $= h - d = \left(\frac{8h}{9}\right)$ metre **Ans.**

18. $d = \frac{1}{2}g(T-t)^2$

$\therefore \quad h = H - d = H - \frac{1}{2}g(T-t)^2$

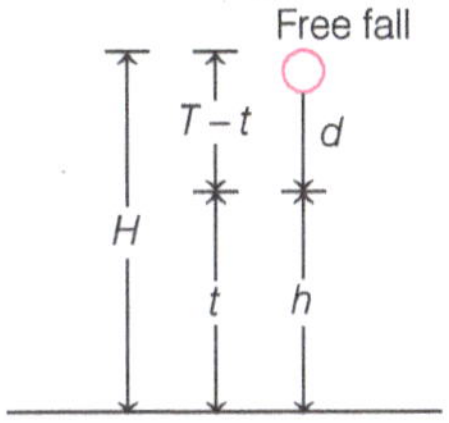

19. $v^2 = u^2 - 2g\left(\frac{h}{2}\right)$

$\therefore \quad (10)^2 = u^2 - gh$

$\therefore \quad u^2 = (100 + gh)$

Now, $\quad h = \frac{u^2}{2g} = \frac{100+gh}{2g}$

$h = 5 + \frac{h}{2}$ (as $2g = 20\text{ m/s}^2$)

$\therefore \quad h = 10\text{ m}$ **Ans.**

20. $1 = \frac{1}{2}g t_1^2$ or $t_1 = \sqrt{\frac{2}{g}}$

$2 = \frac{1}{2}g t^2$ or $t = \sqrt{\frac{4}{g}}$

But, $\quad t_2 = t - t_1 = \sqrt{\frac{4}{g}} - \sqrt{\frac{2}{g}}$

$\therefore \quad \frac{t_1}{t_2} = \frac{\sqrt{2/g}}{\sqrt{4/g} - \sqrt{2/g}}$

$= \frac{\sqrt{2}}{2-\sqrt{2}} = \frac{\sqrt{2}(2+\sqrt{2})}{2}$

$= (\sqrt{2}+1)$

21. $v^2 = u^2 + 2gh$

$\Rightarrow \quad (3u)^2 = (-u)^2 + 2gh \Rightarrow h = \frac{4u^2}{g}$

22. Let h distance is covered in t second

$\Rightarrow \quad h = \frac{1}{2}gt^2$

Distance covered in t^{th} second $= \frac{1}{2}g(2t-1)$

$\Rightarrow \quad \frac{9h}{25} = \frac{g}{2}(2t-1)$

From above two equations, $h = 122.5\text{ m}$

23. $\quad -h = vt - \frac{1}{2}gt^2$

or $\quad gt^2 - 2vt - 2h = 0$

$\therefore \quad t = \frac{2v + \sqrt{4v^2 + 8gh}}{2g}$

$= \frac{v}{g}\left[1 + \sqrt{1 + \frac{2hg}{v^2}}\right]$

24. $t_{ABC} = 10\text{ s}$

$\therefore \quad t_{AB} = 5\text{ s}$

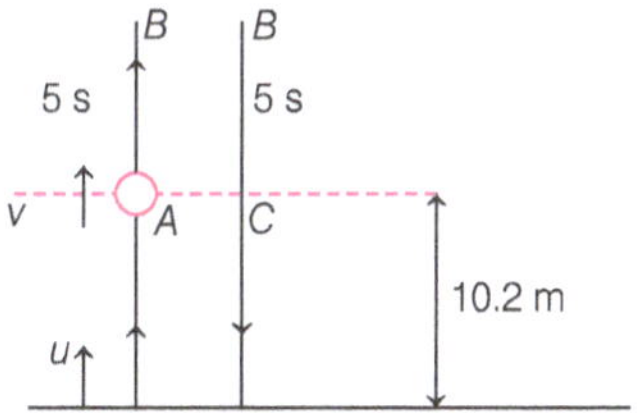

At B, velocity becomes zero. Hence, at A velocity should be 50 m/s.

Now, $\quad (50)^2 = (u^2) - 2 \times 10 \times 10.2$

$\therefore \quad u = 52\text{ m/s}$

25. Retardation during upward motion

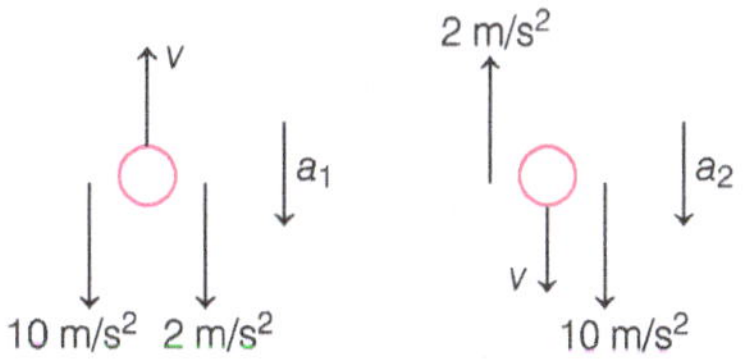

$$a_1 = 10 + 2 = 12\text{ m/s}^2$$

Acceleration during downward motion,

$$a_2 = 10 - 2 = 8\text{ m/s}^2$$

$t = \sqrt{\frac{2s}{a}}$ or $t \propto \frac{1}{\sqrt{a}}$

$\therefore \quad \frac{t_1}{t_2} = \sqrt{\frac{a_2}{a_1}} = \sqrt{\frac{8}{12}} = \sqrt{\frac{2}{3}}$ **Ans.**

26. $s_B = \frac{1}{2} \times 2 \times (1)^2 = 1\,\text{m}$

$$s_A = \frac{1}{2} \times 2 \times (2)^2 = 4\,\text{m}$$

$$\therefore \quad s_A - s_B = 3\,\text{m}$$

27. At the time of overtaking,

$$s_1 = s_2$$

$$\therefore \quad 2ut + \frac{1}{2}at^2 = ut + \frac{1}{2}(2a)t^2$$

$$\therefore \quad t = \frac{2u}{a}$$

$$\therefore \quad s_1 (\text{or } s_2) = (2u)\left(\frac{2u}{a}\right) + \frac{1}{2}(a)\left(\frac{2u}{a}\right)^2$$

$$= \frac{6u^2}{a} \qquad \textbf{Ans.}$$

28. $15 = u + \frac{1}{2}a(2 \times 3 - 1) = u + \frac{5a}{2}$ ...(i)

$23 = u + \frac{1}{2}a(2 \times 5 - 1) = u + \frac{9a}{2}$ ...(ii)

Solving Eqs. (i) and (ii), $a = 4\,\text{ms}^2$, $u = 5\,\text{m/s}$

Displacement in 10 s,

$$s = ut + \frac{1}{2}at^2$$

$$= 5 \times 10 + \frac{1}{2} \times 4 \times 10^2$$

$$= 250\,\text{m}$$

29. Displacements of both should be equal.

or $\quad 8t = \frac{1}{2} \times 4 \times t^2$ or $t = 4\,\text{s}$

30. $s_1 = \frac{1}{2}a_0 t_0^2$

$$v_0 = a_0 t_0$$

$$v^2 = u^2 - 2as$$

$$0^2 = v_0^2 - 2a_0 s_2$$

$$s_2 = \frac{v_0^2}{2a_0} = \frac{(a_0 t_0)^2}{2a_0}$$

$$= \frac{a_0^2 t_0^2}{2a_0} = \frac{1}{2}a_0 t_0^2$$

$\therefore$ Total distance travelled,

$$s = s_1 + s_2 = a_0 t_0^2$$

31. Area of v-t graph = displacement

$$\frac{1}{2}v \times 300 = 3000$$

$$\Rightarrow \quad v = 20\,\text{m/s}$$

Maximum velocity = 20 m/s

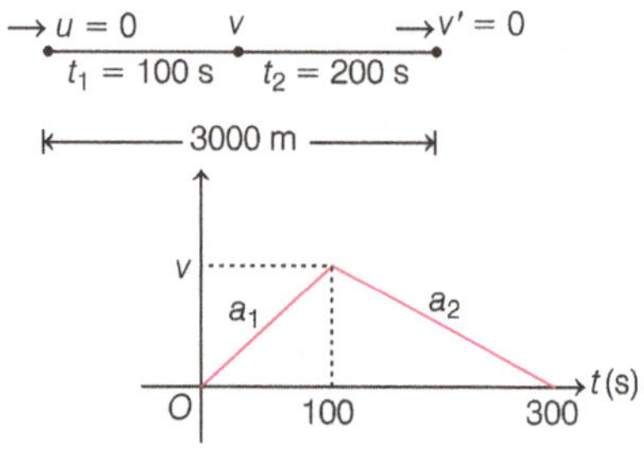

Acceleration, $a_1 = \frac{v}{100} = \frac{20}{100} = 0.2\,\text{m/s}^2$

Retardation, $a_2 = \frac{v}{200} = \frac{20}{200} = 0.1\,\text{m/s}^2$

32. Average velocity in uniformly accelerated motion is given by

$$v_{av} = \frac{s}{t} = \frac{ut + \frac{1}{2}at^2}{t} = u + \frac{1}{2}at$$

Now, $\quad v_1 = u + \frac{1}{2}at_1$

$$v_2 = (u + at_1) + \frac{1}{2}at_2$$

and $\quad v_3 = u + a(t_1 + t_2) + \frac{1}{2}at_3$

$$\therefore \quad (v_1 - v_2) : (v_2 - v_3) = (t_1 + t_2) : (t_2 + t_3)$$

33. $s_1 = s_2$

$$\therefore \quad (40)t - \frac{1}{2} \times 10 \times t^2 = (40)(t-2) - \frac{1}{2} \times 10 \times (t-2)^2$$

Solving this equation, we get

$$t = 5\,\text{s}$$

Then $\quad s_1 = (40)(5) - \frac{1}{2} \times 10 \times (5)^2$

$$= 75\,\text{m} \qquad \textbf{Ans.}$$

34. $v = an \Rightarrow a = \left(\frac{v}{n}\right)$

Now, $\quad s_n = \frac{1}{2}an^2 = \frac{1}{2}\left(\frac{v}{n}\right)(n^2) = \frac{vn}{2}$

and $\quad s_{n-3} = \frac{1}{2}a(n-3)^2$

$$= \frac{1}{2}\left(\frac{v}{n}\right)(n-3)^2$$

$\therefore$ Displacement in last 3 s will be,

$$s = s_n - s_{n-3}$$

$$= \frac{v}{2}\left[n - \frac{(n-3)^2}{n}\right]$$

$$= \left(\frac{6n - 9}{n}\right)\frac{v}{2}$$

35. Let x be the acceleration. Then,

$$a = \frac{1}{2}xp^2 \quad \text{...(i)}$$

$$a + b = +\frac{1}{2}x(p+q)^2 \quad \text{...(ii)}$$

Solving these two equations, we get

$$x = \frac{2b}{(q+2p)q}$$

36. Distance travelled in n^{th} second is, $s_n = u + an - \frac{1}{2}a$

Given, $u = 0$

$$\therefore \quad \frac{s_n}{s_{n+1}} = \frac{an - \frac{1}{2}a}{a(n+1) - \frac{1}{2}a} = \frac{2n-1}{2n+1}$$

37. $v = b\sqrt{x}$

$$\frac{dx}{dt} = bx^{\frac{1}{2}} \qquad \Rightarrow \int_0^x x^{-1/2}dx = \int_0^\tau b\,dt$$

$$\Rightarrow \quad 2\sqrt{x_\tau} = b\tau \qquad \Rightarrow \sqrt{x_\tau} = \frac{b\tau}{2}$$

$$v_\tau = b\frac{b\tau}{2} = \frac{b^2\tau}{2} \qquad \Rightarrow v_\tau = \frac{b^2\tau}{2}$$

38. Let s be the distance between two ends, 'a' be the constant acceleration.

As we know $v^2 - u^2 = 2as$ or $as = \frac{v^2 - u^2}{2}$

Let v_c be velocity at mid-point.

Therefore, $v_c^2 - u^2 = 2a\frac{s}{2}$

$$\Rightarrow \quad v_c^2 = u^2 + as$$

$$\Rightarrow \quad v_c^2 = u^2 + \frac{v^2 - u^2}{2}$$

$$\Rightarrow \quad v_c = \sqrt{\frac{u^2 + v^2}{2}}$$

39. During acceleration, $v = 0 + a_1t_1 = a_1t_1$

During retardation, $0 = v - a_2t_2$

$$\Rightarrow \quad v = a_2t_2$$

$$\therefore \quad a_1t_1 = a_2t_2 \quad \text{or} \quad \frac{t_1}{t_2} = \frac{a_2}{a_1}$$

40. From the third equation of motion

$$v^2 - u^2 = 2as$$

But, $a = \frac{F}{m}$

$$\therefore \quad v^2 = u^2 - 2\left(\frac{F}{m}\right)s$$

$$\Rightarrow \quad v^2 = (1)^2 - (2)\left[\frac{2.5 \times 10^{-2}}{20 \times 10^{-3}}\right]\frac{20}{100}$$

$$\Rightarrow \quad v^2 = 1 - \frac{1}{2} \quad \Rightarrow \quad v = \frac{1}{\sqrt{2}}\text{ m/s}$$

$$\approx 0.7 \text{ m/s}$$

41. Here, $s = \frac{v^2}{2a} \quad \Rightarrow \quad s \propto v^2$

$$\Rightarrow \quad s_2 = s_1\left(\frac{v_2}{v_1}\right)^2 \quad \Rightarrow \quad s_2 = 40\left(\frac{80}{40}\right)^2$$

$$\Rightarrow \quad s_2 = 160 \text{ m}$$

42. Let car starts from A from rest and moves up to point B with acceleration f.

Distance, $AB = s = \frac{1}{2}ft_1^2$

Distance, $BC = (ft_1)t$

Distance, $CD = \frac{u^2}{2a} = \frac{(ft_1)^2}{2(f/2)} = ft_1^2 = 2s$

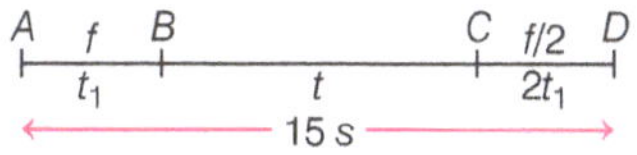

Total distance, $AD = AB + BC + CD = 15s$

$$AD = s + BC + 2s$$

$$\Rightarrow \quad s + f\,t_1t + 2s = 15s$$

$$\Rightarrow \quad ft_1t = 12s \quad \text{...(i)}$$

$$\frac{1}{2}f\,t_1^2 = s \quad \text{...(ii)}$$

Dividing Eq. (i) by Eq. (ii), we get

$$t_1 = \frac{t}{6}$$

$$\Rightarrow \quad s = \frac{1}{2}f\left(\frac{t}{6}\right)^2 = \frac{ft^2}{72}$$

43. In first case :

$$u_1 = u, v_1 = \frac{u}{2}, s_1 = 3 \text{ cm}, a_1 = ?$$

Using $v_1^2 - u_1^2 = 2a_1s_1$ ...(i)

$$\Rightarrow \quad \left(\frac{u}{2}\right)^2 - u^2 = 2 \times a_1 \times 3$$

$$\Rightarrow \quad a_1 = \frac{-u^2}{8}$$

In second case : Assuming the same retardation

$$u_2 = \frac{u}{2},\ v_2 = 0,\ s_2 = ?, a_2 = \frac{-u^2}{8}$$

$$v_2^2 - u_2^2 = 2a_2 \times s_2 \quad \text{...(ii)}$$

$$\therefore \quad 0 - \frac{u^2}{4} = 2\left(\frac{-u^2}{8}\right) \times s_2$$

$$\Rightarrow \quad s_2 = 1 \text{ cm}$$

44. Given, initial velocity of ball, $u = 150$ m/s

$g = 10\,m/s^2$

Using $v = u - gt$

$\therefore \quad v(3) = 150 - 10 \times 3 = 120$

$\therefore \quad v(5) = 150 - 10 \times 5 = 100$

$\therefore \quad \frac{v(3)}{v(5)} = \frac{120}{100} = \frac{x+1}{x} = \frac{6}{5}$

$\Rightarrow \quad x = 5$

45. Maximum height, $h = \frac{u^2}{2g}$

$\Rightarrow \quad u = \sqrt{2gh}$

As, $s = ut + \frac{1}{2}at^2$

$\Rightarrow \quad \frac{h}{3} = t\sqrt{2gh} + \frac{1}{2}(-g)t^2$

$\Rightarrow \quad \frac{gt^2}{2} - t\sqrt{2gh} + \frac{h}{3} = 0$

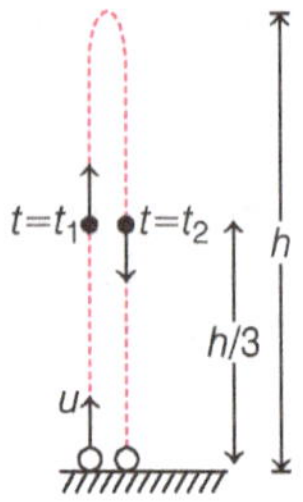

Let t_1 and t_2 be two roots of above equation.

Then, $\frac{t_2}{t_1} = \frac{\sqrt{2gh} + \sqrt{2gh - 4 \times \frac{g}{2} \times \frac{h}{3}}}{\sqrt{2gh} - \sqrt{2gh - 4 \times \frac{g}{2} \times \frac{h}{3}}}$

$= \frac{\sqrt{2gh} + \sqrt{\frac{4gh}{3}}}{\sqrt{2gh} - \sqrt{\frac{4gh}{3}}} = \frac{\sqrt{3}+\sqrt{2}}{\sqrt{3}-\sqrt{2}}$

46. From equation of motion, $s = ut + \frac{1}{2}gt^2$

For first $\frac{h}{2}, \frac{h}{2} = \frac{1}{2}gt_1^2$ ...(i)

Fot total height h,

$$h = \frac{1}{2}g(t_1 + t_2)^2 \quad ...(ii)$$

Dividing Eq. (i) by Eq. (ii), we have

$\frac{1}{2} = \frac{t_1^2}{(t_1+t_2)^2}$

$\Rightarrow \quad \frac{1}{\sqrt{2}} = \frac{t_1}{(t_1+t_2)}$

$\Rightarrow \quad 1 + \frac{t_2}{t_1} = \sqrt{2}$

$\Rightarrow \quad \frac{t_1}{t_2} = \frac{1}{\sqrt{2}-1} \Rightarrow t_2 = (\sqrt{2}-1)\,t_1$

47. From $h = ut + \frac{1}{2}gt^2 \Rightarrow t = \sqrt{\frac{2h}{g}}$

Time taken by mango, $t = \sqrt{\frac{2h}{g}} = \sqrt{\frac{2 \times 19.6}{9.8}} = 2\,s$

Distance $x = vt = \frac{5}{2} \times 2 = 5\,m$

48. Given, initial velocity of hot air balloon, $u = 10\,m/s$

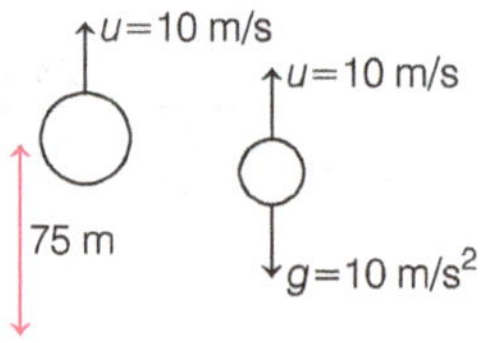

For stone

Using,

$h = ut + \frac{1}{2}gt^2$

$\Rightarrow \quad 75 = -10t + \frac{1}{2}gt^2$

$\Rightarrow \quad 75 = -10t + 5t^2$

$\Rightarrow \quad t^2 - 2t - 15 = 0$

$\Rightarrow \quad t = 5$ sec

Height of balloon

$H = vt + 75$

$\Rightarrow \quad H = 10 \times 5 + 75 = 125\,m$

49.

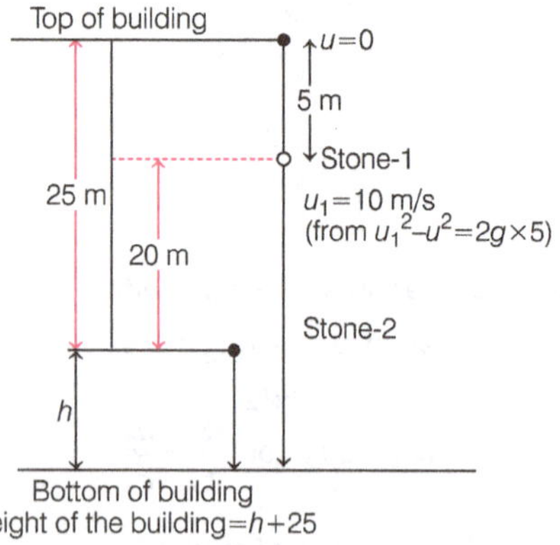

For stone-1, $20 + h = 10t + \frac{1}{2}gt^2$ ...(i)

And for stone-2, $h = \frac{1}{2}gt^2$ ...(ii)

Putting value of h from Eq. (ii) in Eq. (i),

$$20 + \frac{1}{2}gt^2 = 10t + \frac{1}{2}gt^2$$

Therefore, $h = \frac{1}{2}gt^2 = \frac{1}{2} \times 10 \times 2^2$ $(\because t = 2\,s)$

$= 20\,m$

$\therefore$ Height of the building $= h + 25 = 20 + 25 = 45\,m$

50. For upward motion of helicopter,

$$v^2 = u^2 + 2gh$$

$$\Rightarrow \quad v^2 = 0 + 2gh$$

$$\Rightarrow \quad v = \sqrt{2gh}$$

Now, packet will start moving under gravity.

Let 't' be the time taken by the food packet to reach the ground.

$$s = ut + \frac{1}{2}at^2$$

$$\Rightarrow \quad -h = \sqrt{2gh}\,t - \frac{1}{2}gt^2$$

$$\Rightarrow \quad \frac{1}{2}gt^2 - \sqrt{2gh}\,t - h = 0$$

$$\Rightarrow \quad t = \frac{\sqrt{2gh} \pm \sqrt{2gh + 4 \times \frac{g}{2} \times h}}{2 \times \frac{g}{2}}$$

$$\Rightarrow \quad t = \sqrt{\frac{2gh}{g}}\,(1 + \sqrt{2})$$

$$\Rightarrow \quad t = \sqrt{\frac{2h}{g}}\,(1 + \sqrt{2})$$

$$\Rightarrow \quad t = 3.4\sqrt{\frac{h}{g}}$$

51.

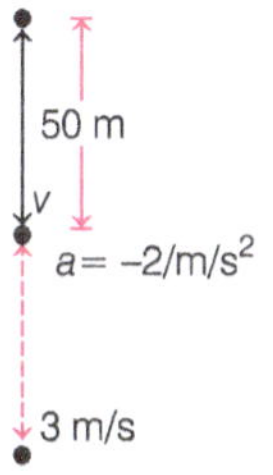

Initial velocity of parachute after bailing out,

$$u = \sqrt{2gh} = \sqrt{2 \times 9.8 \times 50} = \sqrt{980}$$

The velocity at ground,

$$v = 3\,m/s$$

$$s = \frac{v^2 - u^2}{2 \times 2} = \frac{3^2 - 980}{4} \approx 243\,m$$

Initially he has fallen from 50 m.

$\therefore$ Total height from where he bailed out

$= 243 + 50 = 293\,m$

52. From conservation of energy,

potential energy at height h = kinetic energy at ground

Therefore, at height h, potential energy of ball $A = m_A gh$

Kinetic energy at ground $= \frac{1}{2}m_A v_A^2$

So, $m_A gh = \frac{1}{2}m_A v_A^2$ or $v_A = \sqrt{2gh}$

Similarly, $v_B = \sqrt{2gh}$

Therefore, $v_A = v_B$

Note *In this question, it is assumed that both the balls are thrown with same speed.*

53. In the complete journey,

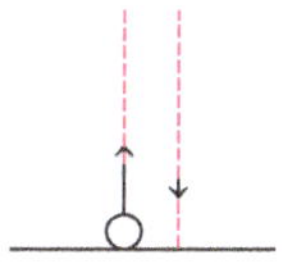

$s = 0$ and $a = \text{constant} = g$ (downwards)

54. At 60 s,

$$h_1 = \frac{1}{2} \times 10 \times (60)^2 = 18000\,m$$

$$v_1 = 10 \times 60 = 600\,m/s$$

After that

$$h_2 = \frac{v_1^2}{2g} = 18000\,m$$

$$\therefore \quad H_{max} = h_1 + h_2$$

$$= 36000\,m = 36\,km$$

Further, $-18000 = 600t - 5t^2$

or $t^2 - 120t - 3600 = 0$

or $$t = \frac{120 + \sqrt{14400 + 14400}}{2}$$

$$= \frac{120 + 120\sqrt{2}}{2}$$

$$= 60 + 60\sqrt{2}$$

$\therefore$ Total time $= 60 + (60 + 60\sqrt{2})$

$= (120 + 60\sqrt{2})\,s$

55. (a) For truck

$$s_1 = \frac{1}{2}a_1 t^2$$

$$60 = \frac{1}{2} \times 2.2 \times t^2$$

$\therefore$ $t = 7.39\,s$ **Ans.**

(b) For automobile

$$s_2 = \frac{1}{2}a_2 t^2 = \frac{1}{2} \times 3.5 \times (7.39)^2$$

$$= 95.5\,m$$

$\therefore$ Initial distance between them

$= s_2 - s_1 = 35.5\,m$ **Ans.**

(c) $v_1 = a_1 t = (2.2)\,(7.39) = 16.2\text{ m/s}$

and $v_2 = a_2 t = (3.5)\,(7.39) = 25.9\text{ m/s}$

56. $s_t = \left[ut + \frac{1}{2}at^2\right] - \left[u(t-1) + \frac{1}{2}a(t-1)^2\right]$

$$= (u)(1) + \frac{1}{2}(a)(2t) - \frac{1}{2}(a)(1)^2$$

The first term is (u) (1), which we are writing only u. Dimensions of u are $[LT^{-1}]$ and dimensions of 1 (which is actually 1 second) are [T]

$\therefore \quad [(u)(1)] = [LT^{-1}]\,[T]$

$$= [L] = s_t$$

Therefore, dimension of (u) (1) are same as the dimensions of s_t. Same logic can be applied with other terms too.

59. $u_{av} = \frac{s}{t} = \frac{ut + \frac{1}{2}at^2}{t} = u + \frac{1}{2}at$ **Ans.**

60. $t = \sqrt{\frac{2h}{g}} = \sqrt{\frac{2\times 125}{10}} = 5\text{s}$

$v_{av} = \frac{s}{t} = \frac{125}{5} = 25\text{ m/s}$ (downwards) **Ans.**

61. (a) $s = ut + \frac{1}{2}at^2 = (2.5\times 2) + \frac{1}{2}(0.5)\,(2)^2$

$= 6\text{ m} = \text{distance}$

(b) $v = u + at$

$7.5 = 2.5 + 0.5\times t \Rightarrow t = 10\text{ s}$

(c) $v^2 = u^2 + 2as$

$\therefore \quad (7.5)^2 = (2.5)^2 + (2)\,(0.5)\,s$

$\Rightarrow \quad s = 50\text{ m} = \text{distance}$

62. (a) $h = \frac{u^2}{2g} = \frac{(50)^2}{2\times 10} = 125\text{ m}$

(b) Time of ascent $= \frac{u}{g} = \frac{50}{10} = 5\text{ s}$

(c) $v^2 = u^2 + 2as = (50)^2 + 2(-10)\left(\frac{125}{2}\right)$

Solving, we get speed $v \approx 35$ m/s

63. Retardation time is 8s (double). Therefore, retardation should be half or 2 m/s^2.

$s_1 =$ acceleration displacement

$= \frac{1}{2}a_1 t_1^2 = \frac{1}{2}\times 4\times (4)^2$

$= 32$ m

$s_2 =$ retardation displacement

$= \frac{1}{2}a_2 t_2^2$

$= \frac{1}{2}\times 2\times (8)^2 = 64\text{m}$

(a) $a_{av} = \frac{v_f - v_i}{t} = \frac{0-0}{12} = 0$ **Ans.**

(b) and (c)

$d = s = s_1 + s_2 = 96$ m

$\therefore$ Average speed = Average velocity

$= \frac{d}{t}$

or $\frac{s}{t} = \frac{96}{12} = 8\text{ m/s}$ **Ans.**

64. At minimum distance, their velocities are same,

$v_A = v_B$ or $u_A + a_A t = u_B + a_B t$

or $(3+t) = (1+2t)$ or $t = 2\text{s}$

Minimum distance, d_{min} = Initial distance – extra displacement of A upto this instant due to its greater speed

$= 10 - (s_A - s_B)$

$= 10 + s_B - s_A$

$= 10 + \left(u_B t + \frac{1}{2}a_B t^2\right) - \left(u_A t + \frac{1}{2}a_A t^2\right)$

$= 10 + \left[(1)(2) + \frac{1}{2}(2)(2)^2\right] - \left[(3)\,(2) + \frac{1}{2}(1)\,(2)^2\right]$

$= 8$ m **Ans.**

65. $s = s_0 + ut + \frac{1}{2}at^2$

At $t = 0, s = s_0 = 2$ m ...(i)

At $t = 10$ s,

$s = 0 = s_0 + ut + \frac{1}{2}at^2$

or $0 = 2 + 10u + 50a$

or $10u + 50a = -2$ ...(ii)

$v = u + at$

$\therefore \quad 0 = u + 6a$ (at $t = 6$ s)

$\therefore \quad u + 6a = 0$ ...(iii)

Solving Eqs. (ii) and (iii), we get

$u = -1.2\text{ m/s}$

and $a = 0.2\text{ m/s}^2$ **Ans.**

Now, $v = u + at$

$\therefore \quad v = (-1.2) + (0.2)\,(10)$

$= 0.8\text{ m/s}$ **Ans.**

66. (a) $v = \alpha\sqrt{x}$ or $v^2 = \alpha^2 x$

Comparing this equation with

$v^2 = u^2 + 2as$

or $u^2 + 2ax$, we get

$u = 0$ and $2a = \alpha^2$

$\Rightarrow \quad a = \frac{\alpha^2}{2}$

$\therefore \quad v = u + at \quad \text{or} \quad v = \frac{\alpha^2}{2} t$

$a = \frac{\alpha^2}{2}$

(b) $s = \frac{1}{2} at^2 = \frac{\alpha^2}{4} t^2$

$\therefore \quad t = \frac{2}{\alpha}\sqrt{s}$

Now, average velocity $= \frac{s}{t} = \frac{\frac{1}{2}at^2}{t} = \frac{1}{2}at$

Substituting the value of a and t, we get

$$v_{av} = \left(\frac{1}{2}\right)\left(\frac{\alpha^2}{2}\right)\left(\frac{2}{\alpha}\right)\sqrt{s} \quad \text{or} \quad v_{av} = \frac{\alpha\sqrt{s}}{2}$$

67. $t_a = \frac{u}{g} = \frac{19.6}{9.8} = 2\text{ s}$

$t_d = (6-2)\text{ s} = \sqrt{\frac{2h_{max}}{g}}$

$\Rightarrow \quad h_{max} = \frac{16 \times 9.8}{2} = \frac{392}{5}$

Module 5.

1. It is not necessary that if $v = 0$, then acceleration is also zero. If a particle is thrown upwards, then at highest point $v = 0$ but $a \neq 0$.

2. $a = 2t \neq$ constant

Therefore, velocity will not increase at a constant rate.

3. $v = 3\alpha t^2 + 2\beta t + \gamma : v_{t=0} = v_i = \gamma$

$a = 6\alpha t + 2\beta : a_{t=0} = a_i = 2\beta$

$\therefore \quad \frac{v_i}{a_i} = \frac{\gamma}{2\beta}$

4. $a = bt$

$\therefore \quad \frac{dv}{dt} = bt$

or $\quad \int_{v_0}^{v} dv = \int_0^t (bt)\, dt$

$\therefore \quad v = v_0 + \frac{bt^2}{2}$

$s = \int_0^t v dt = \int_0^t \left(v_0 + \frac{bt^2}{2}\right) dt$

$= v_0 t + \frac{bt^3}{6}$

5. $\sqrt{x} = t + 3$

$\therefore \quad x = (t+3)^2$

or $\quad v = \frac{dx}{dt} = 2(t+3)$

$\therefore$ v-t equation is linear **Ans.**

6. $v = \frac{dx}{dt} = 32 - 8t^2$

$v = 0 \quad \text{at} \quad t = 2\text{ s}$

$a = \frac{dv}{dt} = -16t$

At $\quad t = 2\text{ s}, \; a = -32\text{ m/s}^2$ **Ans.**

7. $a = \frac{dv}{dt} = 0.2v^2$

$\therefore \quad \int_{10}^{v} -5v^{-2} = \int_0^2 dt$

$\therefore \quad \left[\frac{5}{v}\right]_{10}^{v} = 2$

$\therefore \quad \frac{5}{v} - \frac{5}{10} = 2$

or $\quad v = 2.0\text{ m/s}$ **Ans.**

8. $F = 3t^2 - 32$

$a = \frac{F}{m} = (0.3t^2 - 3.2)$

$\int_{10}^{v} dv = \int_0^5 a dt = \int_0^5 (0.3t^2 - 3.2)\, dt$

$v - 10 = -3.5$

$\therefore \quad v = 6.5\text{ m/s}$ **Ans.**

9. $x = 2 + t - 3t^2, v = \frac{dx}{dt} = 1 - 6t$

Velocity will become zero at time t_0, $0 = 1 - 6t_0$

or $\quad t_0 = \frac{1}{6}\text{ s}$

Since, the given time $t = 1\text{ s}$ is greater than $t_0 = \frac{1}{6}$ s.

Distance > |Displacement|

Displacement, $s = x_f - x_i$

$= (2 + 1 - 3) - (2 + 0 - 0)$

$= -2\text{ m}$

Distance, $d = |s_{0-t_0}| + |s_{t-t_0}|$

$= \frac{u^2}{2|a|} + \frac{1}{2}|a|(t - t_0)^2$

Comparing $v = 1 - 6t$ with $v = u + at$, we have $u = 1\text{ m/s}$ and $a = -6\text{ m/s}^2$

$\therefore \quad \text{Distance} = \frac{(1)^2}{2 \times 6} + \frac{1}{2} \times 6 \times \left(1 - \frac{1}{6}\right)^2$

$= 2.1\text{ m}$

10. We have given,

$v = \alpha t + \beta t^2$

$\Rightarrow \quad \frac{ds}{dt} = \alpha t + \beta t^2 \quad \Rightarrow \quad \int_{s_1}^{s_2} ds = \int_1^2 (\alpha t + \beta t^2)\, dt$

$\Rightarrow \quad s_2 - s_1 = \left[\frac{\alpha t^2}{2} + \frac{\beta t^3}{3}\right]_1^2$

As particle is moving in a straight line.

$\therefore$ Distance = Displacement

$\therefore$ Distance $= \left[\frac{\alpha(4-1)}{2} + \frac{\beta(8-1)}{3}\right] = \frac{3\alpha}{2} + \frac{7\beta}{3}$

11. $x = at + bt^2 - ct^3$

Velocity, $v = \frac{dx}{dt} = \frac{d}{dt}(at + bt^2 - ct^3)$

$= a + 2bt - 3ct^2$

Acceleration, $\frac{dv}{dt} = \frac{d}{dt}(a + 2bt - 3ct^2)$

or $\quad 0 = 2b - 3c \times 2t$

$\therefore \quad t = \left(\frac{b}{3c}\right)$

and $v = a + 2b\left(\frac{b}{3c}\right) - 3c\left(\frac{b}{3c}\right)^2 = a + \frac{2b^2}{3c} - \frac{b^2}{3c} = \left(a + \frac{b^2}{3c}\right)$

12. Given, $\frac{dv}{dt} = -2.5\sqrt{v} \quad \Rightarrow \quad \frac{dv}{\sqrt{v}} = -2.5\, dt$

On integration,

$\int_{6.25}^{0} v^{-1/2}\, dv = -2.5 \int_0^t dt$

$\Rightarrow \quad \left[\frac{v^{+1/2}}{(1/2)}\right]_{6.25}^{0} = -2.5\, [t]_0^t$

$\Rightarrow \quad -2(6.25)^{1/2} = -2.5t$

$\Rightarrow \quad -2 \times 2.5 = -2.5t$

$\Rightarrow \quad t = 2\text{ s}$

13. $x = 0$ at $t = 0$ and $t = \alpha/\beta$. So, the particle returns to starting point at $t = \alpha/\beta$.

$v = \frac{dx}{dt} = 2\alpha t - 3\beta t^2$

At $t = 0$, $v = 0$ i.e. initial velocity of particle is zero.

$v = 0$ at $t = 0$ and $t = \frac{2\alpha}{3\beta}$

Thus, the particle comes to rest after time,

$t = 2\alpha/3\beta$

$a = \frac{dv}{dt} = 2\alpha - 6\beta t$

At $\quad t = 0, a = 2\alpha \neq 0$

14. (a) $x = 0$ at $t = 2\text{s}$

(b) $v = \frac{dx}{dt} = 10t$

and $\quad a = \frac{dv}{dt} = 10$

$v = a$ at 1s

(c) $v = 10t$

$\therefore$ Velocity is positive all the time.

(d) $\quad v = 0$ at $t = 0\text{s}$ **Ans.**

15. (a) $a = \frac{dv}{dt} = 5 - 2t$

At $\quad t = 2\text{s}$

$a = 1\text{ m/s}^2$ **Ans.**

(b) $x = \int_0^3 v\, dt = \int_0^3 (10 + 5t - t^2)\, dt$

$= \left[10t + 2.5t^2 - \frac{t^3}{3}\right]_0^3$

$= (10 \times 3) + (2.5)(3)^2 - \frac{(3)^3}{3}$

$= 43.5\text{ m}$ **Ans.**

16. (a) Acceleration of particle,

$a = \frac{dv}{dt} = (6 + 18t)\text{ cm/s}^2$

At $t = 3\text{ s}$,

$a = (6 + 18 \times 3)\text{ cm/s}^2$

$= 60\text{ cm/s}^2$

(b) Given, $\quad v = (3 + 6t + 9t^2)\text{ cm/s}$

or $\quad \frac{ds}{dt} = (3 + 6t + 9t^2)$

or $\quad ds = (3 + 6t + 9t^2)\, dt$

$\therefore \quad \int_0^s ds = \int_5^8 (3 + 6t + 9t^2)\, dt$

$\therefore \quad s = [3t + 3t^2 + 3t^3]_5^8$

or $\quad s = 1287\text{ cm}$

17. (a) Position, $x = (2t - 3)^2$

Velocity, $\quad v = \frac{dx}{dt} = 4(2t - 3)\text{ m/s}$

and acceleration, $a = \frac{dv}{dt} = 8\ \text{m/s}^2$

At $t = 2\text{ s}$,

$x = (2 \times 2 - 3)^2$

$= 1.0\text{ m}$

$v = 4(2 \times 2 - 3)$

$= 4\text{ m/s}$

and $a = 8\text{ m/s}^2$

(b) At origin, $x = 0$

or $\quad (2t - 3) = 0$

$\therefore \quad v = 4 \times 0 = 0$

18. (a) At $t = 0, x = 2.0$ m **Ans.**

(b) $v = \frac{dx}{dt} = 2t + 6t^2$

At $t = 0, v = 0$ **Ans.**

(c) $a = \frac{dv}{dt} = 2 + 12t$

At $t = 2$ s

$a = 26\, m/s^2$ **Ans.**

19. $v \propto t^{3/4}$

$$a = \frac{dv}{dt} \Rightarrow a \propto t^{-1/4}$$

$$s = \int v dt \Rightarrow s \propto t^{7/4}$$

20. Comparing with $v = u + at$, we have

$u = 40$ m/s

and $a = -10\, m/s^2$

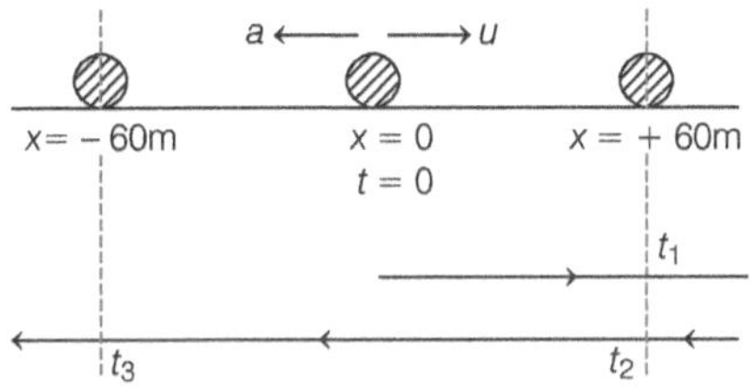

Distance of 60 m from origin may be at $x = +60$ m and $x = -60$ m.

From the figure, we can see that at these two points particle is at three times, t_1, t_2 and t_3.

For $x = +60$ m or t_1 and t_2

$$s = ut + \frac{1}{2}at^2$$

$$\Rightarrow \quad 60 = (+40)t + \frac{1}{2}(-10)t^2$$

Solving this equation, we get

$t_1 = 2$ s and $t_2 = 6$ s **Ans.**

For $x = -60$ m or t_3

$$s = ut + \frac{1}{2}at^2$$

$$\therefore \quad -60 = (+40)t + \frac{1}{2}(-10)t^2$$

Solving this equation, we get positive value of t as

$t_3 = 2(2 + \sqrt{7})$ s **Ans.**

21. Stone comes under gravity.

$\therefore \quad F = mg$

or $\quad a = \frac{F}{m} = g$ **Ans.**

22. Distance x varies with time t as $x^2 = at^2 + 2bt + c$

$$\Rightarrow \quad 2x\frac{dx}{dt} = 2at + 2b$$

$$\Rightarrow \quad x\frac{dx}{dt} = at + b \Rightarrow \frac{dx}{dt} = \frac{(at+b)}{x}$$

$$\Rightarrow \quad x\frac{d^2x}{dt^2} + \left(\frac{dx}{dt}\right)^2 = a$$

$$\Rightarrow \quad \frac{d^2x}{dt^2} = \frac{a - \left(\frac{dx}{dt}\right)^2}{x} = \frac{a - \left(\frac{at+b}{x}\right)^2}{x}$$

$$= \frac{ax^2 - (at+b)^2}{x^3} = \frac{ac - b^2}{x^3}$$

$$\Rightarrow \quad a \propto x^{-3}$$

Hence, $n = 3$

Module 6.

1. $x = \frac{t^2}{2}$

$$\therefore \quad v_x = \frac{dx}{dt} = t$$

$$y = \frac{x^2}{2} = \frac{t^4}{8}$$

$$\therefore \quad v_y = \frac{dy}{dt} = \frac{t^3}{2}$$

At $t = 2$s

$v_x = 2$ m/s

and $v_y = 4$ m/s

$\therefore \quad \mathbf{v} = v_x\,\hat{\mathbf{i}} + v_y\,\hat{\mathbf{j}}$

$= (2\hat{\mathbf{i}} + 4\hat{\mathbf{j}})$ m/s

2. $v_x = \frac{dx}{dt} = 5$

$$v_y = \frac{dy}{dt} = (4t + 1)$$

At 45°, $\quad v_x = v_y$

$\Rightarrow \quad t = 1$ s **Ans.**

3. Comparing with projectile motion we can see that it is like a projectile motion, with $u_x = 4$ m/s, $u_y = 4$ m/s and $a_y = 10\, m/s^2$.

$$x\text{-coordinate} = \text{range} = \frac{2u_x u_y}{a_y}$$

$$= \frac{2 \times 4 \times 4}{10} = 3.2 \text{ m}$$

4. Given, initial position of particle

$$\mathbf{r}_0 = (2\hat{\mathbf{i}} + 4\hat{\mathbf{j}})\text{m}$$

Initial velocity of particle, $\mathbf{u} = (5\hat{\mathbf{i}} + 4\hat{\mathbf{j}})$m/s

Acceleration of particle, $\mathbf{r} = (4\hat{\mathbf{i}} + 4\hat{\mathbf{j}})\text{m/s}^2$

Position of particle at time t is, $\mathbf{r}=\mathbf{r}_0+\mathbf{u}t+\frac{1}{2}\mathbf{a}t^2$

At $t=2$ s, position of particle is,

$$\mathbf{r}=(2\hat{i}+4\hat{j})+(5\hat{i}+4\hat{j})\times 2+\frac{1}{2}(4\hat{i}+4\hat{j})\times 4$$

or $\mathbf{r}=(2+10+8)\hat{i}+(4+8+8)\hat{j}$

$\Rightarrow$ $\mathbf{r}=(20\hat{i}+20\hat{j})$ m

$\therefore$ Distance of particle from origin is,

$$|\mathbf{r}|=\sqrt{(20)^2+(20)^2}=20\sqrt{2}\text{ m}$$

5. Given, $\mathbf{u}=3\hat{i}+4\hat{j}$, $\mathbf{a}=0.4\hat{i}+0.3\hat{j}$, $t=10$ s

From first equation of motion,

$$a=\frac{v-u}{t} \qquad (\because v=u+at)$$

$\Rightarrow$ $v=(0.4\hat{i}+0.3\hat{j})\times 10+(3\hat{i}+4\hat{j})$

$=4\hat{i}+3\hat{j}+3\hat{i}+4\hat{j} \Rightarrow v=7\hat{i}+7\hat{j}$

$\Rightarrow$ $|v|=\sqrt{7^2+7^2}=7\sqrt{2}$ units

6. Coordinates of moving particle at time 't' are $x=\alpha t^3$ and $y=\beta t^3$

$$v_x=\frac{dx}{dt}=3\alpha t^2 \text{ and } v_y=\frac{dy}{dt}=3\beta t^2$$

$\therefore$ $v=\sqrt{v_x^2+v_y^2}=\sqrt{9\alpha^2t^4+9\beta^2t^4}$

$=3t^2\sqrt{\alpha^2+\beta^2}$

7. $x=t^2 \Rightarrow v_x=\frac{dx}{dt}=2t$

$\Rightarrow$ $a_x=\frac{dv_x}{dt}=2$

$\Rightarrow$ $y=t^3-2t$

$\Rightarrow$ $v_y=\frac{dy}{dt}=3t^2-2$

$\Rightarrow$ $a_y=\frac{dv_y}{dt}=6t$

At $t=0$, $v_x=0, v_y=-2, a_x=2$

and $a_y=0$

$\therefore$ $\mathbf{v}=-2\hat{j}$ and $\mathbf{a}=2\hat{i}$ or $\mathbf{v}\perp\mathbf{a}$

At $t=\sqrt{\frac{2}{3}}$, $v_y=0, v_x\neq 0.$

Hence, the particle is moving parallel to x-axis.

8. $v_x=\frac{dx}{dt}=2t-2$

$$a_x=\frac{dv_x}{dt}=2$$

$$v_y=\frac{dy}{dt}=2t-4$$

$$a_y=\frac{dv_y}{dt}=2$$

(a) It crosses y-axis when, $x=0$

$\Rightarrow$ $t=1$ s

At this instant v_y is -2 m/s.

(b) It crosses x-axis when, $y=0$

$\Rightarrow$ $t=2$ s

At this instant v_x is $+2$ m/s

(c) At $t=0$, $v_x=-2$ m/s

and $v_y=-4$ m/s

$\therefore$ Speed $=\sqrt{v_x^2+v_y^2}$

$=2\sqrt{5}$ m/s

(d) At $t=0$, $a_x=2$ m/s^2

and $a_y=2$ m/s^2

$\therefore$ $a=\sqrt{a_x^2+a_y^2}$

$=2\sqrt{2}$ m/s

9. $\mathbf{u}=2\hat{i}$

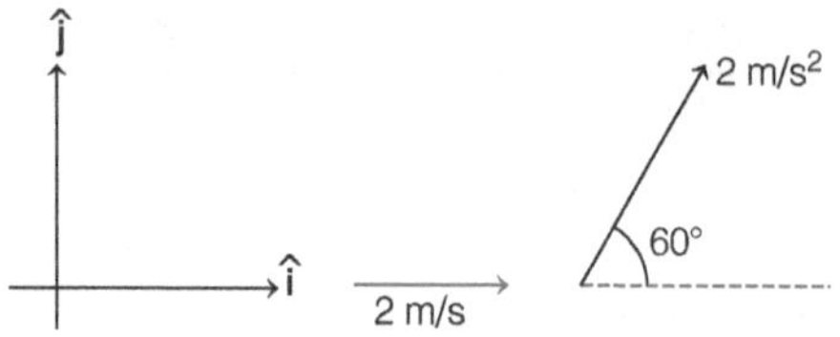

$$\mathbf{a}=(2\cos 60°)\hat{i}+(2\sin 60°)\hat{j}$$

$=(\hat{i}+\sqrt{3}\hat{j})$

$t=2$ s

(i) $\mathbf{v}=\mathbf{u}+\mathbf{a}t=(2\hat{i})+(\hat{i}+\sqrt{3}\hat{j})(2)$

$=4\hat{i}+2\sqrt{3}\hat{j}$

$\therefore$ $|\mathbf{v}|=\sqrt{(4)^2+(2\sqrt{3})^2}$

$=2\sqrt{7}$ m/s **Ans.**

(ii) $\mathbf{s}=\mathbf{u}t+\frac{1}{2}\mathbf{a}t^2=(2\hat{i})(2)+\frac{1}{2}(\hat{i}+\sqrt{3}\hat{j})(2)^2$

$=(6\hat{i}+2\sqrt{3}\hat{j})$

$\therefore$ $|\mathbf{s}|=\sqrt{(6)^2+(2\sqrt{3})^2}$

$=4\sqrt{3}$ m **Ans.**

10. $\mathbf{v}=(2\hat{i}+2t\hat{j})$

(i) $\mathbf{a}=\frac{d\mathbf{v}}{dt}=(2\hat{j})$ m/s^2 = constant

$\therefore$ $\mathbf{v}=\mathbf{u}+\mathbf{a}t$ can be applied.

(ii) $\mathbf{s}=\int_0^1 \mathbf{v}\,dt=\int_0^1(2\hat{i}+2t\hat{j})\,dt=[2t\hat{i}+t^2\hat{j}]_0^1$

$=(2\hat{i}+\hat{j})$ m **Ans.**

11. $\int_{2\hat{i}}^{\mathbf{v}} d\mathbf{v}=\int \mathbf{a}\,dt=\int_0^t(2t\hat{i}+3t^2\hat{j})\,dt$

Solving we get $\mathbf{v} = 2\hat{\mathbf{i}} + t^2\hat{\mathbf{i}} + t^3\hat{\mathbf{j}}$

At $t = 1$ s, $\mathbf{v} = (3\hat{\mathbf{i}} + \hat{\mathbf{j}})$ m/s

$$\int_0^t d\mathbf{r} = \int_0^t \mathbf{v}dt = \int_0^t [(2+t^2)\hat{\mathbf{i}} + t^3\hat{\mathbf{j}}]\,dt$$

$$\Rightarrow \quad \mathbf{r} = \left(2t + \frac{t^3}{3}\right)\hat{\mathbf{i}} + \frac{t^4}{4}\hat{\mathbf{j}}$$

At $t = 1$ s, $\mathbf{r} = \frac{7}{3}\hat{\mathbf{i}} + \frac{1}{4}\hat{\mathbf{j}}$

Therefore, the coordinates are $\left(\frac{7}{3}, \frac{1}{4}\right)$ m.

12. (a) $\mathbf{v}_{av} = \frac{\mathbf{s}}{t} = \frac{\mathbf{r}_f - \mathbf{r}_i}{t}$

$$= \frac{(6\hat{\mathbf{i}} + 4\hat{\mathbf{j}}) - (\hat{\mathbf{i}} + 2\hat{\mathbf{j}})}{4}$$

$$= (1.25\hat{\mathbf{i}} + 0.5\hat{\mathbf{j}})\text{ m/s}$$ **Ans.**

(b) $\mathbf{a}_{av} = \frac{\Delta\mathbf{v}}{t} = \frac{\mathbf{v}_f - \mathbf{v}_i}{t}$

$$= \frac{(2\hat{\mathbf{i}} + 10\hat{\mathbf{j}}) - (4\hat{\mathbf{i}} + 6\hat{\mathbf{j}})}{4}$$

$$= (-0.5\hat{\mathbf{i}} + 4\hat{\mathbf{j}})\text{ m/s}^2$$ **Ans.**

(c) We cannot calculate the distance travelled from the given data.

13. After 2s

$$\mathbf{v}_1 = \mathbf{u} + \mathbf{a}_1 t_1$$
$$= 0 + (2\hat{\mathbf{i}})(2) = (4\hat{\mathbf{i}})\text{ m/s}$$
$$\mathbf{r}_1 = \mathbf{r}_i + \frac{1}{2}\mathbf{a}_1 t_1^2$$
$$= (2\hat{\mathbf{i}} + 4\hat{\mathbf{j}}) + \frac{1}{2}(2\hat{\mathbf{i}})(2)^2$$
$$= (6\hat{\mathbf{i}} + 4\hat{\mathbf{j}})\text{ m}$$

After next 2 s

(a) $\mathbf{v}_2 = \mathbf{v}_1 + \mathbf{a}_2 t_2$

$$= (4\hat{\mathbf{i}}) + (2\hat{\mathbf{i}} - 4\hat{\mathbf{j}})(2) = (8\hat{\mathbf{i}} - 8\hat{\mathbf{j}})\text{ m/s}$$

(b) $\mathbf{r}_2 = \mathbf{r}_1 + \mathbf{v}_1 t_2 + \frac{1}{2}\mathbf{a}_2 t_2^2$

$$= (6\hat{\mathbf{i}} + 4\hat{\mathbf{j}}) + (4\hat{\mathbf{i}})(2) + \frac{1}{2}(2\hat{\mathbf{i}} - 4\hat{\mathbf{j}})(2)^2$$
$$= (18\hat{\mathbf{i}} - 4\hat{\mathbf{j}})\text{ m}$$

$\therefore$ Coordinates are,

$x = 18$ m and $y = -4$ m

14. $x = u_x t + \frac{1}{2}a_x t^2$

$$\therefore \quad 29 = (0)(t) + \frac{1}{2} \times (4.0)t^2$$

$$\therefore \quad t^2 = 14.5 \quad \text{or} \quad t = 3.8 \text{ s}$$

(a) $y = u_y t + \frac{1}{2}a_y t^2$

$$= (8)(3.8) + \frac{1}{2} \times 2 \times 14.5$$
$$= 44.9 \text{ m} \approx 45 \text{ m}$$ **Ans.**

(b) $\mathbf{v} = \mathbf{u} + \mathbf{a}t$

$$= (8\hat{\mathbf{j}}) + (4.0\hat{\mathbf{i}} + 2.0\hat{\mathbf{j}})(3.8)$$
$$= (15.2\hat{\mathbf{i}} + 15.6\hat{\mathbf{j}})$$

$\therefore$ Speed $= |\mathbf{v}| = \sqrt{(15.2)^2 + (15.6)^2}$

$$= 22 \text{ m/s}$$ **Ans.**

15. (a) $u_x = 3$ m/s,

$$a_x = -1.0 \text{ m/s}^2$$

Maximum x-coordinate is attained after time

$$t = \left|\frac{u_x}{a_x}\right| = 3 \text{ s}$$

At this instant $v_x = 0$ and

$$v_y = u_y + a_y t = 0 - 0.5 \times 3 = -1.5 \text{ m/s}$$

$$\therefore \quad \mathbf{v} = (-1.5\hat{\mathbf{j}})\text{ m/s}$$ **Ans.**

(b) $x = u_x t + \frac{1}{2}a_x t^2$

$$= 3 \times 3 + \frac{1}{2}(-1.0)(3)^2 = 4.5 \text{ m}$$

$$y = u_y t + \frac{1}{2}a_y t^2$$

$$= 0 - \frac{1}{2}(0.5)(3)^2 = -2.25 \text{ m}$$

$$\therefore \quad \mathbf{r} = (4.5\hat{\mathbf{i}} - 2.25\hat{\mathbf{j}})\text{ m}$$ **Ans.**

16. (a) Given, $\mathbf{r} = \alpha t\hat{\mathbf{i}} - \beta t^2\hat{\mathbf{j}}$

$$\therefore \quad x = \alpha t \quad \text{or} \quad t = \frac{x}{\alpha}$$

and $\quad y = -\beta t^2$

Substituting the values of t in y, we get

$$y = -\beta\left(\frac{x^2}{\alpha^2}\right)$$

So, the equation of trajectory is,

$$y = -\frac{\beta x^2}{\alpha^2}$$

y - x graph is a parabola as shown below.

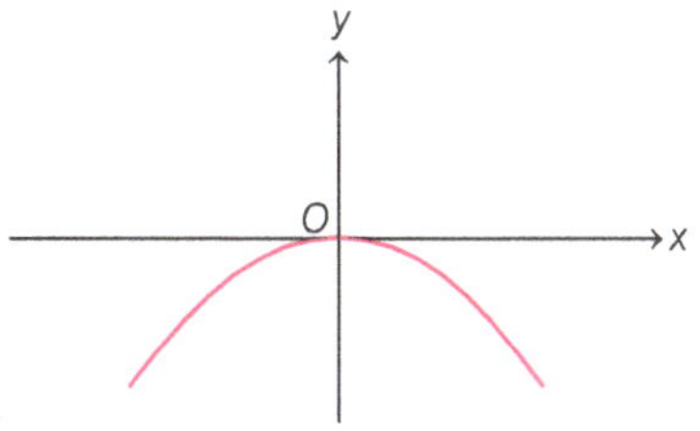

(b) $\mathbf{v} = \frac{d\mathbf{r}}{dt} = \alpha\hat{\mathbf{i}} - 2\beta t\hat{\mathbf{j}}$

and acceleration, $\mathbf{a} = \frac{d\mathbf{v}}{dt} = -2\beta\hat{\mathbf{j}}$

$$|\mathbf{v}| = v = \sqrt{\alpha^2 + 4\beta^2 t^2}$$

and $$|\mathbf{a}| = a = 2\beta$$

(c)

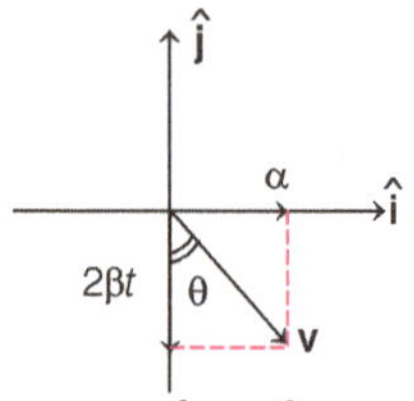

$\mathbf{v} = \alpha\hat{\mathbf{i}} - 2\beta t\hat{\mathbf{j}}$

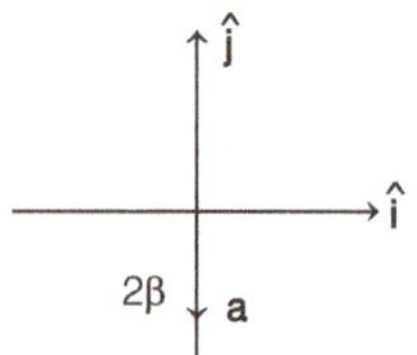

$\mathbf{a} = -2\beta\hat{\mathbf{j}}$

From the above two figures, we can see that θ is the angle between **v** and **a**

$$\tan\theta = \frac{\alpha}{2\beta t}$$

(d) $$\mathbf{v}_{av} = \frac{\mathbf{s}}{t} = \frac{\mathbf{r}_f - \mathbf{r}_i}{t} = \frac{\mathbf{r}_t - \mathbf{r}_0}{t}$$

$$= \frac{(\alpha t\hat{\mathbf{i}} - \beta t^2\hat{\mathbf{j}}) - 0}{t}$$

or $$\mathbf{v}_{av} = \alpha\hat{\mathbf{i}} - \beta t\hat{\mathbf{j}}$$

$$|\mathbf{v}_{av}| = \sqrt{\alpha^2 + \beta^2 t^2}$$

17. (a) $x = \alpha t$

$$\Rightarrow \quad t = \frac{x}{\alpha}$$

Substituting this value of t in the given equation of y,

$$y = \alpha t(1 - \beta t)$$

$$= \alpha\left(\frac{x}{\alpha}\right)\left(1 - \beta\frac{x}{\alpha}\right)$$

$$= x - \frac{\beta x^2}{\alpha}$$

Hence, the desired equation of trajectory of the particle is,

$$y = x - \frac{\beta x^2}{\alpha}$$

y-x graph is a parabola as shown below.

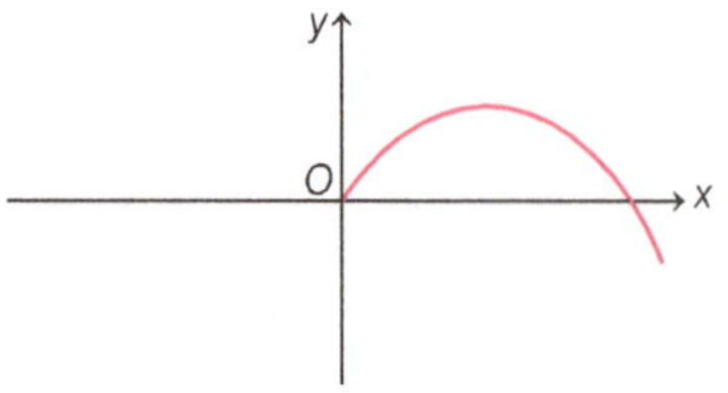

(b) Position vector (w.r.t. origin) of particle at time t is,

$$\mathbf{r} = x\hat{\mathbf{i}} + y\hat{\mathbf{j}}$$

or $$\mathbf{r} = \alpha t\hat{\mathbf{i}} + \alpha t(1 - \beta t)\hat{\mathbf{j}}$$

$$\mathbf{v} = \frac{d\mathbf{r}}{dt} = \alpha\hat{\mathbf{i}} + (\alpha - 2\alpha\beta t)\hat{\mathbf{j}}$$

and acceleration, $a = \frac{d\mathbf{v}}{dt} = (-2\alpha\beta)\hat{\mathbf{j}}$

$$|\mathbf{v}| = \alpha\sqrt{1 + (1 - 2\beta t)^2}$$

$$|\mathbf{a}| = 2\alpha\beta = \text{constant}$$

(c) **a** is along $-\hat{\mathbf{j}}$-direction **v** has one component along $\hat{\mathbf{i}}$-direction and the other component $(\alpha - 2\alpha\beta t)$ along $\hat{\mathbf{j}}$-direction or $(2\alpha\beta t - \alpha)$ component along $-\hat{\mathbf{j}}$-direction or direction of **a**.

Now this vector will make an angle of $\frac{\pi}{4}$ from $-\hat{\mathbf{j}}$ or **a** if both components are same. Or,

$$2\alpha\beta t - \alpha = \alpha$$

$$\therefore \quad t = \frac{1}{\beta} \quad \text{or} \quad t_0 = \frac{1}{\beta}$$

18. (a) $v_x = \frac{dx}{dt} = \alpha\omega\cos\omega t$

$$v_y = \frac{dy}{dt} = \alpha\omega\sin\omega t$$

Speed of particle at time t,

$$v = \sqrt{v_x^2 + v_y^2} = \alpha\omega$$

Since speed is constant. Hence distance = (constant speed) (time)

or $$d = \alpha\omega\tau$$

(b) Velocity vector at time t is,

$$\mathbf{v} = v_x\hat{\mathbf{i}} + v_y\hat{\mathbf{j}}$$

$$= (\alpha\omega\cos\omega t)\hat{\mathbf{i}} + (\alpha\omega\sin\omega t)\hat{\mathbf{j}}$$

$$\mathbf{a} = \frac{d\mathbf{v}}{dt} = (-\alpha\omega^2\sin\omega t)\hat{\mathbf{i}} + (\alpha\omega^2\cos\omega t)\hat{\mathbf{j}}$$

$$\mathbf{v}\cdot\mathbf{a} = -\alpha^2\omega^4\sin\omega t\cos\omega t + \alpha^2\omega^4\sin\omega t\cos\omega t = 0$$

$$\therefore \quad \mathbf{v} \perp \mathbf{a}$$

or the angle between **v** and **a**,

$$\theta = \frac{\pi}{2}$$

Note *Under the given conditions particle executes circular motion with constant speed (called uniform circular motion). In uniform circular motion velocity is tangential to the circle and the acceleration is perpendicular to this towards centre.*

19. Position vector of particle at time t,

$$\mathbf{v} = x\hat{\mathbf{i}} + y\hat{\mathbf{j}} = a\sin\omega t\hat{\mathbf{i}} + b\cos\omega t\hat{\mathbf{j}}$$

$$\mathbf{v} = \frac{d\mathbf{r}}{dt} = a\omega\cos\omega t\hat{\mathbf{i}} - b\omega\sin\omega t\hat{\mathbf{j}}$$

$$\mathbf{a} = \frac{d\mathbf{v}}{dt} = -a\omega^2\sin\omega t\hat{\mathbf{i}} - b\omega^2\cos\omega t\hat{\mathbf{j}}$$

$$= -\omega^2\mathbf{r}$$

$$\mathbf{F} = m\mathbf{a}$$

or $\mathbf{F} = -m\omega^2\mathbf{r}$

Magnitude of force,

$$|\mathbf{F}| = m\omega^2|\mathbf{r}|$$

or $|\mathbf{F}| = m\omega^2\sqrt{x^2 + y^2}$

20. (a) $\sin\omega t = \dfrac{x}{a}$ ...(i)

$\cos\omega t = \dfrac{y}{b}$ ...(ii)

Squaring and adding equations (i) and (ii), we get

$$\frac{x^2}{a^2} + \frac{y^2}{b^2} = 1$$

Note *This is an equation of ellipse.*

(b) $\mathbf{r} = x\hat{\mathbf{i}} + y\hat{\mathbf{j}}$

or $\mathbf{r} = a\sin\omega t\,\hat{\mathbf{i}} + b\cos\omega t\,\hat{\mathbf{j}}$

$$\mathbf{a} = \frac{d^2\mathbf{r}}{dt^2} = -\omega^2[a\sin\omega t\hat{\mathbf{i}} + b\cos\omega t\hat{\mathbf{j}}]$$

or $\mathbf{a} = -\omega^2\mathbf{r}$

Note *At $\omega t = 0$, $x = 0$ and $y = b \rightarrow$ point P.*

At $\omega t = \pi/2$, $x = a$ and $y = 0 \rightarrow$ point Q.

So, in elliptical path particle is moving clockwise as shown below.

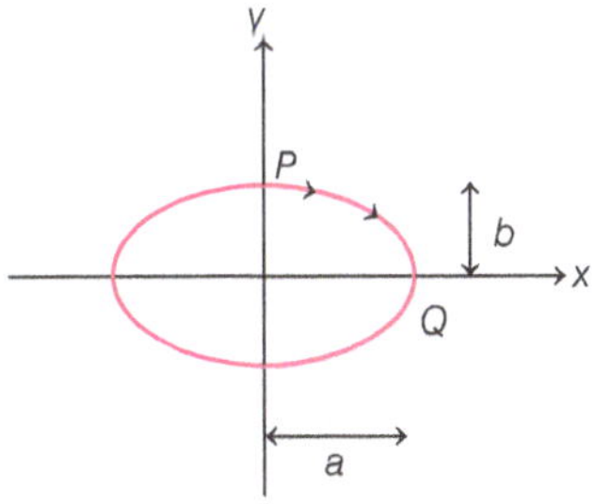

21. Eliminate t and find x - y relations for the trajectories.

22. y-coordinate of particle is zero when

$$4t - 5t^2 = 0$$

$\therefore$ $t = 0$ and 0.8 s

$x = 3t$

At $t = 0, x = 0$

and at $t = 0.8$ s, $x = 2.4$ m **Ans.**

Module 7.

1. Average velocity $= \dfrac{s}{t}$

$$= \frac{\text{area of } v\text{-}t \text{ graph}}{t} = \frac{\left(\frac{1}{2}v_0t_0\right)}{t_0} = \frac{v_0}{2}$$

2. Slope of s-t graph = velocity = positive

At $t = 0$, $s \neq 0$

Further at $t = t_0 : s = 0, v \neq 0$.

3. Area under velocity-time graph gives displacement of body in given time.

Area under acceleration time graph gives change in velocity in the given time.

4. a-t equation from the given graph can be written as

$$a = -2t + 4$$

or $dv = a \cdot dt = (-2t + 4)\,dt$

Integrating we get, $v = -t^2 + 4t$

v-t equation is quadratic. Hence, v-t graph is a parabola.

5. Change in velocity = net area under a - t graph.

6. Motion is first accelerated in positive direction, then uniform in negative direction.

7. Acceleration in second case is more. Hence, slope is more.

8. $a = v\left(\dfrac{dv}{ds}\right) = (10)\left(-\dfrac{20}{30}\right) = -\dfrac{20}{3}\,\text{m/s}^2$

Note *At $s = 15$ m, $v = 10$ m/s.*

9. For uniformly accelerated motion, $v^2 = u^2 + 2as$, i.e. v^2 *versus* s graph is a straight line with intercept u^2 and slope $2a$. Since, intercept is non-zero, so initial velocity is non-zero.

10. Area under v-t graph gives displacement. We can see that area for B is greater than area for A. Hence, B has covered larger distance.

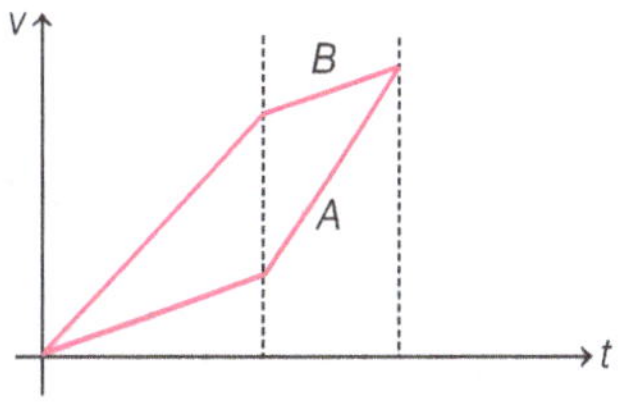

11. Let v_m be the maximum speed,

$$v_m = a_1t_1$$

and $v_m = \sqrt{2a_1l}$

$$t_2 = \frac{2l}{v_m}$$

and $v_m = a_2 t_3 = \sqrt{2a_2(3l)}$

Now, average speed

$$v_{av} = \frac{l + 2l + 3l}{t_1 + t_2 + t_3}$$

$$v_{av} = \frac{6l}{(v_m/a_1) + (2l/v_m) + (v_m/a_2)}$$

$$= \frac{6l}{\left(\frac{v_m}{v_m^2/2l}\right) + \left(\frac{2l}{v_m}\right) + \left(\frac{v_m}{v_m^2/6l}\right)}$$

$$= \frac{6l}{(10l/v_m)} = \frac{3v_m}{5}$$

$$\frac{v_{av}}{v_m} = \frac{3}{5}$$

12. Area of a-t graph gives change in velocity. When net area will become zero, particle will acquire its initial velocity.

13. $a = v \cdot \frac{dv}{ds} = (4)(-\tan 60°)$

$= -4\sqrt{3}\, \text{m/s}^2$.

14. $\frac{dx}{dt} = v = -1 - 2t$

Comparing with $v = u + at$,

we have $u = -1$ m/s and $a = -2\, \text{m/s}^2$

At $t = 0$, $x = 1$ m. Then, u and a both are negative.

Hence, x-coordinate of particle will go on decreasing.

15. Area under acceleration-time graph gives the change in velocity.

v_i is zero. Therefore Δv or $v_f - v_i$ is v_f or v_{max}.

Hence, v_{max} = area of graph $= (1/2) \times 10 \times 11$

$= 55$ m/s

16. The v-x equation from the given graph can be written as

$$v = \left(-\frac{v_0}{x_0}\right)x + v_0 \Rightarrow \frac{dv}{dx} = -\frac{v_0}{x_0}$$

Now, $a = v \cdot \frac{dv}{dx}$

$\Rightarrow$ $a = \left(\frac{v_0}{x_0}\right)^2 x - \frac{v_0^2}{x_0}$ (a linear equation in a and x)

Thus, a-x graph is a straight line with positive slope and negative intercept.

17. Let us take vertically upward direction positive upto highest point velocity decreases at a constant rate, then it increases in negative direction with same rate.

18. a = constant

$v = at$ and $x = \frac{1}{2}at^2$

19.

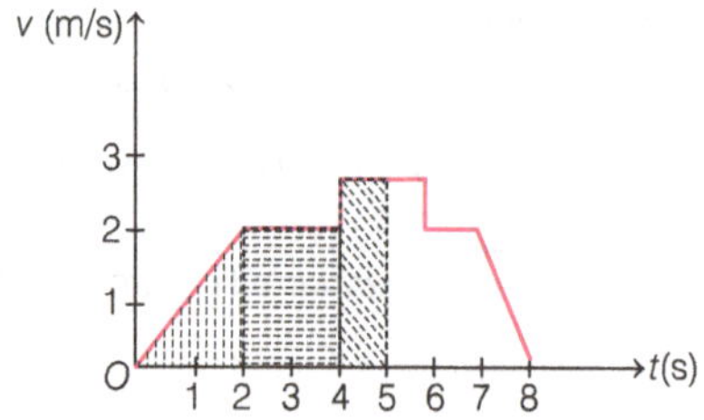

Displacement of particle = Shaded area

$$= \left(\frac{1}{2} \times 2 \times 2\right) + (2 \times 2) + (3 \times 1)$$

$= 2 + 4 + 3 = 9$ m

20. Distance = area under v-t graph with positive value

$$\text{Distance} = \frac{1}{2} \times 10 \times 5 + 5 \times 10 + 5 \times 10 + \frac{1}{2} \times 5 \times 10 + \frac{1}{2} \times 5 \times 20 + \frac{1}{2} \times 5 \times 20 = 250\, \text{m}$$

Displacement = area of graph with sign

$= (200 - 50)$ m $= 150$ m

$$\therefore \quad \frac{\text{Distance}}{\text{Displacement}} = \frac{250}{150} = \frac{5}{3}$$

21.

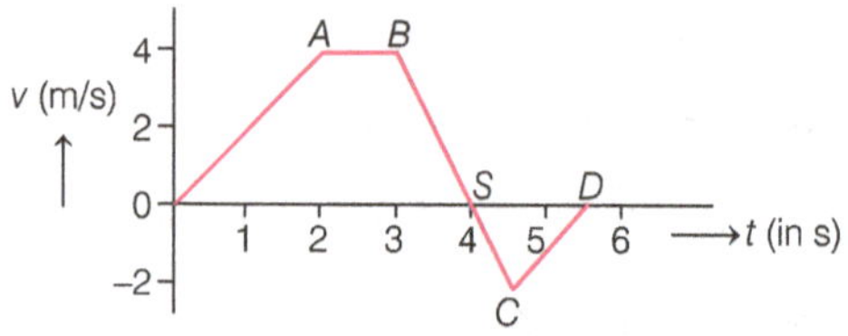

$$OS = 4 + \frac{1}{3} = \frac{13}{3}; \; SD = 2 - \frac{1}{3} = \frac{5}{3}$$

Distance covered by the body = area of v-t graph

= area ($OABS$) + area (SCD)

$$= \frac{1}{2}\left(\frac{13}{3} + 1\right) \times 4 + \frac{1}{2} \times \frac{5}{3} \times 2$$

$$= \frac{32}{3} + \frac{5}{3} = \frac{37}{3}\, \text{m}$$

22. Distance along a line, i.e. displacement (s)

$= t^3$ ($\because s \propto t^3$ given)

By double differentiation of displacement, we get acceleration.

$$v = \frac{ds}{dt} = \frac{dt^3}{dt} = 3t^2$$

and $a = \frac{dv}{dt} = \frac{d\,3t^2}{dt} = 6t$

$a = 6t$ or $a \propto t$

Hence, graph (b) is correct.

23. $\therefore$ $|v_{10}| = (100 + 10 \times 10)$ m/s

$v_{10} = -200$ m/s

and $v_0 = -100$ m/s

From 10s to 20s velocity remains zero.

From $t = 0$ s to 10 s velocity increase in magnitude linearly.

Graph given in option (a) fits correctly.

24. Acceleration (a) is constant.

$\therefore \quad v \propto t$ (straight line graph)

and $\quad x \propto t^2$ (parabolic graph)

25. $a = \frac{F}{m} = \frac{\alpha}{m} t$ ....(i)

or $\quad a \propto t$

i.e. a - t graph is a straight line passing through origin.

If $u = 0$, then integration of Eq. (i) gives,

$$v = \frac{\alpha t^2}{2m}$$

or $\quad v \propto t^2$ ...(ii)

Hence, in this situation (when $u = 0$) v - t graph is a parabola passing through origin.

From Eqs. (i) and (ii), we get

$$v \propto a^2$$

Therefore, curve-2 can show velocity against acceleration.

26. For $t < T$, $v = -$ve

For $t > T$, $v = +$ ve

At $\quad t = T, v = 0$

$\therefore$ Particle changes direction of velocity at $t = T$

$s =$ Net area of v - t graph $= 0$

$a =$ Slope of v - t graph $=$ constant

27. (a) Graph is passing through origin. Hence, option (a) is correct.

(b) From graph, Q starts at $t = 10$ s.

(c) From the given graph, it is clear that particle Q will overtake the particle P at the distance of 30 m and the time difference of 30 s.

28. $\frac{dx}{dt} =$ slope ≥ 0 always increasing

$\frac{dx}{dt} < 0$; and at $t \to \infty$, $\frac{dx}{dt} \to 0$

$\frac{dx}{dt} > 0$ for first half $\frac{dx}{dt} < 0$ for second half.

$\frac{dx}{dt} =$ constant

29. In (a) and (b), if velocity is in the direction of acceleration (or of the same sign as that of acceleration) then speed increases. And if velocity is in opposite direction, then speed decreases.

In (c) slope of s-t graph (velocity) is increasing. Therefore speed is increasing. In (d) slope of s-t graph is decreasing. Therefore, speed is decreasing.

30. In portion AB, we can see that velocity is positive and increasing. Similarly for other parts we can draw the conclusions.

31. (a) Average velocity $= \frac{s}{t} = \frac{\text{Area of } v\text{-}t \text{ graph}}{\text{Time}}$

$$= \frac{20}{4} = 5 \text{ m/s}$$

(b) Average acceleration $= \frac{v_f - v_i}{\text{time}}$

$$= \frac{v_{4\,s} - v_{1\,s}}{4-1} = \frac{0-5}{3} = -\frac{5}{3} \text{ m/s}^2$$

(c) Average speed $= \frac{d}{t} = \frac{|\text{Total area}|}{\text{Time}}$

$$= \frac{20+10}{6} = 5 \text{m/s}$$

(d) Rate of change of speed at 4 s $= |\mathbf{a}|$

$= |$slope of v-t graph $| = 5$ m/s^2

32.

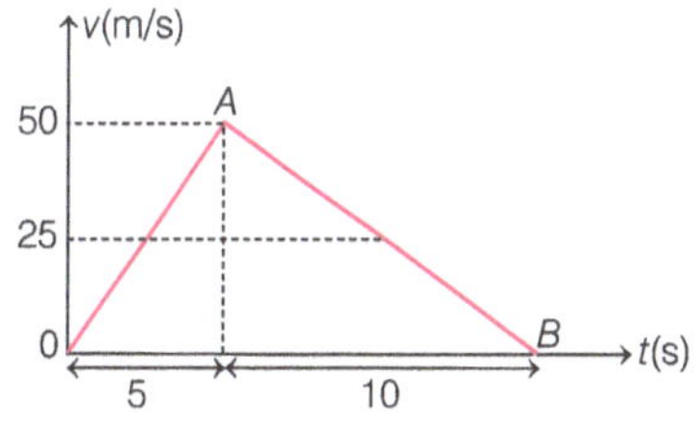

Slope of $OA =$ acceleration

$= +10$ m/s^2

Slope of $AB =$ retardation

$= -5$ m/s^2

Since the particle is not changing the direction.

Hence,

distance $=$ displacement $=$ area under v-t graph.

33. (b) $v =$ Slope of x - t graph

34. (a) Average velocity $= \frac{s}{t} = \frac{x_f - x_i}{t}$

$$= \frac{x_{10\,\text{sec}} - x_{0\,\text{sec}}}{10} = \frac{100}{10} = 10 \text{ m/s}$$

(b) Instantaneous velocity $=$ Slope of x - t graph

35. (a) Velocity $=$ slope of s-t graph

$\therefore$ Sign of velocity $=$ sign of slope of s-t graph

(b) Let us discuss any one of them, let the portion cd, slope of s-t graph ($=$ velocity) of this region is negative but increasing in magnitude. Therefore velocity is negative but increasing. Therefore sign of velocity and acceleration both are negative.

36. Displacement $s =$ net area of v-t graph

$$= 40 + 40 + 40 - 20 = 100\text{ m}$$

Distance $d = |\text{Total area}|$

$$= 40 + 40 + 40 + 20 = 140\text{ m}$$

(a) Average velocity $= \dfrac{s}{t} = \dfrac{100}{14} = \dfrac{50}{7}$ m/s **Ans.**

(b) Average speed $= \dfrac{d}{t} = \dfrac{140}{14} = 10$ m/s **Ans.**

37. (a) $a_{av} = \dfrac{v_f - v_i}{t} = \dfrac{v_{12} - v_6}{12 - 6}$

$$= \frac{(-10) - (20)}{6} = -5\text{ m/s}^2$$ **Ans.**

(b) $\Delta v = v_f - v_i =$ net area of a-t graph

$\therefore$ $v_{14} - v_0 = 40 + 30 + 40 - 20 = 90$

But $v_0 = 0$

$\therefore$ $v_{14} = 90$ m/s **Ans.**

38. Kinetic energy will first increase and then decrease.

$v = gt$ (in downward journey)

$\therefore$ $\text{KE} = \dfrac{1}{2} mv^2 = \dfrac{1}{2} mg^2t^2$

Hence, KE *versus* t graph is a parabola.

39. Speed first increases. Then just after collision, it becomes half and now it decreases.

Pattern of velocity is same (with sign). In the answer, downward direction is taken as positive.

Further, $v = gt$

Hence, v-t graph is straight line.

40. (a) $a = 2t - 2$ (from the graph)

Now, $\int_0^v dv = \int_0^t a\,dt = \int_0^t (2t - 2)\,dt$

$\therefore$ $v = t^2 - 2t$

(b) $s = \int_2^4 v\,dt = \int_2^4 (t^2 - 2t)\,dt = 6.67$ m

41. Note *Let us denote v_x by v, a_x by a and distance s by d.*

a versus t graph

$a = \dfrac{dv}{dt} =$ Slope of v - t graph

Time interval	Slope of v - t graph	a
0-1	1	1
1-3	0	0
3-6	−1	−1
6-7	2	2
7 onwards	0	0

Corresponding a - t graph is as shown below.

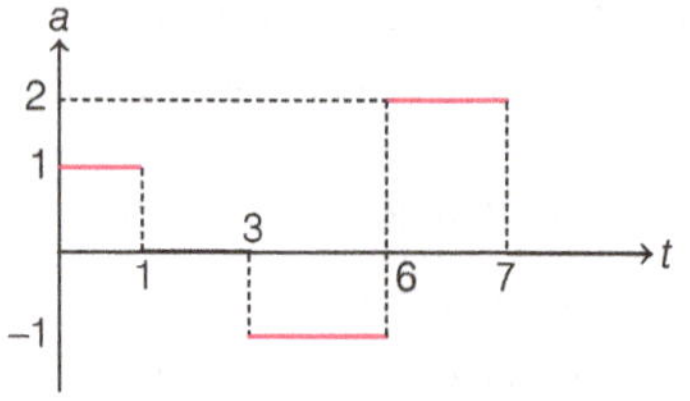

***x*-coordinate *versus t* graph**

Time interval	a	x - t equation	x - t graph
0-1	1	$x = \frac{1}{2}at^2 = \frac{t^2}{2}$	Parabola
1-3	0	$x = x_1 + v_1 t$	Straight line
3-6	−1	$x = x_3 + v_3 t - \frac{1}{2} \times 1 \times t^2$	Parabola
6-7	2	$x = x_6 + v_6 t + \frac{1}{2} \times 2 \times t^2$	Parabola

In the above table,

$x_1 = x$- coordinate at time $t = 1$

$v_1 =$ velocity at time $t = 1$

Similarly x_3, v_3 and so on.

Further area under v-t graph gives displacement. If v is positive then body will be moving towards positive x-direction. Or we can say displacement is positive. But distance travelled is always positive.

Time interval	v	Area	Displacement	Distance
0-1	Positive	0.5	+0.5	0.5
1-3	Positive	2	+2.0	2.0
3-4	Positive	0.5	+0.5	0.5
4-6	Negative	2	−2	2
6-7	Negative	1	−1	1

Note *Total displacement is $0.5 + 2.0 + 0.5 - 2 - 1 = 0$. Hence final x - coordinate at $t = 7$ should also be zero as the journey was started from $x = 0$. Total distance travelled is 6. Corresponding x-coordinate versus time graph is shown below.*

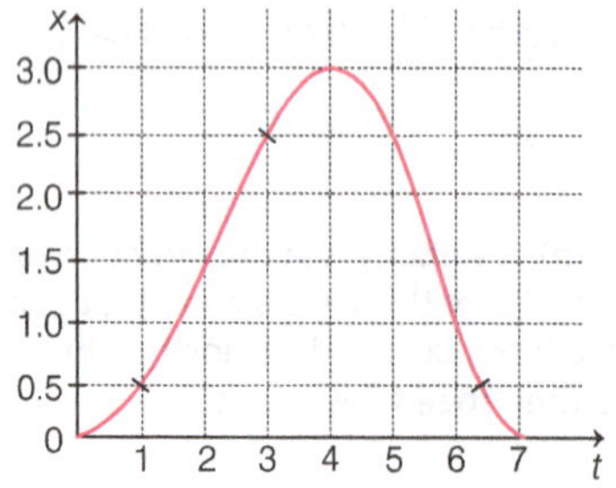

Distance *d* versus time *t*

In displacement we take plus and minus signs. So, displacement may increase and then decrease. But distance keeps on increasing as long as particle is in motion. Nature of graph (parabola or straight line) would be same as we have drawn in displacement graph.

Corresponding *d versus t* graph is as shown below.

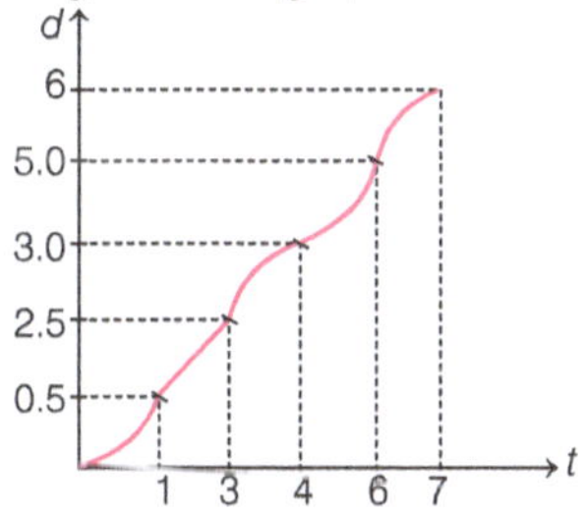

Note *(i) In the question no units were given against v and t. So, we have also not taken any units.*

(ii) From the given graph we can see that v = 0 at t = 0, 4 and 7. Therefore slope of x - t graph should also be zero at these times.

42. $\because\ v = a \sin \omega t$

or $\quad v = \pi \sin \frac{2\pi}{2} t \quad$ or $\quad v = \pi \sin \pi t$

The speed-time graph is

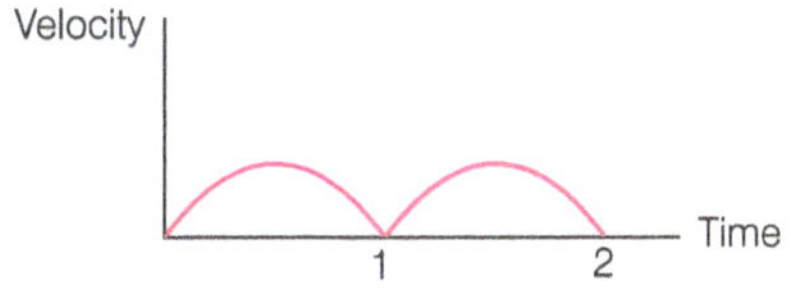

$\because \qquad \text{Speed} = |v|$

$\therefore \quad \text{Distance} = \int_0^2 |v|\, dt = 2\int_0^1 |v|\, dt$

$= \frac{2\pi}{\pi}[-\cos \pi t]_0^t = \frac{2\pi}{\pi} \times 2 = 4\text{ m}$

43. As $v^2 = u^2 + 2ax$

At point 'A',

$$40 = u^2 + 2a \cdot 10 \qquad \text{...(i)}$$

At point 'B',

$$60 = u^2 + 2a \cdot 20 \qquad \text{...(ii)}$$

Subtracting Eq. (ii) from Eq. (i), we get

$$a = 1\text{ m/s}^2$$

44.

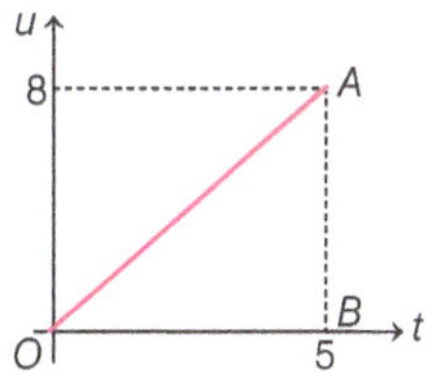

Distance travelled = Area of speed - time graph.

$$= \frac{1}{2} \times 5 \times 8 = 20\text{ m}$$

Module 8.

Introductory Exercise 1

2. $v_A = \text{Slope of } A = \frac{2}{5} = 0.4\text{ m/s}$

$v_B = \text{Slope of } B = \frac{12}{5} = 2.4\text{ m/s}$

$\therefore \quad v_{AB} = v_A - v_B = -2\text{ m/s}$

3. $\quad a_A = a_B = g \qquad$ (downwards)

$\therefore \quad a_{AB} = a_A - a_B = 0 \qquad$ **Ans.**

4.

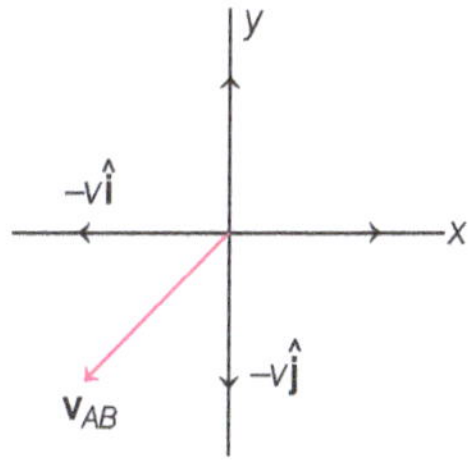

$$\mathbf{v}_{AB} = \mathbf{v}_A - \mathbf{v}_B = -v\hat{\mathbf{j}} - v\hat{\mathbf{i}}$$

$$\mathbf{v}_{AC} = \mathbf{v}_A - \mathbf{v}_C = -v\hat{\mathbf{j}} - v\hat{\mathbf{j}} = -2v\hat{\mathbf{j}}$$

Hence, $\mathbf{v}_{AC}$ is towards negative *y*-direction.

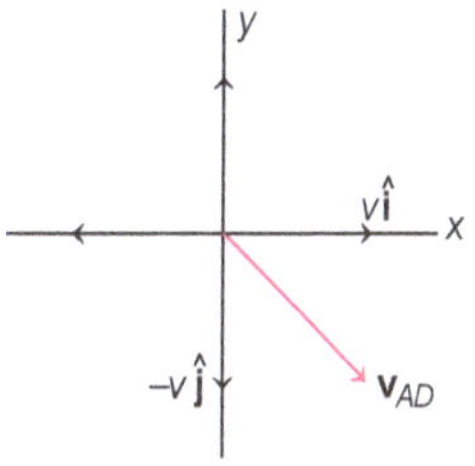

$$\mathbf{v}_{AD} = \mathbf{v}_A - \mathbf{v}_D = -v\hat{\mathbf{j}} - (-v\hat{\mathbf{i}}) = v\hat{\mathbf{i}} - v\hat{\mathbf{j}}$$

Introductory Exercise 2

1. $7.2 = \frac{100}{(v+5)(5/18)}$ or $v = 45\text{ km/h}$

2. $\quad v_A + v_B = 4\text{ m/s}$

and $v_A - v_B = \frac{4.0}{10} = 0.4\text{ m/s}$

Exercise

1. $v_1 = \frac{ds_1}{dt} = (2 - 8t)$

$$v_2 = \frac{ds_2}{dt} = (-2 + 8t)$$

$\therefore \qquad v_{12} = v_1 - v_2 = (4 - 16t)$

v_{12} does not keep on increasing.

2. Reason is the formula of relative velocity. But it does not explain Assertion correctly. The correct explanation of Assertion is due to visual perception of motion. The object appears to be moving faster, when angle subtended by the object on eye (called visual angle) changes fast.

3. $\mathbf{v}_{BA} = \mathbf{v}_B - \mathbf{v}_A = (v_B \cos\theta\,\hat{\mathbf{i}} + v_B \sin\theta)\hat{\mathbf{j}} - v_A\hat{\mathbf{i}}$

$$= (v_B \cos\theta - v_A)\,\hat{\mathbf{i}} + v_B \sin\theta\,\hat{\mathbf{j}}$$

Ship B always remains north of A if $\mathbf{v}_{BA}$ is along north. Or its component along x -axis is zero.

$\therefore \quad v_B \cos\theta - v_A = 0$

or $\quad \dfrac{v_A}{v_B} = \cos\theta$

4. v_{IB} = velocity of insect w.r.t. belt $= v_I - v_B$

$\therefore \quad v_I = v_{IB} + v_B = 4 + 2 = 6\ \text{ms}^{-1}$

5.&6. A and B both are the marks on belt and relative to belt. Relative speed is constant in moving from A to B or B to A. This time is,

$$t = \frac{d}{\text{relative speed}}$$

$$= \frac{60}{4} = 15\ \text{s}$$

7. According to question, train A and train B are running on parallel tracks in the opposite direction.

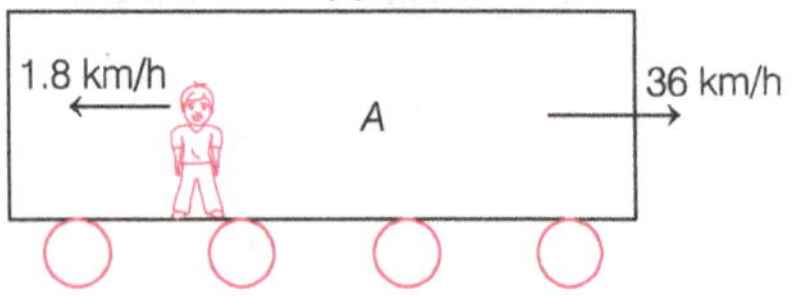

$$v_A = 36\ \text{km/h} = 10\ \text{m/s}$$

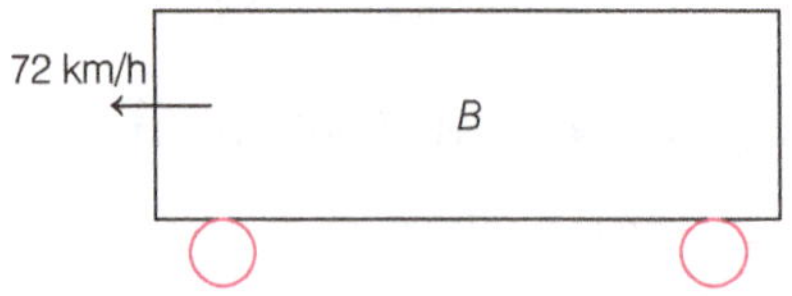

$v_B = -72\ \text{km/h} = -20\ \text{m/s};\ v_{MA} = -1.8\ \text{km/h} = -0.5\ \text{m/s}$

$v_{\text{man},\,B} = v_{\text{man},\,A} + v_{A,\,B}$

$= v_{\text{man},\,A} + v_A - v_B = -0.5 + 10 - (-20)$

$= -0.5 + 30 = 29.5\ \text{m/s}$

8. $\mathbf{r}_B = 0 \quad \text{at} \quad t = 0$

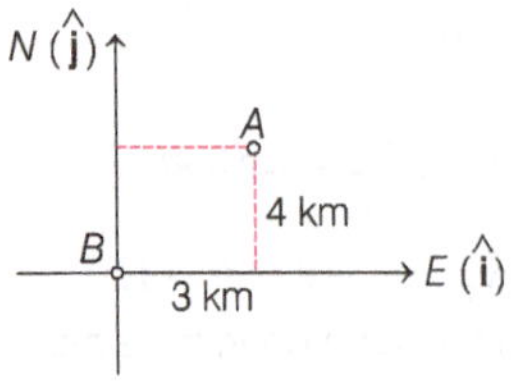

$\mathbf{r}_A = 3\hat{\mathbf{i}} + 4\hat{\mathbf{j}}$

At $\quad t = 0,$

$\mathbf{v}_A = (-20\hat{\mathbf{j}})$

$\mathbf{v}_B = (40\cos 37°)\hat{\mathbf{i}} + (40\sin 37°)\hat{\mathbf{j}}$

$= (32\hat{\mathbf{i}} + 24\hat{\mathbf{j}})$

$\mathbf{v}_{AB} = (-32\hat{\mathbf{i}} - 44\hat{\mathbf{j}})\ \text{km/h}$ **Ans.**

At time $t = 0,$

$\mathbf{r}_{AB} = \mathbf{r}_A - \mathbf{r}_B = (3\hat{\mathbf{i}} + 4\hat{\mathbf{j}})$

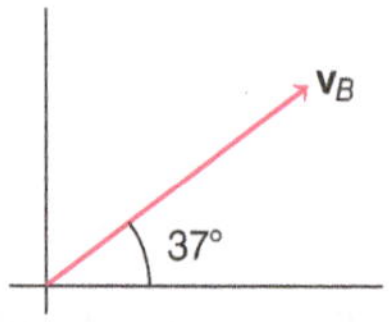

$\therefore$ At time $t = t$,

$\mathbf{r}_{AB} = (\mathbf{r}_{AB}\ \text{at}\ t = 0) + \mathbf{v}_{AB}\,t$

$- = (3 - 32t\hat{\mathbf{i}} + (4 - 44t)\hat{\mathbf{j}}$ **Ans.**

9.

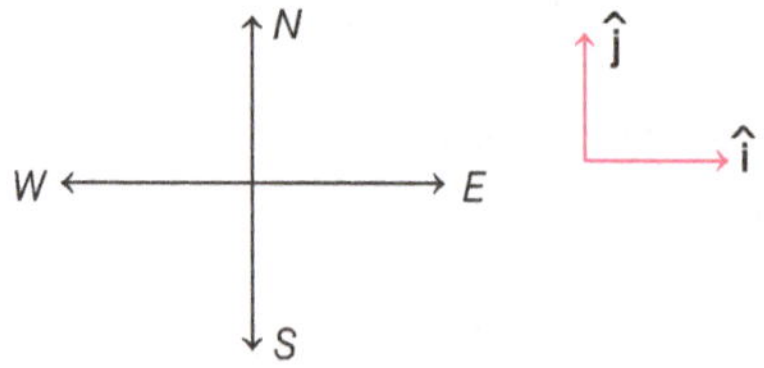

Given, $\mathbf{v}_s = (30\,\hat{\mathbf{i}})\ \text{km/h}$ $\quad (s \rightarrow \text{ship})$

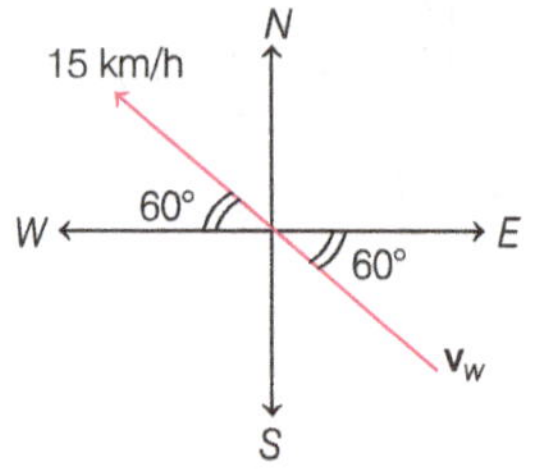

$\mathbf{v}_w = [(15\cos 60°)(-\hat{\mathbf{i}}) + (15\sin 60°)\hat{\mathbf{j}}]\ \text{km/s}$

$= (-7.5\,\hat{\mathbf{i}} + 13\,\hat{\mathbf{j}})\ \text{km/h}$

$\mathbf{v}_{ws} = \mathbf{v}_w - \mathbf{v}_s$

$= (-7.5\,\hat{\mathbf{i}} + 13\,\hat{\mathbf{j}}) - (30\,\hat{\mathbf{i}})$

$= [-37.5\,\hat{\mathbf{i}} + 13\,\hat{\mathbf{j}}]\ \text{km/h}$

$\therefore \quad |\mathbf{v}_{ws}| = \sqrt{(37.5)^2 + (13)^2}$

$\approx 40\ \text{km/h}$

$\therefore \quad |\mathbf{v}_{ws}| \approx 40\text{km/s}$

$\tan\theta = \dfrac{13}{37.5}$

or $\quad \theta = \tan^{-1}\dfrac{13}{37.5}$

$$\Rightarrow \qquad \theta \approx 19°$$

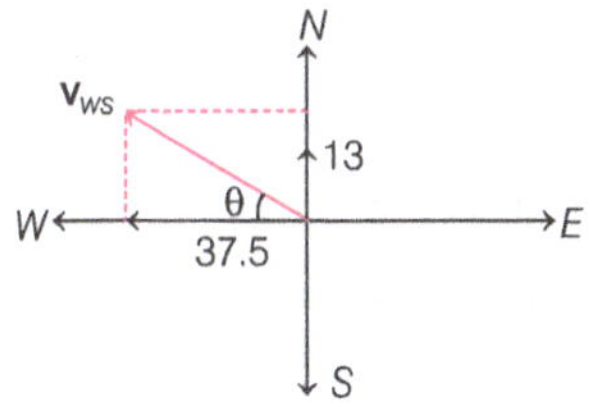

Note *(i) Equator means E-W direction at a particular point.*
(ii) Southwestern wind blows at $\phi = 60°$ to the equator means wind is blowing from S-E direction at 60° as shown in figure.

10.

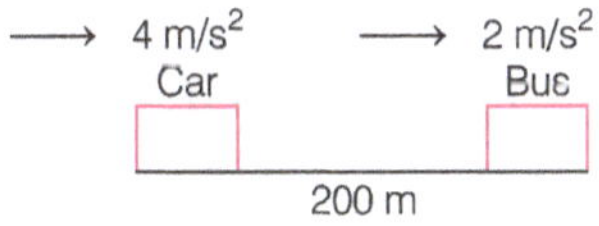

Given, $u_C = u_B = 0$, $a_C = 4\,\text{m/s}^2$, $a_B = 2\,\text{m/s}^2$

Hence, relative acceleration, $a_{CB} = 2\,\text{m/s}^2$

$$\therefore \qquad s = ut + \frac{1}{2}at^2$$

$$\Rightarrow \qquad 200 = \frac{1}{2} \times 2t^2 \qquad [\because u = 0]$$

$$\Rightarrow \qquad t = 10\sqrt{2}\text{ s}$$

11. (a) When the two meet,

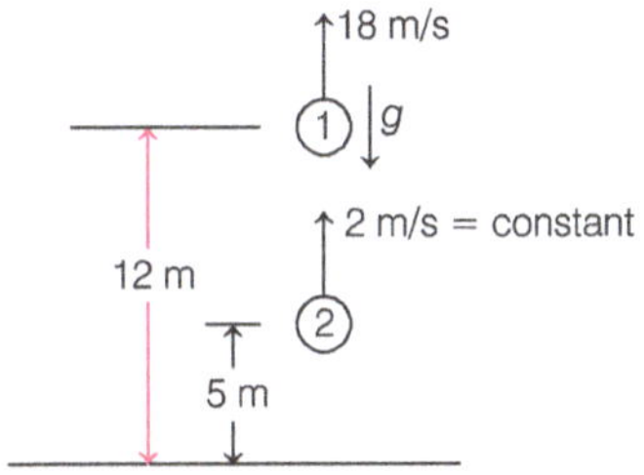

$$s_2 = s_1 + 7$$

or $\quad (2\,t) = (18\,t) - (4.9\,t^2) + 7$

Solving, we get

$$t = 3.65\,s \qquad \textbf{Ans.}$$

$$s_2 = 2 \times 3.65 = 7.3\text{ m}$$

$\therefore$ Height $= 5 + 7.3$

$$= 12.30\text{ m} \qquad \textbf{Ans.}$$

(b) $v_{\text{ball}} = u - gt = 18 - 9.8 \times 3.65$

$$= -17.77\text{ m/s}$$

$\therefore$ Velocity of ball with respect to elevator

= velocity of ball − velocity of elevator

$$= (-17.77) - (2)$$

$$= -19.77\text{ m/s}$$

$$\approx -19.8\text{ m/s} \qquad \textbf{Ans.}$$

Negative sign indicates the downward direction.

12. $t = \dfrac{l_1 + l_2}{v_{\text{Rel}}} \Rightarrow t \propto \dfrac{1}{v_{\text{Rel}}}$

So, $\dfrac{t_1}{t_2} = \dfrac{80+30}{80-30} = \dfrac{110}{50} = \dfrac{11}{5}$

13. Time, $t = \dfrac{\text{distance}}{\text{velocity}}$

$$\therefore \qquad \frac{60l + 4l}{20} - \frac{61l}{30} = 35$$

$$\Rightarrow \qquad l = \frac{1050}{35}$$

Total length of tunnel given,

$$L = 60l = \frac{1050}{35} \times 60$$

$$= 1800\text{ m}$$

14. Velocity of train B with respect to

$$A = \mathbf{v}_B - \mathbf{v}_A$$

$$= 54 - (-90) = 144\text{ km/h}$$

$$= \frac{144 \times 5}{18} = 40\text{ m/s}$$

Time of crossing $= \dfrac{\text{length of train}}{\text{relative velocity}}$

$$\Rightarrow \qquad 8 = \frac{l}{40}$$

$$\therefore \qquad l = 8 \times 40 = 320\text{ m}$$

15. $\mathbf{v}_Q = -20\hat{\mathbf{i}}$

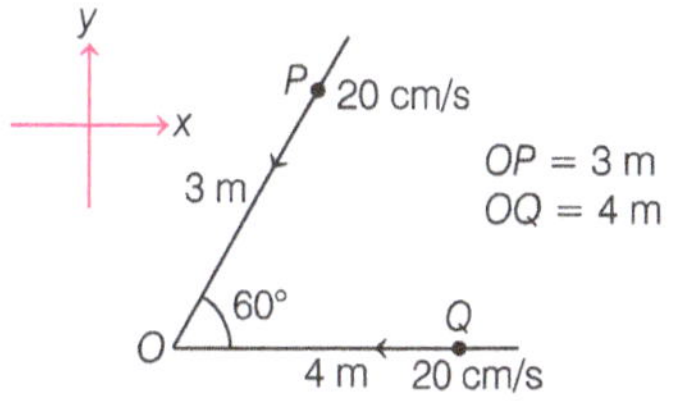

$$\mathbf{v}_P = -20\cos 60\hat{\mathbf{i}} - 20\sin 60°\hat{\mathbf{j}}$$

$$= -10\hat{\mathbf{i}} - 10\sqrt{3}\,\hat{\mathbf{j}}$$

Assuming P to be at rest,

$$\mathbf{v}_{QP} = \mathbf{v}_Q - \mathbf{v}_P$$

$$= -10\hat{\mathbf{i}} + 10\sqrt{3}\,\hat{\mathbf{j}}$$

Now, $\qquad \tan\theta = \dfrac{10\sqrt{3}}{10} = \sqrt{3}$

or $\qquad \theta = 60°$

where, θ is the angle of $\mathbf{v}_{QP}$ from x-axis towards positive y-axis.

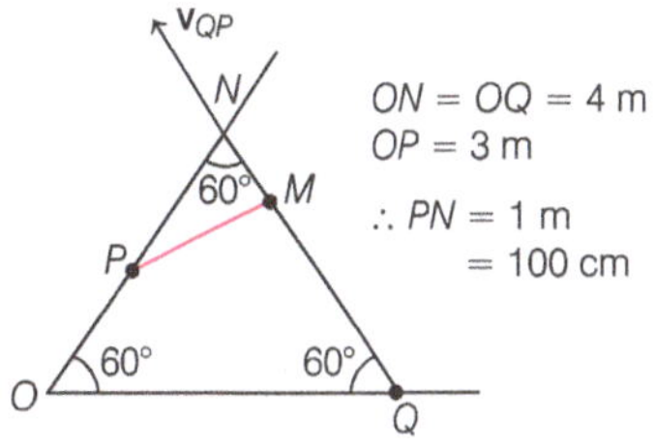

Shortest distance $= PM = PN \sin 60°$

$$= (100)\frac{\sqrt{3}}{2}$$

$$= 50\sqrt{3} \text{ cm}$$

16. Considering the initial position of ship A as origin.

After time t, coordinates of ships A and B are

$$(80 - 10t, 150) \quad \text{and} \quad (30t, 50t)$$

So, distance between A and B after time t is

$$d = \sqrt{(80 - 10t - 30t)^2 + (150 - 50t)^2}$$

$$\Rightarrow \quad d^2 = (80 - 40t)^2 + (150 - 50t)^2$$

Distance is minimum when $\frac{d}{dt}(d^2) = 0$

$$\frac{d}{dt}[(80 - 40t)^2 + (150 - 50t)^2] = 0$$

$$\Rightarrow \quad 2(80 - 40t)(-40) + 2(150 - 50t)(-50) = 0$$

After solving, we get

$$t = 2.6 \text{ h}$$

17.

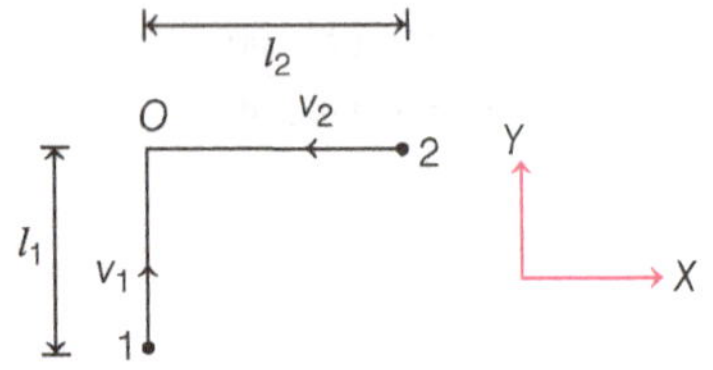

Let us assume particle-2 at rest.

Now, $\mathbf{v}_{12} = \mathbf{v}_1 - \mathbf{v}_2 = v_1\hat{\mathbf{j}} - (-v_2\hat{\mathbf{i}})$

$$= v_2\hat{\mathbf{i}} + v_1\hat{\mathbf{j}}$$

$$|\mathbf{v}_{12}| = \sqrt{v_1^2 + v_2^2}$$

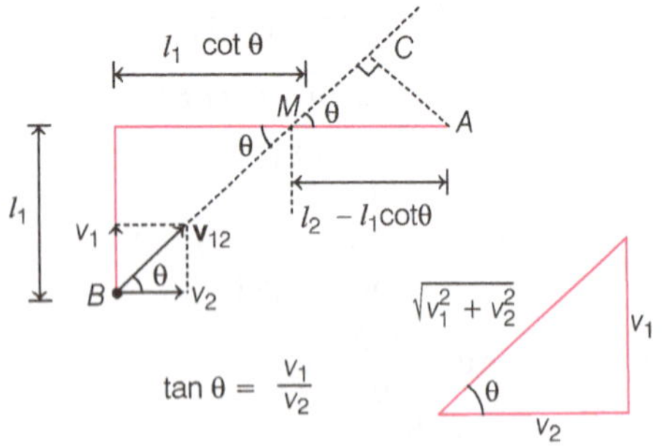

Minimum distance

$$d_{min} = AC = AM \sin\theta$$

$$= (l_2 - l_1 \cot\theta) \sin\theta$$

$$= l_2 \sin\theta - l_1 \cos\theta$$

$$= \frac{l_2 v_1 - l_1 v_2}{\sqrt{v_1^2 + v_2^2}} \left[\sin\theta = \frac{v_1}{\sqrt{v_1^2 + v_2^2}}, \cos\theta = \frac{v_2}{\sqrt{v_1^2 + v_2^2}}\right]$$

Since it's a distance. So, it should be positive.

$$d_{min} = \frac{|l_1 v_2 - l_2 v_1|}{\sqrt{v_1^2 + v_2^2}}$$

Time,

$$t = \frac{BC}{|\mathbf{v}_{12}|} = \frac{BM + MC}{\sqrt{v_1^2 + v_2^2}}$$

$$= \frac{(l_1 \operatorname{cosec}\theta) + (l_2 - l_1 \cot\theta)\cos\theta}{\sqrt{v_1^2 + v_2^2}}$$

$$= \frac{l_1\left(\frac{\sqrt{v_1^2 + v_2^2}}{v_1}\right) + \left[l_2 - l_1 \times \frac{v_2}{v_1}\right]\left[\frac{v_2}{\sqrt{v_1^2 + v_2^2}}\right]}{\sqrt{v_1^2 + v_2^2}}$$

Simplifying this, we get

$$t = \frac{l_1 v_1 + l_2 v_2}{v_1^2 + v_2^2}$$

18. Velocity of rain is at 30° with vertical. So, its horizontal component is $v_R \sin 30° = \frac{v_R}{2}$. When man starts walking with 10 km/h, rain appears vertical. So, horizontal component $\frac{v_R}{2}$ is balanced by his speed of 10 km/h.

Thus,

$$\frac{v_R}{2} = 10$$

or

$$v_R = 20 \text{ km/h}$$

19. Clearly, $\tan 45° = \frac{|\mathbf{v}_{rm}|}{|\mathbf{v}_m|}$

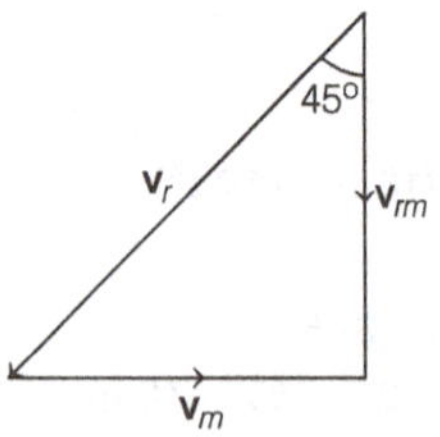

$$\Rightarrow \quad v_{rm} = v_m$$

$$\Rightarrow \quad v_{rm} = 15\sqrt{2} \times \frac{\sqrt{2}}{\sqrt{2}} = \frac{30}{\sqrt{2}} \text{ km/h}$$

20. $t = \dfrac{d}{v_r \cos 30^\circ} = \dfrac{2d}{\sqrt{3}\, v_r}$

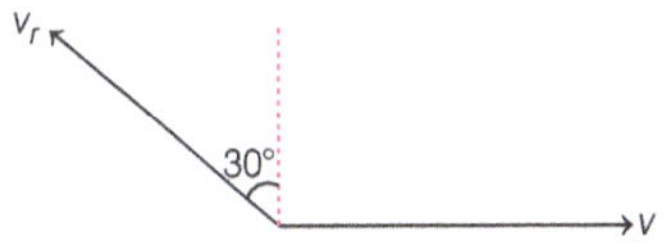

Now drift, $\dfrac{d}{2} = (v - v_r \sin 30^\circ)t = (v - v_r/2)\left(\dfrac{2d}{\sqrt{3}\, v_r}\right)$

$\therefore \quad \sqrt{3}\, v_r = 4v - 2v_r$

$\therefore \quad v_r = \left(\dfrac{4}{2+\sqrt{3}}\right) v$

$= 4(2-\sqrt{3})\, v$

21. $T_0 = \dfrac{2s}{V}$

and $\quad T = \dfrac{s}{V+v} + \dfrac{s}{V-v} = \dfrac{2sV}{V^2 - v^2}$

$\therefore \quad T/T_0 = \dfrac{(2sV/V^2 - v^2)}{2s/V}$

$= \dfrac{V^2}{V^2 - v^2} = \dfrac{1}{1 - v^2/V^2}$

22. To cross the river in shortest time, one has to swim perpendicular to the river current.

23. Let the velocity of the swimmer is $v_s = 4$ km/h and velocity of river is $v_r = 2$ km/h

Also, angle of swimmer with the flow of the river (down stream) is α as shown in the figure below:

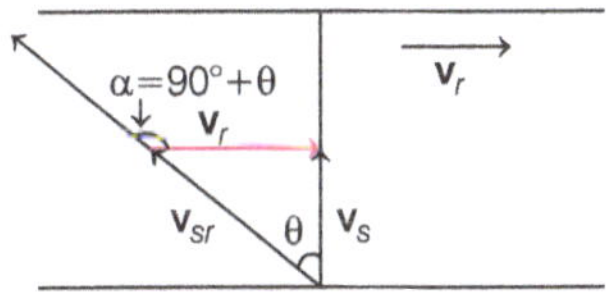

$\sin\theta = \dfrac{v_r}{v_{sr}} = \dfrac{2\text{ km/h}}{4\text{ km/h}} = \dfrac{1}{2}$

$\Rightarrow \quad \theta = 30^\circ$

Clearly, $\quad \alpha = 90^\circ + 30^\circ = 120^\circ$

24. (a) $t = \dfrac{400}{10} = 40$ s **Ans.**

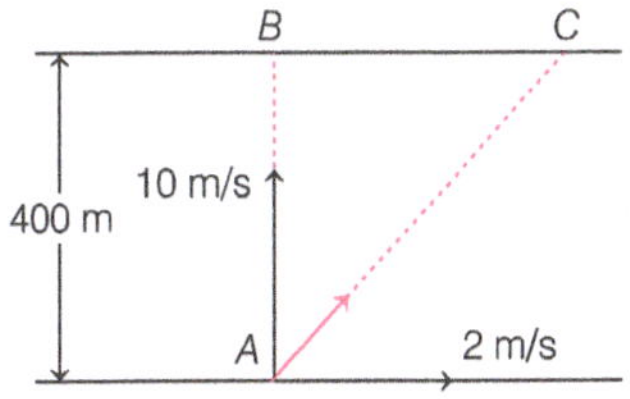

(b) $BC = (2 \text{ m/s})(t)$

$= 2 \times 40 = 80$ m **Ans.**

25. Let v_r = velocity of river

v_{br} = velocity of river in still water and

ω = width of river

Given, $\quad t_{min} = 10$ min $\quad$ or $\quad \dfrac{\omega}{v_{br}} = 10 \quad$ ...(i)

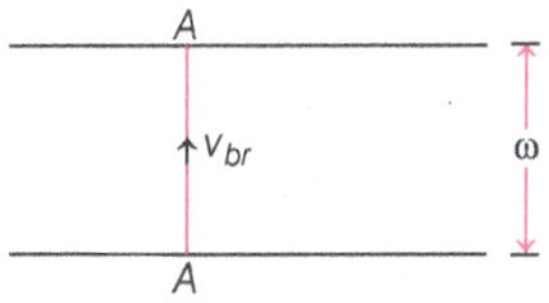

For minimum time

Drift in this case will be,

$x = v_r t$

$\therefore \quad 120 = 10 v_r \quad$ ...(ii)

Shortest path is taken when v_b is along AB. In this case,

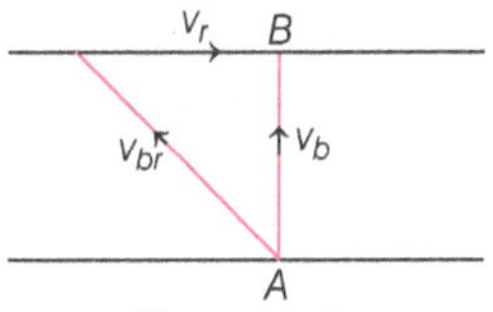

Shortest path

$v_b = \sqrt{v_{br}^2 - v_r^2}$

Now, $\quad 12.5 = \dfrac{\omega}{v_b} = \dfrac{\omega}{\sqrt{v_{br}^2 - v_r^2}} \quad$ ...(iii)

Solving these three equations, we get

$v_{br} = 20$ m/min,

$v_r = 12$ m/min

and $\quad \omega = 200$ m

26.

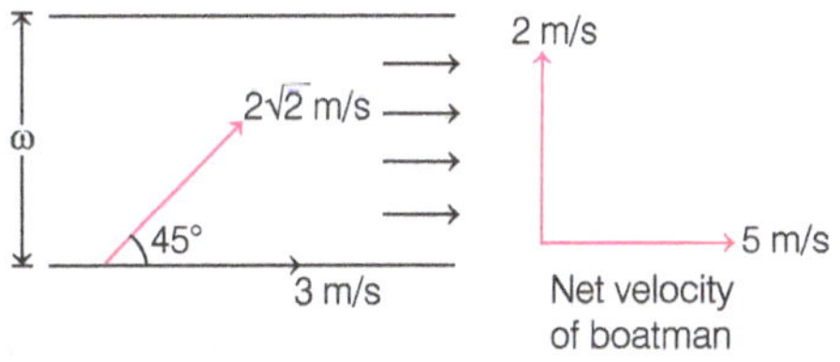

(a) $t = \dfrac{\omega}{2} = \dfrac{20}{2} = 10$ s

(b) Drift $= 5t = 50$ m

27. In the first case,

$BC = vt_1$ and $\omega = ut_1$

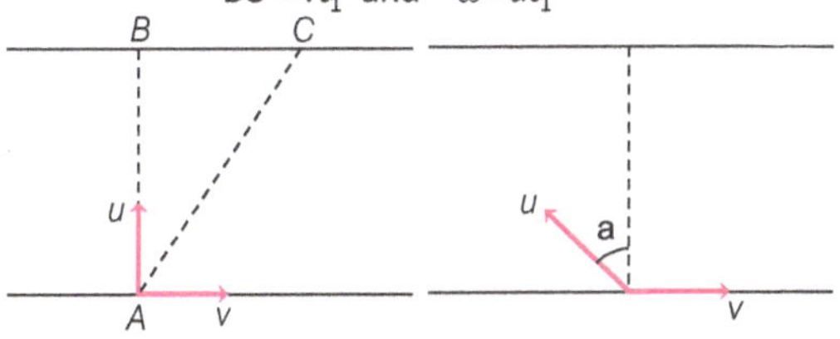

In the second case,

$$u \sin \alpha = v \quad \text{and} \quad \omega = (u \cos \alpha) t_2$$

Solving these four equations with proper substitution, we get

$$\omega = 200 \text{ m}, \ u = 20 \text{ m/min}, \ v = 12 \text{ m/min}$$

and $\alpha = 36°50'$ **Ans.**

28. Let $\omega =$ width of river (in km)

$$|\mathbf{v}_R| = 2.0 \text{ km/h} \Rightarrow |\mathbf{v}_{sw}| = 2.5 \text{ km/h}$$

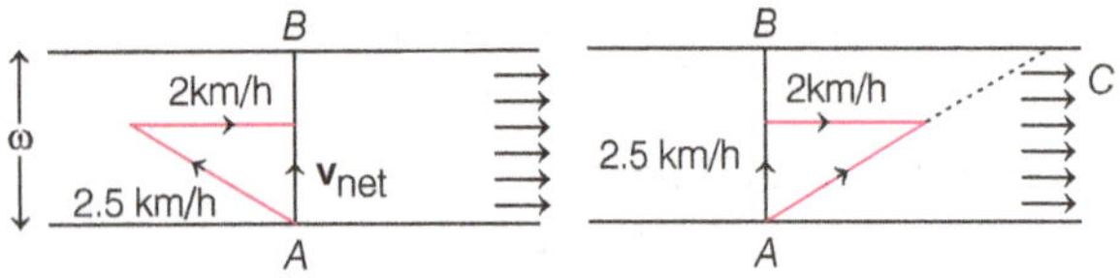

First swimmer   Second swimmer

First swimmer

$$|\mathbf{v}_{net}| = \sqrt{(2.5)^2 - (2)^2}$$

$$= 1.5 \text{ km/h}$$

$$\therefore \quad t_1 = \frac{\omega}{|\mathbf{v}_{net}|} = \left(\frac{\omega}{1.5}\right) \text{h} \quad \text{...(i)}$$

Second swimmer

Time taken to cross the river, $t = \left(\frac{\omega}{2.5}\right)$h

Distance $BC = (2)t = 0.8\,\omega$

Now this distance CB is travelled with velocity u. So further time taken is $\frac{0.8\,\omega}{u}$. Or the total time taken is,

$$t_2 = \frac{\omega}{2.5} + \frac{0.8\,\omega}{u} \quad \text{...(ii)}$$

Given, $t_1 = t_2$

$$\therefore \quad \frac{\omega}{1.5} = \frac{\omega}{2.5} + \frac{0.8\,\omega}{u}$$

Solving this equation, we get

$$u = 3.0 \text{ km/h}$$

29. Let $|\mathbf{v}_R| = v$, then $|\mathbf{v}_{BR}| = 2v$

Minimum drift may be zero. So, net velocity should be perpendicular to the river. Hence the velocity diagram should be as shown below.

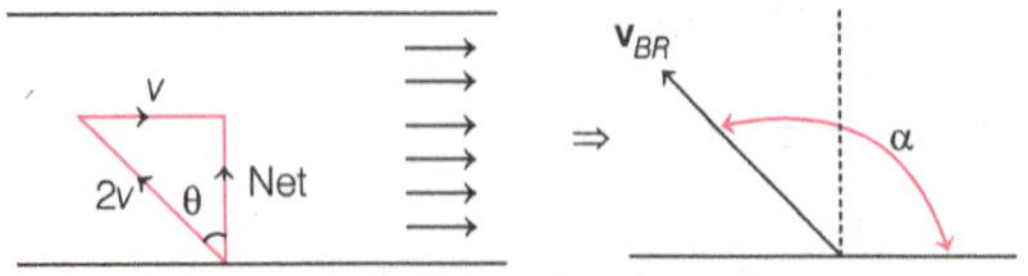

$$\sin \theta = \frac{v}{2v} = \frac{1}{2}$$

$$\therefore \quad \theta = 30°$$

$$\alpha = 90° + \theta = 120°$$

or $\alpha = 120°$

Note *120° is from downstream direction.*

30.

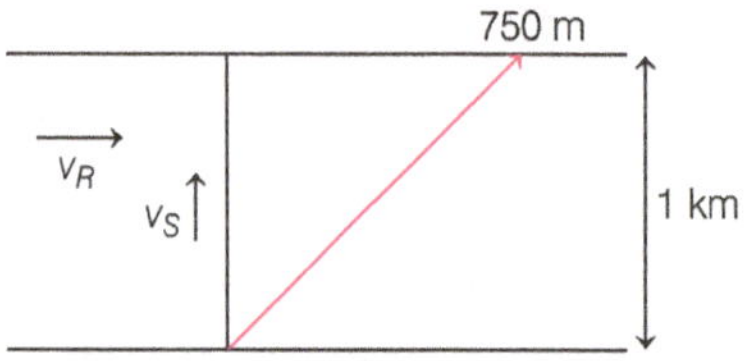

We have,

$$1 \text{ km} = v_S \times t$$

and $750 \text{ m} = v_R \times t$. So, $\frac{1000 \text{ m}}{750 \text{ m}} = \frac{v_S}{v_R}$

$$\therefore \quad \frac{4}{3} = \frac{4}{v_R} \Rightarrow v_R = 3 \text{ km/hr}$$

31.

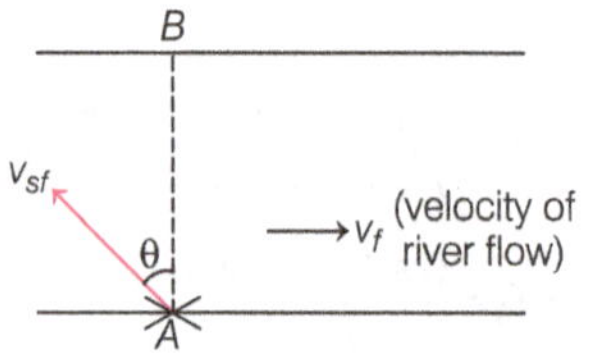

Velocity of swimmer with respect to river flow,

$$v_{sf} = 10 \text{ m/s}$$

Person is swimming at an angle of 120°.

$$\therefore \quad \theta = 120° - 90° = 30°$$

From figure, $\sin \theta = \frac{v_f}{v_{sf}}$

$$\Rightarrow \quad \sin 30° = \frac{x}{10} \Rightarrow \frac{1}{2} = \frac{x}{10}$$

$$\Rightarrow \quad x = \frac{10}{2} = 5 \text{ m/s}$$

32. To reach directly opposite point swimmer should swim with an angle $90° + \theta$.

From figure,

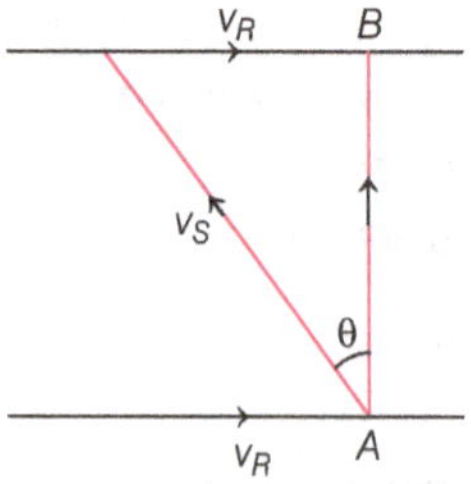

$$\sin \theta = \frac{v_R}{v_S}$$

$$\Rightarrow \quad \sin \theta = \frac{6}{12} = \frac{1}{2}$$

$$\therefore \quad \theta = 30°$$

$\therefore$ Angle with respect to flow $= 90° + \theta = 90° + 30°$

$$= 120°$$

33. Applying sine law in ΔABC, we have

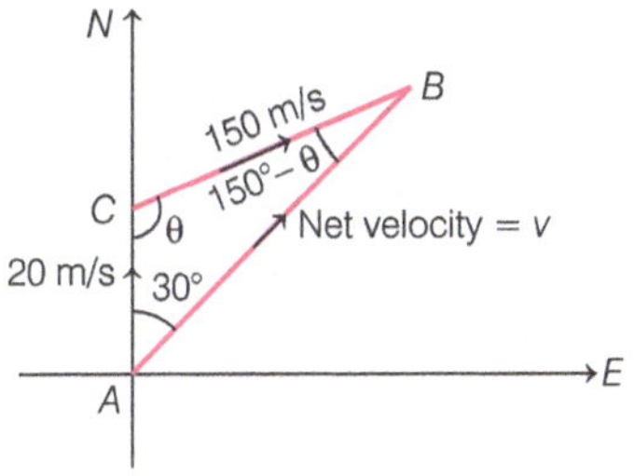

$$\frac{150}{\sin 30°} = \frac{v}{\sin\theta} = \frac{20}{\sin(150° - \theta)}$$

From first and third, we have

$$\sin(150° - \theta) = \frac{20\sin 30°}{150} = \frac{1}{15}$$

$$\therefore \quad 150° - \theta = \sin^{-1}\left(\frac{1}{15}\right) \qquad \textbf{Ans.}$$

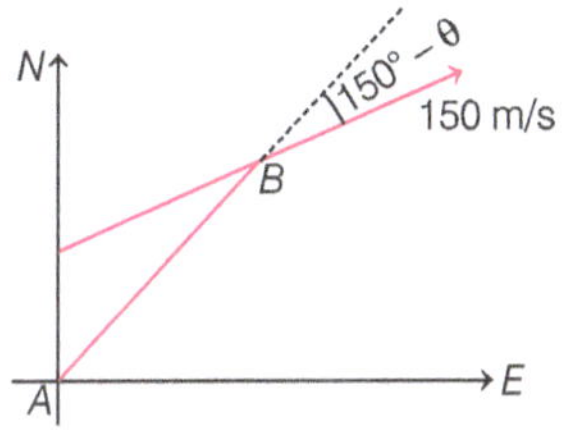

Hence, we can see that direction of 150 m/s is $150° - \theta \left[\text{or } \sin^{-1}\left(\frac{1}{15}\right)\right]$, east of the line AB

(b) $150° - \theta = \sin^{-1}\left(\frac{1}{15}\right) = 3.8°$

$$\therefore \quad \theta = 146.2°$$

From first and second relation,

$$v = \frac{150\sin\theta}{\sin 30°}$$

Substituting the values, we get

$$v = 167 \text{ m/s}$$

$$\therefore \quad t = \frac{AB}{v} = \frac{500 \times 1000}{167 \times 60} \text{ min}$$

$$= 49.9 \text{ min}$$

$$\approx 50 \text{ min} \qquad \textbf{Ans.}$$

34. (a) $\sin\theta = \frac{200}{500} = 0.4$

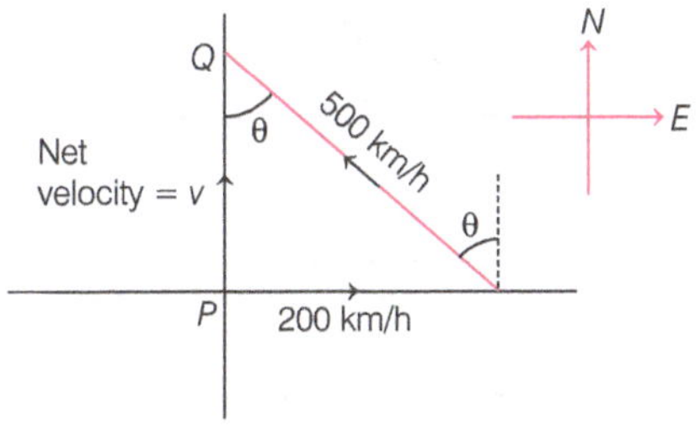

$\therefore \quad \theta = \sin^{-1}(0.4)$, west of north

(b) $v = \sqrt{(500)^2 - (200)^2}$

$$= 100\sqrt{21} \text{ km/h}$$

$$\therefore \quad t = \frac{PQ}{v} = \frac{1000}{100\sqrt{21}} = \frac{10}{\sqrt{21}} \text{h} \qquad \textbf{Ans.}$$

35. By symmetry, we can say that all will meet at the centre of square.

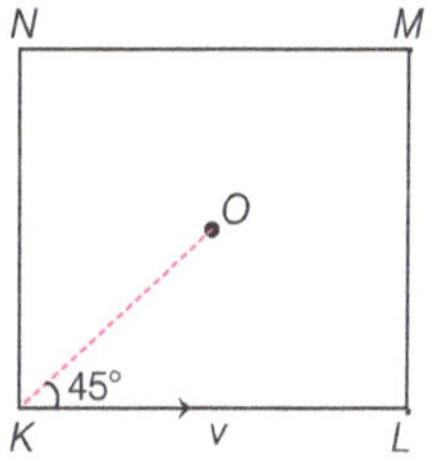

Component of velocity along KO (or LO etc.) at any instant will be $v\cos 45°$ or $\frac{v}{\sqrt{2}}$.

Further distance $KO = \sqrt{2}\left(\frac{d}{2}\right) = \frac{d}{\sqrt{2}}$

$$\therefore \quad \text{Time} = \frac{d/\sqrt{2}}{v/\sqrt{2}} = \frac{d}{v}$$

36. $\frac{x}{3} = \tan\theta$

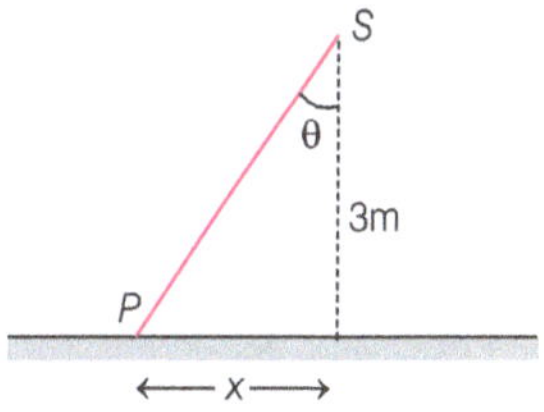

$$\therefore \quad x = 3\tan\theta$$

$$-\left(\frac{dx}{dt}\right) = 3\sec^2\theta\left(-\frac{d\theta}{dt}\right)$$

$$-\frac{dx}{dt} = v_P \quad \text{and} \quad -\frac{d\theta}{dt} = \omega = 0.1 \text{ rad/s},$$

$$\theta = 45°$$

Substituting the values, we get

$$v_P = 3(\sec^2 45°)(0.1) = 0.6 \text{ m/s}$$

37. It should follow the path PQR.

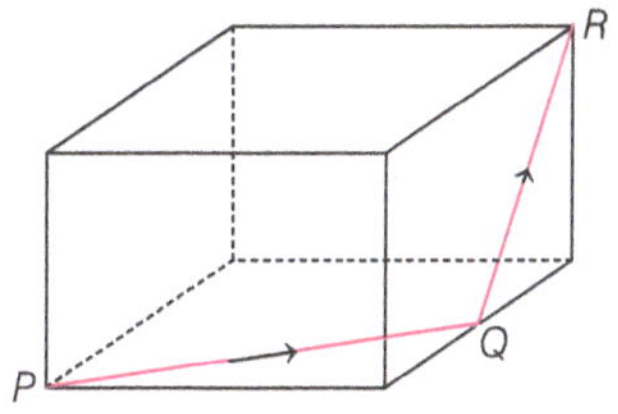

$$PQR = \sqrt{5}\,a$$

$\therefore \quad t = \dfrac{\sqrt{5}\,a}{u}$ **Ans.**

38. $t = \dfrac{80\,\text{m}}{30\,\text{m/s}} = \dfrac{8}{3}\,\text{s}$

Now in vertical direction,

$$t = \frac{2u_r}{a_r} \quad \text{or} \quad u_r = \frac{ta_r}{2}$$

$$= \frac{(8/3)(10)}{2} \qquad (\because a_r = g = 10\,\text{m/s}^2)$$

$$= \frac{40}{3}\,\text{m/s}$$ **Ans.**

39. In the right angle ΔPQR

$$l^2 = c^2 + y^2$$

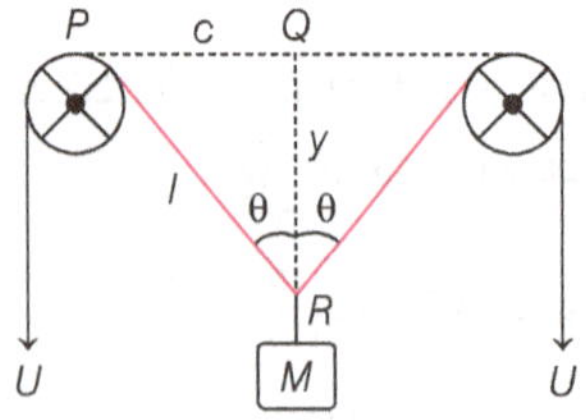

c=constant, l and y are variables

Differentiating this equation with respect to time, we get

$$2l\frac{dl}{dt} = 0 + 2y\frac{dy}{dt} \quad \text{or} \quad \left(-\frac{dy}{dt}\right) = \frac{l}{y}\left(-\frac{dl}{dt}\right)$$

Here, $\quad -\dfrac{dy}{dt} = v_M \quad \Rightarrow \quad \dfrac{l}{y} = \dfrac{1}{\cos\theta}$

and $\quad -dl/dt = U$

Hence, $\quad v_M = \dfrac{U}{\cos\theta}$

40. Initial position of buses A and B and car C are as shown in the figure.

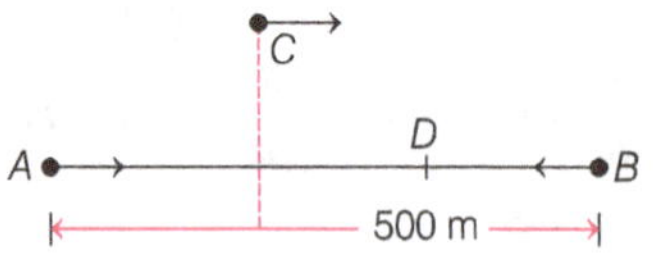

Let A and B cross each other at D.

$$v_{AC} = v_A - v_C = 20 - 15 = 5\,\text{ms}^{-1}$$

and $\quad v_{BC} = v_B - v_C = 20 - (-15)$

$$= 35\,\text{ms}^{-1}$$

Let the time taken by bus B to cover distance BC is t.

$\therefore \quad t = \dfrac{BC}{v_{BC}} = \dfrac{BC}{35}$

For bus A, $\quad AC = v_{AC}t + \dfrac{1}{2}at^2$

$$\Rightarrow \quad 500 - BC = 5\left(\frac{BC}{35}\right) + \frac{1}{2} \times 2\left(\frac{BC}{35}\right)^2$$

$$\Rightarrow \quad (BC)^2 + 1400\,BC - 612500 = 0$$

$$\therefore \quad BC = \frac{-1400 \pm \sqrt{(1400)^2 + 4 \times 612500}}{2}$$

Neglecting negative value, we get

$$BC = 350\,\text{m}$$

41. $AC = 500 - 350 = 150\,\text{m}$

42. $\because$ Time, $t = \dfrac{BC}{35} = \dfrac{350}{35} = 10\,\text{s}$

Unit Exercise (Cumulative)

Level 1

1. $AB = 10 \times 10 = 100\,\text{m}, BC = 20 \times 10 = 200\,\text{m}$

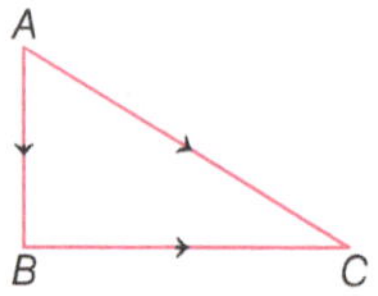

Average velocity $= \dfrac{AC}{t} = \dfrac{100\sqrt{5}}{20} = 5\sqrt{5}\,\text{ms}^{-1}$

Average speed $= \dfrac{AB + BC}{t} = \dfrac{100 + 200}{20} = 15\,\text{ms}^{-1}$

Displacement, $AC = \sqrt{AB^2 + BC^2} = 100\sqrt{5}\,\text{m}$

2. $v_{av} = \dfrac{l_{total}}{t_{total}} = \dfrac{\dfrac{l}{3} + \dfrac{l}{3} + \dfrac{l}{3}}{\dfrac{l}{3v_1} + \dfrac{l}{3v_2} + \dfrac{l}{3v_3}}$

$$= \frac{l}{\dfrac{l}{3}\left(\dfrac{1}{v_1} + \dfrac{1}{v_2} + \dfrac{1}{v_3}\right)} = \frac{1}{\dfrac{1}{3}\left(\dfrac{1}{11} + \dfrac{1}{22} + \dfrac{1}{33}\right)}$$

$$= \frac{33}{1 + \dfrac{1}{2} + \dfrac{1}{3}} = 18\,\text{m/s}$$

3. $\mathbf{a} = \dfrac{d\mathbf{v}}{dt}$. So, direction of **a** and $d\mathbf{v}$ is same.

4. $v = \dfrac{2\pi R}{T} = \dfrac{(2\pi)(1)}{60} = \left(\dfrac{\pi}{30}\right)\text{cm/s}$

In 15 s, velocity vector (of same magnitude) will rotate 90°.

$$\therefore \quad |\Delta\mathbf{v}| = |\mathbf{v}_f - \mathbf{v}_i|$$

$$= \sqrt{v^2 + v^2 - 2vv\cos 90^\circ}$$

$$= \sqrt{2}\,v = \frac{\pi\sqrt{2}}{30}\,\text{cm/s}$$ **Ans.**

5. One taxi leaves each station after every 10 min. As we need to calculate how many taxis will meet each taxi moving from Y to X. We consider one such taxi moving from Y to X and consider all the taxis moving from X to Y during its motion. Now suppose one taxi, taxi A is to leave from station Y towards station X.

The station is shown at $t = 0$ when taxi A leaves from station Y. Taxi 1 will reach station Y in 10 min and taxi 12 is about to leaves from station X.

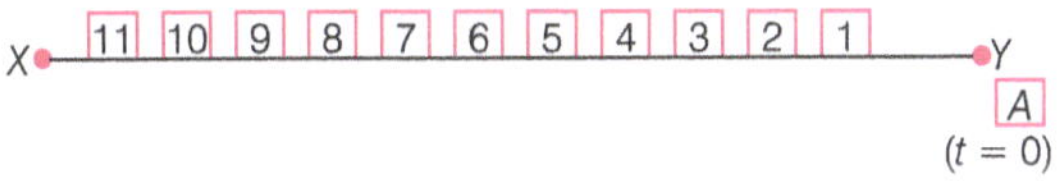

Taxi A will definitely meet these 11 taxis.

Now the situation after 10 min is

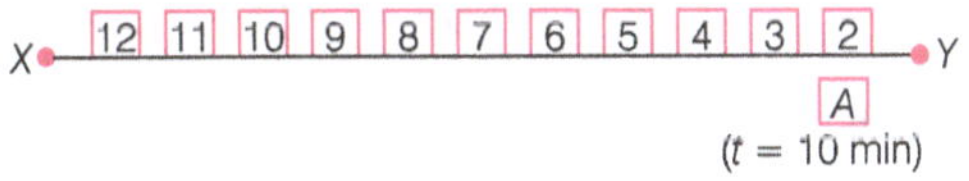

Similarly the situation after 110 min will be

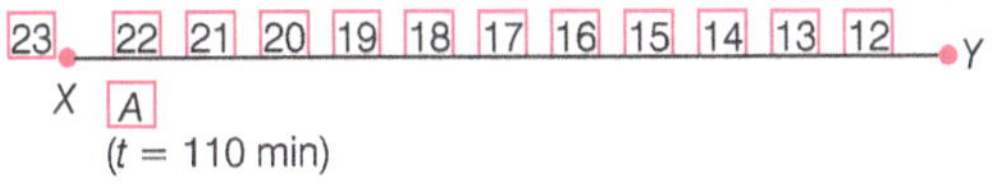

Now as taxi A moves towards X, it meets taxi 23 which will move towards Y.

So, in total taxi A will meet 23 taxis from other side.

6. Retardation is double. Therefore, retardation time will be half.

Let t_0 = acceleration time.

Then, $\frac{t_0}{2} = \text{retardation time}$

$$t_0 + \frac{t_0}{2} = t \quad \therefore \quad t_0 = \frac{2t}{3}$$

Now, $s = s_1 + s_2$

$$= \left(\frac{1}{2}\right)(a)(t_0)^2 + \left(\frac{1}{2}\right)(2a)\left(\frac{t_0}{2}\right)^2$$

$$= \frac{3}{4}at_0^2 = \left(\frac{3}{4}\right)(a)\left(\frac{2t}{3}\right)^2 = \frac{at^2}{3}$$

7. Total height $= 15 + \frac{u^2}{2g}$

$$= 15 + \frac{(10)^2}{2 \times 10}$$

$\Rightarrow$ $h = 20\,\text{m}$

Initial velocity $u = \sqrt{2gh}$

$$= \sqrt{2 \times 10 \times 20}$$

$$= 20\ \text{m/s}$$

Now applying

$v = u + at$, we have

$$v = (+20) + (-10)(3)$$

$$= -10\ \text{m/s}$$

$\therefore$ Velocity is 10 m/s, downwards.

8. In first second distance travelled

$$x = \frac{1}{2} \times g \times t^2 = 5\,\text{m}$$

(for $g = 10\,\text{m/s}^2$) in last second $7x = 35\,\text{m}$

Now, $s_{t\text{th}} = u + at - \frac{1}{2}a$

or $35 = 0 + 10 \times t - \frac{1}{2} \times 10$

or $t = 4\,\text{s}$

9. $u = \sqrt{2g(4h)} = \sqrt{8gh}$

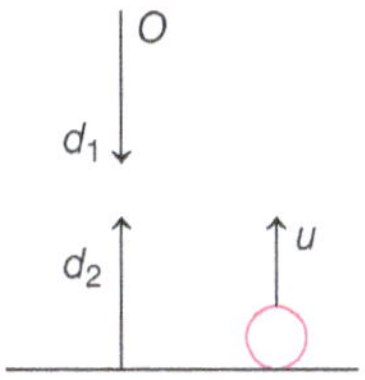

When they cross each other,

$$d_1 + d_2 = h$$

$\therefore$ $\left(\frac{1}{2}gt^2\right) + \left(ut - \frac{1}{2}gt^2\right) = h$

or $t = \frac{h}{u} = \frac{h}{\sqrt{8gh}} = \sqrt{\frac{h}{8g}}$

10.

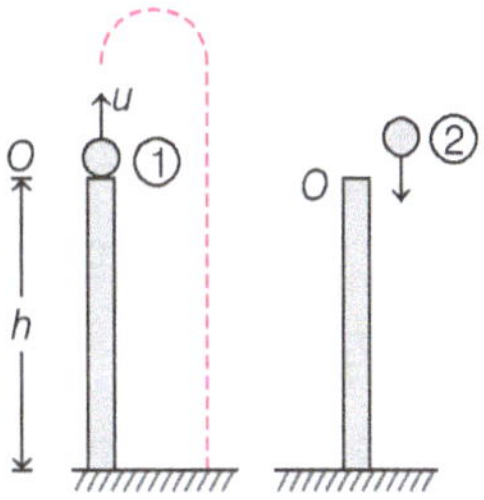

Ball 1 : $-h = u \times 9 - \frac{1}{2}g(9)^2$ ...(i)

Ball 2 : $h = u \times 4 + \frac{1}{2}g \times (4)^2$ ...(ii)

Solving, $h = 180$ m

11. $d = d_1 + d_2$

$$= \frac{v_1^2}{2a_1} + \frac{v_2^2}{2a_2}$$

$$= \frac{(10)^2}{2 \times 2} + \frac{(20)^2}{(2 \times 1)}$$

$$= 225\ \text{m}$$ **Ans.**

12. Taking downward direction as the positive direction,

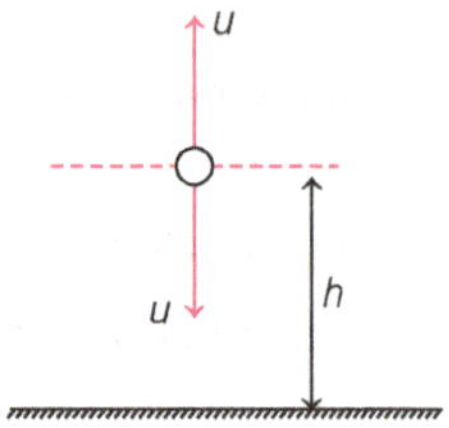

$$+h = -ut_1 + \frac{1}{2}gt_1^2 \quad \text{...(i)}$$

$$+h = ut_2 + \frac{1}{2}gt_2^2 \quad \text{...(ii)}$$

Multiplying Eq. (i) by t_2 and Eq. (ii) by t_1 and adding, we get

$$h(t_1 + t_2) = \frac{1}{2}gt_1t_2(t_1 + t_2)$$

or $$h = \frac{1}{2}gt_1t_2$$

For free fall from rest, $h = \frac{1}{2}gt^2$

$\therefore$ $$t^2 = t_1t_2$$

$$t = \sqrt{t_1t_2}$$

13. **A to B:** $u^2 = 0 + 2g \times 80$

$\Rightarrow$ $u = 40$ m/s

B to C: $v^2 - 2g \times 20 = 0$

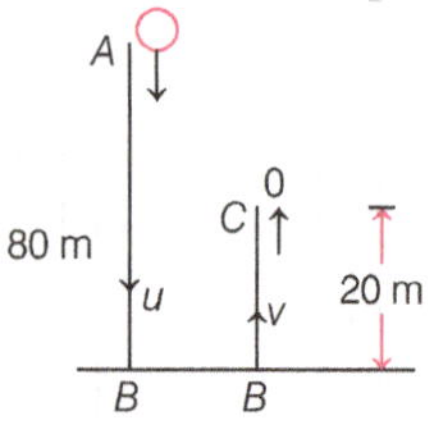

$\Rightarrow$ $v = 20$ m/s

$$\bar{a} = \frac{\Delta v}{\Delta t} = \frac{v + u}{\Delta t}$$

$$= \frac{40 + 20}{0.1}$$

$$= 600 \text{ m/s}^2$$

14. The greatest distance is, when velocities of both are equal.

or $ft = u \Rightarrow t = \frac{u}{f}$

$$s_1 = ut = \frac{u^2}{f}$$

and $$s_2 = \frac{1}{2}ft^2 = \frac{u^2}{2f}$$

$\therefore$ $$s_{max} = s_1 - s_2 = \frac{u^2}{2f}$$

15.

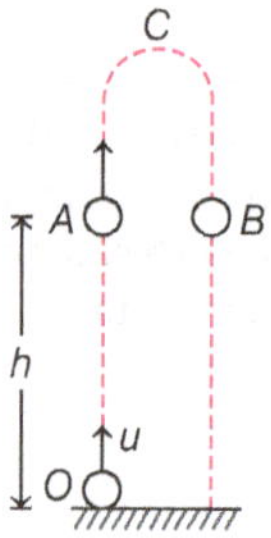

O to A:

$$h = ut_1 - \frac{1}{2}gt_1^2 \quad \text{...(i)}$$

O to B:

$$h = ut_2 - \frac{1}{2}gt_2^2 \quad \text{...(ii)}$$

Solving $u = \frac{g}{2}(t_1 + t_2)$

$$h = \frac{1}{2}gt_1t_2$$

16. $t_{AB} = t_1, t_{ACB} = t_2$

$\therefore$ Total time of flight,

$$T = \frac{2u}{g} = t_1 + t_2 \quad \text{or} \quad u = \frac{g}{2}(t_1 + t_2)$$

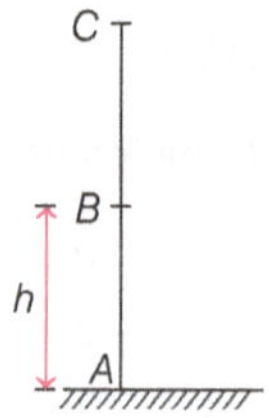

17. Total time of journey is $(t_1 + t_2)$.

Therefore, $$\frac{2u}{g} = t_1 + t_2 \quad \text{...(i)}$$

Further, $$h = ut_1 - \frac{1}{2}gt_1^2$$

or $$h = g\frac{(t_1 + t_2)t_1}{2} - \frac{1}{2}gt_1^2$$

or $$gt_1t_2 = 2h$$

$\therefore$ $$t_1t_2 = \frac{2h}{g}$$

18. Relative velocity $= v_1 + v_2$

Relative retardation $= a_1 + a_2$

$$0 = u^2 - 2as$$

or $$s = d_{max} = \frac{u^2}{2a} = \frac{(v_1 + v_2)^2}{2(a_1 + a_2)}$$

19. $90 \text{ km/h} = 90 \times \frac{5}{18} = 25 \text{ m/s}$

$$108 \text{ km/h} = 108 \times \frac{5}{18} = 30 \text{ m/s}$$

At maximum separation their velocities are same.

$\therefore$ Velocity of motorcycle $= 25$ m/s

or $\quad at = 25$

or $\quad t = 5$ s

But thief has travelled up to 7s.

$s_1 =$ displacement of thief

$= v_1 t_1 = 25 \times 7 = 175$ m

$s_2 =$ displacement of motorcycle

$$= \frac{1}{2} \times a_2 t_2^2$$

$$= \frac{1}{2} \times 5 \times (5)^2 = 62.5 \text{ m}$$

$\therefore$ Maximum separation

$$= s_1 - s_2 = 112.5 \text{ m}$$ **Ans.**

20. $-h = 5t - 5t^2$ and $-2h = -5t - 5t^2$

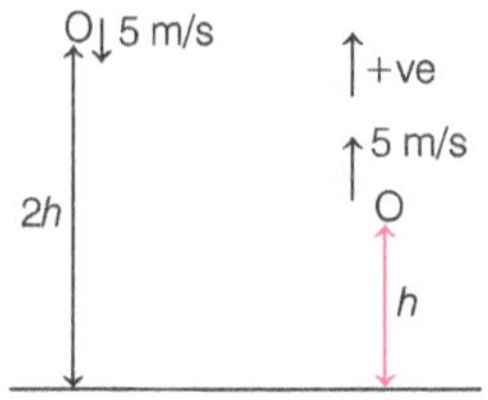

From these two equations,

we get $\quad t = 3$ s

and $h =$ distance between P

and $Q = 30$ m

21. $0 = u^2 - 2a_1(4) - 2a_2(1)$ ...(i)

$$0 = u^2 - 2a_2(2) - 2a_1(2)$$ ...(ii)

From these two equations, we get $a_2 = 2a_1$

22. Time taken by ball to reach highest point, $t = \frac{u}{g}$

Frequency of throw, $n = \frac{1}{t} = \frac{g}{u}$

$\Rightarrow \quad u = \frac{g}{n}$ ...(i)

The maximum height the balls can reach,

$$H_{max} = \frac{u^2}{2g} = \frac{\left(\frac{g}{n}\right)^2}{2g} = \frac{g}{2n^2}$$

23. Let us assume that they meet at $t = t_0$.

For A: $\quad 80 = \frac{1}{2} g t_0^2$ ... (i)

For B: $\quad 80 = u(t_0 - 2) + \frac{1}{2} g (t_0 - 2)^2$ ... (ii)

From Eq. (i), $t_0 = 4$ s

From Eq. (ii), $80 = 2u + 5(2)^2$

$\Rightarrow \quad u = 30$ m/s

24. $\because a = -bv^2$

or $\quad \frac{dv}{dt} = -bv^2$

or $\quad \int_{v_0}^{v} \frac{dv}{v^2} = -b \int_0^t dt$

or $\quad \left[\frac{v^{-2+1}}{-2+1}\right]_{v_0}^{v} = -bt$

or $\quad \left[-\frac{1}{v}\right]_{v_0}^{v} = -bt$

or $\quad \left[\frac{1}{v_0} - \frac{1}{v}\right] = -bt$ or $\frac{1}{v} = \frac{1}{v_0} + bt$

or $\quad \frac{1}{v} = \frac{1 + v_0 bt}{v_0}$ or $v = \frac{v_0}{1 + v_0 bt}$

or $\quad \frac{ds}{dt} = \frac{v_0}{1 + v_0 bt}$ or $\int_0^s ds = \int_0^t \left(\frac{v_0}{1 + v_0 bt}\right) dt$

or $\quad s = \frac{v_0}{v_0 b} [\ln(1 + v_0 bt)]_0^t$

$$= \frac{1}{b} [\ln(1 + v_0 bt) - \ln(1)]$$

$$= \frac{1}{b} [\ln(1 + v_0 bt)]$$

25. $\frac{dv}{dt} = a = -4v + 8$

$\therefore \quad \frac{da}{dt} = -4 \frac{dv}{dt} = -4(-4v + 8) = 16v - 32$

$$\left(\frac{da}{dt}\right)_i = 16 v_i - 32$$

$$= (16)(0) - 32$$

$$= -32 \text{ m/s}^2$$

Further, $\quad \int_0^v \frac{dv}{8 - 4v} = \int_0^t dt$

Solving this equation, we get

$$v = 2(1 - e^{-4t})$$

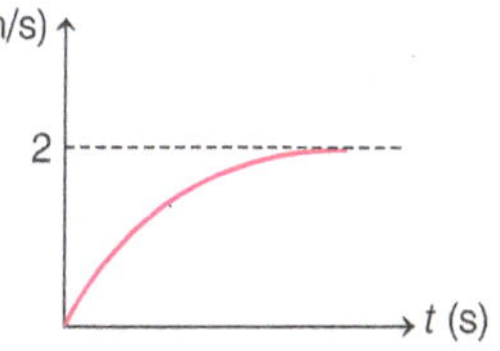

Hence, v-t graph is exponentially increasing graph, terminating at 2 m/s.

26. $a=\frac{F}{m}=\frac{-kx^2}{m}$

$\therefore \quad v\cdot\frac{dv}{dx}=-\frac{kx^2}{m}$

or $\quad \int_0^v v\,dv=\int_a^0 -\frac{kx^2}{m}dx$

$\frac{v^2}{2}=\frac{ka^3}{3m}$

$\therefore \quad v=\sqrt{\frac{2ka^3}{3m}}$ **Ans.**

27. $\because a=\frac{vdv}{dx}$ or $v\,dv=adx$

or $\quad vdv=\frac{10}{v+1}dx$

or $\quad \int_0^{15} v(v+1)\,dv=10\int_0^x dx$

or $\quad \left[\frac{v^3}{3}+\frac{v^2}{2}\right]_0^{15}=10\,x$

or $\quad \frac{1}{10}\left[\frac{15^3}{3}+\frac{15^2}{2}\right]=x$

$\therefore \quad x=123.75$ m

28. Net retardation of the ball when thrown vertically upward is

$$a=-\left(\frac{mg+m\gamma v^2}{m}\right)=-(g+\gamma v^2)=\frac{dv}{dt}$$

$\Rightarrow \quad \frac{dv}{(g+\gamma v^2)}=-dt$

$\Rightarrow \quad \int_{v_0}^{0}\frac{dv}{(\gamma v^2+g)}=\int_0^t -dt$

Solving this integration, we get

$$t=\frac{1}{\sqrt{\gamma g}}\cdot\tan^{-1}\left(\frac{\sqrt{\gamma}}{\sqrt{g}}v_0\right)$$

29. $v_x=\frac{dx}{dt}=(8t-2)$

$\therefore \quad \int_{14}^{x}dx=\int_2^t(8t-2)\,dt$

$x-14=(4t^2-2t)-(12)$

$\therefore \quad x=4t^2-2t+2$ ...(i)

$v_y=\frac{dy}{dt}=2$

$\therefore \quad \int_4^y dy=\int_2^t 2dt$

$\therefore \quad y-4=2t-4$

or $\quad y=2t$

$\therefore \quad t=\frac{y}{2}$

Substituting this value of t in Eq. (i), we have

$x=y^2-y+2$ **Ans.**

30. $y=u_yt$ (as $a_y=0$)

$\Rightarrow \quad t=\frac{y}{u_y}$ ...(i)

$x=u_xt+\frac{1}{2}a_xt^2$

$x=\frac{u_x}{u_y}y-\frac{1}{2}(10)\left(\frac{y}{u_y}\right)^2$

$=\left(\frac{u_x}{u_y}\right)y-\frac{5y^2}{u_y^2}$

Comparing this with the given equation, we get

$u_y=1\frac{\text{m}}{\text{s}}$

and $\quad u_x=\sqrt{3}\frac{\text{m}}{\text{s}}$

$\therefore \quad u=\sqrt{u_x^2+u_y^2}$

$=2\,\text{m/s}$

31. Given, $\mathbf{u}=5\hat{\mathbf{j}}$ m/s

Acceleration, $\mathbf{a}=10\hat{\mathbf{i}}+4\hat{\mathbf{j}}$ and

final coordinate $(20, y_0)$ in time t

$s_x=u_xt+\frac{1}{2}a_xt^2$ $[\because u_x=0]$

$\Rightarrow \quad 20=0+\frac{1}{2}\times10\times t^2$

$\Rightarrow \quad t=2$ s

$s_y=u_y\times t+\frac{1}{2}a_y\,t^2$

$y_0=5\times2+\frac{1}{2}\times4\times2^2$

$=18$ m

32. Given, position vector,

$\mathbf{r}=\cos\omega\hat{\mathbf{i}}+\sin\omega t\,\hat{\mathbf{j}}$

Velocity, $\mathbf{v}=\frac{d\mathbf{r}}{dt}=\omega(-\sin\omega t\hat{\mathbf{i}}+\cos\omega t\hat{\mathbf{j}})$

Acceleration, $\quad \mathbf{a}=\frac{d\mathbf{v}}{dt}=-\omega^2(\cos\omega t\hat{\mathbf{i}}+\sin\omega t\,\hat{\mathbf{j}})$

$\mathbf{a}=-\omega^2\,\mathbf{r}$.

$\therefore$ **a** is antiparallel to **r**.

Also $\mathbf{v}\cdot\mathbf{r}=0$

$\therefore \quad \mathbf{v}\perp\mathbf{r}$

Thus, the particle is performing uniform circular motion.

33. At 4 s

$$u = at = 8\text{ m/s}$$
$$s_1 = \frac{1}{2}at^2 = \frac{1}{2}\times 2\times 4^2 = 16\text{ m}$$

From 4 s to 8 s

$$a = 0,\ v = \text{constant} = 8\text{ m/s}$$
$$\therefore\quad s_2 = vt = (8)(4) = 32\text{ m}$$

From 8 s to 12 s

$$s_3 = s_1 = 16\text{ m}$$
$$\therefore\quad s_{Total} = s_1 + s_2 + s_3 = 64\text{ m}$$ **Ans.**

34. s = net area of v-t graph

At 2 s, net area = 0

$\therefore\quad s = 0$

and the particle crosses its initial position.

35. Deceleration is four times. Therefore, deceleration time should be $\frac{1}{4}$th.

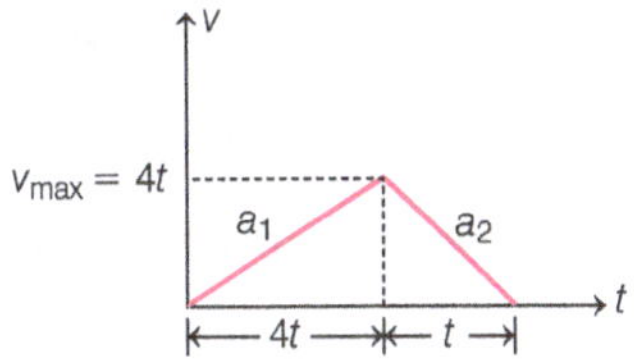

$$v_{max} = (a_1)(4t) = (1)(4t) = 4t$$

Area of v-t graph = displacement

$$\therefore\quad 200 = \frac{1}{2}(5t)(4t)$$

or $\quad t = \sqrt{20}$ s

Total journey time $= 5t = 22.4$ s **Ans.**

36. Slope of v-t graph gives acceleration/retardation.

$|a| = 5\text{ m/s}^2$ for both of them. A has acceleration while B has retardation.

$$|s_A| + |s_B| = 60$$

or $$\left\{\frac{1}{2}\times 5\times t^2\right\} + \left\{20t - \frac{1}{2}\times 5\times t^2\right\} = 60$$

or $$t = 3\text{ s}$$

37. (i) For uniformly accelerated/decelerated motion

$$v^2 = u^2 \pm 2gh$$

i.e. v-h graph will be a parabola (because equation is quadratic).

(ii) Initially velocity is downwards (–ve) and then after collision, it reverses its direction with lesser magnitude, i.e. velocity is upwards (+ve). Graph (a) satisfies both these conditions.

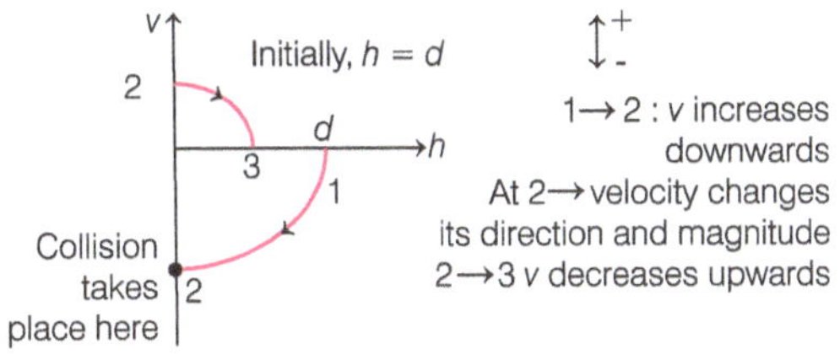

38. In graph (c) initial slope is zero which is not possible, since initial velocity is not zero. In all other three graphs: initial velocity is not zero.

39. Using equation, $a = \frac{v-u}{t}$ and $s = ut + \frac{1}{2}at^2$

Distance travelled by car in 15 s $= \frac{1}{2}\frac{(45)}{15}(15)^2 = \frac{675}{2}$ m

Distance travelled by scooter in 15 s $= 30\times 15 = 450$ m

($\because$ distance = speed × time)

$\therefore$ Difference between distance travelled by car and scooter in 15 s $= 450 - \frac{675}{2} = 112.5$ m

Let car catches scooter in time t;

$\therefore$ Car travelled in t sec = Scooter travelled in t sec

$$\frac{675}{2} + 45(t - 15) = 30t$$
$$\Rightarrow\quad 337.5 + 45t - 675 = 30t$$
$$\Rightarrow\quad 15t = 337.5$$
$$\Rightarrow\quad t = 22.5\text{ s}$$

40. $x_2 - x_1 = vt - \frac{1}{2}at^2$

$x_2 - x_1$ is zero at $t = 0$ and $t = \frac{2v}{a}$.

For $t < \frac{2v}{a}$, $(x_2 - x_1)$ is positive and after that it is negative.

Further, $(x_2 - x_1)$ *versus* t graph is a parabola.

41.

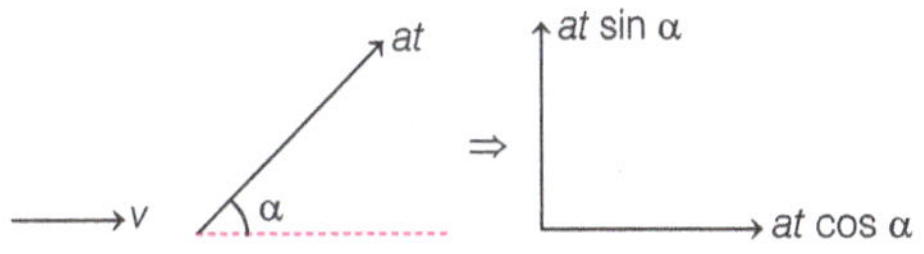

Relative velocity is perpendicular when

$$v = at\cos\alpha$$
$$\Rightarrow\quad t = \frac{v}{a\cos\alpha} = \frac{v}{a}\sec\alpha$$

42.

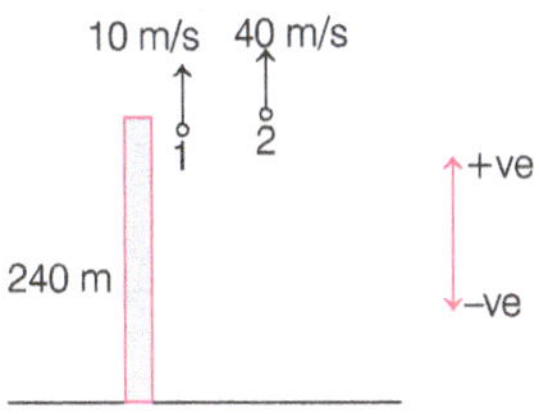

Let us first find, time of collision of two particles with ground :

$$s = ut + \frac{1}{2}at^2$$
$$-240 = 10t_1 - \frac{1}{2}\times 10\times t_1^2$$

Solving, we get positive value of $t_1 = 8$ sec

Similarly,

$$-240 = 40t_2 - \frac{1}{2} \times 10 \times t_2^2$$

Solving this equation, we get positive value of $t_2 = 12$ sec

From 0 to 8 sec, both particles are moving under gravity. Their absolute accelerations are same and equal to **g**. So, relative acceleration is zero or relative motion is uniform. So, relative position will change (or increase) linearly. At 8 sec, first particle strikes with ground. Its acceleration has become zero. The second particle has reached its initial position and is moving downwards so relative position will now decrease. The particle is yet in air. So its acceleration is **g**, or relative acceleration is **g**. Hence, graph is parabola.

43. For collision,

$$\mathbf{r}_A = \mathbf{r}_B \quad \text{(at same instant)}$$

$\therefore \quad (\mathbf{r}_i + \mathbf{v})_A = (\mathbf{r}_i + \mathbf{v})_B$

$\Rightarrow \quad (5\hat{\mathbf{i}} + 10\hat{\mathbf{j}} + 5\hat{\mathbf{k}})t = 30\hat{\mathbf{i}} + (10\hat{\mathbf{i}} + 5\hat{\mathbf{j}} + 5\hat{\mathbf{k}})t$

Equating the coefficients of x

$$5t = 30 + 10t \Rightarrow t = -\text{ve}$$

So, they will never collide.

44. Relative velocity of A with respect to B should be along AB or absolute velocity components perpendicular to AB should be same.

$\therefore \quad \frac{2u}{3} \sin\theta = u \sin 30°$

$\therefore \quad \theta = \sin^{-1}\left(\frac{3}{4}\right)$ **Ans.**

45. Let velocity of rain is

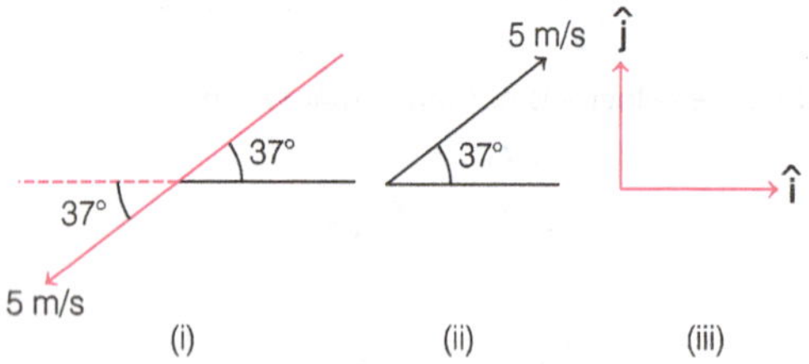

$$\mathbf{v}_R = a\hat{\mathbf{i}} - b\hat{\mathbf{j}}$$

In first case, $\mathbf{v}_M = (-4\hat{\mathbf{i}} - 3\hat{\mathbf{j}})$

$\therefore \quad \mathbf{v}_{RM} = \mathbf{v}_R - \mathbf{v}_M = (a + 4)\hat{\mathbf{i}} + (-b + 3)\hat{\mathbf{j}}$

It appears vertical.

$\therefore \quad a + 4 = 0$ or $a = -4$

In second case, $\mathbf{v}_M = (4\hat{\mathbf{i}} + 3\hat{\mathbf{j}})$

$\therefore \quad \mathbf{v}_{RM} = \mathbf{v}_R - \mathbf{v}_M = (a - 4)\hat{\mathbf{i}} + (-b - 3\hat{\mathbf{j}})$

$$= -8\hat{\mathbf{i}} + (-b - 3\hat{\mathbf{j}})$$

It appears at $\theta = \tan^{-1}\left(\frac{7}{8}\right)$

$\therefore \quad -b - 3 = -7$

$\therefore \quad b = 4$

and speed of rain $= \sqrt{a^2 + b^2}$

$= \sqrt{32}$ m/s **Ans.**

46. The given situation is shown in the diagram. Here v_r be the velocity of rain drop.

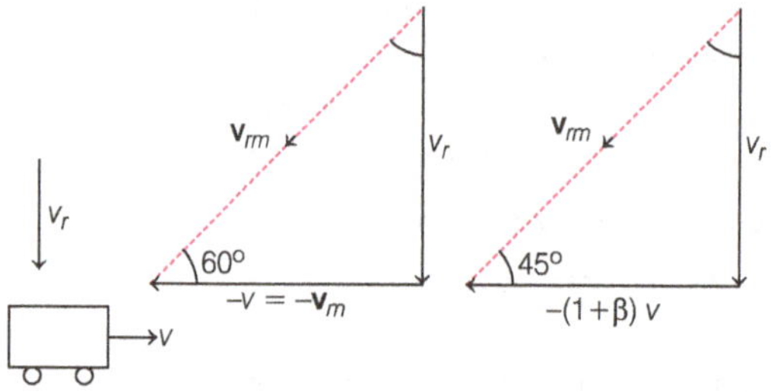

When car is moving with speed v,

$$\tan 60° = \frac{v_r}{v} \quad \text{...(i)}$$

When car is moving with speed $(1 + \beta)\,v$,

$$\tan 45° = \frac{v_r}{(\beta + 1)\,v} \quad \text{...(ii)}$$

Dividing Eq. (i) by Eq. (ii), we get

$$\sqrt{3}v = (\beta + 1)\,v$$

$\Rightarrow \quad \beta = \sqrt{3} - 1 = 0.732$

47. $\frac{dv_y}{dt} = 2t$

$\therefore \quad v_y = t^2$ or $\frac{dy}{dt} = t^2$

or $\quad y = \frac{t^3}{3}$ ...... (i)

and $\quad x = v_0 t \Rightarrow t = \frac{x}{v}$

Substituting in Eq. (i), we have

$$y = \frac{x^3}{3v_0^3}$$ **Ans.**

48. Equation of line RS is $y = -mx + C$

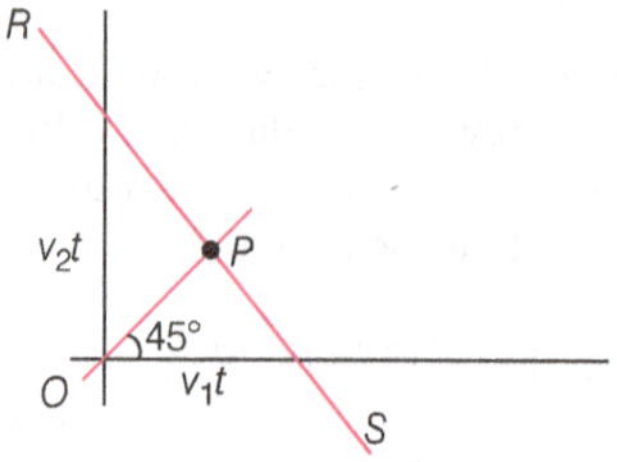

or $\quad y = -\left(\frac{v_2}{v_1}\right)x + v_2 t$

or $\quad v_1 y = -v_2 x + v_1 v_2 t$ ...(i)

Equation of line OP is

$$y = x \quad \text{...(ii)}$$

Point P is the point of intersection. So, we get

$$x_P = y_P = \frac{v_1 v_2 t}{v_1 + v_2}$$

$$\therefore \quad OP = \sqrt{x_P^2 + y_P^2} = \frac{\sqrt{2}\, v_1 v_2 t}{v_1 + v_2}$$

or $$v_3 = \frac{OP}{t} = \frac{\sqrt{2}\, v_1 v_2}{v_1 + v_2}$$

49. Let v be the speed of buses. Distance between two buses $= vT = d$

In the same direction relative velocity is $(v - 20)\,\text{kmh}^{-1}$.

$$d = v_r t$$

$$\Rightarrow \quad vT = (v - 20)\left(\frac{18}{60}\right) \quad \text{...(i)}$$

In the opposite direction, relative velocity is $(v + 20)\,\text{kmh}^{-1}$.

Again using,

$$d = v_r t$$

We get, $$vT = (v + 20)\left(\frac{6}{60}\right) \quad \text{...(ii)}$$

Solving Eqs. (i) and (ii), we get

$$T = \frac{9}{60}\,\text{hr} = 9\,\text{min} \quad \text{and} \quad v = 40\,\text{kmh}^{-1}$$

50. For particle P, motion between AC will be accelerated one while between CB a retarded one. But in any case horizontal component of its velocity will be greater than or equal to v. On the other hand, in case of particle Q, it is always equal to v. Horizontal displacement for both the particles is same. Therefore, $t_P < t_Q$.

51. If $\mathbf{a}$ = constant

Then $|\mathbf{a}|$ is also constant or $\left|\frac{dv}{dt}\right|$ = constant

52. $\mathbf{v}_{av} = \frac{\mathbf{s}}{t}$ and $v_{av} = \frac{d}{t}$

Now, $d \geq |\mathbf{s}|$

$\therefore \quad v_{av} \geq |\mathbf{v}_{av}|$

53. A body having a constant speed can have a varying velocity due to change in the direction of velocity. Thus a body having constant speed can have an acceleration. If velocity and acceleration are in the same direction, then distance is equal to displacement, because then there is no change in direction of motion. The body will continuously travel in one direction only.

54. $a = \frac{dv}{dt}$, if velocity changes, definitely there will be acceleration. If speed changes, then velocity also changes, so definitely there will be acceleration.

Acceleration may be due to change in the direction of velocity only and not magnitude.

If body has acceleration, its speed may change if acceleration is due to change in the magnitude of velocity.

55. If a particle is moving in x- axis

(a) v_{av} = Displacement /Total time

Here, displacement is increasing in $+x$- direction, therefore $v_{av} = +$ ve

(b) Velocity may be negative and acceleration may be positive.

(c) x is decreasing in $-x$- direction, hence particle is moving in $+x$-axis, therefore $v_{av} = +$ ve.

(d) x is increasing with time in positive direction. So velocity is positive because particle is moving in $+x$- direction.

57. A body moving on a circular path with uniform speed must have varying velocity as well as acceleration.

59. $(14)^2 = (2)^2 + 2\,as$

$$\therefore \quad 2\,as = 192 \text{ units}$$

At mid-point, $$v^2 = (2)^2 + 2a\left(\frac{s}{2}\right)$$

$$= 4 + \frac{192}{2} = 100$$

$$\therefore \quad v = 10\ \text{m/s} \qquad \textbf{Ans.}$$

$$XA : AY = 1 : 3$$

$$\therefore \quad XA = \frac{1}{4}s \quad \text{and} \quad AY = \frac{3}{4}s$$

$$v_1^2 = (2)^2 + 2a\left(\frac{s}{4}\right)$$

$$= 4 + \frac{192}{4} = 52$$

$$\therefore \quad v_1 = \sqrt{52} \neq 6\ \text{m/s} \qquad \textbf{Ans.}$$

$$10 = 2 + at_1 \qquad (v = u + at)$$

$$\therefore \quad t_1 = \frac{8}{a} \quad \text{and } 14 = 10 + at_2$$

$$\therefore \quad t_2 = \frac{4}{a} \quad \text{or} \quad t_1 = 2t_2 \qquad \textbf{Ans.}$$

$s_1 = (2t) + \frac{1}{2}a(t^2)$ = distance travelled in first half

$$s_2 = 2(2t) + \frac{1}{2}a(2t)^2$$

$s_3 = s_2 - s_1$ = distance travelled in second half

We can see that,

$$s_1 \neq \frac{s_3}{2}$$

60. $a = -0.5t = \frac{dv}{dt}$

$\therefore \quad \int_{16}^{v} dv = \int_{0}^{t} -0.5t\,dt$

$\therefore \quad v = 16 - 0.25t^2$

$s = \int v\,dt = 16t - \dfrac{0.25t^3}{3}$

$v = 0$ when $16 - 0.25t^2 = 0$

or $\quad t = 8\text{ s}$ **Ans.**

So direction of velocity changes at 8 s. Up to 8 s distance = displacement

$\therefore$ **At 4s :**

$d = 16 \times 4 - \dfrac{0.25 \times (4)^3}{3} = 58.67\text{ m}$ **Ans.**

$s_{8s} = 16 \times 8 - \dfrac{(0.25)(8)^3}{3}$

$= 85.33\text{ m}$

$s_{10\,s} = 16 \times 10 - \dfrac{(0.25)(10)^3}{3}$

$= 76.67\text{ m}$

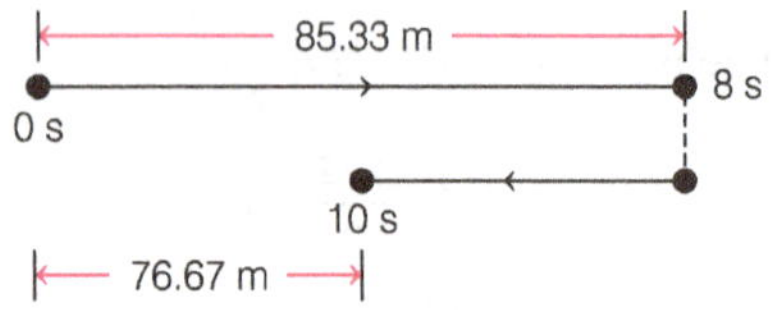

Distance travelled in 10 s,

$d = (85.33) + (85.33 - 76.67)$

$= 94\text{ m}$ **Ans.**

At 10 s :

$v = 16 - 0.25\,(10)^2$

$= -9\text{ m/s}$

$\therefore \quad$ Speed = 9 m/s **Ans.**

61. Comparing with projectile motion,

$$\mathbf{r} = (u_x t)\hat{\mathbf{i}} + \left(u_y t - \frac{1}{2}gt^2\right)\hat{\mathbf{j}}$$

We have, $u_x = 3\text{ m/s}, u_y = 4\text{ m/s}$ and $g = 10\text{ m/s}^2$

$$T = \frac{2u_y}{g} = 0.8\text{ s}$$

(a) $R = u_x T = 2.4\text{ m}$

(b) $u = \sqrt{u_x^2 + u_y^2} = 5\text{ m/s}$

(c) $\theta = \tan^{-1}\left(\dfrac{u_y}{u_x}\right) = \tan^{-1}\left(\dfrac{4}{3}\right)$

(d) $T = 0.8\text{ s}$

62. $x = a\cos(pt) \Rightarrow \quad \cos(pt) = \dfrac{x}{a}$ ...(i)

$y = b\sin(pt)$

$\Rightarrow \quad \sin(pt) = y/b$ ...(ii)

Squaring and adding Eqs. (i) and (ii), we get

$$\frac{x^2}{a^2} + \frac{y^2}{b^2} = 1$$

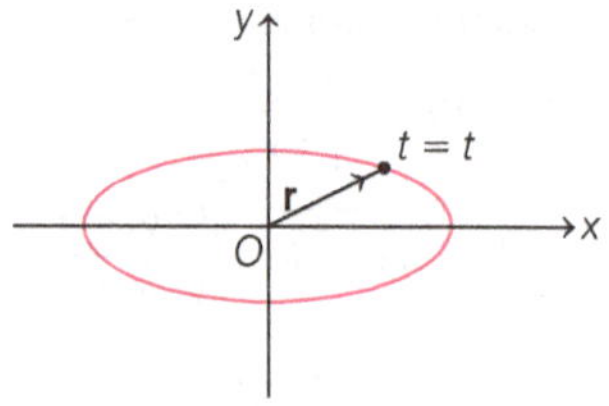

Therefore, path of the particle is an ellipse.

Hence, option (a) is correct.

From the given equations, we can find

$\dfrac{dx}{dt} = v_x = -ap\sin pt$

$\Rightarrow \quad \dfrac{d^2x}{dt^2} = a_x = -ap^2\cos pt$

$\dfrac{dy}{dt} = v_y = bp\cos pt$

and $\quad \dfrac{d^2y}{dt^2} = a_y = -bp^2\sin pt$

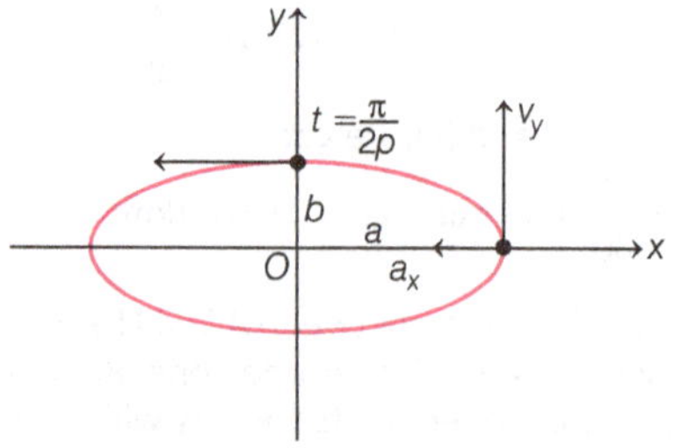

At time $t = \pi/2p$ or $pt = \pi/2$

a_x and v_y become zero (because $\cos\pi/2 = 0$), only v_x and a_y are left.

or we can say that velocity is along negative x-axis and acceleration along y-axis.

Hence, at $t = \pi/2p$, velocity and acceleration of the particle are normal to each other. So, option (b) is also correct.

At $t = t$, position of the particle

$\mathbf{r}(t) = x\hat{\mathbf{i}} + y\hat{\mathbf{j}} = a\cos pt\,\hat{\mathbf{i}} + b\sin pt\,\hat{\mathbf{j}}$

and acceleration of the particle is

$\mathbf{a}(t) = a_x\hat{\mathbf{i}} + a_y\hat{\mathbf{j}} = -p^2[a\cos pt\,\hat{\mathbf{i}} + b\sin pt\,\hat{\mathbf{j}}]$

$= -p^2[x\hat{\mathbf{i}} + y\hat{\mathbf{j}}] = -p^2\,\mathbf{r}(t)$

or, **a** is always antiparallel to **r**.

Therefore, acceleration of the particle is always directed towards origin.

Hence, option (c) is also correct.

At $t = 0$, particle is at $(a,0)$ and at $t = \pi/2p$, particle is at $(0,b)$. Therefore, the distance covered is one-fourth of the elliptical path not a.

Hence, option (d) is wrong.

63. a - s equation corresponding to given graph is,

$$a = 6 - \frac{s}{5}$$

$\therefore$ $$v \cdot \frac{dv}{ds} = \left(6 - \frac{s}{5}\right)$$

or $$\int_0^v v\,dv = \int_0^s \left(6 - \frac{s}{5}\right) ds$$

or $$v = \sqrt{12s - \frac{s^2}{5}}$$

At $s = 10\,\text{m}, v = 10\,\text{m/s}$ **Ans.**

Maximum values of v is obtained when

$\frac{dv}{ds} = 0$ which gives $s = 30\,\text{m}$

$\therefore$ $$v_{max} = \sqrt{12 \times 30 - \frac{(30)^2}{5}}$$

$$= \sqrt{180}\,\text{m/s}$$ **Ans.**

64. $u_A + a_A T = u_B - a_B T$

Putting $T = 4$ s, we get

$$4(a_A + a_B) = u_B - u_A \quad \text{....(i)}$$

Now, $s_A = s_B$

$\therefore$ $$u_A t + \frac{1}{2} a_A t^2 = u_B t - \frac{1}{2} a_B t^2$$

$\therefore$ $$t = 2\frac{(u_B - u_A)}{(a_A + a_B)} = 2 \times 4$$

$$= 8\,\text{s}$$ **Ans.**

65. $s_A = s_B$

$$5t + \frac{1}{2} a_A t^2 = 15t - \frac{1}{2} a_B t^2$$

or $$10 + a_A t = 30 - a_B t$$

$\therefore$ $$(5 + a_A t) - (15 - a_B t) = 10$$

or $$v_A - v_B = 10\,\text{m/s}$$ **Ans.**

66. $8 = 6 + a_A T = 6 + 4a_A$

$\therefore$ $$a_A = 0.5\,\text{m/s}$$

At 10 s,

$$v_A = u_A + a_A t$$

$$= (6) + (0.5)(10)$$

$$= 11\,\text{m/s}$$ **Ans.**

67.& 68.

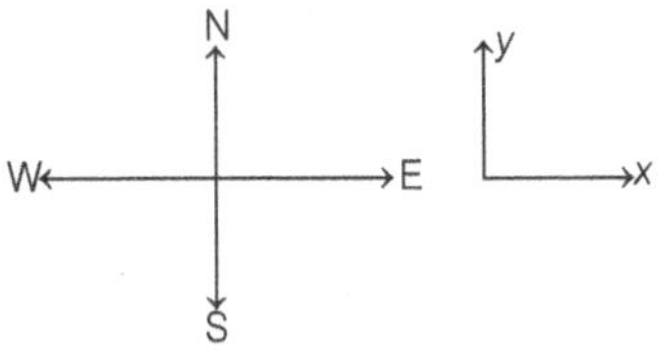

Let $\mathbf{v}_W = x\hat{\mathbf{i}} + y\hat{\mathbf{j}}$

(i) $\mathbf{v}_A = -30\,\hat{\mathbf{j}}$

$\therefore$ $$\mathbf{v}_{WA} = \mathbf{v}_W - \mathbf{v}_A = x\hat{\mathbf{i}} + (y + 30)\,\hat{\mathbf{j}}$$

It appears in x-direction.

$\therefore$ $$y + 30 = 0$$

or $$y = -30\,\text{km/h}$$

(ii) $\mathbf{v}_B = -50\,\hat{\mathbf{j}}$

$\therefore$ $$\mathbf{v}_{WB} = x\hat{\mathbf{i}} + (y + 50)\,\hat{\mathbf{j}}$$

$$= x\hat{\mathbf{i}} + 20\,\hat{\mathbf{j}}$$

It appears at 45°.

$\therefore$ $$x = 20 \text{ km/hr}$$

$\therefore$ $$\mathbf{v}_W = (20\,\hat{\mathbf{i}} - 30\hat{\mathbf{j}})\,\frac{\text{km}}{\text{hr}}$$

69. Velocity of ball with respect to elevator is 15 m/s (up) and elevator has a velocity of 10 m/s (up). Therefore, absolute velocity of ball is 25 m/s (upwards). Ball strikes the floor of elevator, if

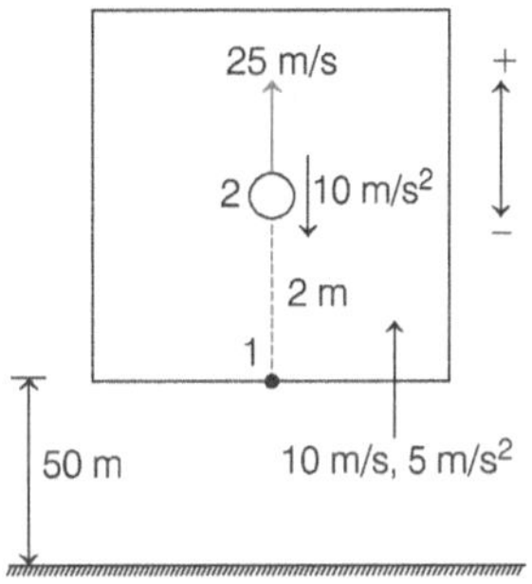

$$s_1 = s_2 + 2$$

$\therefore$ $$10t + 2.5t^2 = 25t - 5t^2 + 2$$

Solving this equation, we get

$$t = 2.13\,\text{s}$$ **Ans.**

Here s_1 and s_2 are the displacements of 1 and 2.

70. If the ball does not collide, then it will reach its maximum height in time,

$$t_0 = \frac{u}{g} = \frac{25}{10} = 2.5\,\text{s}$$

Since, $t < t_0$, therefore as per the question ball is at its maximum height at 2.13 s.

$$h_{max} = 50 + 2 + 25 \times 2.13 - 5 \times (2.13)^2$$

$$= 82.56\,\text{m}$$ **Ans.**

71. $s = 25 \times 2.13 - 5 \times (2.13)^2$

$$= 30.56\,\text{m}$$ **Ans.**

72. At maximum separation, their velocities are same

$\therefore$ $$25 - 10t = 10 + 5t \quad \text{or} \quad t = 1\text{s}$$

Maximum separation $= 2 + s_2 - s_1$

$$= 2 + [25 \times 1 - 5 \times (1)^2] - [10 \times 1 + 2.5(1)^2]$$

$$= 9.5\,\text{m}$$ **Ans.**

73. In 15 s, it will rotate 90° or $\frac{1}{4}$th circle.

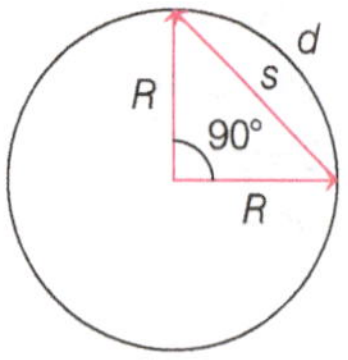

$$\therefore \quad \text{Average speed} = \frac{\text{Total distance}}{\text{Total time}} = \frac{d}{t}$$

$$= \frac{(2\pi R/4)}{15} = \frac{\pi R}{30} = \frac{\pi (2.0)}{30}$$

$$= \left(\frac{\pi}{15}\right) \text{cm/s} \qquad \textbf{Ans.}$$

$$\text{Average velocity} = \frac{\text{Total displacement}}{\text{Total time}} = \frac{s}{t}$$

$$= \frac{\sqrt{2}R}{t} = \frac{(\sqrt{2})(2.0)}{15}$$

$$= \frac{2\sqrt{2}}{15} \text{ cm / s} \qquad \textbf{Ans.}$$

74. (a) Yes, for example, a freely falling particle at its highest point has zero velocity but accelerating downward.

(b) No, if distance is zero, then displacement is also zero.

(c) Yes, if velocity is also in the negative direction.

(d) Yes, if motion takes place continuously in one direction.

75. $\text{Average velocity} = \dfrac{\text{Total displacement}}{\text{Total time}}$

$$\therefore \quad 2.5v = \frac{vt_0 + 2vt_0 + 3vT}{t_0 + t_0 + T}$$

Solving, we get

$$T = 4t_0 \qquad \textbf{Ans.}$$

76.

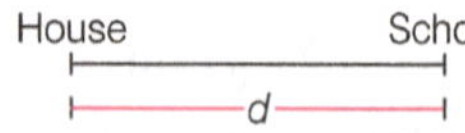

Let distance between house and school $= d$ km

and time required to reach the school be T hour

$$\frac{d}{12} = T + \frac{3}{60} \qquad \text{...(i)}$$

Again,

$$\frac{d}{13} = T - \frac{3}{60} \qquad \text{...(ii)}$$

(speed increases by 1 kmh^{-1})

Subtracting from Eq. (i) to Eq. (ii), we get

$$\frac{d}{12} - \frac{d}{13} = \frac{3}{60} + \frac{3}{60}$$

$$d = 15.6 \text{ km}$$

77. If we start time calculations from point P and apply the equation,

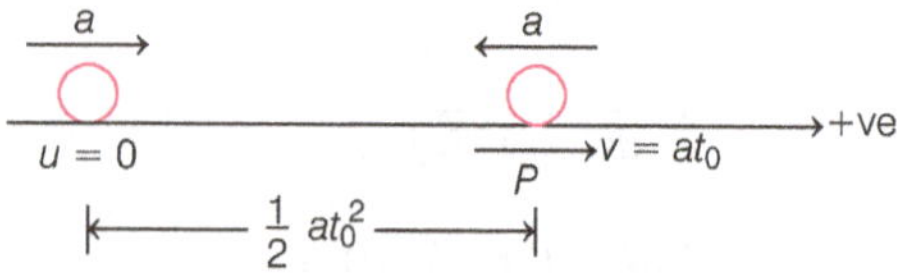

$$s = ut + \frac{1}{2}at^2$$

$$\left(-\frac{1}{2}at_0^2\right) = (+at_0)t + \frac{1}{2}(-a)t^2$$

Solving we get, $\quad t = (\sqrt{2} + 1)t_0$

$$= 2.414\,t_0$$

$$\therefore \quad \text{Total time, } T = t + t_0$$

$$= (3.414)\,t_0 \qquad \textbf{Ans.}$$

78. Let t_0 be the breaking time and a the magnitude of deceleration.

$\because \quad$ 80.5 km/h = 22.36 m/s, 48.3 km/h = 13.42 m/s

In the first case, $56.7 = (22.36 \times t_0) + \dfrac{(22.36)^2}{2a}$ ......(i)

and $\quad 24.4 = (13.42 \times t_0) + \dfrac{(13.42)^2}{2a}$ ...... (ii)

Solving these two equations, we get

$$t_0 = 0.74 \text{ s and } a = 6.2 \text{m / s}^2 \qquad \textbf{Ans.}$$

79. $v = 3t^2 - 6t$

$\Rightarrow \quad v = 0 \quad \text{at} \quad t = 2\text{s}$

For $t < 2$s, velocity is negative. At $t = 2$ s, velocity is zero and for $t > 2$ s, velocity is positive.

$$\therefore \quad s_1 = \int_0^{3.5} v dt = \int_0^{3.5} (3t^2 - 6t)\, dt$$

$$= 6.125 \text{ m}$$

= displacement upto 3.5 s

$$s_2 = \int_0^{2} v dt = \int_0^{2} (3t^2 - 6t)\, dt$$

$$= -4 \text{ m}$$

= displacement upto 2 s

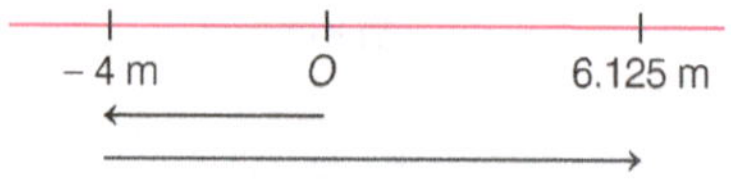

$\therefore \quad d =$ distance travelled in 3.5 s

$= 4 + 4 + 6.125$

$= 14.125$ m **Ans.**

$$\text{Average speed} = \frac{d}{t} = \frac{14.125}{3.5}$$

$= 4.03$ m/s **Ans.**

$$\text{Average velocity} = \frac{s_1}{t} = \frac{6.125}{3.5}$$
$$= 1.75 \text{ m/s}$$ **Ans.**

80. Substituting the values of v_0 and τ, we get

$$v = (10 - 2t)$$

Now comparing this with $v = u + at$, we get

$$u = +10 \text{ cm/s}$$

and $$a = -2 \text{ cm/s}^2$$

At $t = 0$, particle starts from $x = 0$.

So, the situation is as shown below.

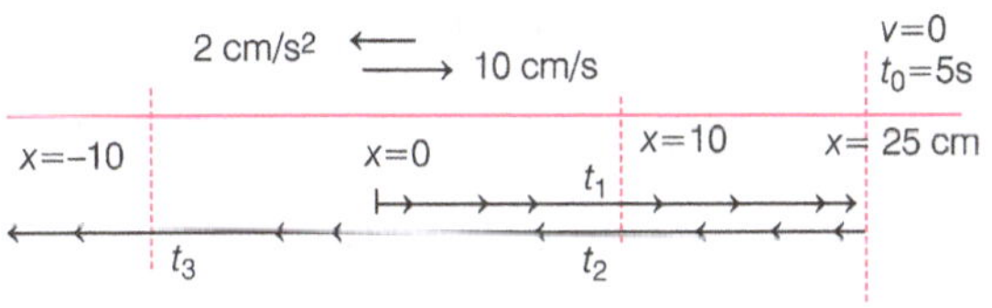

(a) Whatever is displacement at given time it will also become x - coordinate at that time, as particle is at $x = 0$ at $t = 0$.

Further, $a = -2 \text{ cm/s}^2$ is constant. Hence, we can use,

$$x = s = ut + \frac{1}{2}at^2$$

Now, we can find different values of x - coordinates at given times by putting the proper values.

$$x_{6s} = 10 \times 6 - \frac{1}{2} \times 2 \times (6)^2 = 24 \text{ cm or } 0.24 \text{ m}$$

$$x_{10s} = 10 \times 10 - \frac{1}{2} \times 2 \times (10)^2 = 0$$

$$x_{20s} = 10 \times 20 - \frac{1}{2} \times 2 \times (20)^2 = -200 \text{ cm or } -2.0 \text{ m}$$

(b) 10.0 cm from origin means,

$$s = x = \pm 10.0 = 10t - \frac{1}{2} \times 2 \times t^2$$

Solving this equation, we get three positive values of t,

1.1, 9 and 11 s

Note *These are the approximate values.*

(c) Let velocity of particle becomes zero at time t_0, then

$$v = u + at \quad \text{or} \quad 0 = 10 - 2t_0$$

or $$t_0 = 5 \text{ s}$$

Now, upto 5s, velocity is positive, so the particle is moving in positive x - direction. At 5s, velocity is zero and after 5s, velocity is negative, so the particle is moving towards negative x - direction.

Distance in first 4s

$$4\text{s} < t_0$$

$$\therefore \quad d = s = ut + \frac{1}{2}at^2$$

$$= 10 \times 4 - \frac{1}{2} \times 2 \times (4)^2$$

$$= 24 \text{ cm}$$

Distance in first 8s

$$8\text{ s} > t_0$$

$$\therefore \quad d = d_{0\text{-}5} + d_{5\text{-}8} \quad (d = \text{positive})$$

$$= \frac{u^2}{2a} + \frac{1}{2}a(t - t_0)^2 \quad \text{(only magnitude)}$$

$$= \frac{(10)^2}{2 \times 2} + \frac{1}{2} \times 2 \times (8 - 5)^2 = (25 + 9) \text{ cm}$$

$$= 34 \text{ cm}$$

Distance *versus* time graph

From (0 - 5) s, particle is retarded and then accelerated in opposite direction. Hence d *versus* t graph is as shown below.

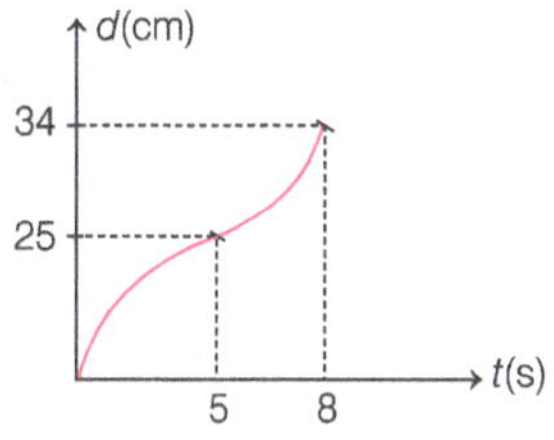

81. Acceleration $a = v \cdot \frac{dv}{ds} = -\alpha\sqrt{v}$

$$\therefore \quad \alpha\, ds = -\sqrt{v}\, dv \quad \text{or} \quad \alpha \int_0^s ds = -\int_{v_0}^0 \sqrt{v}\, dv$$

Solving this equation, we get

$$s = \left(\frac{2}{3\alpha}\right) v_0^{3/2}$$

Further, acceleration, $a = \frac{dv}{dt} = -\alpha\sqrt{v}$

$$\therefore \quad \alpha dt = -\frac{dv}{\sqrt{v}} \quad \text{or} \quad \alpha \int_0^t dt = -\int_{v_0}^0 \frac{dv}{\sqrt{v}}$$

Solving this equation, we get

$$t = \frac{2\sqrt{v_0}}{\alpha}$$

82. (a) Acceleration or retardation,

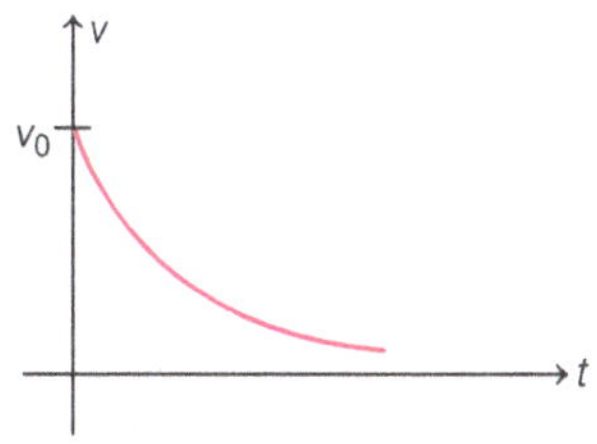

$$a = \frac{F}{m} = -\frac{r}{m}v$$

or $\frac{dv}{dt} = -\frac{r}{m}v$

$\therefore$ $\int_{v_0}^{v} \frac{dv}{v} = -\frac{r}{m}\int_0^t dt$

Solving this equation, we get

$$v = v_0 e^{-\frac{r}{m}t}$$

i.e. velocity decreases exponentially from v_0 to 0.

So, motorboat will continue to move until $t \to \infty$.

(b) $a = -\frac{r}{m}v = v\cdot\frac{dv}{ds}$

$\Rightarrow$ $dv = -\frac{r}{m}ds$

Integrating, we get

$$v = -\frac{r}{m}s + c$$

At $s = 0, v = v_0$

$\Rightarrow$ $c = v_0$

$\therefore$ $v = v_0 - \frac{r}{m}s$

At $v = 0, s = \frac{mv_0}{r}$

$\therefore$ $s_{total} = \frac{mv_0}{r}$

(c) $v = v_0 e^{-\frac{r}{m}t}$

$\Rightarrow$ $\frac{v_0}{\eta} = v_0 e^{-\frac{r}{m}t}$

Solving this equation, we get

$$t = \frac{m}{r}\ln\eta$$

Further, $v = v_0 - \frac{r}{m}s$

$\therefore$ $\frac{v_0}{\eta} = v_0 - \frac{r}{m}s$

or $s = \frac{m}{r}\left(v_0 - \frac{v_0}{\eta}\right)$

Now, average velocity,

$$v_{av} \text{ or } \langle v\rangle = \frac{s}{t}$$

or $\langle v\rangle = \frac{v_0}{\ln\eta}\left(1 - \frac{1}{\eta}\right)$

83. Resistance force $\propto v^2$ means

Retardation $a \propto -v^2$

or $a = -kv^2$ (k = positive constant)

$\therefore$ $v\cdot\frac{dv}{ds} = -kv^2$

or $\frac{dv}{v} = -kds$

Integrating, we get

$$\ln v = -ks + c_1 \quad \text{...(i)}$$

At $s = 0, v = v_0$

$\therefore$ $c_1 = \ln v_0$

Substituting this value in Eq. (i), we get

$$v = v_0 e^{-ks}$$

At $s = h$, velocity is v

$\therefore$ $v = v_0 e^{-kh}$

$\therefore$ $k = \frac{1}{h}\ln\left(\frac{v_0}{v}\right)$

Now again we write

$$a = \frac{dv}{dt} = -kv^2$$

or $\frac{dv}{v^2} = -kdt$

Integrating, we get

$$-\frac{1}{v} = -kt + c_2 \quad \text{...(ii)}$$

At $t = 0, v = v_0$

$\Rightarrow$ $c_2 = -\frac{1}{v_0}$

Substituting this value in Eq. (ii), we get

$$\frac{1}{v_0} - \frac{1}{v} = -kt \quad \text{...(iii)}$$

Now substituting, $v = v$ and $k = \frac{1}{h}\ln\left(\frac{v_0}{v}\right)$ in Eq. (iii), we get

$$\frac{1}{v_0} - \frac{1}{v} = \frac{1}{h}\ln\left(\frac{v_0}{v}\right)t$$

$\therefore$ $t = \frac{h(v_0 - v)}{v_0 v\ln(v_0/v)}$

84. In upward journey, retardation

$$a_1 = -v\frac{dv}{ds} = \frac{mg + kv^2}{m}$$

$$= g + \frac{k}{m}v^2$$

$\therefore$ $\frac{vdv}{g + \frac{k}{m}v^2} = -ds$

Integrating this equation, we get

$$\frac{m}{2k}\ln\left(g + \frac{k}{m}v^2\right) = -s + c \quad \text{...(i)}$$

At $s = 0, v = v_0$

$\therefore$ $c = \frac{m}{2k}\ln\left(g + \frac{k}{m}v_0^2\right)$

Substituting in equation (i), we get

$$\frac{g+\frac{k}{m}v^2}{g+\frac{k}{m}v_0^2}=e^{-\frac{2ks}{m}}$$

At maximum height $s=h$ and $v=0$.

Substituting in above equation, we get

$$e^{\frac{2kh}{m}}=1+\frac{k}{mg}v_0^2$$

$\therefore$ Maximum height,

$$h=\frac{m}{2k}\ln\left(1+\frac{k}{mg}v_0^2\right)$$

In downward journey, net acceleration,

$$a_2=v\cdot\frac{dv}{ds}$$

$$=\frac{mg-kv^2}{m}$$

$$=g-\frac{k}{m}v^2$$

Integrating this equation with limits,

$v=0$ at $s=0$, we can find v - s equation.

Then, putting $s=h$ and $v=v'$, we can find v'.

Finally, we get

$$v'=v_0/\sqrt{1+kv_0^2/mg}$$

Note *Mathematics of this question is complex. So students can skip its mathematical part.*

85. (a) $a=$ slope of v-t graph

(b) $s=r_f-r_i=$ net area of v-t graph

$\therefore\ r_f=r_i+$ net area

$=10+10+20+10-10-10$

$=00$ m

(c) (i) **For $0\le t\le 2$ s**

$u=$ initial velocity $=0$

$s_0=$ initial displacement $=10$ m

$a=$ slope of v-t graph $=+5\ \text{m/s}^2$

$$\therefore\quad s=s_0+ut+\frac{1}{2}at^2=10+2.5t^2$$

(ii) **For $4\text{ s}\le t\le 8$ s**

$u=$ initial velocity $=10$ m/s

$=$ velocity at 4 s

$s_0=(10\text{ m})+$ area of v-t graph upto 4 s

$=10+10+20=40$ m

$a=$ slope of v-t graph $=-5\ \text{m/s}^2$

$$\therefore\quad s=s_0+u(t-4)+\frac{1}{2}a(t-4)^2$$

$$=40+10(t-4)-2.5(t-4)^2$$ **Ans.**

86. $v\,dv=a\,ds$

$$\therefore\quad \int_0^v v\,dv=\int_0^{12\text{ m}} a\,ds$$

$\therefore\ \frac{v^2}{2}=$ area under a-s graph from $s=0$ to $s=12$ m.

$$=2+12+6+4$$

$$=24\ \text{m}^2/\text{s}^2$$

or $$v=\sqrt{48}\ \text{m/s}$$

$$=4\sqrt{3}\ \text{m/s}$$ **Ans.**

87.

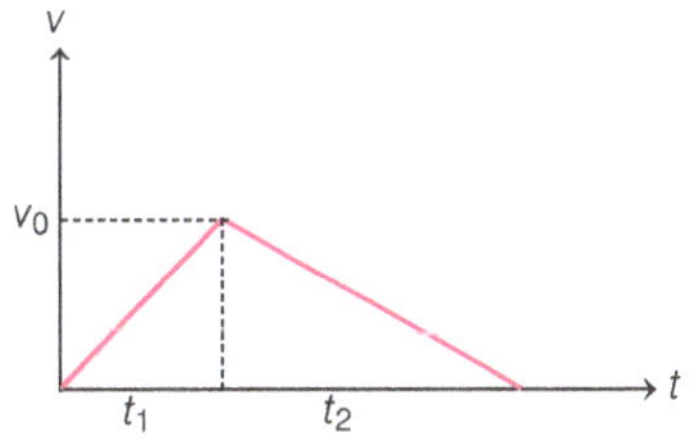

$$\frac{v_0}{t_1}=x\ \Rightarrow\ t_1=\frac{v_0}{x}$$

$$\frac{v_0}{t_2}=y\ \Rightarrow\ t_2=\frac{v_0}{y}$$

$$t_1+t_2=v_0\left(\frac{1}{x}+\frac{1}{y}\right)=4 \qquad \text{...(i)}$$

Further,

area $=$ displacement

$$\therefore\quad \frac{1}{2}v_0\times t=s$$

But numerically, $t=s=4$ units

$$\therefore\quad v_0=2\text{ units}$$

Substituting in Eq. (i), we get

$$\frac{1}{x}+\frac{1}{y}=2$$

88. Since the particle is moving rectilinearly in one direction. Therefore distance and displacement are same.

(a) After 20 s, distance (or displacement) has become constant (or slope of s-t graph has become zero $\Rightarrow v=0$). Hence particle was in motion only upto 20 s.

$$\text{Average velocity}=\frac{\text{Total displacement}}{\text{Total time}}$$

$$=\frac{2.0\text{ m}}{20\text{ s}}=0.1\text{ m/s}$$

$$=10\text{ cm/s}\quad\text{or}\quad v_{av}=10\text{ cm/s}$$

(b) Maximum velocity means, maximum slope (instantaneous velocity).

From the given graph we can see that the middle portion from 10 s to 14 s graph is straight and it has maximum slope.

$\therefore \quad v_{max}$ = Maximum slope from 10-14 s

$$= \frac{(1.4 - 0.4)\text{ m}}{(14 - 10)\text{ s}}$$

$= 0.25$ m/s or 25 cm/s

$\therefore \quad v_{max} = 25$ cm/s

(c) Instantaneous velocity = Slope of s - t graph at that instant and,

$$\text{average velocity} = \frac{\text{total displacement}}{\text{total time}}$$

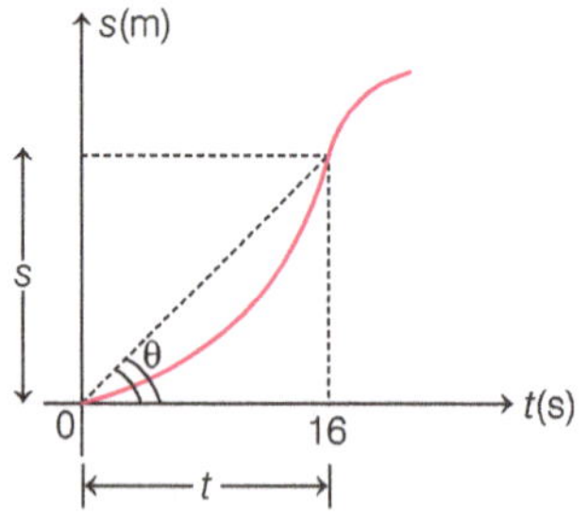

At 16 s, we can see that slope of graph is also equal to $\frac{s}{t}$.

$\therefore \quad t_0 = 16$ s

Note *At 16 s, instantaneous velocity = slope = tan θ =* $\frac{s}{t}$ *and* $\frac{s}{t}$ *is* $\frac{\text{total displacement}}{\text{total time}}$ *or average velocity.*

89. (a)

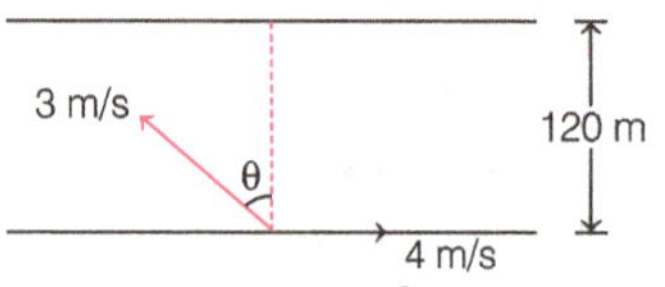

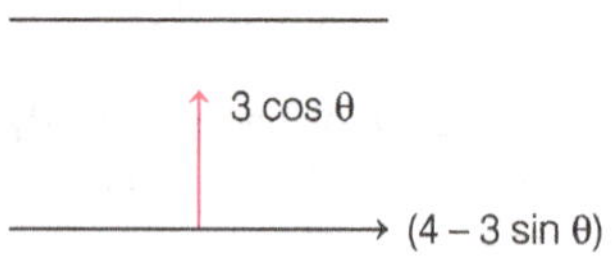

Time to cross the river

$$t_1 = \frac{120}{3\cos\theta} = \frac{40}{\cos\theta} = 40\sec\theta$$

Drift along the river $x = (4 - 3\sin\theta)\left(\frac{40}{\cos\theta}\right)$

$= (160\sec\theta - 120\tan\theta)$

To reach directly opposite, this drift will be covered by walking speed.

Time taken in this,

$$t_2 = \frac{160\sec\theta - 120\tan\theta}{1}$$

$= 160\sec\theta - 120\tan\theta$

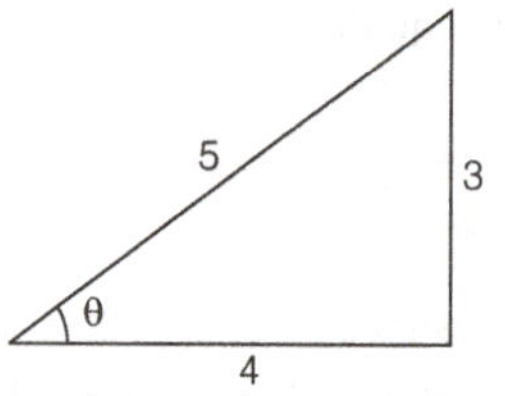

$\therefore$ Total time taken

$t = t_1 + t_2 = (200\sec\theta - 120\tan\theta)$

For t to be minimum, $\frac{dt}{d\theta} = 0$

or $\quad 200\sec\theta\tan\theta - 120\sec^2\theta = 0$

or $\quad \theta = \sin^{-1}(3/5)$

(b) $t_{min} = 200\sec\theta - 120\tan\theta$ (where, $\sin\theta = \frac{3}{5}$)

$= 200 \times \frac{5}{4} - 120 \times \frac{3}{4}$

$= 250 - 90 = 160$ s $= 2$ min 40 s **Ans.**

90. Let $\mathbf{v}_r$ = velocity of river

$\mathbf{v}_{br}$ = velocity of motorboat w.r.t. river

Let us assume that river (or the raft) is at rest. Then the motorboat will move with velocity v_{br}, during downstream. Since, river (or raft) is assumed to be at rest. Therefore, downstream and upstream both times are same or 1 hour each.

In these two hours (1 hour + 1 hour), raft has travelled a distance of 6 km downstream (given) with velocity v_r.

$$\therefore \quad v_r = \frac{\text{Distance}}{\text{Time}} = \frac{6\text{ km}}{2\text{ h}} \quad \text{or} \quad v_r = 3\text{ km/h}$$

91. Given $|\mathbf{v}_{br}| = 1.2|\mathbf{v}_r|$

So, let $|\mathbf{v}_r| = v$, then $|\mathbf{v}_{br}| = 1.2\,v$

Boat *A*

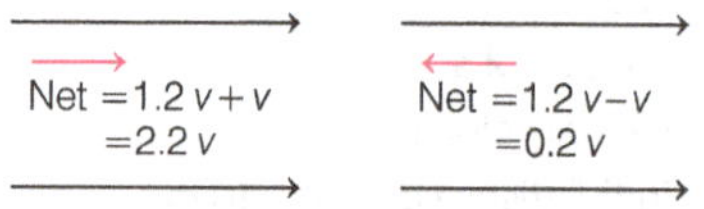

$$\tau_A = \frac{d}{2.2\,v} + \frac{d}{0.2\,v} = \frac{6d}{1.1\,v} \quad \text{...(i)}$$

Boat *B*

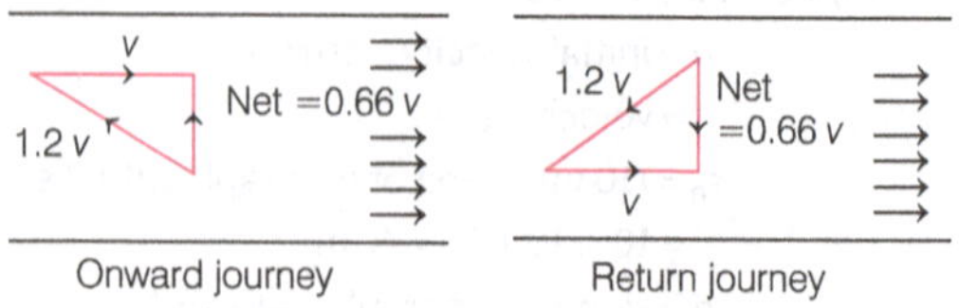

Let d be the distance travelled on one side. Then,

$$\tau_B = \frac{d}{0.66v} + \frac{d}{0.66v} \approx \frac{3d}{v} \quad \text{...(ii)}$$

From equations (i) and (ii), we get

$$\frac{\tau_A}{\tau_B} \approx 1.8$$

92. Let the desired distance is x.

Distance AD = constant = a (say)

$\therefore \quad AC = (a - x)$

$BD = l$ is also constant

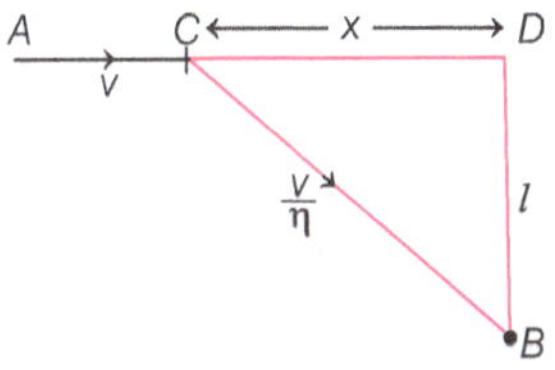

Now, $\quad t_{AB} = t\,(\text{say}) = t_{AC} + t_{BC}$

or $\quad t = \dfrac{AC}{v} + \dfrac{BC}{v/\eta}$

or $\quad t = \dfrac{(a - x)}{v} + \dfrac{\sqrt{x^2 + l^2}}{\dfrac{v}{\eta}}$

For t to be minimum, $\dfrac{dt}{dx} = 0$ and $\dfrac{d^2t}{dx^2}$ should be positive.

Solving the mathematical part of it, we get

$$x = \frac{l}{\sqrt{\eta^2 - 1}}$$

93. $\because \; v^2 = u^2 + 2as$

$\Rightarrow \quad 0^2 = \left(72 \times \dfrac{5}{18}\right)^2 - 2 \times 2 s_0$

$\therefore \quad s_0 = \dfrac{400}{4} = 100 \text{ m}$

Thus, stopping distance

$$= \left(72 \times \frac{5}{18}\right) \times 0.2 + s_0$$

$$= 4 + 100 = 104 \text{ m} = 13x$$

$\therefore \quad x = \dfrac{104}{13} = 8$

94. $\because \quad a = -kv \quad$ or $\quad \dfrac{dv}{dt} = -kv$

or $\quad \displaystyle\int_{v_0}^{v} \frac{dv}{v} = -k \int_0^t dt$

or $\quad [\ln v]_{v_0}^{v} = -kt$

or $\quad \ln v - \ln v_0 = -kt \quad$ or $\quad \ln \dfrac{v}{v_0} = -kt$

or $\quad \dfrac{v}{v_0} = e^{-kt}$

For $\quad t = t_0,\; v = v_0 - v_0 \times \dfrac{50}{100} = \dfrac{v_0}{2}$

$\therefore \quad \dfrac{v_0}{2} = v_0 e^{-kt_0}$

or $\quad \dfrac{1}{2} = e^{-kt_0} \quad$ or $\quad \ln \dfrac{1}{2} = -kt_0 \quad$ or $\quad k = \dfrac{\ln 2}{t_0}$

$\therefore \quad v = v_0 e^{-kt} \quad$ or $\quad v = v_0 e^{-\frac{\ln 2}{t_0} t}$

or $\quad \ln v = \ln v_0 - \dfrac{\ln 2}{t_0} t \quad$ or $\quad \ln \dfrac{v}{v_0} = -\dfrac{\ln 2}{4 \ln 2} t$

or $\quad \ln \dfrac{v}{v_0} = -\dfrac{t}{4} \quad$ or $\quad \dfrac{v}{v_0} = e^{-\frac{t}{4}}$

$\therefore \quad v = v_0 e^{-\frac{t}{4}} = 13.5\, e^{-\frac{4}{4}} = \dfrac{13.5}{e} = \dfrac{13.5}{2.7} = 5 \text{ ms}^{-1}$

95. The acceleration of ball at instant t is

$a = g - bv$

or $\quad \dfrac{dv}{dt} = g - bv \quad$ or $\quad \displaystyle\int_0^v \frac{dv}{g - bv} = \int_0^t dt$

or $\quad -\dfrac{1}{b}[\ln (g - bv)]_0^v = [t]_0^t$

or $\quad [\ln (g - bv) - \ln g] = -bt$

or $\quad \ln \dfrac{g - bv}{g} = -bt$

or $\quad 1 - \dfrac{bv}{g} = e^{-bt} \quad$ or $\quad \dfrac{bv}{g} = 1 - e^{-bt}$

or $\quad v = \dfrac{g}{b}(1 - e^{-bt}) = \dfrac{10}{10}\left(1 - \dfrac{1}{e}\right)$

$$= \left(\frac{e - 1}{e}\right) = \frac{2.7 - 1}{2.7} = \frac{1.7}{2.7} = \frac{17}{27} = \frac{n}{27}$$

$\therefore \quad n = 17$

96. Here, $a = -g - bv \quad$ or $\quad \dfrac{dv}{dt} = -g - bv$

or $\quad \displaystyle\int_{v_0}^{0} \frac{dv}{g + bv} = -\int_0^{t_0} dt$

or $\quad \dfrac{1}{b}[\ln (g + bv)]_{v_0}^{0} = -t_0$

or $\quad \dfrac{1}{b}[\ln g - \ln (g + bv_0)] = -t_0$

or $\quad -\dfrac{1}{b} \ln \dfrac{g + bv_0}{g} = -t_0 \quad$ or $\quad \ln \left(1 + \dfrac{bv_0}{g}\right) = bt_0$

$\therefore \quad 1 + \dfrac{bv_0}{g} = e^{bt_0}$

$\therefore \quad e^{bt_0} = 1 + 2 = 3$

97. As per question, magnitude of velocity of the swimmer is same as that of the river.

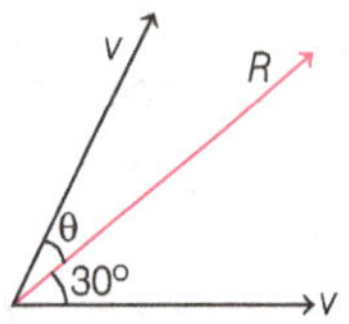

So, resultant would pass exactly midway through them.

$\therefore$ Angle $\theta = 30°$

98. If he starts from home at 6 pm, he reaches at coaching at 6:20 pm. According to the question, he reaches at coaching 8 min before his daily routine. It means he starts at 5:52 pm. and will reach at the coaching at 6:12 pm. Let the distance between home and turning point 'C' is l. The distance between the coaching and home is $L = 20\,v$.

According to the problem,

$$\frac{l}{v} + \frac{l}{v} + \frac{l}{v} + \frac{L-l}{v} = (6:20 - 5:52) + 10 \text{ min}$$

or $$\frac{2l}{v} + \frac{L}{v} = 38 \text{ min}$$

or $$\frac{2l}{v} + \frac{20v}{v} = 38 \text{ min}$$

or $$\frac{2l}{v} = 18 \text{ min} \quad \text{or} \quad \frac{l}{v} = 9 \text{ min}$$

The fraction is $\frac{l}{L} = \frac{9v}{20v} = \frac{9}{20} = \frac{x}{20}$

$\therefore$ $x = 9$

Level 2

99. Area of v-t graph = displacement

$\therefore$ $$1032 = \frac{1}{2}(56 + t_0)(24)$$

or $$t_0 = 30 \text{ s}$$

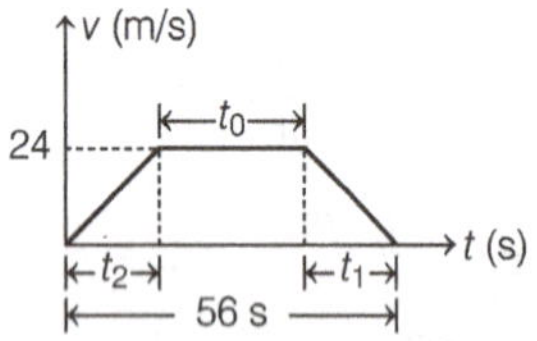

Deceleration time $t_1 = \frac{24}{4} = 6 \text{ s}$

$\therefore$ Acceleration time $t_2 = 56 - t_0 - t_1 = 20 \text{ s}$

$\therefore$ Acceleration $= \frac{24}{20} = 1.2 \text{ m/s}^2$ **Ans.**

100.

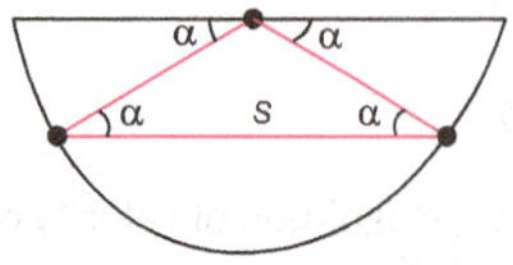

$s = 2r\cos\alpha$

$\therefore$ $$s = \frac{1}{2}a_0 t^2$$

or $$2r\cos\alpha = \frac{1}{2}a_0 t^2$$

$\therefore$ $$t = \sqrt{\frac{4r\cos\alpha}{a_0}}$$

101. $v = \alpha t_1 \Rightarrow t_1 = \frac{v}{\alpha}$

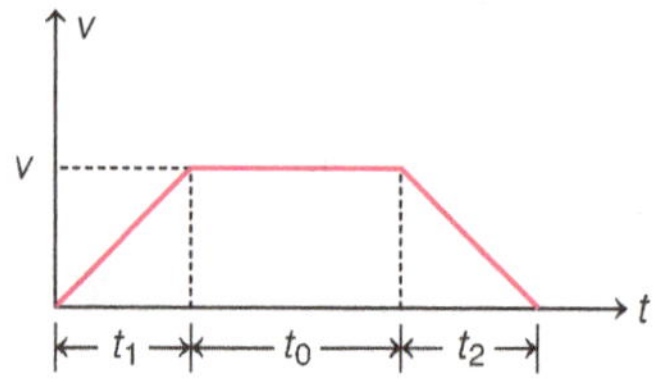

$$v = \beta t_2 \Rightarrow t_2 = \frac{v}{\beta}$$

$\therefore$ $$t_0 = t - t_1 - t_2 = \left(t - \frac{v}{\alpha} - \frac{v}{\beta}\right)$$

Now, $$l = \frac{1}{2}\alpha t_1^2 + v t_0 + \frac{1}{2}\beta t_2^2$$

$$= \frac{1}{2}(\alpha)\left(\frac{v}{\alpha}\right)^2 + v\left(t - \frac{v}{\alpha} - \frac{v}{\beta}\right) + \frac{1}{2}(\beta)\left(\frac{v}{\beta}\right)^2$$

$$= vt - \frac{v^2}{2\alpha} - \frac{v^2}{2\beta}$$

$\therefore$ $$t = \frac{l}{v} + \frac{v}{2}\left(\frac{1}{\alpha} + \frac{1}{\beta}\right)$$

For t to be minimum its first derivation with respect to velocity be zero, i.e.

$$0 = -\frac{l}{v^2} + \frac{\alpha + \beta}{2\alpha\beta}$$

$\therefore$ $$v = \sqrt{\frac{2l\alpha\beta}{\alpha + \beta}}$$

102. $a_1 t_1 = a_2 t_2$

$$v_{max} = a_1 t_1 = a_2 t_2$$

s_1 = Area of v-t graph

$$= \frac{1}{2}(t_1 + t_2)(a_1 t_1)$$

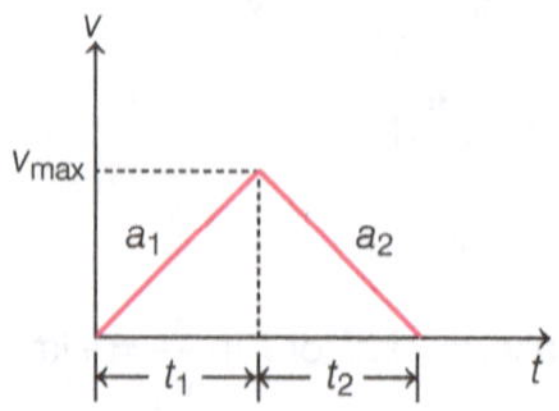

$$v_1 = \frac{s_1}{t_1 + t_2} = \frac{1}{2}a_1 t_1$$

$$s_2 = \frac{1}{2}(t_1 + t_3)\, v_{max}$$

$$= \frac{1}{2}(t_1 + 2t_2)(2a_1 t_1)$$

$$v_2 = \frac{s_2}{(t_1 + t_3)} = a_1 t_1$$

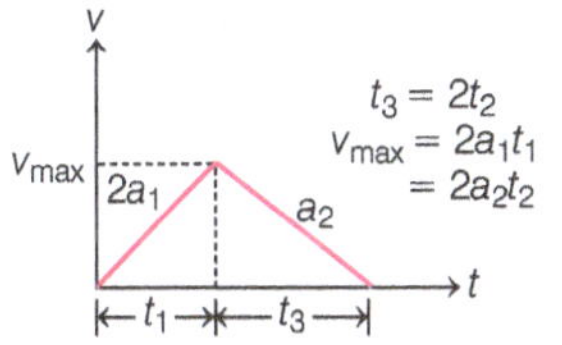

From the four relations we can see that

$v_2 = 2v_1$ and $2s_1 < s_2 < 4s_1$ **Ans.**

103. Since, the body is at rest at $x = 0$ and $x = 1$. Hence, α cannot be positive for all time in the interval $0 \le t \le 1$.

Therefore, first the particle is accelerated and then retarded.

Now, total time $t = 1$ s (given)

Total displacement, $s = 1$ m (given)

$s =$ Area under v-t graph

$\therefore$ Height or $v_{max} = \frac{2s}{t} = 2$ m/s is also fixed.

$$\left[\text{Area or } s = \frac{1}{2} \times t \times v_{max}\right]$$

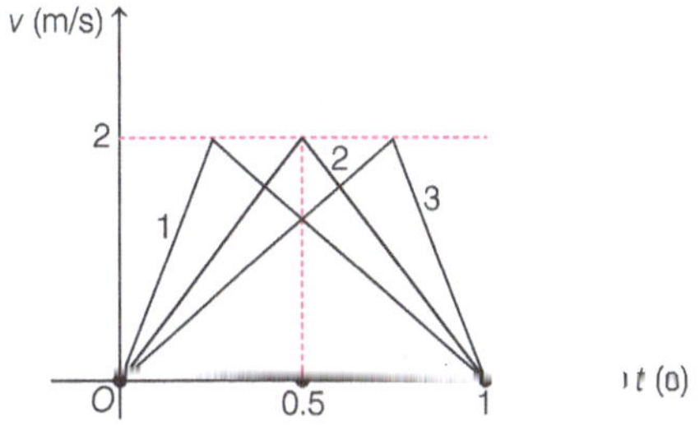

If height and base are fixed, area is also fixed .

In case 2 : Acceleration = Retardation = 4 m/s^2

In case 1 : Acceleration > 4 m /s^2 while Retardation < 4m /s^2

While in case 3 : Acceleration < 4 m /s^2 and Retardation > 4m /s^2

Hence, $|\alpha| \ge 4$ at some point or points in its path.

104. (a) Since, ball A is moving with constant velocity, so its net acceleration is zero.

Let its initial velocity is v_0.

$\therefore$ $a = g - 5v_0$

or $0 = 10 - 5v_0$

$\therefore$ $v_0 = 2 \text{ ms}^{-1}$

(b) Ball B is moving with increasing speed.

$\therefore$ $a > 0$

or $g - 5v > 0$

or $10 - 5v > 0$

or $5v < 10$

or $v < 2 \text{ ms}^{-1}$

$\therefore$ During acceleration, its velocity is lying in the range of $0 \le v \le 2$.

When it gains constant velocity, $a = 0$

$\therefore$ $g - 5v_0 = 0$

$\therefore$ $v_0 = 2 \text{ ms}^{-1}$, this is its final velocity.

(c) Initially ball C is moving with decreasing speed. Let its initial velocity is u.

$\therefore$ $a < 0$

$\therefore$ $5v > 10$

or $g - 5v < 0$

or $v > 2 \text{ ms}^{-1}$

$\therefore$ $v_0 > 2 \text{ ms}^{-1}$

When it gains constant velocity,

$a = 0$

$\therefore$ $g - 5v_0 = 0$

$\therefore$ $v_0 = 2 \text{ ms}^{-1}$

Hence, (a) $\rightarrow$ (p,s), (b) $\rightarrow$ (q,s), (c) $\rightarrow$ (r,s)

105. $a = \frac{4}{v}$ or $\frac{dv}{dt} = \frac{4}{v}$

$$\therefore \int_6^v v\,dv = \int_2^t 4\,dt$$

$$\therefore \frac{v^2}{2} - 18 = 4t - 8$$

$$\therefore v = \sqrt{8t + 20}$$

$$a = \frac{dv}{dt} = \frac{4}{\sqrt{8t + 20}}$$

At $t = 3$ s

$a = 0.603 \text{ m/s}^2$ **Ans.**

106. $\because$ $a = t - x$ or $\frac{da}{dt} = 1 - \frac{dx}{dt}$

or $\frac{da}{dt} = 1 - v$ or $\frac{d^2a}{dt^2} = -\frac{dv}{dt} = -a$

The solution of this differential equation is

$a = a_0 \sin(\omega t + \phi)$ where, $\omega = 1$ unit

At $t = 0, x = 0$

$\Rightarrow$ $a = 0$

$\therefore$ $0 = a_0 \sin(\omega \times 0 + \phi)$

$\therefore$ $\phi = 0$

$\therefore$ $a = a_0 \sin t$ or $\frac{dv}{dt} = a_0 \sin t$

or $\int_0^v dv = a_0 \int_0^t \sin t\, dt$

or $v = a_0[-\cos t]_0^t = -a_0[\cos t - 1]$

$$v = a_0(1-\cos t) = 2\left(1-\cos\frac{\pi}{3}\right)$$
$$= \left(2\times\frac{1}{2}\right) = 1\ \text{ms}^{-1}$$

107. (a) Given $\mathbf{r} = \mathbf{w}t\,(1-\alpha t)$

Here $\mathbf{w}$ = constant vector

$$\mathbf{v} = \frac{d\mathbf{r}}{dt} = \mathbf{w}\,(1-2\alpha t)$$
$$\mathbf{a} = \frac{d\mathbf{v}}{dt} = -2\alpha\mathbf{w} = \text{constant}$$

(b) At initial point, $\mathbf{r} = 0$

From the given equation we can see that,

$\mathbf{r} = 0$ at $t = 0$

and $t = \frac{1}{\alpha}$

Hence, $\Delta t = \frac{1}{\alpha}$

$$\mathbf{v} = \mathbf{w}\,(1-2\alpha t)$$

$\Rightarrow$ $\mathbf{v} = 0$ at $t = \frac{1}{2\alpha}$

From 0 to $t = \frac{1}{2\alpha}$, **v** is parallel to **w** and then from $t = \frac{1}{2\alpha}$ to $t = \frac{1}{\alpha}$, velocity is antiparallel to **w**. So, if we talk about speed, then

$v = w(1-2\alpha t)$ for $t \le \frac{1}{2\alpha}$

and $v = w(2\alpha t - 1)$ for $t \ge \frac{1}{2\alpha}$

Distance $d = \int v\,dt$

$\therefore$ Total distance $= d_1 + d_2$

$$d = \int_0^{1/2\alpha} w(1-2\alpha t)\,dt + \int_{1/2\alpha}^{1/\alpha} w(2\alpha t - 1)\,dt$$

Solving this equation, we get

$$d = \frac{w}{2\alpha}$$

108. (a) $\frac{dv}{dt} = \frac{1}{2\sqrt{v_x^2+v_y^2}}\left(2v_x\frac{dv_x}{dt} + 2v_y\frac{dv_y}{dt}\right)$

$$\frac{dv}{dt} = \frac{(v_x a_x + v_y a_y)}{\sqrt{v_x^2+v_y^2}}$$

(b) $\mathbf{v} = v_x\,\hat{\mathbf{i}} + v_y\,\hat{\mathbf{j}}$ and $\mathbf{a} = a_x\,\hat{\mathbf{i}} + a_y\,\hat{\mathbf{j}}$

$$\mathbf{v}\cdot\mathbf{a} = v_x a_x + v_y a_y$$

Further $v = \sqrt{{v_x}^2 + {v_y}^2}$

$\therefore$ $\frac{\mathbf{v}\cdot\mathbf{a}}{v} = \frac{v_x a_x + v_y a_y}{\sqrt{{v_x}^2+{v_y}^2}} = \frac{dv}{dt} = a_t$

or component of **a** parallel to **v** = tangential acceleration.

109. (a) Let us assume that at time $t = 0$, **F** was along x-axis. Then at time t, **F** is as shown in figure.

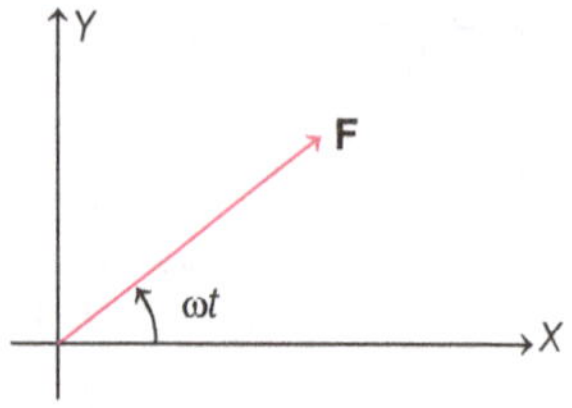

$$\mathbf{F} = F\cos\omega t\,\hat{\mathbf{i}} + F\sin\omega t\,\hat{\mathbf{j}}$$

Linear impulse = change in linear momentum

$\therefore$ $\int_0^t \mathbf{F}\,dt = m\mathbf{v}$

or $\int_0^t (F\cos\omega t\,\hat{\mathbf{i}} + F\sin\omega t\,\hat{\mathbf{j}})\,dt = m\mathbf{v}$

or $\frac{F}{\omega}\sin\omega t\,\hat{\mathbf{i}} + \frac{F}{\omega}(1-\cos\omega t)\,\hat{\mathbf{j}} = m\mathbf{v}$

$\therefore$ $\mathbf{F} = \frac{F}{m\omega}[(\sin\omega t)\hat{\mathbf{i}} + (1-\cos\omega t)\hat{\mathbf{j}}]$

$\therefore$ $|\mathbf{v}| = \frac{F}{m\omega}\sqrt{\sin^2\omega t + (1-\cos\omega t)^2}$

or $|\mathbf{v}| = \frac{2F}{m\omega}\left|\sin\frac{\omega t}{2}\right|$

(b) Speed *versus* time graph is as shown in figure.

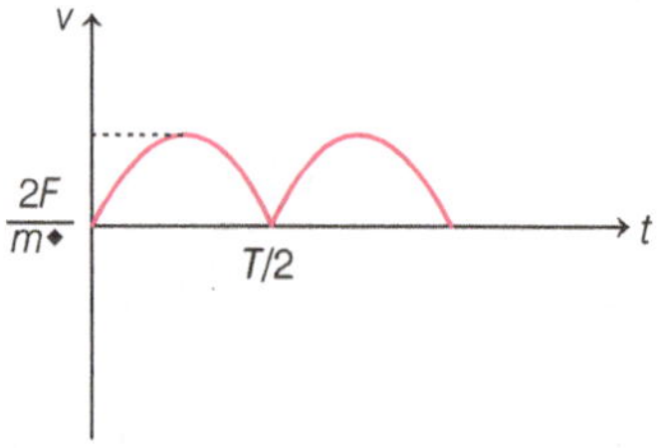

$$\frac{\omega}{2} = \frac{2\pi}{T}$$

$\therefore$ $T = \frac{4\pi}{\omega}$ or $\frac{T}{2} = \frac{2\pi}{\omega}$

Distance $d = \int_0^{T/2 \text{ or } 2\pi/w} \left(\frac{2F}{m\omega}\right)\sin\frac{\omega t}{2}\,dt = \frac{8F}{m\omega^2}$

$\therefore$ $d = \frac{8F}{m\omega^2}$

Note *In the last part answer is given according to average speed.*

$\therefore$ Average speed $= \frac{\text{Total distance}}{\text{Total time}} = \frac{8F/m\omega^2}{2\pi/\omega}$

$\therefore$ Average speed $= \frac{4F}{\pi m\omega}$

110. Let us draw v -t graph of the given situation, area of which will give the displacement and slope the acceleration.

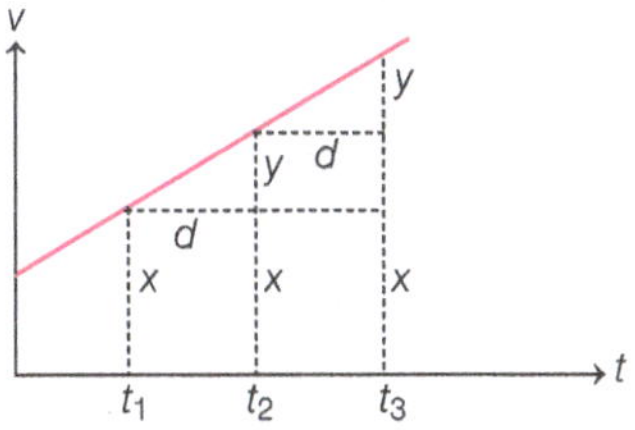

$$s_2 - s_1 = x\,d + \frac{1}{2} yd \quad \text{...(i)}$$

$$s_3 - s_2 = xd + yd + \frac{1}{2} yd \quad \text{...(ii)}$$

Subtracting Eq. (i) from Eq. (II), we have

$$s_3 + s_1 - 2s_2 = yd$$

or $$s_3 + s_1 - 2\sqrt{s_1 s_3} = yd \quad (\because s_2 = \sqrt{s_1 s_3})$$

Dividing by d^2 both sides, we have

$$\frac{(\sqrt{s_1} - \sqrt{s_3})^2}{d^2} = \frac{y}{d}$$

$=$ slope of v -t graph $= a$. **Hence proved.**

111. Area of v-t graph $=$ displacement

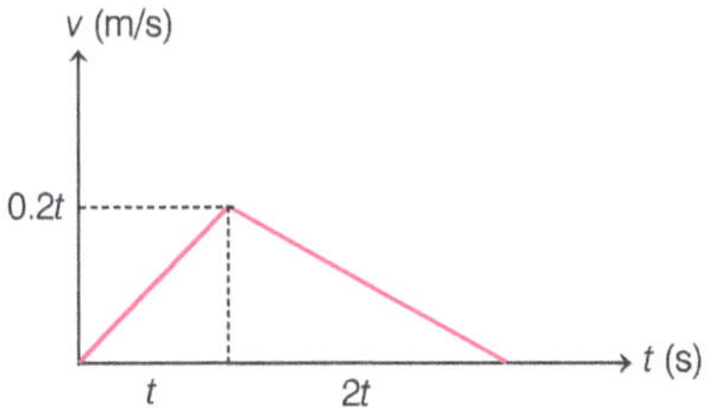

$\therefore$ $$\frac{1}{2}(3t)(0.2\,t) = 14$$

Solving this equation, we get

$$3t = 20.5 \text{ s}$$ **Ans.**

Note *Maximum speed* $0.2\,t$ *is less than* $2.5\ m/s$.

112. For $0 < s \le 60$ m

$$v = \frac{12}{60}s + 3 = 3 + \frac{s}{5}$$

$\therefore$ $$\frac{dv}{dt} = \left(\frac{1}{5}\right) \cdot \frac{ds}{dt}$$

$$= \frac{1}{5}(v) = \frac{1}{5}\left(3 + \frac{s}{5}\right) = \frac{3}{5} + \frac{s}{25} \quad \text{...(i)}$$

or $$a = \frac{3}{5} + \frac{s}{25}$$

i.e. a-s graph is a straight line.

At $s = 0$,

$$a = \frac{3}{5}\text{ m/s}^2 = 0.6 \text{ m/s}^2$$

and at $s = 60$ m, $a = 3.0 \text{ m/s}^2$

For $s > 60$ m,

$v =$ constant

$\therefore$ $a = 0$

Therefore, the corresponding a-s graph is shown in figure.

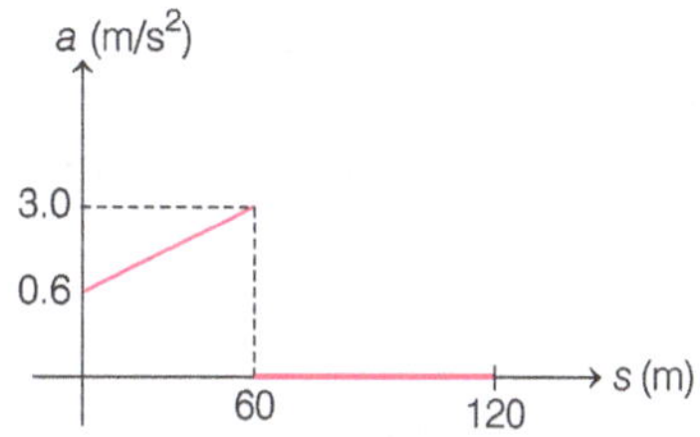

From Eq. (i), $\frac{dv}{dt} = \frac{v}{5}$

or $$\int_3^v \frac{dv}{v} = \frac{1}{5}\int_0^t dt$$

$$\ln\left(\frac{v}{3}\right) = \frac{t}{5}$$

$\therefore$ $$v = 3\,e^{t/5}$$

or $$\int_0^{60} ds = 3\int_0^{t_1} e^{t/5}\,dt$$

$$60 = 15\,(e^{t_1/5} - 1)$$

or $$t_1 = 8.0 \text{ s}$$

Time taken to travel next 60 m with speed 15m/s will be $\frac{60}{15} = 4$ s

$\therefore$ Total time $= 12.0$ s **Ans..**

113. Given $v_{av} = 72 \text{ km/h} = 72 \times \frac{5}{18} \text{ m/s} = 20 \text{ m/s}$

Since acceleration and retardation have same magnitudes $(= 5 \text{ m/s}^2)$.

$\therefore$ Acceleration time $=$ Retardation time $= t$ (say)

Then time of journey with uniform velocity will be $(25 - t - t)$ or $(25 - 2t)$ seconds.

Let us draw v -t diagram of complete journey.

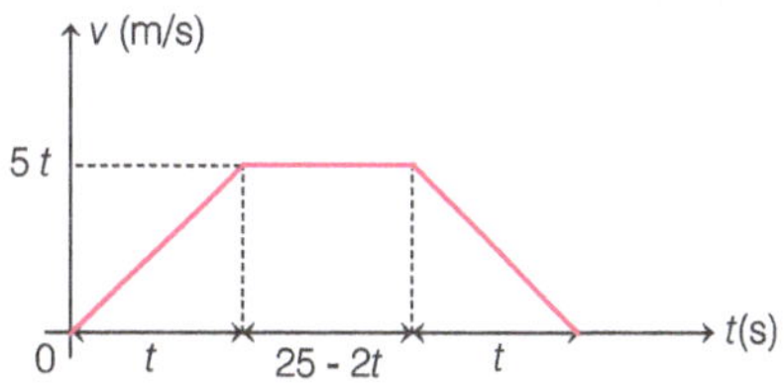

Velocity after t seconds would be

$$v = at \quad (a = 5 \text{ m/s}^2)$$

or $$v = 5t$$

Now, area of v - t graph gives the total displacement and

$$\text{Average velocity} = \frac{\text{Total displacement}}{\text{Total time}}$$

$$= \frac{\text{Total area of } v\text{-}t \text{ graph}}{\text{Total time}}$$

Substituting the values, we get

$$20 = \frac{\frac{1}{2}[(25)+(25-2t)][5t]}{25}$$

Solving this equation, we get, $t = 5$ s

But answer would be $(25-2t)$ or 15 s.

Alternate solution

Total displacement $= s_{0 \to t} + s_{t \to (25-t)} + s_{(25-t) \to t}$

$$= \frac{1}{2}at^2 + (at)(25-2t) + \frac{1}{2}at^2$$

$$= (125t - 5t^2) \qquad (a = 5 \text{ m/s}^2)$$

$$\text{Average velocity} = \frac{\text{Total displacement}}{\text{Total time}}$$

or $\qquad 20 = \dfrac{125t - 5t^2}{25}$

From this equation also, we get $t = 5$ s.

114. (a) Let A stands for trolley and B for ball.

Relative velocity of B with respect to A ($\mathbf{v}_{BA}$) should be along OA for the ball to hit the trolley. Hence, $\mathbf{v}_{BA}$ will make an angle of 45° with positive X-axis.

$$\theta = 45°$$

(b) Let $v =$ absolute velocity of ball.

$$\phi = \frac{4\theta}{3} = \frac{4}{3}(45°) = 60° \to \text{with } x\text{-axis}$$

$\therefore \qquad \mathbf{v}_B = (v\cos\theta)\,\hat{\mathbf{i}} + (v\sin\theta)\,\hat{\mathbf{j}}$

$$= \frac{v}{2}\hat{\mathbf{i}} + \frac{\sqrt{3}\,v}{2}\hat{\mathbf{j}}$$

$$\mathbf{v}_A = (\sqrt{3}-1)\,\hat{\mathbf{j}}$$

$\therefore \qquad \mathbf{v}_{BA} = \dfrac{v}{2}\hat{\mathbf{i}} + \left(\dfrac{\sqrt{3}v}{2} - \sqrt{3} + 1\right)\hat{\mathbf{j}}$

Since, $\mathbf{v}_{BA}$ is at 45°.

$\therefore \qquad \dfrac{v}{2} = \dfrac{\sqrt{3}v}{2} - \sqrt{3} + 1$

Solving, we get

or $\qquad v = 2$ m/s

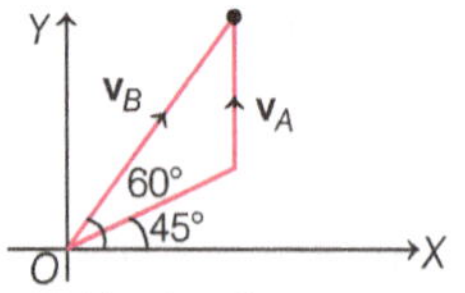

Absolute figure

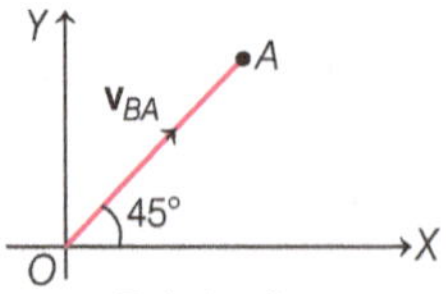

Relative figure

Alternate Solution

As time is same for both, x and y is directly proportional to their speed. We can use sine rule to find,

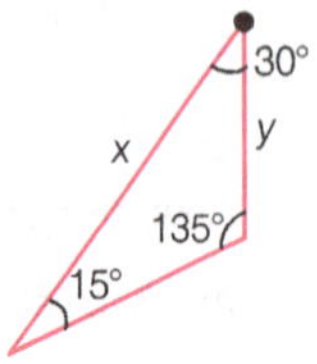

$$x = \frac{2y}{\sqrt{3}-1}$$

$\therefore \qquad v_B = \dfrac{2v_A}{\sqrt{3}-1} = 2 \text{ m/s}$

115. Given that $|\mathbf{v}_{br}| = v_y = \dfrac{dy}{dt} = u \qquad$ ...(i)

$$|\mathbf{v}_r| = v_x = \frac{dx}{dt} = \left(\frac{2v_0}{c}\right)y \qquad \text{...(ii)}$$

From Eqs. (i) and (ii) we have, $\dfrac{dy}{dx} = \dfrac{uc}{2v_0 y}$

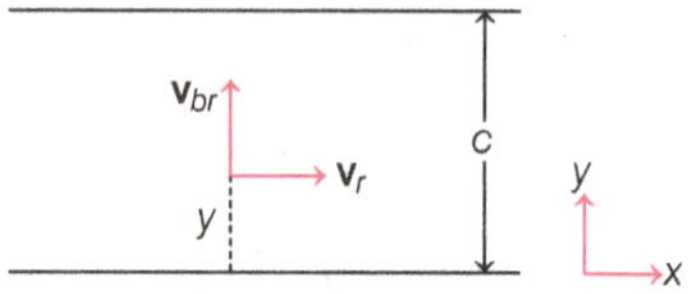

or $\qquad \displaystyle\int_0^y y\,dy = \frac{uc}{2v_0}\int_0^x dx$

or $\qquad y^2 = \dfrac{ucx}{v_0} \qquad$ **Ans.**

At $y = \dfrac{c}{2}, x = \dfrac{cv_0}{4u}$

or $\qquad x_{net} = 2x = \dfrac{cv_0}{2u} \qquad$ **Ans.**

116. Let $\mathbf{v}_{br}$ be the velocity of boatman relative to river, $\mathbf{v}_r$ the velocity of river and $\mathbf{v}_b$ is the absolute velocity of boatman. Then,

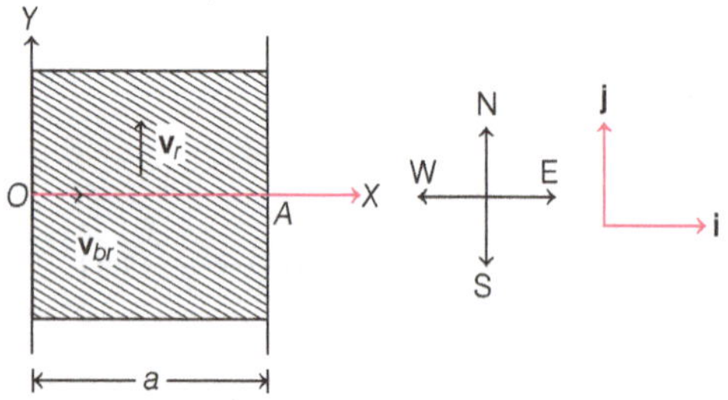

$$\mathbf{v}_b = \mathbf{v}_{br} + \mathbf{v}_r$$

Given, $\quad |\mathbf{v}_{br}| = v \quad$ and $\quad |\mathbf{v}_r| = u$

Now $\qquad u = v_y = \dfrac{dy}{dt} = x(a-x)\dfrac{v}{a^2} \qquad$ ...(i)

and $\qquad v = v_x = \dfrac{dx}{dt} = v \qquad$ ...(ii)

Dividing Eq. (i) by Eq. (ii), we get

$$\frac{dy}{dx} = \frac{x(a-x)}{a^2} \quad \text{or} \quad dy = \frac{x(a-x)}{a^2}dx$$

or $\displaystyle\int_0^y dy = \int_0^x \frac{x(a-x)}{a^2}dx$ or $y = \dfrac{x^2}{2a} - \dfrac{x^3}{3a^2} \qquad$ ...(iii)

This is the desired equation of trajectory.

(b) Time taken to cross the river is

$$t = \frac{a}{v_x} = \frac{a}{v}$$

(c) When the boatman reaches the opposite side, $x = a$ or $v_y = 0$ [from Eq. (i)]

Hence, resultant velocity of boatman is v along positive x-axis or due east.

(d) From Eq. (iii)

$$y = \frac{a^2}{2a} - \frac{a^3}{3a^2} = \frac{a}{6}$$

At $x = a$ (at opposite bank)

Hence, displacement of boatman will be

$$\mathbf{s} = x\,\hat{\mathbf{i}} + y\,\hat{\mathbf{j}} \quad \text{or} \quad \mathbf{s} = a\hat{\mathbf{i}} + \frac{a}{6}\hat{\mathbf{j}}$$

117. (a) Since, the resultant velocity is always perpendicular to the line joining boat and R, the boat is moving in a circle of radius 2ω and centre at R.

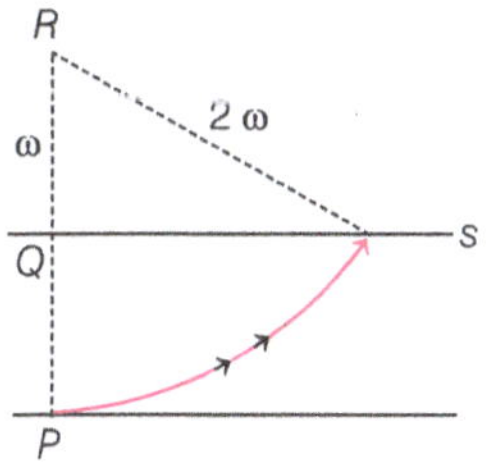

(b) Drifting $= Qs = \sqrt{4\omega^2 - \omega^2} = \sqrt{3}\omega$.

(c) Suppose at any arbitrary time, the boat is at point B.

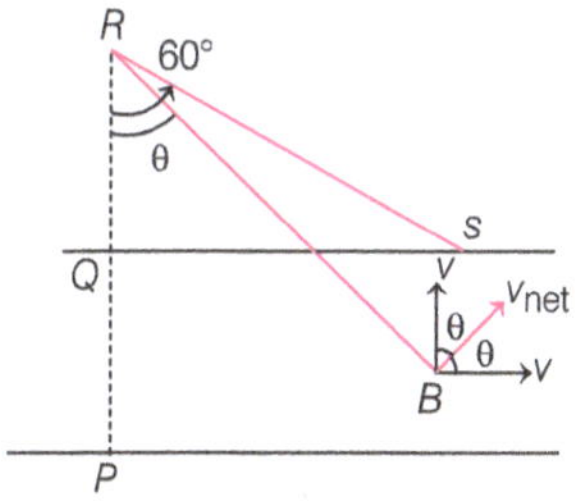

$$v_{net} = 2v\cos\theta$$

$$\frac{d\theta}{dt} = \frac{v_{net}}{2\omega} = \frac{v\cos\theta}{\omega} \quad \text{or} \quad \frac{\omega}{v}\sec\theta\, d\theta = dt$$

$$\therefore \quad \int_0^t dt = \frac{\omega}{v}\int_0^{60^\circ} \sec\theta\, d\theta$$

$$\therefore \quad t = \frac{\omega}{v}[\ln(\sec\theta + \tan\theta)]_0^{60^\circ}$$

or $$t = \frac{1.317\omega}{v}$$ **Ans.**

118. (a) $v_{br} = 4\text{m/s} \quad v_r = 2\text{m/s}$

$$\tan\theta = \frac{BC}{AB} = \frac{|\mathbf{v}_r|}{|\mathbf{v}_{br}|} = \frac{2}{4} = \frac{1}{2}$$

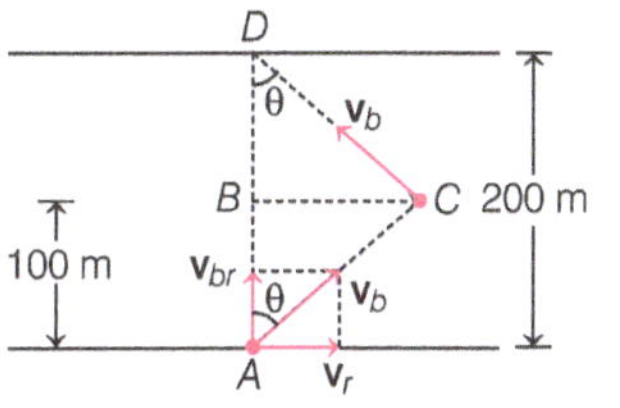

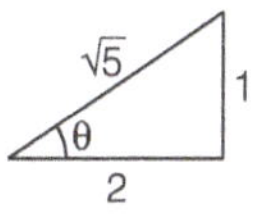

In this case, $\mathbf{v}_b$ should be along CD.

$$\therefore \quad v_r\cos\theta = v_{br}\sin\alpha$$

$$2\left(\frac{2}{\sqrt{5}}\right) = 4\sin\alpha \quad \text{or} \quad \sin\alpha = \frac{1}{\sqrt{5}}$$

$$\therefore \quad \alpha = \theta = \tan^{-1}\left(\frac{1}{2}\right)$$

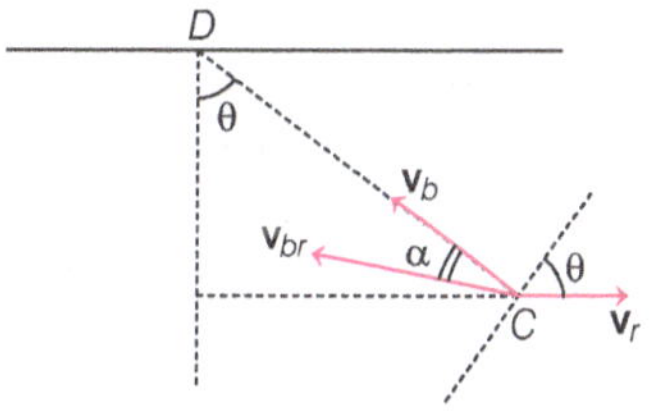

(b) $$t_1 = \frac{200}{|\mathbf{v}_{br}|} = \frac{200}{4} = 50\text{ s}$$

$$DC = DB\sec\theta = (100)\frac{\sqrt{5}}{2} = 50\sqrt{5}\text{ m}$$

$$|\mathbf{v}_b| = |\mathbf{v}_{br}|\cos\alpha - |\mathbf{v}_r|\sin\theta$$

$$= 4\left(\frac{2}{\sqrt{5}}\right) - 2\left(\frac{1}{\sqrt{5}}\right) = \frac{6}{\sqrt{5}}\text{m/s}$$

$$\therefore \quad t_2 = \frac{t_1}{2} + \frac{DC}{|\mathbf{v}_b|} = 25 + \frac{50\sqrt{5}}{\frac{6}{\sqrt{5}}} = \frac{200}{3}\text{ s}$$

or $$\frac{t_2}{t_1} = \frac{4}{3}$$ **Ans.**

119. $\mathbf{v}_b$ = velocity of boatman $= \mathbf{v}_{br} + \mathbf{v}_r$

and $\mathbf{v}_c$ = velocity of child $= \mathbf{v}_r$

$\therefore$ $\mathbf{v}_{bc} = \mathbf{v}_b - \mathbf{v}_c = \mathbf{v}_{br}$

$\mathbf{v}_{bc}$ should be along BC.

i.e. $\mathbf{v}_{br}$ should be along BC,

where $$\tan\alpha = \frac{0.6}{0.8} = \frac{3}{4}$$

or $$\alpha = 37^\circ$$ **Ans.**

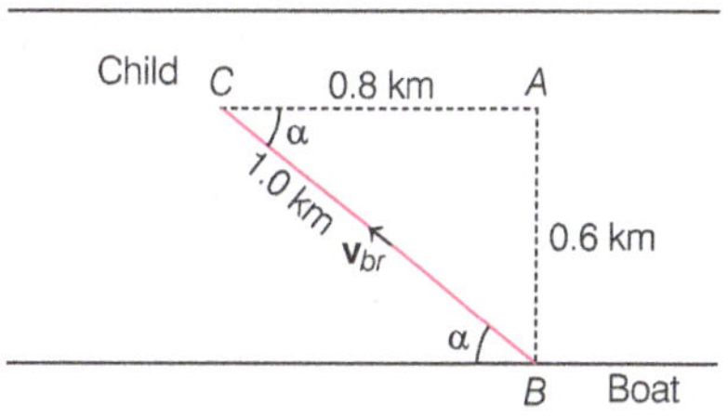

Further $$t = \frac{BC}{|\mathbf{v}_{br}|} = \frac{1}{20}\text{h}$$

$$= 3\text{ min}$$ **Ans.**

120. In order that the moving launch is always on the straight line AB, the components of velocity of the current and of the launch in the direction perpendicular to AB should be equal, i.e.

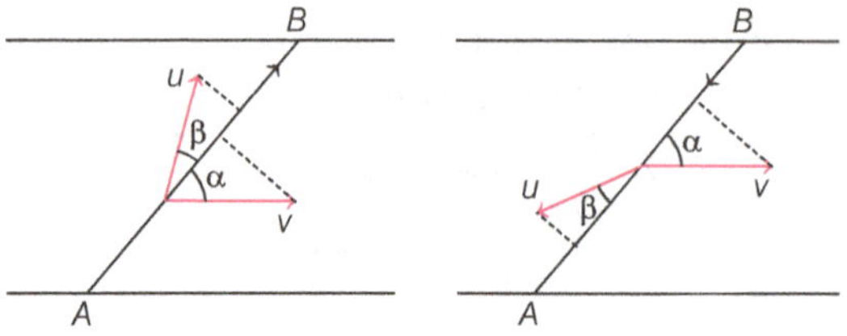

$$u \sin\beta = v \sin\alpha \qquad \text{...(i)}$$

$$S = AB = (u\cos\beta + v\cos\alpha)t_1 \qquad \text{...(ii)}$$

Further $\quad BA = (u\cos\beta - v\cos\alpha)t_2 \qquad \text{...(iii)}$

$$t_1 + t_2 = t \qquad \text{...(iv)}$$

Solving these equations after proper substitution, we get

$u = 8$ m/s and $\beta = 12°$ **Ans.**

121.

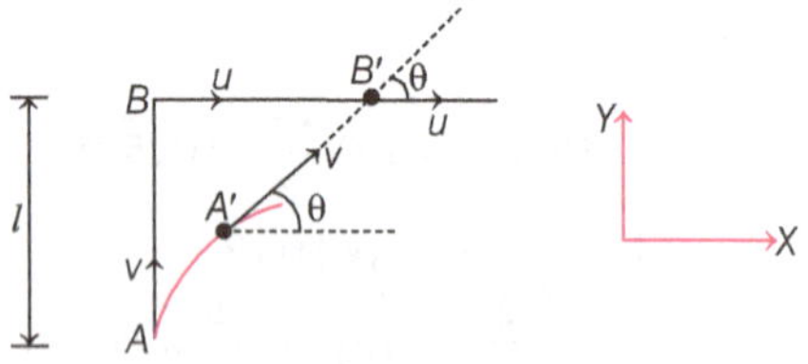

The situation is as shown in figure. In the figure θ is variable.

At some intermediate position suppose A is at point A' and B at B'. Let they collide after time t.

First condition

At the time of collision x -displacement of both the particles will be same.

$$(s_x)_A = (s_x)_B \quad \text{or} \quad \int_0^t (v_x)_A \, dt = ut$$

or $\displaystyle\int_0^t (v\cos\theta)dt = ut$ or $\displaystyle\int_0^t (\cos\theta)dt = \frac{ut}{v}$ ...(i)

Second condition

Component of relative velocity along $A'B'$ is $(v - u\cos\theta)$. This is the rate (variable due to θ) by which distance between them decreases from initial value l to final value zero in time t.

$\therefore \displaystyle\int_0^t (v - u\cos\theta)dt = l$ or $v\displaystyle\int_0^t dt - u\int_0^t (\cos\theta)\,dt = l$ ...(ii)

From Eqs. (i) and (ii), we have

$$vt - \frac{u(ut)}{v} = l$$

$$\therefore \quad t = \frac{vl}{v^2 - u^2}$$

122. Relative velocity of B with respect to A is perpendicular to line PA. Therefore, along line PA, velocity components of A and B should be same.

AP and BM are parallel and AB is perpendicular to both

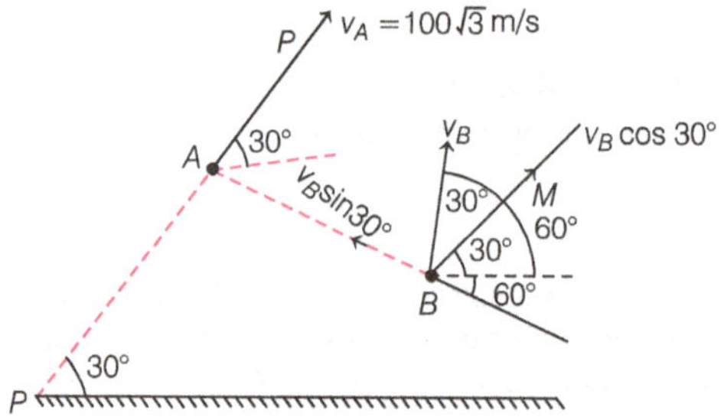

$$\Rightarrow \quad v_A = 100\sqrt{3} = v_B \cos 30°$$

$$\Rightarrow \quad v_B = 200 \text{ m/s}$$

$$t_0 = \frac{500}{v_B \sin 30°} = \frac{500}{200 \times 1/2} = 5 \text{ s}$$

123. Here, $OA = 25$ cm and $OB = 20$ cm

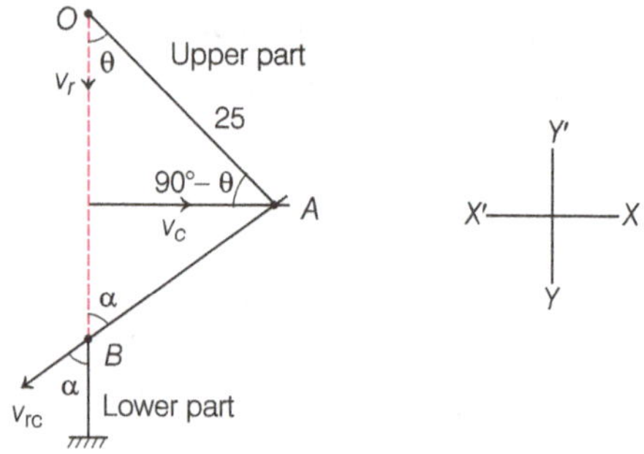

Here, $\mathbf{v}_{rc} = \mathbf{v}_r - \mathbf{v}_c$

$$v_{rc}\cos\alpha\hat{\mathbf{j}} - v_{rc}\sin\alpha\hat{\mathbf{i}} = 20\hat{\mathbf{j}} - 20\sqrt{3}\hat{\mathbf{i}}$$

$$\Rightarrow \quad \tan\alpha = \frac{20\sqrt{3}}{20} \Rightarrow \alpha = 60°$$

$$\therefore \quad \frac{25}{\sin\alpha} = \frac{20}{\sin(90° - \theta + 90° - \alpha)}$$

$$\Rightarrow \quad \frac{25}{\sin 60°} = \frac{20}{\sin(60° + \theta)}$$

$$\Rightarrow \quad \sin(60° + \theta) = \frac{10\sqrt{3}}{25} = \frac{2\sqrt{3}}{5} = \frac{\sqrt{12}}{5}$$

$$\therefore \quad \sin\left(\frac{\pi}{3} + \theta\right) = \frac{\sqrt{12}}{5} = \frac{\sqrt{12}}{n}$$

$$\therefore \quad n = 5$$

UNIT-7

PROJECTILE MOTION

Learning Modules

1. Introduction to Projectile Motion
2. Left Over Topics of Projectile Motion

MODULE 1

Introduction to Projectile Motion

Projectile Motion

Motion of a particle under constant acceleration is either a straight line (one-dimensional) or parabolic (two-dimensional).

Motion is one dimensional under following three conditions :

(i) Initial velocity of the particle is zero.

(ii) Initial velocity of the particle is in the direction of constant acceleration (or parallel to it).

(iii) Initial velocity of the particle is in the opposite direction of constant acceleration (or antiparallel to it).

For small heights acceleration due to gravity (g) is almost constant. The three cases discussed about are as shown in the figure.

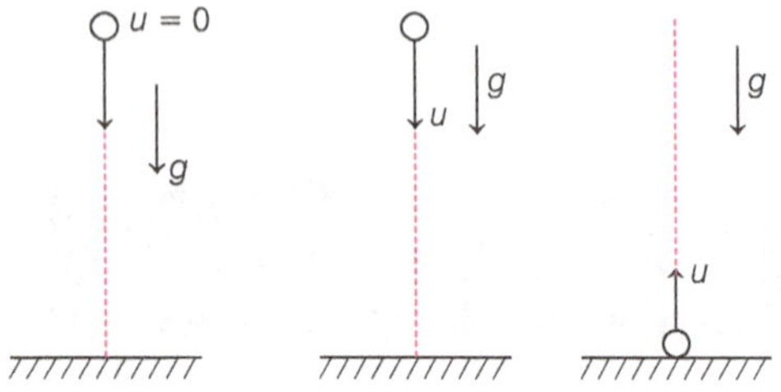

Case-(i) Case-(ii) Case-(iii)

In all other cases when initial velocity is at some angle ($\neq 0°$ or $180°$) with constant acceleration, motion is parabolic as shown below.

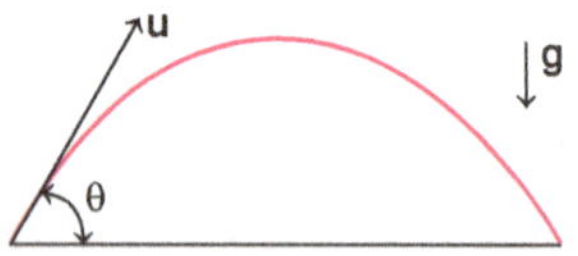

This motion under gravity is called projectile motion.

As we have seen above, projectile motion is a two-dimensional motion (or motion in a plane) with constant acceleration (or acceleration due to gravity for small heights).

The different types of projectile motion are as shown below.

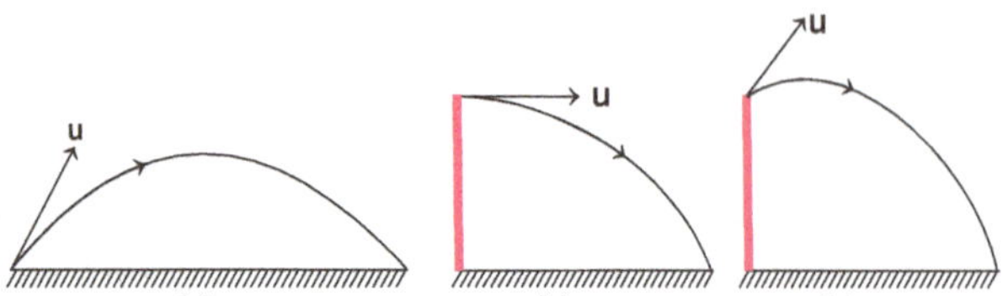

(a) (b) (c)

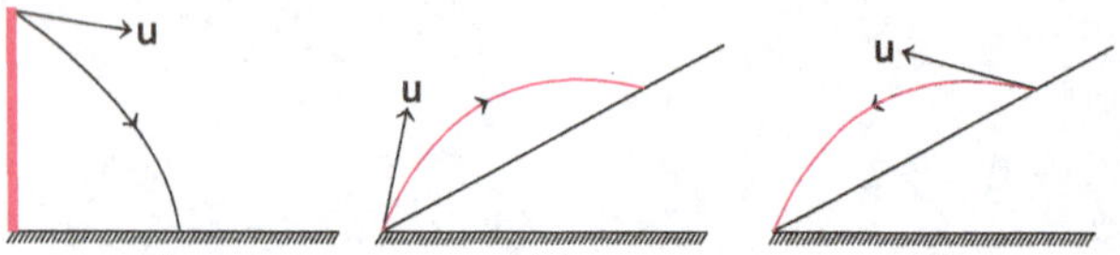

(d) (e) (f)

The plane of the projectile motion is a vertical plane.

Two Methods of Solving a Projectile Motion

Every projectile motion can be solved by either of the following two methods:

Method 1 Projectile motion is a two dimensional motion with constant acceleration. Therefore, we can use the equations

$$\mathbf{v} = \mathbf{u} + \mathbf{a}t \text{ and } \mathbf{s} = \mathbf{u}t + \frac{1}{2}\mathbf{a}t^2$$

For example, in the shown figure

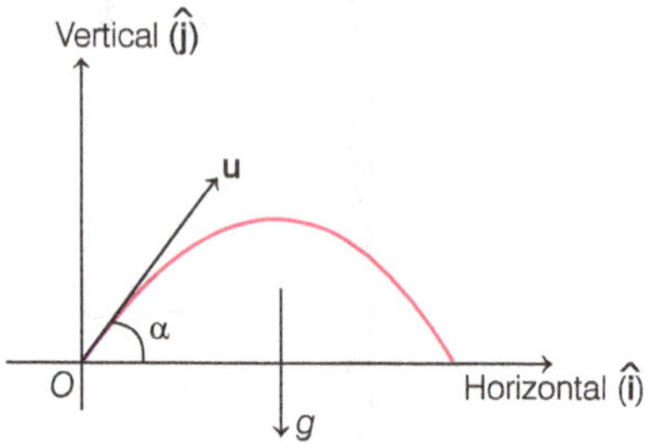

$$\mathbf{u} = u\cos\alpha\,\hat{\mathbf{i}} + u\sin\alpha\,\hat{\mathbf{j}} \quad \text{and} \quad \mathbf{a} = -g\hat{\mathbf{j}}$$

Now, suppose we want to find velocity at time t.

$$\mathbf{v} = \mathbf{u} + \mathbf{a}t = (u\cos\alpha\,\hat{\mathbf{i}} + u\sin\alpha\,\hat{\mathbf{j}}) - gt\hat{\mathbf{j}}$$

or $\quad \mathbf{v} = u\cos\alpha\,\hat{\mathbf{i}} + (u\sin\alpha - gt)\hat{\mathbf{j}}$

Similarly, displacement at time t will be

$$\mathbf{s} = \mathbf{u}t + \frac{1}{2}\mathbf{a}t^2$$

$$= (u\cos\alpha\,\hat{\mathbf{i}} + u\sin\alpha\,\hat{\mathbf{j}})t - \frac{1}{2}gt^2\hat{\mathbf{j}}$$

$$= ut\cos\alpha\,\hat{\mathbf{i}} + \left(ut\sin\alpha - \frac{1}{2}gt^2\right)\hat{\mathbf{j}}$$

Note *In all problems, value of* **a** (= **g**) *will be same only* **u** *will be different.*

Example 1 A particle is projected with a velocity of 50 m/s at 37° with horizontal. Find velocity, displacement and co-ordinates of the particle (w.r.t. the starting point) after 2 s.

(Given, $g = 10\,\text{m/s}^2$, $\sin 37° = 0.6$ and $\cos 37° = 0.8$)

Sol. In the given problem,

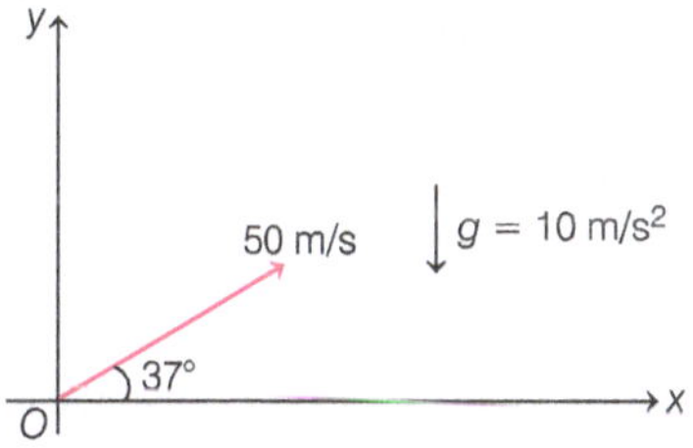

$$\mathbf{u} = (50\cos 37°)\hat{\mathbf{i}} + (50\sin 37°)\hat{\mathbf{j}}$$
$$= (40\hat{\mathbf{i}} + 30\hat{\mathbf{j}})\,\text{m/s}$$
$$\mathbf{a} = (-10\hat{\mathbf{j}})\,\text{m/s}^2$$
$$t = 2\,\text{s}$$
$$\mathbf{v} = \mathbf{u} + \mathbf{a}t$$
$$= (40\hat{\mathbf{i}} + 30\hat{\mathbf{j}}) + (-10\hat{\mathbf{j}})(2)$$
$$= (40\hat{\mathbf{i}} + 10\hat{\mathbf{j}})\,\text{m/s} \qquad \textbf{Ans.}$$
$$\mathbf{s} = \mathbf{u}t + \frac{1}{2}\mathbf{a}t^2$$
$$= (40\hat{\mathbf{i}} + 30\hat{\mathbf{j}})(2) + \frac{1}{2}(-10\hat{\mathbf{j}})(2)^2$$
$$= (80\hat{\mathbf{i}} + 40\hat{\mathbf{j}})\,\text{m}$$

∴ Coordinates of the particle are

$$x = 80\,\text{m}$$
and $$y = 40\,\text{m} \qquad \textbf{Ans.}$$

Example 2 A particle is projected with velocity u at angle θ with horizontal. Find the time when velocity vector is perpendicular to initial velocity vector.

Sol.

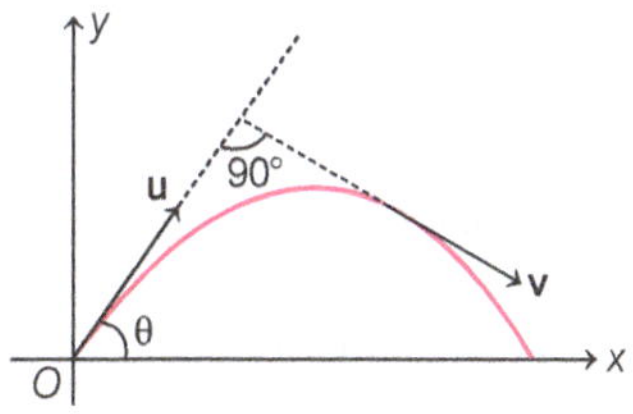

Given, $\mathbf{v} \perp \mathbf{u}$

$$\Rightarrow \quad \mathbf{v}\cdot\mathbf{u} = 0$$
$$\Rightarrow \quad (\mathbf{u} + \mathbf{a}t)\cdot\mathbf{u} = 0 \qquad \text{......(i)}$$

Substituting the proper values in Eq. (i), we have

$$[\{(u\cos\theta)\hat{\mathbf{i}} + (u\sin\theta)\hat{\mathbf{j}}\} + (-g\hat{\mathbf{j}})t]\cdot[(u\cos\theta)\hat{\mathbf{i}} + (u\sin\theta)\hat{\mathbf{j}}] = 0$$
$$\Rightarrow \quad u^2\cos^2\theta + u^2\sin^2\theta - (ug\sin\theta)t = 0$$
$$\Rightarrow \quad u^2(\sin^2\theta + \cos^2\theta) = (ug\sin\theta)t$$

Solving this equation, we get

$$t = \frac{u}{g\sin\theta} = \frac{u\,\text{cosec}\,\theta}{g} \qquad \textbf{Ans.}$$

Alternate method

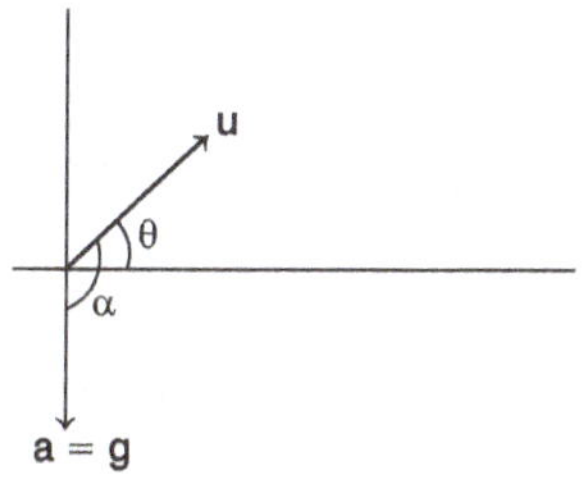

$$\mathbf{a} = \mathbf{g}$$

Angle between $\mathbf{u}$ and $\mathbf{a}$ is $\alpha = 90° + \theta$

Now, Eq. (i) can be written as

$$\mathbf{u}\cdot\mathbf{u} + \mathbf{u}\cdot\mathbf{a}t = 0$$
or $$u^2 + (ug\cos\alpha)\,t = 0$$
or $$u^2 + [ug\cos(90° + \theta)]t = 0$$

Solving this equation, we get

$$t = \frac{u}{g\sin\theta} = \frac{u\,\text{cosec}\,\theta}{g} \qquad \textbf{Ans.}$$

Method 2 In this method, select two mutually perpendicular directions x and y and find the two components of initial velocity and acceleration along these two directions, i.e. find u_x, u_y, a_x and a_y. Now apply the appropriate equation (s) from the following six equations:

$$\left.\begin{aligned} v_x &= u_x + a_x t \\ s_x &= u_x t + \frac{1}{2}a_x t^2 \\ v_x^2 &= u_x^2 + 2u_x s_x \end{aligned}\right] \rightarrow \text{Along } x\text{ - axis}$$

and

$$\left.\begin{aligned} v_y &= u_y + a_y t \\ s_y &= u_y t + \frac{1}{2}a_y t^2 \\ v_y^2 &= u_y^2 + 2a_y s_y \end{aligned}\right] \rightarrow \text{Along } y\text{ - axis}$$

Substitute $v_x, u_x, a_x, s_x, v_y, u_y, a_y$ and s_y with proper signs by choosing one direction as positive and the other as negative along both axes. In most of the problems $s = ut + \frac{1}{2}at^2$ equation is useful for time calculation. Under normal projectile motion, x-axis is taken along horizontal direction and y-axis along vertical direction. In projectile motion along an inclined plane, x-axis is normally taken along the plane and y-axis perpendicular to the plane. Two simple cases are shown below.

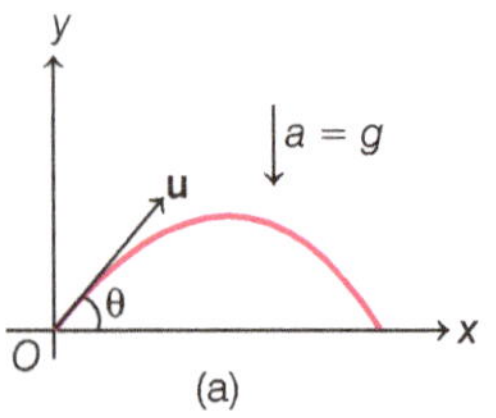

(a)

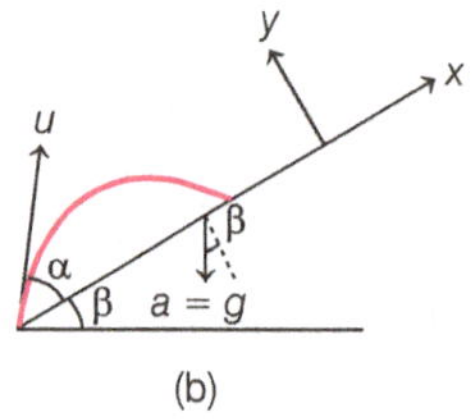

(b)

In Fig. (a),

$$u_x = u\cos\theta, u_y = u\sin\theta, a_x = 0, a_y = -g$$

In Fig.(b),

$$u_x = u\cos\alpha, u_y = u\sin\alpha, a_x = -g\sin\beta, a_y = -g\cos\beta$$

Example 3 A projectile is fired horizontally with velocity of 98 m/s from the top of a hill 490 m high. Find

(a) the time taken by the projectile to reach the ground,

(b) the distance of the point where the particle hits the ground from foot of the hill and

(c) the velocity with which the projectile hits the ground. ($g = 9.8$ m/s^2)

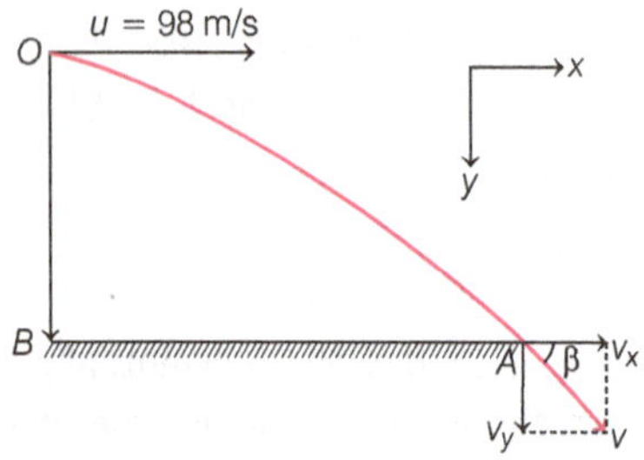

Sol. Here, it will be more convenient to choose *x* and *y* directions as shown in figure.

$u_x = 98$ m/s, $a_x = 0$, $u_y = 0$ and $a_y = g$

(a) At *A*, $s_y = 490$ m . So, applying

$$s_y = u_y t + \frac{1}{2} a_y t^2$$

$$\therefore \quad 490 = 0 + \frac{1}{2}(9.8)t^2$$

$$\therefore \quad t = 10 \text{ s}$$ **Ans.**

(b) $BA = s_x = u_x t + \frac{1}{2} a_x t^2$

or $BA = (98)(10) + 0$

or $BA = 980$ m **Ans.**

(c) $v_x = u_x + a_x t = 98 + 0 = 98$ m/s

$v_y = u_y + a_y t = 0 + (9.8)(10) = 98$ m/s

$$\therefore \quad v = \sqrt{v_x^2 + v_y^2} = \sqrt{(98)^2 + (98)^2} = 98\sqrt{2} \text{ m/s}$$

and $\tan\beta = \frac{v_y}{v_x} = \frac{98}{98} = 1$

$$\therefore \quad \beta = 45°$$

Thus, the projectile hits the ground with velocity $98\sqrt{2}$ m/s at an angle of $\beta = 45°$ with horizontal as shown in figure.

Example 4 A body is thrown horizontally from the top of a tower and strikes the ground after three seconds at an angle of 45° with the horizontal. Find the height of the tower and the speed with which the body was projected. (Take $g = 9.8 \text{m/s}^2$)

Sol. As shown in the figure of Example 3,

$$u_y = 0 \text{ and } a_y = g = 9.8 \text{ m/s}^2, s_y = u_y t + \frac{1}{2} a_y t^2$$

$$s_y = 0 \times 3 + \frac{1}{2} \times 9.8 \times (3)^2 = 44.1 \text{ m}$$

Thus, height of the tower is 44.1 m.

Further, $v_y = u_y + a_y t = 0 + (9.8)(3) = 29.4$ m/s

As the resultant velocity v makes an angle of 45° with the horizontal, so

$$\tan 45° = \frac{v_y}{v_x} \quad \text{or} \quad 1 = \frac{29.4}{v_x}$$

$$v_x = 29.4 \text{ m/s}$$

$$v_x = u_x + a_x t$$

$$\Rightarrow \quad 29.4 = u_x + 0$$

$$\text{or} \quad u_x = 29.4 \text{ m/s}$$

Therefore, the speed with which the body was projected (horizontally) is 29.4 m/s.

INTRODUCTORY Exercise 1

1. Two particles are projected from a tower horizontally in opposite directions with velocities 10 m/s and 20 m/s. Find the time when their velocity vectors are mutually perpendicular. Take $g = 10$ m/s^2.
2. Projectile motion is a 3-dimensional motion. Is this statement true or false?
3. Projectile motion (at low speed) is uniformly accelerated motion. Is this statement true or false?
4. A particle is projected from ground with velocity 50 m/s at 37° from horizontal. Find velocity and displacement after 2 s.
$$\sin 37° = \frac{3}{5}$$
5. A particle is projected from a tower of height 25 m with velocity $20\sqrt{2}$ m/s at 45° in upward direction. Find the time when particle strikes with ground. The horizontal distance from the foot of tower where it strikes. Also find the velocity at the time of collision.

Note *In question numbers 4 and 5, $\hat{i}$ is in horizontal direction and $\hat{j}$ is vertically upwards.*

Answers

1. $\sqrt{2}$ s 2. False 3. True
4. $\mathbf{v} = (40\hat{i} + 10\hat{j})$ m/s, $\mathbf{s} = (80\hat{i} + 40\hat{j})$m
5. $t = 5$ s, $d = 100$ m, $\mathbf{v} = (20\hat{i} - 30\hat{j})\text{ms}^{-1}$

See hints on page 320.

Time of Flight, Maximum Height and Horizontal Range of a Projectile

Figure given below shows a particle projected from the point O with an initial velocity u at an angle α with the horizontal. It goes through the highest point A and falls at B on the horizontal surface passing through O.

The point O is called the **point of projection**, the angle α is called the **angle of projection**, the distance OB is called the **horizontal range** (R) or simply range and the vertical height AC is called the **maximum height** (H). The total time taken by the particle in describing the path OAB is called the **time of flight** (T).

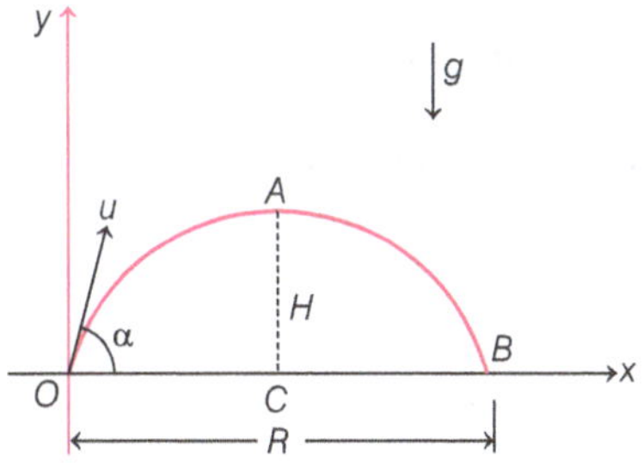

Time of Flight (T)

Refer the figure shown above. Here, x and y-axes are in the directions shown in figure. Axis x is along horizontal direction and axis y is vertically upwards. Thus,

$$u_x = u\cos\alpha,$$
$$u_y = u\sin\alpha,\ a_x = 0$$
and $$a_y = -g$$

At point B, $s_y = 0$. So, applying

$$s_y = u_y t + \frac{1}{2} a_y t^2, \text{ we have}$$
$$0 = (u\sin\alpha)t - \frac{1}{2} g t^2$$
$$\therefore \quad t = 0,\ \frac{2u\sin\alpha}{g}$$

Both $t = 0$ and $t = \frac{2u\sin\alpha}{g}$ correspond to the situation where $s_y = 0$. The time $t = 0$ corresponds to the point O and time $t = \frac{2u\sin\alpha}{g}$ corresponds to point B. Thus, time of flight of the projectile is

$$T = t_{OAB} \quad \text{or} \quad \boxed{T = \frac{2u\sin\alpha}{g}}$$

Maximum Height (H)

At point A vertical component of velocity becomes zero, i.e. $v_y = 0$. Substituting the proper values in

$$v_y^2 = u_y^2 + 2a_y s_y$$
we have, $$0 = (u\sin\alpha)^2 + 2(-g)(H)$$
$$\therefore \quad \boxed{H = \frac{u^2\sin^2\alpha}{2g}}$$

Horizontal Range (R)

Distance OB is the range R. This is also equal to the displacement of particle along x-axis in time $t = T$. Thus, applying $s_x = u_x t + \frac{1}{2} a_x t^2$, we get

$$R = (u\cos\alpha)\left(\frac{2u\sin\alpha}{g}\right) + 0$$
as $$a_x = 0 \quad \text{and} \quad t = T = \frac{2u\sin\alpha}{g}$$
$$\therefore \quad R = \frac{2u^2\sin\alpha\cos\alpha}{g} = \frac{u^2\sin 2\alpha}{g}$$
or $$\boxed{R = \frac{u^2\sin 2\alpha}{g}}$$

Note Points

(i) Range is maximum where $\sin 2\alpha = 1$ *or* $\alpha = 45°$ *and this maximum range is*

$$R_{max} = \frac{u^2}{g} \qquad (\text{at } \alpha = 45°)$$

(ii) For given value of u range at α *and range at* $90° - \alpha$ *are equal, although times of flight and maximum heights may be different. Because*

$$R_{90°-\alpha} = \frac{u^2\sin 2(90° - \alpha)}{g} = \frac{u^2\sin(180° - 2\alpha)}{g} = \frac{u^2\sin 2\alpha}{g} = R_\alpha$$

So, $R_{30°} = R_{60°}$ *or* $R_{20°} = R_{70°}$

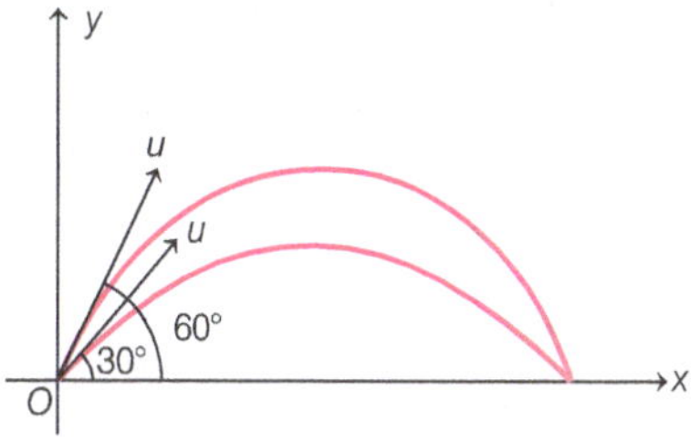

This is shown in the above figure.

Important POINTS

- Formulae of T, H and R can be applied directly between two points lying on same horizontal line.

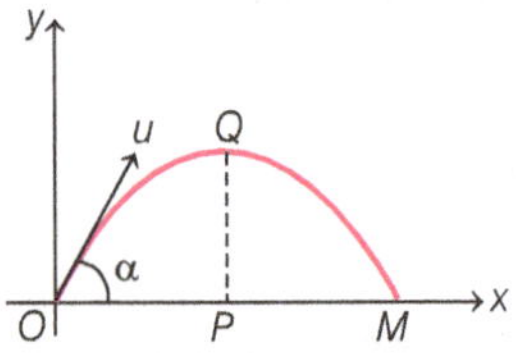

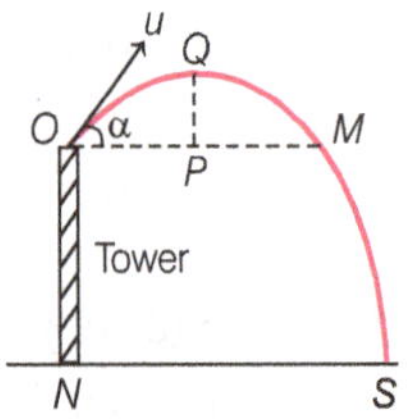

For example, in the two projectile motions shown in figure,

$$t_{OQM} = T = \frac{2u\sin\alpha}{g}, PQ = H = \frac{u^2\sin^2\alpha}{2g}$$

and $$OM = R = \frac{u^2\sin 2\alpha}{g}$$

- As we have seen in the above derivations that $a_x = 0$, i.e. motion of the projectile in horizontal direction is uniform. Hence, horizontal component of velocity $u\cos\alpha$ does not change during its motion.
- Motion in vertical direction is first retarded then accelerated in opposite direction. Because u_y is upwards and a_y is downwards. Hence, vertical component of its velocity first decreases from O to A and then increases from A to B. This is shown in the figure given below.

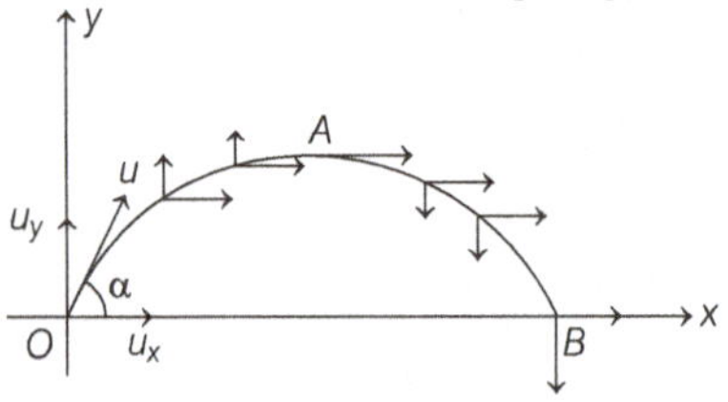

- The coordinates and velocity components of the projectile at time t are

$$x = s_x = u_x t = (u\cos\alpha)t$$

$$y = s_y = u_y t + \frac{1}{2}a_y t^2$$

$$= (u\sin\alpha)t - \frac{1}{2}gt^2$$

$$v_x = u_x = u\cos\alpha$$

and $$v_y = u_y + a_y t = u\sin\alpha - gt$$

Therefore, speed of projectile at time t is $v = \sqrt{v_x^2 + v_y^2}$ and the angle made by its velocity vector with positive x-axis is

$$\theta = \tan^{-1}\left(\frac{v_y}{v_x}\right)$$

- **Equation of trajectory of projectile**

$$x = (u\cos\alpha)t$$

$\therefore$ $$t = \frac{x}{u\cos\alpha}$$

Substituting this value of t in, $y = (u\sin\alpha)t - \frac{1}{2}gt^2$,

we get

$$y = x\tan\alpha - \frac{gx^2}{2u^2\cos^2\alpha}$$

or $$y = x\tan\alpha - \frac{gx^2}{2u^2}\sec^2\alpha$$

$$y = x\tan\alpha - \frac{gx^2}{2u^2}(1 + \tan^2\alpha)$$

The equation is quadratic in x. This is the reason why the path of a projectile is parabola. The above equation can also be written in terms of range (R) as:

$$y = x\left(1 - \frac{x}{R}\right)\tan\alpha$$

Direct Formula Based Problems

Example 5 Find the angle of projection of a projectile for which the horizontal range and maximum height are equal.

Sol. Given, $R = H$

$\therefore$ $$\frac{u^2\sin 2\alpha}{g} = \frac{u^2\sin^2\alpha}{2g} \quad \text{or}$$

$$2\sin\alpha\cos\alpha = \frac{\sin^2\alpha}{2}$$

or $$\frac{\sin\alpha}{\cos\alpha} = 4 \quad \text{or} \quad \tan\alpha = 4$$

$\therefore$ $$\alpha = \tan^{-1}(4)$$ **Ans.**

Example 6 Prove that the maximum horizontal range is four times the maximum height attained by the projectile; when fired at an inclination so as to have maximum horizontal range.

Sol. For $\theta = 45°$, the horizontal range is maximum and is given by

$$R_{max} = \frac{u^2}{g}$$

Maximum height attained

$$H_{max} = \frac{u^2\sin^2 45°}{2g} = \frac{u^2}{4g} = \frac{R_{max}}{4}$$

or $$R_{max} = 4H_{max}$$ **Proved.**

Example 7 For given value of u, there are two angles of projection for which the horizontal range is the same. Show that the sum of the maximum heights for these two angles is independent of the angle of projection.

Sol. There are two angles of projection α and $90° - \alpha$ for which the horizontal range R is same.

Now, $$H_1 = \frac{u^2 \sin^2 \alpha}{2g}$$

and $$H_2 = \frac{u^2 \sin^2 (90° - \alpha)}{2g} = \frac{u^2 \cos^2 \alpha}{2g}$$

Therefore,

$$H_1 + H_2 = \frac{u^2}{2g}(\sin^2 \alpha + \cos^2 \alpha) = \frac{u^2}{2g}$$

Clearly the sum of the heights for the two angles of projection is independent of the angles of projection.

Example 8 Show that there are two values of time for which a projectile is at the same height. Also show mathematically that the sum of these two times is equal to the time of flight.

Sol. For vertical motion of a projectile,

$$y = (u \sin \alpha) t - \frac{1}{2} gt^2$$

or $$\frac{1}{2} gt^2 - (u \sin \alpha) t + y = 0$$

This is a quadratic equation in t. Its roots are

$$t_1 = \frac{u \sin \alpha - \sqrt{u^2 \sin^2 \alpha - 2gy}}{g}$$

and $$t_2 = \frac{u \sin \alpha + \sqrt{u^2 \sin^2 \alpha - 2gy}}{g}$$

$$\therefore \quad t_1 + t_2 = \frac{2u \sin \alpha}{g} = T$$

(time of flight of the projectile)

INTRODUCTORY Exercise 2

1. A particle is projected from ground with velocity $40\sqrt{2}$ m/s at 45°. Find
   (a) velocity and
   (b) displacement of the particle after 2s. ($g = 10$ m/s^2)
2. Under what conditions the formulas of range, time of flight and maximum height can be applied directly in case of a projectile motion?
3. What is the average velocity of a particle projected from the ground with speed u at an angle α with the horizontal over a time interval from beginning till it strikes the ground again?
4. What is the change in velocity in the above question?
5. A particle is projected from ground with initial velocity $u = 20\sqrt{2}$ m/s at $\theta = 45°$. Find

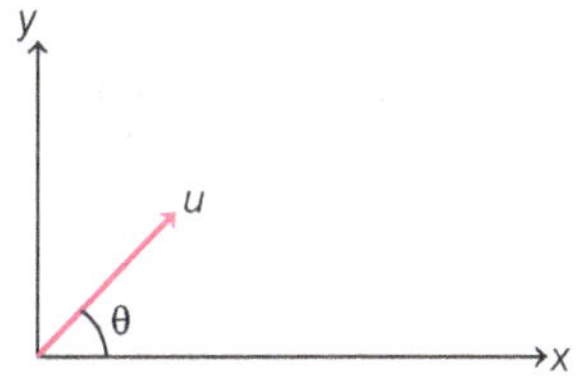

   (a) R, H and T,
   (b) velocity of particle after 1 s
   (c) velocity of particle at the time of collision with the ground (x-axis).
6. A particle is projected from ground at an angle 45° with initial velocity $20\sqrt{2}$ m/s. Find
   (a) change in velocity,
   (b) magnitude of average velocity in a time interval from $t = 0$ to $t = 3$ s.
7. Coach of baseball throws a ball to a player with an initial speed of 20 m/s at an angle of 45° with the horizontal. At the moment the ball is thrown, the player is 50 m from the coach. At what speed and in what direction must the player run to catch the ball at the same height at which it was thrown? ($g = 10$ m/s^2)
8. A bullet fired at an angle of 30° with the horizontal hits the ground 3.0 km away. By adjusting its angle of projection, can one hope to hit a target 5.0 km away? Assume the muzzle speed to be fixed and neglect air resistance.

Answers

1. (a) $20\sqrt{5}$ m/s at angle $\tan^{-1}\left(\frac{1}{2}\right)$ with horizontal (b) 100 m
2. Between two points lying on the same horizontal line
3. $u \cos \alpha$
4. $2u \sin \alpha$, downwards
5. (a) 80 m, 20 m, 4s (b) $(20\hat{i} + 10\hat{j})$ms^{-1} (c) $(20\hat{i} - 20\hat{j})$ms^{-1}
6. (a) 30ms^{-1} (vertically downwards) (b) 20.62 ms^{-1}
7. $\frac{5}{\sqrt{2}}$ ms^{-1}
8. No

See hints on pages 320-321.

Important POINTS

1. In projectile motion speed (and hence kinetic energy) is minimum at highest point. At this point,

   speed $=(\cos\theta)$ times the speed of projection.

   $\therefore$ Kinetic energy $=(\cos^2\theta)$ times the initial kinetic energy

   Here, $\theta=$ angle of projection

2. In projectile motion it is sometimes better to write the equations of H, R and T in terms of u_x and u_y as

$$T=\frac{2u_y}{g},\quad H=\frac{u_y^2}{2g}\quad \text{and}\quad R=\frac{2u_xu_y}{g}$$

3. If a particle is projected vertically upwards, then during upward journey gravity forces (weight) and air drag both are acting downwards. Hence, |retardation| > $|g|$. During its downward journey air drag is upwards while gravity is downwards. Hence, acceleration $<g$.. Therefore we may conclude that the upward journey time is less than the downward journey time.

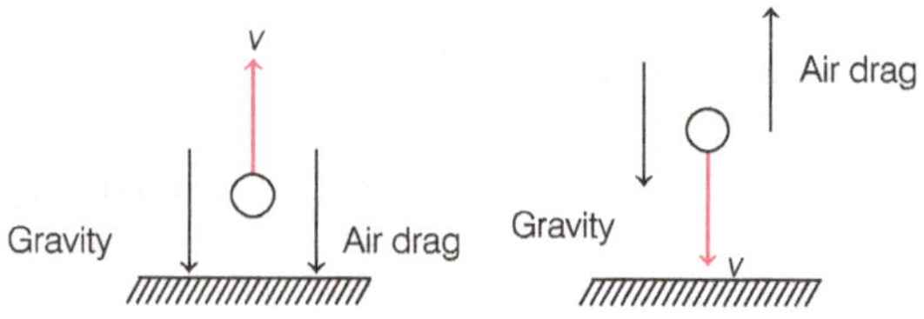

Exercise : In projectile motion, if air drag is taken into consideration than state whether the values of H and R will increase, decrease or remain same.

Ans. Both will decrease.

Typed Problems

Type 1. Based on the concept that horizontal component of velocity remains unchanged.

Ex. 1 A particle is projected from ground with velocity 40m/s at 60° from horizontal.

(a) Find the speed when velocity of the particle makes an angle of 37° from horizontal.

(b) Find the time for the above situation.

(c) Find the vertical height and horizontal distance of the particle from the starting point in the above position. Take $g=10\text{ m/s}^2$.

Sol. In the figure shown,

$$u_x=40\cos 60°=20\text{ m/s}$$
$$u_y=40\sin 60°=20\sqrt{3}\text{ m/s}$$
$$a_x=0 \text{ and } a_y=-10\text{ m/s}^2$$

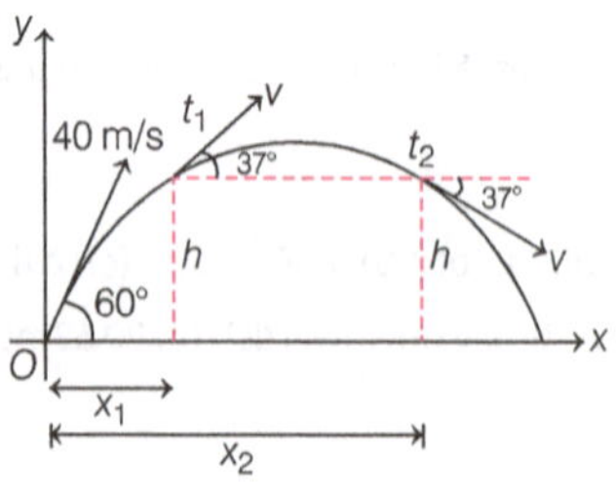

(a) Horizontal component of velocity remains unchanged.

$$v_x=u_x$$

or $\quad v\cos 37°=20$

$\Rightarrow\quad v(0.8)=20$

$\therefore\quad v=25\text{ m/s}$

(b) Using, $v_y=u_y+a_yt \Rightarrow t=\dfrac{v_y-u_y}{a_y}$

For t_1, $\quad t_1=\dfrac{v\sin 37°-20\sqrt{3}}{-10}=\dfrac{(25)(0.6)-20\sqrt{3}}{-10}$

$=1.96$ s.

For t_2, $\quad t_2=\dfrac{-v\sin 37°-20\sqrt{3}}{-10}$

$=\dfrac{-(25)(0.6)-20\sqrt{3}}{-10}$

$=4.96$ s

(c) Vertical height $\quad h=s_y \quad$ (at t_1 or t_2)

Let us calculate at t_1

$\therefore\quad h=s_y=u_yt_1+\dfrac{1}{2}a_yt_1^2$

$=(20\sqrt{3})(1.96)+\dfrac{1}{2}(-10)(1.96)^2$

$=48.7$ m.

Horizontal distances

$x_1=u_xt_1 \quad$ (as $a_x=0$)

$=(20)(1.96)=39.2$m

Similarly, $\quad x_2=u_xt_2=(20)(4.96)$

$=99.2$m

Type 2. Situations where the formulae, *H*, *R* and *T* cannot be applied directly.

Concept

As discussed earlier also, formulae of *H*, *R* and *T* can be applied directly between two points lying on the same horizontal line.

How to Solve?

In any other situation, apply component method (along x and y-axes). In most of the problems it is advisable to first find the time, using the equation, $s = ut + \frac{1}{2}at^2$ in vertical (or y) direction.

Ex.2 In the figures shown, three particles are thrown from a tower of height 40 m as shown in figure. In each case find the time when the particles strike the ground and the distance of this point from foot of the tower.

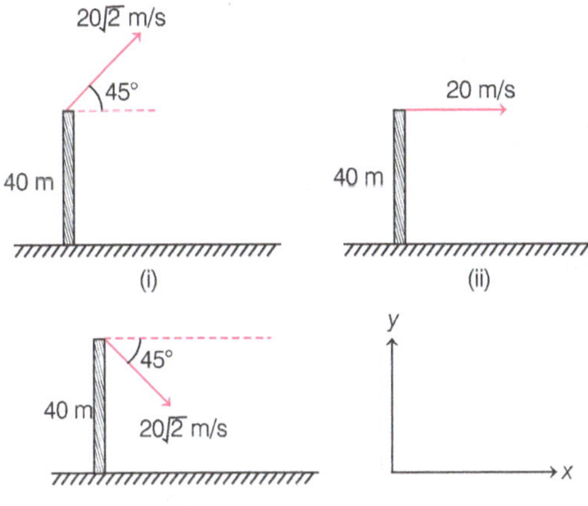

(i) (ii) (iii)

Sol. For time calculation, apply

$$s_y = u_y t + \frac{1}{2}a_y t^2 \quad \text{(in vertical direction)}$$

In all figures, $s_y = -40$ m, $a_y = -10$ m/s^2

In first figure

$u_y = +20\sqrt{2}\sin 45° = +20$ m/s

In second figure, $u_y = 0$

In third figure, $u_y = -20\sqrt{2}\sin 45°$

$= -20$ m/s

For horizontal distance,

$$x = u_x t$$

$u_x = 20$ m/s in all cases

Substituting the proper values and then solving we get,

In first figure

Time $t_1 = 5.46$ s and horizontal distance

$$x_1 = 109.2 \text{ m}$$

In second figure

Time $t_2 = 2.83$ s and the horizontal distance $x_2 = 56.6$ m

In third figure

Time $t_3 = 1.46$ s and the horizontal distance $x_3 = 29.2$ m

Type 3. Situation when a particle is projected horizontally from some height.

Concept

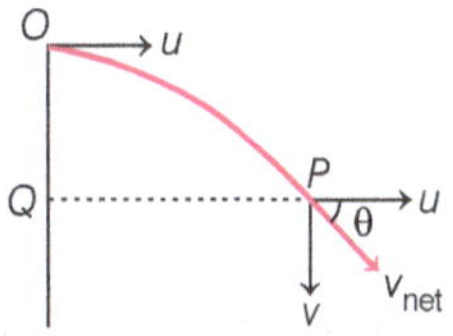

After time t
Fig. (i)

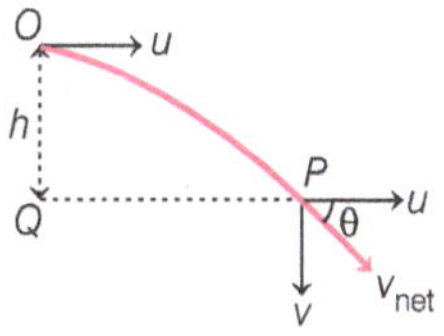

After falling a height h
Fig. (ii)

Suppose a particle is projected from point O with a horizontal velocity u as shown in two figures. Then,

In Fig. (i) or after time t

Suppose the particle is at point P, then

Horizontal component of velocity $= u$

Vertical component of velocity $v = gt$

If $g = 10$ m/s^2, then $v = 10t$ (downwards)

Horizontal distance $QP = ut$ and vertical height fallen

$$OQ = \frac{1}{2}gt^2$$

If $g = 10$ m/s^2, then $OQ = 5t^2$

If Fig. (ii) or after falling a height h

Suppose the particle is at point P, then

Horizontal component of velocity $= u$

Vertical component of velocity $v = \sqrt{2gh}$ (downwards)

Time taken in falling a height h is

$$t = \sqrt{\frac{2h}{g}} \quad \left(\text{as } h = \frac{1}{2}gt^2\right)$$

The horizontal distance $QP = ut$

Note *In both figures, net velocity of the particle at point P is,*

$$v_{net} = \sqrt{v^2 + u^2}$$

and the angle θ of this net velocity with horizontal is

$$\tan\theta = \frac{v}{u} \quad \text{or} \quad \theta = \tan^{-1}\left(\frac{v}{u}\right)$$

Ex. 3 A ball rolls off the edge of a horizontal table top 4 m high. If it strikes the floor at a point 5 m horizontally away from the edge of the table, what was its speed at the instant it left the table? Take $g = 9.8$ m/s^2

Sol. Using $t = \sqrt{\frac{2h}{g}}$

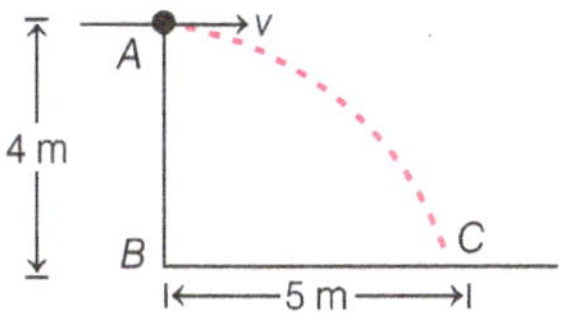

or $$t_{AC} = \sqrt{\frac{2h_{AB}}{g}}$$

$$= \sqrt{\frac{2 \times 4}{9.8}} = 0.9 \text{ s}$$

Further, $BC = vt_{AC}$

or $$v = \frac{BC}{t_{AC}} = \frac{5.0}{0.9} = 5.55 \text{ m/s}$$ **Ans.**

Ex. 4 An aeroplane is flying in a horizontal direction with a velocity 600 km/h at a height of 1960 m. When it is vertically above the point A on the ground, a body is dropped from it. The body strikes the ground at point B. Calculate the distance AB. (Take $g = 9.8 \text{ m/s}^2$)

Sol. From $t = \sqrt{\frac{2h}{g}}$

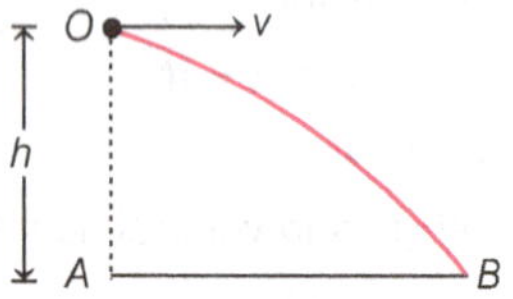

we have, $$t_{OB} = \sqrt{\frac{2h_{OA}}{g}} = \sqrt{\frac{2 \times 1960}{9.8}} = 20 \text{ s}$$

Horizontal distance

$$AB = vt_{OB}$$

$$= \left(600 \times \frac{5}{18} \text{ m/s}\right)(20 \text{ s})$$

$$= 3333.33 \text{ m} = 3.33 \text{ km}$$ **Ans.**

Ex. 5 In the figure shown, find

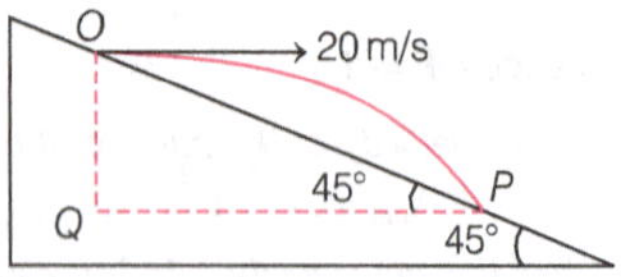

(a) the time of flight of the projectile over the inclined plane

(b) range OP. (Take $g = 10 \text{ m/s}^2$)

Sol. (a) Let the particle strikes the plane at point P at time t, then

$$OQ = \frac{1}{2}gt^2 = 5t^2$$

$$QP = 20t$$

In ΔOPQ, angle OPQ is 45°.

$\therefore$ $OQ = QP$ or $5t^2 = 20t$

$\therefore$ $t = 4 \text{ s}$ **Ans.**

(b) $OP = QP \sin 45° = (20t)(\sqrt{2})$

Substituting $t = 4$ s, we have

$$OP = 80\sqrt{2} \text{ m}$$ **Ans.**

Ex. 6 In the figure shown, find

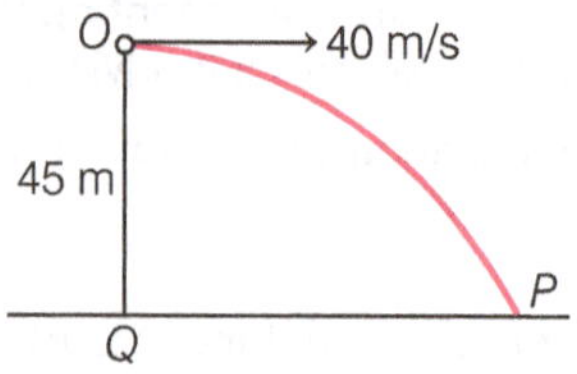

(a) the time when the particle strikes the ground at P

(b) the horizontal distance QP

(c) velocity of the particle at P

Take $g = 10 \text{ m/s}^2$

Sol. (a) $t = \sqrt{\frac{2h}{g}} = \sqrt{\frac{2 \times 45}{10}} = 3 \text{ s}$

(b) Horizontal distance $QP = 40t$

$$= 40 \times 3 = 120 \text{m}$$ **Ans.**

(c) Horizontal component of velocity at $P = 40$ m/s

Vertical compound of velocity

$$= gt = 10t = 10 \times 3 = 30 \text{ m/s}$$ (downwards)

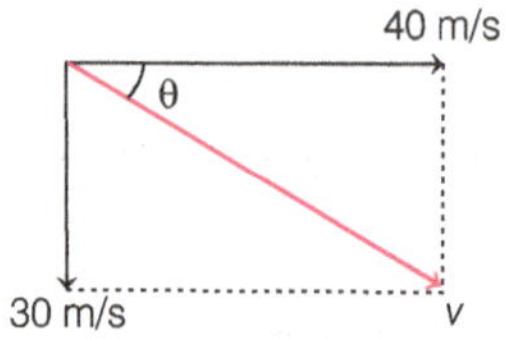

Net velocity $v = \sqrt{(40)^2 + (30)^2} = 50 \text{ m/s}$ **Ans.**

$$\tan\theta = \frac{30}{40} \text{ or } \frac{3}{4}$$

$\therefore$ $$\theta = \tan^{-1}\left(\frac{3}{4}\right) = 37°$$ **Ans.**

Type 4. Based on the trajectory of a projectile.

Concept

We have seen that equation of trajectory of projectile is

$$y = x\tan\theta - \frac{gx^2}{2u^2\cos^2\theta} = x\tan\theta - \frac{gx^2}{2u^2}(1 + \tan^2\theta)$$

In some problems, the given equation of a projectile is compared with this standard equation to find the unknowns.

Ex. 7 A particle moves in the plane xy with constant acceleration 'a' directed along the negative y-axis. The equation of motion of the particle has the form $y = px - qx^2$ where p and q are positive constants. Find the velocity of the particle at the origin of coordinates.

Sol. Comparing the given equation with the equation of a projectile motion,

$$y = x \tan\theta - \frac{gx^2}{2u^2}(1 + \tan^2\theta)$$

We find that $g = a, \tan\theta = p$

and $\frac{a}{2u^2}(1+\tan^2\theta)=q$

$\therefore$ u = velocity of particle at origin

$$=\sqrt{\frac{a(1+\tan^2\theta)}{2q}}=\sqrt{\frac{a(1+p^2)}{2q}}$$

Type 5. Change in velocity components at a height difference of *h*.

Concept

(i) Horizontal component of velocity remains same at all heights.

(ii) Vertical component of velocity decreases in moving up and increases in moving down.

For vertical component we can use,

$$v_f=\sqrt{v_i^2-2gh}$$

Here v_f is the component at greater heights. Similarly,

$$v_f=\sqrt{v_i^2+2gh}$$

Here, v_f is at lower heights.

h = height difference between two points.

Following example will illustrate the concept.

Ex. 8 At a height of 25 m from ground, velocity of a projectile is, $\mathbf{v}=(20\hat{\mathbf{i}}-20\hat{\mathbf{j}})$ m/s. Taking $g=10\text{ m/s}^2$, find T, H, R and angle of projection for this projectile. Here $\hat{\mathbf{i}}$ is in horizontal direction and $\hat{\mathbf{j}}$ in vertically upward direction.

Sol.

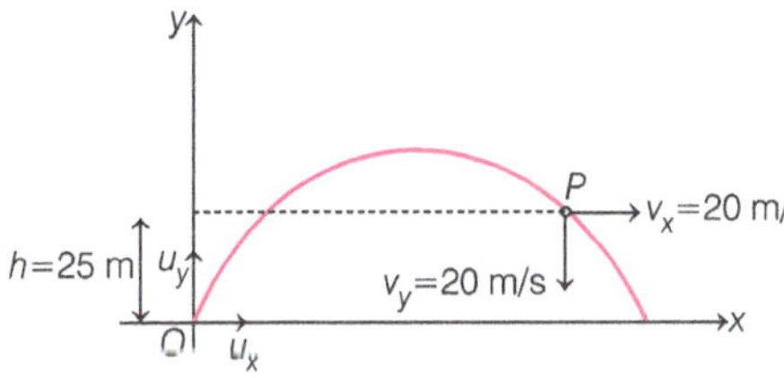

Velocity is given at point P as shown in figure.

$\mathbf{u_x}$: $u_x=v_x=20$ m/s

$\mathbf{u_y}$: $u_y>v_y$

$$\therefore \quad u_y=\sqrt{v_y^2+2gh}$$

$$=\sqrt{(20)^2+(2)(10)(25)}$$

$$=\sqrt{400+500}=\sqrt{900}$$

$$=30\text{ m/s}$$

T: $T=\frac{2u\sin\theta}{g}=\frac{2u_y}{g}$

$$=\frac{2(30)}{10}=6\text{ s}$$

H: $H=\frac{u^2\sin^2\theta}{2g}=\frac{u_y^2}{2g}=\frac{(30)^2}{2\times 10}=45\text{ m}$

R: $R=u_xT=(20)(6)=120$ m

Angle of projection

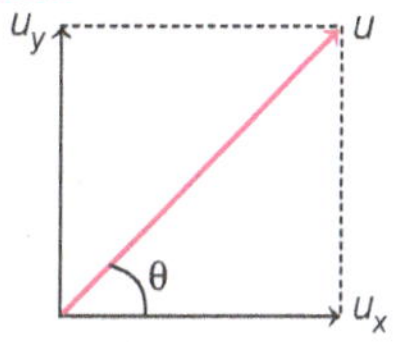

$$\tan\theta=\frac{u_y}{u_x}=\frac{30}{20}=\frac{3}{2}$$

$$\therefore \quad \theta=\tan^{-1}\left(\frac{3}{2}\right)$$

Exercise At a height of 45 m from ground velocity of a projectile is,

$$\mathbf{v}=(30\hat{\mathbf{i}}+40\hat{\mathbf{j}})\text{m/s}$$

Find initial velocity, time of flight, maximum height and horizontal range of this projectile.

Here $\hat{\mathbf{i}}$ and $\hat{\mathbf{j}}$ are the unit vectors in horizontal and vertical directions.

Ans. $10\sqrt{34}$ m/s, 10s, 125m, 300m

Type 6. Collision of two projectiles.

At the time of collision coordinates of particles should be same, i.e.

$$x_1=x_2 \quad \text{and} \quad y_1=y_2$$

Their time of journeys will be same if they start at same time. Otherwise their time of journeys are different.

Ex. 9 Two guns situated on the top of a hill of height 10 m fire one shot each with the same speed $5\sqrt{3}$ m/s at some interval of time. One gun fires horizontally and other fires upwards at an angle of 60° with the horizontal. The shots collide in air at point P ($g=10\text{ m/s}^2$). Find **(1996 JEE)**

(a) the time interval between the firings and

(b) the coordinates of the point P. Take origin of the coordinate system at the foot of the hill right below the muzzle and trajectories in x-y plane.

Sol. $u_{x_1}=u=5\sqrt{3}$ m/s, $u_{y_1}=0$

$$u_{x_2}=u\cos 60°=(5\sqrt{3})\left(\frac{1}{2}\right)=2.5\sqrt{3}\text{ m/s}$$

$$u_{y_2}=u\sin 60°=(5\sqrt{3})\left(\frac{\sqrt{3}}{2}\right)=7.5\text{ m/s}$$

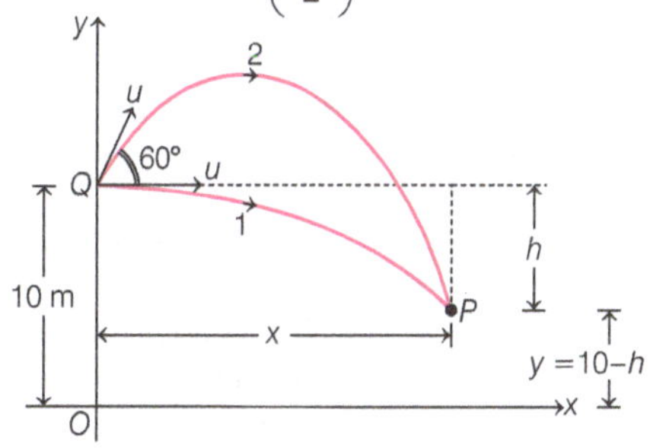

Since, the horizontal displacement of both the shots are equal. The second should be fired early because

$$u_{x_2} < u_{x_1}$$

Now, let first shot takes t_1 time to reach the point P and the second t_2. Using, $x = s_x = u_x t$ and $y = s_y = u_y t + \frac{1}{2} a_y t^2 = u_y t - \frac{1}{2} g t^2$. We have,

$$x_2 = x_1$$

$\therefore$ $$2.5\sqrt{3}\, t_2 = 5\sqrt{3}\, t_1 \quad \text{...(i)}$$

or $$t_2 = 2t_1 \quad \text{...(ii)}$$

$$y_2 = y_1$$

$\therefore$ $$7.5 t_2 - \frac{1}{2} g t_2^2 = 0 - \frac{1}{2} g t_1^2$$

or $$7.5 t_2 - 5t_2^2 = -5t_1^2$$

$\Rightarrow$ $$5t_2^2 - 7.5\, t_2 = 5t_1^2 \quad \text{...(iii)}$$

Substituting $t_2 = 2t_1$ in Eq. (iii), we get

$$5(2t_1)^2 - 7.5\,(2t_1) = 5t_1^2$$

or $$15t_1^2 = 15t_1$$

$\Rightarrow$ $$t_1 = 1\text{ s}$$

and $$t_2 = 2t_1 = 2\text{ s}$$

$$x = 5\sqrt{3}\, t_1 = 5\sqrt{3}\text{ m} \quad \text{[From Eq. (i)]}$$

and $$h = 5t_1^2 = 5\,(1)^2 = 5\text{ m} \quad \text{[From Eq. (iii)]}$$

$\therefore$ $$y = 10 - h = (10 - 5) = 5\text{ m}$$

Hence,

(a) Time interval between the firings $= t_2 - t_1 = (2 - 1)$ s

$$\Delta t = 1\text{ s}$$

(b) Coordinates of point $P = (x, y) = (5\sqrt{3}\text{ m}, 5\text{ m})$

Miscellaneous Examples

Ex. 10 A particle is thrown over a triangle from one end of a horizontal base and after grazing the vertex falls on the other end of the base. If α and β be the base angles and θ the angle of projection, prove that $\tan\theta = \tan\alpha + \tan\beta$.

Sol. The situation is shown in figure.

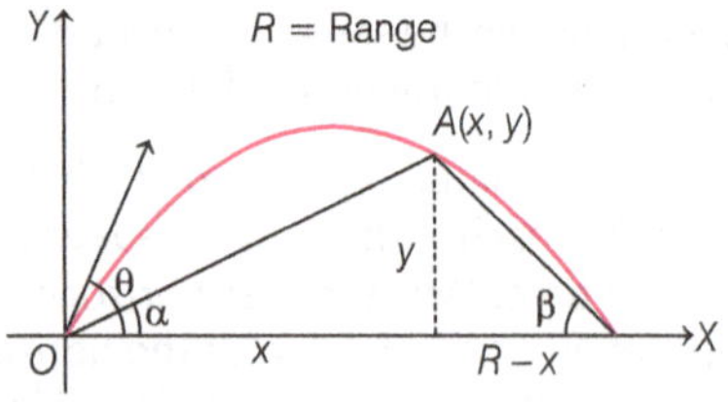

From figure, we have

$$\tan\alpha + \tan\beta = \frac{y}{x} + \frac{y}{R - x}$$

$$\tan\alpha + \tan\beta = \frac{yR}{x(R - x)} \quad \text{...(i)}$$

Equation of trajectory is

$$y = x\tan\theta \left[1 - \frac{x}{R}\right]$$

or, $$\tan\theta = \frac{yR}{x(R - x)} \quad \text{...(ii)}$$

From Eqs. (i) and (ii), we have

$$\tan\theta = \tan\alpha + \tan\beta$$ **Hence Proved.**

Ex. 11 The velocity of a projectile when it is at the greatest height is $\sqrt{2/5}$ times its velocity when it is at half of its greatest height. Determine its angle of projection.

Sol. Suppose the particle is projected with velocity u at an angle θ with the horizontal. Horizontal component of its velocity at all height will be $u\cos\theta$.

At the greatest height, the vertical component of velocity is zero, so the resultant velocity is

$$v_1 = u\cos\theta$$

At half the greatest height during upward motion,

$$y = h/2,\ a_y = -g,\ u_y = u\sin\theta$$

Using $$v_y^2 - u_y^2 = 2a_y y$$

we get, $$v_y^2 - u^2\sin^2\theta = 2(-g)\frac{h}{2}$$

or $$v_y^2 = u^2\sin^2\theta - g \times \frac{u^2\sin^2\theta}{2g} = \frac{u^2\sin^2\theta}{2}$$

$$\left[\because h = \frac{u^2\sin^2\theta}{2g}\right]$$

or $$v_y = \frac{u\sin\theta}{\sqrt{2}}$$

Hence, resultant velocity at half of the greatest height is

$$v_2 = \sqrt{v_x^2 + v_y^2}$$

$$= \sqrt{u^2\cos^2\theta + \frac{u^2\sin^2\theta}{2}}$$

Given, $$\frac{v_1}{v_2} = \sqrt{\frac{2}{5}}$$

$\therefore$ $$\frac{v_1^2}{v_2^2} = \frac{u^2\cos^2\theta}{u^2\cos^2\theta + \frac{u^2\sin^2\theta}{2}} = \frac{2}{5}$$

or $$\frac{1}{1 + \frac{1}{2}\tan^2\theta} = \frac{2}{5}$$

or $$2 + \tan^2\theta = 5 \quad \text{or} \quad \tan^2\theta = 3$$

or $$\tan\theta = \sqrt{3}$$

$\therefore$ $$\theta = 60^\circ$$ **Ans.**

Ex. 12 A car accelerating at the rate of $2\ m/s^2$ from rest from origin is carrying a man at the rear end who has a gun in his hand. The car is always moving along positive x-axis. At $t = 4$ s, the man fires a bullet from the gun and the bullet hits a bird at $t = 8$ s. The bird has a position vector $40\hat{\mathbf{i}} + 80\hat{\mathbf{j}} + 40\hat{\mathbf{k}}$. Find velocity of projection of the bullet. Take the y-axis in the horizontal plane. ($g = 10\ m/s^2$)

Sol. Let velocity of bullet be,

$$\mathbf{v} = v_x\hat{\mathbf{i}} + v_y\hat{\mathbf{j}} + v_z\hat{\mathbf{k}}$$

At $t = 4$ s, x-coordinate of car is

$$x_c = \frac{1}{2}at^2 = \frac{1}{2} \times 2 \times 16 = 16\ \text{m}$$

x-coordinate of bird is $x_b = 40$ m

$\therefore \quad x_b = x_c + v_x(8-4)$

or $\quad 40 = 16 + 4v_x$

$\therefore \quad v_x = 6$ m/s

Similarly, $\quad y_b = y_c + v_y(8-4)$

or $\quad 80 = 0 + 4v_y$

or $\quad v_y = 20$ m/s

and $\quad z_b = z_c + v_z(8-4) - \frac{1}{2}g(8-4)^2$

or $\quad 40 = 0 + 4v_z - \frac{1}{2} \times 10 \times 16$

or $\quad v_z = 30$ m/s

$\therefore$ Velocity of projection of bullet

$$\mathbf{v} = (6\hat{\mathbf{i}} + 20\hat{\mathbf{j}} + 30\hat{\mathbf{k}})\ \text{m/s}$$

Ex. 13 Two inclined planes OA and OB having inclinations 30° and 60° with the horizontal respectively intersect each other at O, as shown in figure. A particle is projected from point P with velocity $u = 10\sqrt{3}$ m/s along a direction perpendicular to plane OA. If the particle strikes plane OB perpendicular at Q. Calculate

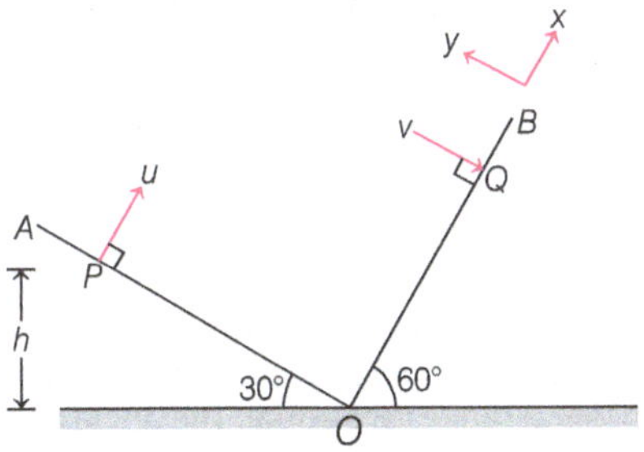

(a) time of flight,

(b) velocity with which the particle strikes the plane OB,

(c) height h of point P from point O,

(d) distance PQ. (Take $g = 10\ m/s^2$)

Sol. Let us choose the x and y directions along OB and OA respectively. Then,

$$u_x = u = 10\sqrt{3}\ \text{m/s},\ u_y = 0$$

$$a_x = -g\sin 60° = -5\sqrt{3}\ \text{m/s}^2$$

and $\quad a_y = -g\cos 60° = -5\ \text{m/s}^2$

(a) At point Q, x-component of velocity is zero. Hence, substituting in

$$v_x = u_x + a_xt$$

$$0 = 10\sqrt{3} - 5\sqrt{3}t$$

or $\quad t = \frac{10\sqrt{3}}{5\sqrt{3}} = 2\text{s}$ **Ans.**

(b) At point Q, $\quad v = v_y = u_y + a_yt$

$\therefore \quad v = 0 - (5)(2) = -10$ m/s

Here, negative sign implies that velocity of particle at Q is along negative y-direction. **Ans.**

(c) Distance $PO = |$ displacement of particle along y-direction $| = |s_y|$

Here, $\quad s_y = u_yt + \frac{1}{2}a_yt^2$

$$= 0 - \frac{1}{2}(5)(2)^2 = -10\text{m}$$

$\therefore \quad PO = 10$ m

Therefore, $\quad h = PO \sin 30°$

$$= (10)\left(\frac{1}{2}\right)$$

or $\quad h = 5$ m **Ans.**

(d) Distance OQ = displacement of particle along x-direction $= s_x$

Here, $\quad s_x = u_xt + \frac{1}{2}a_xt^2$

$$= (10\sqrt{3})(2) - \frac{1}{2}(5\sqrt{3})(2)^2$$

$$= 10\sqrt{3}\ \text{m}$$

or $\quad OQ = 10\sqrt{3}$ m

$\therefore \quad PQ = \sqrt{(PO)^2 + (OQ)^2}$

$$= \sqrt{(10)^2 + (10\sqrt{3})^2}$$

$$= \sqrt{100 + 300} = \sqrt{400}$$

$\therefore \quad PQ = 20$ m **Ans.**

MODULE 1 Exercise

Theory/Formula Based Questions

True/False

1. A projectile fired from the ground follows a parabolic path. The speed of the projectile is minimum at the topmost point of its path. **(1984 JEE)**

Assertion and Reason

Directions : Choose the correct option.

*(a) If both **Assertion** and **Reason** are true and the **Reason** is correct explanation of the **Assertion**.*
*(b) If both **Assertion** and **Reason** are true but **Reason** is not the correct explanation of **Assertion**.*
*(c) If **Assertion** is true, but the **Reason** is false.*
*(d) If **Assertion** is false but the **Reason** is true.*
*(e) If both **Assertion** and **Reason** are false.*

2. **Assertion :** A particle follows only a parabolic path, if acceleration is constant.
Reason : In projectile motion, path is parabolic, as acceleration is assumed to be constant at low heights.

3. **Assertion :** If time of flight in a projectile motion is made two times, its maximum height will become four times.
Reason : In projectile motion $H \propto T^2$, where H is maximum height and T is time of flight.

4. **Assertion :** Two projectiles have maximum heights $4H$ and H respectively. The ratio of their horizontal components of velocities should be 1 : 2 for their horizontal ranges to be the same.
Reason : Horizontal range = horizontal component of velocity × time of flight.

5. **Assertion :** If a particle is projected vertically upwards with velocity u, the maximum height attained by the particle is h_1. The same particle is projected at angle 30° from horizontal with the same speed u. Now, the maximum height is h_2. Then, $h_1 = 4h_2$.
Reason : In first case $v = 0$ at highest point and in the second case $v \neq 0$ at highest point.

6. **Assertion :** When a body is projected at an angle 45°, it's range is maximum.
Reason : For maximum range, the value of $\sin 2\theta$ should be equal to one. **(2023 Main)**

7. **Assertion :** Two identical balls A and B thrown with same velocity u at two different angles with horizontal attained the same range R. If A and B reached the maximum height h_1 and h_2 respectively, then $R = 4\sqrt{h_1h_2}$.
Reason : Product of said heights.
$$h_1h_2 = \left(\frac{u^2\sin^2\theta}{2g}\right)\cdot\left(\frac{u^2\cos^2\theta}{2g}\right)$$
(2022 Main)

Single Correct Option

8. Two bodies are thrown with the same initial velocity at angles θ and $(90° - \theta)$ respectively with the horizontal, then their maximum heights are in the ratio of
(a) 1 : 1
(b) $\sin\theta : \cos\theta$
(c) $\sin^2\theta : \cos^2\theta$
(d) $\cos\theta : \sin\theta$

9. The range of a projectile at an angle θ is equal to half of the maximum range, if thrown at the same speed. The angle of projection θ is given by
(a) 15°
(b) 30°
(c) 60°
(d) data insufficient

10. A ball is projected with a velocity 20 ms^{-1} at some angle with the horizontal. In order to have the maximum range, its velocity at the highest position must be
(a) 10 ms^{-1} (b) 14 ms^{-1} (c) 18 ms^{-1} (d) 16 ms^{-1}

11. A particle has initial velocity $\mathbf{v} = 3\hat{\mathbf{i}} + 4\hat{\mathbf{j}}$ and a constant force $\mathbf{F} = 4\hat{\mathbf{i}} - 3\hat{\mathbf{j}}$ acts on it. The path of the particle is
(a) straight line
(b) parabolic
(c) circular
(d) elliptical

12. A body is projected at an angle 60° with the horizontal with kinetic energy K. When the velocity makes an angle 30° with the horizontal, the kinetic energy of the body will be
(a) $K/2$ (b) $K/3$ (c) $2K/3$ (d) $3K/4$

13. A gun is firing bullets with velocity v_0 by rotating it through 360° in the horizontal plane. The maximum area covered by the bullets is
(a) $\frac{\pi v_0^2}{g}$ (b) $\frac{\pi^2 v_0^2}{g}$ (c) $\frac{\pi v_0^4}{g^2}$ (d) $\frac{\pi^2 v_0^4}{g}$

14. A grasshopper can jump a maximum distance of 1.6 m. It spends negligible time on ground. How far can it go in $10\sqrt{2}$ s?
(a) 45 m (b) 30 m (c) 20 m (d) 40 m

15. Two stones are projected with the same speed but making different angles with the horizontal. Their horizontal ranges are equal. The angle of projection of one is $\frac{\pi}{3}$ and the maximum height reached by it is 102 m. Then, the maximum height reached by the other (in metres) is
(a) 76 (b) 84 (c) 56 (d) 34

16. A ball is projected upwards from the top of a tower with a velocity 50 ms^{-1} making an angle 30° with the horizontal. The height of tower is 70 m. After how many seconds from the instant of throwing, will the ball reach the ground? ($g = 10\ \text{ms}^{-2}$)
(a) 2 s (b) 5 s
(c) 7 s (d) 9 s

17. The range of a projectile when launched at angle θ is same as when launched at angle 2θ. What is the value of θ?
(a) 15° (b) 30°
(c) 45° (d) 60°

18. A stone is projected in air. Its time of flight is 3 s and range is 150 m. Maximum height reached by the stone is (Take, $g = 10\ \text{ms}^{-2}$)
(a) 37.5 m (b) 22.5 m
(c) 90 m (d) 11.25 m

19. The velocity at the maximum height of a projectile is half of its initial velocity of projection (u). Its range on horizontal plane is
(a) $\frac{3u^2}{g}$ (b) $\frac{3}{2}\cdot\frac{u^2}{g}$
(c) $\frac{u^2}{3g}$ (d) $\frac{\sqrt{3}}{2}\cdot\frac{u^2}{g}$

20. Two balls are thrown simultaneously from ground with same velocity of 10 m/s but different angles of projection with horizontal. Both balls fall at same point at a distance of $5\sqrt{3}$ m from point of projection. What is the time interval between the two balls when they strike the ground?
(a) $(\sqrt{3}-1)$ s (b) $(\sqrt{3}+1)$ s
(c) $\sqrt{3}$ s (d) 1 s

21. The maximum height attained by a projectile is increased by 10% by increasing its speed of projection, without changing the angle of projection. The percentage increase in the horizontal range will be
(a) 20% (b) 15%
(c) 10% (d) 5%

22. A ball is thrown at different angles with the same speed u and from the same point. It has the same range in both cases. If y_1 and y_2 be the heights attained in the two cases, then $y_1 + y_2$ is equal to
(a) $\frac{u^2}{g}$ (b) $\frac{2u^2}{g}$
(c) $\frac{u^2}{2g}$ (d) $\frac{u^2}{4g}$

23. A projectile is thrown with an initial velocity of $(a\hat{i} + b\hat{j})\ \text{ms}^{-1}$. If the range of the projectile is twice the maximum height reached by it, then
(a) $a = 2b$ (b) $b = a$
(c) $b = 2a$ (d) $b = 4a$

24. If the instantaneous velocity of a particle projected as shown in figure is given by $\mathbf{v} = a\hat{\mathbf{i}} + (b - ct)\hat{\mathbf{j}}$, where a, b and c are positive constants, the range on the horizontal plane will be

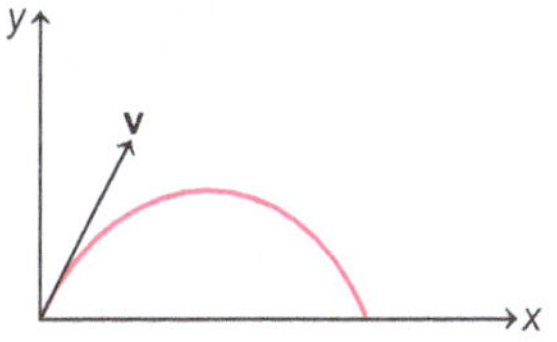

(a) $2ab/c$ (b) ab/c (c) ac/b (d) $a/2bc$

25. A shell is fired from a fixed artillery gun with an initial speed u such that it hits the target on the ground at a distance R from it. If t_1 and t_2 are the values of the time taken by it to hit the target in two possible ways, the product t_1t_2 is **(2019 Main)**
(a) $\frac{R}{4g}$ (b) $\frac{R}{g}$ (c) $\frac{R}{2g}$ (d) $\frac{2R}{g}$

26. A projectile is projected at 30° from horizontal with initial velocity $40\ \text{ms}^{-1}$. The velocity of the projectile at $t = 2$ s from the start will be (Given, $g = 10\ \text{m/s}^2$) **(2023 Main)**
(a) $20\sqrt{3}\ \text{ms}^{-1}$ (b) $40\sqrt{3}\ \text{ms}^{-1}$
(b) $20\ \text{ms}^{-1}$ (d) zero

27. The range of the projectile projected at an angle of 15° with horizontal is 50 m. If the projectile is projected with same velocity at an angle of 45° with horizontal, then its range will be **(2023 Main)**
(a) 50 m (b) $50\sqrt{2}$ m (c) 100 m (d) $100\sqrt{2}$ m

28. Two projectiles are projected at 30° and 60° with the horizontal with the same speed. The ratio of the maximum height attained by the two projectiles respectively is **(2023 Main)**
(a) $2:\sqrt{3}$ (b) $\sqrt{3}:1$ (c) $1:3$ (d) $1:\sqrt{3}$

29. The initial speed of a projectile fired from ground is u. At the highest point during its motion, the speed of projectile is $\frac{\sqrt{3}}{2}u$. The time of flight of the projectile is **(2023 Main)**
(a) $\frac{u}{2g}$ (b) $\frac{u}{g}$ (c) $\frac{2u}{g}$ (d) $\frac{\sqrt{3}u}{g}$

30. The maximum vertical height to which a man can throw a ball is 136 m. The maximum horizontal distance upto which he can throw the same ball is **(2023 Main)**
(a) 192 m (b) 136 m (c) 272 m (d) 68 m

31. Two projectiles P_1 and P_2 thrown with speed in the ratio $\sqrt{3}:\sqrt{2}$, attain the same height during their motion. If P_2 is thrown at an angle of 60° with the horizontal, the angle of projection of P_1 with horizontal will be **(2022 Main)**
(a) 15° (b) 30° (c) 45° (d) 60°

32. The ranges and heights for two projectiles projected with the same initial velocity at angles 42° and 48° with the horizontal are R_1, R_2 and H_1, H_2, respectively. Choose the correct option. **(2021 Main)**

(a) $R_1 > R_2$ and $H_1 = H_2$ (b) $R_1 = R_2$ and $H_1 < H_2$
(c) $R_1 < R_2$ and $H_1 < H_2$ (d) $R_1 = R_2$ and $H_1 = H_2$

33. The position of a projectile launched from the origin at $t = 0$ is given by $\mathbf{r} = (40\,\hat{\mathbf{i}} + 50\,\hat{\mathbf{j}})$ m at $t = 2$ s. If the projectile was launched at an angle θ from the horizontal, then θ is (Take $g = 10\ \text{ms}^{-2}$) **(2014 Main)**

(a) $\tan^{-1}\frac{2}{3}$ (b) $\tan^{-1}\frac{3}{2}$ (c) $\tan^{-1}\frac{7}{4}$ (d) $\tan^{-1}\frac{4}{5}$

Multiple Correct Options

34. Two particles projected from the same point with same speed u at angles of projection α and β strike the horizontal ground at the same point. If h_1 and h_2 are the maximum heights attained by the projectile, R is the range for both and t_1 and t_2 are their time of flight respectively, then

(a) $\alpha + \beta = \frac{\pi}{2}$ (b) $R = 4\sqrt{h_1 h_2}$
(c) $\frac{t_1}{t_2} = \tan\alpha$ (d) $\tan\alpha = \sqrt{\frac{h_1}{h_2}}$

35. At a height of 15 m from ground, velocity of a projectile is $\mathbf{v} = (10\hat{\mathbf{i}} + 10\hat{\mathbf{j}})$. Here, $\hat{\mathbf{j}}$ is vertically upwards and $\hat{\mathbf{i}}$ is along horizontal direction, then (Take, $g = 10\ \text{ms}^{-2}$)

(a) particle was projected at an angle of 45° with horizontal
(b) time of flight of projectile is 4 s
(c) horizontal range of projectile is 100 m
(d) maximum height of projectile from ground is 20 m

36. Which of the following quantities remain constant during projectile motion?

(a) Average velocity between two points
(b) Average speed between two points
(c) $\frac{d\mathbf{v}}{dt}$ (d) $\frac{d^2\mathbf{v}}{dt^2}$

Match the Columns

37. In a projectile motion, given $H = \frac{R}{2} = 20$ m. Here, H is maximum height and R the horizontal range. For the given condition, match the following two columns. (Take, $g = 10\ \text{ms}^{-2}$)

	Column I		Column II
(a)	Time of flight	(p)	1
(b)	Ratio of vertical component of velocity and horizontal component of velocity	(q)	2
(c)	Horizontal component of velocity (in m/s)	(r)	10
(d)	Vertical component of velocity (in m/s)	(s)	None of the above

38. A particle can be thrown at a constant speed at different angles. When it is thrown at 15° with horizontal, it falls at a distance of 10 m from point of projection. For this speed of particle, match the following two columns.

	Column I		Column II
(a)	Maximum horizontal range which can be taken with this speed	(p)	10 m
(b)	Maximum height which can be taken with this speed	(q)	20 m
(c)	Range at 75°	(r)	15 m
(d)	Height at 30°	(s)	None of the above

39. In projectile motion, if vertical component of velocity is increased to two times, keeping horizontal component unchanged, then

	Column I		Column II
(a)	Time of flight	(p)	will remain same
(b)	Maximum height	(q)	will become two times
(c)	Horizontal range	(r)	will become four times
(d)	Angle of projection with horizontal	(s)	None of the above

40. In projectile motion shown in figure.

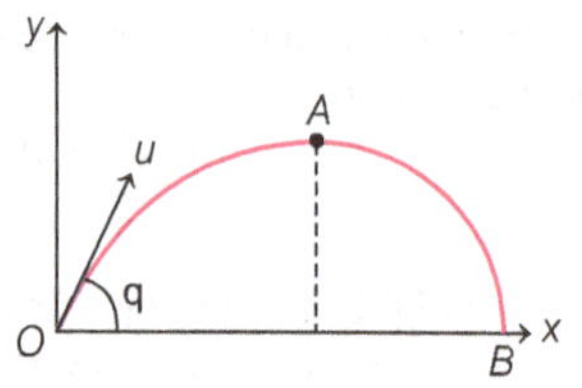

	Column I		Column II
(a)	Change in velocity between O and A	(p)	$u\cos\theta$
(b)	Average velocity between O and A	(q)	$u\sin\theta$
(c)	Change in velocity between O and B	(r)	$2u\sin\theta$
(d)	Average velocity between O and B	(s)	None of the above

41. The trajectories of the motion of three particles are shown in the figure. Match the entries of Column I with the entries of Column II. Neglect air resistance.

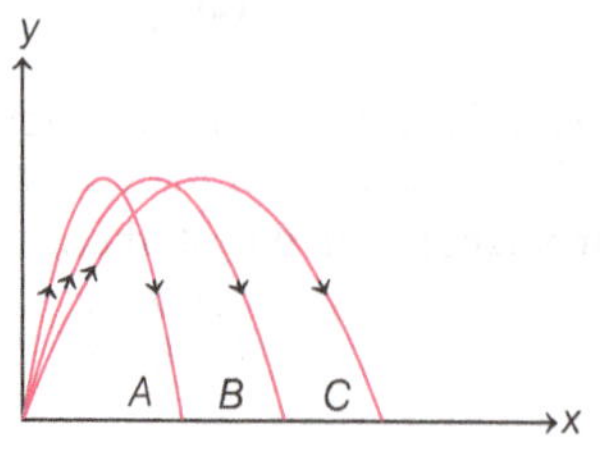

	Column I		Column II
(a)	Time of flight is least for	(p)	A
(b)	Vertical component of velocity is greatest for	(q)	B
(c)	Horizontal component of velocity is greatest for	(r)	C
(d)	Launch speed is least for	(s)	same for all

Integer Type Questions

42. Two bodies are projected from ground with same speeds 40 ms^{-1} at two different angles with respect to horizontal. The bodies were found to have same range. If one of the body was projected at an angle of 60°, with horizontal then sum of the maximum heights, attained by the two projectiles is .... m. (Given $g = 10\ ms^{-2}$) **(2023 Main)**

43. An object is projected in the air with initial velocity u at an angle θ. The projectile motion is such that the horizontal range R is maximum.
Another object is projected in the air with a horizontal range half of the range of first object. The initial velocity remains same in both the cases. The value of the angle of projection, at which the second object is projected, will be .... degree. **(2022 Main)**

44. A ball of mass m is thrown vertically upwards. Another ball of mass $2m$ is thrown at an angle θ with the vertical. Both the balls stay in air for the same period of time. The ratio of the heights attained by the two balls respectively is $\frac{1}{x}$. The value of x is .... **(2022 Main)**

Trajectory Based Questions

Fill in the Blank

45. The trajectory of a projectile in a vertical plane is $y = ax - bx^2$, where a, b are constants, and x and y are respectively, the horizontal and vertical distances of the projectile from the point of projection. The maximum height attained is .... and the angle of projection from the horizontal is .... **(1997 JEE)**

Single Correct Option

46. The equation of a projectile is $y = \sqrt{3}x - \frac{g}{2}x^2$. The angle of its projection is
(a) 90° (b) zero
(c) 60° (d) 30°

47. The equation of motion of a projectile is
$$y = 12x - \frac{3}{4}x^2$$
What is the range of the projectile?
(a) 12 m (b) 16 m
(c) 20 m (d) 24 m

48. The trajectory of a projectile near the surface of the earth is given as $y = 2x - 9x^2$. If it were launched at an angle θ_0 with speed v_0, then (Take, $g = 10\ ms^{-2}$) **(2019 Main)**
(a) $\theta_0 = \sin^{-1}\left(\frac{1}{\sqrt{5}}\right)$ and $v_0 = \frac{5}{3}\ ms^{-1}$
(b) $\theta_0 = \cos^{-1}\left(\frac{2}{\sqrt{5}}\right)$ and $v_0 = \frac{3}{5}\ ms^{-1}$
(c) $\theta_0 = \cos^{-1}\left(\frac{1}{\sqrt{5}}\right)$ and $v_0 = \frac{5}{3}\ ms^{-1}$
(d) $\theta_0 = \sin^{-1}\left(\frac{2}{\sqrt{5}}\right)$ and $v_0 = \frac{3}{5}\ ms^{-1}$

49. A projectile is given an initial velocity of $(\hat{i} + 2\hat{j})$ m/s, where $\hat{i}$ is along the ground and $\hat{j}$ is along the vertical. If $g = 10\ m/s^2$, the equation of its trajectory is **(2013 JEE)**
(a) $y = x - 5x^2$ (b) $y = 2x - 5x^2$
(c) $4y = 2x - 5x^2$ (d) $4y = 2x - 25x^2$

Multiple Correct Options

50. The trajectory of a projectile is $y = x(1 - x)$, where y and x are in metres.

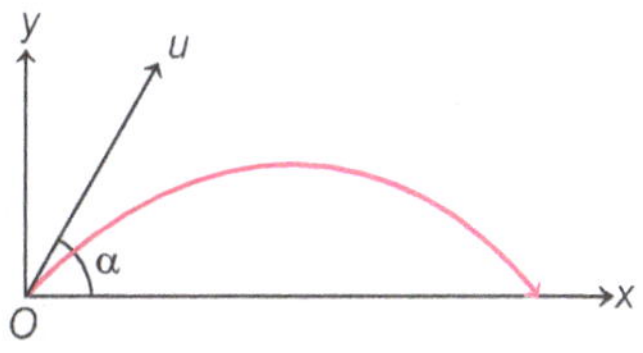

Choose the correct option(s).
(a) The horizontal range is 1 m
(b) The angle of projection is 45°
(c) The horizontal range is 2m
(d) The angle of projection is 30°

Collision of Two Projectiles

Assertion and Reason

See directions on page 302.

51. **Assertion :** Particle A is projected upwards. Simultaneously particle B is projected as shown. Particle A returns to ground in 4 s. At the same time particle B collides with A. Maximum height H attained by B should be 20 m. ($g = 10\ ms^{-2}$)

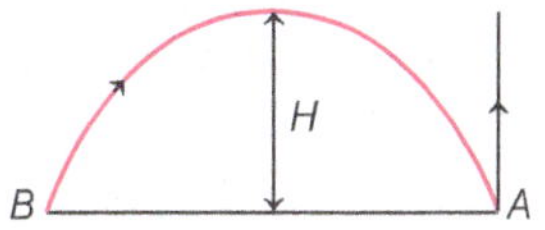

Reason : Speed of projection of both the particles should be same under the given condition.

Subjective Questions

52. A cannon fires successively two shells with velocity $v_0 = 250$ m/s the first at the angle $\theta_1 = 60°$ and the second at the angle $\theta_2 = 45°$ to the horizontal. Neglecting the air drag, find the time interval between firings leading to the collision of the shells. **(Irodov 1.33)**

Miscellaneous Questions

Assertion and Reason

See directions on page 302.

53. **Assertion :** A particle is projected with velocity $\mathbf{u}$ at angle 45° with ground. Let $\mathbf{v}$ be the velocity of particle at time $t (\neq 0)$, then value of $\mathbf{u} \cdot \mathbf{v}$ can be zero.

Reason : Value of dot product is zero when angle between two vectors is 90°.

54. **Assertion :** In projectile motion at any two positions $\dfrac{\mathbf{v}_2 - \mathbf{v}_1}{t_2 - t_1}$ always remains constant.

Reason : The given quantity is average acceleration, which should remain constant as acceleration is constant.

55. **Assertion :** In projectile motion, if particle is projected with speed u, then speed of particle at height h would be $\sqrt{u^2 - 2gh}$.

Reason : If particle is projected with vertical component of velocity u_y, then vertical component at the height h would be $\pm\sqrt{u_y^2 - 2gh}$.

56. **Assertion :** At height 20 m from ground, velocity of a projectile is $\mathbf{v} = (20\hat{\mathbf{i}} + 20\hat{\mathbf{j}})$ m/s. Here, $\hat{\mathbf{i}}$ is in the horizontal direction and $\hat{\mathbf{j}}$ in vertical direction. Then after this moment the particle is at the same height after another 4 s.

Reason : Maximum height of particle from ground is 40 m. (Take $g = 10 \text{ m/s}^2$)

Single Correct Option

57. Two bodies were thrown simultaneously from the same point, one straight up, and the other, at an angle of $\theta = 30°$ to the horizontal. The initial velocity of each body is 20 ms^{-1}. Neglecting air resistance, the distance between the bodies at $t = 1.2$ later is

(a) 20 m (b) 30 m
(c) 24 m (d) 50 m

58. Average velocity of a particle in projectile motion between its starting point and the highest point of its trajectory is (projection speed $= u$, angle of projection from horizontal $= \theta$)

(a) $u \cos\theta$ (b) $\dfrac{u}{2}\sqrt{1 + 3\cos^2\theta}$
(c) $\dfrac{u}{2}\sqrt{2 + \cos^2\theta}$ (d) $\dfrac{u}{2}\sqrt{1 + \cos^2\theta}$

59. A train is moving on a track at 30 ms^{-1}. A ball is thrown from it perpendicular to the direction of motion with 30 ms^{-1} at 45° from horizontal. Find the distance of ball from the point of projection on train to the point, where it strikes the ground.

(a) 90 m (b) $90\sqrt{3}$ m
(c) 60 m (d) $60\sqrt{3}$ m

60. A body is projected at time $t = 0$, from a certain point on a planet's surface with a certain velocity at a certain angle with the planet's surface (assumed horizontal). The horizontal and vertical displacements x and y (in metre) respectively vary with time t in second as, $x = (10\sqrt{3})\, t$ and $y = 10t - t^2$. The maximum height attained by the body is

(a) 75 m (b) 100 m
(c) 50 m (d) 25 m

61. A fixed mortar fires a bomb at an angle of 53° above the horizontal with a muzzle velocity of 80 ms^{-1}. A tank is advancing directly towards the mortar on level ground at a constant speed of 5 m/s. The initial separation (at the instant mortar is fired) between the mortar and tank, so that the tank would be hit is [Take $g = 10 \text{ ms}^{-2}$]

(a) 662.4 m (b) 526.3 m
(c) 486.6 m (d) None of these

62. The equation of motion of a projectile are given by $x = 36\, t$ m and $2y = 96\, t - 9.8\, t^2$ m. The angle of projection from horizontal is

(a) $\sin^{-1}\left(\dfrac{4}{5}\right)$ (b) $\sin^{-1}\left(\dfrac{3}{5}\right)$
(c) $\sin^{-1}\left(\dfrac{4}{3}\right)$ (d) $\sin^{-1}\left(\dfrac{3}{4}\right)$

63. A projectile, thrown with velocity v_0 at an angle α to the horizontal, has a range R. It will strike a vertical wall at a distance $\dfrac{R}{2}$ from the point of projection with a speed of

(a) v_0 (b) $v_0 \sin\alpha$
(c) $v_0 \cos\alpha$ (d) $\sqrt{\dfrac{gR}{2}}$

64. A boy throws a ball with a velocity u at an angle θ with the horizontal. At the same instant, he starts running with uniform velocity to catch the ball before it hits the ground. To achieve this, he should run with a velocity of

(a) $u \cos\theta$ (b) $u \sin\theta$ (c) $u \tan\theta$ (d) $u \sec\theta$

65. A ball is thrown up with a certain velocity at an angle θ to the horizontal. The kinetic energy (KE) of the ball varies with horizontal displacement x as

(a) 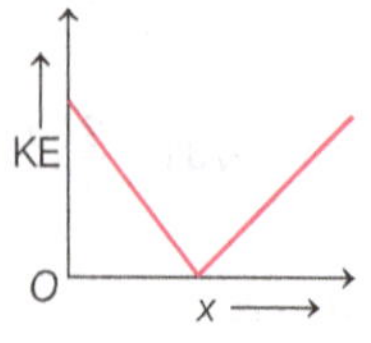

(b) 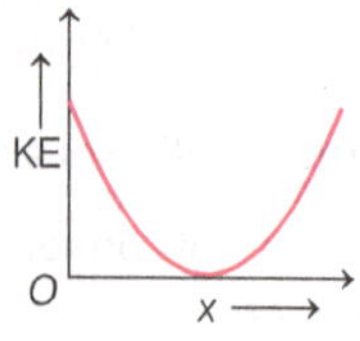

(c) 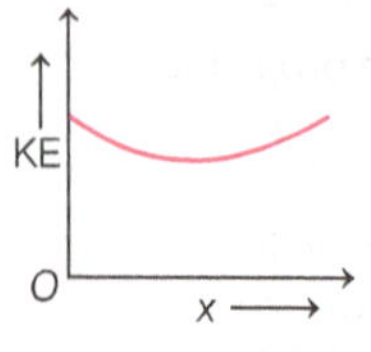

(d) 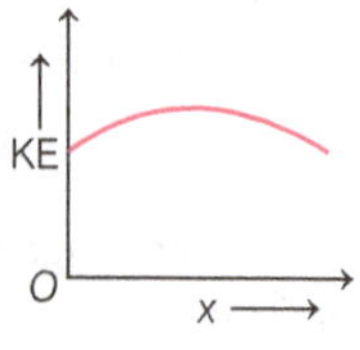

66. A ball is thrown up with a certain velocity at an angle θ to the horizontal. The kinetic energy (KE) of the ball varies with height h as

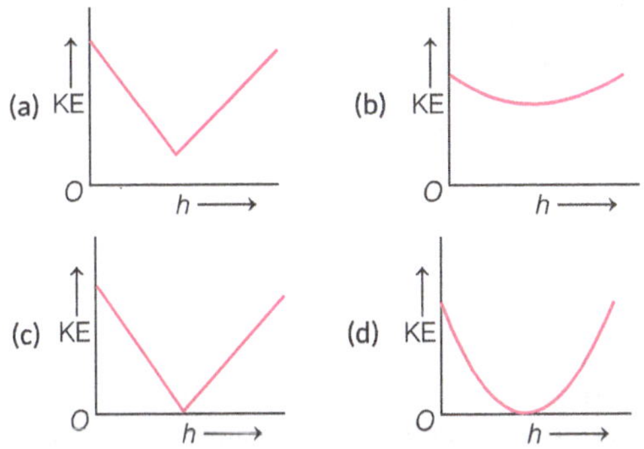

67. A projectile is thrown with a velocity of 10 m/s at an angle of 60° with horizontal. The interval between the moments when speed of the projectile is $5\sqrt{2}$ m/s is (Take, $g = 10\text{ m/s}^2$)

(a) 1 s (b) 3 s (c) 2 s (d) 4 s

68. A body, projected horizontally with a speed u from the top of a tower of height h, reaches the ground at a horizontal distance R from the tower. Another body, projected horizontally from the top of a tower of height $4h$, reaches the ground at horizontal distance $2R$ from the tower. The initial speed of the second body is

(a) u (b) $2u$ (c) $3u$ (d) $4u$

69. A projectile is projected and it takes 9 s to reach in the horizontal plane through the point of projection. In its paths, it passes a point P after 4 s. The height of P above the horizontal plane is

(a) 10 m (b) 50 m (c) 100 m (d) 200 m

70. The ceiling of a hall is 40 m. For maximum horizontal distance, the angle at which the ball can be thrown with a speed of 56 ms^{-1} without hitting the ceiling of the hall is (Take, $g = 9.8\text{ ms}^{-2}$)

(a) 60° (b) 30°
(c) 45° (d) data insufficient

71. A projectile is thrown with velocity u making an angle θ with the vertical. It just crosses the tops of two poles each of height h after 1s and 3s, respectively. The maximum height of projectile is

(a) 9.8 m (b) 19.6 m (c) 39.2 m (d) 4.9 m

72. Two particles A and B are projected from a building. Particle A is projected with speed $2v$ at angle 30° with horizontal and B with speed v at angle 60° with horizontal. Which particle will hit the ground earlier?

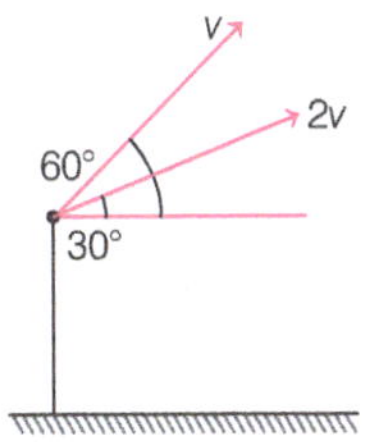

(a) Particle A
(b) Particle B
(c) Both will hit at the same time
(d) Data insufficient

73. A projectile is thrown with velocity v at an angle θ with the horizontal. When the projectile is at a height equal to half of the maximum height, the vertical component of the velocity of the projectile is

(a) $\frac{v\sin\theta}{4}$ (b) $\frac{v\sin\theta}{3}$
(c) $\frac{v\sin\theta}{\sqrt{2}}$ (d) $\frac{v\sin\theta}{\sqrt{3}}$

74. The initial speed of a bullet fired from a rifle is 630 m/s. The rifle is fired at the centre of a target 700 m away at the same level as the target. How far above the centre of the target the rifle must be aimed in order to hit the target? **(2014 Main)**

(a) 1.0 m (b) 4.2 m
(c) 6.1 m (d) 9.8 m

75. A ball is thrown from a point with a speed 'v_0' at an elevation angle of θ. From the same point and at the same instant, a person starts running with a constant speed $\frac{'v_0'}{2}$ to catch the ball. Will the person be able to catch the ball? If yes, what should be the angle of projection θ? **(2004 Main)**

(a) No (b) Yes, 30°
(c) Yes, 60° (d) Yes, 45°

76. A boy playing on the roof of a 10 m high building throws a ball with a speed of 10 m/s at an angle of 30° with the horizontal. How far from the throwing point will the ball be at the height of 10 m from the ground? **(2003 Main)**

$\left(g = 10\text{ ms}^{-2}, \sin 30° = \frac{1}{2}, \cos 30° = \frac{\sqrt{3}}{2}\right)$

(a) 5.20 m (b) 4.33 m (c) 2.60 m (d) 8.66 m

Multiple Correct Options

77. A particle is projected from a point P with a velocity v at an angle θ with horizontal. At a certain point Q, it moves at right angles to its initial direction. Then,

(a) velocity of particle at Q is $v\sin\theta$
(b) velocity of particle at Q is $v\cot\theta$
(c) time of flight from P to Q is $(v/g)\,\text{cosec}\,\theta$
(d) time of flight from P to Q is $(v/g)\sec\theta$

Match the Columns

78. Particle-1 is just dropped from a tower. After 1 s particle-2 is thrown from the same tower horizontally with velocity 10 ms^{-1}. Taking $g = 10\text{ms}^{-2}$, match the following two columns at $t = 2$ s.

	Column I		Column II
(a)	Horizontal displacement between two	(p)	10 SI units
(b)	Vertical displacement between two	(q)	20 SI units
(c)	Magnitude of relative horizontal component of velocity	(r)	$10\sqrt{2}$ SI units
(d)	Magnitude of relative vertical component of velocity	(s)	None of the above

79. Particle-1 is projected from ground (take it origin) at time $t = 0$, with velocity $(30\hat{i} + 30\hat{j})\ ms^{-1}$. Particle-2 is projected from (130 m, 75 m) at time $t = 1$ s with velocity $(-20\ \hat{i} + 20\ \hat{j})\ ms^{-1}$. Assuming $\hat{j}$ to be vertically upward and $\hat{i}$ to be in horizontal direction, match the following two columns at $t = 2$ s.

	Column I		Column II
(a)	Horizontal distance between two	(p)	30 SI units
(b)	Vertical distance between two	(q)	40 SI units
(c)	Relative horizontal component of velocity between two	(r)	50 SI units
(d)	Relative vertical component of velocity between two	(s)	None of the above

Subjective Questions

80. A particle is projected from ground with velocity $20\sqrt{2}$ m/s at 45°. At what time, particle is at height 15 m from ground? (Take, $g = 10\ m/s^2$)

81. A particle is projected at an angle 60° with horizontal with a speed $v = 20$ m/s. Taking $g = 10\ m/s^2$. Find the time after which the speed of the particle remains half of its initial speed.

82. A ball is shot from the ground into the air. At a height of 9.1 m, its velocity is observed to be $\mathbf{v} = 7.6\hat{i} + 6.1\hat{j}$ in metre per second ($\hat{i}$ is horizontal, $\hat{j}$ is upward). Give the approximate answers.
(a) To what maximum height does the ball rise?
(b) What total horizontal distance does the ball travel?
(c) What is the magnitude and
(d) What is the direction of the ball's velocity just before it hits the ground?

83. A particle is projected at an angle of elevation α and after t second, it appears to have an elevation of β as seen from the point of projection. Find the initial velocity of projection.

84. Two particles are simultaneously thrown in horizontal direction from two points on a riverbank, which are at certain height above the water surface. The initial velocities of the particles are $v_1 = 5$ m/s and $v_2 = 7.5$ m/s, respectively. Both particles fall into the water at the same time. First particle enters the water at a point $s = 10$ m from the bank. Determine
(a) the time of flight of the two particles,
(b) the height from which they are thrown,
(c) the point where the second particle falls in water.

Integer Type Questions

85. A fighter jet is flying horizontally at a certain altitude with a speed of 200 ms^{-1}. When it passes directly overhead an anti-aircraft gun, a bullet is fired from the gun, at an angle θ with the horizontal, to hit the jet. If the bullet speed is 400 m/s, the value of θ will be .... °. **(2022 Main)**

86. The projectile motion of a particle of mass 5 g is shown in the figure.

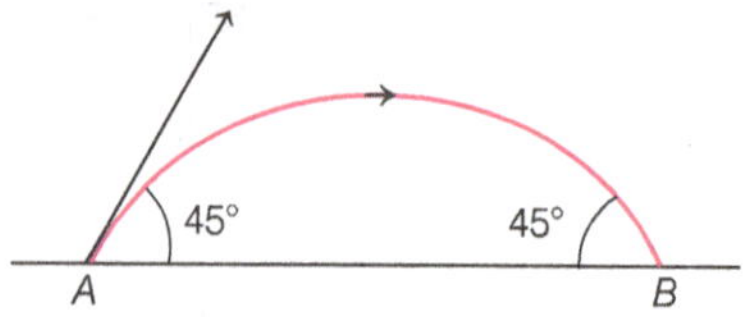

The initial velocity of the particle is $5\sqrt{2}\ ms^{-1}$ and the air resistance is assumed to be negligible. The magnitude of the change in momentum between the points A and B is $x \times 10^{-2}$ $kgms^{-1}$. **(2021 Main)**
The value of x, to the nearest integer, is ....

87. A body is projected from the ground at an angle of 45° with the horizontal. Its velocity after 2s is 20 ms^{-1}. The maximum height reached by the body during its motion is .... m. (Use $g = 10\ ms^{-2}$)

Answers

1. (True) **2.** (d) **3.** (a) **4.** (a or b) **5.** (b)
6. (a) **7.** (a) **8.** (c) **9.** (a) **10.** (b)
11. (b) **12.** (b) **13.** (c) **14.** (d) **15.** (d)
16. (c) **17.** (b) **18.** (d) **19.** (d) **20.** (a)
21. (c) **22.** (c) **23.** (c) **24.** (a) **25.** (d)
26. (a) **27.** (c) **28.** (c) **29.** (b) **30.** (c)
31. (c) **32.** (b) **33.** (c) **34.** (a, b, c, d) **35.** (b, d)
36. (c, d) **37.** (a-s; b-q; c-r; d-s) **38.** (a-q; b-p; c-p; d-s)
39. (a-q; b-r; c-q; d-s) **40.** (a-q; b-s; c-r; d-p)
41. (a-s; b-s; c-r; d-p) **42.** (80) **43.** (15 or 75) **44.** (1)
45. $\left(\frac{a^2}{4b}, \tan^{-1}(a)\right)$ **46.** (c) **47.** (b) **48.** (c)
49. (b) **50.** (a, b) **51.** (c) **52.** ($\Delta t \approx 11$ s)
53. (May be a or b) **54.** (a) **55.** (May be a or b)
56. (b) **57.** (c) **58.** (b) **59.** (a) **60.** (d)
61. (d) **62.** (a) **63.** (c) **64.** (a) **65.** (c)
66. (a) **67.** (a) **68.** (a) **69.** (c) **70.** (b)
71. (b) **72.** (b) **73.** (c) **74.** (c) **75.** (c)
76. (d) **77.** (b, c) **78.** (a-p; b-s; c-p; d-p)
79. (a-r; b-r; c-r; d-s) **80.** 3s and 1s **81.** $\sqrt{3}$ s
82. (a) 11m (b) 23 m (c) 16.6 m/s (d) $\tan^{-1}(2)$, below horizontal
83. $u = gt\dfrac{\cos\beta}{\sin(\alpha - \beta)}$ **84.** (a) 2s (b) 19.6 m (c) 15m
85. (60) **86.** (5) **87.** (20)

See hints on page 321-329.

MODULE 2

Left Over Topics of Projectile Motion

Projectile Motion Along an Inclined Plane

Following two concepts will solve all logical problems of projectile motion along an inclined plane.

(i) s_x: Displacement in horizontal x-direction,

$$s_x = u_x t$$

(ii) s_y: Displacement in vertical y-direction has two components.

(a) $s_1 = u_y t$ (due to initial velocity u_y)

Since, u_y may be zero, upwards or downwards.

So, s_1 may be zero, upwards or downwards.

(b) $s_2 = \frac{1}{2} g t^2$ (due to gravity)

$= 5t^2$ (if we take $g = 10\ \text{m/s}^2$)

Since g is always downwards. So, s_2 is always downwards. It cannot be zero.

Now total $s_y = s_1 + s_2$

Let us take few examples in support of this.

Example 1

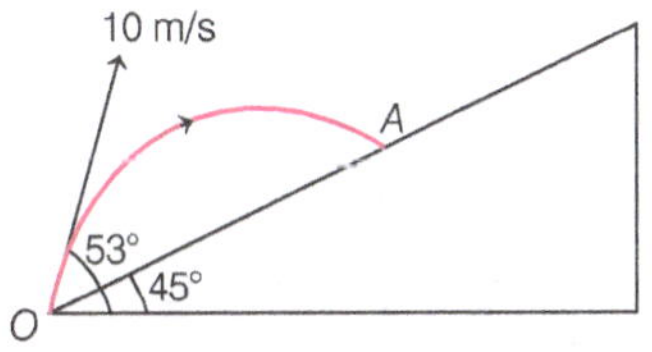

Taking $g = 10\ \text{m/s}^2$, find time of flight from O to A and range along the inclined plane OA.

Sol. Step-1 :

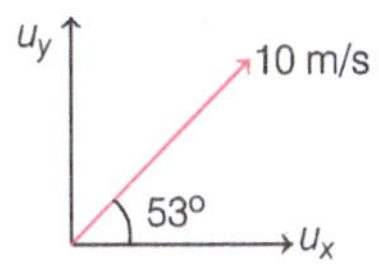

Horizontal component of velocity,

$$u_x = 10 \cos 53° = 10 \times \frac{3}{5} = 6\ \text{m/s}$$

Vertical component of velocity upwards,

$$u_y = 10 \sin 53° = 10 \times \frac{4}{5} = 8\ \text{m/s}$$

Step-2 :

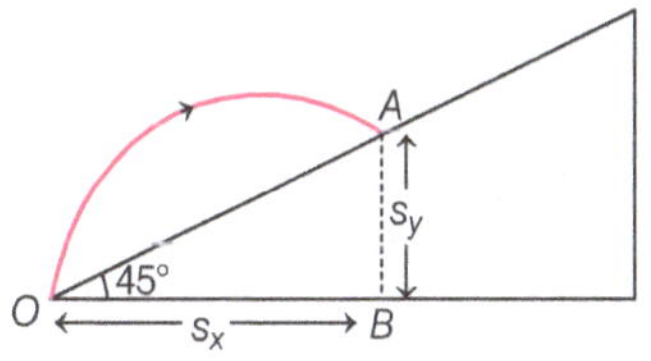

$t = t_{OA}$

$s_x = u_x t = 6t$

s_y: $s_1 = u_y t = 8t$ (upwards); $s_2 = 5t^2$ (downwards)

But net displacement of particle in vertical direction is upwards.

$\therefore \quad s_1 > s_2$

or $\quad s_y = s_1 - s_2 = (8t - 5t^2)$

Step 3 :

In triangle OAB,

$$\tan 45° = \frac{s_y}{s_x}$$

or
$$1 = \frac{8t - 5t^2}{6t}$$

Solving this equation, we get

$t = 0.4\ \text{s}$ **Ans.**

and $\quad OA = s_x \sec 45°$

$= (6t)\sqrt{2}$

$= (6)(0.4)\sqrt{2}$

$= 2.4\sqrt{2}\ \text{m}$ **Ans.**

Example 2

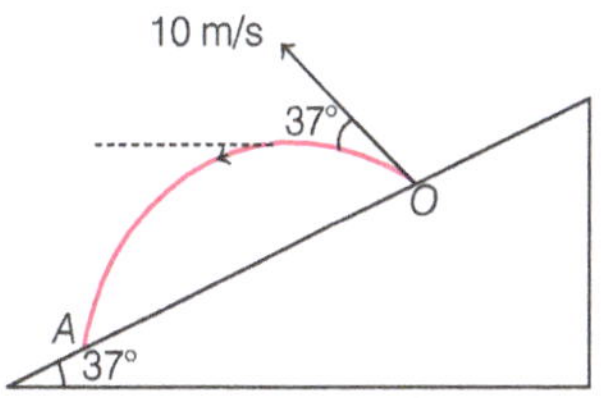

Taking $g = 10\ \text{m/s}^2$, find time of flight t and range along inclined plane OA.

Sol. **Step 1 :**

$$u_x = 10\cos 37° = 10 \times \frac{4}{5} = 8 \text{ m/s}$$

$$u_y = 10\sin 37° = 10 \times \frac{3}{5} = 6 \text{ m/s}$$

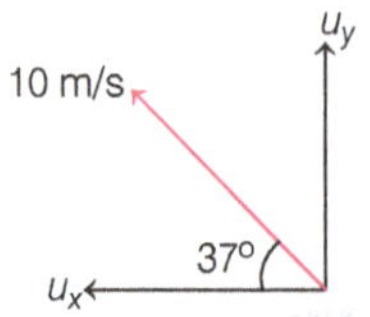

Step 2 :

$$s_x = u_x t = 8t$$

s_y: $s_1 = u_y t = 6t$ (upwards)

$s_2 = 5t^2$ (downwards)

Since net displacement of particle in vertical direction is downwards. So,

$$s_2 > s_1$$

and $s_y = s_2 - s_1$ (downwards)

$= (5t^2 - 6t)$ (downwards)

Step 3 :

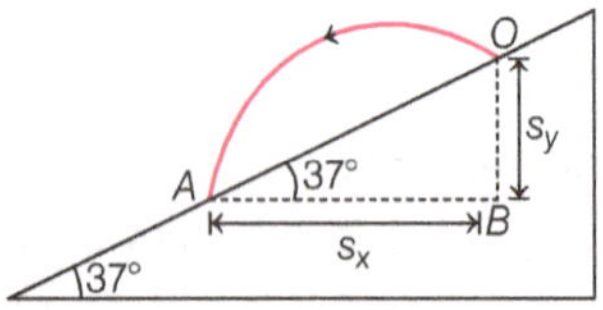

In triangle OAB,

$$\tan 37° = \frac{s_y}{s_x} \quad \therefore \quad \frac{3}{4} = \frac{5t^2 - 6t}{8t}$$

Solving this equation, we get

$$t = 2.4 \text{ s}$$ **Ans.**

and $OA = s_x \sec 37°$

$$= (8t)\left(\frac{5}{4}\right) = 10t$$

$$= (10)(2.4) = 24 \text{ m}$$ **Ans.**

Example 3

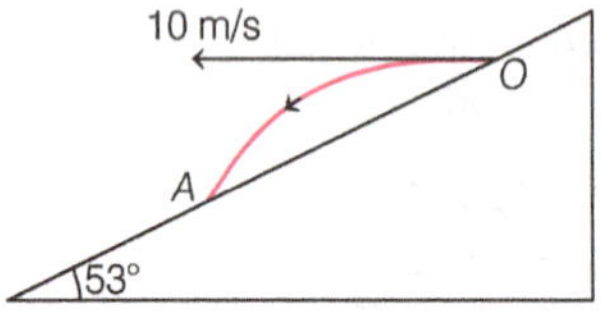

Taking $g = 10 \text{ m/s}^2$, find time of flight t and range along the inclined plane OA.

Sol. **Step 1 :** Here initial velocity is horizontal.

$\therefore$ $u_x = u = 10$ m/s and

$u_y = 0$

Step 2 : $s_x = u_x t = 10t$

s_y: $s_1 = 0$ because $u_y = 0$

$s_2 = 5t^2$ (downwards)

$\therefore$ $s_y = s_1 + s_2$

$= 5t^2$ (downwards)

Step 3 :

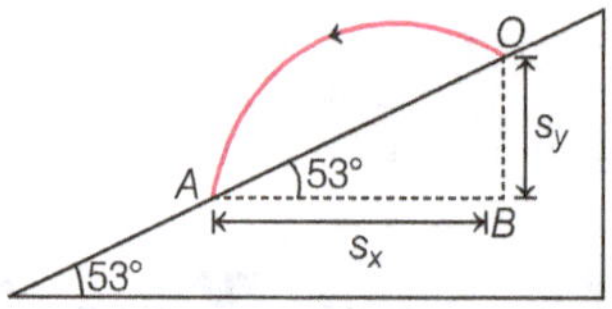

In triangle OAB,

$$\tan 53° = \frac{s_y}{s_x}$$

$$\therefore \quad \frac{4}{3} = \frac{5t^2}{10t}$$

Solving this equation, we get

$$t = \frac{8}{3} \text{ s}$$ **Ans.**

and $OA = s_x \sec 53°$

$$= (10t)\left(\frac{5}{3}\right) = \left(\frac{50}{3}\right)\left(\frac{8}{3}\right)$$

$$= \frac{400}{9} \text{ m}$$ **Ans.**

Example 4

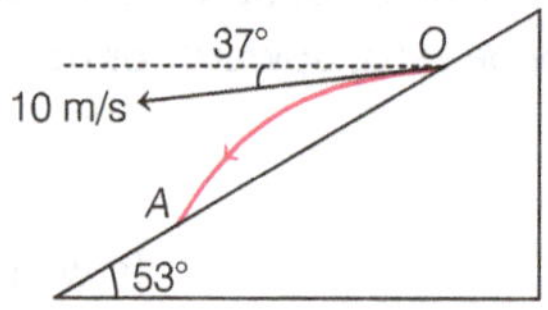

Taking $g = 10 \text{ m/s}^2$, find time of flight t and range along the inclined plane OA.

Sol. **Step 1 :**

$$u_x = 10\cos 37° = 10 \times \frac{4}{5} = 8 \text{ m/s}$$

$$u_y = 10\sin 37° = 10 \times \frac{3}{5} = 6 \text{ m/s}$$

(downwards)

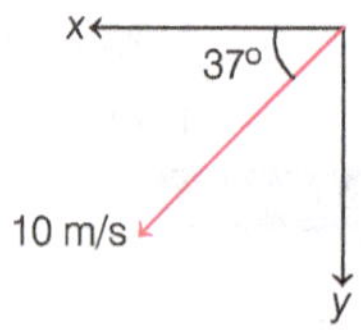

Step 2 : $s_x = u_x t = 8t$

s_y: $s_1 = u_y t = 6t$ (downwards)

$s_2 = 5t^2$ (downwards)

Since s_1 and s_2 both are downwards. So, net

$s_y = s_1 + s_2$

$= (6t + 5t^2)$ (downwards)

Step 3 :

In triangle OAB,

$$\tan 53° = \frac{s_y}{s_x}$$

$$\therefore \quad \frac{4}{3} = \frac{6t + 5t^2}{8t}$$

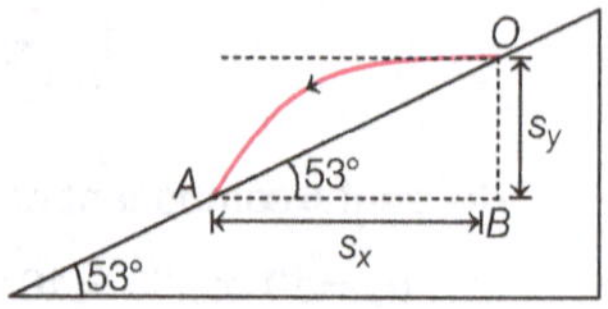

Solving this equation, we get

$$t = \frac{14}{15} \text{ s}$$ **Ans.**

and $OA = s_x \sec 53^\circ = (8t)\left(\frac{5}{3}\right) = \left(\frac{40}{3}\right)t$

$$= \left(\frac{40}{3}\right)\left(\frac{14}{15}\right) = \frac{112}{9}\text{ m}$$ **Ans.**

Note *(i) Basic concepts discussed in above four examples are very useful for solving any type of problem of projectile motion.*

(ii) In different books you will find direct formulae of time of flight T and range R. But in my opinion, the above conceptual approach is better. So, you need not to go in detail of those formulas.

(iii) Formulas are as given below :

(a) Up the plane

$$T = \frac{2u\sin(\alpha - \beta)}{g\cos\beta}$$

$$R = \frac{u^2}{g\cos^2\beta}[\sin(2\alpha - \beta) - \sin\beta]$$

$$R_{max} = \frac{u^2}{g\cos^2\beta}[1 - \sin\beta]$$

at $2\alpha - \beta = \frac{\pi}{2}$

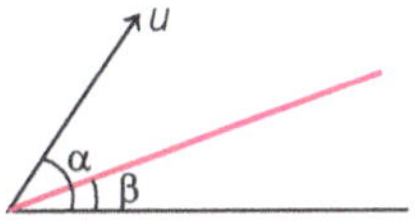

(b) Down the plane

$$T = \frac{2u\sin(\alpha + \beta)}{g\cos\beta}$$

$$R = \frac{u^2}{g\cos^2\beta}[\sin(2\alpha + \beta) + \sin\beta]$$

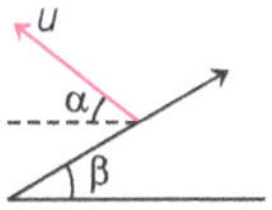

INTRODUCTORY Exercise

1. In the shown figure, find

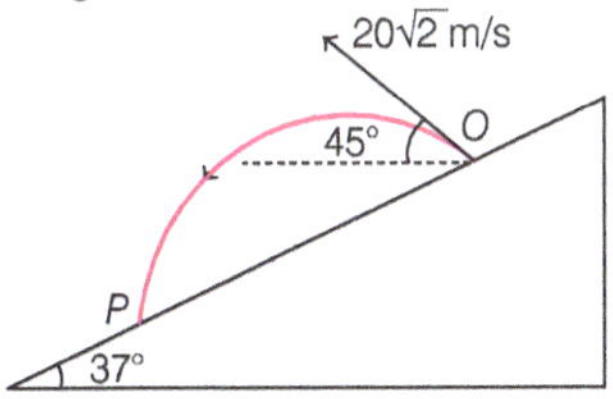

(a) time of flight of the projectile along the inclined plane

(b) range OP

2. In the figure shown, find

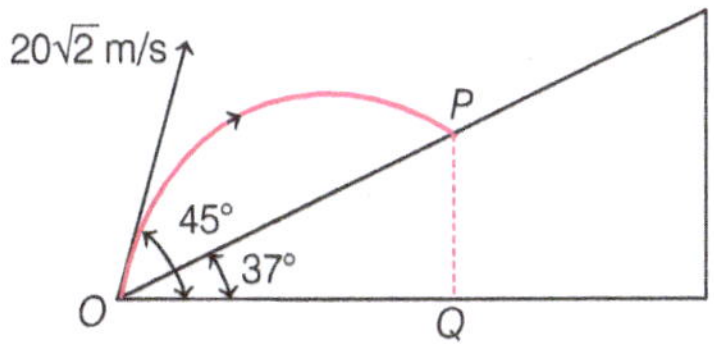

(a) time of flight of the projectile along the inclined plane.

(b) range OP

3. Find time of flight and range of the projectile along the inclined plane as shown in figure. ($g = 10\text{ m/s}^2$)

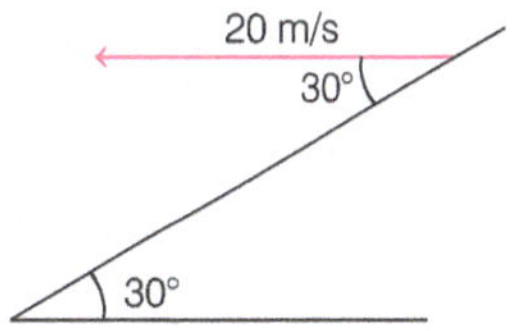

Answers

1. (a) 7s (b) 175 m 2. 1s, 25 m

3. 2.31 s, 53.33 m

See hints on page 329.

Two Projectiles

Let us now discuss the relative motion between two projectiles or the path observed by one projectile of the other. Suppose that two particles are projected from the ground with speeds u_1 and u_2 at angles α_1 and α_2 as shown in Fig. (a) and (b). Acceleration of both the particles is g downwards. So, relative acceleration between them is zero because

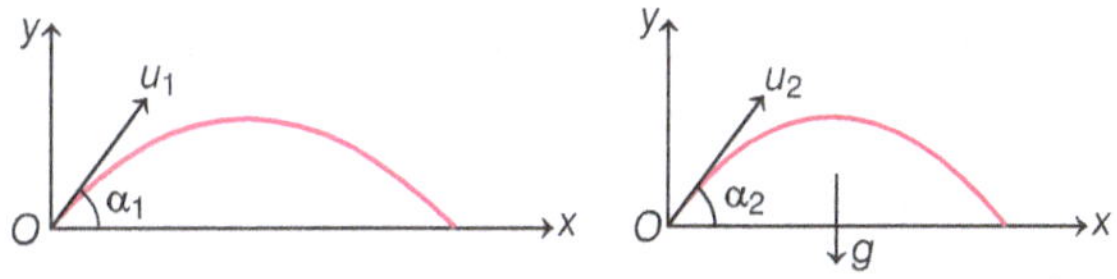

(a) (b)

$$\mathbf{a}_{12} = \mathbf{a}_1 - \mathbf{a}_2 = \mathbf{g} - \mathbf{g} = 0$$

i.e. the relative motion between the two particles is uniform.

Now, $u_{1x} = u_1\cos\alpha_1,\ u_{2x} = u_2\cos\alpha_2$

$u_{1y} = u_1\sin\alpha_1$ and $u_{2y} = u_2\sin\alpha_2$

Therefore, $u_{12x} = u_{1x} - u_{2x} = u_1\cos\alpha_1 - u_2\cos\alpha_2$

and $u_{12y} = u_{1y} - u_{2y} = u_1\sin\alpha_1 - u_2\sin\alpha_2$

u_{12x} and u_{12y} are the x and y-components of relative velocity of 1 with respect to 2.

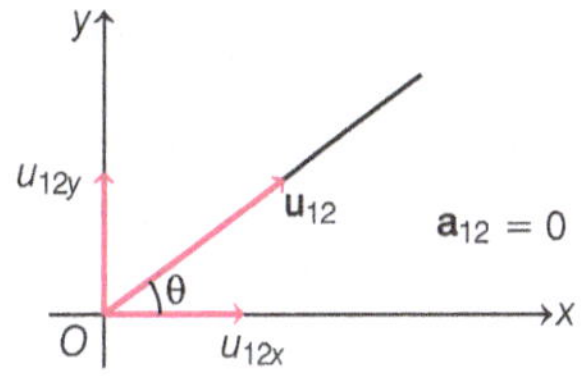

Hence, relative motion of 1 with respect to 2 is a straight line at an angle $\theta = \tan^{-1}\left(\frac{u_{12y}}{u_{12x}}\right)$ from the positive x-axis.

If $u_{12x} = 0$ or $u_1\cos\alpha_1 = u_2\cos\alpha_2$, the relative motion is along y-axis or in vertical direction. Similarly, if $u_{12y} = 0$ or $u_1\sin\alpha_1 = u_2\sin\alpha_2$, the relative motion is along x-axis or in horizontal direction.

Note *Relative acceleration between two projectiles is zero. Relative motion between them is uniform. Therefore, condition of collision of two particles in air is that relative velocity of one with respect to the other should be along line joining them, i.e., if two projectiles A and B collide in mid air, then $\mathbf{v}_{AB}$ should be along AB or $\mathbf{v}_{BA}$ along BA.*

Example 5 Passenger of a train just drops a stone from it. The train was moving with constant velocity. What is path of the stone as observed by (a) the passenger itself (b) a man standing on ground?

Sol. (a) Horizontal component of velocities of passenger and stone are same. Therefore relative velocity in horizontal direction is zero. Hence the relative motion is only in vertical direction.

(b) With respect to the man, stone has both velocity components, horizontal and vertical. Therefore, path of the stone is a projectile.

Example 6 A particle is projected upwards with velocity 20 m/s. Simultaneously another particle is projected with velocity $20\sqrt{2}$ m/s at 45°. ($g = 10\text{ m/s}^2$)

(a) What is acceleration of first particle relative to the second?

(b) What is initial velocity of first particle relative to the other?

(c) What is distance between two particles after 2 s?

Sol. (a) $a_1 = a_2 = g$

$\therefore$ Relative acceleration = 0

(b) $\mathbf{u}_{12} = \mathbf{u}_1 - \mathbf{u}_2$

$= (20\hat{\mathbf{j}}) - (20\hat{\mathbf{i}} + 20\hat{\mathbf{j}})$

$= (-20\hat{\mathbf{i}})$ m/s

or 20 m/s in horizontal direction

(c) $\mathbf{u}_{12} = (-20\hat{\mathbf{i}})$ m/s is constant.

$\therefore$ $|\mathbf{u}_{12}| = 20$ m/s = constant

So, the desired distance is, $d = |\mathbf{u}_{12}|t = 20 \times 2$ m

$= 40$ m **Ans.**

Example 7 A particle A is projected with an initial velocity of 60 m/s at an angle 30° to the horizontal. At the same time a second particle B is projected in opposite direction with initial speed of 50 m/s from a point at a distance of 100 m from A. If the particles collide in air, find (a) the angle of projection α of particle B, (b) time when the collision takes place and (c) the distance of P from A, where collision occurs. ($g = 10\text{ m/s}^2$)

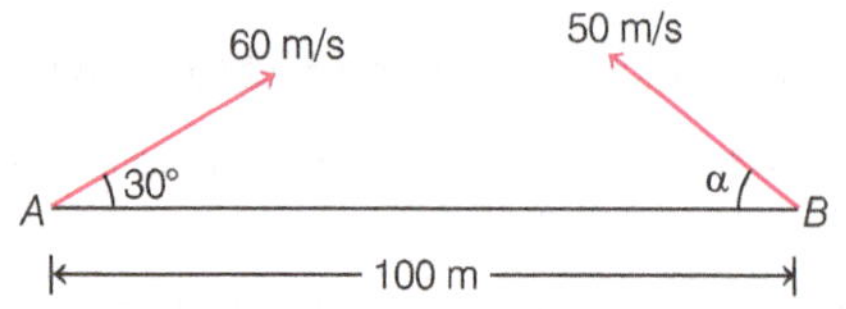

Sol. (a) Taking x and y-directions as shown in figure.

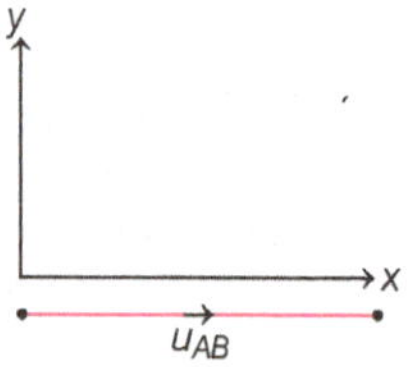

Here, $\mathbf{a}_A = -g\hat{\mathbf{j}}$

$\mathbf{a}_B = -g\hat{\mathbf{j}}$

$u_{Ax} = 60 \cos 30° = 30\sqrt{3}$ m/s

$u_{Ay} = 60 \sin 30° = 30$ m/s

$u_{Bx} = -50 \cos \alpha$

and $u_{By} = 50 \sin \alpha$

Relative acceleration between the two is zero as $\mathbf{a}_A = \mathbf{a}_B$. Hence, the relative motion between the two is uniform. It can be assumed that B is at rest and A is moving with $\mathbf{u}_{AB}$. Hence, the two particles will collide, if $\mathbf{u}_{AB}$ is along AB. This is possible only when

$$u_{Ay} = u_{By}$$

i.e. component of relative velocity along y-axis should be zero.

or $30 = 50 \sin \alpha$

$\therefore$ $\alpha = \sin^{-1}(3/5) = 37°$ **Ans.**

(b) Now, $|\mathbf{u}_{AB}| = |u_{Ax}| + |u_{Bx}|$

$= (30\sqrt{3} + 50 \cos \alpha)$ m/s

$= \left(30\sqrt{3} + 50 \times \frac{4}{5}\right)$ m/s

$= (30\sqrt{3} + 40)$ m/s

Therefore, time of collision is

$$t = \frac{AB}{|\mathbf{u}_{AB}|} = \frac{100}{30\sqrt{3} + 40}$$

or $t = 1.09$ s **Ans.**

(c) Distance of point P from A where collision takes place is

$$d = \sqrt{(u_{Ax}t)^2 + \left(u_{Ay}t - \frac{1}{2}gt^2\right)^2}$$

$$= \sqrt{(30\sqrt{3} \times 1.09)^2 + \left(30 \times 1.09 - \frac{1}{2} \times 10 \times 1.09 \times 1.09\right)^2}$$

or $d = 62.64$ m **Ans.**

Note *For the collision to take place, $t < T_A$ and T_B both. Here T_A and T_B are the times of flight of A and B.*

MODULE 2 Exercise

Projectile Motion Along an Inclined Plane

Single Correct Option

1. A man standing on a hill top projects a stone horizontally with speed v_0 as shown in figure. Taking the coordinates system as shown in the figure. The coordinates of the point, where the stone will hit the hill surface are

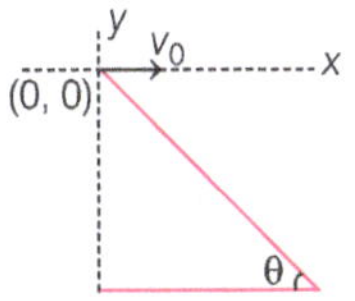

(a) $\left(\frac{2v_0^2 \tan\theta}{g}, -\frac{2v_0^2 \tan^2\theta}{g}\right)$ (b) $\left(\frac{2v_0^2}{g}, -\frac{2v_0^2 \tan^2\theta}{g}\right)$

(c) $\left(\frac{2v_0^2 \tan\theta}{g}, -\frac{2v_0^2}{g}\right)$ (d) $\left(\frac{2v_0^2 \tan^2\theta}{g}, -\frac{2v_0^2 \tan\theta}{g}\right)$

2. A ball is projected from point A with velocity $10\ \text{ms}^{-1}$ perpendicular to the inclined plane as shown in figure. Range of the ball on the inclined plane is

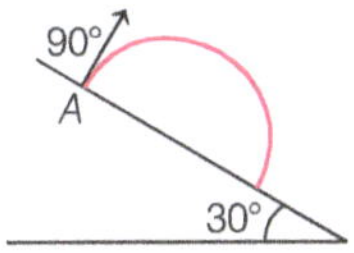

(a) $\frac{40}{3}$ m (b) $\frac{20}{3}$ m (c) $\frac{12}{3}$ m (d) $\frac{60}{3}$ m

3. A plane is inclined at an angle $\alpha = 30°$ with respect to the horizontal. A particle is projected with a speed $u = 2\ \text{ms}^{-1}$, from the base of the plane, making an angle $\theta = 15°$ with respect to the plane as shown in the figure. The distance from the base, at which the particle hits the plane is close to (Take, $g = 10\ \text{ms}^{-2}$) **(2019 Main)**

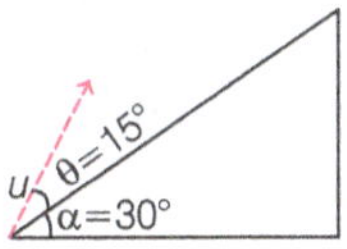

(a) 26 cm (b) 20 cm (c) 18 cm (d) 14 cm

Problems of Two Projectiles

Subjective Questions

4. Two particles A and B are projected from ground towards each other with speeds 10 m/s and $5\sqrt{2}$ m/s at angles 30° and 45° with horizontal from two points separated by a distance of 15 m. Will they collide or not?

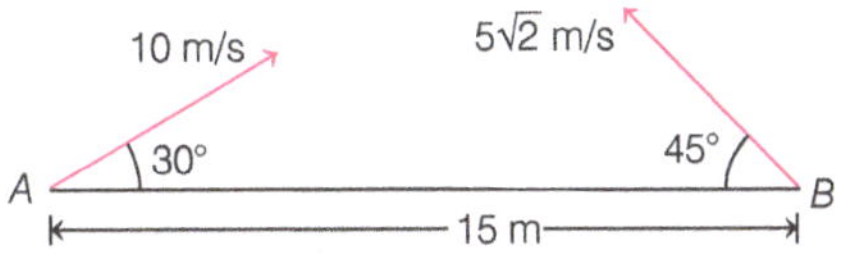

5. Two particles A and B are projected simultaneously in the directions shown in figure with velocities $v_A = 20$ m/s and $v_B = 10$ m/s, respectively. They collide in air after $\frac{1}{2}$ s. Find

(a) the angle θ (b) the distance x.

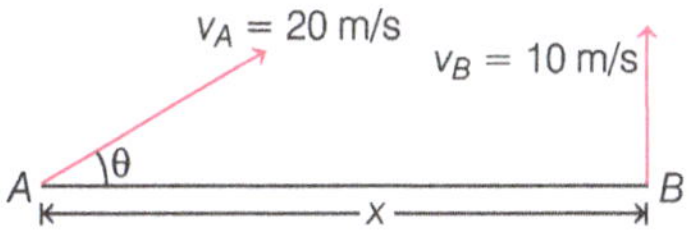

6. Two particles A and B are projected simultaneously in a vertical plane as shown in figure. They collide at time t in air. Write down two necessary equations for collision to take place.

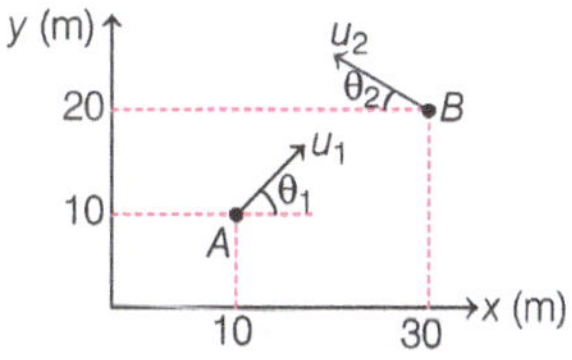

7. Two bodies were thrown simultaneously from the same point; one straight up, and the other at an angle of $\theta = 60°$ to the horizontal. The initial velocity of each body is equal to $v_0 = 25\ \text{ms}^{-1}$. Neglecting the air drag, find the distance between the bodies $t = 1.70$ s later. **(Irodov 1.10)**

Miscellaneous Problems

Single Correct Option

8. Two particles A and B are projected simultaneously from a fixed point of the ground. Particle A is projected on a smooth horizontal surface with speed v, while particle B is projected in air with speed $\frac{2v}{\sqrt{3}}$. If particle B hits the particle A, the angle of projection of B with the horizontal is

(a) 30° (b) 60° (c) 45°

(d) Both (a) and (b)

9. A projectile is given an initial velocity of $(\hat{\mathbf{i}} + 2\hat{\mathbf{j}})$ m/s, where $\hat{\mathbf{i}}$ is along the ground and $\hat{\mathbf{j}}$ is along the vertical. If $g = 10\ \text{m/s}^2$, then the equation of its trajectory is

(a) $y = x - 5x^2$ (b) $y = 2x - 5x^2$ **(2013 Main)**

(c) $4y = 2x - 5x^2$ (d) $4y = 2x - 25x^2$

10. A child stands on the edge of the cliff 10 m above the ground and throws a stone horizontally with an initial speed of $5\ \text{ms}^{-1}$. Neglecting the air resistance, the speed with which the stone hits the ground will be (Given $g = 10\ \text{ms}^{-2}$) **(2023 Main)**

(a) $15\ \text{ms}^{-1}$ (b) $20\ \text{ms}^{-1}$ (c) $30\ \text{ms}^{-1}$ (d) $25\ \text{ms}^{-1}$

Answers

1. (a) **2.** (a) **3.** (b) **4.** (No)

5. (a) 30° (b) $5\sqrt{3}$ m **6.** (i) $(u_1 \cos\theta_1 + u_2 \cos\theta_2)t = 20$

(ii) $(u_1 \sin\theta_1 - u_2 \sin\theta_2)\,t = 10$ **7.** 22 m **8.** (a)

9. (b) **10.** (a)

See hints on pages 329-330.

Unit Exercise (Cumulative)

Level 1

Single Correct Option

1. A particle starts from the origin of coordinates at time $t = 0$ and moves in the x-y plane with a constant acceleration α in the y-direction. Its equation of motion is $y = \beta x^2$. Its velocity component in the x-direction is

(a) variable (b) $\sqrt{\frac{2\alpha}{\beta}}$

(c) $\frac{\alpha}{2\beta}$ (d) $\sqrt{\frac{\alpha}{2\beta}}$

2. A heavy particle is projected with a velocity at an angle with the horizontal into the uniform gravitational field. The slope of the trajectory of the particle varies as

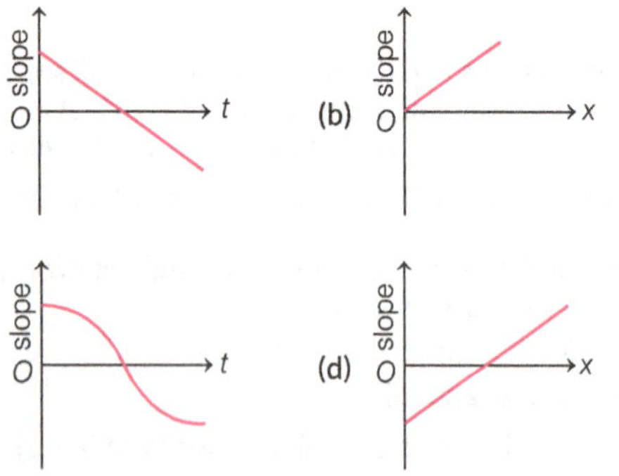

3. A particle moves along the parabolic path $x = y^2 + 2y + 2$ in such a way that y-component of velocity vector remains $5\ ms^{-1}$ during the motion. The magnitude of the acceleration of the particle is

(a) $50\ ms^{-2}$ (b) $100\ ms^{-2}$
(c) $10\sqrt{2}\ ms^{-2}$ (d) $0.1\ ms^{-2}$

4. A particle moves along a parabolic path $y = -9x^2$ in such a way that the x-component of velocity remains constant and has a value $\frac{1}{3}$ m/s. The acceleration of the particle is

(a) $\frac{1}{3}\ m/s^2$ (b) $3\ m/s^2$

(c) $\frac{2}{3}\ m/s^2$ (d) $2\ m/s^2$

5. A ground to ground projectile is at point A at $t = \frac{T}{3}$. It is at point B at $t = \frac{5T}{6}$ and reaches the ground at $t = T$. The difference in heights between points A and B is

(a) $\frac{gT^2}{6}$ (b) $\frac{gT^2}{12}$

(c) $\frac{gT^2}{18}$ (d) None of these

6. A body of mass 1 kg is projected with velocity 50 m/s at an angle of 30° with the horizontal. At the highest point of its path, a force 10 N starts acting on the body for 5 s vertically upward besides gravitational force, what is horizontal range of the body? (Take $g = 10\ m/s^2$)

(a) $125\sqrt{3}$ m (b) $200\sqrt{3}$ m
(c) 500 m (d) $250\sqrt{3}$ m

7. A particle is projected on smooth inclined plane in a direction perpendicular to line of greatest slope with speed $8\ ms^{-1}$. Its speed at $t = 1$ s is

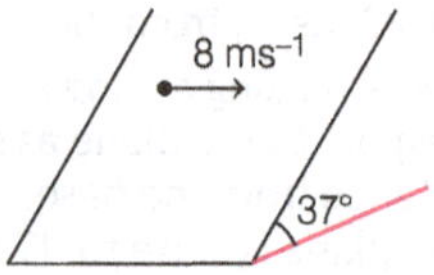

(a) $5\ ms^{-1}$ (b) $8\ ms^{-1}$ (c) $6\ ms^{-1}$ (d) $10\ ms^{-1}$

8. A shell fired from the base of a mountain just clears it. If α is the angle of projection, then the angular elevation of the summit β is

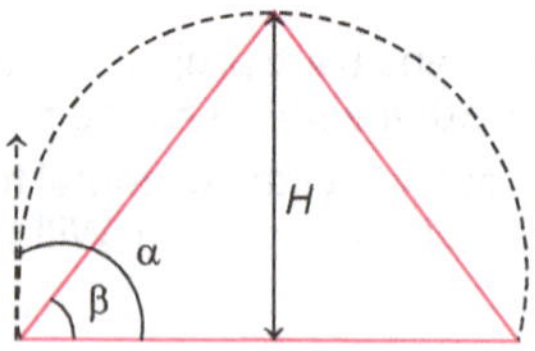

(a) $\frac{\alpha}{2}$ (b) $\tan^{-1}\left(\frac{1}{2}\right)$

(c) $\tan^{-1}\left(\frac{\tan\alpha}{2}\right)$ (d) $\tan^{-1}(2\tan\alpha)$

9. A ball is thrown from the ground to clear a wall 3 m high at a distance of 6 m and falls 18 m away from the wall, the angle of projection of ball is

(a) $\tan^{-1}\left(\frac{3}{2}\right)$ (b) $\tan^{-1}\left(\frac{2}{3}\right)$

(c) $\tan^{-1}\left(\frac{1}{2}\right)$ (d) $\tan^{-1}\left(\frac{3}{4}\right)$

10. A projectile is launched at an angle 'α' with the horizontal with a velocity $20\,\text{ms}^{-1}$. After 10 s, its inclination with horizontal is 'β'. The value of $\tan\beta$ will be ($g = 10\,\text{ms}^{-2}$) **(2022 Main)**

(a) $\tan\alpha + 5\sec\alpha$ (b) $\tan\alpha - 5\sec\alpha$
(c) $2\tan\alpha - 5\sec\alpha$ (d) $2\tan\alpha + 5\sec\alpha$

11. A projectile is projected with velocity of 25 m/s at an angle θ with the horizontal. After t seconds its inclination with horizontal becomes zero. If R represents horizontal range of the projectile, the value of θ will be [Use $g = 10\,\text{m/s}^2$] **(2022 Main)**

(a) $\frac{1}{2}\sin^{-1}\left(\frac{5t^2}{4R}\right)$ (b) $\frac{1}{2}\sin^{-1}\left(\frac{4R}{5t^2}\right)$

(c) $\tan^{-1}\left(\frac{4t^2}{5R}\right)$ (d) $\cot^{-1}\left(\frac{R}{20t^2}\right)$

12. A projectile is projected with speed u at an angle of 60° with horizontal from the foot of an inclined plane. If the projectile hits the inclined plane horizontally, the range on inclined plane will be

(a) $\frac{u^2\sqrt{21}}{2g}$ (b) $\frac{3u^2}{4g}$

(c) $\frac{u^2}{2g}$ (d) $\frac{\sqrt{21}\,u^2}{8g}$

13. A particle is dropped from a height h. Another particle which is initially at a horizontal distance d from the first is simultaneously projected with a horizontal velocity u and the two particles just collide on the ground. Then

(a) $d^2 = \frac{u^2h}{2g}$ (b) $d^2 = \frac{2u^2h}{g}$

(c) $d = h$ (d) $gd^2 = u^2h$

14. A particle is projected at an angle 60° with speed $10\sqrt{3}$ m/s from the point A as shown in the figure. At the same time, the wedge is made to move with speed $10\sqrt{3}$ m/s towards right as shown in the figure. Then, the time after which particle will strike with wedge is

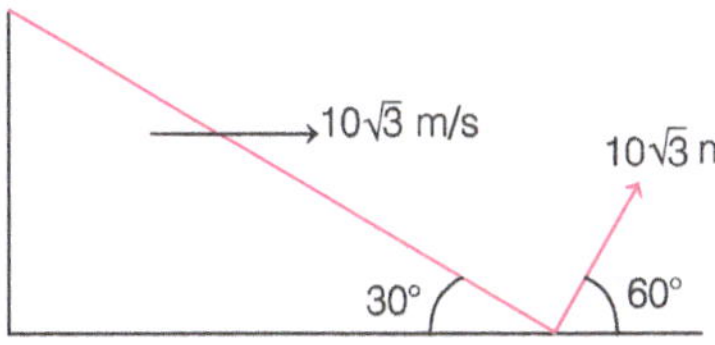

(a) 2 s (b) $2\sqrt{3}$ s

(c) $\frac{4}{\sqrt{3}}$ s (d) None of these

Multiple Correct Options

15. A ball is dropped from a height of 49 m. The wind is blowing horizontally. Due to wind a constant horizontal acceleration is provided to the ball. Choose the correct statement (s). (Take, $g = 9.8\,\text{ms}^{-2}$)

(a) Path of the ball is a straight line
(b) Path of the ball is a curved one
(c) The time taken by the ball to reach the ground is 3.16 s
(d) Actual distance travelled by the ball is more than 49 m

16. In the projectile motion shown is figure, given $t_{AB} = 2$ s, then (Take, $g = 10\,\text{ms}^{-2}$)

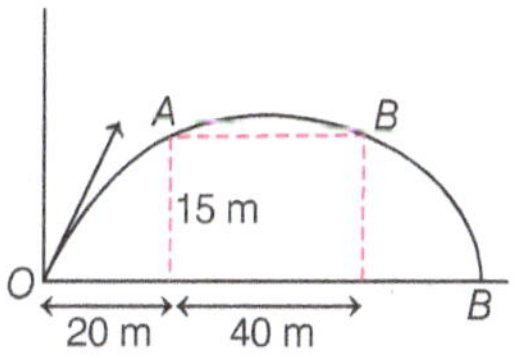

(a) particle is at point B at 3 s
(b) maximum height of projectile is 20 m
(c) initial vertical component of velocity is $20\,\text{ms}^{-1}$
(d) horizontal component of velocity is $20\,\text{ms}^{-1}$

Subjective Questions

17. A cannon and a target are 5.10 km apart and located at the same level. How soon will the shell launched with the initial velocity 240 m/s reach the target in the absence of air drag? **(Irodov 1.32)**

18. Two particles are simultaneously projected in the same vertical plane from the same point with velocities u and v at angles α and β with horizontal. Find the time that elapses when their velocities are parallel.

19. A plank fitted with a gun is moving on a horizontal surface with speed of 4 m/s along the positive x-axis. The z-axis is in vertically upward direction. The mass of the plank including the mass of the gun is 50 kg. When the plank reaches the origin, a shell of mass 10 kg is fired at an angle of 60° with the positive x-axis with a speed of $v = 20$ m/s with respect to the velocity of plank (and gun) just after shell is fired in x-z plane. Find the position vector of the shell at $t = 2$ s after firing it. (Take, $g = 9.8\,\text{m/s}^2$)

20. A body falling freely from a given height H hits an inclined plane in its path at a height h. As a result of this impact the direction of the velocity of the body becomes horizontal. For what value of (h/H), the body will take maximum time to reach the ground? **(1986 JEE)**

21. At time $t = 0$, a small ball is projected from point A with a velocity of 60 m/s at 60° angle with horizontal. Neglect atmospheric resistance and determine the

two times t_1 and t_2 when the velocity of the ball makes an angle of 45° with the horizontal x-axis.

22. A particle is projected along an inclined plane as shown in figure. What is the speed of the particle when it collides at point A? (Take, $g = 10\ \text{m/s}^2$)

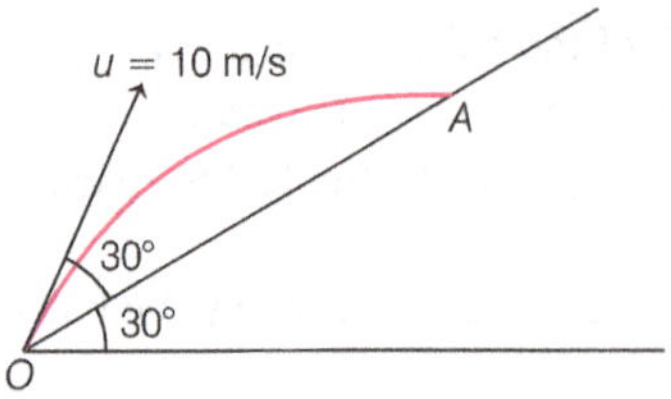

23. In the above problem, what is the component of its velocity perpendicular to the plane when it strikes at A ?

24. A particle is projected from the bottom of an inclined plane of inclination 30° with velocity of 40 m/s at an angle of 60° with horizontal. Find the speed of the particle when its velocity vector is parallel to the plane. (Take, $g = 10\ \text{m/s}^2$)

25. Two particles A and B are projected simultaneously from two towers of heights 10 m and 20 m respectively. Particle A is projected with an initial speed of $10\sqrt{2}$ m/s at an angle of 45° with horizontal, while particle B is projected horizontally with speed 10 m/s. If they collide in air, what is the distance d between the towers?

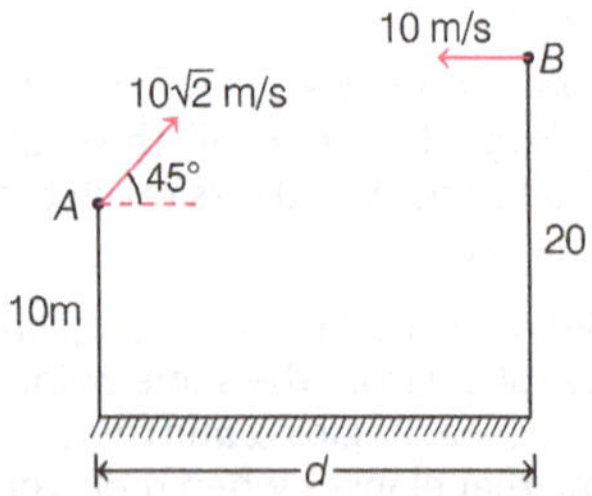

26. Two particles move in a uniform gravitational field with an acceleration g. At the initial moment, the particles were located at one point and moved with velocities $v_1 = 3.0\ \text{ms}^{-1}$ and $v_2 = 4.0\ \text{ms}^{-1}$ horizontally in opposite directions. Find the distance between the particles at the moment when their velocity vectors become mutually perpendicular. **(Irodov 1.11)**

27. A particle is dropped from a point P at time $t = 0$. At the same time another particle is thrown from point O as shown in the figure and it collides with the particle P. Acceleration due to gravity is along the negative y-axis. If the two particles collide 2 s after they start, find the initial velocity v_0 of the particle which was projected from O. Point O is not necessarily on ground.

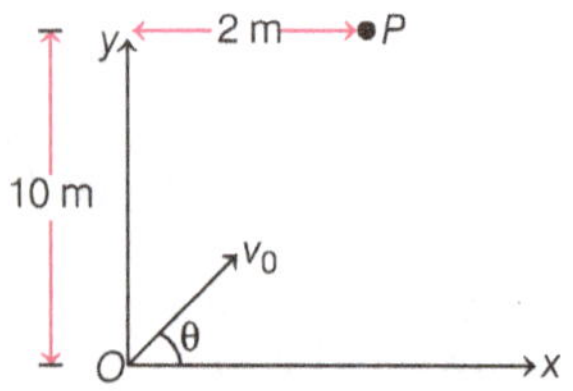

28. Determine the horizontal velocity v_0 with which a stone must be projected horizontally from a point P, so that it may hit the inclined plane perpendicularly. The inclination of the plane with the horizontal is θ and point P is at a height h above the foot of the incline, as shown in the figure.

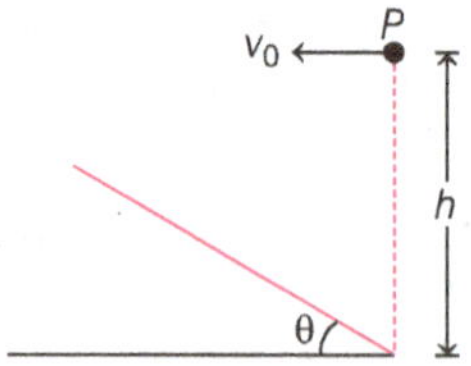

Integer Type Questions

29. If the initial velocity component in horizontal direction of a projectile is unit vector $\hat{i}$ and the equation of trajectory is $y = 5x(1 - x)$. The y-component vector of the initial velocity is .... $\hat{j}$. (Take $g = 10\ \text{m/s}^2$) **(2022 Main)**

30. A train is moving along a straight line with a constant acceleration a. A boy standing in the train throws a ball forward with a speed of 10 m/s, at an angle of 60° to the horizontal. The boy has to move forward by 1.15 m inside the train to catch the ball back at the initial height. The acceleration of the train, in m/s^2, is. .... **(2011 JEE)**

31. A projectile fired at 30° to the ground is observed to be at same height at time 3s and 5s after projection, a during its flight. The speed of projection of the projectile is .... ms^{-1}. (Given $g = 10\ \text{ms}^{-2}$) **(2023 Main)**

32. Two particles A and B are projected from point O with equal speeds. They both hit the point P of an inclined plane of inclination 15°. Particle A is projected at an angle 30° with inclined plane. If the ratio of time of flight of particles A and B is $1 : \sqrt{n}$, find the value of n.

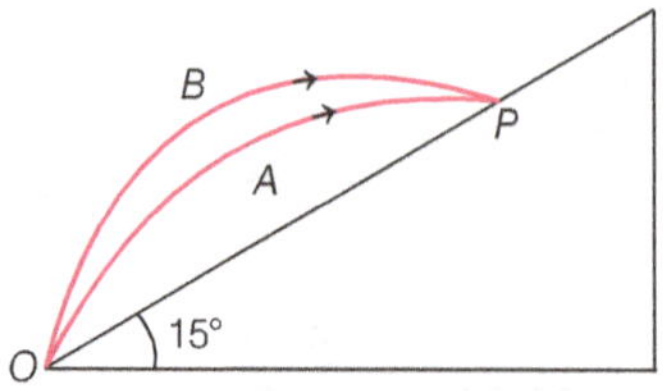

33. The length of inclined plane is 15 m. A stone is projected with speed v_0 from point A to cross through point B. If minimum value of v_0 is $3x$ ms^{-1}, find the value of x.

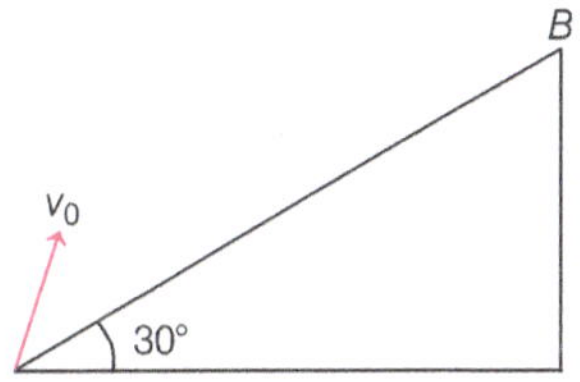

Level 2

Single Correct Option

34. A body of mass 10 kg is projected at an angle of 45° with the horizontal. The trajectory of the body is observed to pass through a point (20, 10). If T is the time of flight, then its momentum vector, at time $t = \frac{T}{\sqrt{2}}$, is : (Take $g = 10\ m/s^2$) **(2022 Main)**

(a) $100\hat{i} + (100\sqrt{2} - 200)\hat{j}$

(b) $100\sqrt{2}\hat{i} + (100 - 200\sqrt{2})\hat{j}$

(c) $100\hat{i} + (100 - 200\sqrt{2})\hat{j}$

(d) $100\sqrt{2}\hat{i} + (100\sqrt{2} - 200)\hat{j}$

35. A stone is projected from a point on the ground so as to hit a bird on the top of a vertical pole of height h and then attain a maximum height $2h$ above the ground. If at the instant of projection the bird flies away horizontally with a uniform speed and if the stone hits the bird while descending, then the ratio of the speed of the bird to the horizontal speed of the stone is

(a) $\frac{\sqrt{2}}{\sqrt{2}+1}$ (b) $\frac{\sqrt{2}}{\sqrt{2}-1}$

(c) $\frac{1}{\sqrt{2}} + \frac{1}{2}$ (d) None of these

Subjective Questions

36. An object A is kept fixed at the point $x = 3$ m and $y = 1.25$ m on a plank P raised above the ground. At time $t = 0$, the plank starts moving along the $+x$-direction with an acceleration $1.5\ m/s^2$. At the same instant, a stone is projected from the origin with a velocity $\mathbf{u}$ as shown. A stationary person on the ground observes the stone hitting the object during its downward motion at an angle of 45° to the horizontal. All the motions are in x-y plane. Find $\mathbf{u}$ and the time after which the stone hits the object. (Take $g = 10\ m/s^2$) **(2000 Adv.)**

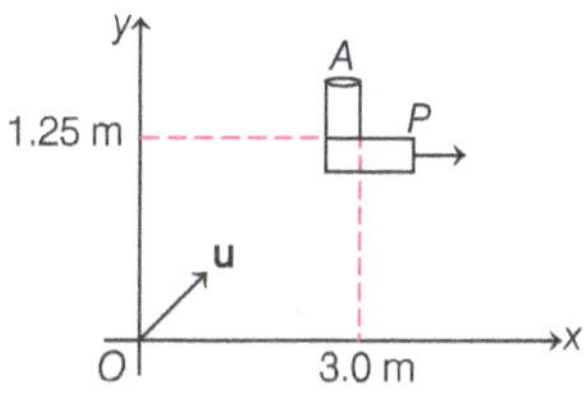

37. A balloon is ascending at the rate $v = 12$ km/h and is being carried horizontally by the wind at $v_w = 20$ km/h. If a bag is dropped from the balloon at the instant $h = 50$ m, determine the time needed for it to strike the ground. Assume that the bag was released from the balloon with the same velocity as the balloon. Also, find the speed with which the bag strikes the ground.

38. An elevator is going up with an upward acceleration of $1\ m/s^2$. At the instant, when its velocity is 2 m/s, a stone is projected upward from its floor with a speed of 2 m/s relative to the elevator, at an elevation of 30°.

(a) Calculate the time taken by the stone to return to the floor.

(b) Sketch the path of the projectile as observed by an observer outside the elevator.

(c) If the elevator was moving with a downward acceleration equal to g, how would the motion be altered?

39. A projectile of mass m is fired into a liquid at an angle θ_0 with an initial velocity v_0 as shown. If the liquid develops a frictional or drag resistance on the projectile which is proportional to its velocity, i.e. $F = -kv$ where k is a positive constant, determine the x and y components of its velocity at any instant. Also find the maximum distance x_{max} that it travels?

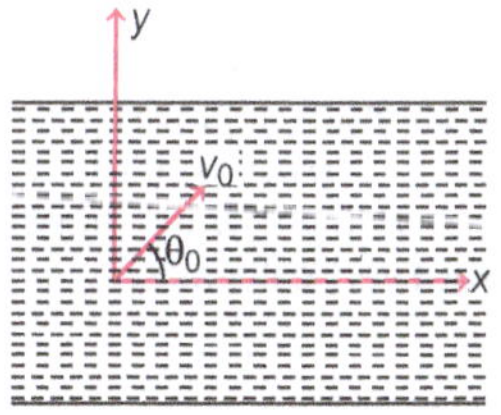

40. A projectile takes off with an initial velocity of 10 m/s at an angle of elevation of 45°. It is just able to clear two hurdles of height 2 m each, separated from each other by a distance d. Calculate d. At what distance from the point of projection is the first hurdle placed? (Take, $g = 10\ m/s^2$)

41. A particle is released from a certain height $H = 400$ m. Due to the wind, the particle gathers the horizontal velocity component $v_x = ay$, where $a = \sqrt{5}\ s^{-1}$ and y is the vertical displacement of the particle from the point of release, then find

(a) the horizontal drift of the particle when it strikes the ground,

(b) the speed with which particle strikes the ground. (Take $g = 10\,\text{m/s}^2$)

42. A particle is projected with velocity $2\sqrt{gh}$, so that it just clears two walls of equal height h which are at a distance of $2h$ from each other. Show that the time of passing between the walls is $2\sqrt{\frac{h}{g}}$.

43. A projectile aimed at a mark, which is in the horizontal plane through the point of projection, falls a cm short of it when the elevation is α and goes b cm far when the elevation is β. Show that, if the speed of projection is same in all the cases the proper elevation is $\frac{1}{2}\sin^{-1}\left[\frac{b\sin 2\alpha + a\sin 2\beta}{a+b}\right]$.

44. A particle is projected from an inclined plane OP_1 from A with velocity $v_1 = 8\ \text{ms}^{-1}$ at an angle 60° with horizontal. Another particle is projected at the same instant from B with velocity $v_2 = 16\ \text{ms}^{-1}$ and perpendicular to the plane OP_2 as shown in figure. After time $10\sqrt{3}$ s, there separation was minimum and found to be 70 m. Then, find distance AB.

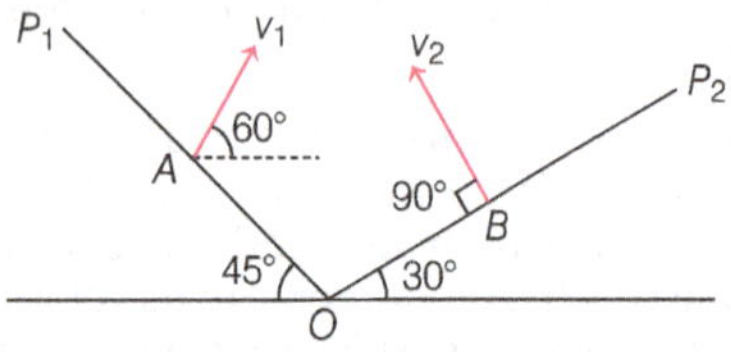

45. A train is moving with a constant speed of 10 m/s in a circle of radius $\frac{16}{\pi}$ m. The plane of the circle lies in horizontal x-y plane. At time $t = 0$, train is at point P and moving in counter-clockwise direction. At this instant, a stone is thrown from the train with velocity 10 m/s relative to train towards negative x-axis at an angle of 37° with vertical z-axis.

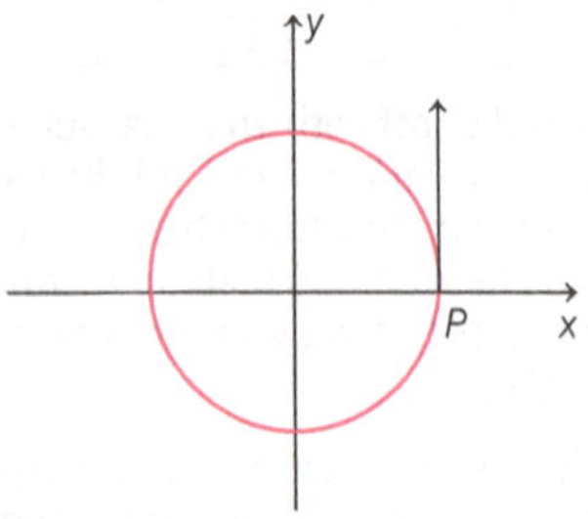

Find

(a) the velocity of particle relative to ground at the highest point of its trajectory.

(b) the coordinates of points on the ground, where it finally falls and that of the highest point of its trajectory. $\left(\text{Take, } g = 10\,\text{m/s}^2, \sin 37° = \frac{3}{5}\right)$

46. A large, heavy box is sliding without friction down a smooth plane of inclination θ. From a point P on the bottom of the box, a particle is projected inside the box. The initial speed of the particle with respect to the box is u and the direction of projection makes an angle α with the bottom as shown in the figure.

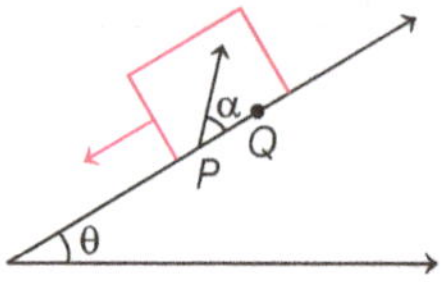

(a) Find the distance along the bottom of the box between the point of projection P and the point Q where the particle lands. (Assume that the particle does not hit any other surface of the box. Neglect air resistance.)

(b) If the horizontal displacement of the particle as seen by an observer on the ground is zero, find the speed of the box with respect to the ground at the instant when the particle was projected. **(1988 Adv.)**

Integer Type Questions

47. A stone targeted at point P which is in the same horizontal plane through the point of projections, falls x_1 distance before the point P when angle of projection is 15°. It goes x_2 distance beyond when the angle of projection is 45°. The proper angle of projection is 30° to hit the point P. If the velocity of projection remains same, find the value of $\frac{(\sqrt{3}-1)x_1}{x_2}$.

48. A particle is projected from the base of a cone with a speed of u at an angle of projection α. The particle grazes the vertex of the cone and strikes again at the base. If θ is half angle of cone and h is its height. Find the value of $\left(\frac{u^2}{gh} - \frac{1}{2}\tan^2\theta\right)$.

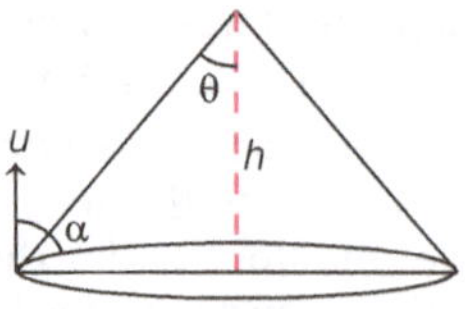

49. Three balls are thrown in the same vertical plane with different speeds and at different angles. The area of triangle formed by the balls at instant t is proportional to t^n. Find the value of n.

50. A projectile is projected from ground with least velocity to cross a wall 3.6 m high and 4.8 m away from the point of projection. The range of projectile is 8.4 m. If angle of projection is α and $\tan\alpha = \frac{n}{4}$, find the value of n.

51. A particle starts from point A and slides along smooth inclined with an acceleration of 6 ms^{-2} and finally strikes at point P on ground. If $x_0 = \frac{36}{n}$ m, find the value of n.

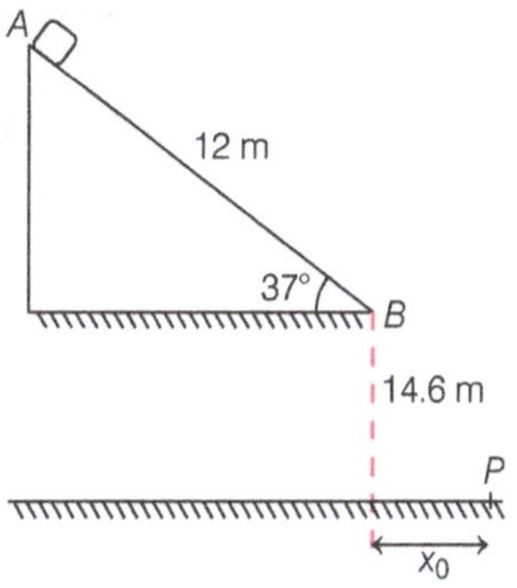

Answers

Level 1

1. (d) **2.** (a) **3.** (a) **4.** (d) **5.** (d) **6.** (d) **7.** (d) **8.** (c) **9.** (b) **10.** (b)
11. (d) **12.** (d) **13.** (b) **14.** (a) **15.** (a,c,d) **16.** (a,b,c,d) **17.** (0.41 min or 0.71 min)
18. $t = \frac{uv\sin(\alpha-\beta)}{g(v\cos\beta - u\cos\alpha)}$ **19.** $(24\hat{i} + 15\hat{k})$ m **20.** $\frac{1}{2}$ **21.** $t_1 = 2.19$ s, $t_2 = 8.20$ s
22. $\frac{10}{\sqrt{3}}$ m/s **23.** 5 m/s **24.** $\frac{40}{\sqrt{3}}$ m/s **25.** 20 m **26.** 2.5 m **27.** $\sqrt{26}$ ms^{-1} at angle $\theta = \tan^{-1}(5)$ with x-axis
28. $v_0 = \sqrt{\frac{2gh}{2+\cot^2\theta}}$ **29.** (5) **30.** (c) **31.** (80) **32.** (2) **33.** (5)

Level 2

34. (d) **35.** (c) **36.** ($\mathbf{u} = (3.75\hat{i} + 6.25\hat{j})$ m/s, $t = 1$ s) **37.** (3.55 s, 32.7 m/s)
38. (a) 0.18 s (c) a straight line with respect to elevator and projectile with respect to ground
39. $v_x = v_0\cos\theta_0 e^{-kt/m}$, $v_y = \frac{m}{k}\left[\left(\frac{k}{m}v_0\sin\theta_0 + g\right)e^{-kt/m} - g\right]$, $x_m = mv_0\cos\theta_0/k$ **40.** (4.47 m, 2.75 m)
41. (a) 2.67 km (b) 0.9 km/s **44.** (250 m) **45.** (a) $(-6\hat{i} + 10\hat{j})$ ms^{-1} (b) (–4.5 m, 16 m, 0), (0.3 m, 8.0 m, 3.2 m)
46. (a) $\frac{u^2\sin 2\alpha}{g\cos\theta}$ (b) $\frac{u\cos(\alpha+\theta)}{\cos\theta}$ **47.** (2) **48.** (2) **49.** (2) **50.** (7) **51.** (5)

See hints on pages 331-341.

Hints & SOLUTIONS

Module 1.

Introductory Exercise 1

1. $\mathbf{v}_1 \cdot \mathbf{v}_2 = 0$

$\Rightarrow (\mathbf{u}_1 + \mathbf{a}_1 t)\cdot(\mathbf{u}_2 + \mathbf{a}_2 t) = 0$

$\Rightarrow (10\hat{\mathbf{i}} - 10t\hat{\mathbf{j}})\cdot(-20\hat{\mathbf{i}} - 10t\hat{\mathbf{j}}) = 0$

$\Rightarrow \quad -200 + 100t^2 = 0$

$\Rightarrow \quad t = \sqrt{2}\ \text{s}$

2. It is two dimensional motion.

3. The uniform acceleration is $\mathbf{g}$.

4. $\mathbf{u} = (40\hat{\mathbf{i}} + 30\hat{\mathbf{j}})\ \text{m/s}, \mathbf{a} = (-10\hat{\mathbf{j}})\ \text{m/s}^2, t = 2\ \text{s}$

Now, $\mathbf{v} = \mathbf{u} + \mathbf{a}t$

and $\mathbf{s} = \mathbf{u}t + \frac{1}{2}\mathbf{a}t^2$

5. $u_x = u_y = 20\ \text{m/s}, a_y = -10\ \text{m/s}^2$

$$s_y = u_y t + \frac{1}{2}a_y t^2$$

$$\Rightarrow \quad -25 = 20t - \frac{1}{2}\times 10 \times t^2$$

Solving this equation, we get the positive value of,

$$t = 5\ \text{s}$$

Now apply, $\mathbf{v} = \mathbf{u} + \mathbf{a}t$ and $s_x = u_x t$

Introductory Exercise 2

1. $\mathbf{u} = 40\hat{\mathbf{i}} + 40\hat{\mathbf{j}}$

$$\mathbf{a} = -10\hat{\mathbf{j}}$$

$$t = 2\ \text{s}$$

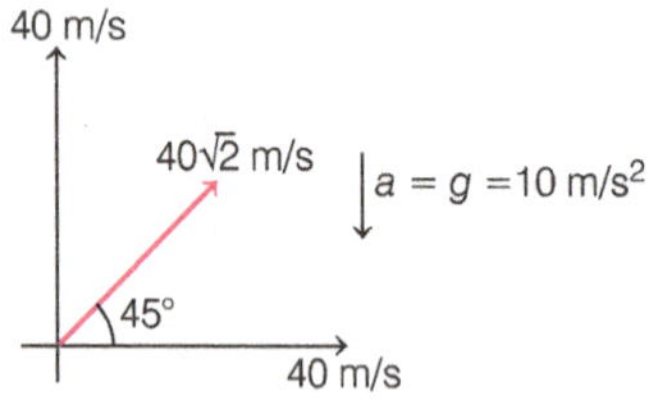

(a) Apply $\mathbf{v} = \mathbf{u} + \mathbf{a}t$

as $\quad \mathbf{a} = \text{constant}$

(b) Apply $\mathbf{s} = \mathbf{u}t + \frac{1}{2}\mathbf{a}t^2$

3. Average velocity

$$= \frac{s}{t} = \frac{R}{T} = \frac{(u_x T)}{T} = u_x$$

$$= u\cos\alpha$$

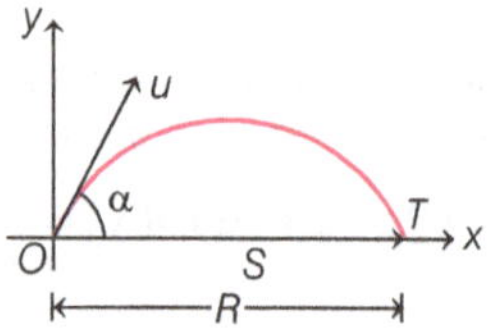

4. $\Delta\mathbf{v} = \mathbf{v}_f - \mathbf{v}_i$

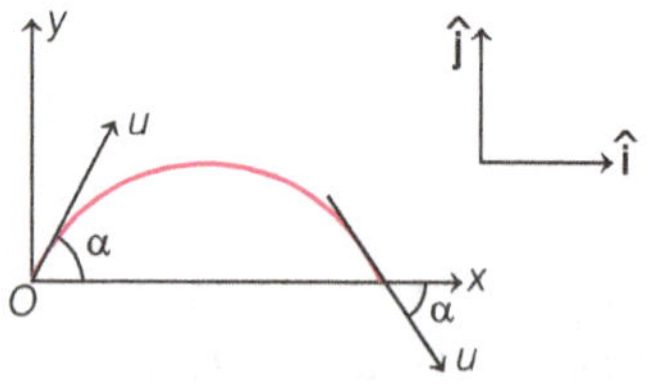

$= (u\cos\alpha\,\hat{\mathbf{i}} - u\sin\alpha\,\hat{\mathbf{j}}) - (u\cos\alpha\,\hat{\mathbf{i}} + u\sin\alpha\,\hat{\mathbf{j}})$

$= (-2u\sin\alpha)\,\hat{\mathbf{j}}$

Therefore, change in velocity is $2u\sin\alpha$ in downward direction.

5. (a) $R = \frac{u^2 \sin 2\theta}{g}, \quad H = \frac{u^2\sin^2\theta}{2g}$

and $\quad T = \frac{2u\sin\theta}{g}$

Here, $\quad u = 20\sqrt{2}\ \text{m/s}$ and $\theta = 45°$

(b) $\mathbf{v} = \mathbf{u} + \mathbf{a}t$

where, $\quad \mathbf{u} = (20\hat{\mathbf{i}} + 20\hat{\mathbf{j}})\ \text{m/s}$

and $\quad \mathbf{a} = (-10\hat{\mathbf{j}})\ \text{m/s}^2$

(c) Horizontal component remains unchanged $(= 20\hat{\mathbf{i}})$ and vertical component is reversed in direction $(= -20\hat{\mathbf{j}})$

$$\mathbf{v} = (20\hat{\mathbf{i}} - 20\hat{\mathbf{j}})\ \text{m/s}$$ **Ans.**

6. (a) Since, acceleration is constant. Therefore,

$$\mathbf{a}_{av} = \mathbf{a} = \mathbf{g} = (-10\hat{\mathbf{j}}) = \frac{\Delta\mathbf{v}}{\Delta t}$$

$\therefore \quad \Delta\mathbf{v} = (-10\hat{\mathbf{j}})(\Delta t) = (-10\hat{\mathbf{j}})(3)$

$= (-10\hat{\mathbf{j}})(3) = (-30\hat{\mathbf{j}})\ \text{m/s}$

$\therefore$ Change in velocity is 30 m/s, vertically downwards. **Ans.**

(b) $\mathbf{v}_{av} = \frac{\mathbf{s}}{t} = \frac{\mathbf{u}t + \frac{1}{2}\mathbf{a}t^2}{t} = \mathbf{u} + \frac{1}{2}\mathbf{a}t$

$= (20\hat{\mathbf{i}} + 20\hat{\mathbf{j}}) + \frac{1}{2}(-10\hat{\mathbf{j}})(3)$

$= (20\hat{\mathbf{i}} + 5\hat{\mathbf{j}})\ \text{m/s}$

$\therefore \quad |\mathbf{v}_{av}| = \sqrt{(20)^2 + (5)^2}$

$= 20.62 \text{ m/s}$ **Ans.**

7. $T = \dfrac{2u\sin\theta}{g} = \dfrac{2\times 20 \times \frac{1}{\sqrt{2}}}{10} = 2\sqrt{2}\text{ s}$

$R = \dfrac{u^2 \sin 2\theta}{g} = \dfrac{(20)^2 \sin 90^\circ}{10}$

$= 40 \text{ m}$

Now, the remaining horizontal distance is $(50-40)\text{ m} = 10\text{ m}$. Let v is the speed of player, then

$vT = 10$

or $\quad v = \dfrac{10}{T} = \dfrac{10}{2\sqrt{2}} = \dfrac{5}{\sqrt{2}}\text{ m/s}$ **Ans.**

8. $R = \dfrac{u^2 \sin 60^\circ}{g} = 3\text{ km}$

$\Rightarrow \quad \dfrac{u^2}{g} = R_{max} = \dfrac{3}{\sin 60^\circ}$

$= 2\sqrt{3}\text{ km}$

Since, $\quad R_{max} < 5\text{ km}$

So, it can't hit the target at 5 km.

Exercise

1. From conservation of mechanical energy, at highest point, potential energy is maximum. Therefore, kinetic energy and hence speed should be minimum.

2. In the cases shown below path is straight line, even if **a** is constant

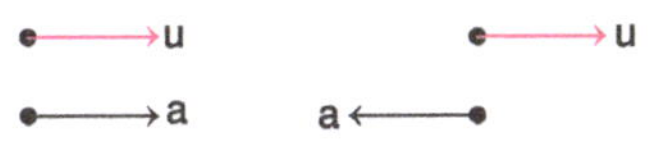

3. $T = \dfrac{2u\sin\theta}{g} \quad \Rightarrow \quad u\sin\theta = \dfrac{gT}{2}$

$H = \dfrac{u^2\sin^2\theta}{2g} = \dfrac{(gT/2)^2}{2g}$

or $\quad H \propto T^2$

4. $H = \dfrac{u^2\sin^2\theta}{2g}$

or $\quad u\sin\theta \propto \sqrt{H}$

$\dfrac{(u\sin\theta)_1}{(u\sin\theta)_2} = \sqrt{\dfrac{4H}{H}} = 2$

$T = \dfrac{2u\sin\theta}{g} \quad \Rightarrow \quad T \propto u\sin\theta$

$\dfrac{T_1}{T_2} = \dfrac{(u\sin\theta)_1}{(u\sin\theta)_2} = \dfrac{2}{1}$

$R = u_x T$

$\therefore \quad \dfrac{R_1}{R_2} = \dfrac{(u_x)_1}{(u_x)_2}\cdot\dfrac{T_1}{T_2} = \left(\dfrac{1}{2}\right)\left(\dfrac{2}{1}\right) = 1$

$\therefore \quad R_1 = R_2$

5. $h_1 = \dfrac{u^2}{2g}, h_2 = \dfrac{u^2\sin^2 30^\circ}{2g} = \dfrac{u^2}{8g}$

6. Range, $R = \dfrac{u^2}{g}\sin 2\theta$

R is maximum for $2\theta = 90^\circ$.

7. For same range, $\theta_1 + \theta_2 = 90^\circ$

$\because \quad h_1 h_2 = \dfrac{u^2\sin^2\theta_1}{2g} \times \dfrac{u^2\sin^2\theta_2}{2g}$

$\therefore \quad h_1 h_2 = \dfrac{u^2\sin^2\theta_1}{2g}\cdot\dfrac{u^2\cos^2\theta_1}{2g} \qquad (\because \theta_2 = 90^\circ - \theta_1)$

$= \left[\dfrac{u^2\sin\theta_1\cos\theta_1}{2g}\right]^2$

$= \left[\dfrac{u^2\sin\theta_1\cos\theta_1}{2g}\times\dfrac{2}{2}\right]^2 = \dfrac{R^2}{16}$

$\Rightarrow \quad R = 4\sqrt{h_1 h_2}$

8. $H_\theta = \dfrac{u^2\sin^2\theta}{2g}$

$H_{90-\theta} = \dfrac{u^2\sin^2(90-\theta)}{2g} = \dfrac{u^2\cos^2\theta}{2g}$

$\therefore \quad \dfrac{H_\theta}{H_{90-\theta}} = \dfrac{\sin^2\theta}{\cos^2\theta}$

9. $R_\theta = \dfrac{R_{45^\circ}}{2}$

$\therefore \quad \dfrac{u^2\sin 2\theta}{g} = \dfrac{1}{2}\left(\dfrac{u^2}{g}\right)$

or $\quad \sin 2\theta = \dfrac{1}{2}$

$\therefore \quad 2\theta = 30^\circ$

or $\quad \theta = 15^\circ$

10. At 45°, range is maximum. At highest point, it has only horizontal component of velocity or $20\cos 45^\circ \approx 14\text{ m/s}$.

11. $\mathbf{F}\cdot\mathbf{v} = 0 \quad \Rightarrow \quad \mathbf{F} \perp \mathbf{v}$

Hence path is parabola.

12. $v_x = u_x$

$\therefore \quad v\cos 30^\circ = u\cos 60^\circ$

or $\quad v = \dfrac{u}{\sqrt{3}}$

Velocity has become $\dfrac{1}{\sqrt{3}}$ times. Therefore, kinetic energy will become 1/3 times.

13. $R_{max} = \frac{v_0^2}{g}$ at $\theta = 45°$

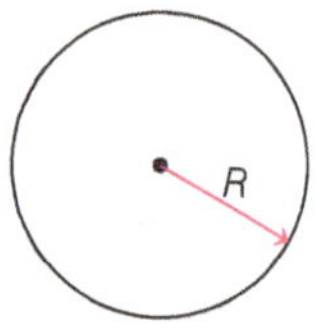

$\therefore \quad A_{max} = \pi R_{max}^2$

14. Maximum range is obtained at 45°

$$\frac{u^2}{g} = 1.6 \quad \text{or} \quad u = 4 \text{ m/s}$$

$$T = \frac{2u \sin 45°}{g} = \frac{2 \times 4 \times (1/\sqrt{2})}{10}$$

$$= 0.4\sqrt{2} \text{ s}$$

Number of jumps in given time,

$$n = \frac{t}{T} = \frac{10\sqrt{2}}{0.4\sqrt{2}} = 25$$

$\therefore$ Total distance travelled $= 1.6 \times 25 = 40$ m **Ans.**

15. $H_1 = \frac{u^2 \sin^2 \theta}{2g}$

$$\therefore \quad 102 = \frac{(u^2) \sin^2 60°}{20}$$

$$\therefore \quad u = 52.2 \text{ m/s}$$

Other stone should be projected at $90° - \theta$ or 30° from horizontal.

$$\therefore \quad H_2 = \frac{u^2 \sin^2 30°}{2g}$$

$$= \frac{(52.2)^2 (1/4)}{20}$$

$$= 34 \text{ m}$$ **Ans.**

16. Using $s = ut + \frac{1}{2}at^2$ in vertical direction

$$\therefore \quad -70 = (50 \sin 30°)t + \frac{1}{2}(-10)t^2$$

On solving this equation, we get

$$t = 7 \text{ s}$$ **Ans.**

17. $2\theta = 90° - \theta$ or $3\theta = 90°$ or $\theta = 30°$

18. $T = \frac{2u_y}{g} = 3$

$$\therefore \quad u_y = 15 \text{ m/s}$$

Now, $H = \frac{u_y^2}{2g} = \frac{(15)^2}{20} = 11.25$ m

19. $u \cos \theta = \frac{u}{2} \Rightarrow \theta = 60°$

Now, $R = \frac{u^2 \sin 2\theta}{g} = \frac{u^2 \sin 120°}{g}$

$$= \frac{\sqrt{3}u^2}{2g}$$

20. $5\sqrt{3} = \frac{(10)^2 \sin 2\theta}{g}$ or $\sin 2\theta = \frac{\sqrt{3}}{2}$

$$\therefore \quad 2\theta = 60° \text{ or } \theta = 30°$$

Two different angles of projection are therefore θ and $90° - \theta$ or 30° and 60°.

$$T_1 = \frac{2u \sin 30°}{g} = 1 \text{ s}$$

$$T_2 = 2\frac{u \sin 60°}{g} = \sqrt{3} \text{ s}$$

$$\therefore \quad \Delta t = T_2 - T_1 = (\sqrt{3} - 1) \text{ s}$$

21. H and R both are proportional to u^2. Hence, percentage increases in horizontal range would also be 10%.

22. $y_1 + y_2 = \frac{u^2 \sin^2 \theta}{2g} + \frac{u^2 \sin^2 (90° - \theta)}{2g} = \frac{u^2}{2g}$

23. $R = 2H$ or $\frac{2u_x u_y}{g} = \frac{2u_y^2}{2g}$

or $\quad 2u_x = u_y$

or $\quad 2a = b$

24. $u_x = a, u_y = b, g = c$

$$\Rightarrow \quad R = \frac{2u_x u_y}{g} = \frac{2ab}{c}$$

25. At θ and $90° - \theta$, ranges are same.

$$R = \frac{u^2 \sin 2\theta}{g}$$

$$t_1 = \frac{2u \sin \theta}{g}$$

$$t_2 = \frac{2u \sin(90° - \theta)}{g} = \frac{2u \cos \theta}{g}$$

From these three equations, we can prove that

$$t_1 t_2 = \frac{2R}{g}$$

26. Given, initial velocity of projectile, $u = 40$ m/s

Angle, $\theta = 30°$

Time of flight

$$T = \frac{2u \sin \theta}{g} = \frac{2 \times 40 \times 1}{10 \times 2} = 4 \text{ s} \quad (\because g = 10 \text{ m/s}^2)$$

It means projectile is at maximum height at $t = 2$ s. At maximum height vertical component of velocity is zero.

Velocity at $t = 2$ s : $v_x = u \cos \theta = 40 \cos 30°$

$$= 20\sqrt{3} \text{ ms}^{-1}$$

27. Range of projectile

$$R=\frac{u^2\sin 2\theta}{g} \quad [\because R\propto \sin(2\theta)]$$

$$\frac{R_1}{R_2}=\frac{\sin(2\theta_1)}{\sin(2\theta_2)}=\frac{\sin(2\times 15)}{\sin(2\times 45)}=\frac{\sin 30^\circ}{\sin 90^\circ}$$

$$\Rightarrow \frac{50}{R_2}=\frac{1}{2} \Rightarrow R_2=100\,\text{m}$$

28. Maximum height in projectile motion

$$H_{max}=\frac{u^2\sin^2\theta}{2g}$$

For two projectiles $\theta_1=30^\circ$ and $\theta_2=60^\circ$

$$\therefore \quad \frac{H_1}{H_2}=\frac{\frac{u^2\sin^2 30^\circ}{2g}}{\frac{u^2\sin^2 60^\circ}{2g}}=\frac{1}{3}$$

29. At highest point, only horizontal component of velocity exists.

So, $u\cos\theta=\frac{\sqrt{3}}{2}u \Rightarrow \cos\theta=\frac{\sqrt{3}}{2}$

$$\Rightarrow \theta=30^\circ \Rightarrow T=\frac{2u\sin\theta}{g}=\frac{2\times u\times\sin 30^\circ}{g}=\frac{u}{g}$$

30. As $v^2=u^2+2as$

$$\Rightarrow \quad 0=u^2-2gH$$

$$\Rightarrow \quad u^2=2gH \Rightarrow u=\sqrt{2\times 10\times 136}$$

$$\therefore \quad R_{max}=\frac{u^2}{g}=\frac{2\times 10\times 136}{10}=272\,\text{m}$$

31. As $H_{max}=\frac{u^2\sin^2\theta}{2g} \Rightarrow u^2\propto\frac{1}{\sin^2\theta}$

$$\Rightarrow u\propto\frac{1}{\sin\theta} \Rightarrow \frac{\sin\theta_2}{\sin\theta_1}=\frac{u_1}{u_2}$$

$$\Rightarrow \sin\theta_1=\frac{u_2}{u_1}\times\sin\theta_2=\frac{\sqrt{2}}{\sqrt{3}}\times\frac{\sqrt{3}}{2}=\frac{1}{\sqrt{2}}$$

So, $\theta_1=\sin^{-1}\left(\frac{1}{\sqrt{2}}\right)=45^\circ$

32. For projectile motion,

Range $R=\frac{u^2\sin 2\theta}{g}$

As range is same for angle of projection θ and $90^\circ-\theta$. So, range is same for 42° and 48°.

Height in projectile motion, $H=\frac{u^2\sin^2\theta}{2g}$

$$\Rightarrow \quad H\propto\sin^2\theta$$

So, H is higher for 48° than H for 42° because $\sin\theta$ increases as θ increases for $0^\circ\le\theta\le 90^\circ$..

33. Horizontal velocity (initial),

$$u_x=\frac{40}{2}=20\,\text{m/s}$$

Vertical velocity (initial), $50=u_y t-\frac{1}{2}gt^2$

$$\Rightarrow \quad u_y\times 2+\frac{1}{2}(-10)\times 4$$

or $\quad 50=2u_y-20$ or $u_y=\frac{70}{2}=35\,\text{m/s}$

$$\therefore \quad \tan\theta=\frac{u_y}{u_x}=\frac{35}{20}=\frac{7}{4}$$

$$\Rightarrow \quad \text{Angle } \theta=\tan^{-1}\frac{7}{4}$$

34. $\alpha+\beta=90^\circ$ or $\beta=90^\circ-\alpha$

$$h_1=\frac{u^2\sin^2\alpha}{2g} \quad \text{and} \quad h_2=\frac{u^2\cos^2\alpha}{2g}$$

$$t_1=\frac{2u\sin\alpha}{g} \quad \text{and} \quad t_2=\frac{2u\cos\alpha}{g}$$

$$R_1=R_2=\frac{2u^2\sin\alpha\cos\alpha}{g}=R$$

From the above expressions we can see that all four options are correct.

35. $u_x=v_x=10\,\text{m/s}$

$$u_y=\sqrt{v_y^2+2gh}=\sqrt{(10)^2+(2)(10)(15)}$$

$$=20\text{ m/s} \qquad \textbf{Ans.}$$

Angle of projection,

$$\theta=\tan^{-1}\left(\frac{u_y}{u_x}\right)=\tan^{-1}(2)$$

$$T=\frac{2u_y}{g}=\frac{(2)(20)}{10}=4\,\text{s}$$

$$R=u_xT=(10)(4)=40\,\text{m}$$

$$H=\frac{u_y^2}{2g}=\frac{(20)^2}{2\times 10}=20\text{ m}$$

36. $\frac{d\mathbf{v}}{dt}=\mathbf{a}=\text{constant}=\mathbf{g}$

$$\frac{d^2\mathbf{v}}{dt^2}=\frac{d\mathbf{a}}{dt}=0=\text{constant}$$

37. $H=\frac{u_y^2}{2g}=20\text{ m}$

$$\therefore \quad u_y=20\,\text{m/s}$$

$$=\frac{2u_y}{g}=4\,\text{s}$$

$$R=u_xT=40\,\text{m}$$

$$\therefore \quad u_x=\frac{40}{T}=10\,\text{m/s}$$

38. $10=\dfrac{u^2\sin 2(15^\circ)}{g} \Rightarrow \dfrac{u^2}{g}=20\,\text{m}$

(a) $R_{max}=\dfrac{u^2}{g}=20\,\text{m}$

(b) $H_{max}=\dfrac{u^2}{2g}$ (when thrown vertically)

$=10\,\text{m}$

(c) $R_{75^\circ}=R_{15^\circ}=10\,\text{m}$

(d) $H_{30^\circ}=\dfrac{u^2\sin^2 30^\circ}{2g}$

$=\dfrac{u^2}{8g}=\dfrac{20}{8}=2.5\,\text{m}$

39. $T=\dfrac{2u_y}{g} \Rightarrow T\propto u_y$

$H=\dfrac{u_y^2}{2g} \Rightarrow H\propto u_y^2$

$R=u_xT$

$\Rightarrow R\propto T$

$\Rightarrow R\propto u_x$

$\Rightarrow \tan\theta=\dfrac{u_y}{u_x}$

$\Rightarrow \tan\theta\propto u_y$

By doubling u_y, $\tan\theta$ will become two times, not θ. From above relations we can see that T and R will become two times and H will become four times.

40. (a) $\mathbf{a}_{av}=\mathbf{a}=(-g\hat{\mathbf{j}})=\dfrac{\Delta\mathbf{v}}{\Delta t}$

$\therefore \Delta\mathbf{v}=(-g\hat{\mathbf{j}})\,\Delta t$

$=(-g\hat{\mathbf{j}})\left(\dfrac{T}{2}\right)$

$=(-g\hat{\mathbf{j}})\left(\dfrac{u\sin\theta}{g}\right)$

$=(-u\sin\theta)\,\hat{\mathbf{j}}$

(b) $v_{av}=\dfrac{s}{t}$

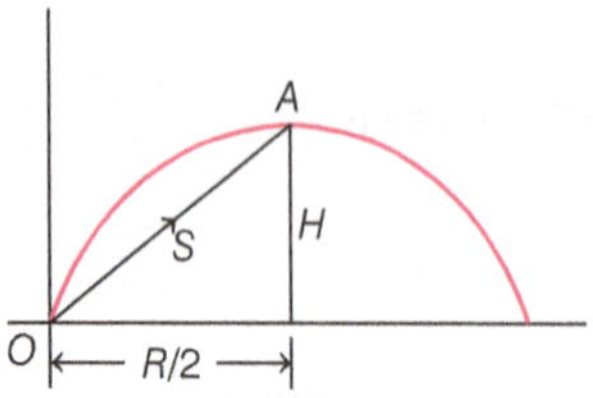

$=\dfrac{\sqrt{(R/2)^2+H^2}}{T/2}$

(c) $\Delta\mathbf{v}=(-g\hat{\mathbf{j}})(\Delta t)$

$=(-g\hat{\mathbf{j}})T$

$=(-g\hat{\mathbf{j}})\left(\dfrac{2u\sin\theta}{g}\right)$

$=(-2u\sin\theta)\,\hat{\mathbf{j}}$

(d) $v_{av}=\dfrac{S}{t}=\dfrac{R}{T}$

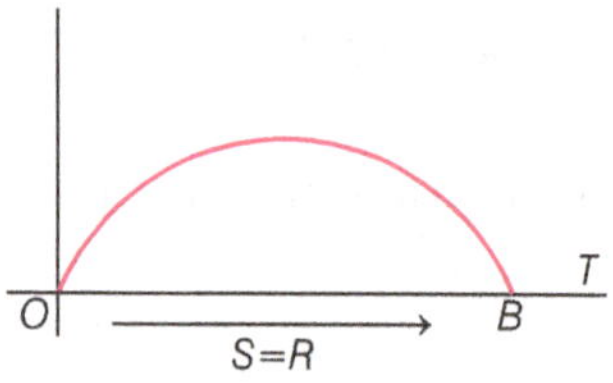

$=\dfrac{u_xT}{T}$

$=u_x=u\cos\theta$

41. $H=\dfrac{u_y^2}{2g}$ or $H\propto u_y^2$

Since H is same. Therefore, u_y is same for all three.

$\therefore T=\dfrac{2u_y}{g}$

or $t\propto u_y$

Since u_y is same. Therefore, T is same for all.

$R=u_xT$

or $R\propto u_x$ (as $T\to$ same)

R for C is maximum. Therefore u_x is greatest for C.

$u=\sqrt{u_y^2+u_x^2}$ ($u_y\to$ same)

R is least for A. Therefore u_x and hence u is least for A.

42. Since range is same.

$\because \theta_1+\theta_2=90^\circ$

$\Rightarrow \theta_2=90^\circ-60^\circ=30^\circ$

$\therefore (H_{max})_1+(H_{max})_2=\dfrac{u^2\sin^2\theta_1}{2g}+\dfrac{u^2\sin^2\theta_2}{2g}$

$=\dfrac{(40)^2}{2\times 10}(\sin^2 60^\circ+\sin^2 30^\circ)$

$=\dfrac{(40)^2}{20}\left(\dfrac{3}{4}+\dfrac{1}{4}\right)=80\,\text{m}$

43. We know that range (R) $=\dfrac{u^2\sin 2\theta}{g}$

So, $R_{max}=\dfrac{u^2\sin 2(45^\circ)}{g}=\dfrac{u^2}{g}$

$\Rightarrow \dfrac{R}{2}=\dfrac{u^2}{2g}=\dfrac{u^2\sin 2\theta}{g}$

$\Rightarrow \sin 2\theta=\dfrac{1}{2} \Rightarrow 2\theta=30^\circ, 150^\circ$

$\Rightarrow \theta=15^\circ, 75^\circ$

44. Here, time of flight is same.

So, vertical component of velocity is same.

Therefore, H_{max} is same.

45. $y = ax - bx^2$

For height (or y) to be maximum.

$$\frac{dy}{dx} = 0 = a - 2bx \quad \text{...(i)}$$

$$\therefore \quad x = \frac{a}{2b}$$

(i) $\therefore$ y_{max} = maximum height

$$H = a(a/2b) - b(a/2b)^2 = a^2/4b$$

(ii) $\left(\frac{dy}{dx}\right)_{(x=0)} = a = \tan\theta_0$

where, θ_0 is the angle of projection.

$$\therefore \quad \theta_0 = \tan^{-1}(a)$$

46. Compare with $y = x\tan\theta - \frac{gx^2}{2u^2\cos^2\theta}$

$$\tan\theta = \sqrt{3}$$

$$\therefore \quad \theta = 60°$$

47. Put $y = 0$,

$$0 = 12x - \frac{3}{4}x^2$$

$$\Rightarrow \quad x = 16\,\text{m}$$

48. Equation of trajectory of the projectile,

$$y = 2x - 9x^2$$

In projectile motion, equation of trajectory is given by

$$y = x\tan\theta_0 - \frac{g\,x^2}{2v_0^2\cos^2\theta_0}$$

Comparing these equations, we get

$$\tan\theta_0 = 2$$

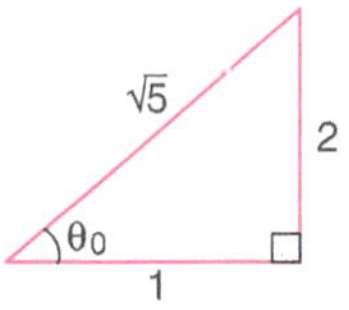

and $\frac{g}{2v_0^2\cos^2\theta_0} = 9$

or $v_0^2 = \frac{g}{9\times 2\cos^2\theta_0}$

$$\therefore \quad v_0^2 = \frac{10\times(\sqrt{5})^2}{9\times 2\times(1)^2} = \frac{10\times 5}{2\times 9}$$

$$\Rightarrow \quad v_0^2 = \frac{25}{9} \text{ or } v_0 = \frac{5}{3}\,\text{m/s}$$

and $\theta_0 = \cos^{-1}\left(\frac{1}{\sqrt{5}}\right)$

49. Given, $\mathbf{u} = \hat{\mathbf{i}} + 2\hat{\mathbf{j}} = u_x\hat{\mathbf{i}} + u_y\hat{\mathbf{j}}$

Then, $u_x = 1 = u\cos\theta$ and $u_y = 2 = u\sin\theta$

$$\therefore \quad \tan\theta = \frac{u\sin\theta}{u\cos\theta} = 2$$

The equation of trajectory of a projectile motion is

$$y = x\tan\theta - \frac{gx^2}{2u^2\cos^2\theta}$$

$$\therefore \quad y = x\times 2 - \frac{10\times x^2}{2(1)^2}$$

$$= 2x - 5x^2$$

50. The equation of trajectory is $y = x\tan\alpha\left(1 - \frac{x}{R}\right)$

$\therefore$ Comparing the equation $y = x(1-x)$

We see that,

$$\tan\alpha = 1 \text{ and } R = 1\,\text{m}$$

$$\therefore \quad \alpha = \frac{\pi}{4}$$

51. For *B*, $T = \frac{2u\sin\theta}{g} = 4$

$$\therefore \quad (u\sin\theta) = 20\,\text{m/s}$$

$$\therefore \quad H = \frac{u^2\sin^2\theta}{2g} = \frac{(20)^2}{20} = 20\,\text{m}$$ **Ans.**

For A, $\frac{2u}{g} = 4$

$$\therefore \quad u = 20\,\text{m/s}$$

Therefore we can see that projection speeds are not same for both.

52. Let time of journey of the two projectiles at the time of collision be t_1 and t_2.

At the time of collisions, their x and y displacements will be same.

Therefore,

$$(s_x)_1 = (s_x)_2$$

or $(u\cos\theta)_1 t_1 = (u\cos\theta)_2 t_2$

$\therefore$ $(250\cos 60°)t_1 = (250\cos 45°)t_2$

or $t_1 = \sqrt{2}\,t_2$ ...(i)

Similarly, $(s_y)_1 = (s_y)_2$

or $(u\sin\theta)_1 t_1 - \frac{1}{2}gt_1^2 = (u\sin\theta)_2 t_2 - \frac{1}{2}gt_2^2$

Substituting the proper values, we get

$$(250\sin 60°)\sqrt{2}t_2 - \frac{1}{2}\times 9.8\times(\sqrt{2}t_2)^2$$

$$= (250\sin 45°)t_2 - \frac{1}{2}\times 9.8\times t_2^2$$

Solving this equation, we get

$$t_2 = 26.41\,\text{s}$$

and $t_1 = \sqrt{2}t_2$ [From equation (i)]

$$= 37.34\,\text{s}$$

Therefore the time interval between the firings,

$$\Delta t = t_1 - t_2$$
$$= 10.94 \text{ s} \approx 11 \text{ s}$$

53.

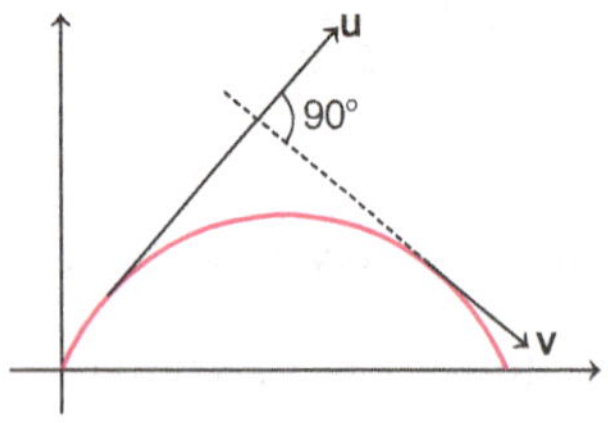

In the figure we can see that **u** and **v** may be mutually perpendicular.

55. $v_x = u_x$

$$v_y = \pm\sqrt{u_y^2 - 2gh}$$

and $v = \sqrt{v_x^2 + v_y^2}$

and $u^2 = u_x^2 + u_y^2$

56. $t = \frac{2u_y}{g} = \frac{2 \times 20}{10} = 4 \text{ s}$

$$H = 20 + \frac{u_y^2}{2g} = 20 + \frac{(20)^2}{2 \times 10}$$
$$= 40 \text{ m}$$

57. Relative acceleration between two is zero. Therefore, relative motion is uniform.

$$\mathbf{u}_{12} = \mathbf{u}_1 + \mathbf{u}_2$$
$$= (20\hat{\mathbf{j}}) - (20 \cos 30° \hat{\mathbf{i}} + 20 \sin 30° \hat{\mathbf{j}})$$
$$= (10\hat{\mathbf{j}} - 10\sqrt{3}\,\hat{\mathbf{i}})$$
$$|\mathbf{u}_{12}| = \sqrt{(10)^2 + (10\sqrt{3})^2} = 20 \text{ m/s}$$

$\therefore \quad d = |\mathbf{u}_{12}|t = 24 \text{ m}$ **Ans.**

58. $S = \sqrt{H^2 + R^2/4}$

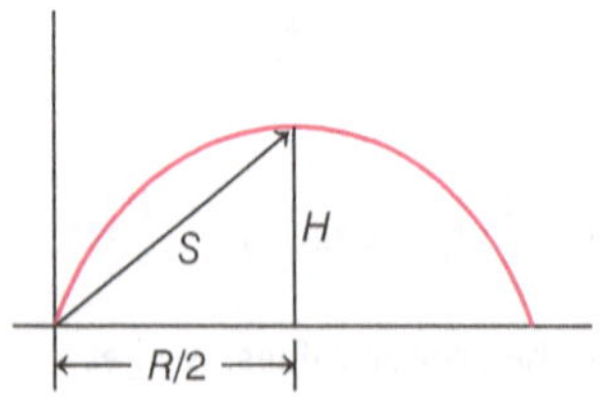

$$\text{Average velocity} = \frac{S}{t} = \frac{S}{T/2} = \frac{2S}{T}$$

59. Velocity of ball in the direction of train is also 30 m/s. So there is no relative motion in this direction. In perpendicular direction,

$$d = R = \frac{u^2 \sin 2\theta}{g}$$
$$= \frac{(30)^2 \sin 90°}{10} = 90 \text{ m}$$ **Ans.**

60. For maximum value of y

$$\frac{dy}{dt} = 10 - 2t = 0$$

$\Rightarrow \quad t = 5 \text{ s}$

$$y_{max} = (10)(5) - (5)^2$$
$$= 25 \text{ m}$$ **Ans.**

61. $T = \frac{2u \sin\theta}{g} = \frac{(2)(80)\sin 53°}{10} = 12.8 \text{ s}$

$$R = \frac{u^2 \sin 2\theta}{g} = \frac{(80)^2 \sin(106°)}{10} = 615.2 \text{ m}$$

Distance travelled by tank,

$$d = (5)T = (5)(12.8) = 64 \text{ m}$$

$\therefore$ Total distance $= (615.2 + 64)$ m

$$= 679.2 \text{ m}$$ **Ans.**

62. $x = 36t$

$\therefore \quad v_x = \frac{dx}{dt} = 36 \text{ m/s}$

$$y = 48t - 4.9t^2$$

$\therefore \quad v_y = \frac{dy}{dt} = 48 - 9.8t$

At $t = 0$, $v_x = 36$ m/s and $v_y = 48$ m/s

So, angle of projection, $\theta = \tan^{-1}\left(\frac{v_y}{v_x}\right)$

$$= \tan^{-1}\left(\frac{4}{3}\right)$$

or $\quad \theta = \sin^{-1}\left(\frac{4}{5}\right)$

63. Projectile will strike at highest point of its path with velocity $v_0 \cos\alpha$.

64. Velocity of boy should be equal to the horizontal component of velocity of ball.

65. At highest point, kinetic energy will be minimum but not zero.

66. $K = K_0 - mgh$, here K = kinetic energy at height h,

K_0 = initial kinetic energy. Variation of K with h is linear. At highest point, kinetic energy is not zero.

67. $v^2 = v_y^2 + v_x^2$

or $\quad (5\sqrt{2})^2 = (u_y - gt)^2 + u_x^2$

or $\quad 50 = (5\sqrt{3} - 10t)^2 + (5)^2$

$\therefore \quad (5\sqrt{3} - 10t) = \pm 5$

$$t_1 = \frac{5\sqrt{3} + 5}{10}$$

and $\quad t_2 = \frac{5\sqrt{3} - 5}{10}$

$\therefore \quad t_1 - t_2 = 1 \text{ s}$

68. $y=\frac{gx^2}{2u^2} \Rightarrow h=\frac{gR^2}{2u^2}, 4h=\frac{g(2R)^2}{2u_1^2}$

$u_1=u$

69. $\frac{2u\sin\theta}{g}=9$

$\Rightarrow \quad u\sin\theta=\frac{9g}{2}$

$(u\cos\theta)_4=x_p$

$\therefore \quad y_p=x_p\tan\theta-g\frac{x_p^2}{2u^2\cos^2\theta}$

On solving, we get

$y_p=10g=100\text{ m}$

70. $H=\frac{u^2\sin^2\theta}{2g}$

$\sin\theta=\frac{\sqrt{2gH}}{u}=\frac{\sqrt{2\times 9.8\times 40}}{56}=\frac{1}{2}$

$\theta=30°$

From 0° to 45°, horizontal range increases. Therefore, the appropriate angle is 30°.

71. $T=t_1+t_2=4\text{ s}=\frac{2u\sin\theta}{g}$

$\therefore \quad u\sin\theta=2g$

$H=\frac{(u\sin\theta)^2}{2g}=2g$

$=19.6\text{ m}$

72. Vertical component of initial velocity of A

$=2v\sin 30°=v=v_1$ (say)

Vertical component of initial velocity of B

$=v\sin 60°=\frac{\sqrt{3}v}{2}=v_2$ (say)

Since $v_2<v_1$. Therefore, B will hit earlier.

73. $H=\frac{(v\sin\theta)^2}{2g}$

At height $h=\frac{H}{2}$ or $\frac{v^2\sin^2\theta}{4g}$, vertical component of velocity is,

$v_y=\sqrt{u_y^2-2gh}$

$=\sqrt{v^2\sin^2\theta-2g\left(\frac{v^2\sin^2\theta}{4g}\right)}$

$=\frac{v\sin\theta}{\sqrt{2}}$

74. Let 't' be the time taken by the bullet to hit the target.

$700\text{ m}=630\text{ ms}^{-1}t$

$\Rightarrow \quad t=\frac{700\text{ m}}{630\text{ ms}^{-1}}=\frac{10}{9}\text{ s}$

For vertical motion,

$h=\frac{1}{2}gt^2 \qquad (\because u=0)$

$=\frac{1}{2}\times 10\times\left(\frac{10}{9}\right)^2=\frac{500}{81}\text{ m}=6.1\text{ m}$

Therefore, the rifle must be aimed 6.1 m above the centre of the target to hit the target.

75. Yes, man will catch the ball, if the horizontal component of velocity becomes equal to the constant speed of man, i.e.

$\frac{v_0}{2}=v_0\cos\theta \text{ or } \theta=60°$

76. Horizontal range is required

$R=\frac{u^2\sin 2\theta}{g}=\frac{(10)^2\sin(2\times 30°)}{10}$

$=5\sqrt{3}=8.66\text{ m}$

77. Horizontal component of velocity remains unchanged.

$\therefore \quad v\cos\theta=v'\cos(90-\theta)$

or $\quad v'=v\cot\theta$

In vertical (y) direction,

$v_y=u_y+a_yt$

$\therefore \quad t=\frac{v_y-u_y}{a_y}$

$=\frac{-v'\sin(90-\theta)-v\sin\theta}{-g}$

$=\frac{(v\cot\theta)\cdot\cos\theta+v\sin\theta}{g}$

$=\frac{v\,\text{cosec}\,\theta}{g}$

78. (a) $\Delta x=(u_x)_2(t-1)$

$=(10)(2-1)$

$=10\text{ m}$

(b) $y_1=\frac{1}{2}gt^2=\frac{1}{2}\times 10\times(2)^2=20\text{m}$

$y_2=\frac{1}{2}g(t-1)^2=5\text{ m}$

$\therefore \quad \Delta y=y_1-y_2=15\text{ m}$

(c) $v_{x_1}=0, v_{x_2}=10\text{ m/s}$

$\therefore \quad v_{x_2}-v_{x_1}=10\text{ m/s}$

(d) $v_{y_1}=gt=10\times 2=20\text{ m/s}$

$v_{y_2}=g(t-1)=10\times 1=10\text{ m/s}$

$\therefore \quad v_{y_1}-v_{y_2}=10\text{ m/s}$

79. (a) $x_1=(u_{x1})t=(30)(2)=60\text{ m}$

$x_2=(130)+(u_{x2})(t-1)$

$=130+(-20)(1)$

$=110\text{ m}$

$\therefore \quad \Delta x = 50\text{ m}$

(b) $y_1 = u_{y1}t - \frac{1}{2}gt^2$

$= (30)(2) - \frac{1}{2}(10)(2)^2$

$= 40\text{ m}$

$y_2 = 75 + u_{y2}(t-1) - \frac{1}{2}g(t-1)^2$

$= 75 + 20 \times 1 - \frac{1}{2} \times 10 \times (1)^2$

$= 90\text{ m}$

$\therefore \quad \Delta y = 50\text{ m}$

(c) $v_{x1} = 30\text{ m/s}$

$v_{x2} = -20\text{ m/s}$

$\therefore \quad v_{x1} - v_{x2} = 50\text{ m/s}$

(d) $v_{y1} = u_{y1} + a_y t$

$= (30) + (-10)(2)$

$= 10\text{ m/s}$

$v_{y2} = u_{y2} + a_y(t-1)$

$= 20 + (-10)(1)$

$= 10\text{ m/s}$

$\therefore \quad v_{y1} - v_{y2} = 0$

80. Vertical component of initial velocity

$$u = 20\sqrt{2}\sin 45° = 20\text{ m/s}$$

Now, apply $s = ut + \frac{1}{2}at^2$ (to find t) in vertical direction,

with

$$s = 15\text{ m}, u = 20\text{ m/s}$$

and $\quad a = -10\text{ m/s}^2$ **Ans.**

81. $v_x = u_x = 20\cos 60° = 10\text{ m/s}$

Given, $\quad v = \frac{u}{2}$

$\therefore \quad 4v^2 = u^2$

or $\quad 4(v_x^2 + v_y^2) = u^2$

$\therefore \quad 4[(10)^2 + v_y^2] = (20)^2$

or $\quad v_y = 0$

Hence, it is the highest point.

$\therefore \quad t = \frac{T}{2} = \frac{u\sin\theta}{g}$

$= \frac{(20)\sin 60°}{10} = \sqrt{3}\text{ s}$ **Ans.**

82. $v_x = u_x = 7.6\text{ m/s}$

$v_y = \sqrt{u_y^2 - 2gh}$

$\therefore \quad u_y = \sqrt{v_y^2 + 2gh}$

$= \sqrt{(6.1)^2 + 2 \times 10 \times 9.1}$

$= 14.8\text{ m/s}$

(a) $H = \frac{u_y^2}{2g} = \frac{(14.8)^2}{2 \times 10} \approx 11\text{ m}$

(b) $R = \frac{2u_x u_y}{g} = \frac{2 \times 7.6 \times 14.8}{10} \approx 23\text{ m}$ **Ans.**

(c) $u = \sqrt{u_x^2 + u_y^2}$

$= \sqrt{(7.6)^2 + (14.8)^2}$

$= 16.6\text{ m/s}$ **Ans.**

(d) $\theta = \tan^{-1}\left(\frac{u_y}{u_x}\right)$

$= \tan^{-1}\left(\frac{14.8}{7.6}\right) \approx \tan^{-1}(2),$ (below horizontal)

Ans.

83. Let v is the velocity at time t,

Using $\quad v_y = u_y + a_y t$

$\therefore \quad v\sin\beta = u\sin\alpha - gt$ ...(i)

$v_x = u_x$

$\therefore \quad v\cos\beta = u\cos\alpha$ ...(ii)

From Eqs. (i) and (ii), we get

$$\left(\frac{u\cos\alpha}{\cos\beta}\right)\sin\beta = u\sin\alpha - gt$$

On solving, we get

$$u = \frac{gt\cos\beta}{\sin(\alpha - \beta)}$$

84. (a) $t = \frac{s_x}{u_x} = \frac{10}{5} = 2\text{ s}$

(b) Vertical components of velocities are zero.

$\therefore \quad h = \frac{1}{2}gt^2 = \frac{1}{2} \times 9.8 \times (2)^2 = 19.6\text{ m}$

(c) $s_x = u_x t = 7.5 \times 2 = 15\text{ m}$

85. To hit the jet both should have same horizontal component of velocity. To hit the jet,

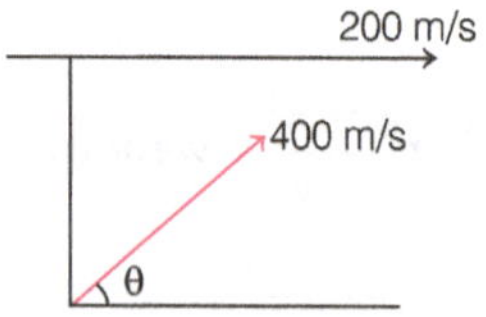

$400\cos\theta = 200$

$\Rightarrow \quad \cos\theta = \frac{1}{2} \Rightarrow \theta = 60°$

86.

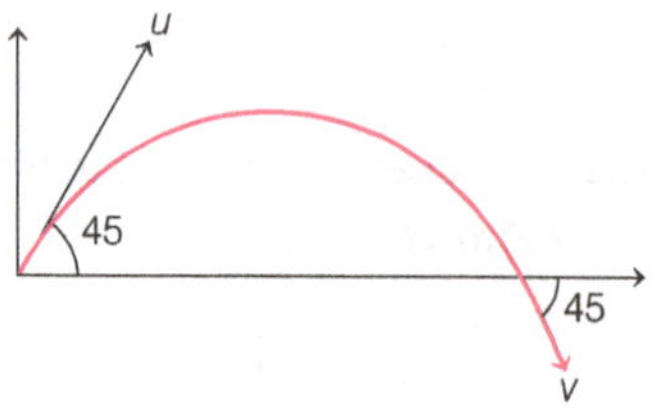

$\therefore \quad |\mathbf{u}| = |\mathbf{v}|$

$\because \quad \mathbf{u} = u\cos 45°\hat{\mathbf{i}} + u\sin 45°\hat{\mathbf{j}}$

and $\quad \mathbf{v} = v\cos 45°\hat{\mathbf{i}} - v\sin 45°\hat{\mathbf{j}}$

$\therefore \quad |\Delta\mathbf{p}| = |m(\mathbf{v} - \mathbf{u})|$

$\Rightarrow \quad |\Delta\mathbf{p}| = 2mu\sin 45°$

$$= 2\times 5\times 10^{-3}\times 5\sqrt{2}\times\frac{1}{\sqrt{2}}$$

$$= 50\times 10^{-3}$$

$$= 5\times 10^{-2}\ \text{kg ms}^{-1}$$

87.

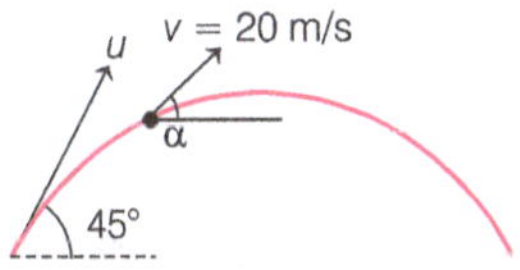

$\therefore \quad v\cos\alpha = u\cos 45° \quad$ ... (i)

and $\quad v\sin\alpha = u\sin 45° - gt \quad$ ...(ii)

Solving for u, we get

$$u = 20\sqrt{2}\ \text{m/s}$$

$$\therefore \quad H = \frac{u^2\sin^2 45°}{20} = 20\ \text{m}$$

Module 2.

Introductory Exercise

1. (a) Horizontal component of initial velocity,

$$u_x = 20\sqrt{2}\cos 45° = 20\ \text{m/s}$$

Let the particle, strikes the inclined plane at P after time t, then horizontal displacement

$$QP = u_x t = 20t$$

In vertical displacement, $u_y t$ or $20\,t$ is upwards and $\frac{1}{2}gt^2$ or $5t^2$ is downwards. But net vertical displacement is downwards. Hence $5t^2$ should be greater than $20\,t$ and therefore,

$$OQ = 5t^2 - 20t$$

In ΔOQP, $\quad \tan 37° = \frac{OQ}{QP}$

or $\quad \frac{3}{4} = \frac{5t^2 - 20t}{20t}$

Solving this equation, we get

$$t = 7\ \text{s}$$

(b) Range, $OP = (PQ)\sec 37°$

$$= (20t)\left(\frac{5}{4}\right)$$

Substituting the value of t, we get

$$OP = 175\ \text{m} \qquad \textbf{Ans.}$$

2. (a) Horizontal component of initial velocity,

$$u_x = 20\sqrt{2}\cos 45° = 20\ \text{m/s}$$

Vertical component of initial velocity

$$u_y = 20\sqrt{2}\sin 45° = 20\ \text{m/s}$$

Let the particle strikes at P after time t, then horizontal displacement

$$OQ = u_x t = 20\,t$$

In vertical displacement, $u_y t$ or $20\,t$ is upwards and $\frac{1}{2}gt^2$ or $5t^2$ is downwards. But net displacement is upwards, therefore $20\,t$ should be greater than $5t^2$ and

$$QP = 20t - 5t^2$$

In ΔOPQ, $\quad \tan 37° = \frac{QP}{OQ}$ or $\frac{3}{4} = \frac{20t - 5t^2}{20t}$

Solving this equation, we get

$$t = 1\ \text{s} \qquad \textbf{Ans.}$$

(b) Range $\quad OP = OQ\sec 37° = (20\,t)\left(\frac{5}{4}\right)$

Substituting $t = 1$ s, we have

$$OP = 25\ \text{m} \qquad \textbf{Ans.}$$

3. $T = \frac{2u\sin(\alpha + \beta)}{g\cos\beta}$

$$T = \frac{2(20)\sin 30°}{(10)\cos 30°} \qquad [\because \alpha = 0°]$$

$$= 2.31\ \text{s}$$

$$R = \frac{(20)^2}{(10)\cos^2 30°}[\sin 30° + \sin 30°]$$

$$= 53.33\ \text{m} \qquad \textbf{Ans.}$$

Here α is the angle of **u** with horizontal and β the angle of inclined plane with horizontal.

Exercise

1. $\frac{AB}{BC} = \frac{\frac{1}{2}gt^2}{v_0 t} = \tan\theta$

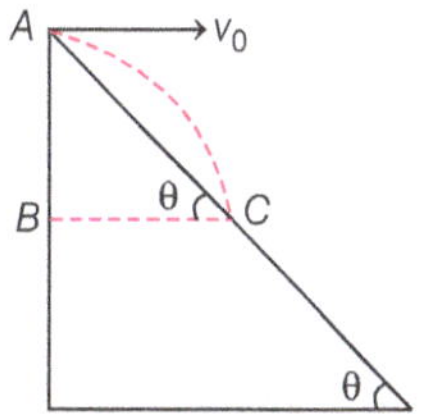

$$t = \frac{2v_0\tan\theta}{g}$$

Now, $\quad x\text{-coordinate} = v_0 t = \frac{2v_0^2\tan\theta}{g}$

and $\quad y\text{-coordinate} = -\frac{1}{2}gt^2 = -\frac{2v_0^2\tan^2\theta}{g}$

2. $R = \dfrac{u^2}{g\cos^2\beta}[\sin(2\alpha+\beta)+\sin\beta]$

$u = 10\,m/s, g = 10\,m/s^2, \beta = 30°$

and $\alpha = 60°$

3. $R = \dfrac{u^2}{g\cos^2\beta}[\sin(2\alpha-\beta)-\sin\beta]$

Here, $u = 2$ m/s, $\alpha = 45°, \beta = 30°$

and $g = 10\,m/s^2$

4. For collision to take place, relative velocity of *A* with respect to *B* should be along *AB* or their vertical components should be same.

Vertical components of *A*

$v_1 = 10\sin 30°$

$= 5\,m/s$

and vertical component of *B*

$v_2 = 5\sqrt{2}\cos 45°$

$= 5\,m/s$

Since, $v_1 = v_2$, so they may collide.

Now, the second condition is,

$R_1 + R_2 \geq d$ $(\because d = 15\,m)$

$\therefore$ $R_1 = \dfrac{(10)^2\sin 60°}{10}$

$= 8.66$ m

$R_2 = \dfrac{(5\sqrt{2})^2\sin 90°}{10}$

$= 5$ m

Since, $R_1 + R_2 < d$, so they will not collide.

5. (a) If they collide in air, then relative velocity of *A* with respect to *B* should be along *AB* or their vertical components should be same.

$\therefore$ $20\sin\theta = 10$

or $\theta = 30°$ **Ans.**

(b) $x = |s_A| + |s_B|$

$= (20\cos 30°)(t) + 0$

$= 20 \times \dfrac{\sqrt{3}}{2} \times \dfrac{1}{2}$

$= 5\sqrt{3}$ m **Ans.**

6. (i) $x_A = x_B$

$\therefore$ $10 + (u_1\cos\theta_1)t = 30 - (u_2\cos\theta_2)t$

$\therefore$ $(u_1\cos\theta_1 + u_2\cos\theta_2)t = 20$

(ii) $y_A = y_B$

$\therefore 10 + (u_1\sin\theta_1)t - \dfrac{1}{2}gt^2 = 20 + (u_2\sin\theta_2)t - \dfrac{1}{2}gt^2$

or $(u_1\sin\theta_1 - u_2\sin\theta_2)t = 10$

7.

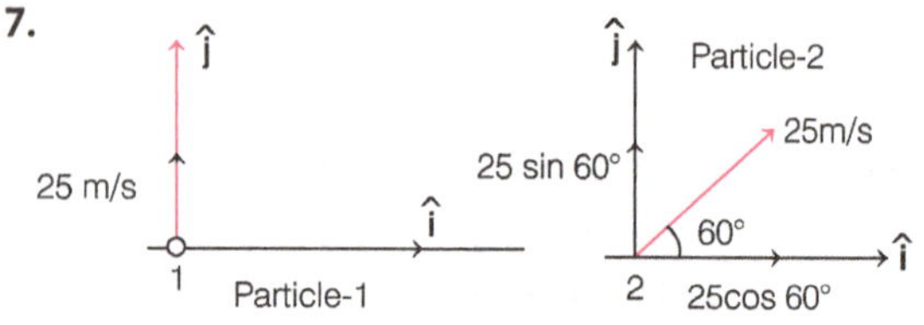

$\mathbf{a}_1 = \mathbf{a}_2 = \mathbf{g}$

$\therefore$ $\mathbf{a}_{relative} = \mathbf{a}_1 - \mathbf{a}_2 = 0$

or, the relative motion is uniform.

$\mathbf{v}_1 = (25\,\hat{\mathbf{j}})$ m/s

$\mathbf{v}_2 = (25\cos 60°)\,\hat{\mathbf{i}} + (25\sin 60°)\,\hat{\mathbf{j}}$

$= (12.5\,\hat{\mathbf{i}} + 21.65\,\hat{\mathbf{j}})$ m/s

$\therefore$ $\mathbf{v}_{12} = \mathbf{v}_1 - \mathbf{v}_2 = [-12.5\,\hat{\mathbf{i}} + 3.35\,\hat{\mathbf{j}}]$ m/s

or $|\mathbf{v}_{12}|$ = Magnitude of relative velocity

$= \sqrt{(-12.5)^2 + (3.35)^2} = 12.94$ m/s

This magnitude is constant as the relative motion is uniform. Hence, distance

$d = |\mathbf{v}_{12}| \times$ time

$= 12.94 \times 1.7 \approx 22$ m

$\therefore$ $d = 22$ m

8. Their horizontal components should be same.

$\therefore$ $\dfrac{2v}{\sqrt{3}}\cdot\cos\theta = v$ or $\theta = 30°$

9. Initial velocity $= (\hat{\mathbf{i}} + 2\hat{\mathbf{j}})$ m/s

Magnitude of initial velocity, $u = \sqrt{u_x^2 + u_y^2} = \sqrt{(1)^2 + (2)^2}$

$= \sqrt{5}$ m/s

Equation of trajectory of projectile is

$y = x\tan\theta - \dfrac{gx^2}{2u^2}(1 + \tan^2\theta)$ $\left[\because \tan\theta = \dfrac{u_y}{u_x} = \dfrac{2}{1} = 2\right]$

Substituting the values, we get

$y = x \times 2 - \dfrac{10(x)^2}{2(\sqrt{5})^2}[1 + (2)^2]$

or $y = 2x - 5x^2$

10. Given, $v_x = 5$ m/s

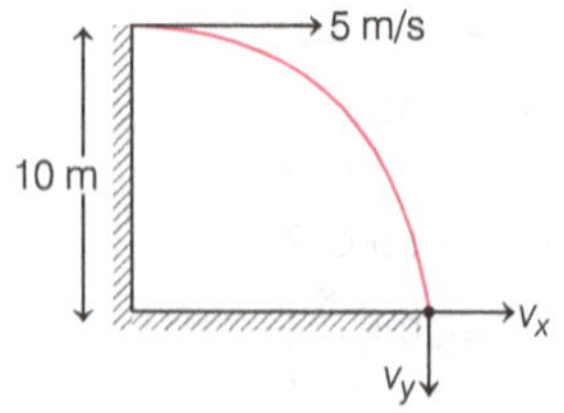

$v_y = \sqrt{2gh} = \sqrt{200}$

$v_{net} = \sqrt{v_x^2 + v_y^2} = \sqrt{(5)^2 + (\sqrt{200})^2}$

$= \sqrt{25 + 200} = 15$ m/s

Unit Exercise (Cumulative)

Level 1

1. $\frac{dy}{dt} = (2\beta x)\cdot\frac{dx}{dt}$

and $\frac{d^2y}{dt^2} = 2\beta\left[\frac{d^2x}{dt^2} + \left(\frac{dx}{dt}\right)^2\right]$

$\frac{d^2y}{dt^2} = \alpha = a_y$

$\frac{d^2x}{dt^2} = a_x = 0$ and $\frac{dx}{dt} = v_x$

$\therefore \quad \alpha = 2\beta\cdot v_x^2$ or $v_x = \sqrt{\frac{\alpha}{2\beta}}$ **Ans.**

2. $y = x\tan\theta - \frac{gx^2}{2u^2\cos^2\theta}$

$\therefore \quad \frac{dy}{dx} = (\tan\theta) - \left(\frac{g}{u^2\cos^2\theta}\right)x$

$\therefore$ $\frac{dy}{dx}$ *versus* x graph is a straight line with negative slope and positive intercept.

3. $a_y = 0 = \frac{d^2y}{dt^2}$

$\frac{dx}{dt} = (2y+2)\cdot\frac{dy}{dt}$

$\frac{d^2x}{dt^2} = (2y+2)\cdot\frac{d^2y}{dt^2} + 2\left(\frac{dy}{dt}\right)^2$

$\therefore \quad a_x = a = (2y+2)(0) + 2(5)^2$

$= 50\,\text{m/s}^2$ **Ans.**

4. Comparing with the trajectory of projectile in which particle is projected from certain height horizontally $(\theta = 0°)$.

$y = x\tan\theta - \frac{gx^2}{2u^2\cos^2\theta}$

Putting $\theta = 0°$ and $g = a = 18u^2 = 2\,\text{m/s}^2$

5. $T = \frac{2u_y}{g}$

$\therefore \quad u_y = \frac{gT}{2} \Rightarrow h_A = u_y t_A - \frac{1}{2}gt_A^2$

$= u_y\left(\frac{2u_y}{3g}\right) - \frac{1}{2}g\left(\frac{2u_y}{3g}\right)^2$

$= \frac{4}{9}\frac{u_y^2}{g} = \left(\frac{4}{9g}\right)\left(\frac{gT}{2}\right)^2 = \frac{gT^2}{9}$

$h_B = u_y\left(\frac{5}{6}\times\frac{2u_y}{g}\right) - \frac{1}{2}\times g\times\left(\frac{5}{6}\times\frac{2u_y}{g}\right)^2$

$= \frac{5}{18}\frac{u_y^2}{g} = \frac{5}{18g}\left(\frac{gT}{2}\right)^2 = \frac{5}{72}gT^2$

$\therefore \quad h_A - h_B = \frac{1}{24}gT^2$

6. For 5 s, weight of the body is balanced by the given force. Hence, it will move in a straight line as shown.

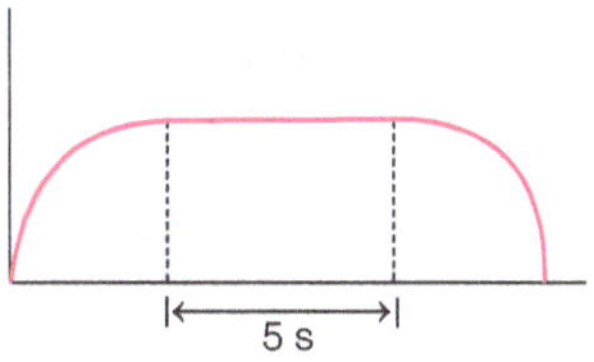

$R = \frac{u^2\sin 2\theta}{g} + (u\cos\theta)(5)$

$= \frac{(50)^2\cdot\sin 60°}{10} + (50\times\cos 30°)(5)$

$= 250\sqrt{3}\,\text{m}$

7.

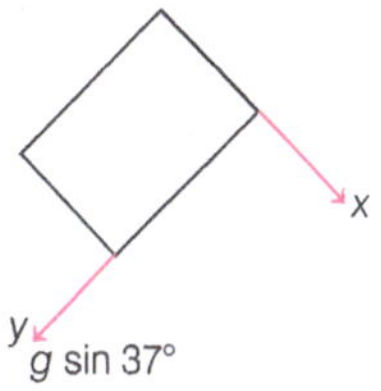

Here, $u_x = 8\,\text{ms}^{-1}$

$a_x = 0$

$\therefore \quad v_x = u_x$

$a_y = g\sin 37°, a_y = 10\times\frac{3}{5} = 6\,\text{ms}^{-2}$

$\therefore \quad v_y = u_y + a_y t = 0 + 6\times 1 = 6\,\text{ms}^{-1}$

$\therefore \quad v = \sqrt{v_x^2 + v_y^2}$

$= \sqrt{8^2 + 6^2} = 10\,\text{ms}^{-1}$

8. $\tan\beta = \frac{H}{R/2} = \frac{2H}{R}$

$= \frac{(2u^2\sin^2\alpha)/2g}{(2u^2\sin\alpha\cos\alpha)/g}$

$= \frac{\tan\alpha}{2}$

$\therefore \quad \beta = \tan^{-1}\left(\frac{\tan\alpha}{2}\right)$ **Ans.**

9. $R = \dfrac{u^2 \sin 2\theta}{g} = 24$ ...(i)

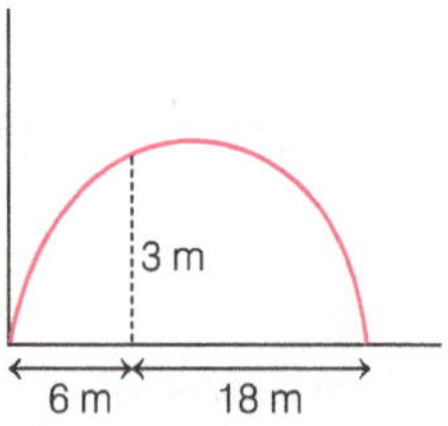

In $y = x \tan\theta - \dfrac{gx^2}{2u^2 \cos^2\theta}$

$3 = 6 \tan\theta - \dfrac{36g}{2u^2 \cos^2\theta}$ ...(ii)

From Eq. (i), $\dfrac{g}{u^2} = \dfrac{\sin 2\theta}{24} = \dfrac{\sin\theta \cos\theta}{12}$

Substituting in Eq. (ii), we have

$3 = 6 \tan\theta - \dfrac{3}{2} \tan\theta$

$= \dfrac{9}{2} \tan\theta$

$\therefore \quad \theta = \tan^{-1}\left(\dfrac{2}{3}\right)$

10. $v_x = 20 \cos\alpha$ $[\because a_x = 0]$

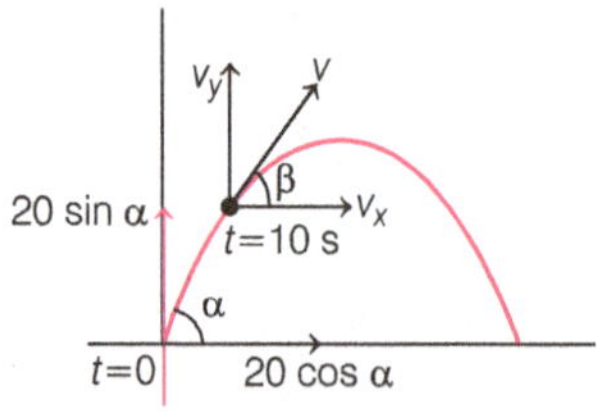

$v_y = 20 \sin\alpha - g \times t$ $[\because a_y = -g]$

$= 20 \sin\alpha - 10 \times 10$

$= 20 \sin\alpha - 100$

So, $\tan\beta = \dfrac{v_y}{v_x} = \dfrac{20 \sin\alpha - 100}{20 \cos\alpha}$

$= \tan\alpha - 5 \sec\alpha$

11. At maximum height inclination with horizontal becomes zero.

So, $t = \dfrac{u \sin\theta}{g}$

$\Rightarrow \quad u = \dfrac{gt}{\sin\theta}$ ...(i)

Now, $R = u \cos\theta \times T$

$\Rightarrow \quad R = u \cos\theta \times 2t$

$\Rightarrow \quad \cos\theta = \dfrac{R}{2ut}$

$\Rightarrow \quad \cos\theta = \dfrac{R \sin\theta}{2tgt}$ [From Eq. (i)]

$\Rightarrow \quad \tan\theta = \dfrac{2gt^2}{R}$

$\Rightarrow \quad \tan\theta = \dfrac{20t^2}{R}$

$\Rightarrow \quad \cot\theta = \dfrac{R}{20t^2}$ **Ans.**

12.

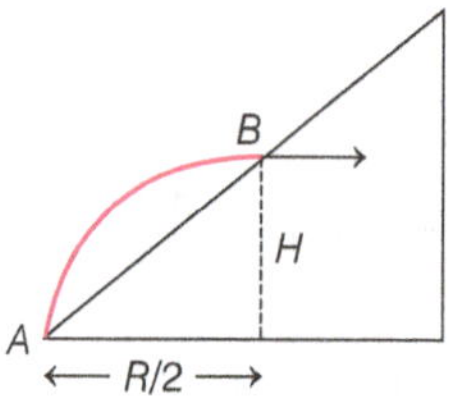

$\dfrac{R}{2} = \dfrac{u^2 \sin 2(60°)}{2g} = \dfrac{\sqrt{3}u^2}{4g}$

$H = \dfrac{u^2 \sin^2 60°}{2g} = \dfrac{3u^2}{8g}$

$\therefore \quad AB = \sqrt{\left(\dfrac{R}{2}\right)^2 + H^2} = \dfrac{\sqrt{21}u^2}{8g}$ **Ans.**

13. $T = \sqrt{\dfrac{2h}{g}}$

$d = (u)T = u\sqrt{\dfrac{2h}{g}}$

$\therefore \quad d^2 = \dfrac{2hu^2}{g}$ **Ans.**

14.

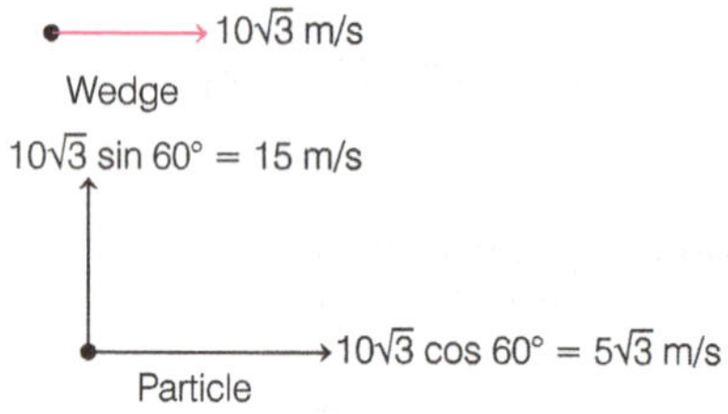

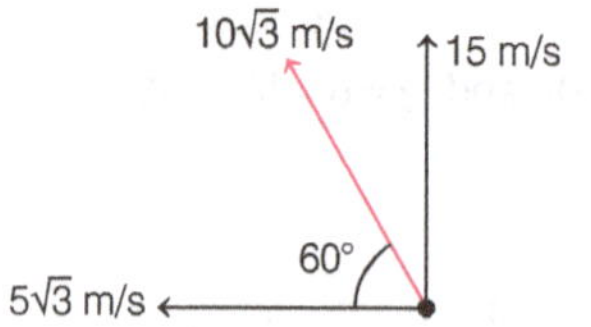

Particle with respect to wedge

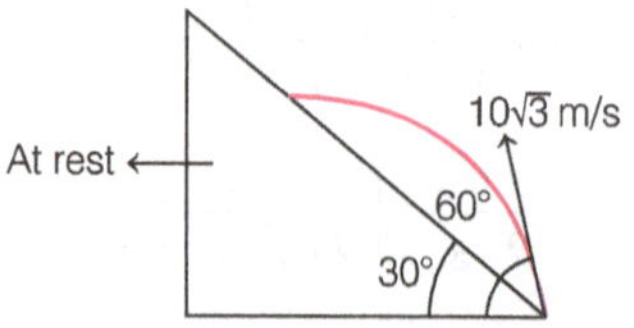

$$T = \frac{2u \sin(\alpha - \beta)}{g \cos \beta}$$

$$= \frac{2(10\sqrt{3}) \cdot \sin(60° - 30°)}{10 \cos 30°} = 2 \text{ s}$$

15. Since $u = 0$, motion of particle is a straight line in the direction of a_{net}.

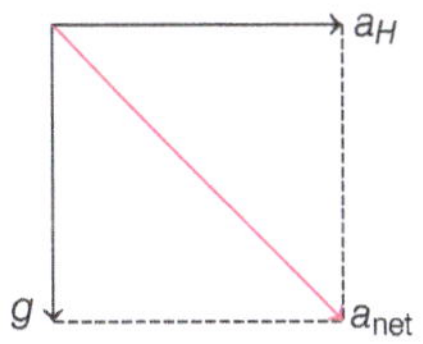

$$t = \sqrt{\frac{2h}{g}} = \sqrt{\frac{2 \times 49}{9.8}} = 3.16 \text{ s}$$ **Ans.**

16. Horizontal component of velocity remains unchanged

$$X_{OA} = 20 \text{ m} = \frac{X_{AB}}{2}$$

$$\therefore \quad t_{OA} = \frac{t_{AB}}{2} = 1 \text{ s}$$

For *AB* projectile

$$T = 2 \text{ s} = \frac{2u_y}{g}$$

$$\therefore \quad u_y = 10 \text{ m/s}$$

$$H = \frac{u_y^2}{2g} = \frac{(10)^2}{2 \times 10} = 5 \text{ m}$$

$\therefore$ Maximum height of total projectile,

$$= 15 + 5 = 20 \text{ m}$$

$$t_{OB} = t_{OA} + t_{AB} = 1 + 2 = 3 \text{ s}$$

For complete projectile

$$T = 2(t_{OA}) + t_{AB}$$

$$= 4 \text{ s} = \frac{2u_y}{g}$$

$$\therefore \quad u_y = 20 \text{ m/s}$$

$$u_x = \frac{AB}{t_{AB}} = \frac{40}{2} = 20 \text{ m/s}$$

17. Let θ be the angle of projection, then

$$T = \frac{2u \sin \theta}{g}$$

$$\Rightarrow \quad \sin \theta = \frac{gT}{2u} \quad \text{...(i)}$$

Further, horizontal range,

$$R = (u \cos \theta) T$$

$$\therefore \quad \cos \theta = \frac{R}{uT} \quad \text{...(ii)}$$

Squaring and adding equations (i) and (ii), we get

$$1 = \frac{g^2 T^2}{(2u)^2} + \frac{R^2}{u^2 T^2} \quad \text{...(iii)}$$

Substituting the values of g, R and u we can find T, which comes out to be,

$$T = 0.41 \text{ min or } 0.71 \text{ min}$$

Alternate method

R and u are given. First find θ. Then,

$$T = \frac{2u \sin \theta}{g}$$

18. $\mathbf{v}_1 = (u \cos \alpha)\hat{\mathbf{i}} + (u \sin \alpha - gt)\hat{\mathbf{j}}$

$\mathbf{v}_2 = (v \cos \beta)\hat{\mathbf{i}} + (v \sin \beta - gt)\hat{\mathbf{j}}$

These two velocity vectors will be parallel when the ratio of coefficients of $\hat{\mathbf{i}}$ and $\hat{\mathbf{j}}$ are equal.

$$\therefore \quad \frac{u \cos \alpha}{v \cos \beta} = \frac{u \sin \alpha - gt}{v \sin \beta - gt}$$

Solving, we get

$$t = \frac{uv \sin(\alpha - \beta)}{g(v \cos \beta - u \cos \alpha)}$$ **Ans.**

19. For shell $u_z = 20 \sin 60° = 17.32 \text{ m/s}$

$$\therefore \quad z = u_z t - \frac{1}{2} g t^2 = (17.32 \times 2) - \left(\frac{1}{2} \times 9.8 \times 4\right)$$

$$\text{or} \quad z = 15 \text{ m} \Rightarrow u_y = 0$$

$$\therefore \quad y = 0$$

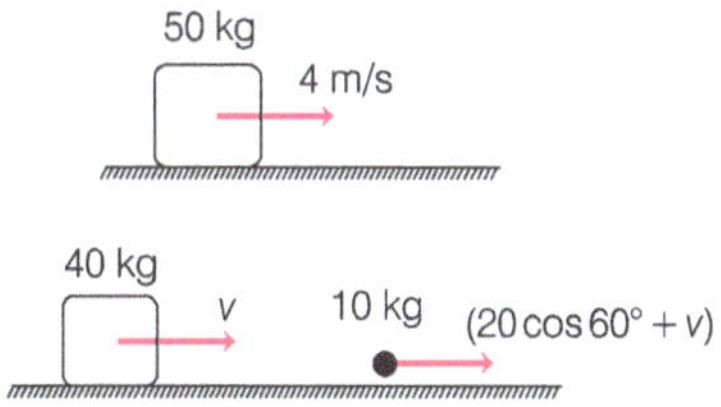

For u_x conservation of linear momentum gives,

$$50 \times 4 = (40)(v) + 10(20 \cos 60° + v)$$

$$\text{or} \quad v = 2 \text{ m/s}$$

$$\therefore \quad u_x = (20 \cos 60°) + 2 = 12 \text{ m/s}$$

$$\therefore \quad x = u_x t = (12)(2) = 24 \text{ m}$$

$$\therefore \quad \mathbf{r} = (24\hat{\mathbf{i}} + 15\hat{\mathbf{k}}) \text{ m}$$ **Ans.**

20. Let t_1 be the time from A to B and t_2 the time from B to C.

Then, $t_1 = \sqrt{\frac{2(H-h)}{g}}$ and $t_2 = \sqrt{\frac{2h}{g}}$

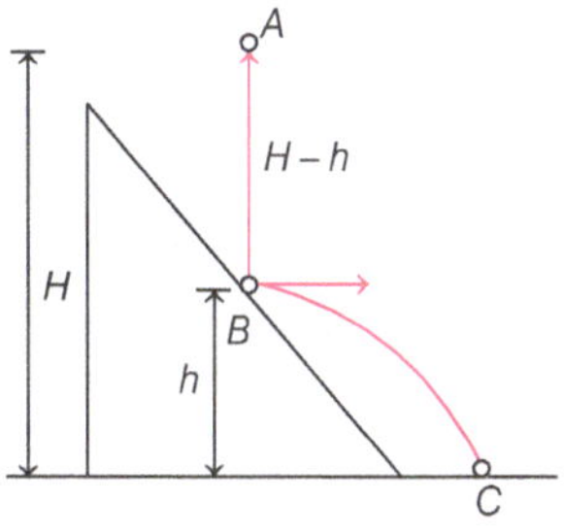

Then, the total time

$$t = t_1 + t_2 = \sqrt{\frac{2}{g}}[\sqrt{H-h} + \sqrt{h}]$$

For t to be maximum $\frac{dt}{dh} = 0$

or $$\sqrt{\frac{2}{g}}\left[\frac{-1}{2\sqrt{H-h}} + \frac{1}{2\sqrt{h}}\right] = 0$$

or $$\frac{1}{\sqrt{h}} = \frac{1}{\sqrt{H-h}}$$

or $$2h = H$$

$\therefore$ $$\frac{h}{H} = \frac{1}{2}$$

21. At 45°

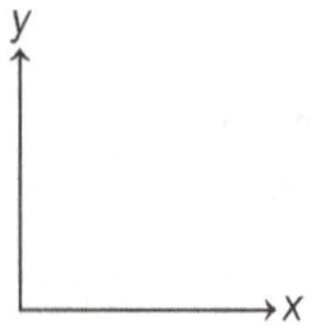

$$v_y = \pm v_x = \pm u_x = \pm 60 \cos 60°$$
$$= \pm 30 \text{ m/s}$$

Now, $$v_y = u_y + a_y t$$

$\therefore$ $$t = \frac{v_y - u_y}{a_y}$$
$$= \frac{(\pm 30) - 60 \sin 60°}{-10}$$

$\therefore$ $t_1 = 2.19\text{s}$ and $t_2 = 8.20\text{s}$ **Ans.**

22. $T = \frac{2u \sin(\alpha - \beta)}{g \cos \beta}$ $(\because \alpha = 60°, \beta = 30°)$

$$= \frac{2 \times 10 \times \sin 30°}{(10) \cos 30°} = \frac{2}{\sqrt{3}} \text{ s}$$

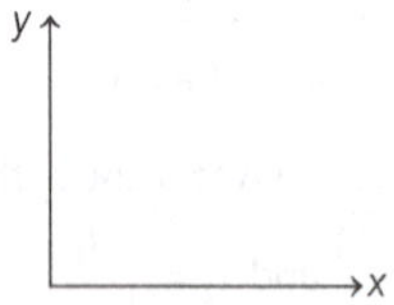

$$v_x = u_x = 10 \cos 60° = 5 \text{ m/s}$$
$$v_y = u_y + a_y t$$
$$= (10 \sin 60°) + (-10)(2/\sqrt{3})$$
$$= 5\sqrt{3} - \frac{20}{\sqrt{3}} = -\frac{5}{\sqrt{3}} \text{ m/s}$$

$\therefore$ $$v = \sqrt{v_x^2 + v_y^2} = \sqrt{25 + \frac{25}{3}}$$
$$= \frac{10}{\sqrt{3}} \text{ m/s}$$

23. $v_y = u_y + a_y t$

$$= (10 \sin 30°) + (-g \cos 30°) T$$

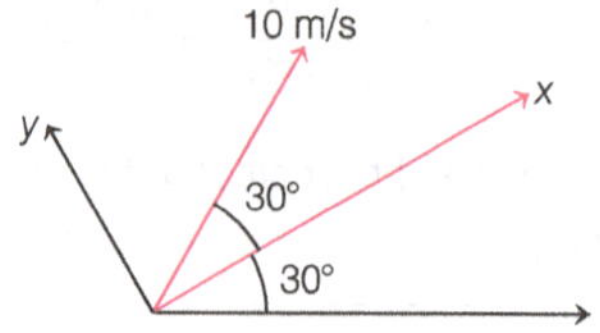

$$= 5 - 10 \times \frac{\sqrt{3}}{2} \times \frac{2}{\sqrt{3}}$$
$$= -5 \text{ m/s}$$

24. Horizontal component of velocity always remains constant

$\therefore$ $$40 \cos 60° = v \cos 30°$$

or $$v = \frac{40}{\sqrt{3}} \text{ m/s}$$ **Ans.**

25. In vertical direction,

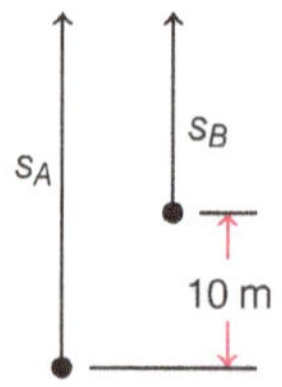

$$s_A = s_B + 10$$

$\therefore$ $$(10)t - \frac{1}{2}gt^2 = -\frac{1}{2}gt^2 + 10$$

$\therefore$ $$t = 1 \text{ s}$$

In horizontal direction,

$$d = |s_A| + |s_B|$$
$$= (10)t + 10t = 20t$$
$$= 20 \text{ m} \quad (\text{as } t = 1 \text{ s})$$

26.

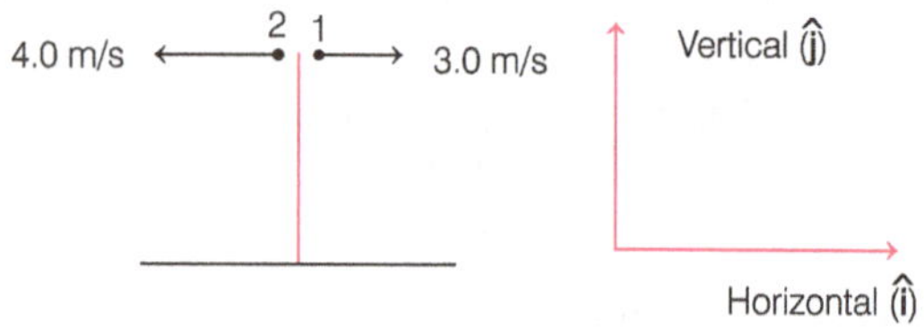

Given, $\mathbf{v}_1 \perp \mathbf{v}_2$

$\therefore$ $$\mathbf{v}_1 \cdot \mathbf{v}_2 = 0$$

$\Rightarrow$ $$(\mathbf{u}_1 + \mathbf{a}_1 t) \cdot (\mathbf{u}_2 + \mathbf{a}_2 t) = 0 \quad \text{...(i)}$$

Here, $$\mathbf{a}_1 = \mathbf{a}_2 = \mathbf{g} = -9.8\hat{\mathbf{j}}$$

Substituting the proper values in equation (i), we get

$$(3\hat{\mathbf{i}} - 9.8t\hat{\mathbf{j}}) \cdot (-4\hat{\mathbf{i}} - 9.8t\hat{\mathbf{j}}) = 0$$

$\Rightarrow$ $$-12 + (9.8t)^2 = 0$$

Solving this equation, we get

$$t = 0.35 \text{ s}$$

Distance

Since, $\mathbf{a}_1 = \mathbf{a}_2 = \mathbf{g}$

$\therefore$ $\mathbf{a}_{rel} = \mathbf{a}_1 - \mathbf{a}_2$

or $\mathbf{a}_2 - \mathbf{a}_1 = 0$

Hence relative motion is uniform.

$$\mathbf{u}_{12} = \mathbf{u}_1 - \mathbf{u}_2 = (3\hat{\mathbf{i}}) - (-4\hat{\mathbf{i}}) = 7\hat{\mathbf{i}}$$

or $|\mathbf{u}_{12}| = 7 \text{ m/s} = \text{constant}$

Hence,

$$d = |\mathbf{u}_{12}| \times t = 7 \times 0.35$$
$$\approx 2.5 \text{ m}$$

$\therefore$ $d = 2.5$ m

27. Let v_x and v_y be the components of v_0 along x and y-directions.

$$(v_x)(2) = 2$$

$\therefore$ $v_x = 1$ m/s

$$v_y(2) = 10$$

or $v_y = 5$ m/s

$\because$ $v_0 = \sqrt{v_x^2 + v_y^2}$

$$= \sqrt{26} \text{ m/s}$$

$$\tan\theta = v_y/v_x = 5/1$$

$\therefore$ $\theta = \tan^{-1}(5)$ **Ans.**

Note *We have seen relative motion between two particles. Relative acceleration between them is zero.*

28. $u_x = v_0 \cos\theta, u_y = v_0 \sin\theta, a_x = -g \sin\theta,\ a_y = g \cos\theta$

At Q, $v_x = 0$

$\therefore$ $u_x + a_x t = 0$

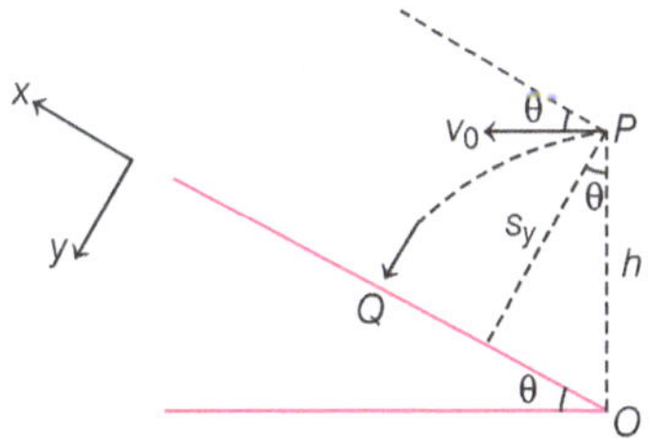

or $t = \dfrac{v_0 \cos\theta}{g \sin\theta}$ ...(i)

$$s_y = h\cos\theta$$

$\therefore$ $u_y t + \dfrac{1}{2} a_y t^2 = h\cos\theta$

$$\therefore (v_0 \sin\theta)\left(\frac{v_0 \cos\theta}{g \sin\theta}\right) + \frac{1}{2}(g\cos\theta)\left(\frac{v_0\cos\theta}{g\sin\theta}\right)^2 = h\cos\theta$$

Solving this equation, we get

$$v_0 = \sqrt{\frac{2gh}{2 + \cot^2\theta}}$$ **Ans**

29. Equation of trajectory of projectile

$$y = 5x(1 - x) = 5x - 5x^2$$

Comparing the above equation with standard equation of projectile $y = x\tan\theta - \dfrac{gx^2}{2u^2\cos^2\theta}$

We get, $\tan\theta = 5$

But $\tan\theta = \dfrac{u\sin\theta}{u\cos\theta} = 5$

$\Rightarrow$ $\dfrac{u\sin\theta}{1} = 5$ ($\because u\cos\theta = 1$ given)

$\Rightarrow$ $u\sin\theta = 5$

30. $t = T = \dfrac{2u\sin\theta}{g}$

$$= \frac{2 \times 10 \times \sin 60^\circ}{10} = \sqrt{3} \text{ s}$$

Displacement of train in time $t = \dfrac{1}{2}at^2$

Displacement of boy with respect to train = 1.15 m

$\therefore$ Displacement of boy with respect to ground

$$= \left(1.15 + \frac{1}{2}at^2\right)$$

Displacement of ball with respect to ground $= (u\cos 60^\circ)t$

To catch the ball back at initial height,

$$1.15 + \frac{1}{2}at^2 = (u\cos 60^\circ)t$$

$\therefore$ $1.15 + \dfrac{1}{2}a(\sqrt{3})^2 = 10 \times \dfrac{1}{2} \times \sqrt{3}$

Solving this equation, we get

$$a = 5 \text{ m/s}^2$$

31. Time of flight, $T = 3 + 5 = 8$ s

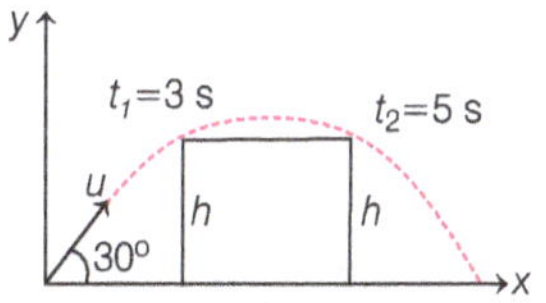

$$T = \frac{2u\sin\theta}{g}$$

$\Rightarrow$ $8 = \dfrac{2u\sin(30^\circ)}{10}$ ($\because \theta = 30^\circ$ given)

$\Rightarrow$ $u = 80$ m/s

32. $R = \dfrac{u^2}{g\cos^2\beta}[\sin(2\alpha - \beta) - \sin\beta]$

$$R_A = R_B$$

$\Rightarrow \quad 2\alpha_A - \beta = \pi - (2\alpha_B - \beta)$

Putting $\quad \alpha_A = 15° + 30° = 45°,$

we get

$$\alpha_B = 60°$$

Now, $\quad T = \dfrac{2u \sin(\alpha - \beta)}{g \cos\beta}$

or $\quad T \propto \sin(\alpha - \beta)$

$\Rightarrow \quad \dfrac{T_A}{T_B} = \dfrac{\sin(\alpha_A - \beta)}{\sin(\alpha_B - \beta)}$

Here, $\alpha_A = 45°, \alpha_B = 60°$ and $\beta = 15°$

33. For minimum value of v_0, the maximum range on inclined plane should be $AB = 15$ m.

The maximum range will be

$$R_{max} = \frac{u^2}{g\cos^2\beta}(1 - \sin\beta)$$

Here, $\quad R_{max} = 15$ m,

$u = v_0,$

$\beta = 30°$

and $\quad g = 10 \text{ m/s}^2$

Level 2

34. Equation of trajectory of projectile

$$y = x\tan\theta - \frac{1}{2}\frac{gx^2}{u^2\cos^2\theta}$$

$$= x\tan 45° - \frac{1}{2}\frac{10x^2}{u^2\cos^2 45°}$$

$\Rightarrow \quad y = x - \dfrac{10x^2}{2u^2\left(\dfrac{1}{2}\right)}$

$\Rightarrow \quad 10 = 20 - \dfrac{(10)(100)}{u^2}$

$\Rightarrow \quad u = 20$

Time period, $T = \dfrac{2u\sin\theta}{g}$

$$= \frac{(2)(20)}{\sqrt{2}(10)} = 2\sqrt{2}$$

$$\mathbf{v} = u\cos\theta\,\hat{\mathbf{i}} + (u\sin\theta - gt)\,\hat{\mathbf{j}}$$

$$= [10\sqrt{2}\,\hat{\mathbf{i}} + (10\sqrt{2} - 10(2)]\,\hat{\mathbf{j}}$$

Momentum $\mathbf{p} = M\mathbf{v} = 100\sqrt{2}\hat{\mathbf{i}} + (100\sqrt{2} - 200)\,\hat{\mathbf{j}}$

35.

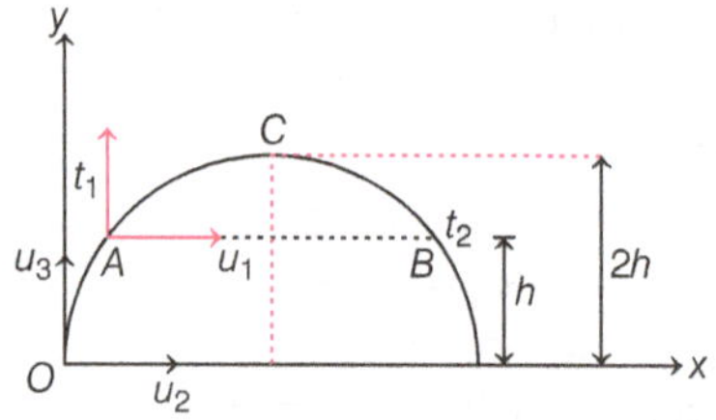

$$2h = \frac{u_3^2}{2g}$$

$\Rightarrow \quad u_3 = 2\sqrt{gh}$ ...(i)

By the time, bird reaches from A to B, stone reaches from **O** to **B**.

$\Rightarrow \quad u_1(t_2 - t_1) = u_2 t_2$

$\Rightarrow \quad \dfrac{u_1}{u_2} = \dfrac{t_2}{t_2 - t_1}$ ...(ii)

Total time of flight,

$$T = t_1 + t_2 = \frac{2u_3}{g} = 4\sqrt{\frac{h}{g}} \quad \text{...(iii)}$$

At A, vertical component of velocity,

$$u_4 = \sqrt{u_3^2 - 2gh}$$

$$= \sqrt{4gh - 2gh}$$

$$= \sqrt{2gh}$$

From A to C, in time $\dfrac{t_2 - t_1}{2}$ this vertical component u_4 also becomes zero.

$\therefore \quad 0 = u_4 - gt$

or $\quad 0 = \sqrt{2gh} - g\left(\dfrac{t_2 - t_1}{2}\right)$

$\therefore \quad t_2 - t_1 = 2\sqrt{2}\left(\sqrt{\dfrac{h}{g}}\right)$ ...(iv)

Solving Eqs. (iii) and (iv), we have

$$\frac{(t_1/t_2) + 1}{1 - (t_1/t_2)} = \sqrt{2}$$

$\therefore \quad \dfrac{t_1}{t_2} = \dfrac{\sqrt{2} - 1}{\sqrt{2} + 1}$

Now, from Eq. (ii), we have

$$\frac{u_1}{u_2} = \frac{1}{1 - \dfrac{t_1}{t_2}} = \frac{1}{1 - \left(\dfrac{\sqrt{2} - 1}{\sqrt{2} + 1}\right)}$$

$$= \frac{\sqrt{2} + 1}{2}$$

36. Let t be the time after which the stone hits the object and θ be the angle which the velocity vector $\mathbf{u}$ makes with horizontal. According to question, we have following three conditions.

Vertical displacement of stone is 1.25 m.

$$1.25 = (u \sin\theta)t - \frac{1}{2}gt^2$$

where, $g = 10\,m/s^2$

or $(u \sin\theta)t = 1.25 + 5t^2$ ..(i)

Horizontal displacement of stone

$= 3 +$ displacement of object A

$\therefore$ $(u \cos\theta)t = 3 + \frac{1}{2}at^2$

where, $a = 1.5\,m/s^2$

or $(u \cos\theta)t = 3 + 0.75t^2$ ...(ii)

Horizontal component of velocity (of stone) = vertical component (because velocity vector is inclined at 45° with horizontal)

$\therefore$ $(u \cos\theta) = gt - (u \sin\theta)$ ...(iii)

(The right hand side is written $gt - u\sin\theta$ because the stone is in its downward motion. Therefore, $gt > u \sin\theta$. In upward motion $u \sin\theta > gt$).

Multiplying Eq. (iii) with t, we can write

$(u \cos\theta)t + (u \sin\theta)t = 10t^2$ ....(iv)

Now Eqs. (iv), (ii) and (i) gives

$$4.25t^2 - 4.25 = 0$$

or $t = 1\,s$

Substituting $t = 1\,s$ in Eqs. (i) and (ii), we get

$u \sin\theta = 6.25$ m/s

or $u_y = 6.25$ m/s

and $u \cos\theta = 3.75$ m/s

or $u_x = 3.75$ m/s

Therefore, $\mathbf{u} = u_x\hat{\mathbf{i}} + u_y\hat{\mathbf{j}}$ m/s

or $\mathbf{u} = (3.75\hat{\mathbf{i}} + 6.25\hat{\mathbf{j}})$ m/s

37. $u_x = 20$ km/h

$$= 20 \times \frac{5}{18} = 5.6\,m/s$$

$u_y = 12$ km/h

$$= 12 \times \frac{5}{18} = 3.3\,m/s$$

Using $s = ut + \frac{1}{2}at^2$ in vertical direction, we have

$$-50 = (3.3)t + \frac{1}{2}(-10)t^2$$

Solving this equation, we get

$t \approx 3.55$ s

At the time of striking with ground,

$v_x = u_x = 5.6$ m/s

$v_y = u_y + a_y t$

$= (3.3) + (-10)(3.55)$

$= 32.2$ m/s

$\therefore$ Speed $= \sqrt{(32.2)^2 + (5.6)^2}$

≈ 32.7 m/s

38. (a) Acceleration of stone is g or $10\,m/s^2$ in downward direction. Acceleration of elevator is $1\,m/s^2$ upwards. Therefore, relative acceleration of stone (with respect to elevator) is $11\,m/s^2$ downwards. Initial velocity is already given relative.

$$T = \frac{2u_r \sin\theta_r}{a_r}$$

$$= \frac{(2)(2)\sin 30°}{11}$$

$= 0.18$ s **Ans.**

(b) $\mathbf{v}_S = \mathbf{v}_{SE} + \mathbf{v}_E$

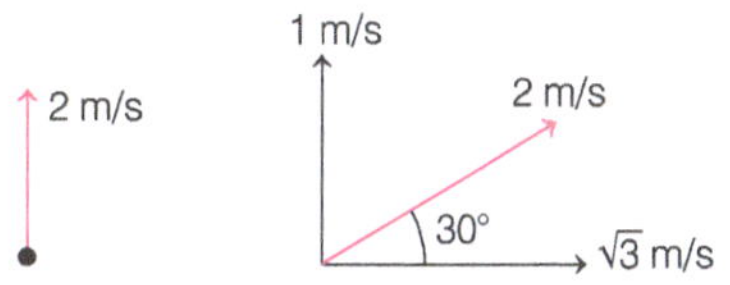

Elevator   Stone relative to elevator

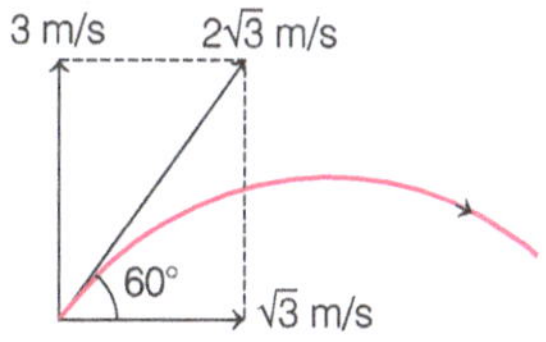

Absolute velocity of stone

(c) In that case, relative acceleration between stone and elevator will be zero. So, with respect to elevator, path is a straight line (uniform) with constant velocity of 20 m/s at 30°. But path, with respect to man on ground will remain unchanged.

39. $a_x = \frac{dv_x}{dt} = -\frac{kv\cos\theta}{m} = -\frac{k}{m}v_x$

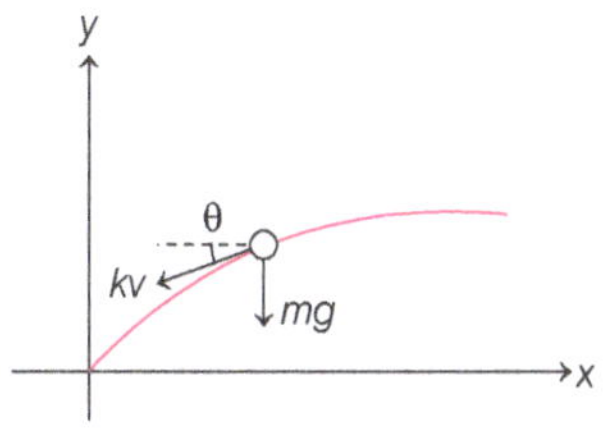

$\therefore$ $$\frac{dv_x}{v_x} = -\frac{k}{m}dt$$

or $$\int_{v_0\cos\theta_0}^{v_x} \frac{dv_x}{v_x} = -\frac{k}{m}\int_0^t dt$$

or $$v_x = v_0 \cos\theta_0 \, e^{-\frac{k}{m}t} \quad ...(i)$$

Similarly,

$$a_y = \frac{dv_y}{dt} = -\frac{kv\sin\theta}{m} - g = -\left(\frac{k}{m}v_y + g\right)$$

or $$\int_{v_0 \sin\theta_0}^{v_y} \frac{dv_y}{\frac{k}{m}v_y + g} = -\int_0^t dt$$

or $$\frac{m}{k}\left[\ln\left(\frac{k}{m}v_y + g\right)\right]_{v_0 \sin\theta_0}^{v_y} = -t$$

or $$\frac{\left(\frac{k}{m}v_y + g\right)}{\left(\frac{k}{m}v_0 \sin\theta_0 + g\right)} = e^{-\frac{k}{m}t}$$

or $$v_y = \frac{m}{k}\left[\left(\frac{k}{m}v_0 \sin\theta_0 + g\right)e^{-\frac{k}{m}t} - g\right] \quad ...(ii)$$

(b) Eq. (i) can be written as

$$\frac{dx}{dt} = v_0 \cos\theta_0 \, e^{-\frac{k}{m}t}$$

or $$\int_0^x dx = v_0 \cos\theta_0 \int_0^t e^{-\frac{k}{m}t} dt$$

or $$x = \frac{mv_0 \cos\theta_0}{k}[1 - e^{-\frac{k}{m}t}]$$

$$x_m = \frac{mv_0 \cos\theta_0}{k}$$

at $t = \infty$ **Ans.**

40. At height 2 m, projectile will be at two times, which are obtained from the equation,

$$2 = (10\sin 45°)t + \frac{1}{2}(-10)t^2$$

or $$2 = 5\sqrt{2}t - 5t^2$$

or $$5t^2 - 5\sqrt{2}t + 2 = 0$$

or $$t_1 = \frac{5\sqrt{2} - \sqrt{50 - 40}}{10} = \frac{5\sqrt{2} - \sqrt{10}}{10}$$

and $$t_2 = \frac{5\sqrt{2} + \sqrt{10}}{10}$$

Now $$d = (10\cos 45°)(t_2 - t_1) = \frac{10}{\sqrt{2}}\left(\frac{2\sqrt{10}}{10}\right) = 4.47 \text{ m}$$

Distance of point of projection from first hurdle

$$= (10\cos 45°)t_1$$

$$= \frac{10}{\sqrt{2}}\left(\frac{5\sqrt{2} - \sqrt{10}}{10}\right)$$

$$= 5 - \sqrt{5}$$

$$= 2.75 \text{ m}$$ **Ans.**

41. (a) Time of descent, $t = \sqrt{\frac{2H}{g}} = \sqrt{\frac{2 \times 400}{10}}$

$$= 8.94 \text{ s}$$

Now $$v_x = ay = \sqrt{5}y$$

or $$\frac{dx}{dt} = \sqrt{5}\left(\frac{1}{2}gt^2\right) = 5\sqrt{5}t^2$$

$\therefore$ $$\int_0^x dx = 5\sqrt{5}\int_0^t t^2 \, dt$$

or horizontal drift

$$x = \frac{5\sqrt{5}}{3}(8.94)^3$$

$$= 2663 \text{ m} \approx 2.67 \text{ km}$$

(b) When particle strikes the ground

$$v_x = \sqrt{5}y = (\sqrt{5})(400) = 400\sqrt{5} \text{ m/s}$$

$$v_y = gt = 89.4 \text{ m/s}$$

$\therefore$ Speed $= \sqrt{v_x^2 + v_y^2} = 899 \text{ m/s}$

$$\approx 0.9 \text{ km/s}$$

42. $u_x^2 + u_y^2 = (2\sqrt{gh})^2 = 4gh$ ...(i)

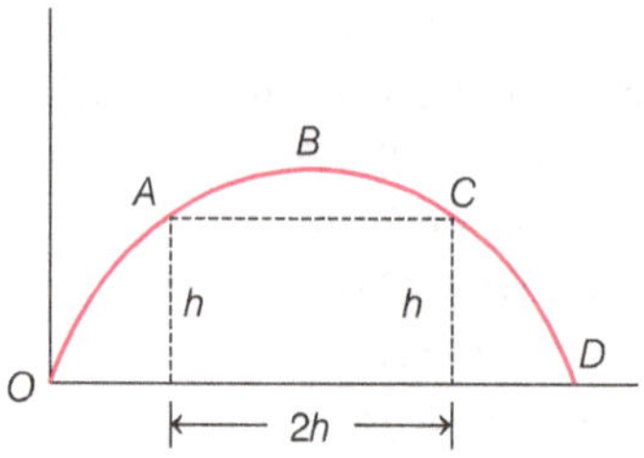

$$v_y = \sqrt{u_y^2 - 2gh} \quad ...(ii)$$

$$v_x = u_x \quad ...(iii)$$

Now for the projectile *ABC*, v_x and v_y are the initial components of velocity.

$\therefore$ $$2h = \text{range} = \frac{2v_x v_y}{g} = \frac{2u_x v_y}{g}$$

or $$u_x = \frac{gh}{v_y} \quad ...(iv)$$

Using Eqs. (ii) and (iv) for rewriting Eq. (i), we have

$$\left(\frac{gh}{v_y}\right)^2 + (v_y^2 + 2gh) = 4gh$$

$\therefore$ $$v_y^4 - (2gh)v_y^2 + g^2h^2 = 0$$

$\therefore \qquad v_y^2 = \dfrac{2gh \pm \sqrt{4g^2h^2 - 4g^2h^2}}{2} = gh$

$\therefore \qquad v_y = \sqrt{gh}$

Now, t_{AC} = time of projectile ABC

$$= \frac{2v_y}{g} = 2\sqrt{\frac{h}{g}} \qquad \textbf{Ans.}$$

43. $R - a = \dfrac{u^2 \sin 2\alpha}{g}$

Multiplying with b, we have

$$bR - ab = \frac{bu^2 \sin 2\alpha}{g} \qquad \text{...(i)}$$

$$R + b = \frac{u^2 \sin 2\beta}{g}$$

Multiplying with a, we have

$$aR + ab = \frac{au^2 \sin 2\beta}{g} \qquad \text{...(ii)}$$

Adding Eqs. (i) and (ii) and by putting

$$R = \frac{u^2 \sin 2\theta}{g},$$

we get the result.

44. $|v_{21x}| = (v_1 + v_2) \cos 60° = 12 \text{ m/s}$

$|v_{21y}| = (v_2 - v_1) \sin 60° = 4\sqrt{3} \text{ m/s}$

$\therefore v_{21} = \sqrt{(12)^2 + (4\sqrt{3})^2} = \sqrt{192} \text{ m/s}$

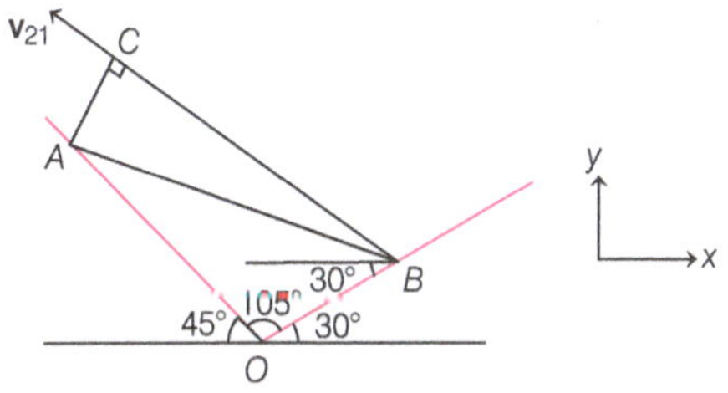

$BC = (v_{21}) t = 240 \text{ m}$

$AC = 70 \text{ m} \qquad$ (Given)

Hence, $AB = \sqrt{(240)^2 + (70)^2}$

$= 250 \text{ m} \qquad$ **Ans.**

45. At $t = 0$, $\mathbf{v}_T = (10\hat{\mathbf{j}}) \text{ m/s}$

$\mathbf{v}_{ST} = 10 \cos 37° \hat{\mathbf{k}} - 10 \sin 37° \hat{\mathbf{i}} = (8\hat{\mathbf{k}} - 6\hat{\mathbf{i}}) \text{ m/s}$

$\therefore \mathbf{v}_S = \mathbf{v}_{ST} + \mathbf{v}_T = (-6\hat{\mathbf{i}} + 10\hat{\mathbf{j}} + 8\hat{\mathbf{k}}) \text{ m/s}$

(a) At highest point, vertical component $(\hat{\mathbf{k}})$ of $\mathbf{v}_S$ will become zero. Hence, velocity of particle at highest point will become $(-6\hat{\mathbf{i}} + 10\hat{\mathbf{j}})$ m/s .

(b) Time of flight, $T = \dfrac{2v_z}{g} = \dfrac{2 \times 8}{10} = 1.6 \text{ s}$

$x = x_i + v_x T$

$= \dfrac{16}{\pi} - 6 \times 1.6 = -4.5 \text{ m}$

$y = (10)(1.6) = 16 \text{ m}$

and $\qquad z = 0$

Therefore, coordinates of particle where it finally lands on the ground are $(-4.5 \text{ m}, 16 \text{ m}, 0)$.

At highest point, $t = \dfrac{T}{2} = 0.8 \text{ s}$

$\therefore \qquad x = \dfrac{16}{\pi} - (6)(0.8) = 0.3 \text{ m}$

$y = (10)(0.8) = 8.0 \text{ m}$

and $\qquad z = \dfrac{v_z^2}{2g} = \dfrac{(8)^2}{20} = 3.2 \text{ m}$

Therefore, coordinates at highest point are (0.3 m, 8.0 m, 3.2 m). **Ans.**

46. (a) Accelerations of particle and block are shown in figure.

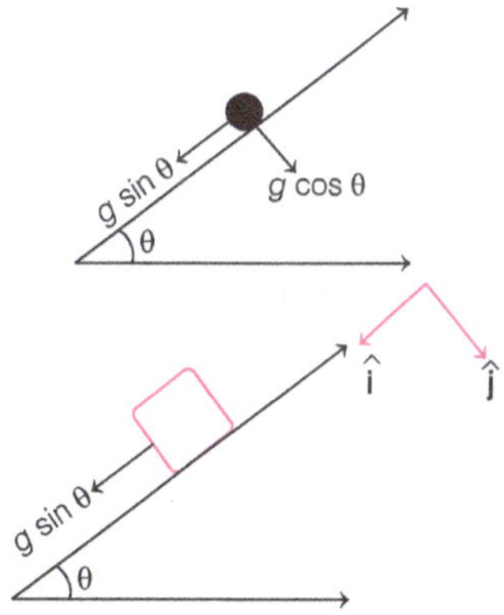

Acceleration of particle with respect to block

= Acceleration of particle − Acceleration of block

$= (g \sin \theta \, \hat{\mathbf{i}} + g \cos \theta \, \hat{\mathbf{j}}) - (g \sin \theta) \hat{\mathbf{i}} = g \cos \theta \, \hat{\mathbf{j}}$

Now, motion of particle with respect to block will be a projectile as shown.

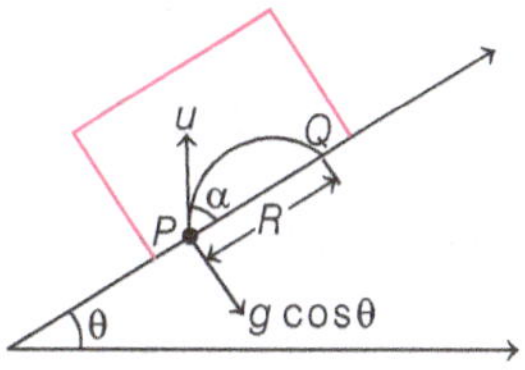

The only difference is, g will be replaced by $g \cos \theta$.

$\therefore \qquad PQ = \text{Range}(R) = \dfrac{u^2 \sin 2\alpha}{g \cos \theta}$

$$\Rightarrow \qquad PQ = \frac{u^2 \sin 2\alpha}{g \cos \theta}$$

(b) Horizontal displacement of particle with respect to ground is zero. This implies that initial velocity with respect to ground is only vertical or there is no horizontal component of the absolute velocity of the particle.

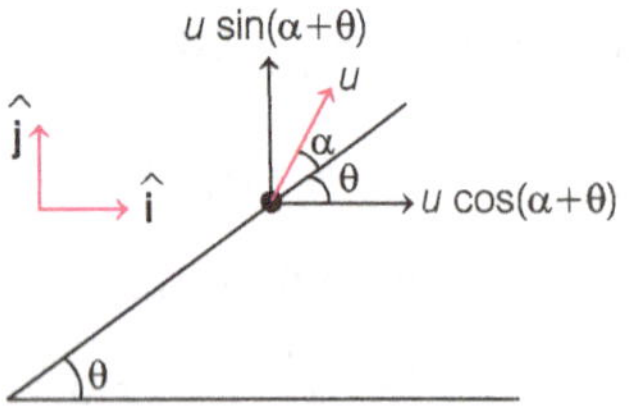

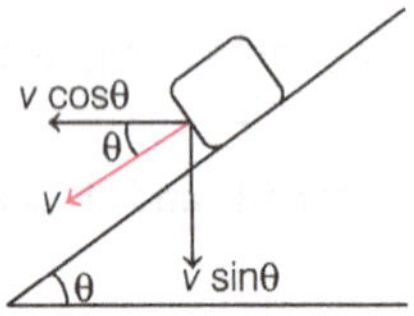

Let v be the velocity of the block down the plane

Velocity of particle with respect to block

$$= u \cos(\alpha + \theta)\,\hat{\mathbf{i}} + u \sin(\alpha + \theta)\,\hat{\mathbf{j}}$$

Velocity of block $= -v\cos\theta\,\hat{\mathbf{i}} - v\sin\theta\,\hat{\mathbf{j}}$

$\therefore$ Velocity of particle with respect to ground

$$= \{u\cos(\alpha+\theta) - v\cos\theta\}\,\hat{\mathbf{i}} + \{u\sin(\alpha+\theta) - v\sin\theta\}\,\hat{\mathbf{j}}$$

Now, as we said earlier that horizontal component of absolute velocity of particle should be zero.

Therefore, $u\cos(\alpha+\theta) - v\cos\theta = 0$

or $\quad v = \dfrac{u\cos(\alpha+\theta)}{\cos\theta}$ (down the plane)

47. $\because \; R = \dfrac{u^2 \sin 60^\circ}{g}$ ...(i)

and $\quad R - x_1 = \dfrac{u^2 \sin 30^\circ}{g}$ ...(ii)

and $\quad R + x_2 = \dfrac{u^2 \sin 90^\circ}{g}$ ...(iii)

From Eqs. (i), (ii) and (iii), we get

$$\frac{x_1}{x_2} = \frac{(\sqrt{3}-1)}{(2-\sqrt{3})}$$

$$\therefore \quad (\sqrt{3}-1)\frac{x_1}{x_2} = \frac{(3+1-2\sqrt{3})}{(2-\sqrt{3})}$$

$$= \frac{2(2-\sqrt{3})}{(2-\sqrt{3})} = 2$$

48. Here, range is $OB = 2h\tan\theta = \dfrac{u^2 \sin 2\alpha}{g}$

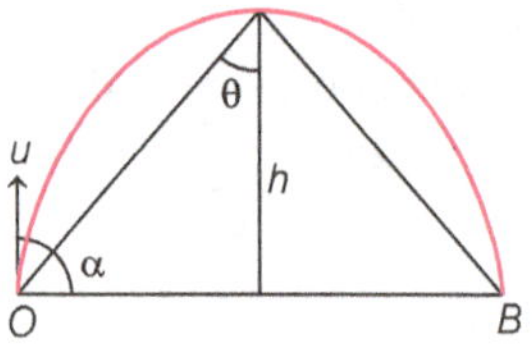

and $\quad h = \dfrac{u^2 \sin^2\alpha}{2g}$

$$\therefore \quad \frac{2h\tan\theta}{h} = \frac{\dfrac{u^2\sin 2\alpha}{g}}{\dfrac{u^2\sin^2\alpha}{2g}}$$

$\therefore \quad \tan\alpha = 2\cot\theta$

or $\quad \sin\alpha = \dfrac{2\cos\theta}{\sqrt{1+4\cot^2\theta}}$

$\because \quad h = \dfrac{u^2\sin^2\alpha}{2g}$

or $\quad \dfrac{u^2}{2g} = \dfrac{h}{\sin^2\alpha} = \dfrac{h}{4\cot^2\theta}(1+4\cot^2\theta)$

or $\quad \dfrac{u^2}{2gh} = \dfrac{1}{4\cot^2\theta}(1+4\cot^2\theta)$

$\quad = \dfrac{1}{4}\tan^2\theta + 1$

or $\quad \dfrac{u^2}{gh} = \dfrac{1}{2}\tan^2\theta + 2$

$\therefore \quad \dfrac{u^2}{gh} - \dfrac{1}{2}\tan^2\theta = 2$

49. Let position vectors of balls are A, B and C at instant t.

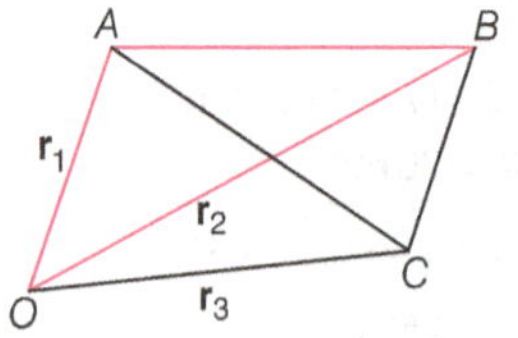

$$\mathbf{r}_1 = (u_1\hat{\mathbf{i}} + v_1\hat{\mathbf{j}})t - \frac{1}{2}gt^2\hat{\mathbf{j}}$$

$$\mathbf{r}_2 = (u_2\hat{\mathbf{i}} + v_2\hat{\mathbf{j}})t - \frac{1}{2}gt^2\hat{\mathbf{j}}$$

$$\mathbf{r}_3 = (u_3\hat{\mathbf{i}} + v_3\hat{\mathbf{j}})t - \frac{1}{2}gt^2\hat{\mathbf{j}}$$

$\therefore \quad \mathbf{BA} = \mathbf{r}_1 - \mathbf{r}_2 = (u_1 - u_2)t\hat{\mathbf{i}} + (v_1 - v_2)t\hat{\mathbf{j}}$

and $\mathbf{BC} = \mathbf{r}_3 - \mathbf{r}_2 = (u_3 - u_2)t\hat{\mathbf{i}} + (v_3 - v_2)t\hat{\mathbf{j}}$

Area of triangle $ABC = \dfrac{1}{2}|\mathbf{BA} \times \mathbf{BC}|$

$$=\frac{1}{2}\begin{vmatrix} \hat{\mathbf{i}} & \hat{\mathbf{j}} & \hat{\mathbf{k}} \\ (u_1-u_2)t & (v_1-v_2)t & 0 \\ (u_3-u_2)t & (v_3-v_2)t & 0 \end{vmatrix}$$

$$=\frac{1}{2}|\hat{\mathbf{i}}\times 0-\hat{\mathbf{j}}(0)+\hat{\mathbf{k}}\{(u_1-u_2)(v_3-v_2)t^2 - (v_1-v_2)(u_3-u_2)t^2|$$

$$=\frac{1}{2}\{(u_1-u_2)(v_3-v_2)-(v_1-v_2)(u_3-u_2)\}t^2$$

$\therefore\ n=2$

50. $R=\dfrac{v_0^2\sin 2\alpha}{g}$ or $8.4=\dfrac{v_0^2\sin 2\alpha}{10}$

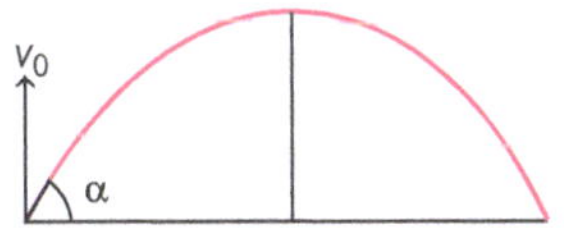

$\Rightarrow\quad v_0^2\sin 2\alpha=84$

$\Rightarrow\quad v_0^2=\dfrac{42}{\sin\alpha\cos\alpha}$

The equation of trajectory is

$$y=x\tan\alpha-\frac{gx^2}{2v_0^2\cos^2\alpha}$$

$$\Rightarrow\quad 3.6=4.8\tan\alpha-\frac{10x^2}{2\times\dfrac{42}{\sin\alpha\cos\alpha}\cos^2\alpha}$$

$$\Rightarrow\quad 3.6=4.8\tan\alpha-\frac{5}{42}(4.8)^2\tan\alpha$$

$$\therefore\quad \tan\alpha=1.75=\frac{7}{4}=\frac{n}{4}$$

$$\therefore\quad n=7$$

51. $v_0^2=2as=2\times 6\times 12$

$$\therefore\quad v_0=12\text{ ms}^{-1}$$

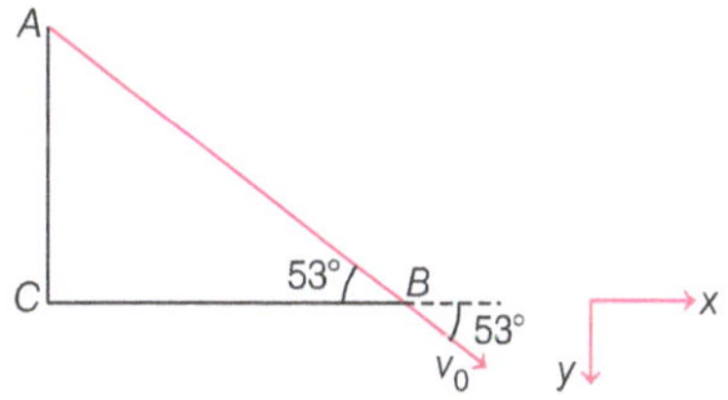

After point B,

$$u_x=12\cos 53°=12\times\frac{3}{5}=7.2\text{ ms}^{-1}$$

and $\quad u_y=12\sin 53°=12\times\dfrac{4}{5}=9.6\text{ ms}^{-1}$

$$\therefore\quad y=u_yt+\frac{1}{2}gt^2=9.6t+\frac{1}{2}\times 10\times t^2$$

or $\quad 14.6=9.6t+5t^2$

$$\therefore\quad t=1\text{ s}$$

$$\therefore\quad x_0=u_xt=7.2\times 1=7.2\text{ m}=\frac{36}{5}$$

$$\therefore\quad n=5$$

UNIT-8

NEWTON'S LAWS OF MOTION

Learning Modules

1. Forces and Free Body Diagram
2. Equilibrium of Forces
3. Newton's Laws of Motion
4. Constraint Equations
5. Pseudo Force
6. Friction

MODULE 1

Forces and Free Body Diagram

Types of Forces

There are basically four forces which are commonly encountered in mechanics.

1. Field Forces

These are the forces in which contact between two objects is not necessary. Gravitational force between two masses and electrostatic force between two charges are two examples of field forces. Weight (**W** $= mg$) of a body comes in this category.

2. Contact Forces

Two bodies in contact exert equal and opposite forces on each other. If the contact is frictionless, the contact force is perpendicular to the common surface and known as **normal reaction.**

If however, the objects are in rough contact and have relative motion (or have a tendency of relative motion) then **frictional force** also acts which oppose such relative motion (or its tendency).

Again each object exerts a frictional force on the other and the two forces are equal and opposite. This force is perpendicular to normal reaction. Thus, the contact force (F) between two objects is made up of two forces.

(i) Normal reaction (N)

(ii) Force of friction (f)

and since these two forces are mutually perpendicular,

$$F = \sqrt{N^2 + f^2}$$

Note *In this book normal reaction at most of the places has been represented by N. But at some places, it is also represented by R. This is because N is confused with the SI unit of force newton.*

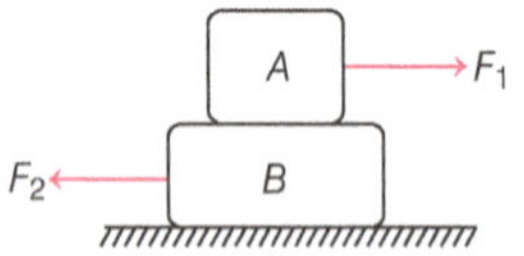

In the above figure , A is being moved to the right while B is being moved leftward. In order to see more clearly which forces act on A and which on B, a second diagram (in the figure shown below) is drawn but they are supposed to be in contact.

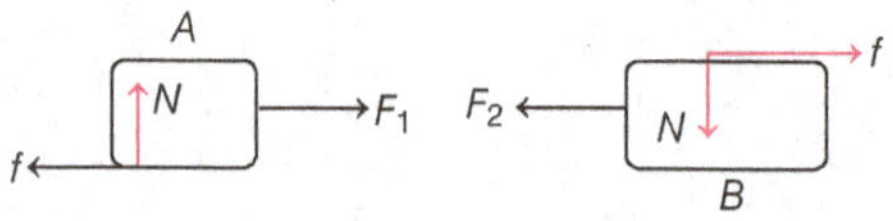

In the above figure, the two normal reactions each of magnitude N are perpendicular to the surface of contact between the blocks and the two frictional forces each of magnitude f act along that surface, in a direction opposing the motion of the block upon which it acts.

Note *(i) Forces on block B from the ground are not shown in the figure.*

(ii) Normal reaction is perpendicular to the common tangent direction and always acts towards the body. It is just like a pressure force ($F = PA$) which is also perpendicular to a surface and acts towards the body.

For example :

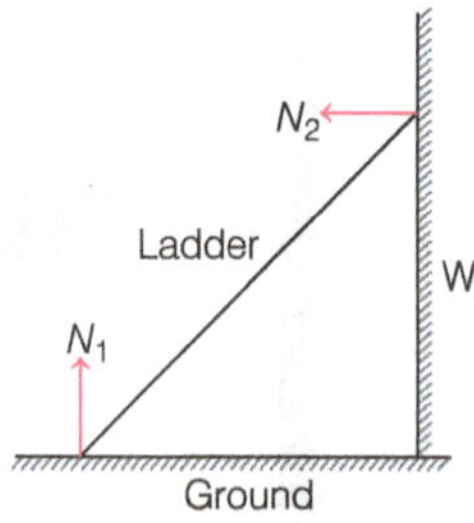

Normal reaction on ladder from ground is N_1 and from wall it is N_2.

3. Attachment to Another Body

Tension (T) in a string and **spring force** ($F = kx$) come in this group. Regarding the tension and string, the following three points are important to remember:

(i) The force exerted on a block (or body) by a rope, cable, chain etc. is called the force of tension T.

(ii) Tension is a pulling force.

If a string is attached with a block then it can apply force on the block only in a direction away from the block (in the form of tension).

String attached with a block

If the block is attached with a rod, then it can apply force on the block in both directions, towards the block (may be called push) or away from the block (called pull)

Rod attached with a block

(iii) We also sometimes say that a string is under tension. Tension in a string (or rope) may be uniform or non-uniform. Now let us see what is the meaning of tension in a string, uniform tension and non-uniform tension.

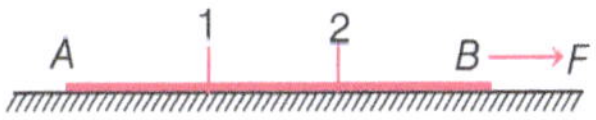

Suppose a rope AB kept on a smooth surface is pulled by a force F, then the rope will come under tension. Now suppose we want to visualize tension at two points (say 1 and 2) then we will have to consider two small sections of the rope at points 1 and 2.

These two sections are stretched by equal and opposite forces T_1 at point-1 and T_2 at point-2. So, tension at point-1 is T_1 and at point-2 is T_2. If tension is uniform then $T_1 = T_2$ and if it is non-uniform then $T_1 \neq T_2$. Note that tension at point-1 is T_1. It is neither zero nor $2T_1$.

For example,

$$T = 10\text{ N}$$
$$\neq 0$$
$$\neq 20\text{ N}$$

(iv) Let us take an example of uniform tension T. It is shown as below :

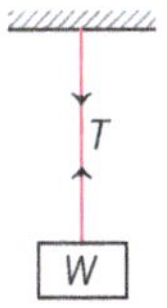

Now, you may be surprised that we are saying, string is stretched and we have shown the arrows as if the string is under compression. For better understanding we will have to see the forces separately.

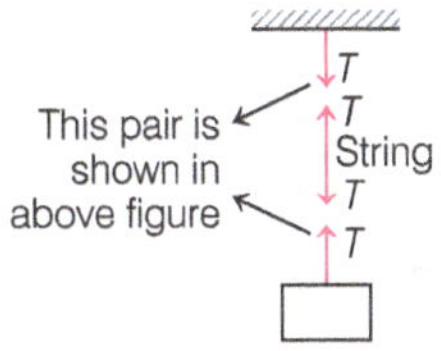

Now, from the above figure we can see that string is under tension.

Example In the figure :

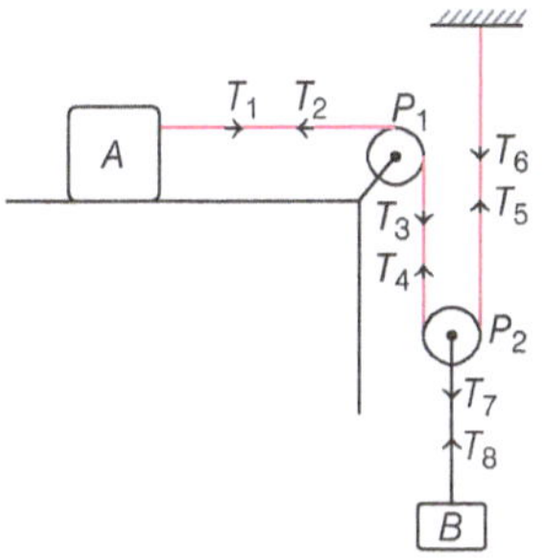

T_1 goes to block A (force applied by string on block A).

T_2 and T_3 to pulley P_1

T_4, T_5 and T_7 to pulley P_2

T_8 to block B and

T_6 to roof

If string and pullies are massless and there is no friction in the pullies, then

$T_1 = T_2 = T_3 = T_4 = T_5 = T_6$ and $T_7 = T_8$

(v) If a string is inextensible then the magnitude of acceleration of any number of masses connected through the string is always same.

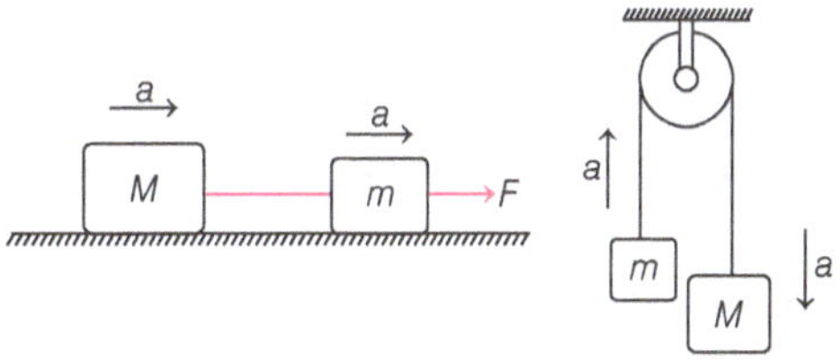

(vi) If a string is massless, the tension in it is same everywhere. However, if a string has a mass and it is accelerated, tension at different points will be different.

(vii) If pulley is massless and frictionless, tension will be same on both sides of the pulley.

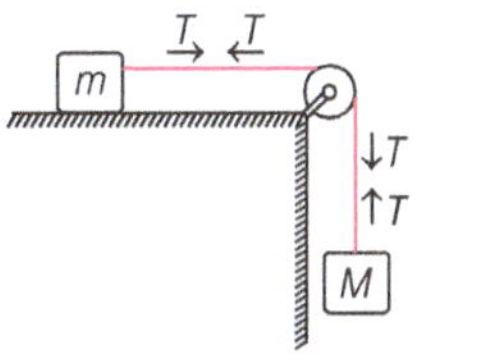

String and pulley are massless and there is no friction between pulley and string

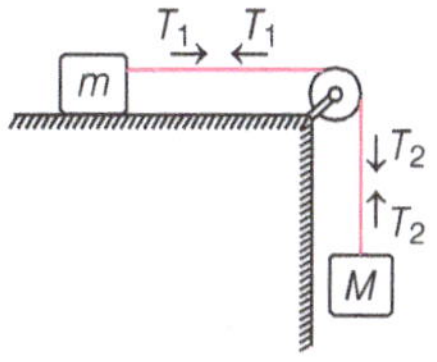

String is massless but pulley is not massless and frictionless

Spring force ($F = kx$) has been discussed in detail in the chapter of work, energy and power.

4. Hinge Force

In the figure shown, there is a hinge force on the rod (from the hinge). There are two methods of finding a hinge force :

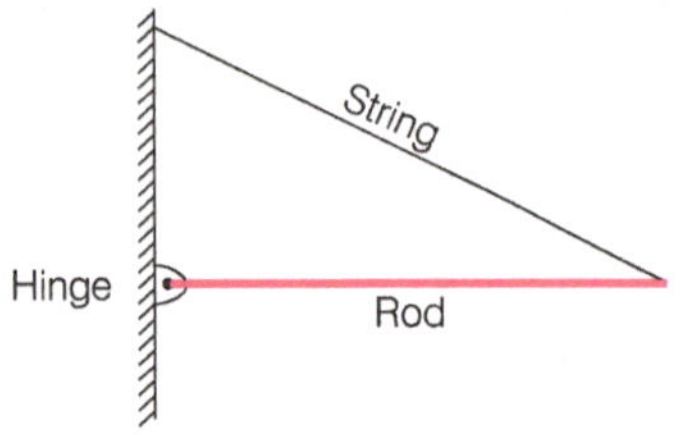

(i) either you find its horizontal (H) and vertical (V) components

(ii) or you find its magnitude and direction.

Later on through problems we will learn how to find hinge force.

Free Body Diagram

Free body diagram (FBD) is nothing but the force diagram.

No system, natural or man made, consists of a single body alone or is complete in itself. A single body or a part of the system can, however be isolated from the rest by appropriately accounting for its effect on the remaining system.

A free body diagram (FBD) consists of a diagrammatic representation of a single body or a sub-system of bodies isolated from its surroundings showing all the forces acting on it.

For example, a book lying on a horizontal surface.

Free body diagram of the book alone would consist of its weight ($W = mg$), acting downwards and the normal reaction (N) exerted on the book by the surface.

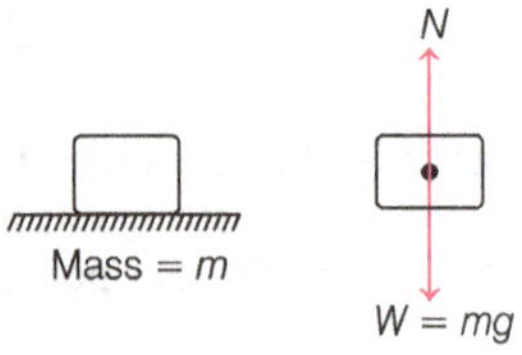

Example 1 A cylinder of weight W is resting on a V-groove as shown in figure. Draw its free body diagram.

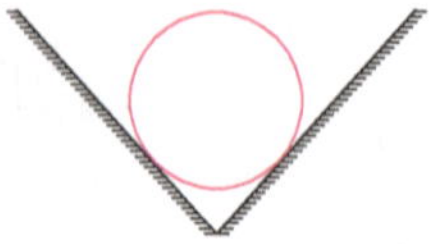

Sol. The free body diagram of the cylinder is as shown in figure.

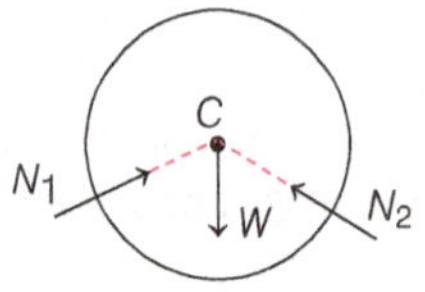

Here, W = weight of cylinder and N_1 and N_2 are the normal reactions from the two inclined surfaces.

Example 2 Three blocks A, B and C are placed one over the other as shown in figure. Draw free body diagrams of all the three blocks.

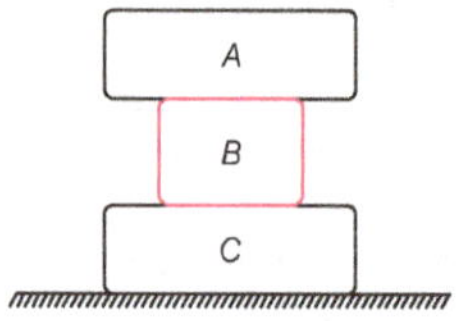

Sol. Free body diagrams of A, B and C are shown below.

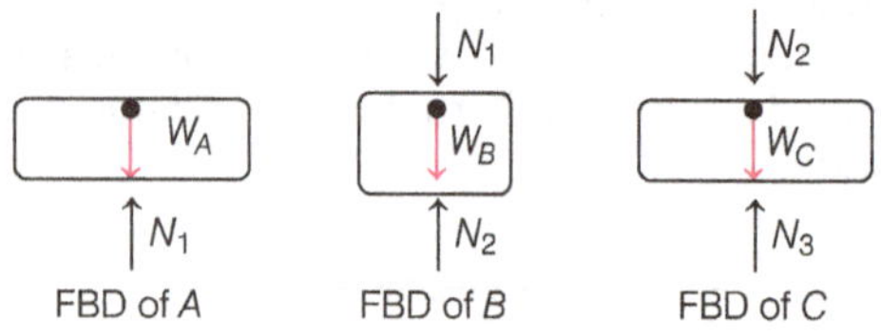

Here, N_1 = normal reaction between A and B

N_2 = normal reaction between B and C

N_3 = normal reaction between C and ground

W_A, W_B and W_C are the weights.

Example 3 A block of mass m is attached with two strings as shown in figure. Draw the free body diagram of the block.

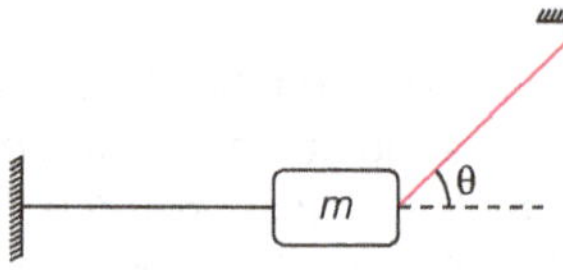

Sol. The free body diagram of the block is as shown in figure.

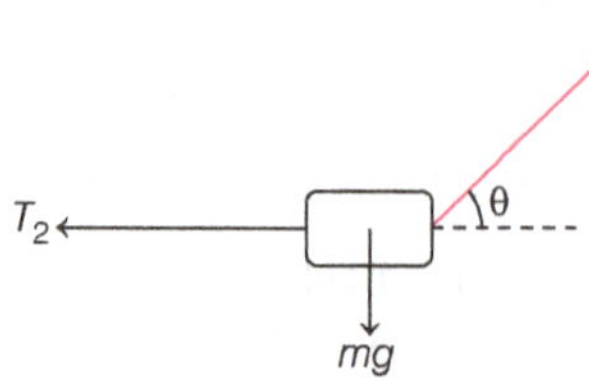

Example 4

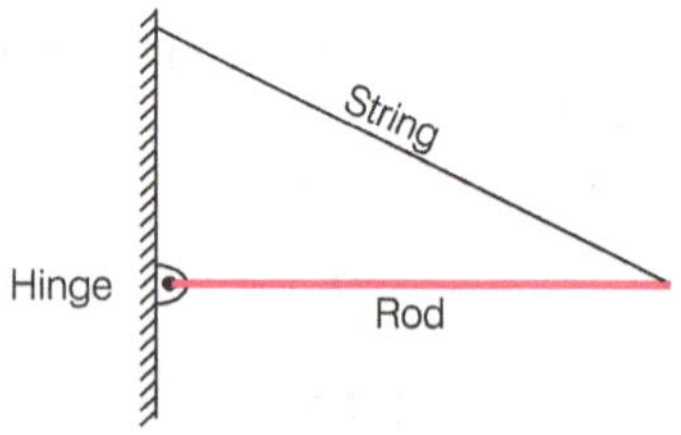

Show free body diagram of the rod as shown in figure.

Sol.

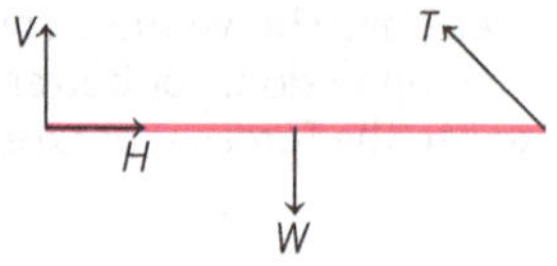

W = weight

T = tension

H and V = horizontal and vertical components of hinge force.

MODULE 1 Exercise

1. The diagram shows a rough plank resting on a cylinder with one end of the plank on rough ground. Neglect friction between plank and cylinder. Draw diagrams to show

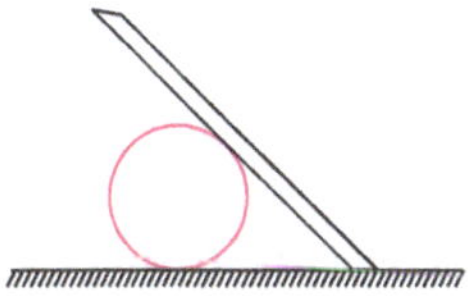

(a) the forces acting on the plank,
(b) the forces acting on the cylinder.

2. Two spheres A and B are placed between two vertical walls as shown in figure. Friction is absent everywhere. Draw the free body diagrams of both the spheres.

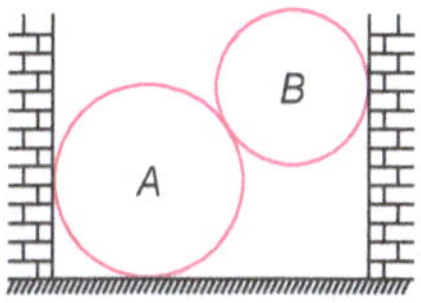

3. A point A on a sphere of weight W rests in contact with a smooth vertical wall and is supported by a string joining a point B on the sphere to a point C on the wall. Draw free body diagram of the sphere.

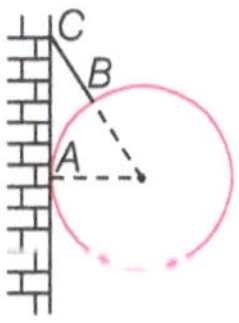

4. A rod AB of weight W_1 is placed over a sphere of weight W_2 as shown in figure. Ground is rough and there is no friction between rod and sphere and sphere and wall. Draw free body diagrams of sphere and rod separately.

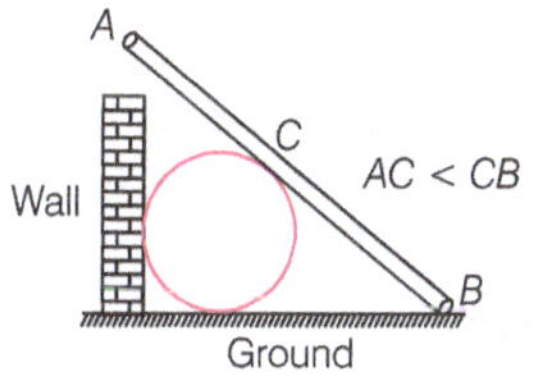

Note *No friction will act between sphere and ground. Think why?*

5. A rod OA is suspended with the help of a massless string AB as shown in figure. Rod is hinged at point O. Draw free body diagram of the rod.

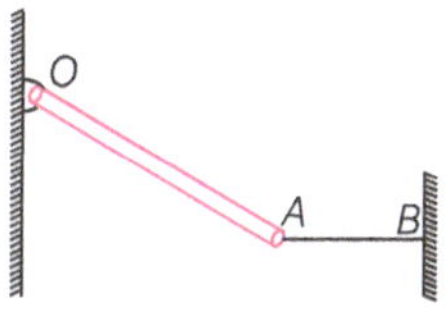

6. A rod AB is placed inside a rough spherical shell as shown in figure. Draw the free body diagram of the rod.

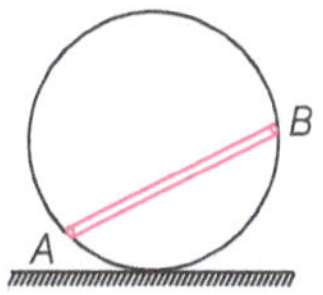

Answers

1 to 6 : See the hints

See hints on page 416.

MODULE 2

Equilibrium of Forces

Equilibrium

Forces which have zero resultant and zero turning effect are said to be in equilibrium. For understanding the equilibrium of forces let us first discuss the resolution of force and moment of a force about some point.

Resolution of a Force

In Unit-5, we have already discussed about the resolution of a vector quantity. Just to remind let us take an example.

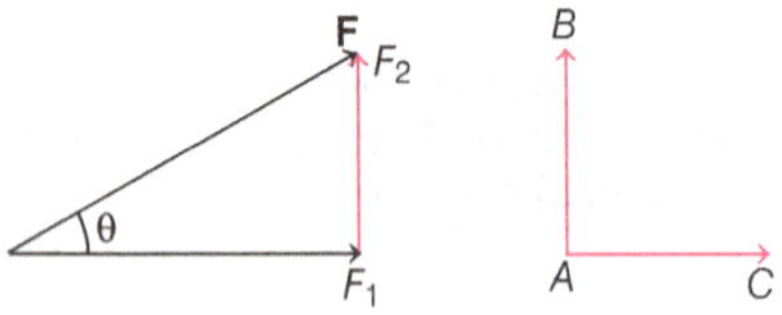

In the above figure,

$F_1 = F\cos\theta =$ component of **F** along AC

$F_2 = F\sin\theta =$ component of **F** perpendicular to AC or along AB

Finding such components is referred to as resolving a force in a pair of perpendicular directions. Note that the component of a force in a direction perpendicular to itself is zero.

For example, if a force of 10 N is applied on an object in horizontal direction then its component along vertical is zero.

Similarly, the component of a force in a direction parallel to the force is equal to the magnitude of the force. For example component of the above force in the direction of force (horizontal) will be 10 N.

In the opposite direction the component is –10 N.

Example 1 Resolve a weight of 10 N in two directions which are parallel and perpendicular to a slope inclined at 30° to the horizontal.

Sol. Component perpendicular to the plane

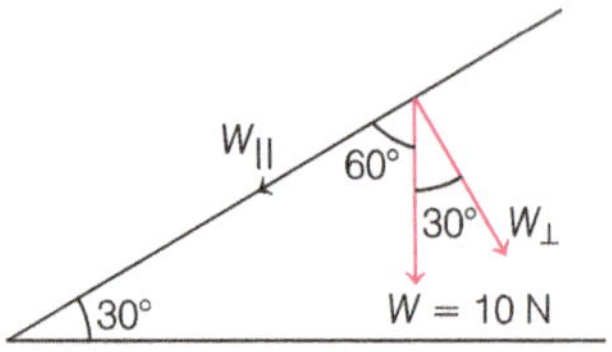

$$W_\perp = W\cos 30^\circ$$
$$= (10)\frac{\sqrt{3}}{2} = 5\sqrt{3}\text{ N}$$

and component parallel to the plane

$$W_{||} = W\sin 30^\circ = (10)\left(\frac{1}{2}\right) = 5\text{ N}$$

Example 2 Resolve horizontally and vertically a force $F = 8$ N which makes an angle of 45° with the horizontal.

Sol. Horizontal component of **F** is

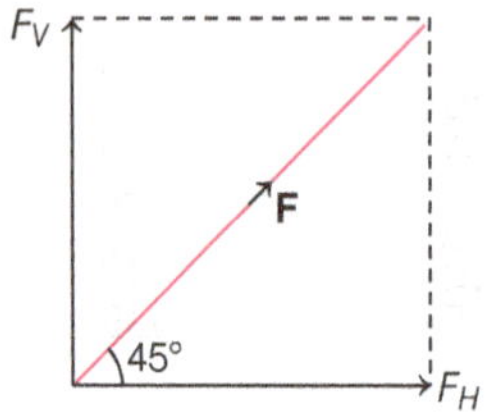

$$F_H = F\cos 45^\circ = (8)\left(\frac{1}{\sqrt{2}}\right)$$
$$= 4\sqrt{2}\text{ N}$$

and vertical component of **F** is $F_V = F\sin 45^\circ$

$$= (8)\left(\frac{1}{\sqrt{2}}\right) = 4\sqrt{2}\text{ N}$$

Example 3 A body is supported on a rough plane inclined at 30° to the horizontal by a string attached to the body and held at an angle of 30° to the plane. Draw a diagram showing the forces acting on the body and resolve each of these forces

(a) horizontally and vertically,

(b) parallel and perpendicular to the plane.

Sol. The forces are

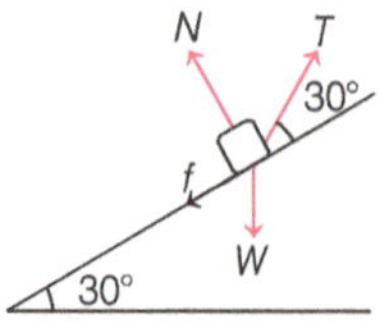

The tension in the string T

The normal reaction with the plane N

The weight of the body W and the friction f

(a) **Resolving horizontally and vertically**

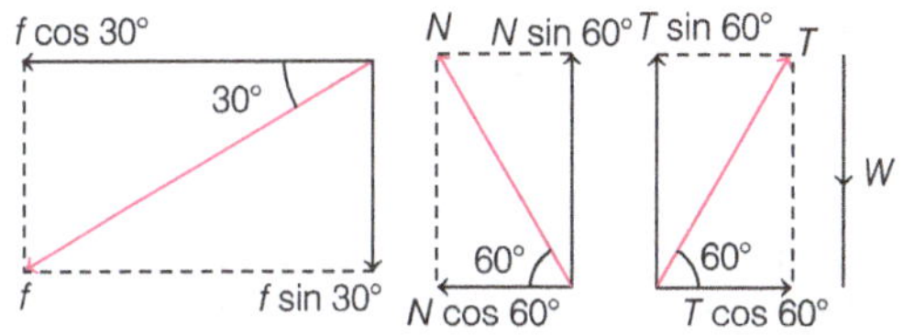

Resolving horizontally and vertically in the directions *OX* and *OY* as shown, the components are

Force	Components	
	Parallel to *OX* (horizontal)	Parallel to *OY* (vertical)
f	$-f\cos 30°$	$-f\sin 30°$
N	$-N\cos 60°$	$N\sin 60°$
T	$T\cos 60°$	$T\sin 60°$
W	0	$-W$

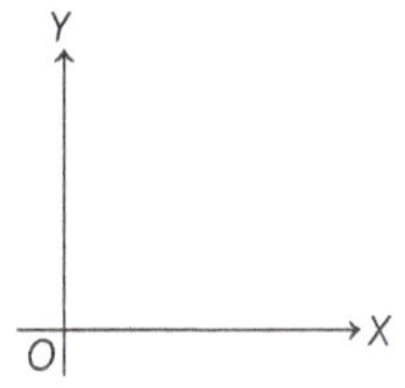

(b) Resolving parallel and perpendicular to the plane

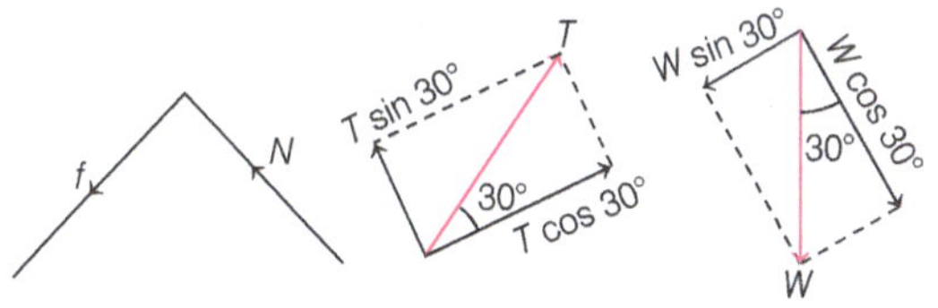

Resolving parallel and perpendicular to the plane in the directions *OX′* and *OY′* as shown, the components are :

Force	Components	
	Parallel to *OX′* (parallel to plane)	Parallel to *OY′* (perpendicular to plane)
f	$-f$	0
N	0	N
T	$T\cos 30°$	$T\sin 30°$
W	$-W\sin 30°$	$-W\cos 30°$

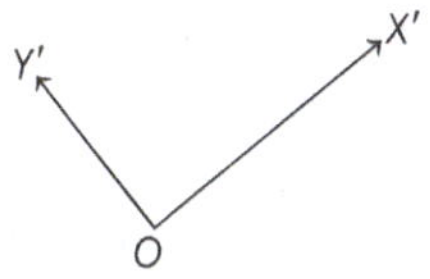

Moment of a Force

The general name given to any turning effect is **torque.** The magnitude of torque, also known as the moment of a force F is calculated by multiplying together the magnitude of the force and its perpendicular distance $r_\perp$ from the point about which we are finding the moment. This is denoted by M or τ (tau).

i.e. $$M = Fr_\perp \quad \text{or} \quad \tau = Fr_\perp$$

At this stage moment (or torque) is either clockwise or anticlockwise.

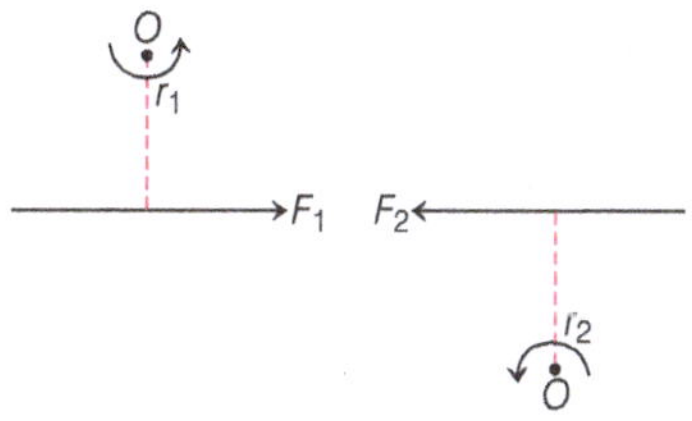

A convenient way to differentiate between clockwise and anticlock-wise torques is to allocate a positive sign to one sense (usually, but not invariably, this is anticlockwise) and negative sign to the other. With this convention, the moments of F_1 and F_2 about point O are $+F_1r_1$ and $-F_2r_2$.

Zero Moment

If the line of action of a force passes through the point O as shown in figure, its perpendicular distance from the axis is zero.

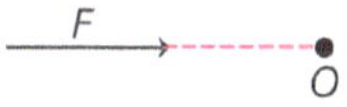

Therefore, its moment about that point is also zero.

Note *Later in the chapter of rotation we will see that torque is a vector quantity.*

Example 4 *ABCD* is a square of side 2 m and *O* is its centre. Forces act along the sides as shown in the diagram. Calculate the moment of each force about

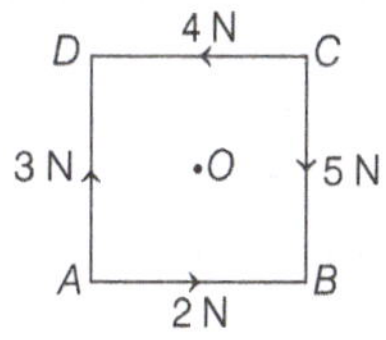

(a) point *A*
(b) point *O*.

Sol. Taking anticlockwise moments as positive we have:

(a)

Magnitude of force	2 N	5 N	4 N	3 N
Perpendicular distance from *A*	0	2 m	2 m	0
Moment about *A*	0	−10 N-m	+8 N-m	0

(b)

Magnitude of force	2 N	5 N	4 N	3 N
Perpendicular distance from O	1 m	1 m	1 m	1 m
Moment about O	+2 N-m	−5 N-m	+4 N-m	−3 N-m

Example 5 Forces act as indicated on a rod *AB* which is pivoted at *A*. Find the anticlockwise moment of each force about the point *A*.

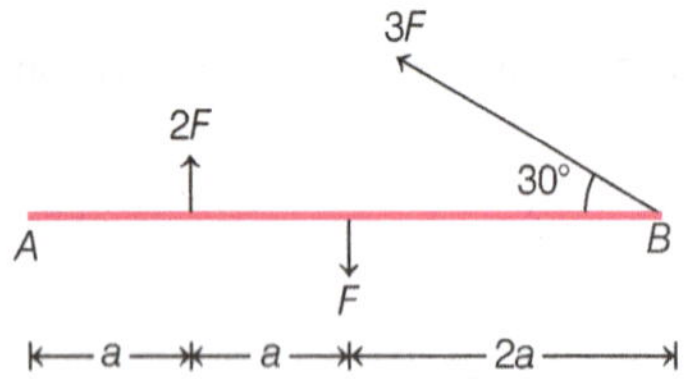

Sol.

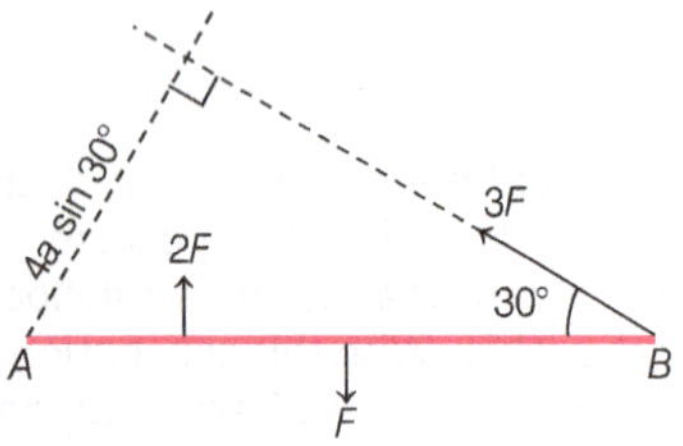

Taking anticlockwise moment as positive we have :

Magnitude of force	2F	F	3F
Perpendicular distance from A	a	$2a$	$4a \sin 30° = 2a$
Moment about A	$+2\,Fa$	$-2\,Fa$	$+6\,Fa$

Coplanar Forces in Equilibrium

When an object is in equilibrium under the action of a set of two or more coplanar forces, each of three factors which comprise the possible movement of the object must be zero, i.e. the object has

(i) no linear movement along any two mutually perpendicular directions *OX* and *OY*.

(ii) no rotation about any axis.

The set of forces must, therefore, be such that

(a) the algebraic sum of the components parallel to *OX* is zero or $\Sigma F_x = 0$

(b) the algebraic sum of the components parallel to *OY* is zero or $\Sigma F_y = 0$

(c) the resultant moment about any point is zero or $\Sigma \tau_{\text{any point}} = 0$

Thus, for the equilibrium of a set of two or more coplanar forces

$$\Sigma F_x = 0$$
$$\Sigma F_y = 0 \quad \text{and} \quad \Sigma \tau_{\text{any point}} = 0$$

Using the above three conditions, we get only three set of equations. So, in a problem number of unknowns should not be more than three.

Example 6 A rod *AB* rests with the end *A* on rough horizontal ground and the end *B* against a smooth vertical wall. The rod is uniform and of weight *W*. If the rod is in equilibrium in the position shown in figure. Find

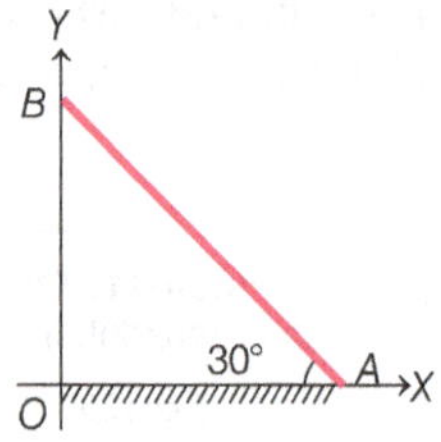

(a) frictional force at *A*
(b) normal reaction at *A*
(c) normal reaction at *B*

Sol. Let length of the rod be $2l$. Using the three conditions of equilibrium. Anticlockwise moment is taken as positive.

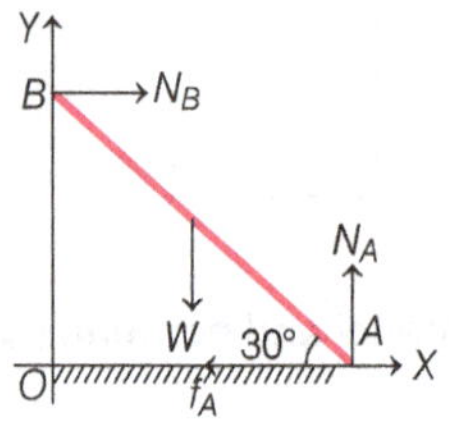

(i) $\Sigma F_X = 0 \Rightarrow \therefore N_B - f_A = 0$

or $N_B = f_A$ ...(i)

(ii) $\Sigma F_Y = 0 \Rightarrow \therefore N_A - W = 0$

or $N_A = W$ ...(ii)

(iii) $\Sigma \tau_O = 0$

$\therefore N_A (2l \cos 30°) - N_B (2l \sin 30°) - W(l \cos 30°) = 0$

or $\sqrt{3}N_A - N_B - \frac{\sqrt{3}}{2} W = 0$ ...(iii)

Solving these three equations, we get

(a) $f_A = \frac{\sqrt{3}}{2} W$ (b) $N_A = W$ (c) $N_B = \frac{\sqrt{3}}{2} W$

Exercise : What happens to N_A, N_B and f_A if

(a) Angle $\theta = 30°$ is slightly increased,

(b) A child starts moving on the ladder from *A* to *B* without changing the angle θ.

Ans. (a) Unchanged, decrease, decrease,
(b) Unchanged, increase, increase

Example 7 The rod shown in figure has a mass of 2 kg and length 4 m. In equilibrium, find the hinge force (or its two components) acting on the rod and tension in the string.

Take $g = 10\ \text{m/s}^2$, $\sin 37° = \frac{3}{5}$ and $\cos 37° = \frac{4}{5}$.

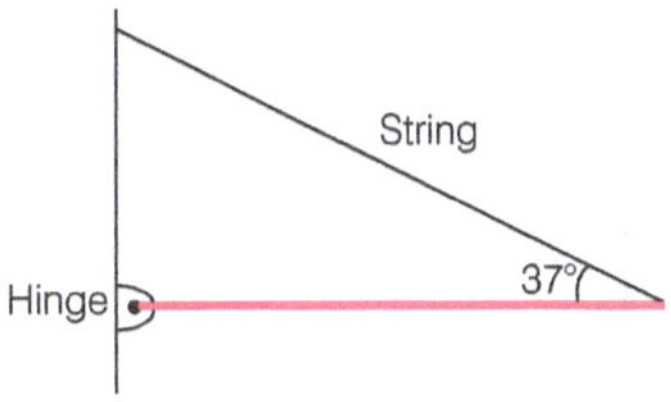

Sol.

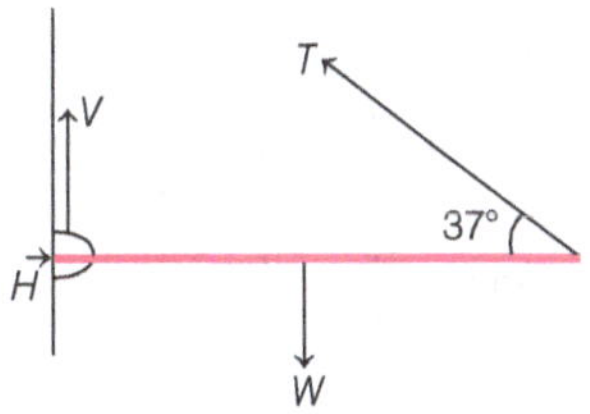

In the figure, only those forces which are acting on the rod have been shown. Here H and V are horizontal and vertical components of the hinge force.

$$W = mg = 20\ \text{N}$$

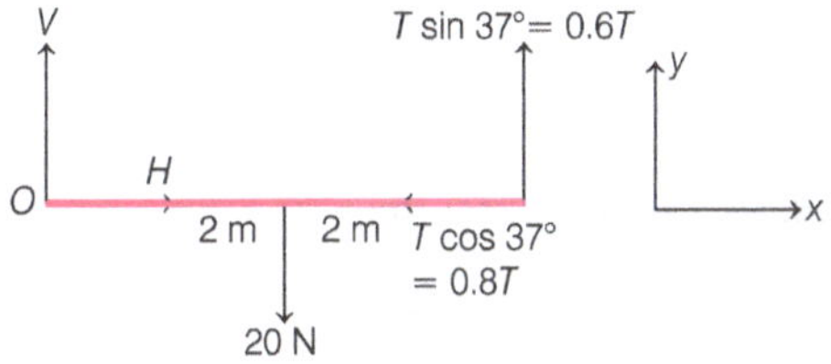

$$\Sigma F_x = 0$$

$\Rightarrow$ $$H - 0.8T = 0 \quad \text{...(i)}$$

$$\Sigma F_y = 0$$

$\Rightarrow$ $$V + 0.6\,T - 20 = 0 \quad \text{...(ii)}$$

$$\Sigma \tau_O = 0$$

$\Rightarrow$ Clockwise torque of 20 N = anticlock-wise torque of 0.6 T.

All other forces (H, V and 0.8 T) pass through O, hence their torques about point O are zero.

$\therefore$ $$20 \times 2 = 0.6T \times 4 \quad \text{...(iii)}$$

Solving these three equations, we get

$$T = 16.67\ \text{N},$$
$$H = 13.33\ \text{N}$$

and $$V = 10\ \text{N}$$ **Ans.**

Hinge force (F)

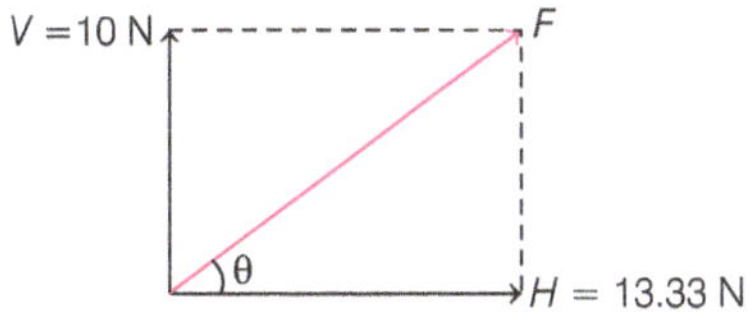

$$F = \sqrt{(13.33)^2 + (10)^2} = 16.67\ \text{N}$$ **Ans.**

$$\tan\theta = \frac{10}{13.33}$$

$\therefore$ $$\theta = \tan^{-1}\left(\frac{10}{13.33}\right) = 37°$$ **Ans.**

Equilibrium of Concurrent Coplanar Forces

If an object is in equilibrium under two or more concurrent coplanar forces the algebraic sum of the components of forces in any two mutually perpendicular directions OX and OY should be zero, i.e. the set of forces must be such that

(i) the algebraic sum of the components parallel to OX is zero, i.e. $\Sigma F_x = 0$.

(ii) the algebraic sum of the components parallel to OY is zero, i.e. $\Sigma F_y = 0$.

Thus, for the equilibrium of two or more concurrent coplanar forces

$$\Sigma F_x = 0$$
$$\Sigma F_y = 0$$

The third condition of zero moment about any point is automatically satisfied if the moment is taken about the point of intersection of the forces. So, here we get only two equations. Thus, number of unknown in any problem should not be more than two.

Example 8 An object is in equilibrium under four concurrent forces in the directions shown in figure. Find the magnitudes of $\mathbf{F}_1$ and $\mathbf{F}_2$.

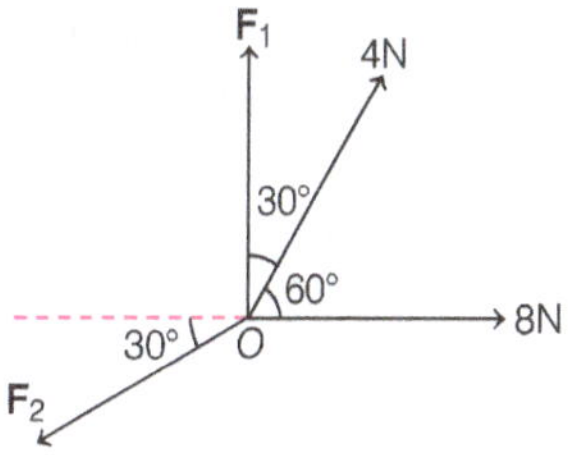

Sol. The object is in equilibrium.

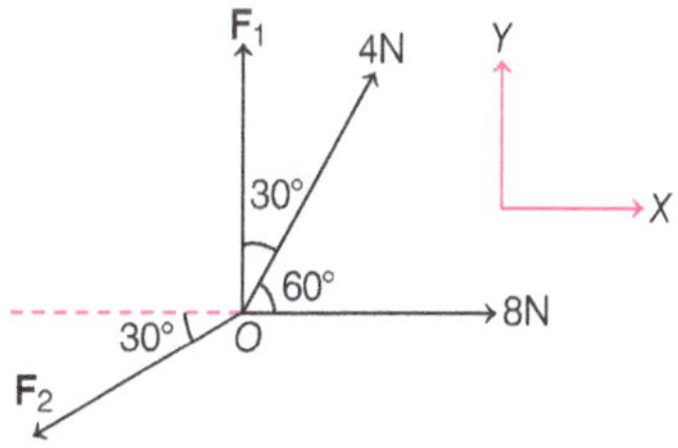

Hence,

(i) $\Sigma F_x = 0$

$\therefore$ $$8 + 4\cos 60° - F_2 \cos 30° = 0$$

or $$8 + 2 - F_2 \frac{\sqrt{3}}{2} = 0$$

or $$F_2 = \frac{20}{\sqrt{3}}\ \text{N}$$

(ii) $\Sigma F_y = 0$

$\therefore \quad F_1 + 4\sin 60° - F_2 \sin 30° = 0$

or $F_1 + \dfrac{4\sqrt{3}}{2} - \dfrac{F_2}{2} = 0$

or $\quad F_1 = \dfrac{F_2}{2} - 2\sqrt{3}$

$\quad = \dfrac{10}{\sqrt{3}} - 2\sqrt{3}$

or $\quad F_1 = \dfrac{4}{\sqrt{3}}$ N

Lami's Theorem

If an object O is in equilibrium under three concurrent forces $\mathbf{F}_1, \mathbf{F}_2$ and $\mathbf{F}_3$ as shown in figure. Then,

$$\frac{F_1}{\sin\alpha} = \frac{F_2}{\sin\beta} = \frac{F_3}{\sin\gamma}$$

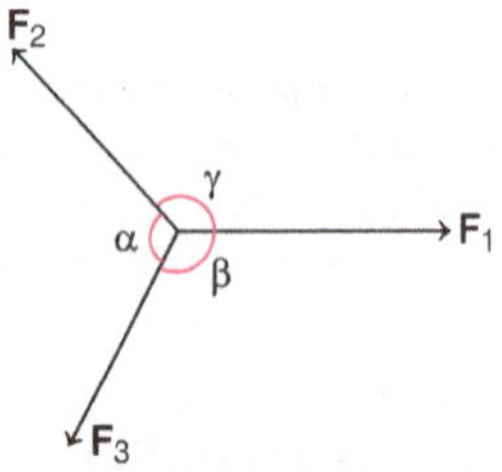

This property of three concurrent forces in equilibrium is known as Lami's theorem and is very useful method of solving problems related to three concurrent forces in equilibrium.

Example 9 One end of a string 0.5 m long is fixed to a point A and the other end is fastened to a small object of weight 8 N. The object is pulled aside by a horizontal force F, until it is 0.3 m from the vertical through A. Find the magnitudes of the tension T in the string and the force F.

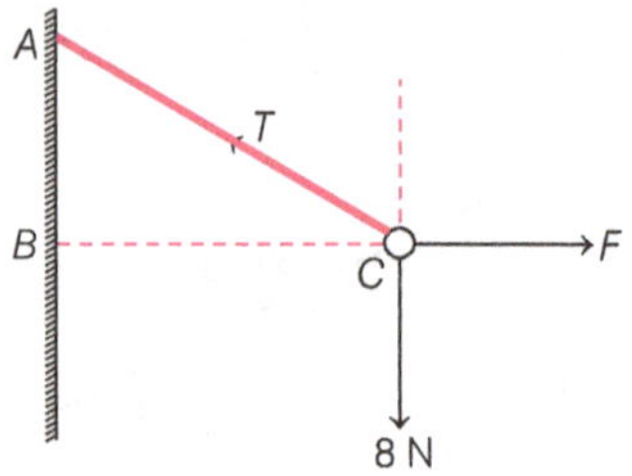

Sol. $AC = 0.5$ m, $\quad BC = 0.3$ m

$\therefore \quad AB = 0.4$ m

and if $\quad \angle BAC = \theta$

Then $\quad \cos\theta = \dfrac{AB}{AC} = \dfrac{0.4}{0.5} = \dfrac{4}{5}$

and $\quad \sin\theta = \dfrac{BC}{AC} = \dfrac{0.3}{0.5} = \dfrac{3}{5}$

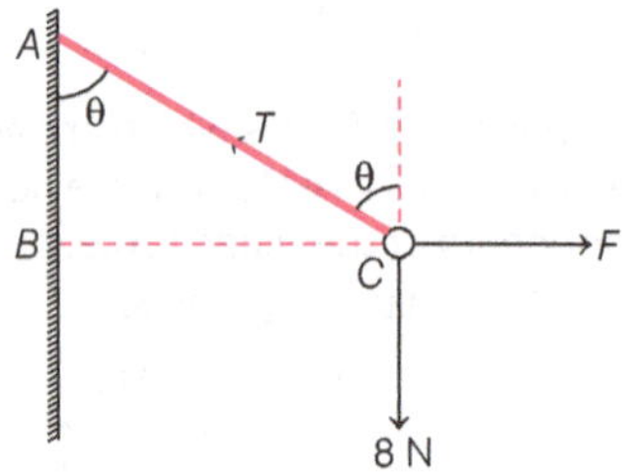

Here, the object is in equilibrium under three concurrent forces. So, we can apply Lami's theorem.

or $\quad \dfrac{F}{\sin(180° - \theta)} = \dfrac{8}{\sin(90° + \theta)} = \dfrac{T}{\sin 90°}$

or $\quad \dfrac{F}{\sin\theta} = \dfrac{8}{\cos\theta} = T$

$\therefore \quad T = \dfrac{8}{\cos\theta} = \dfrac{8}{4/5} = 10$ N

and $\quad F = \dfrac{8\sin\theta}{\cos\theta} = \dfrac{(8)(3/5)}{(4/5)} = 6$ N **Ans.**

MODULE 2 Exercise

Equilibrium of Concurrent Forces

Single Correct Option

1. In the figure, a block of mass 10 kg is in equilibrium. Identify the string in which the tension is zero.

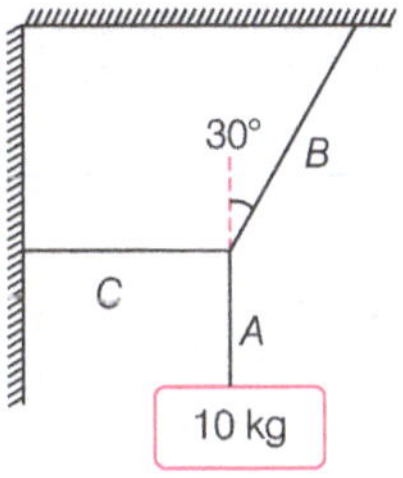

(a) B (b) C
(c) A (d) None of these

2. A metal sphere is hung by a string fixed to a wall. The forces acting on the sphere are shown in figure. Which of the following statements is correct?

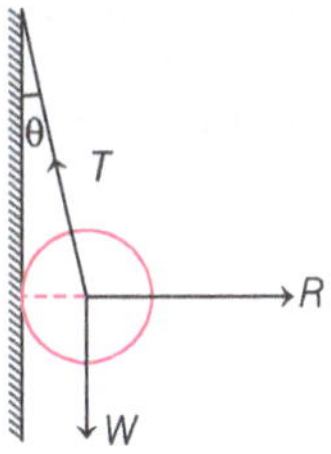

(i) $\mathbf{R} + \mathbf{T} + \mathbf{W} = 0$ (ii) $T^2 = R^2 + W^2$
(iii) $T = R + W$ (iv) $R = W \tan \theta$
(a) (i), (ii), (iii) (b) (ii), (iii), (iv)
(c) (i), (ii), (iv) (d) (i), (ii), (iii), (iv)

3. A smooth rod of mass 1kg is in equilibrium inside a hollow fixed sphere such that $\angle ACB = 120°$. Find the normal reaction on the rod at end A.

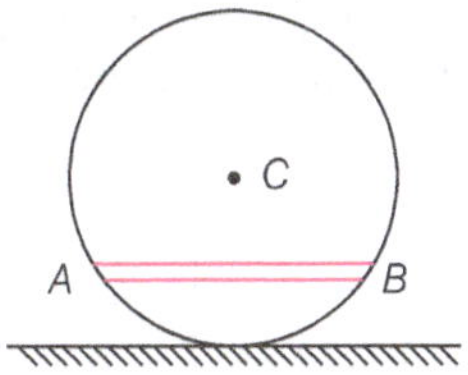

(a) 20 N (b) $10\sqrt{3}$ N
(c) 10 N (d) $5\sqrt{3}$ N

4. A block of mass $\sqrt{3}$ kg is attached to a string whose other end is attached to a wall. An unknown force F is applied so that the string makes an angle of 30° with the wall. The tension T is (Given $g = 10 \text{ ms}^{-2}$)

(2023 Main)

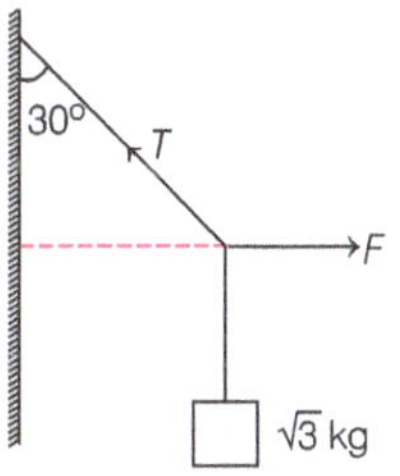

(a) 20 N (b) 25 N
(c) 10 N (d) 15 N

Subjective Questions

5. All the strings shown in figure are massless. Tension in the horizontal string is 30 N. Find W.

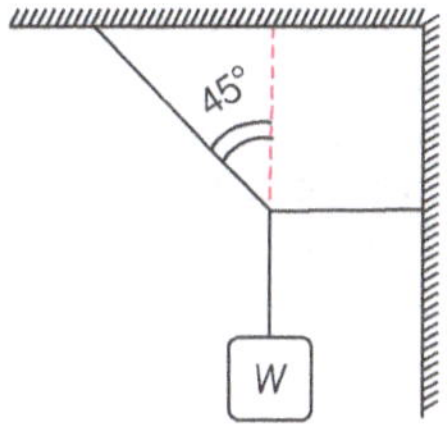

6. The 50 kg homogeneous smooth sphere rests on the 30° incline A and against the smooth vertical wall B. Calculate the contact forces at A and B.

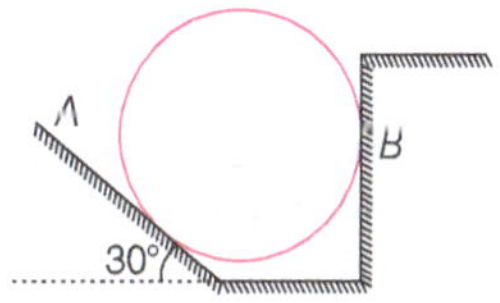

7. In figure, the tension in the diagonal string is 60 N.

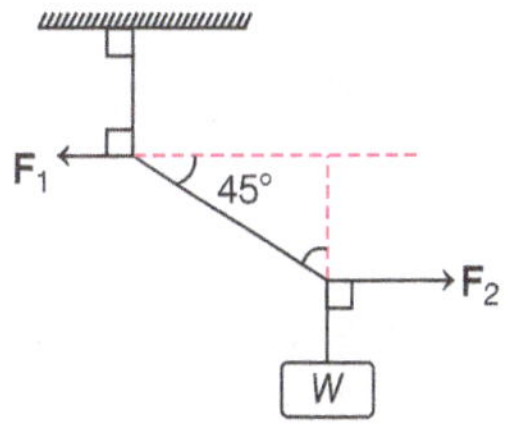

(a) Find the magnitude of the horizontal forces $\mathbf{F}_1$ and $\mathbf{F}_2$ that must be applied to hold the system in the position shown.
(b) What is the weight of the suspended block?

Integer Type Question

8. Four forces are acting at a point P in equilibrium as shown in figure. The ratio of forces F_1 to F_2 is $1 : x$, where $x =$ .... **(2022 Main)**

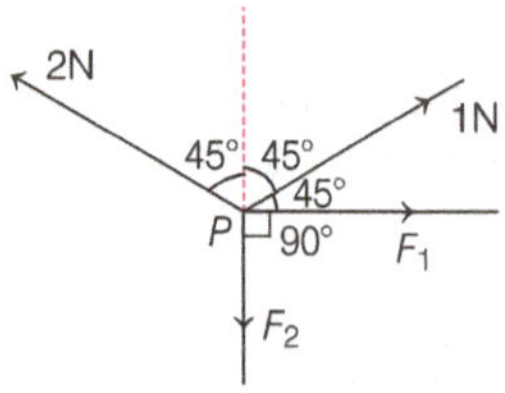

Equilibrium of Non-concurrent Forces

Assertion and Reason

Direction : Choose the correct option.

*(a) If both **Assertion** and **Reason** are true and the **Reason** is correct explanation of the **Assertion**.*

*(b) If both **Assertion** and **Reason** are true but **Reason** is not the correct explanation of **Assertion**.*

*(c) If **Assertion** is true, but the **Reason** is false.*

*(d) If **Assertion** is false but the **Reason** is true.*

*(e) If both **Assertion** and **Reason** are false.*

9. **Assertion :** A massless rod AB is suspended with the help of two strings as shown in figure. Tension on these two strings are T_1 and T_2. A force F is applied at a distance x from end B. If x is decreased then T_1 will decrease and T_2 will increase.

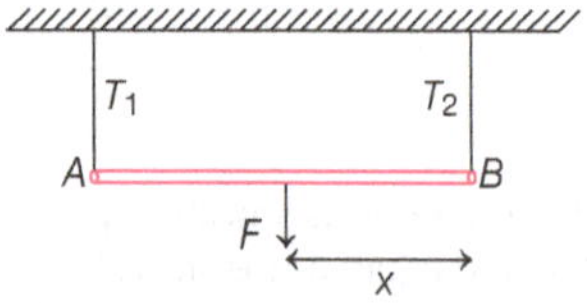

Reason : When $x = 0$, $T_1 = 0$ and $T_2 = F$.

Subjective Questions

10. A sphere of weight $W = 100$ N is kept stationary on a rough inclined plane by a horizontal string AB as shown in figure. Find

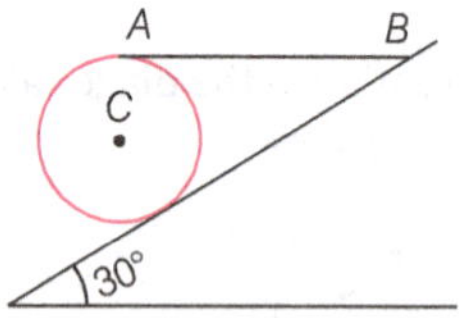

(a) tension in the string,
(b) force of friction on the sphere and
(c) normal reaction on the sphere by the plane.

11. A cube of mass 2 kg is held stationary against a rough wall by a force $F = 40$ N passing through centre C. Find perpendicular distance of normal reaction between wall and cube from point C. Side of the cube is 20 cm. (Take, $g = 10 \text{ m/s}^2$)

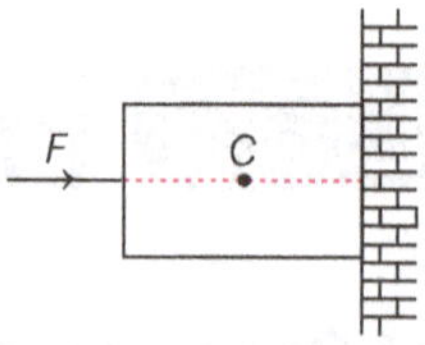

Miscellaneous Problems

Single Correct Option

12. The pulleys and strings shown in the figure are smooth and of negligible mass. For the system to remain in equilibrium, the angle θ should be **(2001 JEE)**

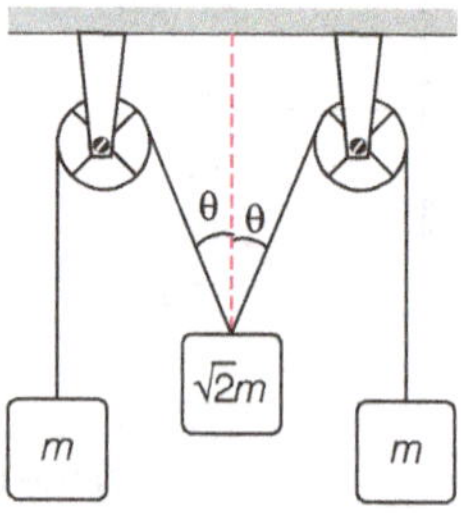

(a) 0° (b) 30°
(c) 45° (d) 60°

13. A string of negligible mass going over a clamped pulley of mass m supports a block of mass M as shown in the figure. The force on the pulley by the clamp is given by **(2001 JEE)**

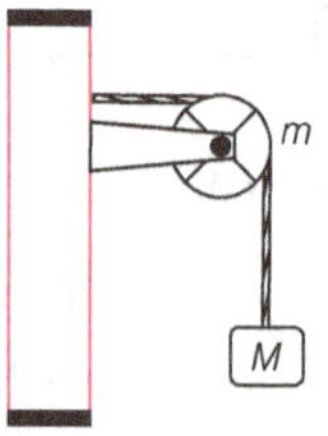

(a) $\sqrt{2}\,Mg$ (b) $\sqrt{2}\,mg$
(c) $g\sqrt{(M+m)^2 + m^2}$ (d) $g\sqrt{(M+m)^2 + M^2}$

14. A block of mass m is at rest under the action of a force F against a wall as shown in figure. Which of the following statement is **incorrect**? **(2005 JEE)**

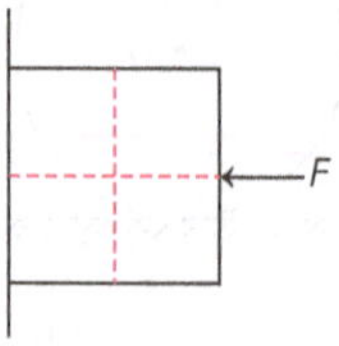

(a) $f = mg$ (where f is the frictional force)
(b) $F = N$ (where N is the normal force)
(c) F will not produce torque
(d) N will not produce torque

15. Two persons of equal heights are carrying a long uniform wooden plank of length l. They are at distance $\frac{l}{4}$ and $\frac{l}{6}$ from nearest end of the rod. The ratio of normal reaction at their heads is

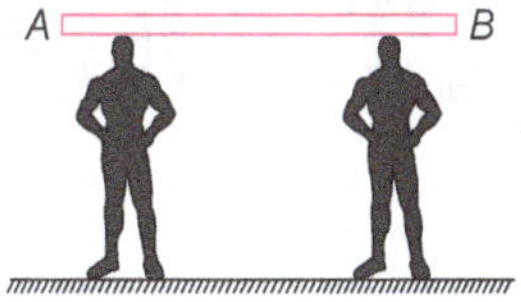

(a) 2 : 3 (b) 1 : 3 (c) 4 : 3 (d) 1 : 2

Subjective Question

16. Write down the components of four forces $\mathbf{F}_1$, $\mathbf{F}_2$, $\mathbf{F}_3$ and $\mathbf{F}_4$ along *ox* and *oy*-directions as shown in figure.

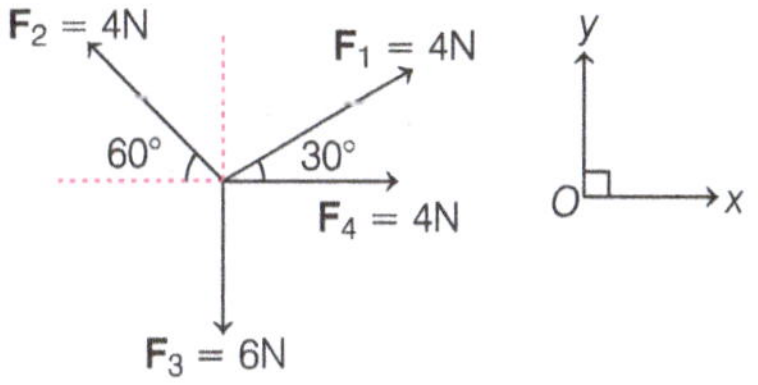

Answers

1. (d) 2. (c) 3. (c) 4. (a) 5. (30N) 6. $N_A = \frac{1000}{\sqrt{3}} N$, $N_B = \frac{500}{\sqrt{3}} N$

7. $F_1 = F_2 = W = 30\sqrt{2}$N 8. (3) 9. (b) 10. (a) 26.8 N (b) 26.8 N (c) 100 N 11. (5 cm) 12. (c) 13. (d)

14. (d) 15. (c) 16. $F_{1x} = 2\sqrt{3}$N, $F_{2x} = -2$N, $F_{3x} = 0$, $F_{4x} = 4$N, $F_{1y} = 2$N, $F_{2y} = 2\sqrt{3}$N, $F_{3y} = -6$N, $F_{4y} = 0$

See hints on pages 416-419.

MODULE 3

Newton's Laws of Motion

It is interesting to read Newton's original version of the laws of motion.

Law I Every body continues in its state of rest or in uniform motion in a straight line unless it is compelled to change that state by forces impressed upon it.

Law II The change of motion is proportional to the magnitude of force impressed and is made in the direction of the straight line in which that force is impressed.

Law III To every action there is always an equal and opposite reaction or the mutual actions of two bodies upon each other are always directed to contrary parts.

The Modern Versions of These Laws are :

1. A body continues in its initial state of rest or motion with uniform velocity unless acted on by an unbalanced external force.
2. The acceleration of a body is inversely proportional of its mass and directly proportional to the resultant external force acting on it. Or

$$\Sigma \mathbf{F} = \mathbf{F}_{net} = m\,\mathbf{a}$$

or

$$\mathbf{a} = \frac{\mathbf{F}_{net}}{m}$$

3. Forces always occur in pairs. If body A exerts a force on body B, an equal but opposite force is exerted by body B on body A.

Problems of Common Acceleration

If all bodies of a system has a common acceleration then that common acceleration is given by

$$a = \frac{\text{Net pulling/pushing force}}{\text{Total mass}} = \frac{\text{NPF}}{\text{TM}}$$

Net pulling/pushing force (NPF) is actually the net force.

Example Suppose two unequal masses m and $2m$ are attached to the ends of a light inextensible string which passes over a smooth massless pulley. We have to find the acceleration of the system. We can assume that the mass $2m$ is pulled downwards by a force equal to its weight, i.e. $2mg$.

Similarly, the mass m is being pulled by a force of mg downwards. Therefore, net pulling force on the system is $2mg - mg = mg$ and total mass being pulled is $2m + m = 3m$.

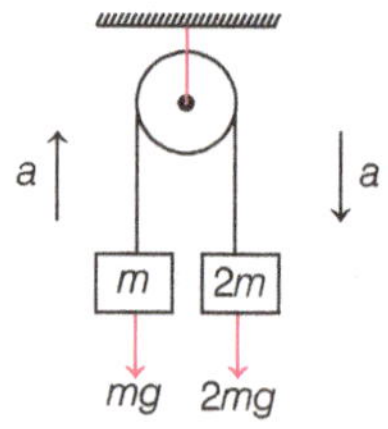

$\therefore$ Acceleration of the system is

$$a = \frac{\text{Net pulling force}}{\text{Total mass to be pulled}} = \frac{mg}{3m} = \frac{g}{3}$$

Note *(i) While finding net pulling force, take the forces (or their components) which are in the direction of motion (or opposite to it) and are single (i.e. they are not forming pair of equal and opposite forces). For example weight (mg) or some applied force F. Tension makes an equal and opposite pair. So, they are not to be included, unless the system in broken at some place and only one tension is taken on the system under consideration.*

(ii) After finding that common acceleration, we will have to draw free body diagrams of different blocks to find normal reaction or tension etc.

*(iii) If **a** is the acceleration of a body, then m**a** force does not act on the body but this much force is required to provide **a** acceleration to the body. The different available forces acting on the body provide this m**a** force or, we can say that vector sum of all forces acting on the body is equal to m**a**. The available forces may be weight, tension, normal reaction, friction or any externally applied force etc.*

Resolution of Forces

In this chapter different situations may be classified in two types:

(i) Permanent rest or body in equilibrium or net force equal to zero or net acceleration equal to zero or moving with constant velocity.

(ii) Accelerated or temporary rest.

In the first situation, forces can be resolved in any direction. Net force (or summation of components of different forces acting on the body) in all directions should be zero.

In the second situation forces are normally resolved along acceleration and perpendicular to it. In the direction perpendicular to acceleration net force is zero and along acceleration net force is ma.

In temporary rest situation, velocity of the body is zero but acceleration is not zero. The direction of acceleration in this case is the direction in which the body is supposed to move just after few seconds. Three situations of temporary rest are shown below.

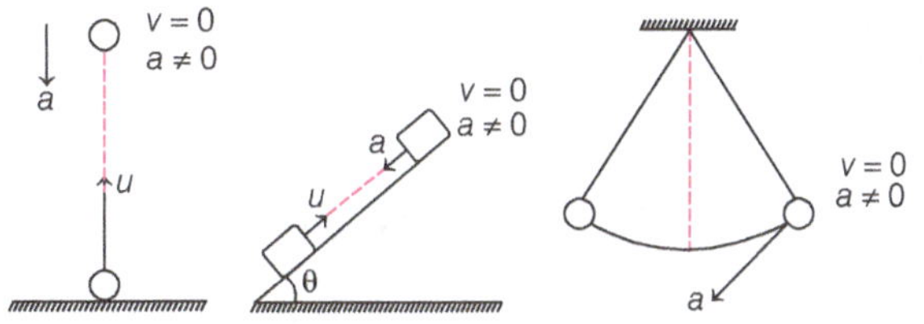

Note *In the second situation also, we can resolve the forces in any direction. In that case, net force along this direction = (mass) (component of acceleration in this direction)*

Typed Problems

Type 1. Problems of common acceleration, tension and normal reaction.

Ex. 1 Two blocks of masses 4 kg and 2 kg are placed side by side on a smooth horizontal surface as shown in the figure. A horizontal force of 20 N is applied on 4 kg block. Find

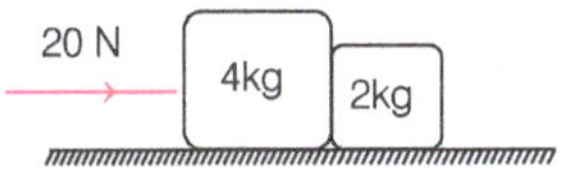

(a) the acceleration of each block.
(b) the normal reaction between two blocks.

Sol. (a) Both the blocks will move with same acceleration (say a) in horizontal direction.

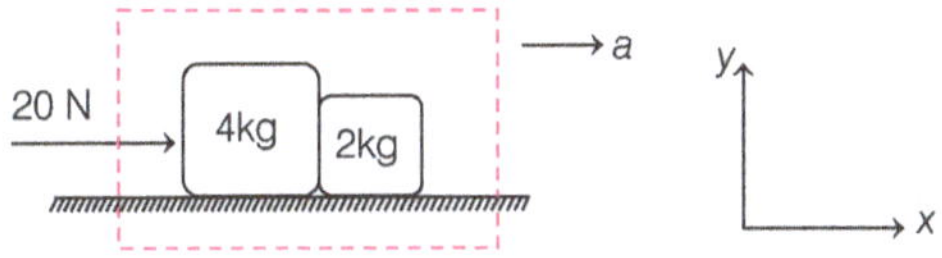

Let us take both the blocks as a system. Net external force on the system is 20 N in horizontal direction.

Using $\Sigma F_x = ma_x$

$$20 = (4+2)a = 6a$$

or $a = \frac{10}{3}\ \text{m/s}^2$ **Ans.**

Alternate Method

$$a = \frac{\text{Net pushing force}}{\text{Total mass}}$$

$$= \frac{20}{4+2} = \frac{10}{3}\ \text{m/s}^2$$

(b) The free body diagram of both the blocks are as shown in figure.

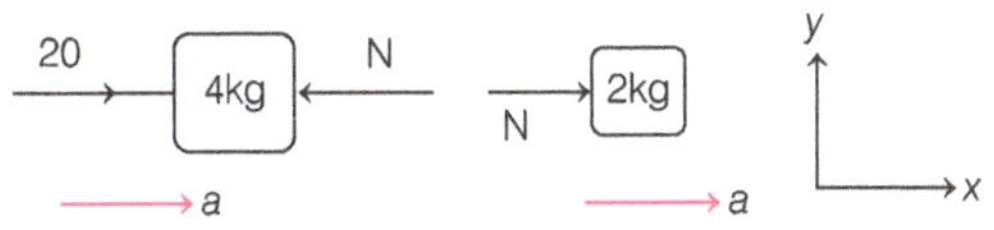

Using $\Sigma F_x = ma_x$

For 4 kg block,

$$20 - N = 4a = 4 \times \frac{10}{3}$$

$$N = 20 - \frac{40}{3} = \frac{20}{3}\ \text{newton}$$ **Ans.**

This can also be solved as under

For 2 kg block,

$$N = 2a = 2 \times \frac{10}{3} = \frac{20}{3}\ \text{newton}$$

Here, N is the normal reaction between the two blocks.

Note *In free body diagram of the blocks we have not shown the forces acting on the blocks in vertical direction, because normal reaction between the blocks and acceleration of the system can be obtained without using $\Sigma F_y = 0$.*

Ex. 2 Three blocks of masses 3 kg, 2 kg and 1 kg are placed side by side on a smooth surface as shown in figure. A horizontal force of 12 N is applied on 3 kg block. Find the net force on 2 kg block.

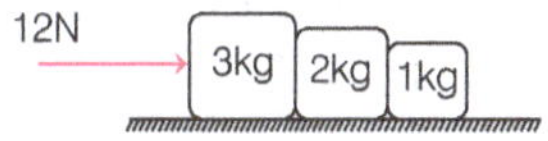

Sol. Since, all the blocks will move with same acceleration (say a) in horizontal direction. Let us take all the blocks as a single system.

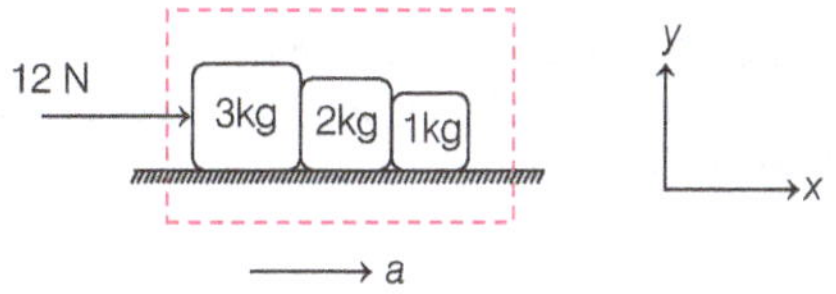

Net external force on the system is 12 N in horizontal direction.

Using $\Sigma F_x = ma_x$

$$12 = (3+2+1)a = 6a$$

or $a = \frac{12}{6} = 2\ \text{m/s}^2$

Alternate Method

$$a = \frac{\text{Net pushing force}}{\text{Total mass}} = \frac{12}{3+2+1} = 2\ \text{m/s}^2$$

Now, let F be the net force on 2 kg block in x-direction, then using $\Sigma F_x = ma_x$ for 2 kg block, we get

$$F = (2)(2) = 4\text{ N} \qquad \textbf{Ans.}$$

Note *Here, net force F on 2 kg block is the resultant of* N_1 *and* $N_2 \, (N_1 > N_2)$
where, N_1 = *normal reaction between 3 kg and 2 kg block,*
and N_2 = *normal reaction between 2 kg and 1 kg block.*
Thus, $F = N_1 - N_2$

Ex. 3 In the arrangement shown in figure. The strings are light and inextensible. The surface over which blocks are placed is smooth. Find

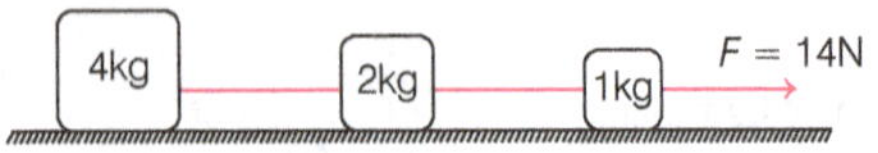

(a) the acceleration of each block.
(b) the tension in each string.

Sol. (a) Let a be the acceleration of each block and T_1 and T_2 be the tensions, in the two strings as shown in figure.

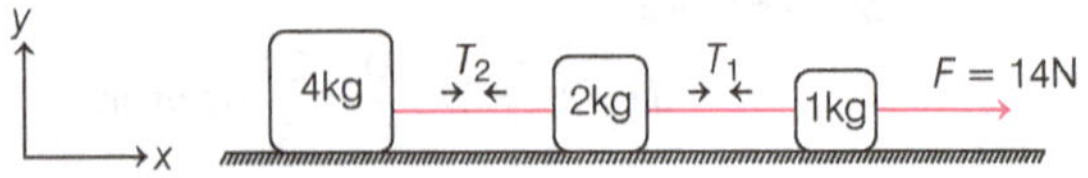

Taking the three blocks and the two strings as the system.

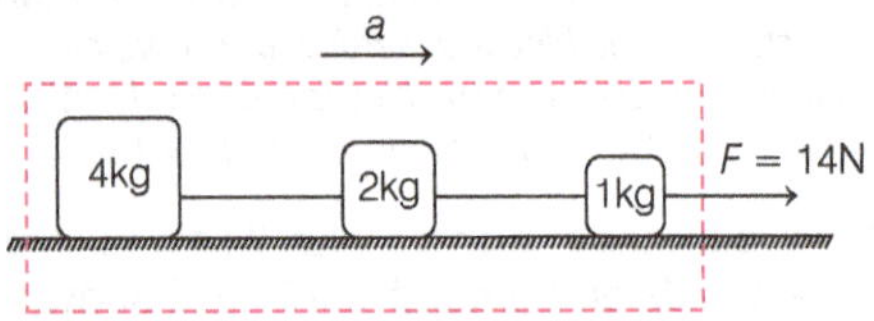

Using $\Sigma F_x = ma_x$

or $\quad 14 = (4 + 2 + 1)a$

or $\quad a = \dfrac{14}{7} = 2\text{ m/s}^2$ **Ans.**

Alternate Method

$$a = \frac{\text{Net pulling force}}{\text{Total mass}} = \frac{14}{4+2+1} = 2\text{ m/s}^2$$

(b) Free body diagram (showing the forces in x-direction only) of 4 kg block and 1kg block are shown in figure.

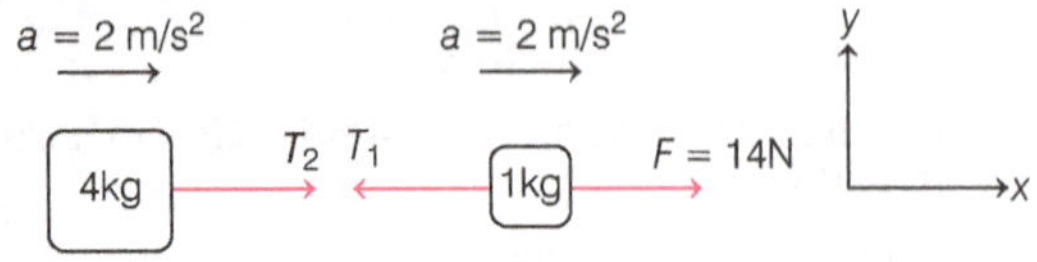

Using $\quad \Sigma F_x = ma_x$

For 1 kg block, $\quad F - T_1 = (1)(a)$

or $\quad 14 - T_1 = (1)(2) = 2$

$\therefore \quad T_1 = 14 - 2 = 12\text{ N}$ **Ans.**

For 4 kg block, $\quad T_2 = (4)(a)$

$\therefore \quad T_2 = (4)(2) = 8\text{ N}$ **Ans.**

Ex. 4 Two blocks of masses 4 kg and 2 kg are attached by an inextensible light string as shown in figure. Both the blocks are pulled vertically upwards by a force $F = 120$ N. Find

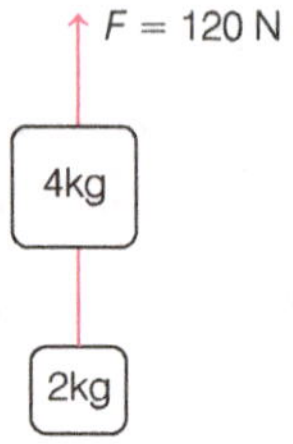

(a) the acceleration of the blocks,
(b) tension in the string. (Take $g = 10\text{ m/s}^2$)

Sol. (a) Let a be the acceleration of the blocks and T the tension in the string as shown in figure.

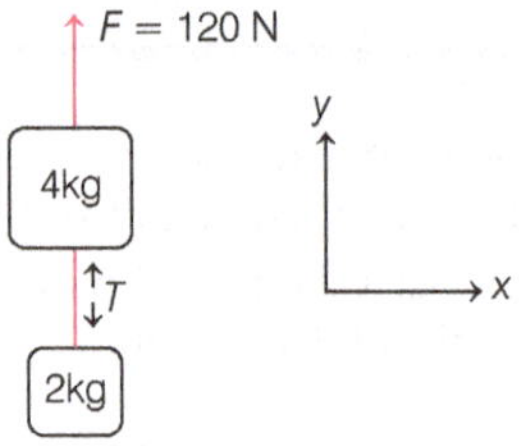

Taking the two blocks and the string as the system (shown below).

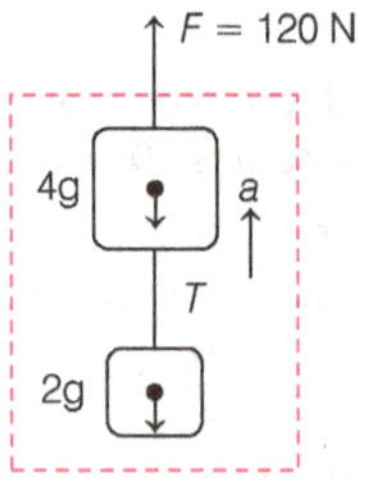

Using $\Sigma F_y = ma_y$, we get

$$F - 4g - 2g = (4 + 2)a$$

or $\quad 120 - 40 - 20 = 6a \quad$ or $\quad 60 = 6a$

$\therefore \quad a = 10\text{ m/s}^2$ **Ans.**

Alternate Method

$$a = \frac{\text{Net pulling force}}{\text{Total mass}} = \frac{120 - 60}{4+2}$$

$$= 10\text{ m/s}^2$$

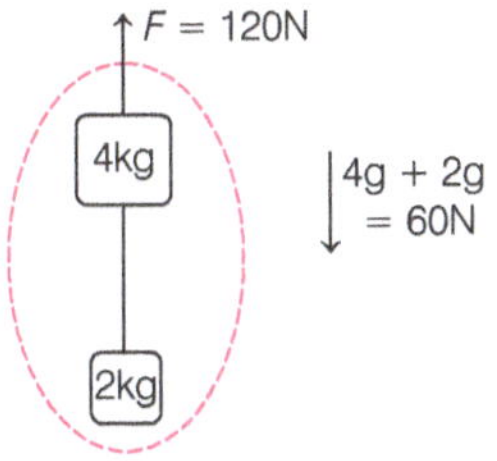

(b) Free body diagram of 2 kg block is as shown in figure.

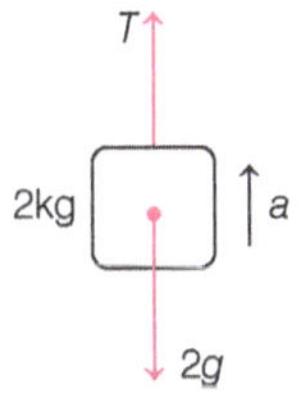

Using $\Sigma F_y = ma_y$

we get, $T - 2g = 2a$

or $T - 20 = (2)(10)$

$\therefore$ $T = 40\text{ N}$ **Ans.**

Ex. 5 In the system shown in figure pulley is smooth. String is massless and inextensible. Find acceleration of the system a, tensions T_1 and T_2. ($g = 10\text{ m/s}^2$)

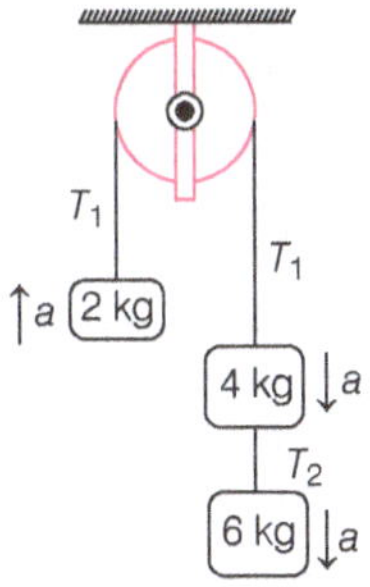

Sol. Here, net pulling force will be

Weight of 4 kg and 6 kg blocks on one side - weight of 2 kg block on the other side. Therefore,

$$a = \frac{\text{Net pulling force}}{\text{Total mass}}$$

$$= \frac{(6 \times 10) + (4 \times 10) - (2)(10)}{6 + 4 + 2}$$

$$= \frac{20}{3}\text{ m/s}^2$$ **Ans.**

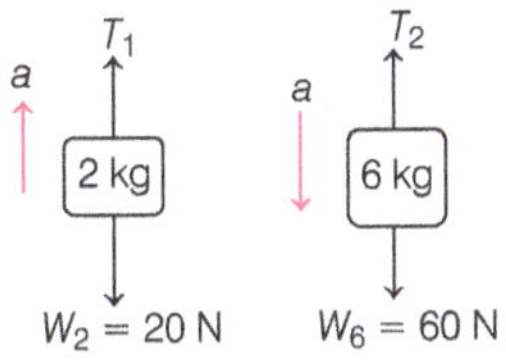

For T_1, let us consider FBD of 2 kg block. Writing equation of motion, we get

$$T_1 - 20 = 2a$$

or $$T_1 = 20 + 2 \times \frac{20}{3}$$

$$= \frac{100}{3}\text{ N}$$ **Ans.**

For T_2, we may consider FBD of 6 kg block. Writing equation of motion, we get

$$60 - T_2 = 6a$$

$\therefore$ $$T_2 = 60 - 6a = 60 - 6\left(\frac{20}{3}\right)$$

$$= 20\text{ N}$$ **Ans.**

Exercise Draw FBD of 4 kg block. Write down the equation of motion for it and check whether the values calculated above are correct or not.

Ex. 6 In the system shown in figure all surfaces are smooth. String is massless and inextensible. Find acceleration a of the system and tension T in the string. ($g = 10\text{ m/s}^2$)

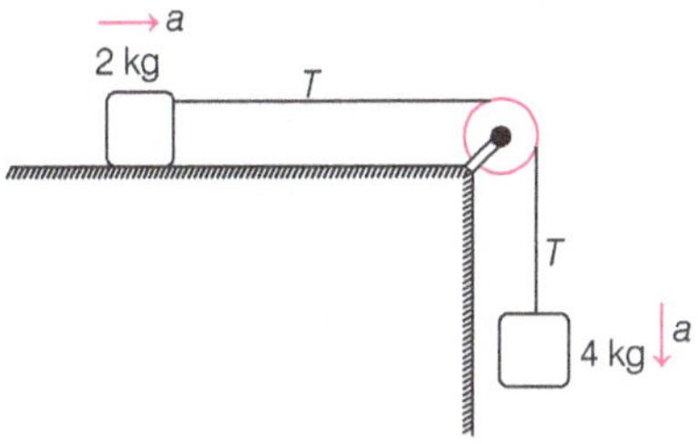

Sol. Here, weight of 2 kg is perpendicular to motion (or a). Hence, it will not contribute in net pulling force. Only weight of 4 kg block will be included.

$\therefore$ $$a = \frac{\text{Net pulling force}}{\text{Total mass}}$$

$$= \frac{(4)(10)}{(4 + 2)} = \frac{20}{3}\text{ m/s}^2$$ **Ans.**

For T, consider FBD of 4 kg block. Writing equation of motion.

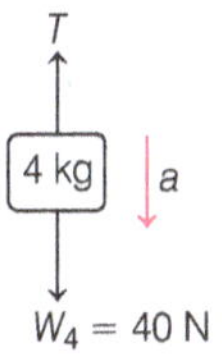

$$40 - T = 4a$$

$\therefore$ $$T = 40 - 4a$$

$$= 40 - 4\left(\frac{20}{3}\right) = \frac{40}{3}\text{ N}$$ **Ans.**

Exercise Draw FBD of 2 kg block and write down equation of motion for it. Check whether the values calculated above are correct or not.

Ex. 7 In the figure, masses of *A*, *B* and *C* are 1 kg, 3 kg and 2 kg respectively. Find

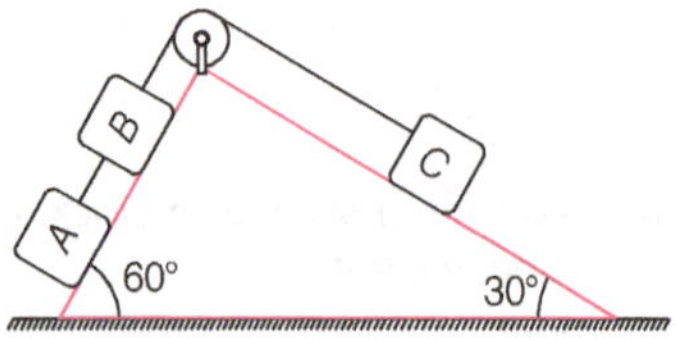

(a) the acceleration of the system and
(b) tensions in the strings.
Neglect friction. ($g = 10 \text{ m/s}^2$)

Sol. (a) In this case net pulling force

$$= m_A g \sin 60° + m_B g \sin 60° - m_C g \sin 30°$$

$$= (1)(10)\frac{\sqrt{3}}{2} + (3)(10)\left(\frac{\sqrt{3}}{2}\right) - (2)(10)\left(\frac{1}{2}\right)$$

$$= 24.64 \text{ N}$$

Total mass being pulled $= 1 + 3 + 2 = 6$ kg

$\therefore$ Acceleration of the system $a = \dfrac{24.64}{6}$

$$= 4.1 \text{ m/s}^2 \qquad \textbf{Ans.}$$

(b) For the tension in the string between *A* and *B*.

FBD of *A*

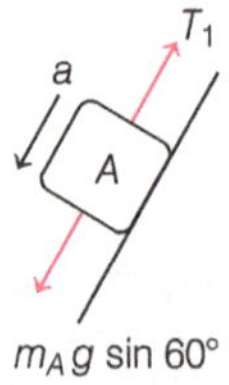

$$m_A g \sin 60° - T_1 = (m_A)(a)$$

$\therefore$ $T_1 = m_A g \sin 60° - m_A a$

$= m_A (g \sin 60° - a)$

$\therefore$ $T_1 = (1)\left(10 \times \frac{\sqrt{3}}{2} - 4.1\right)$

$$= 4.56 \text{ N} \qquad \textbf{Ans.}$$

For the tension in the string between *B* and *C*.

FBD of *C*

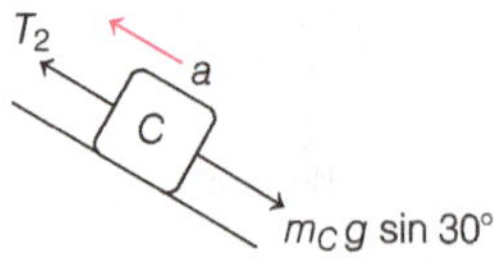

$$T_2 - m_C g \sin 30° = m_C a$$

$\therefore$ $T_2 = m_C (a + g \sin 30°)$

$\therefore$ $T_2 = 2\left[4.1 + 10\left(\frac{1}{2}\right)\right]$

$$= 18.2 \text{ N} \qquad \textbf{Ans.}$$

Note Points

(i) *In Fig. (i), normal reaction at any point P (say between blocks C and D) is given by,*

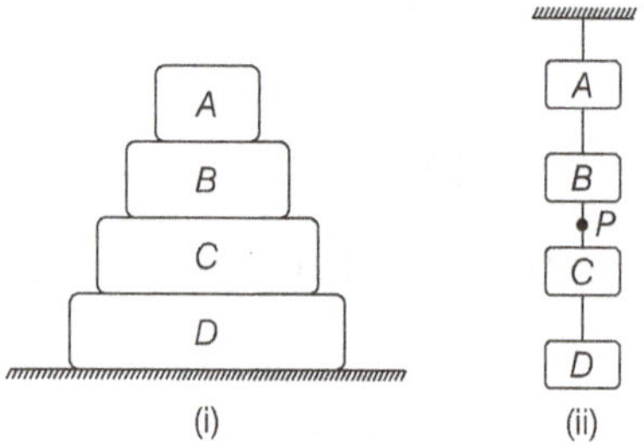

$N = [\Sigma(\text{mass above } P)] \times g_{\text{eff}} = (m_A + m_B + m_C)\, g_{\text{eff}}$

In Fig. (ii), tension at point P is given by,

$T = \Sigma\, [(\text{mass below } P)] \times g_{\text{eff}}$

If the strings are massless then, $T = (m_C + m_D)\, g_{\text{eff}}$

Here, $g_{\text{eff}} = g$ *if acceleration a of system is zero*

$= (g + a)$ if acceleration *a* is upwards

$= (g - a)$ if acceleration *a* is downwards

(ii) *Feeling of weight to a person is due to the normal reaction. Under normal conditions, $N = mg$. Therefore feeling of weight is the actual weight mg. If we are standing on a lift and the lift has an acceleration 'a' upwards then $N = m(g + a)$. Therefore feeling of weight is more than the actual weight mg. Similarly if 'a' is downwards then $N = m(g - a)$ and feeling of weight is less than the actual weight mg.*

Type 2. Problems based on resolution of forces.

Ex. 8 A ball of mass 1 kg is at rest in position *P* by means of two light strings *OP* and *RP*. The string *RP* is now cut and the ball swings to position *Q*. If $\theta = 45°$. Find the tensions in the string *OP* :

(i) When string *RP* was not cut.
(ii) Just after the string *RP* was cut. (Take $g = 10 \text{ m/s}^2$)

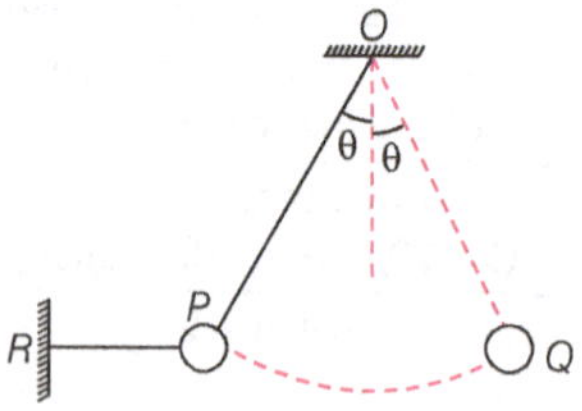

Sol. (i) In the first case, ball is in equilibrium (permanent rest). Therefore, net force on the ball in any direction should be zero.

$\therefore$ (ΣF) *in vertical direction* $= 0$

or $T_1 \cos \theta = mg$

or $T_1 = \dfrac{mg}{\cos \theta}$

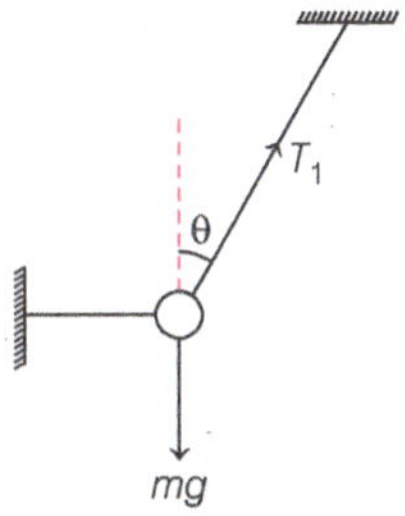

Substituting $m_1 = 1$ kg, $g = 10$ m/s^2 and $\theta = 45°$

we get, $T_1 = 10\sqrt{2}$ N

Note *Here, we have deliberately resolved all the forces in vertical direction because component of the tension in RP in vertical direction is zero. In a direction other than vertical we will have to consider component of tension in RP, which will unnecessarily increase our calculation.*

(ii) In the second case ball is not in equilibrium (temporary rest). After few seconds it will move in a direction perpendicular to OP. Therefore, net force on the ball at P is perpendicular to OP or net force along $OP = 0$.

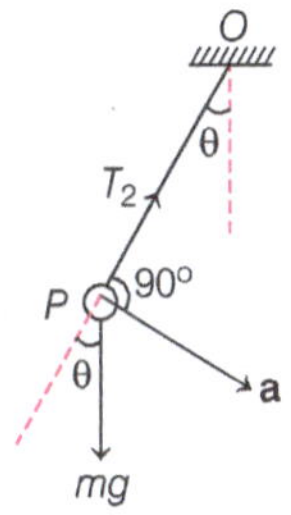

$\therefore \quad T_2 = mg\cos\theta$

Substituting the values, we get

$$T_2 = 5\sqrt{2}\ \text{N}$$

Here, we can see that $T_1 \neq T_2$

Type 3. To find tension at some point (say at P) if it is variable.

How to Solve?

(i) Find acceleration (a common acceleration) of the system by using the equation

$$a = \frac{\text{net pulling or pushing force}}{\text{total mass}}$$

In some cases, 'a' will be given in the question.

(ii) Cut the string at P and divide the system in two parts.

(iii) Make free body diagram of any one part.

(iv) In its FBD make one tension at point P in a direction away from the block with which this part.

(v) Write the equation,

$$F_{net} = ma$$

for this part. You will get tension at P. In this equation m is not the total mass. It is mass of this part only.

Ex. 9 In the given figure mass of string AB is 2 kg. Find tensions at A, B and C, where C is the mid point of string.

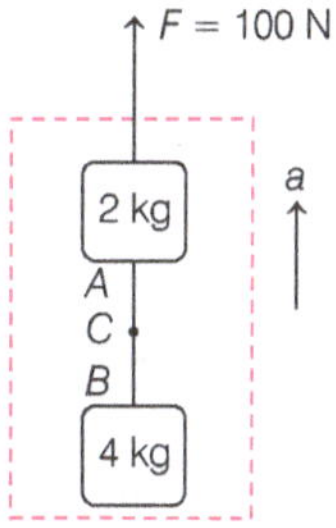

Sol. $a = \dfrac{F - \text{weight of 2kg} - \text{weight of 4kg} - \text{weight of string}}{\text{mass of 2kg} + \text{mass of 4 kg} + \text{mass of string}}$

$$= \frac{100 - 20 - 40 - 20}{2 + 4 + 2} \qquad (g = 10\ \text{m/s}^2)$$

$$= \frac{20}{8} = 2.5\ \text{m/s}^2$$

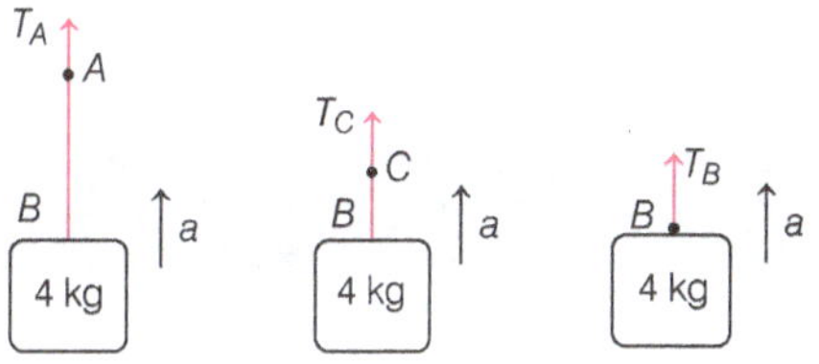

(a) (b) (c)

Refer Fig. (a)

$$T_A - m_{AB}g - 40 = (m_{AB} + 4)a$$

or $\quad T_A - 20 - 40 = (2 + 4)(2.5)$

$T_A = 75$ N **Ans.**

Refer Fig. (b)

$$T_C - m_{BC}g - 40 = (m_{BC} + 4)a$$

or $\quad T_C - 10 - 40 = (1 + 4)(2.5)$

or $\quad T_C = 62.5$ N **Ans.**

Refer Fig. (c) $\quad T_B - 40 = 4a$

or $\quad T_B = 40 + 4 \times 2.5$

or $\quad T_B = 50$ N **Ans.**

Note *Tension at a general point P can be given by:*

$T_P = [(\Sigma \text{ mass below } P) \times g_{eff}]$

Here $g_{eff} = g + a = 12.5\ m/s^2$

For example, $T_B = 4 \times 12.5$

$= 50\ N$

Type 4. Monkey problems.

Following example will illustrate the concept.

Ex. 10 With what minimum and maximum acceleration can a monkey move upwards and downwards if breaking strength of rope is,

(a) $3\,mg$ (b) $\dfrac{mg}{3}$

Here m = mass of monkey

Sol. (a)

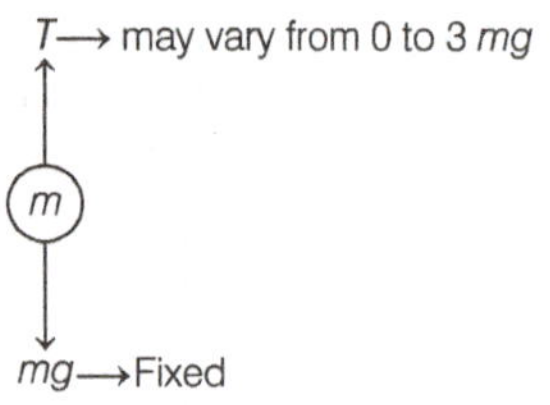

Upward acceleration

$$a_{min} = 0 \text{ when } T = mg$$

$$a_{max} = \frac{T_{max} - mg}{m}$$

$$= \frac{3mg - mg}{m} = 2g$$

Downward acceleration

$a_{min} = 0$ when $T = mg$

$a_{max} = g$ when $T = 0$, a case of free fall.

(b)

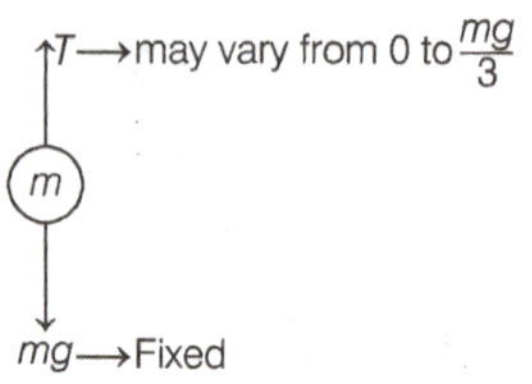

Upward acceleration

Since $T_{max} < mg$. Therefore monkey can't remain at rest. Because, for remaining at rest T should be equal to mg. Similarly it can't be accelerated upwards. Because for upward acceleration T should be greater than mg.

Downward acceleration

$$a_{min} = \frac{mg - T_{max}}{m} = \frac{mg - \frac{mg}{3}}{m} = \frac{2}{3}g$$

$a_{max} = g$ when $T = 0$

MODULE 3 Exercise

True/False

1. The pulley arrangements of figures (a) and (b) are identical. The mass of the rope is negligible. In figure (a), the mass m is lifted up by attaching a mass $2m$ to the other end of the rope. In Fig. (b), m is lifted up by pulling the other end of the rope with a constant downward force $F = 2mg$. The acceleration of m is the same in both cases. **(1984 JEE)**

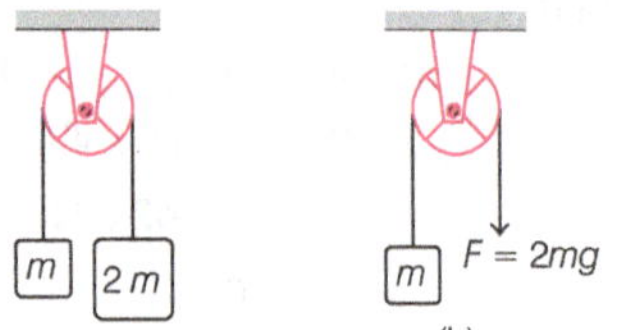

(a) (b)

Assertion and Reason

See directions on page 354.

2. **Assertion :** In the figure shown, tension in string AB always lies between $m_1 g$ and $m_2 g$. ($m_1 \neq m_2$)

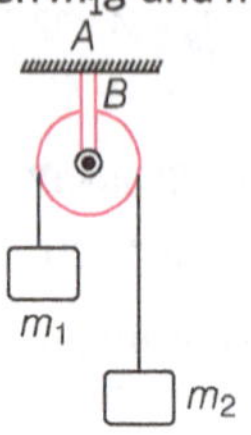

Reason : Tension in massless string is uniform throughout.

3. **Assertion :** Normal reaction on a block can not accelerate the block.

Reason : Normal reaction always acts perpendicular to the surface of contact, no matter what the angle of that surface.

4. **Assertion :** A cloth covers a table. Some dishes are kept on it. The cloth can be pulled out without dislodging the dishes from the table.

Reason : For every action there is an equal and opposite reaction. **(2007 JEE)**

Single Correct Option

5. Following two questions have Statement 1 and Statement 2. Of the four choices given after the Statements, choose the one that best describes the two Statements.

Statement 1 : If you push on a cart being pulled by a horse so that it does not move, the cart pushes you back with an equal and opposite force.

Statement 2 : The cart does not move because the force described in Statement 1 cancel each other. **(2012 Main)**

(a) Statement 1 is true, Statement 2 is true, Statement 2 is the correct explanation of Statement 1.
(b) Statement 1 is false, Statement 2 is true.
(c) Statement 1 is true, Statement 2 is false.
(d) Statement 1 is true, Statement 2 is true, Statement 2 is not the correct explanation of Statement 1.

6. **Statement 1 :** An elevator can go up or down with uniform speed when its weight is balanced with the tension of its cable.

Statement 2 : Force exerted by the floor of an elevator on the foot of a person standing on it is more than his/her weight when the elevator goes down with increasing speed. In the light of the above statements, choose the correct answer from the options given below: **(2023 Main)**

7. As per the given figure, a weightless pulley P is attached on a double inclined frictionless surface. The tension in the string (massless) will be (if $g = 10\ \text{m/s}^2$) **(2023 Main)**

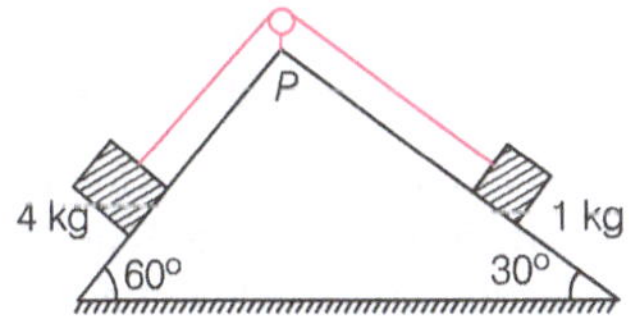

(a) $(4\sqrt{2}+1)$ N (b) $4(\sqrt{3}+1)$ N
(c) $4(\sqrt{3}-1)$ N (d) $(4\sqrt{3}-1)$ N

8. Two bodies of masses $m_1 = 5$ kg and $m_2 = 3$ kg are connected by a light string going over a smooth light pulley on a smooth inclined plane as shown in the figure. The system is at rest. The force exerted by the inclined plane on the body of mass m_1 will be (Take $g = 10\ \text{ms}^{-2}$) **(2022 Main)**

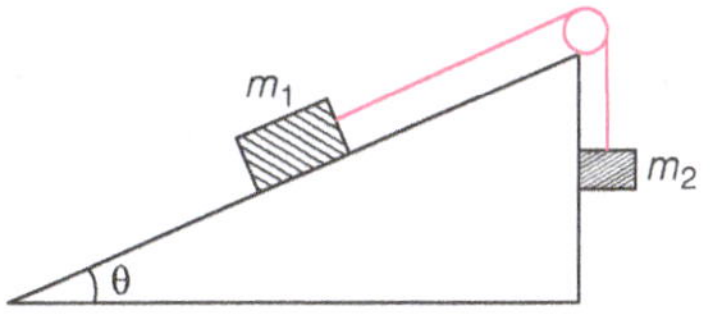

(a) 30 N (b) 40 N
(c) 50 N (d) 60 N

9. Two masses M_1 and M_2 are tied together at the two ends of a light inextensible string that passes over a frictionless pulley. When the mass M_2 is twice that of M_1, the acceleration of the system is a_1. When the mass M_2 is thrice that of M_1, the acceleration of the system is a_2. The ratio $\frac{a_1}{a_2}$ will be **(2022 Main)**

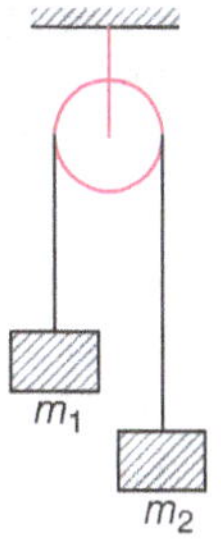

(a) $\frac{1}{3}$ (b) $\frac{2}{3}$ (c) $\frac{3}{2}$ (d) $\frac{1}{2}$

10. Two balls A and B of same size are dropped from the same point under gravity. The mass of A is greater than that of B. If the air resistance acting on each ball is same, then
(a) both the balls reach the ground simultaneously
(b) the ball A reaches earlier
(c) the ball B reaches earlier
(d) nothing can be said

11. At what minimum acceleration should a monkey slide a rope whose breaking strength is $\frac{2}{3}$ rd of its weight ?
(a) $\frac{2g}{3}$ (b) g
(c) $\frac{g}{3}$ (d) zero

12. For the arrangement shown in the figure, the reading of spring balance is

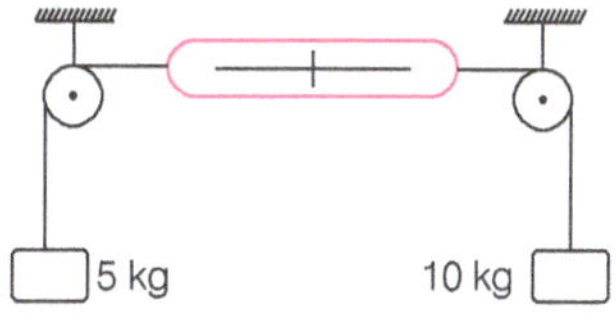

(a) 50 N (b) 100 N
(c) 150 N (d) None of these

13. A force F_1 accelerates a particle from rest to a velocity v. Another force F_2 decelerates the same particle from v to rest, then
(a) F_1 is always equal to F_2
(b) F_2 is greater than F_1
(c) F_2 may be smaller than, greater than or equal to F_1
(d) F_2 cannot be equal to F_1

14. Ten coins are placed on top of each other on a horizontal table. If the mass of each coin is 10 g and acceleration due to gravity is $10\ \text{ms}^{-2}$, what is the magnitude and direction of the force on the 7th coin (counted from the bottom) due to all the coins above it?
(a) 0.3 N downwards
(b) 0.3 N upwards
(c) 0.7 N downwards
(d) 0.7 N upwards

15. A 50 kg boy stands on a platform inside a lift. At bottom, the platform is attached a spring scale. The lift is going down with a constant speed 3 m/s. If the lift is brought to rest by a constant deceleration in a distance of 9 m, what does the scale read during this period? (Take, $g = 9.8\ \text{m/s}^2$)
(a) 500 N (b) 465 N
(c) 515 N (d) Zero

16. If a force F is applied on a block of mass m_1, then it produces an acceleration a_1. The same force is applied to a second block of mass m_2, then it produces an acceleration a_2. If the same force is

applied on a block of mass $(m_1 + m_2)$, then acceleration is

(a) $a_1 + a_2$ (b) $\frac{a_1 + a_2}{2}$

(c) $a_1 - a_2$ (d) $\frac{a_1 a_2}{a_1 + a_2}$

17. One end of a massless rope, which passes over a massless and frictionless pulley P is tied to a hook C while the other end is free. Maximum tension that the rope can bear is 360 N. With what value of maximum safe acceleration (in ms^{-2}) can a man of 60 kg climb up on the rope?

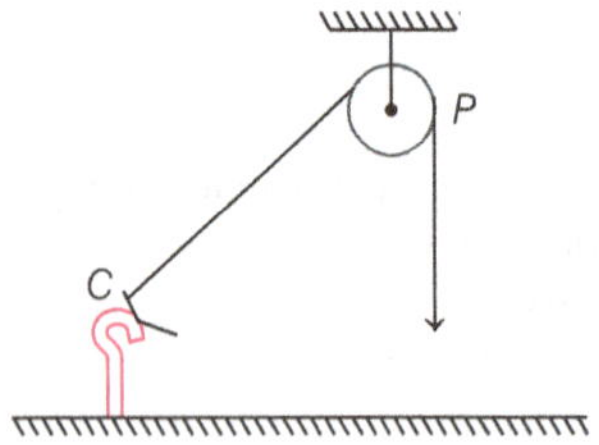

(a) 2 (b) 3
(c) 4 (d) None of these

18. In each case $m_1 = 4$ kg and $m_2 = 3$ kg. If a_1, a_2 and a_3 are the respective accelerations of the block m_1 in given situations, then

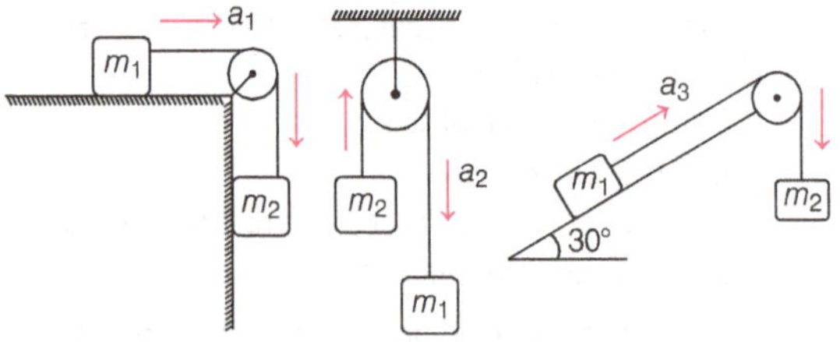

(a) $a_1 > a_2 > a_3$ (b) $a_1 > a_2 = a_3$
(c) $a_1 = a_2 = a_3$ (d) $a_1 > a_3 > a_2$

19. Two blocks A and B each of mass m are placed on a smooth horizontal surface. Two horizontal forces F and $2F$ are applied on the blocks A and B respectively as shown in figure. The block A does not slide on block B. Then, the normal reaction acting between the two blocks is

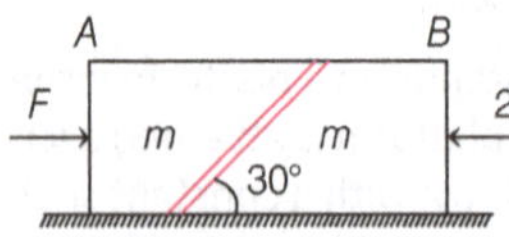

(a) F (b) $\frac{F}{2}$ (c) $\frac{F}{\sqrt{3}}$ (d) $3F$

20. A mass of 3 kg descending vertically downwards supports a mass of 2 kg by means of a light string passing over a pulley. At the end of 5 s, the string breaks. How much high from now the 2 kg mass will go? (Take, $g = 9.8\ m/s^2$)

(a) 4.9 m (b) 9.8 m (c) 19.6 m (d) 2.45 m

21. A ship of mass 3×10^7 kg initially at rest, is pulled by a force of 5×10^4 N through a distance of 3 m. Assuming that the resistance due to water is negligible, the speed of the ship is **(1980 JEE)**

(a) 1.5 m/s (b) 60 m/s
(c) 0.1 m/s (d) 5 m/s

22. A bullet of mass 20 g has an initial speed of $1\ ms^{-1}$, just before it starts penetrating a mud wall of thickness 20 cm. If the wall offers a mean resistance of 2.5×10^{-2} N, the speed of the bullet after emerging from the other side of the wall is close to **(2019 Main)**

(a) $0.3\ ms^{-1}$ (b) $0.4\ ms^{-1}$
(c) $0.1\ ms^{-1}$ (d) $0.7\ ms^{-1}$

23. Three force $F_1 = 10$ N, $F_2 = 8$ N, $F_3 = 6$ N are acting on a particle of mass 5 kg. The forces F_2 and F_3 are applied perpendicular so that particle remains at rest. If the force F_1 is removed, then the acceleration of the particle is **(2023 Main)**

(a) $2\ ms^{-2}$ (b) $0.5\ ms^{-2}$
(c) $4.8\ ms^{-2}$ (d) $7\ ms^{-2}$

24. A body of mass 500 g moves along x-axis such that it's velocity varies with displacement x according to the relation $v = 10\sqrt{x}$ m/s, the force acting on the body is **(2023 Main)**

(a) 166 N (b) 25 N
(c) 125 N (d) 5 N

25. At any instant, the velocity of a particle of mass 500 g is $(2t\,\hat{i} + 3t^2\hat{j})\ ms^{-1}$. If the force acting on the particle at $t = 1$ s is $(\hat{i} + x\hat{j})$ N, then the value of x will be **(2023 Main)**

(a) 3 (b) 4
(c) 6 (d) 2

26. Force acts for 20 s on a body of mass 20 kg, starting from rest, after which the force ceases and then body describes 50 m in the next 10 s. The value of force will be **(2023 Main)**

(a) 40 N (b) 5 N
(c) 20 N (d) 10 N

27. An object of mass 5 kg is thrown vertically upwards from the ground. The air resistance produces a constant retarding force of 10 N throughout the motion. The ratio of time of ascent to the time of descent will be equal to (Use $g = 10\ ms^{-2}$) **(2022 Main)**

(a) 1 : 1 (b) $\sqrt{2} : \sqrt{3}$
(c) $\sqrt{3} : \sqrt{2}$ (d) 2 : 3

28. The position vector of a particle as a function of time t is given by $\mathbf{r} = (10\,t\hat{i} + 15t^2\hat{j} + 7\hat{k})$ m. The direction of net force experienced by the particle is **(2023 Main)**

(a) positive y-axis
(b) positive x-axis
(c) positive z-axis
(d) in x - y plane

29. A person is standing in an elevator. In which situation, he experiences weight loss? **(2022 Main)**
(a) When the elevator moves upward with constant acceleration.
(b) When the elevator moves downward with constant acceleration.
(c) When the elevator moves upward with uniform velocity.
(d) When the elevator moves downward with uniform velocity.

30. Two blocks of masses $M_1 = 20$ kg and $M_2 = 12$ kg are connected by a metal rod of mass 8 kg. The system is pulled vertically up by applying a force of 480 N as shown. The tension at the mid-point of the rod is **(2013 Main)**

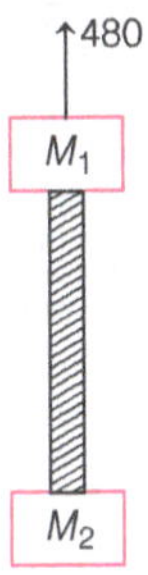

(a) 144 N (b) 96 N (c) 240 N (d) 192 N

31. Two masses $m_1 = 5$ kg and $m_2 = 4.8$ kg tied to a string are hanging over a light frictionless pulley. What is the acceleration of the masses when left free to move? ($g = 9.8\ m/s^2$) **(2004, 02 JEE)**

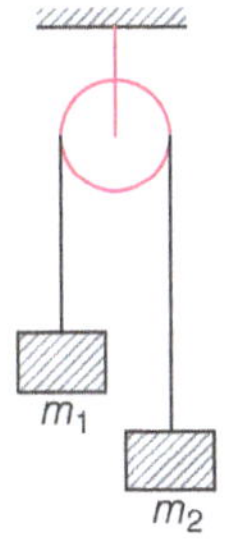

(a) $5\ m/s^2$ (b) $9.8\ m/s^2$
(c) $0.2\ m/s^2$ (d) $4.8\ m/s^2$

32. A spring balance is attached to the ceiling of a lift. A man hangs his bag on the spring and the spring reads 49 N, when the lift is stationary. If the lift moves downward with an acceleration of $5\ m/s^2$, the reading of the spring balance will be **(2003 Main)**
(a) 24 N (b) 74 N (c) 15 N (d) 49 N

33. A block of mass M is pulled along a horizontal frictionless surface by a rope of mass m. If a force P is applied at the free end of the rope, the force exerted by the rope on the block is **(2003 JEE)**
(a) $\frac{Pm}{M+m}$ (b) $\frac{Pm}{M-m}$ (c) P (d) $\frac{PM}{M+m}$

34. A light spring balance hangs from the hook of the other light spring balance and a block of mass M kg hangs from the former one. Then the true statement about the scale reading is **(2003 JEE)**
(a) both the scales read M kg each
(b) the scale of the lower one reads M kg and of the upper one zero
(c) the reading of the two scales can be anything but the sum of the reading will be M kg
(d) both the scales read $M/2$ kg each

35. Two particles start together from a point O and slide down along straight smooth wires inclined at 30° and 60° to the horizontal plane and on the same side of vertical through O. The relative acceleration of second with respect to first will be of magnitude
(a) $\frac{g}{2}$ (b) $\frac{\sqrt{3}g}{2}$ (c) $\frac{g}{\sqrt{3}}$ (d) g

36. If a body loses half of its velocity on penetrating 3 cm in a fixed wooden block, then how much will it penetrate more before coming to rest?
(a) 1cm (b) 2 cm (c) 3 cm (d) 4 cm

Passage Type Questions

Passage (Q.37 to Q.39)

Two smooth blocks A of mass 1 kg and B of mass 2 kg are connected by a light string passing over a smooth pulley as shown. The block B is sliding down with a velocity 2 m/s. A force F is applied on the block A, so that the block B will reverse its direction of motion after 3 s.

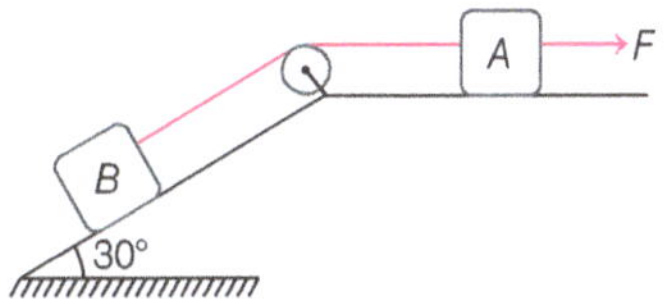

37. Find the acceleration of block A.
(a) $1\ m/s^2$ (b) $\frac{2}{3}\ m/s^2$ (c) $3\ m/s^2$ (d) $\frac{4}{9}\ m/s^2$

38. Find the tension in the string.
(a) 10 N (b) 11.33 N (c) 13.67 N (d) 40 N

39. The value of F is
(a) 9 N (b) 10 N (c) 11 N (d) 12 N

Subjective Questions

40. Three blocks of masses 1 kg, 4 kg and 2 kg are placed on a smooth horizontal plane as shown in figure. Find

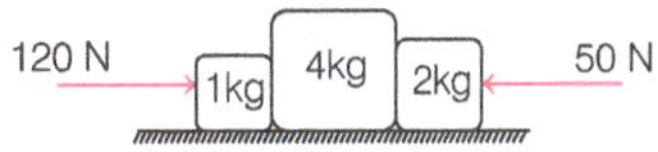

(a) the acceleration of the system
(b) the normal force between 1 kg block and 4 kg block
(c) the net force on 2 kg block.

41. In the arrangement shown in figure, find the ratio of tensions in the strings attached with 4 kg block and that with 1 kg block.

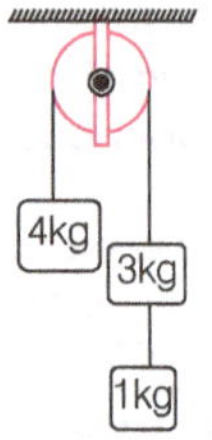

42. Two unequal masses of 1 kg and 2 kg are connected by an inextensible light string passing over a smooth pulley as shown in figure. A force $F = 20$ N is applied on 1 kg block. Find the acceleration of either block. $(g = 10 \text{ m/s}^2)$

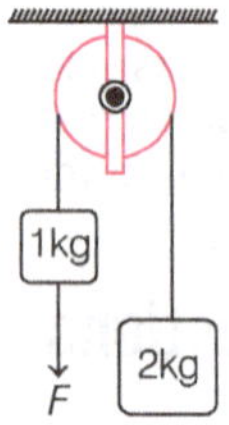

43. In the arrangement shown in figure, what should be the mass of block A, so that the system remains at rest? Neglect friction and mass of strings.

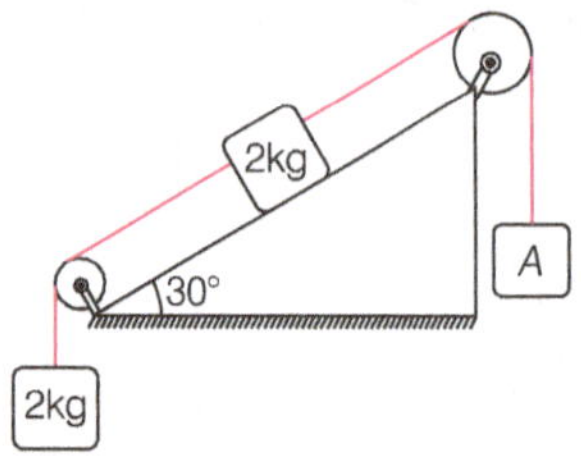

44. Two blocks of masses 2 kg and 4 kg are released from rest over a smooth inclined plane of inclination 30° as shown in figure. What is the normal force between the two blocks?

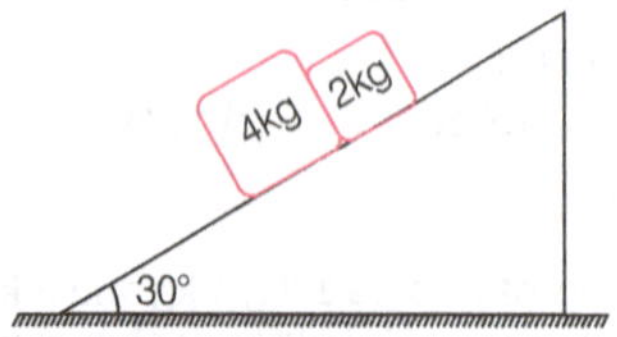

45. What should be the acceleration a of the box shown in figure, so that the block of mass m exerts a force $\frac{mg}{4}$ on the floor of the box?

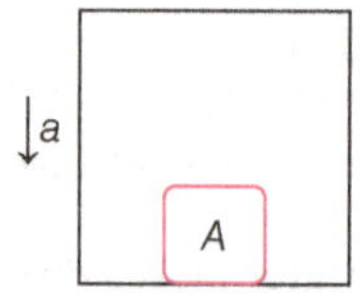

46. In the figure shown, find acceleration of the system and tensions T_1, T_2 and T_3. (Take $g = 10 \text{m/s}^2$)

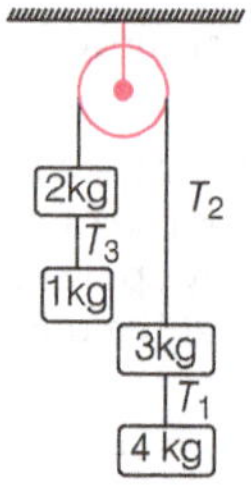

47. In the figure shown, all surfaces are smooth. Find

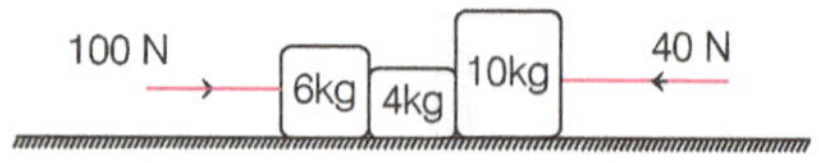

(a) acceleration of all the three blocks,
(b) net force on 6 kg, 4 kg and 10 kg blocks and
(c) force acting between 4 kg and 10 kg blocks.

48. Three blocks of masses $m_1 = 10$ kg, $m_2 = 20$ kg and $m_3 = 30$ kg are on a smooth horizontal table, connected to each other by light horizontal strings. A horizontal force $F = 60$ N is applied to m_3, towards right. Find

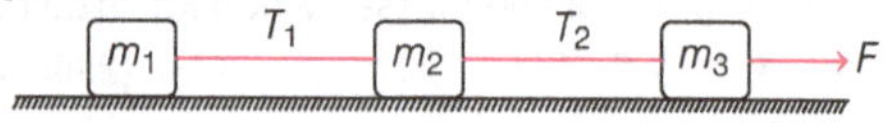

(a) tensions T_1 and T_2 and
(b) tension T_2 if all of a sudden the string between m_1 and m_2 snaps.

49. Two blocks of masses 2.9 kg and 1.9 kg are suspended from a rigid support S by two inextensible wires each of length 1m, as shown in the figure. The upper wire has negligible mass and the lower wire has a uniformly distributed mass of 0.2 kg. The whole system of blocks, wires and support have an upward acceleration of 0.2m/s^2. Acceleration due to gravity is 9.8 m/s^2. **(1989 JEE)**

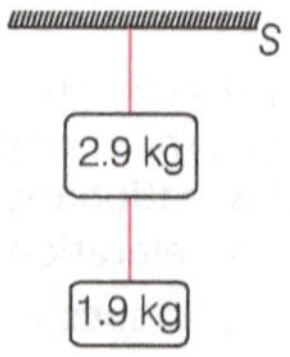

(a) Find the tension at the mid-point of the lower wire.
(b) Find the tension at the mid-point of the upper wire.

50. Two blocks shown in figure are connected by a heavy uniform rope of mass 4 kg. An upward force of 200 N is applied as shown. (Take, $g = 9.8\ m/s^2$)

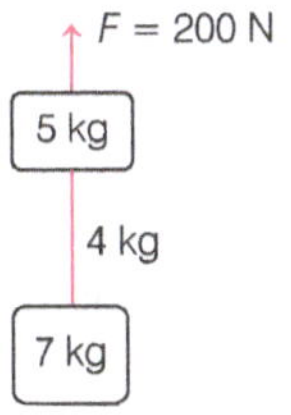

(a) What is the acceleration of the system ?
(b) What is the tension at the top of the rope ?
(c) What is the tension at the mid-point of the rope ?

51. A balloon of mass m starts coming down with a constant acceleration a. Determine the mass to be thrown out for the balloon to move with upward acceleration of the same magnitude. The air drag is to be neglected. **(Irodov 1.59)**

52. A uniform rope of length L and mass M lying on a smooth table is pulled by a constant force F. What is the tension in the rope at a distance l from the end where the force is applied? **(1978 JEE)**

Integer Type Questions

53. A boy pushes a box of mass 2 kg with a force $\mathbf{F} = (20\hat{\mathbf{i}} + 10\hat{\mathbf{j}})$ N on a frictionless surface. If the box was initially at rest, then .... m is displacement along the x-axis after 10 s. **(2021 Main)**

54. A person standing on a spring balance inside a stationary lift measures 60 kg. The weight of that person if the lift descends with uniform downward acceleration of $1.8\ m/s^2$ will be .... N. ($g = 10\ m/s^2$) **(2021 Main)**

Answers

1. False	**2.** (b)	**3.** (d)	**4.** (b)	**5.** (c)	**6.** (b)	**7.** (b)	**8.** (b)	**9.** (b)	**10.** (b)
11. (c)	**12.** (d)	**13.** (c)	**14.** (a)	**15.** (c)	**16.** (d)	**17.** (d)	**18.** (b)	**19.** (d)	**20.** (a)
21. (c)	**22.** (d)	**23.** (a)	**24.** (b)	**25.** (a)	**26.** (b)	**27.** (b)	**28.** (a)	**29.** (b)	**30.** (d)
31. (c)	**32.** (a)	**33.** (d)	**34.** (a)	**35.** (a)	**36.** (a)	**37.** (b)	**38.** (b)	**39.** (d)	

40. (a) $10\ m/s^2$ (b) 110 N (c) 20 N **41.** (4) **42.** $10/3\ ms^{-2}$ **43.** 3 kg **44.** Zero **45.** $3g/4$

46. $4ms^{-2}$, 24N, 42N, 14N **47.** (a) $3\ ms^{-2}$ (b) 18 N, 12 N, 30 N (c) 70 N **48.** (a) 10 N, 30 N (b) 24 N

49. (a) 20 N (b) 50 N **50.** (a) $2.7\ ms^{-2}$ (b) 137.5 N (c) 112.5 N **51.** $M = \dfrac{2ma}{g+a}$ **52.** $F\left(1 - \dfrac{l}{L}\right)$

53. (500) **54.** (492)

See hints on pages 419-425.

MODULE 4

Constraint Equations

In the previous article, we have discussed the cases where different blocks of a system had a common acceleration and that common acceleration was given by :

$$a = \frac{\text{Net pulling/pushing force}}{\text{Total mass}}$$

Now, the question is, if different blocks have different accelerations then what? In those cases, we take help of constraint equations. These equations establish the relation between accelerations (or velocities) of different blocks of a system. Depending upon different kinds of problems, we have divided the constraint equations in following four types. Most of them are directly explained with the help of some example(s) in their support.

Type 1.

Example 1 Using constraint method find the relation between accelerations of 1 and 2.

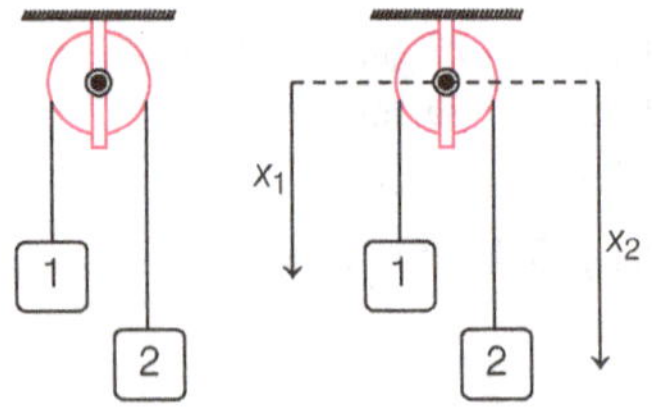

Sol. At any instant of time let x_1 and x_2 be the displacements of 1 and 2 from a fixed line (shown dotted). Here x_1 and x_2 are variables but,

$$x_1 + x_2 = \text{constant}$$

or $x_1 + x_2 = l$ (length of string)

Differentiating with respect to time, we have

$$v_1 + v_2 = 0$$

or $v_1 = -v_2$

Again differentiating with respect to time, we get

$$a_1 + a_2 = 0$$

or $a_1 = -a_2$

or $|a_1| = |a_2|$

This is the required relation between a_1 and a_2, i.e. accelerations of 1 and 2 are equal but in opposite directions.

Note *(i) In the equation $x_1 + x_2 = l$, we have neglected the length of string over the pulley. But that length is also constant.*

(ii) In constraint equation if we get $a_1 = -a_2$, then negative sign does not always represent opposite directions of a_1 and a_2. The real significance of this sign is : x_2 decreases if x_1 increases and vice-versa.

Example 2 Using constraint equations find the relation between a_1 and a_2.

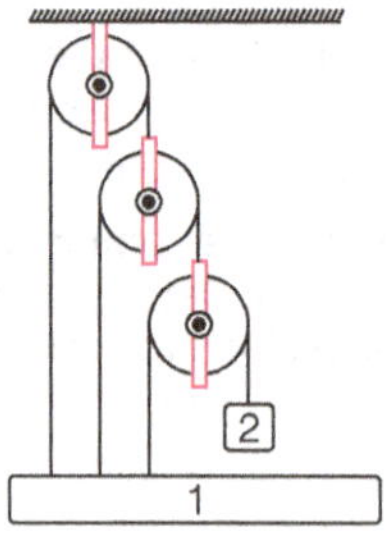

Sol. In the figure shown below, points 1, 2, 3 and 4 are moveable. Let their displacements from a fixed dotted line be x_1, x_2, x_3 and x_4

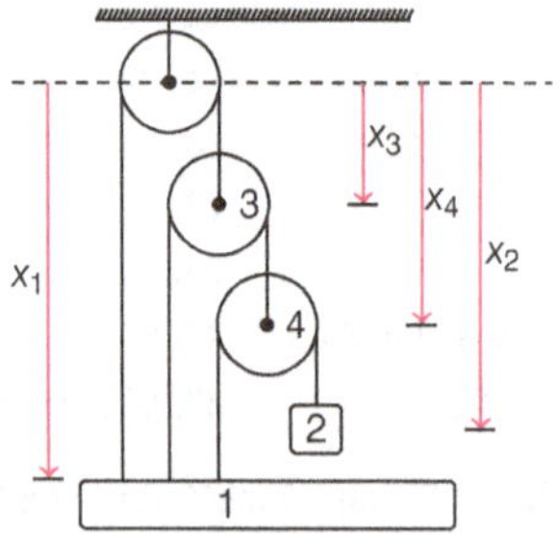

$$x_1 + x_3 = l_1$$
$$(x_1 - x_3) + (x_4 - x_3) = l_2$$
$$(x_1 - x_4) + (x_2 - x_4) = l_3$$

On double differentiating with respect to time, we will get following three constraint relations

$$a_1 + a_3 = 0 \quad \text{...(i)}$$
$$a_1 + a_4 - 2a_3 = 0 \quad \text{...(ii)}$$
$$a_1 + a_2 - 2a_4 = 0 \quad \text{...(iii)}$$

Solving Eqs. (i), (ii) and (iii), we get

$$a_2 = -7a_1 \quad \text{or} \quad |a_2| = 7|a_1|$$

Which is the desired relation between a_1 and a_2.

Note *Pulleys given in this chapter are normally massless. So, net force on the pulley should be zero. Because,*

$$\mathbf{F}_{\text{net}} = ma$$

So, $\mathbf{F}_{\text{net}} = 0$, *if* $m = 0$

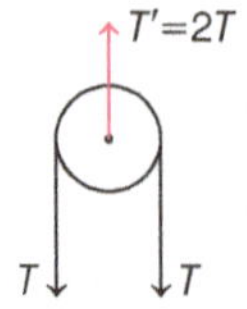

Example 3 In the above example, if two blocks have masses 1 kg and 2 kg respectively then find their accelerations and tensions in different strings.

Sol. Pulleys 3 and 4 are massless. Hence net force on them should be zero. Therefore, if we take T tension in the shortest string, then tension in other two strings will be $2T$ and $4T$.

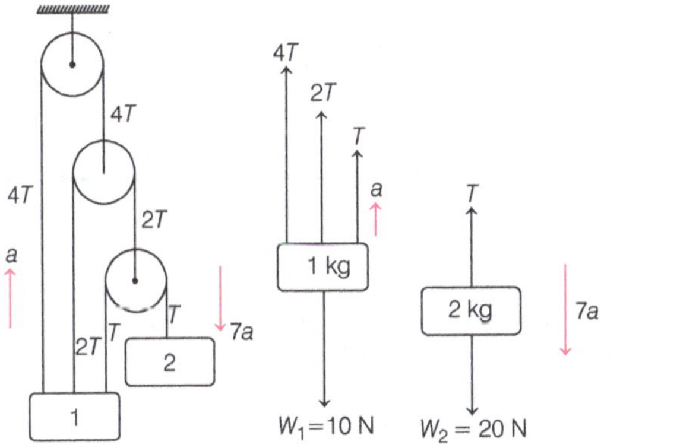

Further, if a is the acceleration of 1 in upward direction, then from the constraint equation $a_2 = -7a_1$, acceleration of 2 will be $7a$ downwards.

Writing the equation, $\mathbf{F}_{net} = m\mathbf{a}$ for the two blocks, we have

$$4T + 2T + T - 10 = 1 \times a$$

or $\quad 7T - 10 = a \quad$ ...(i)

and $\quad 20 - T = 2 \times (7a)$

or $\quad 20 - T = 14a \quad$ ...(ii)

Solving these two equations, we get

$T = 1.62$ N **Ans.**

and $a = 1.31$ m/s^2 **Ans.**

Note *In some problem if 'a' comes out to be negative after calculations then we will change the initially assumed directions of accelerations.*

Type 2.

Example 4 The system shown in figure is released from rest. Find acceleration of different blocks and tension in different strings.

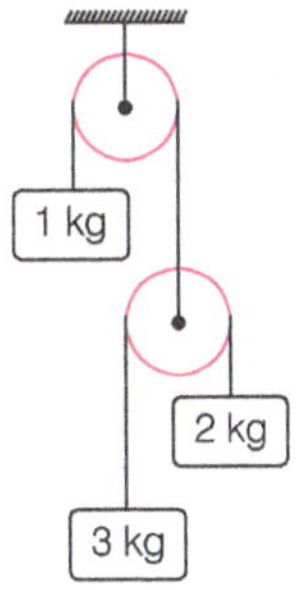

Sol.

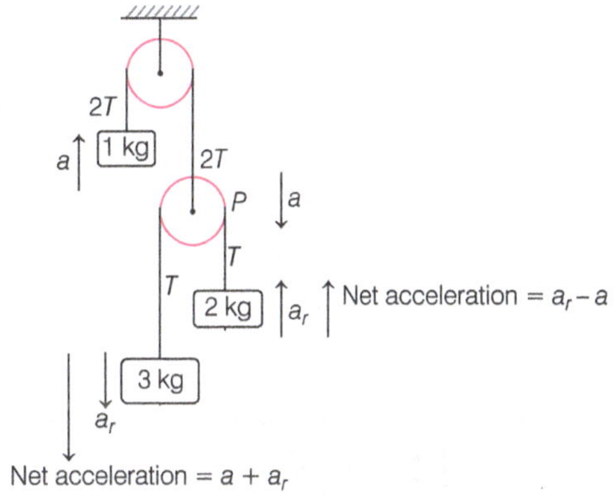

(i) Pulley P and 1 kg mass are attached with the same string. Therefore, if 1 kg mass has an acceleration 'a' in upward direction, then pulley P will have an acceleration 'a' downwards.

(ii) 2 kg and 3 kg blocks are attached with the same string passing over a moveable pulley P. Therefore, their relative accelerations, a_r (relative to pulley) will be same. Their net accelerations (relative to ground) are also shown in figure.

(iii) Pulley P is massless. Hence net force on this pulley should be zero. If T is the tension in the string connecting 2 kg and 3 kg mass, then tension in the upper string will be $2T$.

Now writing the equation, $\mathbf{F}_{net} = m\mathbf{a}$ for three blocks, we have:

1 kg block:

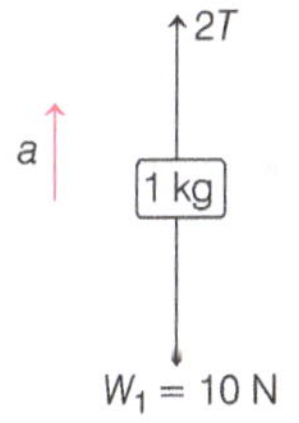

$$2T - 10 = 1 \times a \quad \text{...(i)}$$

2 kg block:

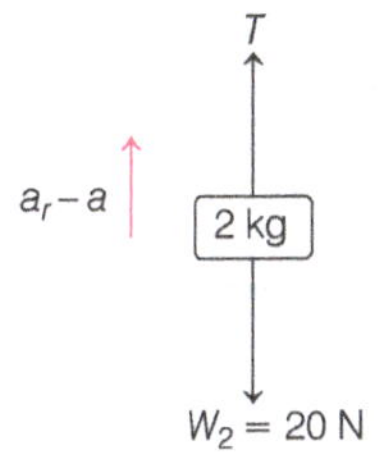

$$T - 20 = 2(a_r - a) \quad \text{...(ii)}$$

3 kg block:

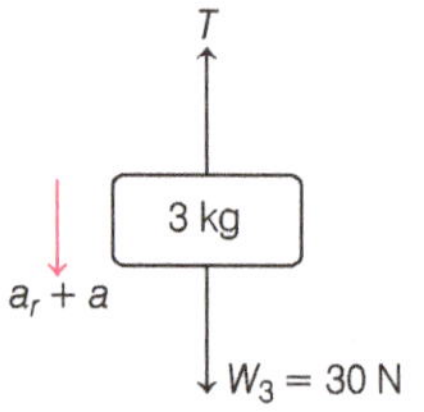

$$30 - T = 3(a_r + a) \quad \text{...(iii)}$$

Solving Eqs. (i), (ii) and (iii), we get

$$T = 8.28\text{ N}, \; a = 6.55\text{ m/s}^2$$

and $\quad a_r = 0.7\text{ m/s}^2$

Now, acceleration of 3kg block is $(a + a_r)$ or 7.25 m/s^2 downwards and acceleration of 2kg is $(a_r - a)$ or -5.85 m/s^2 upwards. Since, this comes out to be negative, hence acceleration of 2 kg block is 5.85 m/s^2 downwards. Acceleration of 1 kg block is a or 6.55 m/s^2 upwards.

Important POINTS

In some cases, acceleration of a block is inversely proportional to tension force acting on the block (or its component in the direction of motion or acceleration). If tension is double (as compared to other block), then acceleration will be half.

Let us take two examples.

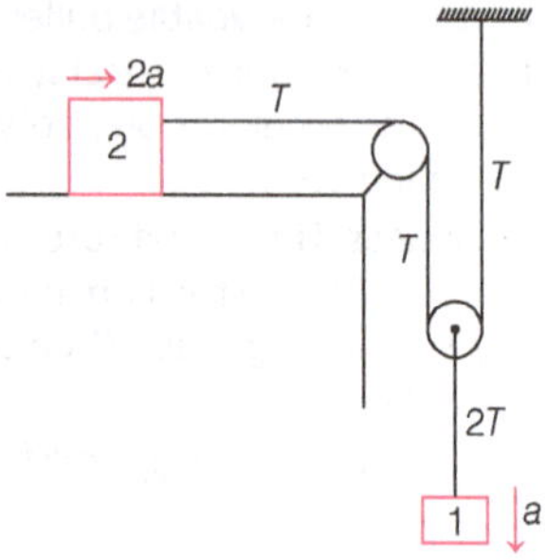

(a)

In Fig. (a): Tension force on block-1 is double $(= 2T)$ than the tension force on block-2 $(= T)$. Therefore, acceleration of block -1 will be half. If block-1 has an acceleration 'a' in downward direction, then block -2 will have an acceleration '$2a$' towards right.

In Fig. (b): Tension force on block-1 is three times $(2T + T = 3T)$ the tension force on block-2 $(= T)$. Therefore acceleration of block-2 will be three times. If block-1 has an acceleration 'a' in upward direction, then acceleration of block-2 will be '$3a$' downwards.

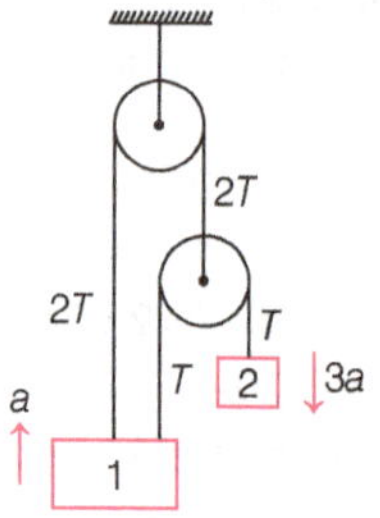

(b)

Type 3. Based on constraint relation between a block (or a plank) and a wedge.

Example 5 In the figure shown, find relation between magnitudes of $\mathbf{a}_A$ and $\mathbf{a}_B$.

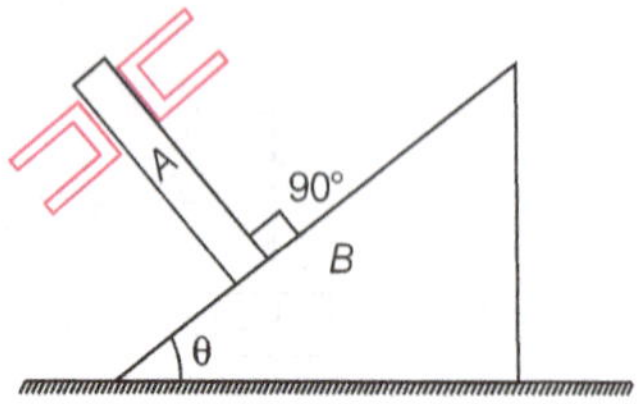

Sol. $x_A = x_B \sin\theta \quad \text{...(i)}$

Here, $\theta = \text{constant}$

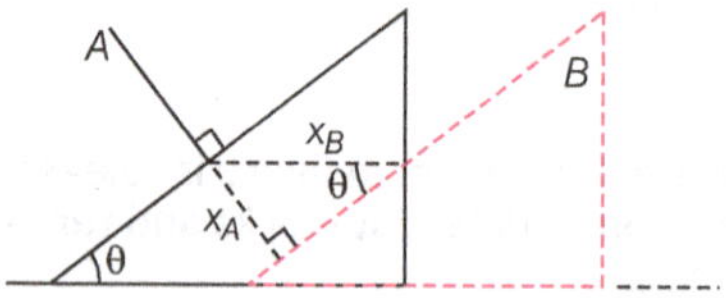

Double differentiating Eq. (i) with respect to time, we get

$$a_A = a_B \sin\theta \quad \textbf{Ans.}$$

Example 6 In the arrangement shown in the figure, the rod of mass m held by two smooth walls, remains always perpendicular to the surface of the wedge of mass M. Assuming all the surfaces to be frictionless, find the acceleration of the rod and that of the wedge.

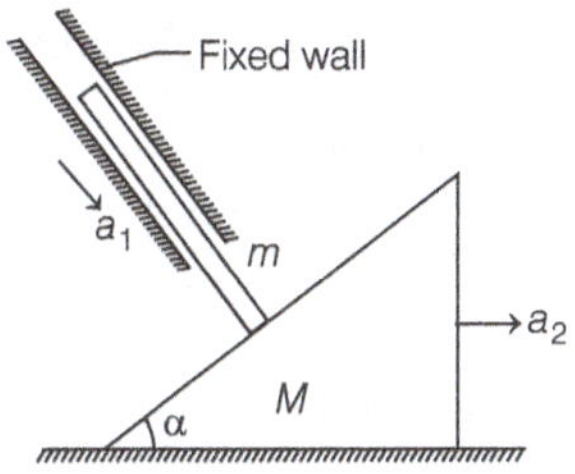

Sol. Let acceleration of m be a_1 (absolute) and that of M be a_2 (absolute).

Writing equations of motion only in the directions of a_1 or a_2.

For m $\quad mg\cos\alpha - N = ma_1 \quad$ ...(i)

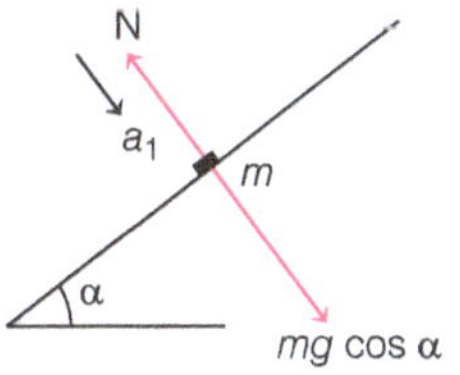

For M, $\quad N\sin\alpha = Ma_2 \quad$ ...(ii)

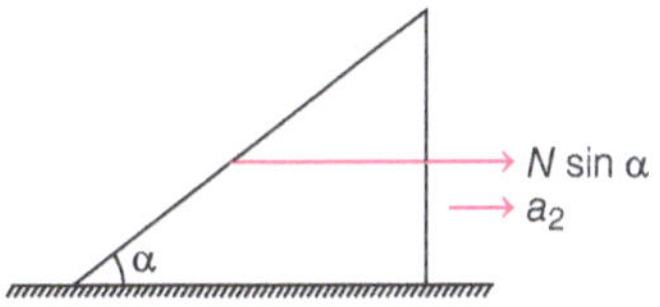

Here, N = normal reaction between m and M

As discussed above, constraint equation can be written as,

$$a_1 = a_2 \sin\alpha \quad \text{...(iii)}$$

Solving above three equations, we get acceleration of rod,

$$a_1 = \frac{mg\cos\alpha\sin\alpha}{\left(m\sin\alpha + \dfrac{M}{\sin\alpha}\right)} \quad \textbf{Ans.}$$

and acceleration of wedge,

$$a_2 = \frac{mg\cos\alpha}{m\sin\alpha + \dfrac{M}{\sin\alpha}} \quad \textbf{Ans.}$$

Type 4. Based on constraint relation which keeps on changing.

Concept

(i) In the constraint relations discussed so far, the relation between different accelerations was fixed.

For example: In the two illustrations discussed above $a_A = a_B\sin\theta$ or $a_1 = a_2\sin\alpha$ but these relations were fixed, as θ or α was constant.

(ii) In some cases, θ or α keeps on changing. Therefore, the constraint relation also keeps on changing.

(iii) In this case, constraint relation between different accelerations becomes very complex. So, normally constraint relation between velocities is only asked.

Example 7 In the arrangement shown in the figure, the ends P and Q of an unstretchable string move downwards with uniform speed U. Pulleys A and B are fixed. Mass M moves upwards with a speed **(1982 JEE)**

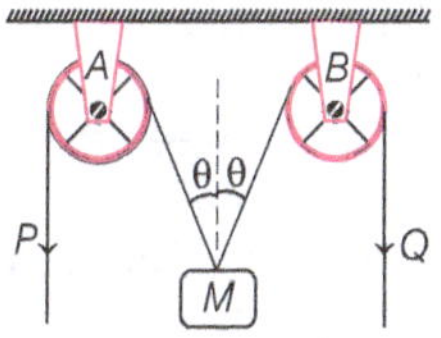

(a) $2U\cos\theta$ (b) $\dfrac{U}{\cos\theta}$

(c) $\dfrac{2U}{\cos\theta}$ (d) $U\cos\theta$

Sol. In the right angle ΔPQR

$$l^2 = c^2 + y^2$$

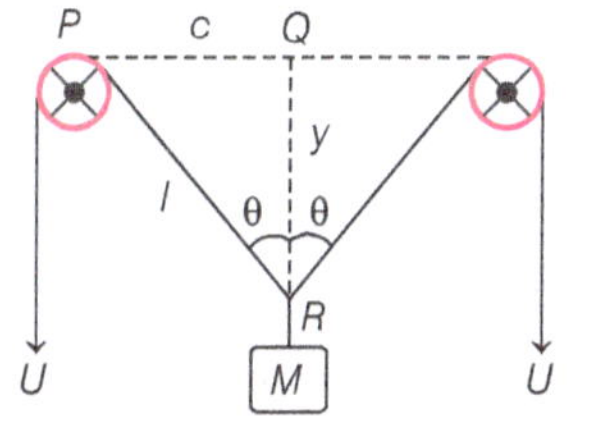

c = constant, l and y are variables

Differentiating this equation with respect to time, we get

$$2l\frac{dl}{dt} = 0 + 2y\frac{dy}{dt}$$

or $$\left(-\frac{dy}{dt}\right) = \frac{l}{y}\left(-\frac{dl}{dt}\right)$$

Here, $$-\frac{dy}{dt} = v_M, \; \frac{l}{y} = \frac{1}{\cos\theta}$$

and $$-dl/dt = U$$

Hence, $$v_M = \frac{U}{\cos\theta}$$

∴ The correct option is (b).

Note *(i) Here θ is variable. Therefore the constraint relation $v_M = \dfrac{U}{\cos\theta}$ is also variable.*

(ii) y and l both are decreasing with time.

∴ $$-\frac{dy}{dt} = v_M$$

and $$-\frac{dl}{dt} = U$$

Example 8 In the adjoining figure, wire PQ is smooth, ring A has a mass of 1 kg and block B has mass of 2 kg. If system is released from rest with $\theta = 60°$, find

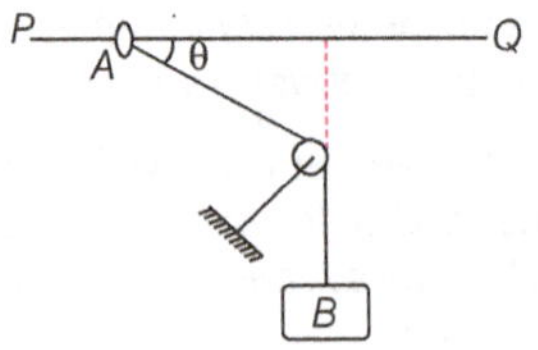

(a) constraint relation between their velocities as a function of θ.
(b) constraint relation between their accelerations just after the release at $\theta = 60°$.
(c) tension in the string and the values of these accelerations at this instant.

Sol. M and Q are two fixed fixed points. Therefore,

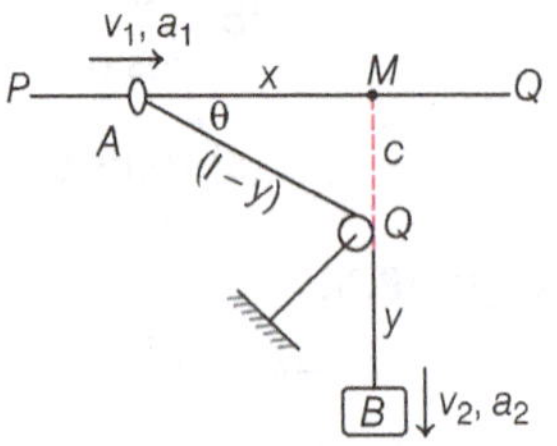

$MQ = \text{constant} = c$

$l = \text{length of string} = \text{constant}.$

(a) In triangle MQA, $(l-y)^2 = x^2 + c^2$

Differentiating w.r.t time, we get

$$2(l-y)\left(-\frac{dy}{dt}\right) = 2x\left(+\frac{dx}{dt}\right) + 0$$

or $$(l-y)\left(+\frac{dy}{dt}\right) = x\left(-\frac{dx}{dt}\right) \quad \text{...(i)}$$

$\therefore$ $$\frac{dy}{dt} = \left(\frac{x}{l-y}\right)\left(-\frac{dx}{dt}\right) \quad \text{...(ii)}$$

y is increasing with time

$\therefore$ $$+\frac{dy}{dt} = v_2$$

x is decreasing with time

$\therefore$ $$-\frac{dx}{dt} = v_1 \quad \text{and} \quad \frac{x}{l-y} = \cos\theta$$

Substituting these values in Eq. (ii), we have

$$v_2 = v_1 \cos\theta \quad \textbf{Ans.}$$

(b) Further differentiating Eq. (i), we have

$$\frac{d^2y}{dt^2}(l-y) - \left(\frac{dy}{dt}\right)^2 = -\left[x \cdot \frac{d^2x}{dt^2} + \left(\frac{dx}{dt}\right)^2\right] \quad \text{...(iii)}$$

Just after the release, $v_1, v_2, \frac{dx}{dt}$ and $\frac{dy}{dt}$ all are zero.

Substituting in Eq. (iii), we have

$$\frac{d^2y}{dt^2} = \left(\frac{x}{l-y}\right)\left(-\frac{d^2x}{dt^2}\right) \quad \text{...(iv)}$$

Here, $$\frac{d^2y}{dt^2} = a_2 \text{ and } -\frac{d^2x}{dt^2} = a_1$$

$$\frac{x}{l-y} = \cos\theta = \cos 60° = \frac{1}{2}$$

Substituting in Eq. (iv), we have

$$a_2 = a_1 \cos\theta = \frac{a_1}{2} \quad \text{...(v)}$$

(c) **For A** Equation is

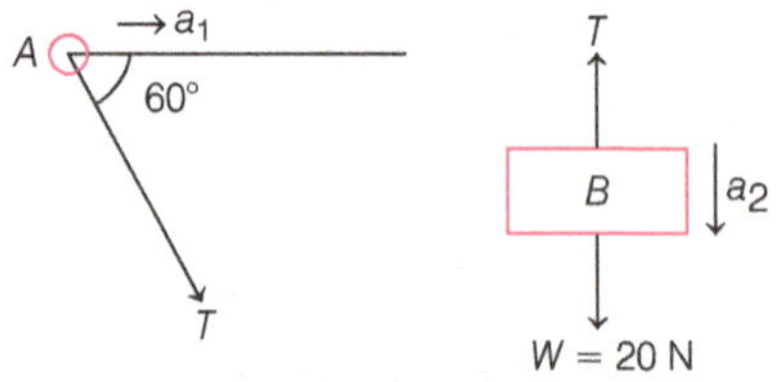

$$T\cos 60° = m_A a_1$$

or $$\frac{T}{2} = (1)a_1 = a_1 \quad \text{...(vi)}$$

For B $$20 - T = m_B a_2$$

$$20 - T = 2a_2 \quad \text{...(vii)}$$

Solving Eqs. (v), (vi) and (vii), we get

$$T = \frac{40}{3}\text{N}$$

$$a_1 = \frac{20}{3}\text{m/s}^2$$

and $$a_2 = \frac{10}{3}\text{m/s}^2$$

Note *(i) From the above example we can see that just after the release, acceleration constraint relation ($a_2 = a_1\cos\theta$) and velocity constraint relation ($v_2 = v_1\cos\theta$) are of same type.*

(ii) We have discussed earlier that in some problems acceleration (or here velocity) is inversely proportional to tension. This concept can be applied here also, if component of tension in the direction of motion is taken into account.

$$v \propto \frac{1}{T} \Rightarrow \frac{v_1}{v_2} = \frac{T_2}{T_1}$$

$$= \frac{T}{T\cos\theta}$$

$$\Rightarrow \quad v_2 = v_1\cos\theta$$

Miscellaneous Examples

Ex. 1 Two blocks of masses $m = 5$ kg and $M = 10$ kg are connected by a string passing over a pulley B as shown. Another string connects the centre of pulley B to the floor and passes over another pulley A as shown. An upward force F is applied at the centre of pulley A. Both the pulleys are massless.

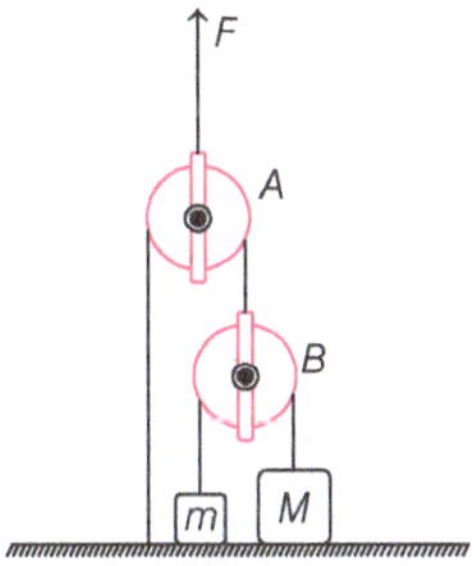

Find the acceleration of blocks m and M, if F is

(a) 100 N

(b) 300 N

(c) 500 N (Take $g = 10 \text{ m/s}^2$)

Sol. Let $T_0 =$ tension in the string passing over A

$T =$ tension in the string passing over B

$2T_0 = F$ and $2T = T_0$

$\therefore \quad T = F/4$

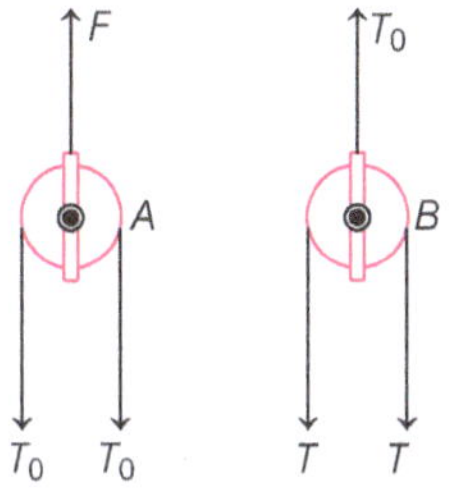

(a) $T = F/4 = 25$ N

Weights of blocks are $mg = 50$ N

and $Mg = 100$ N

As $T < mg$ and Mg both, the blocks will remain stationary on the floor.

(b) $T = F/4 = 75$ N

As $T < Mg$ and $T > mg$, M will remain stationary on the floor, whereas m will move.

Acceleration of m,

$$a = \frac{T - mg}{m} = \frac{75 - 50}{5}$$

$$= 5 \text{ m/s}^2$$ **Ans.**

(c) $T = F/4 = 125$ N

As $T > mg$ and Mg. So, both the blocks will accelerate upwards.

Acceleration of m,

$$a_1 = \frac{T - mg}{m} = \frac{125 - 50}{5}$$

$$= 15 \text{ m/s}^2$$

Acceleration of M,

$$a_2 = \frac{T - Mg}{M} = \frac{125 - 100}{10}$$

$$= 2.5 \text{ m/s}^2$$

Ex. 2 Two blocks A and B of masses 1 kg and 2 kg respectively are connected by a string, passing over a light frictionless pulley. Both the blocks are resting on a horizontal floor and the pulley is held such that string remains just taut.At moment $t = 0$, a force $F = 20t$ newton starts acting on the pulley along vertically upward direction as shown in figure. Calculate

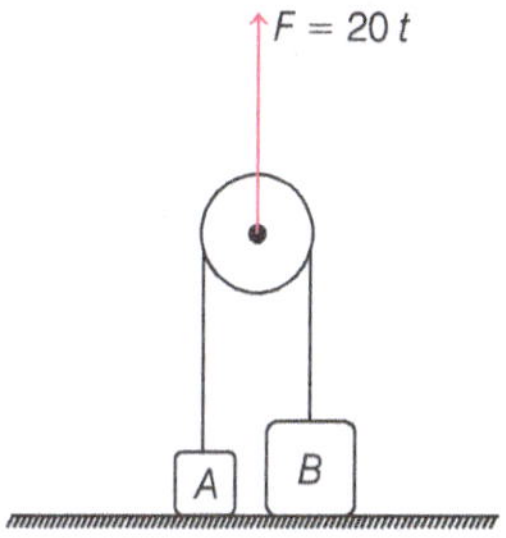

(a) velocity of A when B loses contact with the floor.

(b) height raised by the pulley upto that instant.

(Take, $g = 10 \text{ m/s}^2$)

Sol. (a) Let T be the tension in the string. Then,

$$2T = 20\,t$$

or $$T = 10\,t \text{ newton}$$

Let the block A loses its contact with the floor at time $t = t_1$. This happens when the tension in string becomes equal to the weight of A. Thus,

$$T = mg$$

or $$10\,t_1 = 1 \times 10$$

or $$t_1 = 1 \text{ s} \quad \text{...(i)}$$

Similarly, for block B, we have

$$10t_2 = 2 \times 10$$

or $$t_2 = 2 \text{ s} \quad \text{...(ii)}$$

i.e. the block B loses contact at 2 s. For block A, at time t such that $t \geq t_1$ let a be its acceleration in upward direction. Then,

$$10t - 1 \times 10 = 1 \times a = (dv/dt)$$

or $$dv = 10(t - 1)\,dt \quad \text{...(iii)}$$

Integrating this expression, we get

$$\int_0^v dv = 10 \int_1^t (t - 1)\,dt$$

or $\quad v = 5t^2 - 10t + 5 \quad$ ...(iv)

Substituting $t = t_2 = 2$ s

$$v = 20 - 20 + 5 = 5 \text{ m/s} \quad \text{...(v)}$$

(b) From Eq. (iv), $dy = v \cdot dt = (5t^2 - 10t + 5)\, dt \quad$ ...(vi)

where, y is the vertical displacement of block A at time $t\ (\geq t_1)$.

Integrating, we have

$$\int_{y=0}^{y=h} dy = \int_{t=1}^{t=2} (5t^2 - 10t + 5)\, dt$$

$$h = 5\left[\frac{t^3}{3}\right]_1^2 - 10\left[\frac{t^2}{2}\right]_1^2 + 5\,[t]_1^2 = \frac{5}{3}\text{ m}$$

∴ Height raised by the pulley upto that instant

$$= \frac{h}{2} = \frac{5}{6}\text{ m} \quad \textbf{Ans.}$$

Ex. 3 Find the acceleration of the body of mass m_2 in the arrangement shown in figure. If the mass m_2 is η times great as the mass m_1, and the angle that the inclined plane forms with the horizontal is equal to θ. The masses of the pulleys and threads, as well as the friction, are assmed to be negligible. **(Irodov1.72)**

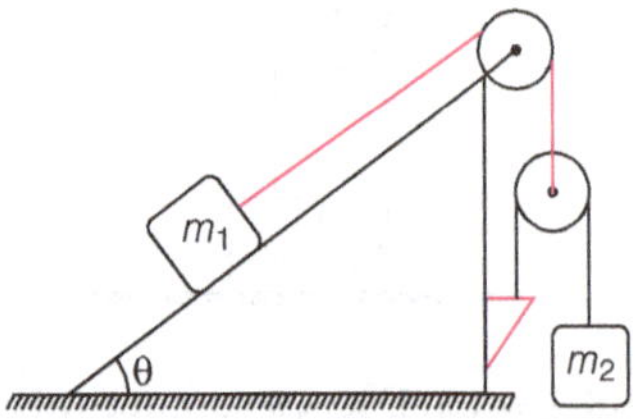

Sol. Here, by constraint relation we can see that the acceleration of m_2 is two times that of m_1. So, let m_1 is moving up the inclined plane with an acceleration a, then the acceleration of mass m_2 is $2a$ downwards. The tensions in different strings are shown in figure.

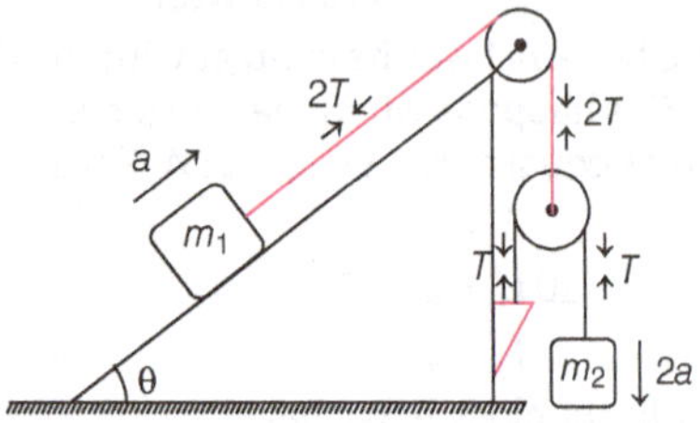

The dynamic equations can be written as

For mass m_1: $\quad 2T - m_1 g \sin\theta = m_1 a \quad$ ...(i)

For mass m_2: $\quad m_2 g - T = m_2(2a) \quad$ ...(ii)

Substituting $m_2 = \eta m_1$ and solving Eqs. (i) and (ii), we get

$$\text{Acceleration of } m_2 = 2a = \frac{2g(2\eta - \sin\theta)}{4\eta + 1} \quad \textbf{Ans.}$$

Ex. 4 In the arrangement shown in figure the mass of the ball is η times as great as that of the rod. The length of the rod is l, the masses of the pulleys and the threads, as well as the friction, are negligible. The ball is set on the same level as the lower end of the rod and then released. How soon will the ball be opposite the upper end of the rod? **(Irodov 1.75)**

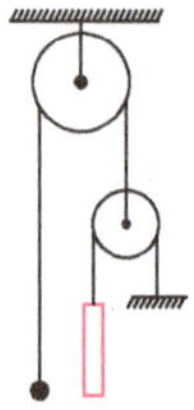

Sol. From constraint relation we can see that the acceleration of the rod is double than that of the acceleration of the ball. If ball is going up with an acceleration a, rod will be coming down with the acceleration $2a$. Thus, the relative acceleration of the ball with respect to rod is $3a$ in upward direction. If it takes time t seconds to reach the upper end of the rod, we have

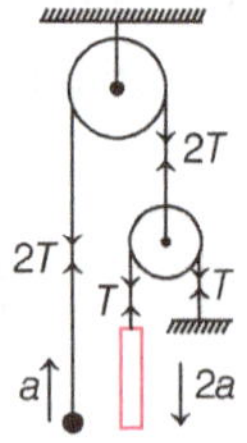

$$t = \sqrt{\frac{2l}{3a}} \quad \text{...(i)}$$

Let mass of ball be m and that of rod is M, the dynamic equations of these are

For rod $\quad Mg - T = M(2a) \quad$ ...(ii)

For ball $\quad 2T - mg = ma \quad$ ...(iii)

Substituting $m = \eta M$ and solving Eqs. (ii) and (iii), we get

$$a = \left(\frac{2 - \eta}{\eta + 4}\right) g$$

From Eq. (i), we have

$$t = \sqrt{\frac{2l(\eta + 4)}{3g(2 - \eta)}} \quad \textbf{Ans.}$$

MODULE 4 Exercise

Single Correct Option

1. A 50 kg man stands on a 25 kg platform. He pulls on the rope which is attached to the platform *via* the frictionless pulleys as shown in the figure. The platform moves upwards at a steady rate, if the force with which the man pulls the rope is

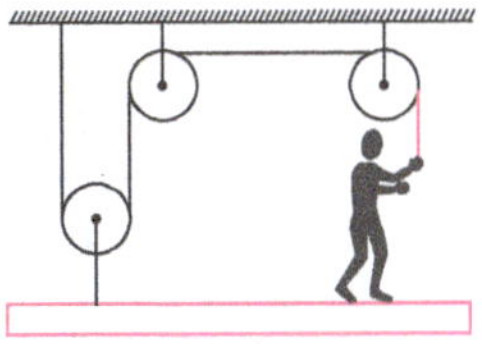

(a) 500 N (b) 250 N
(c) 25 N (d) None of these

2. In the figure shown, a person wants to raise a block lying on the ground to a height *h*. In which case, he has to exert more force. Assume pulleys and strings are light

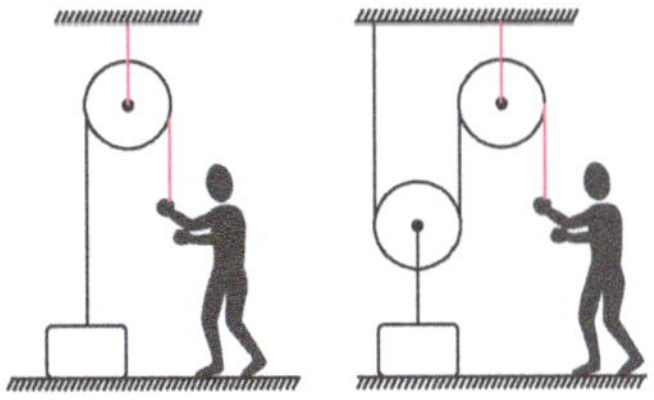

(i) (ii)

(a) Fig. (i) (b) Fig. (ii)
(c) Same in both (d) Cannot be determined

3. A man of mass *m* stands on a platform of equal mass *m* and pulls himself by two ropes passing over pulleys as shown in figure. If he pulls each rope with a force equal to half his weight, his upward acceleration would be

(a) $\frac{g}{2}$ (b) $\frac{g}{4}$
(c) g (d) zero

4. In the arrangement shown in the figure, the pulley has a mass 3*m*. Neglecting friction on the contact surface, the force exerted by the supporting rope *AB* on the ceiling is

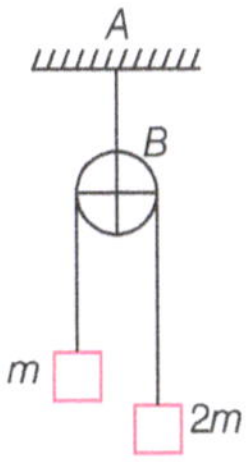

(a) 6 *mg*
(b) 3 *mg*
(c) 4 *mg*
(d) None of the above

5. The acceleration of the 2 kg block, if the free end of string is pulled with a force of 20 N as shown is

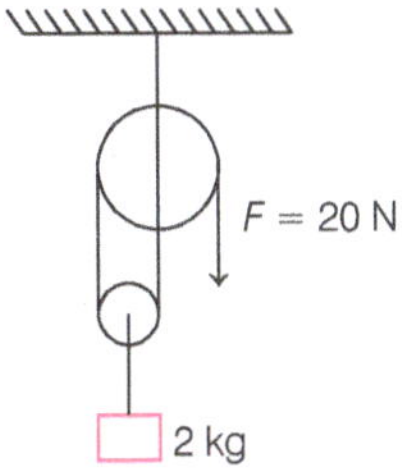

(a) zero (b) $10\,m/s^2$
(c) $5\,m/s^2$ upward
(d) $5\,m/s^2$ downward

6. Two blocks, each having a mass *M*, rest on frictionless surfaces as shown in the figure. If the pulleys are light and frictionless and *M* on the incline is allowed to move down, then the tension in the string will be

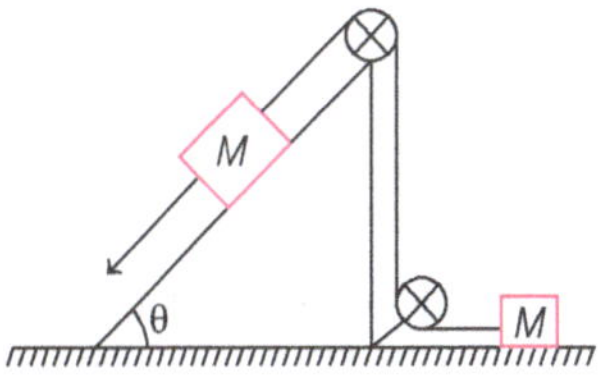

(a) $\frac{2}{3}Mg\sin\theta$ (b) $\frac{3}{2}Mg\sin\theta$ (c) $\frac{Mg\sin\theta}{2}$ (d) $2Mg\sin\theta$

7. Two masses are connected by a string which passes over a pulley accelerating upwards at a rate *A* as shown. If a_1 and a_2 be the accelerations of bodies 1 and 2 respectively, then

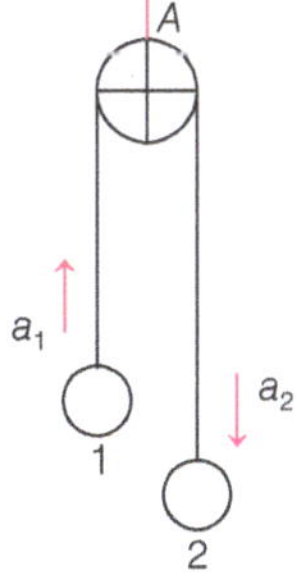

(a) $A = a_1 - a_2$
(b) $A = a_1 + a_2$
(c) $A = \frac{a_1 - a_2}{2}$
(d) $A = \frac{a_1 + a_2}{2}$

8. In the figure shown the block *B* moves with velocity $10\,ms^{-1}$. The velocity of *A* in the position shown is

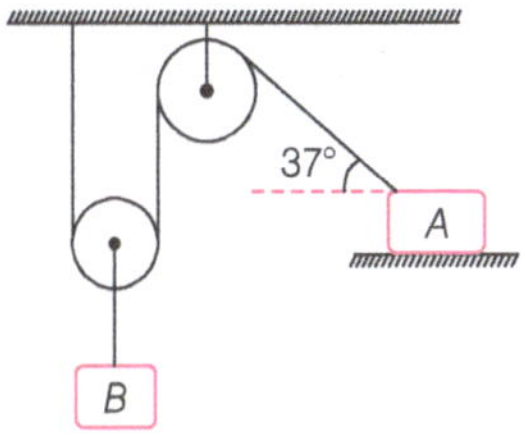

(a) $12.5\,ms^{-1}$ (b) $25\,ms^{-1}$
(c) $8\,ms^{-1}$ (d) $16\,ms^{-1}$

9. Block *B* moves to the right with a constant velocity v_0. The velocity of body *A* relative to *B* is

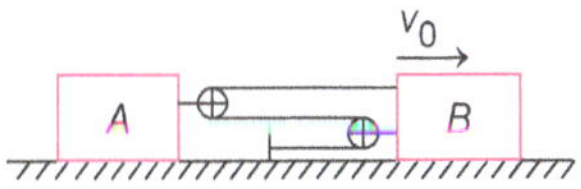

(a) $\frac{v_0}{2}$, towards left (b) $\frac{v_0}{2}$, towards right
(c) $\frac{3v_0}{2}$, towards left (d) $\frac{3v_0}{2}$, towards right

Subjective Questions

10. Make the constraint relation between a_1, a_2 and a_3.

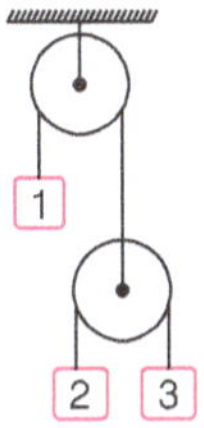

11. Consider the situation shown in figure. Both the pulleys and the string are light and all the surfaces are smooth.

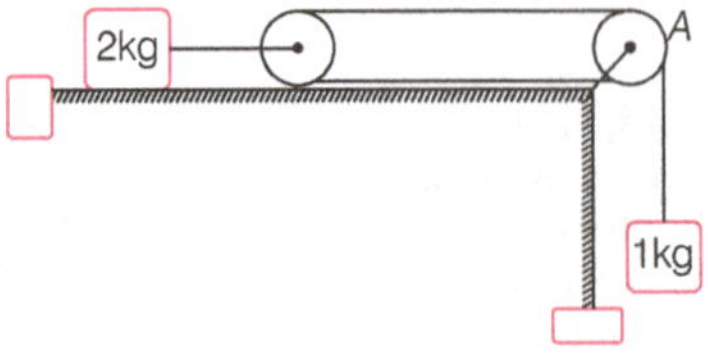

(a) Find the acceleration of 1 kg block.
(b) Find the tension in the string. ($g = 10\,m/s^2$)

12. Calculate the acceleration of either blocks and tension in the string shown in figure. The pulley and the string are light and all surfaces are smooth.

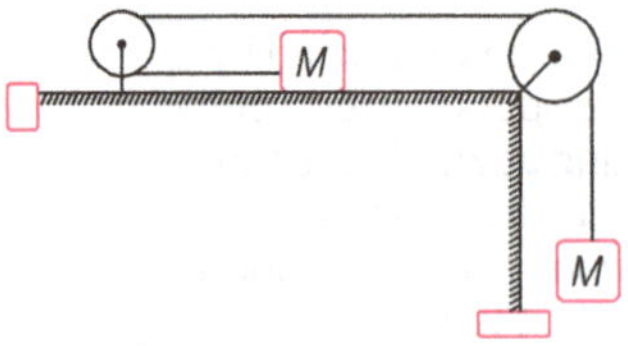

13. In figure assume that there is negligible friction between the blocks and table. Compute the tension in the cord connecting m_2 and the pulley and acceleration of m_2 if $m_1 = 300$ g, $m_2 = 200$ g and $F = 0.40$ N.

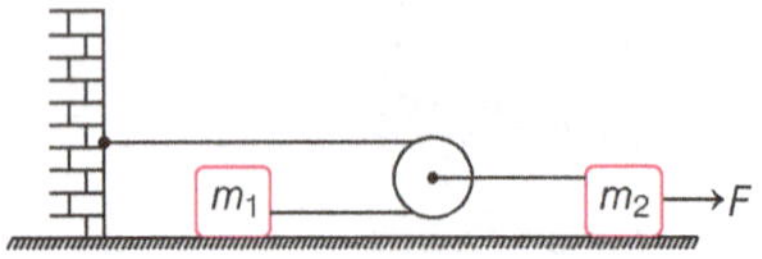

14. In the figure shown, $a_3 = 6\,m/s^2$ (downwards) and $a_2 = 4\,m/s^2$ (upwards). Find acceleration of 1.

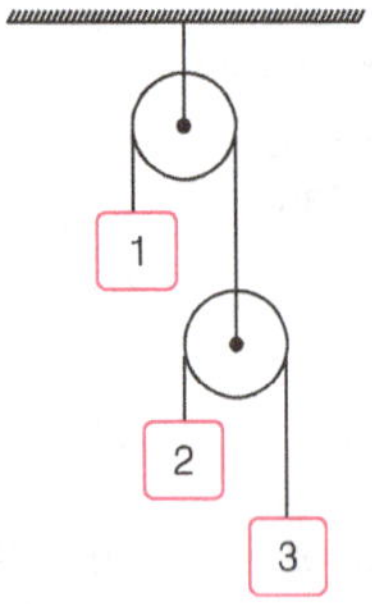

15. Find the acceleration of the block of mass M in the situation shown in the figure. All the surfaces are frictionless.

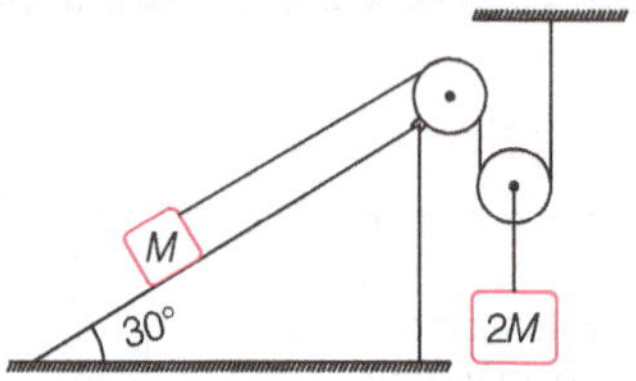

16. A block of mass 200 kg is set into motion on a frictionless horizontal surface with the help of frictionless pulley and a rope system as shown in figure. What horizontal force F should be applied to produce in the block an acceleration of 1 m/s^2 ?

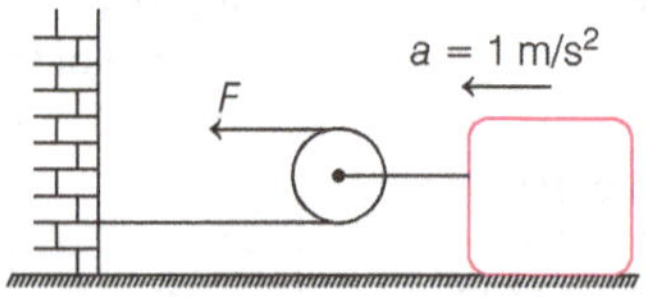

17. In the pulley-block arrangement shown in figure, find the relation between a_A, a_B and a_C.

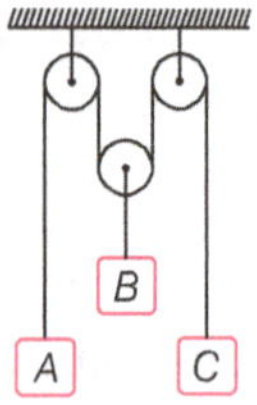

18. Find the acceleration of the blocks A and B in the situation shown in the figure.

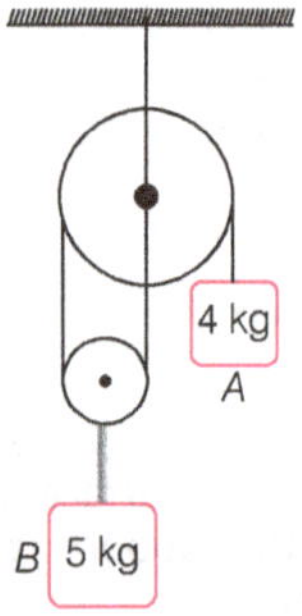

Answers

1. (b) 2. (a) 3. (d) 4. (d) 5. (b) 6. (c) 7. (c) 8. (b) 9. (b)
10. ($a_2 + a_3 + 2a_1 = 0$) 11. (a) $\frac{20}{3}$ m/s^2 (b) 10/3 N 12. $g/2$, $Mg/2$ 13. 12/35 N, 2/7 m/s^2 14. 1 m/s^2
15. $g/3$ 16. 100 N 17. $a_A + 2a_B + a_C = 0$ 18. $2g/7$

See hints on pages 425-427.

MODULE 5

Pseudo Force

Before studying the concept of pseudo force let us first discuss frame of reference.

Frame of reference is the way of observing the things.

Inertial Frame of Reference

A non-accelerating frame of reference is called an inertial frame of reference. A frame of reference moving with a constant velocity is an inertial frame of reference.

Non-Inertial Frame of Reference

An accelerating frame of reference is called a non-inertial frame of reference.

Note *(i) A rotating frame of reference is a non-inertial frame of reference, because it is also an accelerating one.*

(ii) Earth is rotating about its axis of rotation and it is revolving around the centre of sun also. So, it is non-intertial frame of reference. But for most of the cases, we consider it as an inertial frame of reference.

Pseudo Force

Now let us come to the pseudo force. Instead of ground (or inertial frame of reference) when we start watching the objects from a non-inertial (accelerating) frame of reference, its motion conditions of objects are felt differently.

For example, suppose a child is standing inside an accelerating lift. From ground frame of reference this child appears to be accelerating but from lift (non-inertial) frame of reference child appears to be at rest. To justify this changed condition of motion, from equations point of view we have to apply a pseudo force. This pseudo force is given by

$$\mathbf{F}_p = -m\mathbf{a}$$

Here, 'm' is the mass of that body/object which is being observed from non-inertial frame of reference and **a** is the acceleration of frame of reference. Negative sign implies that direction of pseudo force $\mathbf{F}_p$ is opposite to **a**. Hence whenever you make free body diagram of a body from a non-inertial frame, apply all real forces (which are actually acting) on the body plus one pseudo force. Magnitude of this pseudo force is 'ma' and the direction is opposite to **a**.

Example Suppose a block A of mass m is placed on a lift ascending with an acceleration a_0. Let N be the normal reaction between the block and the floor of the lift. Free body diagram of A in ground frame of reference (inertial) is shown in figure.

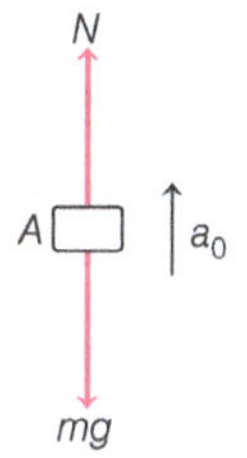

Equation of motion is :

$\therefore \quad N - mg = ma_0$

or $\quad N = m(g + a_0) \quad$ ...(i)

If we draw the free body diagram of A with respect to the elevator (a non-inertial frame of reference) without applying the pseudo force, as shown in figure, we get

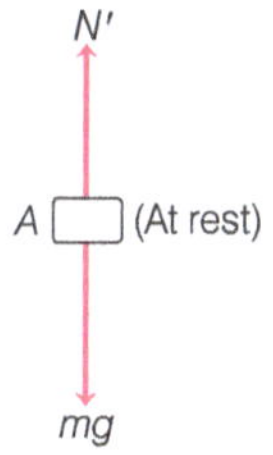

$N' - mg = 0$

or $\quad N' = mg \quad$ ...(ii)

Since, $N' \neq N$. So, we can say that Eq. (ii) is wrong. If we apply a pseudo force in non-inertial frame of reference, N' becomes equal to N as shown in figure. Acceleration of block with respect to elevator is zero.

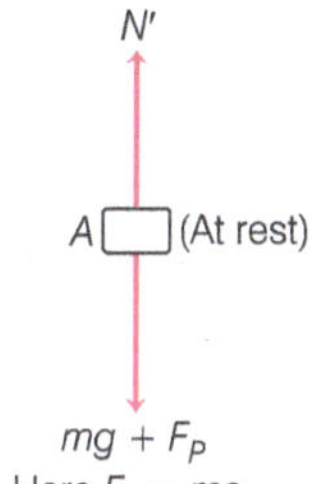

Here $F_P = ma_0$

$\therefore \quad N' - mg - ma_0 = 0$

or $\quad N' = m(g + a_0) \quad$ ...(iii)

$\therefore \quad N' = N$

Example

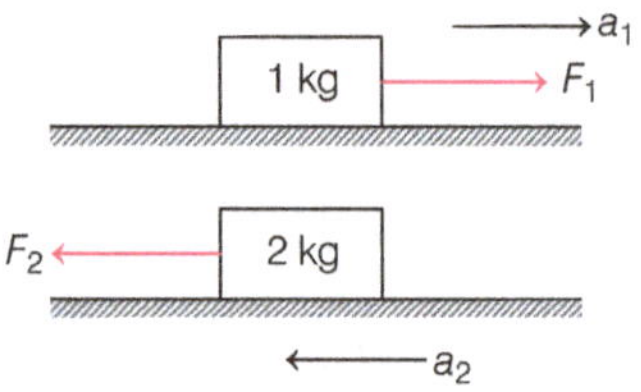

Let, $F_1 = 4\text{N}$ and $F_2 = 6\text{N}$

$\therefore \quad a_1 = \dfrac{F_1}{m_1} = \dfrac{4}{1} = 4\ \text{m/s}^2$

$a_2 = \dfrac{F_2}{m_2} = \dfrac{6}{2} = 3\ \text{m/s}^2$

Both blocks are accelerated. So, both are non-inertial. Hence, we will have to apply pseudo force if we observe one block from the frame of reference of other block.

FBD of 2 kg w.r.t. 1 kg

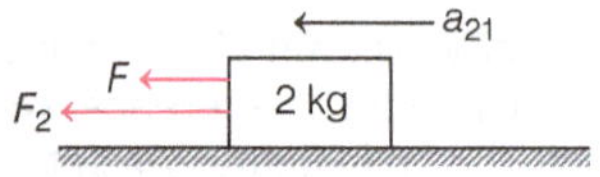

$F = \text{pseudo force} = m_2 a_1$

$= (2)(4) = 8\ \text{N}$

$\therefore \quad a_{21} = \dfrac{F_2 + F}{m_2} = \dfrac{6+8}{2} = 7\ \text{m/s}^2$

From the concept of relative motion we can see that acceleration of 2 w.r.t. 1 is $7\ \text{m/s}^2$ towards left.

FBD of 1 kg w.r.t. 2 kg

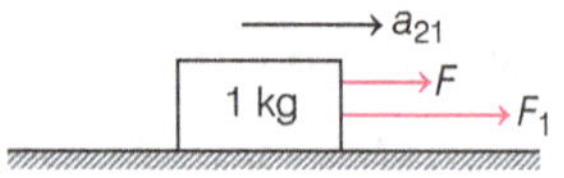

$F = \text{pseudo force} = m_1 a_2$

$= (1)(3) = 3\text{N}$

$a_{12} = \dfrac{F + F_1}{m_1} = \dfrac{3+4}{1} = 7\ \text{m/s}^2$

Again we can see that acceleration of 1 w.r.t. 2 is $7\ \text{m/s}^2$ towards right.

Example 1 All surfaces are smooth in following figure. Find F, such that block remains stationary with respect to wedge.

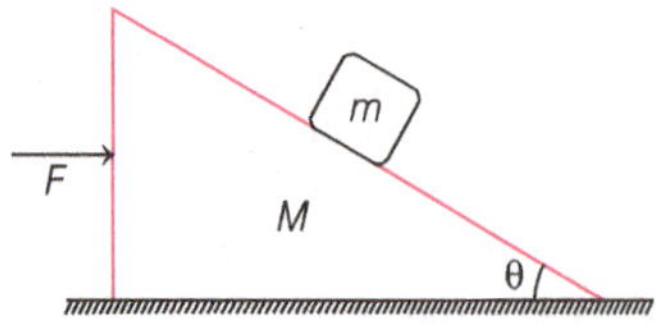

Sol. Acceleration of (block + wedge) $a = \dfrac{F}{(M+m)}$

Method 1.

From Inertial Frame of Reference (Ground)

FBD of block w.r.t. ground (Apply only real forces):

With respect to ground block is moving with an acceleration a.

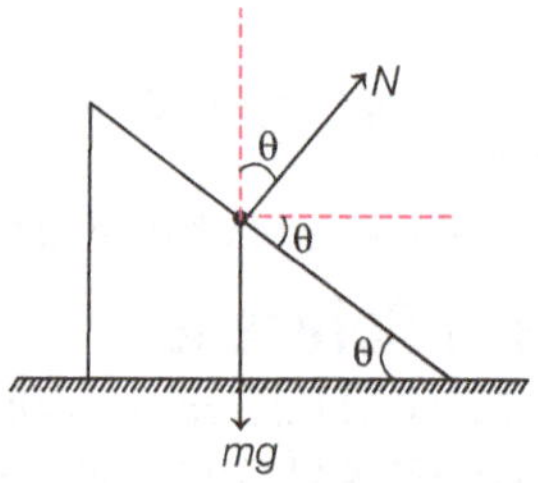

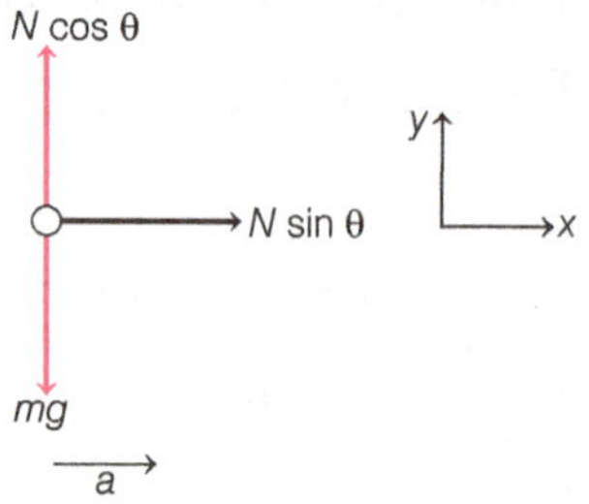

$\therefore \quad \Sigma F_y = 0$

$\Rightarrow \quad N\cos\theta = mg \quad$ ...(i)

and $\quad \Sigma F_x = ma$

$\Rightarrow \quad N\sin\theta = ma \quad$ ...(ii)

From Eqs. (i) and (ii), we get

$a = g\tan\theta$

$\therefore \quad F = (M+m)a$

$= (M+m)\,g\tan\theta$

Method 2.

From Non-inertial Frame of Reference (Wedge)

FBD of block w.r.t. wedge (real forces + pseudo force)

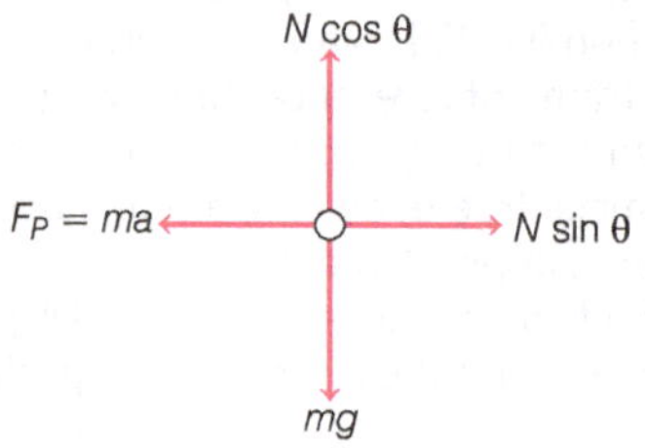

w.r.t. wedge, block is stationary

$\therefore \quad \Sigma F_y = 0 \Rightarrow N\cos\theta = mg \quad$ ...(iii)

$$\Sigma F_x = 0 \Rightarrow N \sin\theta = ma \qquad \text{...(iv)}$$

From Eqs. (iii) and (iv), we will get the same result

i.e. $F = (M + m)g \tan\theta.$

Example 2 A bob of mass m is suspended from the ceiling of a train moving with an acceleration a as shown in figure. Find the angle θ in equilibrium position.

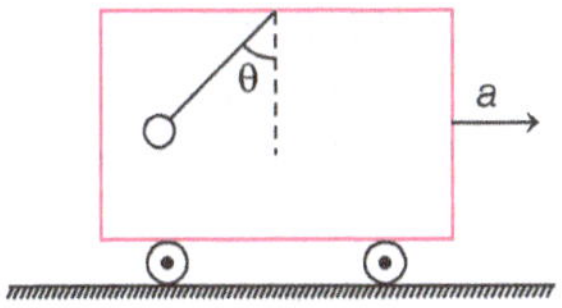

Sol. This problem can also be solved by both the methods.

Method 1.

From Inertial Frame of Reference (Ground)

FBD of bob w.r.t. ground (only real forces)

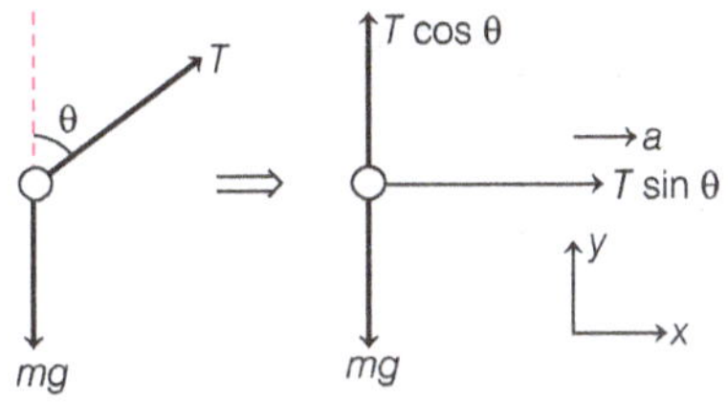

With respect to ground, bob is also moving with an acceleration a.

$$\therefore \quad \Sigma F_x = ma$$

$$\Rightarrow \quad T \sin\theta = ma \qquad \text{...(i)}$$

and $\Sigma F_y = 0 \Rightarrow T\cos\theta = mg$ ...(ii)

From Eqs. (i) and (ii), we get

$$\tan\theta = \frac{a}{g}$$

or

$$\theta = \tan^{-1}\left(\frac{a}{g}\right)$$

Method 2.

From Non-inertial Frame of Reference (Train)

FBD of bob w.r.t. train (real forces + pseudo force):

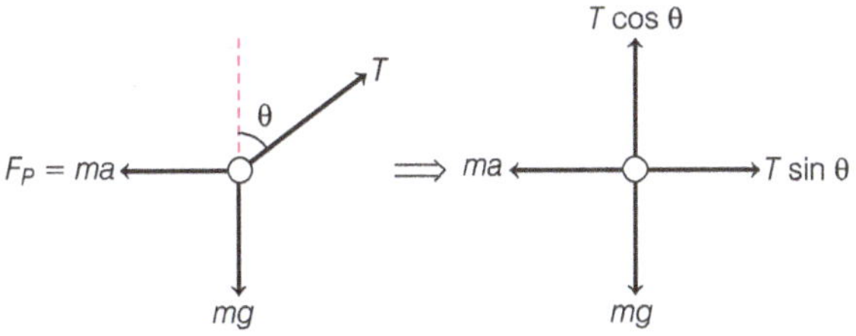

with respect to train, bob is in equilibrium (or at rest)

$$\therefore \quad \Sigma F_x = 0$$

$$\Rightarrow \quad T\sin\theta = ma \qquad \text{...(iii)}$$

$$\therefore \quad \Sigma F_y = 0$$

$$\Rightarrow \quad T\cos\theta = mg \qquad \text{...(iv)}$$

From Eqs. (iii) and (iv), we get the same result, i.e.

$$\theta = \tan^{-1}\left(\frac{a}{g}\right)$$

Example 3 In the arrangement shown in figure the masses m of the bar and M of the wedge, as well as the wedge angle α, are known.

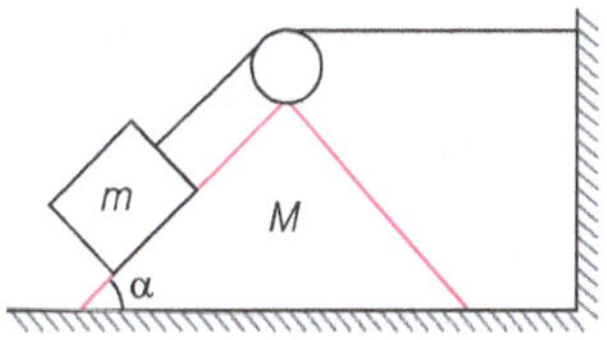

The masses of the pulley and the thread are negligible. The friction is absent. Find the acceleration of the wedge M. **(Irodov 1.82)**

Sol. From constraint relations we can prove that absolute acceleration of M towards right is equal to acceleration of m relative to M (down the plane) and suppose this acceleration is a.

T = tension in string

N = normal reaction between two blocks

FBD of M (only horizontal forces are shown)

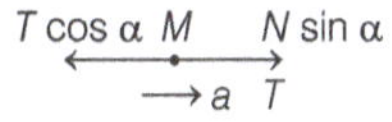

Equation of motion is,

$$T + N\sin\alpha - T\cos\alpha = Ma \qquad \text{...(i)}$$

FBD of m with respect to M, with pseudo force ma towards left

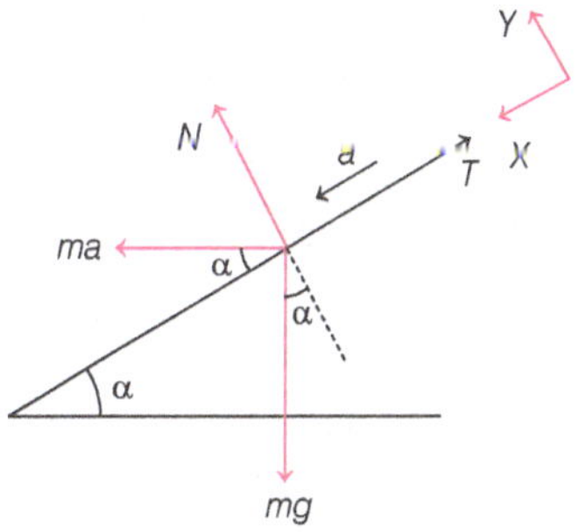

$$\Sigma F_y = 0$$

$$\Rightarrow \quad N + ma\sin\alpha = mg\cos\alpha \qquad \text{...(ii)}$$

$$\Sigma F_x = ma$$

$$\Rightarrow \quad ma\cos\alpha + mg\sin\alpha - T = ma \qquad \text{...(iii)}$$

We have three unknowns T, N and a. Solving these three equations, we get

$$a = \frac{mg\sin\alpha}{M + 2m(1 - \cos\alpha)}$$

MODULE 5 Exercise

Problems Based on Pseudo Force

True/False

1. Pseudo force with respect to a frame moving with constant velocity is zero. Is this statement true or false?
2. Problems of non-intertial frames can be solved only with the concept of pseudo force. Is this statement true or false?

Assertion and Reason

See directions on page 354.

3. **Assertion :** Two frames S_1 and S_2 are non-inertial. Then frame S_2 when observed from S_1 is inertial.

   Reason : A frame in motion is not necessarily a non-inertial frame.

Single Correct Option

4. A smooth inclined plane of length L having inclination θ with the horizontal is inside a lift which is moving down with a retardation a. The time taken by a body to slide down the inclined plane from rest will be

   (a) $\sqrt{\dfrac{2L}{(g+a)\sin\theta}}$ (b) $\sqrt{\dfrac{2L}{(g-a)\sin\theta}}$

   (c) $\sqrt{\dfrac{2L}{a\sin\theta}}$ (d) $\sqrt{\dfrac{2L}{g\sin\theta}}$

5. A block is sliding along an inclined plane as shown in the figure. If the acceleration of chamber is a as shown in the figure. The time required to cover a distance L along the inclined plane is

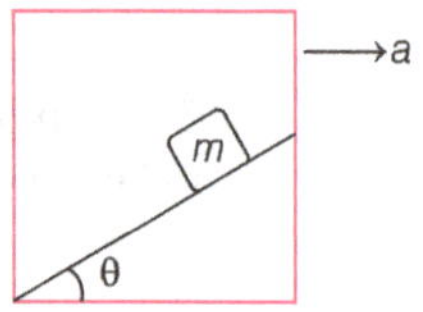

   (a) $\sqrt{\dfrac{2L}{g\sin\theta - a\cos\theta}}$ (b) $\sqrt{\dfrac{2L}{g\sin\theta + a\sin\theta}}$

   (c) $\sqrt{\dfrac{2L}{g\sin\theta + a\cos\theta}}$ (d) $\sqrt{\dfrac{2L}{g\sin\theta}}$

6. A block is kept on a frictionless inclined surface with angle of inclination 'α'. The incline is given an acceleration 'a' to keep the block stationary. Then a is equal to **(2005 JEE)**

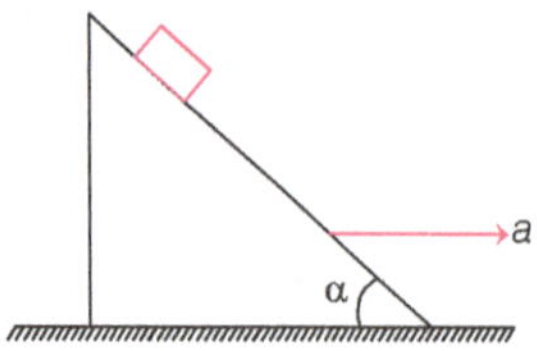

   (a) g cosec α (b) $g/\tan\alpha$

   (c) $g\tan\alpha$ (d) g

Multiple Correct Options

7. A reference frame attached to the earth **(1986 JEE)**

   (a) is an inertial frame by definition

   (b) cannot be an inertial frame because the earth is revolving round the sun

   (c) is an inertial frame because Newton's laws are applicable in this frame

   (d) cannot be an inertial frame because the earth is rotating about its own axis

Subjective Question

8. Two blocks A and B of masses 1kg and 2 kg have accelerations $(2\hat{\mathbf{i}})\text{m/s}^2$ and $(-4\hat{\mathbf{j}})\text{m/s}^2$. Find

   (a) Pseudo force on block A as applied with respect to the block B.

   (b) Pseudo force on block B as applied with respect to the block A.

Miscellaneous Problems

Single Correct Option

9. A man of mass m slides down along a rope which is connected to the ceiling of an elevator with deceleration a relative to the rope. If the elevator is going upward with an acceleration a relative to the ground, then tension in the rope is

   (a) mg (b) $m(g+2a)$ (c) $m(g+a)$ (d) zero

10. A lift is moving down with acceleration a. A man in the lift drops a ball inside the lift. The acceleration of the ball as observed by the man in the lift and a man standing stationary on the ground are respectively **(2002 JEE)**

   (a) g, g (b) $g-a,\ g-a$

   (c) $g-a, g$ (d) a, g

Multiple Correct Options

11. A block of mass 1 kg is at rest relative to a smooth wedge moving leftwards with constant acceleration $a = 5\ \text{ms}^{-2}$. Let N be the normal reaction between the block and the wedge. Then, $(g = 10\text{ms}^{-2})$

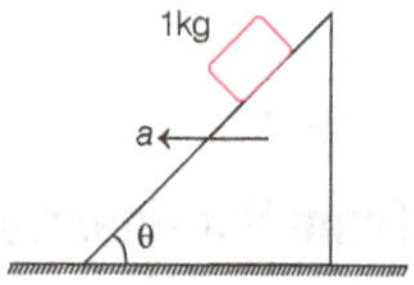

   (a) $N = 5\sqrt{5}$ newton (b) $N = 15$ newton

   (c) $\tan\theta = \dfrac{1}{2}$ (d) $\tan\theta = 2$

Answers

1. (True)	**2.** (False)	**3.** (d)	**4.** (a)	**5.** (c)
6. (c)	**7.** (b, d)	**8.** (a) $(4\hat{j})$N (b) $(-4\hat{i})$N		**9.** (b)
10. (c)	**11.** (a, c)			

See hints on pages 427-428.

MODULE 6

Friction

Regarding the frictional force (f) following points are worthnoting :

1. It is the tangential component of net contact force (F) acting between two bodies in contact.
2. It starts acting when relative motion (between two bodies in contact) is either taking place or there is tendency of relative motion between them. So, friction has a tendency to stop relative motion between two bodies in contact.
3. If there is only tendency of relative motion then static friction acts and if actual relative motion is taking place, then kinetic friction acts.
4. Like any other force of nature friction force also makes a pair of equal and opposite forces acting on two different bodies.
5. Direction of friction force on a given body is opposite to the direction of relative motion (or its tendency) of this body.
6. 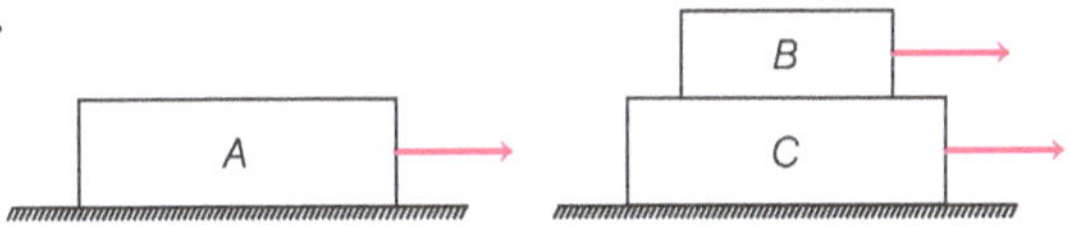

(i) (ii)

   In figure. (i), motion of block A means its relative motion with respect to ground. So, in this case friction between block and ground has a tendency to stop its motion.

   In figure. (ii), relative motion between two blocks B and C means they have different velocities. So, friction between these two blocks has a tendency to make their velocities same.
7. Static friction is self adjusting in nature. This varies from zero to a limiting value f_L. Only that much amount of friction will act which can stop the relative motion.
8. Kinetic friction is constant and it can be denoted by f_k.
9. It is found experimentally that limiting value of static friction f_L and constant value of kinetic friction f_k both are directly proportional to normal reaction N acting between the two bodies.

   $$\therefore \boxed{f_L \text{ or } f_k \propto N \Rightarrow f_L = \mu_s N \text{ and } f_k = \mu_k N}$$

   Here, μ_s = coefficient of static friction and μ_k = coefficient of kinetic friction.

   Both μ_s and μ_k are dimensionless constants which depend on the nature of surfaces in contact. Value of μ_k is usually less than the value of μ_s i.e. constant value of kinetic friction is less than the limiting value of static friction.

Note (i) *In some of the problems, μ_s and μ_k are not given separately but only single value of μ is given. In those cases,*

$$\boxed{f_L = f_k = \mu N}$$

(ii) *If more than two blocks are placed one above the other on a horizontal ground then normal reaction between two blocks will be equal to the weight of the blocks above the common surface.*

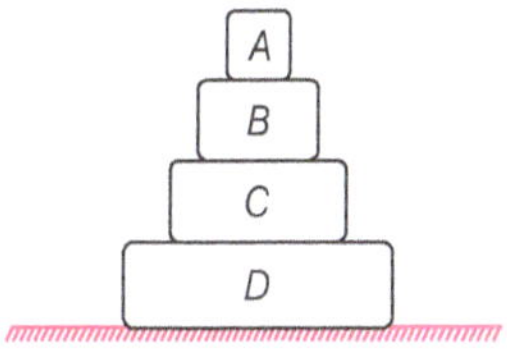

For example, N_1 = *normal reaction between A and B*

$= m_A g$

N_2 = *normal reaction between B and C*

$= (m_A + m_B) g$ *and so on.*

Important POINTS

Friction force is electromagnetic in nature.

The surfaces in contact, however smooth they may appear, actually have imperfections called asperities.

When one surface rests on the other the actual area of contact is very less than the surface area of the face of contact.

The pressure due to the reaction force between the surfaces is very high as the true contact area is very small. Hence, these contact points deform a little and cold welds are formed at these points.

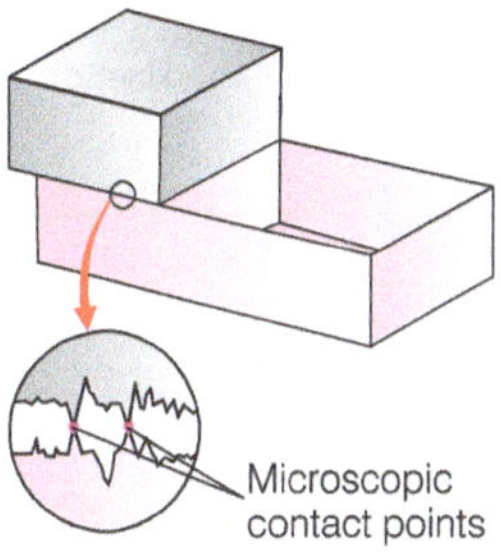

So, in order to start the relative sliding between these surfaces, enough force has to be applied to break these welds. But, once the welds break and the surfaces start sliding over each other, the further formation of these welds is relatively slow and weak and hence a smaller force is enough to keep the block moving with uniform velocity.

This is the reason why limiting value of static friction is greater than the kinetic friction.

Note *By making the surfaces extra smooth, frictional force increases as actual area of contact increases and the two bodies in contact act like a single body.*

Example 1 Suppose a block of mass 1 kg is placed over a rough surface and a horizontal force F is applied on the block as shown in figure. Now, let us see what are the values of force of friction f and acceleration of the block a if the force F is gradually increased. Given that $\mu_s = 0.5, \mu_k = 0.4$ and $g = 10 \text{ m/s}^2$.

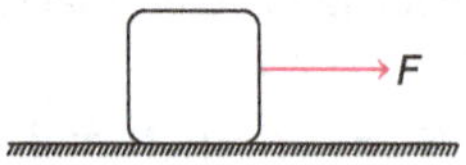

Sol. Free body diagram of the block is

$$\Sigma F_y = 0$$

$\therefore$ $$N - mg = 0$$

or $$N = mg = (1)(10) = 10 \text{ N}$$

$$f_L = \mu_s N = (0.5)(10) = 5 \text{ N}$$

and $$f_k = \mu_k N = (0.4)(10) = 4\text{N}$$

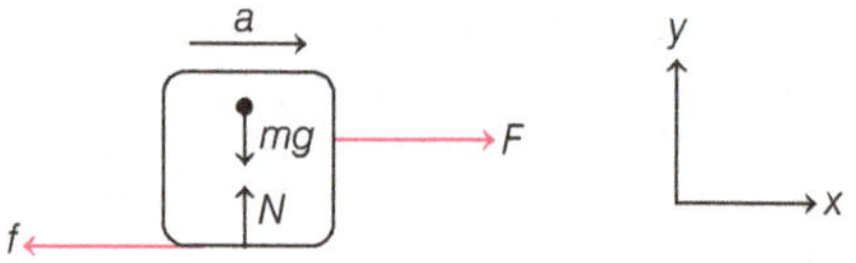

Variation of frictional force f with the applied force F is given below in tabular form.

F	f	$F_{net} = F - f$	Static or kinetic friction	Relative motion or tendency of relative motion	Acceleration of block $a = \frac{F_{net}}{m}$	Diagram
0	0	0	static	Neither tendency nor actual relative motion	0	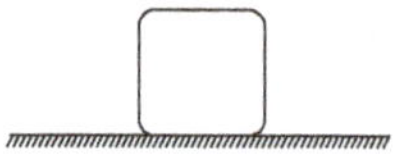
2 N	2 N	0	static	Tendency	0	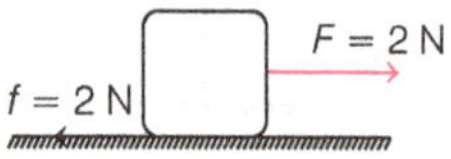
4 N	4 N	0	static	Tendency	0	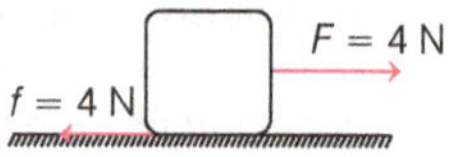
5 N	5 N	0	static	Tendency	0	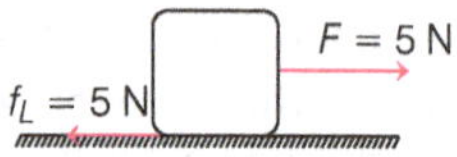
6 N	4N	2 N	kinetic	Actual relative motion	2 m/s^2	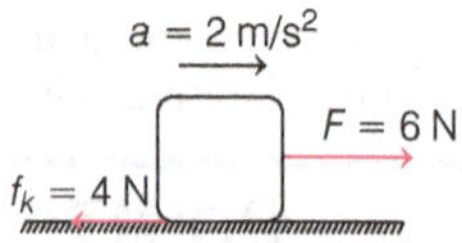
8 N	4 N	4 N	kinetic	Actual relative motion	4 m/s^2	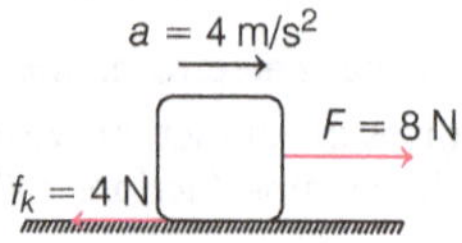

Graphically, this can be explained as below.

Note that $f = F$ till $F \le f_L$. Therefore, slope of line OA will be 1 $(y = mx)$ or angle of line OA with F-axis is 45°.

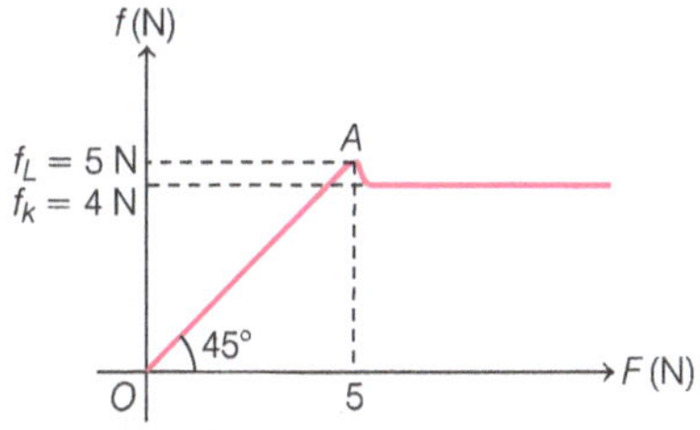

Here, $a = 0$ for $F \le 5$ N

and $a = \frac{F - f_K}{m} = \frac{F - 4}{1} = F - 4$ for $F > 5$ N

Thus, a-F graph is as shown in the figure given below.

When F is slightly increased from 5 N, acceleration of block increases from 0 to 1 m/s^2. Think why?

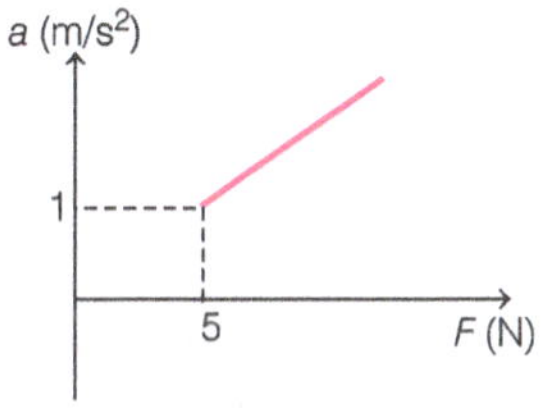

Note *Henceforth, we will take coefficient of friction as* μ *unless* μ_s *and* μ_k *are given separately.*

Angle of Friction (λ)

In limiting case the two forces are the normal reaction N and frictional force μN.

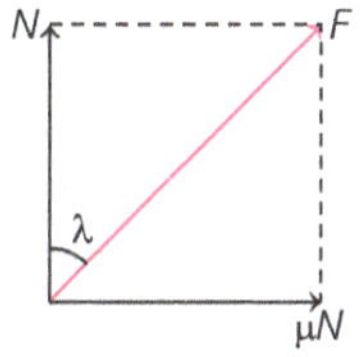

The resultant of these two forces is the net contact force F and suppose it makes an angle λ with the normal reaction, then

$$\tan \lambda = \frac{\mu N}{N} = \mu \quad \text{or} \quad \lambda = \tan^{-1}(\mu) \qquad \text{...(i)}$$

This angle λ is called the angle of friction.

Angle of Repose (α)

Suppose a block of mass m is placed on an inclined plane whose inclination θ can be increased or decreased. Let μ be the coefficient of friction between the block and the plane.

At some general value θ,

Normal reaction $N = mg \cos \theta$

Limiting friction $f_L = \mu N = \mu mg \cos \theta$

and the driving force

$F = mg \sin \theta$ (Down the plane)

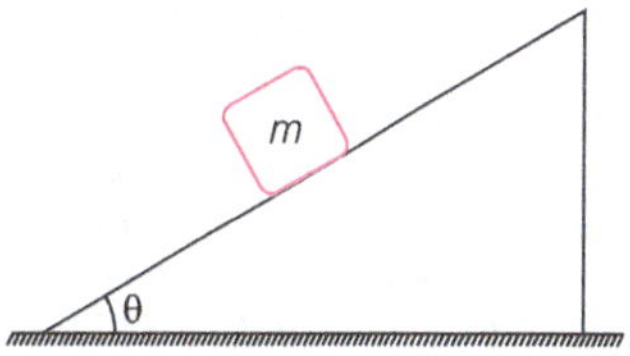

From these three equations we see that, when θ is increased from 0° to 90°, normal reaction N and hence, the limiting friction f_L is decreased while the driving force F is increased. There is a critical angle called angle of repose (α) at which these two forces are equal.

Now, if θ is further increased, then the driving force F becomes more than the limiting friction f_L and the block starts sliding.

Thus, $f_L = F$ at $\theta = \alpha$

or $\mu mg \cos \alpha = mg \sin \alpha$

or $\tan \alpha = \mu$

or $\alpha = \tan^{-1}(\mu)$ ...(ii)

From Eqs. (i) and (ii), we see that angle of friction (λ) is numerically equal to the angle of repose.

or $\lambda = \alpha$

From the above discussions we can conclude that

If $\theta < \alpha$, $F < f_L$ the block is stationary.

If $\theta = \alpha$, $F = f_L$ the block is on the verge of sliding.

and if $\theta > \alpha, F > f_L$ the block slides down with an acceleration

$$a = \frac{F - f_L}{m} = g(\sin \theta - \mu \cos \theta)$$

Variation of N, f_L and F with θ, is shown graphically in the figure shown below.

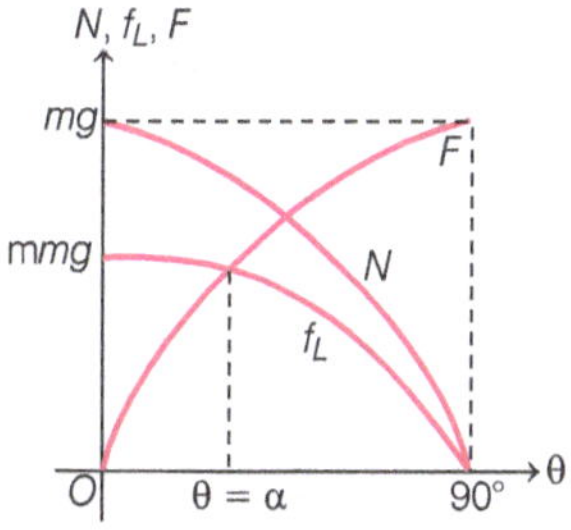

$N = mg \cos \theta$

or $N \propto \cos \theta$

$f_L = \mu mg \cos \theta$

or $f_L \propto \cos \theta$

$F = mg \sin \theta$

or $F \propto \sin \theta$

Normally $\mu < 1$,

So, $f_L < N$.

Note *In the problems of friction, normal reaction N has very important role. Because limiting value of friction and constant value of kinetic friction depend upon N. Following are given some of the basic rules for finding N correctly.*

$N = mg$ on horizontal ground or $N = mg\cos\theta$ on inclined ground as long as the external forces (other than weight and normal reaction) are either zero or tangential to the surface. If the external force is inclined to the horizontal surface (or inclined plane), then normal reaction either increases or decreases depending on the direction of F.

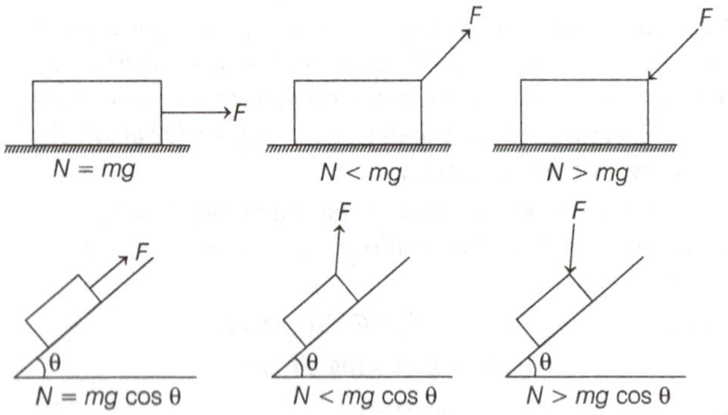

Example 2 A block of mass 1 kg rests on rough contact with a plane inclined at 30° to the horizontal and is just about to slip. Find the coefficient of friction between the plane and the particle.

Sol.

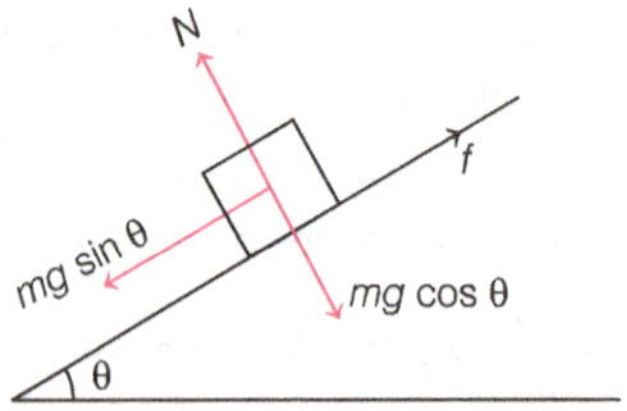

Weight mg has two components $mg\sin\theta$ and $mg\cos\theta$. Block is at rest.

$$\therefore \quad N = mg\cos\theta \quad \ldots\text{(i)}$$

$$f = mg\sin\theta \quad \ldots\text{(ii)}$$

Block is about to slip.

$$\therefore \quad f = f_L = \mu N \quad \ldots\text{(iii)}$$

Solving these three equations, we get

$$\mu = \tan\theta = \tan 30^\circ$$

$$= \frac{1}{\sqrt{3}} \quad \textbf{Ans.}$$

Note *The given angle is also angle of repose α.*

$$\therefore \quad \mu = \tan\theta = \tan\alpha = \tan 30^\circ = \frac{1}{\sqrt{3}}$$

Example 3 In the given figure, the coefficient of friction between wedge (of mass M) and block (of mass m) is μ.

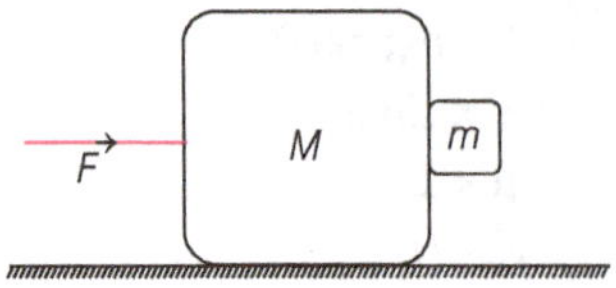

Find the minimum horizontal force F required to keep the block stationary with respect to wedge.

Sol. This problem can be solved with or without using the concept of pseudo force. Let us solve the problem by both the methods.

a = acceleration of (wedge + block) in horizontal direction

$$= \frac{F}{M+m}$$

Method 1.

From Inertial Frame of Reference (Ground)

FBD of block with respect to ground (only real forces are to be applied) is as shown in figure. With respect to ground block is moving with an acceleration a. Therefore,

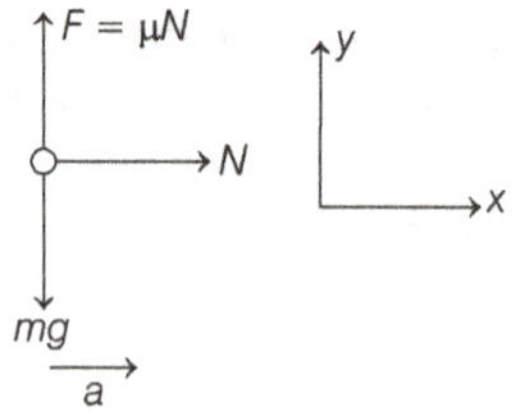

$$\Rightarrow \quad \Sigma F_y = 0 \quad \text{and} \quad \Sigma F_x = ma$$

$$\Rightarrow \quad mg = \mu N \quad \text{and} \quad N = ma$$

From these two equations, we get

$$a = \frac{g}{\mu}$$

$$\therefore \quad F = (M+m)a = (M+m)\frac{g}{\mu}$$

Method 2.

From Non-inertial Frame of Reference (Wedge)

FBD of m with respect to wedge (real + pseudo force) is as shown in figure. With respect to wedge block is stationary.

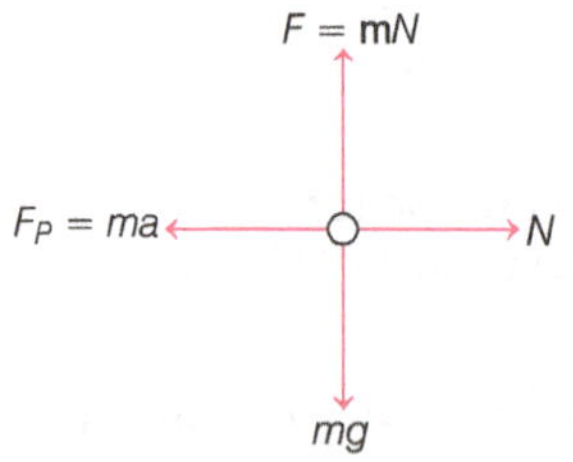

Here, F_P = pseudo force

$$\Sigma F_x = 0 \quad \text{and} \quad \Sigma F_y = 0$$

$$\therefore \quad mg = \mu N \quad \text{and} \quad N = ma$$

From these two equations, we again get

$$a = \frac{g}{\mu}$$

$$\therefore \quad F = (M+m)a$$

$$= (M+m)\frac{g}{\mu}$$

Example 4 A 6 kg block is kept on an inclined rough surface as shown in figure. Find the force F required to

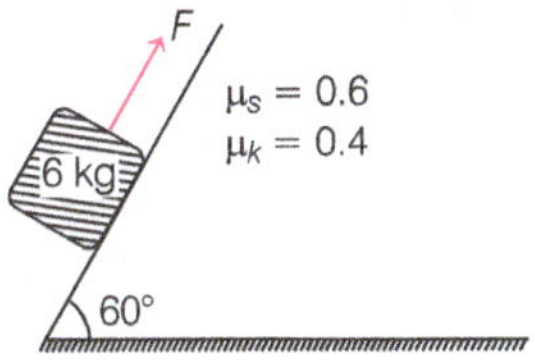

(a) keep the block stationary,

(b) move the block downwards with constant velocity and

(c) move the block upwards with an acceleration of $4\ \text{m/s}^2$. (Take $g = 10\ \text{m/s}^2$)

Sol. $N = mg\cos 60° = (6)(10)\left(\frac{1}{2}\right) = 30\ \text{N}$

$$\mu_s N = 18\ \text{N}$$

$$\Rightarrow \quad \mu_k N = 12\ \text{N}$$

Driving force $F_0 = mg\sin 60° = (6)(10)\left(\frac{\sqrt{3}}{2}\right) = 52\ \text{N}$

(a) Force needed to keep the block stationary is

$$F_1 = F_0 - \mu_s N \quad \text{(upwards)}$$
$$= 52 - 18$$
$$= 34\ \text{N} \quad \text{(upwards)} \qquad \textbf{Ans.}$$

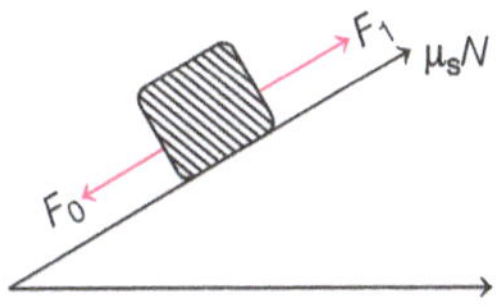

(b) If the block moves downwards with constant velocity ($a = 0, F_{net} = 0$), then kinetic friction will act in upward direction.

∴ Force needed,

$$F_2 = F_0 - \mu_k N \quad \text{(upwards)}$$
$$= 52 - 12$$
$$= 40\ \text{N} \quad \text{(upwards)} \qquad \textbf{Ans.}$$

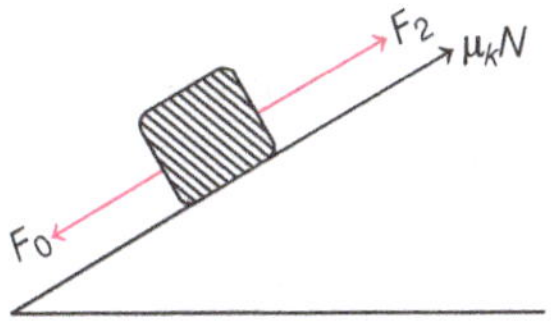

(c) In this case, kinetic friction will act in downward direction

$$F_3 - F_0 - \mu_k N = ma$$

or $\quad F_3 - 52 - 12 = ma = (6)(4)$

∴ $\quad F_3 = 88\ \text{N} \quad \text{(upwards)} \qquad$ **Ans.**

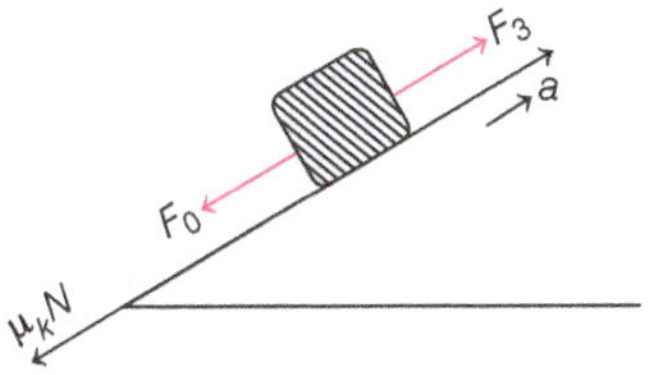

Example 5 A block of mass m is at rest on a rough wedge as shown in figure. What is the force exerted by the wedge on the block?

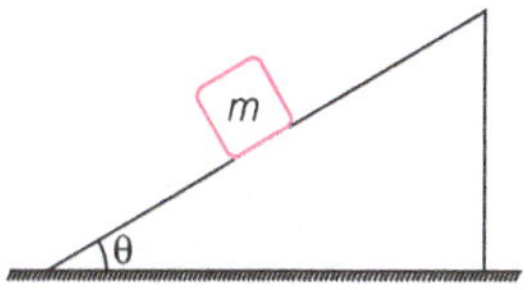

Sol. Since, the block is permanently at rest, it is in equilibrium. Net force on it should be zero. In this case, only two forces are acting on the block

(i) Weight $= mg$ (downwards).

(ii) Contact force (resultant of normal reaction and friction force) applied by the wedge on the block.

For equilibrium, these two forces should be equal and opposite.

Therefore, force exerted by the wedge on the block is mg (upwards).

Note *This result can also be obtained in a different manner. The normal force on the block is $N = mg\cos\theta$ and the friction force on the block is $f = mg\sin\theta$ (not $\mu\ mg\cos\theta$)*

These two forces are mutually perpendicular.

∴ $\quad$ *Net contact force* $= \sqrt{N^2 + f^2}$

$$= \sqrt{(mg\cos\theta)^2 + (mg\sin\theta)^2} = mg$$

INTRODUCTORY Exercise

1. In the two figures shown here, find acceleration of block and force of friction on it in each case.

$\mu_s = 0.6, \mu_k = 0.4$ (a) $\qquad$ $\mu_s = 0.6, \mu_k = 0.4$ (b)

2. In the figure shown here, angle of repose is 45°. Find force of friction, net force and acceleration of the block when

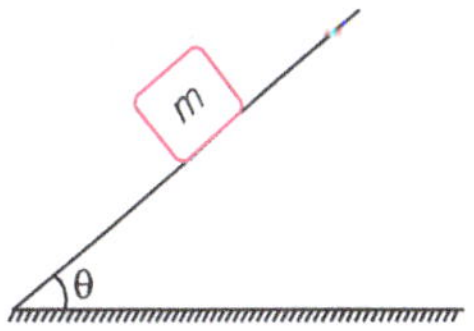

(a) $\theta = 30°$ (b) $\theta = 45°$ and (c) $\theta = 60°$

Answers

1. Zero, 20 N (b) $6\ \text{ms}^{-2}$, 8N

2. (a) $mg/2$, 0,0 (b) $mg/\sqrt{2}$, 0,0 (c) $mg/2$, $\left(\frac{\sqrt{3}-1}{2}\right)mg$, $\left(\frac{\sqrt{3}-1}{2}\right)g$

See hints on page 428.

Typed Problems

Type 1. To find whether the block will move or not under different forces acting on the block.

How to Solve?

- The rough surface may be horizontal, inclined or vertical.
- Resolve the forces along the surface and perpendicular to the surface.
- In most of the cases, acceleration perpendicular to the surface is zero.
- So net force perpendicular to the surface should be zero. By putting net force perpendicular to the surface equal to zero we will get normal reaction N.
- After finding, N, calculate $\mu_s N, \mu_k N$ or μN.
- Calculate net force along the plane and call it the driving force F.

Now, if $F \leq \mu_s N$

Then $f = F$ (f = Force of friction)

and $F_{net} = 0$ or $a = 0$

If $F > \mu_s N$

Then, $f = \mu_k N$ and $F_{net} = F - f$

or $a = \frac{F_{net}}{m}$ (in the direction of F)

Ex. 1 In the figure shown,

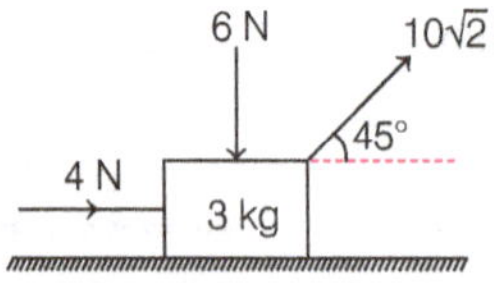

$\mu_s = 0.6, \mu_k = 0.4$

(a) find the force of friction acting on the block.

(b) state whether the block will move or not. If yes then with what acceleration?

Sol. Resolving the force in horizontal (along the plane) and in vertical (perpendicular to the plane) directions (except friction)

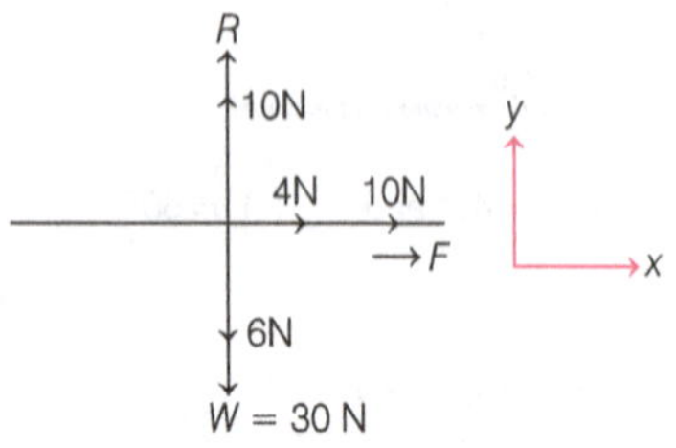

Here, R is the normal reaction.

$$\Sigma F_y = 0$$

$$\Rightarrow \quad R = 26\text{N}$$

$$\mu_s R = 0.6 \times 26 = 15.6 \text{ N}$$

$$\mu_k R = 0.4 \times 26 = 10.4 \text{ N}$$

$$\Sigma F_x = \text{net driving force } F = 14\text{N}$$

(a) Since, $F < \mu_s R$

$\therefore$ Force of friction $f = F$ or 14 N

(b) Since, $F < \mu_s R$, the block will not move and acceleration,

$$a = 0$$

Ex. 2 In the figure shown,

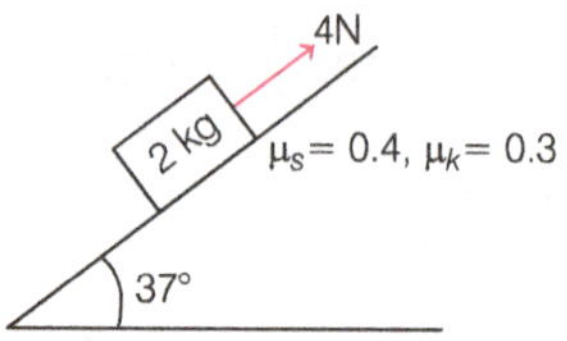

(a) find the force of friction acting on the block.

(b) state whether the block will move or not. If yes then with what acceleration? (Take $g = 10 \text{ m/s}^2$)

Sol. Resolving the forces along the plane and perpendicular to the plane (except friction).

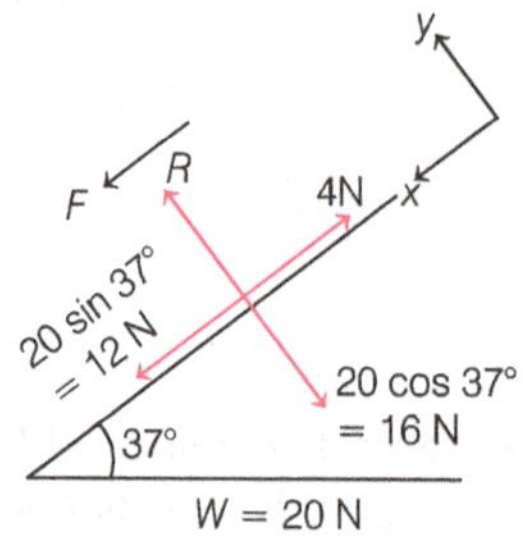

Here, R is the normal reaction.

$$\Sigma F_y = 0 \quad \Rightarrow \quad R = 16\text{N}$$

$$\mu_s R = 0.4 \times 16 = 6.4\text{N}$$

$$\mu_k R = 0.3 \times 16 = 4.8\text{N}$$

$$\Sigma F_x = \text{net driving force } F = (12 - 4)\text{N} = 8\text{N}$$

(a) Since, $F > \mu_s R$, therefore kinetic friction or 4.8 N will act in opposite direction of F.

(b) Since, $F > \mu_s R$, the block will move in the direction of F (or downwards) with an acceleration,

$$a = \frac{F_{net}}{m} = \frac{F - f}{m} = \frac{8 - 4.8}{2}$$

$$= 1.6 \text{ m/s}^2$$

This acceleration is in the direction of F (or downwards).

Ex. 3 A block of mass 1 kg is pushed against a rough vertical wall with a force of 20 N, coefficient of static friction being $\frac{1}{4}$. Another horizontal force of 10 N is applied on the block in a direction parallel to the wall. Will the block move? If yes, then in which direction? If no, find the frictional force exerted by the wall on the block. ($g = 10 \text{ m/s}^2$)

Sol. Normal reaction on the block from the wall will be

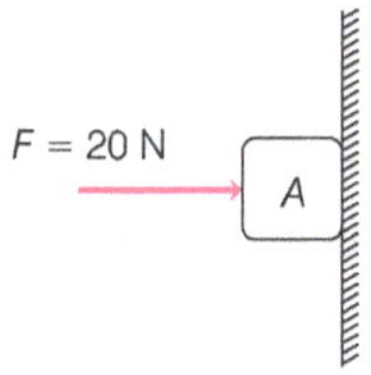

$$R = F = 20\text{ N}$$

($F_{net} = 0$, perpendicular to the wall)

Therefore, limiting friction

$$f_L = \mu R = \left(\frac{1}{4}\right)(20) = 5\text{ N}$$

Weight of the block is $W = mg = (1)(10) = 10\text{ N}$

A horizontal force of 10 N is applied to the block. Both weight and this force are along the wall in mutually perpendicular directions. The resultant of these two forces will be $10\sqrt{2}$ N in the direction shown in figure. Since, this resultant is greater than the limiting friction. The block will move in the direction of $\mathbf{F}_{net}$ with acceleration

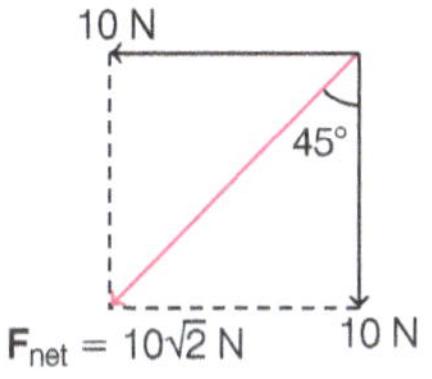

$$a = \frac{\mathbf{F}_{net} - f_L}{m} = \frac{10\sqrt{2} - 5}{1} = 9.14\text{ m/s}^2$$

Type 2. To draw acceleration *versus* time graph.

Ex. 4 In the figure shown, F is in newton and t in seconds. Take $g = 10\text{ m/s}^2$.

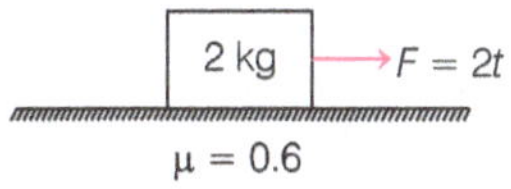

(a) Plot acceleration of the block versus time graph.

(b) Find force of friction at, $t = 2$ s and $t = 8$ s.

Sol. (a) Normal reaction, $R = mg = 20$ N. Limiting value of friction, $f_1 = \mu R = 0.6 \times 20 = 12$ N

The applied force $F (= 2t)$ crosses this limiting value of friction at 6 s. Therefore, up to 6 s block remains stationary and after 6 s it starts moving. After 6 s, friction force becomes constant at 12 N but the applied force keeps on increasing. Therefore, acceleration keeps on increasing.

For $t \le 6$ s

$$f = F = 2t \qquad \text{...(i)}$$

$$F_{net} = F - f = 0$$

$$\therefore \quad a = \frac{F_{net}}{m} = 0$$

For $t > 6$ s

$$F = 2t$$

$$f = 12\text{N} = f_1 \qquad \text{...(ii)}$$

$$F_{net} = F - f = 2t - 12$$

$$\therefore \quad a = \frac{F_{net}}{m} = \frac{2t-12}{2} = t - 6$$

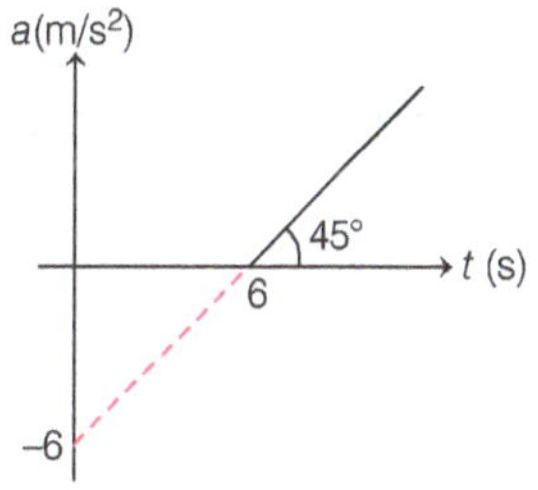

$\therefore$ a-t graph is a straight line with slope = 1 and intercept = –6.

Corresponding a-t graph is as shown in figure.

(b) At $t = 2$ s, $f = 4$ N [from Eq. (i)]

At $t = 8$ s, $f = 12$ N [from Eq. (ii)]

Ex. 5 Repeat the above problem, if instead of μ we are given μ_s and μ_k, where $\mu_s = 0.6$ and $\mu_k = 0.4$.

Sol. (a) $R = mg = 20$N

$$\mu_s R = 0.6 \times 20 = 12\text{N}$$

$$\mu_k R = 0.4 \times 20 = 8\text{N}$$

Upto 6 s, situation is same but after 6 s, a constant kinetic friction of 8N will act. At 6 s, friction will suddenly change from 12N($= \mu_s R$) to 8N($= \mu_k R$) and direction of friction is opposite to its motion. Therefore, at 6 s it will start with an initial acceleration.

$$a_i = \frac{\text{decrease in friction}}{\text{mass}} = \frac{12-8}{2} = 2\text{ m/s}^2$$

For $t \le 6$s

$$f = F = 2t \qquad \text{...(i)}$$

$$F_{net} = F - f = 0$$

$$\therefore \quad a = \frac{F_{net}}{m} = 0$$

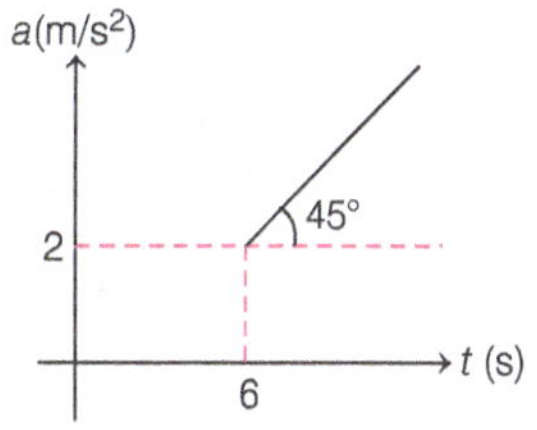

For $t \ge 6$s

$$F = 2t$$

$$f = \mu_k R = 8\text{N} \qquad \text{...(ii)}$$

$$F_{net} = F - f = 2t - 8$$

$$\therefore \qquad a = \frac{F_{net}}{m} = \frac{2t-8}{2} = (t-4)$$

Further, a-t graph is a straight line of slope $= 1$ and intercept $= -4$. Corresponding a-t graph is as shown in figure.

At $t = 6$ s, we can see that, $a_i = 2\ \text{m/s}^2$

(b) At $t = 2$s, $f = 4$N [from Eq. (i)]

At $t = 8$ s, $f = 8$ N [from Eq. (ii)]

Ex. 6 Two blocks A and B of masses 2 kg and 4 kg are placed one over the other as shown in figure. A time varying horizontal force $F = 2t$ is applied on the upper block as shown in figure. Here t is in second and F is in newton. Draw a graph showing accelerations of A and B on y-axis and time on x-axis. Coefficient of friction between A and B is $\mu = \frac{1}{2}$ and the horizontal surface over which B is placed is smooth. ($g = 10\ \text{m/s}^2$)

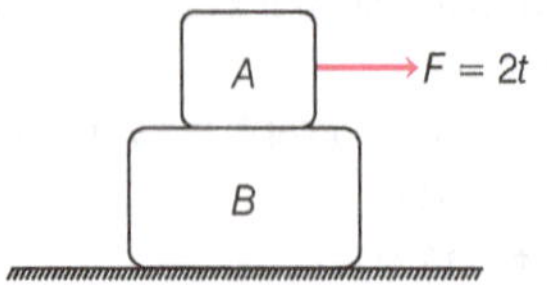

Concept

In the given example, block A will move due to the applied force but block B moves due to friction (between A and B). But there is a limiting value of friction between them. Therefore, there is a limiting value of acceleration (of block B). Up to this acceleration both the blocks move as a single block with a common acceleration, but after that acceleration of B will become constant (as friction acting on this block will become constant). But acceleration of A will keep on increasing as a time increasing force is acting on it.

Sol. Limiting friction between A and B is

$$f_L = \mu m_A g = \left(\frac{1}{2}\right)(2)\ (10) = 10\ \text{N}$$

Block B moves due to friction only. Therefore, maximum acceleration of B can be

$$a_{max} = \frac{f_L}{m_B} = \frac{10}{4} = 2.5\ \text{m/s}^2$$

Thus, both the blocks move together with same acceleration till the common acceleration becomes $2.5\ \text{m/s}^2$. After that acceleration of B will become constant while that of A will go on increasing. To find the time when acceleration of both the blocks becomes $2.5\ \text{m/s}^2$ (or when slipping will start between A and B) we will write

$$2.5 = \frac{F}{(m_A + m_B)} = \frac{2t}{6}$$

$$\therefore \qquad t = 7.5\ \text{s}$$

Hence, for $t \le 7.5$ s

$$a_A = a_B = \frac{F}{m_A + m_B} = \frac{2t}{6} = \frac{t}{3}$$

Thus, a_A *versus* t or a_B *versus* t graph is a straight line passing through origin of slope $\frac{1}{3}$.

For $t \ge 7.5$ s,

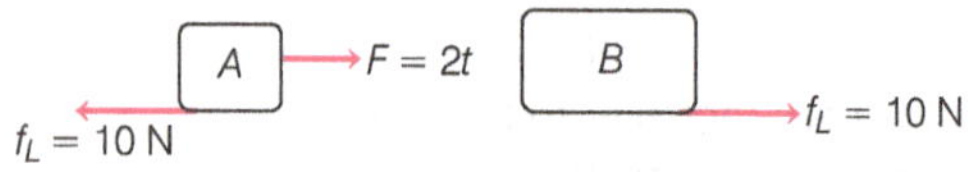

$$a_B = 2.5\ \text{m/s}^2 = \text{constant}$$

and $\qquad a_A = \dfrac{F - f_L}{m_A}$

or $\qquad a_A = \dfrac{2t - 10}{2}$

or $\qquad a_A = t - 5$

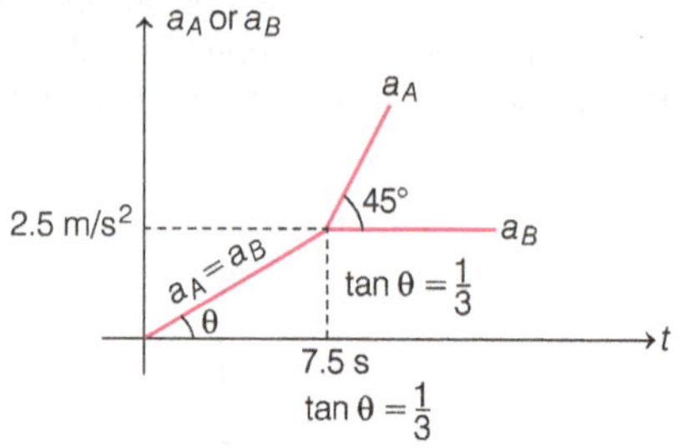

Thus, a_A *versus* t graph is a straight line of slope 1 and intercept -5. While a_B *versus* t graph is a straight line parallel to t-axis. The corresponding graph is as shown in above figure.

Type 3. When two blocks in contact are given different velocities and after some time their velocities become equal due to friction.

Concept

In the figure shown, if $v_1 > v_2$ (or $v_1 \ne v_2$) then there is a relative motion between the two blocks.

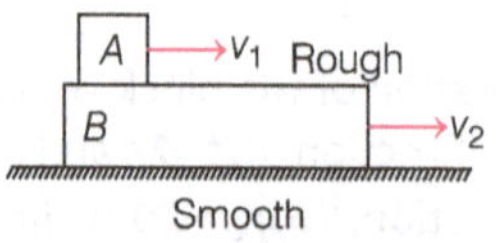

As $v_1 > v_2$, relative motion of A is towards right and relative motion of B is towards left. Since, relative motion is there, so kinetic friction (or limiting value of friction) will act in the opposite direction of relative motion. This friction (and acceleration due to this force) with decrease the velocity of A and increase the velocity of B. After some time when their velocities become equal, frictional force between them becomes zero and they continue to be moving with that common velocity (as the ground is smooth).

How to Solve?

- Find value of kinetic friction or limiting value of friction ($f=\mu_k N$ or μN) between the two blocks and then accelerations of these blocks $\left(=\frac{f}{m}\right)$. Then write $v_1=v_2$, as their velocities become same when relative motion is stopped.

or $$u_1+a_1t=u_2+a_2t \quad \text{...(i)}$$

Substituting the proper values of u_1, a_1, u_2 and a_2 in Eq. (i), we can find the time when the velocities become equal.

Ex. 7 Coefficient of friction between two blocks shown in figure is $\mu=0.6$. The blocks are given velocities in the directions shown in figure. Find

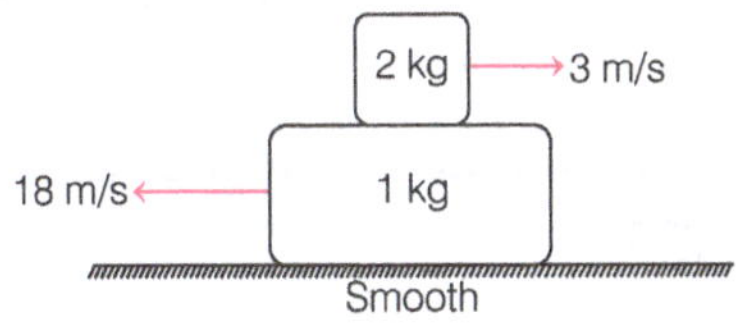

(a) the time when relative motion between them is stopped.

(b) the common velocity of the two blocks.

(c) the displacements of 1 kg and 2 kg blocks up to that instant. (Take $g=10\text{m/s}^2$)

Note *Assume that lower block is sufficiently long and upper block does not fall from it.*

Sol. Relative motion of 2 kg block is towards right. Therefore, maximum friction on this block will act towards left

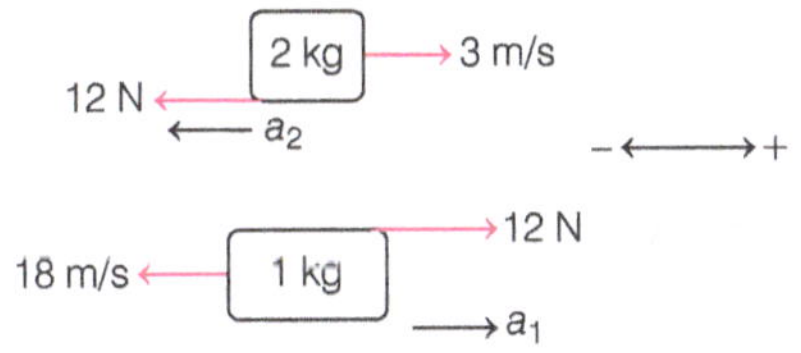

$$f=\mu N=(0.6)(2)(10)=12\text{ N}$$

$$a_2=-\frac{12}{2}=-6\text{ m/s}^2$$

$$a_1=\frac{12}{1}=12\text{ m/s}^2$$

(a) Relative motion between them will stop when

$$v_1=v_2$$

or $$u_1+a_1t=u_2+a_2t$$

or $$-18+12t=3-6t \quad \text{...(i)}$$

Solving we get, $t=\frac{7}{6}\text{ s}$ **Ans.**

(b) Substituting value of t in Eq. (i) either on RHS or on LHS, we have

common velocity $=-4\text{ m/s}$ **Ans.**

(c) $$s_1=u_1t+\frac{1}{2}a_1t^2$$

$$=(-18)\left(\frac{7}{6}\right)+\frac{1}{2}(12)\left(\frac{7}{6}\right)^2$$

$$=-12.83\text{ m}$$ **Ans.**

$$s_2=u_2t+\frac{1}{2}a_2t^2$$

$$=(3)\left(\frac{7}{6}\right)+\frac{1}{2}(-6)\left(\frac{7}{6}\right)^2$$

$$=-0.58\text{ m}$$ **Ans.**

Type 4. Acceleration or retardation of a car.

Concept

A car accelerates or retards due to friction. On a horizontal road maximum available friction is μN or μmg (as $N=mg$). Therefore, maximum acceleration or retardation of a car on a horizontal road is

$$a_{max}=\frac{f_{max}}{m}=\frac{\mu mg}{m}=\mu g$$

On an inclined road maximum value of friction is μN or $\mu mg\cos\theta$ (as $N=mg\cos\theta$). Now $mg\sin\theta$ is a force which is always downwards but depending upon the requirement the frictional force varying from 0 to $\mu mg\cos\theta$ can be applied in upward or downward direction by the application of brakes or accelerator.

Ex. 8 On a horizontal rough road, value of coefficient of friction is $\mu=0.4$. Find the minimum time in which a distance of 400 m can be covered. The car starts from rest and finally comes to rest. Take $g=10\text{ m/s}^2$

Sol. Maximum friction on horizontal rough road, $f_{max}=\mu mg$

$\therefore$ Maximum acceleration or retardation of the car may be

$$a_{max}\text{ or }a=\frac{f_{max}}{m}=\frac{\mu mg}{m}=\mu g$$

$$=0.4\times 10=4\text{ m/s}^2$$

Let, the car accelerates and retards for time t with 4m/s^2.

Then, $$\frac{1}{2}at^2+\frac{1}{2}at^2=400\text{m}$$

or $$at^2=400\text{m}$$

or $$4t^2=400$$

or $$t=10\text{s}$$

Therefore, the minimum time is 20 s (10 s of acceleration and 10 s of retardation). **Ans.**

Ex. 9 A car is moving up the plane. Angle of inclination is θ and coefficient of friction is μ.

(a) What is the condition so that car can be accelerated? If this condition is satisfied then find

(b) maximum acceleration of the car.

(c) minimum retardation of the car.

(d) maximum retardation of the car.

Sol. (a) $mg\sin\theta$ in all conditions is downwards but direction of friction may be upwards or downwards. We will have to press accelerator for upward friction and brakes for downward friction.

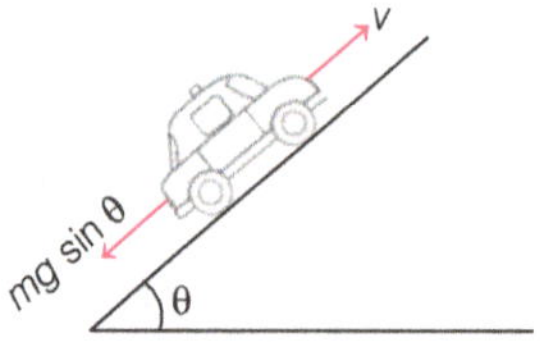

To accelerate the car friction should be upwards. Therefore, car can be accelerated if maximum upward friction $> mg\sin\theta$

or $\mu\, mg\cos\theta > mg\sin\theta$

or $\mu > \tan\theta$

(b) Maximum acceleration

$$= \frac{\text{maximum upwwards force}}{\text{mass}}$$

$$= \frac{\mu\, mg\cos\theta - mg\sin\theta}{m}$$

$$= (\mu\, g\cos\theta - g\sin\theta)$$

(c) Minimum retardation will be zero, when upward friction $= mg\sin\theta$

(d) Maximum retardation

$$= \frac{\text{maximum downward force}}{m}$$

$$= \frac{mg\sin\theta + \mu\, mg\cos\theta}{m}$$

$$= (g\sin\theta + \mu\, g\cos\theta)$$

This is the case, when maximum friction force acts in downward direction.

Miscellaneous Examples

Ex. 10 In the figure shown, angle of plane θ is increased from 0° to 90°. Plot force of friction 'f' *versus* θ graph.

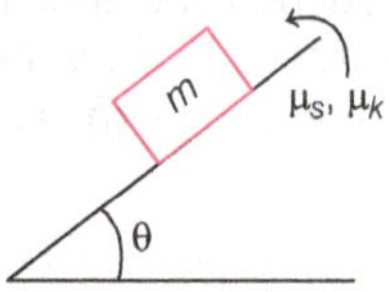

Sol. Normal reaction, $N = mg\cos\theta$

Limiting value of static friction, $f_L = \mu_s N = \mu_s\, mg\cos\theta$

Constant value of kinetic friction, $f_K = \mu_k N = \mu_k mg\cos\theta$

Driving force down the plane, $F = mg\sin\theta$

Now block remains stationary until F becomes equal to f_L

or $mg\sin\theta = \mu_s\, mg\cos\theta$ or $\tan\theta = \mu_s$

or $\theta = \tan^{-1}(\mu_s) = \theta_r$ (say)

After this, block starts moving and kinetic friction will act. Thus,

For $\theta \le \tan^{-1}(\mu_s)$ or θ_r

$$f = F = mg\sin\theta \quad \text{or} \quad f \propto \sin\theta$$

At, $\theta = 0°$, $f = 0$ and at $\theta = \tan^{-1}(\mu_s)$ or θ_r

or $f = mg\sin\theta_r$ or $\mu_s mg\cos\theta_r$

For $\theta > \tan^{-1}(\mu_s)$ or θ_r

$$f = f_k = \mu_k mg\cos\theta \quad \text{or} \quad f \propto \cos\theta$$

At $\theta = \tan^{-1}(\mu_s)$ or θ_r

$f = \mu_k mg\cos\theta_r$ and at $\theta = 90°$

$f = 0$

Corresponding f *versus* θ graph is as shown in figure. In the figure, OP is sine graph and MN is cos graph,

$$f_1 = mg\sin\theta_r = \mu_s\, mg\cos\theta_r$$

$$f_2 = \mu_k\, mg\cos\theta_r$$

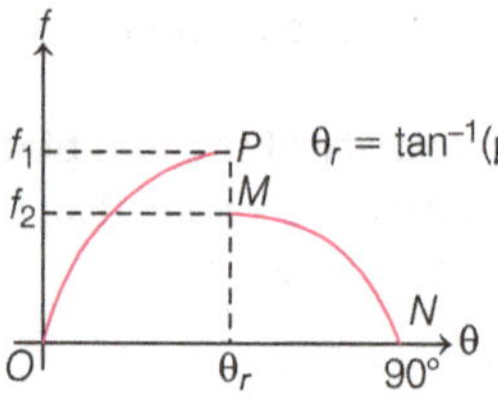

Note *Points P and M will coincide if* $\mu_s = \mu_k = \mu$ *(say).*

Ex. 11 Figure shows two blocks in contact sliding down an inclined surface of inclination 30°. The friction coefficient between the block of mass 2.0 kg and the incline is $\mu_1 = 0.20$ and that between the block of mass 4.0 kg and the incline is $\mu_2 = 0.30$. Find the acceleration of 2.0 kg block. ($g = 10\ \text{m/s}^2$)

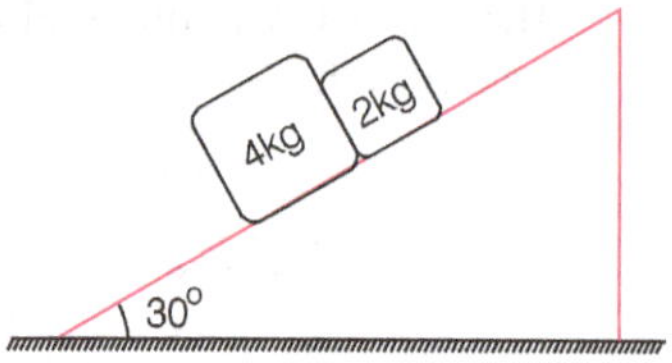

Sol. Since, $\mu_1 < \mu_2$, acceleration of 2 kg block down the plane will be more than the acceleration of 4 kg block, if they are allowed to move separately. But, as the 2.0 kg block is behind the 4.0 kg block both of them will move with same acceleration say a. Taking both the blocks as a single system:

Force down the plane on the system $= (4+2)\, g\sin 30°$

$$= (6)(10)\left(\frac{1}{2}\right) = 30\ \text{N}$$

Force up the plane on the system

$$= \mu_1(2)(g)\cos 30° + \mu_2(4)(g)\cos 30°$$

$$= (2\mu_1 + 4\mu_2)\, g\cos 30°$$

$$= (2\times 0.2 + 4\times 0.3)(10)(0.86)$$

$$\approx 13.76\ \text{N}$$

$\therefore$ Net force down the plane is $F = 30 - 13.76 = 16.24$ N

$\therefore$ Acceleration of both the blocks down the plane will be

$$a = \frac{F}{4+2} = \frac{16.24}{6} = 2.7 \text{ m/s}^2 \quad \textbf{Ans.}$$

Ex. 12 Figure shows a man standing stationary with respect to a horizontal conveyor belt that is accelerating with 1 ms^{-2}. What is the net force on the man? If the coefficient of static friction between the man's shoes and the belt is 0.2, up to what maximum acceleration of the belt can the man continue to be stationary relative to the belt? Mass of the man $= 65$ kg. ($g = 9.8 \text{ m/s}^2$)

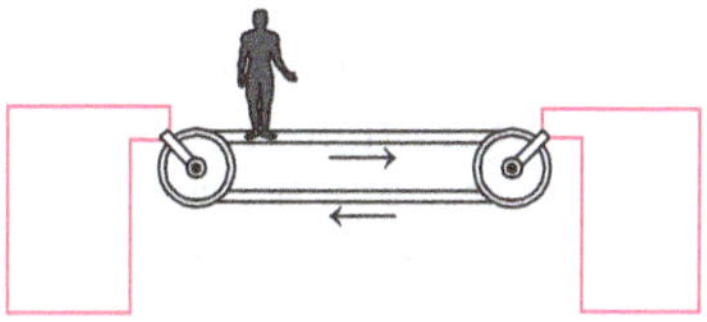

Sol. As the man is standing stationary w.r.t. the belt,

$\therefore$ Acceleration of the man = Acceleration of the belt

$$= a = 1 \text{ ms}^{-2}$$

Mass of the man, $m = 65$ kg

Net force on the man $= ma = 65 \times 1 = 65$ N **Ans.**

Given coefficient of friction,

$$\mu = 0.2$$

$\therefore$ Limiting friction, $f_L = \mu mg$

If the man remains stationary w.r.t. the belt and maximum acceleration of the belt is a_0, then

$$ma_0 = f_L = \mu mg$$

$$\therefore \quad a_0 = \mu g = 0.2 \times 9.8 = 1.96 \text{ ms}^{-2} \quad \textbf{Ans.}$$

Ex. 13 Figure shows a small block A of mass m kept at the left end of a plank B of mass $M = 2m$ and length l. The system can slide on a horizontal road. The system is started towards right with the initial velocity v. The friction coefficients between the road and the plank is $\frac{1}{2}$ and that between the plank and the block is 1/4. Find

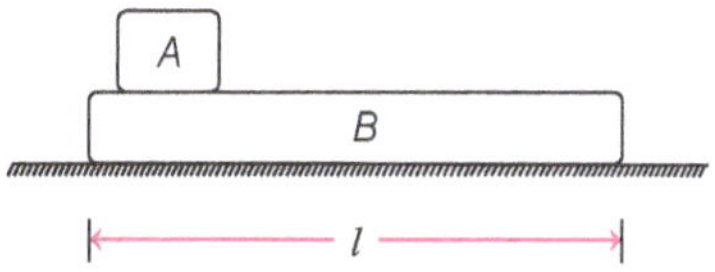

(a) the time elapsed before the block separates from the plank.

(b) displacement of block and plank relative to ground till that moment.

Sol. There will be relative motion between block and plank and plank and road. So at each surface limiting friction will act. The direction of friction forces at different surfaces are as shown in figure.

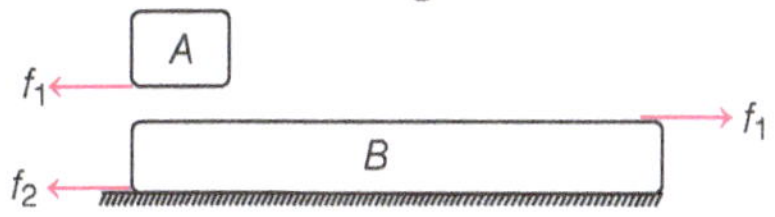

Here, $f_1 = \left(\frac{1}{4}\right)(mg)$ and Retardation of A is $a_1 = \frac{f_1}{m} = \frac{g}{4}$

and retardation of B is $a_2 = \frac{f_2 - f_1}{2m} = \frac{5}{8}g$

Since, $a_2 > a_1$

Relative acceleration of A with respect to B is

$$a_r = a_2 - a_1 = \frac{3}{8}g$$

Initial velocity of both A and B is v. So, there is no relative initial velocity. Hence,

(a) Applying $s = \frac{1}{2}at^2$ or $l = \frac{1}{2}a_r t^2 = \frac{3}{16}gt^2$

$$\therefore \quad t = 4\sqrt{\frac{l}{3g}} \quad \textbf{Ans.}$$

(b) Displacement of block $s_A = u_A t - \frac{1}{2}a_A t^2$

or $s_A = 4v\sqrt{\frac{l}{3g}} - \frac{1}{2}\cdot\frac{g}{4}\cdot\left(\frac{16l}{3g}\right) \quad \left(\because a_A = a_1 = \frac{g}{4}\right)$

or $s_A = 4v\sqrt{\frac{l}{3g}} - \frac{2}{3}l$

Displacement of plank

$$s_B = u_B t - \frac{1}{2}a_B t^2$$

or $s_B = 4v\sqrt{\frac{l}{3g}} - \frac{1}{2}\left(\frac{5}{8}g\right)\left(\frac{16l}{3g}\right) \quad \left(\because a_B = a_2 = \frac{5}{8}g\right)$

or $s_B = 4v\sqrt{\frac{l}{3g}} - \frac{5}{3}l$ **Ans.**

Note *We can see that $s_A - s_B = l$. Which is quite obvious because block A has moved a distance l relative to plank.*

Ex. 14 Consider the situation shown in figure. The block B moves on a frictionless surface, while the coefficient of friction between A and the surface on which it moves is 0.2. Find the acceleration with which the masses move and also the tension in the strings. (Take $g = 10 \text{ m/s}^2$)

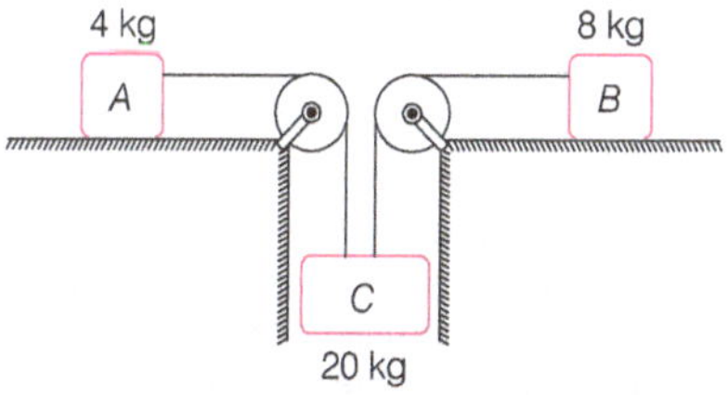

Sol. Let a be the acceleration with which the masses move and T_1 and T_2 be the tensions in left and right strings. Kinetic friction on mass A is $\mu m_A g = 8$N. Then equations of motion of masses A, B and C are

For mass A $\quad T_1 - 8 = 4a \quad$ ...(i)

For mass B $\quad T_2 = 8a \quad$ ...(ii)

For mass C $\quad 200 - T_1 - T_2 = 20a \quad$ ...(iii)

Adding the above three equations, we get $32a = 192$

or $\quad a = 6 \text{ m/s}^2$

From Eqs. (i) and (ii), we have

$$T_2 = 48 \text{ N} \quad \text{and} \quad T_1 = 32 \text{ N}$$

MODULE 6 Exercise

True/False

1. When a person walks on a rough surface, the frictional force exerted by the surface on the person is opposite to the direction of his motion. **(1981 JEE)**

Fill in the Blank

2. A block of mass 1 kg lies on a horizontal surface in a truck. The coefficient of static friction between the block and the surface is 0.6. If the acceleration of the truck is 5 m/s^2, the frictional force acting on the block is .... N. **(1984 JEE)**

Assertion and Reason

See directions on page 354.

3. **Assertion :** A block is kept at rest on a rough ground as shown. Two forces F_1 and F_2 are acting on it. If we increase either of the two forces F_1 or F_2, force of friction acting on the block will increase.

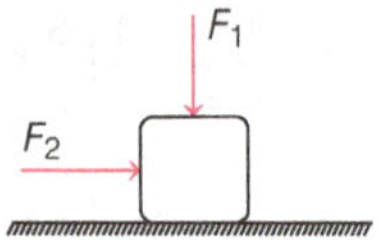

Reason : By increasing F_1, normal reaction from ground will increase.

4. **Assertion :** In the system of two blocks of equal masses as shown, the coefficient of friction between the blocks (μ_2) is less than coefficient of friction (μ_1) between lower block and ground.

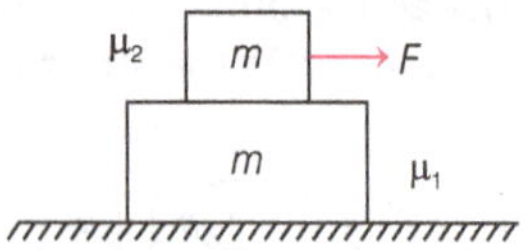

For all values of force F applied on upper block lower block remains at rest.

Reason : Frictional force on lower block due to upper block is not sufficient to overcome the frictional force on lower block due to ground.

5. **Assertion :** In the figure shown, block of mass m is stationary with respect to lift. Force of friction acting on the block is greater than $mg \sin\theta$.

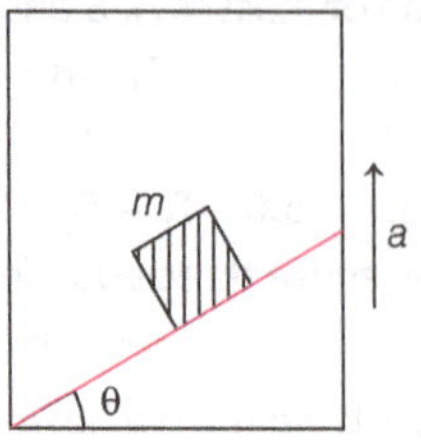

Reason : If lift moves with constant velocity then force of friction is equal to $mg \sin\theta$.

6. **Assertion :** Assume that earth consists of only neutrons. The normal reaction on a man standing on the surface of earth is zero.

Reason : Normal reaction is an electromagnetic force.

7. **Assertion :** If a man of 50 kg balances himself in a horizontal position by pushing his hands and back against two vertical walls, the frictional force on the man by each wall is 250 N.

Reason : The weight of man is balanced by frictional forces by walls on the man.

8. **Assertion :** It is easier to pull a heavy object than to push it on a level ground.

Reason : The magnitude of frictional force depends on the nature of the two surfaces in contact. **(2008 JEE)**

Single Correct Option

9. A particle is placed at rest inside a hollow hemisphere of radius R. The coefficient of friction between the particle and the hemisphere is $\mu = \frac{1}{\sqrt{3}}$. The maximum height up to which the particle can remain stationary is

(a) $\frac{R}{2}$ (b) $\left(1-\frac{\sqrt{3}}{2}\right)R$

(c) $\frac{\sqrt{3}}{2}R$ (d) $\frac{3R}{8}$

10. In the figure shown, the frictional coefficient between table and block is 0.2. Find the ratio of tensions in the right and left strings.

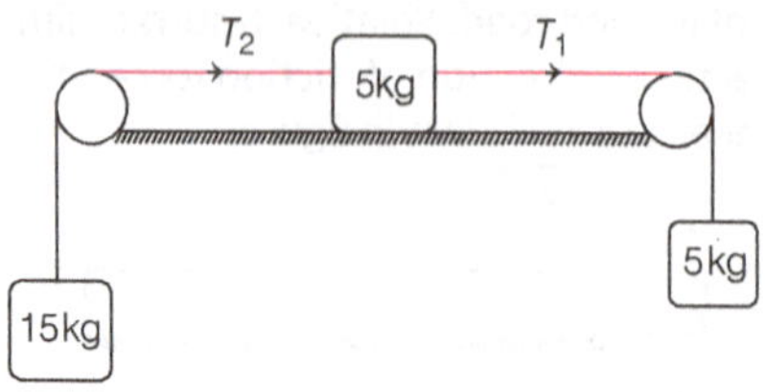

(a) 17 : 24 (b) 34 : 12 (c) 2 : 3 (d) 3 : 2

11. A block rests on a rough inclined plane making an angle of 30° with horizontal. The coefficient of static friction between the block and inclined plane is 0.8. If the frictional force on the block is 10 N, the mass of the block (in kg) is (Take, $g = 10 \text{ m/s}^2$)
(a) 2.0 (b) 4.0 (c) 1.6 (d) 2.5

12. A block of mass 4 kg is placed on a rough horizontal plane. A time dependent horizontal force $F = kt$ acts on the block. Here $k = 2 \text{Ns}^{-1}$. The frictional force between the block and plane at time $t = 2$ s is ($\mu = 0.2$)
(a) 4 N (b) 8 N (c) 12 N (d) 10 N

13. If a ladder weighing 250 N is placed against a smooth vertical wall having coefficient of friction between it and floor 0.3, then what is the maximum force of friction available at the point of contact between the ladder and the floor?
(a) 75 N (b) 50 N (c) 35 N (d) 25 N

14. A rope of length L and mass M is being pulled on a rough horizontal floor by a constant horizontal force $F = Mg$. The force is acting at one end of the rope in the same direction as the length of the rope. The coefficient of kinetic friction between rope and floor is $\frac{1}{2}$. Then, the tension at the mid-point of the rope is
(a) $\frac{Mg}{4}$ (b) $\frac{2Mg}{5}$ (c) $\frac{Mg}{8}$ (d) $\frac{Mg}{2}$

15. A block A of mass 4 kg is kept on ground. The coefficient of friction between the block and the ground is 0.8. The external force of magnitude 30 N is applied parallel to the ground. The resultant force exerted by the ground on the block is ($g = 10 \text{ m/s}^2$)
(a) 40 N (b) 30 N (c) zero (d) 50 N

16. A block A of mass 2 kg rests on another block B of mass 8 kg which rests on a horizontal floor. The coefficient of friction between A and B is 0.2 while that between B and floor is 0.5. When a horizontal force F of 25 N is applied on the block B, the force of friction between A and B is
(a) 3 N (b) 4 N (c) 2 N (d) zero

17. A body of mass 10 kg lies on a rough inclined plane of inclination $\theta = \sin^{-1}\left(\frac{3}{5}\right)$ with the horizontal. When the force of 30 N is applied on the block parallel to and upward the plane, the total force by the plane on the block is nearly along

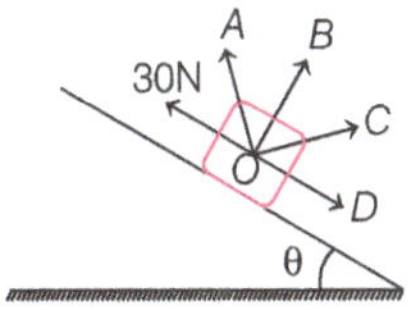

(a) OA (b) OB (c) OC (d) OD

18. A time varying horizontal force $F = at$ acts on a block of mass m kept on a smooth horizontal surface. An identical block is kept on the first block. The coefficient of friction between the blocks is μ. The time after which the relative sliding between the blocks will start is
(a) $\frac{2mg}{a}$ (b) $\frac{2\mu mg}{a}$
(c) $\frac{\mu mg}{a}$ (d) $2\mu mga$

19. A box of mass 8 kg is placed on a rough inclined plane of inclination θ. Its downward motion can be prevented by applying an upward pull F. It can be made to slide upwards by applying a force $2F$. The coefficient of friction between the box and the inclined plane is
(a) $\frac{1}{3}\tan\theta$ (b) $3\tan\theta$
(c) $\frac{1}{2}\tan\theta$ (d) $2\tan\theta$

20. For the arrangement shown in the figure, the tension in the string is given by $\left(\sin 37° = \frac{3}{5}\right)$

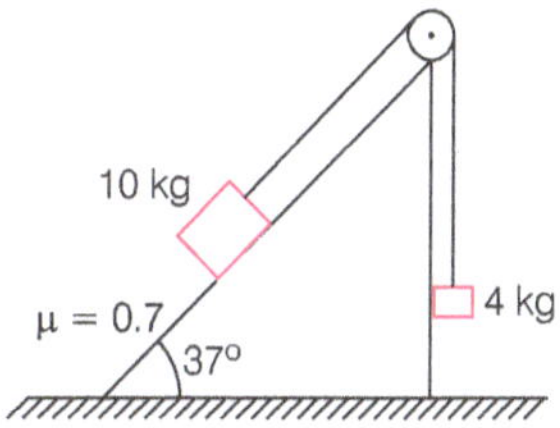

(a) 30 N (b) 40 N
(c) 60 N (d) 30 N

21. A uniform rope of length l lies on a table. If the coefficient of friction is μ, then the maximum length l_1 of the hanging part of the rope which can hang from the edge of the table without sliding down is
(a) l/μ (b) $l/(\mu + 1)$
(c) $\mu l/(\mu + 1)$ (d) $\mu l/(\mu - 1)$

22. For the arrangement shown in figure, the coefficient of friction between the two blocks is μ. If both the blocks are identical and moving, then the acceleration of each block is

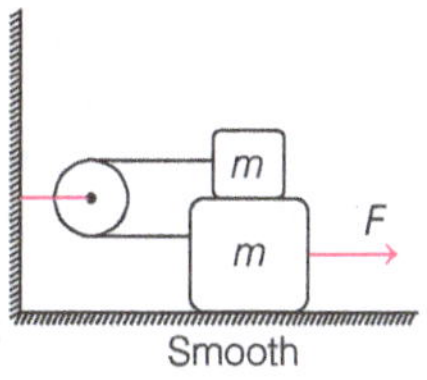

(a) $\frac{F}{2m} - 2\mu g$ (b) $\frac{F}{2m}$
(c) $\frac{F}{2m} - \mu g$ (d) zero

23. In the arrangement shown in figure, there is a friction force between the blocks of masses m and $2m$. The mass of the suspended block is m. The block of mass m is stationary with respect to block of mass $2m$. The minimum value of coefficient of friction between m and $2m$ is

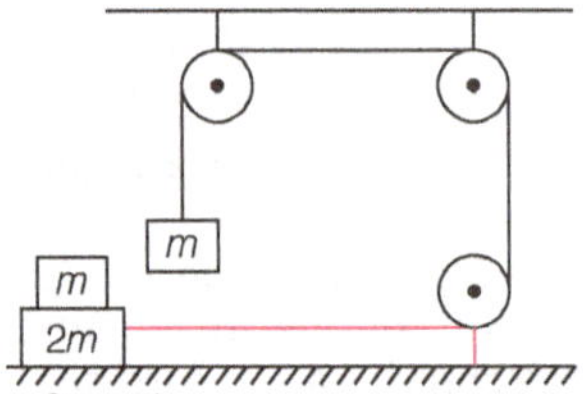

(a) $\frac{1}{2}$ (b) $\frac{1}{\sqrt{2}}$ (c) $\frac{1}{4}$ (d) $\frac{1}{3}$

24. Two blocks of masses $M_1 = 4$ kg and $M_2 = 6$ kg are connected by a string of negligible mass passing over a frictionless pulley as shown in the figure below. The coefficient of friction between the block M_1 and the horizontal surface is 0.4. When the system is released, the masses M_1 and M_2 start accelerating. What additional mass m should be placed over M_1, so that the masses $(M_1 + m)$ slide with a uniform speed?

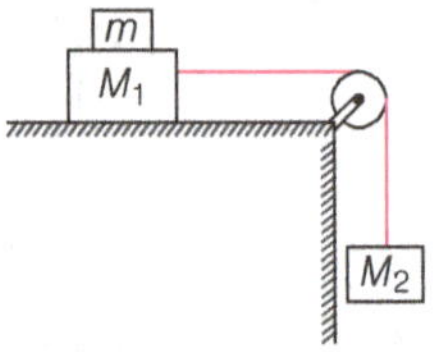

(a) 12 kg (b) 11 kg (c) 10 kg (d) 15 kg

25. A block of mass 2 kg rests on a rough inclined plane making an angle of 30° with the horizontal. The coefficient of static friction between the block and the plane is 0.7. The frictional force on the block is **(1980 JEE)**

(a) 9.8 N (b) $0.7 \times 9.8 \times \sqrt{3}$ N
(c) $9.8 \times \sqrt{3}$ N (d) 0.7×9.8 N

26. During paddling of a bicycle, the force of friction exerted by the ground on the two wheels is such that it acts **(1990 JEE)**
(a) in the backward direction on the front wheel and in the forward direction on the rear wheel
(b) in the forward direction on the front wheel and in the backward direction on the rear wheel
(c) in the backward direction on both the front and the rear wheels
(d) in the forward direction on both the front and the rear wheels

27. A block of mass 0.1 kg is held against a wall applying a horizontal force of 5 N on the block. If the coefficient of friction between the block and the wall is 0.5, the magnitude of the frictional force acting on the block is **(1994 JEE)**

(a) 2.5 N (b) 0.98 N (c) 4.9 N (d) 0.49 N

28. An insect crawls up a hemispherical surface very slowly (see the figure). The coefficient of friction between the surface and the insect is $\frac{1}{3}$. If the line joining the centre of the hemispherical surface to the insect makes an angle α with the vertical, the maximum possible value of α is given **(2001 JEE)**

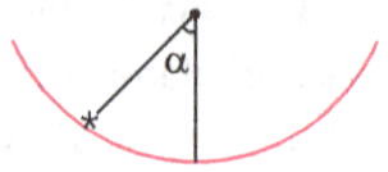

(a) $\cot \alpha = 3$ (b) $\tan \alpha = 3$
(c) $\sec \alpha = 3$ (d) $\text{cosec}\, \alpha = 3$

29. What is the maximum value of the force F such that the block shown in the arrangement, does not move? **(2003 JEE)**

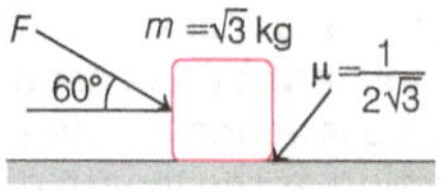

(a) 20 N (b) 10 N
(c) 12 N (d) 15 N

30. Given in the figure are two blocks A and B of weight 20 N and 100 N, respectively. These are being pressed against a wall by a force F as shown in figure. If the coefficient of friction between the blocks is 0.1 and between block B and the wall is 0.15, the frictional force applied by the wall in block B is **(2015 Main)**

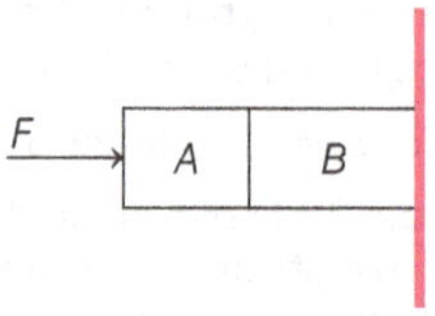

(a) 100 N (b) 120 N
(c) 80 N (d) 150 N

31. Two masses $m_1 = 5$ kg and $m_2 = 10$ kg connected by an inextensible string over a frictionless pulley, are moving as shown in the figure. The coefficient of friction of horizontal surface is 0.15. The minimum weight m that should be put on top of m_2 to stop the motion is **(2018 Main)**

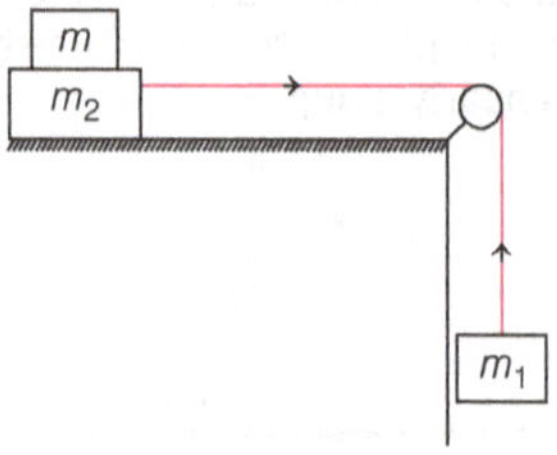

(a) 23.33 kg (b) 18.3 kg
(c) 27.3 kg (d) 43.3 kg

32. A block of mass 5 kg is (i) pushed in case (A) and (ii) pulled in case (B), by a force $F = 20$ N, making an angle of 30° with the horizontal, as shown in the figures. The coefficient of friction between the block and the floor is $\mu = 0.2$. The difference between the accelerations of the block, in case (B) and case (A) will be (Take, $g = 10\,\text{ms}^{-2}$) **(2019 Main)**

(A) (B)

(a) $0.4\,\text{ms}^{-2}$ (b) $3.2\,\text{ms}^{-2}$
(c) $0.8\,\text{ms}^{-2}$ (d) $0\,\text{ms}^{-2}$

33. A block of mass 10 kg is kept on a rough inclined plane as shown in the figure. A force of 3 N is applied on the block. The coefficient of static friction between the plane and the block is 0.6. What should be the minimum value of force F, such that the block does not move downward? (Take, $g = 10\,\text{ms}^{-2}$) **(2019 Main)**

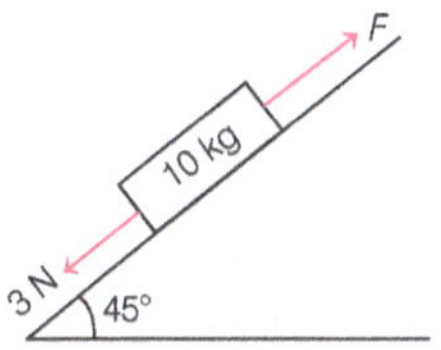

(a) 32 N (b) 25 N
(c) 23 N (d) 18 N

Multiple Correct Options

34. Two blocks each of mass 1 kg are placed as shown. They are connected by a string which passes over a smooth (massless) pulley.

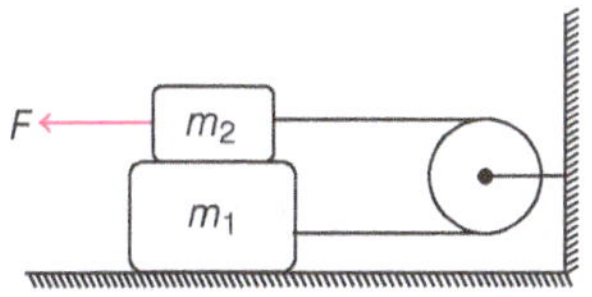

There is no friction between m_1 and the ground. The coefficient of friction between m_1 and m_2 is 0.2. A force F is applied to m_2. Which of the following statements is/are correct?
(a) The system will be in equilibrium if $F \leq 4$ N
(b) If $F > 4$ N tension in the string will be 4 N
(c) If $F > 4$ N the frictional force between the blocks will be 2 N
(d) If $F = 6$ N tension in the string will be 3 N

35. The velocity-time graph of the figure shows the motion of a wooden block of mass 1 kg which is given an initial push at $t = 0$ along a horizontal table.

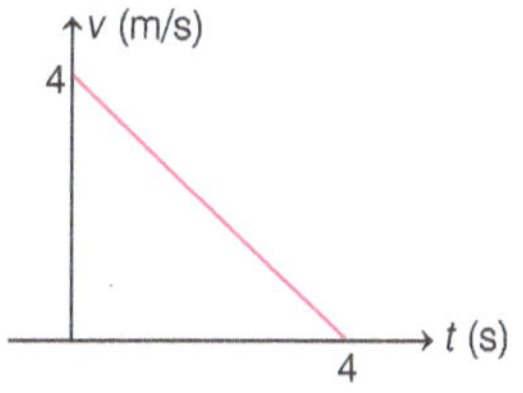

(a) The coefficient of friction between the block and the table is 0.1
(b) The coefficient of friction between the block and the table is 0.2
(c) If the table was half of its present roughness, the time taken by the block to complete the journey is 4 s
(d) If the table was half of its present roughness, the time taken by the block to complete the journey is 8 s

36. A man pulls a block of mass equal to himself with a light string. The coefficient of friction between the man and the floor is greater than that between the block and the floor
(a) if the block does not move, then the man also does not move
(b) the block can move even when the man is stationary
(c) if both move, then the acceleration of the block is greater than the acceleration of man
(d) if both move, then the acceleration of man is greater than the acceleration of block

Match the Columns

37. Angle θ is gradually increased as shown in figure. For the given situation, match the following two columns. (Take, $g = 10\,\text{ms}^{-2}$)

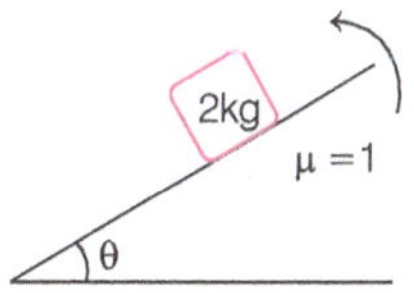

	Column I		Column II
(a)	Force of friction when $\theta = 0°$	(p)	10 N
(b)	Force of friction when $\theta = 90°$	(q)	$10\sqrt{3}$ N
(c)	Force of friction when $\theta = 30°$	(r)	$\frac{10}{\sqrt{3}}$ N
(d)	Force of friction when $\theta = 60°$	(s)	None of the above

38. Match the following two columns regarding fundamental forces of nature.

	Column I		Column II
(a)	Force of friction	(p)	field force
(b)	Normal reaction	(q)	contact force
(c)	Force between two neutrons	(r)	electromagnetic force
(d)	Force between two protons	(s)	nuclear force

39. In the figure shown, match the following two columns. (Take, $g = 10 \text{ms}^{-2}$)

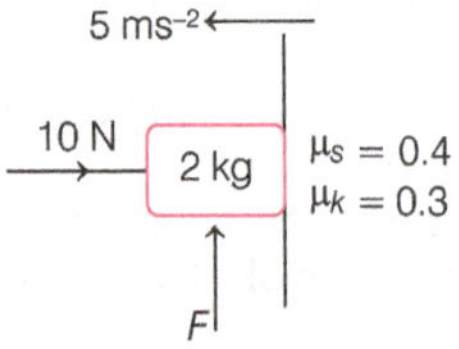

	Column I		Column II
(a)	Normal reaction	(p)	5 N
(b)	Force of friction when $F = 15$ N	(q)	10 N
(c)	Minimum value of F for stopping the block moving down	(r)	15 N
(d)	Maximum value of F for stopping the block moving up	(s)	None of the above

40. Match the Column I with Column II.

Note *Applied force is parallel to plane.*

	Column I		Column II
(a)	If friction force is less than applied force, then friction may be	(p)	static
(b)	If friction force is equal to the force applied, then friction may be	(q)	kinetic
(c)	If a block is moving on ground, then friction is	(r)	limiting
(d)	If a block kept on ground is at rest, then friction may be	(s)	no conclusion can be drawn

41. If the system is released from rest, then match the following two columns.

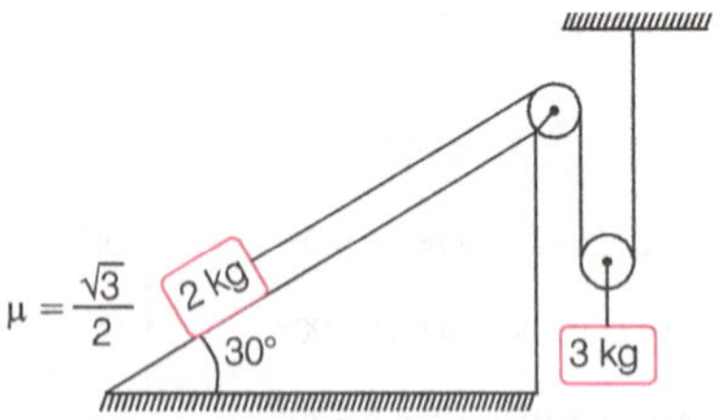

	Column I		Column II
(a)	Acceleration of 2 kg mass	(p)	2 SI unit
(b)	Acceleration of 3 kg mass	(q)	5 SI unit
(c)	Tension in the string connecting 2 kg mass	(r)	Zero
(d)	Frictional force on 2 kg mass	(s)	None of the above

42. Two blocks A and B of masses 2 kg and 3 kg are placed on horizontal surface. The coefficient of friction between block A and horizontal surface is 0.5 while the block B is smooth. A force $F = (\sqrt{2}\ t)$ N is applied on block A as shown in the figure.

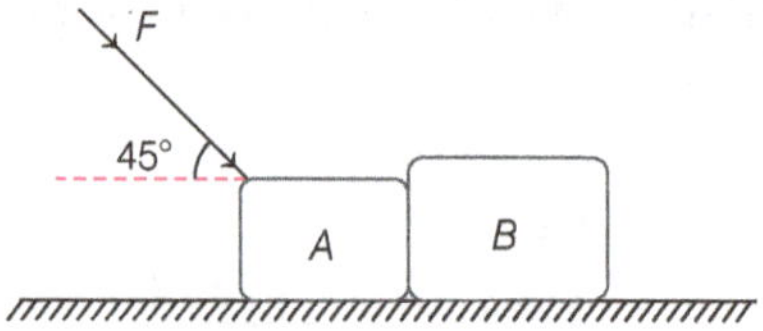

Match the Column I with Column II.

	Column I		Column II
(a)	$t = 4$ s	(p)	The acceleration of block B is zero.
(b)	$t = 20$ s	(q)	Friction on block A is more than 20 N.
(c)	$t = 40$ s	(r)	The acceleration of block A is 2 m/s^2.
(d)	$t = 60$ s	(s)	Normal contact force between blocks A and B is non-zero.

Subjective Questions

43. A 4 m long ladder weighing 25 kg rests with its upper end against a smooth wall and lower end on rough ground. What should be the minimum coefficient of friction between the ground and the ladder for it to be inclined at 60° with the horizontal without slipping? (Take, $g = 10 \text{ m/s}^2$)

44. A block is placed on an inclined plane as shown in figure. What must be the frictional force between block and incline, if the block is not to slide along the incline when the incline is accelerating to the right at $3 \text{ m/s}^2 \left(\sin 37° = \frac{3}{5}\right)$? (Take, $g = 10 \text{ m/s}^2$)

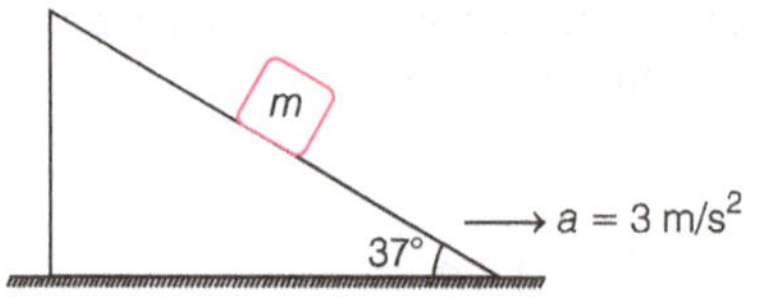

45. A conveyor belt is moving with constant speed of 6 m/s. A small block is just dropped on it. Coefficient of friction between the two is $\mu = 0.3$. Find

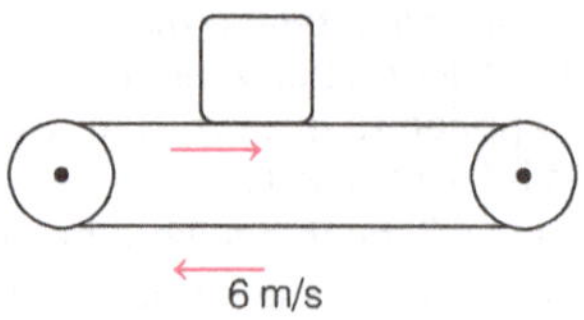

(a) the time when relative motion between them will stop.

(b) displacement of block up to that instant. (Take, $g = 10 \text{m/s}^2$)

46. A 30 kg mass is initially at rest on the floor of a truck. The coefficient of static friction between the mass and the floor of truck is 0.3 and coefficient of kinetic

friction is 0.2. Initially the truck is travelling due east at constant speed. Find the magnitude and direction of the friction force acting on the mass, if :
(Take, $g = 10 \text{ m/s}^2$)

(a) The truck accelerates at 1.8 m/s^2 eastward

(b) The truck accelerates at 3.8 m/s^2 westward.

47. Determine the acceleration of the 5 kg block A. Neglect the mass of the pulley and cords. The block B has a mass of 10 kg. The coefficient of kinetic friction between block B and the surface is $\mu_k = 0.1$. (Take, $g = 10 \text{m/s}^2$)

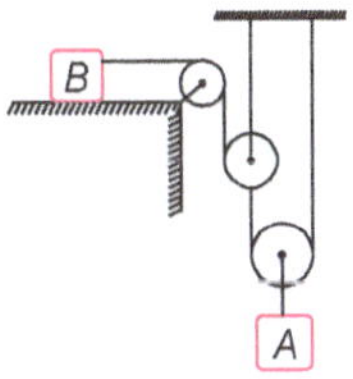

48. The upper portion of an inclined plane of inclination α is smooth and the lower portion is rough. A particle slides down from rest from the top and just comes to rest at the foot. If the ratio of smooth length to rough length is $m : n$, find the coefficient of friction.

49. The conveyor belt is designed to transport packages of various weights. Each 10 kg package has a coefficient of kinetic friction $\mu_k = 0.15$. If the speed of the conveyor belt is 5 m/s, and then it suddenly stops, determine the distance the package will slide before coming to rest. (Take, $g = 9.8 \text{ m/s}^2$)

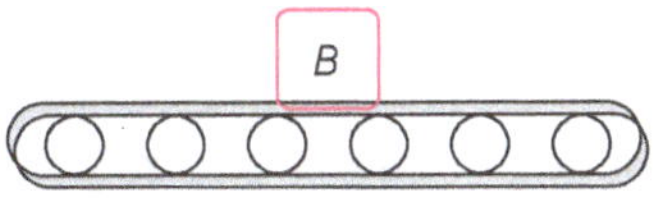

50. In the arrangement of figure the masses m_0, m_1 and m_2 of bodies are equal say m, the masses of the pulley and the threads are negligible, and there is no friction in the pulley. Find the acceleration a with which the body m_0 comes down, and the tension of the thread binding together the bodies m_1 and m_2, if the coefficient of friction between these bodies and the horizontal surface is equal to μ. Consider possible cases. **(Irodov 1.60)**

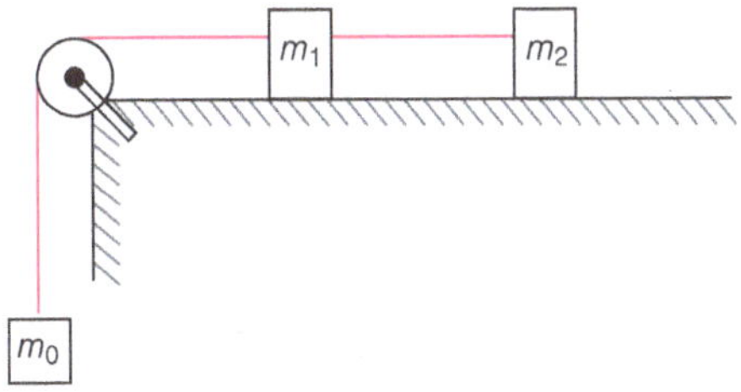

51. Two touching bars 1 and 2 are placed on an inclined plane forming an angle α with the horizontal. The masses of the bars are equal to m_1 and m_2, and the coefficients of friction between the inclined plane and these bars are equal to μ_1 and μ_2 respectively, with $\mu_1 > \mu_2$. Find:

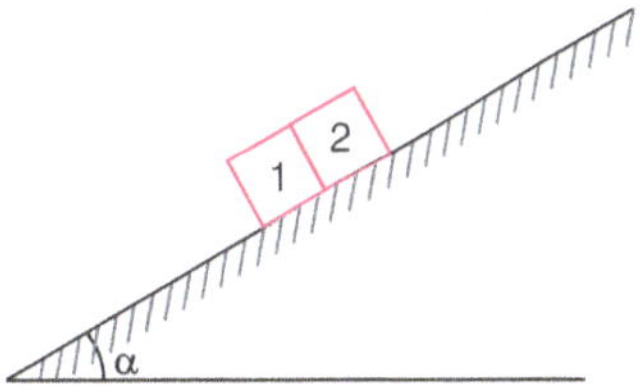

(a) the force of interaction of the bars in the process of motion

(b) the minimum value of the angle α at which the bars start sliding down. **(Irodov 1.61)**

52. A small body was launched up an inclined plane set at an angle $\alpha = 15°$ against the horizontal. Find the coefficient of friction, if the time of the ascent of the body is $\eta = 2.0$ times less than the time of its descent. **(Irodov 1.62)**

53. The following parameters of the arrangement of figure are available; the angle α which the inclined plane forms with the horizontal, and the coefficient of friction μ between the body m_1 and the inclined plane. The masses of the pulley and the threads, as well as the friction in the pulley, are negligible. Assuming both bodies to be motionless at the initial moment, find the mass ratio m_2/m_1 at which the body m_2

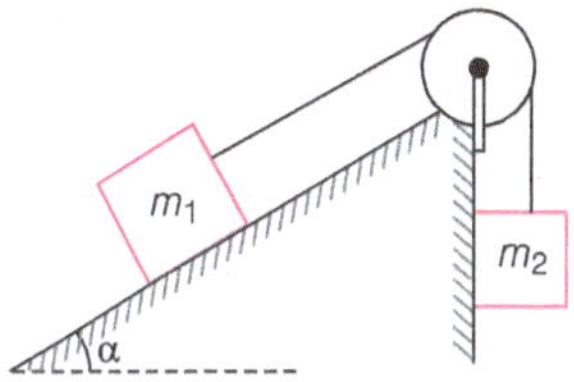

(a) starts coming down

(b) starts going up

(c) is at rest. **(Irodov 1.63)**

54. The inclined plane of the above figure forms an angle $\alpha = 30°$ with the horizontal. The mass ratio $m_2/m_1 = 2/3$. The coefficient of friction between the body m_1 and the inclined plane is equal to $\mu = 0.10$. The masses of the pulley and the threads are negligible. Find the magnitude and the direction of acceleration of the body m_2 when the formerly stationary system of masses starts moving. **(Irodov 1.64)**

55. At the moment $t = 0$ the force $F = kt$ is applied to a small body of mass m resting on a smooth horizontal plane (k is a constant). The permanent direction of this forms an angle α with the horizontal. Find :

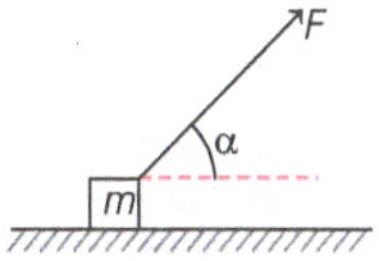

(a) the velocity of the body at the moment of its breaking off the plane

(b) the distance traversed by the body upto this moment.

56. Prism 1 with bar 2 of mass m placed on it gets a horizontal acceleration a directed to the left. At what maximum value of this acceleration will the bar be stationary relative to the prism, if the coefficient of friction between them $\mu < \cot\alpha$? **(Irodov 1.80)**

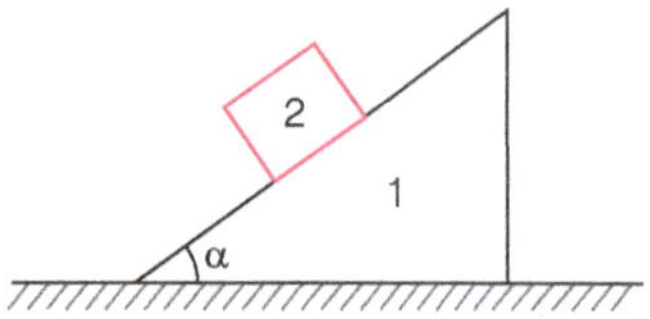

57. A body of mass m rests on a horizontal plane with the friction coefficient μ. At the moment $t = 0$, a horizontal force is applied to it, which varies with time as $\mathbf{F} = \alpha t$, where α is a constant vector.

Find the distance traversed by the body during the first t seconds after the force acting began. **(Irodov 1.103)**

58. A block of mass 2 kg slides on an inclined plane which makes an angle of 30° with the horizontal. The coefficient of friction between the block and the surface is $\sqrt{3}/\sqrt{2}$. What force along the plane should be applied to the block so that it moves (a) down and (b) up without any acceleration?
(Take $g = 10\ \text{m/s}^2$) **(1978 JEE)**

59. Two blocks A and B of equal masses are released from an inclined plane of inclination 45° at t=0. Both the blocks are initially at rest. The coefficient of kinetic friction between the block A and the inclined plane is 0.2 while it is 0.3 for block B. Initially the block A is $\sqrt{2}$ m behind the block B. When and where their front faces will come in a line?
(Take g=10m/s^2) **(2004 JEE)**

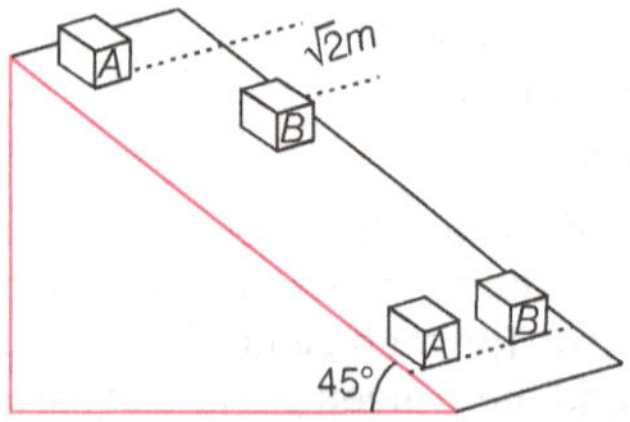

60. A small bar starts sliding down an inclined plane forming angle α with the horizontal. The friction coefficient depends on the distance x covered as $\mu = kx$, where k is a constant. Find the distance covered by the bar till it stops, and its maximum velocity over this distance. **(Irodov 1.102)**

61. In the arrangement shown in figure, the mass of the rod M exceeds the mass m of the ball. The ball has an opening permitting it to slide along the thread with some friction. The mass of the pulley and the friction in its axle are negligible. At the initial moment the ball was located opposite the lower end of the rod. When set free, both bodies began moving with constant accelerations. Find the friction force between the ball and the thread if t seconds after the beginning of motion the ball got opposite the upper end of the rod. The rod length equals l. **(Irodov 1.74)**

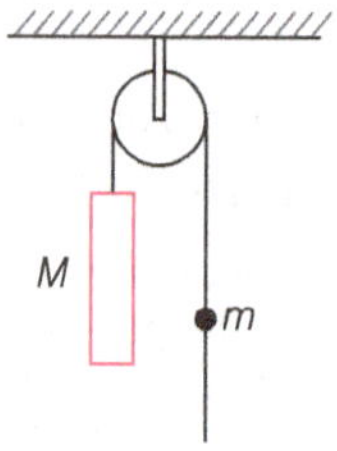

Answers

1. (False) **2.** (5) **3.** (d) **4.** (a) **5.** (b)
6. (a) **7.** (a) **8.** (c) **9.** (b) **10.** (a)
11. (a) **12.** (a) **13.** (a) **14.** (d) **15.** (d)
16. (d) **17.** (a) **18.** (b) **19.** (a) **20.** (b)
21. (c) **22.** (c) **23.** (c) **24.** (b) **25.** (a)
26. (a) **27.** (b) **28.** (a) **29.** (a) **30.** (b)
31. (a) **32.** (c) **33.** (a) **34.** (a, c, d) **35.** (a, d)
36. (a, b, c) **37.** (a-s; b-s; c-p; d-p) **38.** (a-q,r; b-q,r; c-s; d-p,s)
39. (a-s; b-p; c-s; d-s) **40.** (a-q; b-p,q,r; c-q; d-p,r)
41. (a-r; b-r; c-s; d-q) **42.** (a-p; b-p; c-q,r,s; d-q,s) **43.** (0.288)
44. $\left(\frac{9}{25}mg\right)$ **45.** (a) 2s (b) 6 m
46. (a) 54 N (eastwards) (b) 60 N (westwards) **47.** 2/33 m/s^2
48. $\mu = \left(\frac{m+n}{n}\right)\tan\alpha$ **49.** 8.5 m
50. For $\mu \geq \frac{1}{2}$, $a = 0$, $T = mg$.

For $\mu < \frac{1}{2}$, $a = \left(\frac{1-2\mu}{3}\right)g$, $T_2 = \left(\frac{1+\mu}{3}\right)mg$

51. (a) $N = \frac{(\mu_1 - \mu_2)m_1 m_2 g\cos\alpha}{(m_1 + m_2)}$ (b) $\tan\alpha_{min} > \frac{\mu_1 m_1 + \mu_2 m_2}{m_1 + m_2}$
52. 0.16
53. (a) $\frac{m_2}{m_1} > (\sin\alpha + \mu\cos\alpha)$ (b) $\frac{m_2}{m_1} < (\sin\alpha - \mu\cos\alpha)$

(c) $(\sin\alpha - \mu\cos\alpha) < \frac{m_2}{m_1} < (\sin\alpha + \mu\cos\alpha)$

54. (a) $0.05g$ **55.** (a) $v = \frac{mg^2\cos\alpha}{2k\sin^2\alpha}$ (b) $s = \frac{m^2g^3\cos\alpha}{6k^2\sin^3\alpha}$
56. $g\left[\frac{1+\mu\cot\alpha}{\cot\alpha - \mu}\right]$
57. $s = \frac{\alpha}{6m}(t - t_0)^3$, where $t_0 = \mu mg/\alpha$; At $t \leq t_0$, the distance is $s = 0$
58. (a) 11.21 N (b) 31.21 N **59.** 2s, $8\sqrt{2}$m
60. $d = \frac{2}{k}\tan\alpha$, $v_{max} = \sqrt{\frac{g}{k}\sin\alpha\tan\alpha}$
61. $f = 2l\,mM/(M-m)t^2$

See hints on pages 428-437.

Unit Exercise (Cumulative)

Level 1

Assertion and Reason

For directions see on page 354.

1. **Assertion :** Moment of concurrent forces about any point is constant.

   Reason : If vector sum of all the concurrent forces is zero, then moment of all the forces about any point is also zero.

2. **Assertion :** A block A is just placed inside a smooth box B as shown in figure. Now, the box is given an acceleration $\mathbf{a} = (3\hat{\mathbf{j}} - 2\hat{\mathbf{i}})\ \text{ms}^{-2}$. Under this acceleration block A cannot remain in the position shown.

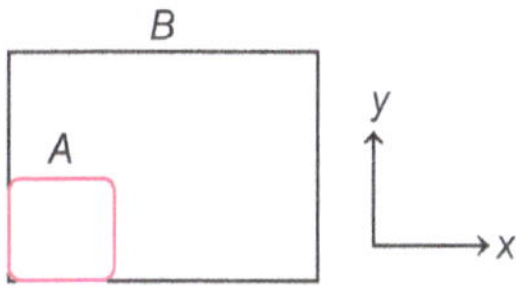

   Reason : Block will require $m\mathbf{a}$ force for moving with acceleration $\mathbf{a}$.

Single Correct Option

3. A weightless rod is acted upon by upward parallel forces of 4 N and 2 N magnitudes at ends A and B, respectively. The total length of the rod $AB = 3$ m. To keep the rod in equilibrium a force of 6 N should act in the following manner
   (a) downward at any point between A and B
   (b) downward at the mid-point of AB
   (c) downward at a point C such that AC = 1 m
   (d) downward at a point C such that BC = 1 m

4. In the figure, two identical particles each of mass m are tied together with an inextensible string. This is pulled at its centre with a constant force F. If the whole system lies on a smooth horizontal plane, then the acceleration of each particle towards each other is

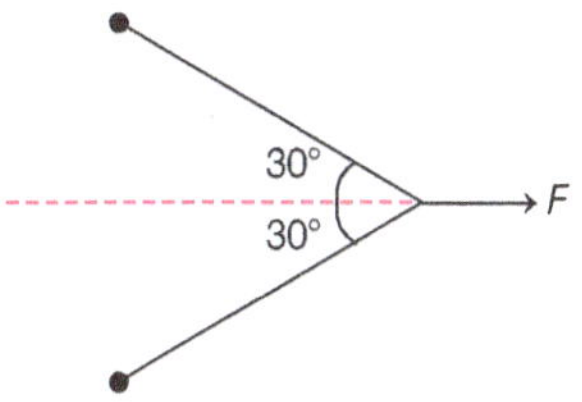

   (a) $\frac{\sqrt{3}}{2}\frac{F}{m}$ (b) $\frac{1}{2\sqrt{3}}\frac{F}{m}$ (c) $\frac{2}{\sqrt{3}}\frac{F}{m}$ (d) $\sqrt{3}\frac{F}{m}$

5. A smooth plank C of mass 1 kg and length 8 m is resting on smooth horizontal surface. The gap between fixed pulley P and plank is larger than the length of plank. A smooth block D is placed on the plank. The masses of blocks A and B are 1 kg and 3 kg, respectively.

   Find the time elapsed before the block D separates from the plank.

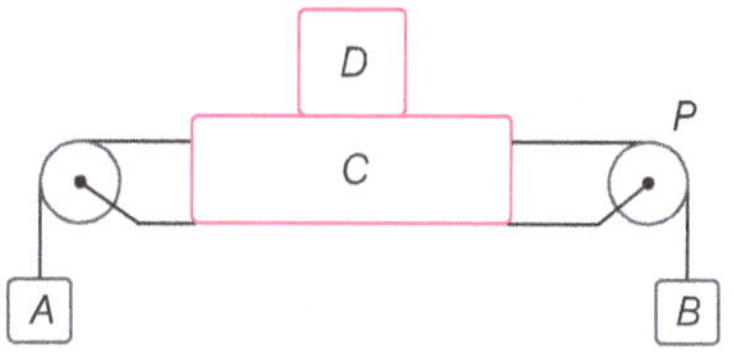

   (a) 1 s
   (b) 2 s
   (c) 3 s
   (d) Insufficient information

Note *Neglect size of block D and assume that initially it is at one end of plank C.*

6. A sphere of mass 1 kg rests at one corner of a cube. The cube is moved with a velocity $\mathbf{v} = (8t\hat{\mathbf{i}} - 2t^2\hat{\mathbf{j}})$, where t is time in second. The force by sphere on the cube at $t = 1$ s is ($g = 10\ \text{ms}^{-2}$) [Figure shows vertical plane of the cube]

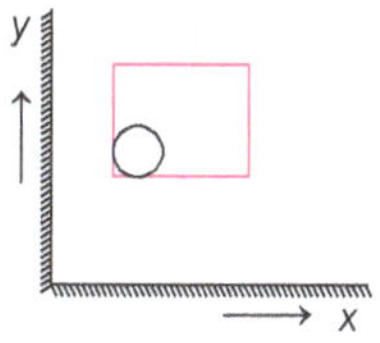

   (a) 8 N (b) 10 N (c) 20 N (d) 6 N

7. Mr. X of mass 80 kg enters a lift and selects the floor he wants. The lift now accelerates upwards at 2 ms^{-2} for 2 s and then moves with constant velocity. As the lift approaches his floor, it decelerates at the same rate as it previously accelerates. If the lift cables can safely withstand a tension of 2×10^4 N and the lift itself has a mass of 500 kg, how many Mr. X's could it safely carry at one time?
   (a) 22 (b) 14 (c) 18 (d) 12

8. Two blocks A and B, each of same mass are attached by a thin inextensible string through an ideal pulley. Initially block B is held in position as shown in figure. Now, the block B is released. Block A will slide to right and hit the pulley in time t_A. Block B will swing and hit the surface in time t_B. Assume the surface as frictionless, then

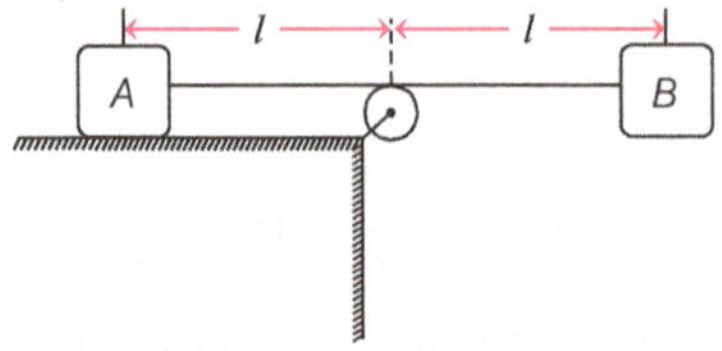

(a) $t_A > t_B$ (b) $t_A < t_B$
(c) $t_A = t_B$ (d) data insufficient

9. A pendulum of mass m hangs from a support fixed to a trolley. The direction of the string (i.e. angle θ) when the trolley rolls up a plane of inclination α with acceleration a is

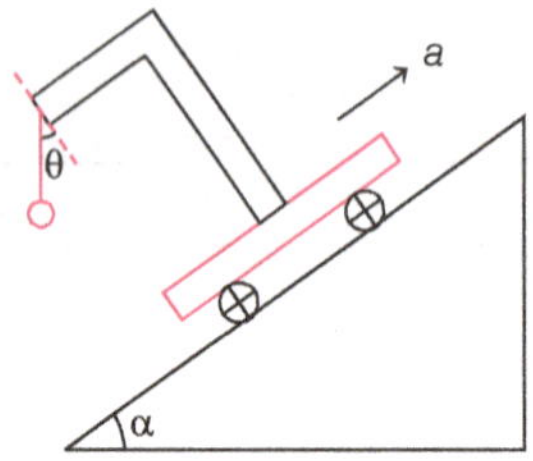

(a) 0 (b) $\tan^{-1}\alpha$
(c) $\tan^{-1}\left(\dfrac{a+g\sin\alpha}{g\cos\alpha}\right)$ (d) $\tan^{-1}\dfrac{a}{g}$

10. A man of mass m has fallen into a ditch of width d. Two of his friends are slowly pulling him out using a light rope and two fixed pulleys as shown in figure. Both the friends exert force of equal magnitudes F.

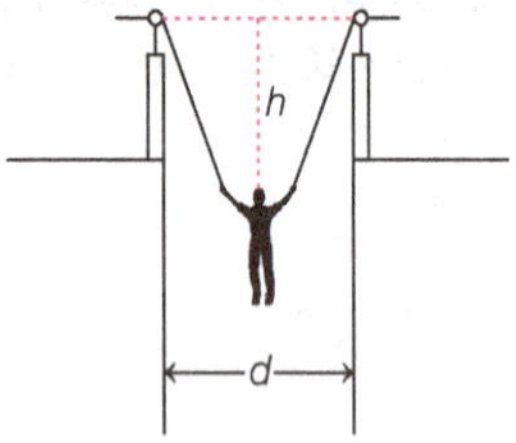

When the man is at a depth h, the value of F is
(a) $\dfrac{mg}{4h}\sqrt{d^2+4h^2}$ (b) hmg
(c) dmg (d) $\dfrac{mg}{2h}\sqrt{h^2+d^2}$

11. System shown in figure is in equilibrium and at rest. The spring and string are massless, now the string is cut. The acceleration of mass $2m$ and m just after the string is cut will be **(2006 JEE)**

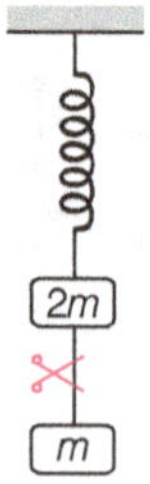

(a) $g/2$ upwards, g downwards
(b) g upwards, $g/2$ downwards
(c) g upwards, $2g$ downwards
(d) $2g$ upwards, g downwards

12. Two particles of mass m each are tied at the ends of a light string of length $2a$. The whole system is kept on a frictionless horizontal surface with the string held tight so that each mass is at a distance a from the centre P (as shown in the figure). Now, the mid-point of the string is pulled vertically upwards with a small but constant force F. As a result, the particles move towards each other on the surface. The magnitude of acceleration when the separation between them becomes $2x$, is **(2007 JEE)**

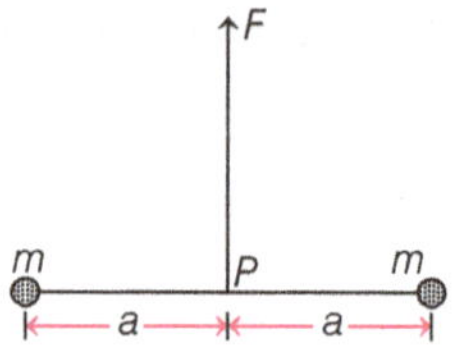

(a) $\dfrac{F}{2m}\dfrac{a}{\sqrt{a^2-x^2}}$ (b) $\dfrac{F}{2m}\dfrac{x}{\sqrt{a^2-x^2}}$
(c) $\dfrac{F}{2m}\dfrac{x}{a}$ (d) $\dfrac{F}{2m}\dfrac{\sqrt{a^2-x^2}}{x}$

13. The force F which is applied to 1 kg block initially at rest varies linearly with time as shown in the figure. Find velocity of the block at $t = 4$ s.

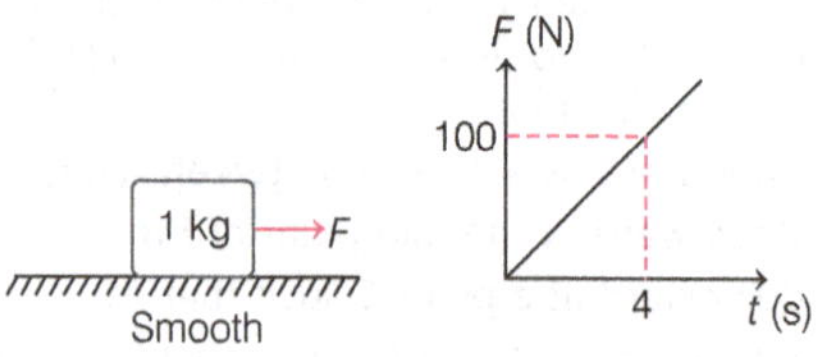

(a) 100 m/s (b) 200 m/s (c) 20 m/s (d) 4 m/s

14. If the acceleration of block A is 5 m/s^2, then acceleration of block B is

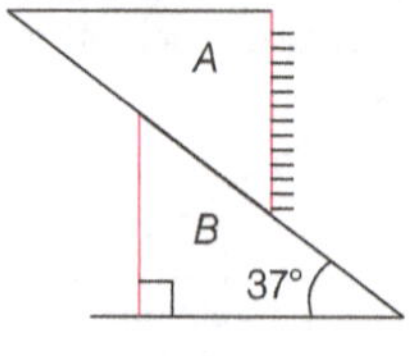

(a) $5\,\text{m/s}^2$ (b) $10\,\text{m/s}^2$ (c) zero (d) $6.67\,\text{m/s}^2$

15. In the arrangement shown in the figure, the rod R is restricted to move in the vertical direction with acceleration a_1 and the block B can slide down the fixed wedge with acceleration a_2. The correct relation between a_1 and a_2 is given by

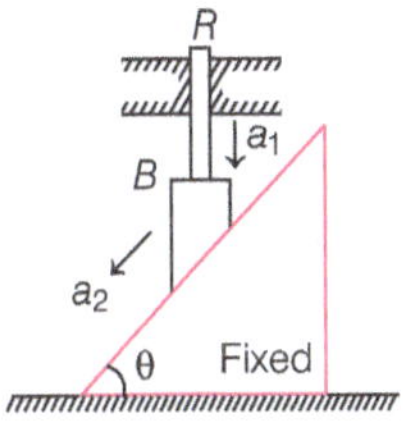

(a) $a_2 = a_1 \sin\theta$ (b) $a_2 \sin\theta = a_1$
(c) $a_2 \cos\theta = a_1$ (d) $a_2 = a_1 \cos\theta$

16. All the surfaces and pulleys are frictionless in the shown arrangement. Pulleys P and Q are massless. The force applied by clamp on pulley P is

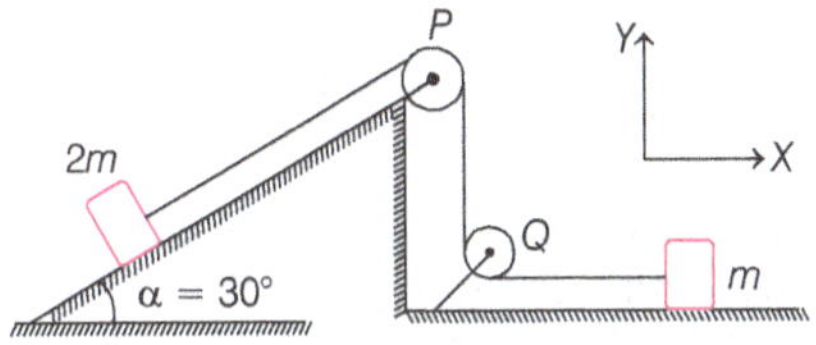

(a) $\frac{mg}{6}(-\sqrt{3}\,\hat{i} - 3\,\hat{j})$ (b) $\frac{mg}{6}(\sqrt{3}\,\hat{i} + 3\,\hat{j})$
(c) $\frac{mg}{6}\sqrt{2}$ (d) None of these

17. In the system shown in figure, all surfaces are smooth. Rod is moved by external agent with acceleration 9 ms^{-2} vertically downwards. Force exerted on the rod by the wedge will be

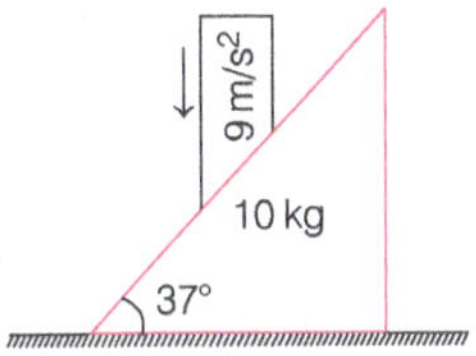

(a) 120 N (b) 200 N (c) 160 N (d) 180 N

18. A pivoted beam of negligible mass has a mass suspended from one end and an Atwood's machine suspended from the other. The frictionless pulley has negligible mass and dimension. Gravity is directed downward and $M_2 = 3M_3$, $l_2 = 3l_1$. Find the ratio M_1/M_2 which will ensure that the beam has no tendency to rotate just after the masses are released.

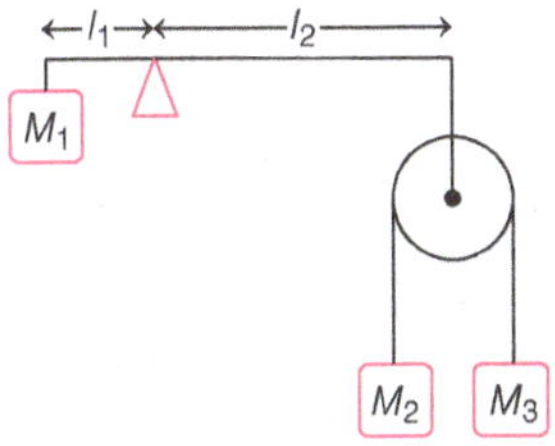

(a) $\frac{M_1}{M_2} = 2$ (b) $\frac{M_1}{M_2} = 3$
(c) $\frac{M_1}{M_2} = 4$ (d) None of these

19. In the figure, pulleys are smooth and strings are massless, $m_1 = 1$ kg and $m_2 = \frac{1}{3}$ kg. To keep m_3 at rest, mass m_3 should be

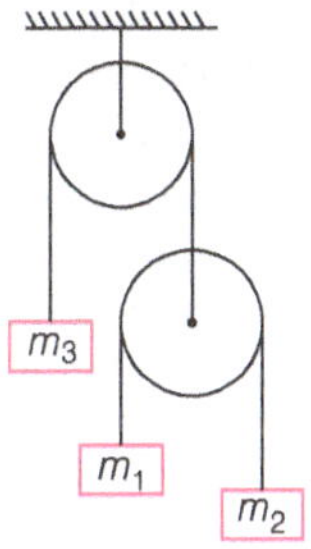

(a) 1 kg
(b) $\frac{2}{3}$ kg
(c) $\frac{1}{4}$ kg
(d) 2 kg

20. In the arrangement shown in figure a_1, a_2, a_3 and a_4 are the accelerations of masses m_1, m_2, m_3 and m_4, respectively. Which of the following relation is true for this arrangement? **(2022 Main)**

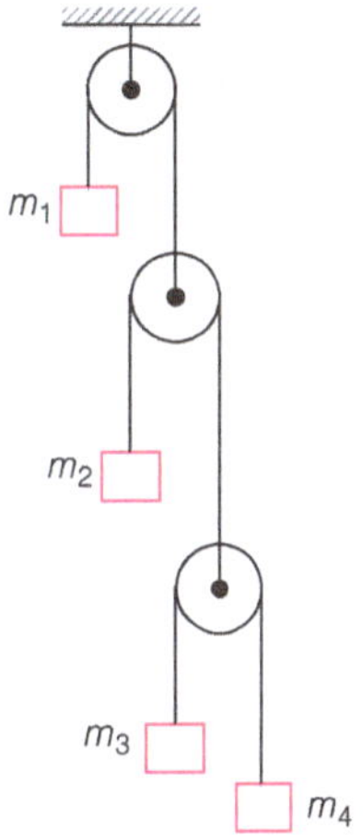

(a) $4a_1 + 2a_2 + a_3 + a_4 = 0$ (b) $a_1 + 4a_2 + 3a_3 + a_4 = 0$
(c) $a_1 + 4a_2 + 3a_3 + 2a_4 = 0$ (d) $2a_1 + 2a_2 + 3a_3 + a_4 = 0$

21. The boxes of masses 2 kg and 8 kg are connected by a massless string passing over smooth pulleys. Calculate the time taken by box of mass 8 kg to strike the ground starting from rest. (Use $g = 10\ m/s^2$) **(2021 Main)**

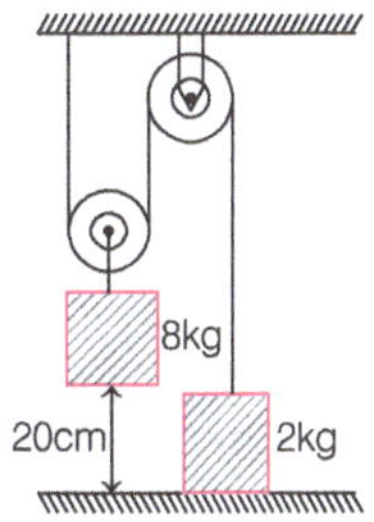

(a) 0.34 s (b) 0.2 s (c) 0.25 s (d) 0.4 s

22. A smooth rod of length l is kept inside a trolley at an angle θ as shown in the figure. What should be the acceleration a of the trolley, so that the rod remains in equilibrium with respect to it?

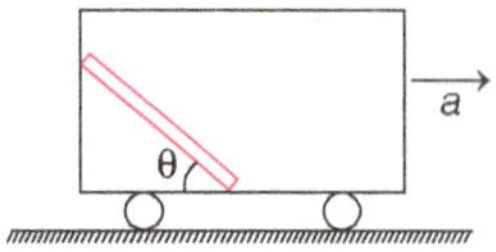

(a) $g \tan\theta$ (b) $g \cos\theta$ (c) $g \sin\theta$ (d) $g \cot\theta$

23. A block of mass m slides on the wooden wedge, which in turn slides backward on the horizontal surface. The acceleration of the block with respect to the wedge is : (Given $m = 8$ kg, $M = 16$ kg)

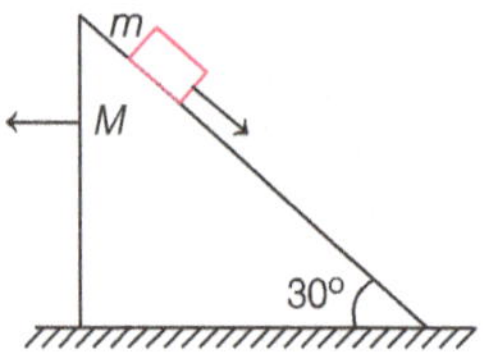

Assume all the surfaces shown in the figure to be frictionless. **(2021 Main)**

(a) $\frac{4}{3}g$ (b) $\frac{6}{5}g$ (c) $\frac{3}{5}g$ (d) $\frac{2}{3}g$

24. Consider the shown arrangement. Assume all surfaces to be smooth. If N represents magnitudes of normal reaction between block and wedge, then acceleration of M along horizontal is equal to

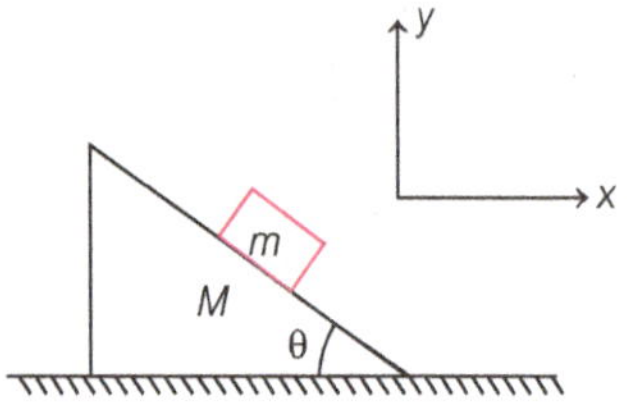

(a) $\frac{N\sin\theta}{M}$ along +ve x-axis

(b) $\frac{N\cos\theta}{M}$ along −ve x-axis

(c) $\frac{N\sin\theta}{M}$ along −ve x-axis

(d) $\frac{N\sin\theta}{m+M}$ along −ve x-axis

25. In the above problem, normal reaction between ground and wedge will have magnitude equal to

(a) $N\cos\theta + Mg$ (b) $N\cos\theta + Mg + mg$
(c) $N\cos\theta - Mg$ (d) $N\sin\theta + Mg + mg$

26. A block of mass m is resting on a wedge of angle θ as shown in the figure. With what minimum acceleration a should the wedge move, so that the mass m falls freely?

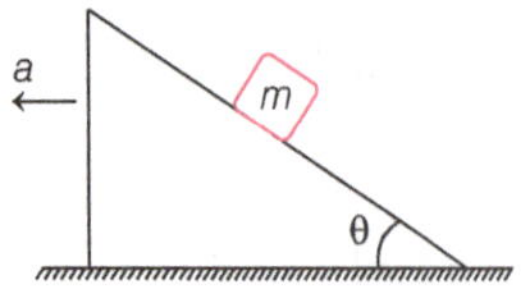

(a) g (b) $g \cos\theta$
(c) $g \cot\theta$ (d) $g \tan\theta$

27. A block is kept on a smooth inclined plane of angle of inclination θ that moves with a constant acceleration, so that the block does not slide relative to the inclined plane. If the inclined plane stops, the normal contact force offered by the plane on the block changes by a factor

(a) $\tan\theta$ (b) $\tan^2\theta$
(c) $\cos^2\theta$ (d) $\cot\theta$

28. In the figure shown the acceleration of A is $\mathbf{a}_A = (15\hat{\mathbf{i}} + 15\hat{\mathbf{j}})$. Then, the acceleration of B is (A remains in contact with B)

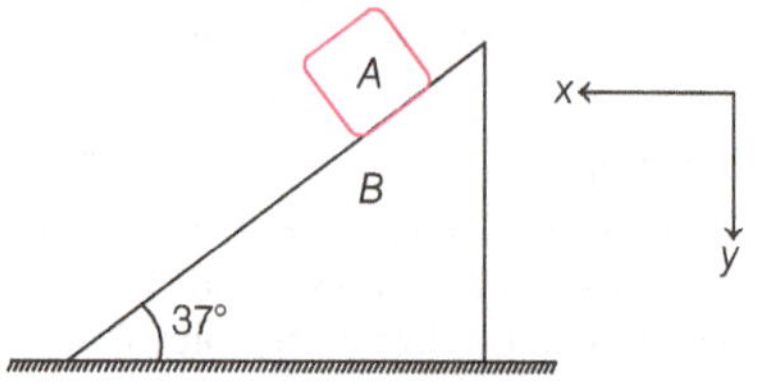

(a) $5\hat{\mathbf{i}}$ (b) $-15\hat{\mathbf{i}}$
(c) $-10\hat{\mathbf{i}}$ (d) $-5\hat{\mathbf{i}}$

29. A ball connected with string is released at an angle 45° with the vertical as shown in the figure. Then, the acceleration of the box at this instant will be (mass of the box is equal to mass of ball)

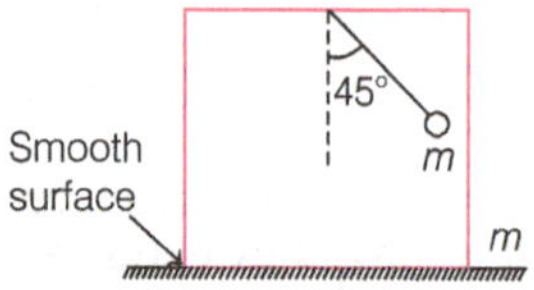

(a) $\frac{g}{4}$ (b) $\frac{g}{3}$ (c) $\frac{g}{2}$ (d) g

30. Three masses $M = 100$ kg, $m_1 = 10$ kg and $m_2 = 20$ kg are arranged in a system as shown in figure. All the surfaces are frictionless and strings are inextensible and weightless. The pulleys are also weightless and frictionless. A force F is applied on the system so that the mass m_2 moves upward with an acceleration of $2\ \text{ms}^{-2}$. The value of F is (Take $g = 10\ \text{ms}^{-2}$) **(2022 Main)**

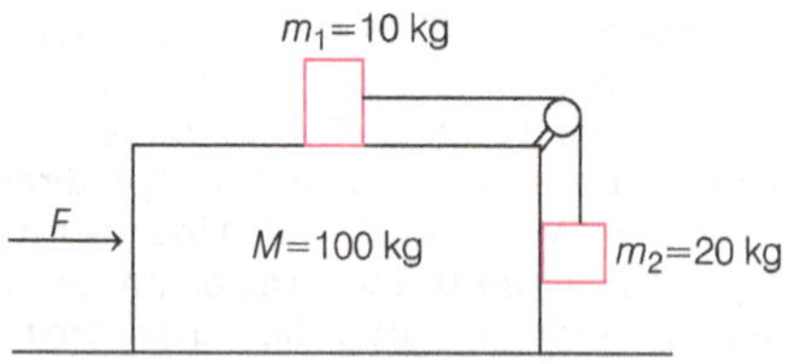

(a) 3360 N (b) 3380 N
(c) 3120 N (d) 3240 N

31. A smooth light rod not touching the rough floor is inserted between two identical blocks. A horizontal force F is applied to upper end of rod. The contact force between block A and rod is F_1 and contact force between block B and rod is F_2. Then,

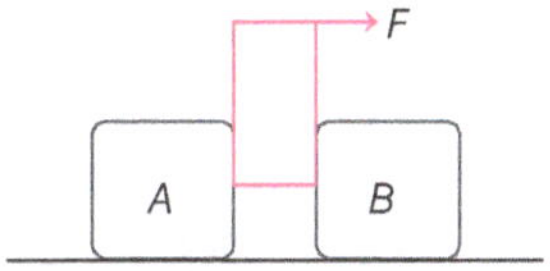

(a) $F_1 = F_2$ (b) $F_1 > F_2$
(c) $F_1 < F_2$ (d) $F_1 = 0$ and $F_2 > 0$

32. A time varying force of $5t$ is applied on a block of mass 1 kg. The magnitude of normal reaction on the block by horizontal ground is N_1 and magnitude of normal reaction on block due to vertical wall is N_2. N_1 *versus* N_2 graph is

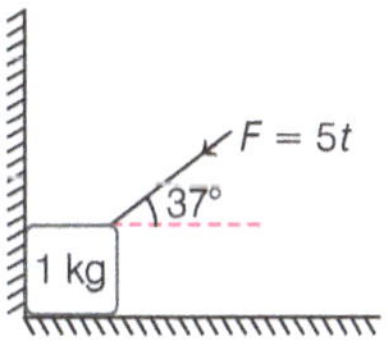

(a)
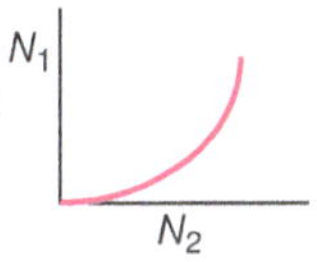

(b)
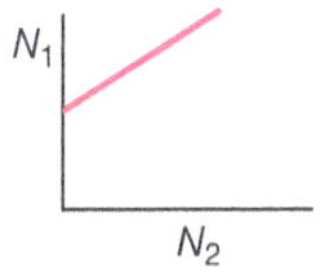

(c)
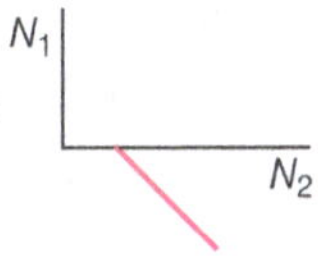

(d) None of these

33. What is the largest mass of C in kg that can be suspended without moving blocks A and B ? The static coefficient of friction for all plane surfaces of contact is 0.3. Mass of block A is 50kg and block B is 70kg. Neglect friction in the pulleys.

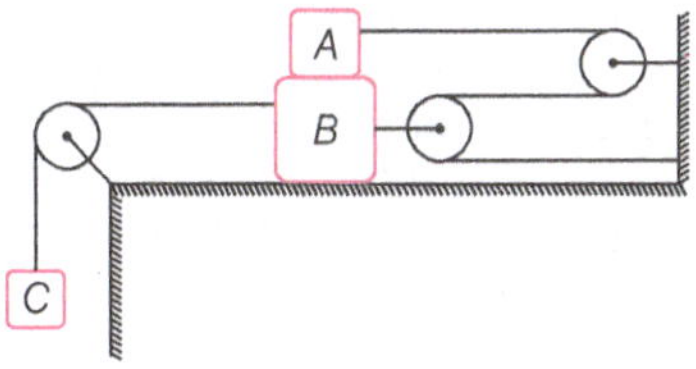

(a) 120 kg (b) 92 kg
(c) 81 kg (d) None of these

34. Two blocks of masses m_1 and m_2 are placed in contact with each other on a horizontal platform as shown in figure. The coefficient of friction between m_1 and platform is 2μ and that between block m_2 and platform is μ. The platform moves with an acceleration a. The normal reaction between the blocks is

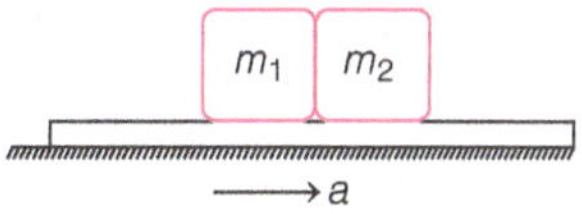

(a) zero in all cases (b) zero only if $m_1 = m_2$
(c) non zero only if $a > 2\mu g$ (d) non zero only if $a > \mu g$

35. The minimum value of μ between the two blocks for no slipping is

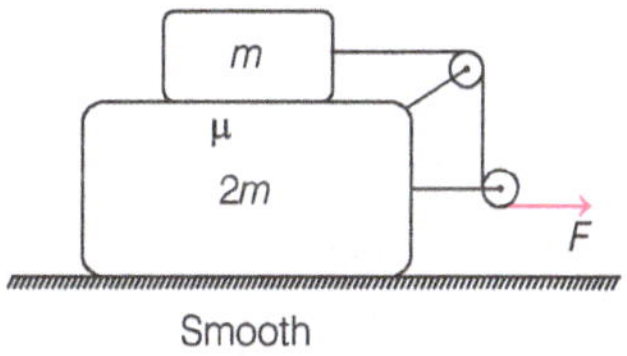

(a) $\dfrac{F}{mg}$ (b) $\dfrac{F}{3mg}$ (c) $\dfrac{2F}{3mg}$ (d) $\dfrac{4F}{3mg}$

36. Two blocks A and B are separated by some distance and tied by a string as shown in the figure. The force of friction in both the blocks at $t = 2$ s is

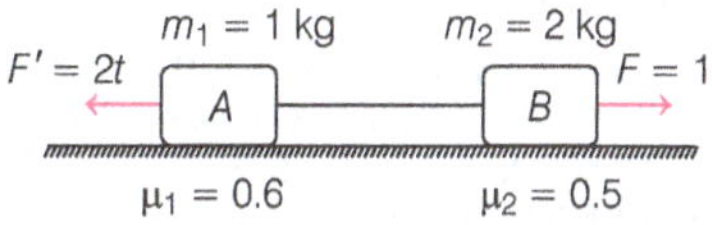

(a) 4N(→), 5 N(←) (b) 2N(→), 5N(←)
(c) 0 N(→), 10 N(←) (d) 1N(←), 10 N(←)

37. Two blocks of masses 2 kg and 4 kg are connected by a light string and kept on horizontal surface. A force of 16 N is acted on 4 kg block horizontally as shown in figure. Besides, it is given that coefficient of friction between 4 kg and ground is 0.3 and between 2 kg block and ground is 0.6. Then, frictional force between 2 kg block and ground is

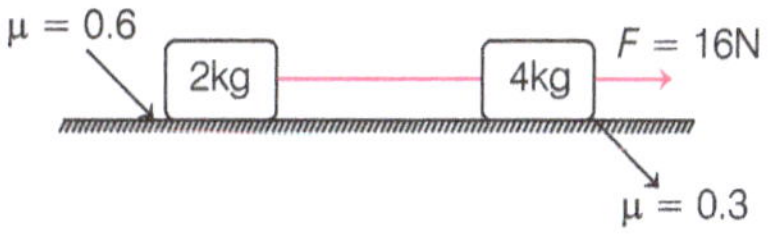

(a) 12 N (b) 6 N (c) 4 N (d) zero

38. A horizontal force $F = \dfrac{mg}{3}$ is applied on the upper surface of a uniform cube of mass m and side a which is resting on a rough horizontal surface having $\mu = \dfrac{1}{2}$.

The distance between lines of action of mg and normal reaction is

(a) $\dfrac{a}{2}$ (b) $\dfrac{a}{3}$
(c) $\dfrac{a}{4}$ (d) None of these

39. Three blocks are kept as shown in figure. Acceleration of 20 kg block with respect to ground is

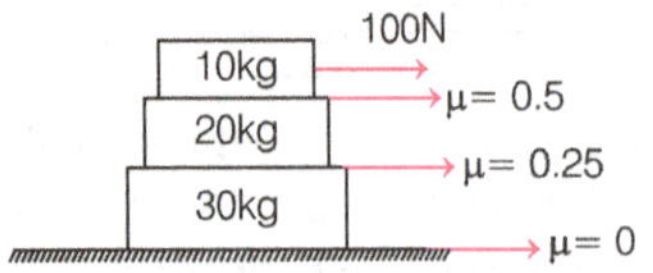

(a) 5 ms^{-2} (b) 2 ms^{-2}
(c) 1 ms^{-2} (d) None of these

40. If force F is increasing with time and at $t=0, F=0$. Where will slipping first start?

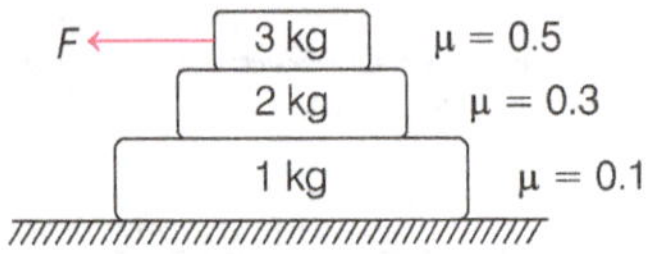

(a) Between 3 kg and 2 kg
(b) Between 2 kg and 1 kg
(c) Between 1 kg and ground
(d) Both (a) and (b)

41. A plank of mass 2 kg and length 1 m is placed on horizontal floor. A small block of mass 1 kg is placed on top of the plank, at its right extreme end. The coefficient of friction between plank and floor is 0.5 and that between plank and block is 0.2. If a horizontal force = 30 N starts acting on the plank to the right, the time after which the block will fall off the plank is
(Take, $g = 10 ms^{-2}$)

(a) $\left(\frac{2}{3}\right)$ s (b) 1.5 s (c) 0.75 s (d) $\left(\frac{4}{3}\right)$ s

42. If coefficient of friction between all surfaces is 0.4, then find the minimum force F to have equilibrium of the system. (Take, $g = 10 m/s^2$)

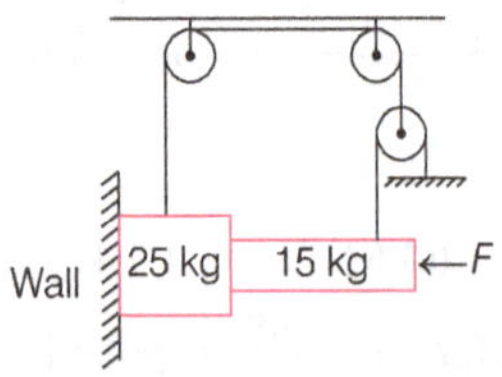

(a) 62.5 N (b) 150 N (c) 135 N (d) 50 N

43. The rear side of a truck is open and a box of mass 20 kg is placed on the truck 4 m away from the open end. Coefficient of friction between truck and box is $\mu = 0.15$ and $g = 10 m/s^2$. The truck starts from rest with an acceleration of $2 m/s^2$ on a straight road. The box will fall off the truck when it is at a distance from the starting point equal to

(a) 4 m (b) 8 m (c) 16 m (d) 32 m

44. In the situation shown in the figure, find the minimum value of coefficient of friction, so that all blocks move together.

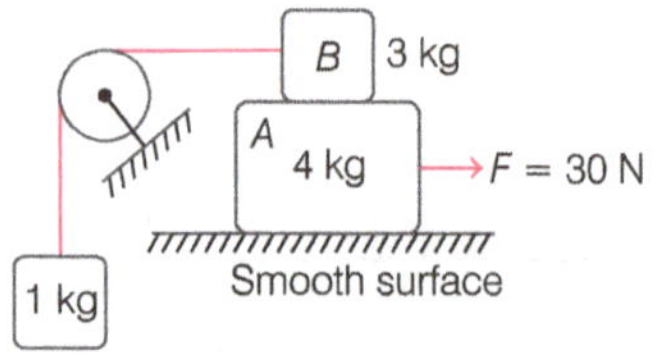

(a) 1 (b) $\frac{1}{2}$ (c) $\frac{1}{3}$ (d) $\frac{2}{3}$

45. In the situation shown in the figure, the coefficient of static friction for all surfaces of contact is 0.5. Find the maximum value of mass of block A, so that the blocks B, C and D move together.

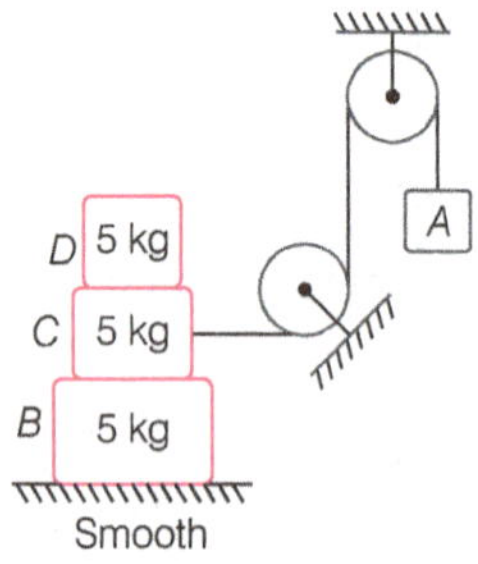

(a) 10 kg (b) 15 kg
(c) 20 kg (d) 25 kg

46. A block of mass m is placed on a surface with a vertical cross-section given by $y = \frac{x^3}{6}$. If the coefficient of friction is 0.5, the maximum height above the ground at which the block can be placed without slipping is **(2014 Main)**

(a) $\frac{1}{6}$ m (b) $\frac{2}{3}$ m (c) $\frac{1}{3}$ m (d) $\frac{1}{2}$ m

47. Two blocks A and B of masses $m_A = 1$ kg and $m_B = 3$ kg are kept on the table as shown in figure. The coefficient of friction between A and B is 0.2 and between B and the surface of the table is also 0.2. The maximum force F that can be applied on B horizontally, so that the block A does not slide over the block B is (Take, $g = 10 m/s^2$) **(2019 Main)**

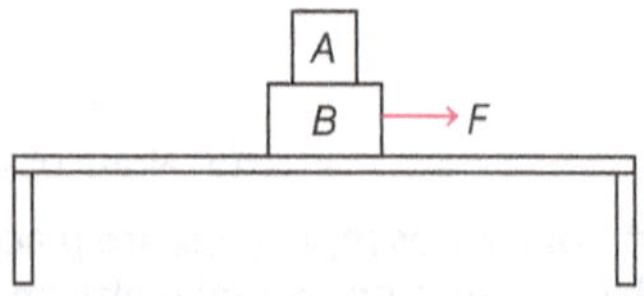

(a) 12 N (b) 16 N
(c) 8 N (d) 40 N

48. A block kept on a rough inclined plane as shown in the figure, remains at rest upto a maximum force 2 N down the inclined plane. The maximum external force up the inclined plane that does not move the block is 10 N. The coefficient of static friction between the block and the plane is (Take, $g = 10 m/s^2$) **(2019 Main)**

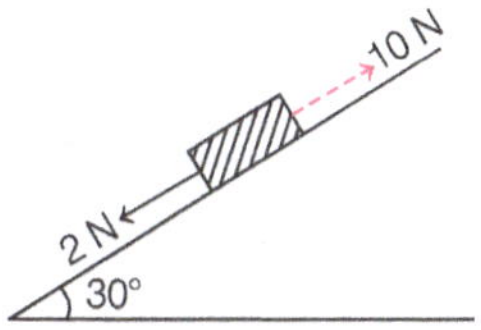

(a) $\frac{2}{3}$ (b) $\frac{\sqrt{3}}{2}$ (c) $\frac{\sqrt{3}}{4}$ (d) $\frac{1}{2}$

Multiple Correct Options

49. Two particles A and B, each of mass m are kept stationary by applying a horizontal force $F = mg$ on particle B as shown in figure. Then,

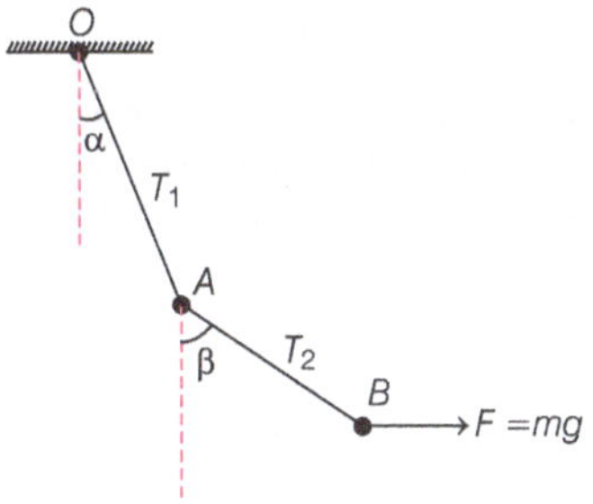

(a) $\tan\beta = 2\tan\alpha$ (b) $2T_1 = 5T_2$
(c) $\sqrt{2}\,T_1 = \sqrt{5}T_2$ (d) $\alpha = \beta$

50. As shown in the figure, A is a man of mass 60 kg standing on a block B of mass 40 kg kept on ground. The coefficient of friction between the feet of the man and the block is 0.3 and that between B and the ground is 0.2. If the person pulls the string with 125 N force, then

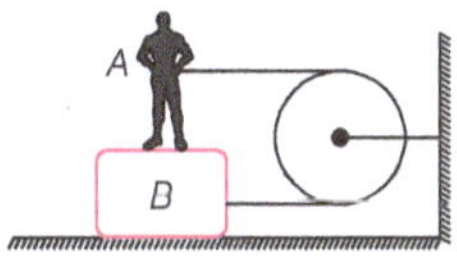

(a) B will slide on ground
(b) A and B will move with acceleration 0.5 ms^{-2}
(c) the force of friction acting between A and B will be 125 N
(d) the force of friction acting between B and ground will be 250 N

51. For the given situation shown in figure, choose the correct options (Take, $g = 10 ms^{-2}$)

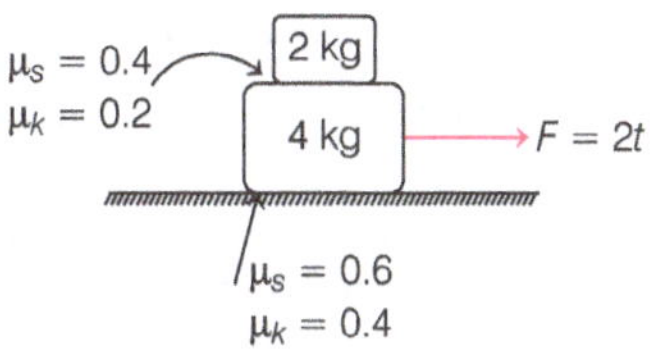

(a) At $t = 1$ s, force of friction between 2 kg and 4 kg is 2 N
(b) At $t = 1$ s, force of friction between 2 kg and 4 kg is zero
(c) At $t = 4$ s, force of friction between 4 kg and ground is 8 N
(d) At $t = 15$ s, acceleration of 2 kg is 1 ms^{-2}

52. Force acting on a block *versus* time graph is as shown in figure. Choose the correct options. (Take, $g = 10 ms^{-2}$)

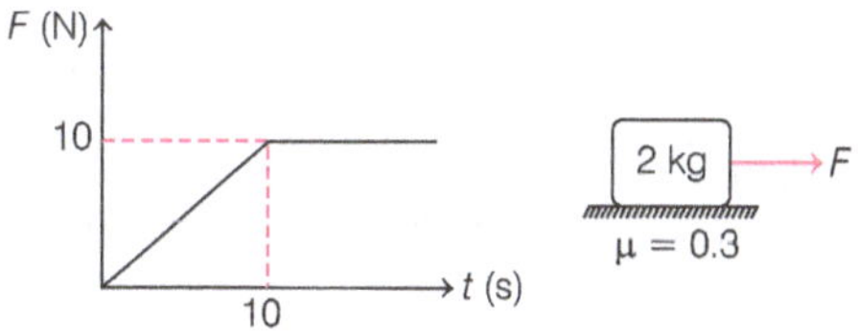

(a) At $t = 2$ s, force of friction is 2 N
(b) At $t = 8$ s, force of friction is 6 N
(c) At $t = 10$ s, acceleration of block is 2 ms^{-2}
(d) At $t = 12$ s, velocity of block is 8 ms^{-1}

53. For the situation shown in figure, mark the correct options.

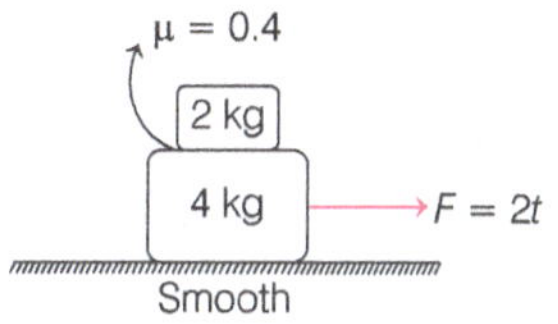

(a) At $t = 3$ s, pseudo force on 4 kg block applied from 2 kg is 4 N in forward direction
(b) At $t = 3$ s, pseudo force on 2 kg block applied from 4 kg is 2 N in backward direction
(c) Pseudo force does not make an equal and opposite pairs
(d) Pseudo force also makes a pair of equal and opposite forces

Passage Based Questions

Passage 1 (Q.54 to Q.55)

A lift with a mass 1200 kg is raised from rest by a cable with a tension 1350 kg-f. After sometime, the tension drops to 1000 kg-f and the lift comes to rest at a height of 25 m above its initial point. (Take, 1 kg-f = 9.8 N)

54. What is the height at which the tension changes?
(a) 10.8 m (b) 12.5 m
(c) 14.3 m (d) 16 m

55. What is the greatest speed of lift?
(a) 9.8 ms^{-1} (b) 7.5 ms^{-1}
(c) 5.92 ms^{-1} (d) None of these

Passage 2 (Q.56 to Q.60)

A man wants to slide down a block of mass m which is kept on a fixed inclined plane of inclination 30° as shown in the figure. Initially the block is not sliding.

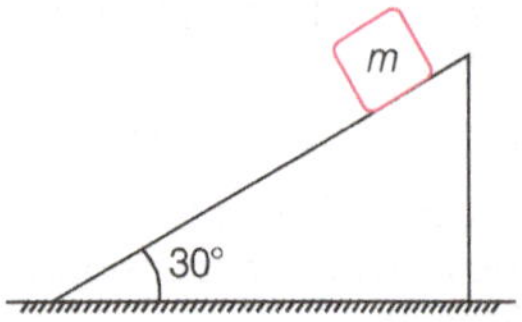

To just start sliding the man pushes the block down the incline with a force F. Now, the block starts accelerating. To move it downwards with constant speed, the man starts pulling the block with same force. Surfaces are such that ratio of maximum static friction to kinetic friction is 2. Now, answer the following questions.

56. What is the value of F ?

(a) $\frac{mg}{4}$ (b) $\frac{mg}{6}$ (c) $\frac{mg\sqrt{3}}{4}$ (d) $\frac{mg}{2\sqrt{3}}$

57. What is the value of μ_s, the coefficient of static friction?

(a) $\frac{4}{3\sqrt{3}}$ (b) $\frac{2}{3\sqrt{3}}$ (c) $\frac{3}{3\sqrt{3}}$ (d) $\frac{1}{2\sqrt{3}}$

58. If the man continues pushing the block by a force F, its acceleration would be

(a) $\frac{g}{6}$ (b) $\frac{g}{4}$ (c) $\frac{g}{2}$ (d) $\frac{g}{3}$

59. If the man wants to move the block up the incline, what minimum force is required to start the motion?

(a) $\frac{2}{3}mg$ (b) $\frac{mg}{2}$ (c) $\frac{7mg}{6}$ (d) $\frac{5mg}{6}$

60. What minimum force is required to move it up the incline with constant speed?

(a) $\frac{2}{3}mg$ (b) $\frac{mg}{2}$ (c) $\frac{7mg}{6}$ (d) $\frac{5mg}{6}$

Passage 3 (Q.61 to Q.62)

Blocks A and B shown in the figure are connected with a bar of negligible weight. A and B each has mass 170 kg, the coefficient of friction between A and the plane is 0.2 and that between B and the plane is 0.4. (Take, $g = 10\ ms^{-2}$)

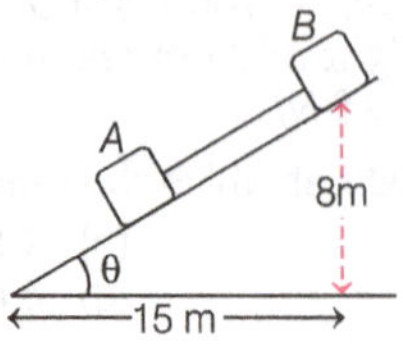

61. What is the total force of friction between the blocks and the plane?

(a) 900 N (b) 700 N
(c) 600 N (d) 300 N

62. What is the force acting on the connecting bar?

(a) 150 N (b) 100 N
(c) 75 N (d) 125 N

Match the Columns

63.

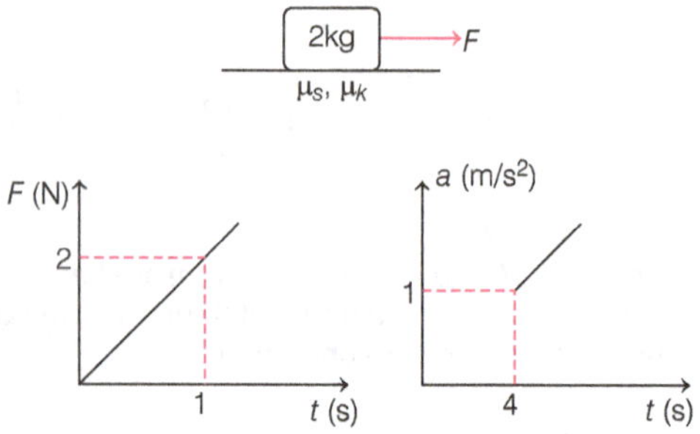

Force acting on a block *versus* time and acceleration *versus* time graph are as shown in figure. Taking value of $g = 10\text{ms}^{-2}$, match the following two columns.

	Column I		Colum n II
(a)	Coefficient of static friction	(p)	0.2
(b)	Coefficient of kinetic friction	(q)	0.3
(c)	Force of friction (in newton) at $t = 0.1$ s	(r)	0.4
(d)	Value of $\frac{a}{10}$, where a is acceleration of block (in m/s^2) at $t = 8$ s	(s)	0.5

64. There is no friction between blocks B and C. But ground is rough. Pulleys are smooth and massless and strings are light. For $F = 10$ N, whole system remains stationary. Match the following two columns. ($m_B = m_C = 1$ kg and $g = 10\ \text{ms}^{-2}$)

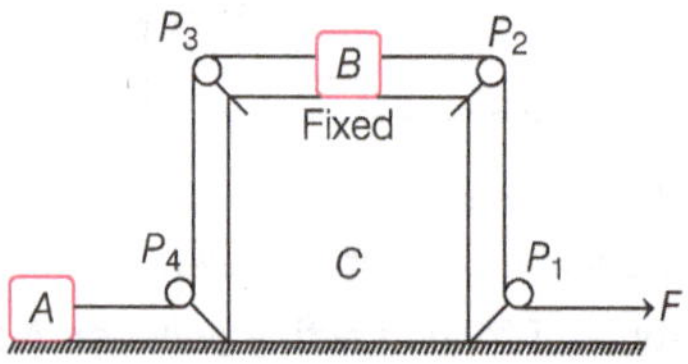

	Column I		Column II
(a)	Force of friction between A and ground	(p)	10 N
(b)	Force of friction between C and ground	(q)	20 N
(c)	Normal reaction on C from ground	(r)	5 N
(d)	Tension in string between P_3 and P_4	(s)	None of the above

65. For the situation shown in figure in Column I, the statements regarding friction forces are mentioned, while in Column II some information related to friction forces are given. Match the entries of Column I with the entries of Column II. (Take, $g = 10\text{ms}^{-2}$)

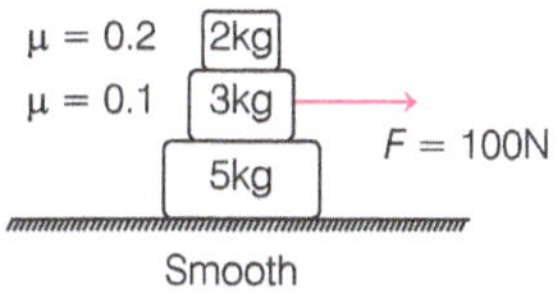

	Column I		Column II
(a)	Total friction force on 3 kg block is	(p)	towards right
(b)	Total friction force on 5 kg block is	(q)	towards left
(c)	Friction force on 2 kg block due to 3 kg block is	(r)	zero
(d)	Friction force on 3 kg block due to 5 kg block is	(s)	non-zero

66. A block of mass $m_1 = 1$ kg and another mass $m_2 = 2$ kg are placed together (see figure) on an inclined plane with angle of inclination θ. Various values of θ are given in List I. The coefficient of friction between the block m_1 and the plane is always zero. The coefficient of static and dynamic friction between the block m_2 and the plane are equal to $\mu = 0.3$.

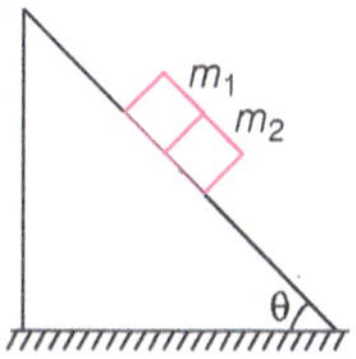

In List II expressions for the friction on the block m_2 are given. Match the correct expression of the friction in List II with the angles given in List I, and choose the correct option. The acceleration due to gravity is denoted by g.
[Useful information $\tan(5.5°) \approx 0.1$;
$\tan(11.5°) \approx 0.2$; $\tan(16.5°) \approx 0.3$]

	List I		List II
P.	$\theta = 5°$	1.	$m_2 g \sin\theta$
Q.	$\theta = 10°$	2.	$(m_1 + m_2) g \sin\theta$
R.	$\theta = 15°$	3.	$\mu m_2 g \cos\theta$
S.	$\theta = 20°$	4.	$\mu (m_1 + m_2) g \cos\theta$

(2014 Adv.)

Codes
(a) P-1, Q-1, R-1, S-3
(b) P-2, Q-2, R-2, S-3
(c) P-2, Q-2, R-2, S-4
(d) P-2, Q-2, R-3, S-3

Subjective Questions

67. Two beads of equal masses m are attached by a string of length $\sqrt{2}a$ and are free to move in a smooth circular ring lying in a vertical plane as shown in figure. Here, a is the radius of the ring. Find the tension and acceleration of B just after the beads are released to move.

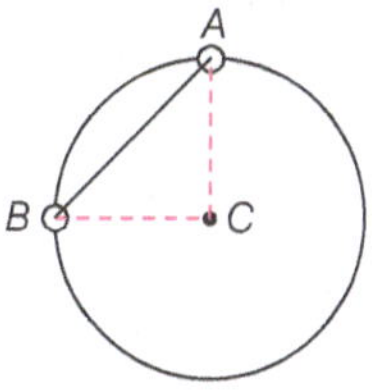

68. Two unequal masses of 1 kg and 2 kg are connected by a string going over a clamped light smooth pulley as shown in figure. The system is released from rest. The larger mass is stopped for a moment 1.0 s after the system is set in motion. Find the time elapsed before the string is tight again.

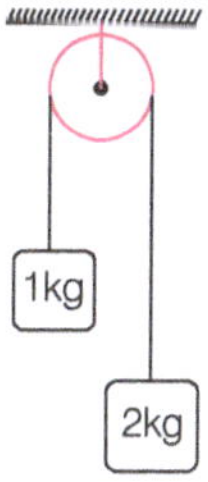

69. A 20 kg monkey has a firm hold on a light rope that passes over a frictionless pulley and is attached to a 20 kg bunch of bananas. The monkey looks upward, sees the bananas and starts to climb the rope to get them.
(a) As the monkey climbs, do the bananas move up, move down or remain at rest ?
(b) As the monkey climbs, does the distance between the monkey and the bananas decrease, increase or remain constant ?
(c) The monkey releases her hold on the rope. What happens to the distance between the monkey and the bananas while she is falling ?
(d) Before reaching the ground, the monkey grabs the rope to stop her fall. What do the bananas do?

70. A small marble is projected with a velocity of 10 m/s in a direction 45° from the y-direction on the smooth inclined plane. Calculate the magnitude v of its velocity after 2s. (Take, $g = 10\text{ m/s}^2$)

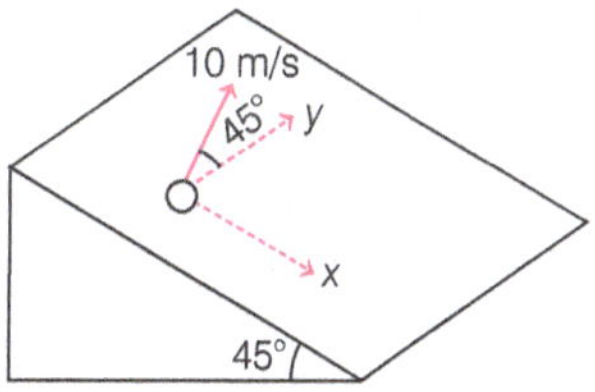

71. If block A of the pulley system is moving downward with a speed of 1 m/s while block C is moving up at 0.5 m/s, determine the speed of block B.

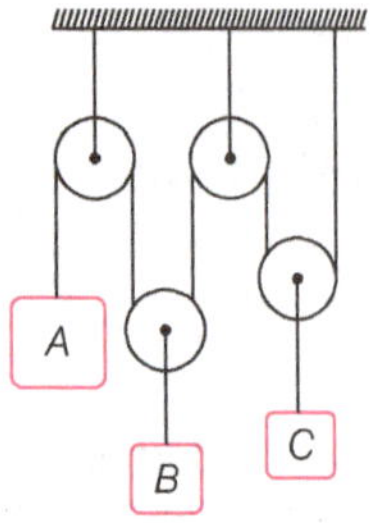

72. In the arrangement shown in figure the bodies have masses m_0, m_1 and m_2 the friction is absent, the masses of the pulleys and the threads are negligible. Find the acceleration of the body m_1. Look into possible cases. **(Irodov 1.73)**

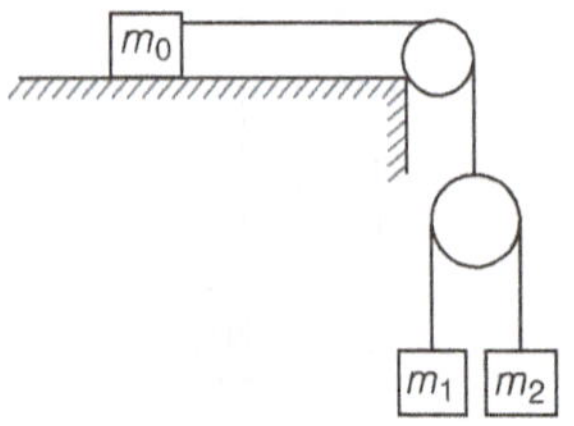

73. Find the accelerations of rod A and wedge B in the arrangement shown in figure, if the ratio of the mass of the wedge to that of the rod equals η and the friction between all contact surfaces is negligible. **(Irodov 1.77)**

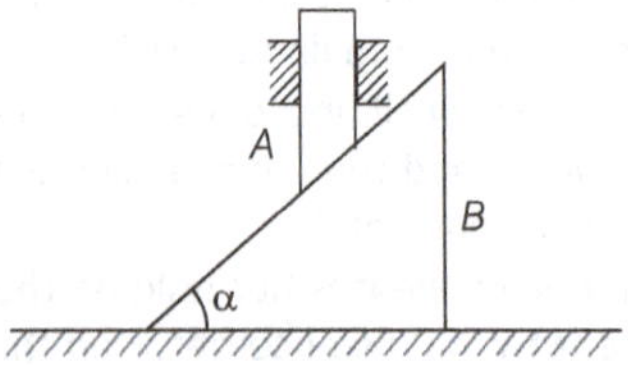

74. A pulley fixed to the ceiling of an elevator car carries a thread whose ends are attached to the loads of masses m_1 and m_2. The car starts going up with an acceleration $\mathbf{a}_0$. Assuming the masses of the pulley and the thread, as well as the friction, to be negligible. Find: **(Irodov 1.71)**

(a) the acceleration of the load m_1 relative to the elevator shaft and relative to the car

(b) the force exerted by the pulley on the ceiling of the car.

75. The collar A is free to slide along the smooth shaft B mounted in the frame. The plane of the frame is vertical. Determine the horizontal acceleration a of the frame necessary to maintain the collar in a fixed position on the shaft. (Take, $g = 9.8\ \text{m/s}^2$)

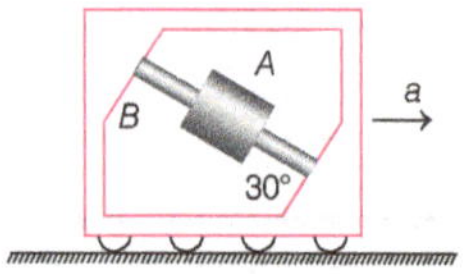

76. In the figure shown all surfaces are frictionless. What force F must by applied to M_1 to keep M_3 free from rising or falling?

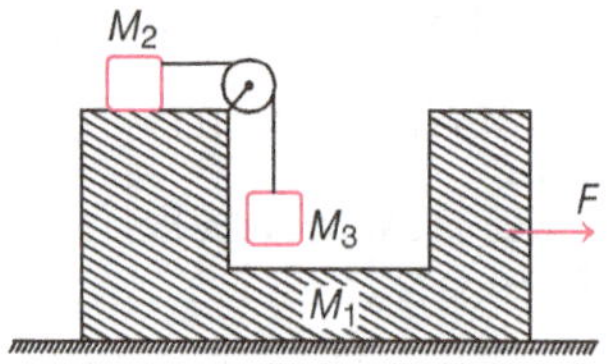

77. In figure $m_1 = 1$ kg and $m_2 = 4$ kg. Find the mass M of the hanging block which will prevent the smaller block from slipping over the triangular block. All the surfaces are frictionless and the strings and the pulleys are light.

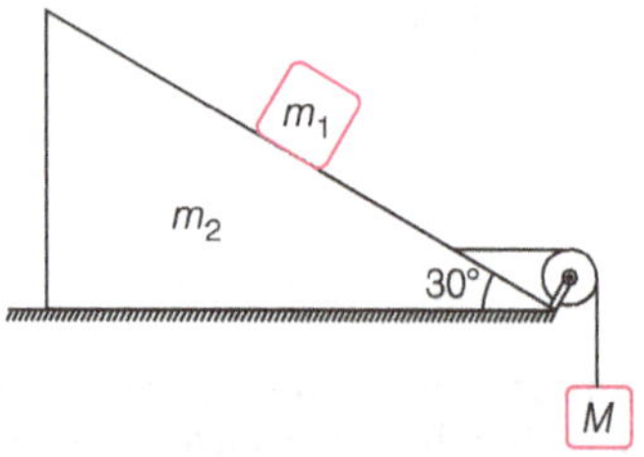

78. A 6 kg block B rests as shown on the upper surface of a 15 kg wedge A. Neglecting friction, determine immediately after the system is released from rest (a) the acceleration of A (b) the acceleration of B relative to A. (Take, $g = 10\ \text{m/s}^2$)

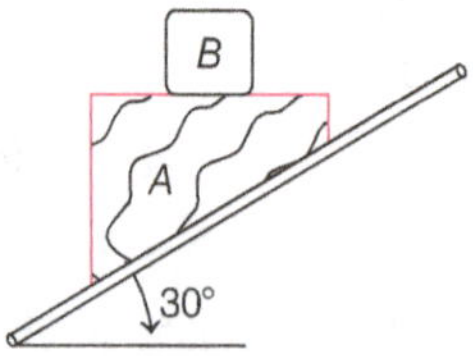

79. Block B has a mass m and is released from rest when it is on top of wedge A, which has a mass $3m$. Determine the tension in cord CD while B is sliding down A. Neglect friction.

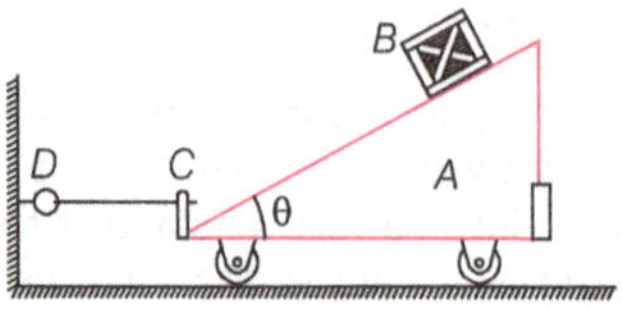

80. Determine the normal force the 10 kg crate A exerts on the smooth cart B, if the cart is given an acceleration of $a = 2\ \text{m/s}^2$ down the plane. Also, find the acceleration of the crate. Set $\theta = 30°$. (Take, $g = 10\ \text{m/s}^2$)

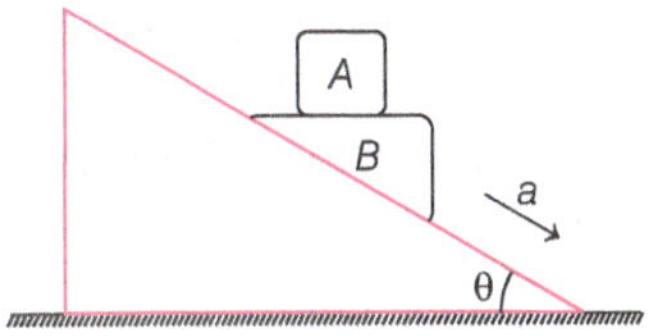

81. In the shown arrangement, both pulleys and the string are massless and all the surfaces are frictionless. Find the acceleration of the wedge.

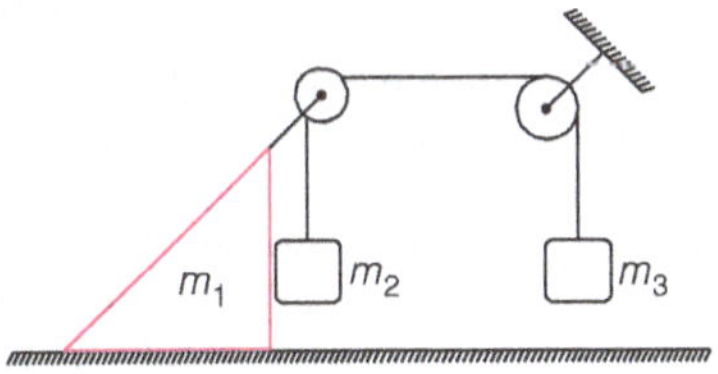

82. Neglect friction. Find accelerations of m, $2m$ and $3m$ as shown in the figure. The wedge is fixed.

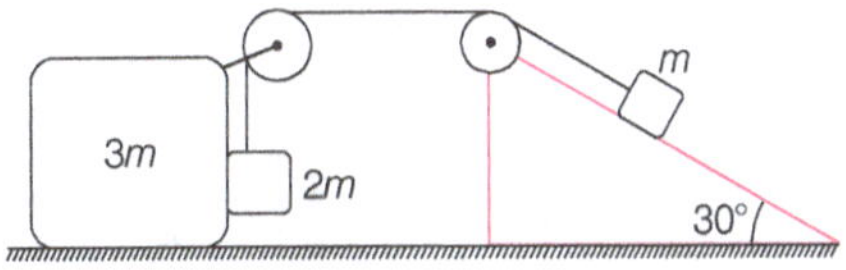

83. At the bottom edge of a smooth vertical wall, an inclined plane is kept at an angle of 45°. A uniform ladder of length l and mass M rests on the inclined plane against the wall such that it is perpendicular to the incline.

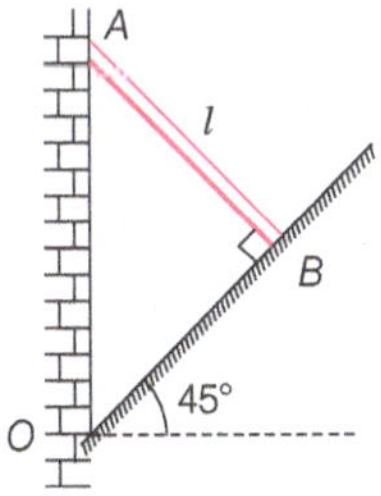

(a) If the plane is also smooth, which way will the ladder slide.

(b) What is the minimum coefficient of friction necessary so that the ladder does not slip on the incline?

84. Coefficients of friction between the flat bed of the truck and crate are $\mu_s = 0.85$ and $\mu_k = 0.8$. The coefficient of kinetic friction between the truck tires and the road surface is 0.9. If the truck stops from an initial speed of 15 m/s with maximum braking (wheels skidding). Determine where on the bed the crate finally comes to rest. (Take, $g = 10\ \text{m/s}^2$)

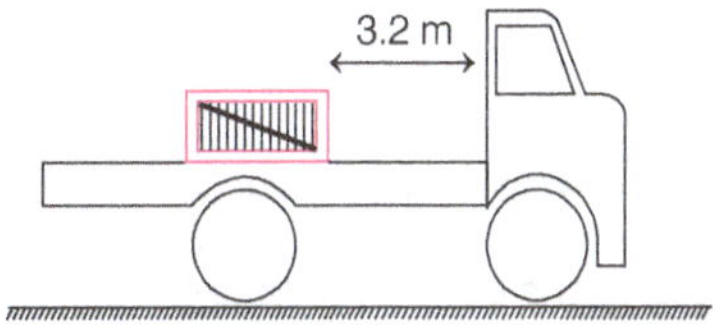

85. The 10 kg block is moving to the left with a speed of 1.2 m/s at time $t = 0$. A force F is applied as shown in the graph. After 0.2 s, the force continues at the 10 N level. If the coefficient of kinetic friction is $\mu_k = 0.2$. Determine the time t at which the block comes to a stop. (Take, $g = 10\ \text{m/s}^2$)

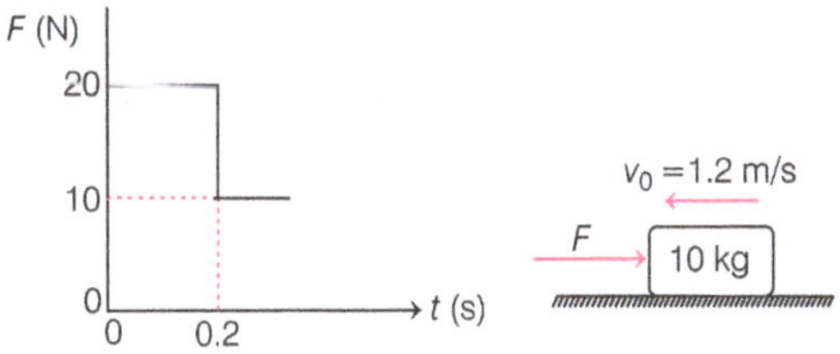

86. The 10 kg block is resting on the horizontal surface when the force F is applied to it for 7 s. The variation of F with time is shown. Calculate the maximum velocity reached by the block and the total time t during which the block is in motion. The coefficients of static and kinetic friction are both 0.50. (Take, $g = 9.8\ \text{m/s}^2$)

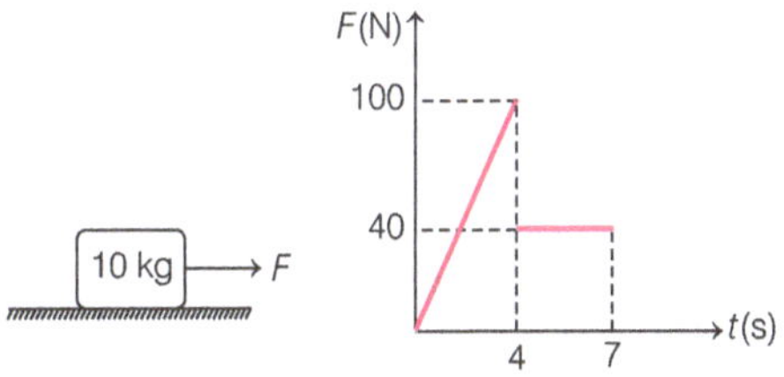

87. In figure, a crate slides down an inclined right-angled trough. The coefficient of kinetic friction between the crate and the trough is μ_k. What is the acceleration of the crate in terms of μ_k, θ and g?

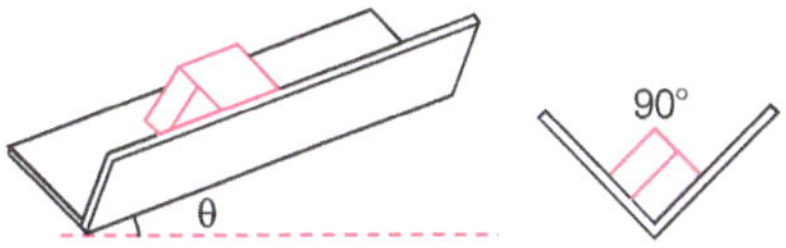

88. A heavy chain with a mass per unit length ρ is pulled by the constant force F along a horizontal surface consisting of a smooth section and a rough section. The chain is initially at rest on the rough surface with $x = 0$. If the coefficient of kinetic friction between the chain and the rough surface is μ_k, determine the velocity v of the chain when $x = L$. The force F is greater than $\mu_k \rho g L$ in order to initiate the motion.

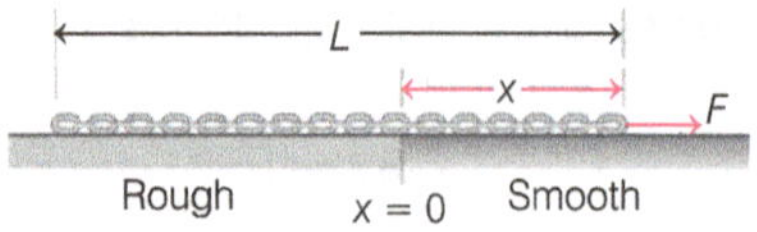

89. The figure shows an L shaped body of mass M placed on smooth horizontal surface. The block A is connected to the body by means of an inextensible string, which is passing over a smooth pulley of negligible mass. Another block B of mass m is placed against a vertical wall of the body. Find the minimum value of the mass of block A so that block B remains stationary relative to the wall. Coefficient of friction between the block B and the vertical wall is μ.

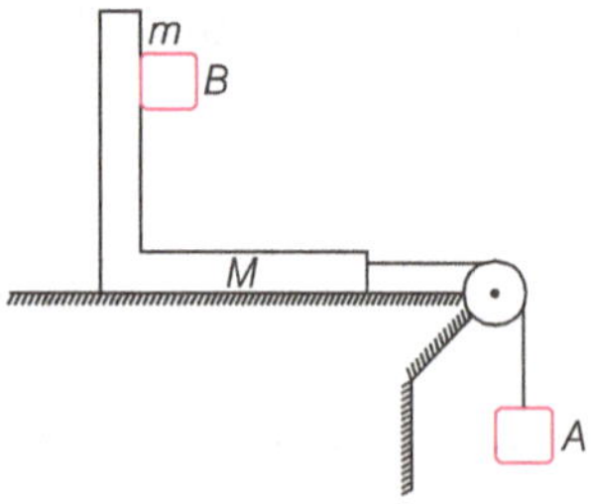

90. A small body A starts sliding down from the top of a wedge whose base is equal to $l = 2.10$ m. The coefficient of friction between the body and the wedge surface is $\mu = 0.140$. At what value of the angle α will the time of sliding be the least? What will it be equal to? **(Irodov 1.66)**

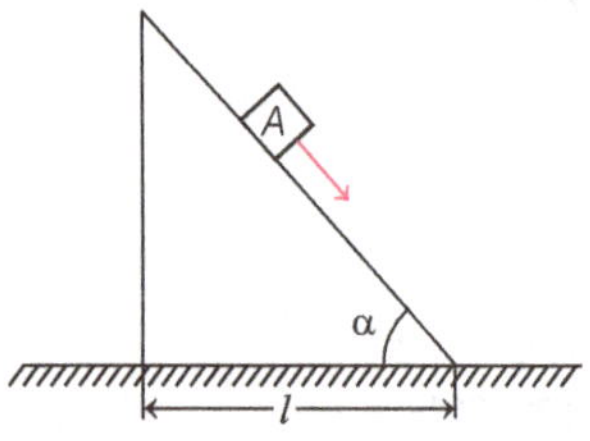

91. A bar of mass m is pulled by means of a thread up an inclined plane forming an angle α with the horizontal. The coefficient of friction is equal to μ. Find the angle β which the thread must form with the inclined plane for the tension of the thread to be minimum. What is it equal to? **(Irodov 1.67)**

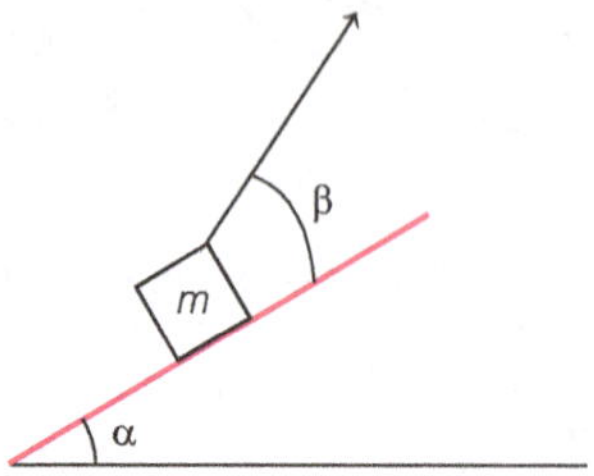

92. A bar of mass m resting on a smooth horizontal plane starts moving due to the force $F = mg/3$ of constant magnitude. In the process of its rectilinear motion the angle α between the direction of this force and the horizontal varies as $\alpha = ks$, where k is a constant and s is the distance traversed by the bar from its initial position. Find the velocity of the bar as a function of the angle α. **(Irodov 1.69)**

93. A horizontal plane with the coefficient of friction μ supports two bodies: a bar and an electric motor with a battery on a block. A thread attached to the bar is wound on the shaft of the electric motor. The distance between the bar and the electric motor is equal to l. When the motor is switched on, the bar whose mass is twice as great as that of the other body, starts moving with a constant acceleration a. How soon will the bodies collide? **(Irodov 1.70)**

94. In the arrangement shown in figure the masses of the wedge M and the body m are known. The appreciable friction exists only between the wedge and the body m, the friction coefficient being equal to μ. The masses of the pulley and the thread are negligible. Find the acceleration of the body m relative to the horizontal surface on which the wedge slides. **(Irodov 1.78)**

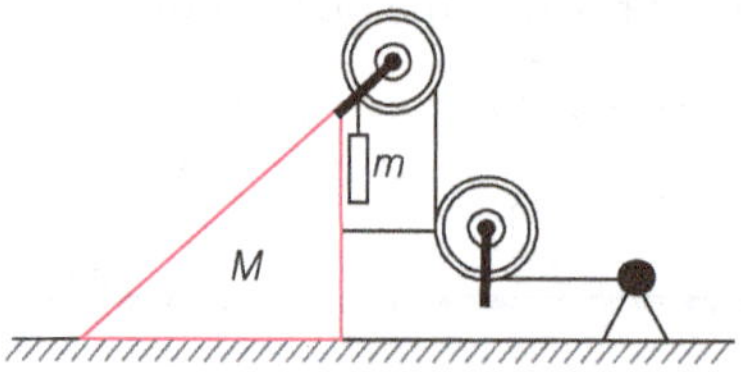

95. What is the minimum acceleration with which bar A should be shifted horizontally to keep bodies 1 and 2 stationary relative to the bar? The masses of the bodies are equal, and the coefficient of friction between the bar and the bodies is equal to μ. The masses of the pulley and the threads are negligible, the friction in the pulley is absent. **(Irodov 1.79)**

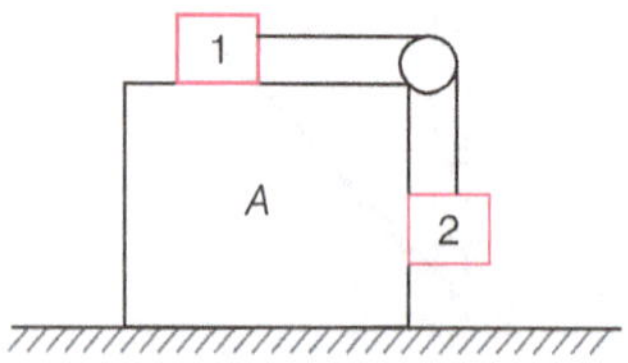

96. In the figure, the blocks A, B and C have masses 3 kg, 4 kg and 8 kg respectively. The coefficient of sliding friction between any two surfaces is 0.25. A is held at rest by a massless rigid rod fixed to the wall, while B and C are connected by a light flexible cord passing around a fixed frictionless pulley. Find the force F necessary to drag C along the horizontal surface to the left at a constant speed. Assume that the arrangement shown

in the figure. i.e. B on C and A on B, is maintained throughout.(Take $g = 10\ m/s^2$) **(1978 JEE)**

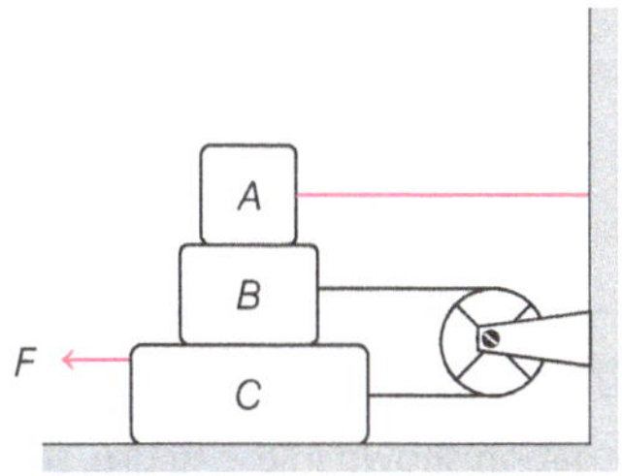

97. Two blocks connected by a massless string slides down an inclined plane having an angle of inclination of 37°. The masses of the two blocks are $M_1 = 4$ kg and $M_2 = 2$ kg respectively and the coefficients of friction of M_1 and M_2 with the inclined plane are 0.75 and 0.25 respectively. Assuming the string to be taut, find (a) the common acceleration of two masses and (b) the tension in the string. ($\sin 37° = 0.6$, $\cos 37° = 0.8$). (Take $g = 9.8\ m/s^2$) **(1979 JEE)**

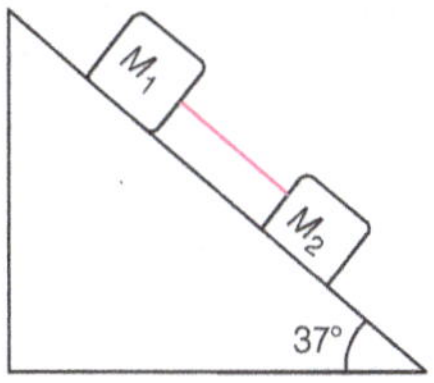

98. Masses M_1, M_2 and M_3 are connected by strings of negligible mass which passes over massless and frictionless pulleys P_1 and P_2 as shown in figure.

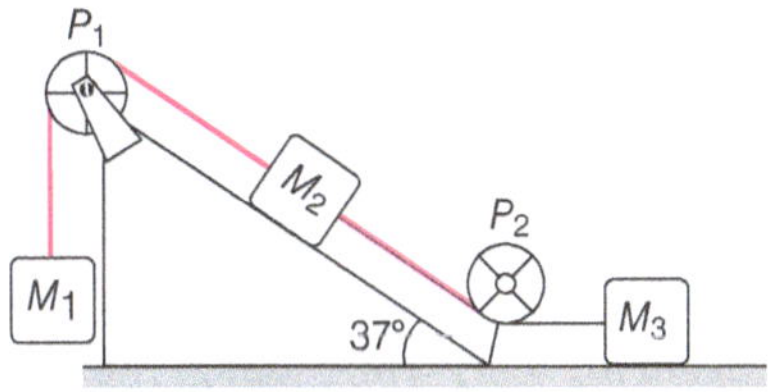

The masses move such that the portion of the string between P_1 and P_2 is parallel to the inclined plane and the portion of the string between P_2 and M_3 is horizontal. The masses M_2 and M_3 are 4.0 kg each and the coefficient of kinetic friction between the masses and the surfaces is 0.25. The inclined plane makes an angle of 37° with the horizontal.

If the mass M_1 moves downwards with a uniform velocity, find **(1981 JEE)**

(a) the mass of M_1,

(b) the tension in the horizontal portion of the string.

(Take $g = 9.8\ m/s^2$, $\sin 37° \cong 3/5$)

99. A block of mass m rests on a horizontal floor with which it has a coefficient of static friction μ. It is desired to make the body move by applying the minimum possible force F. Find the magnitude of F and the direction in which it has to be applied. **(1987 JEE)**

100. Block A of mass m and block B of mass $2m$ are placed on a fixed triangular wedge by means of a massless, inextensible string and a frictionless pulley as shown in figure. The wedge is inclined at 45° to the horizontal on both sides. The coefficient of friction between block A and the wedge is 2/3 and that between block B and the wedge is 1/3. If the blocks A and B are released from rest, find **(1997 JEE)**

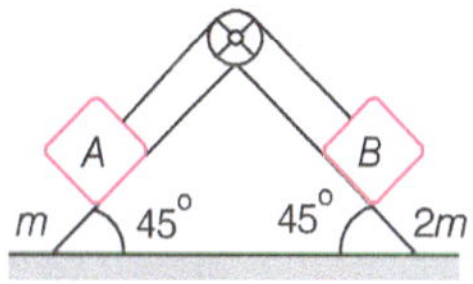

(a) the acceleration of A,

(b) tension in the string and

(c) the magnitude and direction of the force of friction acting on A.

101. A circular disc with a groove along its diameter is placed horizontally. A block of mass 1 kg is placed as shown. The coefficient of friction between the block and all surfaces of groove in contact is $\mu = 2/5$. The disc has an acceleration of 25 m/s². Find the acceleration of the block with respect to disc. Given $\cos\theta = 4/5$. **(2006 JEE)**

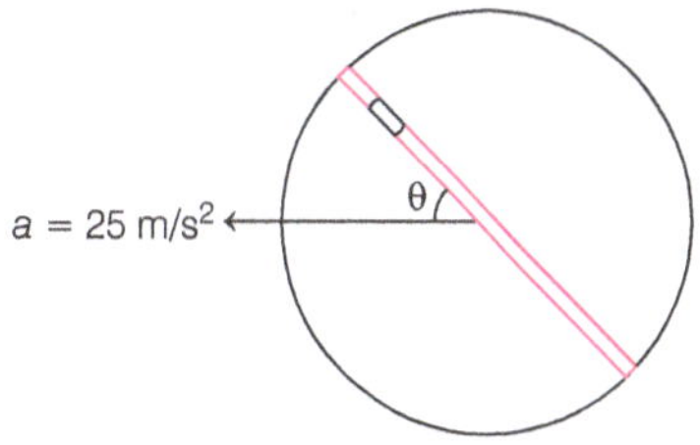

Level 2

Single Correct Option

102. Three identical smooth cylinders each of mass $\sqrt{3}$ kg are placed as shown in the figure. Find normal reaction between cylinder A and cylinder B.

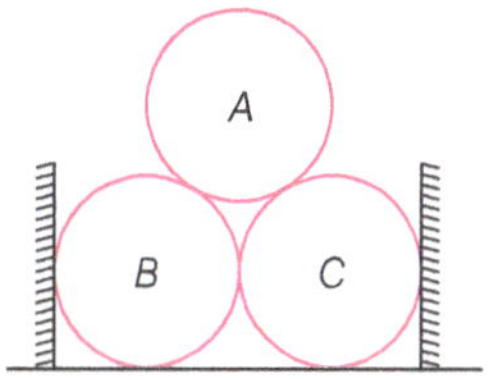

(a) 10 N (b) 20 N
(c) $10\sqrt{3}$ N (d) Cannot be calculated

103. A massive rod of length 1 m placed straight on a smooth horizontal surface is pulled longitudinally by a force F of 10N as shown in the figure.

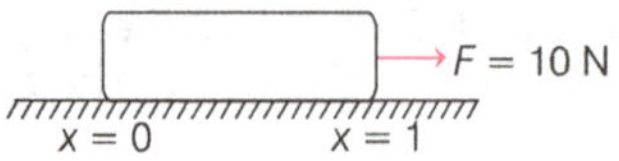

The tension in rod varies as $T = 10\sqrt{x}$. The linear mass density of rod is

(a) $\lambda \propto x$
(b) $\lambda \propto \frac{1}{\sqrt{x}}$
(c) $\lambda \propto \sqrt{x}$
(d) $\lambda \propto x^2$

104. In the figure, block moves downwards with velocity v_1, the wedge moves rightwards with velocity v_2. The correct relation between v_1 and v_2 is

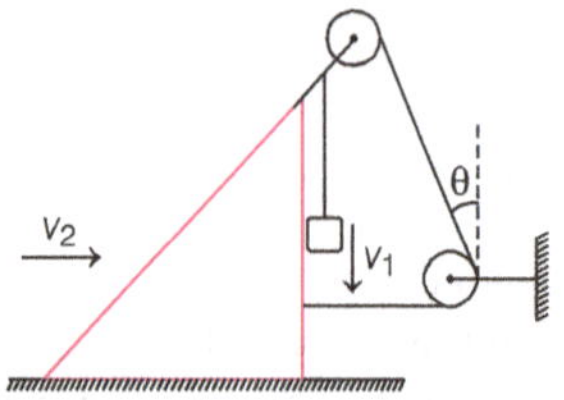

(a) $v_2 = v_1$
(b) $v_2 = v_1 \sin\theta$
(c) $2v_2 \sin\theta = v_1$
(d) $v_2(1 + \sin\theta) = v_1$

105. In the figure, the minimum value of a at which the cylinder starts rising up the inclined surface is

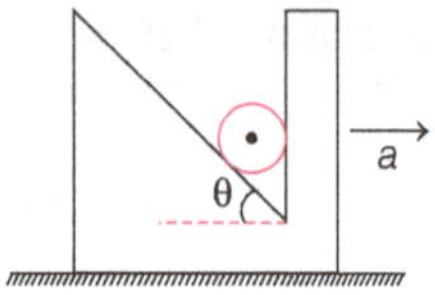

(a) $g \tan\theta$
(b) $g \cot\theta$
(c) $g \sin\theta$
(d) $g \cos\theta$

106. A sphere of radius R is in contact with a wedge. The point of contact is $\frac{R}{5}$ from the ground as shown in the figure. Wedge is moving with velocity 20 ms^{-1} towards left, then the velocity of the sphere at this instant will be

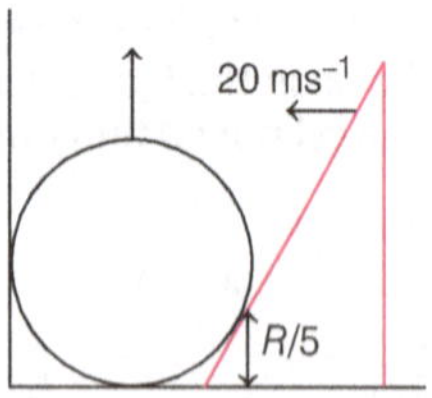

(a) 20 ms^{-1}
(b) 15 ms^{-1}
(c) 16 ms^{-1}
(d) 12 ms^{-1}

107. In the figure it is given that the velocity of lift is 2 ms^{-1} while string is winding on the motor shaft with velocity 2 ms^{-1} and shaft A is moving downward with velocity 2 ms^{-1} with respect to lift, then find out the velocity of block B.

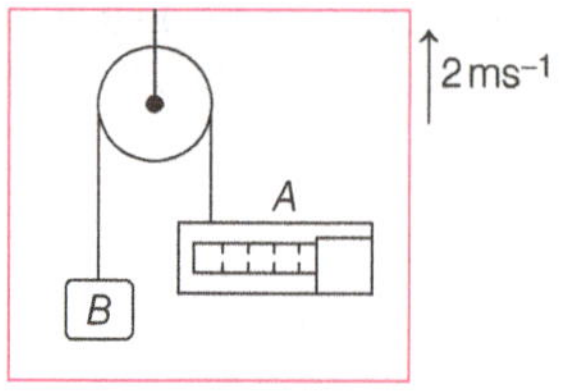

(a) 2 $\text{ms}^{-1}\uparrow$
(b) 2 $\text{ms}^{-1}\downarrow$
(c) 4 $\text{ms}^{-1}\uparrow$
(d) None of these

108. For the situation shown in the figure, find the speed of cylinder at the instant shown in figure.

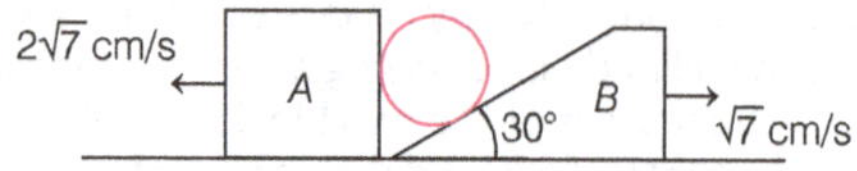

(a) $\sqrt{21}$ cm/s
(b) $\sqrt{17}$ cm/s
(c) 7 cm/s
(d) $2\sqrt{7}$ cm/s

109. A car begins from rest at time $t = 0$, and then accelerates along a straight track during the interval $0 < t \le 2$ s and thereafter with constant velocity as shown in the graph. A coin is initially at rest on the floor of the car. At $t = 1$ s, the coin begins to slip and it stops slipping at $t = 3$ s. The coefficient of static friction between the floor and the coin is
(Take, $g = 10\ \text{m/s}^2$)

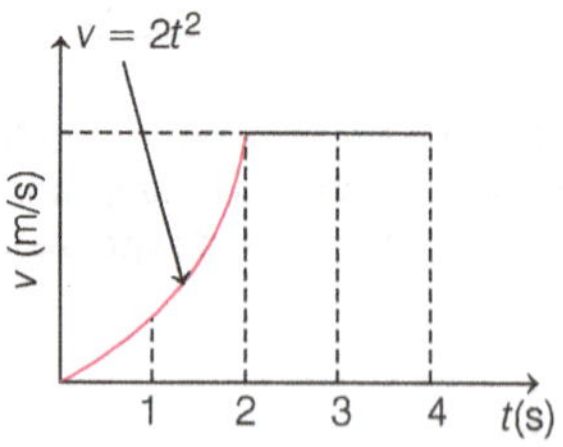

(a) 0.2
(b) 0.3
(c) 0.4
(d) 0.5

110. In the figure $m_A = m_B = m_c = 60$ kg. The coefficient of friction between C and ground is 0.5, B and ground is 0.3, A and B is 0.4. C is pulling the string with the maximum possible force without moving. Then, the tension in the string connected to A will be

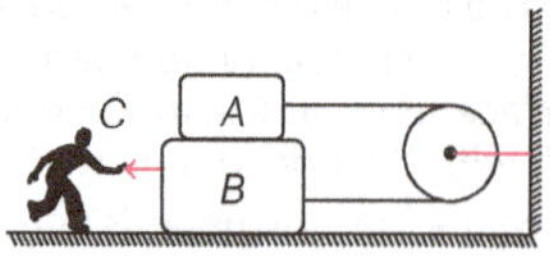

(a) 120 N
(b) 60 N
(c) 100 N
(d) zero

111. Three identical billiard balls each of mass 1 kg are placed as shown in the figure. The upper ball is smooth. Find minimum coefficient of static friction between lower balls and ground, so that balls remain stationary.

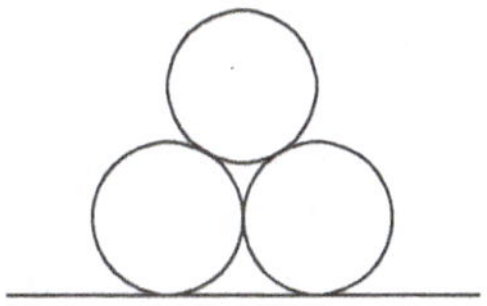

(a) $\frac{1}{\sqrt{3}}$ (b) $\frac{1}{3\sqrt{3}}$ (c) $\frac{2}{\sqrt{3}}$ (d) 0.5

112. A block of mass m is on an inclined plane of angle θ. The coefficient of friction between the block and the plane is μ and $\tan\theta > \mu$. The block is held stationary by applying a force P parallel to the plane. The direction of force pointing up the plane is taken to be positive. As P is varied from $P_1 = mg(\sin\theta - \mu\cos\theta)$ to $P_2 = mg(\sin\theta + \mu\cos\theta)$, the frictional force f *versus* P graph will look like **(2010 JEE)**

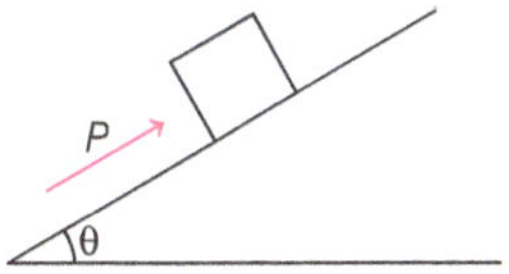

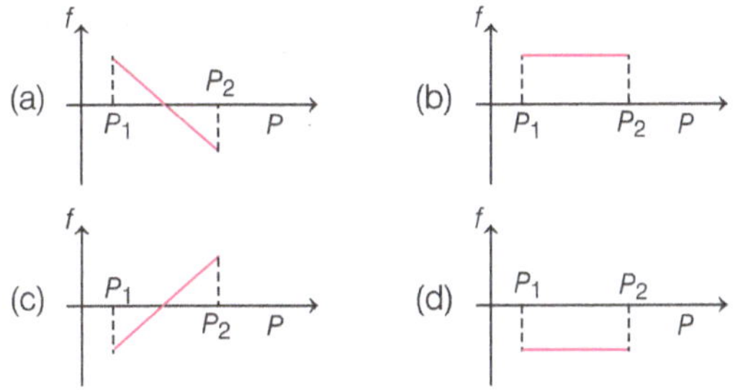

Multiple Correct Options

113. A small block of mass of 0.1 kg lies on a fixed inclined plane PQ which makes an angle θ with the horizontal. A horizontal force of 1 N acts on the block through its centre of mass as shown in the figure. The block remains stationary if (Take $g = 10\ \text{m/s}^2$) **(2012 JEE)**

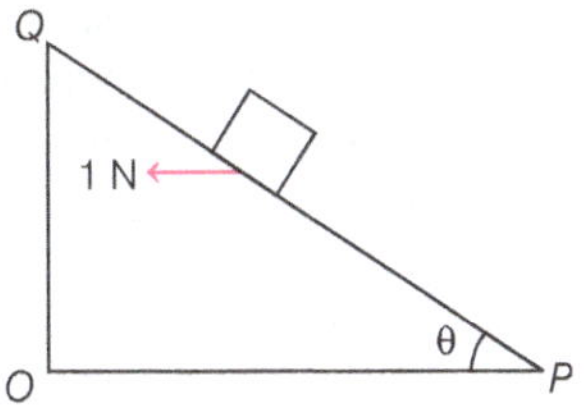

(a) $\theta = 45°$.

(b) $\theta > 45°$ and a frictional force acts on the block towards P

(c) $\theta > 45°$ and a frictional force acts on the block towards Q

(d) $\theta < 45°$ and a frictional force acts on the block towards Q

114. In the figure, a ladder of mass m is leaning against a wall. It is in static equilibrium making an angle θ with the horizontal floor. The coefficient of friction between the wall and the ladder is μ_1 and that between the floor and the ladder is μ_2. The normal reaction of the wall on the ladder is N_1 and that of the floor is N_2. If the ladder is about to slip, then **(2014 Adv.)**

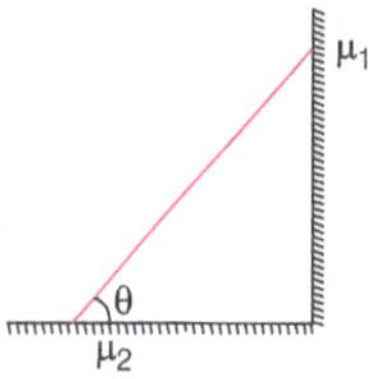

(a) $\mu_1 = 0, \mu_2 \neq 0$ and $N_2 \tan\theta = \frac{mg}{2}$

(b) $\mu_1 \neq 0, \mu_2 = 0$ and $N_1 \tan\theta = \frac{mg}{2}$

(c) $\mu_1 \neq 0, \mu_2 \neq 0$ and $N_2 = \frac{mg}{1+\mu_1\mu_2}$

(d) $\mu_1 = 0, \mu_2 \neq 0$ and $N_1 \tan\theta = \frac{mg}{2}$

Subjective Questions

115. A plank of mass M is placed on a rough horizontal surface and a constant horizontal force F is applied on it. A man of mass m runs on the plank. Find the range of acceleration of the man so that the plank does not move on the surface. Coefficient of friction between the plank and the surface is μ. Assume that the man does not slip on the plank.

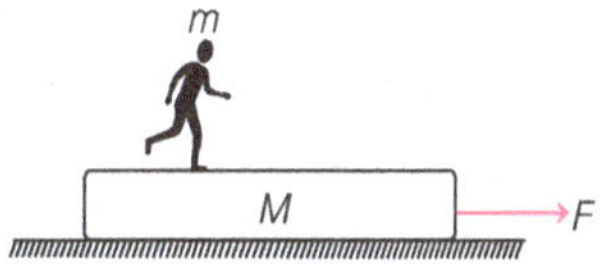

116. A package is at rest on a conveyor belt which is initially at rest. The belt is started and moves to the right for 1.3 s with a constant acceleration of $2\ \text{m/s}^2$. The belt then moves with a constant deceleration a_2 and comes to a stop after a total displacement of 2.2 m. Knowing that the coefficients of friction between the package and the belt are $\mu_s = 0.35$ and $\mu_k = 0.25$, determine (a) the deceleration a_2 of the belt, (b) the displacement of the package relative to the belt as the belt comes to a stop.
(Take, $g = 10\ \text{m/s}^2$)

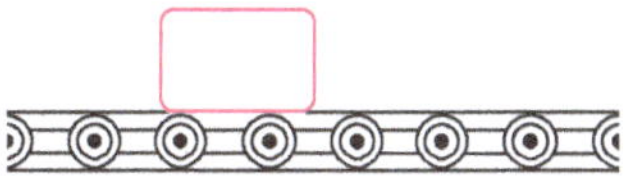

117. A small block of mass m is projected on a larger block of mass $10\,m$ and length l with a velocity v as shown in the figure. The coefficient of friction between the two blocks is μ_2 while that between the lower block and the ground is μ_1. Given that $\mu_2 > 11\,\mu_1$.

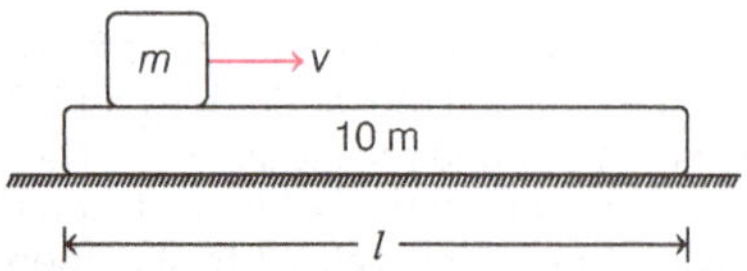

(a) Find the minimum value of v, such that the mass m falls off the block of mass $10\,m$.

(b) If v has this minimum value, find the time taken by block m to do so.

118. A particle of mass m and velocity v_1 in positive y-direction is projected on to a belt that is moving with uniform velocity v_2 in x-direction as shown in figure. Coefficient of friction between particle and belt is μ. Assuming that the particle first touches the belt at the origin of fixed x-y coordinate system and remains on the belt, find the coordinates (x, y) of the point where sliding stops.

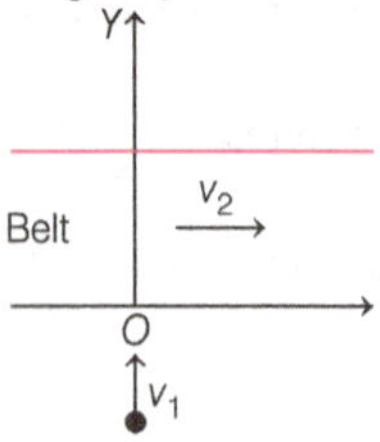

119. A fixed pulley carries a weightless thread with masses m_1 and m_2 at its ends. There is friction between the thread and the pulley. It is such that the thread starts slipping when the ratio $\dfrac{m_2}{m_1} = \eta_0$. Find the friction coefficient. **(Irodov 1.93)**

120. In the figure masses m_1, m_2 and M are 20 kg, 5 kg and 50 kg respectively. The coefficient of friction between M and ground is zero. The coefficient of friction between m_1 and M and that between m_2 and ground is 0.3. The pulleys and the strings are massless. The string is perfectly horizontal between P_1 and m_1 and also between P_2 and m_2.
The string is perfectly vertical between P_1 and P_2. An external horizontal force F is applied to the mass M. (Take, $g = 10\,m/s^2$) **(2000 JEE)**

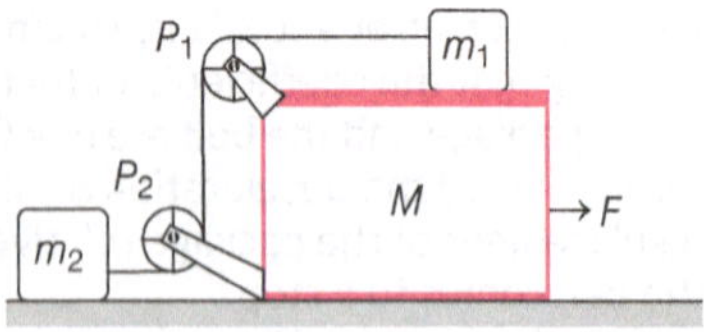

(a) Draw a free body diagram of mass M, clearly showing all the forces.

(b) Let the magnitude of the force of friction between m_1 and M be f_1 and that between m_2 and ground be f_2. For a particular force F it is found that $f_1 = 2f_2$. Find f_1 and f_2. Write equations of motion of all the masses. Find F, tension in the string and accelerations of the masses.

121. Two identical ladders are arranged as shown in the figure. Mass of each ladder is M and length L. The system is in equilibrium. Find direction and magnitude of frictional force acting at A or B. **(2005 JEE)**

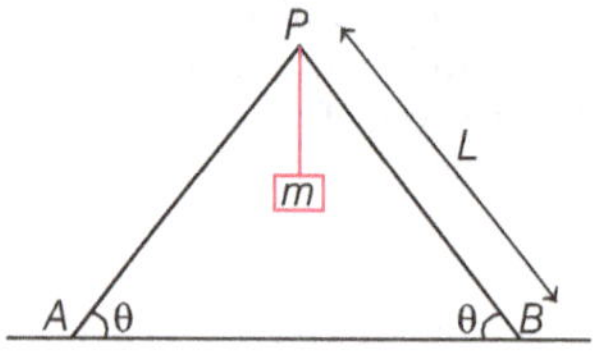

Integer Type Questions

122. Three smooth identical cylinders A, B and C are placed on smooth inclined plane as shown in the figure. The minimum value of $\tan\theta$ that prevent the arrangement from collapse is $\dfrac{1}{3\sqrt{n}}$. Find the value of n.

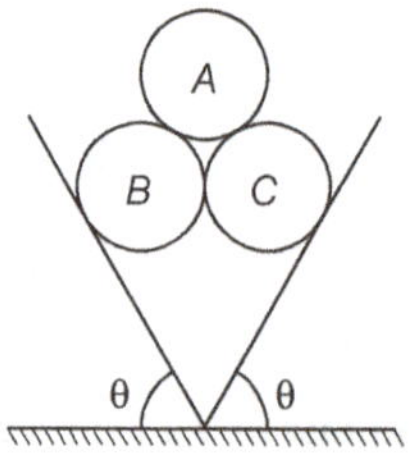

123. A block is at rest in gravity free space. It is acted by four forces, 3 N in East direction, 1 N in South direction, 1 N in downward direction and an unknown force $\mathbf{F}_0$. Due to these forces, the block starts to move in South-East direction. The minimum possible value of F_0 is $\sqrt{n}$ newton. Find the value of n.

124. A force $F = bt - cs$, (here $b = 1$ N/s and $c = 1$ N/m) is acting on a block of mass 1 kg placed on a smooth horizontal surface. During motion, the acceleration of the block is $a = a_0 \sin\omega t$. Find the value of ω.

Answers

Level 1

1. (d) **2.** (b) **3.** (c) **4.** (b) **5.** (b) **6.** (b) **7.** (b) **8.** (b) **9.** (c) **10.** (a)

11. (a) **12.** (b) **13.** (b) **14.** (d) **15.** (b) **16.** (b) **17.** (b) **18.** (b) **19.** (a) **20.** (a)

21. (d) **22.** (d) **23.** (d) **24.** (c) **25.** (a) **26.** (c) **27.** (c) **28.** (d) **29.** (b) **30.** (a)

31. (c) **32.** (b) **33.** (c) **34.** (d) **35.** (c) **36.** (d) **37.** (c) **38.** (b) **39.** (c) **40.** (c)

41. (a) **42.** (a) **43.** (c) **44.** (d) **45.** (b) **46.** (a) **47.** (b) **48.** (b) **49.** (a, c) **50.** (c, d)

51. (b, c) **52.** (a, b, c, d) **53.** (b, c) **54.** (c) **55.** (c) **56.** (b) **57.** (a) **58.** (d) **59.** (c)

60. (d) **61.** (a) **62.** (a) **63.** (a-r; b-q; c-p; d-s) **64.** (a-p; b-s; c-q; d-p) **65.** (a-q,s; b-p, s; c-p, s; d-q, s)

66. (d) **67.** $mg/\sqrt{2}, g/2$ **68.** $\frac{1}{3}$ s **69.** (a) Move up (b) Constant (c) Constant (d) Stop **70.** 10 ms^{-1}

71. (Zero) **72.** $\left[\frac{4m_1m_2 + m_0(m_1 - m_2)}{4m_1m_2 + m_0(m_1 + m_2)}\right]g$ **73.** $a_A = \frac{g}{1 + \eta\cot^2\alpha}$, $a_B = \frac{g}{\tan\alpha + \eta\cot\alpha}$

74. (a) $a_r = \left(\frac{m_1 - m_2}{m_1 + m_2}\right)(g + a_0)$, $a_{m_1} = \frac{(m_1 - m_2)g - 2m_2a_0}{(m_1 + m_2)}$ (b) $F = 4m_1m_2(g + a_0)/(m_1 + m_2)$ **75.** 5.66 m/s^2

76. $(M_1 + M_2 + M_3)\frac{M_3}{M_2}g$ **77.** 6.83 kg **78.** (a) 6.36 m/s^2 (b) 5.5 m/s^2 **79.** $\left(\frac{mg}{2}\sin 2\theta\right)$ **80.** 90 N, 1 m/s^2

81. $\frac{2m_1m_3g}{(m_2 + m_3)(m_1 + m_2) + m_2m_3}$ **82.** $\frac{13}{34}g, \frac{\sqrt{397}g}{34}, \frac{3}{17}g$ **83.** (a) Clockwise (b) 1/3

84. Displacement of crate relative to truck is 1.56 m **85.** $t = 0.33$s **86.** 5.2 m/s, 5.55 s **87.** $g(\sin\theta - \sqrt{2}\mu_k\cos\theta)$

88. $\sqrt{\frac{2F}{\rho} - \mu_k gL}$ **89.** $m_A = \frac{M + m}{\mu - 1}$ but $\mu > 1$ **90.** $\alpha \approx 49°$, $t_{min} = 1.0$ s

91. $\beta = \tan^{-1}(\mu)$, $T_{min} = \mu mg\cos\alpha + \frac{mg\sin\alpha}{\sqrt{1 + \mu^2}}$ **92.** $v = \sqrt{\frac{2g\sin\alpha}{3k}}$ **93.** $t = \sqrt{\frac{2l}{3a + \mu g}}$ **94.** $\frac{\sqrt{2}g}{2 + \mu + \frac{M}{m}}$

95. $g\left(\frac{1 - \mu}{1 + \mu}\right)$ **96.** 80 N **97.** (a) 1.3 m/s^2 (b) 5.2 N **98.** (a) 4.2 kg (b) 9.8 N

99. $F_{min} = \mu mg/\sqrt{1 + \mu^2}$ at angle $\theta = \tan^{-1}(\mu)$ from horizontal

100. (a) Zero (b) $2\sqrt{2}mg/3$ (c) $f_A = mg/3\sqrt{2}$ (down the plane) **101.** 10 m/s^2

Level 2

102. (a) **103.** (b) **104.** (d) **105.** (a) **106.** (b) **107.** (d) **108.** (c) **109.** (c) **110.** (d)

111. (b) **112.** (a) **113.** (a, c) **114.** (c, d) **115.** $\frac{F}{m} - \frac{\mu(M + m)}{m}g \le a \le \frac{F}{m} + \frac{\mu(M + m)}{m}g$

116. (a) 6.63 m/s^2 (b) 0.33 m **117.** (a) $v_{min} = \sqrt{\frac{22(\mu_2 - \mu_1)gl}{10}}$ (b) $t = \sqrt{\frac{20l}{11g(\mu_2 - \mu_1)}}$

118. $x = \frac{v_2\sqrt{v_1^2 + v_2^2}}{2\mu g}$, $y = \frac{v_1\sqrt{v_1^2 + v_2^2}}{2\mu g}$ **119.** $\mu = \frac{\ln(\eta_0)}{\pi}$

120. (b) $f_1 = 30$N, $f_2 = 15$ N, $F = 60$N, $T = 18$N and $a = 3/5\ m/s^2$ **121.** $f = \left(\frac{M + m}{2}\right)g\cot\theta$ **122.** (3) **123.** (3)

124. 1 unit

See hints on pages 437-461.

Hints & SOLUTIONS

Module 1.

1. W_1 = weight of cylinder

W_2 = weight of plank

N_1 = normal reaction between cylinder and plank

N_2 = normal reaction on cylinder from ground

N_3 = normal reaction on plank from ground

f_1 = force of friction on cylinder from ground

f_2 = force of friction on plank from ground

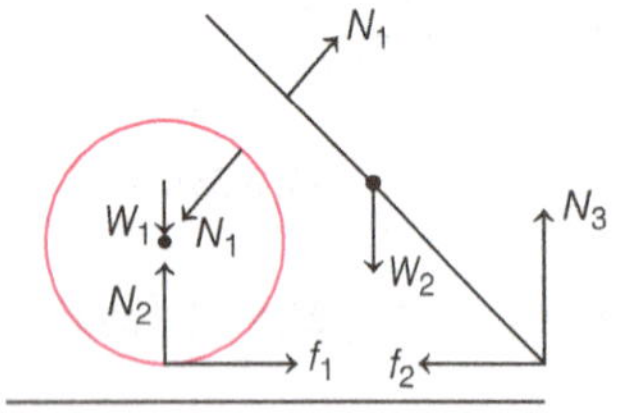

2. N = normal reactions

W = weights

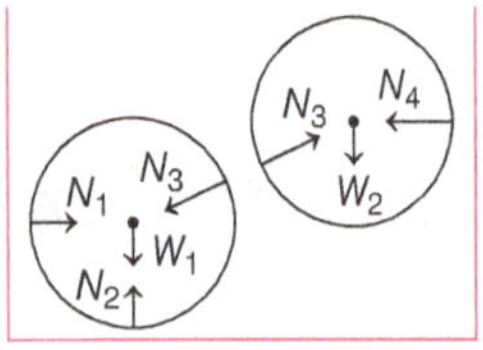

3.

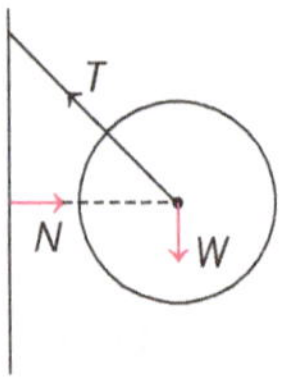

4.

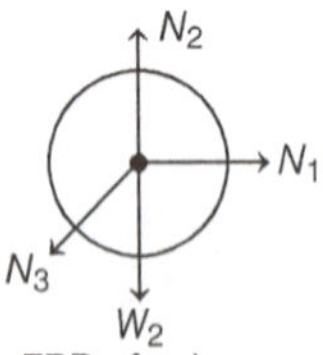

FBD of sphere
(i)

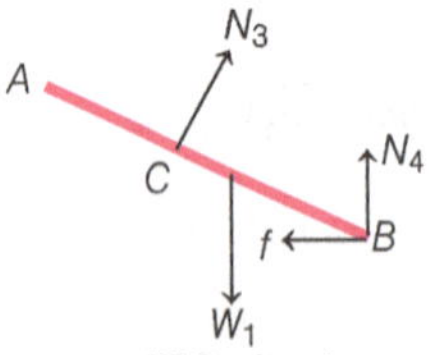

FBD of rod
(ii)

In the figure

N_1 = normal reaction between sphere and wall

N_2 = normal reaction between sphere and ground

N_3 = normal reaction between sphere and rod and

N_4 = normal reaction between rod and ground

f = force of friction between rod and ground

No friction will act between sphere and ground because horizontal component of normal reaction from rod (on sphere) will be balanced by the horizontal normal reaction from the wall.

5.

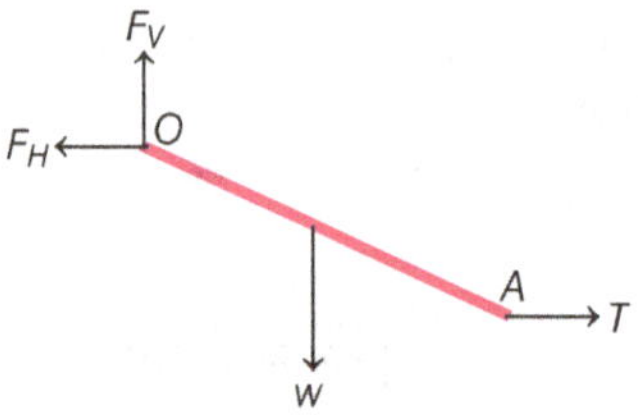

FBD of rod

In the figure

T = tension in the string, W = weight of the rod,

F_V = vertical force exerted by hinge on the rod,

F_H = horizontal force exerted by hinge on the rod

6. In the figure

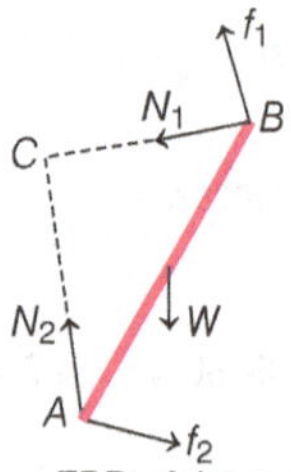

FBD of the rod

N_1 = normal reaction at B, f_1 = force of friction at B,

N_2 = normal reaction at A, f_2 = force of friction at A,

W = weight of the rod.

Module 2.

1. $T_A = 10\,g = 100\,\text{N}$

$$T_B \cos 30° = T_A$$

$$\therefore \quad T_B = \frac{200}{\sqrt{3}}\,\text{N}$$

$$T_B \sin 30° = T_C$$

$$\therefore \quad T_C = \frac{100}{\sqrt{3}}\,\text{N}$$

2. (i) Sphere is in equilibrium, therefore vector sum of all three forces acting on sphere will be zero.

$$\therefore \quad \mathbf{T} = -(\mathbf{W} + \mathbf{R})$$

(ii) $\mathbf{T} = -(\mathbf{W} + \mathbf{R})$

or $|\mathbf{T}| = -|(\mathbf{W} + \mathbf{R})|$

$\therefore \quad T = \sqrt{W^2 + R^2}$ as angle between **W** and **R** is 90°.

(iv)

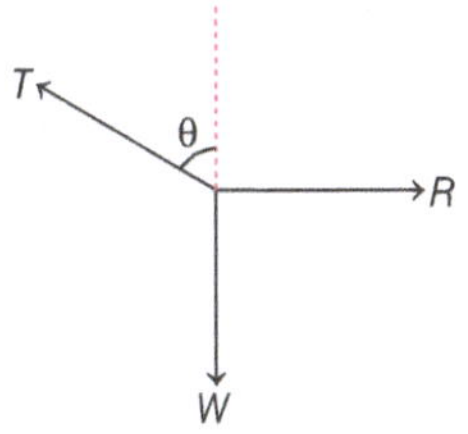

$$T\sin\theta = R \qquad \text{...(i)}$$

$$T\cos\theta = W \qquad \text{...(ii)}$$

Dividing Eq. (i) by Eq. (ii), we get

$$\tan\theta = \frac{R}{W}$$

or $$R = W\tan\theta$$

3. From free body diagram of rod,

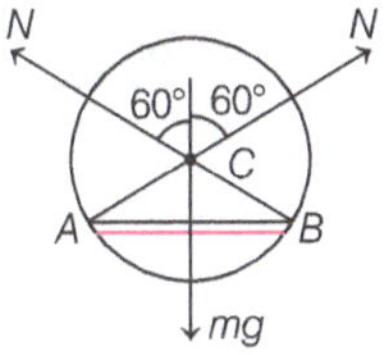

$$2N\cos 60° = mg$$

$\therefore$ $$N = \frac{mg}{2\cos 60°}$$

$\Rightarrow$ $$N = \frac{1\times 10}{2\times\frac{1}{2}} \qquad (\because g = 10\,\text{m/s}^2)$$

$$= 10\text{ N}$$

4.

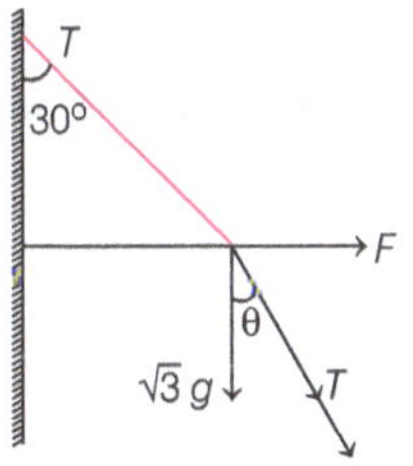

From the free body diagram shown above,

$$\cos\theta = \frac{\sqrt{3}g}{T}$$

$\therefore$ $$\frac{\sqrt{3}}{2} = \frac{\sqrt{3}g}{T} \qquad (\because \theta = 30°)$$

$\Rightarrow$ $$T = 20\text{ N}$$

5. $T_1\cos 45° = W$ and $T_1\sin 45° = 30\text{ N}$

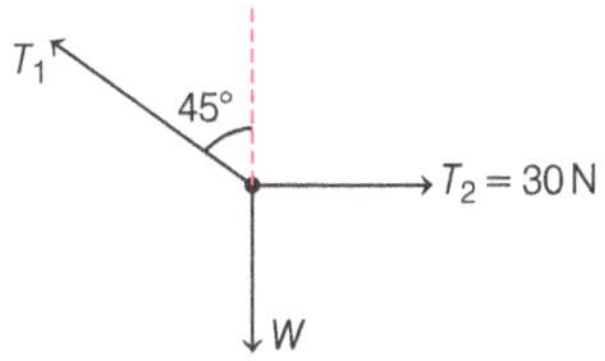

$\therefore$ $$W = 30\text{ N}$$ **Ans.**

6. $N_A\cos 30° = 500\text{ N}$ ...(i)

$$N_A\sin 30° = N_B \qquad \text{...(ii)}$$

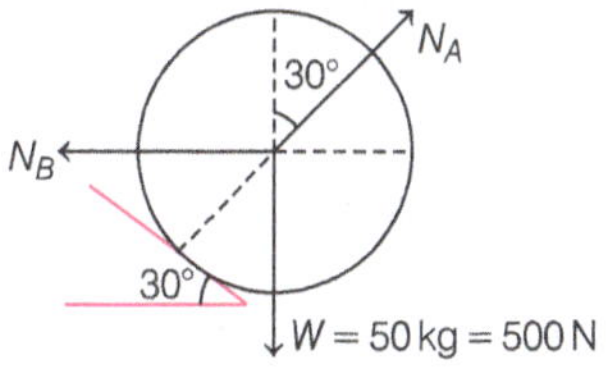

On solving these two equations, we get

$$N_A = \frac{1000}{\sqrt{3}}\text{ N}$$

and $$N_B = \frac{500}{\sqrt{3}}\text{ N}$$ **Ans.**

7. (a) **At P**

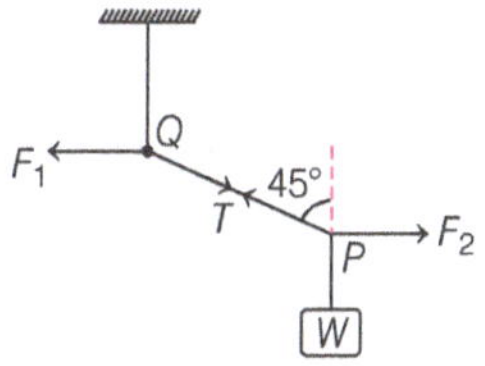

$$F_2 = T\cos 45° = \frac{T}{\sqrt{2}} \qquad \text{...(i)}$$

$$W = T\cos 45° = \frac{T}{\sqrt{2}} \qquad \text{...(ii)}$$

At Q $$F_1 = \frac{T}{\sqrt{2}} \qquad \text{...(iii)}$$

From these three equations, we can see that

$$F_1 = F_2 = W = \frac{T}{\sqrt{2}}$$

$$= \frac{60}{\sqrt{2}} = 30\sqrt{2}\text{ N}$$ **Ans.**

8. Along horizontal

$$F_1 + 1\cos 45° = 2\sin 45°$$

$$F_1 = \frac{2}{\sqrt{2}} - \frac{1}{\sqrt{2}} = \frac{1}{\sqrt{2}}$$

Along vertical

$$F_2 = 1\sin 45° + 2\sin 45°$$

$$F_2 = 3\sin 45° = \frac{3}{\sqrt{2}}$$

So, $\frac{F_1}{F_2} = \frac{1}{3}$, therefore $x = 3$

9. Consider moment of forces about the point of application of force.

10. $W = 100\text{ N}$

Σ(forces in horizontal direction) $= 0$

$\therefore \quad T + f\cos 30° = N\sin 30° \quad$ ...(i)

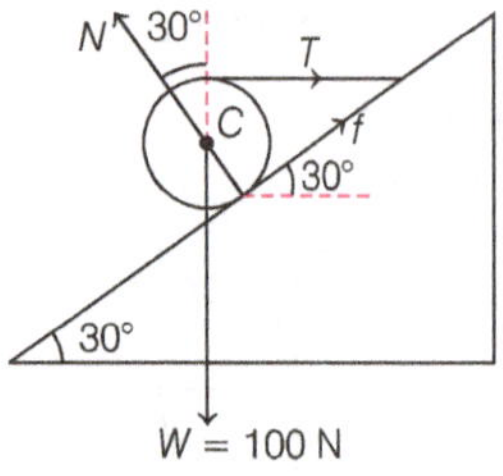

Σ(Forces in vertical direction) $= 0$

$\therefore \quad N\cos 30° + f\sin 30° = 100 \quad$ ...(ii)

Σ(moment of all forces about C) $= 0$

$\therefore \quad (T)(R) = f(R)$

or $\quad T = f \quad$ ...(iii)

On solving these three equations, we get

$T = f = 26.8\,\text{N}$

and $\quad N = 100\,\text{N} \quad$ **Ans.**

11. $N = F = 40\,\text{N}$

Net moment about $C = 0$

$\therefore$ Anti-clockwise moment of f

= clockwise moment of N

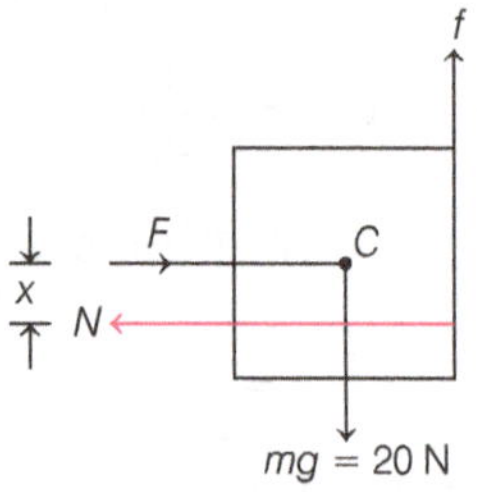

$$\therefore \quad (20)\left(\frac{20}{2}\right) = (40)\cdot x$$

or $\quad x = 5\text{ cm} \quad$ **Ans.**

12. Free body diagram of m is

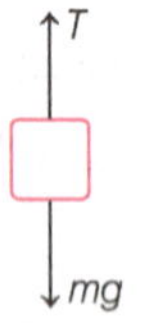

$T = mg \quad$ ...(i)

Free body diagram of mass $\sqrt{2}\,m$ is

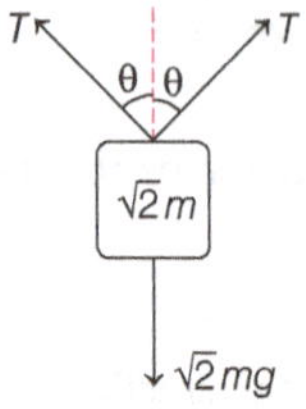

$2T\cos\theta = \sqrt{2}\,mg \quad$ ...(ii)

Dividing Eq. (ii) by Eq. (i), we get

$$\cos\theta = \frac{1}{\sqrt{2}} \text{ or } \theta = 45°$$

13. Free body diagram of pulley is shown in figure. Pulley is in equilibrium under four forces. Three forces as shown in figure and the fourth, which is equal and opposite to the resultant of these three forces, is the force applied by the clamp on the pulley (say F).

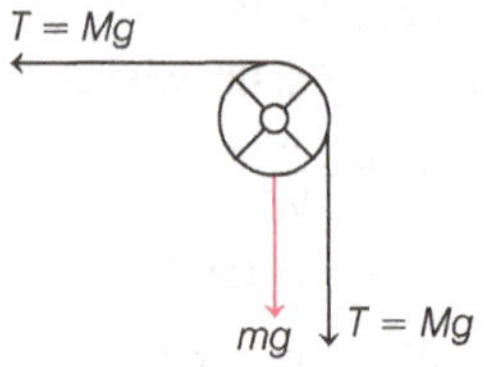

Resultant R of these three forces is

$$R = \left(\sqrt{(M+m)^2 + M^2}\right)g$$

Therefore, the force F is equal and opposite to R as shown in figure.

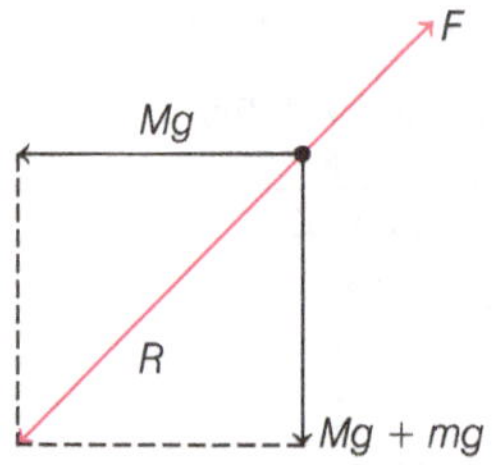

$$\therefore \quad F = \left(\sqrt{(M+m)^2 + M^2}\right)g$$

14. This is the equilibrium of coplanar forces. Hence,

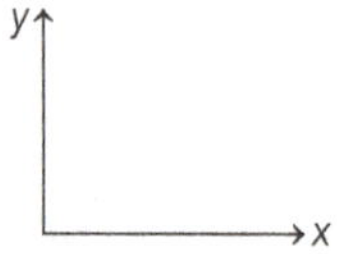

$\Sigma F_x = 0$

$\therefore \quad F = N$

$\therefore \quad \Sigma F_y = 0,\ f = mg$

$\Sigma \tau_c = 0$

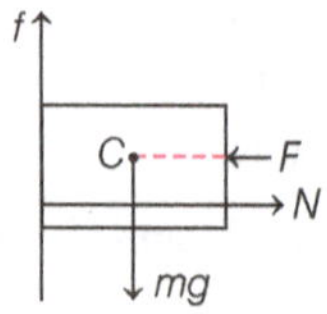

τ_F and τ_{mg} are already zero, as they are passing through C.

$\therefore \quad \tau_N + \tau_f = 0$

Since, $\quad \tau_f \neq 0$

$\therefore \quad \tau_N \neq 0$

Note *τ_f is clockwise. Therefore τ_N should be anticlockwise. Hence, N should act below point C.*

15. $x = \frac{l}{2} - \frac{l}{4} = \frac{l}{4}$

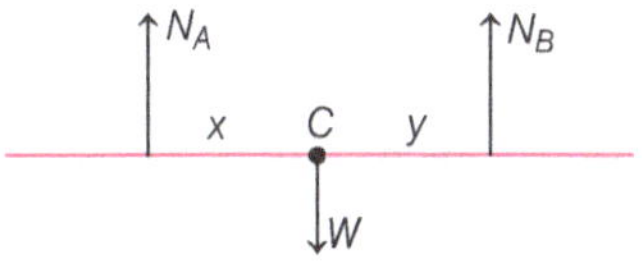

$$y = \frac{l}{2} - \frac{l}{6} = \frac{l}{3}$$

Σ (moments of all forces about point C) $= 0$

$\therefore \quad N_A x = N_B y$

or $\quad \frac{N_A}{N_B} = \frac{y}{x} = \frac{4}{3}$ **Ans.**

16.

Force	F_x	F_y
F_1	$4\cos 30° = 2\sqrt{3}$ N	$4\sin 30° = 2$ N
F_2	$-4\cos 60° = -2$ N	$4\sin 60° = 2\sqrt{3}$ N
F_3	0	-6 N
F_4	4 N	0

Module 3.

1. In these cases, $a = \frac{\text{Net pulling force}}{\text{Total mass}}$

Net pulling force in both the cases is

$$2mg - mg = mg$$

But mass (to be pulled) in case (a) is $3m$ and in case (b) is m. Therefore, $a_1 = \frac{mg}{3m} = \frac{g}{3}$

[in case (a)]

and $\quad a_2 = \frac{mg}{m} = g \quad$ [in case (b)]

or $\quad a_1 < a_2$

2. **If $m_1 > m_2$**

$$a = \frac{\text{Net pulling force}}{\text{Total mass}} = \frac{(m_1 - m_2)g}{m_1 + m_2}$$

FBD of m_1

$$m_1 g - T = m_1 a = \frac{m_1(m_1 - m_2)g}{m_1 + m_2}$$

$$\therefore \quad T = \frac{2m_1 m_2}{(m_1 + m_2)} g$$

$$= m_1 g\left[\frac{2m_2}{m_1 + m_2}\right] = m_2 g\left[\frac{2m_1}{m_1 + m_2}\right]$$

$$= m_1 g\left[\frac{m_2 + m_2}{m_1 + m_2}\right] = m_2 g\left[\frac{m_1 + m_1}{m_1 + m_2}\right]$$

Since $m_1 > m_2, T < m_1 g$ but $T > m_2 g$

Similarly,

we can prove that

$$T < m_2 g \text{ and } T > m_1 g$$

if $\quad m_2 > m_1$.

Therefore under all conditions T lies between $m_1 g$ and $m_2 g$.

3. Suppose a block is placed inside an elevator moving upward with an acceleration a_0

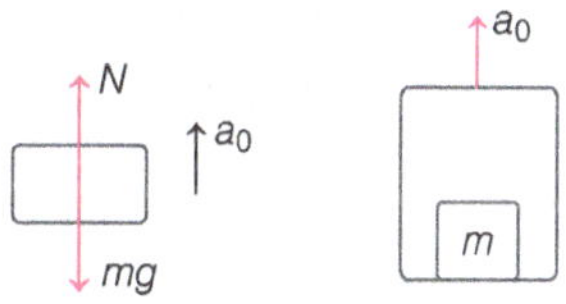

From free body diagram of block,

$$N - mg = ma_0$$

$$\therefore \quad N = mg + ma_0$$

Thus, normal reaction may accelerate a block. Hence, Statement I is wrong.

Normal reaction is +ve component of net contact force perpendicular to the surface of contact. Hence, Statement II is the correct.

4. The cloth can be pulled out without dislodging the dishes from the table due to law of inertia, which is Newton's first law. While, the statement II is true, but it is Newton's third law.

5. If you push a cart with some force, then according to Newton's third law, the cart will exert an equal and opposite force on you. So the cart push you with same amount of force in opposite direction. So, Statement 1 is correct. The action-reaction pairs mentioned in statement one cannot cancel each other because action and reaction forces act on two different bodies. So, they cannot cancel each other. So Statement 2 is wrong.

6. From FBD of lift,

$$T_1 - T_2 + mg, m = \text{mass of lift}$$

$$\Rightarrow \quad T_1 - T_2 = mg$$

$$\Rightarrow \quad T_{net} = mg$$

So, statement I is true.

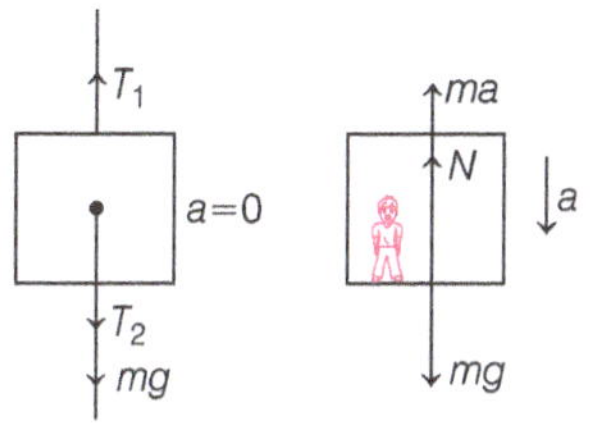

From FBD of person,

$$N + ma = mg$$

$$\Rightarrow \quad N = mg - ma$$

$$\Rightarrow \quad N < mg$$

So, statement II is false.

7. For 4 kg block,

$$4g \sin 60° - T = 4a \quad ...(i)$$

For 1 kg block,

$$T - 1g \sin 30° = 1a$$

$$4(T - g \sin 30°) = 4a \quad ...(ii)$$

From Eqs. (i) and (ii), we get

$$4(T - g \sin 30°) = 4g \sin 60° - T$$

$$\Rightarrow \quad 5T = 20\sqrt{3} + 20$$

$$\Rightarrow \quad T = (4\sqrt{3} + 4)\,\text{N}$$

8. For equilibrium condition, $m_2 g = m_1 g \sin \theta$

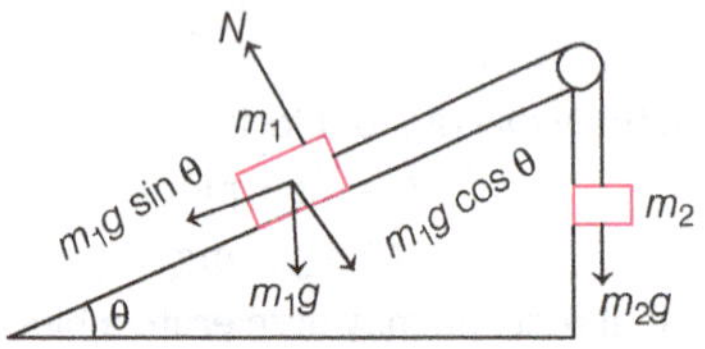

$$\sin \theta = \frac{m_2}{m_1} = \frac{3}{5}$$

$$\cos \theta = \frac{4}{5}$$

Normal force (N) on $m_1 = 5g \cos \theta$

$$= 5 \times 10 \times \frac{4}{5} = 40\,\text{N}$$

9. $a = \dfrac{M_2 g - M_1 g}{M_1 + M_2}$

When $M_2 = 2M_1 \Rightarrow a_1 = \dfrac{2M_1 g - M_1 g}{3M_1} = \dfrac{g}{3}$

When $M_2 = 3M_1 \Rightarrow a_2 = \dfrac{3M_1 g - M_1 g}{4M_1} = \dfrac{g}{2}$

The ratio $\dfrac{a_1}{a_2} = \dfrac{\frac{g}{3}}{\frac{g}{2}} = \dfrac{2}{3}$

10. $a = \dfrac{mg - F}{m} = g - \dfrac{F}{m}$

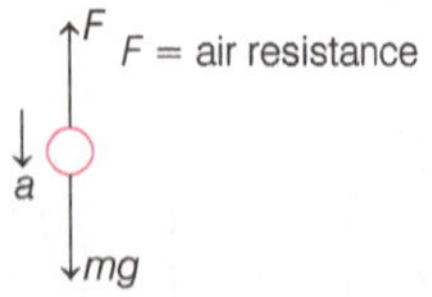

$$m_A > m_B$$

$\therefore \quad a_A > a_B$ and A will reach earlier

11. $a = \dfrac{mg - T}{m} = g - \dfrac{T}{m}$

$$a_{min} = g - \frac{T_{max}}{m}$$

$$= g - \frac{\frac{2}{3}mg}{m} = \frac{g}{3}$$

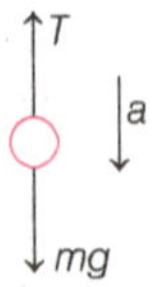

12. $a = \dfrac{\text{Net pulling force}}{\text{Total mass}} = \dfrac{10 \times 10 - 5 \times 10}{10 + 5} = \dfrac{10}{3}\,\text{m/s}^2$

For 5 kg, $\quad T - 5 \times 10 = 5 \times a = \dfrac{50}{3}$

$$\therefore \quad T = \frac{200}{3}\,\text{N}$$

This is also the reading of spring balance.

13. During acceleration,

$$v = a_1 t_1 = \frac{F_1}{m} t_1$$

$$\therefore \quad F_1 = \frac{mv}{t_1} \quad ...(i)$$

During retardation,

$$0 = v - a_2 t_2 = v - \frac{F_2}{m} t_2$$

$$\therefore \quad F_2 = \frac{mv}{t_2} \quad ...(ii)$$

If $t_1 = t_2$, then $F_1 = F_2$

$t_1 < t_2$, then $F_1 > F_2$

and $t_1 > t_2$, then $F_1 < F_2$

14. Three coins are above the 7th coin. Therefore, force from 3 coins above it will be

$3mg = 3(10 \times 10^{-3})(10) = 0.3\,\text{N}$ (in downward direction)

15. $0 = (3)^2 - 2(a)(9)$

$$\therefore \quad a = \frac{1}{2} = 0.5\,\text{m/s}^2 \quad \text{(upwards)}$$

$$N = m(g + a)$$

$$= 50\,(9.8 + 0.5) = 515\,\text{N}$$

16. $a_1 = \dfrac{F}{m_1}$ and $a_2 = \dfrac{F}{m_2}$

and $\quad a = \dfrac{F}{m_1 + m_2} = \dfrac{F}{\frac{F}{a_1} + \frac{F}{a_2}}$

$$= \frac{a_1 a_2}{a_1 + a_2}$$

17. For climbing up $T > mg$

$$T_{max} = 360\,\text{N} \quad \text{(Given)}$$

and $\quad mg = 600\,\text{N}$

So, it's not possible to climb up safely with some acceleration.

18. $a_1 = \dfrac{m_2 g}{m_1 + m_2} = \dfrac{30}{7}\,\text{m/s}^2$

$$a_2 = \frac{(m_1 - m_2)g}{m_1 + m_2} = \frac{10}{7}\,\text{m/s}^2$$

$$a_3 = \frac{m_2 g - m_1 g \sin 30°}{m_1 + m_2}$$

$$= \frac{10}{7}\,\text{m/s}^2$$

$\therefore \quad a_1 > a_2 = a_3$ **Ans.**

19. Acceleration,

$$a = \frac{2F - F}{m + m} = \frac{F}{2m} \quad \text{(towards left)}$$

Horizontal forces on B gives the equation,

$$2F - N \sin 30^\circ = m \cdot a$$

or $$2F - \frac{N}{2} = m\left(\frac{F}{2m}\right)$$

$\therefore \quad N = 3F$ **Ans.**

20. Acceleration of system before breaking the string

$$a = \frac{\text{Net pulling force}}{\text{Total mass}}$$

$$= \frac{3g - 2g}{5} = \frac{g}{5}$$

After 5 s velocity of system,

$$v = at = \frac{g}{5} \times 5 = g \text{ m/s}$$

Now, $$h = \frac{v^2}{2g} = \frac{g^2}{2g} = \frac{g}{2} = 4.9 \text{ m}$$

21. As, $a = \frac{F}{m} = \frac{5 \times 10^4}{3 \times 10^7} = \frac{5}{3} \times 10^{-3} \text{ m/s}^2$

So, $v = \sqrt{2as} = \sqrt{2 \times \frac{5}{3} \times 10^{-3} \times 3}$

$$= 0.1 \text{ m/s}$$

22. Given, resistance offered by the wall

$$= F = -2.5 \times 10^{-2} \text{ N}$$

So, deceleration of bullet,

$$a = \frac{F}{m} = \frac{-2.5 \times 10^{-2}}{20 \times 10^{-3}} = -\frac{5}{4} \text{ms}^{-2}$$

Now, using the equation of motion,

$$v^2 - u^2 = 2as$$

We have, $$v^2 = 1 + 2\left(-\frac{5}{4}\right)(20 \times 10^{-2})$$

$\Rightarrow \quad v^2 = \frac{1}{2}$

$\therefore \quad v = \frac{1}{\sqrt{2}} \approx 0.7 \text{ ms}^{-1}$

23. Given, mass of the particle, $m = 5$ kg

As the particle is at rest. So resultant of $\mathbf{F}_2$ and $\mathbf{F}_3$ should be opposite to $\mathbf{F}_1$.

$$\mathbf{F}_{net} = \sqrt{\mathbf{F}_2^2 + \mathbf{F}_3^2}$$

$$= \sqrt{8^2 + 6^2} = 10$$

$\therefore$ Acceleration, $a = \frac{F_{net}}{m} = \frac{10}{5} = 2 \text{ m/s}^2$

24. Given, mass of body, $m = \frac{500}{1000} = 0.5$ kg

Velocity, $v = 10\sqrt{x}$

$\Rightarrow \quad v^2 = 100x$

$\therefore \quad 2v\frac{dv}{dx} = 100$

$\Rightarrow \quad a = \frac{vdv}{dx} = 50 \text{ m/s}^2$

$\therefore$ Force, $F = ma = 0.5 \times 50 = 25$ N

25. Mass of particle,

$$m = 500 \text{ g} = 0.5 \text{ kg}$$

Velocity of particle,

$$\mathbf{v} = 2t\hat{\mathbf{i}} + 3t^2\hat{\mathbf{j}}$$

$\because \quad \mathbf{a} = \frac{d\mathbf{v}}{dt} = 2\hat{\mathbf{i}} + 6t\hat{\mathbf{j}}$

At $t = 1$ s, $\mathbf{a} = 2\hat{\mathbf{i}} + 6\hat{\mathbf{j}}$

Force acting on the particle,

$$\mathbf{F} = m\mathbf{a} = 0.5(2\hat{\mathbf{i}} + 6\hat{\mathbf{j}}) = \hat{\mathbf{i}} + 3\hat{\mathbf{j}}$$

$$\mathbf{F} = \hat{\mathbf{i}} + x\hat{\mathbf{j}}$$

Hence $x = 3$

26.

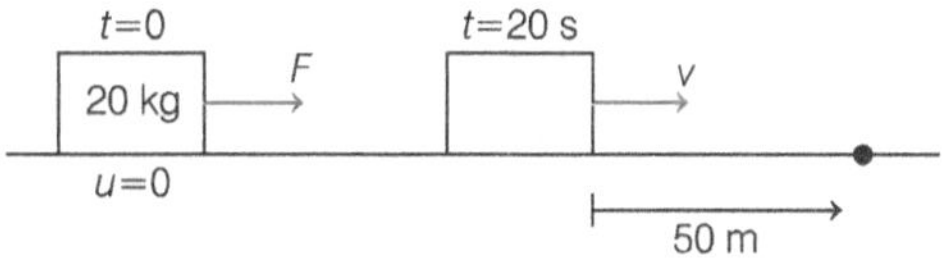

Assume surface to be frictionless.

Then, $50 = v \times 10 \Rightarrow v = 5$ m/s

As $\quad v = u + at \Rightarrow v = 0 + a \times 20$

$\Rightarrow \quad 5 = a \times 20 \Rightarrow a = \frac{1}{4} \text{m/s}^2$

So, $F = ma = 20 \times \frac{1}{4} = 5$ N

27. While going upward, $a_1 = -(g + a)$

$$= -\left(10 + \frac{10}{5}\right) = -12 \text{ m/s}^2$$

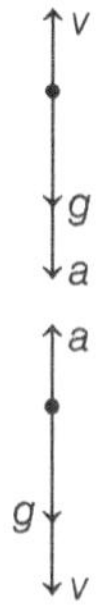

Now, $v = u + a_1t_1$

$\Rightarrow \quad u = -a_1t_1$

and $s = ut_1 + \frac{1}{2}a_1t_1^2$

$$s = ut_1 - \frac{1}{2} \times 2 \times t_1^2$$

$$= -at_1^2 - 6t_1^2 = 12t_1^2 - 6t_1^2 = 6t_1^2$$

While coming downward,

$$a_2 = g - a = 10 - \frac{10}{5} = 8 \text{ m/s}^2$$

Here, $u = 0$

So, $s = \frac{1}{2}a_2t_2^2 = \frac{1}{2} \times 8 \times t_2^2 = 4t_2^2$

$\because \quad s = 6t_1^2 = 4t_2^2$

$\Rightarrow \quad \dfrac{t_1}{t_2} = \sqrt{\dfrac{4}{6}} = \sqrt{\dfrac{2}{3}}$

28. $\mathbf{r} = 10t\hat{\mathbf{i}} + 15t^2\hat{\mathbf{j}} + 7\hat{\mathbf{k}}; \mathbf{v} = \dfrac{d\mathbf{r}}{dt} = 10\hat{\mathbf{i}} + 30t\hat{\mathbf{j}}$

$\mathbf{a} = \dfrac{d\mathbf{v}}{dt} = 30\hat{\mathbf{j}} \qquad (\because \mathbf{F} = m\mathbf{a})$

So, net force is along $+y$-direction.

29. When elevator does downward, then a pseudo force acts on object in upward direction, due to which effective weight of object decreases.

30. $a = \dfrac{480}{20+12+8} = 12\,\text{m/s}^2$

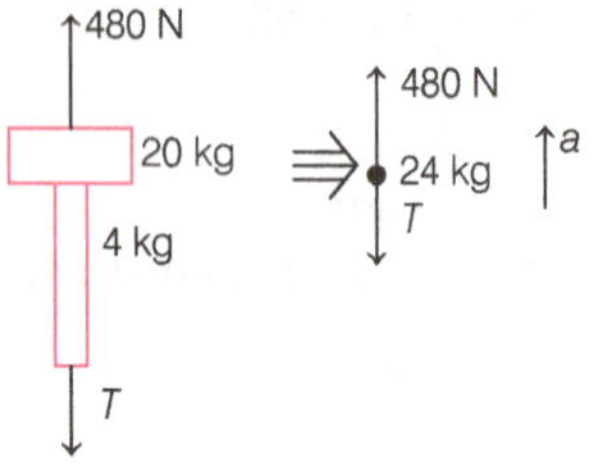

$480 - T = 24a$

$480 - T = 24 \times 12$

$480 - T = 288$

$T = 192\,\text{N}$

31. $a = \dfrac{\text{Net pulling force}}{\text{Total mass}}$

$= \dfrac{(5-4.8)\times 9.8}{5+4.8}$

$= 0.2\,\text{m/s}^2$

32. When lift is stationary, $W_1 = mg$ ...(i)

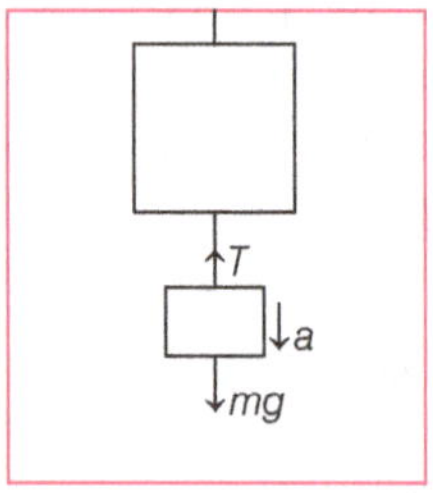

When the lift descends with acceleration a,

$W_2 = m(g-a)$

$W_2 = \dfrac{49}{10}(10-5) = 24.5\,\text{N} \approx 24\text{N}$

33. Taking the rope and the blocks as a system, we get

$P = (m+M)\,a$

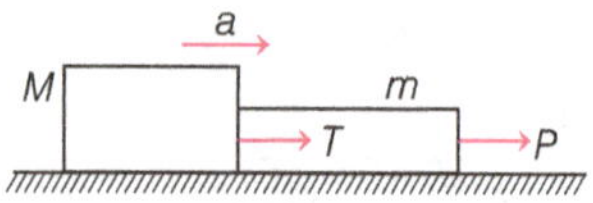

$\therefore$ Acceleration produced,

$a = \dfrac{P}{m+M}$

Taking the block as a system,

Force on the block, $F = Ma$

$\therefore \quad F = \dfrac{MP}{m+M}$

34. As springs are massless

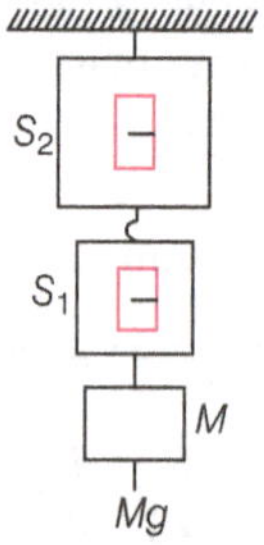

So, tension in spring 1

= tension in spring 2

$= Mg$

And, reading of spring balance is equal to tension of spring, so both will show same reading.

35. $a_1 = g\sin 30° = g/2$

$a_2 = g\sin 60°$

$= \sqrt{3}\,g/2$

$a_r = |\mathbf{a}_1 - \mathbf{a}_2|$

Angle between $\mathbf{a}_1$ and $\mathbf{a}_2$ is 30°.

$\therefore \quad a_r = \sqrt{a_1^2 + a_2^2 - 2a_1 a_2 \cos 30°}$

$= g/2$

36. Assuming the resistance force or retardation to be constant.

$\left(\dfrac{v}{2}\right)^2 = v^2 - 2as_1 \qquad \text{... (i)}$

$0 = \left(\dfrac{v}{2}\right)^2 - 2as_2 \qquad \text{... (ii)}$

Solving these two equations, we get

$s_2 = \dfrac{s_1}{3} = 1\,\text{cm}$

37. When direction of motion will reverse at the instant, the velocity becomes zero.

$\because \quad v = u + at$

or $\quad 0 = 2 + a \times 3$

$$\therefore \qquad a = -\frac{2}{3}\ \text{m/s}^2$$

Thus, acceleration of A is $\frac{2}{3}$ m/s^2 in rightward direction.

38.

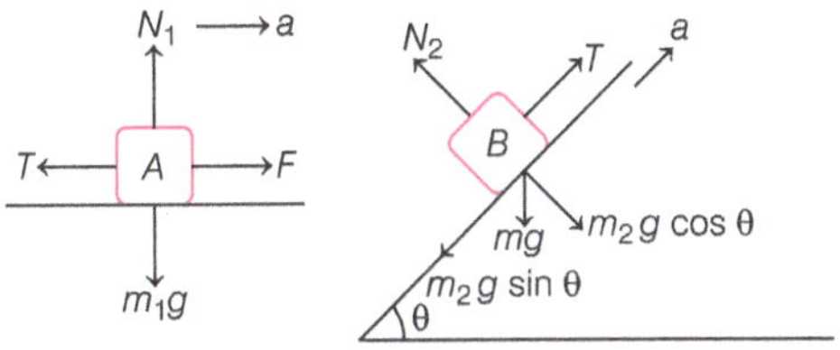

From FBD of A,

$$F - T = m_1 a = 1 \times \frac{2}{3} \qquad \text{...(i)}$$

From FBD of B,

$$T - m_2 g \sin\theta = m_2 a \qquad \text{...(ii)}$$

$$= 2 \times \frac{2}{3} = \frac{4}{3}$$

or $\quad T - 10 = \frac{4}{3}$

$$\therefore \qquad T = 10 + \frac{4}{3} = \frac{34}{3}$$

$$= 11.33\ \text{N}$$

39. From Eq. (i), we get

$$F - T = \frac{2}{3}$$

$$\therefore \qquad F = \frac{2}{3} + 11.33 = 12\ \text{N}$$

40. (a) $a = \dfrac{\text{Net pushing force}}{\text{Total mass}}$

$$= \frac{120 - 50}{1 + 4 + 2}$$

$$= 10\ \text{m/s}^2$$

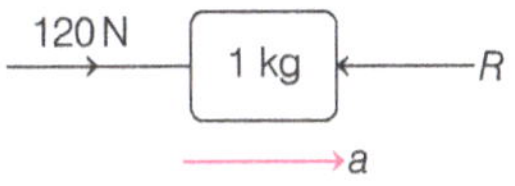

(b) $120 - R = 1 \times a = 10$

$\therefore \qquad R = 110\ \text{N}$

(c) $F_{\text{net}} = ma = (2)(10) = 20\ \text{N}$

41. $T_4 = 4g$ and $T_1 = (1)g$ as $a = 0$

$$\therefore \qquad \frac{T_4}{T_1} = 4 \qquad \textbf{Ans.}$$

42. $a = \dfrac{\text{Net pulling force}}{\text{Total mass}}$

$$= \frac{F + 10 - 20}{1 + 2} \qquad (\because F = 20\ \text{N})$$

$$= \frac{10}{3}\ \text{m/s}^2 \qquad \textbf{Ans.}$$

43. $m_A g = (2)(g) + (2)g \sin 30°$ as $a = 0$

$$\therefore \qquad m_A = 3\ \text{kg} \qquad \textbf{Ans.}$$

44. Since surface is smooth, acceleration of both is $g \sin\theta = 10 \sin 30° = 5\ \text{m/s}^2$, down the plane.

The component of mg down the plane $(= mg \sin\theta)$ provides this acceleration. So, normal reaction will be zero.

45. $N = \dfrac{mg}{4}$ is given

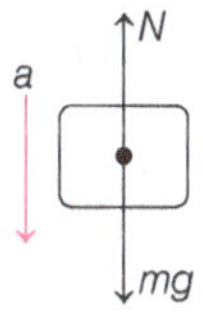

$$mg - \frac{mg}{4} = ma$$

$$\therefore \qquad a = \frac{3g}{4}$$

46. $a = \dfrac{\text{Net pulling force}}{\text{Total mass}}$

$$= \frac{(3 + 4 - 2 - 1)g}{3 + 4 + 2 + 1} = 4\ \text{m/s}^2 \qquad \textbf{Ans.}$$

4 kg $\quad 40 - T_1 = 4a \quad \therefore\ T_1 = 24\ \text{N}$ **Ans.**

3 kg $\quad T_1 + 30 - T_2 = 3a \quad \therefore\ T_2 = 42\ \text{N}$ **Ans.**

1 kg $\quad T_3 - 10 = (1)(a) \quad \therefore\ T_3 = 14\ \text{N}$ **Ans.**

47. (a) $a = \dfrac{\text{Net pushing force}}{\text{Total mass}}$

$$= \frac{100 - 40}{6 + 4 + 10} = 3\ \text{m/s}^2$$

(b) Net force on any block $= ma$

(c) $N - 40 = 10a = 30$

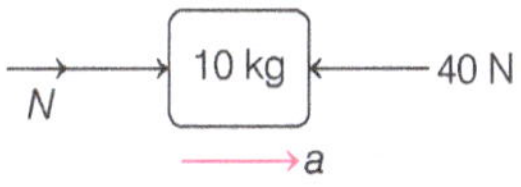

$$\therefore \qquad N = 70\ \text{N} \qquad \textbf{Ans.}$$

48. (a) $T_1 = 10a, T_2 - T_1 = 20a, 60 - T_2 = 30a$

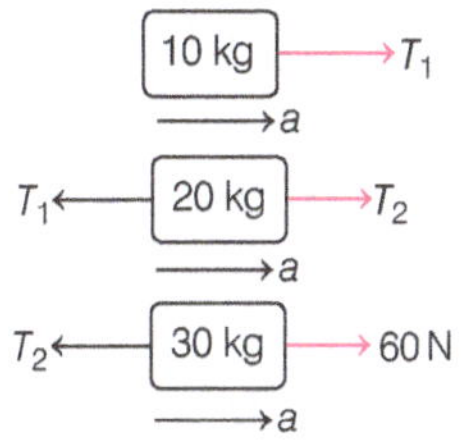

$$a = \frac{\text{Net pulling force}}{\text{Total mass}} = \frac{60}{10 + 20 + 30}$$

$$= 1\ \text{m/s}^2$$

On solving, we get $\quad T_1 = 10\ \text{N}, T_2 = 30\ \text{N}$ **Ans.**

(b) $T_1 = 0, T_2 = 20a, 60 - T_2 = 30a$

$$a = \frac{\text{Net pulling force}}{\text{Total mass}} = \frac{60}{20+30}$$
$$= 1.2 \text{ m/s}^2$$

On solving, we get

$$T_2 = 24 \text{ N}$$ **Ans.**

49. (a) $T_1 - 2g = 2a$

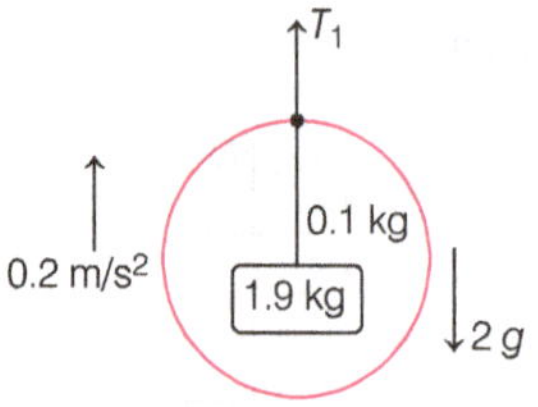

$\therefore$ $T_1 = 2(g+a)$
$= 2(9.8 + 0.2)$
$= 20 \text{ N}$

(b) $T_2 - 5g = 5a$

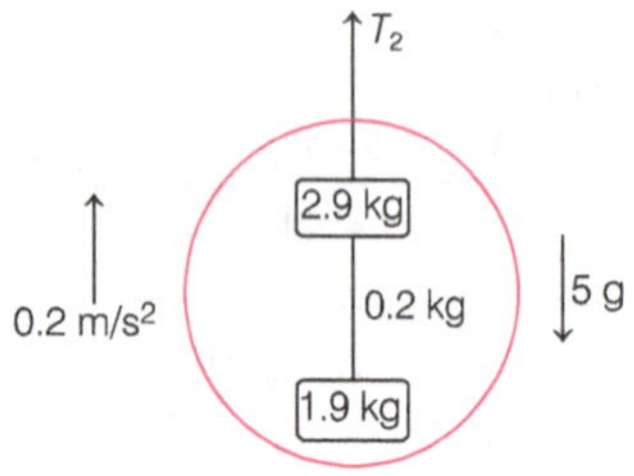

$\therefore$ $T_2 = 5(g+a)$
$= 5(9.8 + 0.2)$
$= 50 \text{ N}$

50. (a) $a = \dfrac{\text{Net pulling force}}{\text{Total mass}}$

$$= \frac{200 - 16 \times 9.8}{16} = 2.7 \text{ m/s}^2$$

(b) $200 - 49 - T_1 = 5a$

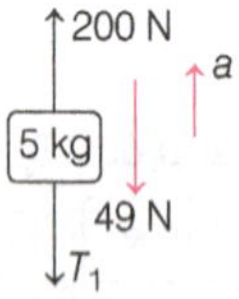

$\therefore$ $T_1 = 200 - 5 \times 2.7 - 49$
$= 137.5 \text{ N}$

(c) $T_2 - 9g = 9a$

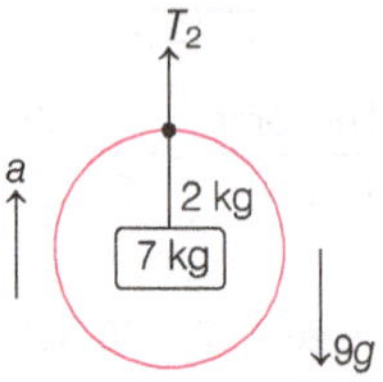

$\therefore$ $T_2 = 9(g+a)$
$= 9(9.8 + 2.7)$
$= 112.5 \text{ N}$

51.

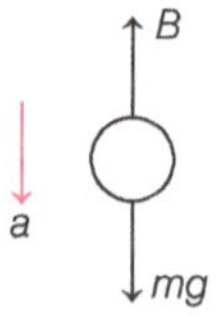

B = Buoyant force from air

$$mg - B = ma \quad \text{...(i)}$$

Let M mass is thrown out, then

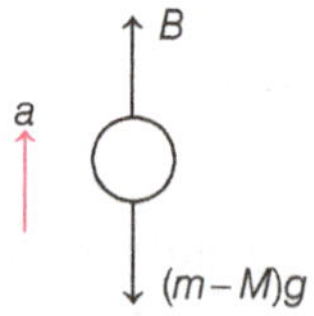

$$B - (m-M)g = (m-M)a \quad \text{...(ii)}$$

Adding Eqs. (i) and (ii), we get

$$Mg = 2ma - Ma$$

$$\therefore \quad M = \frac{2ma}{g+a}$$

Note *Buoyant force depends upon the volume and here volume is not changing. Therefore buoyant force will remain unchanged.*

52. Acceleration of rope, $a = F/M$

Now, to find tension at point C, a distance l from point B, we can write equation of motion of any one part (AC or CB), both moving with acceleration a.

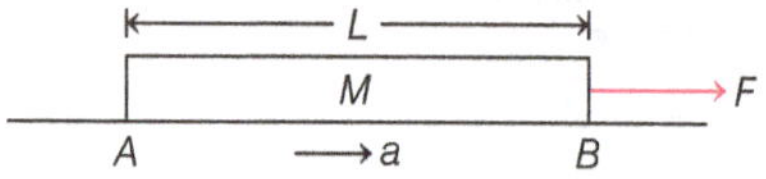

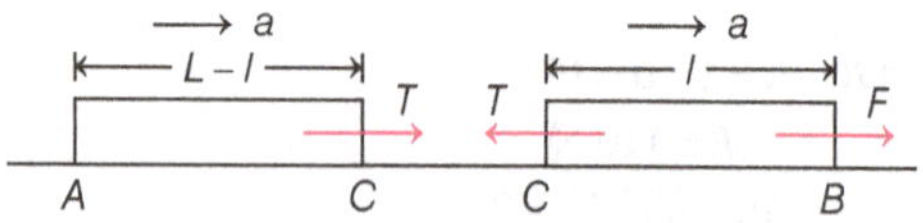

Equation of motion of part AC is

$$T = (\text{mass of } AC) \times (\text{acceleration})$$
$$= \frac{M}{L}(L-l)\left(\frac{F}{M}\right) = F\left(1 - \frac{l}{L}\right)$$

53. $\mathbf{a} = \dfrac{\mathbf{F}}{m} = 10\hat{\mathbf{i}} + 5\hat{\mathbf{j}}$

Displacement of the box along x-axis,

$$x = \frac{1}{2}a_x t^2 = \frac{1}{2} \times 10 \times 100$$
$$= 500 \text{ m}$$

54. Mass of the person, $M = 60\,\text{kg}$

Tension in the rope of the lift when it moves downward with acceleration a.

$$T = M(g-a)$$
$$= 60(10-1.8)$$
$$= 492\,\text{N}$$

Module 4.

1. Steady rate means, net force $= 0$

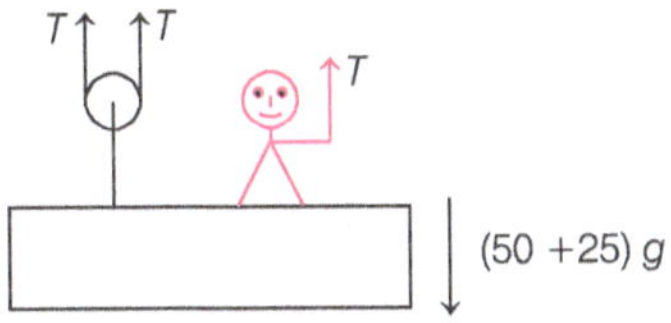

$\therefore \quad 3T = 75\,g = 750$

or $\quad T = 250\,\text{N}$ **Ans.**

2. If force applied by man is F, then in first figure, force transferred to the block is F, while in second figure, force transferred to the block is $2F$.

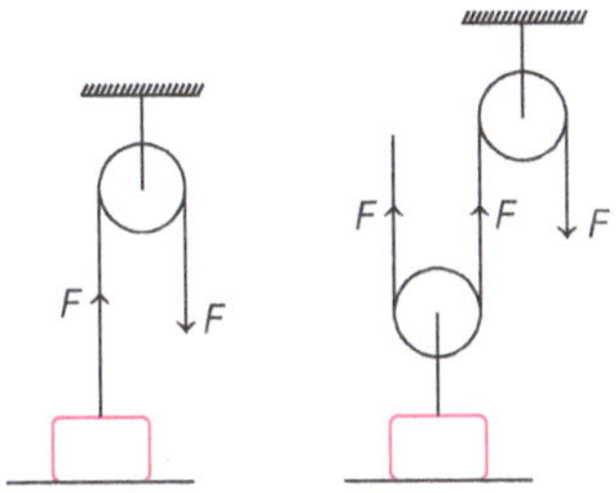

(i) (ii)

3. Total upward force $= 4\left(\dfrac{mg}{2}\right) = 2\,mg$

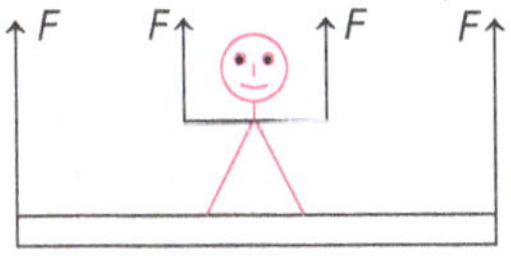

Total downward force $= (m+m)\,g = 2\,mg$

$\therefore \quad$ Net force $= 0$

$$F = \frac{mg}{2}$$

4. Acceleration of system,

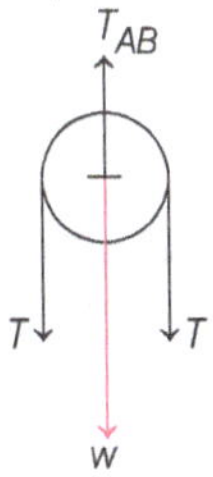

$$a = \frac{\text{Net pulling force}}{\text{Total mass}} = \frac{2mg - mg}{3m} = \frac{g}{3}$$

Now, from equation of motion of m,

$$T - mg = ma = \frac{mg}{3}$$

$$\therefore \quad T = \frac{4mg}{3}$$

For equilibrium of pulley,

$$T_{AB} = 2T + \text{weight of pulley}$$
$$= \frac{8mg}{3} + 3mg = \frac{17mg}{3}$$

5. Upward force on 2 kg block in upward direction will be 40 N $(= 2F)$ in the form of tension.

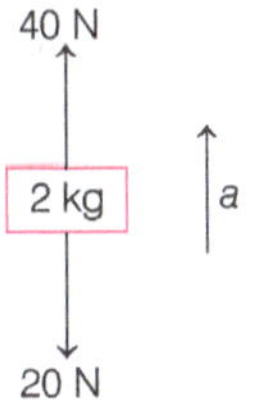

$$\therefore \quad a = \frac{40-20}{2} = 10\,\text{m/s}^2$$

6. Acceleration of system, $a = \dfrac{\text{Net pulling force}}{\text{total mass}}$

$$= \frac{Mg\sin\theta}{2M}$$

$$a = \frac{1}{2}g\sin\theta$$

Now, the block on ground is moving due to tension.

Hence, $\quad T = Ma = \dfrac{Mg\sin\theta}{2}$

7. Relative to pulley let relative acceleration is a_r (clockwise). Then :

$$a_1 = a_r + A \qquad \text{...(i)}$$
$$a_2 = a_r - A \qquad \text{...(ii)}$$

(i) – (ii) gives,

$$A = \frac{a_1 - a_2}{2}$$ **Ans.**

8. $v_1 = v_2 = 20\,\text{m/s}$

$$v_2 = v_A \cos 37° \quad \text{or} \quad 20 = 0.8\,v_A$$

$$\Rightarrow \quad v_A = 25\,\text{m/s}$$

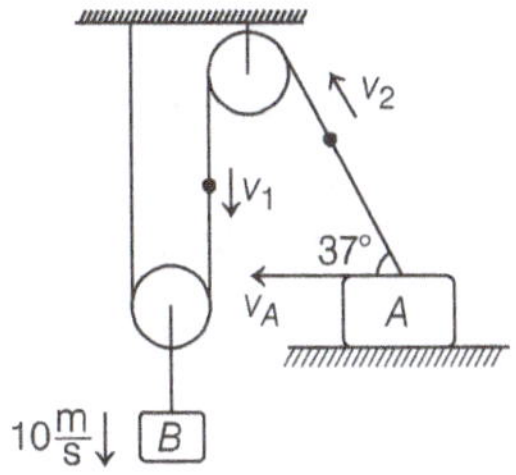

9. $T_B = 3T$ and $T_A = 2T$

$\therefore \quad v_A = \frac{T_B}{T_A} \cdot v_B = \frac{3}{2} v_0$ (towards right)

$\therefore \quad v_{AB} = \frac{3v_0}{2} - v_0 = \frac{v_0}{2}$ (towards right)

In such cases, velocity and acceleration are in inverse ratio of tensions.

10. Let a_1 is upwards. Then acceleration of movable pulley will be a_1 downwards. Further, let a_r is relative acceleration of 2 and 3 (w.r.t. movable pulley) in clockwise direction. Then taking upward direction as positive, we have

$a_2 = a_r - a_1$ ...(i)

$a_3 = -a_r - a_1$ ...(ii)

Adding Eq. (i) with Eq. (ii), we get

$$2a_1 + a_2 + a_3 = 0$$

11. **2 kg** $2T = 2(a)$ ...(i)

1 kg $10 - T = 1(2a)$ ...(ii)

On solving these two equations, we get

$$T = \frac{10}{3}\,\text{N}$$

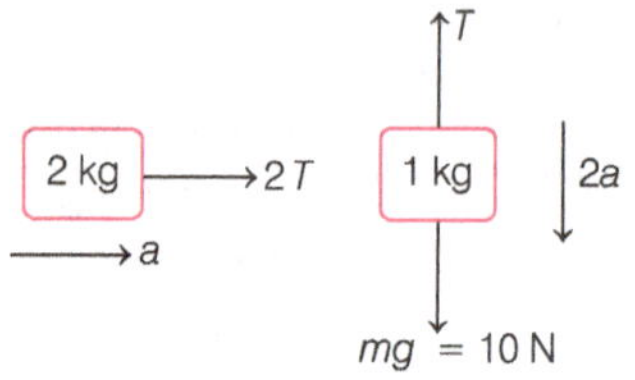

and $\quad 2a = 2T = \frac{20}{3}\,\text{m/s}^2$

= acceleration of 1 kg block

12.

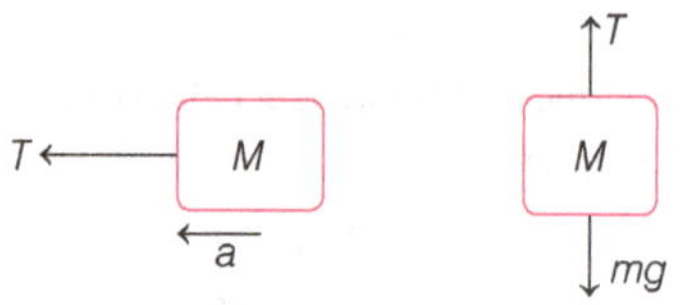

$T = Ma$ ...(i)

$Mg - T = Ma$ ...(ii)

On solving these two equations, we get

$$a = \frac{g}{2}$$

and $\quad T = \frac{Mg}{2}$ **Ans.**

13. $T = m_1(2a) = (0.3)(2a) = 0.6a$ ...(i)

$F - 2T = m_2 a$

$\therefore \quad 2T = 0.4 - 0.2a$ ...(ii)

On solving these two equations, we get

$a = \frac{2}{7}\,\text{m/s}^2$ **Ans.**

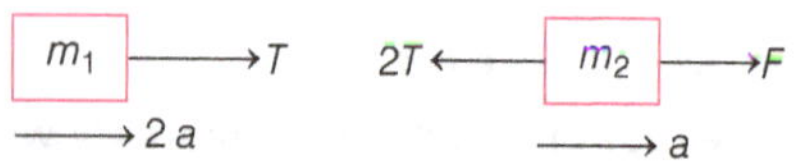

and $\quad 2T = \frac{12}{35}\,\text{N}$ **Ans.**

14. $a_3 = a + a_r = 6$ ...(i)

$a_2 = a_r - a = 4$ ...(ii)

On solving these two equations, we get

$$a = 1\,\text{m/s}^2$$

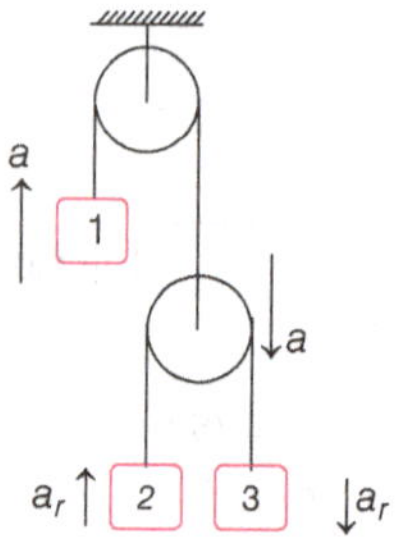

15. $T - Mg \sin 30° = Ma$ ...(i)

$2Mg - 2T = 2M\left(\frac{a}{2}\right)$ ...(ii)

Solving these equations, we get

$a = \frac{g}{3}$ **Ans.**

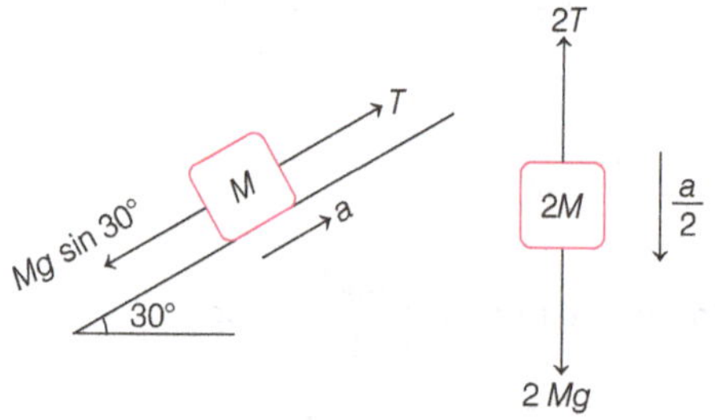

16. As shown in figure, when force F is applied at the end of the string, the tension in the lower part of the string is also F.

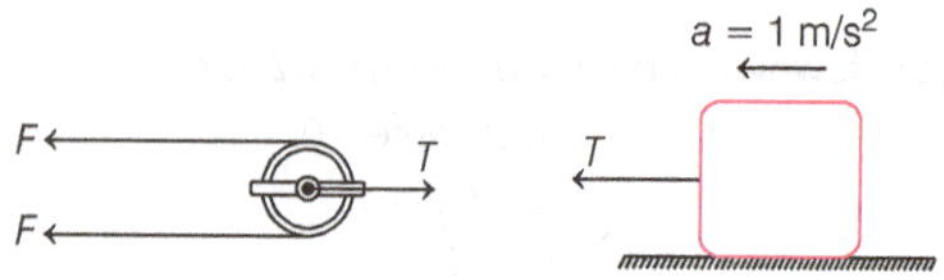

If T is the tension in string connecting the pulley and the block, then

$T = 2F$

But $\quad T = ma = (200)(1) = 200\,\text{N}$

$\therefore \quad 2F = 200\,\text{N}$

or $\quad F = 100\,\text{N}$

17. $X_A + 2X_B + X_C$

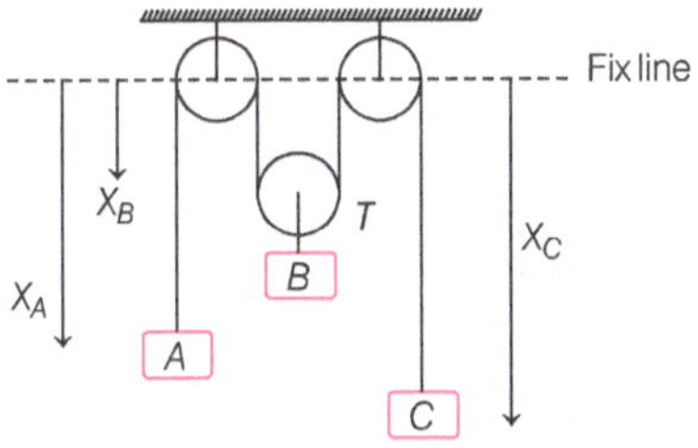

= constant on double differentiating with respect to time, we get

$$a_A + 2a_B + a_C = 0$$ **Ans.**

18. $2T - 50 = 5a$ ...(i)

$$40 - T = 4(2a)$$ ...(ii)

On solving these equations, we get

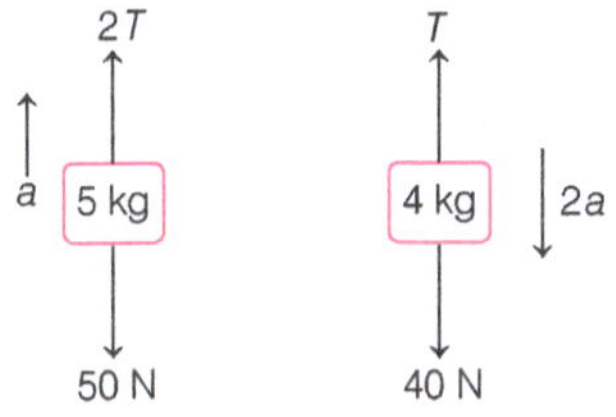

$$a = \frac{10}{7}\,\text{m/s}^2$$

or $\frac{g}{7}$ **Ans.**

and $2a = \frac{20}{7}\,\text{m/s}^2$ or $\frac{2g}{7}$ **Ans.**

Module 5.

1. Constant velocity means acceleration of frame is zero. Hence, it is an inertial frame.

3. A frame moving with constant velocity $(a = 0)$ is inertial.

4. Acceleration of block with respect to lift

Pseudo force = (downward)

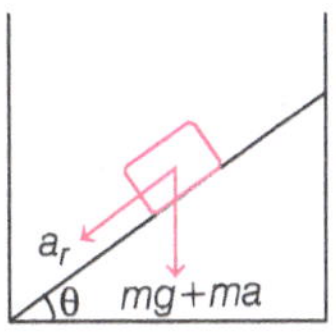

$$a_r = \frac{m(g+a)\sin\theta}{m} = (g+a)\sin\theta$$

$$s = \frac{1}{2}a_r t^2$$

$$\therefore \quad t = \sqrt{\frac{2s}{a_r}} = \sqrt{\frac{2L}{(g+a)\sin\theta}}$$

5. FBD of *m* w.r.t. chamber

Relative acceleration along the inclined plane

$$a_r = \frac{ma\cos\theta + mg\sin\theta}{m}$$

$$= (a\cos\theta + g\sin\theta)$$

$$t = \sqrt{\frac{2s}{a_r}} = \sqrt{\frac{2L}{a\cos\theta + g\sin\theta}}$$ **Ans.**

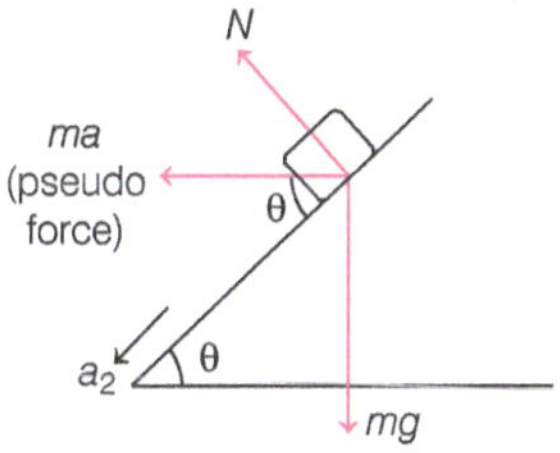

6. When the incline is given an acceleration '*a*' towards the right, the block experience a pseudo force *ma* towards left.

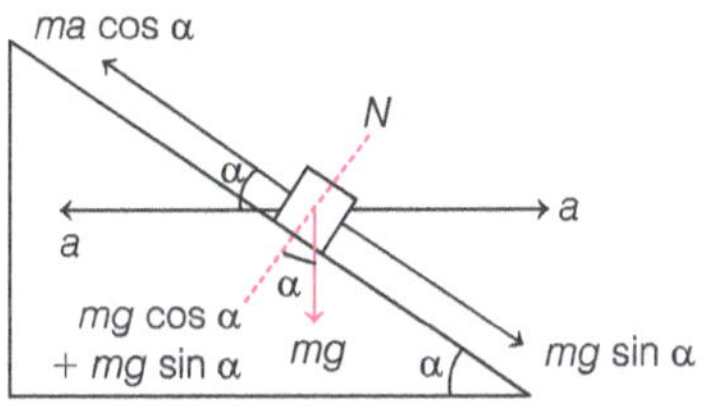

For block to remain stationary, net force along the incline should be zero.

$$mg\sin\alpha = ma\cos\alpha \Rightarrow a = g\tan\alpha$$

7. A rotating/revolving frame is accelerating and hence non-inertial.

8. (a) $\mathbf{F}_A = -m_A\mathbf{a}_B = (4\hat{\mathbf{j}})\,\text{N}$

(b) $\mathbf{F}_B = -m_B\mathbf{a}_A = (-4\hat{\mathbf{i}})\,\text{N}$

9. a_{mr} = acceleration of man relative to rope

$$= a_m - a_r$$

$$\therefore \quad a_m = a_{mr} + a_r$$

$$= (a) + (a)$$

$$= 2a$$

Now, $$T - mg = m(a_m) = m(2a)$$

$$\therefore \quad T = m(g + 2a)$$ **Ans.**

Note *Man slides down with a deceleration a relative to the rope. So, $a_{mr} = +a$ not $-a$.*

10. Case I : For the man standing in the lift, the acceleration of the ball

$$\mathbf{a}_{bm} = \mathbf{a}_b - \mathbf{a}_m \Rightarrow a_{bm} = g - a$$

Case II : The man standing on the ground, the acceleration of the ball

$$\mathbf{a}_{bm} = \mathbf{a}_b - \mathbf{a}_m$$

$$\Rightarrow \quad \mathbf{a}_{bm} = g - 0 = g$$

11.

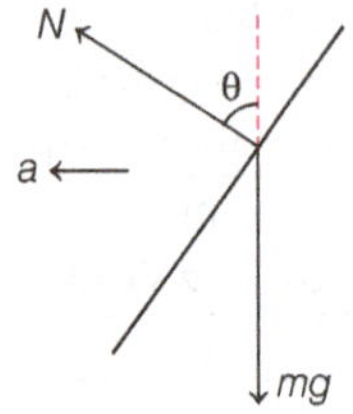

$$N\cos\theta = mg = 10 \qquad \text{...(i)}$$

$$N\sin\theta = ma = 5 \qquad \text{...(ii)}$$

On solving these two equations, we get

$$N = 5\sqrt{5}\ \text{N} \quad \text{and} \quad \tan\theta = \frac{1}{2}$$

Module 6.

Introductory Exercise

1. **Figure (a)** $N = mg = 40\ \text{N}$

$$\mu_s N = 24\ \text{N}$$

Since, $F < \mu_s N$

Block will remain stationary and

$$f = F = 20\ \text{N}$$

Figure (b) $N = mg = 20\ \text{N}$

$$\mu_s N = 12\ \text{N} \quad \text{and} \quad \mu_K N = 8\ \text{N}$$

Since, $F > \mu_s N$, block will slide and kinetic friction (=8 N) will act.

$$a = \frac{F - f}{m} = \frac{20 - 8}{2} = 6\ \text{m/s}^2$$

2. If $\theta \le$ angle of repose, the block is stationary, $a = 0, F_{net} = 0$ and $f = mg\sin\theta$.

If $\theta >$ angle of repose, the block will move,

$$f = \mu mg\cos\theta$$

$$a = \frac{mg\sin\theta - \mu mg\cos\theta}{m}$$

$$= g\sin\theta - \mu g\cos\theta \text{ and } F_{net} = ma$$

Further, $\mu = \tan\alpha$ (α = angle of repose = 45°)

$$= \tan 45° = 1$$

Exercise

1. The frictional force exerted by the surface on the person is in the direction of his motion, since the legs pushes the surface backward direction.

2. The block will move due to friction.

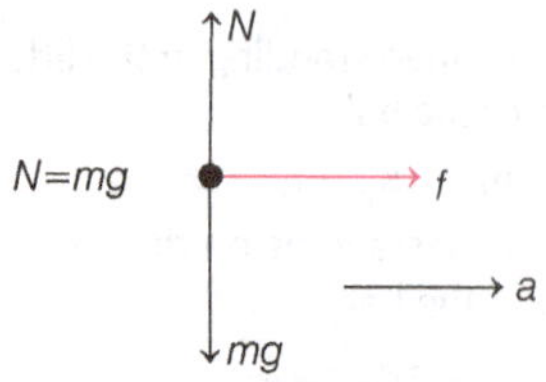

The frictional force (f) is therefore

$$f = ma = (1)(5) = 5\ \text{N}$$

Note $f_{max} = \mu N = \mu mg$

$$= 0.6 \times 1 \times 9.8 = 5.88\ \text{N}$$

Since $f < f_{max}$, therefore 5N friction will act.

3. By increasing F_1 limiting value of friction will increase. But it is not necessary that actual value of friction acting on block will increase.

4. $(f_2)_{max} = \mu_2 mg$

$$(f_1)_{max} = \mu_1 (2m) g$$

Since $(f_1)_{max} > (f_2)_{max}$

Lower block will not move at all.

5. $f - mg\sin\theta = ma\sin\theta$

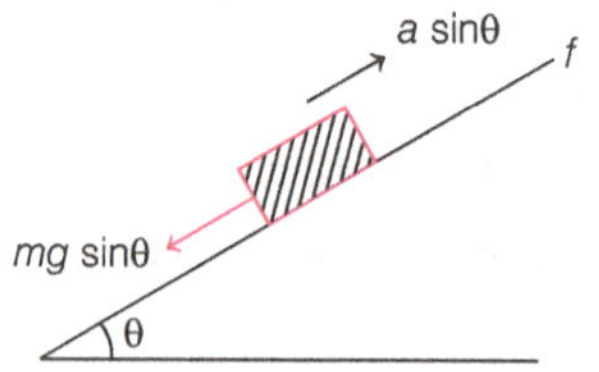

$$\therefore \quad f = m(g + a)\sin\theta > mg\sin\theta$$

7. Free body diagram of man,

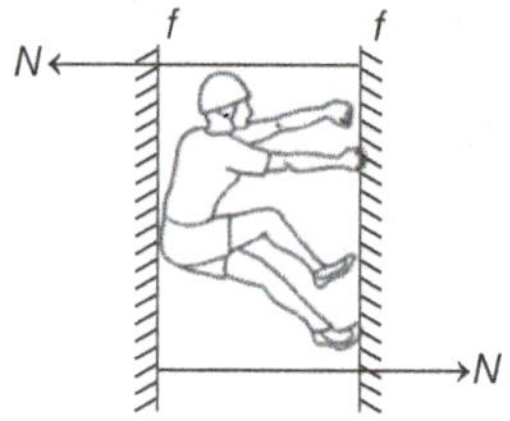

$$f + f = mg$$

$$2f = 50 \times 10$$

$$\Rightarrow \quad f = 250\ \text{N}$$

8.

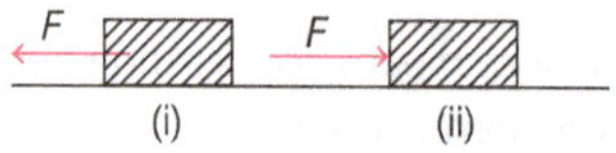

In figure (i), we are pulling the block and in figure (ii), we are pushing the block. In both cases efforts will be the same.

9. Angle of repose $\theta = \tan^{-1}(\mu)$

$$= \tan^{-1}\left(\frac{1}{\sqrt{3}}\right) = 30°$$

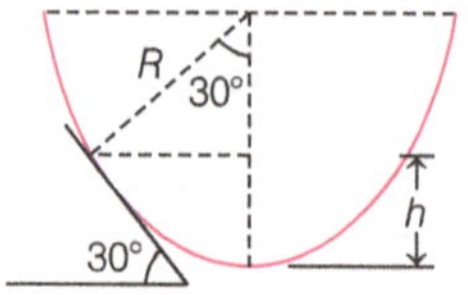

So, particle may be placed maximum up to 30°, as shown in figure

$$h = R - R\cos 30° = \left(1 - \frac{\sqrt{3}}{2}\right)R$$

10. Net pulling force $F_1 = 15g - 5g$

$$F_1 = 10g = 100\text{ N}$$

Net stopping force $F_2 = 0.2 \times 5 \times 10 = 10\text{ N}$

Since $F_1 > F_2$, therefore the system will move (anti-clockwise) with an acceleration,

$$a = \frac{F_1 - F_2}{15 + 5 + 5} = \frac{90}{25}$$

$$= 3.6\text{ m/s}^2$$

15 kg block

$$15 \times 10 - T_2 = 15 \times a = 15 \times 3.6$$

$\therefore$ $T_2 = 96\text{ N}$

5 kg block

$$T_1 - 5 \times 10 = 5 \times a = 5 \times 3.6$$

$\therefore$ $T_1 = 68\text{ N}$

$$\frac{T_1}{T_2} = \frac{68}{96} = \frac{17}{24}$$

11. Since, the block is resting (not moving)

$\therefore$ $f = mg\sin\theta \neq \mu_S\, mg\cos\theta$

or $m = \dfrac{f}{g\sin\theta} = \dfrac{10}{10 \times \sin 30°}$

$= 2\text{ kg}$ **Ans.**

12. $N = mg = 40\text{ N}$

$$\mu N = 0.2 \times 40 = 8\text{ N}$$

At $t = 2\text{ s}, F = 4\text{ N}$

Since, $F < \mu N$

$\therefore$ $f = F = 4\text{ N}$

13. In the figure, N_2 is always equal to W or 250 N

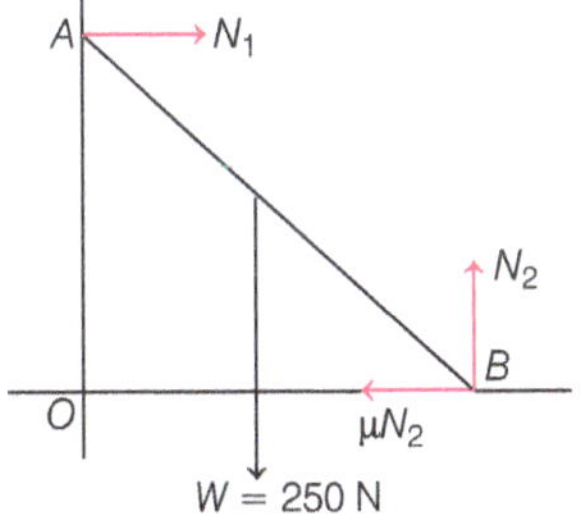

$\therefore$ Maximum value of friction available at B is μN_2 or 75 N.

14. $a = \dfrac{F - f}{M} = \dfrac{Mg - \mu mg}{M} = \dfrac{g}{2}$

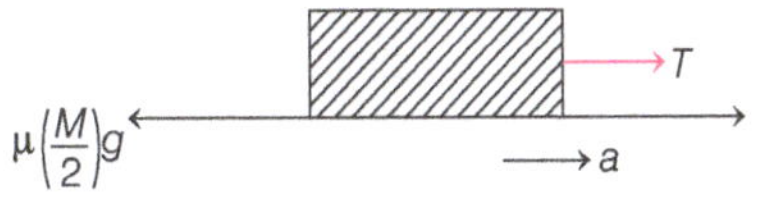

Now, at the mid-point,

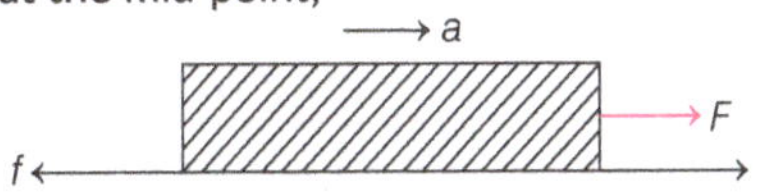

$$T - \mu\left(\frac{M}{2}\right)g = \frac{M}{2}$$

$$a = \left(\frac{M}{2}\right)\left(\frac{g}{2}\right)$$

Putting $\mu = \dfrac{1}{2}$, we get

$T = \dfrac{Mg}{2}$ **Ans.**

15. $N = mg = 40\text{ N}$

$$\mu N = 32\text{ N}$$

Applied force $F = 30\text{ N} < \mu N$

Therefore, friction force,

$$f = F = 30\text{ N}$$

$\therefore$ Net contact force from ground on block

$= \sqrt{N^2 + f^2} = 50\text{ N}$ **Ans.**

16. Maximum value of friction between B and ground

$$= \mu N = \mu\,(m_A + m_B)\,g$$
$$= (0.5)\,(2 + 8)\,(10) = 50\text{ N}$$

Since, applied force $F = 25$ N is less than 50 N. Therefore, system will not move and force of friction between A and B is zero.

17. $mg\sin\theta = (10)\,(10)\,(3/5) = 60\text{ N}$

This 60 N > 30 N. Therefore, friction force f will act in upward direction.

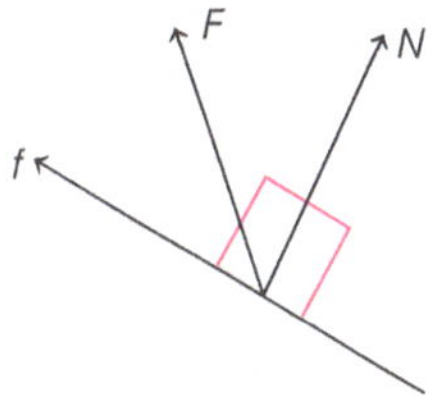

$F = \sqrt{N^2 + f^2}$ = net force by plane on the block.

18. Maximum friction between two,

$$f_{max} = \mu N = \mu\, mg$$

Upper block moves only due to friction. Therefore, its maximum acceleration may be

$$a_{max} = \frac{f_{max}}{m} = \mu g$$

Relative motion between them will start when common acceleration becomes μg.

$\therefore$ $\mu g = \dfrac{\text{Net force}}{\text{Total mass}} = \dfrac{at}{2m}$

$\therefore$ $t = \dfrac{2\mu\, mg}{a}$ **Ans.**

19. $F = mg\sin\theta - \mu mg\cos\theta$ ... (i)

$2F = mg\sin\theta + \mu mg\cos\theta$ ... (ii)

or $2mg\sin\theta - 2\mu mg\cos\theta = mg\sin\theta + \mu mg\cos\theta$

or $$\mu = \frac{1}{3}\tan\theta$$

20. Net pulling force on the system,

$$F = 10g\sin 37° - 4g = 20\text{ N}$$

Maximum force of friction

$$f_{max} = \mu mg\cos 37°$$
$$= 0.7 \times 10 \times 10 \times \frac{4}{5}$$
$$= 56\text{ N}$$

Since, $F < f_{max}$, system will not move. Equilibrium of 4 kg gives $T = 40$N.

21. In critical case, weight of hanging part = force of friction on the part of rope lying on table.

$$\therefore \quad \frac{m}{l}\cdot l_1 g = \mu\frac{m}{l}(l - l_1)g$$

Solving this, we get

$$l_1 = \left(\frac{\mu}{1+\mu}\right)l$$

22. $T - \mu mg = ma$

$$\therefore \quad T = \mu mg + ma$$

$F - T - \mu mg = ma$

$$\therefore \quad F - \mu mg - ma - \mu mg = ma$$

or $$a = \frac{F}{2m} - \mu g$$ **Ans.**

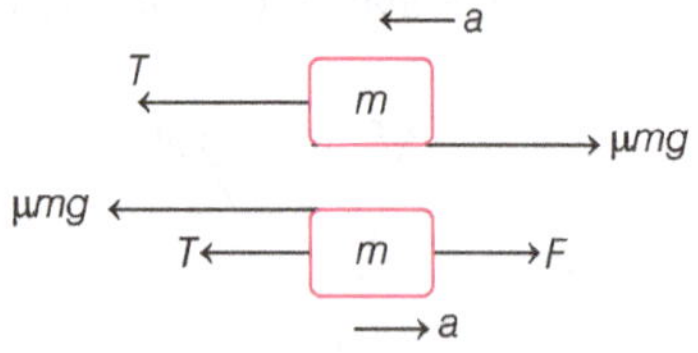

23. Maximum acceleration due to friction of mass m over mass $2m$ can be μg. Now, for the whole system

$$a = \frac{\text{Net pulling force}}{\text{Total mass}}$$

$$\therefore \quad \mu g = \frac{mg}{4m} \text{ or } \mu = \frac{1}{4}$$

24. Constant velocity means net acceleration or net force is zero. Hence,

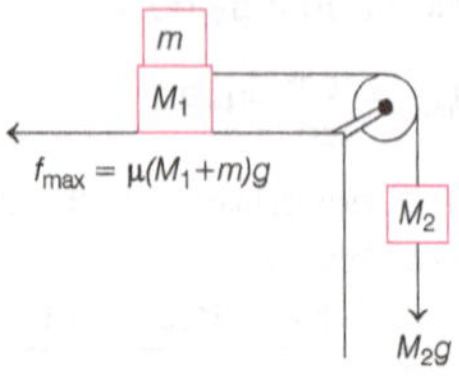

$$\mu(M_1 + m)g = M_2 g$$

or $$0.4(4 + m) = 6$$

$$\therefore \quad m = 11\text{ kg}$$

25. Since, $\mu mg\cos\theta > mg\sin\theta$. So block is in rest.

$\therefore$ Force of friction is $f = mg\sin\theta$

$$= 2 \times 9.8 \times \frac{1}{2} = 9.8\text{ N}$$

27. $N = 5$ N

$$(f)_{max} = \mu N = (0.5)(5) = 2.5\text{ N}$$

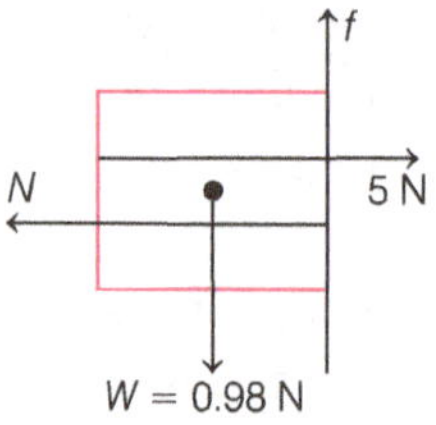

For vertical equilibrium of the block,

$$f = mg = 0.98\text{ N} < (f)_{max}$$

28. Equilibrium of insect gives

$$N = mg\cos\alpha \quad \text{...(i)}$$
$$\mu N = mg\sin\alpha \quad \text{...(ii)}$$

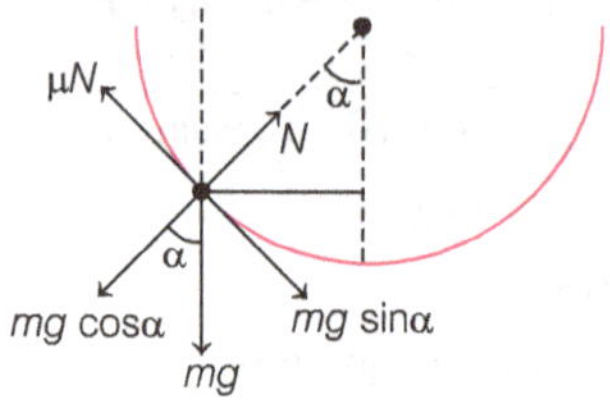

From Eqs. (i) and (ii), we get

$$\cot\alpha = \frac{1}{\mu} = 3$$

Note *Beyond angle of repose insect will slip.*

29. Free body diagram (FBD) of the block (shown by a dot) is shown in figure.

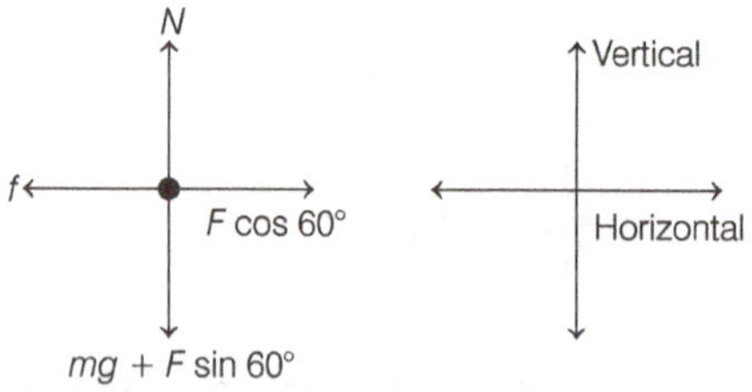

For vertical equilibrium of the block

$$N = mg + F\sin 60°$$
$$= \sqrt{3}g + \sqrt{3}\frac{F}{2} \quad \text{...(i)}$$

For no motion, force of friction

$$f \geq F\cos 60°$$

or $$\mu N \geq F\cos 60°$$

or $$\frac{1}{2\sqrt{3}}\left(\sqrt{3}g + \frac{\sqrt{3}F}{2}\right) \geq \frac{F}{2}$$

or $g \geq \frac{F}{2}$ or $F \leq 2g$ or 20 N

Therefore, maximum value of F is 20 N.

30.

Note *It is not given in the question, but assuming that both blocks are in equilibrium. The free body diagram of two blocks is as shown below,*

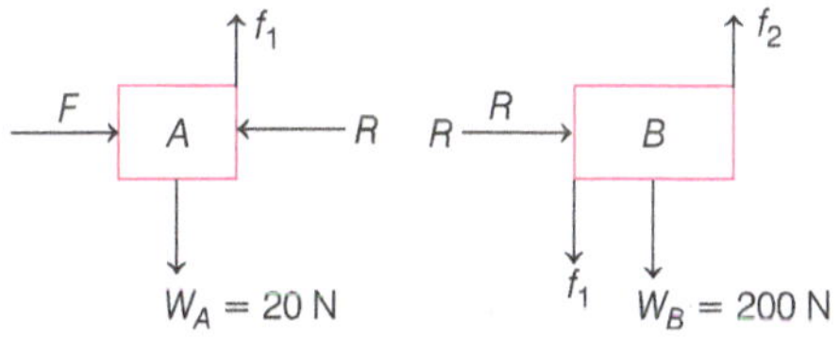

Reaction force, $R =$ applied force F

For vertical equilibrium of A;

$$f_1 = \text{friction between two blocks} = W_A = 20\text{ N}$$

For vertical equilibrium of B;

$$f_2 = \text{friction between block } B \text{ and wall}$$
$$= W_B + f_1 = 100 + 20 = 120\text{ N}$$

31.

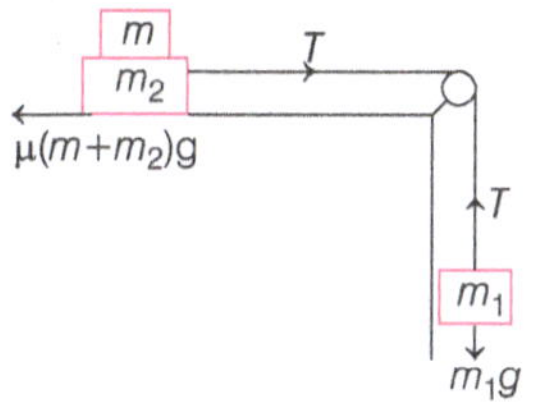

Substituting, $m_1 = 5\text{ kg}$ and $m_2 = 10\text{ kg}$

We get, $\mu(10 + m)g \geq 5g$

$$10 + m \geq \frac{5}{0.15}$$

$\therefore$ $m \geq 23.33\text{ kg}$

32. Case I

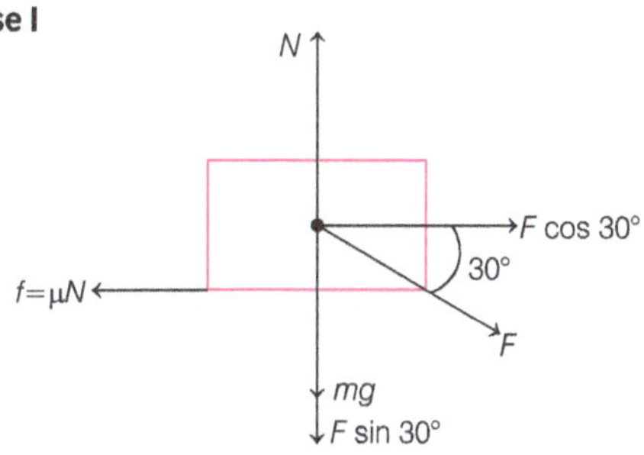

$$N = mg + F\sin 30° = 5 \times 10 + 20 \times \frac{1}{2} = 60\text{ N}$$

$$f = \mu N = 0.2 \times 60 = 12\text{ N}$$

$$F_{\text{net}} = ma_1 = F\cos 30° - f$$

$\Rightarrow$ $ma_1 = 20 \times \frac{\sqrt{3}}{2} - 12$

$\therefore$ $a_1 = \frac{10\sqrt{3} - 12}{5} \approx 1\text{ ms}^{-2}$

Case II

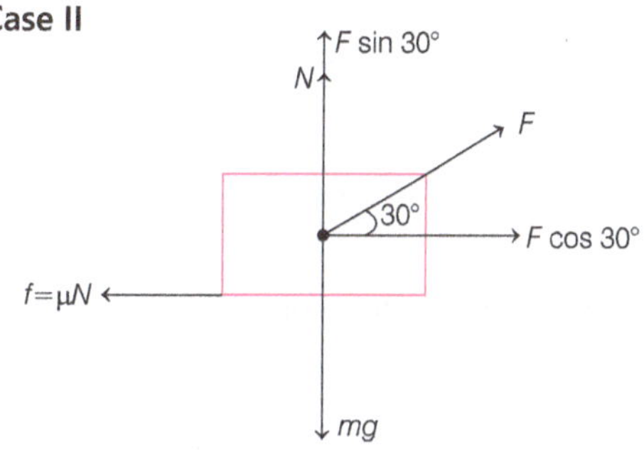

$$F_{\text{net}} = F\cos 30° - f = F\cos 30° - \mu N$$

$\Rightarrow$ $F_{\text{net}} = F\cos 30° - \mu(mg - F\sin 30°)$

$$a_2 = \frac{F_{\text{net}}}{m} = \frac{F\cos 30° - \mu(mg - F\sin 30°)}{m}$$

$$= \frac{20 \times \frac{\sqrt{3}}{2} - 0.2\left(5 \times 10 - 20 \times \frac{1}{2}\right)}{5}$$

$$= \frac{10\sqrt{3} - 8}{5}$$

$\Rightarrow$ $a_2 \approx 1.8\text{ ms}^{-2}$

So, difference $= a_2 - a_1 = 1.8 - 1 = 0.8\text{ ms}^{-2}$

33.

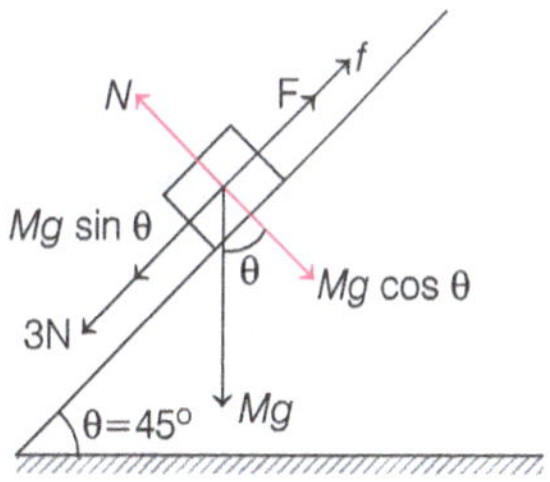

$$3 + Mg\sin\theta = F + f$$

Here, $f = \mu N$

$\therefore$ $3 + Mg\sin\theta = F + \mu N$ ...(i)

or $Mg\cos\theta = N$ ...(ii)

Solving these equations, we get

$$F \simeq 32\text{ N}$$

34. Maximum value of friction between two blocks

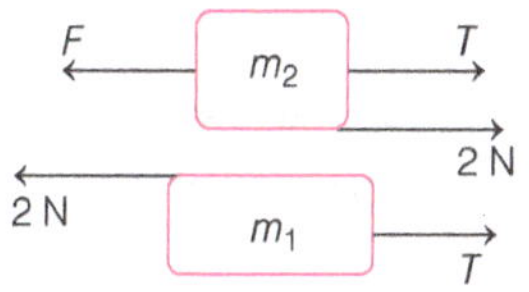

$$f_{\max} = 0.2 \times 1 \times 10 = 2\text{N}$$

In critical case,

$$T = 2\text{ N}$$
$$F = T + 2 = 4\text{ N}$$

$\therefore$ System is in equilibrium, if $f \leq 4\text{ N}$ **Ans.**

For $F > 4$ N

$$F - (T + 2) = m_2 a = (1)(a) \quad \text{...(i)}$$

$$T-2=m_1 a=(1)(a) \quad \text{...(ii)}$$

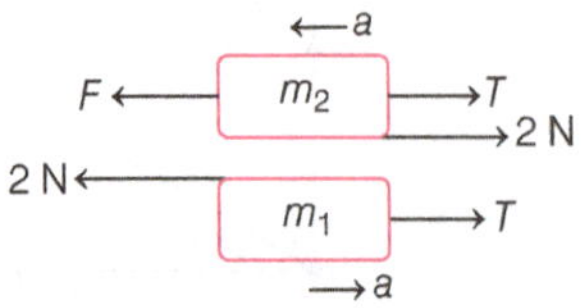

On solving these two equations, we get

$$T=\frac{F}{2}$$

When $F=6\,\text{N}, T=3\,\text{N}$ **Ans.**

35. $a=$ slope of v-t graph

$$=-1\,\text{m/s}^2$$

$\therefore$ Retardation $=1\,\text{m/s}^2=\frac{\mu mg}{m}=\mu g$

or $\mu=\frac{1}{g}=\frac{1}{10}=0.1$

If μ is half, then retardation a is also half. So using

$$v=u-at$$

or $0=u-at$

or $t=\frac{u}{a}$

or $t\propto\frac{1}{a}$

we can see that t will be two times.

36. Since, $\mu_1>\mu_2$

$\therefore$ $(f_1)_{max}>(f_2)_{max}$

Further if both move,

$$a=\frac{T-\mu mg}{m}$$

μ of block is less. Therefore, its acceleration is more.

37. (a) At $\theta=0°$, driving force $F=0$

$\therefore$ friction $=0$

(b) At $\theta=90°, N=0$

$\therefore$ Maximum friction $=\mu N=0$

or friction $=0$

(c) Angle of repose,

$$\theta_r=\tan^{-1}(\mu)=45°$$

Since $\theta<\theta_r$, block is at rest and

$$f=mg\sin\theta$$
$$=2\times10\sin30°$$
$$=10\,\text{N}$$

(d) $\theta>\theta_r$. Therefore block will be moving

$$f=\mu mg\cos\theta$$
$$=(1)(2)(10)\cos60°$$
$$=10\,\text{N}$$

39. (a) $N-10=ma=5\times2$

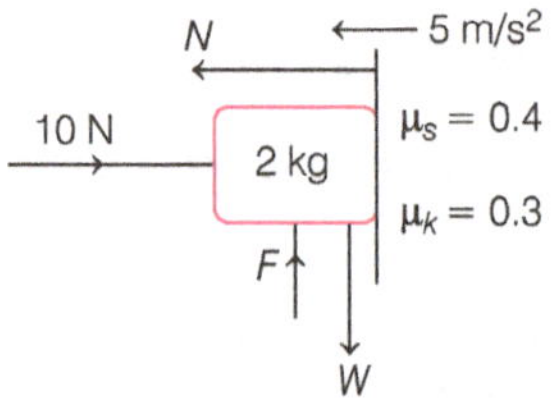

$\therefore$ $N=20\,\text{N} \Rightarrow \mu_s N=8\,\text{N}$

$\mu_k N=6\,\text{N} \Rightarrow W=mg$

$=20\,\text{N}$

(b) When $F=15\,\text{N}$

$W-F=5\,\text{N}$ (downwards)

This is less than $\mu_s N$

$\therefore$ $f=5\,\text{N}$ (upwards)

(c) $F=W-\mu_s N=20-8=12\,\text{N}$

(d) $F=W+\mu_s N=20+8=28\,\text{N}$

41. $\mu mg\cos\theta=15\,\text{N}$

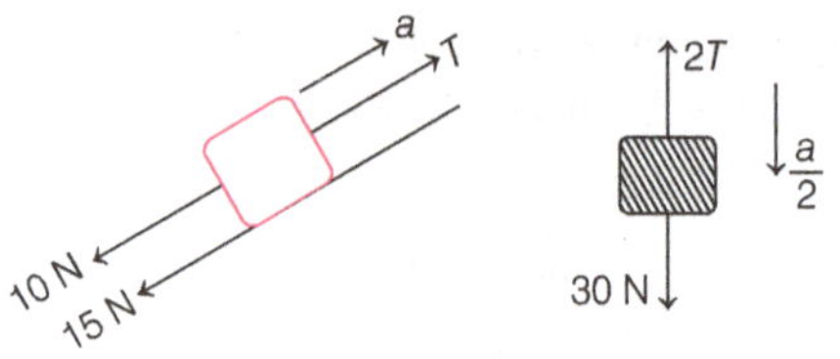

$$T-10-15=2a \quad \text{...(i)}$$

$$30-2T=3\left(\frac{a}{2}\right) \quad \text{...(ii)}$$

Solving these two equations, we get

$$a=-3.63\,\text{m/s}^2$$

So, if we take the other figure,

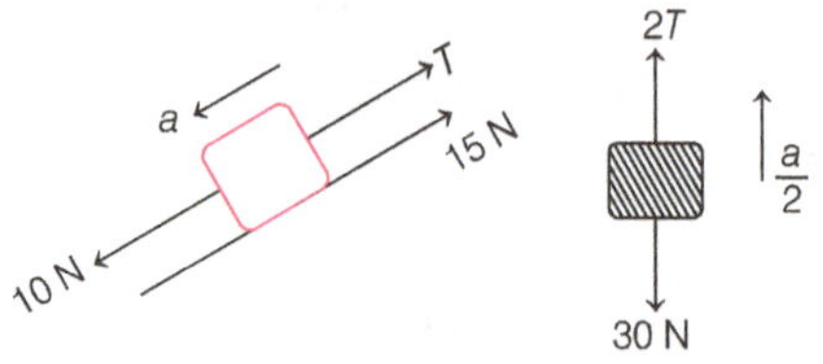

This figure is not feasible. Because for 'a' to be down the plane,

$$10\,\text{N}>T+15$$

which is not possible.

$\therefore$ $a=0$

and free body diagrams are as shown below.

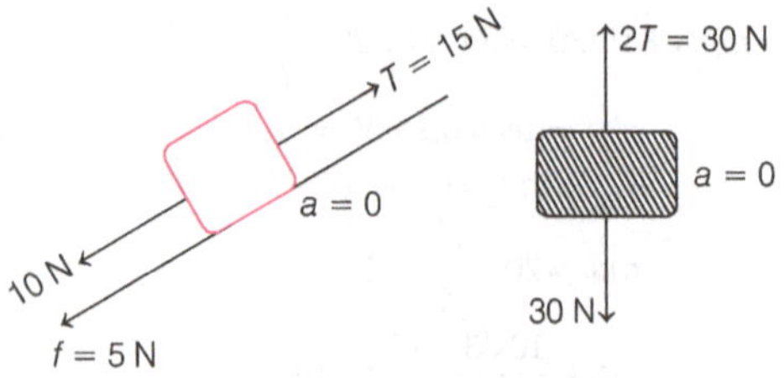

42. The maximum friction on block A is

$$f_{1\,ma} = \mu(20 + t)$$
$$= 10 + \frac{t}{2}$$

(i) When $t = 4\,\text{s}$, $f_{max} = 12\,\text{N}$

The horizontal force on block A is $F_A = t = 4\,\text{N}$, thus static friction acts on the block A.

(ii) When $t = 20\,\text{s}$

$$\because \quad f_{max} = 20 + \frac{20}{2} = 30\,\text{N}$$

$\because$ Horizontal component of F is 20 N and maximum friction is 30 N. Static friction acts on the block A.

(iii) When $t = 40\,\text{s}$

$$f_{max} = 10 + 20 = 30\,\text{N}$$

In this case, limiting friction acts on the block.

$$\because \quad f = 30\,\text{N}$$
$$a = \frac{40 - 30}{5} = 2\,\text{m/s}^2$$

(iv) When $t = 60\,\text{s}$

$$f_{max} = 10 + 30 = 40\,\text{N}$$

The horizontal component of F is

$$F \cos 45° = t = 60\,\text{N}$$

Kinetic friction comes into play.

$$\therefore \quad f = 40\,\text{N}$$
and
$$a = \frac{60 - 40}{5} = 4\,\text{m/s}^2$$

Free body diagram of block A is,

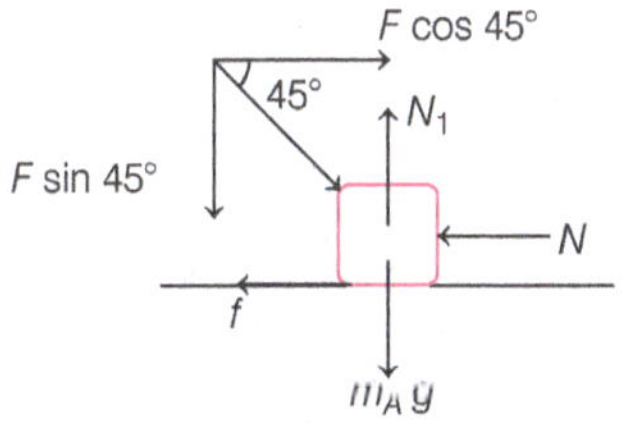

Free body diagram of block B is,

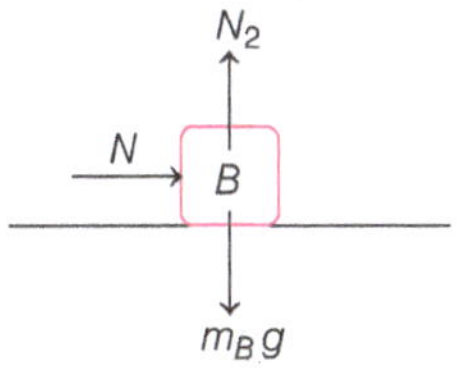

43. In figure, AB is a ladder of weight w which acts at its centre of gravity G.

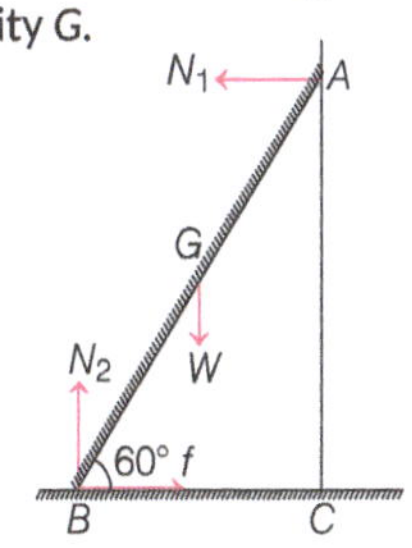

$$\angle ABC = 60°$$
$$\therefore \quad \angle BAC = 30°$$

Let N_1 be the reaction of the wall and N_2 the reaction of the ground.

Force of friction f between the ladder and the ground acts along BC.

For horizontal equilibrium,

$$f = N_1 \qquad \text{...(i)}$$

For vertical equilibrium,

$$N_2 = W \qquad \text{...(ii)}$$

Taking moments about B, we get for equilibrium,

$$N_1(4 \cos 30°) - W(2 \cos 60°) = 0 \qquad \text{...(iii)}$$

Here, $W = 250\,\text{N}$

Solving these three equations, we get

$$f = 72.17\,\text{N}$$
and
$$N_2 = 250\,\text{N}$$
$$\therefore \quad \mu = \frac{f}{N_2} = \frac{72.17}{250} = 0.288 \qquad \textbf{Ans.}$$

44.

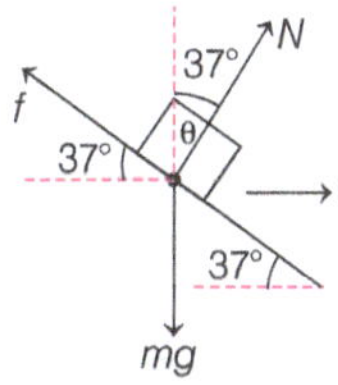

$$N \cos 37° + f \sin 37° = mg \qquad \text{...(i)}$$
$$N \sin 37° - f \cos 37° = ma \qquad \text{...(ii)}$$

On solving these two equations, we get

$$f = 3.6\text{m}$$
$$= \frac{9}{25} mg \qquad \textbf{Ans.}$$

45. $a = \dfrac{f}{m} = \mu g = 3\,\text{m/s}^2$

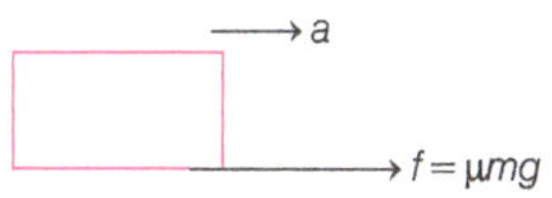

(a) Relative motion will stop when velocity of block also becomes 6 m/s by the above acceleration.

$$v = at$$
$$\therefore \quad t = \frac{v}{a} = \frac{6}{3} = 2\,\text{s} \qquad \textbf{Ans.}$$

(b) $s = \dfrac{1}{2}at^2 = \dfrac{1}{2}(3)(2)^2 = 6\,\text{m}$ **Ans.**

46. (a) When the truck accelerates eastward force of friction on mass is eastwards.

$$f_{\text{required}} = \text{mass} \times \text{acceleration}$$
$$= 30 \times 1.8 = 54\,\text{N}$$

Since it is less than $\mu_s mg$

$\therefore \quad f = 54\text{ N} \quad$ (eastwards)

(b) When the truck accelerates westwards, force of friction is westwards.

$f_{required} = \text{mass} \times \text{acceleration} = 30 \times 3.8 = 114\text{ N}$

Since it is greater than $\mu_s mg$. Hence

$f = f_k = \mu_k mg = 60\text{ N}$ (westwards) **Ans.**

47. Suppose T be the tension in the string attached to block B. Then tension in the string connected to block A would be $4T$.

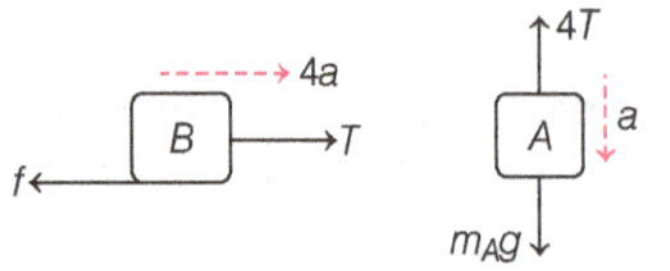

Similarly, if a be the acceleration of block A (downwards), then acceleration of block B towards right will be $4a$.

Equations of motion are

For block A, $\quad m_A g - 4T = m_A a$

or $\quad 50 - 4T = 5a \quad$ ...(i)

For block B, $\quad T - f = 10\,(4a)$

or $\quad T - (0.1)(10)(10) = 40a$

or $\quad T - 10 = 40a \quad$ ...(ii)

Solving Eqs. (i) and (ii), we get

$$a = \frac{2}{33}\text{ m/s}^2$$ **Ans.**

48. $2a_1 s_1 = 2a_2 s_2$

or $$\frac{a_1}{a_2} = \frac{s_2}{s_1} = \frac{n}{m}$$

or $$\frac{g \sin\alpha}{\mu g \cos\alpha - g\sin\alpha} = \frac{n}{m}$$

Solving it, we get

$$\mu = \left(\frac{m+n}{n}\right)\tan\alpha$$ **Ans.**

49. Retardation $a = \mu_k g = 0.15 \times 9.8 = 1.47\text{ m/s}^2$

Distance travelled before sliding stops is,

$$s = \frac{v^2}{2a} = \frac{(5)^2}{2 \times 1.47} \approx 8.5\text{ m}$$ **Ans.**

50. $m_0 = m_1 = m_2 = m$ (say)

Net pulling force is $m_0 g$ or $mg = \text{NPF}$ (say). Friction will act on m_1 and m_2.

Limiting value of friction,

$$f_L = \mu m_1 g + \mu m_2 g = 2\mu mg$$

Possible case-1 → $f_L \geq \text{NPF}$

or $\quad 2\mu mg \geq mg \quad$ or $\quad \mu \geq \frac{1}{2}$

Then acceleration $a = 0$. In this case, force diagram is as shown below.

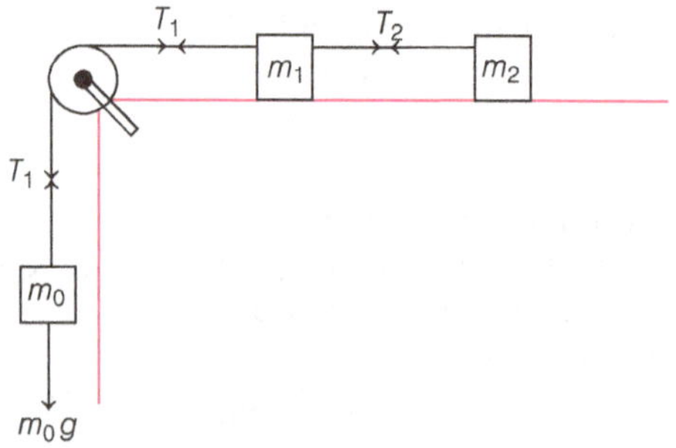

In this case, $T_1 = T_2 = m_0 g$

and friction between the blocks will be zero.

Possible case - 2 → $f_L < \text{NPF}$

or $\quad 2\mu mg < mg \quad$ or $\quad \mu < \frac{1}{2}$

In this case common acceleration of all three blocks will be,

$$a = \frac{\text{NPF} - f_L}{\text{Total mass}} = \frac{mg - 2\mu mg}{3m}$$

or $$a = \left(\frac{1-2\mu}{3}\right)g$$

FBD of m_2

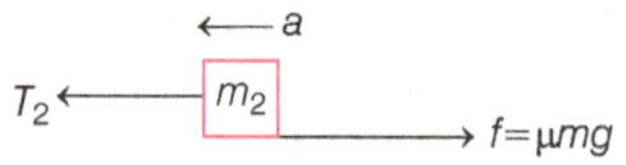

$$T_2 - \mu mg = m_2 a = m\left(\frac{1-2\mu}{3}\right)g$$

$$\therefore \quad T_2 = \left(\frac{1+\mu}{3}\right)mg$$

Here we can write m as m_0, m_1 or m_2.

51. (a) According to the answer given, it has been assumed that both blocks are moving downwards with common acceleration given by,

$$a = \frac{\text{Net pulling or pushing force}}{\text{Total mass}}$$

$$= \frac{(m_1+m_2)g\sin\alpha - \mu_1 m_1 g\cos\alpha - \mu_2 m_2 g\cos\alpha}{(m_1+m_2)}$$

FBD of m_2

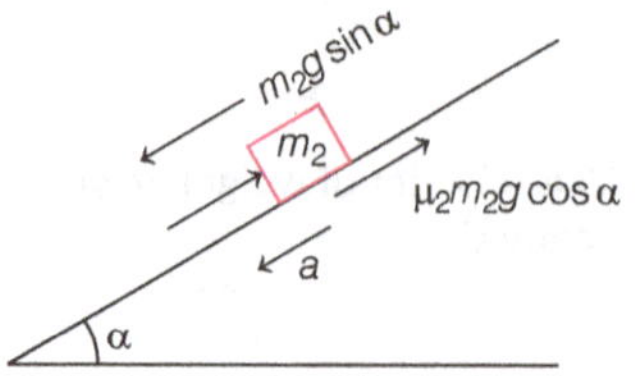

Equation of motion would be,

$$m_2 g\sin\alpha - N - \mu_2 m_2 g\cos\alpha = m_2 a$$

Substituting the value of a and finding the normal reaction N between two blocks, we get

$$N = \frac{(\mu_1 - \mu_2) m_1 m_2 g \cos\alpha}{m_1 + m_2}$$

(b) Bars start sliding down, if

$$(m_1 + m_2)g \sin\alpha > \mu_1 m_1 g \cos\alpha + \mu_2 m_2 g \cos\alpha$$

or $$\tan\alpha_{min} > \frac{\mu_1 m_1 + \mu_2 m_2}{m_1 + m_2}$$

52. In upward journey,

$$\text{retardation } a_1 = \frac{mg \sin\alpha + \mu mg \cos\alpha}{m} = g \sin\alpha + \mu g \cos\alpha$$

In downward journey,

$$\text{acceleration, } a_2 = \frac{mg \sin\alpha - \mu mg \cos\alpha}{m} = g \sin\alpha - \mu g \cos\alpha$$

Further, $s = \frac{1}{2}at^2$ or $t = \sqrt{\frac{2s}{a}}$

or $t \propto \frac{1}{\sqrt{a}}$ $(s \to \text{same})$

$\therefore$ $$\frac{t_1}{t_2} = \sqrt{\frac{a_2}{a_1}}$$

$$\frac{1}{2} = \sqrt{\frac{g \sin\alpha - \mu g \cos\alpha}{g \sin\alpha + \mu g \cos\alpha}}$$

Substituting the proper values, we get

$$\mu = 0.16$$

53. (a) m_2 will start sliding down, if

$$m_2 g > (m_1 g \sin\alpha) + (\mu m_1 g \cos\alpha)$$

or $$\frac{m_2}{m_1} > (\sin\alpha + \mu \cos\alpha)$$

(b) m_2 will start moving up, if

$$(m_1 g \sin\alpha) > (m_2 g) + (\mu m_1 g \cos\alpha)$$

or $$m_2 g < m_1 g \sin\alpha - \mu m_1 g \cos\alpha$$

or $$\frac{m_2}{m_1} < (\sin\alpha - \mu \cos\alpha)$$

(c) From (a) and (b), we can see that between these two ranges of $\frac{m_2}{m_1}$, body m_2 will remain stationary.

Thus the range is,

$$(\sin\alpha - \mu \cos\alpha) < \frac{m_2}{m_1} < (\sin\alpha + \mu \cos\alpha)$$

54. Taking the values from above solution,

$$\sin\alpha + \mu \cos\alpha = \sin 30° + (0.1)\cos 30° = 0.58$$

$$\frac{m_2}{m_1} = \frac{2}{3} = 0.67$$

Since $\frac{m_2}{m_1} > (\sin\alpha + \mu \cos\alpha)$

$\therefore$ m_2 will be moving downwards.

$$\text{Acceleration, } a = \frac{m_2 g - m_1 g \sin\alpha - \mu m_1 g \cos\alpha}{m_1 + m_2}$$

$$= \left[\frac{\frac{m_2}{m_1} - \sin\alpha - \mu \cos\alpha}{1 + \frac{m_2}{m_1}}\right] g$$

$$= \left[\frac{\frac{2}{3} - \sin 30° - 0.1 \cos 30°}{1 + \frac{2}{3}}\right] g$$

$$= 0.05g$$

55. (a) Block will break off the plane when

$$F \sin\alpha = mg$$

or $$kt \sin\alpha = mg$$

or $$t = \frac{mg}{k \sin\alpha} = t_0 \text{ (say)} \quad \text{...(i)}$$

For $t < t_0$

Acceleration of block,

$$a = \frac{F \cos\alpha}{m} = \frac{kt \cos\alpha}{m}$$

Integrating a with respect to time with $v = 0$ at $t = 0$, we get

$$v(t) = kt^2 \cos\alpha / 2m \quad \text{...(ii)}$$

Substituting $t = t_0 = \frac{mg}{k \sin\alpha}$ from Eq. (i) into Eq. (ii), we get

$$v = \frac{mg^2 \cos\alpha}{2k \sin^2\alpha}$$

(b) Integrating $v(t)$ with respect to time with values $s = 0$ at $t = 0$, we get

$$s(t) = \frac{kt^3 \cos\alpha}{6m} \quad \text{...(iii)}$$

Substituting $t = t_0 = \frac{mg}{k \sin\alpha}$ from Eq. (i) into Eq. (iii), we get

$$s = \frac{m^2 g^3 \cos\alpha}{6k^2 \sin^3\alpha}$$

56. At maximum acceleration bar 2 will have a tendency to move upwards. So maximum friction will act downwards.

FBD of 2 with respect to 1, with pseudo force *ma*

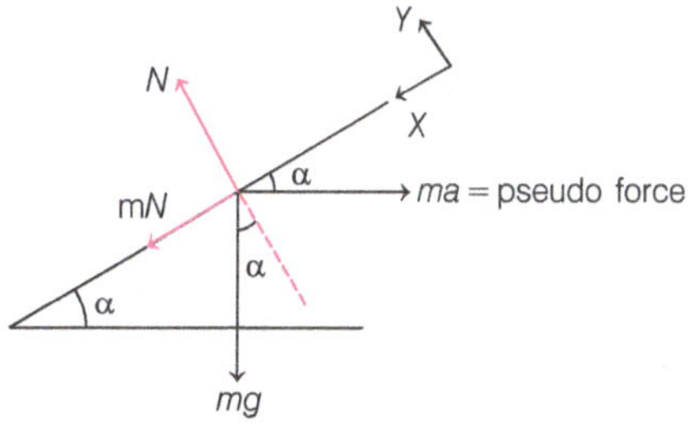

$$\Sigma F_x = 0$$

$\Rightarrow$ $$\mu N + mg\sin\alpha = ma\cos\alpha \quad \text{...(i)}$$

$$\Sigma F_y = 0$$

$$N = mg\cos\alpha + ma\sin\alpha \quad \text{...(ii)}$$

Solving these two equations, we get

$$a = g\left[\frac{1+\mu\cot\alpha}{\cot\alpha - \mu}\right]$$

Note *In the question, it is already given that* $\mu < \cot\alpha$.

57. Block will start moving (say at time t_0) when the applied force is greater than or equal to the limiting friction, or

$$F = \alpha t_0 = \mu N = \mu mg$$

$\therefore$ $$t_0 = \frac{\mu mg}{\alpha}$$

Thus, distance/displacement for $t \le t_0$ is zero.

After $t > t_0$, net acceleration,

$$a = \frac{\text{Applied force} - \text{Friction}}{\text{Mass}}$$

$$= \frac{\alpha t - \mu mg}{m}$$

or $$a = \frac{dv}{dt} = \frac{\alpha}{m}t - \mu g$$

$\Rightarrow$ $$dv = \left(\frac{\alpha}{m}t - \mu g\right)dt$$

Integrating this equation, we get

$$v = \frac{\alpha t^2}{2m} - \mu gt + c_1 \quad \text{...(i)}$$

At $t = t_0, v = 0$

$\therefore$ $$c_1 = \left(\mu g t_0 - \frac{\alpha t_0^2}{2m}\right)$$

Substituting in Eq. (i), we get

$$v = \frac{ds}{dt} = \left(\frac{\alpha t^2}{2m} - \mu gt\right) + \left(\mu g t_0 - \frac{\alpha t_0^2}{2m}\right)$$

or $$ds = \left[\left(\frac{\alpha t^2}{2m} - \mu gt\right) + \left(\mu g t_0 - \frac{\alpha t_0^2}{2m}\right)\right]dt$$

Further integrating, we get

$$s = \frac{\alpha t^3}{6m} - \frac{\mu g t^2}{2} + \left(\mu g t_0 - \frac{\alpha t_0^2}{2m}\right)t + c_2 \quad \text{...(ii)}$$

At $t = t_0, s = 0$

We get, $$c_2 = \frac{\alpha t_0^3}{3m} - \frac{\mu g t_0^2}{2}$$

Substituting in Eq. (ii), we get

$$s = \frac{\alpha t^3}{6m} - \frac{\mu g t^2}{2} + \left(\mu g t_0 - \frac{\alpha t_0^2}{2m}\right)t + \left(\frac{\alpha t_0^3}{3m} - \frac{\mu g t_0^2}{2}\right)$$

Further, $t_0 = \dfrac{\mu mg}{\alpha} \Rightarrow \mu g = \dfrac{\alpha t_0}{m}$

Substituting in above equation, we have

$$s = \frac{\alpha t^2}{6m} - \left(\frac{\alpha t_0}{2m}\right)t^2 + \frac{\alpha t_0^2}{2m}t - \frac{\alpha t_0^3}{6m}$$

$$= \frac{\alpha}{6m}[t^3 - t_0^3 - 3tt_0(t - t_0)]$$

$$= \frac{\alpha}{6m}(t - t_0)^3$$

Thus, $s = 0$ for $t \le 0$

and $$s = \frac{\alpha}{6m}(t - t_0)^3 \text{ for } t > 0$$

where, $$t_0 = \frac{\mu mg}{\alpha}$$

58. $mg\sin\theta = (2)(10)\left(\frac{1}{2}\right) = 10\text{ N} = F_1$ (say)

$$\mu\, mg\cos\theta = \left(\sqrt{\frac{3}{2}}\right)(2)(10)\left(\frac{\sqrt{3}}{2}\right)$$

$$= 21.21\text{ N} = F_2 \text{ (say)}$$

(a) Force required to move the block down the plane with constant velocity.

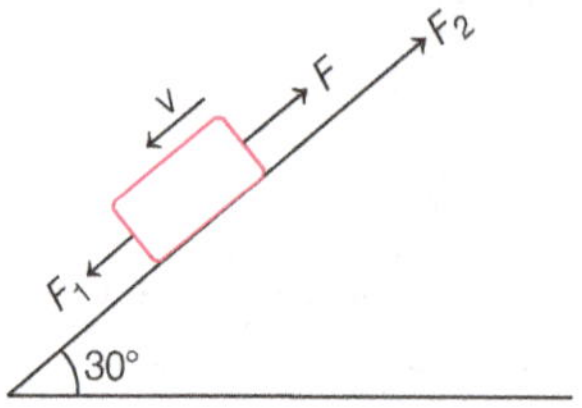

F_1 will be acting downwards, while F_2 upwards.

Since $F_2 > F_1$, force required

$$F = F_2 - F_1 = 11.21\text{N}$$

(b) Force required to move the block up the plane with constant velocity.

F_1 and F_2 both will be acting downwards.

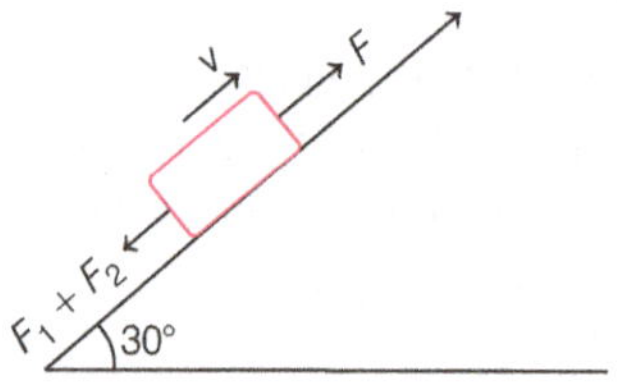

$\therefore$ $$F = F_1 + F_2 = 31.21\text{ N}$$

59. Acceleration of A down the plane,

$$a_A = g\sin 45° - \mu_A g\cos 45°$$

$$= (10)\left(\frac{1}{\sqrt{2}}\right) - (0.2)(10)\left(\frac{1}{\sqrt{2}}\right)$$

$$= 4\sqrt{2}\text{ m/s}^2$$

Similarly acceleration of B down the plane,

$$a_B = g\sin 45° - \mu_B g\cos 45°$$

$$= (10)\left(\frac{1}{\sqrt{2}}\right) - (0.3)(10)\left(\frac{1}{\sqrt{2}}\right)$$

$$= 3.5\sqrt{2}\text{ m/s}^2$$

The front face of A and B will come in a line when

$$s_A = s_B + \sqrt{2}$$

or $$\frac{1}{2}a_A t^2 = \frac{1}{2}a_B t^2 + \sqrt{2}$$

$$\frac{1}{2}\times 4\sqrt{2}\times t^2 = \frac{1}{2}\times 3.5\sqrt{2}\times t^2 + \sqrt{2}$$

Solving this equation, we get $t = 2$ s

Further, $$s_A = \frac{1}{2}a_A t^2 = \frac{1}{2}\times 4\sqrt{2}\times (2)^2 = 8\sqrt{2}\text{ m}$$

Hence, both the blocks will come in a line after A has travelled a distance $8\sqrt{2}$ m down the plane.

60. Acceleration,

$$a = \frac{mg\sin\alpha - \mu mg\cos\alpha}{m}$$

$$a = (g\sin\alpha - kxg\cos\alpha) \qquad (\because \mu = kx)$$

$\therefore$ $$v\cdot\frac{dv}{dx} = (g\sin\alpha - kxg\cos\alpha)$$

or $$v\cdot dv = (g\sin\alpha - kxg\cos\alpha)dx$$

Integrating, we get

$$\frac{v^2}{2} = gx\sin\alpha - \frac{kgx^2\cos\alpha}{2} + c \qquad \text{...(i)}$$

At $x = 0, v = 0 \Rightarrow c = 0$

From Eq. (i), we get

$$v = \sqrt{2gx\sin\alpha - kgx^2\cos\alpha} \qquad \text{...(ii)}$$

$$v = 0 \text{ at } x = 0$$

and $$x = \frac{2}{k}\tan\alpha$$

$\therefore$ Distance covered by the bar till it stops is,

$$d = \frac{2}{k}\tan\alpha$$

From Eq. (ii), we can see that v will be maximum when

$$\frac{d}{dx}[2gx\sin\alpha - kgx^2\cos\alpha] = 0$$

or $$2g\sin\alpha - 2kgx\cos\alpha = 0$$

or $$x = \frac{\tan\alpha}{k}$$

Substituting this value in Eq. (ii), we get

$$v_{max} = \sqrt{\frac{g}{k}\sin\alpha\tan\alpha}$$

61. Let friction on m is f upwards.

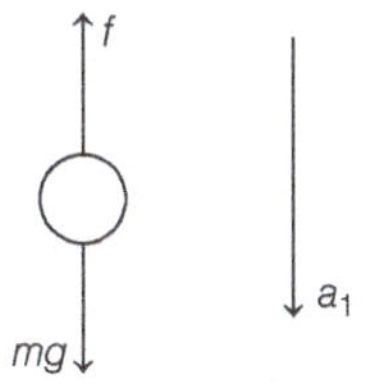

FBD of m

$$a_1 = \frac{mg - f}{m} \qquad \text{...(i)}$$

FBD of M and rope

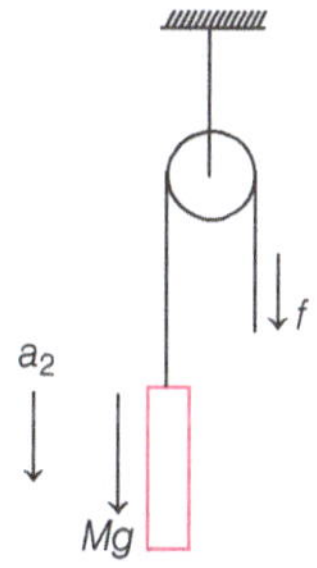

$$a_2 = \frac{Mg - f}{M} \qquad \text{...(ii)}$$

For the ball to get on other end of rod,

$$a_2 > a_1$$

$\therefore$ Relative acceleration,

$$a_r = a_2 - a_1$$

Now, $l = \frac{1}{2}a_r t^2$

or $$l = \frac{1}{2}(a_2 - a_1)t^2$$

or $$l = \frac{1}{2}\left[\frac{Mg - f}{M} - \frac{mg - f}{m}\right]t^2$$

[From Eqs. (i) and (ii)]

Solving this equation, we get

$$f = \frac{2lmM}{(M - m)t^2}$$

Unit Exercise (Cumulative)

Level 1

1. If vector sum of concurrent forces is zero, then all forces can be assumed a pair of two equal and opposite forces acting at one point.

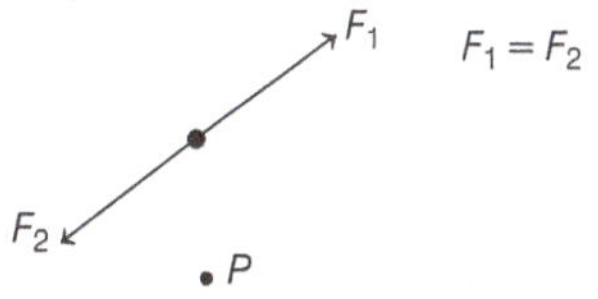

Now, if we take moment about any general point P, then moment of F_1 is clockwise and moment of F_2 (of same magnitude) is anti-clockwise. Therefore net moment is zero.

2. For $(-2\hat{i})\text{ m/s}^2$ component of acceleration, a horizontal force in – ve x-direction is required which can be provided only by the right vertical wall of the box.

3. Net moment about point C should be zero. Or

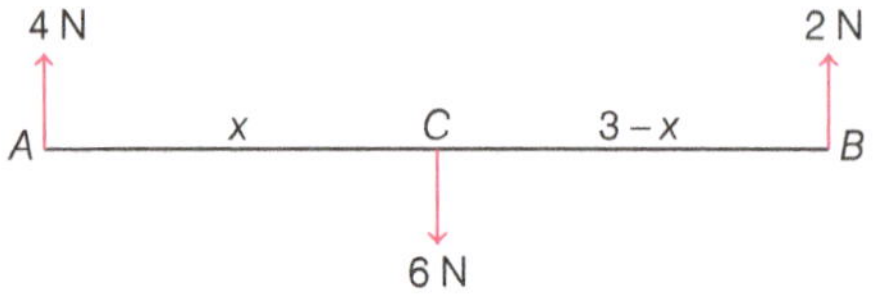

$$4x = 2(3 - x)$$

$\therefore \quad x = 1\text{ m}$

4. At point P,

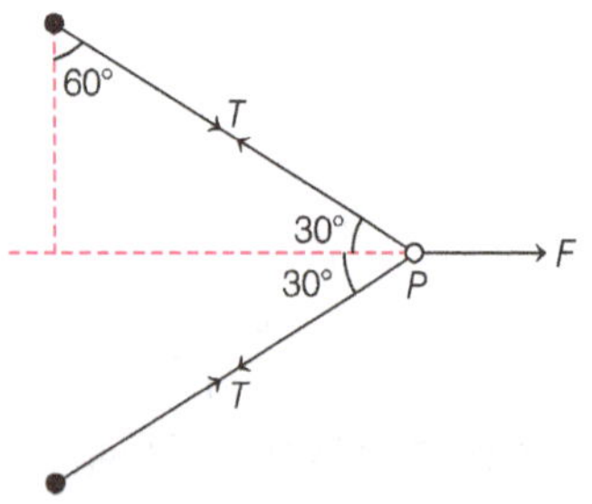

$$F = 2T\cos 30°$$

$\therefore \quad T = \dfrac{F}{\sqrt{3}}$

Acceleration of particle towards each other

$$= \frac{\text{Component of } T \text{ towards each other particle}}{m}$$

$$= \frac{T\cos 60°}{m} = \frac{(F/\sqrt{3})(1/2)}{m}$$

$$= \frac{F}{2\sqrt{3}\,m}$$

5. $a = \dfrac{m_B g - m_A g}{m_A + m_B + m_C} = \dfrac{20}{5} = 4\text{ m/s}^2$

$\therefore \quad s = \dfrac{1}{2}at^2 \quad \text{or} \quad \sqrt{\dfrac{8\times 2}{4}} = t$

$\therefore \quad t = 2\text{ s}$

6. $\mathbf{a} = \dfrac{d\mathbf{v}}{dt} = (8\hat{\mathbf{i}} - 4t\hat{\mathbf{j}})$

At 1 s $\quad \mathbf{F}_{\text{net}} = m\mathbf{a} = (1)(8\hat{\mathbf{i}} - 4\hat{\mathbf{j}}) = (8\hat{\mathbf{i}} - 4\hat{\mathbf{j}})$

$$= \mathbf{W} + \mathbf{F}$$

where, $\mathbf{F}$ = force on cube

$\therefore \quad \mathbf{F} = (8\hat{\mathbf{i}} - 4\hat{\mathbf{j}}) - \mathbf{W}$

$$= (8\hat{\mathbf{i}} - 4\hat{\mathbf{j}}) - (-10\hat{\mathbf{j}})$$

$$= (8\hat{\mathbf{i}} + 6\hat{\mathbf{j}})$$

or $\quad |\mathbf{F}| = \sqrt{(8)^2 + (6)^2}$

$= 10\text{ N}$ **Ans.**

7. $T - (nm + M)g = (nm + M)a$

$\therefore \quad n(mg + ma) = T - Mg - Ma$

or $\quad n = \dfrac{T - M(g + a)}{m(g + a)}$

$$n_{\max} = \frac{T_{\max} - M(g + a)}{m(g + a)}$$

$$= \frac{2\times 10^4 - 500(10 + 2)}{80(10 + 2)}$$

$$= 14.58$$

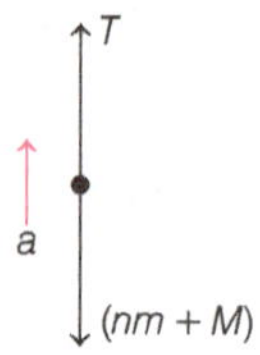

But answer will be 14. **Ans.**

8. Horizontal displacement of both is same (= l). Horizontal force on A is complete T. But horizontal force on B is not complete T. It is component of T. So, horizontal acceleration of B will be less.

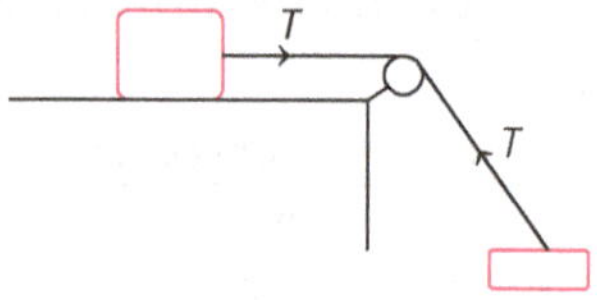

$\therefore \quad t_B > t_A$

9. $T\sin\theta - mg\sin\alpha = ma$

$$T\cos\theta = mg\cos\alpha$$

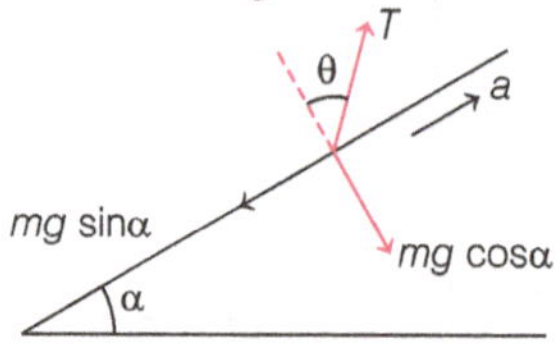

From these two equations, we get

$$\tan\theta = \frac{a + g\sin\alpha}{g\cos\alpha}$$

$$\theta = \tan^{-1}\left(\frac{a + g\sin\alpha}{g\cos\alpha}\right)$$

10. $2F\cos\theta = mg$

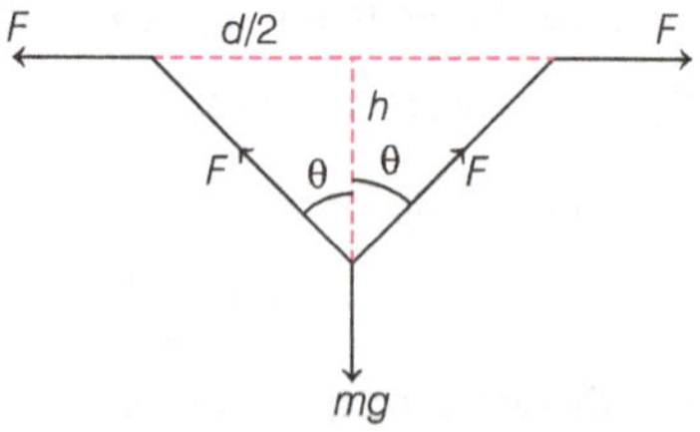

$\therefore \quad F = \dfrac{mg}{2\cos\theta} = \dfrac{mg\sqrt{h^2 + \dfrac{d^2}{4}}}{2h} = \dfrac{mg}{4h}\sqrt{d^2 + 4h^2}$

11. Initially under equilibrium of mass m:

$$T = mg$$

Now, the string is cut. Therefore, $T = mg$, force is decreased on mass m upwards and downwards on mass $2m$.

$\therefore \quad a_m = \dfrac{mg}{m} = g \quad$ (downwards)

and $\quad a_{2m} = \dfrac{mg}{2m} = \dfrac{g}{2} \quad$ (upwards)

12. At point P, $2T\cos\theta = F$

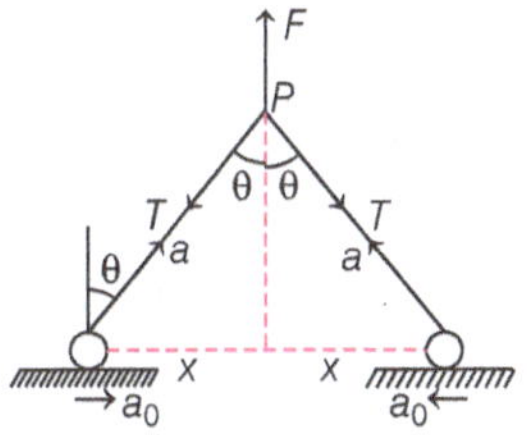

So, $T = \frac{F}{2}\sec\theta$

Acceleration of particle

$$a_0 = \frac{\text{horizontal component of tension}}{\text{mass}} = \frac{T\sin\theta}{m}$$

$$= \frac{F\tan\theta}{2m} = \frac{F}{2m}\frac{x}{\sqrt{a^2 - x^2}}$$

13. The area of force – Time graph = Change in momentum = Impulse

$\therefore \quad p_f - p_i = \frac{1}{2} \times 100 \times 4 = 200$

or $\quad mv - 0 = 200$

$\therefore \quad v = 200\ \text{m/s}$

14. Here, $a_B\cos 53° = a_A\cos 37°$

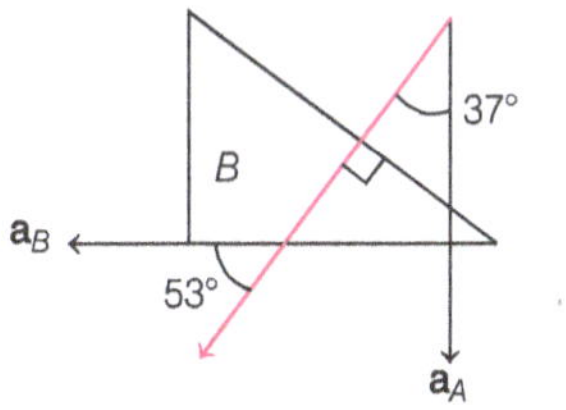

$$a_B \times \frac{3}{5} = 5 \times \frac{4}{5}$$

$\Rightarrow \quad a_B = \frac{20}{3}\ \text{m/s}^2$

$= 6.67\ \text{m/s}^2$

15. $\frac{a_1}{a_2} = \sin\theta$

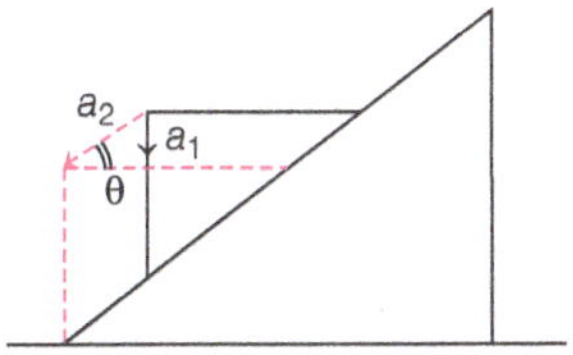

$\therefore \quad a_1 = a_2\sin\theta$

16. $a = \frac{\text{Net pulling force}}{\text{Total mass}}$

$$= \frac{2mg\sin 30°}{2m + m} = \frac{g}{3}$$

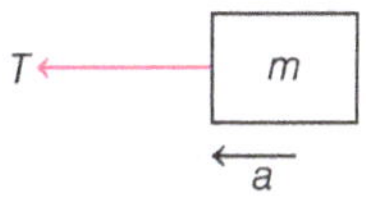

FBD of m

$T = ma = mg/3$

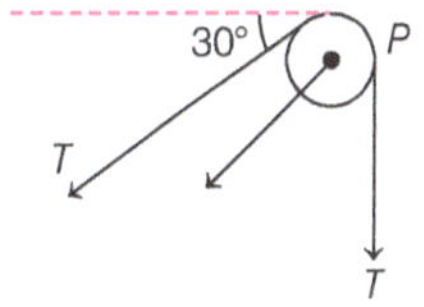

Resultant of tensions

$$\mathbf{R} = -(T\cos 30°)\,\hat{\mathbf{i}} - (T\sin 30° + T)\,\hat{\mathbf{j}}$$

Putting $T = mg/3$

$$\mathbf{R} = -\frac{\sqrt{3}}{6}mg\,\hat{\mathbf{i}} - \frac{mg}{2}\,\hat{\mathbf{j}}$$

Since, pulley P is in equilibrium. Therefore,

$$\mathbf{F} + \mathbf{R} = 0$$

where, $\mathbf{F}$ = force applied by clamp on pulley

$\therefore \quad \mathbf{F} = -\mathbf{R} = \frac{mg}{6}(\sqrt{3}\,\hat{\mathbf{i}} + 3\,\hat{\mathbf{j}})$ **Ans.**

17. $\frac{9}{a} = \tan 37° = \frac{3}{4} \Rightarrow a = 12\ \text{m/s}^2$

Let N = normal reaction between rod and wedge. Then $N\sin 37°$ will provide the necessary ma force to the wedge

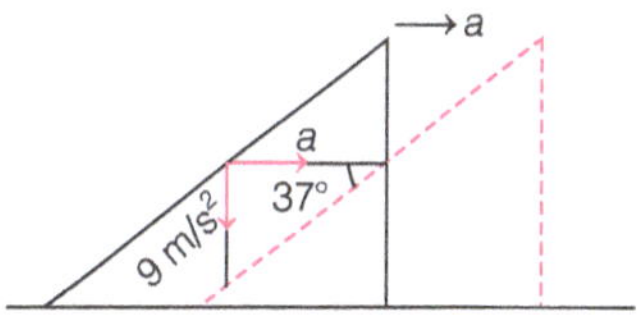

$\therefore \quad N\sin 37° = ma = (10)(12) = 120$

$\therefore \quad N = \frac{120}{\sin 37°} = \frac{120}{0.6}$

$= 200\ \text{N}$ **Ans.**

18. For M_2 and M_3

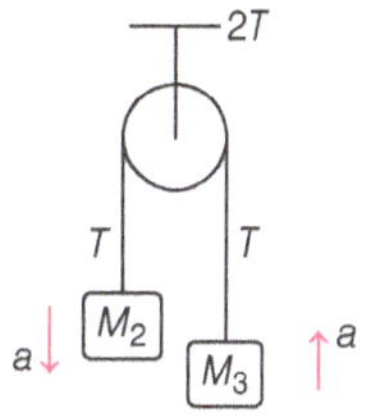

$$a = \frac{M_2g - M_3g}{M_2 + M_3} = \frac{3M_3g - M_3g}{3M_3 + M_3} = \frac{g}{2}$$

Now, FBD of M_2 gives the equation,

$$M_2g - T = M_2 \cdot a = \frac{M_2g}{2}$$

$\therefore \quad T = \frac{M_2 g}{2}$ or $2T = M_2 g$

Now, taking moments of forces about support point

$$M_1 g\,(l_1) = (2T)\,l_2 = (M_2 g)\,(3l_1)$$

$\therefore \quad \frac{M_1}{M_2} = 3$ **Ans.**

19. m_3 is at rest. Therefore,

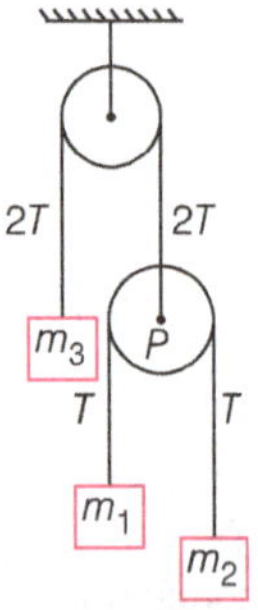

$$2T = m_3 g \quad \text{...(i)}$$

Further if m_3 is at rest, then pulley P is also at rest.
Writing equations of motion,

$$m_1 g - T = m_1 a \quad \text{...(ii)}$$

$$T - m_2 g = m_2 a \quad \text{...(iii)}$$

Solving Eqs. (ii) and (iii), we get

$$m_3 = 1 \text{ kg}$$

20. $\Sigma \mathbf{T} \cdot \mathbf{a} = 0$

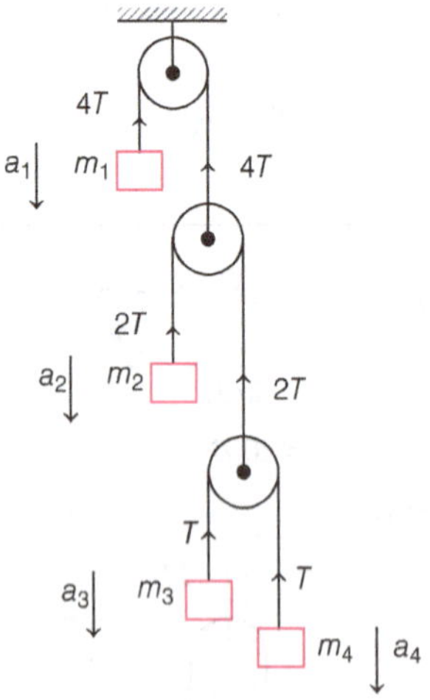

$\Rightarrow \quad -4Ta_1 - 2Ta_2 - Ta_3 - Ta_4 = 0$

$\Rightarrow \quad 4a_1 + 2a_2 + a_3 + a_4 = 0$

21. From free body diagram

$$80 - 2T = 8a \quad \text{...(i)}$$

$$T - 20 = 4a \quad \text{...(ii)}$$

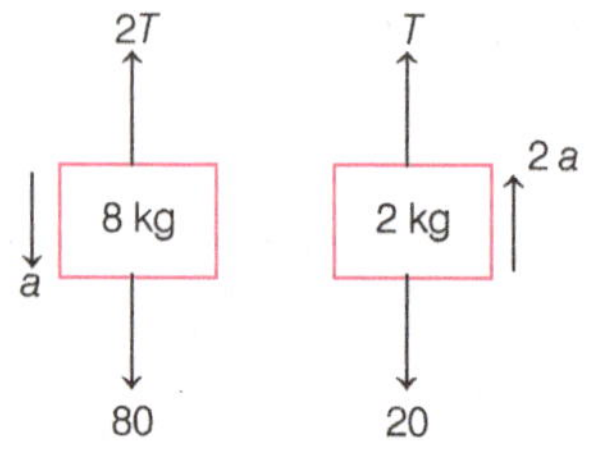

Multiplying Eq. (ii) by 2 and adding with Eq. (i), we get

$$(8 + 8)a = 40$$

$\Rightarrow \quad a = \frac{40}{16} = \frac{10}{4} \text{ m/s}^2$

Using $s = \frac{1}{2}at^2 \quad \Rightarrow \quad t^2 = \frac{2s}{a}$

$\Rightarrow \quad \frac{0.2 \times 2 \times 4}{10} = t^2$

$\Rightarrow \quad t = 0.4 \text{ s}$

22. FBD of rod w.r. t. trolley

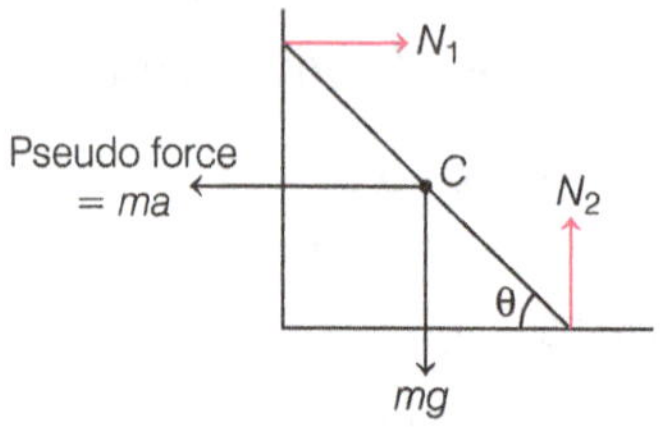

$$N_1 = ma \quad \text{...(i)}$$

$$N_2 = mg \quad \text{...(ii)}$$

Net torque about point $C = 0$

$\therefore \quad N_1\left(\frac{l}{2}\sin\theta\right) = N_2\left(\frac{l}{2}\cos\theta\right)$

or $\quad (ma)(\sin\theta) = (mg)(\cos\theta)$

$\Rightarrow \quad a = g\cot\theta$ **Ans.**

23. Let a_w be the acceleration of wedge and a_b be the acceleration of block w.r.t. wedge.
For the wedge w.r.t. ground

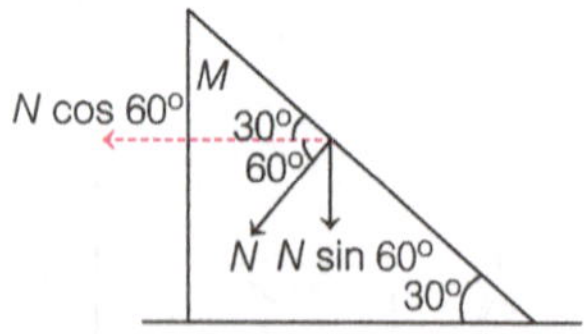

$$N\cos 60° = Ma_w$$

$\Rightarrow \quad N \times \frac{1}{2} = 16\,a_w$

$\Rightarrow \quad N = 32a_w \quad \text{...(i)}$

For block w.r.t. wedge

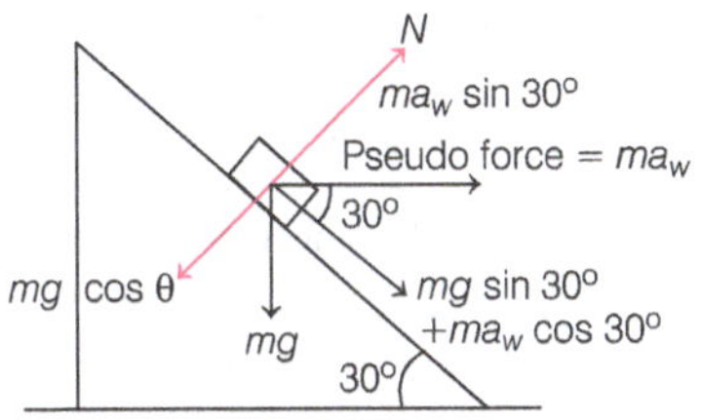

Balancing vertical forces

$$N + ma_w \sin 30° = mg \cos 30°$$

$$\Rightarrow \quad N = 8g \cos 30° - 8a_w \sin 30°$$

$$\Rightarrow \quad 32a_w = 4\sqrt{3}g - 4a_w \quad \text{[Using Eq. (i)]}$$

$$\Rightarrow \quad a_w = \frac{\sqrt{3}}{9}g$$

Along incline plane

$$\Rightarrow \quad m\,g \sin 30° + ma_w \cos 30° = ma_b = 8a_b$$

$$\Rightarrow \quad a_b = \frac{8 \times g \times \frac{1}{2} + 8 \times \frac{\sqrt{3}}{9} \times g \cdot \frac{\sqrt{3}}{2}}{8}$$

$$\Rightarrow \quad a_b = g \times \frac{1}{2} + \frac{\sqrt{3}}{9}g \cdot \frac{\sqrt{3}}{2} = \frac{2g}{3}$$

24. Wedge moves due to horizontal component of normal reaction.

Thus, $a = \frac{N_H}{M} = \frac{N \sin \theta}{M}$ (along $-$ve x-axis)

25. Net force on M in vertical direction should be zero.

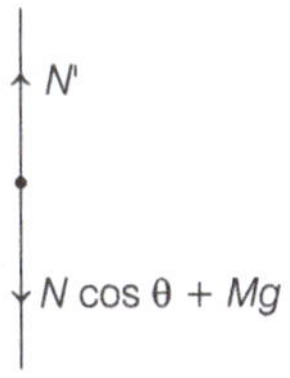

In vertically downwards direction, two forces $N \cos \theta$ and Mg are acting. Therefore N', the normal reaction from ground should be equal to $N \cos \theta + Mg$.

26. $\frac{a}{g} = \cot \theta$

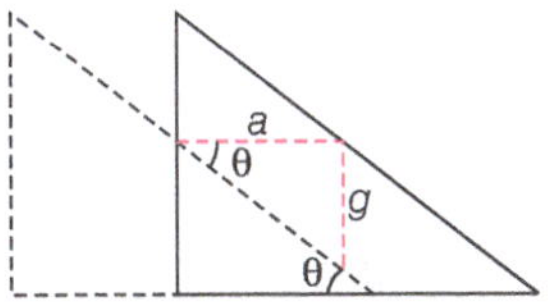

$\therefore \quad a = g \cot \theta$ **Ans.**

27. In vertical direction, net force = 0

$$\therefore \quad N_1 \cos \theta = mg \quad \text{or} \quad N_1 = \frac{mg}{\cos \theta} \quad \text{...(i)}$$

Under normal condition, normal reaction is,

$$N_2 = mg \cos \theta \quad \text{...(ii)}$$

$$\therefore \quad \frac{N_2}{N_1} = \cos^2 \theta \quad \textbf{Ans.}$$

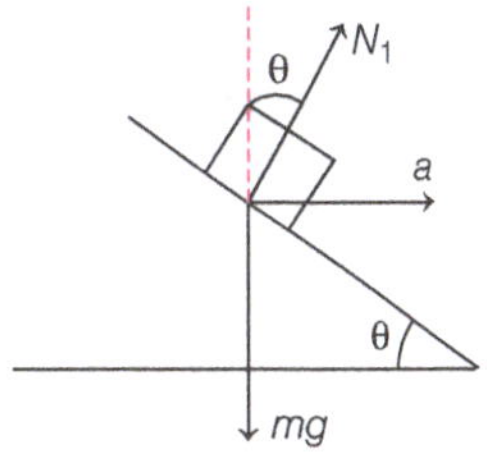

28. Let $\mathbf{a}_B = a\hat{\mathbf{i}}$

Then, $\mathbf{a}_{AB} = \mathbf{a}_A - \mathbf{a}_B$
$= (15 - a)\,\hat{\mathbf{i}} + 15\hat{\mathbf{j}}$

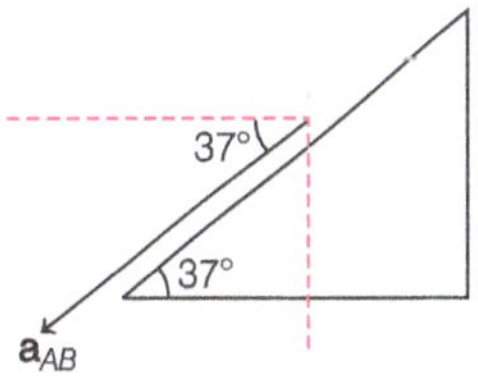

Since, $\mathbf{a}_{AB}$ is along the plane as shown in figure.

$$\therefore \quad \tan 37° = \frac{3}{4} = \frac{15}{15 - a}$$

Solving this equation, we get $a = -5$

or $\quad \mathbf{a}_B = (-5\hat{\mathbf{i}})$

29. Let acceleration of box at this instant is 'a' (towards right).

FBD of ball w.r.t. box

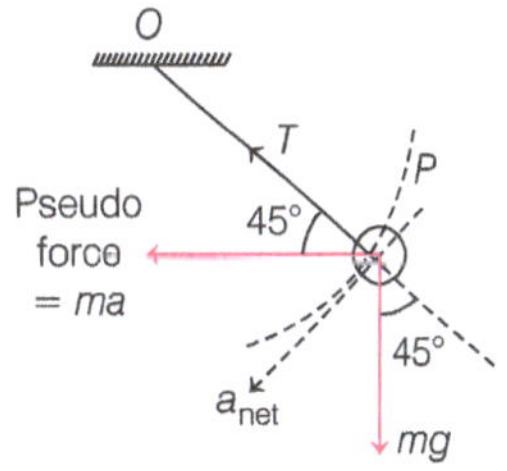

Net force in OP direction is zero.

$$\therefore \quad T + \frac{ma}{\sqrt{2}} = \frac{mg}{\sqrt{2}} \quad \text{...(i)}$$

FBD of box w.r.t. ground

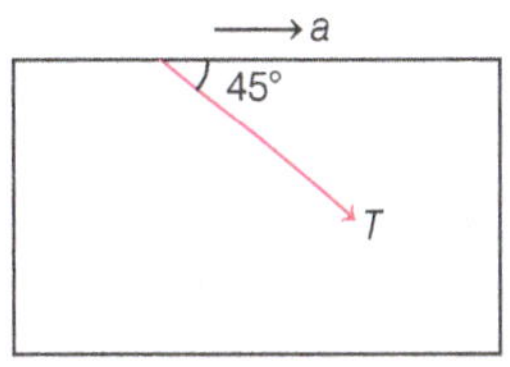

$$T \cos 45° = ma$$

or $\quad T = \sqrt{2}\,ma$

Substituting in Eq. (i), we get

$$a = \frac{g}{3} \qquad \textbf{Ans.}$$

30. Let a_1 be the acceleration of 100 kg block.

FBD of 100 kg block w.r.t. ground

$$F - T - N_1 = 100a_1 \qquad \text{... (i)}$$

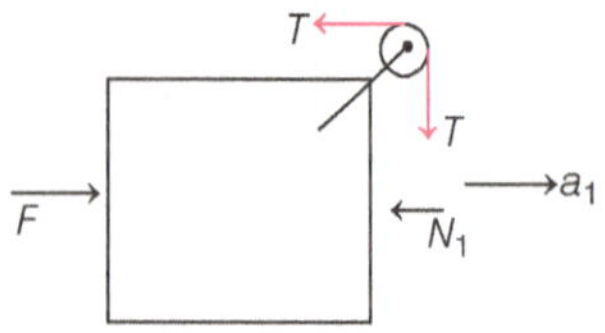

FBD of 20 kg block w.r.t. 100 kg

$$T - 20g = 20(2) \Rightarrow T = 40 + 200$$

$$\Rightarrow \quad T = 240 \qquad \text{...(ii)}$$

$$N_1 = 20a_1 \qquad \text{...(iii)}$$

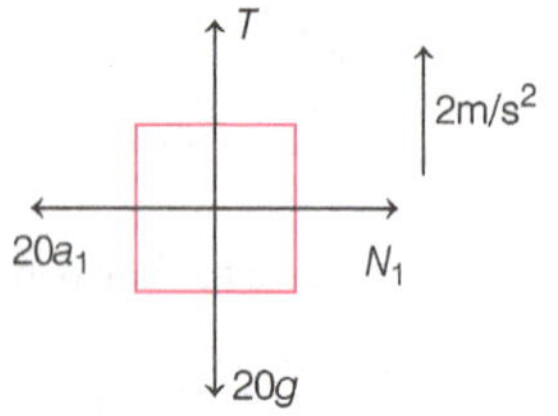

FBD of 10 kg block w.r.t. 100 kg

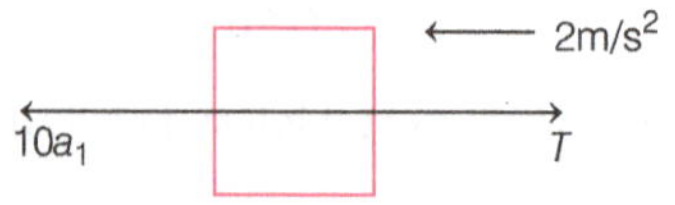

$$10a_1 - 240 = 10(2)$$

$$\Rightarrow \quad a_1 = 26 \text{ m/s}^2$$

$$\because \quad F = 240 + 20(26) + 100 \times 26$$

$$\Rightarrow \quad F = 3360 \text{ N}$$

31. From free body diagram of light rod,

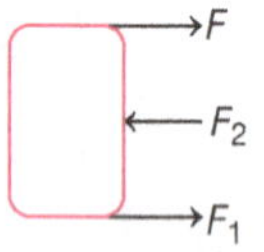

$$\therefore \quad F + F_1 = F_2$$

$$\therefore \quad F_1 < F_2$$

32. From free body diagram of the block,

$$N_2 = 5t \cos 37° = 4t$$

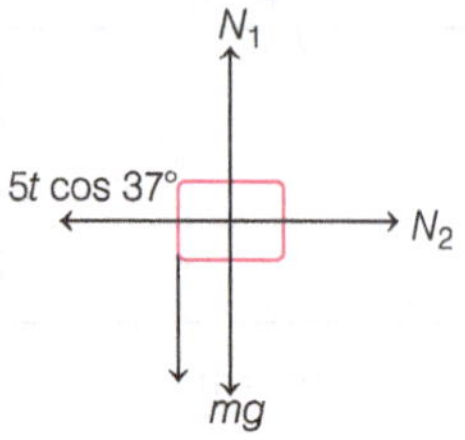

$$N_1 = mg + 5t \sin 37° = 10 + 3t$$

$$\therefore \quad N_1 = 10 + \frac{3}{4}N_2$$

33. Maximum value of friction between A and B

$$(f_1)_{max} = \mu N_1 = \mu m_A g$$
$$= 0.3 \times 50 \times 10$$
$$= 150 \text{ N}$$

Maximum value of friction between B and ground

$$(f_2)_{max} = \mu N_2 = \mu (m_A + m_B) g$$
$$= (0.3)(120)(10) = 360 \text{ N}$$

Force diagram is as shown below

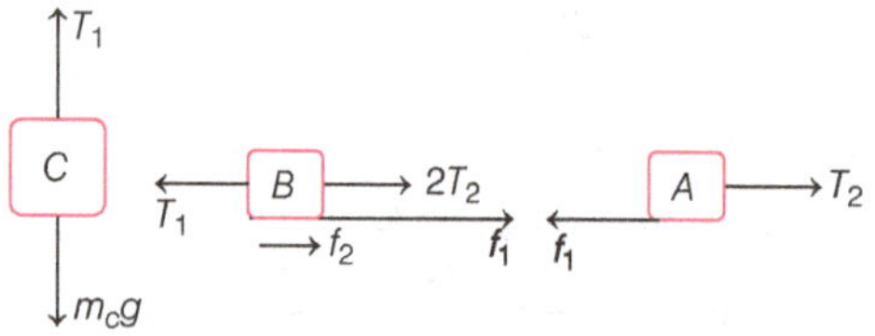

$$T_2 = (f_1)_{max} = 150 \text{ N}$$

$$T_1 = 2T_2 + (f_1)_{max} + (f_2)_{max}$$
$$= 300 + 150 + 360 = 810 \text{ N}$$

$$T_1 = m_c g$$

$$\therefore \quad m_c = \frac{T_1}{g} = \frac{810}{10} = 81 \text{ kg} \qquad \textbf{Ans.}$$

34. Maximum friction available to m_2 is

$$(f_{max}) = \mu m_2 g$$

Therefore, maximum acceleration which can be provided to m_2 by friction, (without the help of normal reaction from m_1) is

$$a_{max} = \frac{f_{max}}{m_2} = \mu g$$

If $a > \mu g$, normal reaction from m_1 (on m_2) is non zero.

35. If they do not slip, then net force on system $= F$

$\therefore$ Acceleration of system $a = \dfrac{F}{3m}$

$$T = F$$

FBD of m $\quad F - \mu mg = ma = \dfrac{F}{3}$

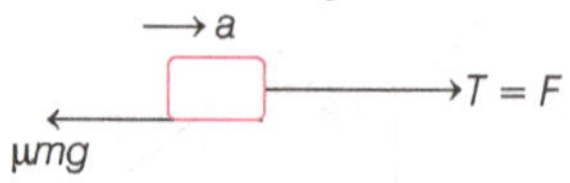

$$\therefore \quad \mu mg = \frac{2F}{3} \quad \text{or} \quad \mu = \frac{2F}{3mg} \qquad \textbf{Ans.}$$

36. $(f_1)_{max} = \mu_1 m_1 g = 6 \text{ N}$

$$(f_2)_{max} = \mu_2 m_2 g = 10 \text{ N}$$

At $t = 2$ s, $F' = 4$ N

Net pulling force,

$$F_1 = F - F' = 11 \text{ N}$$

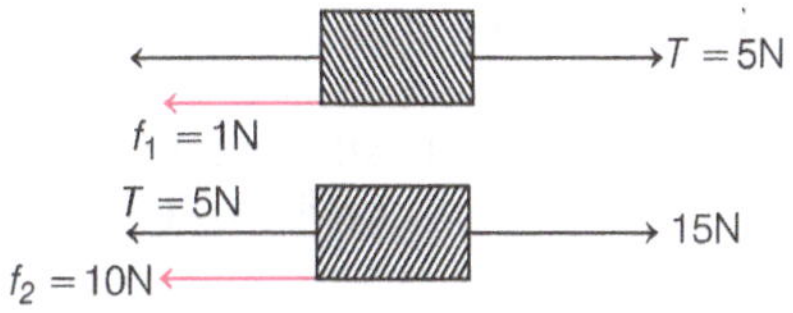

Total maximum resisting force,

$$F_2 = (f_1)_{max} + (f_2)_{max} = 16\text{ N}$$

Since $F_1 < F_2$, system will not move and free body diagrams of the two block are as shown in figure.

37. $(f_{2\text{ kg}})_{max} = \mu_2 m_2 g$

$$= 0.6 \times 2 \times 10 = 12\text{ N}$$

$$(f_{4\text{ kg}})_{max} = \mu_4 m_4 g$$

$$= 0.3 \times 4 \times 10 = 12\text{ N}$$

Net pulling force $F = 16\text{ N}$ and

Net resistive force

$$F' = (f_{2\text{ kg}})_{max} + (f_{4\text{ kg}})_{max}$$

$$= 24\text{ N}$$

Since, $F < F'$, system will not move and free body diagrams of two blocks are as shown below.

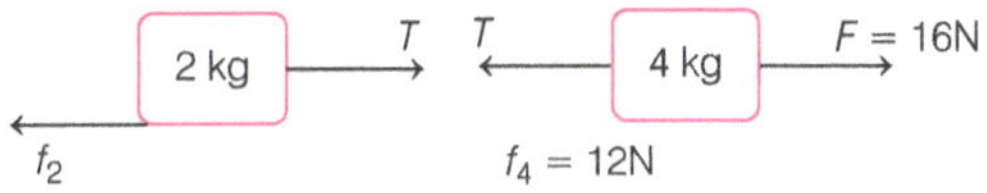

4 kg $T + 12 = 16$

$\Rightarrow$ $T = 4\text{ N}$

2 kg $f_2 = T = 4\text{ N}$ **Ans.**

38. Σ(moments of all forces about point C) = 0

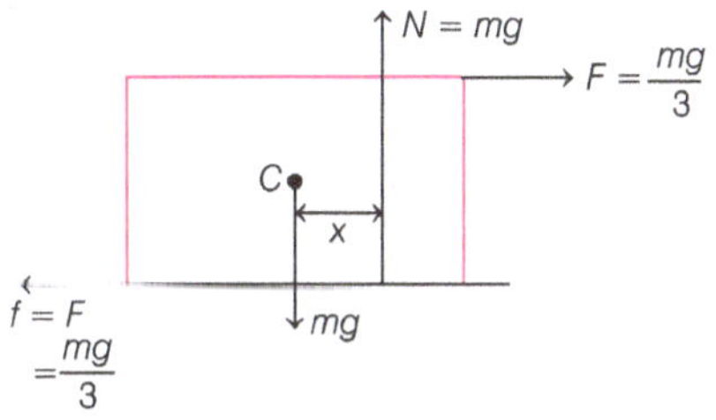

$\therefore$ $Nx = F \cdot \frac{a}{2} + f \cdot \frac{a}{2}$

or $(mg)\,x = \frac{mg}{3}\left(\frac{a}{2}\right) + \frac{mg}{3}\left(\frac{a}{2}\right)$

or $x = \frac{a}{3}$ **Ans.**

39. Maximum value of friction between 10 kg and 20 kg is

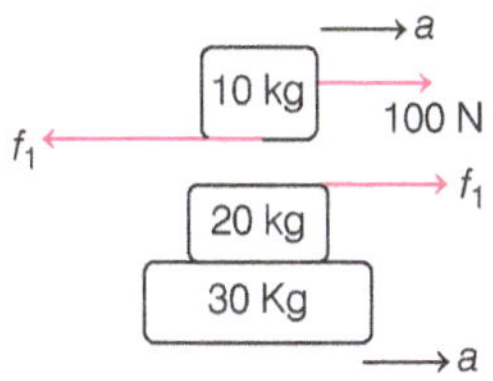

$$(f_1)_{max} = 0.5 \times 10 \times 10 = 50\text{ N}$$

Maximum value of friction between 20 kg and 30 kg is

$$(f_2)_{max} = (0.25)(10 + 20)(10) = 75\text{ N}$$

Now, let us first assume that 20 kg and 30 kg move as a single block with 10 kg block. So, let us first calculate the requirement of f_1 for this

$$100 - f_1 = 10a$$

$$f_1 = 50a$$

On solving these two equations, we get

$$f_1 = 83.33\text{ N}$$

Since, it is greater than $(f_1)_{max}$, so there is slip between 10 kg and other two blocks and 50 N will act here.

Now let us check whether there is slip between 20 kg and 30 kg or not. For this we will have to calculate requirement of f_2 for no slip condition.

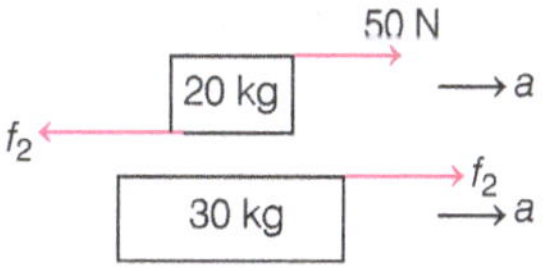

$$50 - f_2 = 20a$$

and $f_2 = 30a$

On solving these two equations, we get

$$f_2 = 30\text{ N}$$

and $a = 1\text{ m/s}^2$

Since, f_2 is less than $(f_2)_{max}$, so there is no slip between 20 kg and 30 kg and both move together with same acceleration of 1 m/s^2.

40. $f_1 \rightarrow$ force of friction between 2 kg and 3 kg

$$(f_1)_{max} = 0.5 \times 3 \times 10$$

$$= 15\text{ N}$$

$f_2 \rightarrow$ force of friction between 2 kg and 1 kg

$$(f_2)_{max} = 0.3 \times 5 \times 10 = 15\text{ N}$$

$f_3 \rightarrow$ force of friction between 1 kg and ground

$$(f_3)_{max} = 0.1 \times 6 \times 10 = 6\text{ N}$$

When $F > 6\text{ N}$ system will start moving with a common acceleration

$$a = \frac{F - 6}{3 + 2 + 1} = \left(\frac{F}{6} - 1\right)\text{m/s}^2$$

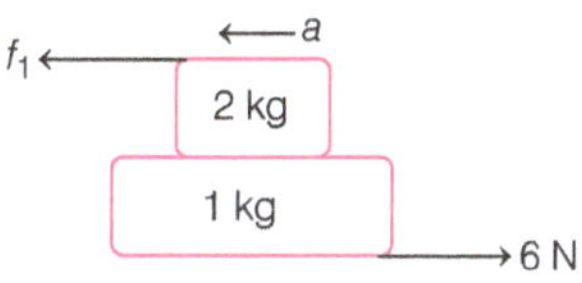

$$f_1 - 6 = (3)\,a = \frac{F}{2} - 3$$

$\therefore$ $f_1 = \left(6 + \frac{F}{2} - 3\right)$

$$=\left(\frac{F}{2}+3\right)$$

Since, F is slightly greater than 6 N

$\therefore \quad f_1 < 15\text{ N} \quad \text{or} \quad < (f_1)_{max}$

$\therefore$ No slipping will occur here

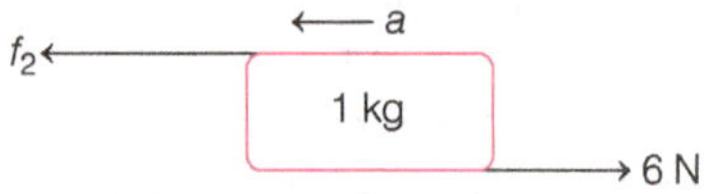

$$f_2 - 6 = (1)(a) = \frac{F}{6} - 1$$

$$\therefore \quad f_2 = \frac{F}{6} + 5$$

Again $f_2 < (f_2)_{max}$. So no slip will take place here also.

41. $(f_1)_{max}$ = between 1 kg and 2 kg

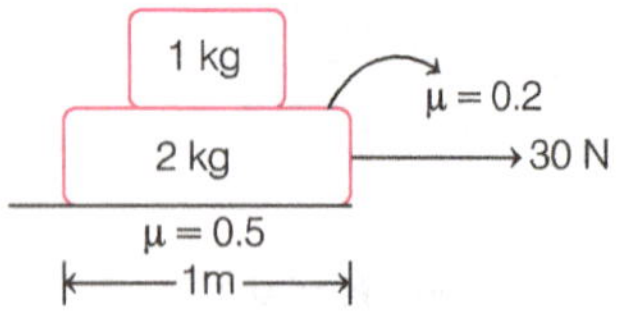

$$= 0.2 \times 1 \times 10 = 2\text{ N}$$

$(f_2)_{max}$ = between 2 kg and ground

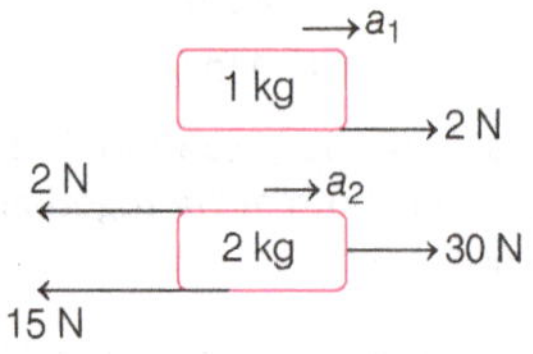

$$= 0.5 \times 3 \times 10 = 15\text{ N}$$

$$a_1 = \frac{2}{1} = 2\text{ m/s}^2$$

$$a_2 = \frac{30 - 15 - 2}{2} = 6.5\text{ m/s}^2$$

$$a_r = a_2 - a_1 = 4.5\text{ m/s}^2$$

$$t = \sqrt{\frac{2S_r}{a_r}} = \sqrt{\frac{2 \times 1}{4.5}} = \frac{2}{3}\text{ s} \qquad \textbf{Ans.}$$

42. $2T = 250$

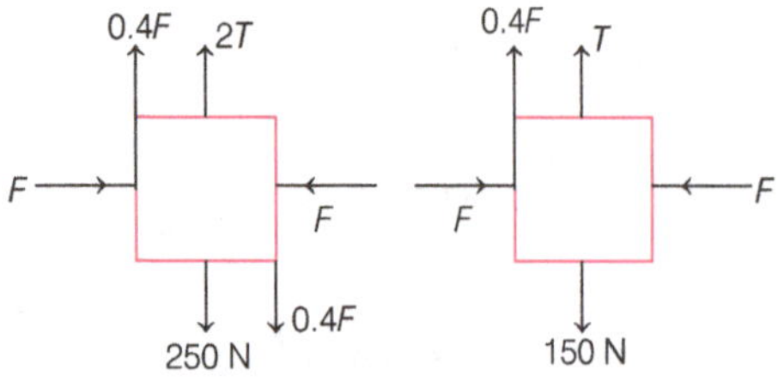

$$\therefore \quad T = 125\text{ N}$$

$$T + 0.4F = 150$$

$$\therefore \quad F = 62.5\text{ N}$$

43. Maximum acceleration of the box can be μg or 1.5 m/s^2, while acceleration of truck is 2 m/s^2. Therefore, relative acceleration of the box will be $a_r = 0.5\text{ m/s}^2$ (backward). It will fall off the truck in a time.

$$t = \sqrt{\frac{2l}{a_r}} \qquad \left(\because s = \frac{1}{2}at^2\right)$$

$$= \sqrt{\frac{2 \times 4}{0.5}} = 4\text{s}$$

Displacement of truck upto this instant is

$$s_t = \frac{1}{2}a_r t^2$$

$$= \frac{1}{2} \times 2 \times (4)^2 = 16\text{m}$$

44. Here, $F - 10 = 8a$ (where, a is the acceleration of the system)

or $\quad 30 - 10 = 8a$

or $\quad a = \frac{20}{8} = \frac{5}{2}\text{ m/s}^2$

Since, blocks are moving together,

Free body diagram of A,

For block A, $F - f = 4a$

or $\quad 30 - 4 \times \frac{5}{2} = f \le f_{max}$

or $\quad 20 \le 30\mu \quad$ or $\quad 30\mu \ge 20$

$\Rightarrow \quad \mu \ge \frac{2}{3} \Rightarrow \mu_{min} = \frac{2}{3}$

45. Free body diagram of block A is as shown below,

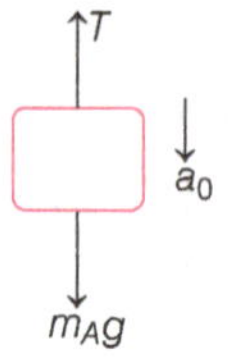

$$m_A g - T = m_A a_0 \qquad \text{...(i)}$$

For all the blocks moving together, common acceleration is a_0.

$$\Rightarrow \quad T = 15a_0 \qquad \text{...(ii)}$$

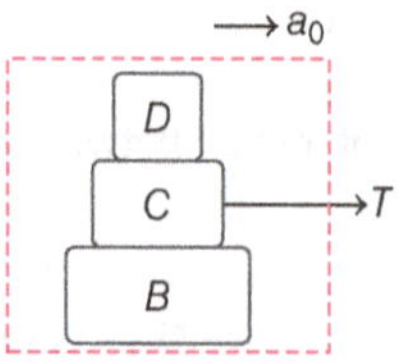

From Eqs. (i) and (ii), we get

$$m_A g - 15a_0 = m_A a_0$$

or $$a_0 = \frac{m_A g}{(15 + m_A)}$$

Free body diagram of block D in the frame of reference of block C.

For all blocks moving together,

$$f_1 \le (f_1)_{max}$$

or $$m_D a_0 \le \mu m_D g$$

or $$a_0 \le \mu g$$

$\Rightarrow$ $$\frac{m_A g}{(15 + m_A)} \le \mu g$$

$$m_A \le (0.5)(15 + m_A)$$

$$m_A - 0.5 m_A \le 0.5 \times 15$$

$\therefore$ $$m_A \le 15$$

$\therefore$ $$m_{A\,max} = 15 \text{ kg}$$

46.

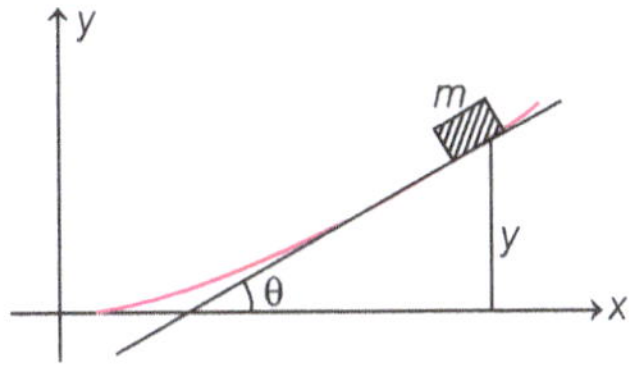

$$\tan\theta = \frac{dy}{dx} = \frac{d\left(\frac{x^3}{6}\right)}{dx} = \frac{x^2}{2}$$

At limiting case θ should be angle of repose and tan of this angle of repose is μ.

$\therefore$ $$\mu = \tan\theta \Rightarrow 0.5 = \frac{x^2}{2}$$

$\Rightarrow$ $$x^2 = 1$$

$\Rightarrow$ $$x = \pm 1$$

Now, putting the value of x in $y = \frac{x^3}{6}$, we get

When $x = 1$	When $x = -1$
$y = \frac{(1)^3}{6} = \frac{1}{6}$	$y = \frac{(-1)^3}{6} = \frac{-1}{6}$

So, the maximum height above the ground at which the block can be placed without slipping is 1/6 m.

47. Acceleration a of system of blocks A and B is

$$a = \frac{\text{Net force}}{\text{Total mass}} = \frac{F - f_1}{m_A + m_B}$$

where, f_1 = friction between B and the surface

$$= \mu(m_A + m_B)g$$

So, $$a = \frac{F - \mu(m_A + m_B)g}{(m_A + m_B)} \quad \text{...(i)}$$

Here, $\mu = 0.2$, $m_A = 1$ kg, $m_B = 3$ kg, $g = 10 \text{ ms}^{-2}$

Substituting the above values in Eq. (i), we have

$$a = \frac{F - 0.2(1+3) \times 10}{1+3}$$

$\Rightarrow$ $$a = \frac{F - 8}{4} \quad \text{...(ii)}$$

Block A moves due to friction between A and B (say f_2).

We consider the limiting case, $m_A a = f_2$

$\Rightarrow$ $$m_A a = \mu(m_A)g$$

$\Rightarrow$ $$a = \mu g = 0.2 \times 10$$

$$= 2 \text{ ms}^{-2} \quad \text{...(iii)}$$

Putting the value of a from Eq. (iii) into Eq. (ii), we get

$$\frac{F - 8}{4} = 2$$

$\Rightarrow$ $$F = 16 \text{ N}$$

48.

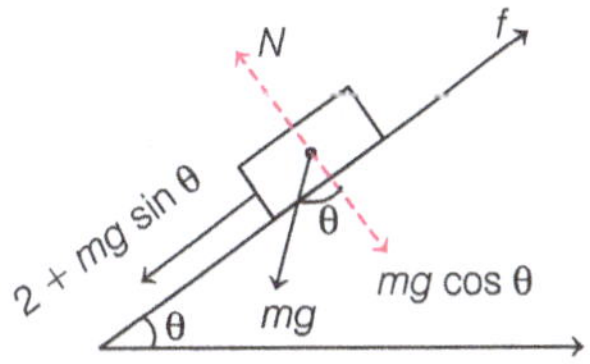

$$N = mg\cos\theta \text{ and } f = \mu N$$

or $$2 + mg\sin\theta = \mu\, mg\cos\theta \quad \text{...(i)}$$

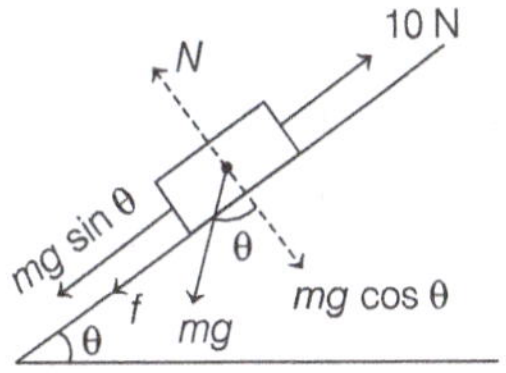

$$mg\sin\theta + f = 10 \text{ N}$$

Here, $$f = \mu N = \mu mg\cos\theta$$

or $$mg\sin\theta + \mu mg\cos\theta = 10 \quad \text{...(ii)}$$

Now, solving Eqs. (i) and (ii), we get

$$\mu = \frac{\sqrt{3}}{2}$$

49. Resultant of mg and mg is $\sqrt{2}\, mg$.

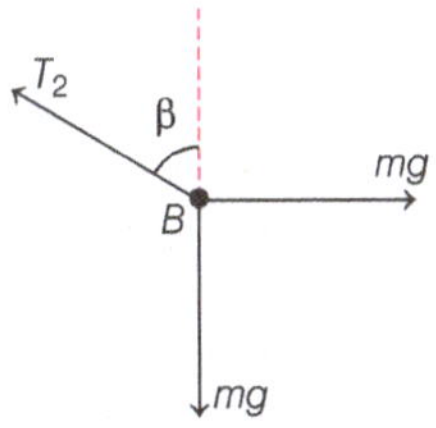

Therefore, T_2 should be equal and opposite of this.

or $$T_2 = \sqrt{2}\, mg \quad \text{...(i)}$$

Further, $$T_2\cos\beta = mg \quad \text{...(ii)}$$

and $$T_2\sin\beta = mg \quad \text{...(iii)}$$

or $$\sin\beta = \cos\beta \Rightarrow \beta = 45°$$

$$T_1 \cos \alpha = mg + T_2 \cos \beta$$
$$= mg + \sqrt{2}\, mg \left(\frac{1}{\sqrt{2}}\right)$$

or $\quad T_1 \cos \alpha = 2\, mg \quad$ ...(iv)

$$T_1 \sin \alpha = T_2 \sin \beta = \sqrt{2}\, mg \left(\frac{1}{\sqrt{2}}\right)$$

$\therefore \quad T_1 \sin \alpha = mg \quad$ ...(v)

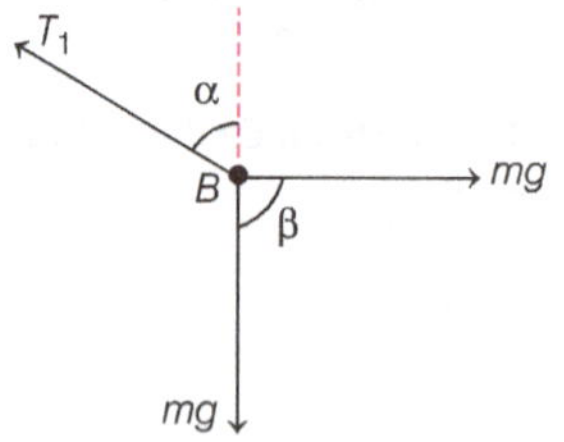

From Eqs. (iv) and (v), we get

$$\tan \alpha = \frac{1}{2} \quad \text{and} \quad T_1 = \sqrt{5}\, mg$$
$$\tan \beta = \tan 45° = 1 \quad \text{and} \quad T_2 = \sqrt{2}\, mg$$
$$\therefore \quad \tan \beta = 2 \tan \alpha \quad \text{and} \quad \sqrt{2}\, T_1 = \sqrt{5}\, T_2$$

50. Maximum force of friction between A and B

$$(f_1)_{max} = 0.3 \times 60 \times 10 = 180 \text{ N}$$

Maximum force of friction between B and ground

$$(f_2)_{max} = 0.3 \times (60 + 40)\, g = 300 \text{ N}$$

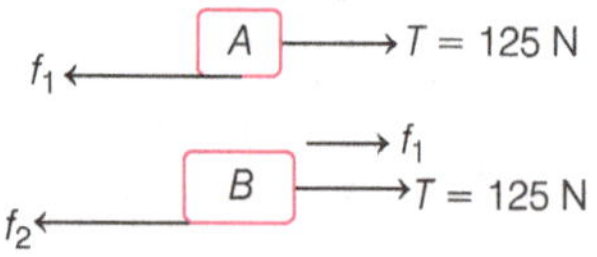

Both are stationary

$$f_1 = T = 125 \text{ N}$$
$$f_2 = T + f_1 = 250 \text{ N}$$

51. $f_1 \rightarrow$ force of friction between 2 kg and 4 kg

$f_2 \rightarrow$ force of friction between 4 kg and ground

$$(f_{s_1})_{max} = 0.4 \times 2 \times 10 = 8 \text{N}$$
$$F_{k_1} = 0.2 \times 2 \times 10 = 4 \text{N}$$
$$(f_{s_2})_{max} = 0.6 \times 6 \times 10 = 36 \text{N}$$
$$F_{k_2} = 0.4 \times 6 \times 10 = 24 \text{N}$$

At $t = 1$ s, $F = 2\text{N} < 36$ N, therefore system remains stationary and force of friction between 2 kg and 4 kg is zero.

At $t = 4$ s, $F = 8$ N < 36 N. Therefore system is again stationary and force of friction on 4 kg from ground is 8 N.

At $t = 15$ s, $F = 30$ N < 36 N and system is stationary.

52. $f_{max} = 0.3 \times 2 \times 10 = 6 \text{N}$

At $t = 2$s, $F = 2\text{N} < f_{max}$

$\therefore \quad f = F = 2\text{N}$

At $t = 8$ s, $F = 8 \text{ N} > f_{max}$

$\therefore \quad f = 6 \text{N}$

At $t = 10$ s, $F = 10 \text{ N} > f_{max}$

$\therefore \quad f = 6 \text{ N}$

$$a = \frac{F - f}{m} = \frac{10 - 6}{2}$$
$$= 2 \text{ m/s}^2$$
$$F = f_{max} = 6 \text{ N at } 6 \text{ s}$$

For $6 \text{ s} \le t \le 10 \text{ s}$

$$a = \frac{F - f}{m} = \frac{t - 6}{2}$$
$$= 0.5t - 3$$
$$\int_0^v dv = \int a\, dt = \int_6^{10} (0.5t - 3)\, dt$$
$$v = 4 \text{ m/s}$$

After 10 s

$$a = \frac{F - f}{m} = \frac{10 - 6}{2}$$
$$= 2 \text{ m/s}^2 = \text{constant}$$
$$\therefore \quad v' = v + at$$
$$= 4 + 2(12 - 10)$$
$$= 8 \text{ m/s}$$

53. Maximum force of friction between 2 kg and 4 kg

$$= 0.4 \times 2 \times 10 = 8 \text{ N}$$

2 kg moves due to friction. Therefore its maximum acceleration may be

$$a_{max} = \frac{8}{2} = 4 \text{ m/s}^2$$

Slip will start when their combined acceleration becomes 4 m/s^2

$$\therefore \quad a = \frac{F}{m} \quad \text{or} \quad 4 = \frac{2t}{6}$$

or $\quad t = 12$ s

At $t = 3$ s

$$a_2 = a_4 = \frac{F}{m} = \frac{2t}{6} = \frac{2 \times 3}{6} = 1 \text{ m/s}^2$$

Both a_2 and a_4 are towards right. Therefore, pseudo forces F_1 (on 2 kg from 4 kg) and F_2 (on 4 kg from 2 kg) are towards left

$$F_1 = (2)(1) = 2 \text{N}$$
$$\Rightarrow \quad F_2 = (4)(1) = 4 \text{ N}$$

From here, we can see that F_1 and F_2 do not make a pair of equal and opposite forces.

54. Acceleration, $a_1 = \dfrac{1350 \times 9.8 - 1200 \times 9.8}{1200}$

$$= 1.225 \text{ m/s}^2$$

Retardation, $a_2 = \dfrac{1200\, g - 1000\, g}{1200}$

$$=1.63\ m/s^2$$

$$h_1 + h_2 = 25 \quad ...(i)$$

$$v = \sqrt{2a_1 h_1} \text{ or } \sqrt{2a_2 h_2}$$

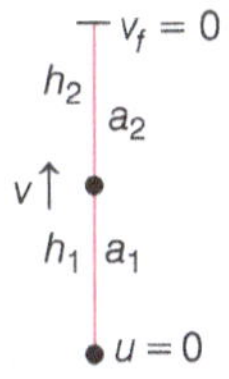

or $$2a_1 h_1 = 2a_2 h_2$$

$\therefore$ $$\frac{h_1}{h_2} = \frac{a_2}{a_1} \quad ...(ii)$$

$$= \frac{1.63}{1.225} = 1.33$$

Solving these equations, we get

$$h_1 = 14.3\ m$$ **Ans.**

55. $v = \sqrt{2a_1 h_1}$

$$= \sqrt{2 \times 1.225 \times 14.3}$$

$$= 4.92\ m/s$$ **Ans.**

56. Let $\mu_k = \mu$, then $\mu_s = 2\mu$

According to first condition,

$$F + mg \sin\theta = \mu_s mg \cos\theta = 2\mu\, mg \cos\theta \quad ...(i)$$

According to second condition,

$$mg \sin\theta = F + \mu_k mg \cos\theta$$

$$= F + \mu\, mg \cos\theta \quad ...(ii)$$

Putting $\theta = 30°$, we get

$$F + mg/2 = 2\mu\, mg\left(\frac{\sqrt{3}}{2}\right)$$

or $$\sqrt{3}\mu\, mg = F + 0.5\, mg \quad ...(iii)$$

$$\frac{mg}{2} = F + \mu\, mg\left(\frac{\sqrt{3}}{2}\right)$$

or $$0.5\sqrt{3}\mu\, mg = 0.5\, mg - F \quad ...(iv)$$

Dividing Eq. (iii) and (iv), we get

$$F = \frac{mg}{6}$$ **Ans.**

57. Substituting the value of F in Eq. (iii), we have

$$\mu = \frac{2}{3\sqrt{3}} = \mu_k$$

$\therefore$ $$\mu_s = 2\mu = \frac{4}{3\sqrt{3}}$$

58. $$a = \frac{F + mg \sin\theta - \mu_k mg \cos\theta}{m}$$

$$= \frac{(mg/6) + (mg/2) - \left(\frac{2}{3\sqrt{3}}\right) mg\left(\frac{\sqrt{3}}{2}\right)}{m}$$

$$= \frac{g}{3}$$ **Ans.**

59. $F' = mg \sin\theta + \mu_s mg \cos\theta$

$$= \frac{mg}{2} + \frac{4}{3\sqrt{3}} mg\left(\frac{\sqrt{3}}{2}\right) = \frac{7mg}{6}$$ **Ans.**

60. $F'' = mg \sin\theta + \mu_k mg \cos\theta$

$$= (mg/2) + \left(\frac{2}{3\sqrt{3}}\right) mg\left(\frac{\sqrt{3}}{2}\right) = \frac{5mg}{6}$$ **Ans.**

61. $\tan\theta = \frac{8}{15}$

$$(f_A)_{max} = 0.2 \times 170 \times 10 \times \cos\theta$$

$$= 300\ N$$

$$(f_B)_{max} = 0.4 \times 170 \times 10 \times \cos\theta$$

$$= 600\ N$$

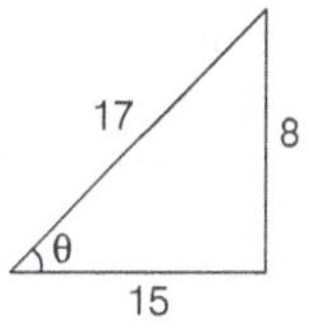

Now, $(m_A + m_B) g \sin\theta = (340)(10) \sin\theta = 1600\ N$

Since, this is greater than $(f_A)_{max} + (f_B)_{max}$, therefore blocks slides downward and maximum force of friction will act on both surfaces.

$\therefore$ $$f_{total} = (f_A)_{max} + (f_B)_{max}$$

$$= 900\ N$$ **Ans.**

62. $$a = \frac{(m_A + m_B) g \sin\theta - f_{total}}{m_A + m_B}$$

$$= \frac{1600 - 900}{340} = 2.06\ m/s^2$$

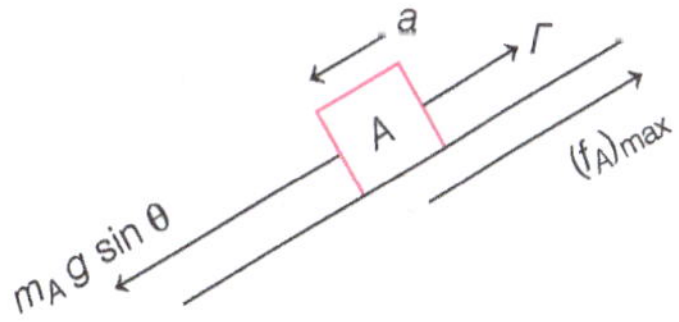

F = force on connecting bar

$$m_A g \sin\theta - F - (f_A)_{max} = m_A a$$

$\therefore$ $$F = m_A g \sin\theta - (f_A)_{max} - m_A a$$

$$= 170 \times 10 \times \sin\theta - 300 - 170 \times 2.06$$

$$\approx 150\ N$$ **Ans.**

63. $F = 2t$

$$\mu_s mg = 20\mu_s$$

$$\mu_k mg = 20\mu_k$$

(a) Motion starts at 4 s

$\therefore$ $$F = \mu_s mg$$

$\Rightarrow$ $$(2)(4) = 20\mu_s$$

$\therefore \quad \mu_s = 0.4$

(b) At 4 s, when motion starts,

$$a = \frac{F - \mu_k mg}{m}$$

$$\therefore \quad 1 = \frac{8 - 20\mu_k}{2}$$

Solving we get, $\mu_k = 0.3$

(c) At $t = 0.1$ s, when motion has not started,

$$f = F = 2 \times 0.1$$
$$= 0.2 \text{ N}$$

(d) **At 8 s**

$$a = \frac{F - \mu_k mg}{m}$$
$$= \frac{2 \times 8 - 0.3 \times 20}{2}$$
$$= 5 \text{ m/s}^2$$

$$\therefore \quad \frac{a}{10} = 0.5 \text{ m/s}^2 \qquad \textbf{Ans.}$$

64. Net pulling force F = net resisting frictional force at $A = 10$ N

(b) $f_c = 0$

(c) $N_c = (m_B + m_C)g = 20$ N

(d) $T = F = 10$ N (everywhere)

65. Ground is smooth. So all blocks will move towards right 2 kg and 5 kg blocks due to friction.

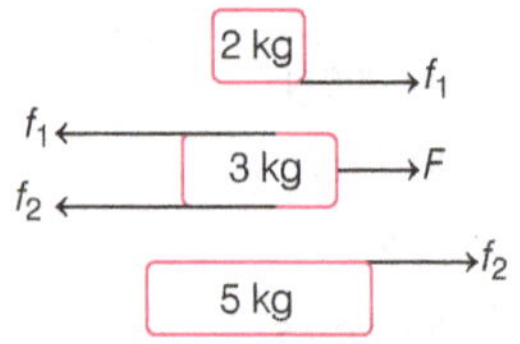

66. Block will not slip if

$$(m_1 + m_2) g \sin\theta \le \mu m_2 g \cos\theta$$

$$\Rightarrow \quad 3 \sin\theta \le \left(\frac{3}{10}\right)(2) \cos\theta$$

$$\tan\theta \le 1/5$$

$$\Rightarrow \quad \theta \le 11.5°$$

(P) $\theta = 5°$ friction is static

$$f = (m_1 + m_2) g \sin\theta$$

(Q) $\theta = 10°$ friction is static

$$f = (m_1 + m_2) g \sin\theta$$

(R) $\theta = 15°$ friction is kinetic

$$f = \mu m_2 g \cos\theta$$

(S) $\theta = 20°$ friction is kinetic

$$\Rightarrow f = \mu m_2 g \cos\theta$$

67.

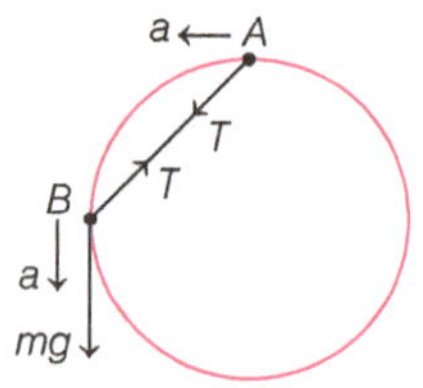

For **A** $\quad T \cos 45° = ma$

or $\quad T = \sqrt{2}\, ma \qquad$ ...(i)

For **B** $\quad mg - T\cos 45° = ma$

$\therefore \quad mg - ma = ma$

or $\quad a = \dfrac{g}{2}$

Substituting in Eq. (i), we get

$$T = \frac{mg}{\sqrt{2}}$$

68. $a = \dfrac{\text{Net pulling force}}{\text{Total mass}}$

$$= \frac{2g - (1)g}{2 + 1} = \frac{g}{3} = \frac{10}{3} \text{ m/s}^2$$

After 1 s, $\quad v = at = \dfrac{10}{3}$ m/s

At this moment string slacks ($T = 0$)

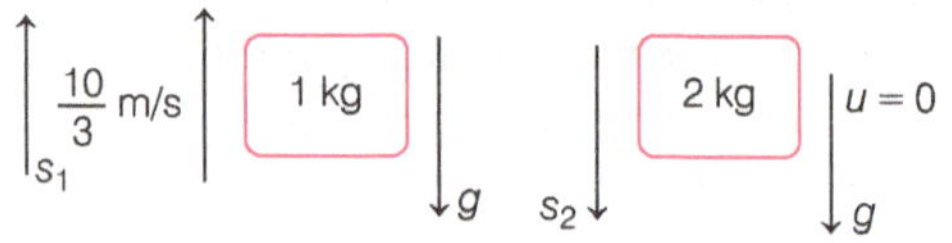

String is again tight when

$$s_1 = s_2$$

$$\therefore \quad \frac{10}{3}t - \frac{1}{2}gt^2 = \frac{1}{2}gt^2 \qquad (g = 10 \text{ m/s}^2)$$

On solving, we get

$$t = \frac{1}{3} \text{ s} \qquad \textbf{Ans.}$$

69. Force diagram on both sides is always similar. Therefore motion of both sides is always similar. For example, if monkey accelerates upwards, then $T > 20g$. But same T is on RHS also.

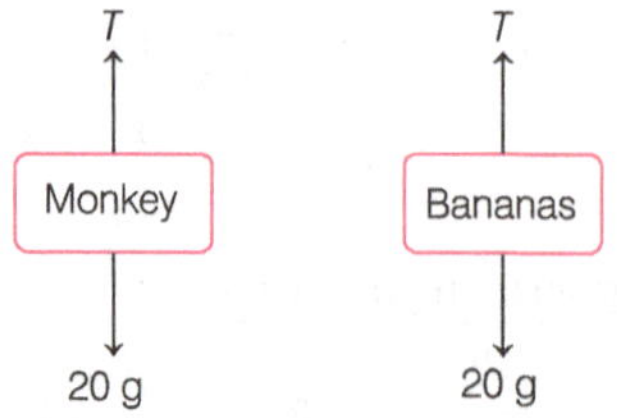

Therefore, bananas also accelerate upwards.

70. It is just like a projectile motion with g to be replaced by $g\sin 45°$.

After 2 s,

$$v=\sqrt{v_x^2+v_y^2}$$

$$=\sqrt{\left(10\sin 45°-\frac{g}{\sqrt{2}}\times 2\right)+(10\cos 45°)^2}$$

$$=10\,\text{m/s}$$ **Ans.**

71. Let B and C both move upwards (alongwith their pulleys) with speeds v_B and v_C then we can see that, A will move downward with speed, $2v_B+2v_C$. So, with sign we can write,

$$v_B=\frac{v_A}{2}-v_c$$

Substituting the values we have, $v_B=0$ **Ans.**

72.

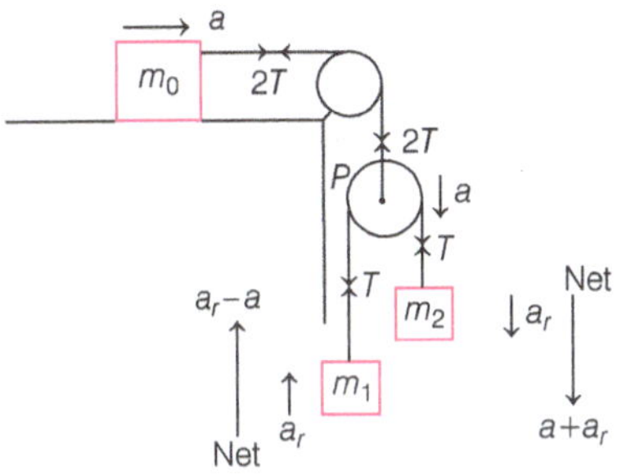

Let $a_{m_0}=a_P=a$, in the directions shown in figure.

With respect to pulley P,

let $a_{m_1}=a_{m_2}=a_r$, in the directions shown in figure.

Then net accelerations of m_1 and m_2 are a_r-a and a_r+a in the directions shown.

Equations of motion of m_0, m_1 and m_2 will be,

$$2T=m_0a \quad \text{...(i)}$$

$$T-m_1g=m_1(a_r-a) \quad \text{...(ii)}$$

$$m_2g-T=m_2(a_r+a) \quad \text{...(iii)}$$

Solving Eqs. (i), (ii) and (iii), we can find T, a and a_r and then a_{m_1} or a_r-a. Answer is already given.

Possible cases

Net acceleration of m_1 may be upwards, downwards or zero. In physics, whenever you get this type of situation (means three possible cases) you get an answer like,

$$a_{m_1}=k(\alpha-\beta)$$

Here k, α and β are constants.

Now three possible cases are :

(i) $a_{m_1}=0$ if $\alpha=\beta$

(ii) $a_{m_1}=$ positive if $\alpha>\beta$ and

(iii) $a_{m_1}=$ negative if $\alpha<\beta$

73. Let $m_A=m$

Then $m_B=\eta m$

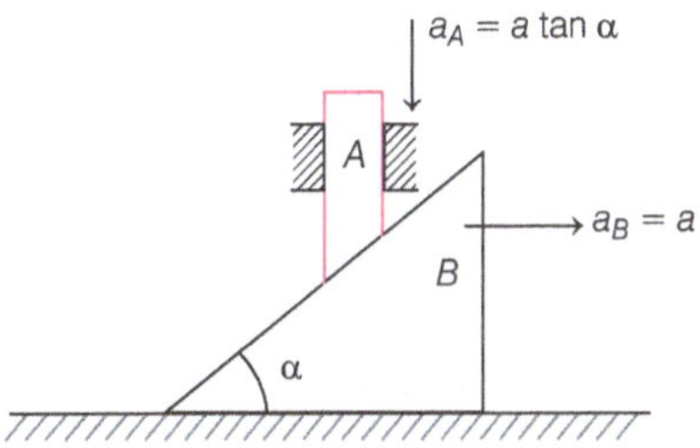

From constraint relations, we can prove that,

$$\frac{a_A}{a_B}=\tan\alpha$$

So, if $a_B=a$ (in horizontal direction)

Then, $a_A=a\tan\alpha$ (in vertical direction)

Let N be the normal reaction between A and B.

Vertical component of N comes out to be $N\cos\alpha$ and the horizontal component $N\sin\alpha$.

FBD of A (only vertical forces)

$$mg-N\cos\alpha=ma\tan\alpha \quad \text{...(i)}$$

FBD of B (only horizontal forces)

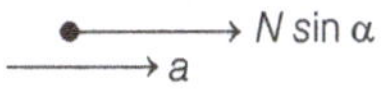

$$N\sin\alpha=\eta ma \quad \text{...(ii)}$$

Solving these two equations, we get

Acceleration of B,

$$a_B=a=\frac{g}{\tan\alpha+\eta\cot\alpha}$$

and $$a_A=a\tan\alpha=\frac{g}{1+\eta\cot^2\alpha}$$

74. (a)

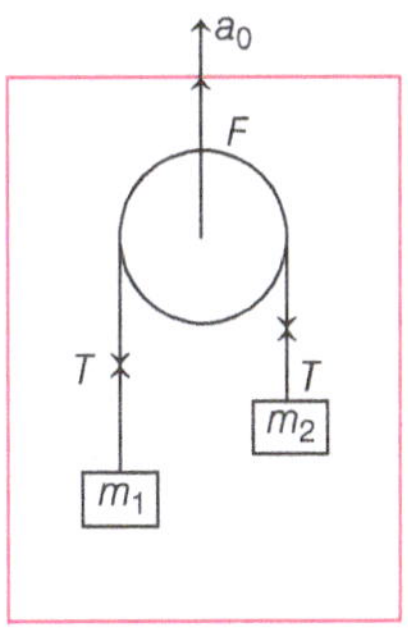

Relative acceleration of m_1 with respect to pulley is equal and opposite to relative acceleration of m_2 with respect to pulley and suppose this is a_r.

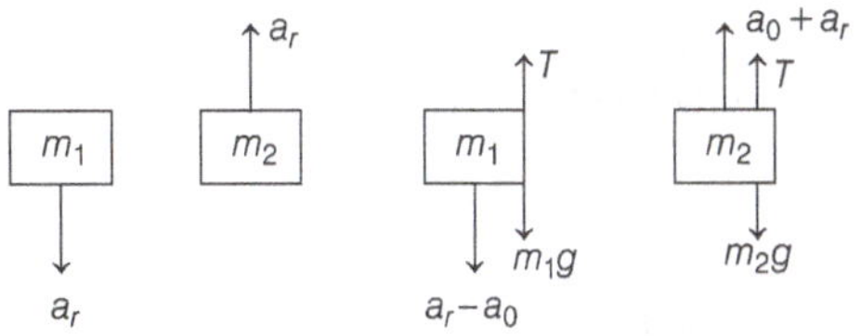

Relative Absolute

FBD of m_1

$$m_1 g - T = m_1(a_r - a_0) \quad \text{...(i)}$$

FBD of m_2

$$T - m_2 g = m_2(a_r + a_0) \quad \text{...(ii)}$$

Solving Eqs. (i) and (ii), we get

$$a_r = \left(\frac{m_1 - m_2}{m_1 + m_2}\right)(g + a_0)$$

Note *(i) This a_r is acceleration of m_1 relative to the car (or pulley) which is accelerated upwards with acceleration a_0. This is a non-inertial frame.*

(ii) You can also solve this part by making use of pseudo force. This is your home assignment.

Acceleration of m_1 with respect to elevator shaft (or earth)

This is absolute acceleration of m_1 which is $a_r - a_0$, downwards.

Thus,

$$a_{m_1} = a_r - a_0 = \left(\frac{m_1 - m_2}{m_1 + m_2}\right)(g + a_0) - a_0$$

or $$a_{m_1} = \frac{(m_1 - m_2)g - 2m_2 a_0}{(m_1 + m_2)} \quad \text{(downwards)}$$

(b) Eq. (i) gives

$$T = m_1(g - a_r + a_0)$$

Substituting the values, we get

$$T = \frac{2m_1 m_2 (g + a_0)}{m_1 + m_2}$$

Now force exerted by the pulley on the ceiling of the car will be,

$$F = 2T \quad \text{or} \quad F = \frac{4m_1 m_2 (g + a_0)}{m_1 + m_2}$$

Note *Answers of Irodov are given in vector form and here we have taken the magnitudes of the answers. So they seem to be different. This is your home assignment to think why they are same.*

75. FBD of A with respect to frame is shown in figure.

A is in equilibrium under three concurrent forces shown in figure, so applying Lami's theorem

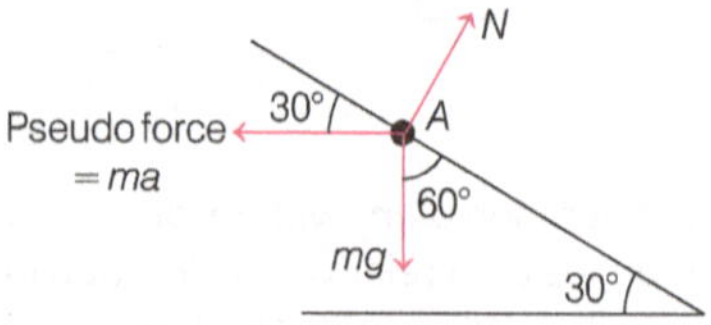

$$\frac{ma}{\sin(90 + 60)} = \frac{mg}{\sin(90 + 30)}$$

$$\therefore \quad a = \frac{g \cos 60°}{\cos 30°} = 5.66 \text{ m/s}^2$$

76. FBD of M_2 and M_3 in accelerated frame of reference is shown in figure.

Note *Only the necessary forces have been shown.*

Mass M_3 will neither rise nor fall if net pulling force is zero.

i.e. $$M_2 a = M_3 g$$

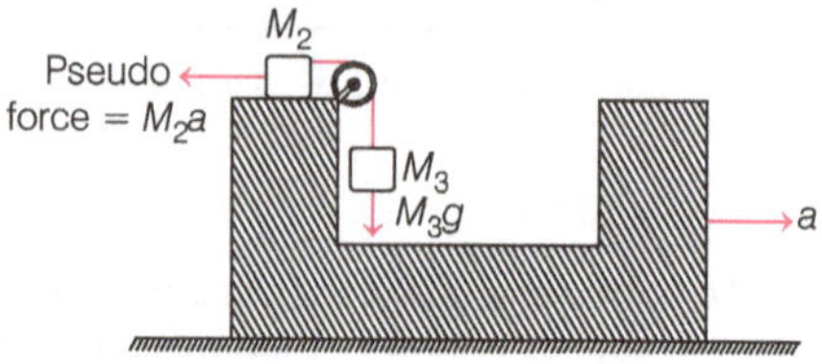

or $$a = \frac{M_3}{M_2} g$$

$$\therefore \quad F = (M_1 + M_2 + M_3)a$$

$$= (M_1 + M_2 + M_3)\frac{M_3}{M_2} g \quad \textbf{Ans.}$$

77. $a = \dfrac{\text{Net pulling force}}{\text{Total mass}}$

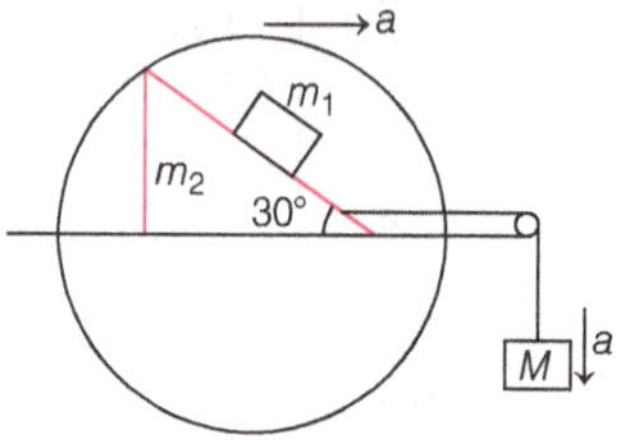

$$a = \frac{Mg}{m_1 + m_2 + M} \quad \text{...(i)}$$

For m_1

$$N \cos 30° = m_1 g \quad \text{...(ii)}$$

$$N \sin 30° = m_1 a \quad \text{...(iii)}$$

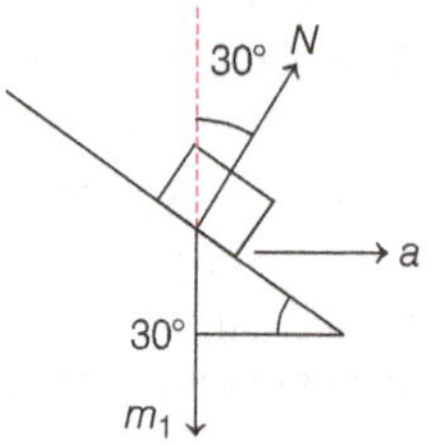

From Eqs. (ii) and (iii), we get

$$a = g \tan 30° = \frac{g}{\sqrt{3}}$$

Substituting this value in Eq. (i), we get

$$M = 6.83 \text{ kg} \quad \textbf{Ans.}$$

78. Block B will fall vertically downwards and A along the plane.

Writing the equations of motion.

For block B, $$m_B g - N = m_B a_B$$

or $$60 - N = 6a_B \quad \text{...(i)}$$

$$(N + m_A g) \sin 30° = m_A a_A$$

or $\quad (N + 150) = 30\, a_A \quad$ ...(ii)

Further $\quad a_B = a_A \sin 30°$

or $\quad a_A = 2a_B \quad$ ...(iii)

Solving these three equations, we get

(a) $a_A = 6.36\ \text{m/s}^2$ **Ans.**

(b) $a_{BA} = a_A \cos 30° = 5.5\ \text{m/s}^2$ **Ans.**

79. Normal reaction between A and B would be $N = mg\cos\theta$. Its horizontal component is $N\sin\theta$. Therefore, tension in cord CD is equal to this horizontal component.

Hence, $\quad T = N\sin\theta = (mg\cos\theta)(\sin\theta)$

$$= \frac{mg}{2}\sin 2\theta$$ **Ans.**

80. Free body diagram of crate A w.r.t. ground is shown in figure.

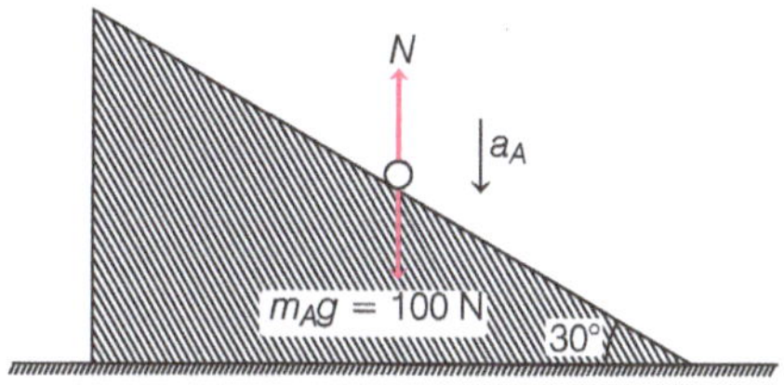

Equation of motion is

$$100 - N = 10a_A \quad ...(i)$$

$$a_A = a\sin 30° = (2)\left(\frac{1}{2}\right)$$

or $\quad a_A = 1\ \text{m/s}^2$

Substituting in Eq. (i), we get

$$N = 90\ \text{N}$$

81. FBD of m_1 (showing only the horizontal forces)

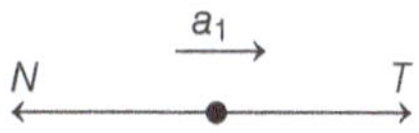

Equation of motion for m_1 is

$$T - N = m_1a_1 \quad ...(i)$$

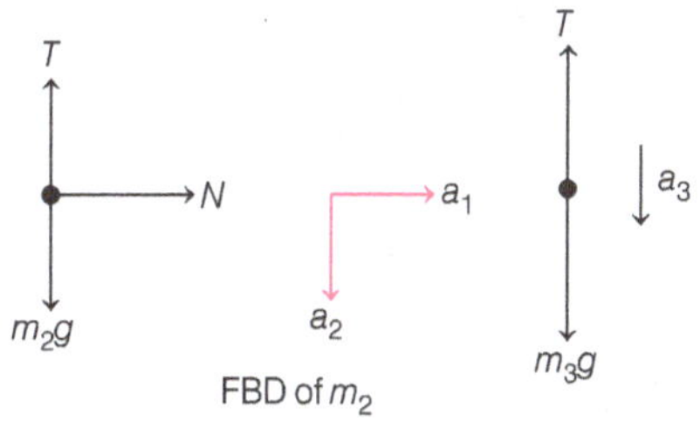

FBD of m_2

Equations of motion for m_2 are

$$N = m_2a_1 \quad ...(ii)$$

and $\quad m_2g - T = m_2a_2 \quad$ ...(iii)

Equation of motion for m_3 are

$$m_3g - T = m_3a_3 \quad ...(iv)$$

Further from constraint equation we can find the relation,

$$a_1 = a_2 + a_3 \quad ...(v)$$

We have five unknowns a_1, a_2, a_3, T and N solving, we get

$$a_1 = \frac{2m_1m_3\,g}{(m_2 + m_3)(m_1 + m_2) + m_2m_3}$$ **Ans.**

82. Writing equations of motion,

$$T - N = 3ma_1 \quad ...(i)$$

$$N = 2ma_1 \quad ...(ii)$$

$$2mg - T = 2ma_2 \quad ...(iii)$$

$$T - \frac{mg}{2} = ma_3 \quad ...(iv)$$

From constraint equation,

$$a_1 = a_2 - a_3 \quad ...(v)$$

We have five unknowns. Solving the above five equations, we get

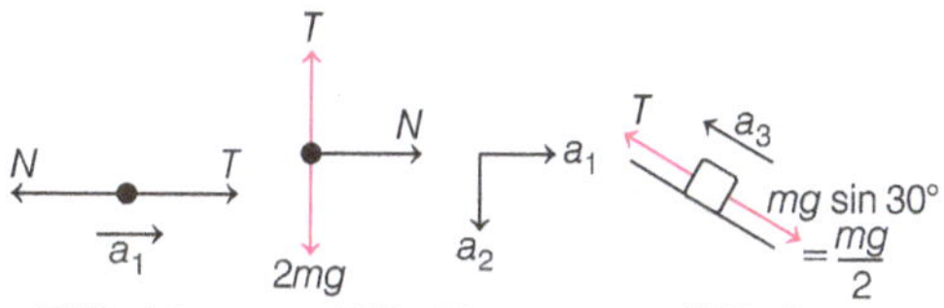

FBD of $3m$ $\quad$ FBD of $2m$ $\quad$ FBD of m

$$a_1 = \frac{3}{17}g,\ a_2 = \frac{19}{34}g$$

and $\quad a_3 = \frac{13}{34}g$

Acceleration of $m = a_3 = \frac{13}{34}g$

Acceleration of $2m = \sqrt{a_1^{\,2} + a_2^{\,2}} = \frac{\sqrt{397}}{34}g$

and acceleration of $3m = a_1 = \frac{3}{17}g$

83. (a) N_2 and mg pass through G. N_1 has clockwise moment about G, so the ladder has a tendency to slip by rotating clockwise and the force of friction (f) at B is then up the plane.

(b) $\Sigma M_A = 0$

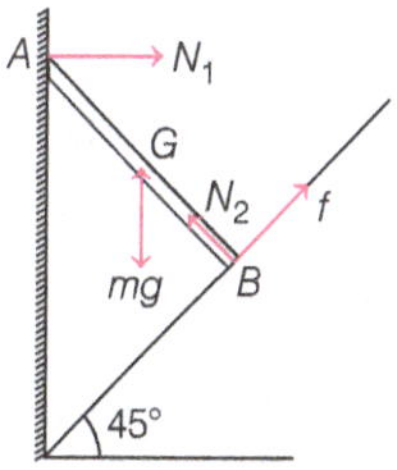

$\therefore \quad fl = mg\left(\frac{l}{2}\sin 45°\right) \quad$ ...(i)

$$\Sigma F_V = 0$$

$\therefore \quad mg = N_2 \cos 45° + f \sin 45°$ ...(ii)

From Eqs. (i) and (ii), we get

$$N_2 = \frac{3}{2\sqrt{2}} mg$$

and $\quad f = \frac{mg}{2\sqrt{2}}$

or $\quad \mu_{min} = \frac{f}{N_2} = \frac{1}{3}$ **Ans.**

84. Since μ_s between crate and truck is less than 0.9. Therefore, retardation of truck and crate will be different.

Retardation of truck,

$$a_1 = 0.9g = 9 \text{ m/s}^2$$

Retardation of crate,

$$a_2 = 0.8g = 8 \text{ m/s}^2$$

Displacement of truck, before it finally stops is

$$s_1 = \frac{u^2}{2a_1} = \frac{(15)^2}{2 \times 9}$$

$$= 12.5 \text{ m}$$

and displacement of crate

$$s_2 = \frac{u^2}{2a_2} = \frac{(15)^2}{2 \times 8}$$

$$\approx 14.06 \text{ m}$$

$\therefore \quad s_2 - s_1 = 1.56 \text{ m}$

85. $\mu_k mg = 0.2 \times 10 \times 10 = 20 \text{ N}$

For $t \le 0.2$ s

Retardation, $a_1 = \frac{F + \mu_k mg}{m}$

$$= \frac{20 + 20}{10} = 4 \text{ m/s}^2$$

At the end of 0.2 s,

$$v = u - a_1 t$$

$$v = 1.2 - 4 \times 0.2$$

$$= 0.4 \text{ m/s}$$

For $t > 0.2$ s

Retardation, $a_2 = \frac{10 + 20}{10} = 3 \text{ m/s}^2$

Block will come to rest after time

$$t_0 = \frac{v}{a_2} = \frac{0.4}{3} = 0.13 \text{ s}$$

$\therefore$ Total time $= 0.2 + 0.13$

$= 0.33$ s **Ans.**

86. Block will start moving at, $F = \mu mg$

or $\quad 25t = (0.5)(10)(9.8)$

$= 49$ N

$\therefore \quad t = 1.96$ s

Velocity is maximum at the end of 4s.

$\therefore \quad \frac{dv}{dt} = \frac{25t - 49}{10}$

$= 2.5t - 4.9$

$\therefore \quad \int_0^{v_{max}} dv = \int_{1.96}^{4} (2.5t - 4.9)\, dt$

$\therefore \quad v_{max} = 5.2 \text{ m/s}$ **Ans.**

For 4 s < t < 7 s

Net retardation $a_1 = \frac{49 - 40}{10} = 0.9 \text{ m/s}^2$

$\therefore \quad v = v_{max} - a_1 t_1$

$= 5.2 - 0.9 \times 3$

$= 2.5 \text{ m/s}$

For t > 7 s

Retardation $a_2 = \frac{49}{10} = 4.9 \text{ m/s}^2$

$\therefore \quad t = \frac{v}{a_2} = \frac{2.5}{4.9} = 0.51 \text{ s}$

$\therefore$ Total time $= (4 - 1.96) + (7 - 4) + (0.51)$

$= 5.55$ s **Ans.**

87. $\sqrt{2} N = mg \cos \theta$

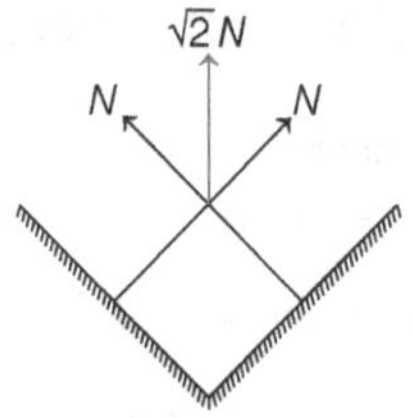

$\therefore \quad N = \frac{mg \cos \theta}{\sqrt{2}}$

$$a = \frac{mg \sin \theta - 2\mu_k N}{m}$$

$$= g \sin \theta - \sqrt{2} \mu_k g \cos \theta$$

$= g (\sin \theta - \sqrt{2} \mu_k \cos \theta)$ **Ans.**

88. $v \cdot \frac{dv}{dx} = \frac{\text{Net force}}{\text{mass}} = \frac{F - \mu_k \rho (L - x) g}{\rho L}$

$\therefore \quad \int_0^v v\, dv = \int_0^L \frac{F - \mu_k \rho (L - x) g}{\rho L} dx$

$\therefore \quad \frac{v^2}{2} = \frac{F}{\rho} - \mu_k gL + \frac{\mu_k gL}{2}$

$\therefore \quad v = \sqrt{\frac{2F}{\rho} - \mu_k gL}$ **Ans.**

89. $a = \frac{m_A g}{m_A + M + m}$

For the equilibrium of B,

$$mg = \mu N = \mu (ma) = \frac{\mu m m_A g}{m_A + M + m}$$

$\therefore \quad m_A = \dfrac{(M+m)m}{(\mu-1)m}$

$m_A = \dfrac{(M+m)}{\mu-1}$ **Ans.**

Note $m_A > 0 \quad \therefore \quad \mu > 1$

90. $s = \dfrac{1}{2}at^2$ or $t = \sqrt{\dfrac{2s}{a}}$ ...(i)

Here, $s = \dfrac{l}{\cos\alpha}$

and $a = \dfrac{mg\sin\alpha - \mu mg\cos\alpha}{m}$

$= (g\sin\alpha - \mu g\cos\alpha)$

Substituting in Eq. (i), we get

$t = \sqrt{\dfrac{2l}{\cos\alpha[g\sin\alpha - \mu g\cos\alpha]}}$ ...(ii)

t will be minimum when the variable term $\cos\alpha(g\sin\alpha - \mu g\cos\alpha)$ is maximum or, its first derivative with respect to α is zero.

I am leaving this mathematical part for the students. The value of α comes out to be,

$\alpha \approx 49°$

Now, substituting this value of α in Eq. (ii) with values of μ, g and l, we get

$t_{min} = 1.0$ s

91. To just pull the block upwards with almost zero acceleration,

$T\cos\beta = \mu N + mg\sin\alpha$

$= \mu(mg\cos\alpha - T\sin\beta) + mg\sin\alpha$

$\therefore \quad T = \dfrac{\mu mg\cos\alpha + mg\sin\alpha}{\cos\beta + \mu\sin\beta}$ ...(i)

Here β or denominator is variable. Therefore, for T to be minimum

$\dfrac{d}{d\beta}(\cos\beta + \mu\sin\beta) = 0$

or $-\sin\beta + \mu\cos\beta = 0$

$\therefore \quad \tan\beta = \mu$

or $\beta = \tan^{-1}(\mu)$

At this value of β value of T will be minimum.

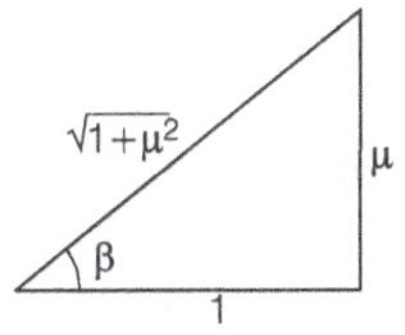

Substituting $\cos\beta = \dfrac{1}{\sqrt{1+\mu^2}}$

and $\sin\beta = \dfrac{\mu}{\sqrt{1+\mu^2}}$

From Eq. (i), we get

$T_{min} = \dfrac{\mu mg\cos\alpha + mg\sin\alpha}{\sqrt{1+\mu^2}}$

92. Acceleration, $a = \dfrac{F\cos\alpha}{m}$

$= \dfrac{(mg/3)\cos\alpha}{m}$

$= \dfrac{g\cos\alpha}{3} = \dfrac{g\cos(ks)}{3}$

$\therefore \quad v \cdot \dfrac{dv}{ds} = \dfrac{g\cos(ks)}{3}$

or $vdv = \dfrac{g\cos(ks)}{3}ds$

Integrating, we get

$\dfrac{v^2}{2} = \dfrac{g\sin(ks)}{3k} + c$

At $s = 0, \ v = 0 \Rightarrow c = 0$

$\therefore \quad \dfrac{v^2}{2} = \dfrac{g\sin(ks)}{3k}$

or $v = \sqrt{\dfrac{2g\sin(ks)}{3k}}$

Putting $ks = \alpha$, we get

$v = \sqrt{\dfrac{2g\sin\alpha}{3k}}$

93.

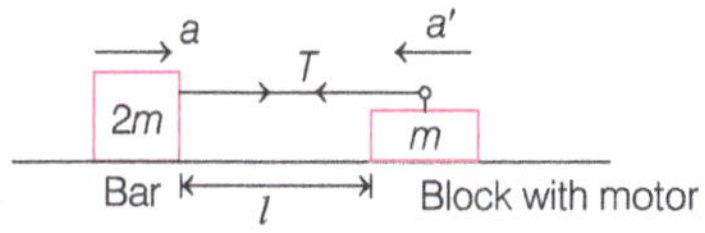

$a = \dfrac{T - \mu(2mg)}{2m}$

$= \dfrac{T}{2m} - \mu g$

$\Rightarrow \quad \dfrac{T}{m} = 2[a + \mu g]$

Now, $a' = \dfrac{T - \mu mg}{m} = \dfrac{T}{m} - \mu g$

$= (2a + 2\mu g) - \mu g$

$= (2a + \mu g)$

Relative acceleration between two,

$a_r = a + a' = (3a + \mu g)$

From $s_r = \dfrac{1}{2}a_r t^2$,

$t = \sqrt{\dfrac{2s_r}{a_r}}$ or $t = \sqrt{\dfrac{2l}{3a + \mu g}}$

94. From the constraint relation we can prove that vertical acceleration of m relative to M is equal to absolute acceleration of M towards right and suppose this acceleration is a.

FBD of M (showing only horizontal forces)

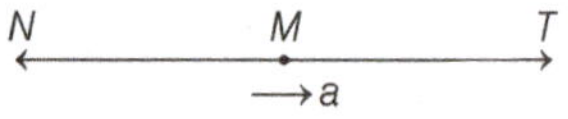

$N =$ normal reaction
$T =$ tension
Equation of motion is,

$$T - N = Ma \quad \text{...(i)}$$

FBD of *m*

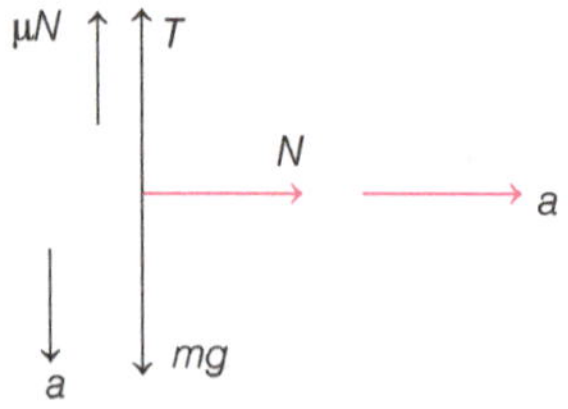

Equations of motion are,

$$N = ma \quad \text{...(ii)}$$

$$mg - T - \mu N = ma \quad \text{...(iii)}$$

We have three unknowns T, N and a. Solving these three equations, we get

$$a = \frac{g}{2 + \mu + \frac{M}{m}}$$

Net acceleration of m will be,

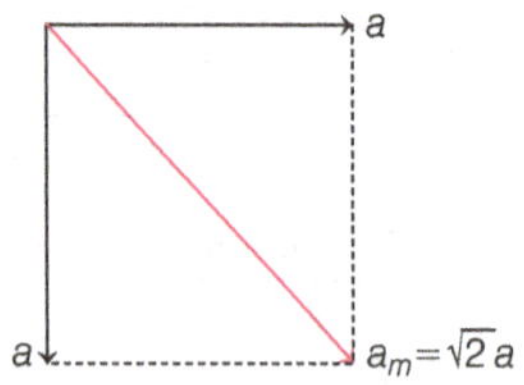

$$a_m = \sqrt{2}\,a = \frac{\sqrt{2}g}{2 + \mu + \frac{M}{m}}$$

95. Let this minimum acceleration is a.

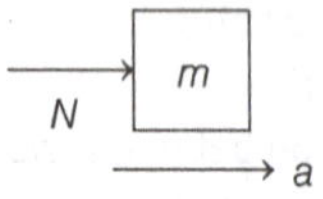

FBD of body-2 (only horizontal forces)

$$N = ma$$

FBD of 1 and 2 with respect to A with pseudo forces : (Forces are shown only along the string)

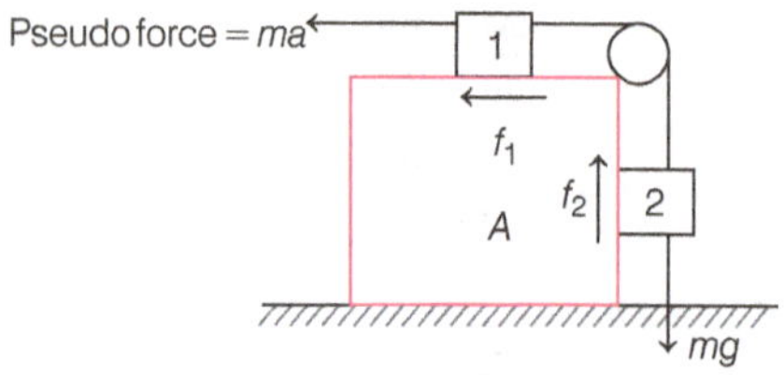

$f_1 =$ limiting friction between 1 and A

$$= \mu N_1$$

$$= \mu mg$$

$f_2 =$ limiting friction between 2 and A

$$= \mu N_2 \text{ or } \mu N$$

$$= \mu ma$$

Now net pulling force $= 0$

$\therefore \quad mg = ma + f_1 + f_2$

$\Rightarrow \quad mg = ma + \mu mg + \mu ma$

Solving this equation, we get

$$a = g\left(\frac{1-\mu}{1+\mu}\right)$$

96. Maximum force of friction between A and $B = \mu\, m_A g$

or $\quad f_1 = (0.25)(3)(10) = 7.5\text{ N}$

Maximum force of friction between B and C

$$= \mu\,(m_A + m_B)\,g$$

or $\quad f_2 = (0.25)(3+4)(10) = 17.5\text{ N}$

Maximum force of friction between C and ground

$$f_3 = \mu\,(m_A + m_B + m_C)\,g$$

$$= (0.25)(3+4+8)(10) = 37.5\text{ N}$$

Block C and hence block B are moving in opposite directions with constant velocities and block A is at rest. Hence, net force on all three blocks should be zero. Free body diagrams have been shown below (Only horizontal forces are shown)

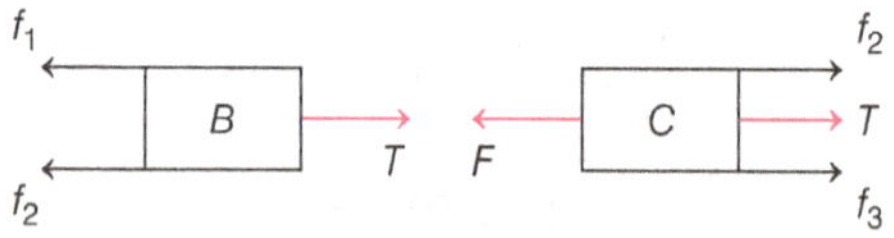

For equilibrium of B, $T = f_1 + f_2 = 25\text{ N}$

For equilibrium of C, $F = T + f_2 + f_3 = 80\text{ N}$

97. Maximum force of friction between M_1 and inclined plane

$$f_1 = \mu_1 M_1 g \cos\theta = (0.75)(4)(9.8)(0.8) = 23.52\text{ N}$$

$$M_1 g \sin\theta = (4)(9.8)(0.6) = 23.52\text{ N} = F_1 \quad \text{(say)}$$

Maximum force of friction between M_2 and inclined plane

$$f_2 = \mu_2 M_2 g \cos\theta$$

$$=(0.25)(2)(9.8)(0.8)=3.92\,\text{N}$$

$$M_2 g \sin\theta=(2)(9.8)(0.6)=11.76\,\text{N}=F_2 \quad \text{(say)}$$

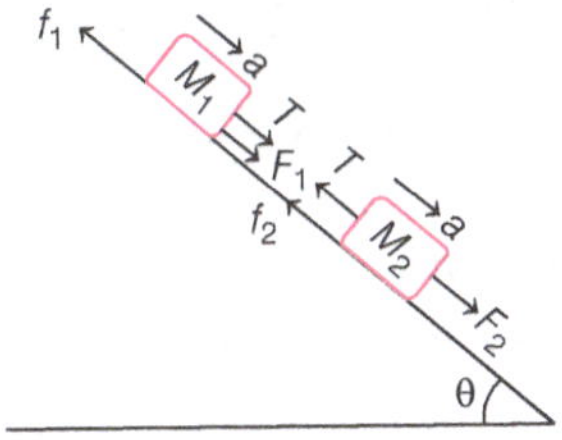

Both the blocks will be moving downwards with same acceleration a. Different forces acting on two blocks are as shown in above figure.

Equation of motion of M_1

$$T+F_1-f_1=M_1 a$$

Substituting the values of F_1, f_1 and M_1, we get

or $$T=4a \quad \text{...(i)}$$

Equation of motion of M_2

$$F_2-T-f_2=M_2 a$$

or $$7.84-T=2a \quad \text{...(ii)}$$

Solving Eqs. (i) and (ii), we get

$$a=1.3\,\text{m/s}^2 \quad \text{and} \quad T=5.2\,\text{N}$$

98. Constant velocity means net acceleration of the system is zero. Or net pulling force on the system is zero. While calculating the pulling force, tension forces are not taken into consideration. Therefore,

(a) $$M_1 g=M_2 g\sin 37°+\mu M_2 g\cos 37°+\mu M_3 g$$

or $$M_1=M_2\sin 37°+\mu M_2\cos 37°+\mu M_3$$

Substituting the values, we get

$$M_1=(4)\left(\frac{3}{5}\right)+(0.25)(4)\left(\frac{4}{5}\right)+(0.25)(4)$$

$$=4.2\,\text{kg}$$

(b) Since, M_3 is moving with uniform velocity

$$T=\mu M_3 g=(0.25)(4)(9.8)=9.8\,\text{N}$$

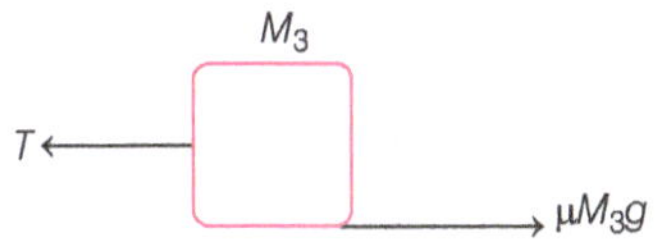

99. Let F is applied at angle θ as shown in figure.

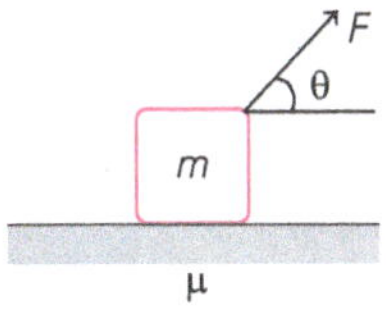

Normal reaction in this case will be,

$$N=mg-F\sin\theta$$

The limiting friction is therefore

$$f_L=\mu N=\mu(mg-F\sin\theta)$$

For the block to move,

$$F\cos\theta=f_L=\mu(mg-F\sin\theta)$$

or $$F=\frac{\mu mg}{\cos\theta+\mu\sin\theta} \quad \text{...(i)}$$

For F to be minimum, denominator should be maximum.

or $$\frac{d}{d\theta}(\cos\theta+\mu\sin\theta)=0$$

or $$-\sin\theta+\mu\cos\theta=0$$

or $$\tan\theta=\mu$$

or $$\theta=\tan^{-1}(\mu)$$

Substituting this value of θ in Eq. (i), we get

$$F_{\min}=mg\sin\theta$$

$$\tan\theta=\mu \Rightarrow \sin\theta=\frac{\mu}{\sqrt{1+\mu^2}}$$

$$\Rightarrow \quad F_{\min}=\frac{\mu mg}{\sqrt{1+\mu^2}}$$

100. (a) Acceleration of block

Maximum friction force that can be obtained at A is

$$(f_{\max})_A=\mu_A(mg\cos 45°)$$

$$=\frac{2}{3}(mg/\sqrt{2})=\frac{\sqrt{2}mg}{3}$$

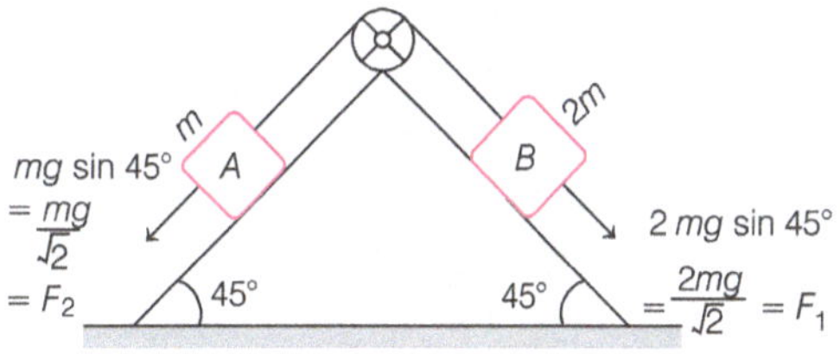

Similarly, $$(f_{\max})_B=\mu_B(2mg\cos 45°)$$

$$=\frac{1}{3}(2mg/\sqrt{2})=\frac{\sqrt{2}mg}{3}$$

Therefore, maximum value of friction that can be obtained on the system is

$$(f_{\max})=(f_{\max})_A+(f_{\max})_B=\frac{2\sqrt{2}\,mg}{3} \quad \text{...(i)}$$

Net pulling force on the system is

$$F=F_1-F_2=\frac{2mg}{\sqrt{2}}-\frac{mg}{\sqrt{2}}=\frac{mg}{\sqrt{2}} \quad \text{...(ii)}$$

From Eqs. (i) and (ii), we can see that

Net pulling force $< f_{\max}$. Therefore, the system will not move or the acceleration of block A will be zero.

(b) and (c) Tension in the string and friction at A

Net pulling force on the system (block A and B)

$$F=F_1-F_2=mg/\sqrt{2}$$

Therefore, total friction force on the blocks should also be equal to $\frac{mg}{\sqrt{2}}$

or $$f_A + f_B = F = mg/\sqrt{2}$$

Now, since the blocks will start moving from block B first (if they move), therefore f_B will reach its limiting value first and if still some force is needed, it will be provided by f_A.

Here, $$(f_{\max})_B < F$$

Therefore, f_B will be in its limiting value and rest will be provided by f_A.

Hence, $f_B = (f_{\max})_B = \frac{\sqrt{2}mg}{3}$ and $f_A = F - f_B$

$$= \frac{mg}{\sqrt{2}} - \frac{\sqrt{2}mg}{3} = \frac{mg}{3\sqrt{2}}$$

The FBD of the whole system is be as shown in the figure.

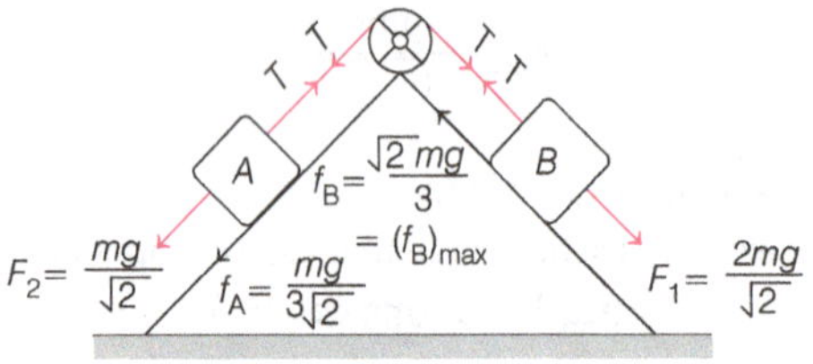

Therefore, friction on A is

$$f_A = mg/3\sqrt{2} \quad \text{(down the plane)}$$

Now, for tension T in the string, we may consider either equilibrium of A or B.

Equilibrium of A gives

$$T = F_2 + f_A = \frac{mg}{\sqrt{2}} + \frac{mg}{3\sqrt{2}} = \frac{4mg}{3\sqrt{2}} \text{ or } T = \frac{2\sqrt{2}mg}{3}$$

Similarly, equilibrium of B gives $T + f_B = F_1$

or $$T = F_1 - f_B = \frac{2mg}{\sqrt{2}} - \frac{\sqrt{2}mg}{3} = \frac{4mg}{3\sqrt{2}}$$

or $$= \frac{2\sqrt{2}\,mg}{3}$$

Therefore, tension in the string is $\frac{2\sqrt{2}\,mg}{3}$.

101. Let us draw free boy diagram of the block with respect to disc (with pseudo force ma)

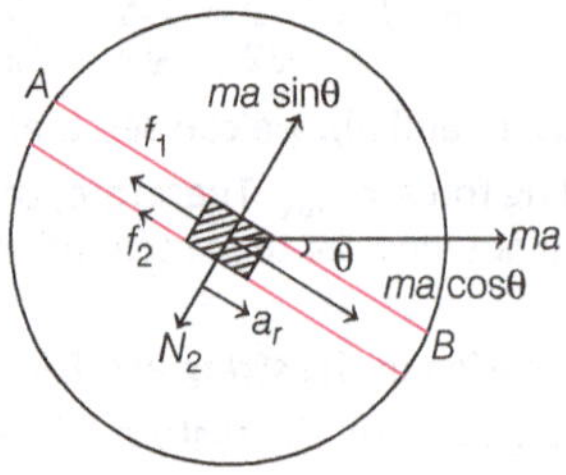

$N_1 = mg$ = normal reaction on block from ground

$N_2 = ma\sin\theta$ = normal reaction on block from AB side of groove

Now, a_r = acceleration of block w.r.t. disc

$$= \frac{ma\cos\theta - f_1 - f_2}{m} \quad (\because f_1 = \mu N_1 \text{ and } f_2 = \mu N_2)$$

$$\therefore \quad a_r = \frac{ma\cos\theta - \mu N_1 - \mu N_2}{m}$$

Substituting the values of N_1 and N_2, we get

$$a_r = a\cos\theta - \mu g - \mu a\sin\theta$$

$$= 25 \times \frac{4}{5} - \frac{2}{5} \times 10 - \frac{2}{5} \times 25 \times \frac{3}{5}$$

$$= 10\text{ m/s}^2$$

Level 2

102. mg can be resolved into two forces of equal magnitude F inclined at an angle 60° with each other.

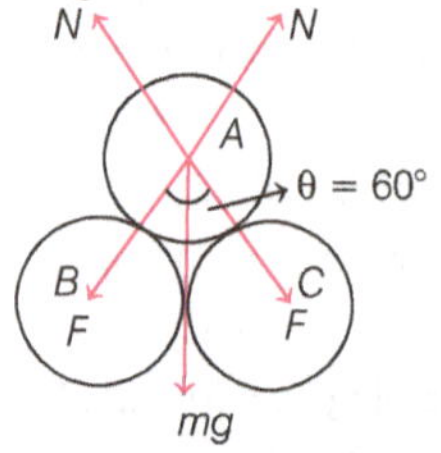

$$\therefore \quad \sqrt{F^2 + F^2 + 2F^2\cos 60^\circ} = \sqrt{3}g$$

$$\Rightarrow \quad 3F^2 = 3g^2$$

$$F = g$$

or $$F = 10\text{ N}$$

Hence, reaction of B on $A = 10$ N.

103. Here, $T = ma$, $10\sqrt{x} = ma$

$$\Rightarrow \quad m = \frac{10}{a}\sqrt{x}$$

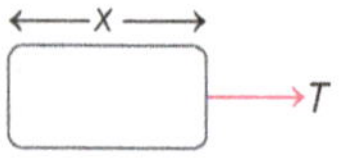

or $$\frac{dm}{dx} = \frac{10}{a}\frac{d(\sqrt{x})}{dx}$$

or $$\frac{dm}{dx} = \frac{10}{a}\frac{1}{2\sqrt{x}} \text{ or } \frac{dm}{dx} = \frac{5}{a\sqrt{x}}$$

$$\Rightarrow \quad \lambda \propto \frac{1}{\sqrt{x}}$$

104.

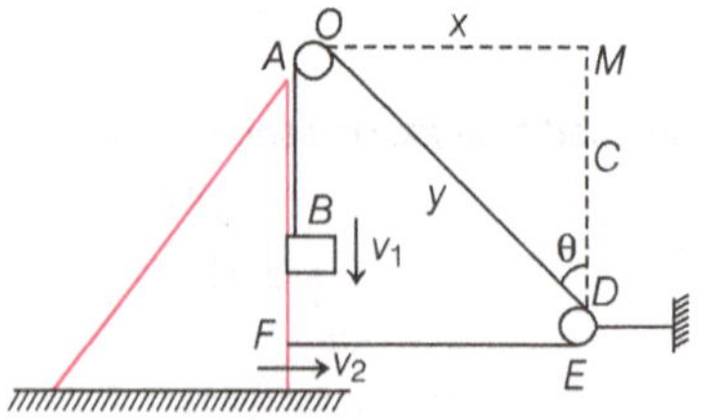

Rate of increase of AB = Rate of decrease of EF and OD

$$\therefore \quad v_1 = v_2 + v_2\sin\theta$$

or $$v_1 = v_2(1 + \sin\theta)$$

Note $y^2 = x^2 + C^2$ (C = constant)

$\therefore \quad 2y\left(\frac{dy}{dt}\right) = 2x\left(\frac{dx}{dt}\right)$

$\therefore \quad \left(-\frac{dy}{dt}\right) = \frac{x}{y}\left(-\frac{dx}{dt}\right)$

or rate of decrease of OD $= (\sin\theta)(v_2)$

105. On the cylinder

If N = normal reaction between cylinder and inclined plane

$N\sin\theta$ = horizontal component of N

$= ma$ ....(i)

$N\cos\theta$ = vertical component of N

$= mg$ ...(ii)

Dividing Eq. (i) by Eq. (ii), we get

$$\tan\theta = \frac{a}{g}$$

$\therefore \quad a = g\tan\theta$ **Ans.**

106. $\cos\theta = \frac{4R/5}{R} = 0.8$

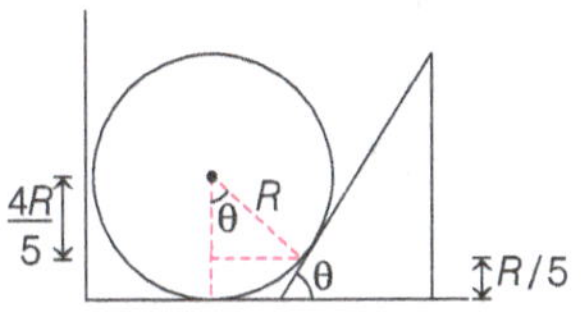

$\therefore \quad \theta = 37°$

$$\frac{v_1}{v_2} = \tan\theta = \tan 37° = \frac{3}{4}$$

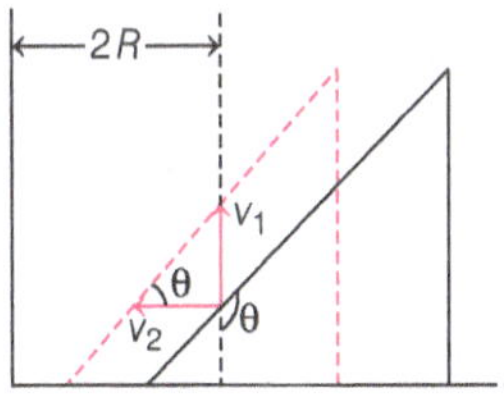

$\therefore \quad v_1 = \frac{3}{4}v_2 = \frac{3}{4}\times 20$

$= 15\ \text{m/s}$ **Ans.**

107. $v_{AL} = v_A - v_L$

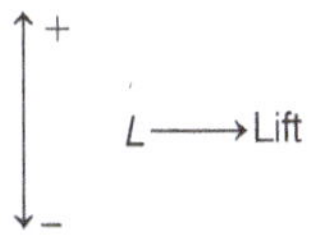

$\therefore \quad v_A = v_{AL} + v_L = (-2) + 2 = 0$

Let z = length of string at some instant. Then,

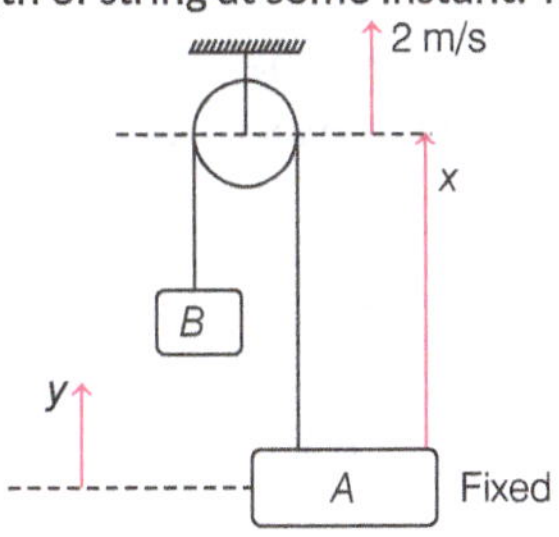

$-\frac{dz}{dt} = 2\ \text{m/s}$ (Given)

Now, $y = x - (z - x) = 2x - z$

$\therefore \quad \frac{dy}{dt} = 2\frac{dx}{dt} + \left(-\frac{dz}{dt}\right)$

$= 2(2\ \text{m/s}) + 2\ \text{m/s}$

$= 6\ \text{m/s} = v_B$ **Ans.**

108. For maintaining contact of cylinder with block A,

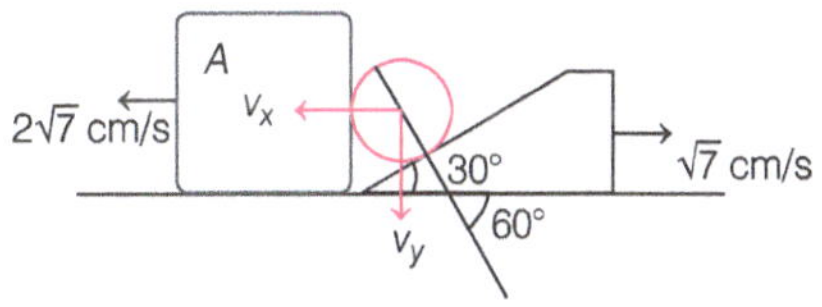

For maintaining contact of cylinder with block B,

$$\sqrt{7}\cos 60° = v_y\cos 30° - v_x\sin 30°$$

$$\frac{\sqrt{7}}{2} = \frac{\sqrt{3}}{2}v_y - \sqrt{7}$$

$\therefore \quad v_y = \sqrt{21}$

$\therefore$ The speed of cylinder is

$$v = \sqrt{v_x^2 + v_y^2} = \sqrt{(2\sqrt{7})^2 + (\sqrt{21})^2}$$

$= 7\ \text{cm/s}$

109. $v = 2t^2$

$\therefore \quad a = \frac{dv}{dt} = 4t$

At $t = 1\ \text{s}$,

$a = 4\ \text{m/s}^2$ ...(i)

Limiting value of static friction,

$$f_L = \mu_S mg$$

$\therefore$ Maximum value of acceleration of coin which can be provided by the friction,

$a_{max} = \frac{f_L}{m} = \mu_S g$ ...(ii)

Equating Eqs. (i) and (ii), we get

$4 = \mu_S(10)$

$\therefore \quad \mu_S = 0.4$ **Ans.**

110. Maximum force of friction between C and ground is

$$(f_C)_{max} = (0.5)(60)(10) = 300\ \text{N}$$

Since, it is pulling the blocks by the maximum force (without moving). Therefore, the applied force is $F = 300\ \text{N}$

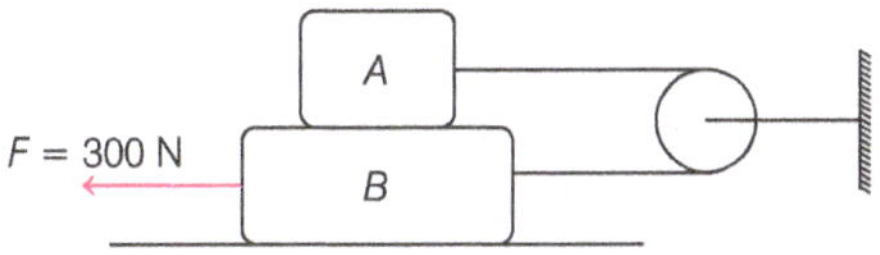

$(f_{AB})_{max} = 0.4\times 60\times 10 = 240\ \text{N}$

$(f_{BG})_{max} = 0.3\times 120\times 10 = 360\ \text{N}$

Since, $(f_{BG})_{max}$ is greater than 300 N, blocks will not move. Free body diagrams of block are as shown below.

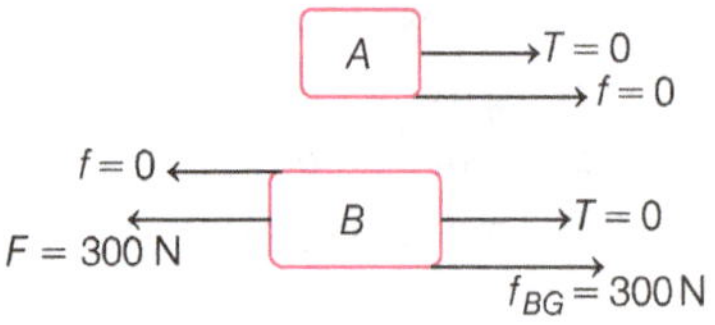

111. For equilibrium of upper ball,

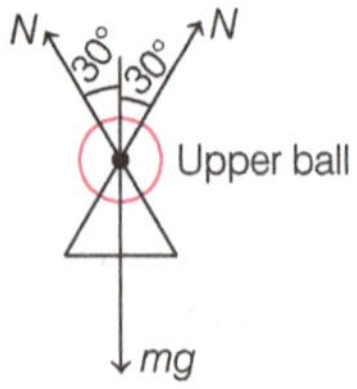

$$2N\cos 30° = mg$$

$$\therefore \qquad N = \frac{mg}{2\cos 30°} = \frac{mg}{\sqrt{3}}$$

From free body diagram of leftward ball,

$N_0 = N\sin 60° + mg$

(resolving the forces vertically)

$N\cos 60° = f \le \mu N_0$

(resolving the forces horizontally)

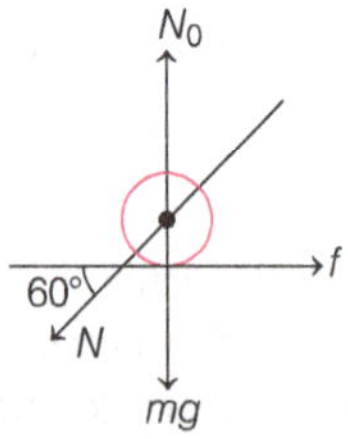

$$\text{or} \qquad \frac{N}{2} \le \mu\left(\frac{\sqrt{3}}{2}N + mg\right)$$

$$\text{or} \qquad \frac{mg}{2\sqrt{3}} \le \mu\frac{\sqrt{3}}{2} \times \frac{mg}{\sqrt{3}} + \mu mg$$

$$\text{or} \qquad \mu \ge \frac{1}{3\sqrt{3}}$$

$$\Rightarrow \qquad \mu_{min} = \frac{1}{3\sqrt{3}}$$

112. When $P = mg(\sin\theta - \mu\cos\theta)$

$f = \mu\, mg\cos\theta$ (upwards)

When $P = mg\sin\theta;\ f = 0$

and when $P = mg(\sin\theta + \mu\cos\theta)$

$f = \mu mg\cos\theta$ (downwards)

Hence, friction is first positive, then zero and then negative.

113. $W = mg = 0.1 \times 10 = 1$ N

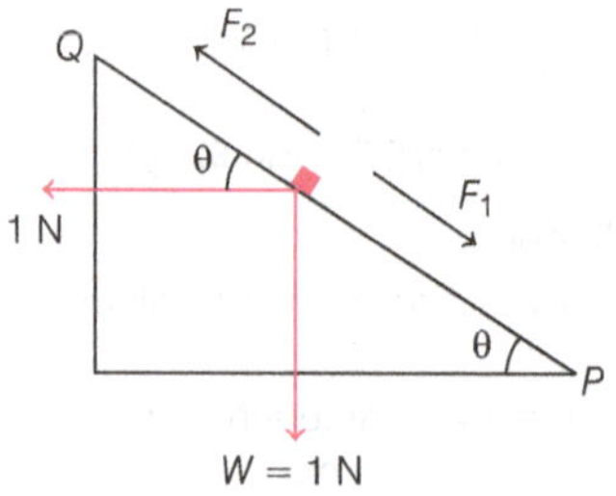

F_1 = component of weight $= 1 \cdot \sin\theta = \sin\theta$

F_2 = component of applied force $= 1 \cdot \cos\theta = \cos\theta$

Now, at $\theta = 45°: F_1 = F_2$ and block remains stationary without the help of friction.

For $\theta > 45°$, $F_1 > F_2$, so friction will act towards Q.

For $\theta < 45°$, $F_2 > F_1$ and friction will act towards P.

114. μ_2 can never be zero for equilibrium.

When $\mu_1 = 0$, we have

$$N_1 = \mu_2 N_2 \qquad \text{...(i)}$$

$$N_2 = mg \qquad \text{...(ii)}$$

$$\tau_B = 0$$

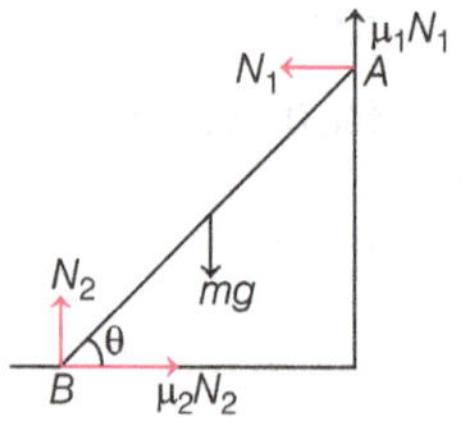

$$\Rightarrow \qquad mg\frac{L}{2}\cos\theta = N_1 L\sin\theta$$

$$\Rightarrow \qquad N_1 = \frac{mg\cot\theta}{2}$$

$$\Rightarrow \qquad N_1\tan\theta = \frac{mg}{2}$$

When $\mu_1 \ne 0$, we have

$$\mu_1 N_1 + N_2 = mg \qquad \text{...(iii)}$$

$$\mu_2 N_2 = N_1 \qquad \text{...(iv)}$$

Solving these two equations, we get

$$N_2 = \frac{mg}{1 + \mu_1\mu_2}$$

115. Here f_1 = force of friction between man and plank and f_2 = force of friction between plank and surface.

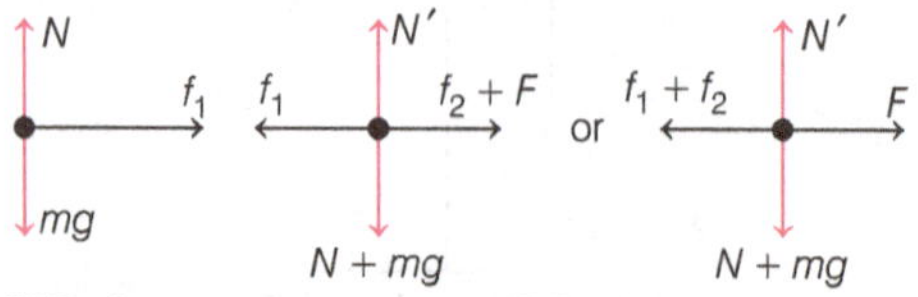

FBD of man FBD of plank

For the plank not to move

$$F-(f_2)_{max} \le f_1 \le F+(f_2)_{max}$$

or $F-\mu(M+m)g \le ma \le F+\mu(M+m)g$

or a should lie between $\frac{F}{m}-\frac{\mu(M+m)g}{m}$

and $\frac{F}{m}+\frac{\mu(M+m)g}{m}$ **Ans.**

116. (a) $v=a_1t_1=2.6\,\text{m/s}$

$$s_1=\frac{1}{2}a_1t_1^2=\frac{1}{2}\times 2\times(1.3)^2=1.69\,\text{m}$$

$$s_2=(2.2-1.69)=0.51\,\text{m}$$

Now, $s_2=\frac{v^2}{2a_2}$

$\therefore$ $a_2=\frac{v^2}{2s_2}=\frac{(2.6)^2}{2\times 0.51}$

$$=6.63\,\text{m/s}^2$$

and $t_2=\frac{v}{a_2}=0.4\,\text{s}$

(b) Acceleration of package will be $2\,\text{m/s}^2$ while retardation will be $\mu_k g$ or $2.5\,\text{m/s}^2$ not $6.63\,\text{m/s}^2$.

For the package,

$$v=a_1t_1=2.6\,\text{m/s}$$

$\Rightarrow$ $s_1=\frac{1}{2}a_1t_1^2=1.69\,\text{m}$

$$s_2=vt_2-\frac{1}{2}a_2' t_2^2$$

$$=2.6\times 0.4-\frac{1}{2}\times 2.5\times(0.4)^2$$

$$=0.84\,\text{m}$$

$\therefore$ Displacement of package w.r.t. belt

$$=(0.84-0.51)\,\text{m}$$

$$=0.33\,\text{m}$$ **Ans.**

Alternate Solution For last 0.4 s

$$|a_r|=6.63-2.5=4.13\,\text{m/s}^2$$

$\therefore$ $s_r=\frac{1}{2}|a_r|t^2=\frac{1}{2}\times 4.13\times(0.4)^2$

$$=0.33\,\text{m}$$

117. (a) Force of friction at different contacts are shown in figure.

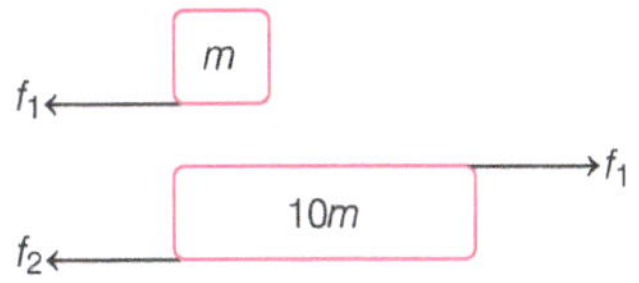

Here, $f_1=\mu_2 mg$

and $f_2=\mu_1(11\,mg)$

Given that $\mu_2>11\mu_1$

$\therefore$ $f_1>f_2$

Retardation of upper block

$$a_1=\frac{f_1}{m}=\mu_2 g$$

Acceleration of lower block

$$a_2=\frac{f_1-f_2}{m}=\frac{(\mu_2-11\mu_1)g}{10}$$

Relative retardation of upper block

$$a_r=a_1+a_2 \quad \text{or} \quad a_r=\frac{11}{10}(\mu_2-\mu_1)g$$

Now, $0=v_{min}^2-2a_rl$

$\therefore$ $v_{min}=\sqrt{2a_rl}$

$$=\sqrt{\frac{22(\mu_2-\mu_1)gl}{10}}$$ **Ans.**

(b) $0=v_{min}-a_rt$

or $t=\frac{v_{min}}{a_r}=\sqrt{\frac{20l}{11(\mu_2-\mu_1)g}}$ **Ans.**

118. $v_r=\sqrt{v_1^2+v_2^2}$

Retardation $a=\mu g$

$\therefore$ Time when slipping will stop is $t=\frac{v_r}{a}$

or $t=\frac{\sqrt{v_1^2+v_2^2}}{\mu g}$ ...(i)

$$s_r=\frac{v_r^2}{2a}=\frac{v_1^2+v_2^2}{2\mu g}$$

$$x_r=-s_r\cos\theta=-\left(\frac{v_1^2+v_2^2}{2\mu g}\right)\left(\frac{v_2}{\sqrt{v_1^2+v_2^2}}\right)$$

$$=\frac{-v_2\sqrt{v_1^2+v_2^2}}{2\mu g}$$

$$y_r=s_r\sin\theta=\left(\frac{v_1^2+v_2^2}{2\mu g}\right)\left(\frac{v_1}{\sqrt{v_1^2+v_2^2}}\right)$$

$$=\frac{v_1\sqrt{v_1^2+v_2^2}}{2\mu g}$$

In time t, belt will move a distance $s=v_2t$

or $\frac{v_2\sqrt{v_1^2+v_2^2}}{\mu g}$ in x-direction.

Hence, coordinate of particle,

$$x=x_r+s=\frac{v_2\sqrt{v_1^2+v_2^2}}{2\mu g}$$

and $y=y_r=\frac{v_1\sqrt{v_1^2+v_2^2}}{2\mu g}$ **Ans.**

119.

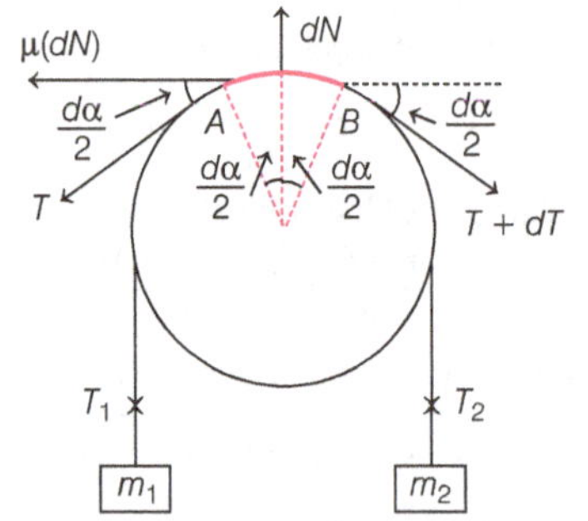

We have taken a small element AB of massless string.

At $\frac{m_2}{m_1} = \eta_0$, string will just start sliding. So, we have taken maximum friction $\mu(dN)$ and the portion is in equilibrium.

Net force towards centre = 0.

$$\therefore \quad dN = T\sin\frac{d\alpha}{2} + (T + dT)\sin\frac{d\alpha}{2}$$

For small angle, $\sin\frac{d\alpha}{2} \approx \frac{d\alpha}{2}$ and then neglecting the term, $(dT)\left(\frac{d\alpha}{2}\right)$, we get

$$dN = Td\alpha \qquad \text{...(i)}$$

Net force in tangential direction = 0

$$\therefore \quad \mu(dN) + T\cos\frac{d\alpha}{2} = (T + dT)\cos\frac{d\alpha}{2}$$

For small angles,

$$\cos\frac{d\alpha}{2} \approx 1$$

$$\therefore \quad \mu dN + T = T + dT$$

$$\text{or} \quad \mu dN = dT \qquad \text{...(ii)}$$

From Eqs. (i) and (ii), we get

$$\mu(Td\alpha) = dT$$

$$\text{or} \quad \int_{T_1}^{T_2} \frac{dT}{T} = \mu\int_0^{\pi} d\alpha$$

$$\text{or} \quad \ln\frac{T_2}{T_1} = \mu\pi$$

$$\mu = \frac{\ln(T_2/T_1)}{\pi} \qquad \text{...(iii)}$$

Now, $T_1 = m_1 g$ and $T_2 = m_2 g$

$$\therefore \quad \frac{T_2}{T_1} = \frac{m_2}{m_1} = \eta_0$$

Substituting in Eq. (iii), we get

$$\mu = \frac{\ln(\eta_0)}{\pi}$$

120. Given, $m_1 = 20\,\text{kg}$, $m_2 = 5\,\text{kg}$, $M = 50\,\text{kg}$, $\mu = 0.3$ and $g = 10\,\text{m/s}^2$

(a) Free body diagram of mass M is given as below:

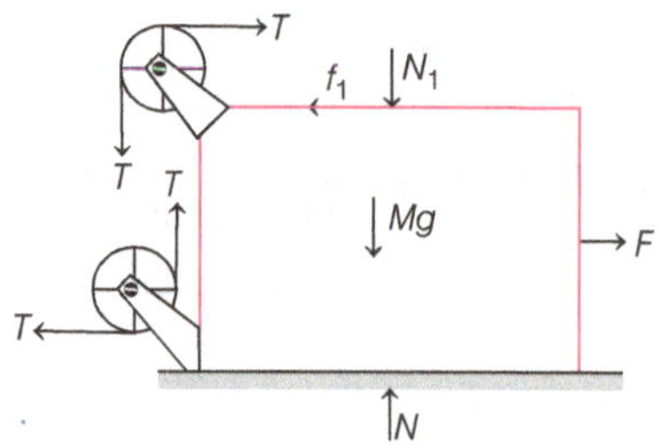

(b) The maximum value of f_1 is

$$(f_1)_{\max} = (0.3)(20)(10) = 60\,\text{N}$$

The maximum value of f_2 is

$$(f_2)_{\max} = (0.3)(5)(10) = 15\,\text{N}$$

Forces on m_1 and m_2 in horizontal direction are as shown below

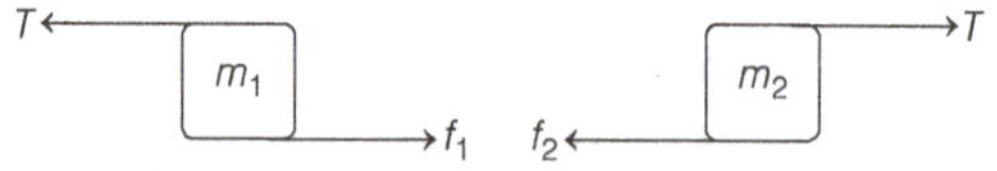

Now, there are only two possibilities :

(1) Either both m_1 and m_2 will remain stationary (w.r.t. ground) or

(2) Both m_1 and m_2 will move (w.r.t. ground)

First case is possible when

$$T \le (f_1)_{\max} \quad \text{or} \quad T \le 60\,\text{N}$$

$$\text{and} \quad T \le (f_2)_{\max} \quad \text{or} \quad T \le 15\,\text{N}$$

These conditions will be satisfied when $T \le 15\,\text{N}$ say $T = 14\,\text{N}$, then $f_1 = f_2 = 14\,\text{N}$

Therefore, the condition $f_1 = 2f_2$ will not be satisfied. Thus, m_1 and m_2 both cannot remain stationary.

In the second case, when m_1 and m_2 both move

$$f_2 = (f_2)_{\max} = 15\,\text{N}$$

$$\therefore \quad f_1 = 2f_2 = 30\,\text{N}$$

Now, since $f_1 < (f_1)_{\max}$, there is no relative motion between m_1 and M i.e. all the masses move with same acceleration, say a.

$$\therefore \quad f_2 = 15\,\text{N} \quad \text{and} \quad f_1 = 30\,\text{N}$$

Free body diagrams and equations of motion are as shown below

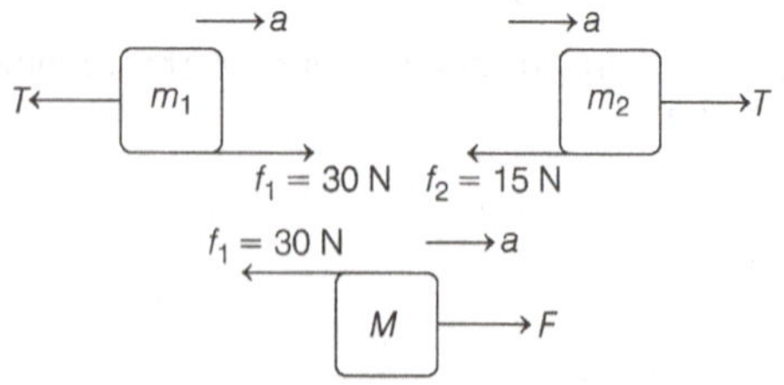

For m_1, $\quad 30 - T = 20a \qquad$ ...(i)

For m_2, $\quad T - 15 = 5a \qquad$ ...(ii)

For M, $\quad F - 30 = 50a \qquad$ ...(iii)

Solving these three equations, we get

$$F = 60 \text{ N}$$

$$T = 18 \text{ N}$$

and $$a = \frac{3}{5} \text{ m/s}^2$$

121. Let us first take equilibrium of whole system,

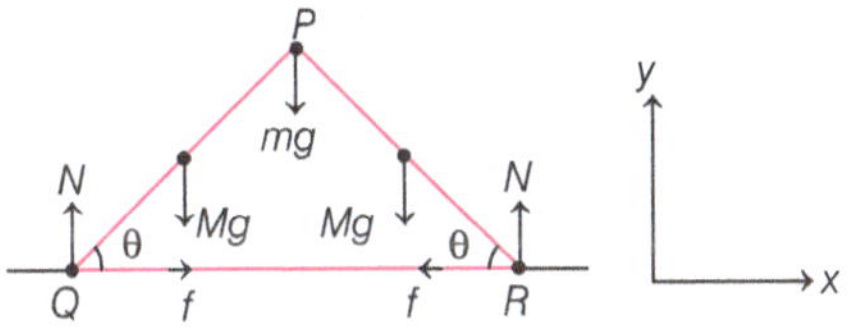

$$\Sigma F_y = 0 \Rightarrow Mg + Mg + mg = 2N$$

or $$N = \frac{(2M + m)g}{2} \quad \text{...(i)}$$

Now, if we take equilibrium of rod PQ alone and put net torque about P equal to zero, then we will get the equation,

clockwise torque of N = anticlockwise torque of Mg and f

$$\Rightarrow N(l\cos\theta) = Mg\left(\frac{l}{2}\cos\theta\right) + f(l\sin\theta)$$

or $$f = N\cot\theta - \frac{Mg}{2}\cot\theta$$

$$= \frac{(2M + m)g}{2}\cot\theta - \frac{Mg}{2}\cot\theta$$

$$= \left(\frac{M + m}{2}\right)g\cot\theta$$

122. From free body diagram of cylinder A,

$$2N\cos 30° = mg$$

$$\therefore \quad N = \frac{mg}{\sqrt{3}}$$

From free body diagram of cylinder B,

$$N_0 \cos\theta = mg + N\cos 30°,$$

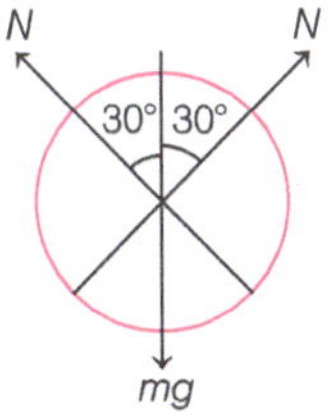

$$N_0 \sin\theta = N\sin 30°$$

$$\tan\theta = \frac{N\sin 30°}{mg + N\cos 30°}$$

Putting the value of N, $\tan\theta = \frac{1}{3\sqrt{3}}$

$$\therefore \quad n = 3$$

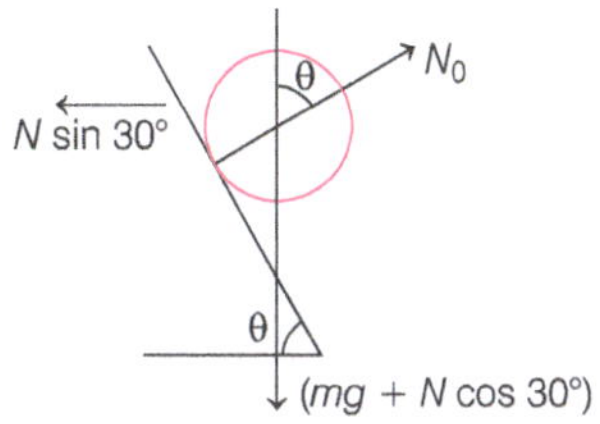

Note *At minimum value of θ normal reaction between B and C will be zero.*

123. $\mathbf{F}_{net} = \mathbf{F}_{East} + \mathbf{F}_{South} + \mathbf{F}_{downn} + \mathbf{F}_0$

$$= 3\hat{\mathbf{i}} + (-1\hat{\mathbf{j}}) + (-1\hat{\mathbf{k}}) + \mathbf{F}_0$$

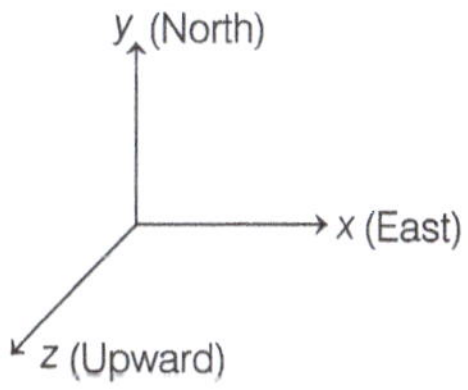

Since, the block moves in South-East direction,

$$\mathbf{F}_{net} = x\hat{\mathbf{i}} - x\hat{\mathbf{j}} + 0\hat{\mathbf{k}} \quad [\because \tan\theta = -1]$$

or $$x\hat{\mathbf{i}} - x\hat{\mathbf{j}} = 3\hat{\mathbf{i}} - \hat{\mathbf{j}} - \hat{\mathbf{k}} + \mathbf{F}_0$$

or $$\mathbf{F}_0 = (x - 3)\hat{\mathbf{i}} + (1 - x)\hat{\mathbf{j}} + \hat{\mathbf{k}}$$

$$|F_0| = \sqrt{(x - 3)^2 + (1 - x)^2 + (1)^2}$$

or $$|F_0|^2 = (x - 3)^2 + (1 - x)^2 + 1$$

For minimum F_0, $|F_0|^2$ must be minimum.

Differentiating $|F_0|^2$ w.r.t. x, we get

$$\frac{dy}{dx} = 2(x - 3) - 2(1 - x) \quad [\text{let } |F_0|^2 = y]$$

For minima, $\frac{dy}{dx} = 0$

or $2(x - 3) - 2(1 - x) = 0$

or $4x + 8 = 0$

or $x = 2$

$$\mathbf{F}_{0(\min)} = (2 - 3)\hat{\mathbf{i}} + (1 - 2)\hat{\mathbf{j}} + \hat{\mathbf{k}}$$

$$= (-\hat{\mathbf{i}} - \hat{\mathbf{j}} + \hat{\mathbf{k}}) \text{ N}$$

$$\therefore \quad \mathbf{F}_{0(\min)} = \sqrt{(-1)^2 + (-1)^2 + 1^2}$$

$$= \sqrt{3} = \sqrt{n}$$

$$\therefore \quad n = 3$$

124. Here, $a = \frac{F}{m} = \frac{bt - cs}{m} = t - s$

or $$\frac{da}{dt} = 1 - v$$

or $$\frac{d^2a}{dt^2} = \frac{-dv}{dt} = -a$$

The solution of equation is $a = a_0 \sin\omega t$.

When, $\omega = 1$ unit.

UNIT-9

WORK, ENERGY AND POWER

Learning Modules

1. Introduction to Work Done
2. Conservative Force and Its Potential Energy
3. Kinetic Energy and Work-Energy Theorem
4. Law of Conservation of Mechanical Energy
5. Types of Equilibrium
6. Power of a Force

MODULE 1

Introduction to Work Done

Introduction

In our daily life, 'work' has many different meanings. For example, Ram is working in a factory. The machine is in working order. Let us work out a plan for the next year, etc. In physics however, the term 'work' has a special meaning. In physics, work is always associated with a force and a displacement. Consider a person holding a weight at a distance 'h' off the floor as shown in figure.

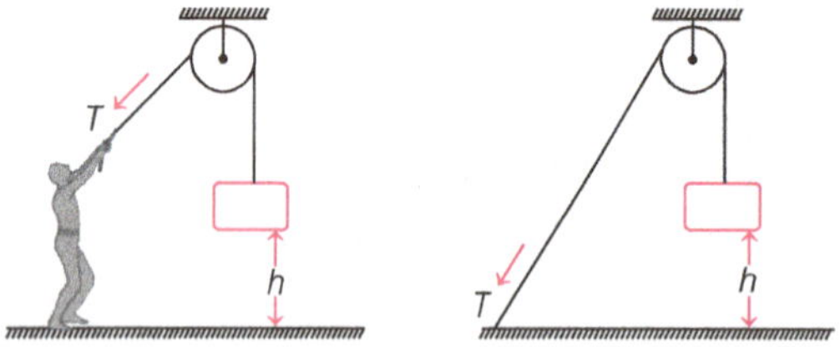

No work is done by the man holding the weight at a fixed position. The same task could be accomplished by tying the rope to a fixed point.

In everyday usage, we might say that the man is doing work, but in our scientific definition, no work is done by a force acting on a stationary object. We could eliminate the effort of holding the weight by merely tying the string to some object and the weight could be supported with no help from us.

Work Done

There are mainly four methods of finding work done by a force:

(i) Work done by a constant force.

(ii) Work done by a variable force.

(iii) Work done by the area under force and displacement graph.

(iv) General method of finding work done.

1. Work done by a Constant Force

Work done by a constant force is given by

$$W = \mathbf{F} \cdot \mathbf{s} = Fs\cos\theta \qquad (\mathbf{F} = \text{force}, \mathbf{s} = \text{displacement})$$

= (magnitude of force) (component of displacement in the direction of force)

= (magnitude of displacement) (component of force in the direction of displacement)

Here, θ is the angle between **F** and **s**.

Thus, work done is the dot product of **F** and **s**.

Special Cases

(i) If $\theta = 0°$,

$W = Fs\cos 0° = Fs$

(ii) If $\theta = 90°$,

$W = Fs\cos 90° = 0$

(iii) If $\theta = 180°$,

$W = Fs\cos 180° = -Fs$

Important POINTS

(i) Work done by a force may be **positive**, **negative** or even **zero** also, depending upon the angle (θ) between the force vector **F** and displacement vector **s**. Work done by a force is zero when $\theta = \mathbf{90°}$, it is **positive** when $\mathbf{0°} \le \theta < \mathbf{90°}$ and **negative** when $\mathbf{90°} < \theta \le \mathbf{180°}$.

(ii) Work depends on frame of reference. With change of frame of reference, inertial force does not change while displacement may change. So, the work done by a force will be different in different frames. For example, if a person is pushing a box inside a moving train, then work done as seen from the frame of reference of train is $\mathbf{F} \cdot \mathbf{s}$ while as seen from the ground it is $\mathbf{F} \cdot (\mathbf{s} + \mathbf{s}_0)$. Here $\mathbf{s}_0$, is the displacement of train relative to ground.

(iii) Suppose a body is displaced from point A to point B, then

$$\mathbf{s} = \mathbf{r}_B - \mathbf{r}_A = (x_B - x_A)\hat{\mathbf{i}} + (y_B - y_A)\hat{\mathbf{j}} + (z_B - z_A)\hat{\mathbf{k}}$$

Here, (x_A, y_A, z_A) and (x_B, y_B, z_B) are the coordinates of points A and B.

Example 1 A body is displaced from $A = (2\text{ m}, 4\text{ m}, -6\text{ m})$ to $\mathbf{r}_B = (6\hat{\mathbf{i}} - 4\hat{\mathbf{j}} + 2\hat{\mathbf{k}})$ m under a constant force $\mathbf{F} = (2\hat{\mathbf{i}} + 3\hat{\mathbf{j}} - \hat{\mathbf{k}})$ N. Find the work done.

Sol. $\mathbf{r}_A = (2\hat{\mathbf{i}} + 4\hat{\mathbf{j}} - 6\hat{\mathbf{k}})$ m

$$\therefore \quad \mathbf{s} = \mathbf{r}_B - \mathbf{r}_A = (6\hat{\mathbf{i}} - 4\hat{\mathbf{j}} + 2\hat{\mathbf{k}}) - (2\hat{\mathbf{i}} + 4\hat{\mathbf{j}} - 6\hat{\mathbf{k}}) = 4\hat{\mathbf{i}} - 8\hat{\mathbf{j}} + 8\hat{\mathbf{k}}$$

$$W = \mathbf{F} \cdot \mathbf{s} = (2\hat{\mathbf{i}} + 3\hat{\mathbf{j}} - \hat{\mathbf{k}}) \cdot (4\hat{\mathbf{i}} - 8\hat{\mathbf{j}} + 8\hat{\mathbf{k}}) = 8 - 24 - 8 = -24\text{ J}$$ **Ans.**

Note *Work done is negative. Therefore angle between* **F** *and* **s** *is obtuse.*

Example 2 A block of mass $m = 2$ kg is pulled by a force $F = 40$ N upwards through a height $h = 2$m. Find the work done on the block by the applied force F and its weight mg. ($g = 10\text{ m/s}^2$)

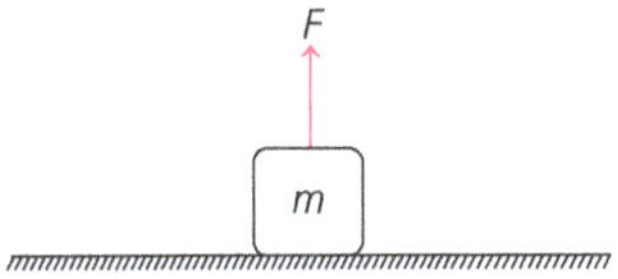

Sol. Weight $mg = (2)(10) = 20$ N

Work done by the applied force $W_F = Fh\cos 0°$

As the angle between force and displacement is 0°

or $\quad W_F = (40)(2)(1) = 80$ J **Ans.**

Similarly, work done by its weight

$$W_{mg} = (mg)(h)\cos 180°$$

or $\quad W_{mg} = (20)(2)(-1) = -40$ J **Ans.**

Example 3 Two unequal masses of 1 kg and 2 kg are attached at the two ends of a light inextensible string passing over a smooth pulley as shown in figure. If the system is released from rest, find the work done by string on both the blocks in 1 s. (Take $g = 10\text{ m/s}^2$)

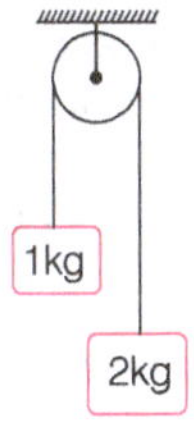

Sol. Net pulling force on the system is

$$F_{\text{net}} = 2g - 1g = 20 - 10 = 10\text{ N}$$

Total mass being pulled

$$m = (1 + 2) = 3\text{ kg}$$

Therefore, acceleration of the system will be

$$a = \frac{F_{\text{net}}}{m} = \frac{10}{3}\text{ m/s}^2$$

Displacement of both the blocks in 1 s is

$$s = \frac{1}{2}at^2 = \frac{1}{2}\left(\frac{10}{3}\right)(1)^2 = \frac{5}{3}\text{ m}$$

Free body diagram of 2 kg block is shown in figure (b).

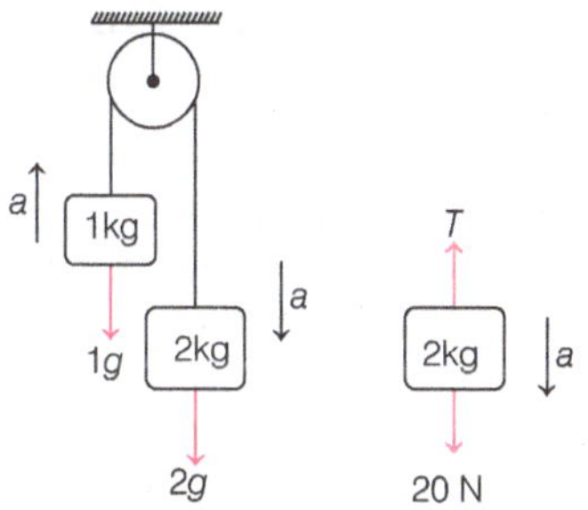

(a) (b)

Using $\Sigma F = ma$, we get

$$20 - T = 2a = 2\left(\frac{10}{3}\right)$$

or $$T = 20 - \frac{20}{3} = \frac{40}{3}\text{ N}$$

$\therefore$ Work done by string (tension) on 1 kg block in 1 s is

$$W_1 = (T)(s)\cos 0° = \left(\frac{40}{3}\right)\left(\frac{5}{3}\right)(1) = \frac{200}{9}\text{ J}$$ **Ans.**

Similarly, work done by string on 2 kg block in 1 s will be

$$W_2 = (T)(s)(\cos 180°) = \left(\frac{40}{3}\right)\left(\frac{5}{3}\right)(-1) = -\frac{200}{9}\text{ J}$$ **Ans.**

2. Work Done by a Variable Force

Until now we have considered the work done by a force which is constant both in magnitude and direction. Let us now consider a force which acts always in one direction but whose magnitude may keep on varying. We can choose the direction of the force as x-axis. Further, let us assume that the force is also a function of x or say $F(x)$ is known to us. Now, we are interested in finding the work done by this force in moving a body from x_1 to x_2.

Work done in a small displacement from x to $x + dx$ will be

$$dW = F \cdot dx$$

Now, the total work can be obtained by integration of the above elemental work from x_1 to x_2 or

$$W = \int_{x_1}^{x_2} dW = \int_{x_1}^{x_2} F \cdot dx$$

Note *In this method of finding work done, you need not to worry for the sign of work done. If we put proper limits in integration then sign of work done automatically comes.*

It is important to note that $\int_{x_1}^{x_2} F\,dx$ is also the area under F-x graph between $x = x_1$ to $x = x_2$.

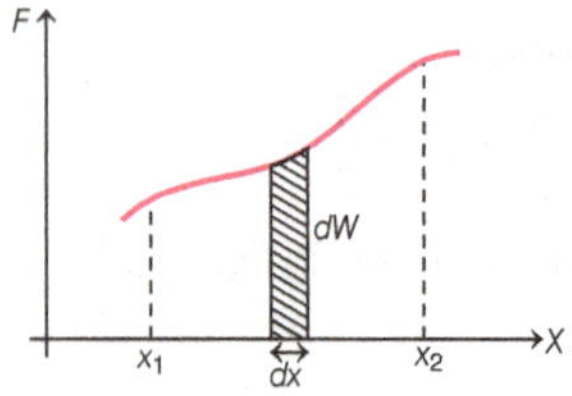

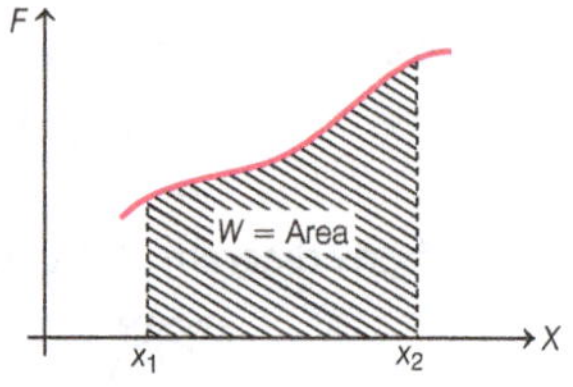

Spring Force

An important example of the above idea is a spring that obeys Hooke's law. Consider the situation shown in figure. One end of a spring is attached to a fixed vertical support and the other end to a block which can move on a horizontal table.

Let $x = 0$ denote the position of the block when the spring is in its natural length. When the block is displaced by an amount x (either compressed or elongated) a restoring force (F) is applied by the spring on the block.

The direction of this force F is always towards its mean position ($x = 0$) and the magnitude is directly proportional to x or

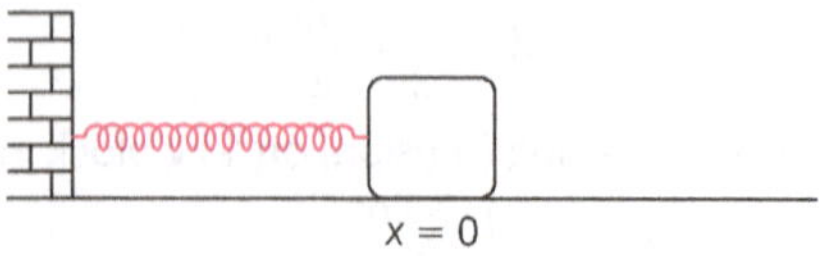

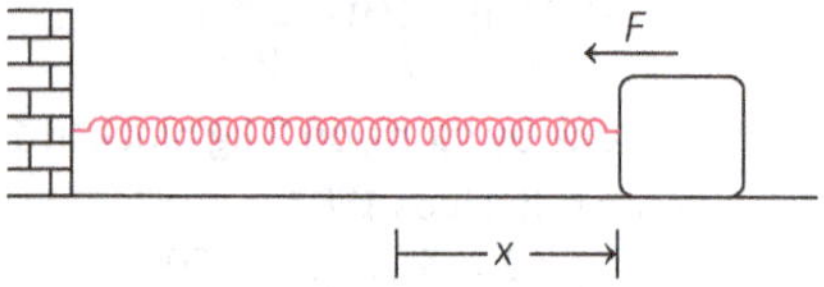

$$F \propto x \qquad \text{(Hooke's law)}$$

$$\therefore \qquad F = -kx \qquad \text{...(i)}$$

Here, k is a constant called **force constant of spring** and depends on the nature of spring.

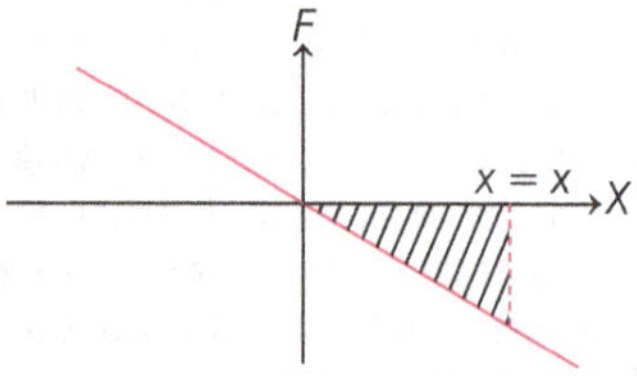

From Eq. (i) we see that F is a variable force and F-x graph is a straight line passing through origin with slope $= -k$. Negative sign in Eq. (i) implies that the spring force F is directed in a direction opposite to the displacement x of the block.

Let us now find the work done by this force F when the block is displaced from $x = 0$ to $x = x$. This can be obtained either by integration or the area under F-x graph.

Thus, $\quad W = \int dW = \int_0^x F dx = \int_0^x -kx\, dx = -\frac{1}{2} kx^2$

Here, work done is negative because force is in opposite direction of displacement.

Similarly, if the block moves from $x = x_1$ to $x = x_2$. The limits of integration are x_1 and x_2 and the work done is

$$W = \int_{x_1}^{x_2} -kx\, dx = \frac{1}{2} k (x_1^2 - x_2^2)$$

Note *Force constant of the spring is inversely proportional to its length.*

Example 4 A force $F = (2 + x)$ acts on a particle in x-direction where F is in newton and x in metre. Find the work done by this force during a displacement from $x = 1.0$ m to $x = 2.0$ m.

Sol. As the force is variable, we shall find the work done in a small displacement from x to $x + dx$ and then integrate it to find the total work. The work done in this small displacement is

$$dW = F\, dx = (2 + x)\, dx$$

Thus, $$W = \int_{1.0}^{2.0} dW = \int_{1.0}^{2.0} (2 + x)\, dx$$

$$= \left[2x + \frac{x^2}{2}\right]_{1.0}^{2.0} = 3.5 \text{ J} \qquad \textbf{Ans.}$$

Example 5 A force $F = -\frac{k}{x^2}$ $(x \neq 0)$ acts on a particle in x-direction. Find the work done by this force in displacing the particle from. $x = +a$ to $x = +2a$. Here, k is a positive constant.

Sol. $$W = \int F\, dx = \int_{+a}^{+2a} \left(\frac{-k}{x^2}\right) dx = \left[\frac{k}{x}\right]_{+a}^{+2a}$$

$$= -\frac{k}{2a} \qquad \textbf{Ans.}$$

Note *It is important to note that work comes out to be negative which is quite obvious as the force acting on the particle is in negative x-direction* $\left(F = -\frac{k}{x^2}\right)$ *while displacement is along positive x-direction. (from* $x = a$ *to* $x = 2a$*)*

3. Work Done by Area Under F-s or F-x Graph

As we have discussed above

$$W = \int F dx = \text{area under } F\text{-}x \text{ graph}$$

So, work done by a force can be obtained from the area under F-x graph. Unlike the integration method of finding work done in which sign of work done automatically comes after integration, in this method area of the graph will only give us the magnitude of work done. If force and displacement have same sign, work done will be positive and if both have opposite signs, work done is negative.

Let us take an example.

Example 6 A force F acting on a particle varies with the position x as shown in the figure. Find the work done by this force in displacing the particle from

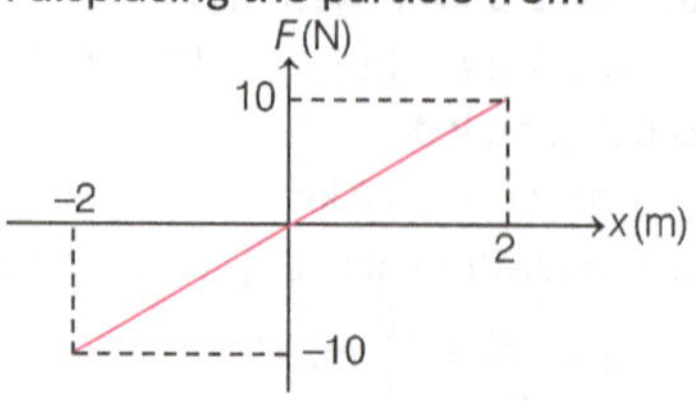

(a) $x = -2$ m to $x = 0$ (b) $x = 0$ to $x = 2$ m

Sol. (a) From $x = -2$m to $x = 0$, displacement of the particle is along positive x-direction while force acting on the particle is along negative x-direction (you can see from the graph). Therefore, work done is negative and given by the area under F-x graph with projection along x-axis.

$\therefore \qquad W = -\frac{1}{2}(2)(10) = -10$ J **Ans.**

(b) From $x = 0$ to $x = 2$ m, displacement of the particle and force acting on the particle both are along positive x-direction. Therefore, work done is positive and given by the area under F-x graph,

or $\qquad W = \frac{1}{2}(2)(10) = 10$ J **Ans.**

INTRODUCTORY Exercise

1. A block is displaced from (1m, 4m, 6m) to $(2\hat{\mathbf{i}} + 3\hat{\mathbf{j}} - 4\hat{\mathbf{k}})$ m under a constant force $\mathbf{F} = (6\hat{\mathbf{i}} - 2\hat{\mathbf{j}} + \hat{\mathbf{k}})$ N. Find the work done by this force.
2. A block of mass 2.5 kg is pushed 2.20 m along a frictionless horizontal table by a constant force 16 N directed 45° above the horizontal. Determine the work done by
   (a) the applied force,
   (b) the normal force exerted by the table,
   (c) the force of gravity and
   (d) determine the total work done on the block
3. A block is pulled a distance x along a rough horizontal table by a horizontal string. If the tension in the string is T, the weight of the block is w, the normal reaction is N and frictional force is F. Write down expressions for the work done by each of these forces.
4. A bucket tied to a string is lowered at a constant acceleration of $g/4$. If mass of the bucket is m and it is lowered by a distance l then find the work done by the string on the bucket.
5. A 1.8 kg block is moved at constant speed over a surface for which coefficient of friction $\mu = \frac{1}{4}$. It is pulled by a force F acting at 45° with horizontal as shown in figure. The block is displaced by 2 m. Find the work done on the block by (a) the force F (b) friction (c) gravity.

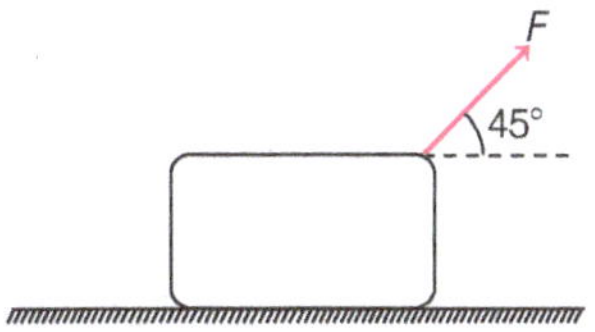

6. A block is constrained to move along x-axis under a force $F = -2x$. Here, F is in newton and x in metre. Find the work done by this force when the block is displaced from $x = 2$ m to $x = -4$ m.
7. A block is constrained to move along x-axis under a force $F = \frac{4}{x^2}$ $(x \neq 0)$. Here, F is in newton and x in metre. Find the work done by this force when the block is displaced from $x = 4$ m to $x = 2$ m.
8. Force acting on a particle varies with displacement as shown in figure. Find the work done by this force on the particle from $x = -4$ m to $x = +4$ m.

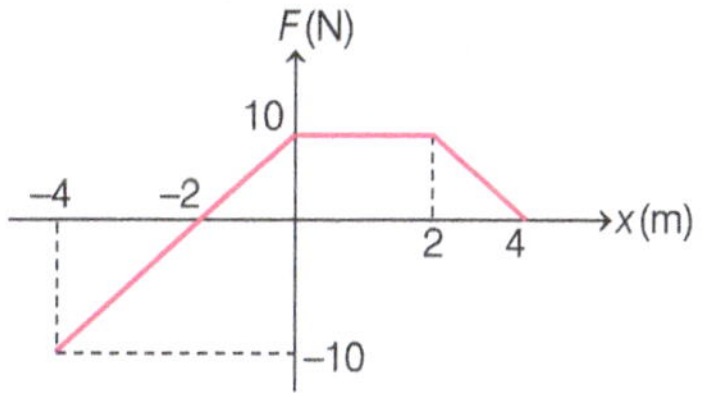

9. A particle is subjected to a force F_x that varies with position as shown in figure. Find the work done by the force on the body as it moves
   (a) from $x = 10.0$ m to $x = 5.0$ m,
   (b) from $x = 5.0$ m to $x = 10.0$ m,
   (c) from $x = 10.0$ m to $x = 15.0$ m,
   (d) what is the total work done by the force over the displacement from $x = 0$ to $x = 15.0$ m ?

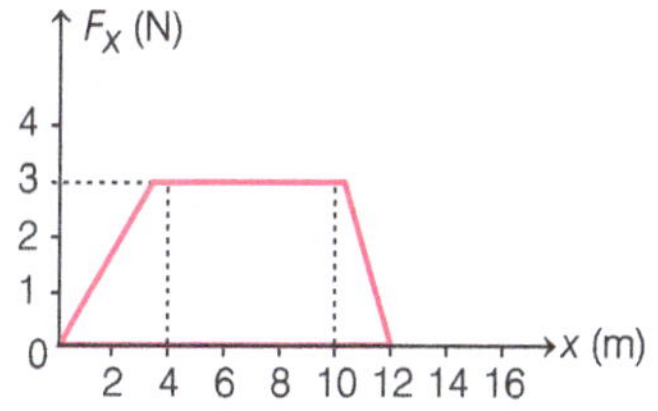

10. A child applies a force **F** parallel to the x-axis to a block moving on a horizontal surface. As the child controls the speed of the block, the x-component of the force varies with the x-coordinate of the block as shown in figure. Calculate the work done by the force **F** when the block moves
    (a) from $x = 0$ to $x = 3.0$ m
    (b) from $x = 3.0$ m to $x = 4.0$ m
    (c) from $x = 4.0$ m to $x = 7.0$ m
    (d) from $x = 0$ to $x = 7.0$ m

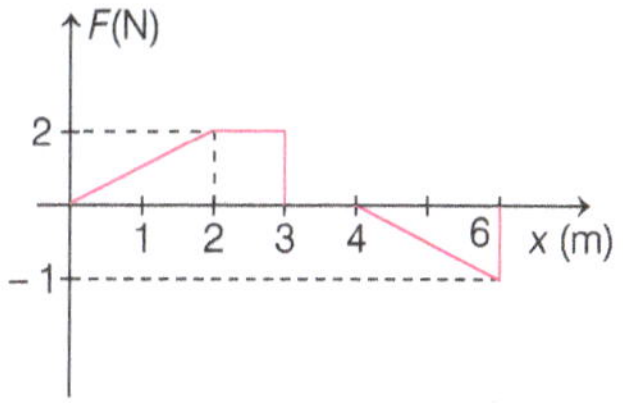

Answers

1. -2 J **2.** (a) 24.89 J (b) Zero (c) Zero (d) 24.9 J

3. Tx, 0, 0, $-Fx$ **4.** $\frac{-3}{4}mgl$

5. (a) 7.2 J (b) -7.2 J (c) Zero

6. -12 J **7.** -1 J **8.** 30 J

9. (a) -15J (b) $+15$J (c) 3 J (d) 27 J

10. (a) 4.0 J (b) Zero (c) -1.0 J (d) 3.0 J

See hints on pages 513-514.

4. General method of finding work done

General expression of work done is,

$$W = \int dW = \int \mathbf{F} \cdot d\mathbf{r} \quad \text{or} \quad \int \mathbf{F} \cdot d\mathbf{s}$$

Here **F** will be given and $d\mathbf{r}$ is a standard vector given by,

$$d\mathbf{r} = dx\hat{\mathbf{i}} + dy\hat{\mathbf{j}} + dz\hat{\mathbf{k}}$$

Note *In the following three cases work done is path independent.*
*(i) **F** is a constant force.*
*(ii) **F** is of the type* $\mathbf{F} = f_1(x)\hat{\mathbf{i}} + f_2(y)\hat{\mathbf{j}} + f_3(z)\hat{\mathbf{k}}$
(iii) $\mathbf{F} \cdot d\mathbf{r}$ *is in the form d (function of x, y and z) e.g. d(xy), so that*

$$W_{A \to B} = \int_A^B \mathbf{F} \cdot d\mathbf{r} = \int_A^B d(xy) = [xy]_A^B$$

In all other cases, we will have to mention the path. Along different paths work done will be different.

Example 7 A body is displaced from origin to (2m, 4m) under the following two forces :

(a) $\mathbf{F} = (2\hat{\mathbf{i}} + 6\hat{\mathbf{j}})$N, a constant force

(b) $\mathbf{F} = (2x\hat{\mathbf{i}} + 3y^2\hat{\mathbf{j}})$ N

Find work done by the given forces in both cases.

Sol. (a) $\mathbf{F} = (2\hat{\mathbf{i}} + 6\hat{\mathbf{j}})$N

$$d\mathbf{r} = (dx\hat{\mathbf{i}} + dy\hat{\mathbf{j}} + dz\hat{\mathbf{k}})$$

$$\therefore \quad \mathbf{F} \cdot d\mathbf{r} = 2dx + 6dy$$

$$W = \int_{(0,0)}^{(2\text{m},4\text{m})} \mathbf{F} \cdot d\mathbf{r} = \int_{(0,0)}^{(2\text{m},4\text{m})} (2dx + 6dy)$$

$$= [2x + 6y]_{(0,0)}^{(2\text{ m},4\text{ m})} = (2 \times 2 + 6 \times 4)$$

$$= 28 \text{ J}$$ **Ans.**

Note *Here, **F** is constant, so the work done is path independent.*

(b) $\mathbf{F} = (2x\hat{\mathbf{i}} + 3y^2\hat{\mathbf{j}})$N

$$d\mathbf{r} = (dx\hat{\mathbf{i}} + dy\hat{\mathbf{j}} + dz\hat{\mathbf{k}})$$

$$\therefore \quad \mathbf{F} \cdot d\mathbf{r} = (2x\,dx + 3y^2dy)$$

$$W = \int_{(0,0)}^{(2\text{m},4\text{m})} \mathbf{F} \cdot d\mathbf{r} = \int_{(0,0)}^{(2\text{m},4\text{m})} (2x\,dx + 3y^2dy)$$

$$= [x^2 + y^3]_{(0,0)}^{(2\text{m},4\text{m})} = (2)^2 + (4)^3$$

$$= 68 \text{ J}$$ **Ans.**

Note *Here the given force is of type* $\mathbf{F} = f_1(x)\hat{\mathbf{i}} + f_2(y)\hat{\mathbf{j}}$
So, the work done is path independent.

Example 8 A force $\mathbf{F} = -k(y\hat{\mathbf{i}} + x\hat{\mathbf{j}})$ (where k is a positive constant) acts on a particle moving in the x-y plane. Starting from the origin, the particle is taken along the positive x-axis to the point $(a,0)$ and then parallel to the y-axis to the point (a,a). The total work done by the force **F** on the particle is **(JEE 1998)**

(a) $-2ka^2$ (b) $2ka^2$ (c) $-ka^2$ (d) ka^2

Sol. $d\mathbf{r} = dx\hat{\mathbf{i}} + dy\hat{\mathbf{j}} + dz\hat{\mathbf{k}}$

and $\mathbf{F} = -k(y\hat{\mathbf{i}} + x\hat{\mathbf{j}})$

$$\therefore \quad dW = \mathbf{F} \cdot d\mathbf{r} = -k(ydx + xdy) = -kd(xy)$$

$$\therefore \quad W = \int_{(0,0)}^{(a,a)} dW = -k\int_{(0,0)}^{(a,a)} d(xy)$$

$$= -k[xy]_{(0,0)}^{(a,a)}$$

$$W = -ka^2$$

$\therefore$ The correct option is (c).

Alternate Method

While moving from (0, 0) to $(a, 0)$ along positive x-axis, $y = 0$

$\therefore$ $\mathbf{F} = -kx\hat{\mathbf{j}}$ i.e. force is in negative y-direction while the displacement is in positive x-direction. Therefore, $W_1 = 0$ (Force $\perp$ displacement).

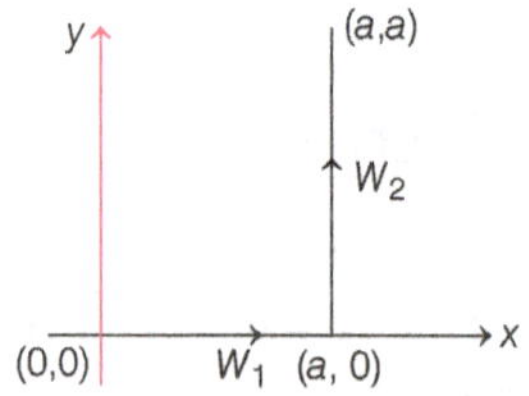

Then, it moves from $(a, 0)$ to (a, a) along a line parallel to y-axis $(x = +a)$. During this

$$\mathbf{F} = -k(y\hat{\mathbf{i}} + a\hat{\mathbf{j}})$$

The first component of force, $-ky\hat{\mathbf{i}}$ will not contribute any work, because this component is along negative x-direction $(-\hat{\mathbf{i}})$ while displacement is in positive y-direction $(a,0)$ to (a, a).

The second component of force i.e. $-ka\hat{\mathbf{j}}$ will perform negative work as :

$$\mathbf{F} = -ka\hat{\mathbf{j}} \text{ and } \mathbf{s} = a\hat{\mathbf{j}}$$

$$\therefore \quad W_2 = \mathbf{F} \cdot \mathbf{s}$$

$$\text{or} \quad W_2 = (-ka)(a) = -ka^2$$

$$\therefore \quad W = W_1 + W_2 = -ka^2$$

Note *For the given force, work done is path independent. It depends only on initial and final positions.*

Example 9 A body is displaced from origin to (1m, 1m) by a force $\mathbf{F} = (2y\hat{\mathbf{i}} + 3x^2\hat{\mathbf{j}})$ along two paths

(a) $x = y$ (b) $y = x^2$

Find the work done along both paths.

Sol. $\mathbf{F} = (2y\hat{\mathbf{i}} + 3x^2\hat{\mathbf{j}})$

$$d\mathbf{r} = (dx\hat{\mathbf{i}} + dy\hat{\mathbf{j}})$$

$$\Rightarrow \quad \mathbf{F} \cdot d\mathbf{r} = (2y\,dx + 3x^2\,dy)$$

We cannot integrate $\mathbf{F} \cdot d\mathbf{r}$ or $(2y\,dx + 3x^2\,dy)$ directly to find the work done. But along the given paths we can change this expression.

(a) Along the path $x = y$,

$$(2y\,dx + 3x^2dy) = (2x\,dx + 3y^2dy)$$

$$\therefore \quad W_1 = \int_{(0,0)}^{(1\text{m},1\text{m})} \mathbf{F} \cdot d\mathbf{r} = \int_{(0,0)}^{(1\text{m},1\text{m})} (2x\,dx + 3y^2dy)$$

$$= [x^2 + y^3]_{(0,0)}^{(1m, 1m)}$$

$$= (1)^2 + (1)^3 = 2\text{ J}$$ **Ans.**

(b) Along the path $y = x^2$

$$(2y\,dx + 3x^2 dy) = (2x^2\,dx + 3y\,dy)$$

$$\therefore \quad W_2 = \int_{(0,0)}^{(1m,1m)} \mathbf{F}\cdot d\mathbf{r} = \int_{(0,0)}^{(1m,1m)} (2x^2 dx + 3y\,dy)$$

$$= \left[\frac{2}{3}x^3 + \frac{3}{2}y^2\right]_{(0,0)}^{(1m, 1m)}$$

$$= \frac{2}{3}(1)^3 + \frac{3}{2}(1)^2$$

$$= \frac{13}{6}\text{ J}$$ **Ans.**

Note *We can see that $W_1 \neq W_2$ or work done is path dependent in this case.*

Important POINTS

Following method will explain you whether $\mathbf{F}\cdot d\mathbf{r}$ is in the form of d (function of x, y and z) or not.

(i) Let $\mathbf{F}\cdot d\mathbf{r}$ is $(ydx + xdy)$ then this expression is $d(xy)$.

Reason is :

First term : $y\int dx = xy$

Second term : $x\int dy = xy$

Since both are same $(= xy)$. So, it is $d(xy)$.

(ii) Let $\mathbf{F}\cdot d\mathbf{r}$ is, $(3x^2y^2dx + 2x^3ydy)$. Then,

$(3x^2y^2dx + 2x^3ydy) = d(x^3y^2)$

First term : $3y^2\int x^2dx = y^2x^3$

Second term : $2x^3\int ydy = x^3y^2$

Since both are same $(= x^3y^2)$. So,

$(3x^2y^2dx + 2x^3ydy) = d(x^3y^2)$

If both terms are not same then $\mathbf{F}\cdot d\mathbf{r}$ is not in the desired form.

Method of Finding Work Done in Complex Problems

Following example will illustrate the method.

Example 10 A smooth track in the form of a quarter-circle of radius 6 m lies in the vertical plane. A ring of weight 4 N moves from P_1 and P_2 under the action of forces $\mathbf{F}_1, \mathbf{F}_2$ and $\mathbf{F}_3$. Force $\mathbf{F}_1$ is always towards P_2 and is always 20 N in magnitude; force $\mathbf{F}_2$ always acts horizontally and is always 30 N in magnitude; force $\mathbf{F}_3$ always acts tangentially to the track and is of magnitude $(15 - 10\,s)$ N, where s is in metre. Find the work done by $\mathbf{F}_1, \mathbf{F}_2, \mathbf{F}_3$ and weight in moving the ring from P_1 to P_2.

Sol. Work done by $\mathbf{F}_1$:

$$W_1 = \int_{P_1}^{P_2} F_1 \cos\theta\, ds$$

ds is tangential.

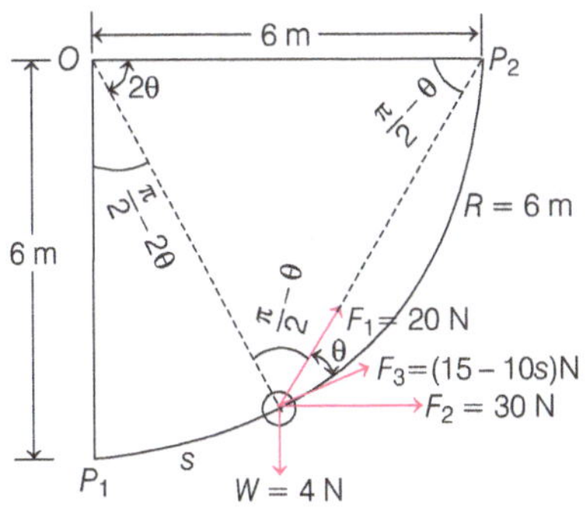

From figure, $s = R\left(\frac{\pi}{2} - 2\theta\right) = 6\left(\frac{\pi}{2} - 2\theta\right)$

or $ds = -12\,d\theta$

and $F_1 = 20.$

At P_1, $2\theta = \frac{\pi}{2}$

$\Rightarrow$ $\theta = \frac{\pi}{4}$

At P_2, $2\theta = 0$

$\Rightarrow$ $\theta = 0$

Hence, $W_1 = -240\int_{\pi/4}^{0} \cos\theta\, d\theta$

$$= 240 \sin\frac{\pi}{4} = 120\sqrt{2}\text{ J}$$

Work done by $\mathbf{F}_3$:

$$W_3 = \int F_3\, ds = \int_0^{3\pi} (15 - 10s)\, ds \quad \left(P_1P_2 = R\frac{\pi}{2} = \frac{6\pi}{2} = 3\pi\right)$$

$$= [15s - 5s^2]_0^{3\pi} = -302.76\text{J} \approx -302.8\text{ J}$$

Work done by $\mathbf{F}_2$ and W :

To calculate the work done by $\mathbf{F}_2$ and by W, it is convenient to take the projection of the displacement in the direction of the force. Because both forces are constant (in magnitude as well as direction). Thus,

$$W_2 = F_2(OP_2) = 30(6) = 180\text{ J}$$

$$W_4 = (-W)(P_1O) = (-4)(6) = -24\text{ J} \qquad (W = \text{weight})$$

MODULE 1 Exercise

Assertion and Reason

Direction : Choose the correct option.

(a) If both **Assertion** and **Reason** are true and the **Reason** is correct explanation of the **Assertion**.

(b) If both **Assertion** and **Reason** are true but **Reason** is not the correct explanation of **Assertion**.

(c) If **Assertion** is true, but the **Reason** is false.

(d) If **Assertion** is false but the **Reason** is true.

(e) If both **Assertion** and **Reason** are false.

1. **Assertion :** All surfaces shown in figure are smooth. Block *A* comes down along the wedge *B*. Work done by normal reaction (between *A* and *B*) on *B* is positive while on *A* it is negative.

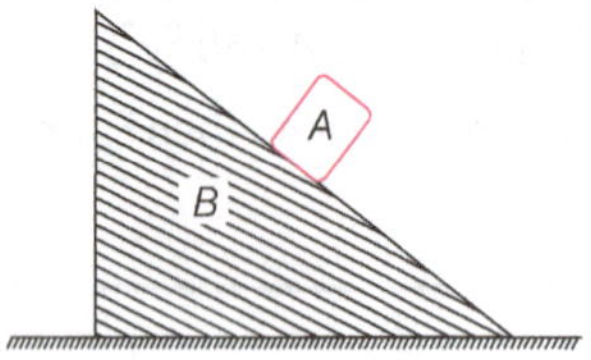

Reason : Angle between normal reaction and net displacement of *A* is greater than 90° while between normal reaction and net displacement of *B* is less than 90°.

2. **Assertion :** Consider a person of mass 80 kg who is climbing a ladder. In climbing up a vertical distance of 5 m, the contact force exerted by ladder on person's feet does 4000 J of work. [Take, $g = 10 \text{m/s}^2$]

Reason : Work done by a force *F* is defined as the dot product of force and the displacement of point of application of force.

Single Correct Option

3. A block of mass *m* is pulled along a horizontal surface by applying a force at an angle θ with the horizontal. If the block travels with a uniform velocity and has a displacement *d* and the coefficient of friction is μ, then the work done by the applied force is

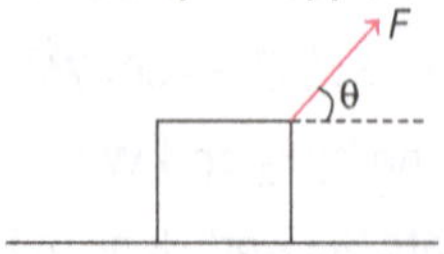

(a) $\dfrac{\mu mgd}{\cos\theta + \mu \sin\theta}$ (b) $\dfrac{\mu mgd \cos\theta}{\cos\theta + \mu \sin\theta}$

(c) $\dfrac{\mu mgd \sin\theta}{\cos\theta + \mu \sin\theta}$ (d) $\dfrac{\mu mgd \cos\theta}{\cos\theta - \mu \sin\theta}$

4. The net work done by the tension in the figure when the bigger block of mass *M* touches the ground is

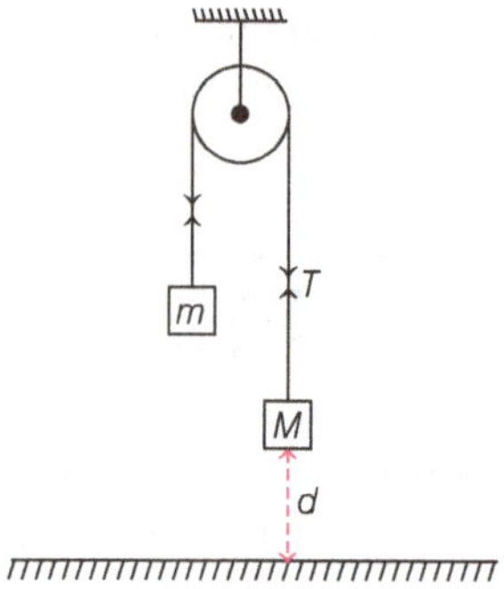

(a) $+Mgd$ (b) $-(M+m)gd$

(c) $-mgd$ (d) zero

5. A block of mass 10 kg is released on a fixed wedge inside a cart which is moving with constant velocity 10 ms^{-1} towards right. There is no relative motion between block and cart. Then, work done by normal reaction on block in two seconds from ground frame will be

(Take, $g = 10 \text{ ms}^{-2}$)

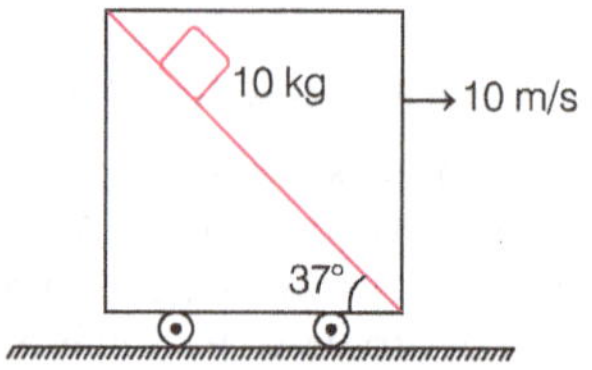

(a) 1320 J (b) 960 J

(c) 1200 J (d) 240 J

6. When a rubber band is stretched by a distance *x*, it exerts a restoring force of magnitude $F = ax + bx^2$, where *a* and *b* are constants. The work done in stretching the unstretched rubber band by *L* is

(a) $aL^2 + bL^3$ (b) $\dfrac{1}{2}(aL^2 + bL^3)$

(c) $\dfrac{aL^2}{2} + \dfrac{bL^3}{3}$ (d) $\dfrac{1}{2}\left(\dfrac{aL^2}{2} + \dfrac{bL^3}{3}\right)$

7. The pointer reading *versus* load graph for a spring balance is as shown

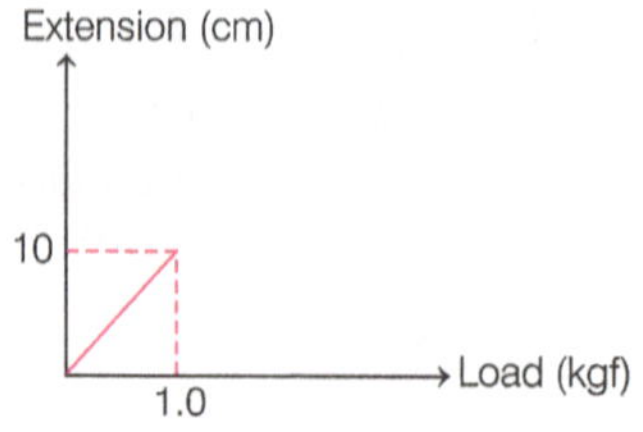

The spring constant is

(a) $\frac{15\,\text{kgf}}{\text{cm}}$ (b) $\frac{5\,\text{kgf}}{\text{cm}}$

(c) $\frac{0.1\,\text{kgf}}{\text{cm}}$ (d) $\frac{10\,\text{kgf}}{\text{cm}}$

8. A block of mass m is kept on a platform which starts from rest with constant acceleration $\frac{g}{2}$ upwards as shown in figure. Work done by normal reaction on block in time t is

(2019 Main)

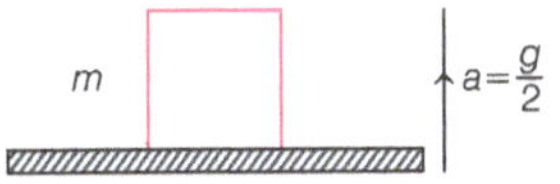

(a) $\frac{mg^2t^2}{8}$ (b) $\frac{3mg^2t^2}{8}$ (c) 0 (d) $-\frac{mg^2t^2}{8}$

9. Identify the correct statements from the following:

(A) Work done by a man in lifting a bucket out of a well by means of a rope tied to the bucket is negative.

(B) Work done by gravitational force in lifting a bucket out of a well by a rope tied to the bucket is negative.

(C) Work done by friction on a body sliding down an inclined plane is positive.

(D) Work done by an applied force on a body moving on a rough horizontal plane with uniform velocity is zero.

(E) Work done by the air resistance on an oscillating pendulum is negative.

Choose the correct answer from the options given below:

(2023 Main)

(a) B and E only (b) A and C only

(c) B, D and E only (d) B and D only

10. Consider a force $\mathbf{F} = -x\hat{\mathbf{i}} + y\hat{\mathbf{j}}$. The work done by this force in moving a particle from point A (1, 0) to B (0, 1) along the line segment is (all quantities are in SI units)

(2020 Main)

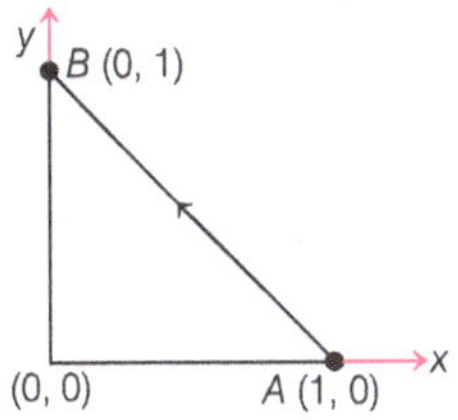

(a) 2 J (b) $\frac{1}{2}$ J (c) 1 J (d) $\frac{3}{2}$ J

11. The force required to stretch a spring varies with the distance as shown in the figure. If the experiment is performed with the above spring of half the length, the line OA will

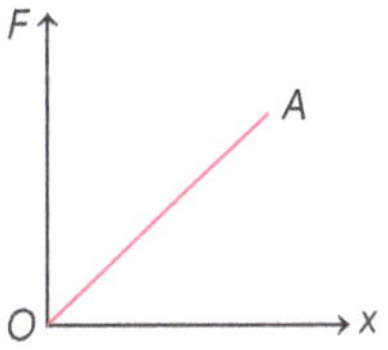

(a) shift towards F-axis

(b) shift towards x-axis

(c) remain as it is

(d) become double in length

Multiple Correct Options

12. Block A has no relative motion with respect to wedge fixed to the lift as shown in figure during motion 1 or motion 2. Then,

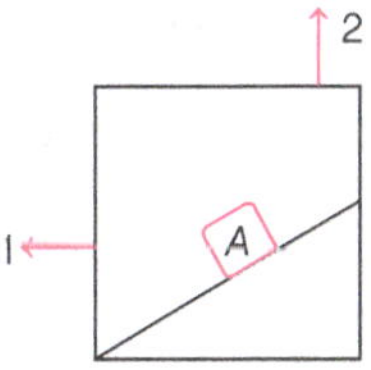

(a) work done by gravity on block A in motion 2 is less than in motion 1

(b) work done by normal reaction on block A in both the motions will be positive

(c) work done by force of friction in motion 1 may be positive

(d) work done by force of friction in motion 1 may be negative

Passage Based Questions

Passage (Q.13 to Q.15)

Consider the situation shown in the figure. The system is released from rest and block A travels a distance 5 m in downward direction. (Take, $g = 10\,\text{m/s}^2$)

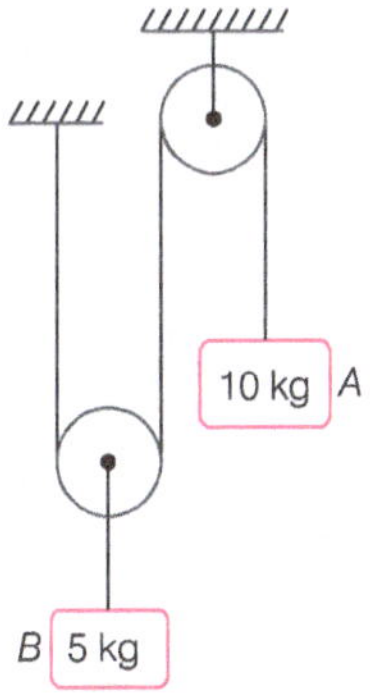

13. Find the work done by tension on the block B.

(a) Zero (b) 166.67 J

(c) 375 J (d) −116.67 J

14. Find the total work done by the tension on the system (block A + block B).

(a) Zero (b) 116.67 J (c) 375 J (d) −375 J

15. Find net work done on the system.

(a) Zero (b) 116.67 J (c) 375 J (d) −116.67 J

Match the Columns

16. A body is displaced from $x = 4$ m to $x = 2$ m along the x-axis. For the forces mentioned in Column I, match the corresponding work done is Column II.

	Column I		Column II
(a)	$\mathbf{F} = 4\hat{\mathbf{i}}$	(p)	positive
(b)	$\mathbf{F} = (4\hat{\mathbf{i}} - 4\hat{\mathbf{j}})$	(q)	negative
(c)	$\mathbf{F} = -4\hat{\mathbf{i}}$	(r)	zero
(d)	$\mathbf{F} = (-4\hat{\mathbf{i}} - 4\hat{\mathbf{j}})$	(s)	$\lvert W \rvert = 8$ units

17. In Column I, some statements are given related to work done by a force on an object while in Column II the sign and information about value of work done is given. Match the entries of Column I with the entries of Column II.

	Column I		Column II
(a)	Work done by friction force on a block as it slides down a rigid fixed incline with respect to ground.	(p)	Positive
(b)	In the above case, work done by friction force on incline with respect to ground.	(q)	Negative
(c)	Work done by a man in lifting a bucket out of a well by means of a rope tied to the bucket with respect to ground.	(r)	Zero
(d)	Total work done by friction force in (a) with respect to ground.	(s)	may be positive, negative or zero.

Subjective Questions

18. A particle is pulled a distance l up a rough plane inclined at an angle α to the horizontal by a string inclined at an angle β to the plane $(\alpha + \beta < 90°)$. If the tension in the string is T, the normal reaction between the particle and the plane is N, the frictional force is F and the weight of the particle is w. Write down expressions for the work done by each of these forces.

19. A small block of mass 1 kg is kept on a rough inclined wedge of inclination 45° fixed in an elevator. The elevator goes up with a uniform velocity $v = 2$ m/s and the block does not slide on the wedge. Find the work done by the force of friction on the block in 1 s. (Take, $g = 10\ \text{m/s}^2$)

20. A block of mass 2 kg is released from rest on a rough inclined ground as shown in figure. Find the work done on the block by

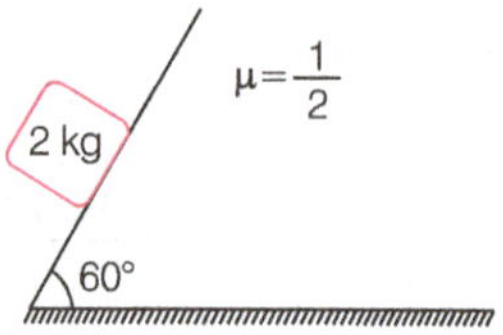

(a) gravity, (b) force of friction

when the block is displaced downwards along the plane by 2 m. (Take, $g = 10\ \text{m/s}^2$)

21. A particle has shifted along some trajectory in the plane xy from point 1 whose radius vector $\mathbf{r}_1 = \hat{\mathbf{i}} + 2\hat{\mathbf{j}}$ to point 2 with the radius vector $\mathbf{r}_2 = 2\hat{\mathbf{i}} - 3\hat{\mathbf{j}}$. During that time the particle experienced the action of certain forces, one of which being $\mathbf{F} = 3\hat{\mathbf{i}} + 4\hat{\mathbf{j}}$. Find the work performed by the force $\mathbf{F}$. (Here r_1, r_2 and F are given in SI units) **(Irodov 1.118)**

22. A light inextensible string that goes over a smooth fixed pulley as shown in the figure connects two blocks of masses 0.36 kg and 0.72 kg. Taking $g = 10\ \text{ms}^{-2}$, find the work done (in joule) by string on the block of mass 0.36 kg during the first second after the system is released from rest. **(2004 JEE)**

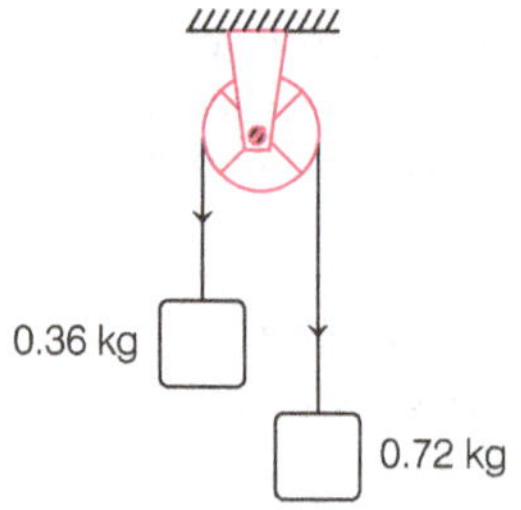

Integer Type Questions

23. A force $\mathbf{F} = (5 + 3y^2)$ acts on a particle in the y-direction, where $\mathbf{F}$ is newton and y is in metre. The work done by the force during a displacement from $y = 2\text{m}$ to $y = 5\text{m}$ is .... J. **(2023 Main)**

24. Two persons A and B perform same amount of work in moving a body through a certain distance d with application of forces acting at angles 45° and 60° with the direction of displacement respectively. The ratio of force applied by person A to the force applied by person B is $\frac{1}{\sqrt{x}}$. The value of x is .... **(2021 Main)**

Answers

1. (a) **2.** (d) **3.** (b) **4.** (d) **5.** (b) **6.** (c) **7.** (c) **8.** (b) **9.** (a) **10.** (c)
11. (a) **12.** (a, b, c, d) **13.** (b) **14.** (a) **15.** (c) **16.** (a -q, s; b -q, s; c -p, s; d -p,s)
17. (a -q; b -r; c -p; d -q) **18.** $Tl \cos\beta$, 0, $-Fl$, $wl \sin\alpha$ **19.** 10 J **20.** (a) 34.6 J (b) −10J **21.** −17 J **22.** 8 J
23. (132) **24.** (2)

See hints on pages 514-516.

MODULE 2

Conservative Force and Its Potential Energy

Conservative Forces

The forces in which work is path independent are known as **conservative forces.**

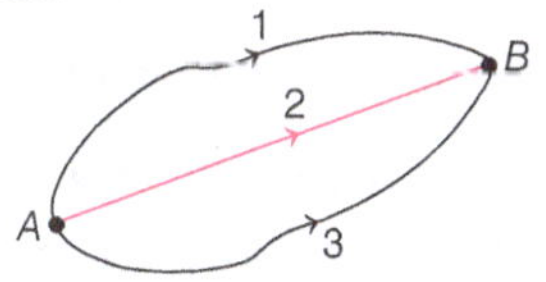

Thus, if a particle or an object is displaced from position A to position B through three different paths under a conservative force field. Then,

$$W_1 = W_2 = W_3$$

Further, it can be shown that work done in a closed path is zero under a conservative force field. ($W_{AB} = -W_{BA}$ or $W_{AB} + W_{BA} = 0$).

Gravitational force, Coulomb's force and spring force are few examples of conservative forces. On the other hand, if the work is path dependent or $W_1 \neq W_2 \neq W_3$, the force is called a **non-conservative.** Frictional force and viscous force are non-conservative in nature. Work done in a closed path is not zero in a non-conservative force field.

Note (i) *The word potential energy is defined only for conservative forces like gravitational force, electrostatic force and spring force etc.*

(ii) *If only conservative forces act on a system, then mechanical energy (kinetic energy + potential energy) of the system remains constant.*

We can differentiate the conservative and non-conservative forces in a better way by making a table as given below.

S.No	Conservative Forces	Non-conservative Forces
1.	Work done is path independent	Work done is path dependent
2.	Work done in a closed path is zero	Work done in a closed path is not zero
3.	The word potential energy is defined for conservative forces	The word potential energy is not defined for non-conservative forces.
4.	Examples are: gravitational force, electrostatic force, spring force etc.	Examples are: frictional force, viscous force etc.
5.	Mechanical energy under conservative forces remains constant.	Mechanical energy does not remain constant.

Important POINTS

1. Gravitational force is a conservative force. Its work done is path independent. For small heights it only depends on the height difference 'h' between two points. Work done by gravitational force is $\pm$***mgh***, in moving the mass 'm' from one point to another point.

   This is $+$***mgh*** if the **mass is moving downwards** (as the force mg and displacement both are downwards, in the same direction) and $-$***mgh*** if the **mass is moving upwards**.

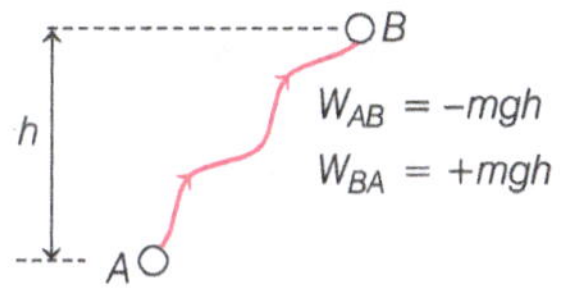

2. The magnetic field (and therefore the magnetic force) is neither conservative nor non-conservative.
3. Electric field (and therefore electric force) is produced either by static charge or by time varying magnetic field. First is conservative.
4. In Module -1 of this unit we have seen that there are certain cases where work done is path independent. For example, work done by a constant force is path independent. But it does not mean if force is constant then it is conservative in nature.

   Let us take one example. If a block is kept on a rough horizontal table. Now block is displaced from A to B along a straight line, then force of friction is constant (in magnitude as well as direction). But friction is not a conservative force. Because if we displace the block along a curved path then direction of friction will keep on changing.

   Similarly, if we displace the block from A to B along different paths then again direction of friction will change and work done will be different along different paths. Similarly work done by friction will not be zero along a closed path.

   Therefore a constant force is said to be a conservative force if :

   (i) It is constant along any path.

   Even in a closed path also it is constant (in magnitude as well as direction) at every point.

   (ii) Mechanical energy should remain constant.

Potential Energy

The energy possessed by a body or system by virtue of its position or configuration is known as the potential energy. For example, a block attached to a compressed or elongated spring possesses some energy called elastic potential energy. This block has a capacity to do work.

Similarly, a stone when released from a certain height also has energy in the form of gravitational potential energy. Two charged particles kept at certain distance have electrostatic potential energy.

Regarding the potential energy it is important to note that it is defined for a conservative force field only. For non-conservative forces it has no meaning. The change in potential energy (dU) of a system corresponding to a conservative force **F** is given by

$$dU = -\mathbf{F}\cdot d\mathbf{r} = -dW \quad \text{or} \quad \int_i^f dU = -\int_{r_i}^{r_f} \mathbf{F}\cdot d\mathbf{r}$$

or

$$\Delta U = U_f - U_i = -\int_{r_i}^{r_f} \mathbf{F}\cdot d\mathbf{r}$$

For defining absolute potential energy at some point we have to choose a reference point where potential energy is assumed to be zero.

We generally choose the reference point at infinity and assume potential energy to be zero there, i.e. if we take $r_i = \infty$ (infinite) and $U_i = 0$ then we can write

$$U = -\int_{\infty}^{r} \mathbf{F}\cdot d\mathbf{r} = -W$$

or potential energy of a body or system is the negative of work done by the conservative forces in bringing it from infinity to the present position.

Regarding the potential energy it is worth noting that:

1. Potential energy can be defined only for conservative forces and it should be considered to be a property of the entire system rather than assigning it to any specific particle.
2. Potential energy depends on frame of reference.
3. If conservative force F and potential energy associated with this force U are functions of single variable r or x then :

$$F = -\frac{dU}{dr} \quad \text{or} \quad -\frac{dU}{dx}$$

Now, let us discuss three different types of potential energies which we usually come across.

Elastic Potential Energy

We have already discussed about the spring force. We have seen there that the work done by the spring force (of course conservative for an ideal spring) is $-\frac{1}{2}kx^2$ when the spring is stretched or compressed by an amount x from its unstretched position. Thus,

$$U = -W = -\left(-\frac{1}{2}kx^2\right) \quad \text{or} \quad U = \frac{1}{2}kx^2$$

(k = spring constant)

Note that elastic potential energy is always positive.

In the above expression x is either compression or elongation.

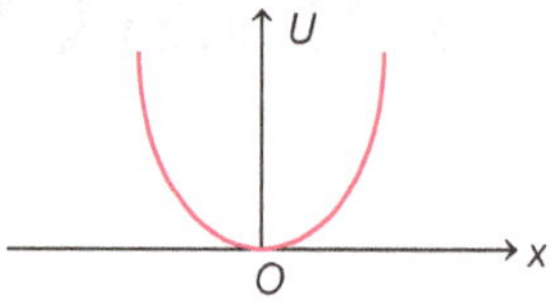

At O, $x = 0$

$\therefore \quad U = 0$. So, at O spring is in natural length.

If we take $x > 0$ as elongation then $x < 0$ will be compression.

Gravitational Potential Energy

The gravitational potential energy of two particles of masses m_1 and m_2 separated by a distance r is give by

$$U = -G\frac{m_1 m_2}{r}$$

Here, G = Universal gravitation constant

$$= 6.67 \times 10^{-11} \frac{\text{N-m}^2}{\text{kg}^2}$$

If a body of mass m is raised to a height h from the surface of earth, the change in potential energy of the system (earth + body) comes out to be

$$\Delta U = \frac{mgh}{\left(1 + \frac{h}{R}\right)} \qquad (R = \text{radius of earth})$$

or $\quad \Delta U \approx mgh \quad$ if $\quad h << R$

Thus, the potential energy of a body at height h, i.e. mgh is really the change in potential energy of the system for $h << R$ w.r.t. ground and if we take ground as the reference point then absolute potential energy at height h is mgh. So, be careful while using $U = mgh$, that h should not be too large. We will discuss this point in detail in the chapter of Gravitation.

Electric Potential Energy

The electric potential energy of two point charges q_1 and q_2 separated by a distance r in vacuum is given by

$$U = \frac{1}{4\pi\varepsilon_0}\cdot\frac{q_1 q_2}{r}$$

Here, $\quad \frac{1}{4\pi\varepsilon_0} = 9.0 \times 10^9 \frac{\text{N-m}^2}{\text{C}^2}$ = a constant

Important POINTS

1. Elastic potential energy is either zero or positive, but gravitational and electric potential energy may be zero, positive or negative.
2. For increase or decrease in gravitational potential energy of a particle (for small heights) we write,

$$|\Delta U| = mgh$$

Here, h is the change in height of particle. In case of a rigid body, h is measured from centre of mass of the rigid body.

3. Change in potential energy is equal to the negative of work done by the conservative force ($\Delta U = W$). If work done by the conservative force is negative, change in potential energy will be positive or potential energy of the system will increase.

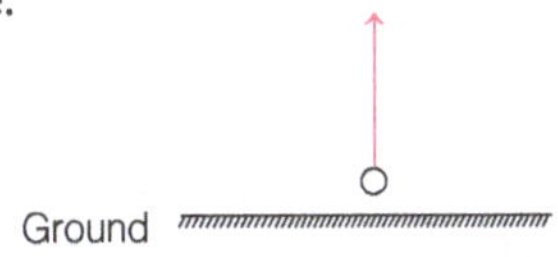

Suppose a ball is taken from the ground to some height, work done by gravity is negative, i.e. change in potential energy should be positive or potential energy of the ball will increase.

$F = -\dfrac{dU}{dr}$, represents that conservative forces always act in a direction where potential energy of the system is decreased.

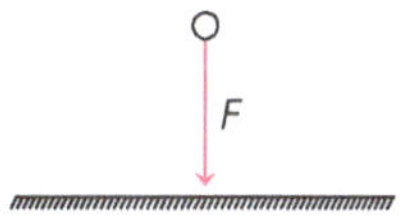

4. If a ball is dropped from a certain height. The force on it (its weight) acts in a direction in which its potential energy decreases.

Example 1 A chain of mass 'm' and length 'l' is kept in three positions as shown below. Assuming $h=0$ on the ground find potential energy of the chain in all three cases.

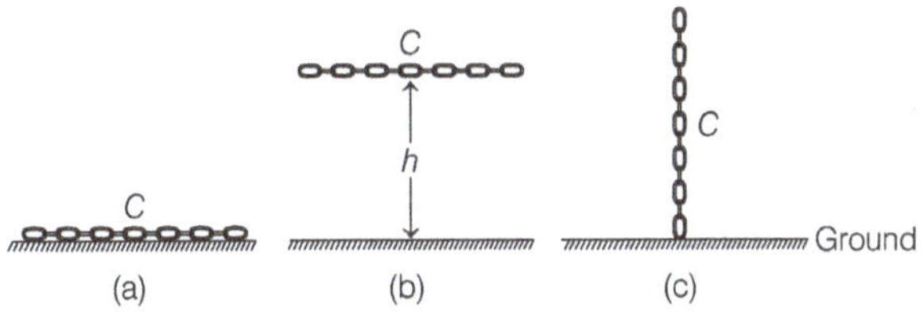

Sol. For finding potential energy of chain we will have to see its centre of mass height 'h' from ground.

In figure (a): $h_c = 0 \Rightarrow$ Potential energy, $U = 0$

In figure (b): $h_c = h \Rightarrow$ Potential energy, $U = mgh$

In figure (c): $h_c = \dfrac{l}{2} \Rightarrow$ Potential energy, $U = mg\dfrac{l}{2}$

Example 2 Potential energy of a body in position A is -40J. Work done by conservative force of this potential energy in moving the body from A to B is -20J. Find potential energy of the body in position B.

Sol. Work done by a conservative force is given by

$$W_{A\to B} = -\Delta U = -(U_B - U_A) = U_A - U_B$$

$$\Rightarrow \quad U_B = U_A - W_{A\to B}$$
$$= (-40) - (-20)$$
$$= -20\,\text{J} \qquad \textbf{Ans.}$$

Note *Potential energy from A to B is increasing (from -40J to -20 J). So, work done by conservative force should be negative (-20 J).*

INTRODUCTORY Exercise

1. If work done by a conservative force is positive then select the correct option(s).
(a) potential energy will decrease
(b) potential energy may increase or decrease
(c) kinetic energy will increase
(d) kinetic energy may increase or decrease.
2. Work done by a conservative force in bringing a body from infinity to A is 60 J and to B is 20 J. What is the difference in potential energy between points A and B, i.e. $U_B - U_A$.

Answers

1. (a, d) 2. 40 J

See hints on page 516.

Relation between Conservative Force (F) and its Potential Energy (U)

$$F = -\frac{dU}{dx} \qquad \text{...(i)}$$

$$\therefore \quad \int dU = -\int F dx \qquad \text{...(ii)}$$

If U-x function is given, we can make F-x function by simple differentiation by using Eq. (i).

If F-x function is given, then U-x function can be made by integration by using Eq. (ii).

In this case, some limit of U (or value of U at some given value of x) should be known to us to make complete

U-x function. Otherwise an unknown constant of integration will be there in U-x equation. If no limit is given in the question and we have to select the most appropriate answer then we can take $U=0$ at $x=0$.

Note *If U is a function of more than one variables (say x, y and z), then*

$$F = -\left[\frac{\partial U}{\partial x}\hat{i} + \frac{\partial U}{\partial y}\hat{j} + \frac{\partial U}{\partial z}\hat{k}\right]$$

Here, $\frac{\partial U}{\partial x}$ is partial differentiation of U with respect to x.

Means, differentiate U with respect to x by assuming y and z to be constant. Similarly $\frac{\partial U}{\partial y}$ and $\frac{\partial U}{\partial z}$.

Example 3 Spring potential energy is given by,

$$U = \frac{1}{2}kx^2$$

Here k is force constant of spring and x is either compression or elongation. Using this expression find the spring force F.

Sol. As we know,

$$F = -\frac{dU}{dx}$$

$$\therefore \quad F = -\frac{d}{dx}\left(\frac{1}{2}kx^2\right) = -\frac{1}{2}k\frac{d}{dx}(x^2)$$

$$= -\frac{1}{2}k(2x) = -kx$$

Example 4 A particle, which is constrained to move along x-axis, is subjected to a force in the same direction which varies with the distance x of the particle from the origin as $F(x) = -kx + ax^3$. Here, k and a are positive constants. For $x \geq 0$, the functional form of the potential energy $U(x)$ of the particle is **(2002 JEE)**

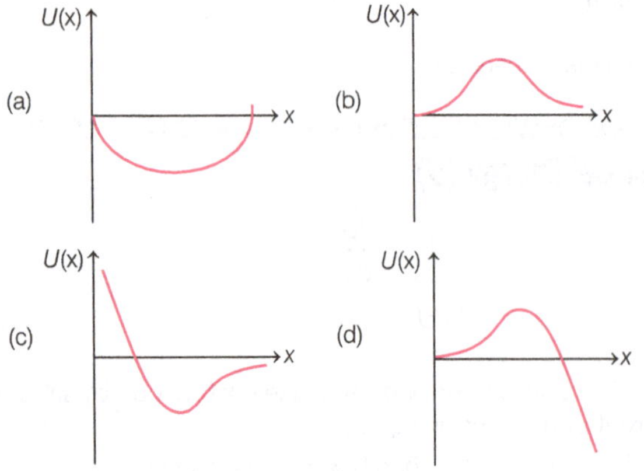

Sol. $F = -\frac{dU}{dx}$

$$\therefore \quad dU = -F \cdot dx$$

$$\text{or} \quad U(x) = -\int_0^x (-kx + ax^3)\,dx$$

$$U(x) = \frac{kx^2}{2} - \frac{ax^4}{4}$$

$$U(x) = 0 \text{ at } x = 0 \text{ and } x = \sqrt{\frac{2k}{a}}$$

$$U(x) = \text{negative for } x > \sqrt{\frac{2k}{a}}$$

Further, $F = 0$ at $x = 0$. Therefore slope of U-x graph should be zero at $x = 0$, because $F = -\frac{dU}{dx}$.

Hence, the correct answer is (d).

Note *In this example, we have assumed a limit : $U = 0$ at $x = 0$.*

Example 5 A particle is placed at the origin and a force $F = kx$ is acting on it (where k is a positive constant). If $U(0) = 0$, the graph of $U(x)$ *versus* x will be (where, U is the potential energy function) **(2004 JEE)**

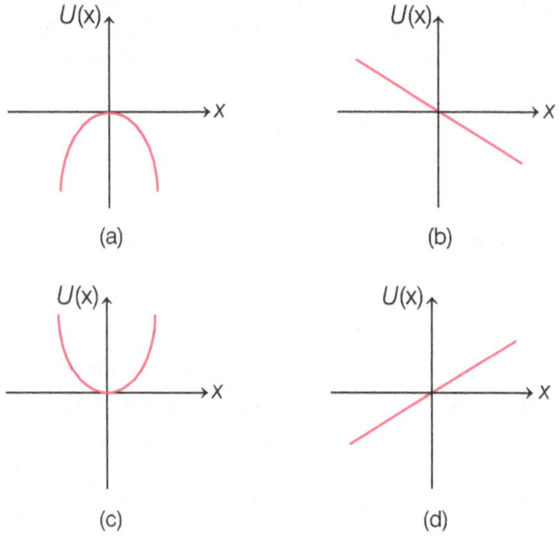

Sol. From $F = -\frac{dU}{dx}$

$$\int_0^{U(x)} dU = -\int_0^x F dx = -\int_0^x (kx)\,dx$$

$$\therefore \quad U(x) = -\frac{kx^2}{2}$$

$$\text{as} \quad U(0) = 0$$

Therefore, the correct option is (a).

MODULE 2 Exercise

Assertion-Reason

See directions on page 470.

1. **Assertion :** The work done in bringing a body down from the top to the base along a frictionless inclined plane is the same as the work done in bringing it down from the vertical side.

   Reason : The gravitational force on the body along the inclined plane is the same as that along the vertical side.

Single Correct Option

2. Two identical cylindrical vessels with their bases at the same level, each contains a liquid of density ρ. The height of the liquid in one vessel is h_1 and that in the other h_2. The area of either bases is A. The work done by the gravity in equalizing the levels when the vessels are interconnected is

   (a) $A\rho g\left[\frac{(h_1-h_2)}{2}\right]$ (b) $A\rho g\left[\frac{(h_1+h_2)}{2}\right]^2$

   (c) $A\rho g\left[\frac{(h_1-h_2)}{4}\right]$ (d) $A\rho g\left[\frac{(h_1-h_2)}{2}\right]^2$

3. A ball is released from the top of a tower. The ratio of work done by force of gravity in first, second and third second of the motion of ball is

   (a) 1:2:3 (b) 1:4:9 (c) 1:3:5 (d) 1:9:25

4. A particle is moving in a conservative force field from point A to point B. U_A and U_B are the potential energies of the particle at points A and B and W_C is the work done by conservative forces in the process of taking the particle from A and B. Then,

   (a) $W_C = U_B - U_A$ (b) $W_C = U_A - U_B$
   (c) $U_A > U_B$ (d) $U_B > U_A$

5. If W_1, W_2 and W_3 represent the work done in moving a particle from A to B along three different paths 1, 2 and 3 respectively (as shown) in the gravitational field of a point mass m. Find the correct relation between W_1, W_2 and W_3. **(2003 JEE)**

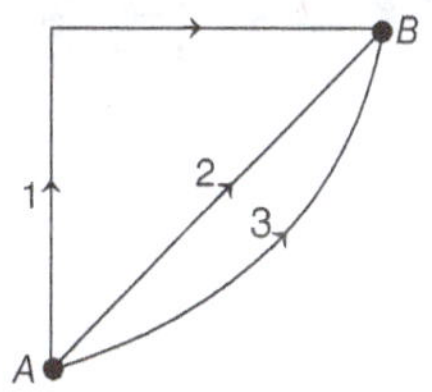

   (a) $W_1 > W_2 > W_3$ (b) $W_1 = W_2 = W_3$
   (c) $W_1 < W_2 < W_3$ (d) $W_2 > W_1 > W_3$

Multiple Correct Options

6. One end of a light spring of spring constant k is fixed to a wall and the other end is tied to a block placed on a smooth horizontal surface. In a displacement, the work done by the spring is $+\left(\frac{1}{2}\right)kx^2$. The possible cases are

   (a) The spring was initially compressed by a distance x and was finally in its natural length
   (b) It was initially stretched by a distance x and finally was in its natural length
   (c) It was initially in its natural length and finally in a compressed position
   (d) It was initially in its natural length and finally in a stretched position

Subjective Question

7. Two equal masses are attached to the two ends of a spring of force constant k. The masses are pulled out symmetrically to stretch the spring by a length $2x_0$ over its natural length. Find the work done by the spring on each mass.

Answers

1. (c) **2.** (d) **3.** (c) **4.** (b) **5.** (b) **6.** (a,b) **7.** $-k\,x_0^2$

See hints on pages 516-517.

MODULE 3

Kinetic Energy and Work-Energy Theorem

Kinetic Energy

Kinetic energy (KE) is the capacity of a body to do work by virtue of its motion. If a body of mass m has a velocity v, its kinetic energy is equivalent to the work which an external force would have to do to bring the body from rest up to its velocity v.

The numerical value of the kinetic energy can be calculated from the formula.

$$KE = \frac{1}{2} mv^2$$

This can be derived as follows:

Consider a constant force F which acting on a mass m initially at rest. This force provides the mass m a velocity v.

If in reaching this velocity, the particle has been moving with an acceleration a and has been given a displacement s, then

$$F = ma$$

$$v^2 = 2as \Rightarrow$$

$$s = \frac{v^2}{2a}$$

Work done by the constant force $= Fs$

or $$W = (ma)\left(\frac{v^2}{2a}\right) = \frac{1}{2} mv^2$$

But the kinetic energy of the body is equivalent to the work done in giving the body this velocity.

Hence, $KE = \frac{1}{2} mv^2$

Regarding the kinetic energy the following two points are important to note:

1. Since, both m and v^2 are always positive. KE is always positive and does not depend on the direction of motion of the body.
2. Kinetic energy depends on the frame of reference. For example, the kinetic energy of a person of mass m sitting in a train moving with speed v is zero in the frame of train but $\frac{1}{2} mv^2$ in the frame of earth.

Work-Energy Theorem

This theorem is a very important tool that relates work with kinetic energy. According to this theorem:

Work done by all the forces (conservative or non-conservative, external or internal) acting on a particle or an object is equal to the change in kinetic energy of it.

$\therefore$ $$W_{net} = \Delta KE = K_f - K_i$$

Let, $\mathbf{F}_1, \mathbf{F}_2 ...$ be the individual forces acting on a particle. The resultant force is $\mathbf{F} = \mathbf{F}_1 + \mathbf{F}_2 + ...$ and the work done by the resultant force is

$$W = \int \mathbf{F} \cdot d\mathbf{r} = \int (\mathbf{F}_1 + \mathbf{F}_2 + ...) \cdot d\mathbf{r}$$

$$= \int \mathbf{F}_1 \cdot d\mathbf{r} + \int \mathbf{F}_2 \cdot d\mathbf{r} + ...$$

where, $\int \mathbf{F}_1 \cdot d\mathbf{r}$ is the work done on the particle by $\mathbf{F}_1$ and so on. Thus, work energy theorem can also be written as: work done by the resultant force which is also equal to the sum of the work done by the individual forces is equal to change in kinetic energy.

Regarding the work-energy theorem it is worthnoting that :

(i) If W_{net} is positive then $K_f - K_i$ = positive, i.e. $K_f > K_i$ or kinetic energy will increase and if W_{net} is negative then kinetic energy will decrease.

(ii) This theorem can be applied to non-inertial frames also. In a non-inertial frame it can be written as work done by all the forces (including the pseudo forces) = change in kinetic energy in non-inertial frame. Let us take an example.

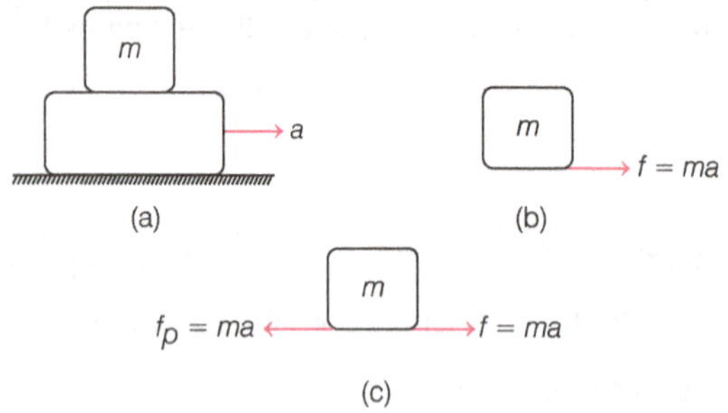

Refer Figure (a)

A block of mass m is kept on a rough plank moving with an acceleration a. There is no relative motion between the block and the plank. Hence, force of friction on block is $f = ma$ in forward direction.

Refer Figure (b)

Horizontal force on the block has been shown from ground (inertial) frame of reference.

If the plank moves a distance s on the ground, the block will also move the same distance s (as there is no slipping between the two). Hence, work done by friction on the block (w.r.t. ground) is

$$W_f = fs = mas$$

From work-energy principle if v is the speed of block (w.r.t. ground), then

$$\text{KE} = W_f$$

or $$\frac{1}{2}mv^2 = mas$$

or $$v = \sqrt{2as}$$

Thus, velocity of block relative to ground is $\sqrt{2as}$.

Refer Figure (c)

Free body diagram of the block has been shown from accelerating or non-inertial frame (plank).

Here, $f_p =$ pseudo force $= ma$

Work done by all the forces,

$$W = W_f + W_p = mas - mas = 0$$

From work-energy theorem,

$$\frac{1}{2}mv_r^2 = W = 0$$

or $$v_r = 0$$

Thus, velocity of block relative to plank is zero.

Note *Work-energy theorem is very useful in finding the work-done by a force whose exact nature is not known to us or to find the work done of a variable force whose exact variation is not known to us.*

Example 1 An object of mass 5 kg falls from rest through a vertical distance of 20 m and attains a velocity of 10 m/s. How much work is done by the resistance of the air on the object ? ($g = 10 \text{ m/s}^2$)

Sol. Applying work-energy theorem,

work done by all the forces = change in kinetic energy

or $$W_{mg} + W_{air} = \frac{1}{2}mv^2$$

$$\therefore \quad W_{air} = \frac{1}{2}mv^2 - W_{mg} = \frac{1}{2}mv^2 - mgh$$

$$= \frac{1}{2} \times 5 \times (10)^2 - (5) \times (10) \times (20)$$

$$= -750 \text{ J}$$ **Ans.**

Example 2 An object of mass m is tied to a string of length l and a variable force F is applied on it which brings the string gradually at angle θ with the vertical. Find the work done by the force F.

Note *Gradually means* $(K.E)_{initial} = (K.E)_{final} = 0$

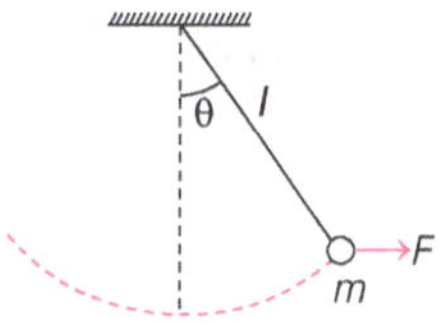

Sol. In this case, three forces are acting on the object:

1. tension (T)
2. weight (mg) and
3. applied force (F)

Using work-energy theorem

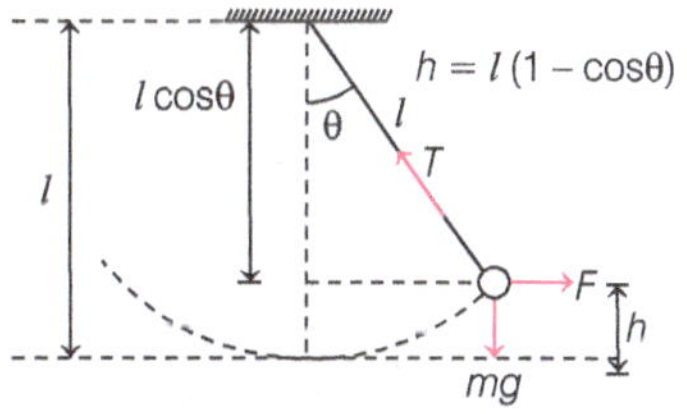

$$W_{net} = \Delta\text{KE}$$

or $$W_T + W_{mg} + W_F = 0 \quad \text{...(i)}$$

as $$\Delta\text{KE} = 0$$

because $$K_i = K_f = 0$$

Further, $W_T = 0$, as tension is always perpendicular to displacement.

$$W_{mg} = -mgh$$

or $$W_{mg} = -mgl\,(1 - \cos\theta)$$

Substituting these values in Eq. (i), we get

$$W_F = mgl\,(1 - \cos\theta)$$ **Ans.**

Note *Here, the applied force F is variable. So, if we do not apply the work energy theorem we will first find the magnitude of F at different locations and then integrate* $dW \; (= \mathbf{F} \cdot d\mathbf{r})$ *with proper limits.*

Example 3 A body of mass m was slowly hauled up the hill as shown in the figure by a force F which at each point was directed along a tangent to the trajectory. Find the work performed by this force, if the height of the hill is h, the length of its base is l and the coefficient of friction is μ.

(Irodov 1.121)

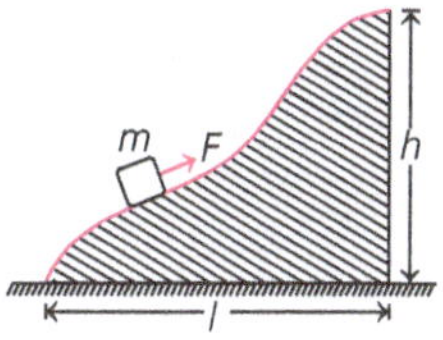

Sol. Four forces are acting on the body:

1. weight (mg)
2. normal reaction (N)
3. friction (f) and
4. the applied force (F)

Using work-energy theorem

$$W_{net} = \Delta KE$$

or $\quad W_{mg} + W_N + W_f + W_F = 0 \quad$ ...(i)

Here, $\Delta KE = 0$, because $K_i = 0 = K_f$

$$W_{mg} = -mgh$$

$$W_N = 0$$

(as normal reaction is perpendicular to displacement at all points)

W_f can be calculated as under

$$f = \mu N = \mu \, mg \cos\theta$$

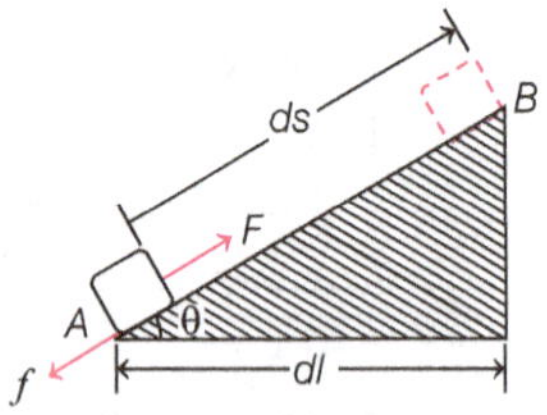

$\therefore \quad (dW_{AB})_f = -f\,ds$

$= -(\mu\, mg \cos\theta)\,ds$

$= -\mu\, mg\,(dl) \quad$ (as $ds \cos\theta = dl$)

$\therefore \quad W_f = -\mu\, mg\, \Sigma dl = -\mu\, mgl$

Substituting these values in Eq. (i), we get

$$W_F = mgh + \mu mgl \quad \textbf{Ans.}$$

Note *(i) Here again, if we want to solve this problem without using work-energy theorem we will first find magnitude of applied force* **F** *at different locations and then integrate $dW (= \mathbf{F} \cdot d\mathbf{r})$ with proper limits.*

(ii) In the above example we have obtained a very important result which can be used in some other questions too.

The result is :

$$\boxed{W_f = -\mu mgl}$$

if the forces are only tangential or when, $N = mg\cos\theta$.

Example 4 The displacement x of a particle moving in one dimension, under the action of a constant force is related to time t by the equation

$$t = \sqrt{x} + 3$$

where, x is in metre and t in second. Calculate: (a) the displacement of the particle when its velocity is zero, (b) the work done by the force in the first 6 s. **(1980 JEE)**

Sol. As $t = \sqrt{x} + 3$

i.e. $\quad x = (t-3)^2 \quad$ ...(i)

So, $\quad v = (dx/dt) = 2(t-3) \quad$ ...(ii)

(a) v will be zero when $2(t-3) = 0$ i.e. $t = 3$

Substituting this value of t in Eq. (i),

$$x = (3-3)^2 = 0$$

i.e. when velocity is zero, displacement is also zero. **Ans.**

(b) From Eq. (ii), $\quad (v)_{t=0} = 2(0-3)$

$= -6$ m/s

and $\quad (v)_{t=6} = 2(6-3)$

$= 6$ m/s

So, from work-energy theorem

$$w = \Delta KE = \frac{1}{2} m[v_f^2 - v_i^2]$$

$$= \frac{1}{2} m[6^2 - (-6)^2] = 0$$

i.e. work done by the force in the first 6 s is zero. **Ans.**

MODULE 3 Exercise

Single Correct Option

1. A block of mass 5 kg is raised from the bottom of a lake to a height of 3 m without change in kinetic energy at any instant. If the density of the block is 3000 kg m^{-3}, then the work done by the external force is equal to
(a) 100 J (b) 150 J (c) 50 J (d) 75 J

Note *Upthrust = (volume immersed) (density of liquid) (g)*

2. A force $\mathbf{F} = (3t\hat{\mathbf{i}} + 5\hat{\mathbf{j}})$N acts on a body due to which its displacement varies as $\mathbf{s} = (2t^2\hat{\mathbf{i}} - 5\hat{\mathbf{j}})$ m. Work done by this force in 2 second is
(a) 32 J (b) 24 J (c) 46 J (d) 20 J

3. A block of mass m is directly pulled up slowly on a smooth inclined plane of height h and inclination θ with the help of a string parallel to the incline. Which of the following statement is **incorrect** for the block when it moves up from the bottom to the top of the incline?

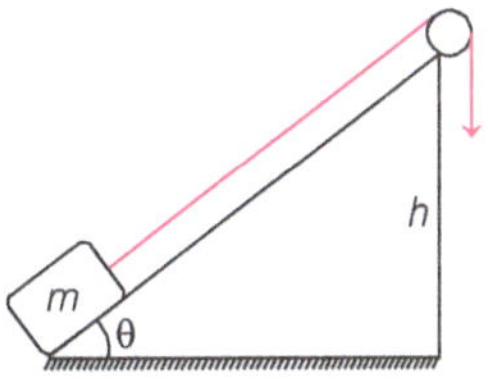

(a) Work done by the normal reaction force is zero
(b) Work done by the string is mgh
(c) Work done by gravity is mgh
(d) Net work done on the block is zero

4. A block of mass 10 kg is moving in x-direction with a constant speed of 10 m/s. It is subjected to a retarding force $F = -0.1x$ J/m during its travel from $x = 20$ m to $x = 30$ m. Its final kinetic energy will be
(a) 475 J (b) 450 J (c) 275 J (d) 250 J

5. A particle at rest on a frictionless table is acted upon by a horizontal force which is constant in magnitude and direction. A graph is plotted of the work done on the particle W, against the speed of the particle v. If there are no frictional forces acting on the particle, the graph will look like

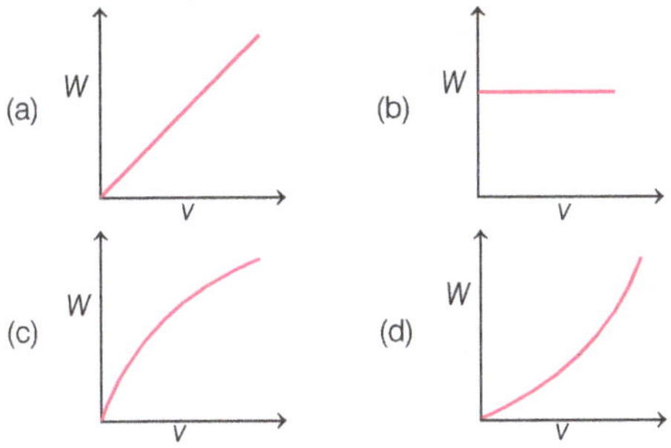

6. A body moves from rest with a constant acceleration. Which one of the following graphs represents the variation of its kinetic energy K with the distance travelled x?

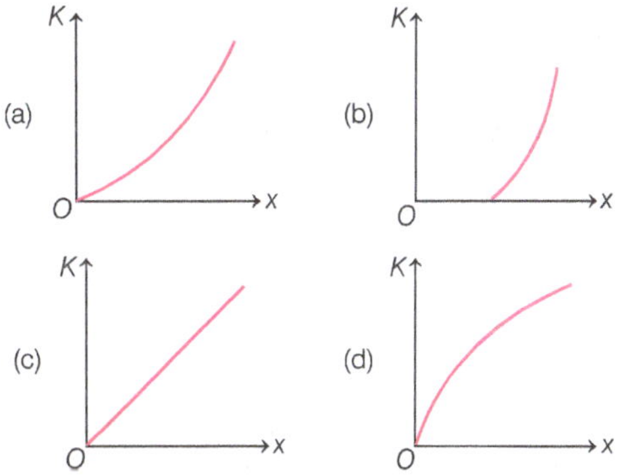

7. v-t graph of an object of mass 1 kg is shown. Select the **wrong** statement.

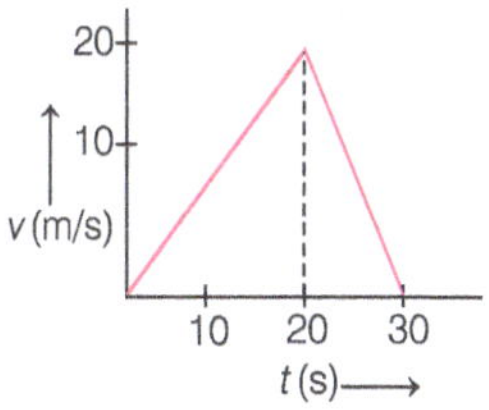

(a) Work done on the object in 30 s is zero
(b) The average acceleration of the object is zero
(c) The average velocity of the object is zero
(d) The average force on the object is zero

8. Kinetic energy of a particle moving in a straight line varies with time t as $K = 4t^2$. The force acting on the particle
(a) is constant
(b) is increasing
(c) is decreasing
(d) first increases and then decreases

9. A bullet moving with a speed of 100 m/s can just penetrate two planks of equal thickness. Then, the number of such planks penetrated by the same bullet when the speed of bullet is doubled.
(a) 4 (b) 8 (c) 6 (d) 10
Hint : Assume same retardation.

10. A bead of mass $\frac{1}{2}$ kg starts from rest from A to move in a vertical plane along a smooth fixed quarter ring of radius 5 m, under the action of a constant horizontal force $F = 5$ N as shown. The speed of bead as it reaches the point B is (Take, $g = 10 \text{ms}^{-2}$)

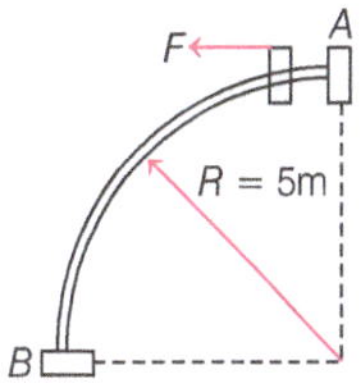

(a) 14.14 ms^{-1} (b) 7.07 ms^{-1}
(c) 4 ms^{-1} (d) 25 ms^{-1}

11. Equal net forces act on two different blocks A and B of masses m and $4m$, respectively. For same displacement, identify the correct statement.
(a) Their kinetic energies are in the ratio $\frac{K_A}{K_B}=\frac{1}{4}$
(b) Their speeds are in the ratio $\frac{v_A}{v_B}=\frac{1}{1}$
(c) Work done on the blocks are in the ratio $\frac{W_A}{W_B}=\frac{1}{1}$
(d) All of the above

12. A particle moves in a straight line with retardation proportional to its displacement. Its loss of kinetic energy in any displacement x is proportional to
(a) x^2 (b) e^x (c) x (d) $\log_e x$

13. A force acts on a 2 kg object, so that its position is given as a function of time as $x=3t^2+5$. What is the work done by this force in first 5 seconds? **(2019 Main)**
(a) 850 J (b) 900 J (c) 950 J (d) 875 J

14. A particle moves in one dimension from rest under the influence of a force that varies with the distance travelled by the particle as shown in the figure. The kinetic energy of the particle after it has travelled 3 m is **(2019 Main)**

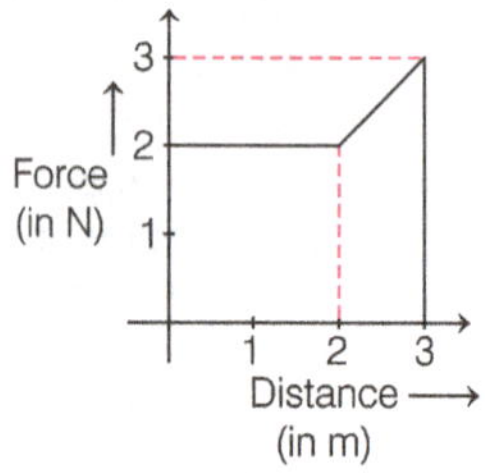

(a) 4 J (b) 2.5 J
(c) 6.5 J (d) 5 J

15. A particle which is experiencing a force, is given by $\mathbf{F}=3\hat{\mathbf{i}}-12\hat{\mathbf{j}}$, undergoes a displacement of $\mathbf{d}=4\hat{\mathbf{i}}$. If the particle had a kinetic energy of 3 J at the beginning of the displacement, what is its kinetic energy at the end of the displacement? **(2019 Main)**
(a) 9 J (b) 15J (c) 12 J (d) 10 J

Subjective Questions

16. A ball of mass 100 gm is projected upwards with velocity 10 m/s. It returns back with 6 m/s. Find work done by air resistance.

17. Velocity-time graph of a particle of mass 2 kg moving in a straight line is as shown in figure. Find the work done by all the forces acting on the particle.

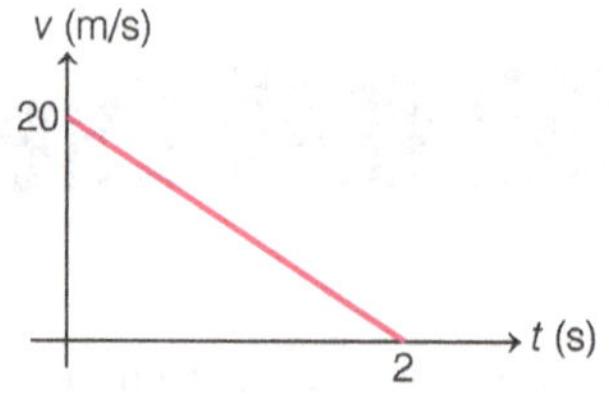

18. Is work-energy theorem valid in a non-inertial frame?

19. A particle of mass m moves on a straight line with its velocity varying with the distance travelled according to the equation $v=\alpha\sqrt{x}$, where α is a constant. Find the total work done by all the forces during a displacement from $x=0$ to $x=b$.

20. A 5 kg mass is raised a distance of 4 m by a vertical force of 80 N. Find the final kinetic energy of the mass if it was originally at rest. ($g=10\text{ m/s}^2$)

21. An object of mass m has a speed v_0 as it passes through the origin. It is subjected to a retarding force given by $F_x=-Ax$. Here, A is a positive constant. Find its x-coordinate when it stops.

22. A block of mass M is hanging over a smooth and light pulley through a light string. The other end of the string is pulled by a constant force F. The kinetic energy of the block increases by 40J in 1 s. State whether the following statements are true or false
(a) The tension in the string is Mg
(b) The work done by the tension on the block is 40 J
(c) The tension in the string is F
(d) The work done by the force of gravity is 40J in the above 1s

23. A helicopter lifts a 72 kg astronaut 15 m vertically from the ocean by means of a cable. The acceleration of the astronaut is $\frac{g}{10}$. How much work is done on the astronaut by (Take, $g=9.8\text{ m/s}^2$)
(a) the force from the helicopter and
(b) the gravitational force on her ?
(c) What are the kinetic energy and
(d) the speed of the astronaut just before she reaches the helicopter ?

24. A block of mass 30 kg is being brought down by a chain. If the block acquires a speed of 40 cm/s in dropping down 2 m. Find the work done by the chain during the process. ($g=10\text{ m/s}^2$)

Answers

1. (a) **2.** (a) **3.** (c) **4.** (a) **5.** (d) **6.** (c) **7.** (c) **8.** (a) **9.** (b) **10.** (a)
11. (c) **12.** (a) **13.** (b) **14.** (c) **15.** (b) **16.** – 3.2 J **17.** – 400 J **18.** Yes **19.** $\frac{1}{2}m\alpha^2 b$
20. 120 J **21.** $v_0\sqrt{\frac{m}{A}}$ **22.** (a) False (b) False (c) True (d) False
23. (a) 11642 J (b) – 10584 J (c) 1058 J (d) 5.42 m/s **24.** – 597.6 J

See hints on pages 517-519.

MODULE 4

Law of Conservation of Mechanical Energy

Suppose, only conservative forces are acting on a system of particles and potential energy U is defined corresponding to these forces. There are either no other forces or the work done by them is zero. We have

$$U_f - U_i = -(\text{work done by conservative forces}) = -W$$

Further, $W = K_f - K_i$ (from work-energy theorem)

then $U_f - U_i = -(K_f - K_i)$

or $U_f + K_f = U_i + K_i$ ...(i)

The sum of the potential energy and the kinetic energy is called the total mechanical energy. We see from Eq. (i), that the **total mechanical energy of a system remains constant, if only conservative forces are acting on a system of particles and the work done by all other forces is zero.** This is called the conservation of mechanical energy.

The total mechanical energy is not constant, if non-conservative forces such as friction is also acting on the system. However, the work energy theorem, is still valid. Thus, we can apply

$$W_c + W_{nc} + W_{ext} = K_f - K_i$$

Here, $W_c = -(U_f - U_i)$

So, we get $W_{nc} + W_{ext} = (K_f + U_f) - (K_i + U_i)$

or $W_{nc} + W_{ext} = E_f - E_i = \Delta E$

Here, $E = K + U$ is the total mechanical energy.

Work done by Different Forces

Work done by conservative forces is equal to minus of change in potential energy

$$W_c = -\Delta U = -(U_f - U_i) = U_i - U_f$$

Work done by all the forces is equal to change in kinetic energy.

$$W_{all} = \Delta K = K_f - K_i$$

Work done by the forces other than the conservative forces (non-conservative + external forces) is equal to change in mechanical energy

$$W_{nc} + W_{ext} = \Delta E = E_f - E_i = (K_f + U_f) - (K_i + U_i)$$

Note *Third type of work done is also sometimes called :*

(i) only W_{ext}

(ii) work done by the forces other than conservative

(iii) work done by you

(iv) or sometimes written as : find the work done.

Further, in this case if no information is given regarding the change in kinetic energy then we can take it zero. In that case,

$$W_{ext} = \Delta U = U_f - U_i$$

Example 1 A body is displaced from position A to position B. Kinetic and potential energies of the body at positions A and B are

$$K_A = 50\text{ J}, U_A = -30\text{ J}, K_B = 10\text{ J} \text{ and } U_B = 20\text{ J}.$$

Find work done by

(a) conservative forces

(b) all forces

(c) forces other than conservative forces.

Sol. (a) $W_c = -\Delta U = -(U_f - U_i)$

$= U_i - U_f = U_A - U_B$

$= -30 - 20 = -50\text{ J}$ **Ans.**

(b) $W_{all} = \Delta K = K_f - K_i$

$= K_B - K_A$

$= 10 - 50 = -40\text{ J}$ **Ans.**

(c) Work done by the forces other than conservative

$= \Delta E = E_f - E_i$

$= (K_f + U_f) - (K_i + U_i)$

$= (K_B + U_B) - (K_A + U_A)$

$= (10 + 20) - (50 - 30)$

$= 10\text{ J}$ **Ans.**

Note *(i) In the above example work done by conservative force is negative (= −50 J). Therefore potential energy should increase and we can see that,*

$U_f > U_i$ *as* $U_B > U_A$

(ii) Suppose a particle is released from point A with u = 0.

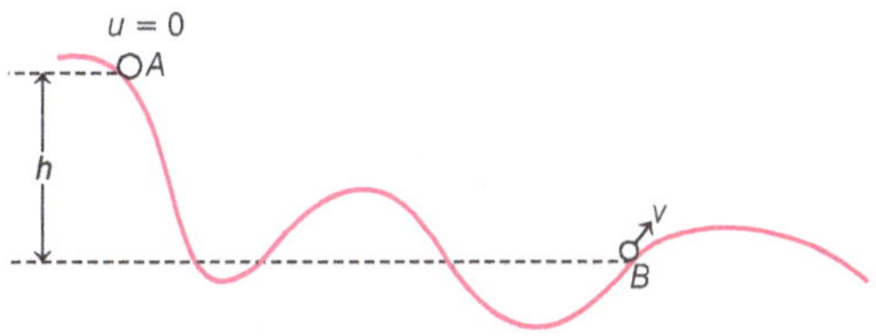

Friction is absent everywhere. Then velocity at B will be

$v = \sqrt{2gh}$

(irrespective of the track it follows from A to B)

Here, $h = h_A - h_B$

Typed Problems

Type 1. Based on conservation of mechanical energy.

Concept

If only conservative forces are acting on a system then its mechanical energy remains conserved. If all surfaces are smooth, then also mechanical energy will be conserved.

How to Solve?

If mechanical energy is conserved then it is possible that some part of the energy may be decreasing while the other part may be increasing. Now the energy conservation equation can be written in following two ways :

First method Magnitude of decrease of energy = magnitude of increase of energy.

Second method $E_i = E_f$

i.e. write down total initial mechanical energy on one side and total final mechanical energy on the other side. While writing gravitational potential energy we choose some reference point (where $h = 0$), but throughout the question this reference point should not change. Let us take a simple example.

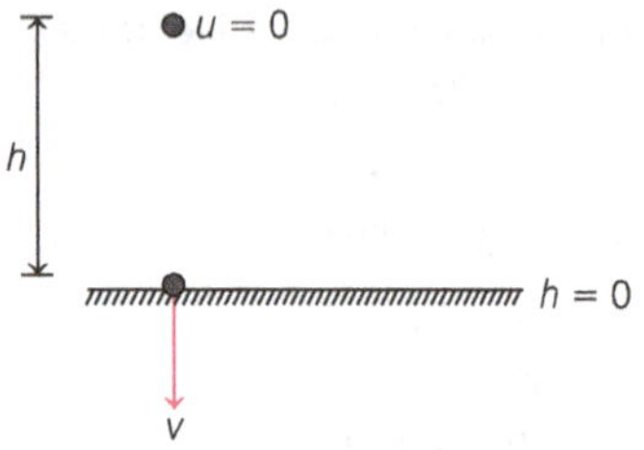

A ball of mass 'm' is released from a height h as shown in figure. The velocity of particle at the instant when it strikes the ground can be found using energy conservation principle by following two methods.

Method-1 Decrease in gravitational potential energy = increase in kinetic energy

$$\therefore \quad mgh = \frac{1}{2}mv^2$$

$$\text{or} \quad v = \sqrt{2gh}$$

Method-2 $E_i = E_f$

$$\text{or} \quad K_i + U_i = K_f + U_f$$

$$\text{or} \quad 0 + mgh = \frac{1}{2}mv^2 + 0$$

$$\Rightarrow \quad v = \sqrt{2gh}$$

Ex. 1 In the figure shown, all surfaces are smooth and force constant of spring is 10 N/m. Block of mass 2 kg is not attached with the spring. The spring is compressed by 2m and then released. Find the maximum distance 'd' travelled by the block over the inclined plane. Take $g = 10 \text{m/s}^2$.

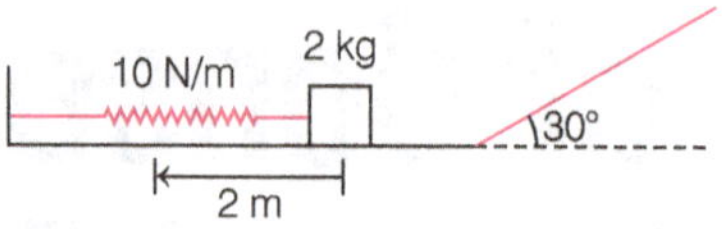

Sol. In the final position, block will stop for a moment and then it will return back.

In the initial position system has only spring potential energy $\frac{1}{2}kx^2$ and in the final position it has only gravitational potential energy.

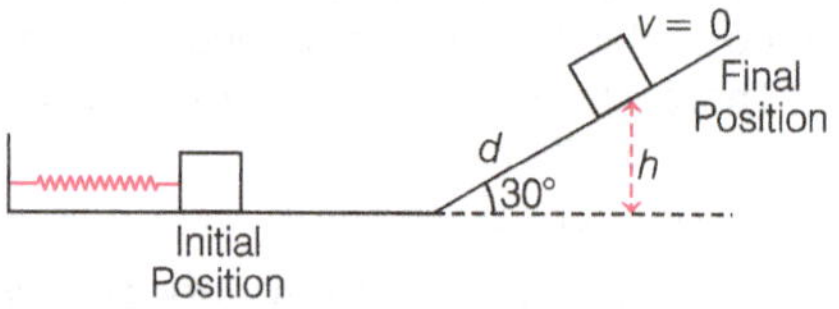

Since, all surfaces are smooth, therefore mechanical energy will remain conserved.

$$\Rightarrow \quad E_i = E_f \quad \text{or} \quad \frac{1}{2}kx^2 = mgh = mg\left(\frac{d}{2}\right)$$

$$\text{where} \quad h = d \sin 30° = \frac{d}{2}$$

$$\Rightarrow \quad d = \frac{kx^2}{mg}$$

Substituting the values we have,

$$d = \frac{(10)(2)^2}{(2)(10)} = 2 \text{ m}$$ **Ans.**

Ex. 2 A smooth narrow tube in the form of an arc AB of a circle of centre O and radius r is fixed so that A is vertically above O and OB is horizontal. Particles P of mass m and Q of mass $2m$ with a light inextensible string of length $(\pi r/2)$ connecting them are placed inside the tube with P at A and Q at B and released from rest. Assuming the string remains taut during motion, find the speed of particles when P reaches B.

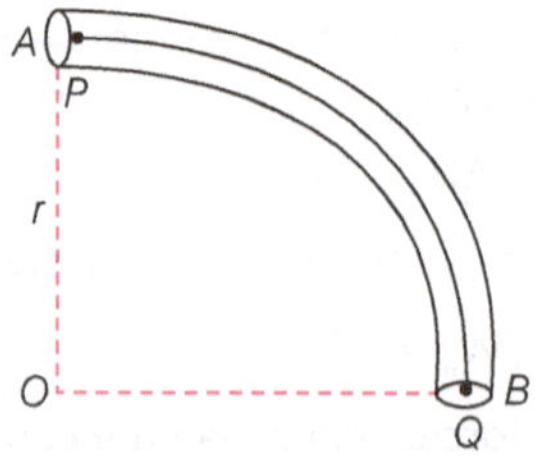

Sol. All surfaces are smooth. Therefore, mechanical energy of the system will remain conserved.

$\therefore$ Decrease in PE of both the blocks = increase in KE of both the blocks

$$\therefore \quad (mgr) + (2mg)\left(\frac{\pi r}{2}\right) = \frac{1}{2}(m + 2m)v^2$$

$$\text{or} \quad v = \sqrt{\frac{2}{3}(1 + \pi)gr}$$ **Ans.**

Ex. 3 One end of a light spring of natural length d and spring constant k is fixed on a rigid wall and the other is attached to a smooth ring of mass m which can slide without friction on a vertical rod fixed at a distance d from the wall. Initially the spring makes an angle of 37° with the horizontal as shown in figure. When the system is released from rest, find the speed of the ring when the spring becomes horizontal. ($\sin 37° = 3/5$)

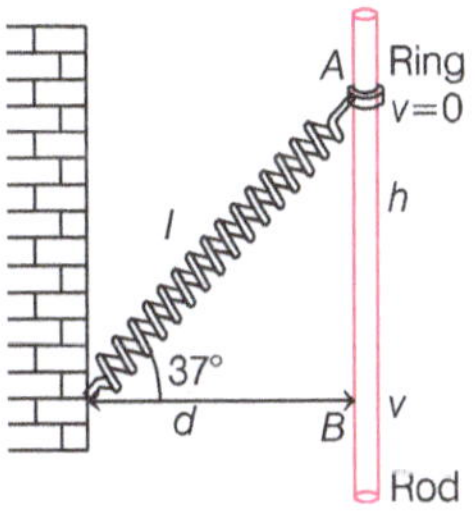

Sol. If l is the stretched length of the spring, then from figure

$$\frac{d}{l} = \cos 37° = \frac{4}{5}, \quad \text{i.e.} \quad l = \frac{5}{4}d$$

So, the stretch $x = l - d = \frac{5}{4}d - d = \frac{d}{4}$

and $h = l \sin 37° = \frac{5}{4}d \times \frac{3}{5} = \frac{3}{4}d$

Now, taking point B as the reference level ($h = 0$) and applying law of conservation of mechanical energy between A and B,

$$E_A = E_B$$

or $mgh + \frac{1}{2}kx^2 = \frac{1}{2}mv^2$ [At B, $h = 0$ and $x = 0$]

or $\frac{3}{4}mgd + \frac{1}{2}k\left(\frac{d}{4}\right)^2 = \frac{1}{2}mv^2$ [at A, $h = \frac{3}{4}d$ and $x = \frac{1}{4}d$]

Solving this equation, we get

$$v = d\sqrt{\frac{3g}{2d} + \frac{k}{16\,m}}$$ **Ans.**

Type 2. Based on the position of equilibrium and momentary rest.

Concept

In the figure shown, block attached with the spring is released from rest at natural length of the spring at A.

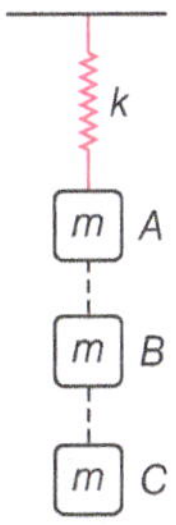

At this position a constant force mg is acting on the block in downward direction. So, block starts moving downwards and spring is stretched. So, a variable spring force kx starts acting on the block in upward direction which keeps on increasing with extension 'x'. In position B, also called equilibrium position,

$$kx = mg$$

$\therefore$ $F_{net} = 0$

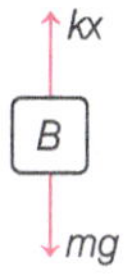

But the block does not stop here. Rather, it has maximum velocity in this position. After crossing B the block retards, as $kx > mg$ and net force is upwards. At point C (the maximum extension) block stops for a moment ($v=0$) and then it returns back.

The block starts oscillating between A and C.

Thus,

From	kx and mg	Direction of net force	Direction of velocity	Speed
A to B	$mg > kx$	Downwards	Downwards	Increasing
B to C	$kx > mg$	Upwards	Downwards	Decreasing
C to B	$kx > mg$	Upwards	Upwards	Increasing
B to A	$mg > kx$	Downwards	Upwards	Decreasing

Note *(i) Points A and C are the points of momentary rest, where $v=0$ but $\mathbf{F}_{net} \neq 0$. So the maximum extension (at point C) can be obtained by energy conservation principle but not by putting $\mathbf{F}_{net} = 0$.*

(ii) Point B is the point of equilibrium where $\mathbf{F}_{net} = 0$ and speed is maximum. This point can be obtained by putting $\mathbf{F}_{net} = 0$ and after that, maximum speed can be obtained by energy conservation principle.

Ex. 4 In the figure shown in the above concept, find

(a) Equilibrium extension $x_0 (=AB)$

(b) Maximum extension $x_m (=AC)$

(c) Maximum speed at point B.

Sol. (a) At point B,

$$F_{net} = 0$$

$\Rightarrow$ $kx_0 = mg$

$\Rightarrow$ $x_0 = \frac{mg}{k}$ **Ans.**

(b) **From A to C**

Decrease in gravitational potential energy

= increase in spring potential energy.

$\therefore$ $mgx_m = \frac{1}{2}kx_m^2$ $(AC = x_m)$

$$\Rightarrow \quad x_m = \frac{2mg}{k} \quad (v_A = v_C = 0) \qquad \textbf{Ans.}$$

(c) **From A to B**

Decrease in gravitational potential energy = increase in (spring potential energy + kinetic energy)

$$\Rightarrow \quad mg\,x_0 = \frac{1}{2}kx_0^2 + \frac{1}{2}mv_{max}^2$$

Substituting the value of $x_0 = \frac{mg}{k}$ in the above equation, we get

$$v_{max} = \left(\sqrt{\frac{m}{k}}\right)g \qquad \textbf{Ans.}$$

Ex. 5 Consider the situation shown in figure. Mass of block A is m and that of block B is 2 m. The force constant of spring is k. Friction is absent everywhere. System is released from rest with the spring unstretched. Find

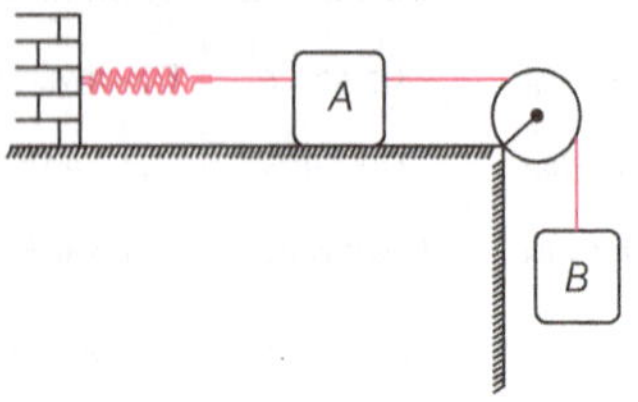

(a) the maximum extension of the spring x_m

(b) the speed of block A when the extension in the spring is $x = \frac{x_m}{2}$

(c) net acceleration of block B when extension in the spring is $x = \frac{x_m}{4}$

Sol. (a) At maximum extension of the spring

$$v_A = v_B = 0 \qquad \text{(momentarily)}$$

Therefore, applying conservation of mechanical energy:

decrease in gravitational potential energy of block B = increase in elastic potential energy of spring.

$$\text{or} \quad m_B g x_m = \frac{1}{2}kx_m^2$$

$$\text{or} \quad 2\,mgx_m = \frac{1}{2}kx_m^2$$

$$\therefore \quad x_m = \frac{4\,mg}{k} \qquad \textbf{Ans.}$$

(b) At $x = \frac{x_m}{2} = \frac{2\,mg}{k}$

$$v_A = v_B = v \text{ (say)}$$

Then, decrease in gravitational potential energy of block B = increase in elastic potential energy of spring + increase in kinetic energy of both the blocks.

$$\therefore \quad m_B g x = \frac{1}{2}kx^2 + \frac{1}{2}(m_A + m_B)v^2$$

$$\text{or} \quad (2m)(g)\left(\frac{2\,mg}{k}\right) = \frac{1}{2}k\left(\frac{2\,mg}{k}\right)^2 + \frac{1}{2}(m+2m)v^2$$

Solving this equation, we get

$$\therefore \quad v = 2g\sqrt{\frac{m}{3k}} \qquad \textbf{Ans.}$$

(c) At $x = \frac{x_m}{4} = \frac{mg}{k}$

or $kx = mg$

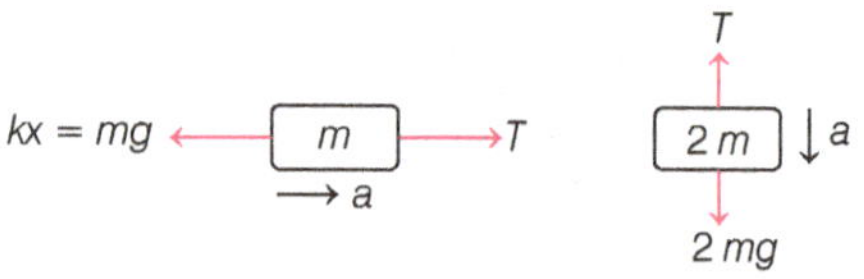

$$a = \frac{\text{Net pulling force}}{\text{Total mass}} = \frac{2\,mg - mg}{3\,m}$$

$$= \frac{g}{3} \text{ (downwards)} \qquad \textbf{Ans.}$$

Type 3. Problems with friction.

Concept

Mechanical energy does not remain constant if friction is there and work done by friction is not zero. So, initial mechanical energy is more than the final mechanical energy and the difference goes in the work done against friction.

How to Solve?

The problem can be solved by the following simple equation:

Initial mechanical energy − final mechanical energy = work done against friction.

Here, work done against friction is equal to ($\mu\,mg\,d$) if the block is moving on horizontal ground and this is equal to ($\mu d\,mg\cos\theta$) if the block is moving on an inclined plane. In these expressions 'd' is the distance travelled over the rough ground (not the displacement). If μ_s and μ_k two coefficients of friction are given separately, then we will have to take μ_k.

Ex. 6

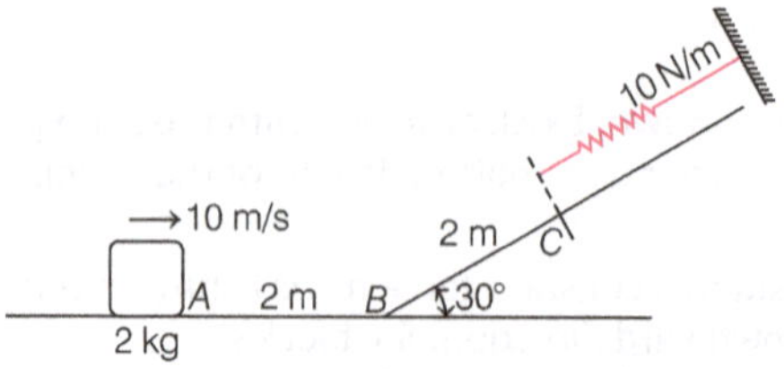

In the figure shown, $AB = BC = 2$m. Friction coefficient everywhere is $\mu = 0.2$. Find the maximum compression of the spring.

Sol. Let x be the maximum compression.

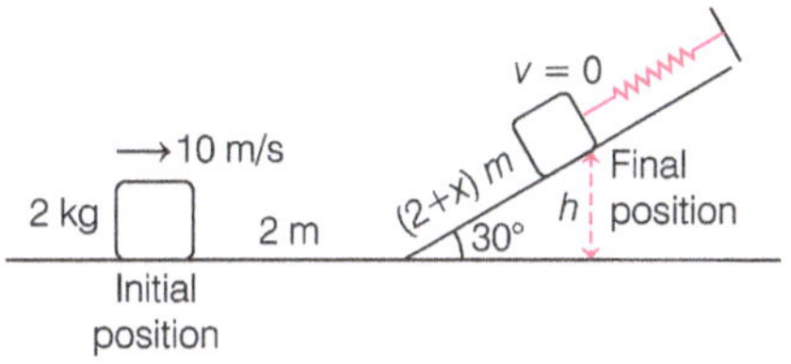

$$h = (2+x)\sin 30° = (1+0.5x)m$$

The block has travelled $d_1 = 2m$ on rough horizontal ground and $d_2 = (2+x)m$ on rough inclined ground. In the initial position, block has only kinetic energy and in the final position we have spring and gravitational potential energy. So, applying the equation.

$E_i - E_f$ = work done against friction

$$\left(\frac{1}{2}mv^2\right) - \left(\frac{1}{2}kx^2 + mgh\right) = \mu\, mg\, d_1 + (\mu\, mg \cos\theta)d_2$$

$$\Rightarrow \quad \frac{1}{2}\times 2\times (10)^2 + \frac{1}{2}\times 10\times x^2 + 2\times 10\times (1+0.5x)$$

$$= 0.2\times 2\times 10\times 2 + (0.2\times 2\times 10\times \cos 30°)(2+x)$$

Solving this equation, we get

$$x = 2.45\text{m}$$ **Ans.**

Ex. 7 A small block slides along a track with elevated ends and a flat central part as shown in figure. The flat portion BC has a length $l = 3.0$ m. The curved portions of the track are frictionless. For the flat part, the coefficient of kinetic friction is $\mu_k = 0.20$, the particle is released at point A which is at height $h = 1.5$ m above the flat part of the track. Where does the block finally comes to rest?

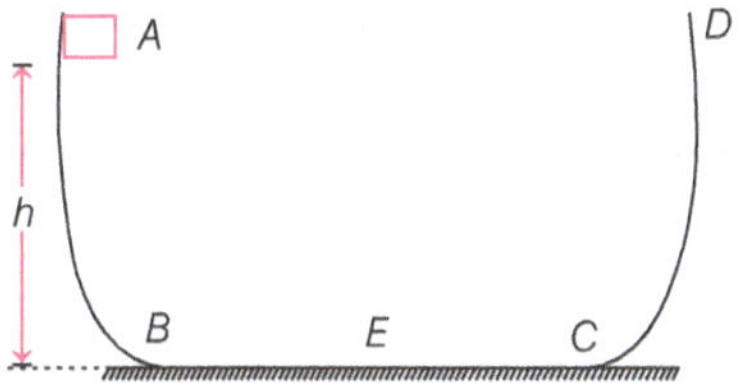

Sol. As initial mechanical energy of the block is mgh and final is zero, so

loss in mechanical energy $= mgh$.

This mechanical energy is lost in doing work against friction in the flat part only.

So, loss in mechanical energy = work done against friction

or $$mgh = \mu mgd$$

i.e. $$d = \frac{h}{\mu} = \frac{1.5}{0.2} = 7.5\text{m}$$

After starting from B, the block will reach C and then will rise up till the remaining KE at C is converted into potential energy. It will then again descend and at C will have the same value as it had when ascending, but now it will move from C to B. The same will be repeated and finally the block will come to rest at E such that

$$BC + CB + BE = 7.5$$

or $$3 + 3 + BE = 7.5$$

i.e. $$BE = 1.5$$

So, the block comes to rest at the centre of the flat part.

Ex. 8 A 0.5 kg block slides from the point A on a horizontal track with an initial speed 3 m/s towards a weightless horizontal spring of length 1 m and force constant 2 N/m. The part AB of the track is frictionless and the part BC has the coefficient of static and kinetic friction as 0.22 and 0.20 respectively. If the distances AB and BD are 2 m and 2.14 m respectively, find the total distance through which the block moves before it comes to rest completely. ($g = 10$ m/s^2) **(1983 JEE)**

Sol. As the track AB is frictionless, the block moves this distance without loss in its initial

$$KE = \frac{1}{2}mv^2 = \frac{1}{2}\times 0.5\times 3^2 = 2.25\text{J}$$

Path BD is rough, so

work done against friction

$$= \mu_k mgd = 0.2\times 0.5\times 10\times 2.14 = 2.14\text{ J}$$

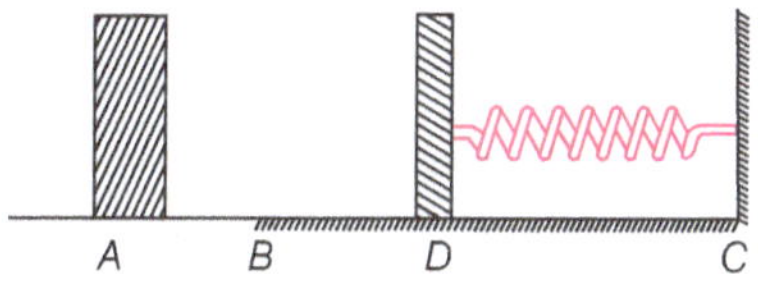

So, at D the KE of the block is $= 2.25 - 2.14 = 0.11$ J

Now, if the spring is compressed by x, then

$$0.11 = \frac{1}{2}\times k\times x^2 + \mu_k mgx$$

i.e. $$0.11 = \frac{1}{2}\times 2\times x^2 + 0.2\times 0.5\times 10x$$

or $x^2 + x - 0.11 = 0$

which on solving gives positive value of $x = 0.1$ m

After moving the distance $x = 0.1$ m the block comes to rest.

Now, the compressed spring exerts a force,

$$F = kx = 2\times 0.1 = 0.2\text{ N}$$

on the block while limiting frictional force between block and track is

$$f_L = \mu_s mg = 0.22\times 0.5\times 10 = 1.1\text{ N}$$

Since, $F < f_L$. The block will not move back. So, the total distance moved by the block

$$= AB + BD + 0.1 = 2 + 2.14 + 0.1$$

$$= 4.24\text{ m}$$ **Ans.**

Miscellaneous Examples

Ex. 9 A small mass m starts from rest and slides down the smooth spherical surface of R. Assume zero potential energy at the top. Find

(a) the change in potential energy,

(b) the kinetic energy,

(c) the speed of the mass as a function of the angle θ made by the radius through the mass with the vertical.

Sol. In the figure, $h = R(1 - \cos\theta)$

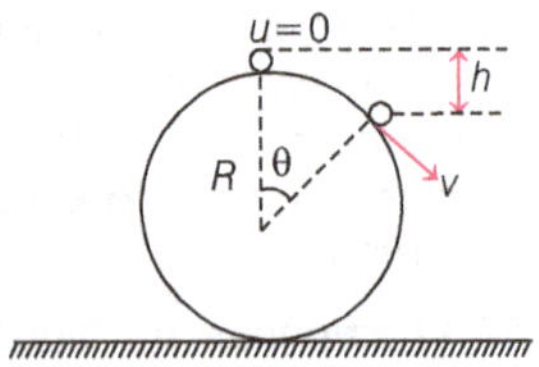

(a) As the mass comes down, potential energy will decrease. Hence,

$$\Delta U = -mgh = -mgR(1 - \cos\theta)$$

(b) Magnitude of decrease in potential energy = increase in kinetic energy

$\therefore$ Kinetic energy $= mgh$

$$= mgR(1 - \cos\theta) \qquad \textbf{Ans.}$$

(c) $$\frac{1}{2}mv^2 = mgR(1 - \cos\theta)$$

$\therefore$ $$v = \sqrt{2gR(1 - \cos\theta)} \qquad \textbf{Ans.}$$

Ex. 10 A small disc A slides down with initial velocity equal to zero from the top of a smooth hill of height H having a horizontal portion. What must be the height of the horizontal portion h to ensure the maximum distance s covered by the disc? What is it equal to? **(Irodov 1.135)**

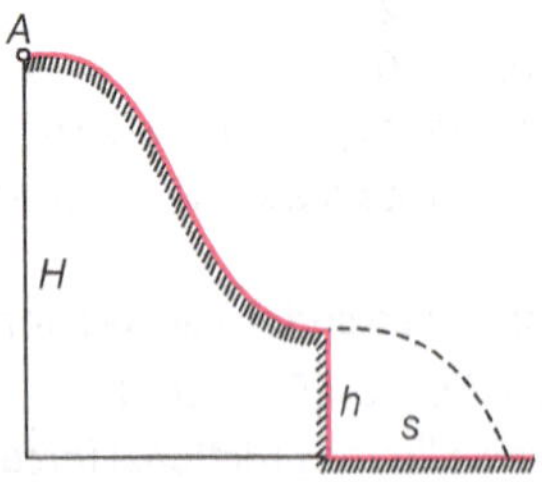

Sol. Velocity at the time of leaving the horizontal position,

$$v = \sqrt{2g(H - h)}$$

Since v is horizontal, so time is free fall time, or

$$t = \sqrt{\frac{2h}{g}}$$

Now, $$s = vt = 2\sqrt{h(H - h)} \qquad \text{...(i)}$$

For s to be maximum, $\dfrac{ds}{dh} = 0$.

By putting $\dfrac{ds}{dh} = 0$, we get

$$h = \frac{H}{2}$$

So, s is maximum at

$$h = \frac{H}{2}$$

Putting $h = \dfrac{H}{2}$ in equation (i), we get

$$s_{max} = H$$

Note *(i) At $h = \dfrac{H}{2}$, we can check that $\dfrac{d^2s}{dh^2}$ is negative.*

(ii) This problem is similar to a problem of fluid mechanics. This we will see in the chapter of fluid mechanics.

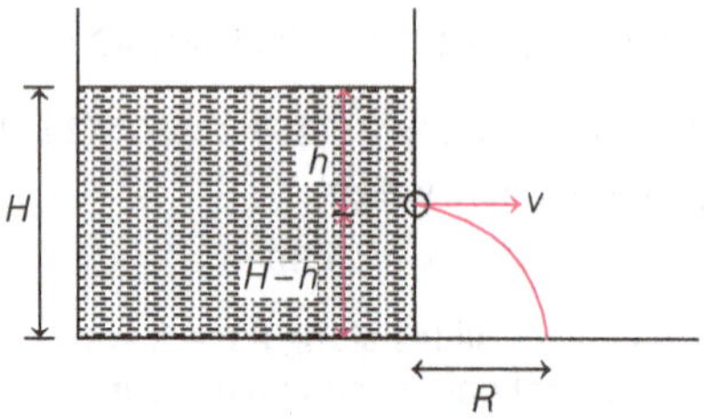

$$R = 2\sqrt{h(H - h)}$$

$$R_{max} = H \text{ at } h = \frac{H}{2}$$

Ex. 11 A small block of mass m slides down a smooth hill of height h without initial velocity and gets onto a plank of mass M lying on a smooth horizontal plane at the base of hill figure. Due to friction between the block and the plank, block slows down and finally moves as one piece with the plank. (a) Find out total work performed by the friction forces in this process. (b) can it be stated that the result obtained does not depend on the choice of the reference frame.

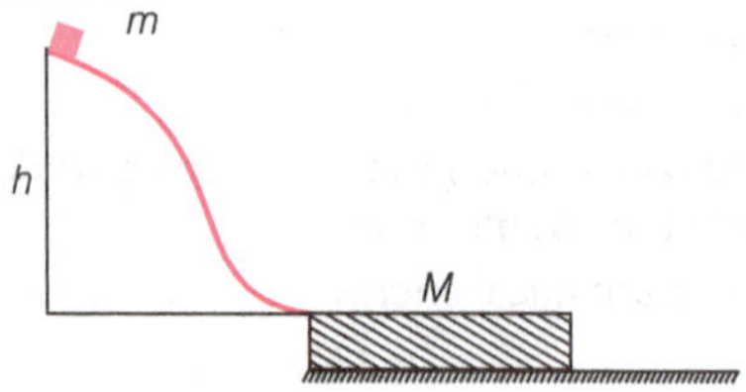

Sol. (a) When the block slides down and comes onto the plank, then

$$mgh = \frac{1}{2}mv^2$$

$\therefore$ $$v = \sqrt{(2gh)} \qquad \text{...(i)}$$

Let v_1 be the common velocity of both, then from law of conservation of linear momentum (we will discuss it later),

$$mv = (M + m)\,v_1$$

$$\therefore \quad v_1 = \frac{mv}{(M + m)} \qquad \text{...(ii)}$$

Now, change in KE $= (K)_f - (K)_i =$ (work done)$_{\text{friction}}$

$$\therefore \quad \frac{1}{2}(M + m)v_1^2 - \frac{1}{2}mv^2 = (\text{work done})_{\text{friction}}$$

$$\text{or} \quad W_{\text{fr}} = \frac{1}{2}(M + m)\left[\frac{m\,v}{M + m}\right]^2 - \frac{1}{2}m\,v^2$$

$$= \frac{1}{2}mv^2\left[\frac{m}{M + m} - 1\right]$$

$$\text{as} \quad \frac{1}{2}mv^2 = mgh$$

$$\therefore \quad W_{\text{fr}} = -mgh\left[\frac{M}{M + m}\right] \qquad \textbf{Ans.}$$

(b) In part (a), we have calculated work done from the ground frame of reference. Now, let us take plank as the reference frame.

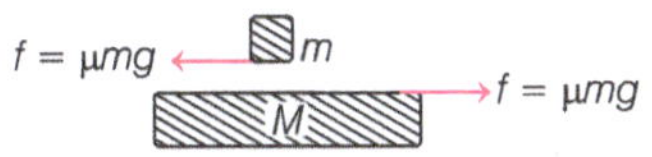

$$\text{Acceleration of plank } a_0 = \frac{f}{M} = \frac{\mu\, mg}{M}$$

Free body diagram of block with respect to plank is shown in figure.

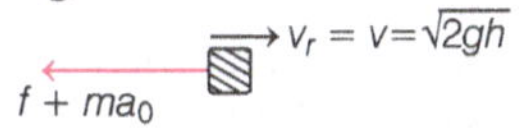

Here, $ma_0 =$ pseudo force.

$\therefore$ Retardation of block w.r.t. plank,

$$a_r = \frac{f + ma_0}{m} = \frac{\mu mg + \dfrac{\mu m^2 g}{M}}{m}$$

$$= \mu g + \frac{\mu mg}{M}$$

$$= \left(\frac{M + m}{M}\right)\mu g$$

The block will stop after travelling a distance S_r relative to plank, where

$$s_r = \frac{v_r^2}{2a_r} = \frac{Mgh}{(M + m)\mu g} \qquad (0 = v_r^2 - 2a_r s_r)$$

$\therefore$ Work done by friction in this frame of reference

$$W_{fr} = -fs_r = -(\mu mg)\left[\frac{Mgh}{(M + m)\mu g}\right]$$

$$= -\frac{M\,mgh}{(M + m)}$$

which is same as part (a).

Note *Work done by friction in this problem does not depend upon the frame of reference, otherwise in general work depends upon reference frame.*

Ex. 12 Two blocks A and B are connected to each other by a string and a spring. The string passes over a frictionless pulley as shown in figure. Block B slides over the horizontal top surface of a stationary block C and the block A slides along the vertical side of C, both with the same uniform speed. The coefficient of friction between the surfaces of the blocks is 0.2. The force constant of the spring is 1960 Nm^{-1}. If the mass of block A is 2 kg, calculate the mass of block B and the energy stored in the spring ($g = 9.8\ \text{m/s}^2$) **(1982 JEE)**

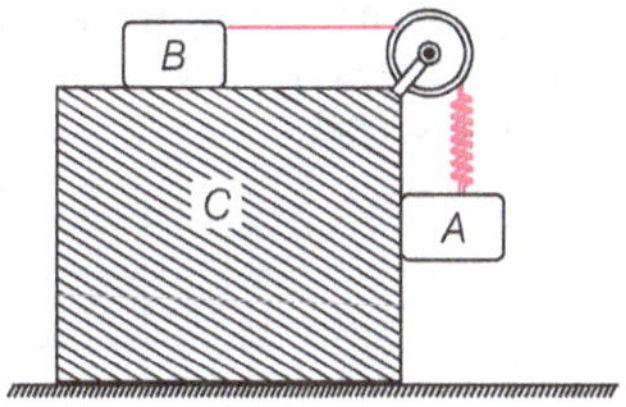

Sol. Let m be the mass of B. From free-body diagram of B :

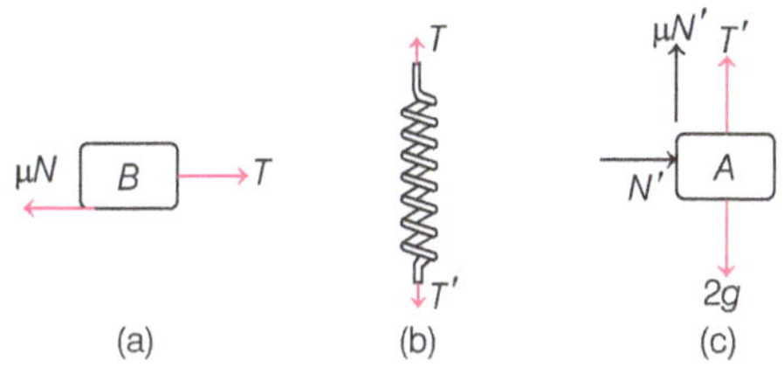

$$T - \mu N = m_B a = m \times 0 = 0$$

where, $T =$ tension of the string and $N = mg$

$$\therefore \quad T = \mu mg$$

From the free-body diagram of the spring

$$T - T' = 0$$

where, T' is the force exerted by A on the spring or

$$T = T' = \mu mg$$

From the free-body diagram of A

$$2g - (T' + \mu N') = m_A a = 2 \times 0 = 0$$

where, N' is the normal reaction of the vertical wall of C on A and $N' = m_A a = 2 \times 0$

(as there is no horizontal acceleration of A)

$$\therefore \quad 2g = T' = \mu mg \quad \text{or} \quad m = \frac{2g}{\mu g} = \frac{2}{0.2} = 10\ \text{kg} \qquad \textbf{Ans.}$$

Tensile force on the spring $= T$

$$\text{or} \quad T' = \mu mg = 0.2 \times 10 \times 9.8$$

$$= 19.6\ \text{N}$$

Now, in a spring

tensile force = force constant × extension

$$\therefore \quad 19.6 = 1960\,x$$

$$\text{or} \quad x = \frac{1}{100}\ \text{m}$$

$$\text{or} \quad U\ (\text{energy of a spring}) = \frac{1}{2}kx^2$$

$$= \frac{1}{2} \times 1960 \times \left(\frac{1}{100}\right)^2 = 0.098\ \text{J} \qquad \textbf{Ans.}$$

MODULE 4 Exercise

Conservation of Mechanical Energy

Single Correct Option

1. A body with mass 1 kg moves in one direction in the presence of a force which is described by the potential energy graph. If the body is released from rest at $x = 2$ m, than its speed when it crosses $x = 5\,\text{m}$ is (Neglect dissipative forces)

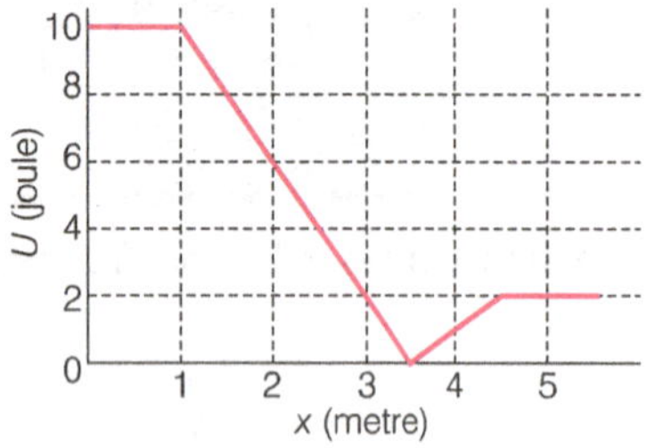

(a) $2\sqrt{2}\,\text{ms}^{-1}$ (b) $1\,\text{ms}^{-1}$ (c) $2\,\text{ms}^{-1}$ (d) $3\,\text{ms}^{-1}$

2. In the figure, m_1 and m_2 $(< m_1)$ are joined together by a pulley. When the mass m_1 is released from the height h above the floor, it strikes the floor with a speed

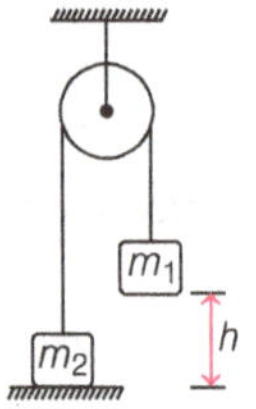

(a) $\sqrt{2gh\left(\dfrac{m_1 - m_2}{m_1 + m_2}\right)}$ (b) $\sqrt{2gh}$

(c) $\sqrt{\dfrac{2m_2gh}{m_1 + m_2}}$ (d) $\sqrt{\dfrac{2m_1gh}{m_1 + m_2}}$

3. A ball of mass 12 kg and another of mass 6 kg are dropped from a 60 feet tall building. After a fall of 30 feet each, towards earth, their kinetic energies will be in the ratio of

(a) $\sqrt{2} : 1$ (b) $1 : 4$ (c) $2 : 1$ (d) $1 : \sqrt{2}$

4. If v be the instantaneous velocity of the body dropped from the top of a tower, when it is located at height h, then which of the following remains constant?

(a) $gh + v^2$ (b) $gh + \dfrac{v^2}{2}$ (c) $gh - \dfrac{v^2}{2}$ (d) $gh - v^2$

5. The figures shows a particle sliding on a frictionless track, which terminates in a straight horizontal section. If the particle starts slipping from the point A, how far away from the track will the particle hit the ground? (Take $g = 10\,\text{m/s}^2$)

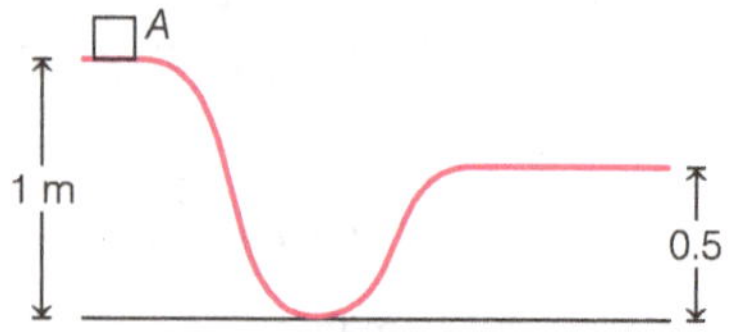

(a) 1 m (b) 2 m (c) 3 m (d) 4 m

6. A block of mass m is attached to two unstretched springs of spring constants k each as shown. The block is displaced towards right through a distance x and is released. The speed of the block as it passes through the mean position will be

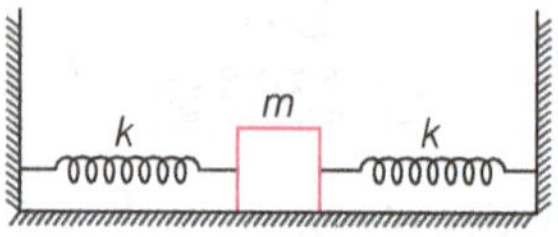

(a) $x\sqrt{\dfrac{m}{2k}}$ (b) $x\sqrt{\dfrac{2k}{m}}$ (c) $x\dfrac{m}{k}$ (d) $x\dfrac{2k}{m}$

7. A smooth chain AB of mass m rests against a surface in the form of a quarter of a circle of radius R. If it is released from rest, the velocity of the chain after it comes over the horizontal part of the surface is

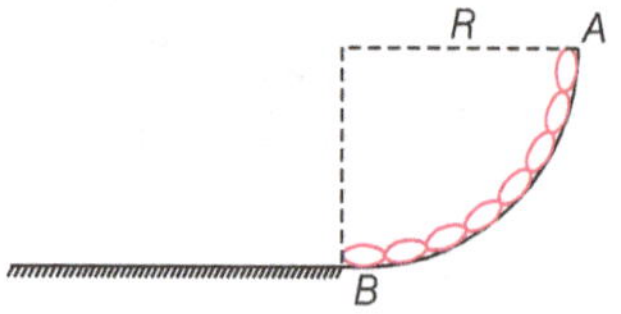

(a) $\sqrt{2gR}$ (b) $\sqrt{gR}$

(c) $\sqrt{2gR\left(1 - \dfrac{2}{\pi}\right)}$ (d) $\sqrt{2gR\,(2 - \pi)}$

8. A uniform chain of length πr lies inside a smooth semicircular tube AB of radius r. Assuming a slight disturbance to start the chain in motion, the velocity with which it will emerge from the end B of the tube will be

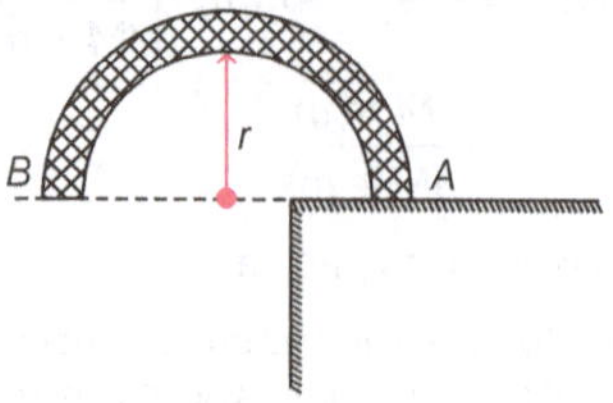

(a) $\sqrt{gr\left(1+\frac{2}{\pi}\right)}$ (b) $\sqrt{2gr\left(\frac{2}{\pi}+\frac{\pi}{2}\right)}$

(c) $\sqrt{gr\,(\pi+2)}$ (d) $\sqrt{\pi gr}$

9. A uniform flexible chain of mass m and length l hangs in equilibrium over a smooth horizontal pin of negligible diameter. One end of the chain is given a small vertical displacement so that the chain slips over the pin. The speed of chain when it leaves the pin is

(a) $\sqrt{\frac{gl}{2}}$ (b) $\sqrt{gl}$ (c) $\sqrt{2gl}$ (d) $\sqrt{3gl}$

10. A block (B) is attached to two unstretched springs S_1 and S_2 with spring constants k and $4k$, respectively. The other ends are attached to two supports M_1 and M_2 are not attached to the walls. The springs and supports have negligible mass. There is no friction anywhere.

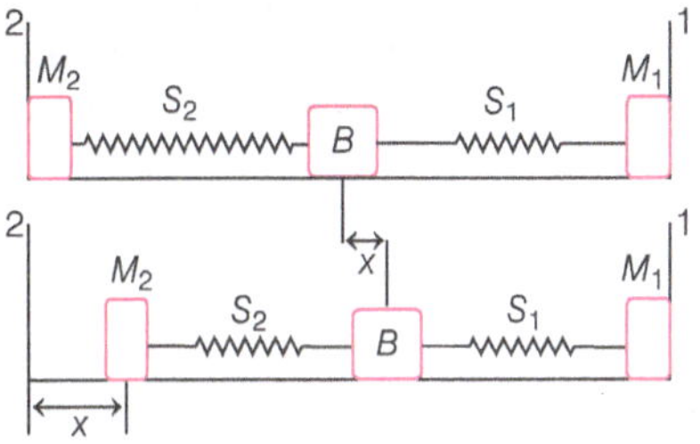

The block B is displaced towards wall 1 by a small distance x and released. The block returns and moves a maximum distance y towards wall 2. Displacements x and y are measured with respect to the equilibrium position of the block B. The ratio $\frac{y}{x}$ is **(2008 JEE)**

(a) 4 (b) 2 (c) $\frac{1}{2}$ (d) $\frac{1}{4}$

Subjective Questions

11. As shown in figure a smooth rod is mounted just above a table top. A 10 kg collar, which is able to slide on the rod with negligible friction is fastened to a spring whose other end is attached to a pivot at O. The spring has negligible mass, a relaxed length of 10 cm and a spring constant of 500 N/m. The collar is released from rest at point A. (a) What is its velocity as it passes point B ? (b) Repeat for point C.

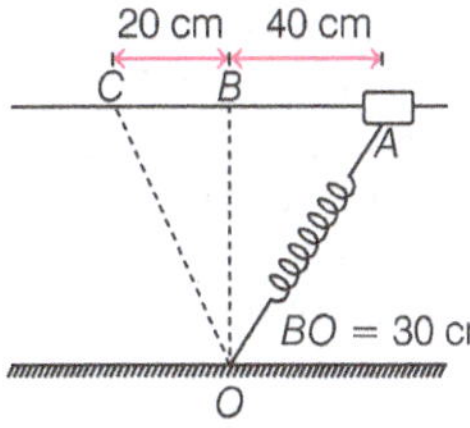

12. A 4 kg block is on a smooth horizontal table. The block is connected to a second block of mass 1 kg by a massless flexible taut cord that passes over a frictionless pulley. The 1 kg block is 1 m above the floor. The two blocks are released from rest. With what speed does the 1 kg block hit the ground ?

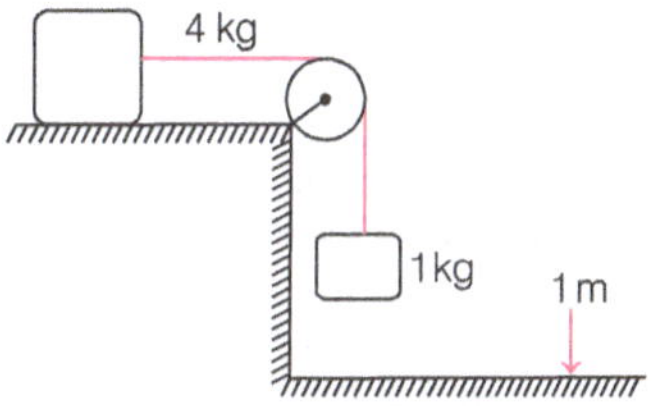

13. Block A has a weight of 300 N and block B has a weight of 50 N. Determine the distance that A must descend from rest before it obtains a speed of 2.5 m/s. Neglect the mass of the cord and pulleys. (Take $g = 9.8$ m/s^2)

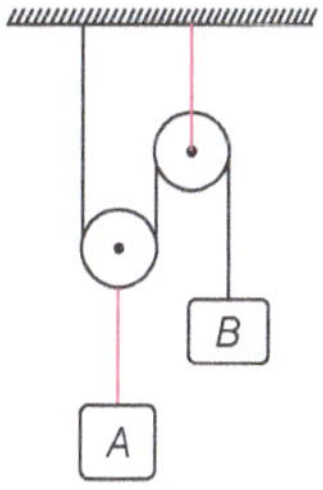

14. The flexible bicycle type chain of length $\frac{\pi r}{2}$ and mass per unit length ρ is released from rest with $\theta = 0°$ in the smooth circular channel and falls through the hole in the supporting surface. Determine the velocity v of the chain as the last link leaves the slot.

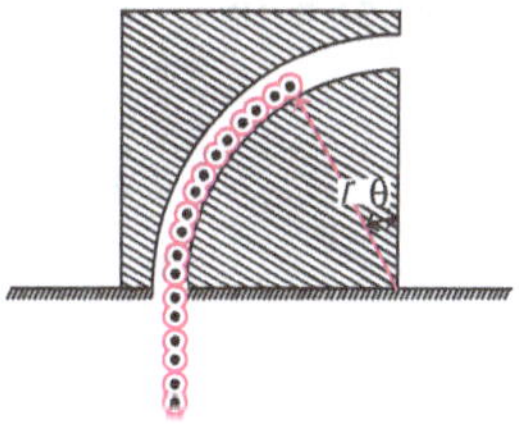

15. A block of mass m moving at a speed v compresses a spring through a distance x before its speed is halved. Find the spring constant of the spring.

Problems Based on Frictional Forces

Assertion and Reason

See directions on page 470.

16. **Assertion :** A block of mass m starts moving on a rough horizontal surface with a velocity v. It stops due to friction between the block and the surface after moving through a ceratin distance. The surface is now tilted to an angle of 30° with the horizontal and the same block is made to go up on the surface with the same initial velocity v. The block stops after moving certain distance. The decrease in the mechanical energy in the second situation is smaller than that in the first situation.

Reason : The coefficient of friction between the block and the surface decreases with the increase in the angle of inclination.

Single Correct Option

17. A block of mass 1 kg slides down a curved track which forms one quadrant of a circle of radius 1 m as shown in figure. The speed of block at the bottom of the track is $v = 2\,\text{ms}^{-1}$. The work done by the force of friction is

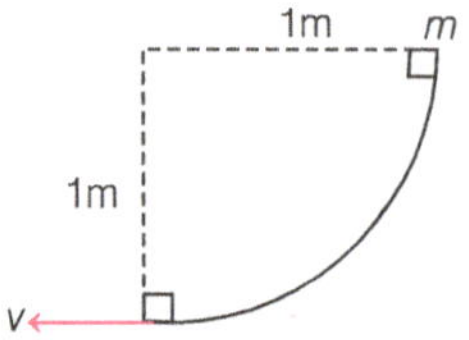

(a) + 4 J (b) − 4 J (c) −8 J (d) +8 J

18. A 2 kg block slides on a horizontal floor with a speed of 4 m/s. It strikes a uncompressed spring and compresses it till the block becomes motionless. The kinetic friction force is 15 N and spring constant is 10000 N/m. The spring compresses by

(a) 5.5 cm (b) 2.5 cm (c) 11.0 cm (d) 8.5 cm

19. A pendulum of length 2 m is left from *A*. When it reaches *B*, it loses 10% of its total energy due to air resistance. The velocity at *B* is

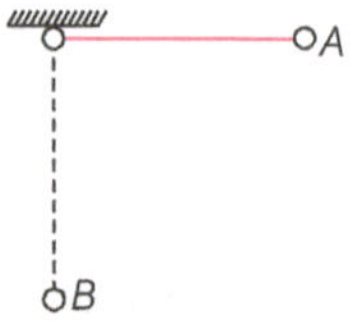

(a) 6 m/s (b) 1 m/s
(c) 2 m/s (d) 8 m/s

20. A particle moves on a rough horizontal ground with some initial velocity say v_0. If $\left(\frac{3}{4}\right)$th of its kinetic energy is lost due to friction in time t_0, then coefficient of friction between the particle and the ground is

(a) $\frac{v_0}{2gt_0}$ (b) $\frac{v_0}{4gt_0}$
(c) $\frac{3v_0}{4gt_0}$ (d) $\frac{v_0}{gt_0}$

21. A block of mass *m* is attached to one end of a mass less spring of spring constant *k*. The other end of the spring is fixed to a wall. The block can move on a horizontal rough surface. The coefficient of friction between the block and the surface is μ. Then, the compression of the spring for which maximum extension of the spring becomes half of maximum compression is

(a) $\frac{2mg\,\mu}{k}$ (b) $\frac{mg\,\mu}{k}$
(c) $\frac{4\,mg\,\mu}{k}$ (d) None of these

22. A particle of mass 1 g executes an oscillatory motion on the concave surface of a spherical dish of radius 2 m placed on a horizontal plane. If the motion of the particle begins from a point on the dish at a height of 1 cm from the horizontal plane and the coefficient of friction is 0.01, the total distance covered by the particle before it comes to rest, is approximately

(a) 2.0 m (b) 10.0 m (c) 1.0 m (d) 20.0 m

Subjective Questions

23. Two masses $m_1 = 10$ kg and $m_2 = 5$ kg are connected by an ideal string as shown in the figure. The coefficient of friction between m_1 and the surface is $\mu = 0.2$. Assuming that the system is released from rest. Calculate the velocity of blocks when m_2 has descended by 4 m. (Take, $g = 10\,\text{m/s}^2$)

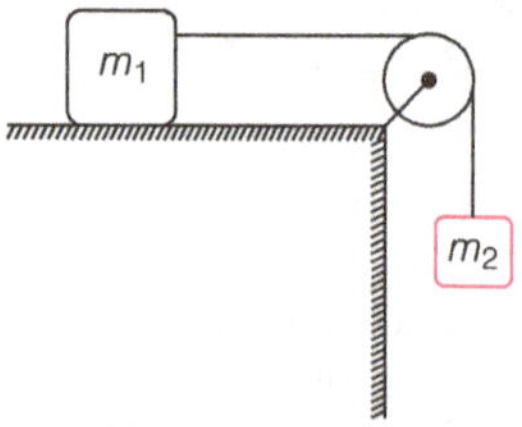

24. In the arrangement shown in figure $m_A = 4.0$ kg and $m_B = 1.0$ kg. The system is released from rest and block *B* is found to have a speed 0.3 m/s after it has descended through a distance of 1 m. Find the coefficient of friction between the block and the table. Neglect friction elsewhere. (Take, $g = 10\,\text{m/s}^2$)

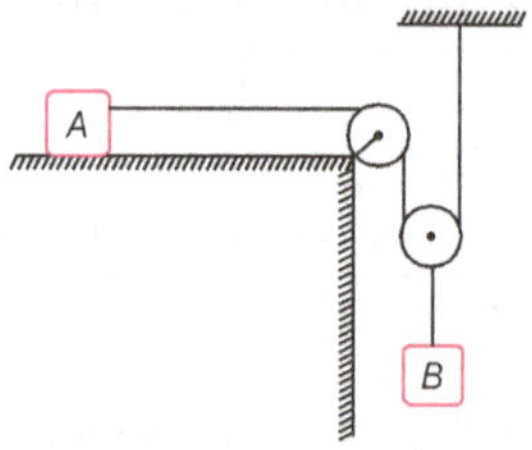

25. Block *A* has a weight of 300 N and block *B* has a weight of 50 N. If the coefficient of kinetic friction between the incline and block *A* is $\mu_k = 0.2$. Determine the speed of block *A* after it moves 1 m down the plane, if initially it is at rest. Neglect the mass of the cord and pulleys.

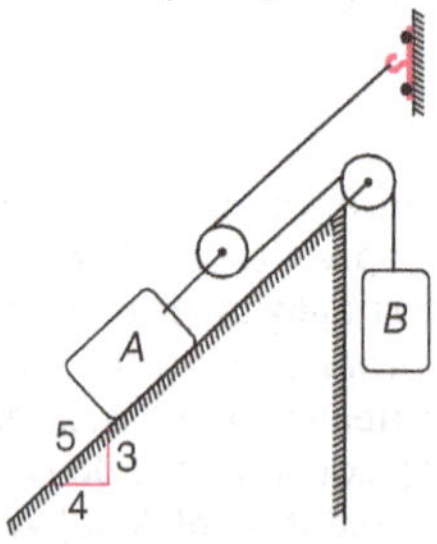

26. A disc of mass $m = 50$ g slides with the zero initial velocity down an inclined plane set at an angle $\alpha = 30°$ to the horizontal; having traversed the distance $l = 50$ cm along the horizontal plane, the disc stops. Find the work performed by the friction forces over the whole distance, assuming the friction coefficient $\mu = 0.15$ for both inclined and horizontal planes. **(Irodov 1.122)**

27. A body of mass *m* is pushed with the initial velocity v_0 up an inclined plane set at an angle α to the horizontal. The friction coefficient is equal to μ. What distance will the body cover before it stops and what

work do the friction forces perform over this distance? **(Irodov 1.134)**

28. A body of mass 2 kg is being dragged with a uniform velocity of 2 m/s on a rough horizontal plane. The coefficient of friction between the body and the surface is 0.20, $J = 4.2$ J/cal and $g = 9.8$ m/s^2. Calculate the amount of heat generated in 5 s. **(1980 JEE)**

29. A lead bullet just melts when stopped by an obstacle. Assuming that 25 percent of the heat is absorbed by the obstacle, find the velocity of the bullet if its initial temperature is 27°C.
(Melting point of lead $= 327°$C, specific heat of lead $= 0.03$ cal/g-°C, latent heat of fusion of lead $= 6$ cal/g, $J = 4.2$ J/cal) **(1981 JEE)**

Equilibrium, Momentary Rest, Maximum Speed

Single Correct Option

30. Two light vertical springs with spring constants k_1 and k_2 are separated by a distance l. Their upper ends are fixed to the ceiling and their lower ends to the points A and B of a light horizontal rod AB. A vertical downward force F is applied at point C on the rod. AB will remain horizontal in equilibrium, if the distance AC is

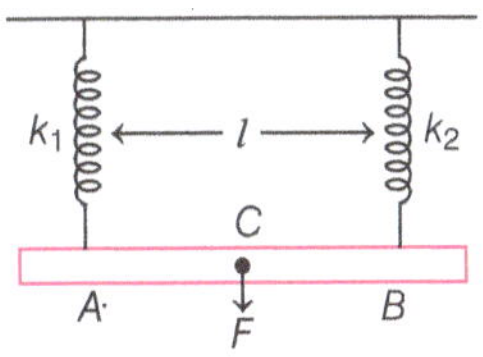

(a) $\frac{lk_1}{k_2}$ (b) $\frac{lk_1}{k_2 + k_1}$ (c) $\frac{lk_2}{k_1}$ (d) $\frac{lk_2}{k_1 + k_2}$

31. A system of wedge and block as shown in figure, is released with the spring in its natural length. All surfaces are frictionless. Maximum elongation in the spring will be

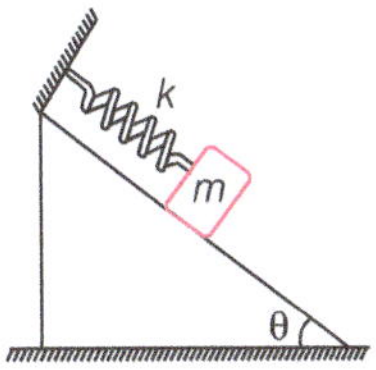

(a) $\frac{2\,mg\sin\theta}{k}$ (b) $\frac{mg\sin\theta}{k}$
(c) $\frac{4\,mg\sin\theta}{k}$ (d) $\frac{mg\sin\theta}{2k}$

32. An ideal massless spring S can be compressed 1.0 m in equilibrium by a force of 100 N. This same spring is placed at the bottom of a frictionless inclined plane which makes an angle $\theta = 30°$ with the horizontal. A 10 kg mass m is released from the rest at the top of the inclined plane and is brought to rest momentarily after compressing the spring by 2.0 m. The distance through which the mass moved before coming to rest is

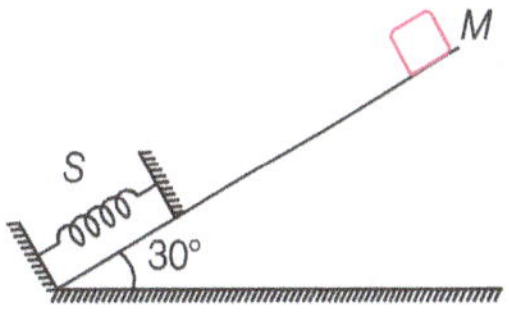

(a) 8 m (b) 6 m (c) 4 m (d) 5 m

33. In the above problem, what is the speed of mass just before it touches the spring?
(a) $\sqrt{20}$ ms^{-1} (b) $\sqrt{30}$ ms^{-1} (c) $\sqrt{10}$ ms^{-1} (d) $\sqrt{40}$ ms^{-1}

34. A body of mass m is released from a height h on a smooth inclined plane that is shown in the figure. The following can be true about the velocity of the block knowing that the wedge is fixed.

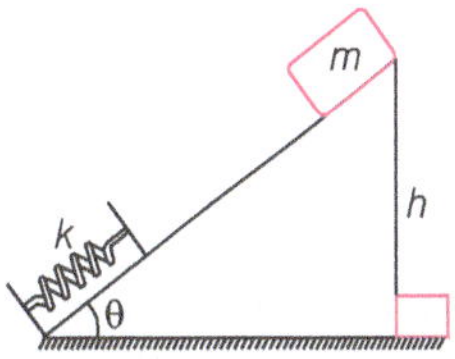

(a) v is highest when it just touches the spring
(b) v is highest when it compresses the spring by some amount
(c) v is highest when the spring comes back to natural position
(d) v is highest at the maximum compression

35. A block of mass m moving with velocity v_0 on a smooth horizontal surface hits the spring of constant k as shown. The maximum compression in the spring is

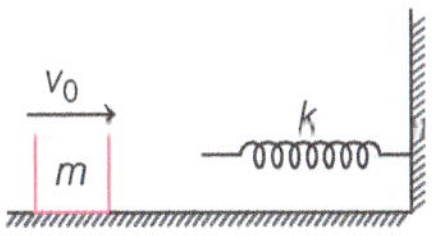

(a) $\sqrt{\frac{2m}{k}} \cdot v_0$ (b) $\sqrt{\frac{m}{k}} \cdot v_0$ (c) $\sqrt{\frac{m}{2k}} \cdot v_0$ (d) $\frac{m}{2k} \cdot v_0$

36. The figure shows a smooth curved track terminating in a smooth horizontal part. A spring of spring constant 400 N/m is attached at one end to a wedge fixed rigidly with the horizontal part. A 40 g mass is released from rest at a height of 5 m on the curved track. The maximum compression of the spring will be

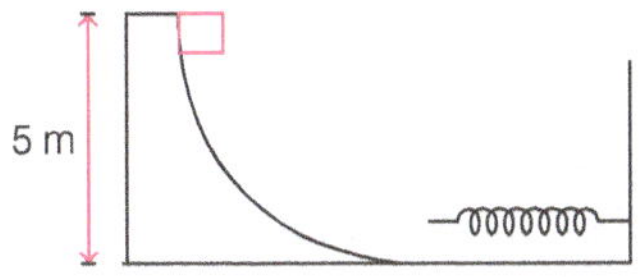

(a) 10 cm (b) 2 cm (c) 3 cm (d) 4 cm

37. A block of mass 1 kg is connected with a fixed wall through a light spring of constant k = 100 N/m. The surfaces are frictionless. Initially, spring is in relaxed position. A horizontal force F of 50 N is applied on the block, find maximum speed of the block during the motion of the block.

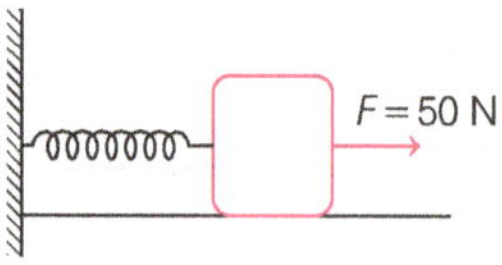

(a) 1 m/s (b) 2 m/s (c) 5 m/s (d) 10 m/s

38. Two blocks each of mass 1 kg are connected by a light spring of spring constant k = 100 N/m. When spring is in natural length, each block is projected in opposite directions with speed of 10 m/s. The work done by spring on each block upto maximum compression is

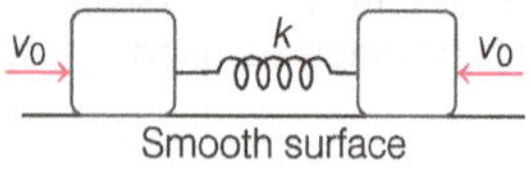

(a) 100 J (b) – 100 J (c) 50 J (d) – 50 J

Match the Columns

39. A block attached with a spring is released from A. Position B is the mean position and the block moves up to point C. Match the following two columns.

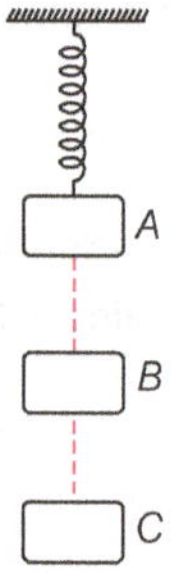

	Column I		Column II
(a)	From A to B decrease in gravitational potential energy is........, the increase in spring potential energy.	(p)	less than
(b)	From A to B increase in kinetic energy of block is........., the decrease in gravitational potential energy.	(q)	more than
(c)	From B to C decrease in kinetic energy of block is..... the increase in spring potential energy.	(r)	equal to
(d)	From B to C decrease in gravitational potential energy is ...., the increase in spring potential energy.		

Subjective Question

40. A block of mass 1 kg is attached to three unstretched springs of spring constants $k_1 = 40$ N/m, $k_2 = 15$ N/m and $k_3 = 45$ N/m as shown in the given figure.

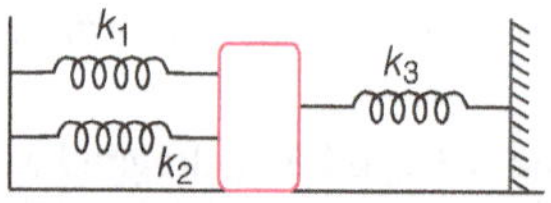

The block is displaced towards left on the horizontal smooth surface through a distance 10 cm and is released. Find maximum speed during its motion in m/s.

Work Done by Different Forces

Assertion and Reason

See directions on page 470.

41. **Assertion :** If work done by conservative forces is positive, kinetic energy will increase.

Reason : Because potential energy will decrease.

42. **Assertion :** Displacement-time graph of a particle moving in a straight line is shown in figure. Work done by all the forces between time interval t_1 and t_2 is definitely zero.

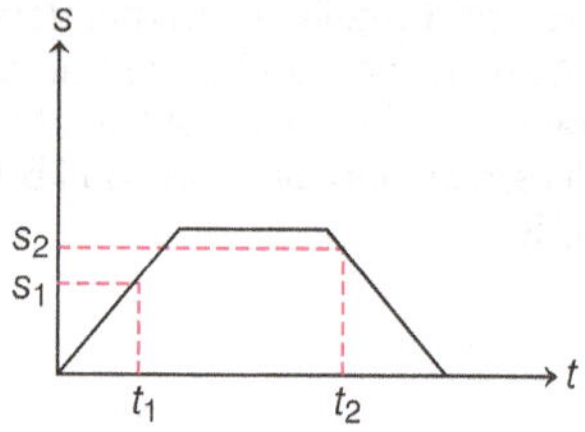

Reason : Work done by all the forces is equal to change in kinetic energy.

Single Correct Option

43. A person pulls a bucket of water from a well of depth h. If the mass of uniform rope is m and that of the bucket full of water is M, then work done by the person is

(a) $\left(M+\frac{m}{2}\right)gh$ (b) $\frac{1}{2}(M+m)gh$

(c) $(M+m)\,gh$ (d) $\left(\frac{M}{2}+m\right)gh$

44. The velocity of a particle decreases uniformly from 20 ms^{-1} to zero in 10 s as shown in figure. If the mass of the particle is 2 kg, then identify the correct statement.

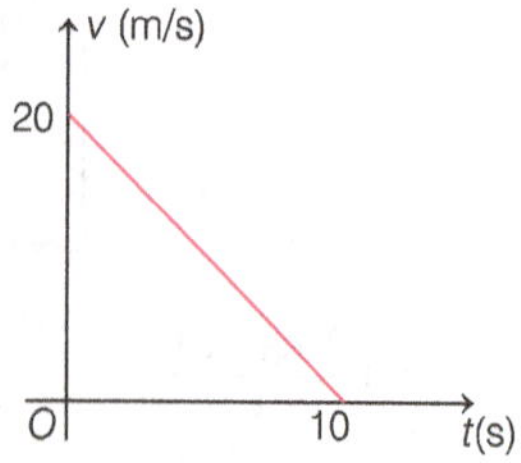

(a) The net force acting on the particle is opposite to the direction of motion

(b) The work done by friction force is –400 J

(c) The magnitude of friction force acting on the particle is 4 N

(d) All of the above

45. A spring of spring constant 5×10^3 N/m is stretched initially by 5 cm from the unstretched position. The work required to further stretch the spring by another 5 cm is

(a) 6.25 N-m (b) 12.50 N-m
(c) 18.75 N-m (d) 25.00 N-m

46. A mass M is lowered with the help of a string by a distance x at a constant acceleration $\frac{g}{2}$. The magnitude of work done by the string will be

(a) Mgx (b) $\frac{1}{2}Mgx^2$ (c) $\frac{1}{2}Mgx$ (d) Mgx^2

47. A spring of force constant k is first stretched by a distance a from its natural length and then further by a distance b. The work done in stretching the part b is

(a) $\frac{1}{2}ka(a-b)$ (b) $\frac{1}{2}ka(a+b)$
(c) $\frac{1}{2}kb(a-b)$ (d) $\frac{1}{2}kb(2a+b)$

48. A uniform chain of length L and mass M is lying on a smooth table and one-third of its length is hanging vertically down over the edge of the table. If g is acceleration due to gravity, the work required to pull the hanging part onto the table is **(1985 JEE)**

(a) MgL (b) $MgL/3$ (c) $MgL/9$ (d) $MgL/18$

Multiple Correct Options

49. Identify the correct statement :

(a) Work done by all the conservative forces is equal to the decrease in potential energy
(b) Work done by all the forces except, the conservative forces is equal to the change in mechanical energy
(c) Work done by all the forces is equal to the change in kinetic energy
(d) Work done by all the forces is equal to the change in potential energy

50. If kinetic energy of a body is increasing then

(a) work done by conservative forces must be positive
(b) work done by conservative forces may be positive
(c) work done by conservative forces may be zero
(d) work done by non-conservative forces may be zero

51. In position A, kinetic energy of a particle is 60 J and potential energy is –20 J. In position B, kinetic energy is 100 J and potential energy is 40 J. Then, in moving the particle from A to B,

(a) work done by conservative forces is –60 J
(b) work done by external forces is 40 J
(c) net work done by all the forces is 40 J
(d) net work done by all the forces is 100 J

Subjective Question

52. A locomotive of mass m starts moving so that its velocity varies according to the law $v = \alpha\sqrt{s}$, where α is a constant and s is the distance covered. Find the total work performed by all the forces which are acting on the locomotive during the first t seconds after the beginning of motion. **(Irodov 1.119)**

Miscellaneous Problems

Single Correct Option

53. Initially the system shown in figure is in equilibrium. At the moment, the string is cut the downward acceleration of blocks A and B are respectively a_1 and a_2. The magnitudes of a_1 and a_2 are

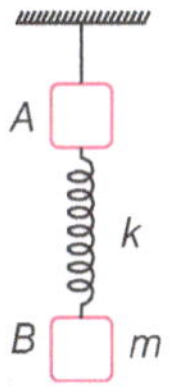

(a) zero and zero (b) $2g$ and zero
(c) g and zero (d) None of these

54. The minimum stopping distance of a car moving with velocity v is x. If the car is moving with velocity $2v$, then the minimum stopping distance will be

(a) $2x$ (b) $4x$
(c) $3x$ (d) $8x$

55. A projectile is fired from the origin with a velocity v_0 at an angle θ with the x-axis. The speed of the projectile at an altitude h Is

(a) $v_0 \cos\theta$ (b) $\sqrt{v_0^2 - 2gh}$
(c) $\sqrt{v_0^2 \sin^2\theta - 2gh}$ (d) None of these

56. A ball is dropped at $t = 0$ from some height on a smooth elastic surface. Identify the graph which correctly represents the variation of kinetic energy K with time t.

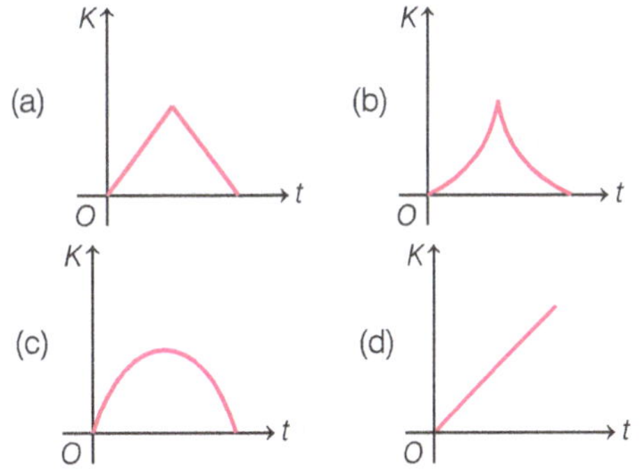

57. A spring of force constant k is cut in two parts at its one-third length. When both the parts are stretched by same amount. The work done in the two parts will be

(a) equal in both
(b) greater for the longer part
(c) greater for the shorter part
(d) data insufficient

Note *Spring constant of a spring is inversely proportional to length of spring.*

58. An open knife of mass m is dropped from a height h on a wooden floor. If the blade penetrates up to the depth d into the wood, the average resistance offered by the wood to the knife edge is

(a) $mg\left(1+\frac{h}{d}\right)$ (b) $mg\left(1+\frac{h}{d}\right)^2$
(c) $mg\left(1-\frac{h}{d}\right)$ (d) $mg\left(1+\frac{d}{h}\right)$

59. Two springs have force constants k_A and k_B such that $k_B = 2k_A$. The four ends of the springs are stretched by the same force. If energy stored in spring A is E, then energy stored in spring B is

(a) $\frac{E}{2}$ (b) $2E$ (c) E (d) $4E$

60. A body of mass 100 g is attached to a hanging spring whose force constant is 10 N/m. The body is lifted until the spring is in its unstretched state and then released. Calculate the speed of the body when it strikes the table 15 cm below the release point

(a) 1 m/s (b) 0.866 m/s
(c) 0.225 m/s (d) 1.5 m/s

61. The kinetic energy of a projectile at its highest position is K. If the range of the projectile is four times the height of the projectile, then the initial kinetic energy of the projectile is

(a) $\sqrt{2}K$ (b) $2K$ (c) $4K$ (d) $2\sqrt{2}K$

62. A juggler keeps four balls in air. He throws each ball vertically upwards with the same speed at equal interval of time. The maximum height attained by each ball is 20m. Find kinetic energy of first ball when fourth ball is in his hand. Assume that the mass of each ball is 10 g.

(a) 0.5 J (b) 1 J
(c) 1.5 J (d) 2.5 J

63. A rod of mass M hinged at O is kept in equilibrium with a spring of stiffness k as shown in figure. The potential energy stored in the spring is

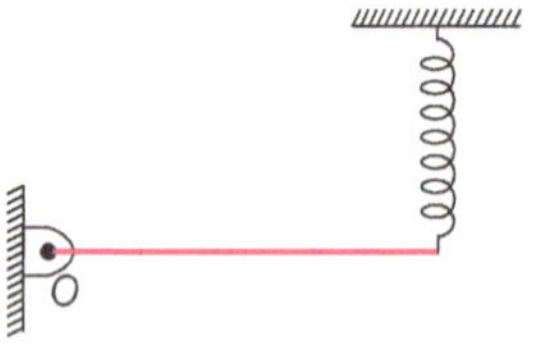

(a) $\frac{(mg)^2}{4k}$ (b) $\frac{(mg)^2}{2k}$ (c) $\frac{(mg)^2}{8k}$ (d) $\frac{(mg)^2}{k}$

64. Two masses of 1 g and 4 g are moving with equal kinetic energies. The ratio of the magnitudes of their momenta is **(1980 JEE)**

(a) 4 : 1 (b) $\sqrt{2}$: 1 (c) 1 : 2 (d) 1 : 16

65. A spring of force constant k is cut into two pieces such that one piece is double the length of the other. Then, the long piece will have a force constant of **(1999 JEE)**

(a) (2/3) k (b) (3/2) k (c) 3 k (d) 6 k

Subjective Questions

66. The potential energy of a two particle system separated by a distance r is given by $U(r) = \frac{A}{r}$, where A is a constant. Find the radial force F_r, that each particle exerts on the other.

67. A small block of ice with mass 0.120 kg is placed against a horizontal compressed spring mounted on a horizontal table top that is 1.90 m above the floor. The spring has a force constant $k = 2300$ N/m and is initially compressed 0.045 m. The mass of the spring is negligible. The spring is released and the block slides along the table, goes off the edge and travels to the floor. If there is negligible friction between the ice and the table, what is the speed of the block of ice when it reaches the floor. (Take, $g = 9.8$ m/s^2)

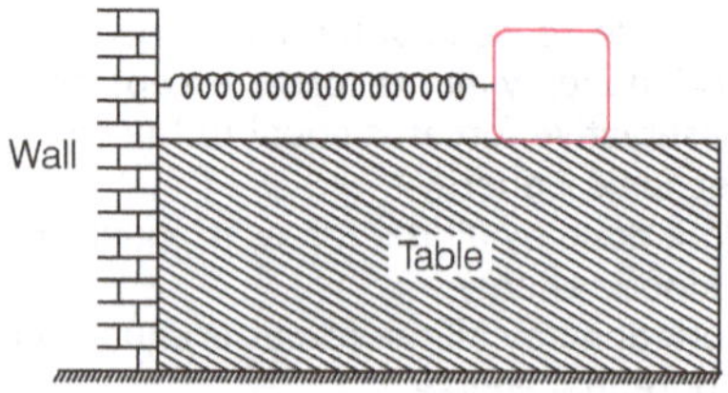

Answers

1. (a) **2.** (a) **3.** (c) **4.** (b) **5.** (a) **6.** (b) **7.** (c) **8.** (b) **9.** (a) **10.** (c)

11. (a) 2.45 m/s (b) 2.15 m/s **12.** 2 m/s **13.** 0.797 m **14.** $v = \sqrt{rg\left(\frac{\pi}{2} + \frac{4}{\pi}\right)}$ **15.** $k = \frac{3v^2 m}{4x^2}$ **16.** (c)

17. (c) **18.** (a) **19.** (a) **20.** (a) **21.** (c) **22.** (c) **23.** 4 m/s **24.** 0.115 **25.** 1.13 m/s

26. – 0.05 J **27.** $s = \frac{v_0^2}{2(g\sin\alpha + \mu g\cos\alpha)}$, $W_f = \frac{-mv_0^2\mu}{2g(1+\tan\alpha)}$ **28.** 9.33 cal **29.** 409.8 m/s **30.** (d) **31.** (a)

32. (c) **33.** (a) **34.** (b) **35.** (b) **36.** (a) **37.** (c) **38.** (d) **39.** (a-q; b-p; c-p; d-p) **40.** 1 m/s

41. (d) **42.** (d) **43.** (a) **44.** (a) **45.** (c) **46.** (c) **47.** (d) **48.** (d) **49.** (b , c)

50. (b, c, d) **51.** (a, c) **52.** $W_{Total} = m\alpha^4 t^2/8$ **53.** (b) **54.** (b) **55.** (b) **56.** (b) **57.** (c) **58.** (a)

59. (a) **60.** (b) **61.** (b) **62.** (a) **63.** (c) **64.** (c) **65.** (b) **66.** A/r^2 **67.** 8.72 m/s

See hints on pages 519-525.

MODULE 5

Types of Equilibrium

A body is said to be in translatory equilibrium, if net force acting on the body is zero, i.e.

$$F_{net} = 0$$

If the forces are conservative

$$F = -\frac{dU}{dr}$$

and for equilibrium $F = 0$.

So, $$-\frac{dU}{dr} = 0 \quad \text{or} \quad \frac{dU}{dr} = 0$$

i.e. at equilibrium position slope of U-r graph is zero or the potential energy is maximum, minimum or constant.

Equilibrium are of three types, i.e. the situation where $F = 0$ and $\frac{dU}{dr} = 0$ can be obtained under three conditions.

These are : **stable equilibrium**, **unstable equilibrium** and **neutral equilibrium**. These three types of equilibrium can be better explained from the given three figures.

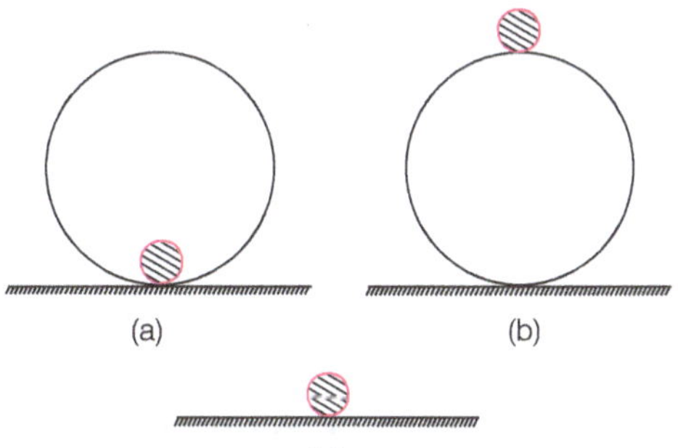

(a) (b) (c)

Three identical balls are placed in equilibrium positions as shown in figures (a), (b) and (c) respectively.

In Fig. (a), ball is placed inside a smooth spherical shell. This ball is in stable equilibrium position. In Fig. (b), the ball is placed over a smooth sphere. This is in unstable equilibrium position. In Fig. (c), the ball is placed on a smooth horizontal ground. This ball is in neutral equilibrium position.

The table given below explains what is the difference and what are the similarities between these three equilibrium positions in the language of physics.

Differences and similarities between stable equilibrium, unstable equilibrium and neutral equilibrium

S. No.	Stable Equilibrium	Unstable Equilibrium	Neutral Equilibrium
1.	Net force is zero.	Net force is zero.	Net force is zero.
2.	$\frac{dU}{dr} = 0$ or slope of U-r graph is zero.	$\frac{dU}{dr} = 0$ or slope of U-r graph is zero.	$\frac{dU}{dr} = 0$ or slope of U-r graph is zero.
3.	When displaced from its equilibrium position a net restoring force starts acting on the body which has a tendency to bring the body back to its equilibrium position.	When displaced from its equilibrium position, a net force starts acting on the body which moves the body in the direction of displacement or away from the equilibrium position.	When displaced from its equilibrium position the body has neither the tendency to come back nor to move away from the original position. It is again in equilibrium.
4.	Potential energy in equilibrium position is minimum as compared to its neighbouring points. or $\frac{d^2U}{dr^2}$ = positive	Potential energy in equilibrium position is maximum as compared to its neighbouring points. or $\frac{d^2U}{dr^2}$ = negative	Potential energy remains constant even if the body is displaced from its equilibrium position. or $\frac{d^2U}{dr^2} = 0$
5.	When displaced from equilibrium position the centre of gravity of the body goes up.	When displaced from equilibrium position the centre of gravity of the body comes down.	When displaced from equilibrium position the centre of gravity of the body remains at the same level.

Important POINTS

1. If we plot graphs between F and r or U and r, F will be zero at equilibrium position while U will be maximum, minimum or constant depending on the type of equilibrium. This all is shown in figure.

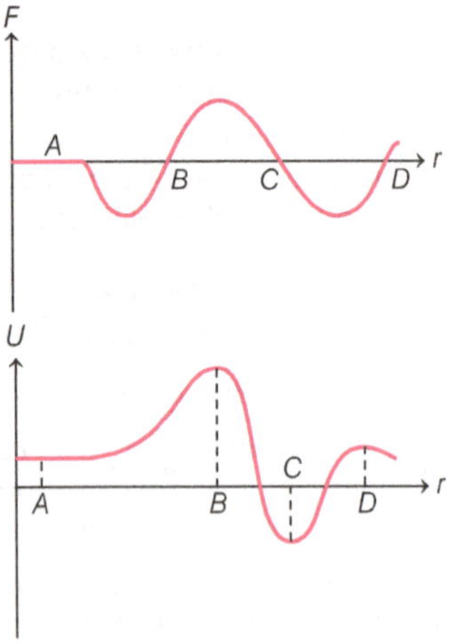

At point A, $F = 0$, $\frac{dU}{dr} = 0$, but U is constant. Hence, A is neutral equilibrium position.

At points B and D, $F = 0$, $\frac{dU}{dr} = 0$ but U is maximum. Thus, these are the points of unstable equilibrium.

At point C, $F = 0$, $\frac{dU}{dr} = 0$, but U is minimum. Hence, point C is stable equilibrium position.

2. Oscillations of a body take place about stable equilibrium position. For example, bob of a pendulum oscillates about its lowest point which is also the stable equilibrium position of the bob. Similarly, in Fig. (b), the ball will oscillate about its stable equilibrium position.

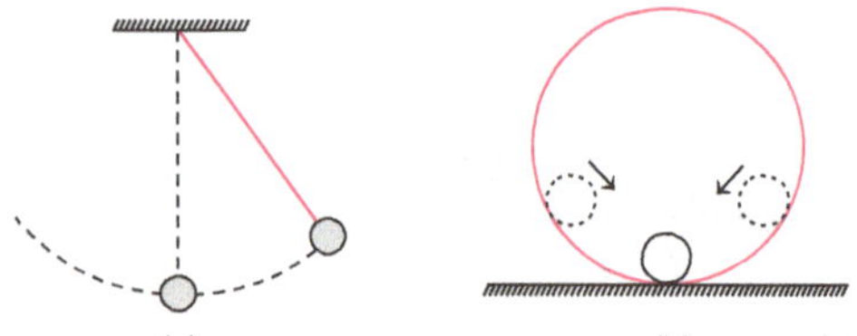

3. If a graph between F and r is as shown in figure, then $F = 0$, at $r = r_1$, $r = r_2$ and $r = r_3$. Therefore, at these three points, body is in equilibrium. But these three positions are three different types of equilibriums. For example

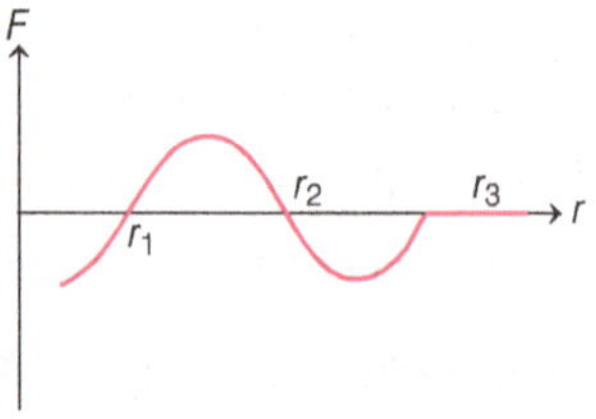

At $r = r_1$, body is in unstable equilibrium. This is because, if we displace the body slightly rightwards (positive direction), force acting on the body is also positive, i.e. away from $r = r_1$ position.

At $r = r_2$, body is in stable equilibrium. Because if we displace the body rightwards (positive direction) force acting on the body is negative (or leftwards) or the force acting is restoring in nature.

At $r = r_3$, equilibrium is neutral in nature. Because if we displace the body rightwards or leftwards force is again zero.

Example 1 For the potential energy curve shown in figure.

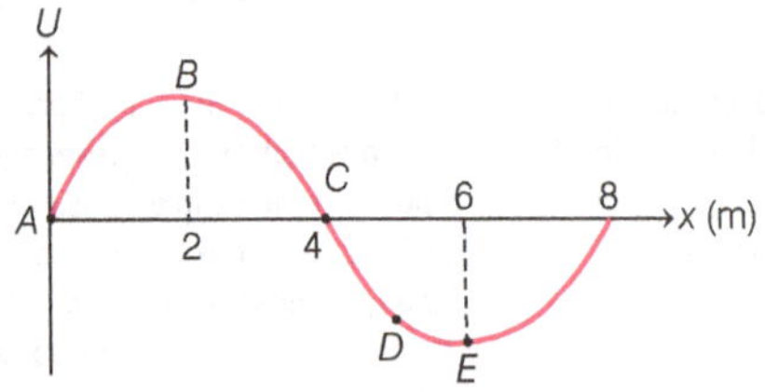

(a) Find directions of force at points A, B, C, D and E.

(b) Find positions of stable, unstable and neutral equilibriums.

Sol. (a) $F = -\frac{dU}{dr}$

or $-\frac{dU}{dx} = -$(slope of U - x graph)

Point	Slope of U - x graph	F = (Slope of U - x graph)
A	Positive	Negative
B	Positive	Negative
C	Negative	Positive
D	Negative	Positive
E	Zero	Zero

(b) At $x = 6$m, potential energy is minimum. So. it is stable equilibrium position.

At $x = 2$m, potential energy is maximum. So, it is unstable equilibrium position.

There is no point, where potential energy is constant. So, we don't have any point of neutral equilibrium position.

Example 2 The potential energy of a conservative force field is given by

$$U = ax^2 - bx$$

where, a and b are positive constants. Find the equilibrium position and discuss whether the equilibrium is stable, unstable or neutral.

Sol. In a conservative field $F = -\frac{dU}{dx}$

$\therefore \quad F = -\frac{d}{dx}(ax^2 - bx) = b - 2ax$

For equilibrium $F = 0$

or $\quad b - 2ax = 0 \quad \therefore \quad x = \frac{b}{2a}$

From the given equation we can see that $\frac{d^2U}{dx^2} = 2a$ (positive), i.e. U is minimum.

Therefore, $x = \frac{b}{2a}$ is the stable equilibrium position.

Example 3 A single conservative force $F(x)$ acts on a 1.0 kg particle that moves along the x-axis. The potential energy $U(x)$ is given by:

$$U(x) = 20 + (x - 2)^2$$

where, x is in metres. At $x = 5.0$ m the particle has a kinetic energy of 20 J.

(a) What is the mechanical energy of the system?

(b) Make a plot of $U(x)$ as a function of x for $-10\text{m} \le x \le 10\text{ m}$, and on the same graph draw the line that represents the mechanical energy of the system.
Use part (b) to determine.

(c) The least value of x and

(d) The greatest value of x between which the particle can move.

(e) The maximum kinetic energy of the particle and

(f) The value of x at which it occurs.

(g) Determine the equation for $F(x)$ as a function of x.

(h) For what (finite) value of x does $F(x) = 0$?

Sol. (a) Potential energy at $x = 5.0$m is

$$U = 20 + (5 - 2)^2 = 29\text{J}$$

$\therefore$ Mechanical energy

$$E = K + U = 20 + 29 = 49\text{J}$$

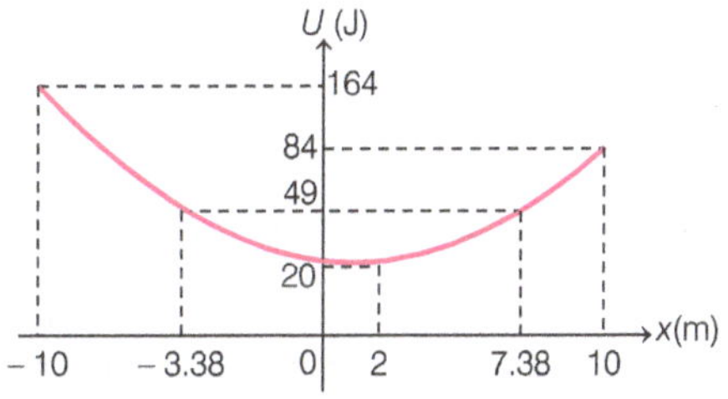

(b) At $x = 10$m, $U = 84$ J at $x = -10$ m, $U = 164$ J
and at $x = 2$ m, $U = \text{minimum} = 20$ J

(c) and (d) Particle will move between the points where its kinetic energy becomes zero or its potential energy is equal to its mechanical energy.

Thus, $\quad 49 = 20 + (x - 2)^2$

or $\quad (x - 2)^2 = 29$

or $\quad x - 2 = \pm\sqrt{29} = \pm 5.38$ m

$\therefore \quad x = 7.38$ m and -3.38 m

or the particle will move between $x = -3.38$ m and $x = 7.38$ m

(e) and (f) Maximum kinetic energy is at $x = 2$ m, where the potential energy is minimum and this maximum kinetic energy is,

$$K_{max} = E - U_{min} = 49 - 20 = 29 \text{ J}$$

(g) $F = -\frac{dU}{dx} = -2(x - 2) = 2(2 - x)$

(h) $F(x) = 0$, at $x = 2.0$ m
where potential energy is minimum (the position of stable equilibrium).

MODULE 5 Exercise

Single Correct Option

1. The force acting on a body moving along x-axis varies with the position of the particle shown in the figure. The body is in stable equilibrium at

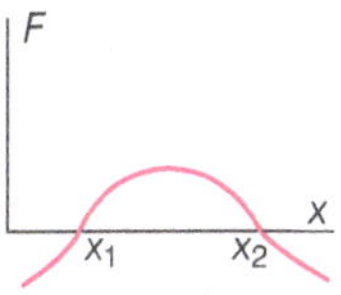

(a) $x = x_1$ (b) $x = x_2$ (c) both x_1 and x_2 (d) neither x_1 nor x_2

2. The given plot shows the variation of U, the potential energy of interaction between two particles with the distance separating them r.

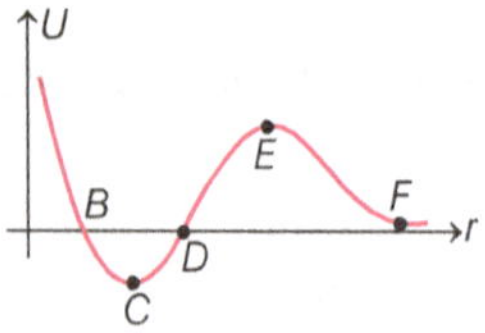

1. B and D are equilibrium points
2. C is a point of stable equilibrium
3. The force of interaction between the two particles is attractive between points C and D and repulsive between D and E
4. The force of interaction between particles is repulsive between points E and F.

Which of the above statements are correct?

(a) 1 and 2 (b) 1 and 4 (c) 2 and 4 (d) 2 and 3

3. The potential energy function for a diatomic molecule is $U(x) = \dfrac{a}{x^{12}} - \dfrac{b}{x^6}$. In stable equilibrium, the distance between the particles is

(a) $\left(\dfrac{2a}{b}\right)^{1/6}$ (b) $\left(\dfrac{a}{b}\right)^{1/6}$ (c) $\left(\dfrac{b}{2a}\right)^{1/6}$ (d) $\left(\dfrac{b}{a}\right)^{1/6}$

Multiple Correct Options

4. The potential energy of a particle is given by $U = 100 - 5x + 100x^2$, where U and x are in SI units. If mass of the particle is 0.1 kg, then magnitude of it's acceleration

(a) At 0.05 m from the origin is 50 ms^{-2}
(b) At 0.05 m from the mean position is $100\,ms^{-2}$
(c) At 0.05 m from the origin is 150 ms^{-2}
(d) At 0.05 m from the mean position is $200\,ms^{-2}$

Match the Columns

5. Two positive charges $+q$ each are fixed at points $(-a, 0)$ and $(a, 0)$. A third charge $+Q$ is placed at origin. Corresponding to small displacement of $+Q$ in the direction mentioned in Column I, match the corresponding equilibrium of Column II.

	Column I		Column II
(a)	Along positive x-axis	(p)	stable equilibrium
(b)	Along positive y-axis	(q)	unstable equilibrium
(c)	Along positive z-axis	(r)	neutral equilibrium
(d)	Along the line $x = y$	(s)	no equilibrium

Subjective Questions

6. Potential energy of a particle moving along x-axis is given by

$$U = \left(\frac{x^3}{3} - 4x + 6\right)$$

Here, U is in joule and x in metre. Find position of stable and unstable equilibrium.

7. A single force acting on a particle moving along x-axis is as shown in figure. Find points of stable and unstable equilibrium.

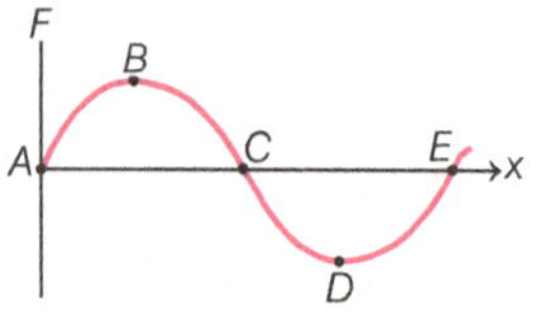

8. Two point charges $+q$ and $+q$ are fixed at $(a, 0, 0)$ and $(-a, 0, 0)$. A third point charge $-q$ is at origin. State whether its equilibrium is stable, unstable or neutral if it is slightly displaced :

(a) along x-axis. (b) along y-axis.

9. Potential energy of a particle along x-axis, varies as, $U = -20 + (x-2)^2$, where U is in joule and x in metre. Find the equilibrium position and state whether it is stable or unstable equilibrium.

10. Force acting on a particle constrained to move along x-axis is $F = (x - 4)$. Here, F is in newton and x in metre. Find the equilibrium position and state whether it is stable or unstable equilibrium.

11. A single conservative force F_x acts on a 2 kg particle that moves along the x-axis. The potential energy is given by $U = (x-4)^2 - 16$.

Here, x is in metre and U in joule. At $x = 6.0$ m, kinetic energy of particle is 8 J. Find

(a) total mechanical energy
(b) maximum kinetic energy
(c) values of x between which particle moves
(d) the equation of F_x as a function of x
(e) the value of x at which F_x is zero

Answers

1. (b) **2.** (c) **3.** (a) **4.** (a, b, c)

5. (a-p; b-q; c-q; d-s)

6. $x = 2$m is position of stable equilibrium. $x = -2$m is position of unstable equilibrium.

7. Points A and E are unstable equilibrium positions. Point C is stable equilibrium position.

8. (a) Unstable (b) Stable **9.** $x = 2$m, stable

10. $x = 4$m, unstable

11. (a) -4 J (b) 12 J (c) $x = (4 \pm 2\sqrt{3})$ m (d) $F_x = (8 - 2x)$ (e) $x = 4$ m

See hints on pages 526-527.

MODULE 6

Power of a Force

Power of a force is the rate of work done by this force. Now, power may be of two types:

(i) Instantaneous power (P_i or P)

(ii) Average power (P_{av})

Instantaneous Power

The rate of doing work done by a force at a given instant is called instantaneous power of this force.

Thus,

$$P = \frac{dW}{dt} = \frac{\mathbf{F}\cdot d\mathbf{r}}{dt} \qquad (\text{as } dW = \mathbf{F}\cdot d\mathbf{r})$$

$$= \mathbf{F}\cdot\mathbf{v} \qquad \left(\text{as } \frac{d\mathbf{r}}{dt} = \mathbf{v}\right)$$

$$= Fv\cos\theta$$

$$\therefore \quad \boxed{P = \frac{dW}{dt} = \mathbf{F}\cdot\mathbf{v} = Fv\cos\theta}$$

Here, θ is the angle between **F** and **v**. Hence power of a force is the dot product of this force and instantaneous velocity. So, power is a scalar quantity.

If angle between **F** and **v** is acute then dot product is positive (or $\cos\theta$ is positive). So, power is positive. If angle is 90°, then dot product, $\cos\theta$ and hence power are zero.

If θ is obtuse, then dot product, $\cos\theta$ and hence power are negative.

Average Power

The ratio of total work done and total time taken in doing that work is defined as the average power.

Thus $\boxed{P_{av} = \frac{W_{Total}}{t_{total}}}$

Example 1 A ball of mass 1 kg is dropped from a tower. Find power of gravitational force at time $t = 2$ s. Take $g = 10\text{ m/s}^2$.

Sol. At $t = 2$ s, velocity of the ball,

$$v = gt = 10\times 2 = 20\text{ m/s} \qquad \text{(downwards)}$$

Gravitational force on the ball,

$$F = mg = 1\times 10 = 10\text{ N} \qquad \text{(downwards)}$$

$\therefore$ Instantaneous power, $P = \mathbf{F}\cdot\mathbf{v} = Fv\cos\theta$

$$= (10)(20)\cos 0°$$

$$= 200\text{ W} \qquad \textbf{Ans.}$$

Example 2 A particle of mass m is lying on smooth horizontal table. A constant force F tangential to the surface is applied on it. Find

(a) average power over a time interval from $t = 0$ to $t = t$,

(b) instantaneous power as function of time t.

Sol. (a) $a = \frac{F}{m} = \text{constant}$

$$v = at = \frac{F}{m}t \Rightarrow P_{av} = \frac{W}{t} = \frac{\frac{1}{2}mv^2}{t}$$

$$= \frac{\left(\frac{1}{2}\right)(m)\left(\frac{Ft}{m}\right)^2}{t} = \frac{F^2t}{2m} \qquad \textbf{Ans.}$$

(b) $P_i = Fv\cos 0° = Fv = (F)\left(\frac{Ft}{m}\right) = \frac{F^2t}{m}$ **Ans.**

Note *In circular motion, centripetal force (we will discuss it in next unit) acts towards the centre. This force is perpendicular to small displacement d**s** and velocity **v**. Hence, work done by it is zero and power of this force is also zero.*

MODULE 6 Exercise

Fill in the Blank

1. A body is moved along a straight line by a machine delivering constant power. The distance moved by the body in time t is proportional to.......... **(1984 JEE)**

Assertion and Reason

See directions on page 470.

2. **Assertion :** Power of a constant force is also constant.

   Reason : Net constant force will always produce a constant acceleration.

Single Correct Option

3. A particle is projected at $t = 0$ from a point on the ground with certain velocity at an angle with the horizontal. The power of gravitation force is plotted against time. Which of the following is the best representation?

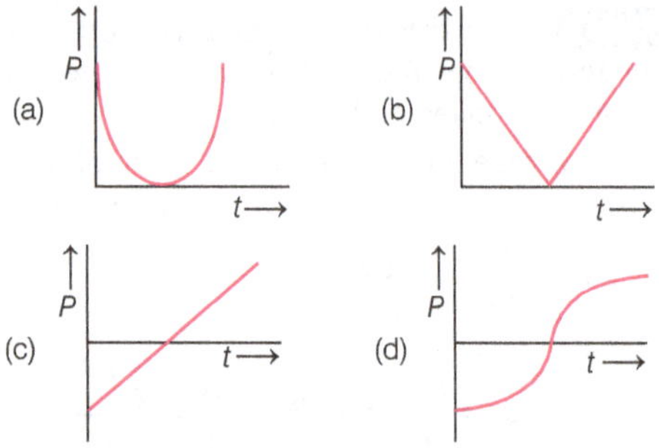

4. A block of mass m slides along the track with kinetic friction μ. A man pulls the block through a rope which makes an angle θ with the horizontal as shown in the figure. The block moves with constant speed v. Power delivered by man is

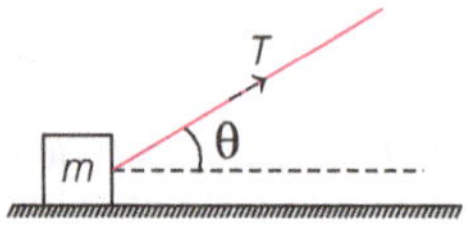

(a) Tv (b) $Tv \cos\theta$
(c) $(T\cos\theta - \mu mg)\,v$ (d) zero

5. A pump is required to lift 800 kg of water per minute from a 10 m deep well and eject it with speed of 20 m/s. The required power (in watts) of the pump will be
(a) 6000 (b) 4000
(c) 5000 (d) 8000

6. A particle of mass m moves from rest under the action of a constant force F which acts for two seconds. The maximum power attained is
(a) $2Fm$ (b) $\frac{F^2}{m}$
(c) $\frac{2F}{m}$ (d) $\frac{2F^2}{m}$

7. A body moves under the action of a constant force along a straight line. The instantaneous power developed by this force with time t is correctly represented by

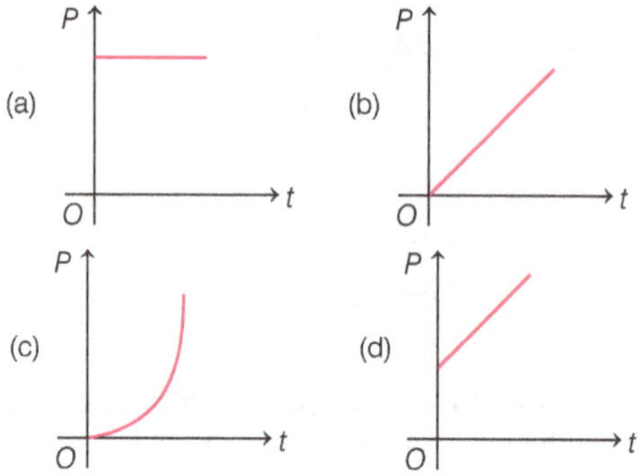

8. A car of mass m is accelerating on a level smooth road under the action of a single force F. The power delivered to the car is constant and equal to P. If the velocity of the car at an instant is v, then after travelling how much distance it becomes double?

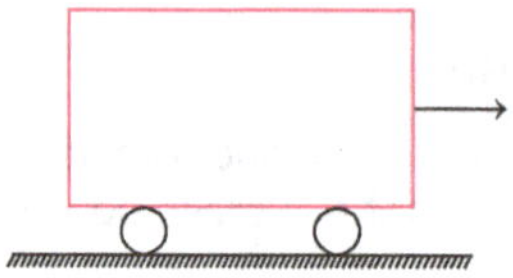

(a) $\frac{7mv^3}{3P}$ (b) $\frac{4mv^3}{3P}$
(c) $\frac{mv^3}{P}$ (d) $\frac{18mv^3}{7P}$

Subjective Questions

9. A block of mass 1 kg starts moving with constant acceleration $a = 4\,\text{m/s}^2$. Find
   (a) average power of the net force in a time interval from $t = 0$ to $t = 2$ s,
   (b) instantaneous power of the net force at $t = 4$ s.

10. A constant power P is applied on a particle of mass m. Find kinetic energy, velocity and displacement of particle as function of time t.

11. A time varying power $P = 2t$ is applied on a particle of mass m. Find
(a) kinetic energy and velocity of particle as function of time,
(b) average power over a time interval from $t = 0$ to $t = t$.

12. A body of mass m slides down a plane inclined at an angle α. The coefficient of friction is μ. Find the rate at which kinetic plus gravitational potential energy is dissipated at any time t.

13. A body of mass m is thrown at an angle α to the horizontal with the initial velocity v_0. Find the mean power developed by gravity over the whole time of motion of the body, and the instantaneous power of gravity as a function of time. **(Irodov 1.125)**

Integer Type Question

14. A particle of mass 0.2 kg is moving in one dimension under a force that delivers a constant power 0.5 W to the particle. If the initial speed (in ms^{-1}) of the particle is zero, the speed (in ms^{-1}) after 5 s is.......... **(2013 JEE)**

Answers

1. $t^{3/2}$ **2.** (d) **3.** (c) **4.** (b) **5.** (b) **6.** (d) **7.** (b) **8.** (a)

9. (a) 16 W (b) 64 W **10.** $K = Pt$, $v = \sqrt{\frac{2Pt}{m}}$, $S = \sqrt{\frac{8P}{9m}}\, t^{3/2}$ **11.** (a) $K = t^2$, $v = \sqrt{\frac{2}{m}}\, t$ (b) $P_{av} = t$

12. $\mu mg^2 \cos\alpha(\sin\alpha - \mu\cos\alpha)t$ **13.** $\langle P \rangle = 0$, $P = mg(gt - v_0 \sin\alpha)$ **14.** (5)

See hints on pages 527-528.

Unit Exercise (Cumulative)

Level 1

Assertion and Reason

See direction on page 470.

1. **Assertion :** A plank A is placed on a rough surface over which a block B is placed. In the shown situation, elastic cord is unstretched. Now a gradually increasing force F is applied slowly on A until the relative motion between the block and plank starts.

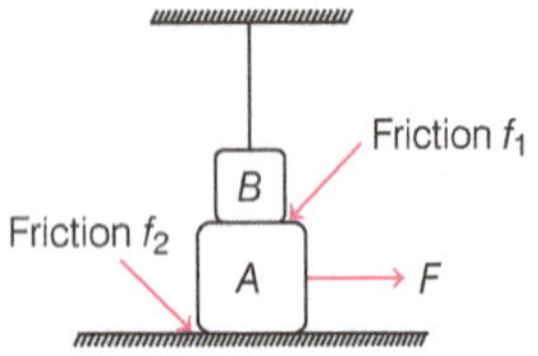

At this moment, cord is making an angle θ with the vertical. Work done by force F is equal to energy lost against friction f_2.

Reason : Work done by static friction f_1 on the system as a whole is zero.

Single Correct Option

2. A particle moves under the action of a force $\mathbf{F} = 20\hat{\mathbf{i}} + 15\hat{\mathbf{j}}$ along a straight line $3y + \alpha x = 5$, where, α is a constant. If the work done by the force F is zero, then the value of α is

(a) $\frac{4}{9}$ (b) $\frac{9}{4}$ (c) 3 (d) 4

3. A plank of mass 10 kg and a block of mass 2 kg are placed on a horizontal plane as shown in the figure. There is no friction between plane and plank. The coefficient of friction between block and plank is 0.5. A force of 60 N is applied on plank horizontally. In first 2 s, the work done by friction on the block is

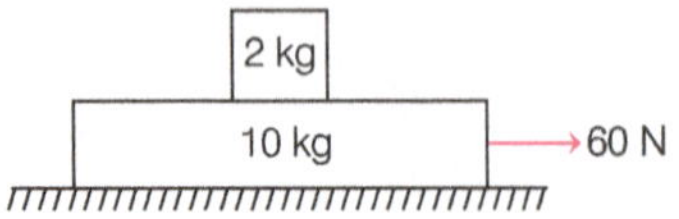

(a) -100 J (b) 100 J
(c) zero (d) 200 J

4. A block of mass 1 kg is projected from point A along irregular rough inclined surface and reaches at the point B as shown in the figure. The coefficient of friction between the block and the inclined plane is 0.5. Find work done by the friction on the block, if $AC = 1$ m.

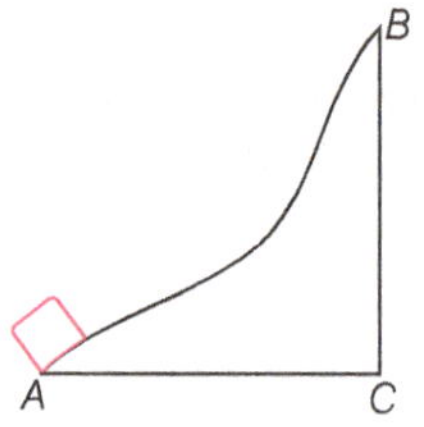

(a) -1 J (b) -2 J (c) -4 J (d) -5 J

5. A block A of mass 45 kg is placed on another block B of mass 123 kg. Now, block B is displaced by external agent by 50 cm horizontally towards right. During the same time, block A just reaches to the left end of block B. Initial and final positions are shown in figures. The work done by friction on block A in ground frame is

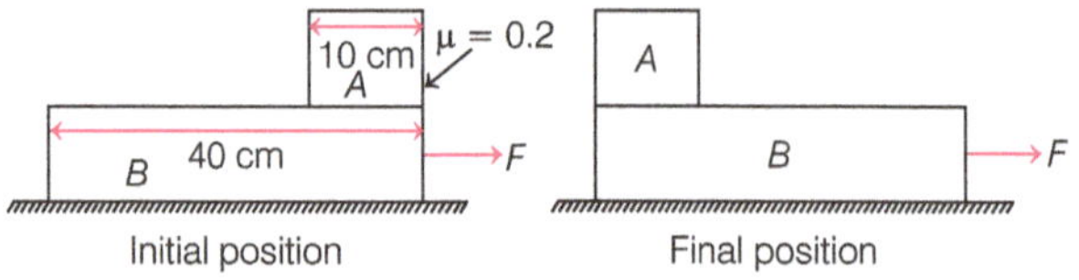

(a) -18 J (b) 18 J (c) 36 J (d) -36 J

6. A person pushes a box on a rough horizontal platform surface. He applies a force of 200 N over a distance of 15 m. Therefore, he gets progressively tired and his applied force reduces linearly with distance to 100 N. The total distance through which the box has been moved is 30 m. What is the work done by the person during the total movement of the box? **(2020 Main)**

(a) 3280 J (b) 2780 J (c) 5690 J (d) 5250 J

7. The potential energy function of a particle in the x-y plane is given by $U = k(x + y)$, where k is a constant. The work done by the conservative force in moving a particle from (1, 1) to (2, 3) is

(a) $-3k$ (b) $+3k$
(c) k (d) None of these

8. A body of mass $m = 10^{-2}$kg is moving in a medium and experiences a frictional force $F = -kv^2$. Its initial speed is $v_0 = 10\text{ms}^{-1}$. If after 10 s, its energy is $\frac{1}{8}mv_0^2$, the value of k will be **(2017 Main)**

(a) 10^{-3} kgs^{-1} (b) 10^{-4} kgm^{-1}
(c) 10^{-1} kgm^{-1}s^{-1} (d) 10^{-3} kgm^{-1}

9. The potential energy of a 1 kg particle free to move along the x-axis is given by

$$U = \left(\frac{x^4}{4} - \frac{x^2}{2}\right) \text{J}$$

The total mechanical energy of the particle is 2 J. Then, the maximum speed (in m/s) is

(a) $\sqrt{2}$ (b) $1/\sqrt{2}$
(c) 2 (d) $3/\sqrt{2}$

10. A body is moving down an inclined plane of slope 37°. The coefficient of friction between the body and the plane varies as $\mu = 0.3\,x$, where x is the distance traveled down the plane by the body. The body will have maximum speed. $\left(\text{Take, } \sin 37° = \frac{3}{5}\right)$

(a) at $x = 1.16$ m
(b) at $x = 2$ m
(c) at bottommost point of the plane
(d) at $x = 2.5$m

11. A small mass slides down an inclined plane of inclination θ with the horizontal. The coefficient of friction is $\mu = \mu_0 x$, where x is the distance through which the mass slides down and μ_0 a positive constant. Then, the distance covered by the mass before it stops is

(a) $\frac{2}{\mu_0}\tan\theta$ (b) $\frac{4}{\mu_0}\tan\theta$ (c) $\frac{1}{2\mu_0}\tan\theta$ (d) $\frac{1}{\mu_0}\tan\theta$

12. A block A of mass M rests on a wedge B of mass $2M$ and inclination θ. There is sufficient friction between A and B so that A does not slip on B. If there is no friction between B and ground, the compression in spring is

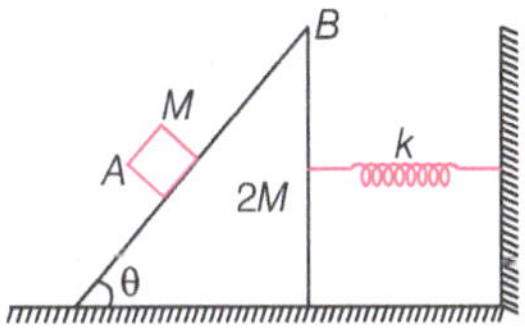

(a) $\frac{Mg\cos\theta}{k}$ (b) $\frac{Mg\cos\theta\sin\theta}{k}$
(c) $\frac{Mg\sin\theta}{k}$ (d) zero

13. A point particle of mass m, moves along the uniformly rough track PQR as shown in the figure. The coefficient of friction, between the particle and the rough track equals μ.

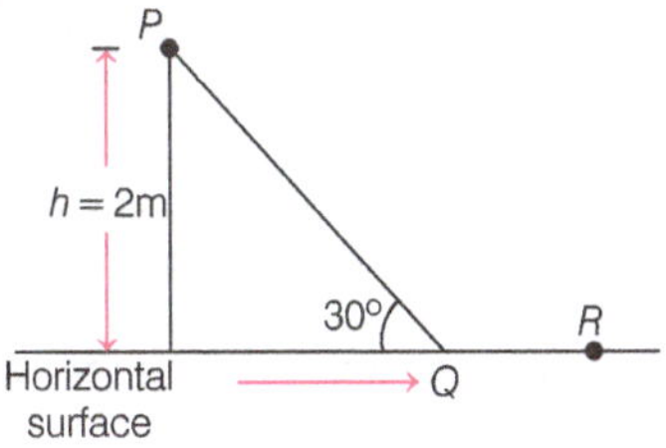

The particle is released from rest, from the point P and it comes to rest at a point R. The energies lost by the ball, over the parts PQ and QR of the track, are equal to each other, and no energy is lost when particle changes direction from PQ to QR. The values of the coefficient of friction μ and the distance x (=QR) are respectively close to **(2016 Main)**

(a) 0.2 and 6.5 m
(b) 0.2 and 3.5 m
(c) 0.29 and 3.5 m
(d) 0.29 and 6.5 m

14. System shown in figure is in equilibrium. Find the magnitude of net change in the string tension between two masses just after, when one of the springs is cut. Mass of both the blocks is same and equal to m and spring constant of both the springs is k

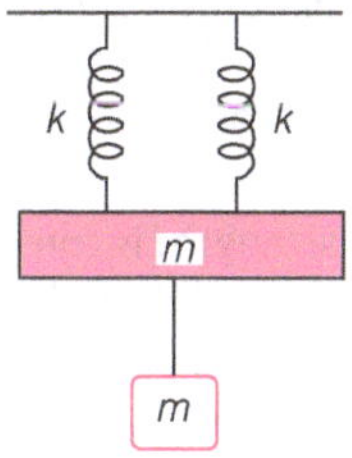

(a) $\frac{mg}{2}$ (b) $\frac{mg}{4}$ (c) $\frac{mg}{3}$ (d) $\frac{3mg}{2}$

15. A block of mass m is connected with the wall of a toy car through a light spring of constant k. The toy car is moving with constant acceleration a_0 (neglect friction). The block is stationary with respect to car. Now, car is abruptly stopped. Then, maximum velocity of the block after stopping the car is

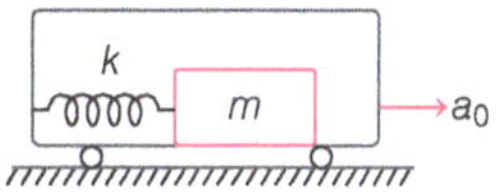

(a) zero (b) $\sqrt{\frac{k}{m}}a_0$ (c) $\sqrt{\frac{m}{k}}a_0$ (d) $\sqrt{\frac{m}{2k}}a_0$

16. In the diagram shown, the blocks A and B are of the same mass M and the mass of the block C is M_1. Friction is present only under the block A. The whole system is suddenly released from the state of rest. The minimum coefficient of friction to keep the block A in the state of rest is equal to

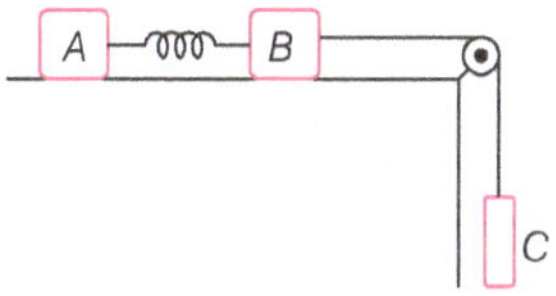

(a) $\frac{M_1}{M}$ (b) $\frac{2M_1}{M}$
(c) $\frac{M_1}{2M}$ (d) None of these

17. A vertical spring is fixed to one of its end and a massless plank fitted to the other end. A block is released from a height h as shown. Spring is in relaxed position. Then, choose the correct statement.

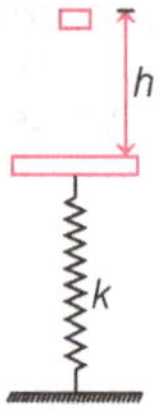

(a) The maximum compression of the spring does not depend on h
(b) The maximum kinetic energy of the block does not depend on h
(c) The compression of the spring at maximum KE of the block does not depend on h
(d) The maximum compression of the spring does not depend on k

18. A spring of natural length l is compressed vertically downward against the floor so that its compressed length becomes $\frac{l}{2}$. On releasing, the spring attains its natural length. If k is the stiffness constant of spring, then the work done by the spring on the floor is

(a) zero (b) $\frac{1}{2}kl^2$ (c) $\frac{1}{2}k\left(\frac{l}{2}\right)^2$ (d) kl^2

19. A block tied between two identical springs is in equilibrium. If upper spring is cut, then the acceleration of the block just after cut is 5 ms^{-2}. Now, if instead of upper spring, lower spring is cut, then the acceleration of the block just after the cut will be (Take, $g = 10\text{ m/s}^2$)

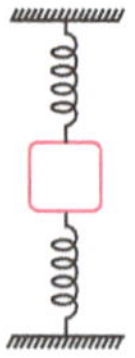

(a) 1.25 ms^{-2} (b) 5 ms^{-2}
(c) 10 ms^{-2} (d) 2.5 ms^{-2}

20. The rate of doing work by a force acting on a particle of mass m moving along x-axis depends on position x of particle and is equal to $2x$. The velocity of particle is given by the expression

(a) $\left[\frac{3x^2}{m}\right]^{1/3}$ (b) $\left[\frac{3x^2}{2m}\right]^{1/3}$
(c) $\left[\frac{2mx}{9}\right]^{1/3}$ (d) $\left[\frac{mx^2}{3}\right]^{1/2}$

21. Two blocks are connected to an ideal spring of stiffness 200 N/m. At a certain moment, the two blocks are moving in opposite directions with speeds 4 ms^{-1} and 6 ms^{-1}, and the instantaneous elongation of the spring is 10 cm. The rate at which the spring energy $\left(\frac{kx^2}{2}\right)$ is increasing is

(a) 500 J/s (b) 400 J/s (c) 200 J/s (d) 100 J/s

22. A pendulum of mass 1 kg and length $l = 1$ m is released from rest at angle $\theta = 60°$. The power delivered by all the forces acting on the bob at angle $\theta = 30°$ will be (Take, $g = 10\text{ m/s}^2$)

(a) 13.4 W (b) 20.4 W
(c) 24.6 W (d) zero

Multiple Correct Options

23. The acceleration of a particle with speed v is $\mathbf{a} = \mathbf{b} \times \mathbf{v}$, where $\mathbf{b}$ is a constant vector which is neither parallel nor anti-parallel to velocity. Then,
(a) the velocity of the particle remains constant
(b) the work done on the particle is zero
(c) the speed of the particle remains constant
(d) the work done on the particle is non-zero

24. In the shown figure, block A is of mass m and block B is of mass $2m$. The spring has force constant k. All the surfaces are smooth and the system is released from rest with spring unstretched.

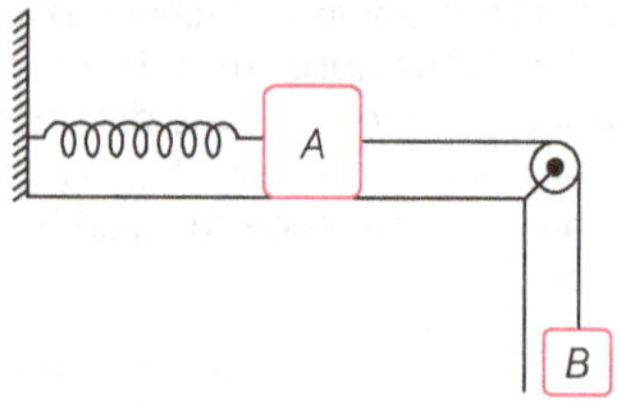

(a) The maximum extension of the spring is $\frac{4mg}{k}$
(b) The speed of block A when extension in spring is $\frac{2mg}{k}$, is $2g\sqrt{\frac{m}{3k}}$
(c) Net acceleration of block B when the extension in the spring is maximum, is $\frac{2}{3}g$
(d) Tension in the thread for extension of $\frac{2mg}{k}$ in spring is mg

25. Two identical blocks each of mass m are connected by a light spring of constant $k = 100$ N/m. Initially, spring is compressed upto 0.2 m and then released. Which of the following statements is/are correct?

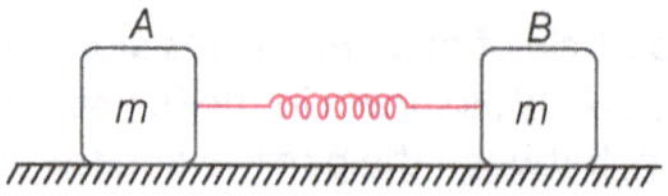

(a) The work done by the spring on the block A when the spring comes in natural length is 1 J
(b) The work done by the spring on the block B when the spring comes in natural length is 1 J
(c) The work done by the spring on the system (block A + block B) is 2 J
(d) The work done by the spring on the system (block A + block B) is -2 J

26. The potential energy of a particle of mass 5 kg moving in xy-plane is given as $U = (7x + 24y)$ joule, x and y being in metre. Initially at $t = 0$, the particle is at the origin (0, 0) moving with a velocity of $(8.6\hat{i} + 23.2\hat{j})\ \text{ms}^{-1}$. Then,
(a) the velocity of the particle at $t = 4$ s, is $5\ \text{ms}^{-1}$
(b) the acceleration of the particle is $5\ \text{ms}^{-2}$
(c) the direction of motion of the particle initially (at $t = 0$) is at right angles to the direction of acceleration
(d) the path of the particle is circle

Passage Based Questions

Passage (Q.27 to Q.28)

The figure shows the variation of potential energy of a particle as a function of x, the x-coordinate of the region. It has been assumed that potential energy depends only on x. For all other values of x, U is zero, i.e. for $x < -10$ and $x > 15$, $U = 0$.

Based on above information answer the following questions:

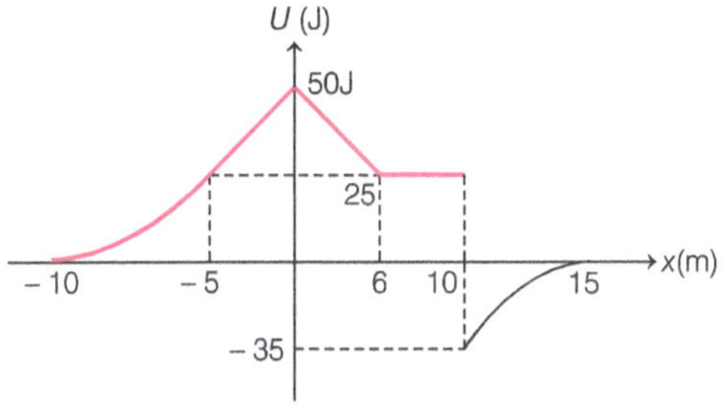

27. If total mechanical energy of the particle is 25 J, then it can be found in the region
(a) $-10 < x < -5$ and $6 < x < 15$
(b) $-10 < x < 0$ and $6 < x < 10$
(c) $-5 < x < 6$
(d) $-10 < x < 10$

28. If total mechanical energy of the particle is -40 J, then it can be found in region
(a) $x < -10$ and $x > 15$
(b) $-10 < x < -5$ and $6 < x < 15$
(c) $10 < x < 15$
(d) It is not possible

Match the Columns

29. Two light springs S_1 and S_2 of spring constants $k_1 = 100$ N/m and $k_2 = 200$ N/m are connected in series. A man is keeping the springs stretched by applying a constant force of 50 N. Match the Column I with Column II.

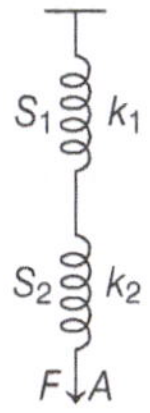

	Column I		Column II
(a)	How much work has been done by the spring S_1 on the spring S_2?	(p)	18.75 J
(b)	How much work has been done by the spring S_2 on the spring S_1?	(q)	– 6.25 J
(c)	How much total work has been done by the spring S_1 on both the spring S_2 and the man?	(r)	12.5 J
(d)	How much work is done by the man on the spring S_2?	(s)	–12.5 J

Subjective Questions

30. A smooth sphere of radius R is made to translate in a straight line with a constant acceleration $a = g$. A particle kept on the top of the sphere is released from there at zero velocity with respect to the sphere. Find the speed of the particle with respect to the sphere as a function of angle θ as it slides down.

31. A 1.5 kg block is initially at rest on a horizontal frictionless surface when a horizontal force in the positive direction of x-axis is applied to the block. The force is given by $\mathbf{F}(x) = (2.5 - x^2)\hat{i}$ N, where, x is in metre and the initial position of the block is $x = 0$.
(a) What is the kinetic energy of the block as it passes through $x = 2.0$ m ?
(b) What is the maximum kinetic energy of the block between $x = 0$ and $x = 2.0$ m ?

32. The system is released from rest with the spring initially stretched 75 mm. Calculate the velocity v of the block after it has dropped 12 mm. The spring has a stiffness of 1050 N/m. Neglect the mass of the small pulley.

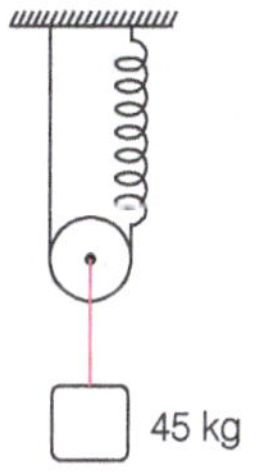

33. Find the speed of both the blocks at the moment of the block m_2 hits the wall AB, after the blocks are released from rest. Given that $m_1 = 0.5$ kg and $m_2 = 2$ kg. (Take, $g = 10\ \text{m/s}^2$) **(JEE 1985)**

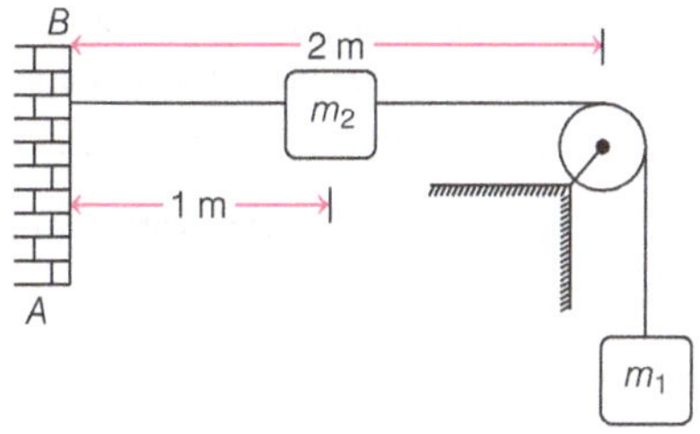

34. In the arrangement shown in figure the mass of body 1 is $\eta = 4.0$ times as great as that of body 2. The height $h = 20$ cm. The masses of the pulleys and the threads, as well as the friction, are negligible. At a certain moment body 2 is released and the arrangement set in motion. What is the maximum height that body 2 will go up to? **(Irodov 1.76)**

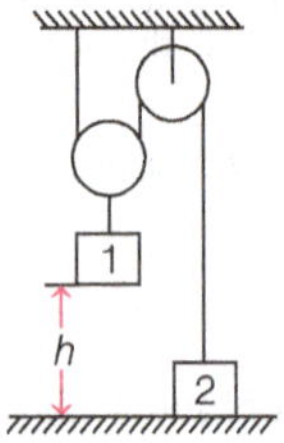

35. Figure shows a 3.5 kg block accelerated by a compressed spring whose spring constant is 640 N/m. After leaving the spring at the spring's relaxed length, the block travels over a horizontal surface, with a coefficient of kinetic friction of 0.25, for a distance of 7.8 m before stopping.
(Take, $g = 9.8 \text{ m/s}^2$)

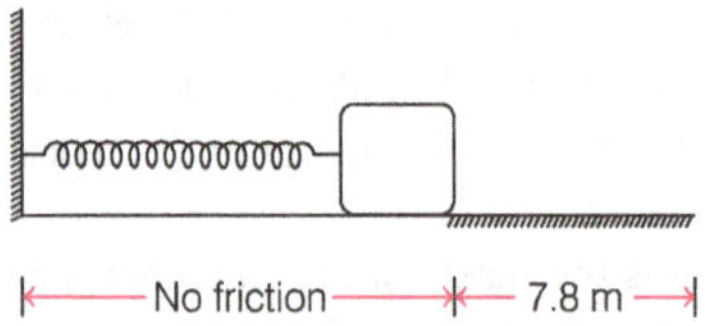

(a) What is the increase in the thermal energy of the block-floor system ?
(b) What is the maximum kinetic energy of the block?
(c) Through what distance is the spring compressed before the block begins to move ?

36. A block of mass M slides along a horizontal table with speed v_0. At $x = 0$, it hits a spring with spring constant k and begins to experience a friction force. The coefficient of friction is variable and is given by $\mu = bx$, where b is a positive constant. Find the loss in mechanical energy when the block has first come momentarily to rest.

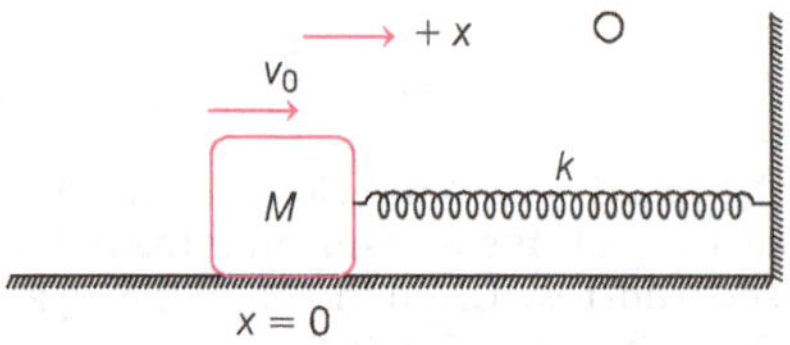

37. A chain of mass $m = 0.80$ kg and length $l = 1.5$ m rests on a rough surfaced table so that one of its ends hangs over the edge. The chain starts sliding off the table provided the overhanging part equals $\eta = \frac{1}{3}$ of the chain length. What will be the total work performed by the friction forces acting on the chain by the moment it slides completely off the table? **(Irodov 1.124)**

38. In the figures (a) and (b) AC, DG and GF are fixed inclined planes, $BC = EF = x$ and $AB = DE = y$. A small block of mass M is released from the point A. It slides down AC and reaches C with a speed v_C. The same block is released from rest from the point D. It slides down DGF and reaches the point F with speed v_F. The coefficients of kinetic frictions between the block and both the surfaces AC and DGF are μ. Calculate v_C and v_F. **(1980 JEE)**

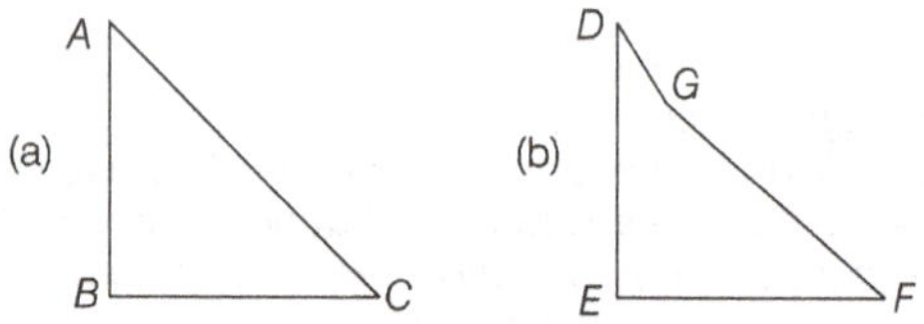

39. In the figure, block A is released from rest when the spring is in its natural length. For the block B of mass m to leave contact with the ground at some stage what should be the minimum mass of block A?

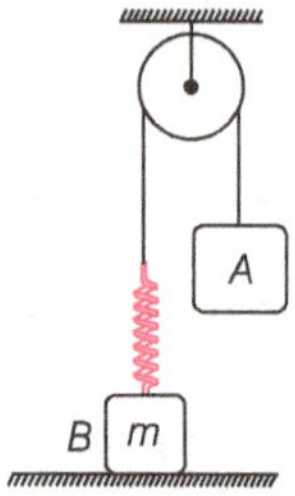

40. A block of mass m is attached with a massless spring of force constant K. The block is placed over a rough inclined surface for which the coefficient of friction is $\mu = \frac{3}{4}$. Find the minimum value of M required to move the block up the plane. (Neglect mass of string and pulley. Ignore friction in pulley).

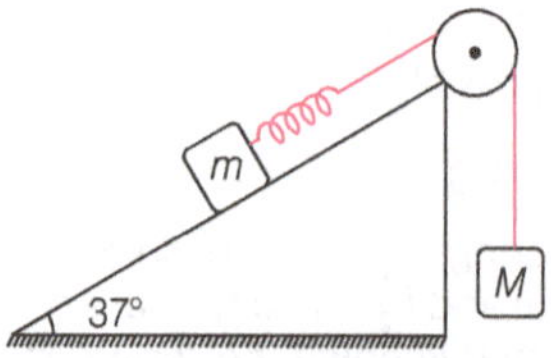

41. In the figure shown masses of the blocks A, B and C are 6kg, 2 kg and 1 kg, respectively. Mass of the spring is negligibly small and its stiffness is 1000 N/m. The coefficient of friction between the block A and the table is $\mu = 0.8$. Initially block C is held such that spring is in relaxed position. The block is released from rest. Find (Take, $g = 10\,m/s^2$)

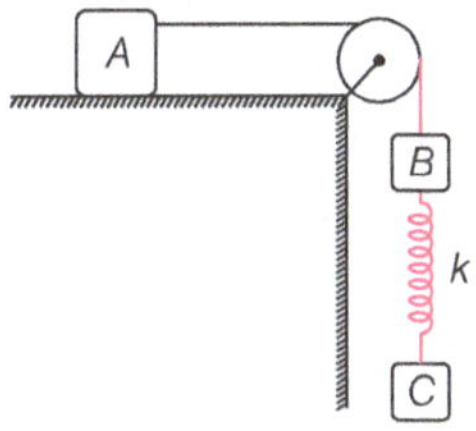

(a) the maximum distance moved by the block C.
(b) the acceleration of each block, when elongation in the spring is maximum.

42. Two bars of masses m_1 and m_2 connected by a non-deformed light spring rest on a horizontal plane. The coefficient of friction between the bars and the surface is equal to μ. What minimum constant force has to be applied in the horizontal direction to the bar of mass m_1 in order to shift the other bar? **(Irodov 1.123)**

43. A sphere of mass m held at a height $2R$ between a wedge of same mass m and a rigid wall, is released from rest. Assuming that all the surfaces are frictionless. Find the speed of both the bodies when the sphere hits the ground.

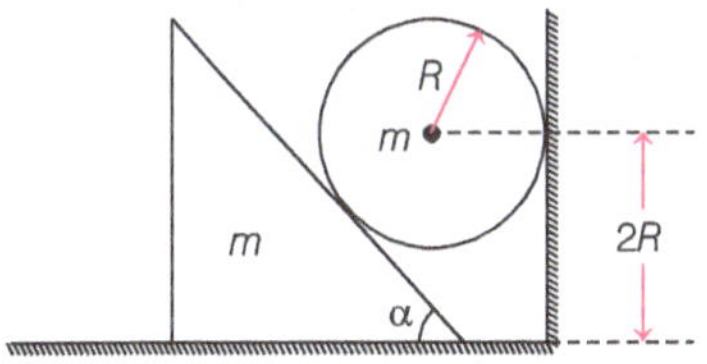

44. A 0.5 kg block is attached to a spring with length 0.6 m and force constant $k = 40.0\,N/m$. The mass of the spring is negligible. You pull the block to the right along the surface with a constant horizontal force $F = 20.0\,N$.
(a) What is the block's speed when the block reaches point B, which is 0.25 m to the right of point A?
(b) When the block reaches point B, you let go off the block. In the subsequent motion, how close does the block get to the wall, where the left end of the spring is attached? Neglect size of block and friction.

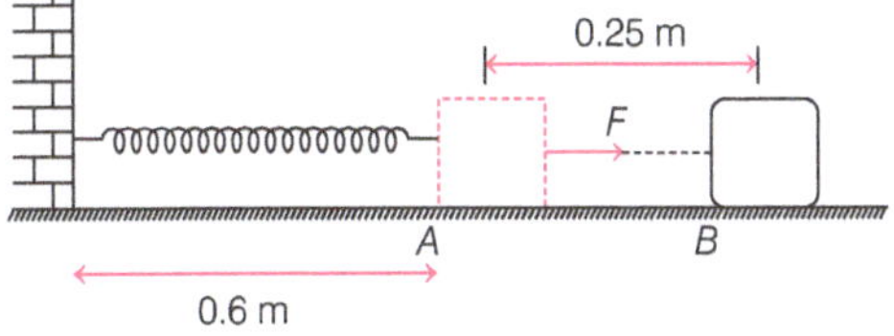

45. A baseball having a mass of 0.4 kg is thrown such that the force acting on it varies with time as shown in the first graph. The corresponding velocity-time graph is shown in the second graph. Determine the power applied as a function of time and the work done till $t = 0.3$ s.

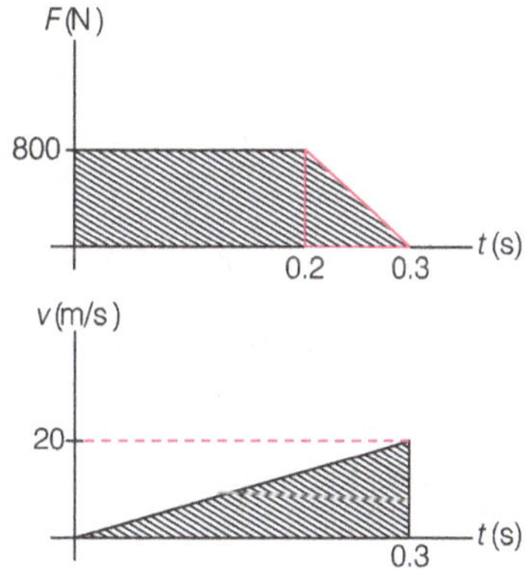

46. A small body of mass m is located on a horizontal plane at the point O. The body acquires a horizontal velocity v_0. Find:
(a) the mean power developed by the friction force during the whole time of motion, if the friction coefficient $\mu = 0.27$, $m = 1.0$ kg and $v_0 = 1.5\,ms^{-1}$
(b) the maximum instantaneous power developed by the friction force, if the friction coefficient varies as $\mu = \alpha x$, where α is a constant, and x is the distance from the point O. **(Irodov 1.127)**

Integer Type Questions

47. Consider an elliptically shaped rail PQ in the vertical plane with $OP = 3$ m and $OQ = 4$ m. A block of mass 1 kg is pulled along the rail from P to Q with a force of 18 N, which is always parallel to line PQ (see figure). Assuming no frictional losses, the kinetic energy of the block when it reaches Q is $(n \times 10)$ J. The value of n is (take acceleration due to gravity $= 10\,ms^{-2}$) **(2014 Adv.)**

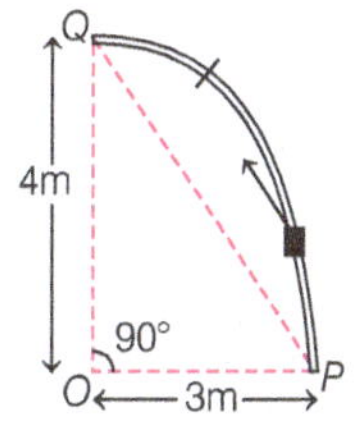

48. Two blocks A and B of equal masses ($m = 10$ kg) are connected by a light spring of spring constant $k = 150$ N/m. The system is in equilibrium. The minimum value of initial downward velocity v_0 of the

block B for which the block A bounce up is $\frac{20}{\sqrt{3n}}$ m/s.

Find the value of n.

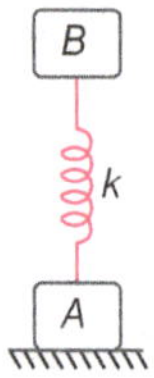

49. A cricket ball of mass 0.15 kg is thrown vertically up by a bowling machine, so that it rises to a maximum height of 20 m after leaving the machine. If the part pushing the ball applies a constant force F on the ball and moves horizontally a distance of 0.2 m, while launching the ball, the value of F (in N) is (Take, $g = 10\ \text{ms}^{-2}$) .... **(2015 JEE)**

Level 2

Single Correct Option

50. A force $\left(\mathbf{F} = \frac{-ky}{\sqrt{x^2+y^2}}\hat{\mathbf{i}} + \frac{kx}{\sqrt{x^2+y^2}}\hat{\mathbf{j}}\right)$ N is applied on a block, whose position is $P(x, y)$. The block moves on a circular path of radius R. Find the work done by the force $\mathbf{F}$ in a complete rotation.
(a) $2\pi Rk$ (b) πRk
(c) $4\pi Rk$ (d) Zero

51. The work done on a particle of mass m by a force, $K\left[\frac{x}{(x^2+y^2)^{3/2}}\hat{\mathbf{i}} + \frac{y}{(x^2+y^2)^{3/2}}\hat{\mathbf{j}}\right]$ (K being a constant of appropriate dimensions), when the particle is taken from the point $(a, 0)$ to the point $(0, a)$ along a circular path of radius a about the origin in the x-y plane is **(2013 Adv.)**
(a) $\frac{2K\pi}{a}$ (b) $\frac{K\pi}{a}$
(c) $\frac{K\pi}{2a}$ (d) 0

52. The potential energy ϕ in joule of a particle of mass 1 kg moving in x-y plane obeys the law, $\phi = 3x + 4y$. Here, x and y are in metres. If the particle is at rest at (6m, 8m) at time 0, then the work done by conservative force on the particle from the initial position to the instant when it crosses the x-axis is
(a) 25 J (b) −25 J
(c) 50 J (d) −50 J

Multiple Correct Options

53. A particle of mass m is initially at rest at the origin. It is subjected to a force and starts moving along the x-axis. Its kinetic energy K changes with time as $\frac{dK}{dt} = \gamma t$, where γ is a positive constant of appropriate dimensions. Which of the following statements is (are) true? **(2018 Adv.)**
(a) The force applied on the particle is constant
(b) The speed of the particle is proportional to time
(c) The distance of the particle from the origin increases linearly with time
(d) The force is conservative

54. The force on a particle of mass m placed on a smooth horizontal surface is $F = (2x^2 - 4)$ N directed along positive x-axis. The particle is projected with minimum speed from $x = 2$ m towards mean position along x-axis for crossing the origin, then the kinetic energy at
(a) the origin must be zero (b) the origin is non-zero
(c) $x = \sqrt{2}$ must be zero (d) $x = 2$ m may be zero

55. In previous problem, the particle is projected with minimum speed from $x = 2$ m towards mean position along x-axis for crossing the origin, then
(a) the minimum initial kinetic energy is $\frac{8}{3}(\sqrt{2}-1)$
(b) the minimum kinetic energy at origin is just greater than $\frac{8\sqrt{2}}{3}$ J
(c) the minimum kinetic energy at $x = \sqrt{2}$ m is just greater than or equal to zero
(d) the minimum kinetic energy at origin is 4J

56. A disc of mass $3m$ and a disc of mass m are connected by a massless spring of stiffness k. The heavier disc is placed on the ground with the spring vertical and lighter disc on top. From its equilibrium position, the upper disc is pushed down by a distance δ and released. Then,
(a) if $\delta > \frac{3\,mg}{k}$, the lower disc will bounce up
(b) if $\delta = \frac{2\,mg}{k}$, maximum normal reaction from ground on lower disc $= 6\,mg$
(c) if $\delta = \frac{2\,mg}{k}$, maximum normal reaction from ground on lower disc $= 4\,mg$
(d) if $\delta > \frac{4\,mg}{k}$, the lower disc will bounce up

Match the Columns

57. A particle of unit mass is moving along the x-axis under the influence of a force and its total energy is

conserved. Four possible forms of the potential energy of the particle are given in Column I (a and U_0 are constants). Match the potential energies in Column I to the corresponding statements in Column II **(2015 Adv.)**

	Column I		Column II
A.	$U_1(x) = \frac{U_0}{2}\left[1-\left(\frac{x}{a}\right)^2\right]^2$	p.	The force acting on the particle is zero at $x = a$
B.	$U_2(x) = \frac{U_0}{2}\left(\frac{x}{a}\right)^2$	q.	The force acting on the particle is zero at $x = 0$
C.	$U_3(x) = \frac{U_0}{2}\left(\frac{x}{a}\right)^2 \exp\left[-\left(\frac{x}{a}\right)^2\right]$	r.	The force acting on the particle is zero at $x = -a$
D.	$U_4(x) = \frac{U_0}{2}\left[\frac{x}{a} - \frac{1}{3}\left(\frac{x}{a}\right)^3\right]$	s.	The particle experiences an attractive force towards $x = 0$ in the region $\lvert x\rvert < a$
		t.	The particle with total energy $\frac{U_0}{4}$ can oscillate about the point $x = -a$.

Subjective Questions

58. In the situation shown in figure, two blocks A and B, each having mass $m = \sqrt{2}$ kg are connected by a combination of light string and a light spring of constant k. The planes on which blocks A and B are placed are rough having coefficient of friction 0.5. Find the minimum value of constant force F required to move the block B up the inclined plane (Take, $g = 10 \text{m/s}^2$).

Assume that initially spring is relaxed.

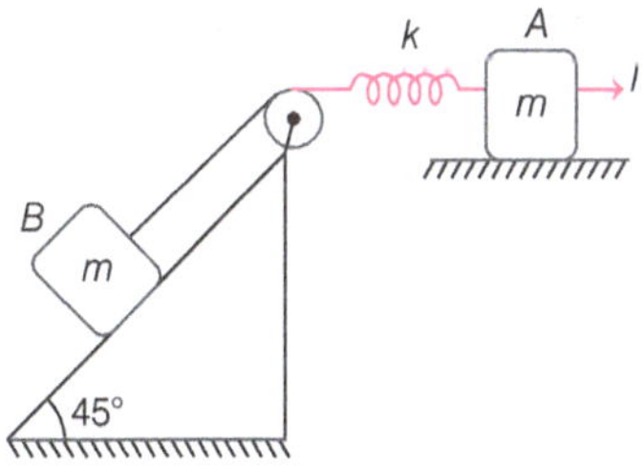

59. The block shown in the figure is acted upon by a spring with spring constant k and a weak frictional force of constant magnitude f. The block is pulled a distance x_0 from equilibrium position and then released. It oscillates many times and ultimately comes to rest.

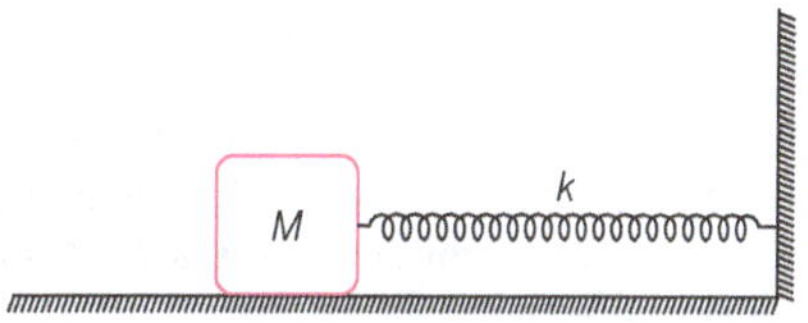

(a) Show that the decrease of amplitude is the same for each cycle of oscillation.

(b) Find the number of cycles the mass oscillates before coming to rest.

60. A body of mass m is pulled upwards from the Earth's surface by applying a force **F** varying with the height of ascent y as $\mathbf{F} = 2(ay - 1)\, m\mathbf{g}$, where a is a positive constant. Find the work performed by this force and the increment of the body's potential energy in the gravitational field of the Earth over the first half of the ascent. **(Irodov 1.130)**

61. A small bar A resting on a smooth horizontal plane is attached by threads to a point P and by means of a weightless pulley, to a weight B possessing the same mass as the bar itself.

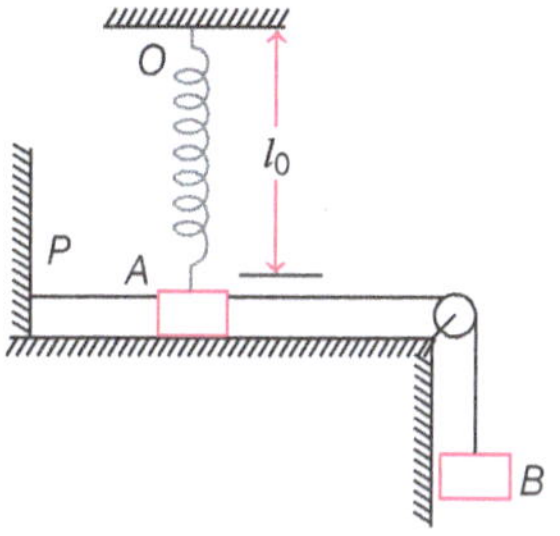

Besides, the bar is also attached to a point O by means of light non-deformed spring of length $l_0 = 50$ cm and stiffness $k = 5\, mg/l_0$, where m is the mass of the bar. The thread PA having been burned, the bar starts moving. Find its velocity at the moment when it is breaking off the plane. **(Irodov 1.140)**

62. A horizontal plane supports a plank with a bar of mass $m = 1.0$ kg placed on it and attached by a light elastic non-deformed cord of length $l_0 = 40$ cm to a point O. The coefficient of friction between the bar and the plank is equal to $\mu = 0.20$. The plank is slowly shifted to the right until the bar starts sliding over it. It occurs at the moment when the cord deviates from the vertical by an angle $\theta = 30°$. Find the work that has been performed by that moment by the friction force acting on the bar in the reference frame fixed to the plane. **(Irodov 1.141)**

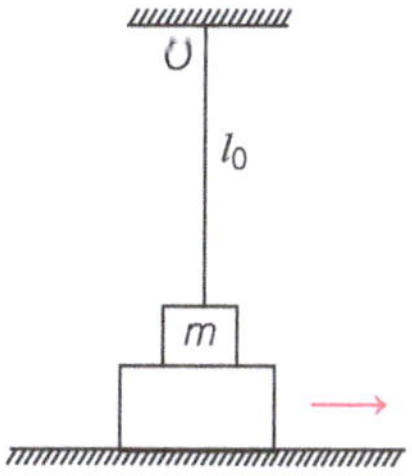

63. The potential energy of a particle in a certain field has the form $U = \frac{a}{r^2} - \frac{b}{r}$, where a and b are positive constants, r is the distance from the centre of the field. Find :
(a) the value of r_0 corresponding to the equilibrium position of the particle; examine whether this position is steady
(b) the maximum magnitude of the attraction force; draw the plots $U(r)$ and $F_r(r)$ (the projection of the force on the radius vector r). **(Irodov 1.131)**

64. A particle moving in a straight line is acted upon by a force which works at a constant rate and changes its velocity from u to v over a distance x. Prove that the time taken in it is

$$\frac{3}{2}\frac{(u+v)\,x}{u^2+v^2+uv}$$

Answers

Level 1

1. (d) **2.** (d) **3.** (b) **4.** (d) **5.** (b) **6.** (d) **7.** (a) **8.** (b) **9.** (d) **10.** (d)
11. (a) **12.** (d) **13.** (c) **14.** (a) **15.** (c) **16.** (b) **17.** (c) **18.** (a) **19.** (b) **20.** (a)
21. (c) **22.** (a) **23.** (b,c) **24.** (a,b) **25.** (a,b,c) **26.** (a.b) **27.** (a) **28.** (d) **29.** (a-s; b-r; c-q; d-p)
30. $v_r = \sqrt{2gR\,(1+\sin\theta-\cos\theta)}$ **31.** (a) 2.33 J (b) 2.635 J **32.** 0.37 m/s **33.** $v_1 = 3.03$ m/s, $v_2 = 3.39$ m/s
34. $h_{max} = 0.6$ m **35.** (a) 66.88 J (b) 66.88 J (c) 45.7 cm **36.** $\Delta E = v_0^2 bM^2g/2(k+bMg)$ **37.** – 1.3 J
38. $v_C = v_F = \sqrt{2gy - 2\mu gx}$ **39.** $m_A = m/2$ **40.** $M = \frac{3}{5}$ m **41.** 0.02m, 10 m/s^2
42. $F = \left(m_1 + \frac{m_2}{2}\right)\mu g$ **43.** $v_{wedge} = \sqrt{2gR}\cos\alpha$, $v_{sphere} = \sqrt{2gR}\sin\alpha$ **44.** (a) 3.87 m/s (b) 0.10 m
45. $(160t - 533t^2)$ kW, 1.69 kJ **46.** (a) $P_{av} \approx -2$ W (b) $P_{max} = -\frac{1}{2}mv_0^2\sqrt{\alpha g}$ **47.** (5) **48.** (5) **49.** (150)

Level 2

50. (a) **51.** (d) **52.** (c) **53.** (a,b,d) **54.** (b,c) **55.** (a,b,c) **56.** (b,d)
57. (A-p,q,r,t; B-q,s; C-p,q,r,s; D-p,r,t) **58.** 14.57 N **59.** (a) $\frac{2f}{k}$ (b) $\frac{1}{4}\left(\frac{kx_0}{f}-1\right)$ **60.** $W = \frac{3mg}{4a}$, $\Delta U = \frac{mg}{2a}$
61. 1.7 m/s **62.** 0.09 J **63.** (a) $r_0 = \frac{2a}{b}$, steady (b) $F_{max} = \frac{b^3}{27a^2}$

See hints on pages 528-539.

Hints & SOLUTIONS

Module 1.

Introductory Exercise

1. $W = \mathbf{F} \cdot \mathbf{s} = \mathbf{F} \cdot (\mathbf{r}_f - \mathbf{r}_i)$

$= (6\hat{\mathbf{i}} - 2\hat{\mathbf{j}} + \hat{\mathbf{k}}) \cdot [(2\hat{\mathbf{i}} + 3\hat{\mathbf{j}} - 4\hat{\mathbf{k}}) - (\hat{\mathbf{i}} + 4\hat{\mathbf{j}} + 6\hat{\mathbf{k}})]$

$= -2\text{ J}$

2. (a) $W_F = Fs\cos 45°$

$= (16)(2.2)\left(\dfrac{1}{\sqrt{2}}\right) = 24.89\text{J}$ **Ans.**

(b) $W_N = Ns\cos 90° = 0$ **Ans.**

(c) $W_{mg} = (mg)(s)\cos 90° = 0$ **Ans.**

(d) Only three forces are acting. So, total work done is summation of all above work done.

3. $W_T = (T)(x)\cos 0° = Tx$

$W_W = (W)(x)\cos 90° = 0$

$W_N = (N)(x)\cos 90° = 0$

$W_F = (F)(x)\cos 180° = -Fx$

4. $mg - T = ma = \dfrac{mg}{4}$

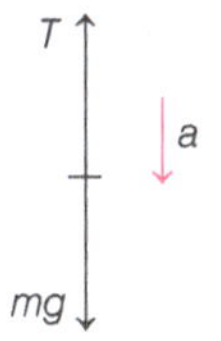

$\Rightarrow \quad T = \dfrac{3mg}{4}$

$\therefore \quad W_T = (T)(l)(\cos 180°)$

$= -\dfrac{3}{4}mgl$

5. $N = mg - F\sin 45° = 18 - \dfrac{F}{\sqrt{2}}$

Moving with constant speed means net force $= 0$

$F\cos 45° = \mu N = \dfrac{1}{4}\left(18 - \dfrac{F}{\sqrt{2}}\right)$

$\therefore \quad \dfrac{4F}{\sqrt{2}} = 18 - \dfrac{F}{\sqrt{2}}$

$\therefore \quad F = \dfrac{18\sqrt{2}}{5}\text{N}$

(a) $W_F = Fs\cos 45°$

$= \left(\dfrac{18\sqrt{2}}{5}\right)(2)\left(\dfrac{1}{\sqrt{2}}\right) = 7.2\text{ J}$

(b) $W_f = (\mu N)(s)\cos 180°$

$= \left(\dfrac{1}{4}\right)\left(18 - \dfrac{F}{\sqrt{2}}\right)(2)(-1)$

$= -7.2\text{ J}$

(c) $W_{mg} = (mg)(s)\cos 90° = 0$

6. $W = \int_2^{-4} F dx = \int_2^{-4} (-2x)\, dx$

$= [-x^2]_2^{-4} = -[16 - 4] = -12\text{ J}$ **Ans.**

7. $W = \int_4^2 F dx = \int_4^2 \dfrac{4}{x^2} dx$

$= -4\left[\dfrac{1}{x}\right]_4^2 = -4\left[\dfrac{1}{2} - \dfrac{1}{4}\right] = -1\text{ H}$ **Ans.**

8. $W =$ area under F-x graph

From $X = -4$ to $X = -2$

$F = -\text{ve}$ (from graph)

$s = +\text{ve}$

$\therefore \quad W_1 = -\dfrac{1}{2} \times 2 \times 10 = -10\text{ J}$

From -2 to 4

$F = +\text{ve}$

$s = +\text{ve}$

$\therefore \quad W_2 = +\dfrac{1}{2}(6 + 2)(10)$

$= +40\text{ J}$

$\therefore \quad W_T = W_1 + W_2 = 30\text{ J}$ **Ans.**

9. (a) From $x = 10\text{m}$ to $x = 5\text{m}$

$s = -\text{ve}$

$F = +\text{ve}$ (from graph)

$\therefore \quad W_1 = -\text{Area}$

$= -5 \times 3$

$= -15\text{ J}$ **Ans.**

(b) From $x = 5\text{ m}$ to $x = 10\text{m}$

$s = +\text{ve}$

and $F = +\text{ve}$

$\therefore \quad W_2 = +\text{Area} = 5 \times 3$

$= 15\text{ J}$ **Ans.**

(c) From $x = 10\text{m}$ to $x = 15\text{m}$

$s = +\text{ve}$

$F = +\text{ve}$

$\therefore \quad W_3 = \text{Area}$

$= 3\text{J}$ **Ans.**

(d) From $x = 0$ to $x = 15\text{m}$

$s = +\text{ve}$ and $F = +\text{ve}$

$\therefore \quad W_4 = +\text{Area}$

$= \dfrac{1}{2} \times 3 \times (12 + 6) = 27\text{ J}$ **Ans.**

10. (a) From $x = 0$ to $x = 3.0$ m

$$s = +\text{ve}$$
$$F = +\text{ve}$$
$$\therefore \quad W_1 = +\text{Area} = +4\text{ J}$$

(b) From $x = 3$ m to $x = 4$ m

$$F = 0 \Rightarrow W_2 = 0$$

(c) From $x = 4$m to $x = 7$m

$$s = +\text{ve and } F = -\text{ve}$$
$$\therefore \quad W_3 = +\text{Area} = -1\text{ J}$$

(d) From $x = 0$ to $x = 7$ m

$$W = W_1 + W_2 + W_3 = 3\text{ J}$$

Exercise

1.

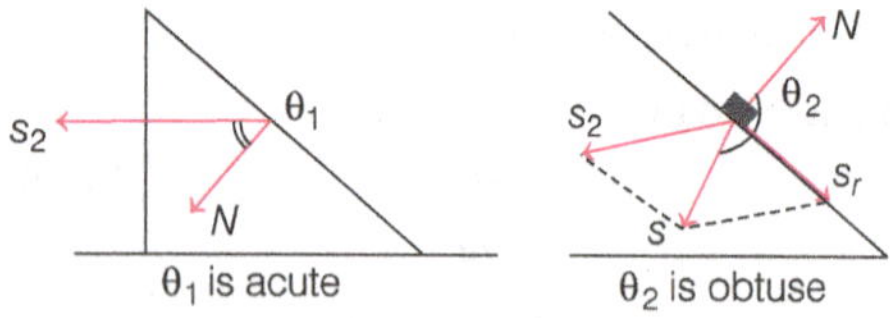

2. $W = 0$. That feet remains stationary, which is in contact with the ladder.

3. $N = mg - F\sin\theta$

Block moves with uniform velocity. Hence, net force = 0

or $\quad F\cos\theta = \mu N = \mu(mg - F\sin\theta)$

$$\therefore \quad F = \frac{\mu mg}{\cos\theta + \mu\sin\theta}$$

$$W = Fd\cos\theta = \frac{\mu\, mgd\cos\theta}{\cos\theta + \mu\sin\theta}$$

4. Work done by tension on M is negative (force and displacement are in opposite directions). But, work done by tension on m is positive. Net work done will be zero.

5. Constant velocity means net force = 0

Using Lami's theorem in the figure,

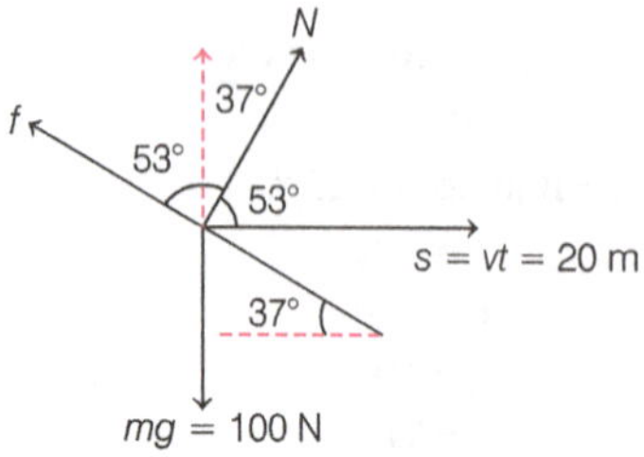

We have, $\quad \dfrac{N}{\sin(180° - 53°)} = \dfrac{100}{\sin 90°}$

$$\therefore \quad N = 100\sin 53° = 80\text{ N}$$

Now, $\quad W_N = Ns\cos 53°$

$$= (80)(20)(0.6)$$

$$= 960\text{ J} \qquad \textbf{Ans.}$$

6. We know that, change in potential energy of a system corresponding to a conservative internal force as

$$U_f - U_i = -W = -\int_i^f \mathbf{F}\cdot d\mathbf{r}$$

Given, $\quad F = ax + bx^2$

We know that, work done in stretching the rubber band by L is $|dW| = |Fdx|$

$$|W| = \int_0^L (ax + bx^2)\,dx$$

$$= \left[\frac{ax^2}{2}\right]_0^L + \left[\frac{bx^3}{3}\right]_0^L$$

$$= \left[\frac{aL^2}{2} - \frac{a\times(0)^2}{2}\right] + \left[\frac{b\times L^3}{3} - \frac{b\times(0)^3}{3}\right]$$

$$= |W| = \frac{aL^2}{2} + \frac{bL^3}{3}$$

7. $F = kx$ or k = slope of F-x graph $\quad$ (F along y-axis)

Here, F is along x-axis.

So, $\quad k = \dfrac{1.0}{10} = 0.1\,\dfrac{\text{kgf}}{\text{cm}}$

8. Normal reaction force on the block is

$$N - mg = ma = \frac{mg}{2}$$

$$\therefore \quad N = m\left(g + \frac{g}{2}\right) = \frac{3mg}{2}$$

Now, in time 't' displacement of block,

$$s = \frac{1}{2}at^2 = \frac{1}{2}\left(\frac{g}{2}\right)t^2 = \frac{gt^2}{4}$$

$\therefore$ Work done = Force × Displacement

$$\Rightarrow \quad W = \frac{3mg}{2}\times\frac{gt^2}{4}$$

$$= \frac{3mg^2t^2}{8}$$

9. Work done by a force is positive if displacement is in direction of force and work done by a force is negative if displacement is in direction opposite to that of force.

10. Work done, $W = \int \mathbf{F}\cdot d\mathbf{s} = (-x\hat{\mathbf{i}} + y\hat{\mathbf{j}})\cdot(dx\hat{\mathbf{i}} + dy\hat{\mathbf{j}})$

$$\Rightarrow W = -\int_1^0 xdx + \int_0^1 ydy = \left(0 + \frac{1}{2}\right) + \frac{1}{2} = 1\text{ J}$$

11. $K = \dfrac{F}{x}$ = slope of F - x graph.

$$K \propto \frac{1}{l}$$

Length is reduced to half. Therefore, K will become two times. Slope will increase.

12. (a) Work done by gravity in motion 1 is zero ($\theta = 90°$) and in motion 2 is negative ($\theta = 180°$).

(b) In both cases, angle between N and S is acute.

(c) and (d) : Depending on the value of acceleration in motion 1, friction may act up the plane or down the plane. Therefore, angle between friction and displacement may be obtuse or acute. So, work done by friction may be negative or positive.

13. Here, $s_A = 2s_B$

$v_A = 2v_B$

and $a_A = 2a_B$

FBD

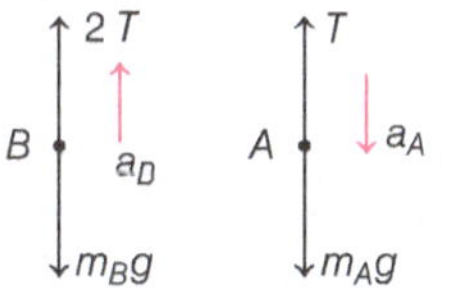

Also, $100 - T = 10a_A = 20a_B$ ...(i)

and $2T - 50 = 5a_B$ ...(ii)

From Eqs. (i) and (ii), we get

$$200 - 2T = 40a_B$$
$$-50 + 2T = 5a_B$$

$\Rightarrow$ $150 = 45\,a_B$

$\therefore$ $a_B = \dfrac{150}{45} = \dfrac{10}{3}$ m/s^2

$\therefore$ $a_A = \dfrac{20}{3}$ m/s^2

$\because$ $2T - 50 = 5a_B = \dfrac{50}{3}$

or $2T = \dfrac{200}{3}$

or $T = \dfrac{100}{3}$

$W_B = 2T \times s_B = \dfrac{200}{3} \times 2.5$

$= \dfrac{500}{3}$ J $= 166.67$ J

14. $W_A = -Ts_A = -\dfrac{100}{3} \times 5$

$= -\dfrac{500}{3}$ J

$= -166.67$ J

$W_T = W_A + W_B = 0$

15. $W_g = 100 \times 5 - 50 \times 2.5 = 375$ J

$W = W_T + W_g = 0 + 375 = 375$ J

16. $\mathbf{s} = \mathbf{r}_f - \mathbf{r}_i = (+2\hat{\mathbf{i}}) - (+4\hat{\mathbf{i}}) = -2\hat{\mathbf{i}}$

Now, apply $W = \mathbf{F} \cdot \mathbf{s}$

17. (a)

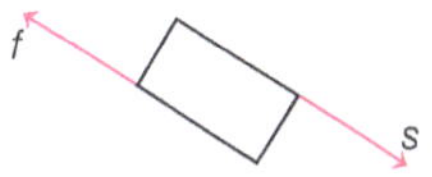

$W_f = fs \cos 180° =$ negative

(b)

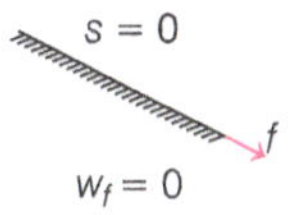

(c) and (d) : No solution is required.

18. $W_T = (T)(l) \cos \beta$

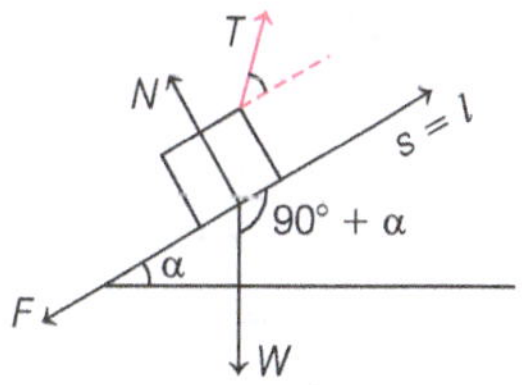

$W_N = (N)(l) \cos 90° = 0$

$W_w = (W)(l) \cos (90 + \alpha) = -Wl \sin \alpha$

$W_f = (F)(l) \cos 180° = -Fl$

19. $W_f = fs \cos 45°$

Here, $f = mg \sin \theta$, because uniform velocity means net acceleration $= 0$ or net force $= 0$

$\therefore$ $W_f = (mg \sin \theta)(s)(\cos 45°)$

$= (1)(10)(\sin 45°)(s)(\cos 45°)$

If $s = vt = 2 \times 1 = 2$ m

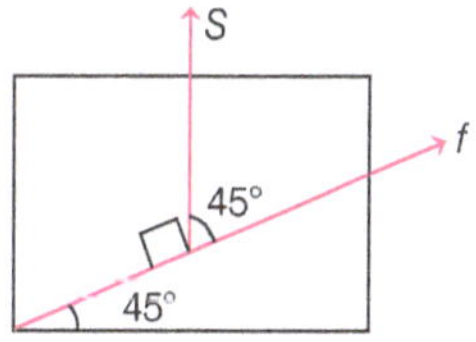

$W_f = (1)(10)\left(\dfrac{1}{\sqrt{2}}\right)(2)\left(\dfrac{1}{\sqrt{2}}\right) = 10$ J **Ans.**

20. (a) $W_{mg} = (mg)(s) \cos 30°$

$= (20)(2)(\sqrt{3}/2)$

$= 34.6$ J **Ans.**

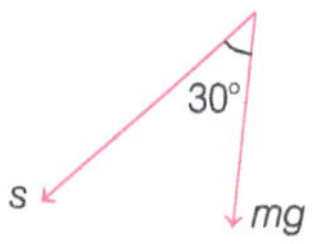

(b) $W_f = fs \cos 180°$

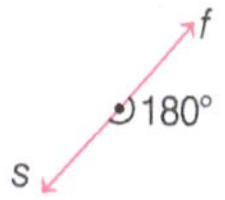

$= (\mu mg \cos\theta)(s)(-1)$

$= -\left(\frac{1}{2}\right)(20)(\cos 60^\circ)(2)$

$= -10$ J **Ans.**

21. Work done by a constant force,

$$W = \mathbf{F} \cdot \mathbf{s}$$

where, $\mathbf{F} = (3\hat{\mathbf{i}} + 4\hat{\mathbf{j}})$ N

and $\mathbf{s} = \mathbf{r}_f - \mathbf{r}_i = \mathbf{r}_2 - \mathbf{r}_1$

$= (\hat{\mathbf{i}} - 5\hat{\mathbf{j}})$ m

$\therefore$ $W = \mathbf{F} \cdot \mathbf{s}$

$= (3 - 20)$ J

or $W = -17$ J

22. $a = \frac{\text{Net pulling force}}{\text{Total mass}} = \frac{0.72g - 0.36g}{0.72 + 0.36} = \frac{g}{3}$

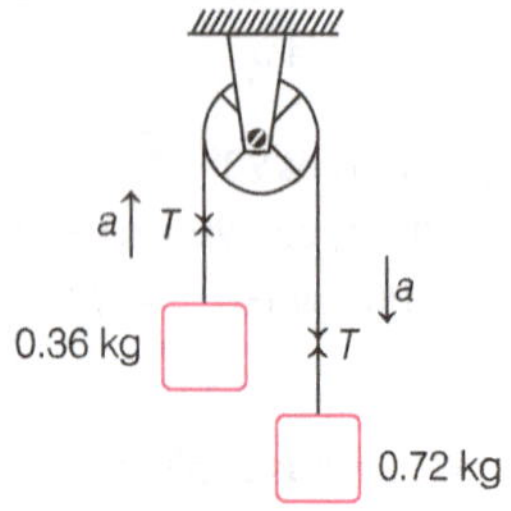

$$s = \frac{1}{2}at^2 = \frac{1}{2}\left(\frac{g}{3}\right)(1)^2 = \frac{g}{6}$$

Equation of motion of 0.36 kg mass is,

$$T - 0.36g = 0.36a = (0.36)\frac{g}{3}$$

$\therefore$ $T = 0.48g$

Now, $W_T = Ts \cos 0^\circ$ (on 0.36 kg mass)

$= (0.48g)\left(\frac{g}{6}\right)(1)$

$= 0.08(g^2) = (0.08)(10^2)$

$= 8$ J

23. We know that,

$$W = \int F dy = \int_2^5 (5 + 3y^2)\, dy$$

$= [5y + y^3]_2^5 = [(25 + 125) - (10 + 8)]$

$= 150 - 18 = 132$ J

24. Work done by A = Work done by B

$$F_A\, d \cos 45^\circ = F_B\, d \cos 60^\circ$$

$\Rightarrow$ $F_A \times \frac{1}{\sqrt{2}} = F_B \times \frac{1}{2}$

$\Rightarrow$ $\frac{F_A}{F_B} = \frac{\sqrt{2}}{2} = \frac{1}{\sqrt{2}}$

$\Rightarrow$ $x = 2$

Module 2.

Introductory Exercise

1. $\Delta U = -W$

So, if work done by conservative force is positive then ΔU is negative or potential energy will decrease. But there is no straight forward rule regarding the kinetic energy.

2. $U_A = -60$ J

$U_B = -20$ J

$\therefore$ $U_B - U_A = 40$ J

Exercise

1. Work done in moving an object against gravitational force (conservative force) depends only on the initial and final position of the object, not upon the path taken. But gravitational force on the body along the inclined plane is not same as that along the vertical and it varies with the angle of inclination.

2.

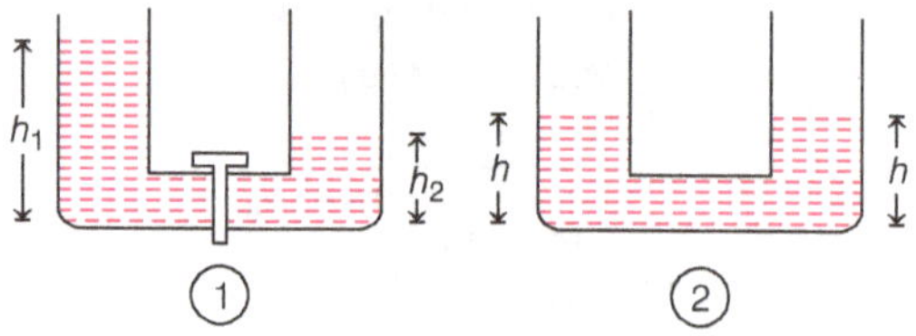

Since, the volume remains same.

$$Ah_1 + Ah_2 = Ah + Ah$$

$\Rightarrow$ $h = \frac{h_1 + h_2}{2}$

$$U_1 = \frac{[(\rho Ah_1)gh_1 + (\rho Ah_2)gh_2]}{2}$$

$$U_2 = \frac{(\rho Ah)gh}{2} \times 2$$

$$W_{gr} = -(U_2 - U_1) = U_1 - U_2$$

$$= \frac{\rho Ag}{2}[(h_1^2 + h_2^2) - 2h^2]$$

$$= \frac{\rho Ag}{2}\left[(h_1^2 + h_2^2) - 2\left(\frac{h_1 + h_2}{2}\right)^2\right]$$

$$= \rho Ag\left[\frac{h_1 - h_2}{2}\right]^2$$

3. Ratio of displacement is

$$\frac{1}{2}g(1)^2 : \left(\frac{1}{2}(g)(2)^2 - \frac{1}{2}g(1)^2\right) : \left(\frac{1}{2}g(3)^2 - \frac{1}{2}g(2)^2\right)$$

or 1 : 3 : 5.

Therefore, ratio of work done will also be 1 : 3 : 5.

As $W = mgs$ or $W \propto s$

4. Work done by conservative force $= -\Delta U$

5. Work done is equal to change in potential energy. Potential energy is a state function. So change in potential energy from A to B will be same.

Hence, $W_1 = W_2 = W_3$

6. Spring force is always towards mean position. If displacement is also towards mean position, F and s will be of same sign and work done will be positive.

7. Total work done $= -\frac{1}{2} k (2x_0)^2 = -2kx_0^2$

Work done on one mass

$$= \frac{-2kx_0^2}{2} = -kx_0^2$$

Module 3.

1. Upthrust = (Volume immersed) (density of liquid) g

$$= \left(\frac{5}{3000}\right)(1000)(10)$$

$$= \frac{50}{3} \text{N}$$

Weight = 50 N

$\therefore$ Applied force (upwards) = weight − upthrust

$$= 50 - \frac{50}{3} = \frac{100}{3} \text{N}$$

$$W = Fs = \frac{100}{3} \times 3$$

$$= 100 \text{ J} \quad \textbf{Ans.}$$

2. $\mathbf{v} = \frac{ds}{dt} = (4t)\hat{\mathbf{i}}$

$$P = \mathbf{F} \cdot \mathbf{v} = 12t^2$$

$$\therefore \quad W = \int_0^2 P dt = \int_0^2 (12) t^2 dt$$

$$= 32 \text{ J} \quad \textbf{Ans.}$$

3. (a) $W_N = Ns \cos 90° = 0$

(b) $W_T = Ts \cos 0° = (mg \sin \theta)\left(\frac{h}{\sin \theta}\right) = mgh$

(c) $W_{mg} = (mg)(h) \cos 180° = -mgh$

(d) $W_{\text{Total}} = \Delta K = K_f - K_i = 0$

As block is moved slowly or $K_f = K_i$

4. (a) $K_f - K_i = W = \int F dx$

$$\therefore \quad K_f = K_i + \int_{20}^{30} (-0.1x)\, dx$$

$$= \frac{1}{2} \times 10 \times (10)^2 - \left[0.1 \frac{x^2}{2}\right]_{20}^{30}$$

$$= 475 \text{ J} \quad \textbf{Ans.}$$

5. From work-energy theorem,

W = change in kinetic energy

or $W = \frac{1}{2} mv^2$

$\therefore$ W - v graph is a parabola.

6. $K = \frac{1}{2} mv^2$ ...(i)

$$v^2 = 2as \quad \text{...(ii)}$$

From Eqs. (i) and (ii), we get

$$K = mas$$

or $K \propto s$

7. Average velocity $= \frac{\text{Total displacement}}{\text{Time}}$

$$= \frac{\text{Area under } v\text{-}t \text{ graph}}{\text{Time}}$$

Since, area $\neq 0$

$\therefore$ Average velocity $\neq 0$

8. $k = 4t^2$ or $v^2 \propto t^2$

$\therefore$ $v \propto t$

v varies linearly with time when acceleration or force is constant.

9. $\frac{1}{2} m(100)^2 = R(2t)$

Here, R is the retarding force and t, thickness of one plank.

$$\frac{1}{2} m(200)^2 = R(nt)$$

$$n = 8$$

10. $W_{\text{All}} = \frac{1}{2} mv^2$

$$\therefore \quad W_F + W_{mg} + W_N = \frac{1}{2} mv^2$$

$$\therefore \quad (5 \times 5) + \left(\frac{1}{2} \times 10 \times 5\right) + 0 = \frac{1}{2} \times \frac{1}{2} \times v^2$$

$$\therefore \quad v = 14.14 \text{ m/s} \quad \textbf{Ans.}$$

11. $W = Fs$

F and s are same. Therefore,

$$\frac{W_A}{W_B} = \frac{1}{1}$$

From work-energy theorem,

$$\frac{K_A}{K_B} = \frac{W_A}{W_B} = \frac{1}{1}$$

or $\frac{1}{2} m_A v_A^2 = \frac{1}{2} m_B v_B^2$

$$\therefore \quad \frac{v_A}{v_B} = \sqrt{\frac{m_B}{m_A}} = \frac{2}{1}$$

12. $a \propto -x$

$$a = -\lambda x \qquad (\lambda = \text{constant})$$

$$v\frac{dv}{dx} = -\lambda x$$

$$\int_{v_1}^{v_2} v\,dv = -\lambda \int_0^x x\,dx$$

$$\frac{|v^2|_{v_1}^{v_2}}{2} = -\frac{\lambda x^2}{2}$$

$$\frac{1}{2}mv_2^2 - \frac{1}{2}mv_1^2 = -\frac{\lambda m x^2}{2}$$

$$(\text{Loss in KE}) \propto x^2$$

13. $x = (3t^2 + 5)$ m

Therefore, velocity $(v) = \dfrac{dx}{dt}$

or $\quad v = 6t$ m/s ...(i)

$$v_i = 6 \times 0 = 0$$

$$v_f = 6 \times 5 = 30 \text{ m/s}$$

Using Work-energy theorem,

$$W = \Delta K \cdot E = \frac{1}{2}m(v_f^2 - v_i^2)$$

$$= \frac{1}{2}m \times (30^2 - 0^2)$$

$$= \frac{1}{2} \times 2 \times 900 = 900 \text{ J}$$

14.

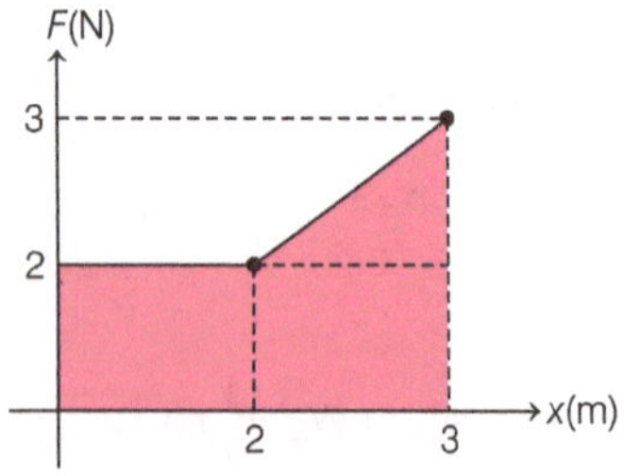

Work done on the particle = Shaded area

$$= 2 \times 2 + \frac{1}{2} \times 1 \times 1 + 2 \times 1 = 6.5 \text{ J}$$

Now, from work-energy theorem,

$$\Delta W = K_f - K_i$$

$\Rightarrow \quad K_f = \Delta W = 6.5$ J $\qquad [\because K_i = 0]$

15. $\quad W = \mathbf{F} \cdot \mathbf{d}$

$\Rightarrow \quad W = (3\hat{\mathbf{i}} - 12\hat{\mathbf{j}}) \cdot (4\hat{\mathbf{i}})$

$\Rightarrow \quad W = 12$ J ...(i)

Now, using work-energy theorem, we get

work done (W) = change in kinetic energy (ΔK)

or $\quad W = K_2 - K_1$ ...(ii)

Comparing Eqs. (i) and (ii), we get

$$K_2 - K_1 = 12 \text{ J}$$

or $\quad K_2 = K_1 + 12$ J

Given, initial kinetic energy, $K_1 = 3$ J

$\therefore$ Final kinetic energy, $K_2 = 3\text{ J} + 12\text{ J} = 15$ J

16. From work-energy theorem,

$$W_{\text{net}} = W_{mg} + W_{\text{air}} = \frac{1}{2}m(v_f^2 - v_i^2)$$

$$\Rightarrow \quad 0 + W_{\text{air}} = \frac{1}{2} \times 0.1[(6)^2 - (10)^2] = -3.2 \text{ J}$$

17. $W_{\text{All}} = \Delta KE = K_f - K_i$

$$= \frac{1}{2}m(v_f^2 - v_i^2) = \frac{1}{2} \times 2 \times (0 - 20^2)$$

$$= -400 \text{ J} \qquad \textbf{Ans.}$$

19. $v_{x=0} = 0 \quad : \quad v_{x=b} = \alpha\sqrt{b}$

$$\therefore \quad W_{\text{All}} = \Delta KE = K_f - K_i$$

$$= \frac{1}{2}m(v_f^2 - v_i^2) = \frac{1}{2} \times m[(\alpha\sqrt{b})^2 - 0]$$

$$= \frac{1}{2}m\alpha^2 b \qquad \textbf{Ans.}$$

20. $W_F = Fs\cos 0° = 80 \times 4 \times 1 = 320$ J

$$W_{mg} = (mg)(s)\cos 180°$$

$$= (50)(4)(-1)$$

$$= -200 \text{ J}$$

$$K_f = W_{\text{All}} = 120 \text{ J} \qquad \textbf{Ans.}$$

21. $K_f - K_i = W_F = \int F dx$

$$0 - \frac{1}{2}mv_0^2 = \int_0^x -Ax\,dx$$

$$\therefore \quad \frac{1}{2}mv_0^2 = \frac{Ax^2}{2}$$

$$\text{or} \quad x = v_0\sqrt{\frac{m}{A}} \qquad \textbf{Ans.}$$

22. $T = F$, (b) $W_{\text{All forces}} = 40$J

23. $T - mg = ma$

$\therefore \quad T = m(g + a) = 72\,(9.8 + 0.98)$

$$= 776.16 \text{ N}$$

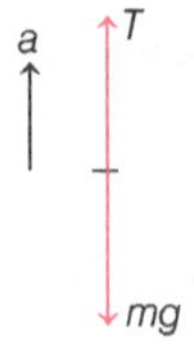

(a) $W_T = Ts\cos 0°$

$$= (776.16)(15)$$

$$= 11642 \text{ J} \qquad \textbf{Ans.}$$

(b) $W_{mg} = (mg)(s)\cos 180°$

$$= (72 \times 9.8 \times 15)(-1)$$

$$= -10584 \text{ J} \qquad \textbf{Ans.}$$

(c) $K = W_{\text{Total}} = 11642 - 10584$

$$= 1058 \text{ J} \qquad \textbf{Ans.}$$

(d) $K = \frac{1}{2}mv^2$

$$\therefore \quad v = \sqrt{\frac{2K}{m}} = \sqrt{\frac{2 \times 1058}{72}}$$

$$= 5.42 \text{ m/s} \qquad \textbf{Ans.}$$

24. $W_{\text{all}} = \Delta K = K_f - K_i = K_f \qquad (\text{as } K_i = 0)$

$$\therefore \quad W_{mg} + W_{\text{chain}} = K_f$$

or $W_{\text{chain}} = K_f - W_{mg} = \frac{1}{2}mv_f^2 - mgh$

$$= \frac{1}{2} \times 30 \times (0.4)^2 - 30 \times 10 \times 2$$

$$= -597.6 \text{ J} \qquad \textbf{Ans.}$$

Module 4.

1. $K_i + U_i = K_f + U_f$

$$\therefore \quad 0 + 6 = \frac{1}{2} \times 1 \times v^2 + 2$$

$$\therefore \quad v = 2\sqrt{2} \text{ m/s}$$

2. $E_i = E_f$

$$0 = m_2gh - m_1gh + \frac{1}{2}(m_1 + m_2)v^2$$

$$\therefore \quad v = \sqrt{2gh\left(\frac{m_1 - m_2}{m_1 + m_2}\right)} \qquad \textbf{Ans.}$$

3. KE = decrease in potential energy = mgh

or $\quad \text{KE} \propto m$

or $\quad \frac{(\text{KE})_1}{(\text{KE})_2} = \frac{12}{6} = \frac{2}{1}$

4. $mgh + \frac{1}{2}mv^2 = \text{constant} = mgH \qquad (H = \text{initial height})$

or $\quad gh + \frac{v^2}{2} = \text{constant}$

5.

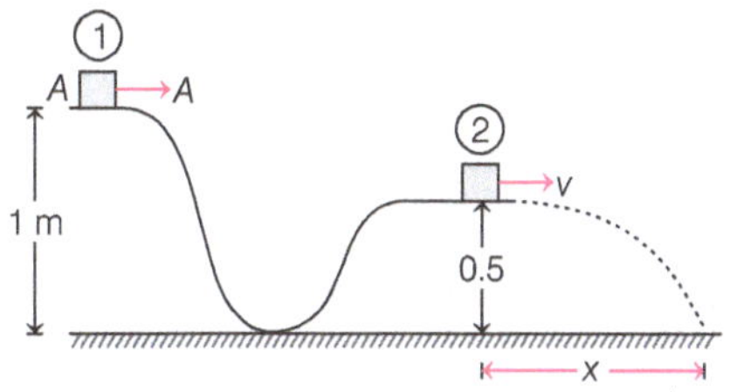

$$mg(1) = \frac{1}{2}mv^2 + mg(0.5)$$

$$v^2 = 2g(1 - 0.5) = g = 10$$

$$\Rightarrow \quad v = \sqrt{10} \text{ m/s}$$

$$t = \sqrt{\frac{2h}{g}} = \sqrt{\frac{2 \times 0.5}{10}} = \sqrt{\frac{1}{10}}$$

Now, $x = vt = 1 \text{ m}$

6.

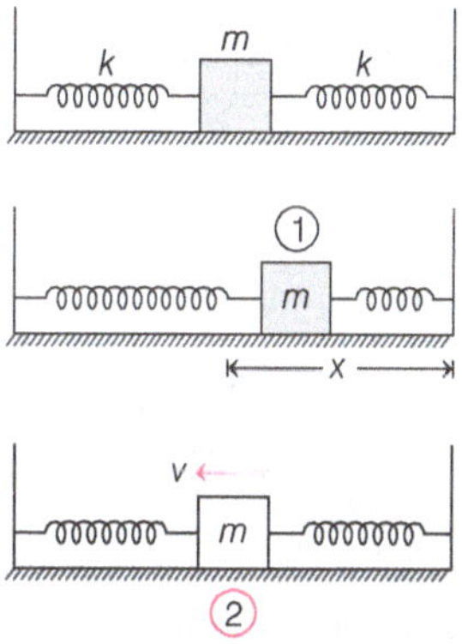

$$\frac{1}{2}kx^2 + \frac{1}{2}kx^2 = \frac{1}{2}mv^2$$

$$kx^2 = \frac{1}{2}mv^2$$

$$v = \sqrt{\frac{2k}{m}}x$$

7. $dm = \left(\frac{m}{\pi/2}\right)d\theta = \left(\frac{2m}{\pi}\right)d\theta$

$$h = R(1 - \cos\theta)$$

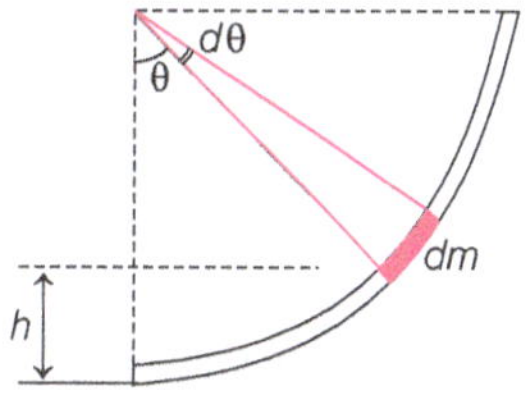

$$dU_i = (dm)gh = \frac{2mgR}{\pi}(1 - \cos\theta)\,d\theta$$

$$\therefore \quad U_i = \int_0^{\pi/2} dU_i = \frac{2mgR}{\pi}\left(\frac{\pi}{2} - 1\right)$$

$$= mgR\left(1 - \frac{2}{\pi}\right)$$

Now, $\quad U_i + K_i = U_f + K_f$

$$\therefore \quad mgR\left(1 - \frac{2}{\pi}\right) = 0 + \frac{1}{2}mv^2$$

or $\quad v = \sqrt{2gR\left(1 - \frac{2}{\pi}\right)} \qquad$ **Ans.**

8. $dm = \left(\dfrac{m}{\pi}\right) d\theta$

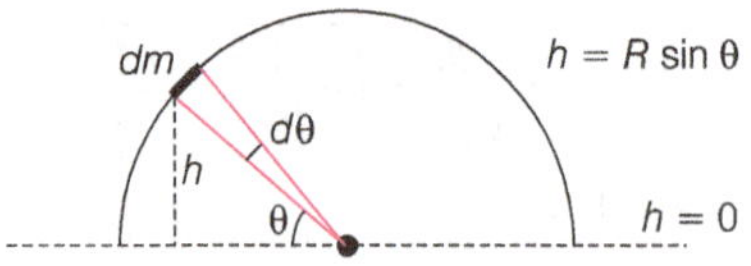

$$\therefore \quad dU = (dm)\,gh = \left(\frac{m}{\pi}\,d\theta\right) gr\sin\theta$$

$$U_i = \int_0^{\pi} dU = \frac{2mgr}{\pi}$$

Now, $\quad K_i + U_i = K_f + U_f$

$$\therefore \quad 0 + \frac{2mgr}{\pi} = \frac{1}{2}mv^2 - mg\left(\frac{\pi r}{2}\right)$$

$$\therefore \quad v = \sqrt{2gr\left(\frac{2}{\pi} + \frac{\pi}{2}\right)} \qquad \textbf{Ans.}$$

9. Decrease in gravitational potential energy = increase in kinetic energy.

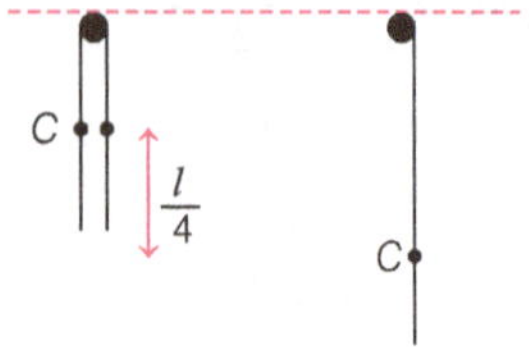

Initially centre of mass of chain was at distance $\dfrac{l}{4}$ below the pin and in final position, it is at distance $\dfrac{l}{2}$ below the pin. Hence, centre of mass has descended $\dfrac{l}{4}$.

$$\therefore \quad mg\frac{l}{4} = \frac{1}{2}mv^2 \quad \text{or} \quad v = \sqrt{\frac{gl}{2}}$$

10. From energy conservation equation,

$$\frac{1}{2}kx^2 = \frac{1}{2}(4k)y^2$$

$$\therefore \quad \frac{y}{x} = \frac{1}{2}$$

11. (a) $K_A + U_A = K_B + U_B$

$$\therefore \quad 0 + \frac{1}{2}\times 500\times(0.5-0.1)^2 = \frac{1}{2}\times 10\times v^2 + \frac{1}{2}\times 500\times(0.3-0.1)^2$$

On solving, we get

$$v = 2.45 \text{ m/s} \qquad \textbf{Ans.}$$

(b) $CO = \sqrt{(BO)^2 + (BC)^2}$

$$= \sqrt{(30)^2 + (20)^2}$$

$$\approx 36 \text{ cm} = 0.36 \text{ m}$$

Again applying the equation,

$$K_A + U_A = K_C + U_C$$

$$0 + \frac{1}{2}\times 500\times(0.5-0.1)^2 = \frac{1}{2}\times 10\times v^2 + \frac{1}{2}\times 500\,(0.36-0.1)^2$$

On solving, we get

$$v = 2.15 \text{ m/s} \qquad \textbf{Ans.}$$

12. Decrease in potential energy of 1 kg = increase in kinetic energy of both

$$\Rightarrow \quad 1\times 10\times 1 = \frac{1}{2}\times(4+1)\times v^2$$

$$\therefore \quad v = 2 \text{ m/s} \qquad \textbf{Ans.}$$

13. If A descends X, then B will ascend $2x$. Further, if speed of A at this instant is 2.5 m/s, then speed of B at this instant will be 5 m/s. Now,
Decrease in potential energy of A = increase in potential energy of B + increase in kinetic energy of both

$$\therefore \quad (300)\,x = (50)\,(2x) + \frac{1}{2}\left(\frac{300}{9.8}\right)(2.5)^2 + \frac{1}{2}\left(\frac{50}{9.8}\right)(5.0)^2$$

Solving, we get

$$x = 0.797 \text{ m} \qquad \textbf{Ans.}$$

14.

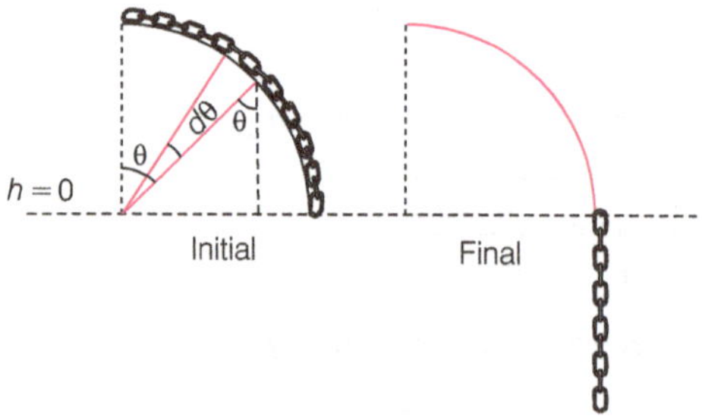

Initial PE,

$$U_i = \int_{\theta=0^\circ}^{\theta=\pi/2} (r\,d\theta)(\rho)(g)(r\cos\theta)$$

$$= (\rho g r^2)[\sin\theta]_0^{\pi/2} = \rho g r^2$$

Final PE,

$$U_f = \left(\frac{\pi r}{2}\times\rho\right)(g)\left(-\frac{\pi r/2}{2}\right) = -\frac{\pi^2 r^2 \rho g}{8}$$

$$\Delta U = r^2 \rho\, g\left(1 + \frac{\pi^2}{8}\right)$$

$$\Delta U = KE$$

$$\text{or} \quad r^2\rho g\left(1 + \frac{\pi^2}{8}\right) = \frac{1}{2}\left(\frac{\pi r}{2}\right)(\rho)\,v^2$$

$$\Rightarrow \quad v = \sqrt{4rg\left(\frac{1}{\pi} + \frac{\pi}{8}\right)}$$

$$\text{or} \quad v = \sqrt{rg\left(\frac{\pi}{2} + \frac{4}{\pi}\right)} \qquad \textbf{Ans.}$$

15. $E_i = E_f$

$\therefore \quad \frac{1}{2}mv^2 = \frac{1}{2}m\left(\frac{v}{2}\right)^2 + \frac{1}{2}kx^2$

or $\quad k = \frac{3v^2m}{4x^2}$ **Ans.**

16. Decrease in mechanical energy in first case

$$= E_i - E_f = \frac{1}{2}mv^2 - 0 = \frac{1}{2}mv^2$$

Decrease in mechanical energy in second case

$$= E_i - E_f = \frac{1}{2}mv^2 - mgh$$

Further, μ does not depend on angle of inclination.

17. Work done by friction $= E_f - E_i$

$$= \frac{1}{2} \times 1 \times (2)^2 - 1 \times 10 \times 1$$

$$= -8 \text{ J}$$ **Ans.**

18. According to work-energy theorem,

Loss in kinetic energy = Work done against friction + Potential energy of spring

$\Rightarrow \quad \frac{1}{2}mv^2 = fx + \frac{1}{2}kx^2$

where, x is the compression of the spring.

$\Rightarrow \frac{1}{2} \times 2 \times (4)^2 = 15x + \frac{1}{2} \times 10000x^2$

$\Rightarrow \quad 5000x^2 + 15x - 16 = 0$

$\therefore \quad x = 0.055$ m

$= 5.5$ cm

19. $\frac{1}{2}mv^2 = (0.9)(mgh)$

$\therefore \quad v = \sqrt{1.8gh} = \sqrt{1.8 \times 10 \times 2}$

$= 6$ m/s

20. $\frac{3}{4}$ th is lost. Hence, left is $\frac{1}{4}$ th, or

$$v^2 = \frac{v_0^2}{4}$$

$\therefore \quad v = \frac{v_0}{2} = v_0 - at_0 = v_0 - \mu g t_0$

or $\quad \mu = \frac{v_0}{2gt_0}$

21. $E_i - E_f =$ Work done against friction

$\therefore \quad \frac{1}{2}kx^2 - \frac{1}{2}K\left(\frac{x}{2}\right)^2 = \mu mg\left(x + \frac{x}{2}\right)$

$\therefore \quad x = \frac{4\mu mg}{k}$ **Ans.**

22. Since, the particle is released from a small height, θ (angle of radius with vertical) will be very small. Force of friction throughout the journey can be assumed to be μmg. Particle will finally come to rest when whole of its energy ($= mgh$) is lost in the work done against friction. Let particle stops after travelling a distance d. Then,

$$\mu mgd = mgh$$

or $\quad d = \frac{h}{\mu} = \frac{10^{-2}}{0.01} = 1.0$ m

23. $v_{m_1} = v_{m_2} = v$ (say)

$$d_{m_1} = h_{m_2} = 4\text{m}$$

Using the equation,

$E_i - E_f =$ Work done against friction

$$0 - \left[\frac{1}{2} \times (10+5)(v^2) - 5 \times 10 \times 4\right]$$

$$= 0.2 \times 10 \times 10 \times 4$$

Solving, we get

$$v = 4 \text{ m/s}$$

24. From constraint relations, we can see that

$$v_A = 2v_B$$

Therefore, $\quad v_A = 2(0.3) = 0.6$ m/s

as $\quad v_B = 0.3$ m / s (given)

Applying $\quad W_{nc} = \Delta U + \Delta K$, we get

$$-\mu m_A g s_A = -m_B g s_B + \frac{1}{2}m_A v_A^2 + \frac{1}{2}m_B v_B^2$$

Here, $\quad s_A = 2s_B = 2$ m as $s_B = 1$ m (given)

$\therefore \quad -\mu(4.0)(10)(2) = -(1)(10)(1) + \frac{1}{2}(4)(0.6)^2 + \frac{1}{2}(1)(0.3)^2$

or $\quad -80\mu = -10 + 0.72 + 0.045$

or $\quad 80\mu = 9.235$

or $\quad \mu = 0.115$ **Ans.**

25. If A moves 1 m down the plane and its speed is v, then B will move 2m upwards and its speed will be $2v$.

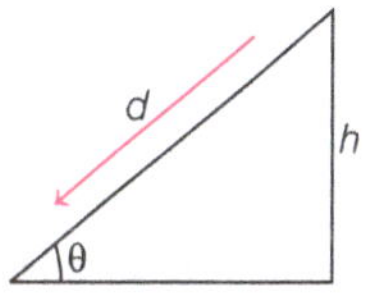

Using the equation,

$E_i - E_f =$ Work done against friction

$\therefore \quad 0 - \left[\frac{1}{2} \times 30 \times v^2 + \frac{1}{2} \times 5 \times (2v)^2 + 5 \times 10 \times 2 - 30 \times 10 \times \frac{3}{5}\right]$

$$= 0.2 \times 30 \times 10 \times \frac{4}{5} \times 1$$

Solving this equation, we get

$$v = 1.13 \text{ m/s}$$ **Ans.**

Note $\quad h = d \sin\theta = (1)\left(\frac{3}{5}\right) m = \frac{3}{5}m$

$$m_A = \frac{w_A}{g} = 30 \text{ kg and } m_B = \frac{w_B}{g} = 5 \text{ kg}$$

26.

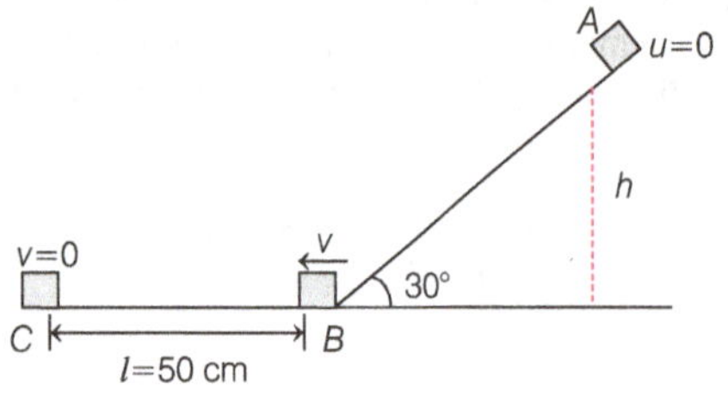

From B to C

Retardation $a_1 = \dfrac{\text{Maximum friction}}{\text{Mass}}$

$$= \frac{\mu mg}{m} = \mu g = 0.15 \times 9.81$$

$$= 1.47\ m/s^2$$

$$v^2 = 2a_1 l = 2 \times 1.47 \times 0.5$$

$$= 1.47\ m^2/s^2 \quad ...(i)$$

From A to B

Acceleration $a_2 = \dfrac{mg \sin\alpha - \mu mg \cos\alpha}{m}$

$$= g \sin\alpha - \mu g \cos\alpha$$

$$= 9.81\,[\sin 30° - 0.15 \cos 30°]$$

$$= 3.63\ m/s^2$$

$\therefore \quad v^2 = 2a_2 s = 2 \times 3.63 \times h(\text{cosec}\, 30°)$

or $\quad v^2 = 14.52\,h \quad ...(ii)$

Equating equations (i) and (ii), we get

$$h \approx 0.1\ m$$

Total work done by friction from A to C

Friction is an external force.

So, total work done by friction

$$= \text{change in mechanical energy}$$

$$= -mgh = -0.05 \times 9.81 \times 0.1$$

$$= -0.049\ J \approx -0.05\ J$$

$\therefore \quad W_f = -0.05\ J$

27. Distance or displacement covered before the body stops

Retardation, $a = \dfrac{mg \sin\alpha + \mu mg \cos\alpha}{m}$

$$= (g \sin\alpha + \mu g \cos\alpha)$$

Using $v^2 = u^2 - 2as$, we have

$$0 = v_0^2 - 2(g \sin\alpha + \mu g \cos\alpha)\, s$$

$$\therefore \quad s = \frac{v_0^2}{2(g \sin\alpha + \mu g \cos\alpha)}$$

Work done by friction force $f = \mu mg \cos\alpha$

$$W_f = Fs \cos 180°$$

(angle between F and s is 180°)

$$= -\frac{(\mu mg \cos\alpha) v_0^2}{2(g \sin\alpha + \mu g \cos\alpha)}$$

or $\quad W_f = -\dfrac{mv_0^2 \mu}{2g\,(1 + \tan\alpha)}$

28. $s = vt = 2 \times 5 = 10\text{m}$

Q = work done against friction

$$= \mu mgs = 0.2 \times 2 \times 9.8 \times 10$$

$$= 39.2\ J = 9.33\ \text{cal}$$

29. Heat energy required to just melt the bullet.

$$Q = Q_1 + Q_2$$

Here, $\quad Q_1 = ms\Delta\theta$

$$= (m \times 10^3)\,(0.03 \times 4.2)\,(327 - 27)$$

$$= (3.78 \times 10^4\ m)$$

$$Q_2 = mL = (m \times 10^3)\,(6 \times 4.2)$$

$$= (2.52 \times 10^4\ m)$$

$\therefore \quad Q = (6.3 \times 10^4\ m)$

If v be the speed of bullet, then 75% of $\frac{1}{2}mv^2$ should be equal to Q. Thus,

$$0.75 \times \frac{1}{2} \times m \times v^2 = 6.3 \times 10^4\ m$$

$\Rightarrow \quad v = 409.8\ m/s$

30. Σ (Moments about C) = 0

$\therefore \quad (k_1 x)\,AC = (k_2 x)\,BC$

$\Rightarrow \quad \dfrac{AC}{BC} = \dfrac{k_2}{k_1} \quad ...(i)$

$$AC + BC = l \quad ...(ii)$$

Solving these two equations, we get

$$AC = \left(\frac{k_2}{k_1 + k_2}\right) l \qquad \textbf{Ans.}$$

31. Decrease in gravitational potential energy of block

= increase in spring potential energy

$\therefore \quad mg\,(x_m \sin\theta) = \dfrac{1}{2} k x_m^2$

$\therefore \quad x_m = \dfrac{2\,mg \sin\theta}{k} \qquad$ **Ans.**

32. From $F = kx$

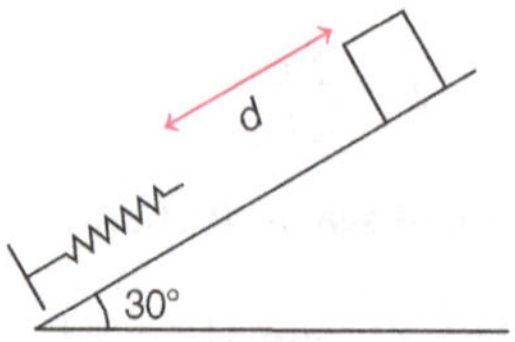

$$k = \frac{F}{x} = \frac{100}{1.0} = 100\ N/m$$

Decrease in gravitational potential energy = increase in spring potential energy

$$\Rightarrow \quad 10 \times 10 \times (d+2) \sin 30° = \frac{1}{2} \times 100 \times (2)^2$$

Solving, we get $\quad d = 2\,\text{m}$

∴ Total distance covered before coming momentarily to rest

$$= d + 2 = 4\,\text{m}$$ **Ans.**

33. $F = kx$

$$\therefore \quad k = \frac{F}{x} = \frac{100}{1}$$

$$= 100\ \text{N/m}$$

$$E_i = E_f$$

$$\therefore \quad \frac{1}{2} \times 10 \times v^2 = \frac{1}{2} \times 100 \times (2)^2 - (10)(10)(2 \sin 30°)$$

Solving, we get

$$v = \sqrt{20}\ \text{ms}^{-1}$$ **Ans.**

34. Speed (and hence the kinetic energy) will increase as long as $mg \sin\theta > kx$.

35. $E_i = E_f$

$$\frac{1}{2} mv_0^2 = \frac{1}{2} kx_{max}^2 \quad \Rightarrow \quad x_{max} = \sqrt{\frac{m}{k}}\, v_0$$

36. $mgh = \frac{1}{2} kx^2$

$$\frac{40}{1000} \times 10 \times 5 = \frac{1}{2} \times 400\, x^2$$

$$x = \frac{1}{10}\,\text{m} = 10\,\text{cm}$$

37. At maximum speed, acceleration of the block should be zero.

$$\therefore \quad F = kx$$

$$\Rightarrow \quad x = \frac{F}{k}$$

According to work-energy theorem,

$$W = \Delta K = \frac{1}{2} mv_{max}^2 - 0$$

or $$W_S + W_F = \frac{1}{2} mv_{max}^2$$

or $$-\frac{1}{2} kx^2 + Fx = \frac{1}{2} mv_{max}^2$$

or $$-\frac{1}{2} k \left(\frac{F}{k}\right)^2 + F \times \left(\frac{F}{k}\right) = \frac{1}{2} mv_{max}^2$$

or $$-\frac{F^2}{2k} + \frac{F^2}{k} = \frac{1}{2} mv_{max}^2$$

or $$\frac{F^2}{2k} = \frac{1}{2} mv_{max}^2$$

$$\therefore \quad v_{max} = \sqrt{\frac{F^2}{mk}} = \frac{F}{\sqrt{mk}}$$

$$= \frac{50}{\sqrt{1 \times 100}} = \frac{50}{10}$$

$$= 5\ \text{m/s}$$

38. At the maximum compression, velocity of each blocks are zero.

∴ Gain in PE $= \Delta U =$ Loss in KE

$$= \frac{1}{2} mv_0^2 + \frac{1}{2} mv_0^2 - 0 = mv_0^2$$

∵ Spring force is conservative in nature.

∴ The work done by spring is

$$W_s = -\Delta U = -mv_0^2$$

∴ The work done on each block by spring is

$$W = \frac{W_s}{2} = -\frac{1}{2} mv_0^2$$

$$= -\frac{1}{2} \times 1 \times 100 = -50\,\text{J}$$

39. (a) From A to B speed (or kinetic energy) will be increasing. Therefore, net potential energy should decrease.

(b) From A to B, a part of decrease in gravitational potential energy goes in increasing the kinetic energy and rest goes in increasing the potential energy of spring.

(c) and (d) : From B to C kinetic and gravitational potential energy are decreasing and spring potential energy is increasing.

∴ (decrease in kinetic energy) + (decrease in gravitational potential energy) = increase in spring potential energy.

40. The equivalent spring constant is

$$k_{eq} = k_1 + k_2 + k_3$$

$$= 40 + 15 + 45 = 100\,\text{N/m}$$

When block will achieve maximum speed, then springs come in relaxed positions.

According to mechanical energy conservation principle,

$$\frac{1}{2} k_{eq}\, x_0^2 = \frac{1}{2} mv_0^2$$

or $$100 \times (0.1)^2 = 1 \times v_0^2$$

$$\therefore \quad v_0 = 1\,\text{m/s}$$

41. In the figure, work done by conservative force (gravity force) is positive, potential energy is decreasing. But kinetic energy may increase, decrease or remain constant, depending on the value of F.

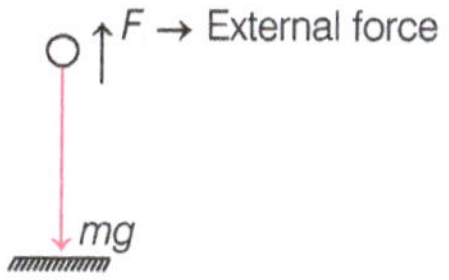

42. $W_{All} = \Delta K = K_f - K_i$

$$= \frac{1}{2} m (v_f^2 - v_i^2)$$

$v_f = v_{t_2}$ = Slope of s-t graph at t_2

$v_i = v_{t_1}$ = Slope of s-t graph at t_1

These two slopes are not necessarily equal.

43. $W = Mgh + mg\dfrac{h}{2} = \left(M + \dfrac{m}{2}\right)gh$

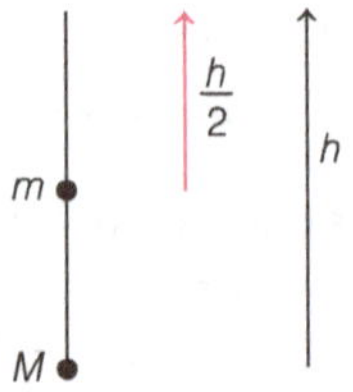

44. Velocity is decreasing. Therefore, acceleration (or net force) is opposite to the direction of motion.

(b) and (c): some other forces (other than friction) may also act which retard the motion.

45. $W = U_f - U_i = \dfrac{1}{2}K(X_f^2 - X_i^2)$

$= \dfrac{1}{2} \times 5 \times 10^3 [(0.1)^2 - (0.05)^2]$

= 18.75 N - m **Ans.**

46. $Mg - T = \dfrac{Mg}{2}$ or $T = \dfrac{Mg}{2}$

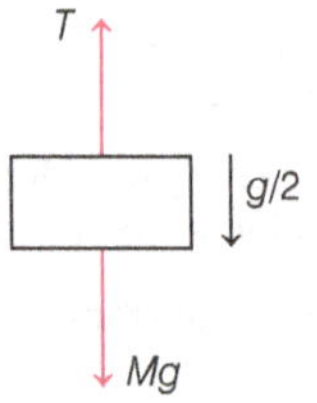

or magnitude of work done $= T \times x = \dfrac{Mgx}{2}$

47. $W = \Delta E = \dfrac{1}{2}k(a+b)^2 - \dfrac{1}{2}ka^2$

$= \dfrac{1}{2}k(b^2 + 2ab)$

$= \dfrac{1}{2}kb(b + 2a)$

48. Mass of hanging portion is $M/3$ (one-third) and centre of mass C, is at a distance $h = \dfrac{L}{6}$ below the table top.

Therefore, the required work done is change in potential energy,

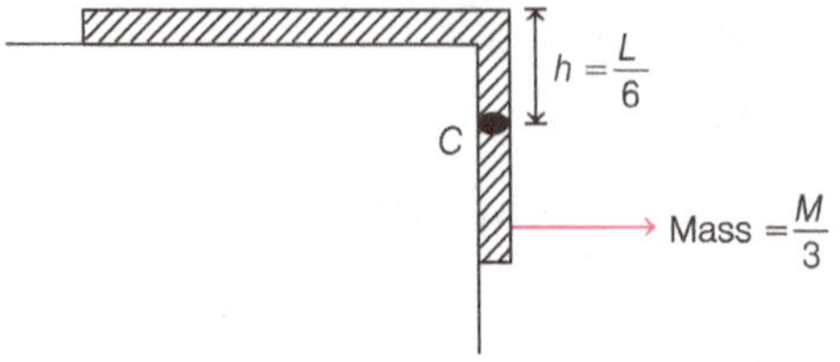

$$W = mgh = \left(\frac{M}{3}\right)(g)\left(\frac{L}{6}\right) = \frac{MgL}{18}$$

49. Work done by conservative force $= -\Delta U$

Work done by all the forces $= \Delta K$

Work done by forces other than conservative forces $= \Delta E$

51. Work done by conservative forces $= U_i - U_f$

Work done by external forces $= E_f - E_i$

and net work done by all the forces $= K_f - K_i$

52. $v = \alpha\sqrt{s}$

or $v^2 = \alpha^2 s$

Comparing with,

$v^2 = 2as$ (a = acceleration)

We get, $u = 0$ and $2a = \alpha^2$

$\Rightarrow$ $a = \alpha^2/2$

At time t, $v = at = \dfrac{\alpha^2}{2}t$

From work-energy theorem,

total work done = change in kinetic energy

$$= \frac{1}{2}mv^2 = \frac{1}{2}m\left(\frac{\alpha^2 t}{2}\right)^2$$

or $W_{\text{Total}} = \dfrac{m\alpha^4 t^2}{8}$

53. $T = 2mg$

As soon as string is cut T (on A) suddenly becomes zero. Therefore, a force of $2mg$ acting on upward direction on A suddenly becomes zero.

So, net force on it will become $2mg$ downwards.

$\therefore$ $a_1 = \dfrac{2mg}{m} = 2g$ (downwards)

Spring force does not become instantly zero. So, acceleration of B will not change abruptly.

or $a_2 = 0$

54. Let retarding force is F.

Then, $Fx = \dfrac{1}{2}mv^2$ ...(i)

and $F(x') = \dfrac{1}{2}m(2v)^2$ ...(ii)

Solving these two equations, we get

$x' = 4x$ **Ans.**

55. $K_i + U_i = K_f + U_f$

$\therefore$ $\dfrac{1}{2}mv_0^2 + 0 = \dfrac{1}{2}mv^2 + mgh$

$\therefore$ $v = \sqrt{v_0^2 - 2gh}$ **Ans.**

56. $K = \frac{1}{2}mv^2 = \frac{1}{2}m(gt^2)$

i.e. $K \propto t^2$

i.e.during downward motion. K-t graph is a parabola passing through origin with K increasing with time. Then, in upward journey K will decrease with time.

57. $k \propto \frac{1}{l}$

l of shorter part is less, therefore value of k is more

$$W = \frac{1}{2}kx^2$$

$\therefore$ $W_{\text{shorter part}}$ will be more.

58. Decrease in potential energy = Work done against friction

$\therefore$ $mg(h+d) = F \cdot d$

Here, F = average resistance

$\Rightarrow$ $F = mg\left(1 + \frac{h}{d}\right)$ **Ans.**

59. $F = kx \Rightarrow x = \frac{F}{k}$

Now, $U = \frac{1}{2}kx^2 = \frac{1}{2}k\left(\frac{F}{k}\right)^2$

or $U \propto \frac{1}{k}$

k_B is double. Therefore, U_B will be half.

60. $E_i = E_f$

$$0 = \frac{1}{2} \times 10 \times (0.15)^2 - 0.1 \times 10 \times 0.15 + \frac{1}{2} \times 0.1 \times v^2$$

$\therefore$ $v = 0.866 \text{ m/s}$ **Ans.**

61. At highest point,

$$\frac{1}{2}mu_x^2 = K$$

$\therefore$ $u_x = \sqrt{\frac{2K}{m}}$

$R = 4H$

$$\frac{2u_x u_y}{g} = \frac{4u_y^2}{2g}$$

$\therefore$ $u_y = u_x = \sqrt{\frac{2K}{m}}$

Now, $K_i = \frac{1}{2}mu^2 = \frac{1}{2}m(u_x^2 + u_y^2)$

$$= \frac{1}{2}m\left(\frac{2K}{m} + \frac{2K}{m}\right)$$

$= 2K$ **Ans.**

62. Let the time of descent of each ball is t_0.

$\therefore$ $h = \frac{1}{2}gt_0^2$ or $\frac{20 \times 2}{10} = t_0^2$

$\therefore$ $t_0 = 2 \text{ s}$

$\therefore$ Each ball remains in air for four second. Thus, the time interval between balls is 1 s.

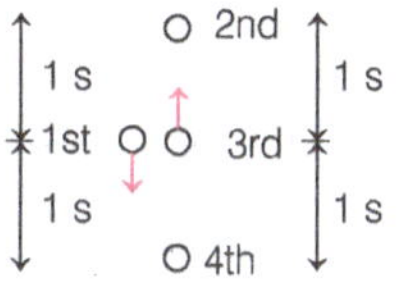

The speed of first ball is

$$v = gt = 10 \times 1 = 10 \text{ m/s}$$

$\therefore$ The kinetic energy of the first ball is $KE = \frac{1}{2}mv_0^2$

$$= \frac{1}{2} \times 10 \times 10^{-3} \times 100 = 0.5 \text{ J}$$

63. $\sum$(Moment about O) = 0

$\therefore$ $(kx)l = mg\left(\frac{l}{2}\right)$ or $x = \frac{mg}{2k}$

$U = \frac{1}{2}kx^2 = \frac{(mg)^2}{8k}$ **Ans.**

64. $p = \sqrt{2Km}$ or $p \propto \sqrt{m}$

$$\frac{m_1}{m_2} = \frac{1}{4}$$

$\therefore$ $\frac{p_1}{p_2} = \frac{1}{2}$

65. $l_1 = 2l_2 \Rightarrow l_1 = \frac{2}{3}l$

Force constant $k \propto \frac{1}{\text{length of spring}}$

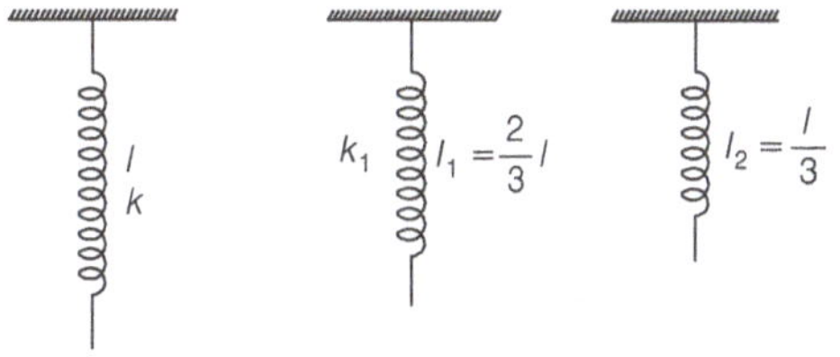

In second figure length is $\frac{2}{3}$ times as compared to the first figure.

$\therefore$ $k_1 = \frac{3}{2}k$

66. $F = -\frac{dU}{dr} = \frac{A}{r^2}$

67. From conservation of mechanical energy,

$E_i = E_f$ or $\frac{1}{2}kx_i^2 + mgh_i = \frac{1}{2}mv_f^2$

$\therefore$ $v_f = \sqrt{2gh_i + \frac{k}{m}x_i^2}$

Substituting the values, we have

$$v_f = \sqrt{2 \times 9.8 \times 1.9 + \frac{2300}{0.12}(0.045)^2}$$

$= 8.72 \text{ m/s}$ **Ans.**

Module 5.

1. $F = 0$ at $x = x_2$. When displaced from x_2 in negative direction, force is positive, i.e. in the opposite direction of displacement. Similarly, when displaced in positive direction, force is negative.

2. At C, potential energy is minimum. So, it is stable equilibrium position.
Further,
$$F = -\frac{dU}{dr} = -(\text{Slope of } U\text{-}r \text{ graph})$$
Negative force means attraction and positive force means repulsion.

3. $F = -\frac{dU}{dx} = \frac{12a}{x^{13}} - \frac{6b}{x^7}$
At equilibrium, $F = 0$
or $$x = \left(\frac{2a}{b}\right)^{1/6}$$
At this value of x, we can see that $\frac{d^2U}{dx^2}$ is positive. So, potential energy is minimum or equilibrium is stable.

4. $F = -\frac{dU}{dx} = 5 - 200x$
At origin, $x = 0$
$\therefore$ $F = 5\text{N}$
$$a = \frac{F}{m} = \frac{5}{0.1} = 50\text{ m/s}^2$$
Mean position is at $F = 0$
or at, $x = \frac{5}{200} = 0.025$ m
$$a = \frac{F}{m} = \frac{5 - 200x}{0.1} = (50 - 2000x) \quad \text{...(i)}$$
At 0.05 m from the origin,
$x = +0.05$ m or $x = -0.05$ m
Substituting in Eq. (i), we have
$$|a| = 150\text{ m/s}^2$$
or $$= 50\text{ m/s}^2$$
At 0.05 m from the mean position means,
$$x = 0.075$$
or $$x = -0.025\text{ m}$$
Substituting in Eq. (i), we have
$$|a| = 100\text{ m/s}^2$$ **Ans.**

5. (a) Net force is towards the mean position $x = 0$ where, $F = 0$ when displaced from this position. Therefore, equilibrium is stable.

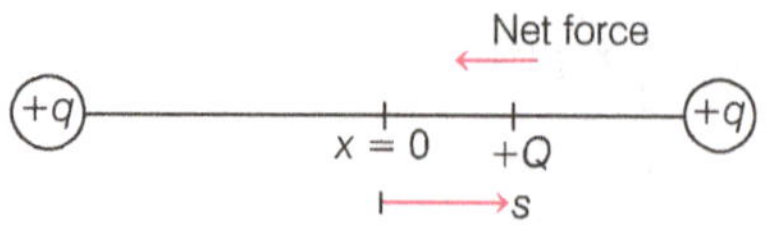

(b) Net force is away from the mean position. Therefore, equilibrium is unstable.

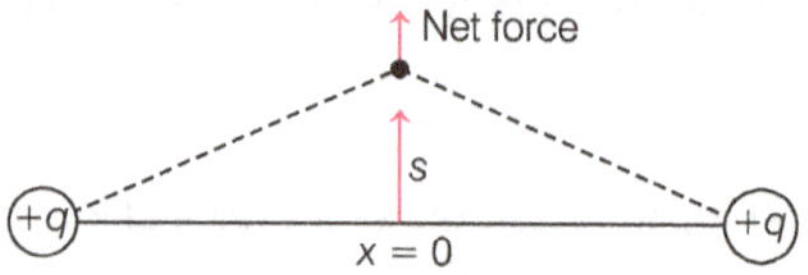

(c) Same logic can be applied as was applied in part (b).
(d) Net force is neither towards $x = 0$ nor away from $x = 0$. Therefore, equilibrium is none of the three.

6. $U = \frac{x^3}{3} - 4x + 6$
$$F = -\frac{dU}{dx} = -x^2 + 4$$
$$F = 0 \quad \text{at} \quad x = \pm 2\text{m}$$

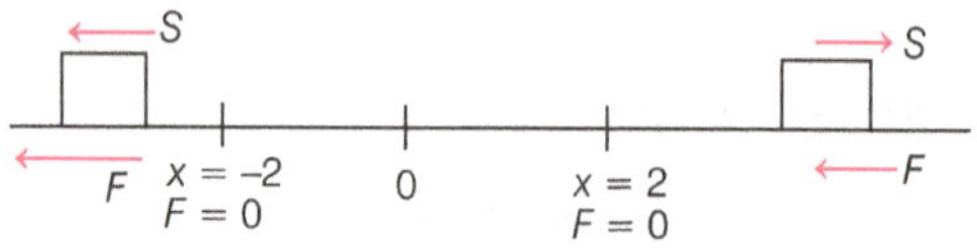

For $x > 2$ m, $F = -$ve i.e. displacement is in positive direction and force is negative. Therefore $x = 2$ is stable equilibrium position.
For $x < -2$ m, $F = -$ve
i.e. force and displacement are in negative directions. Therefore, $x = -2$ m is unstable equilibrium position.

7. At A, $x = 0$ and $F = 0$
For $x > 0$, $F = +$ ve i.e. force is in the direction of displacement. Hence A is unstable equilibrium position.
Same concept can be applied with E also.
At point C, $F = 0$.
For $x > x_C$, $F = -$ ve
Displacement is positive and force is negative (in opposite direction of displacement). Therefore, C point is stable equilibrium point.

8. (a) At $x = 0$, $F = 0$. At P, attraction on $-q$
or $F_1 >$ attraction F_2

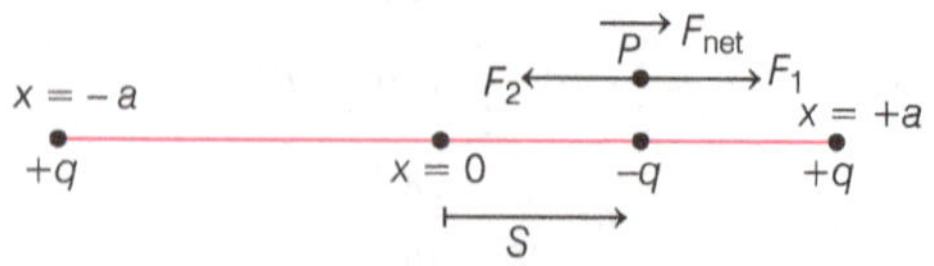

$\therefore$ Net force is in the direction of displacement. So, equilibrium is unstable.

(b) F_{net} is in opposite direction of S. Therefore equilibrium is stable.

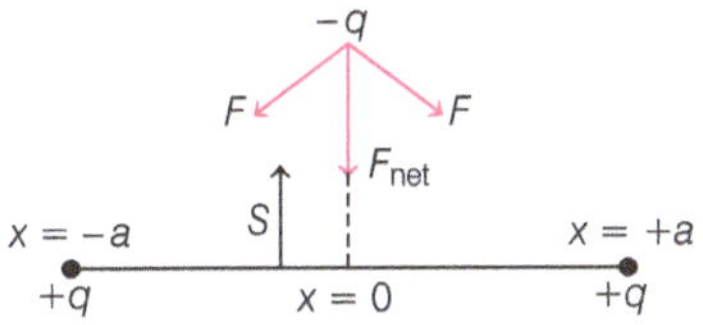

9. $U = \text{minimum} = -20\ \text{J}$

at $x = 2\text{m}$

$\therefore$ $x = 2\text{m}$ is stable equilibrium position **Ans.**

10. $F = (x - 4)$

$$F = 0 \quad \text{at} \quad x = 4\ \text{m}$$

When $x > 4\,\text{m}, F = +\text{ve}$

When displaced from $x = 4$ m (towards positive direction) force also acts in the same direction.

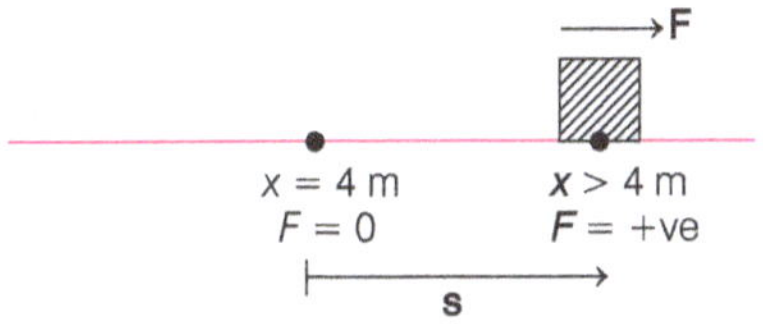

Therefore, equilibrium is unstable.

11. (a) At $x = 6\text{m}$,

$$U = (6-4)^2 - 16 = -12\ \text{J}$$

$$K = 8\text{J}$$

$\therefore$ $E = U + K = -4\ \text{J}$ **Ans.**

(b) $U_{min} = -16\text{J}$ at $x = 4\text{m}$

$\therefore$ $K_{max} = E - U_{min}$

$= -4 + 16$

$= 12\ \text{J}$ **Ans.**

(c) $K = 0$

$\therefore$ $U = E$

or $(x-4)^2 - 16 = -4$

or $x = (4 \pm 2\sqrt{3})\ \text{m}$ **Ans.**

(d) $F_x = -\dfrac{dU}{dx} = (8 - 2x)$ **Ans.**

(e) $F_x = 0$ at $x = 4$ m **Ans.**

Module 6.

1. $P = \text{constant}$

$\therefore$ Work done in time t,

$$W = Pt$$

From work-energy theorem, net work done is change in kinetic energy. Therefore,

$$\frac{1}{2}mv^2 = Pt \quad \text{or} \quad v \propto t^{1/2}$$

Integrating, we get $s \propto t^{3/2}$

2. $F = \text{constant}$

$\therefore$ $a = \dfrac{F}{m} = \text{constant}$

$$v = at = \frac{F}{m}t$$

$$P = F \cdot v = \frac{F^2}{m}t$$

$\therefore$ $P \neq \text{constant}$

But $P \propto t$

3. $P = \mathbf{F} \cdot \mathbf{v}$

$$= (-mg\,\hat{\mathbf{j}}) \cdot [u_x \hat{\mathbf{i}} + (u_y - gt)\,\hat{\mathbf{j}}]$$

$$= (-mgu_y) + mg^2t$$

i.e. P *versus* t graph is a straight line with negative intercept and positive slope.

4. $P = \mathbf{F} \cdot \mathbf{v} = Fv\cos\theta = Tv\cos\theta$ **Ans.**

5. $P = \dfrac{W}{t} = \dfrac{mgh + \frac{1}{2}mv^2}{t}$

$$= \frac{(800 \times 10 \times 10) + \frac{1}{2} \times 800 \times (20)^2}{60}$$

$= 4000$ W **Ans.**

6. $a = \dfrac{F}{m}$

$$v = at = \left(\frac{F}{m}\right)(2)$$

$P = F \cdot v = \dfrac{2F^2}{m}$ **Ans.**

7. $a = \dfrac{F}{m}$

$$v = at = \frac{F}{m}t$$

$$P = F \cdot v = \left(\frac{F^2}{m}\right)t$$

or $P \propto t$

i.e. P-t graph is a straight line passing through origin.

8. $P = Fv = \text{constant}$

$\therefore$ $F = \dfrac{P}{v}$

$\therefore$ $mv\left(\dfrac{dv}{ds}\right) = \dfrac{P}{v}$

$\therefore$ $\displaystyle\int_0^s ds = \frac{m}{P}\int_v^{2v} v^2 dv$

Solving, we get $s = \dfrac{7mv^3}{3P}$ **Ans.**

9. (a) $P_{av} = \frac{W}{t} = \frac{\frac{1}{2}mv^2}{t} = \frac{\frac{1}{2}m(at)^2}{t}$

$= \frac{1}{2}ma^2t = \frac{1}{2} \times 1 \times (4)^2 (2)$

$= 16\ W$ **Ans.**

(b) $P_i = Fv = (ma)(at)$

$= ma^2t = (1)(4)^2(4)$

$= 64\ W$ **Ans.**

10. (i) $W = Pt = \frac{1}{2}mv^2$

(ii) $v = \sqrt{\frac{2Pt}{m}}$

(iii) Integrating the velocity, we will get displacement

$\therefore \quad S = \sqrt{\frac{2P}{m}}\left(\frac{t^{3/2}}{3/2}\right) = \sqrt{\frac{8P}{9m}}\, t^{3/2}$ **Ans.**

11. (a) $KE = W = \int P dt = \int 2t \cdot dt = t^2$

$\frac{1}{2}mv^2 = t^2$

$\therefore \quad v = \sqrt{\frac{2}{m}}\, t$ **Ans.**

(b) $P_{av} = \frac{W}{t} = \frac{t^2}{t} = t$ **Ans.**

12. Rate at which kinetic plus gravitational potential energy is dissipated at time t is actually the magnitude of power of frictional force at time t.

$|P_f| = f \cdot v = (\mu mg \cos\alpha)(at)$

$= (\mu mg \cos\alpha)[(g \sin\alpha - \mu g \cos\alpha)t]$

$= \mu mg^2 \cos\alpha(\sin\alpha - \mu\cos\alpha)t$ **Ans.**

13. Mean power by gravity

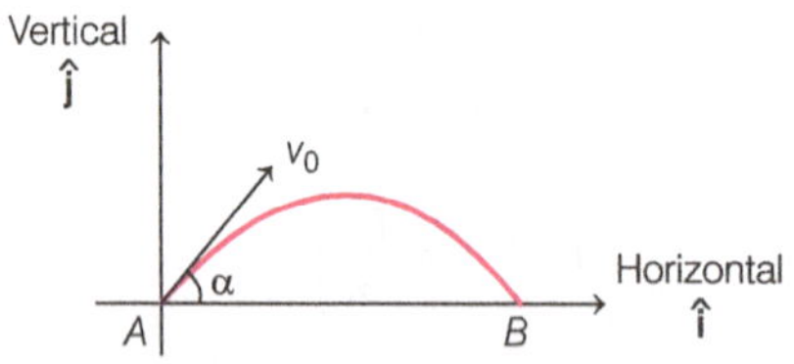

From A to B, total work done by gravity is zero because net vertical displacement is zero. Therefore,

Mean or average power $= \frac{\text{Total work done}}{\text{Total time}}$

$= 0$

$\therefore \quad \langle P \rangle = 0$

Power of gravity force as a function of time t

$P = \mathbf{F} \cdot \mathbf{v}$

$= \mathbf{F} \cdot (\mathbf{u} + \mathbf{a}t)$

$= [(mg(-\hat{\mathbf{j}})] \cdot [(v_0 \cos\alpha\hat{\mathbf{i}} + v_0 \sin\alpha\hat{\mathbf{j}}) + (-g\hat{\mathbf{j}})t]$

$= mg\,(gt - v_0 \sin\alpha)$

$\therefore \quad P = mg(gt - v_0 \sin\alpha)$

14. From work-energy theorem, $W = \frac{1}{2}mv^2$

Since, power is constant. Therefore, $W = Pt$

$\therefore \quad Pt = \frac{1}{2}mv^2$

$\therefore \quad v = \sqrt{\frac{2Pt}{m}}$

$= \sqrt{\frac{2 \times 0.5 \times 5}{0.2}} = 5\ m/s$

Unit Exercise (Cumulative)

Level 1

1. $(W_{f_1})_{\text{on } A} = -f_1 s$

$(W_{f_1})_{\text{on } B} = +f_1 s$

$\therefore$ Net work done by $f_1 = 0$

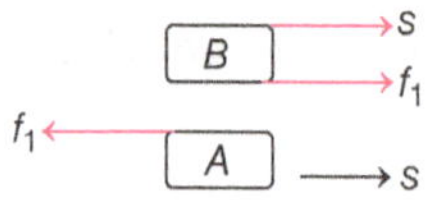

2. Let θ_1 is the angle of **F** with positive x-axis.

$\therefore \quad \tan\theta_1 = \frac{F_y}{F_x} = \frac{15}{20} = \frac{3}{4} = m_1$ (say)

Slope of given line, $m_2 = -\frac{\alpha}{3}$

$W = 0$ if $\mathbf{F} \perp \mathbf{s}$ or $m_1 m_2 = -1$

$\therefore \quad \left(\frac{3}{4}\right)\left(-\frac{\alpha}{3}\right) = -1$

$\therefore \quad \alpha = 4$ **Ans.**

3. Maximum acceleration of 2 kg block due to friction can be μg or $5\ m/s^2$.

Combined acceleration, if both move together with same acceleration would be

$a = \frac{60}{12} = 5 m/s^2$

Since, both accelerations are equal, upper block will move with acceleration $5 m/s^2$ due to friction.

In first two seconds, $s = \frac{1}{2}at^2 = \frac{1}{2} \times 5 \times 4 = 10 m$

and force of friction, $f = ma = 10\ N$

$\therefore \quad W_f = fs\cos 0° = 100\ J$

4. $dW = -(\mu\, mg \cos\theta)\, ds = -\mu mg\,(ds \cos\theta)$

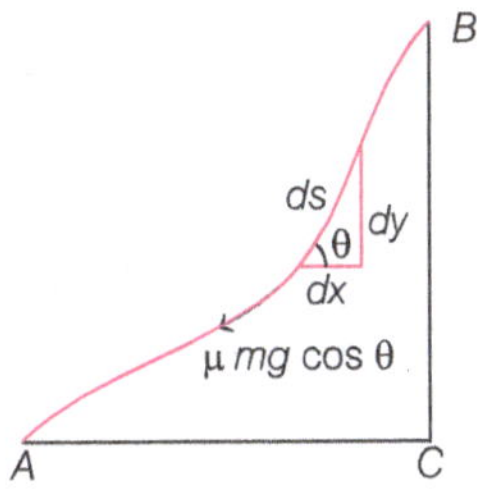

$$= -\mu\, mg\, dx = -\mu\, mg \int_0^1 dx \qquad (\because\ ds\cos\theta = dx)$$

$$= -0.5 \times 1 \times 10 \times 1 = -5\text{ J}$$

5. There is slip, so maximum friction will act on A in the direction of motion (or towards right)

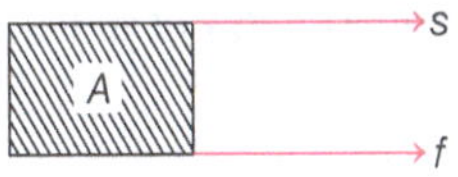

$$f = \mu mg = 0.2 \times 45 \times 10 = 90\text{ N}$$

$$s = 40 - 10 - 10 = 20\text{ cm} = 0.2\text{ m}$$

$\therefore \quad W = fs = 18\text{ J}$ **Ans.**

6. The given situation can be drawn graphically as shown in figure.

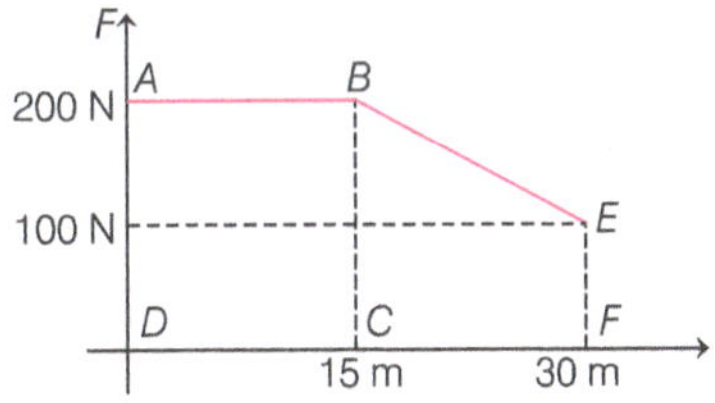

Work done = Area under F - x graph

= Area of rectangle $ABCD$ + Area of trapezium $BCFE$

$$W = (200 \times 15) + \frac{1}{2}(100 + 200) \times 15 = 3000 + 2250$$

$\Rightarrow W = 5250\text{ J}$

7. Work done by conservative force

$$= -\Delta U$$
$$= U_i - U_f$$
$$= [k(1+1)] - [k(2+3)]$$
$$= -3k$$ **Ans.**

8. Given, force, $\quad F = -kv^2$

$\therefore$ Acceleration, $a = \dfrac{-k}{m}v^2$

or $\quad \dfrac{dv}{dt} = \dfrac{-k}{m}v^2$

$\Rightarrow \quad \dfrac{dv}{v^2} = -\dfrac{k}{m}.dt$

Now, with limits, we have

$$\int_{10}^{v} \frac{dv}{v^2} = -\frac{k}{m}\int_0^t dt$$

$$\Rightarrow \quad \left(-\frac{1}{v}\right)_{10}^{v} = -\frac{k}{m}t$$

$$\Rightarrow \quad \frac{1}{v} = 0.1 + \frac{kt}{m}$$

$$\Rightarrow \quad v = \frac{1}{0.1 + \frac{kt}{m}} = \frac{1}{0.1 + 1000k}$$

$$\Rightarrow \quad \frac{1}{2} \times m \times v^2 = \frac{1}{8} \times v_0^2 \times m$$

$$\Rightarrow \quad v = \frac{v_0}{2} = 5\ \text{m/s}$$

$$\Rightarrow \quad \frac{1}{0.1 + 1000\,k} = 5$$

$$\Rightarrow \quad 1 = 0.5 + 5000\,k$$

$$\Rightarrow \quad k = \frac{0.5}{5000}$$

$$\Rightarrow \quad k = 10^{-4}\text{kg/m}$$

9. $U = \dfrac{x^4}{4} - \dfrac{x^2}{2}$

$$F = -\frac{dU}{dx} = -[x^3 - x] = 0$$

$$\Rightarrow \quad x(x^2 - 1) = 0$$

$$x = 0, x = \pm 1$$

$$\Rightarrow \quad \frac{d^2U}{dx} = 3x^2 - 1$$

At $x = \pm 1, \dfrac{d^2U}{dx} = +\text{ve}$, i.e.

at $x = \pm 1$, PE is minimum

$$U_{min} = -\frac{1}{4}$$

$$E = K_{max} + U_{min}$$

$$\Rightarrow \quad 2 = \frac{1}{2} \times 1 \times v_{max}^2 - \frac{1}{4}$$

$$\Rightarrow \quad v_{max} = \frac{3}{\sqrt{2}}\text{m/s}$$

10. $F_{net} = mg\sin\theta - \mu mg\cos\theta$

$$= mg\sin\theta - 0.3\,xmg\cos\theta \qquad \text{...(i)}$$

At maximum speed $F_{net} = 0$. Because after this net force will become negative and speed will decrease.

From Eq. (i), $F_{net} = 0$ at

$$x = \frac{\tan\theta}{0.3} = \frac{3/4}{0.3}$$

$$= 2.5\text{ m}$$ **Ans.**

11. $F_{net} = mg\sin\theta - \mu mg\cos\theta$

$$= mg\sin\theta - \mu_0 xg\cos\theta$$

$$a = \frac{F_{net}}{m} = g \sin\theta - \mu_0 xg \cos\theta$$

$\therefore \quad v \cdot \dfrac{dv}{dx} = g \sin\theta - \mu_0 xg \cos\theta$

or $\quad \displaystyle\int_0^0 v\,dv = \int_0^{x_m} (g \sin\theta - \mu_0 xg \cos\theta)\, dx$

Solving this equation, we get

$$x_m = \frac{2}{\mu_0} \tan\theta \qquad \textbf{Ans.}$$

12. On M, horizontal components of N and f are balanced (as Mg is vertical). Hence, on $2M$ also they will be balanced.

$\therefore$ Horizontal Kx force on $2M$ should be zero.

13. As energy loss is same, thus

$\mu\, mg \cos\theta \cdot (PQ) = \mu mg \cdot (QR)$

$\therefore \quad QR = (PQ)\cos\theta = (2\,\text{cosec}\,30°)\cos\theta$

$\Rightarrow \quad QR = 4 \times \dfrac{\sqrt{3}}{2}$

$= 2\sqrt{3} \approx 3.5$ m

Further, decrease in potential energy = loss due to friction

$\therefore \quad mgh = (\mu mg \cos\theta) d_1 + (\mu mg) d_2$

$m \times 10 \times 2 = \mu \times m \times 10 \times \dfrac{\sqrt{3}}{2} \times 4 + \mu \times m \times 10 \times 2\sqrt{3}$

$\Rightarrow \quad 4\sqrt{3}\,\mu = 2$

$\Rightarrow \quad \mu = \dfrac{1}{2\sqrt{3}}$

$= 0.288 = 0.29$

14. $T_i = mg \qquad$ ...(i)

$2kx = 2mg$

$\therefore \quad kx = mg$

One kx force (acting in upward direction) is suddenly removed. So, net downward force on system will be kx or mg. Therefore, net downward acceleration of system,

$$a = \frac{mg}{2m} = \frac{g}{2}$$

Free body diagram of lower block gives the equation,

$$mg - T_f = ma = \frac{mg}{2}$$

$\therefore \quad T_f = \dfrac{mg}{2} \qquad$ ..(ii)

From these two equations, we get

$$\Delta T = \frac{mg}{2} \qquad \textbf{Ans.}$$

15. In the frame of car,

$\therefore \quad ma_0 = kx_0$

$\therefore \quad x_0 = \dfrac{ma_0}{k}$

When car stops, the maximum speed is achieved by the block when spring comes in natural length. Thus, loss in spring potential energy appears as kinetic energy of the block

$$\frac{1}{2}kx_0^2 = \frac{1}{2}mv_{max}^2$$

$\Rightarrow \quad v_{max} = \sqrt{\dfrac{m}{k}}\, a_0$

16. Let X_m is maximum elongation of spring. Then,

increase in potential energy of spring = decrease in potential energy of C.

$\therefore \quad \dfrac{1}{2}kX_m^2 = M_1 g X_m$

or $\quad kX_m$ = maximum spring force

$= 2M_1 g = \mu_{min} Mg$

$\therefore \quad \mu_{min} = \dfrac{2M_1}{M} \qquad$ **Ans.**

17. After falling on plank downward force on block is mg and upward force is kx. Kinetic energy will increase when $mg > kx$ and it will decrease when $kx > mg$. Therefore, it is maximum when

$$kx = mg$$

or $\quad x = \dfrac{mg}{k}$

and this does not depend on h.

18. Displacement of floor = 0

19. In both cases sudden changes in force by cutting the spring would be kx.

$\therefore \quad a = \dfrac{\Delta F}{m} = \dfrac{kx}{m} \qquad$ (in both case)

In one case, it is downwards and in other case, it is upwards.

20. $P = 2x = Fv = mav = m\left(v\dfrac{dv}{dx}\right)v$

$\therefore \quad v^2 dv = \dfrac{1}{m} 2x\, dx$

Integrating, we get

$$\frac{v^3}{3} = \frac{x^2}{m}$$

$\therefore \quad v = \left(\dfrac{3x^2}{m}\right)^{\frac{1}{3}}$

21. Increase in potential energy per unit time = decrease in kinetic energy of both

$$= -\frac{d}{dt}\left(\frac{1}{2}m_1v_1^2 + \frac{1}{2}m_2v_2^2\right)$$

$$= v_1\left(-m_1\frac{dv_1}{dt}\right) + v_2\left(-m_2\frac{dv_2}{dt}\right)$$

$$= v_1(-m_1a_1) + v_2(-m_2a_2)$$

or $$\frac{dU}{dt} = v_1(-F_1) + v_2(-F_2) \qquad \text{...(i)}$$

Here, $-F_1 = -F_2 = kx = 200 \times 0.1 = 20\text{ N}$

Substituting in Eq. (i), we have

$$\frac{dU}{dt} = (4)(20) + (6)(20)$$

$$= 200\text{ J/s}$$ **Ans.**

22. Power of tension = 0

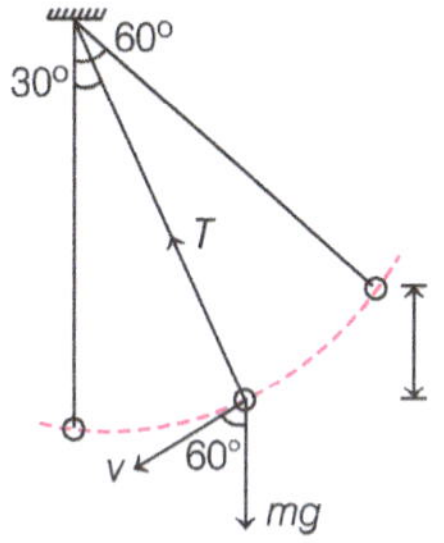

Power of $mg = (mg)(v)\cos 60°$

Here, $v = \sqrt{2gh}$

and $h = l(\cos 30° - \cos 60°)$

23. $\mathbf{a} = \mathbf{b} \times \mathbf{v} \Rightarrow \mathbf{a}$ (acceleration) is always perpendicular to velocity or we can say that force **F** is perpendicular to velocity. Hence, the work done on the particle by the force is zero, its change in kinetic energy is also zero and the speed of the particle remains constant.

Options (b) and (c) are correct.

24. (a) Decrease in potential energy of B = increase in spring potential energy

$$\therefore \quad 2mg\,x_m = \frac{1}{2}kx_m^2$$

$$\therefore \quad x_m = \frac{4mg}{k}$$ **Ans.**

(b) $E_i = E_f$

$$0 = \frac{1}{2}(m + 2m)v^2 + \frac{1}{2} \times k \times \left(\frac{2mg}{k}\right)^2 - (2mg)\left(\frac{2mg}{k}\right)$$

$$\therefore \quad v = 2g\sqrt{\frac{m}{3k}}$$

(c) $a = \dfrac{kx_m - 2mg}{2m}$ (upwards)

$$= \frac{k\left(\frac{4mg}{k}\right) - 2mg}{2m} = g$$

(d) $T - 2mg = ma$ ...(i)

$2mg - T = 2ma$ ...(ii)

Solving these two equations, we get

$$a = 0$$

and $$T = 2mg$$

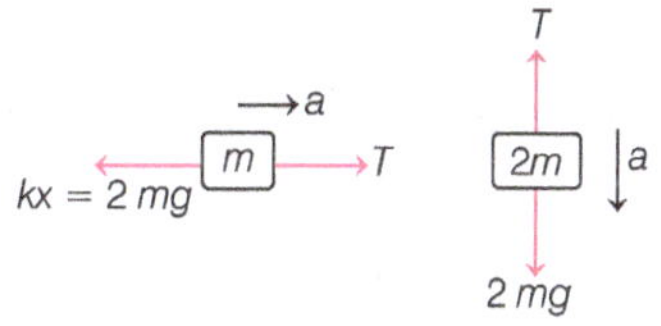

25. Total work done is

$$W = W_A + W_B = \frac{1}{2}kx_i^2 - \frac{1}{2}kx_f^2$$

$$= 2 - 0 = 2\text{ J}$$

$$\therefore \quad W_A = W_B = 1\text{ J}$$

26. $\mathbf{F} = -\left[\frac{\partial U}{\partial X}\hat{\mathbf{i}} + \frac{\partial U}{\partial y}\hat{\mathbf{j}}\right] = (-7\hat{\mathbf{i}} - 24\hat{\mathbf{j}})\text{ N}$

$$\mathbf{a} = \frac{\mathbf{F}}{m} = \left(-\frac{7}{5}\hat{\mathbf{i}} - \frac{24}{5}\hat{\mathbf{j}}\right)\text{m/s}$$

$$|\mathbf{a}| = \sqrt{\left(\frac{7}{5}\right)^2 + \left(\frac{24}{5}\right)^2} = 5\text{ m/s}^2$$ **Ans.**

Since, $\mathbf{a}$ = constant, we can apply,

$$\mathbf{v} = \mathbf{u} + \mathbf{a}t$$

$$= (8.6\hat{\mathbf{i}} + 23.2\hat{\mathbf{j}}) + \left(-\frac{7}{5}\hat{\mathbf{i}} - \frac{24}{5}\hat{\mathbf{j}}\right)(4)$$

$$= (3\hat{\mathbf{i}} + 4\hat{\mathbf{j}})\text{ m/s}$$

$$|\mathbf{v}| = \sqrt{(3)^2 + (4)^2}$$

$$= 5\text{ m/s}$$ **Ans.**

27. $U = E - K = 25 - K$

Since, $K \geq 0$

$\therefore \quad U \leq 25\text{ J}$

28. $U = E - K = -40 - K$

Since, $K \geq 0$

$\therefore \quad U \leq -40\text{ J}$

29. $W_1 = \dfrac{-F^2}{2k_1}$

$$= \frac{-50 \times 50}{2 \times 100} = -12.5\text{ J}$$

(b) $W_2 = \dfrac{F^2}{2k_1} = 12.5\text{ J}$

(c) $W_3 = \dfrac{-F^2}{2k_2}$

$$=-\frac{50\times 50}{2\times 200}=-6.25\text{ J}$$

(d) $W_4=\dfrac{F^2}{2}\left(\dfrac{1}{k_1}+\dfrac{1}{k_2}\right)$

$$=\frac{2500}{2}\left(\frac{1}{100}+\frac{1}{200}\right)$$

$$=\frac{1250}{200}\times 3=18.75\text{ J}$$

Hence, (a) → (s), (b) → (r), (c) → (q), (d) → (p)

30. FBD of particle w.r.t. sphere

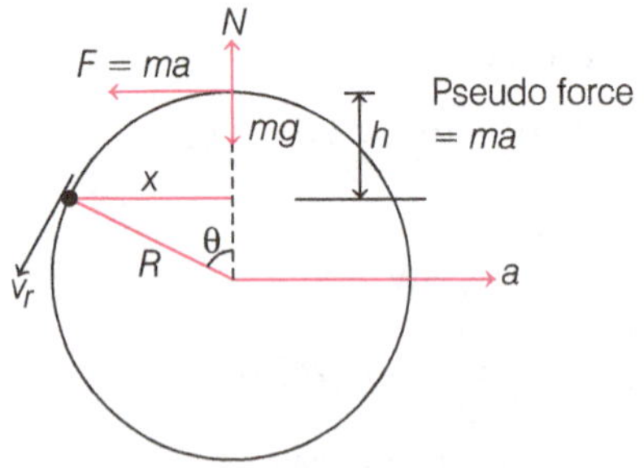

$$K_f=\frac{1}{2}mv_r^2=W_{\text{All}}$$

or $\quad \dfrac{1}{2}mv_r^2=W_N+W_F+W_{mg}$

$$=0+Fx+mgh$$

$$=(ma)(R\sin\theta)+mg\,[R\,(1-\cos\theta)]$$

∴ $\quad v_r=\sqrt{2gR\,(1+\sin\theta-\cos\theta)}$ **Ans.**

31. (a) $K=W=\int F dx=\int_0^2 (2.5-x^2)\,dx$

$$=2.33\text{ J}$$ **Ans.**

(b) Maximum kinetic energy of the block is at a point, where force changes its direction.

or $\quad F=0$

at $\quad X=\sqrt{2.5}$

$$=1.58\text{ m}$$

∴ $\quad K_{max}=\int_0^{1.58}(2.5-X^2)\,dx$

$$=2.635\text{ J}$$ **Ans.**

32. If block drops 12 mm, then spring will further stretch by 24 mm. Now,

$$E_i=E_f$$

∴ $\dfrac{1}{2}\times 1050\times(0.075)^2=-45\times 10\times 0.012$

$$+\frac{1}{2}\times 45\times v^2+\frac{1}{2}\times 1050\times(0.099)^2$$

Solving, we get $v=0.37$ m/s **Ans.**

33. From conservation of energy,

$$m_2gh_2=m_1gh_1+\frac{1}{2}m_1v_1^2+\frac{1}{2}m_2v_2^2$$

Here, $\quad v_1=v_2\cos\theta$

∴ $\quad 2\times 10\times 1=(0.5)(10)(\sqrt{5}-1)+\dfrac{1}{2}\times 0.5$

$$\times v_2^2\times\left(\frac{2}{\sqrt{5}}\right)^2+\frac{1}{2}\times 2\times v_2^2$$

∴ $\quad 20=6.18+0.2\,v_2^2+v_2^2$

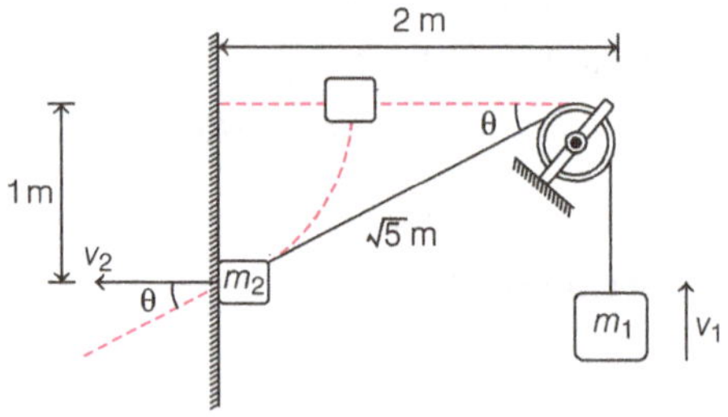

$$v_2=3.39\text{ m/s}$$ **Ans.**

and $\quad v_1=v_2\cos\theta=\dfrac{2}{\sqrt{5}}\times 3.39$

or $\quad v_1=3.03$ m/s **Ans.**

34. From constraint relations we can show that, if 1 moves h downwards then 2 will move $2h$ upwards. If velocity of 1 at the time of striking ground is v then velocity of 2 will be $2v$ upwards. So, let us first find v from energy conservation.

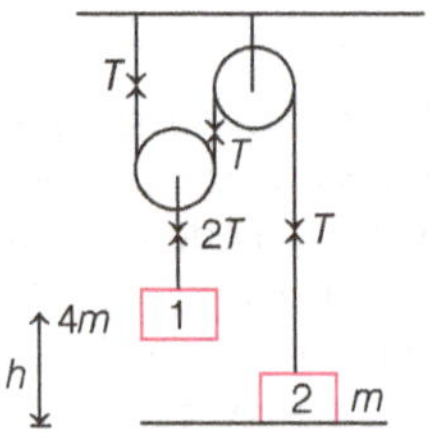

Decrease in potential energy of $4m$

= increase in potential energy of m

+ increase in kinetic energy of both.

$$\Rightarrow (4mgh)=mg(2h)+\frac{1}{2}m(2v)^2+\frac{1}{2}\times 4m\times v^2$$

Solving this equation, we get

$$v_2=2v=\sqrt{2gh}$$

After 1 strikes with ground, string slacks.

Body 2 comes under gravity. So further height attained by 2 is,

$$H=\frac{v_2^2}{2g}=h$$

So, total or maximum height attained by 2 is,

$$h_{max}=H+2h=3h=60\text{ cm}$$

or $\qquad h_{max} = 0.6$ m

35. (a) Thermal energy = Work done against friction

$$= \mu_K \, mgd$$
$$= (0.25)\,(3.5)\,(9.8)\,(7.8) \text{ J}$$
$$= 66.88 \text{ J} \qquad \textbf{Ans.}$$

(b) Maximum kinetic energy

$$= \text{work done against friction}$$
$$= 66.88 \text{ J} \qquad \textbf{Ans.}$$

(c) $\frac{1}{2} k x_m^2$ = maximum kinetic energy

$$\frac{1}{2} \times 640 \times x_m^2 = 66.88$$
$$x_m = 0.457 \text{ m}$$
$$= 45.7 \text{ cm} \qquad \textbf{Ans.}$$

36. Net retarding force $= kx + bMgx$

$\therefore$ Net retardation $= \left(\frac{k + bMg}{M}\right) \cdot x$

So, we can write

$$v \cdot \frac{dv}{dx} = -\left(\frac{k + bMg}{M}\right) \cdot x$$

or
$$\int_{v_0}^{0} v \cdot dv = -\left(\frac{k + bMg}{M}\right) \int_0^x x\,dx$$

or
$$x = \sqrt{\frac{M}{k + bMg}}\, v_0$$

Loss in mechanical energy,

$$\Delta E = \frac{1}{2} M v_0^2 - \frac{1}{2} k x^2$$

or
$$\Delta E = \frac{1}{2} M v_0^2 - \frac{k}{2}\left(\frac{M}{k + bMg}\right) v_0^2$$

or
$$\Delta E = \frac{v_0^2}{2}\left[M - k\left(\frac{M}{k + bMg}\right)\right]$$

$$= \frac{v_0^2 b M^2 g}{2(k + bMg)} \qquad \textbf{Ans.}$$

37. Friction coefficient μ

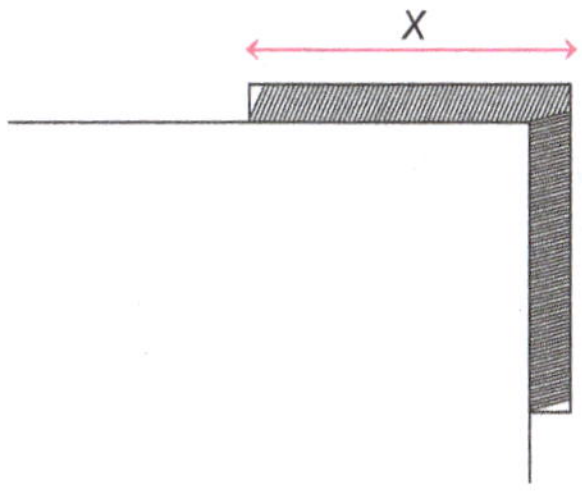

At $\frac{1}{3}$rd length overhanging portion,

pulling force = limiting friction

$$\therefore \quad \frac{mg}{3} = \mu\left(\frac{2}{3} mg\right)$$

$$\therefore \quad \mu = \frac{1}{2}$$

Work done by friction f

At some intermediate position when x length is on the table $\left(\frac{2l}{3} < x < l\right)$:

$$f_x = m_x \mu g = \left(\frac{m}{l} x\right)\left(\frac{1}{2}\right) g = \left(\frac{mg}{2l}\right) x$$

$$dW_f = (f_x)(-dx) \cos 180°$$
$$= \left(\frac{mg}{2l}\right) x\,dx$$

$$\therefore \quad W_f = \int_{x = \frac{2l}{3}}^{x = 0} dW_f = -\frac{1}{9} mgl$$

$$= -\frac{1}{9} \times 0.8 \times 9.81 \times 1.5$$
$$= -1.308 \text{ J}$$

$$\therefore \quad W_f = -1.3 \text{ J}$$

Note *x is decreasing, so we have taken $-dx$.*

38. In both the cases, work done by friction will be $-\mu Mgx$.

$$\therefore \quad \frac{1}{2} M v_C^2 = \frac{1}{2} M v_F^2$$
$$= Mgy - \mu Mgx$$

$$\therefore \quad v_C = v_F = \sqrt{2gy - 2\mu gx}$$

39. Let x_{max} = maximum extension of spring

Decrease in potential energy of A

= Increase in elastic potential energy of spring

$$\therefore \quad m_A g x_{max} = \frac{1}{2} k x_{max}^2$$

$$\therefore \quad x_{max} = \frac{2 m_A g}{k}$$

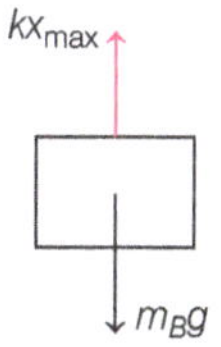

To just lift the block B,

$$k x_{max} = m_B g$$

$$\therefore \quad 2 m_A g = m_B g = mg$$

$$\therefore \quad m_A = \frac{m}{2} \qquad \textbf{Ans.}$$

40. Let X_m is maximum extension of spring. Then,
decrease in potential energy of M = increase in elastic potential energy of spring

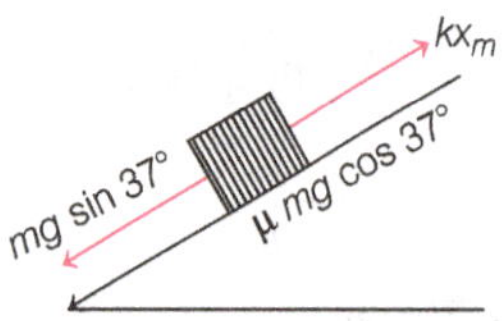

$\therefore \quad \frac{1}{2}kX_m^2 = MgX_m$

or $\quad X_m = \frac{2Mg}{k}$

Now, $\quad kX_m = mg \sin 37° + \mu mg \cos 37°$

or $\quad 2Mg = (mg)\left(\frac{3}{5}\right) + \left(\frac{3}{4}\right) mg \left(\frac{4}{5}\right)$

$\therefore \quad M = \frac{3}{5}m$ **Ans.**

41. $\mu m_A g = 0.8 \times 6 \times 10 = 48\ \text{N}$

$(m_B + m_C)g = (1+2) \times 10 = 30\ \text{N}$

Since, $\quad (m_B + m_C)g < \mu m_A g, a_A = a_B = 0$

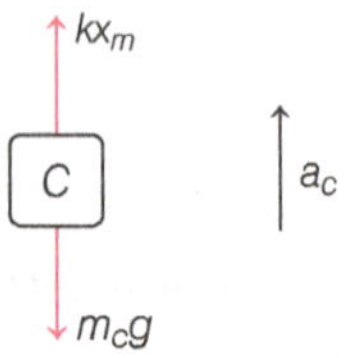

From conservation of energy principle, we can prove that maximum distance moved by C or maximum extension in the spring would be

$$x_m = \frac{2m_C g}{k} = \frac{2 \times 1 \times 10}{1000}$$

$= 0.02\ \text{m}$

At maximum extension,

$$a_C = \frac{kx_m - m_C g}{m_C}$$

Substituting the values, we have

$a_C = 10\ \text{m/s}^2$ **Ans.**

42.

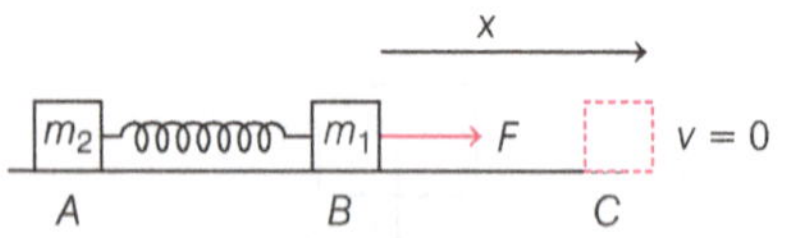

Let force constant of the spring is k. m_2 will shift when

$kx > \mu m_2 g \quad (x = \text{elongation})$

In limiting case let us take

$kx = \mu m_2 g \quad$ ...(i)

B to C

Work done by external forces (F and friction) = change in mechanical energy.

Thus,

$Fx - \mu m_1 gx = \frac{1}{2}kx^2 \quad$ ...(ii)

Solving equations (i) and (ii), we get

$$F = \left(m_1 + \frac{m_2}{2}\right)\mu g$$

43. If speed of sphere is v downwards, then speed of wedge at this instant will be $v \cot \alpha$ in horizontal direction.

Now, Decrease in potential energy of sphere

= Increase in kinetic energy of both

$\therefore \quad mgR = \frac{1}{2}mv^2 + \frac{1}{2}m(v \cot \alpha)^2$

$= \frac{1}{2}mv^2 \operatorname{cosec}^2 \alpha$

$\therefore \quad v = \sqrt{2gR} \sin \alpha$ = speed of sphere

and speed of wedge $= v \cot \alpha$

$= \sqrt{2gR} \cos \alpha$ **Ans.**

44. From work-energy theorem,

Work done by all forces = Change in kinetic energy

$\therefore \quad Fx - \frac{1}{2}kx^2 = \frac{1}{2}mv^2$

$\therefore \quad v = \sqrt{\frac{2Fx - kx^2}{m}}$

Substituting the values, we have

$$v = \sqrt{\frac{2 \times 20 \times 0.25 - 40 \times 0.25 \times 0.25}{0.5}}$$

$= \sqrt{15}$ m/s

= 3.87 m/s **Ans.**

(b) From conservation of mechanical energy,

$E_i = E_f \quad$ or $\quad \frac{1}{2}mv_i^2 + \frac{1}{2}kx_i^2 = \frac{1}{2}kx_f^2$

or $\quad x_f = \sqrt{\frac{mv_i^2}{k} + x_i^2} = \sqrt{\frac{0.5 \times 15}{40} + (0.25)^2}$

$= 0.5\ \text{m}$ (compression)

$\therefore$ Distance of block from the wall

$= (0.6 - 0.5)\ \text{m}$

= 0.10 m **Ans.**

45. For $t \le 0.2\ \text{s}$

$F = 800\ \text{N}$ and $v = \left(\frac{20}{0.3}\right)t$

$\therefore \quad P = Fv = (53.3t)\ \text{kW}$ **Ans.**

For $\quad t > 0.2\ \text{s}$

$$F = 800 - \left(\frac{800}{0.1}\right)(t - 0.2)$$

and $\quad v = \frac{20}{0.3}t$

$\therefore \quad P = Fv = (160t - 533t^2)\ \text{kW}$

$$W = \int_0^{0.2} (53.3t)\,dt + \int_{0.2}^{0.3} (160t - 533t^2)\,dt$$

= 1.69 kJ **Ans.**

46. (a) Retardation, $a = \frac{\mu mg}{m} = \mu g = 0.27 \times 9.81$

$$= 2.65 \text{ m/s}^2$$

Let t be the time when body comes to rest. Then,

$$0 = v_0 - at = 1.5 - 2.65t$$

$\therefore$ $t = 0.57$ s

Mean power or average power

$$= \frac{\text{Total work done (by friction)}}{\text{Total time}}$$

$$= \frac{\text{Change in kinetic energy (from work - energy theorem)}}{\text{Total time}}$$

$$= \frac{-\frac{1}{2}mv_0^2}{t} = \frac{-\frac{1}{2} \times 1.0 \times (1.5)^2}{0.57} = -1.97 \text{ W}$$

or $P_{av} \approx -2$ W

(b) Retardation, $a = -\frac{\mu mg}{m} = -\mu g = -(\alpha x g)$

or $v \cdot \frac{dv}{dx} = -\alpha x g$

$\therefore$ $v \cdot dv = -(\alpha x g)\, dx$

Integrating, we get

$$\frac{v^2}{2} = \frac{-\alpha g x^2}{2} + c$$

At $x = 0, v = v_0$

$\therefore$ $c = \frac{v_0^2}{2}$

$\therefore$ $\frac{v^2}{2} = \frac{v_0^2}{2} - \frac{\alpha g x^2}{2}$

or $v = \sqrt{v_0^2 - \alpha g x^2}$

Instantaneous power of friction,

$$P = fv \cos 180° \quad (f = \text{limiting friction})$$

$$= -(\mu mg)\, v$$

$$= -(\alpha mgx)\sqrt{v_0^2 - \alpha g x^2}$$

$$|P| = \alpha mgx \sqrt{v_0^2 - \alpha g x^2}$$

For maximum value of $|P|$, its first derivative with respect to x should be zero.

I am leaving the mathematical part for the students. Finally, we get

$$P_{max} = -\frac{1}{2} m v_0^2 \sqrt{\alpha g}$$

47. From work-energy theorem

Work done by all forces = change in kinetic energy

or $W_F + W_{mg} = K_f - K_i$

$18 \times 5 + (1 \times 10)(-4) = K_f$

$90 - 40 = K_f$ or $K_f = 50 \text{ J} = 5 \times 10$ J

48. Let for bouncing of the block A, the elongation in the spring is x_1.

$kx_1 > m_A g$ or $kx_1 > 100$

$\therefore$ $x_1 > \frac{100}{150}$ or $x_1 > \frac{2}{3}$

$\therefore$ $x_{1\,min} = \frac{2}{3}$

For $x_{1\,min}$, the velocity of block B will be zero.

According to conservation principle of mechanical energy,

$$U_i + K_i = U_f + K_f$$

$$\frac{1}{2}kx_0^2 + \frac{1}{2}mv_{0\,min}^2 = mg(x_0 + x_{1\,min}) + \frac{1}{2}kx_{1\,min}^2$$

Here, $x_0 = \frac{mg}{k}$

By putting the value,

$$v_{0_{min}} = \frac{20}{\sqrt{15}} \text{ m/s}$$

$\therefore$ $n = 5$

49. From work-energy theorem

$$F(0.2) - mg(20) = 0$$

$$F = mg\frac{(20)}{0.2} = mg(100)$$

$$= 0.15 \times 10 \times 100$$

$\therefore$ $F = 150$ N

Level 2

50. $\tan\theta = \frac{F_y}{F_x} = -\frac{x}{y}$

or $\frac{dy}{dx} = -\frac{x}{y}$

or $\int y\, dy = -\int x\, dx$

or $x^2 + y^2 = c$ is a circular curve.

Thus, the force is along tangent to the path of the block.

The magnitude of the force is

$$F = \sqrt{F_x^2 + F_y^2} = k$$

$\therefore$ $W = \int_c F\, ds = k \int_c ds$

$$= k\, 2\pi R = 2\pi Rk$$

51. $\mathbf{r} = \mathbf{OP} = x\hat{\mathbf{i}} + y\hat{\mathbf{j}}$

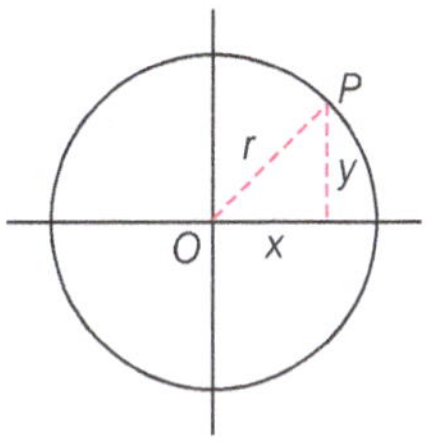

$$\mathbf{F} = \frac{K}{(x^2 + y^2)^{3/2}}(x\hat{\mathbf{i}} + y\hat{\mathbf{j}})$$

$$= \frac{K}{r^3}(\mathbf{r})$$

Since, $\mathbf{F}$ is along $\mathbf{r}$ or in radial direction.

Therefore, work done is zero.

52. $\mathbf{F} = -\left(\frac{\partial \phi}{\partial x}\hat{\mathbf{i}} + \frac{\partial \phi}{\partial y}\hat{\mathbf{j}}\right] = (-3\hat{\mathbf{i}} - 4\hat{\mathbf{j}})$

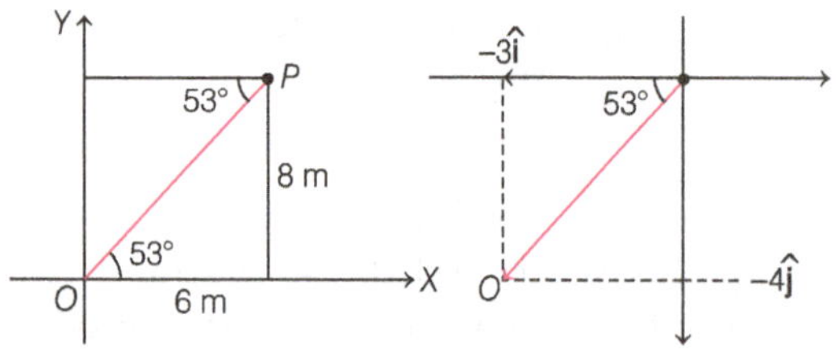

Since, particle was initially at rest. So, it will move in the direction of force.

We can see that initial velocity is in the direction of PO. So, the particle will cross the X-axis at origin.

$$K_i + U_i = K_f + U_f$$

$\therefore \quad 0 + (3\times 6 + 4\times 8) = K_f + (3\times 0 + 4\times 0)$

or $\quad K_f = 50\text{ J}$ **Ans.**

53. $K = \dfrac{1}{2}mv^2 \Rightarrow \dfrac{dK}{dt} = mv\dfrac{dv}{dt}$

Given, $\dfrac{dK}{dt} = \gamma t \quad \Rightarrow mv\dfrac{dv}{dt} = \gamma t$

$\Rightarrow \quad \int_0^v v\,dv = \int_0^t \dfrac{\gamma}{m} t\,dt$

$\Rightarrow \quad \dfrac{v^2}{2} = \dfrac{\gamma}{m}\dfrac{t^2}{2} \Rightarrow v = \sqrt{\dfrac{\gamma}{m}}t$

$\Rightarrow \quad a = \dfrac{dv}{dt} = \sqrt{\dfrac{\gamma}{m}}$

$\therefore \quad F = ma = \sqrt{\gamma m} = \text{constant}$

$\therefore \quad v = \dfrac{ds}{dt} = \sqrt{\dfrac{\gamma}{m}}t \Rightarrow s = \sqrt{\dfrac{\gamma}{m}}\dfrac{t^2}{2}$

54. If $F = 0$, $x = \sqrt{2}$ m

When $x > \sqrt{2}$ m, force is directed away from the origin.

When $0 \le x < \sqrt{2}$ m, force is directed towards origin.

If the particle just cross $x = \sqrt{2}$ m, then it starts to accelerate. For minimum value of initial speed, the speed of particle should be zero at $x = \sqrt{2}$ m. This means that KE at $x = \sqrt{2}$ must be zero and it is non-zero at origin.

55. $W = \int F\,dx = \int_2^{\sqrt{2}} (2x^2 - 4)\,dx$

$$= \left[\frac{2x^3}{3} - 4x\right]_2^{\sqrt{2}}$$

$$= \left(\frac{4\sqrt{2}}{3} - 4\sqrt{2}\right) - \left(\frac{16}{3} - 8\right)$$

$$= \frac{-8\sqrt{2}}{3} - \left(\frac{-8}{3}\right) = \frac{-8}{3}(\sqrt{2} - 1)$$

(a) $\because W = \Delta K$ or $\dfrac{-8}{3}(\sqrt{2} - 1) = 0 - T_i$

$\therefore \quad T_i = \dfrac{8}{3}(\sqrt{2} - 1)\text{ J}$

(b) Also, $W = \left[\dfrac{2x^3}{3} - 4x\right]_{\sqrt{2}}^{0}$

$$= -\left(\frac{4\sqrt{2}}{3} - 4\sqrt{2}\right) = \frac{8\sqrt{2}}{3}\text{ J}$$

But $\quad W = \Delta K$

or $\quad \dfrac{8\sqrt{2}}{3} = K_f - K_i = 0$

$\therefore \quad T_f = \dfrac{8\sqrt{2}}{3}\text{ J}$

(c) $T(x = \sqrt{2}) \ge 0$

56. At equilibrium

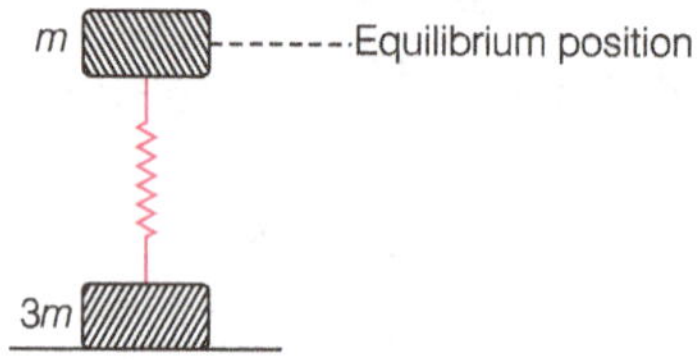

$$k\delta_0 = mg$$

or $\quad \delta_0 = \dfrac{mg}{k}$

where, $\delta_0 = \text{compression}$

(b) $\delta_{\text{Total}} = \delta + \delta_0 = \dfrac{3mg}{k}$

$F_{max} = k\delta_{max} = k\left(\dfrac{3mg}{k}\right) = 3mg$ (downward)

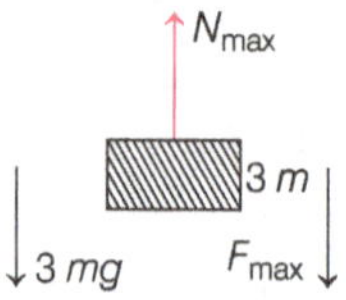

$\therefore \quad N_{max} = 3mg + F_{max} = 6mg$ **Ans.**

(d) If $\delta > \dfrac{4mg}{k}$, then upper block will move a distance

$x > \dfrac{4mg}{k} - \delta_0$ or $x > \dfrac{3mg}{k}$ from natural length.

Hence in this case, extension

$$x > \frac{3mg}{k}$$

or $\quad F = kx > 3mg$ (upwards on lower block)

So, lower block will bounce up.

57. (A) $F_x = \dfrac{-dU}{dx} = -\dfrac{2U_0}{a^3}[x - a][x][x + a]$

$$F = 0 \text{ at } x = 0,\ x = a,\ x = -a$$

and $\quad U = 0$ at $x = -a$

and $\quad x = a$

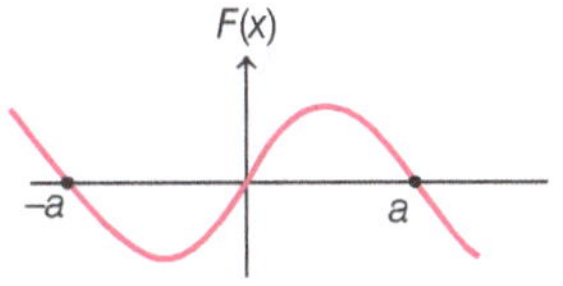

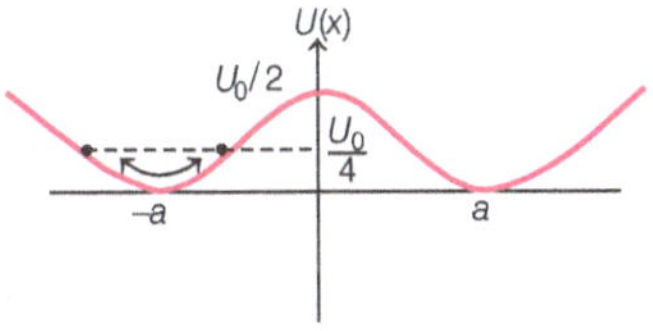

(B) $F_x = -\frac{dU}{dx} = -U_0\left(\frac{x}{a}\right)$

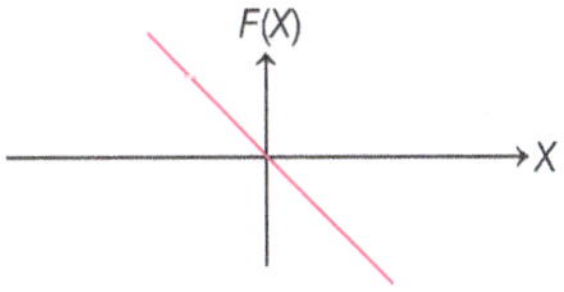

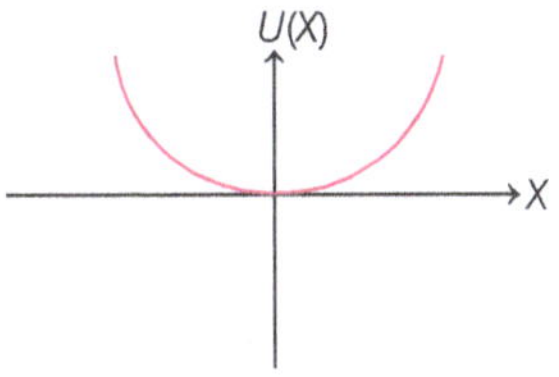

(C) $F_x = -\frac{dU}{dx} = U_0 \frac{e^{-x^2/x^2}}{a^3}[x][x-a][x+a]$

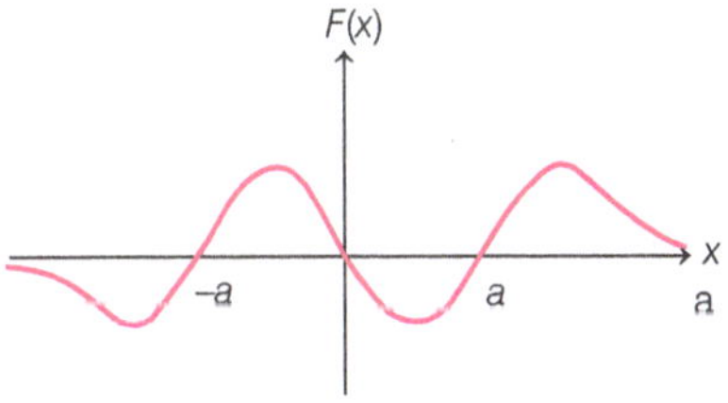

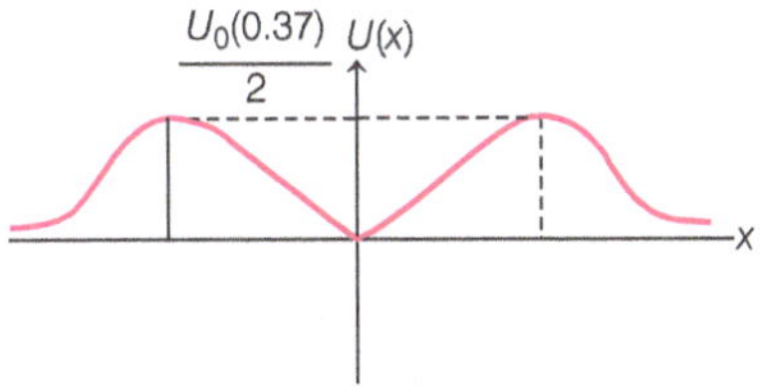

(D) $F_x = -\frac{dU}{dx} = -\frac{U_0}{2a^3}[(x-a)(x+a)]$

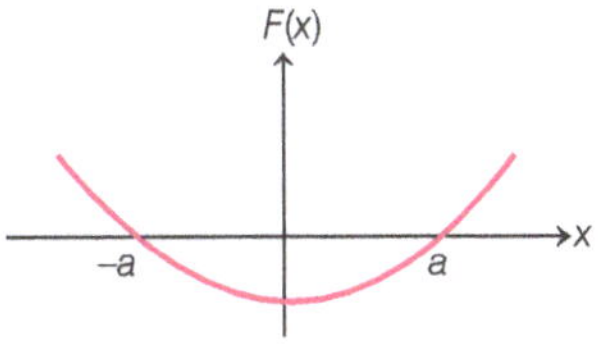

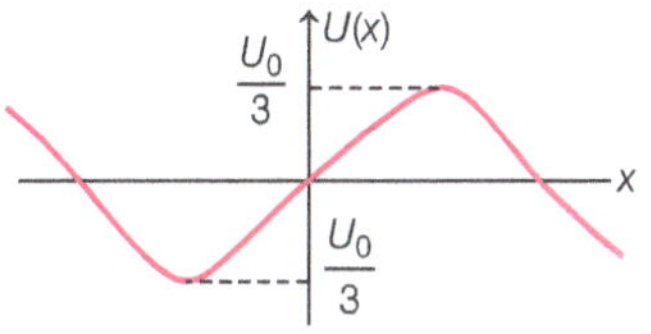

58. For shifting of B, $kx_0 > mg\sin\theta + \mu\, mg\cos\theta$

or $kx_0 > \sqrt{2}\times 10\times\frac{1}{\sqrt{2}} + 0.5\times\sqrt{2}\times 10\times\frac{1}{\sqrt{2}}$

or $\quad kx_0 > 15$

$\therefore \quad x_{0\,\min} = \frac{15}{k}$

For minimum value of F, the block B will be in the condition of just shift, when the block A is at its extreme position ($v = 0$). Applying work-energy theorem for the block A,

$$W = \Delta K = K_f - K_i = 0 - 0 = 0 \quad \text{or} \quad W_F + W_S + W_f = 0$$

or $\quad Fx_{0\,\min} - \frac{1}{2}kx^2_{0\,\min} - \mu\, mg\, x_{0\,\min} = 0$

or $\quad F - \frac{k}{2}x_{0\,\min} - \mu\, mg = 0$

or $\quad F - \frac{k}{2}\times\frac{15}{k} - \mu\, mg = 0$

or $\quad F - 7.5 - 0.5\times\sqrt{2}\times 10 = 0$

$\therefore \quad F = 7.5 + 5\sqrt{2} = 14.57\text{ N}$

59. (a) From energy conservation principle,

Work done against friction = decrease in elastic PE

or $\quad f(x_0 + a_1) = \frac{1}{2}k(x_0^2 - a_1^2)$ or $x_0 - a_1 = \frac{2f}{k}$ $\quad$ ...(i)

From Eq. (i), we see that decrease of amplitude $(x_0 - a_1)$ is $\frac{2f}{k}$, which is constant and same for each cycle of oscillation.

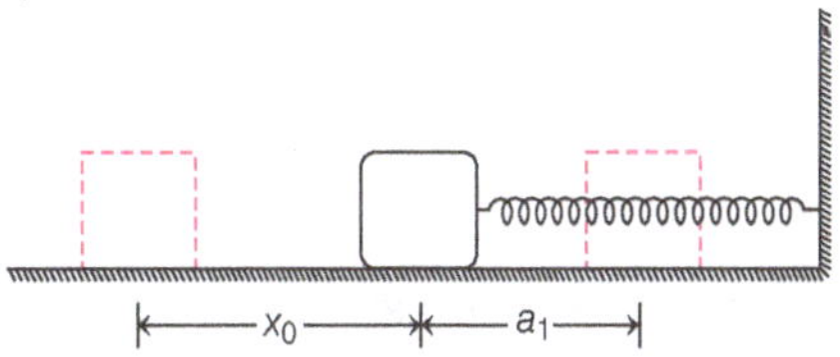

(b) The block will come to rest when $ka = f$

or $\quad a = \frac{f}{k}$ $\quad$ ...(A)

In the similar manner, we can write

$a_1 - a_2 = \frac{2f}{k}$ $\quad$ ...(i)

$a_2 - a_3 = \frac{2f}{k}$ $\quad$ ...(ii)

... ...

$a_{n-1} - a_n = \frac{2f}{k}$ $\quad$ ...(n)

Adding Eqs. (i), (ii),... etc., we get

$$x_0 - a_n = n\left(\frac{2f}{k}\right)$$

or $$a_n = x_0 - n\left(\frac{2f}{k}\right) \quad ...(B)$$

Equating Eqs. (A) and (B), we get

$$\frac{k}{f} = x_0 - n\left(\frac{2f}{k}\right)$$

or $$n = \frac{x_0 - \frac{k}{f}}{\frac{2f}{k}} = \frac{kx_0}{2f} - \frac{1}{2}$$

Number of cycles,

$$m = \frac{n}{2} = \frac{kx_0}{4f} - \frac{1}{4} = \frac{1}{4}\left(\frac{kx_0}{f} - 1\right) \quad \textbf{Ans.}$$

60. $\mathbf{F = 2(ay - 1)\,mg}$

On ground, $y = 0$

$\therefore \quad F = -2mg$

g is downwards, so **F** is upwards. Or we can say, $F = 2mg$ upwards.

Now as y increases, this force decreases. Thus, $F = 2mg\,(1 - ay)$ is acting upwards.

Suppose h is the maximum height upto which body rises (where velocity becomes zero).

So work done by applied force is equal to change in mechanical energy or change in only potential energy (as Δ KE $= 0$). Thus,

$$\int_0^h F dy = \Delta U = mgh$$

or $$\int_0^h 2mg\,(1 - ay)\,dy = mgh$$

or $$2h - 2a\frac{h^2}{2} = h$$

or $$h = \frac{1}{a}$$

Work performed by the applied force in first half, $\frac{h}{2}$ or $\frac{1}{2a}$

$$W = \int_0^{1/2a} 2mg\,(1 - ay)\,dy$$

Solving, we get

$$W = \frac{3mg}{4a}$$

Increment in potential energy

$$\Delta U = mg\frac{h}{2} = mg\left(\frac{1}{2a}\right) \quad \text{or} \quad \Delta U = \frac{mg}{2a}$$

61.

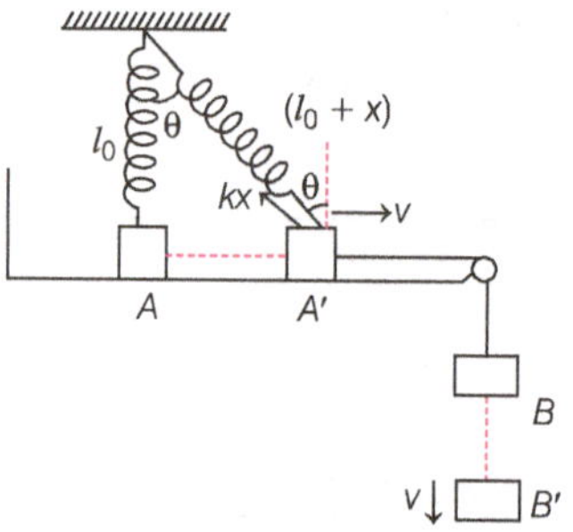

$$AA' = BB'$$

Force constant $k = \dfrac{5mg}{l_0}$

Let A breaks off in position A' where extension in spring is x. At the time of breaking off,

$$kx\cos\theta = mg \quad ...(i)$$

$$\cos\theta = \frac{l_0}{l_0 + x} \quad ...(ii)$$

$$AA' = BB' = (l_0 + x)\sin\theta \quad ...(iii)$$

Suppose at this moment velocity of blocks is v.

Applying energy conservation principle, we have decrease in potential energy of B = increase in kinetic energy of both + increase in potential energy of spring

$$\therefore \quad mg(BB') = 2 \times \frac{1}{2}mv^2 + \frac{1}{2}kx^2 \quad ...(iv)$$

We have four unknowns x, θ, BB' and v.

Solving these four equations with given values, we get

$$v = 1.7 \text{ m/s}$$

62. Let k be the force constant of the elastic cord and let x be the extension in the cord when bar starts sliding on the plank.

Friction coefficient $= \mu = 0.20$.

Writing the equations of motion for the bar at the time of sliding in horizontal and vertical directions, we have

$$\mu N = kx\sin\theta \quad ...(i)$$

$$mg = (kx\cos\theta + N) \quad ...(ii)$$

and $$\cos\theta = \frac{l_0}{l_0 + x} \quad ...(iii)$$

Plank is shifted slowly, so we can assume that velocity is almost zero everytime or its acceleration is also zero. So, it is an intertial frame.

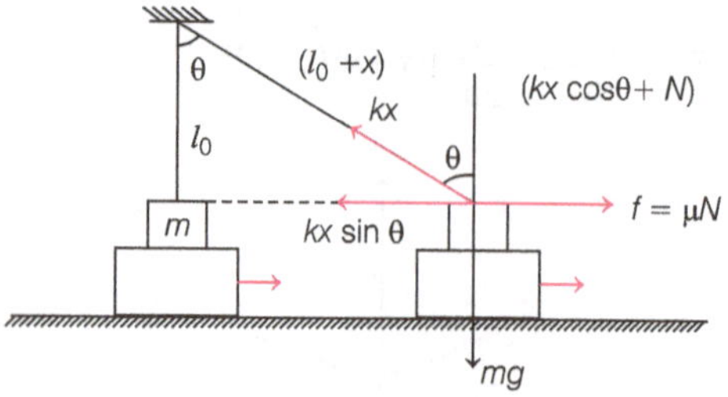

Let us take cord and bar (of mass m) as a system.

Work done by external force friction on this system $=$ Change in mechanical energy of this system $=$ Change in potential energy as $\Delta KE = 0$

or $$W_f = \frac{1}{2}kx^2 \qquad \text{...(iv)}$$

We have four unknowns x, N, θ and W_f.

Solving these four equations, we can find

$$W_f = 0.09 \text{ J}$$

Note *Friction f is acting on the bottom of bar m, for better understanding, we have shown it on the top of it.*

63. (a) $U = \frac{a}{r^2} - \frac{b}{r}$

$$F = -\frac{dU}{dr} = \frac{2a}{r^3} - \frac{b}{r^2}$$

At equilibrium position,

$$F = 0$$

$$\therefore \quad \frac{2a}{r^3} - \frac{b}{r^2} = 0 \quad \text{or} \quad r = \frac{2a}{b} = r_0 \qquad \text{(Given)}$$

Thus $$r_0 = \frac{2a}{b}$$

For $r > r_0$

$$\frac{b}{r^2} > \frac{2a}{r^3}$$

$\therefore$ Force will be negative or restoring in nature. Thus equilibrium is steady or stable.

(b) In central force, field force will be attractive, if it is negative and force is negative for $r > r_0$. We have to find maximum attractive force or magnitude wise maximum negative force. Or we can say we have to find minimum value of F.

$$F = \frac{2a}{r^3} - \frac{b}{r^2} \qquad \text{...(i)}$$

For F to be minimum,

$$\frac{dF}{dr} = 0 \quad \text{or} \quad -\frac{6a}{r^4} + \frac{2b}{r^3} = 0$$

$$\therefore \quad r = \frac{3a}{b} \Rightarrow \frac{d^2F}{dr^2} = \frac{24a}{r^5} - \frac{6b}{r^4}$$

At $r = \frac{3a}{b}, \frac{d^2F}{dr^2}$ is positive.

Therefore F is minimum.

From equation (i), value of F at $r = \frac{3a}{b}$ is

$$F = -\frac{b^3}{27a^2}$$

Hence magnitude of maximum attractive force,

$$F_{max} = \frac{b^3}{27a^2}$$

***U - r* and *F - r* Graphs**

(i) At $r = \frac{a}{b}, U = 0$

(ii) At $r = \frac{2a}{b}, \frac{dU}{dr}$ or $F = 0$

(iii) For $r < \frac{a}{b}, U =$ positive

(iv) For $r > \frac{a}{b}, U =$ negative

(v) At $r = \frac{2a}{b}, U$ is minimum

(vi) For $r < \frac{2a}{b}, F$ is positive

(vii) For $r > \frac{2a}{b}, F$ is negative

(viii) At $r = \frac{3a}{b}, F$ is minimum

(ix) For $r \to 0, U \to +\infty$ and $F \to +\infty$

(x) For $r \to \infty, U$ and $F \to 0$

Keeping all these points in mind U - r and F - r graphs are as shown below.

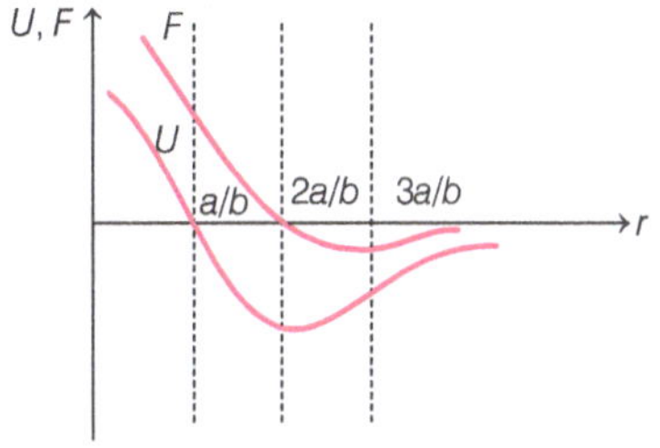

Magnitude of maximum negative value of $F = \frac{b^3}{27a^2}$

at $r = \frac{3a}{b}$

Magnitude of maximum negative value of $U = \frac{b^2}{4a^2}$

at $r = \frac{2a}{b}$.

64. From work-energy principle, $W = \Delta KE$

$$\therefore \quad Pt = \frac{1}{2}m(v^2 - u^2) \qquad (P = \text{power})$$

or $$t = \frac{m}{2P}(v^2 - u^2) \qquad \text{...(i)}$$

Further $$F \cdot v = P$$

$$\therefore \quad m \cdot \frac{dv}{ds} \cdot v^2 = P \quad \text{or} \quad \int_u^v v^2\, dv = \frac{P}{m}\int_0^x ds$$

$$\therefore \quad (v^3 - u^3) = \frac{3P}{m} \cdot x \quad \text{or} \quad \frac{m}{P} = \frac{3x}{v^3 - u^3}$$

Substituting in Eq. (i), we get

$$t = \frac{3x(u+v)}{2(u^2 + v^2 + uv)}$$ **Hence proved.**

UNIT-10

CIRCULAR MOTION

Learning Modules

1. Kinematics of Circular Motion
2. Dynamics of Circular Motion
3. Centrifugal Force
4. Vertical Circular Motion

MODULE 1

Kinematics of Circular Motion

Introduction

Circular motion is a two dimensional motion or motion in a plane. This plane may be horizontal, inclined or vertical. But in most of the cases, this plane is horizontal.

In circular motion, direction of velocity continuously keeps on changing. Therefore, even though speed is constant, velocity certainly keeps on changing. So body is accelerated.

Later on we will see that this is a variable acceleration. So, we cannot apply the equations $\mathbf{v} = \mathbf{u} + \mathbf{a}t$ etc. directly.

Key Points

(i) In general, in any curvilinear motion direction of instantaneous velocity is tangential to the path, while acceleration may have any direction. Resolve the acceleration in two perpendicular directions, one parallel to velocity and the other normal to velocity. The first component is a_t while the other is a_n

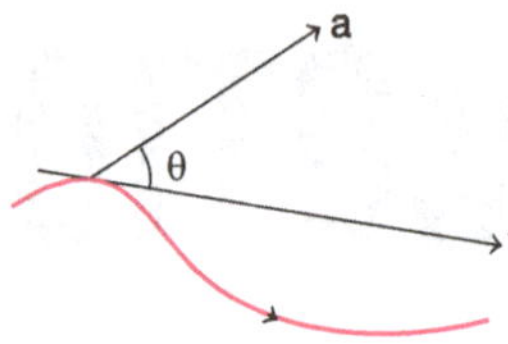

Thus, $a_t =$ component of $\mathbf{a}$ along $\mathbf{v} = a\cos\theta = \dfrac{\mathbf{a}\cdot\mathbf{v}}{v}$

$$= \frac{dv}{dt} = \frac{d|\mathbf{v}|}{dt} = \text{rate of change of speed}$$

and $a_n =$ component of $\mathbf{a}$ perpendicular to $\mathbf{v} = a\sin\theta$

$$= \sqrt{a^2 - a_t^2} = \frac{v^2}{R}$$

Here, v is the speed of particle at that instant and R is called the radius of curvature to the curvilinear path at that point.

(ii) Acceleration means change in velocity. Velocity (a vector quantity) can be changed either but changing its magnitude or direction. This is the reason why we break $\mathbf{a}$ into $\boldsymbol{a}_t$ and $\boldsymbol{a}_n$.

Here $\boldsymbol{a}_t$ is responsible for **change in speed** (or magnitude of velocity) and $\boldsymbol{a}_n$ is responsible for **change in direction** of velocity vector. This is also called **radial (a_r)** or **centripetal acceleration (a_c)**.

(iii) In $\boldsymbol{a}_t = \boldsymbol{a}\cos\theta$, if θ **is acute**, a_t will be **positive** and speed will increase. If θ is **obtuse** a_t will be **negative** and speed will decrease. If θ **is 90°**, a_t is **zero** and speed will remain constant.

Kinematics of Circular Motion

Velocity

In circular motion, a particle has two velocities :

(i) Angular velocity

(ii) Linear velocity

Angular Velocity

Suppose a particle P is moving in a circle of radius r and centre O.

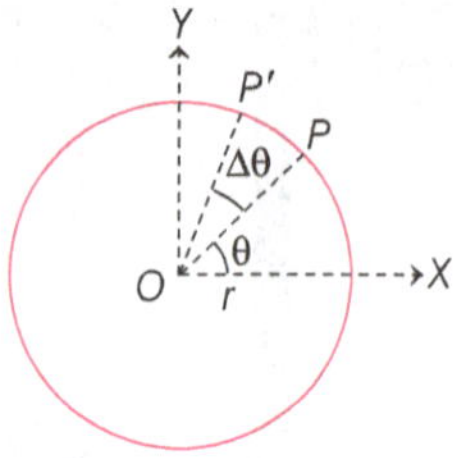

The position of the particle P at a given instant may be described by the angle θ between OP and OX. This angle θ is called the **angular position** of the particle. As the particle moves on the circle, its angular position θ changes. Suppose the point rotates an angle $\Delta\theta$ in time Δt. The rate of change of angular position is known as the **angular velocity (ω)**. Thus,

$$\omega = \lim_{\Delta t \to 0} \frac{\Delta\theta}{\Delta t} = \frac{d\theta}{dt}$$

Here, ω is the angular speed or magnitude of angular velocity. Angular velocity is a vector quantity. Direction of ω is perpendicular to plane of circle and given by screw law.

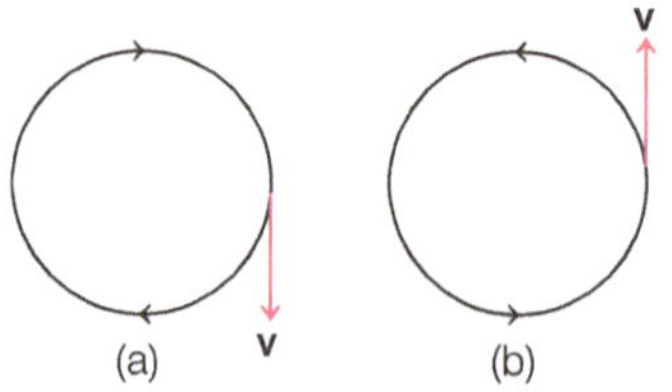

(a) (b)

In the Fig. (a), when the particle is rotating clockwise, direction of ω is perpendicular to paper inwards or in ⊗ direction.

In Fig. (b), when the particle is rotating in anticlockwise direction, direction of ω is perpendicular to paper outwards or in ⊙ direction.

Linear Velocity

Linear velocity is as usual, $v = \frac{ds}{dt}$ or $\frac{dr}{dt}$

Direction of linear velocity is tangential to the circle as shown in figure.

Magnitude of linear velocity is called linear speed v. Thus,

$$v = |\mathbf{v}| = \left|\frac{ds}{dt} \text{ or } \frac{dr}{dt}\right|$$

Relation between linear speed and angular speed

In the figure (in angular velocity) linear distance *PP′* travelled by the particle in time Δt is

$$\Delta s = r\Delta\theta \quad \text{or} \quad \lim_{\Delta t \to 0} \frac{\Delta s}{\Delta t} = r\left(\lim_{\Delta t \to 0} \frac{\Delta\theta}{\Delta t}\right)$$

or $$\frac{ds}{dt} = r\frac{d\theta}{dt} \quad \text{or} \quad v = r\omega$$

Acceleration

Like velocity, a particle in circular motion has two accelerations:

(i) Angular acceleration (ii) Linear acceleration

Angular Acceleration

The rate of change of angular velocity is called the **angular acceleration** (α).

Thus, $$\alpha = \frac{d\omega}{dt} = \frac{d^2\theta}{dt^2}$$

Angular acceleration is also a vector quantity. Direction of α is also perpendicular to plane of circle, either parallel or antiparallel to ω. If angular speed of the particle is increasing, then α is parallel to ω and if angular speed is decreasing, then α is antiparallel to ω. Angular acceleration is zero if angular speed (or angular velocity) is constant.

Linear Acceleration

In circular motion, linear speed of the particle may or may not be constant but direction of linear velocity continuously keeps on changing. So, velocity is continuously changing. Therefore, acceleration cannot be zero. But of course we can resolve the linear acceleration into two components:

(i) tangential acceleration (a_t)

(ii) radial or centripetal acceleration (a_r)

Component of linear acceleration in tangential direction is called tangential acceleration (a_t). This component is responsible for change in linear speed. This is the rate of change of speed. Thus,

$$a_t = \frac{dv}{dt} = \frac{d|\mathbf{v}|}{dt}$$

If speed of the particle is constant, then a_t is zero. If speed is increasing, then this is positive and in the direction of linear velocity. If speed is decreasing, then this component is negative and in the opposite direction of linear velocity.

Tangential component of the linear acceleration and angular acceleration have following relation:

$$a_t = \frac{dv}{dt} = \frac{d(r\omega)}{dt} = r\frac{d\omega}{dt}$$

$$= r\alpha \qquad \left(\text{as } \frac{d\omega}{dt} = \alpha\right)$$

$\therefore$ $$a_t = r\alpha$$

Component of linear acceleration in radial direction (towards centre) is called radial or normal or centripetal acceleration. This component is responsible for change in direction of linear velocity. So, this component can never be zero, as the direction continuously keeps on changing. Value of this component is

$$a_r = \frac{v^2}{r} = r\omega^2 \qquad (\text{as } v = r\omega)$$

These two components are mutually perpendicular. So, the net linear acceleration is the vector sum of these two, as shown in figure.

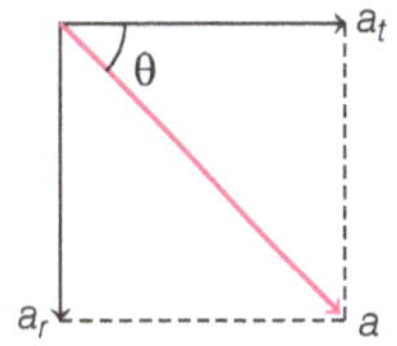

$$a = \sqrt{a_t^2 + a_r^2} = \sqrt{\left(\frac{dv}{dt}\right)^2 + \left(\frac{v^2}{r}\right)^2}$$

or $$a = \sqrt{(r\alpha)^2 + (r\omega^2)^2}$$

and $$\tan\theta = \frac{a_r}{a_t} \quad \text{or} \quad \theta = \tan^{-1}\left(\frac{a_r}{a_t}\right)$$

Conclusion

In circular motion, acceleration of the particle has two components

(i) tangential acceleration $a_t = \frac{dv}{dt} = r\alpha$

(ii) normal or radial or centripetal acceleration

$$a_n \text{ or } a_r = \frac{v^2}{r} = r\omega^2$$

a_t and a_n are two perpendicular components of **a**. Hence, we can write $a = \sqrt{a_t^2 + a_n^2}$

Since, circular motion, is a 2-D motion we can write

$$a = \sqrt{a_x^2 + a_y^2} = \sqrt{\left(\frac{dv}{dt}\right)^2 + \left(\frac{v^2}{r}\right)^2}$$

Here, $v = \sqrt{v_x^2 + v_y^2}$

or $v^2 = v_x^2 + v_y^2$

Three Types of Circular Motion

For better understanding, we can classify the circular motion in following three types:

(i) **Uniform circular motion** in which v and ω are constant.

In this motion,

$$v \text{ or } |\mathbf{v}| = \text{constant}$$

$$\Rightarrow \quad a_t = 0$$

$$\Rightarrow \quad \omega = \text{constant}$$

$$\Rightarrow \quad \alpha = 0$$

$$a = a_r = \frac{v^2}{r} = r\omega^2$$

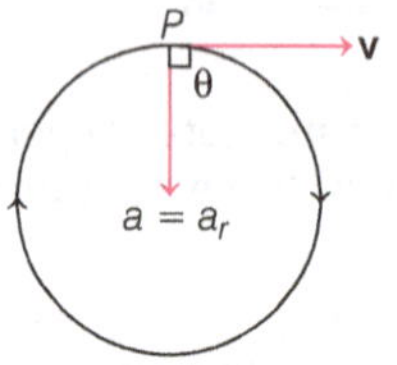

θ = 90°

a is towards centre, **v** is tangential and according to the shown figure, ω is perpendicular to paper inwards or in ⊗ direction.

(ii) **Accelerated circular motion** in which v and ω are increasing. So, a_t is in the direction of **v** and α is in the direction of ω.

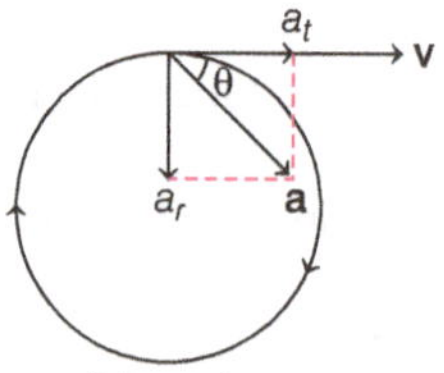

θ is acute

In the figure shown, α and ω both are perpendicular to paper inwards.

Further,

$$a = \sqrt{a_t^2 + a_r^2} \quad \text{and} \quad \tan\theta = \frac{a_r}{a_t}$$

(iii) **Retarded circular motion** in which v and ω are decreasing. So, a_t is in the opposite direction of **v** and α is in the opposite direction of ω.

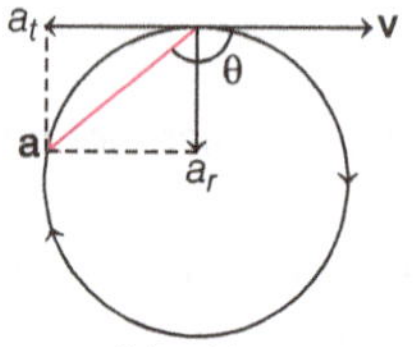

θ is obtuse

In the figure shown, ω is **perpendicular to paper** inwards in ⊗ direction and α is perpendicular to paper outwards in ⊙ direction.

Note *In the above figures, θ is the angle between **v** and **a**.*

Important POINTS

- Relation between angular velocity vector ω, velocity vector **v** and position vector of the particle with respect to centre **r** is given by

$$\mathbf{v} = \boldsymbol{\omega} \times \mathbf{r}$$

- In circular motion, if angular acceleration α is constant then we can apply the following equations directly:

$$\omega = \omega_0 + \alpha t$$

$$\Rightarrow \quad \omega^2 = \omega_0^2 + 2\alpha\theta$$

$$\text{and} \quad \theta = \omega_0 t + \frac{1}{2}\alpha t^2$$

Here, ω_0 is the initial angular velocity and ω, the angular velocity at time t. Similarly, θ is the angle rotated by position vector of the particle (with respect to centre).

- If angular acceleration is not constant, then we will have to take help of differentiation or integration. The basic equations are

$$\omega = \frac{d\theta}{dt}$$

$$\text{and} \quad \alpha = \frac{d\omega}{dt} = \omega\frac{d\omega}{d\theta}$$

$$\therefore \quad \int d\theta = \int \omega\, dt, \int d\omega = \int \alpha\, dt$$

$$\text{and} \quad \int \omega\, d\omega = \int \alpha\, d\theta$$

- Vectors lying in the plane of circular motion are : **v**, **a**, a_t and a_r and vectors perpendicular to the plane of circular motion are ω and α.

Example 1 A particle moves in a circle of radius 0.5 m at a speed that uniformly increases. Find the angular acceleration of particle if its speed changes uniformly from 2.0 m/s to 4.0 m/s in 4.0 s.

Sol. The tangential acceleration of the particle is

$$a_t = \frac{dv}{dt}$$

or $$\frac{\Delta v}{\Delta t} = \frac{4.0 - 2.0}{4.0} = 0.5 \text{ m/s}^2$$

The angular acceleration is $\alpha = \frac{a_t}{r}$

$$= \frac{0.5}{0.5} = 1 \text{ rad/s}^2$$ **Ans.**

Example 2 Speed of a particle moving in a circle of radius $r = 2\text{m}$ varies with time t as $v = t^2$, where t is in second and v in m/s. Find the radial, tangential and net acceleration at $t = 2$s.

Sol. Linear speed of particle at $t = 2$ s is

$$v = (2)^2 = 4 \text{ m/s}$$

∴ Radial acceleration, $a_r = \frac{v^2}{r} = \frac{(4)^2}{2} = 8 \text{ m/s}^2$

The tangential acceleration is $a_t = \frac{dv}{dt} = 2t$

∴ Tangential acceleration at $t = 2$ s is

$$a_t = (2)(2) = 4 \text{ m/s}^2$$

∴ Net acceleration of particle at $t = 2$ s is

$$a = \sqrt{(a_r)^2 + (a_t)^2}$$

$$= \sqrt{(8)^2 + (4)^2}$$

or $$a = \sqrt{80} \text{ m/s}^2$$

Example 3 In circular motion, what are the possible values (zero, positive or negative) of the following :

(a) $\omega \cdot \mathbf{v}$ (b) $\mathbf{v} \cdot \mathbf{a}$ (c) $\omega \cdot \alpha$

Sol. (a) **v** lies in the plane of circle and ω is always perpendicular to this plane.

∴ $\mathbf{v} \perp \omega$ (always)

Hence, $\omega \cdot \mathbf{v}$ is always zero.

(b) **v** and **a** both lie in the plane of circle and the angle between these two vectors may be acute (when speed is increasing) obtuse (when speed is decreasing) or 90° (when speed is constant).

Hence, $\mathbf{v} \cdot \mathbf{a}$ may be positive, negative or zero.

(c) ω and α are either parallel ($\theta = 0°$ between ω and α) or antiparallel ($\theta = 180°$). In uniform circular motion, α has zero magnitude. Hence, $\omega \cdot \alpha$ may be positive, negative or zero.

INTRODUCTORY Exercise

1. Is the acceleration of a particle in uniform circular motion constant or variable?
2. Which of the following quantities may remain constant during the motion of an object along a curved path?
   (i) Velocity
   (ii) Speed
   (iii) Acceleration
   (iv) Magnitude of acceleration
3. A particle moves in a circle of radius 1.0 cm with a speed given by $v = 2t$, where v is in cm/s and t in seconds.
   (a) Find the radial acceleration of the particle at $t = 1$ s.
   (b) Find the tangential acceleration at $t = 1$ s.
   (c) Find the magnitude of net acceleration at $t = 1$ s.
4. A particle is moving with a constant speed in a circular path. Find the ratio of average velocity to its instantaneous velocity when the particle rotates an angle $\theta = \left(\frac{\pi}{2}\right)$.
5. A particle is moving with a constant angular acceleration of 4 rad/s^2 in a circular path. At time $t = 0$, particle was at rest. Find the time at which the magnitudes of centripetal acceleration and tangential acceleration are equal.
6. A particle rotates in a circular path of radius 54 m with varying speed $v = 4t^2$. Here v is in m/s and t in second. Find angle between velocity and acceleration at $t = 3$ s.
7. Figure shows the total acceleration and velocity of a particle moving clockwise in a circle of radius 2.5 m at a given instant of time. At this instant, find :

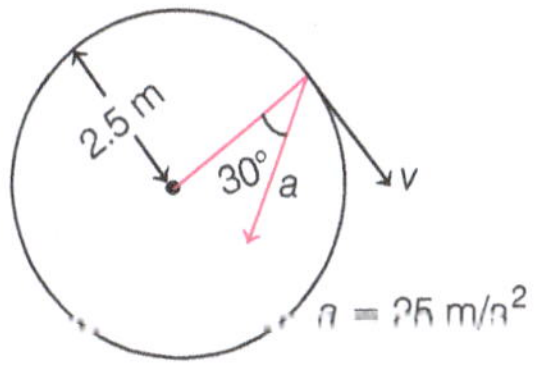

   (a) the radial acceleration,
   (b) the speed of the particle and
   (c) its tangential acceleration.

Answers

1. Variable
2. Speed, acceleration, magnitude of acceleration
3. (a) 4.0 cm/s^2 (b) 2.0 cm/s^2 (c) $2\sqrt{5}$ cm/s^2
4. $2\sqrt{2}/\pi$ 5. 1/2 s 6. 45°
7. (a) 21.65 ms^{-2} (b) 7.35 ms^{-1} (c) 12.5 ms^{-2}

See hints on page 581.

Miscellaneous Examples

Ex. 1 If a point moves along a circle with constant speed, prove that its angular speed about any point on the circle is half of that about the centre.

Sol. Let, O be a point on a circle and P be the position of the particle at any time t, such that

$$\angle POA = \theta$$

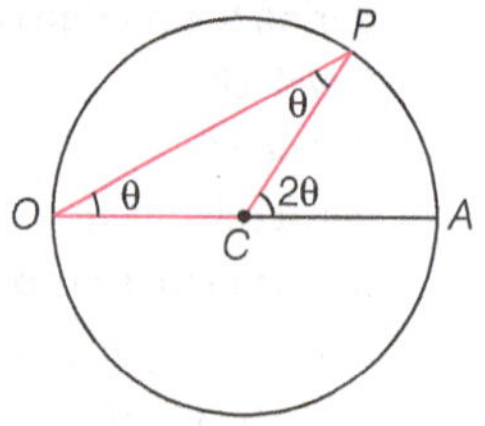

Then, $\angle PCA = 2\theta$

Here, C is the centre of the circle.

Angular velocity of P about O is $\omega_O = \dfrac{d\theta}{dt}$

and angular velocity of P about C is,

$$\omega_C = \frac{d}{dt}(2\theta) = 2\frac{d\theta}{dt}$$

or $\omega_C = 2\omega_O$ **Proved.**

Ex. 2 A particle is projected with a speed u at an angle θ with the horizontal. What is the radius of curvature of the parabola traced out by the projectile at a point where the particle velocity makes an angle $\dfrac{\theta}{2}$ with the horizontal.

Sol. Let v be the velocity at the desired point. Horizontal component of velocity remains unchanged. Hence,

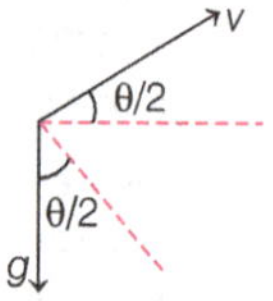

$$v\cos\frac{\theta}{2} = u\cos\theta$$

$$\therefore \quad v = \frac{u\cos\theta}{\cos\dfrac{\theta}{2}} \quad \ldots \text{(i)}$$

Radial acceleration is the component of acceleration perpendicular to velocity or

$$a_n = g\cos\left(\frac{\theta}{2}\right)$$

$$\therefore \quad \frac{v^2}{R} = g\cos\left(\frac{\theta}{2}\right) \quad \ldots \text{(ii)}$$

Substituting the value of v from Eq. (i) in Eq. (ii), we have radius of curvature

$$R = \frac{\left[\dfrac{u\cos\theta}{\cos\left(\dfrac{\theta}{2}\right)}\right]^2}{g\cos\left(\dfrac{\theta}{2}\right)} = \frac{u^2\cos^2\theta}{g\cos^3\left(\dfrac{\theta}{2}\right)}$$ **Ans.**

Ex. 3 A point moves along a circle with a speed $v = kt$, where $k = 0.5\ \text{m/s}^2$. Find the total acceleration of the point at the moment when it has covered the nth fraction of the circle after the beginning of motion, where $n = \dfrac{1}{10}$.

Sol. $v = \dfrac{ds}{dt} = kt$ or $\int_0^s ds = k\int_0^t t\,dt \quad \therefore \quad s = \dfrac{1}{2}kt^2$

For completion of nth fraction of circle,

$$s = 2\pi rn = \frac{1}{2}kt^2 \quad \text{or} \quad t^2 = (4\pi nr)/k \quad \ldots\text{(i)}$$

$$\text{Tangential acceleration} = a_t = \frac{dv}{dt} = k \quad \ldots\text{(ii)}$$

$$\text{Normal acceleration} = a_n = \frac{v^2}{r} = \frac{k^2t^2}{r} \quad \ldots\text{(iii)}$$

Substituting the value of t^2 from Eq. (i), we have

$$a_n = 4\pi nk$$

$$\therefore \quad a = \sqrt{(a_t^2 + a_n^2)} = [k^2 + 16\pi^2n^2k^2]^{1/2}$$

$$= k\,[1 + 16\pi^2n^2]^{1/2}$$

$$= 0.50\,[1 + 16\times(3.14)^2\times(0.10)^2]^{1/2}$$

$$= 0.8\ \text{m/s}^2$$ **Ans.**

Ex. 4 In a two dimensional motion of a body, prove that tangential acceleration is nothing but component of acceleration along velocity.

Sol. Let velocity of the particle be,

$$\mathbf{v} = v_x\hat{\mathbf{i}} + v_y\hat{\mathbf{j}}$$

$$\text{Acceleration} \quad \mathbf{a} = \frac{dv_x}{dt}\hat{\mathbf{i}} + \frac{dv_y}{dt}\hat{\mathbf{j}}$$

Component of $\mathbf{a}$ along $\mathbf{v}$ will be,

$$\frac{\mathbf{a}\cdot\mathbf{v}}{|\mathbf{v}|} = \frac{v_x\dfrac{dv_x}{dt} + v_y\cdot\dfrac{dv_y}{dt}}{\sqrt{v_x^2 + v_y^2}} \quad \ldots\text{(i)}$$

Further, tangential acceleration of particle is rate of change of speed.

$$\text{or} \quad a_t = \frac{dv}{dt} = \frac{d}{dt}\left(\sqrt{v_x^2 + v_y^2}\right)$$

$$\text{or} \quad a_t = \frac{1}{2\sqrt{v_x^2 + v_y^2}}\left[2v_x\cdot\frac{dv_x}{dt} + 2v_y\frac{dv_y}{dt}\right]$$

$$\text{or} \quad a_t = \frac{v_x\cdot\dfrac{dv_x}{dt} + v_y\cdot\dfrac{dv_y}{dt}}{\sqrt{v_x^2 + v_y^2}} \quad \ldots\text{(ii)}$$

From Eqs. (i) and (ii), we can see that

$$a_t = \frac{\mathbf{a}\cdot\mathbf{v}}{|\mathbf{v}|}$$

or Tangential acceleration = component of acceleration along velocity. **Hence proved.**

MODULE 1 Exercise

Assertion and Reason

Direction : Choose the correct option.

(a) If both **Assertion** and **Reason** are true and the **Reason** is correct explanation of the **Assertion**.
(b) If both **Assertion** and **Reason** are true but **Reason** is not the correct explanation of **Assertion**.
(c) If **Assertion** is true, but the **Reason** is false.
(d) If **Assertion** is false but the **Reason** is true.
(e) If both **Assertion** and **Reason** are false.

1. **Assertion :** A particle is rotating in a circle with constant speed as shown. Between points *A* and *B*, ratio of average acceleration and average velocity is angular velocity of particle about point *O*.

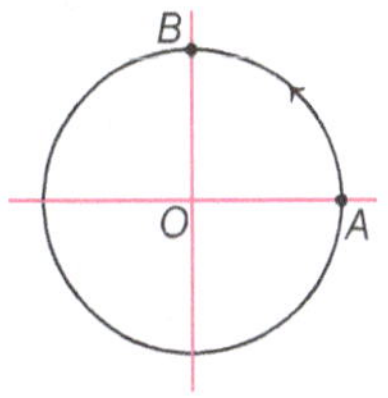

Reason : Since speed is constant, angular velocity is also constant.

2. **Assertion :** Velocity and acceleration of a particle in circular motion at some instant are: $\mathbf{v} = (2\hat{\mathbf{i}})\ \text{ms}^{-1}$ and $\mathbf{a} = (-\hat{\mathbf{i}} + 2\hat{\mathbf{j}})\ \text{ms}^{-2}$, then radius of circle is 2 m.

Reason : Speed of particle is decreasing at a rate of $1\ \text{ms}^{-2}$.

Single Correct Option

3. In a clock, what is the time period of meeting of the minute hand and the second hand?
(a) 59 s (b) $\frac{60}{59}$ s (c) $\frac{59}{60}$ s (d) $\frac{3600}{59}$ s

4. A particle is moving in a circular path with a constant speed. If θ is the angular displacement, then starting from $\theta = 0°$, the maximum and minimum change in the linear momentum will occur when value of θ is
(a) 45° and 90° (b) 90° and 180°
(c) 180° and 360° (d) 90° and 270°

5. A point moves along a circle having a radius 20 cm with a constant tangential acceleration $5\ \text{cm/s}^2$. How much time is needed after motion begins for the normal acceleration of the point to be equal to tangential acceleration?
(a) 1 s (b) 2 s (c) 3 s (d) 4 s

6. Two particles revolve concentrically in a horizontal plane in the same direction. The time required to complete one revolution for particle *A* is 3 min, while for particle *B* is 1 min. The time required for *A* to complete one revolution relative to *B* is
(a) 2 min (b) 1 min
(c) 1.5 min (d) 1.25 min

7. Three particles *A*, *B* and *C* move in a circle in anti-clockwise direction with speeds $1\ \text{ms}^{-1}$, $2.5\ \text{ms}^{-1}$ and $2\ \text{ms}^{-1}$, respectively. The initial positions of *A*, *B* and *C* are as shown in figure. The ratio of distance travelled by *B* and *C* by the instant *A*, *B* and *C* meet for the first time is

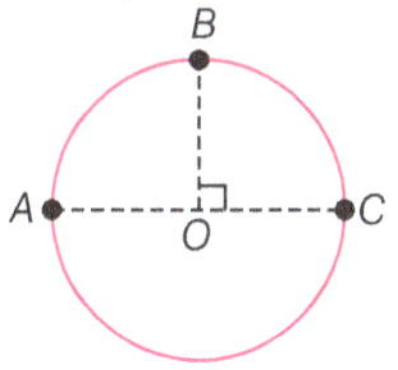

(a) 3 : 2 (b) 5 : 4
(c) 3 : 5 (d) data insufficient

8. The ratio of angular speeds of minute hand and hour hand of a watch is
(a) 1 : 12 (b) 6 : 1 (c) 12 : 1 (d) 1 : 6

9. A particle moves in a circular path of radius *R* with an angular velocity $\omega = a - bt$, where *a* and *b* are positive constants and *t* is time. The magnitude of the acceleration of the particle after time $\frac{2a}{b}$ is
(a) $\frac{a}{R}$ (b) a^2R
(c) $R(a^2 + b)$ (d) $R\sqrt{a^4 + b^2}$

10. A circular disc of radius *R* is rotating about its axis *O* with a uniform angular velocity ω as shown in the figure. The magnitude of the relative velocity of point *A* relative to point *B* on the disc is

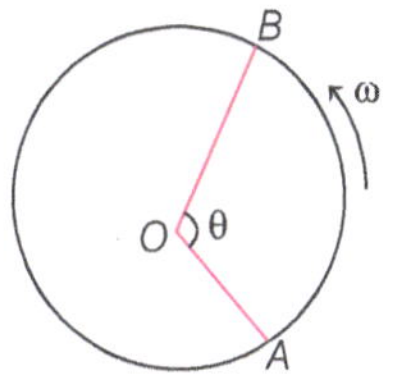

(a) zero
(b) $R\omega \sin\left(\frac{\theta}{2}\right)$
(c) $2R\omega \sin\left(\frac{\theta}{2}\right)$
(d) $\sqrt{3}\, R\omega \sin\left(\frac{\theta}{2}\right)$

11. A particle of mass *m* describes a circle of radius (*r*). The centripetal acceleration of the particle is $\frac{4}{r^2}$. The momentum of the particle is
(a) $\frac{2m}{r}$ (b) $\frac{2m}{\sqrt{r}}$ (c) $\frac{4m}{r}$ (d) $\frac{4m}{\sqrt{r}}$

12. For a particle in a uniform circular motion, the acceleration **a** at a point *P* (*R*, θ) on the circle of radius *R* is (here, θ is measured from the *x*-axis)
(a) $-\frac{v^2}{R}\cos\theta\,\hat{\mathbf{i}} + \frac{v^2}{R}\sin\theta\,\hat{\mathbf{j}}$ (b) $-\frac{v^2}{R}\sin\theta\,\hat{\mathbf{i}} + \frac{v^2}{R}\cos\theta\,\hat{\mathbf{j}}$
(c) $-\frac{v^2}{R}\cos\theta\,\hat{\mathbf{i}} - \frac{v^2}{R}\sin\theta\,\hat{\mathbf{j}}$ (d) $\frac{v^2}{R}\hat{\mathbf{i}} + \frac{v^2}{R}\hat{\mathbf{j}}$

13. Two particles A and B are moving on two concentric circles of radii R_1 and R_2 with equal angular speed ω. At $t = 0$, their positions and direction of motion are shown in the figure. The relative velocity $\mathbf{v}_A - \mathbf{v}_B$ at $t = \frac{\pi}{2\omega}$ is given by **(2019 Main)**

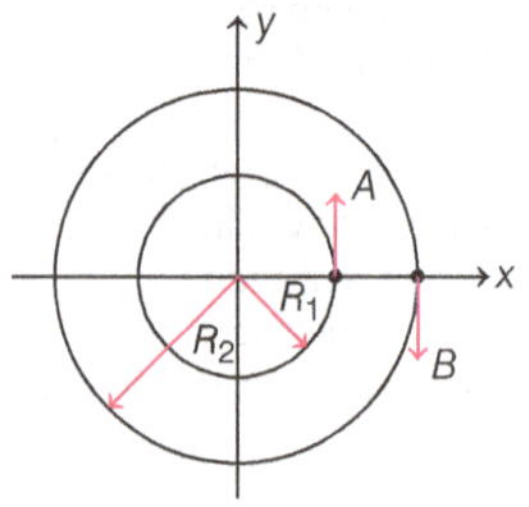

(a) $\omega(R_1 + R_2)\hat{\mathbf{i}}$ (b) $-\omega(R_1 + R_2)\hat{\mathbf{i}}$
(c) $\omega(R_1 - R_2)\hat{\mathbf{i}}$ (d) $\omega(R_2 - R_1)\hat{\mathbf{i}}$

14. A particle is moving along a circular path with a constant speed of 10 ms^{-1}. What is the magnitude of the change in velocity of the particle, when it moves through an angle of 60° around the centre of the circle? **(2019 Main)**

(a) $10\sqrt{2}$ m/s (b) 10 m/s (c) $10\sqrt{3}$ m/s (d) Zero

15. A point P moves in a counter-clockwise direction on a circular path as shown in the figure. The movement of P is such that it sweeps out a length $s = t^3 + 5$, where s is in metre and t is in second. The radius of the path is 20 m. The acceleration of P when $t = 2$ s is nearly, is

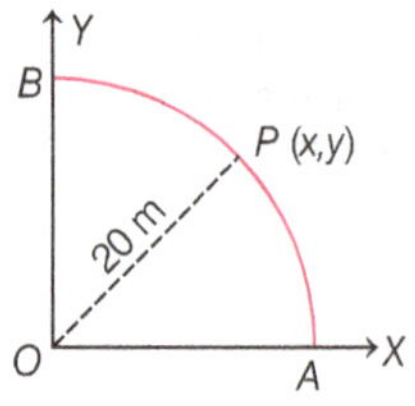

(a) 13 ms^{-2} (b) 12 ms^{-2} (c) 7.2 ms^{-2} (d) 14 ms^{-2}

Multiple Correct Options

16. A jeep runs around a curve of radius 0.3 km at a constant speed of 60 ms^{-1}. The jeep covers a curve of 60° arc. Then,

(a) change in velocity of jeep is 60 ms^{-1}
(b) instantaneous acceleration of jeep is 12 ms^{-2}
(c) average acceleration of jeep is approximately 11.5 ms^{-2}
(d) instantaneous and average acceleration are same in this case

Match the Columns

17. A particle is rotating in a circle of radius $R = \left(\frac{2}{\pi}\right)$ m, with constant speed 1 ms^{-1}. Match the following two columns for the time interval when it completes $\frac{1}{4}$th of the circle.

Column I		Column II (SI units)	
(a)	Average speed	(p)	$\frac{\sqrt{2}}{\pi}$
(b)	Average velocity	(q)	$2\frac{\sqrt{2}}{\pi}$
(c)	Average acceleration	(r)	$\sqrt{2}$
(d)	Displacement	(s)	1

Subjective Questions

18. A particle is projected with a speed u at an angle θ with the horizontal. Consider a small part of its path near the highest position and take it approximately to be a circular arc. What is the radius of this circle? This radius is called the radius of curvature of the curve at that point.

19. A particle is projected with velocity $20\sqrt{2}$ m/s at 45° with horizontal. After 1 s, find tangential and normal acceleration of the particle. Also, find radius of curvature of the trajectory at that point.
(Take, $g = 10$ m/s^2)

20. A particle A moves along a circle of radius $R = 50$ cm so that its radius vector r relative to the point O rotates with the constant angular velocity $\omega = 0.40$ rad s^{-1}. Find the modulus of the velocity of the particle, and the modulus and direction of its total acceleration. **(Irodov 1.43)**

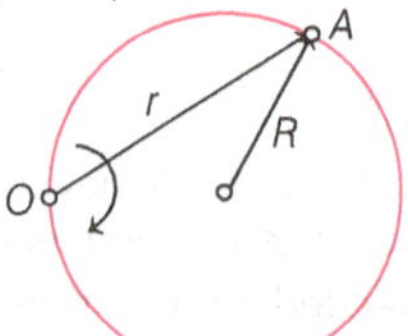

Answers

1. (b) 2. (b) 3. (d) 4. (c) 5. (b) 6. (c) 7. (b) 8. (c) 9. (d) 10. (c)
11. (b) 12. (c) 13. (d) 14. (b) 15. (d) 16. (a, b, c) 17. (a-s; b-q; c-r; d-q) 18. $(u\cos\theta)^2/g$
19. $-2\sqrt{5}$ m/s^2, $4\sqrt{5}$ m/s^2, $25\sqrt{5}$ m 20. 0.4 m/s, 0.32 m/s^2

See hints on pages 581-583.

MODULE 2

Dynamics of Circular Motion

In the previous topic, we have learnt that linear acceleration of a particle in circular motion has two components, tangential and radial (or centripetal). So, normally we resolve the forces acting on the particle in two directions:

(i) tangential (ii) radial

In **tangential direction**, net force on the particle is $\boldsymbol{ma_t}$ and in **radial direction** net force is $\boldsymbol{ma_r}$.

In uniform circular motion, tangential acceleration is zero. Hence, net force in tangential direction is zero and in radial direction

$$F_{\text{net}} = ma_r = \frac{mv^2}{r} = mr\,\omega^2$$

as, $a_r = \dfrac{v^2}{r} = r\,\omega^2$

This net force (towards centre) is also called **centripetal force**.

In most of the cases plane of our uniform circular motion will be horizontal and one of the tangent is in vertical direction also. So, in this case we resolve the forces in:

(i) horizontal radial direction

(ii) vertical tangential direction

In vertical tangential direction net force is zero (as $a_t = 0$) and in horizontal radial direction (towards centre) net force is $\dfrac{mv^2}{r}$ or $mr\,\omega^2$.

Note *Centripetal force $\dfrac{mv^2}{r}$ or $mr\,\omega^2$ (towards centre) does not act on the particle but this much force is required to the particle for rotating it in a circle (as it is accelerated due to change in direction of velocity). The real forces acting on the particle provide this centripetal force or we can say that vector sum of all the forces acting on the particle is equal to $\dfrac{mv^2}{r}$ or $mr\,\omega^2$ (in case of uniform circular motion). The real forces acting on the particle may be, friction force, weight, normal reaction, tension etc.*

Given below are some examples of uniform circular motion where plane of circular motion is horizontal. So, we have resolved the forces in vertical tangential direction $(\Sigma F_t = 0)$ and horizontal radial direction towards centre $\left(\Sigma F_r = ma_r = \dfrac{mv^2}{r} = mr\omega^2\right)$.

Conical Pendulum

If a small particle of mass m tied to a string is whirled in a horizontal circle, as shown in figure. The arrangement is called the 'conical pendulum'. In case of conical pendulum, the vertical component of tension balances the weight in tangential direction, while its horizontal component provides the necessary centripetal force in radial direction (towards centre). Thus,

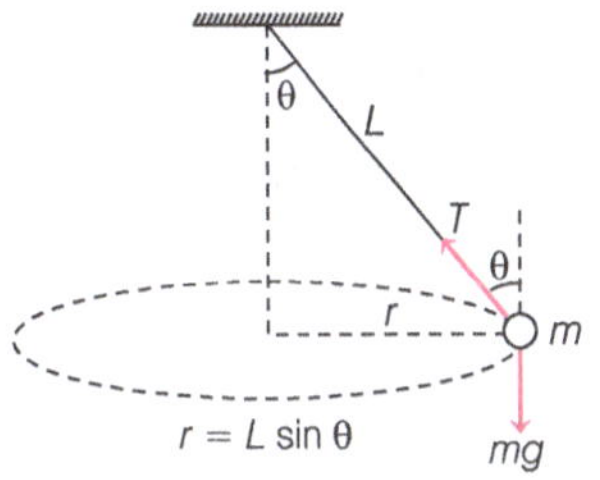

$$T \sin\theta = \frac{mv^2}{r} \quad \text{...(i)}$$

and $$T \cos\theta = mg \quad \text{...(ii)}$$

From these two equations, we can find

$$v = \sqrt{rg \tan\theta}$$

$\therefore$ Angular speed $$\omega = \frac{v}{r} = \sqrt{\frac{g \tan\theta}{r}}$$

So, the time period of pendulum is

$$T = \frac{2\pi}{\omega} = 2\pi\sqrt{\frac{r}{g\tan\theta}}$$

$$= 2\pi\sqrt{\frac{L\cos\theta}{g}} \quad (\text{as } r = L\sin\theta)$$

or $$T = 2\pi\sqrt{\frac{L\cos\theta}{g}}$$

Motion of a Particle Inside a Smooth Cone

A particle of mass m is rotating inside a smooth cone in horizontal circle of radius r as shown in figure. constant speed of the particle is suppose v.

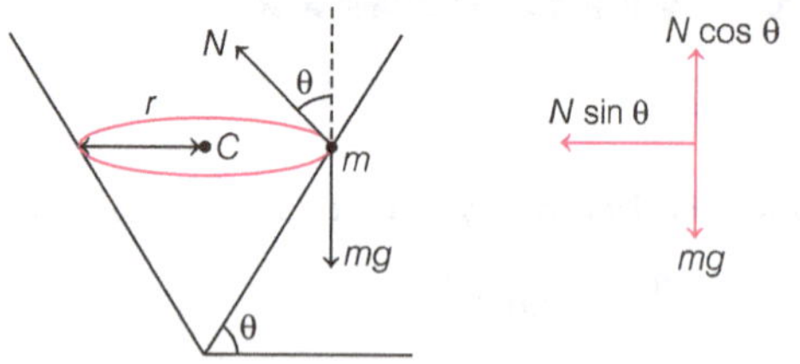

Only two forces are acting on the particle in the shown directions:

(i) normal reaction N

(ii) weight mg

We have resolved these two forces in vertical tangential direction and horizontal radial direction. In vertical tangential direction,

net force is zero.

$$\therefore \qquad N\cos\theta = mg \qquad \text{...(i)}$$

In horizontal radial direction (towards centre), net force is $\frac{mv^2}{r}$.

$$\therefore \qquad N\sin\theta = \frac{mv^2}{r} \qquad \text{...(ii)}$$

From these two equations, we can find that,

$$v = \sqrt{rg\tan\theta}$$

Death Well or Rotor

In case of 'death well' a person drives a bicycle on a vertical surface of a large wooden well while in case of a rotor, at a certain angular speed of rotor a person hangs resting against the wall without any support from the bottom. In death well walls are at rest and person revolves while in case of rotor person is at rest and the walls rotate.

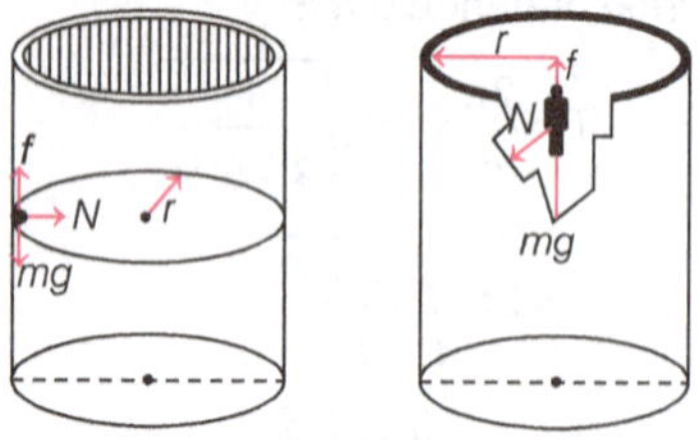

(A) Death well (B) Rotor

In both cases, friction force balances the weight of person while reaction provides the centripetal force for circular motion, i.e.

$$f = mg$$

and $$N = \frac{mv^2}{r} = mr\omega^2 \qquad (\because v = r\omega)$$

A Cyclist Bends Towards Centre on a Circular Path

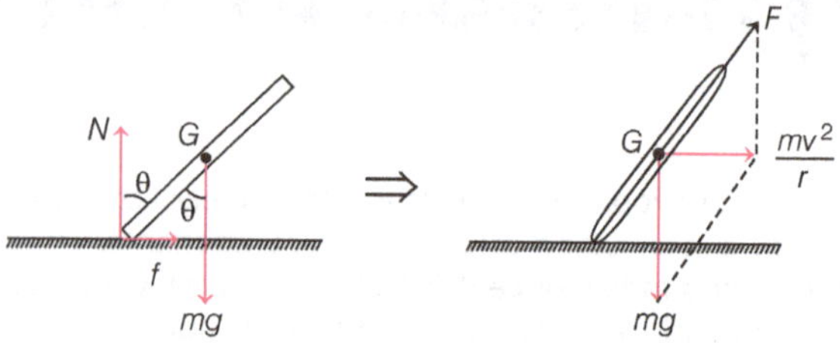

In the figure, F is the resultant of N and f.

$$\therefore \qquad F = \sqrt{N^2 + f^2}$$

When the cyclist is inclined to the centre of the circular path, the resultant of N, f and mg is directed horizontally to the centre of the circular path of the cycle. This resultant force imparts a centripetal acceleration to the cyclist.

Resultant of N and f, i.e. F should pass through G, the centre of gravity of cyclist (for complete equilibrium, rotational as well as translational). Hence,

$$\tan\theta = \frac{f}{N}, \text{ where } f = \frac{mv^2}{r} \text{ and } N = mg$$

$$\therefore \qquad \tan\theta = \frac{v^2}{rg}$$

Circular Turning of Roads

When vehicles go through turnings, they travel along a nearly circular arc. There must be some force which will provide the required centripetal acceleration. If the vehicles travel in a horizontal circular path, this resultant force is also horizontal. The necessary centripetal force is being provided to the vehicles by following three ways:

1. By friction only.
2. By banking of roads only.
3. By friction and banking of roads both.

In real life, the necessary centripetal force is provided by friction and banking of roads both. Now, let us write equations of motion in each of the three cases separately and see what are the constraints in each case.

1. By Friction Only

Suppose a car of mass m is moving at a speed v in a horizontal circular arc of radius r. In this case, the necessary centripetal force to the car will be provided by force of friction f (in horizontal direction) acting towards centre.

Thus, $$f = \frac{mv^2}{r}$$

Further, limiting value of f is μN.

or $$f_L = \mu N = \mu mg \qquad \text{(in vertical direction } N = mg\text{)}$$

Therefore, for a safe turn without sliding

$$\frac{mv^2}{r} \le f_L \quad \text{or} \quad \frac{mv^2}{r} \le \mu mg$$

or
$$\mu \ge \frac{v^2}{rg} \quad \text{or} \quad v \le \sqrt{\mu rg}$$

Here, two situations may arise. If μ and r are known to us, the speed of the vehicle should not exceed $\sqrt{\mu rg}$ and if v and r are known to us, the coefficient of friction should be greater than $\frac{v^2}{rg}$.

Important POINTS

(i) If a car is taking a turn on rough horizontal road then two speeds are critical. These two speeds are :

v_1 = maximum speed beyond which car skids outwards $= \sqrt{\mu rg}$ (already discussed)

and v_2 = maximum speed beyond which car topples about outer wheel. Derivation of v_2 is given below.

While moving in a circular track normal reaction on the outer wheels (N_1) is more than the normal reaction on inner wheels (N_2).

or $$N_1 > N_2$$

This can be proved as below.

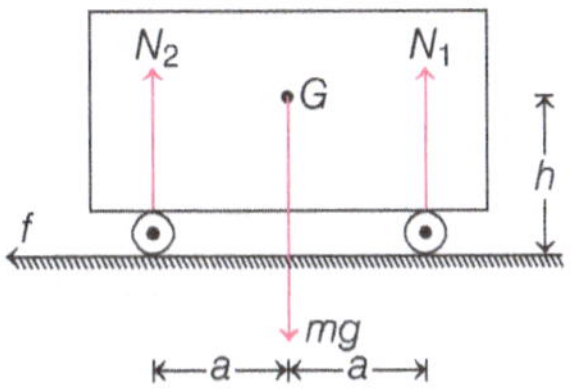

Distance between two wheels $= 2a$

Height of centre of gravity of car from road $= h$

For translational equilibrium of car

$$N_1 + N_2 = mg \quad \text{...(i)}$$

and $$f = \frac{mv^2}{r} \quad \text{...(ii)}$$

and for rotational equilibrium of car, net moment of all forces about centre of gravity should be zero.

or $$N_1(a) = N_2(a) + f(h) \quad \text{...(iii)}$$

From Eq. (iii), we can see that

$$N_2 = N_1 - \left(\frac{h}{a}\right)f = N_1 - \left(\frac{mv^2}{r}\right)\left(\frac{h}{a}\right) \quad \text{...(iv)}$$

or $$N_2 < N_1$$

From Eq. (iv), we see that N_2 decreases as v is increased.

In critical case, $N_2 = 0$

and $N_1 = mg$ [From Eq. (i)]

$\therefore$ $N_1(a) = f(h)$ [From Eq. (iii)]

or $$(mg)(a) = \left(\frac{mv^2}{r}\right)(h)$$

or $$v = \sqrt{\frac{gra}{h}}$$

Now, if $v > \sqrt{\frac{gra}{h}}$, $N_2 < 0$, and the car topples outwards.

Therefore, for a safe turn without toppling $v \le \sqrt{\frac{gra}{h}}$.

(ii) Thus $v_1 = \sqrt{\mu rg}$ and $v_2 = \sqrt{\frac{gra}{h}}$. While designing a road or a car we should keep v_1 and v_2 as high as possible (within financial and other constraints). $v_1 \propto \sqrt{\mu}$. So, to increase v_1, coefficient of friction between road and tyre should be high. Further, $v_2 \propto \sqrt{\frac{a}{h}}$. So, to keep v_2 large, a should be more and h should be less.

This you can see in the design of a racing car.

Loaded sugarcane trucks overturn at low speed because h in that case is more. So, value of v_2 is less.

While driving a car, actual speed should be less than both v_1 and v_2, so that car neither skids nor topples.

Note *You might have seen that if the speed of the car is too high, car starts skidding outwards. With this, radius of the circle increases or the required centripetal force is reduced (as centripetal force* $\propto \frac{1}{r}$*).*

2. By Banking of Roads Only

Friction is not always reliable at circular turns if high speeds and sharp turns are involved. To avoid dependence on friction, the roads are banked at the turn so that the outer part of the road is some what lifted compared to the inner part.

Let us assume that road is smooth (not practically possible) and centripetal force is provided only by banking of roads.

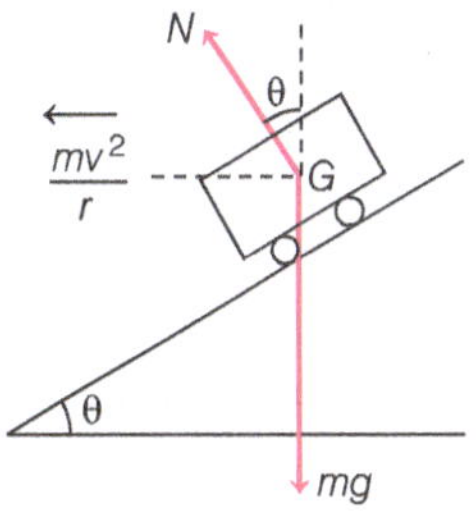

Resolving the forces in horizontal and vertical directions, we have

$$N \sin\theta = \frac{mv^2}{r} \quad ...(i)$$

and $$N \cos\theta = mg \quad ...(ii)$$

From these two equations, we get

$$\tan\theta = \frac{v^2}{rg} \quad \text{or} \quad v = \sqrt{rg \tan\theta}$$

Note *This is the speed at which car does not slide down even if track is smooth. If track is smooth and speed is less than $\sqrt{rg \tan\theta}$, vehicle will move down so that r gets decreased as ($v \propto \sqrt{r}$) and if speed is more than this vehicle will move up.*

3. By Friction and Banking of Road Both

If a vehicle is moving on a circular road which is rough and banked also, then **three forces** (weight, normal reaction and friction) may act on the vehicle. Of these three forces the first force, i.e. **weight (mg)** is fixed both in magnitude and direction. The direction of second force, i.e. **normal reaction** N is also fixed (perpendicular to the road) while the direction of the third force, i.e. **friction** f can be either inwards or outwards, while its magnitude can be varied from zero to a maximum limit ($f_L = \mu N$).

So, the magnitude of normal reaction N and direction plus magnitude of friction f are so adjusted that the resultant of the three forces mentioned above is $\frac{mv^2}{r}$ towards the centre. Of these m and r are also constant. Therefore, magnitude of N and direction plus magnitude of friction mainly depend on the speed of the vehicle v. Although situation varies from problem to problem yet, we can see that

(i) Friction f is upwards if the vehicle is at rest or $v = 0$. Because in this case the component of weight $mg \sin\theta$ is balanced by f.

(ii) Friction f is downwards if $v > \sqrt{rg \tan\theta}$

(iii) Friction f is upwards if $v < \sqrt{rg \tan\theta}$

(iv) Friction f is zero if $v = \sqrt{rg \tan\theta}$

Let us now see how the force of friction and normal reaction change as speed is gradually increased.

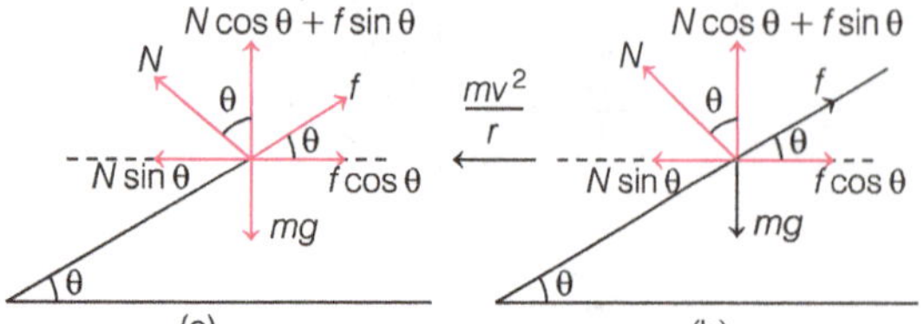

In Fig. (a) When the car is at rest force of friction is upwards. We can resolve the forces in any two mutually perpendicular directions. Let us resolve them in horizontal and vertical directions.

$$\Sigma F_H = 0 \quad \therefore \quad N \sin\theta - f \cos\theta = 0 \quad ...(i)$$

$$\Sigma F_V = 0 \quad \therefore \quad N \cos\theta + f \sin\theta = mg \quad ...(ii)$$

In Fig. (b) Now, the car is given a small speed v, so that a centripetal force $\frac{mv^2}{r}$ is now required in horizontal direction towards centre. So, Eq. (i) will now become,

$$N \sin\theta - f \cos\theta = \frac{mv^2}{r} \quad ...(iii)$$

or we can say that in the first case $N \sin\theta$ and $f \cos\theta$ are equal while in the second case their difference is $\frac{mv^2}{r}$. *This can occur in following three ways:*

(i) N increases while f remains same.

(ii) N remains same while f decreases or

(iii) N increases and f decreases.

But only third case is possible, i.e. N will increase and f will decrease. This is because

Eq. (ii), $N \cos\theta + f \sin\theta = mg$ = constant is still has to be valid.

So, to keep $N \cos\theta + f \sin\theta$ to be constant N should increase and f should decrease (as θ = constant).

Now, as speed goes on increasing, force of friction first decreases. Becomes zero at $v = \sqrt{rg \tan\theta}$ and then starts acting in downward direction, so that its horizontal component $f \cos\theta$ with $N \sin\theta$ now provides the required centripetal force.

Note Points

(i) In uniform circular motion

$$T = \frac{2\pi}{\omega} = \frac{1}{f} \text{ or } \frac{1}{n} = \frac{2\pi r}{v}$$

$$\therefore \quad \omega = 2\pi f \text{ or } 2\pi n$$

Here f or n = frequency (in Hz) or number of revolutions per second.

(ii) If a car is taking a turn on rough inclined road then again two speeds are critical. Let us call them v_3 and v_4. Here,

v_3 = minimum speed below which car skids downwards and

v_4 = maximum speed beyond which car skids upwards.

Now let us discuss v_3 and v_4.

v_3: Again there are two cases :

Case 1 : If $\theta <$ angle of repose. Then $v_3 = 0$.

Case 2 : If $\theta >$ angle of repose. In Eqs. (ii) and (iii) discussed above. Put $f = \mu N$ and find the value of v. This is actually v_3.

v_4: In Eqs. (ii) and (iii) discussed above put $f = \mu N$ and change the direction of f. Now find v. This is v_4.

(iii) In the following two figures, surface is smooth. So, only two forces N and mg are acting. But direction of acceleration are different.

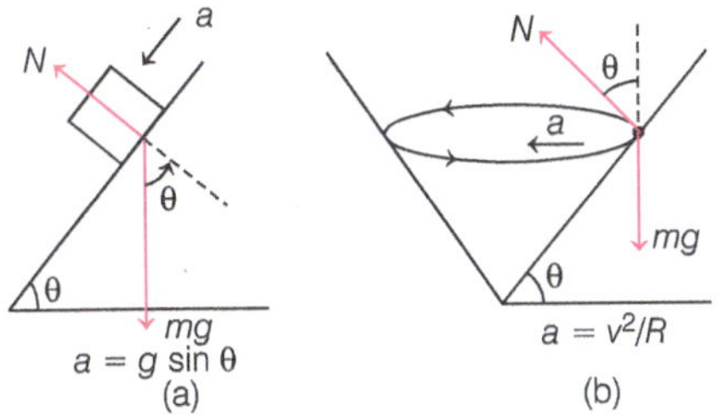

Net force perpendicular to acceleration should be zero. So, in the first figure.

$$N = mg\cos\theta$$

and in the second figure, $N\cos\theta = mg$

Example 1 A small block of mass 100 g moves with uniform speed in a horizontal circular groove, with vertical side walls of radius 25 cm. If the block takes 2.0 s to complete one round, find the normal contact force by the side wall of the groove.

Sol. The speed of the block is

$$v = \frac{2\pi r}{T} = \frac{2\pi \times (25\text{ cm})}{2.0\text{ s}}$$

$$= 78.5\text{ cm/s} = 0.785\text{ m/s}$$

The acceleration of the block is

$$a = \frac{v^2}{r} = \frac{(0.785\text{ m/s})^2}{0.25\text{ m}} = 2.464\text{ m/s}^2$$

towards the centre. The only force in this direction is the normal contact force due to the side walls. Thus, from Newton's second law, this force is

$$N = ma = (0.100\text{ kg})(2.464\text{ m/s}^2)$$

$$= 0.246\text{ N}$$ **Ans.**

Example 2 A fighter plane is pulling out for a dive at a speed of 900 km/h. Assuming its path to be a vertical circle of radius 2000 m and its mass to be 16000 kg, find the force exerted by the air on it at the lowest point. (Take, $g = 9.8\text{ m/s}^2$)

Sol. At the lowest point in the path, the acceleration is vertically upward (towards the centre) and its magnitude is v^2/r.

The forces on the plane are :

(a) weight Mg downward and

(b) force F by the air upward.

Hence, Newton's second law of motion gives

$$F - Mg = Mv^2/r$$

or $$F = M(g + v^2/r)$$

Here, $$v = 900\text{ km/h} = \frac{9 \times 10^5}{3600}\text{ m/s}$$

$$= 250\text{ m/s}$$

$\therefore$ $$F = 16000\left(9.8 + \frac{62500}{2000}\right)\text{N}$$

$$= 6.56 \times 10^5\text{ N} \quad \text{(upward)}$$

Example 3 Three particles, each of mass m are situated at the vertices of an equilateral triangle of side a. The only forces acting on the particles are their mutual gravitational forces. It is desired that each particle moves in a circle while maintaining the original mutual separation a. Find the initial velocity that should be given to each particle and also the time period of the circular motion. $\left(F = \dfrac{Gm_1m_2}{r^2}\right)$.

Sol. $$r = \frac{a}{2}\sec 30° = \frac{a}{\sqrt{3}}$$

$$F = \frac{Gmm}{a^2}$$

$$F_{net} = \sqrt{3}F = \left(\frac{Gmm}{a^2}\right)(\sqrt{3})$$

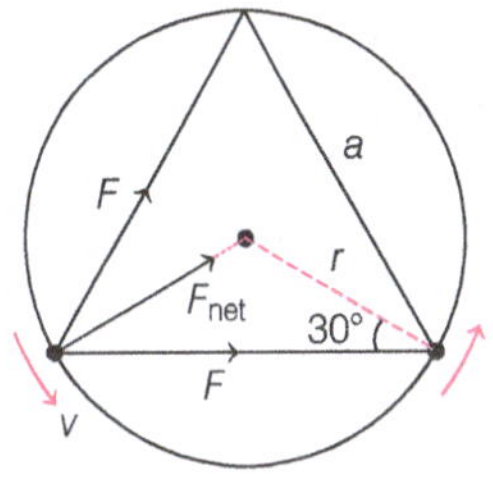

This will provide the necessary centripetal force.

$\therefore$ $$\frac{mv^2}{r} = \frac{\sqrt{3}\,Gm^2}{a^2}$$

or $$\frac{mv^2}{(a/\sqrt{3})} = \frac{\sqrt{3}\,Gm^2}{a^2}$$

$\Rightarrow$ $$v = \sqrt{\frac{Gm}{a}}$$ **Ans.**

$$T = \frac{2\pi r}{v} = \frac{2\pi(a/\sqrt{3})}{\sqrt{Gm/a}}$$

$$= 2\pi\sqrt{\frac{a^3}{3Gm}}$$ **Ans.**

Example 4 (a) How many revolutions per minute must the apparatus shown in figure make about a vertical axis so that the cord makes an angle of 45° with the vertical?

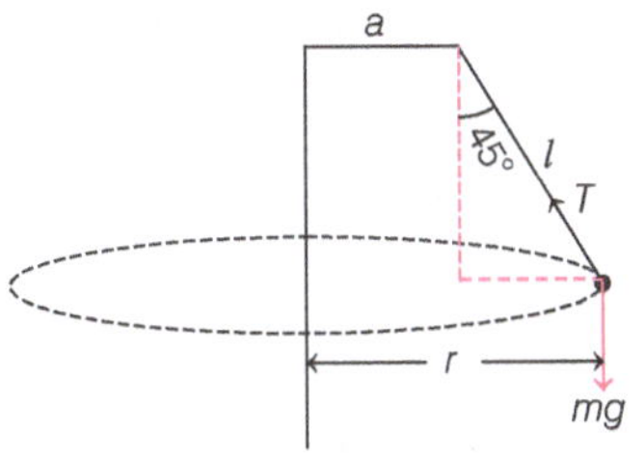

(b) What is the tension in the cord then? Given, $l = \sqrt{2}$ m, $a = 20$ cm and $m = 5.0$ kg?

Sol. (a) $r = a + l \sin 45°$

$$= (0.2) + (\sqrt{2})\left(\frac{1}{\sqrt{2}}\right) = 1.2 \text{ m}$$

Now, $T \cos 45° = mg$ ...(i)

and $T \sin 45° = mr\omega^2$ ...(ii)

From Eqs. (i) and (ii), we have $\omega = 2n\pi = \sqrt{\frac{g}{r}}$

$$\therefore \quad n = \frac{1}{2\pi}\sqrt{\frac{g}{r}}$$

$$= \frac{60}{2\pi}\sqrt{\frac{9.8}{1.2}} \text{ rpm}$$

$$= 27.3 \text{ rpm}$$ **Ans.**

(b) From Eq. (i), we have

$$T = \sqrt{2}\, mg = (\sqrt{2})(5.0)(9.8)$$

$$= 69.3 \text{ N}$$ **Ans.**

Example 5 A particle of mass m starts moving in a circular path of constant radius r, such that its centripetal acceleration a_c is varying with time t as $a_c = k^2rt^2$, where k is a constant. What is the power delivered to the particle by the forces acting on it ? **(1994 JEE)**

Sol. As $a_c = (v^2/r)$ so $(v^2/r) = k^2rt^2$

$\Rightarrow \quad v^2 = k^2r^2t^2$

$\therefore$ Kinetic energy $K = \frac{1}{2}mv^2 = \frac{1}{2}mk^2r^2t^2$

Now, from work-energy theorem

$$W = \Delta K = \frac{1}{2}mk^2r^2t^2 - 0 \qquad [\text{as at } t = 0, K = 0]$$

So, $P = \frac{dW}{dt} = \frac{d}{dt}\left(\frac{1}{2}mk^2r^2t^2\right) = mk^2r^2t$ **Ans.**

Alternate solution : Given that $a_c = k^2rt^2$, so that

$$F_c = ma_c = mk^2rt^2$$

Now, as $a_c = (v^2/r)$,

so $(v^2/r) = k^2rt^2$

or $v = krt$

So, that $a_t = (dv/dt) = kr$

i.e. $F_t = ma_t = mkr$

Now, as $\mathbf{F} = \mathbf{F}_c + \mathbf{F}_t$

So, $P = \frac{dW}{dt} = \mathbf{F} \cdot \mathbf{v} = (\mathbf{F}_c + \mathbf{F}_t) \cdot \mathbf{v}$

In circular motion, $\mathbf{F}_c$ is perpendicular to $\mathbf{v}$ while $\mathbf{F}_t$ parallel to it, so

$$P = F_t v \qquad [\text{as } \mathbf{F}_c \cdot \mathbf{v} = 0]$$

$$\therefore \quad P = mk^2r^2t$$ **Ans.**

Example 6 A turn of radius 20 m is banked for the vehicle of mass 200 kg going at a speed of 10 m/s. Find the direction and magnitude of frictional force acting on a vehicle if it moves with a speed

(a) 5 m/s

(b) 15 m/s

Assume that friction is sufficient to prevent slipping. ($g = 10$ m/s^2)

Sol. (a) The turn is banked for speed $v = 10$ m/s

Therefore, $\tan\theta = \frac{v^2}{rg} = \frac{(10)^2}{(20)(10)} = \frac{1}{2}$

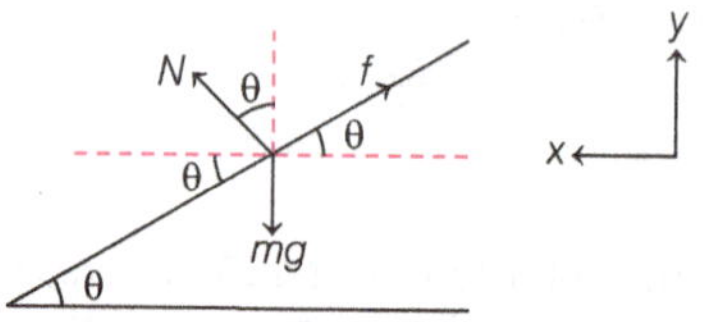

Now, as the given speed is less than 10 m/s, force of friction f acts upwards.

Using the equations $\Sigma F_x = \frac{mv^2}{r}$

and $\Sigma F_y = 0$, we get

$$N \sin\theta - f \cos\theta = \frac{mv^2}{r} \qquad \text{...(i)}$$

$$N \cos\theta + f \sin\theta = mg \qquad \text{...(ii)}$$

Substituting, $\theta = \tan^{-1}\left(\frac{1}{2}\right)$, $v = 5$ m/s, $m = 200$ kg

and $r = 20$ m, in the above equations,

we get $f = 300\sqrt{5}$ N (upwards)

(b) In this case force of friction f will act downwards.

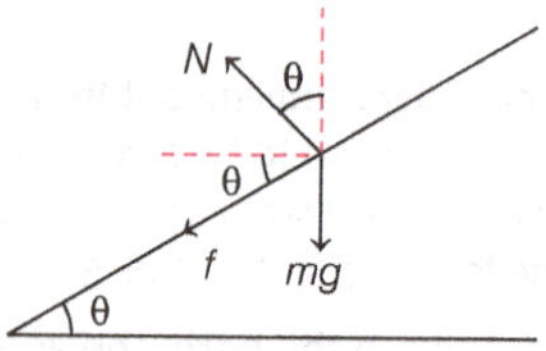

Using $\Sigma F_x = \frac{mv^2}{r}$

and $\Sigma F_y = 0$, we get

$$N \sin\theta + f \cos\theta = \frac{mv^2}{r} \qquad \text{...(iii)}$$

$$N \cos\theta - f \sin\theta = mg \qquad \text{...(iv)}$$

Substituting $\theta = \tan^{-1}\left(\frac{1}{2}\right)$, $v = 15$ m/s,

$m = 200$ kg

and $r = 20$ m in the above equations, we get

$f = 500\sqrt{5}$ N (downwards)

Example 7 An automobile enters a turn of radius r. If the road is banked at an angle of 45° and the coefficient of friction is $\frac{1}{2}$, the minimum speed with which the automobile can negotiate the turn without skidding is

(a) $\sqrt{\frac{rg}{2}}$ (b) $\frac{\sqrt{rg}}{2}$

(c) $\sqrt{rg}$ (d) $\sqrt{2rg}$

Sol. $\tan\theta = \tan 45° = 1$

$$\tan\theta_r = \mu = \frac{1}{2} \qquad (\theta_r = \text{angle of repose})$$

$$\tan\theta > \tan\theta_r \quad \Rightarrow \quad \theta > \theta_r$$

Therefore for minimum speed (v_3, case -2, discussed in theory), maximum friction ($= \mu N$) will act upwards.

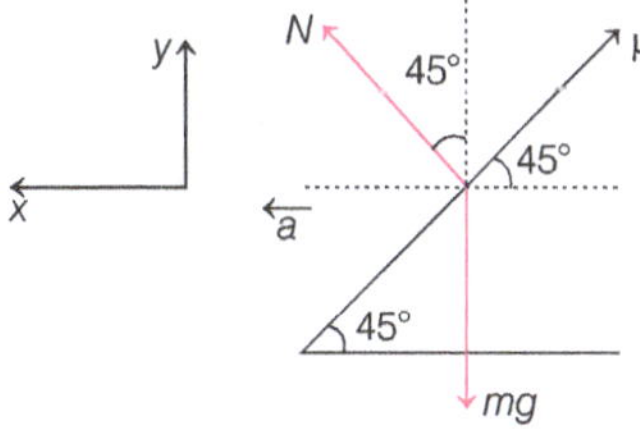

$$\Sigma F_y = 0$$

$$\therefore \quad N\cos 45° = \frac{N}{\sqrt{2}} = mg \qquad ...(i)$$

$$\Sigma F_x = ma$$

$$\therefore \quad N\sin 45° - \mu N\cos 45° = \frac{N}{\sqrt{2}} - \frac{\mu N}{\sqrt{2}} = \frac{mv_{min}^2}{r} \qquad ...(ii)$$

$$\mu = \frac{1}{2} \qquad ...(iii)$$

Solving these three equations, we get

$$v_{min} = \sqrt{\frac{rg}{2}}$$

$\therefore$ The correct option is (a).

Example 8 The kinetic energy K of a particle moving along a circle of radius R depends on the distance covered s as $K = \beta s^2$. The force acting on the particle is

(Irodov 1.120, Similar type)

(a) $\frac{2\beta s^2}{R}$ (b) $2\beta s\left(1 + \frac{s^2}{R^2}\right)^{1/2}$

(c) $\beta s\left(1 + \frac{s^2}{R^2}\right)^{1/2}$ (d) None of these

Sol. $\frac{1}{2}mv^2 = \beta s^2 \Rightarrow mv^2 = 2\beta s^2$

or $$F_n = \frac{mv^2}{R} = \frac{2\beta s^2}{R} \qquad ...(i)$$

Further, $$v = \sqrt{\frac{2\beta}{m}} \cdot s$$

or $$a_t = \frac{dv}{dt} = \sqrt{\frac{2\beta}{m}} \cdot \frac{ds}{dt} = \sqrt{\frac{2\beta}{m}} \cdot v$$

$$= \sqrt{\frac{2\beta}{m}} \cdot \sqrt{\frac{2\beta}{m}} \cdot s = \frac{2\beta s}{m}$$

$$\therefore \quad F_t = ma_t = 2\beta s \qquad ...(ii)$$

$$\therefore \quad F_{net} = \sqrt{F_n^2 + F_t^2}$$

$$= 2\beta s\sqrt{1 + \frac{s^2}{R^2}}$$

$\therefore$ The correct option is (b).

Example 9 A body moves on a horizontal circular road of radius r, with a tangential acceleration a_t. The coefficient of friction between the body and the road surface is μ. It begins to slip when its speed is v. Then,

(i) $v^2 = \mu rg$

(ii) $\mu g = \left(\frac{v^4}{r^2} + a_t\right)$

(iii) $\mu^2 g^2 = \left(\frac{v^4}{r^2} + a_t^2\right)$

(iv) The force of friction makes an angle $\tan^{-1}(v^2/a_t r)$ with the direction of motion at the point of slipping.

Choose the correct option.

(a) (i), (ii) (b) (ii), (iii) (c) (i), (iv) (d) (iii), (iv)

Sol. Here road is horizontal (not banked). So, centripetal and tangential both horizontal acceleration components (or force components) will be provided by friction. Further, in vertical direction net force is zero.

So, $N = mg$

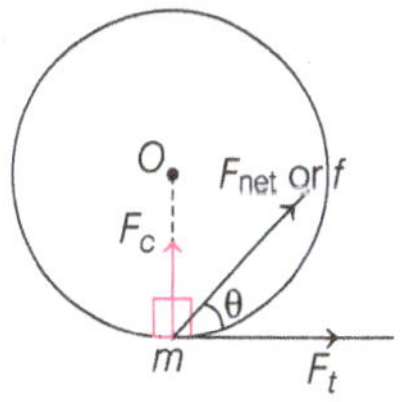

f = frictional force

$$F_{net} \le \mu N$$

$$\therefore \quad \sqrt{F_c^2 + F_t^2} \le \mu mg$$

or $$F_c^2 + F_t^2 = \mu^2 m^2 g^2$$

$$\Rightarrow \quad \left(\frac{mv^2}{r}\right)^2 + (ma_t)^2 \le (\mu^2 m^2 g^2)$$

$$\frac{v^4}{r^2} + a_t^2 \le \mu^2 g^2 \qquad ...(i)$$

$$\tan\theta = \frac{F_c}{F_t} = \frac{mv^2/r}{ma_t} = \frac{v^2}{a_t r}$$

$$\theta = \tan^{-1}\left(\frac{v^2}{a_t r}\right) \quad ...(ii)$$

∴ The correct option is (d).

Example 10 A turn of radius 20 m is banked for the vehicles going at a speed of 36 km/h. If the coefficient of static friction between the road and the tyre is 0.4. What are the possible speeds of a vehicle, so that it neither slips down nor skids up ? (Take, $g = 9.8 \text{ m/s}^2$)

Sol. Banking angle, $\theta = \tan^{-1}\left(\frac{v^2}{Rg}\right)$

$$36 \text{ km/h} = 10 \text{ m/s}$$

$$\therefore \quad \theta = \tan^{-1}\left(\frac{100}{20 \times 9.8}\right) = 27°$$

Angle of repose,

$$\theta_r = \tan^{-1}(\mu) = \tan^{-1}(0.4) = 21.8°$$

Since $\theta > \theta_r$, vehicle cannot remain in the given position with $v = 0$.

At rest, it will slide down. To find minimum speed, so that vehicle does not slip down, maximum friction will act up the plane.

To find maximum speed, so that the vehicle does not skid up, maximum friction will act down the plane.

Minimum Speed

Equation of motion are,

$$\Sigma F_y = 0$$

$$\therefore \quad N\cos\theta + \mu N \sin\theta = mg \quad ...(i)$$

$$\Sigma F_x = ma$$

$$\therefore \quad N\sin\theta - \mu N\cos\theta = \frac{m}{R}v_{min}^2 \quad ...(ii)$$

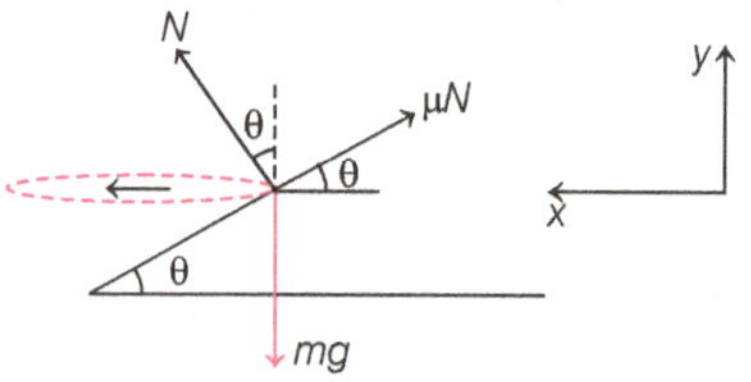

Solving these two equations, we get

$$v_{min} = 4.2 \text{ m/s} \quad \textbf{Ans.}$$

Maximum Speed

Equations of motion are,

$$N\cos\theta - \mu N\sin\theta = mg \quad ...(iii)$$

$$N\sin\theta + \mu N\cos\theta = \frac{m}{R}v_{max}^2 \quad ...(iv)$$

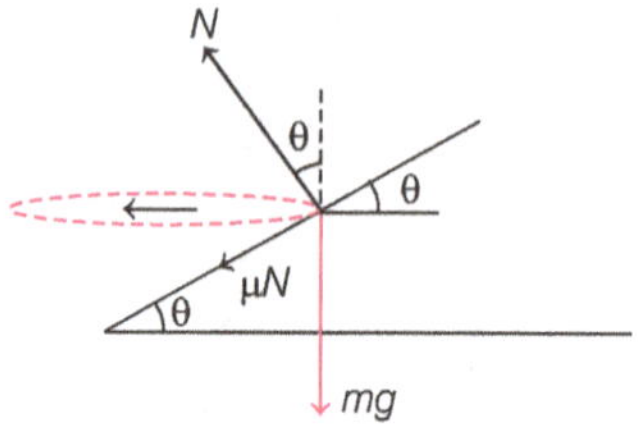

Solving these two equations, we have

$$v_{max} = 15 \text{ m/s}$$

MODULE 2 Exercise

Assertion and Reason

See directions on page 547.

1. **Assertion :** A car moving on a horizontal rough road with velocity v can be stopped in a minimum distance d. If the same car, moving with same speed v takes a circular turn, then minimum safe radius can be $2d$.

   Reason : $d = \frac{v^2}{2\mu g}$ and minimum safe radius $= \frac{v^2}{\mu g}$.

2. **Assertion :** A particle of mass m takes uniform horizontal circular motion inside a smooth funnel as shown. Normal reaction in this case is not $mg\cos\theta$.

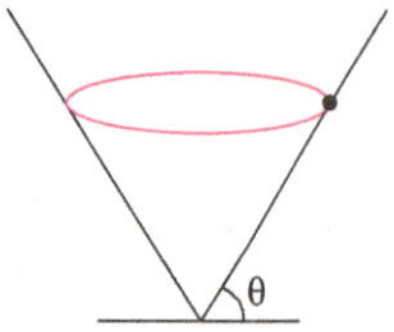

   Reason : Acceleration of particle is not along the surface of funnel.

Single Correct Option

3. An insect of mass $m = 3$ kg is inside a vertical drum of radius 2 m that is rotating with an angular velocity of 5 rad s^{-1}. The insect doesn't fall off. Then, the minimum coefficient of friction required is

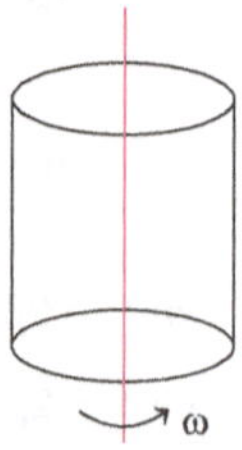

(a) 0.5 (b) 0.4
(c) 0.2 (d) None of these

4. A car is moving on a circular level road of curvature 300 m. If the coefficient of friction is 0.3 and acceleration due to gravity is $10\ \mathrm{m/s^2}$, the maximum speed of the car can be
(a) 90 km/h (b) 81 km/h
(c) 108 km/h (d) 162 km/h

5. A string of length 1 m is fixed at one end with a bob of mass 100 g and the string makes $\left(\frac{2}{\pi}\right)$ revs^{-1} around a vertical axis through a fixed point. The angle of inclination of the string with vertical is
(a) $\tan^{-1}\left(\frac{5}{8}\right)$ (b) $\tan^{-1}\left(\frac{3}{5}\right)$
(c) $\cos^{-1}\left(\frac{3}{5}\right)$ (d) $\cos^{-1}\left(\frac{5}{8}\right)$

6. In the previous question, the tension in the string is
(a) $\frac{5}{8}$ N (b) $\frac{8}{5}$ N (c) $\frac{50}{8}$ N (d) $\frac{80}{5}$ N

7. A small particle of mass 0.36 g rests on a horizontal turn-table at a distance 25 cm from the axis of spindle. The turn-table is accelerated at a rate of $\alpha = \frac{1}{3}\ \mathrm{rad\ s^{-2}}$. The frictional force that the table exerts on the particle 2 s after the start is
(a) 40 μN (b) 30 μN (c) 50 μN (d) 60 μN

8. A coin, placed on a rotating turn-table slips, when it is placed at a distance of 9 cm from the centre. If the angular velocity of the turn-table is tripled, it will just slip, if its distance from the centre is
(a) 27 cm (b) 9 cm
(c) 3 cm (d) 1 cm

9. A mass of 100 g is tied to one end of a string 2 m long. The body is revolving in a horizontal circle making a maximum of 200 rev per min. The other end of the string is fixed at the centre of the circle of revolution. The maximum tension that the string can bear is (approximately)
(a) 8.76 N (b) 8.94 N
(c) 89.42 N (d) 87.64 N

10. Toy cart tied to the end of an unstretched string of length *a*, when revolved moves in a horizontal circle of radius 2*a* with a time period *T*. Now, the toy cart is speeded up until it moves in a horizontal circle of radius 3*a* with a period *T′*. If Hooke's law ($F = kx$) holds, then
(a) $T' = \sqrt{\frac{3}{2}}T$ (b) $T' = \left(\frac{\sqrt{3}}{2}\right)T$
(c) $T' = \left(\frac{3}{2}\right)T$ (d) $T' = T$

11. A string of length *l* fixed at one end carries a mass *m* at the other end. The strings makes $\frac{2}{\pi}$ rev/s around the axis through the fixed end as shown in the figure. The tension in the string is

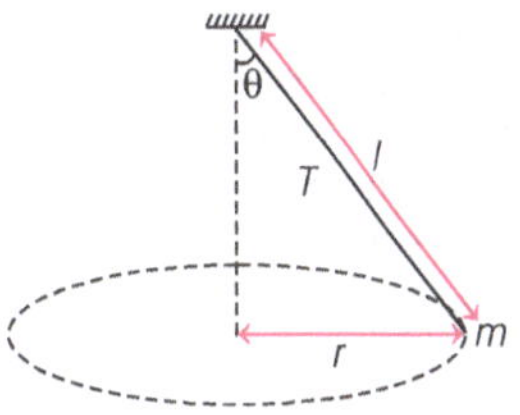

(a) 16 *ml* (b) 4 *ml*
(c) 8 *ml* (d) 2 *ml*

12. A mass *m* is fixed at $x = a$ on a parabolic wire with its axis vertical and vertex at the origin as shown in the figure. The equation of parabola is $x^2 = 4ay$. The wire frame is rotating with constant angular velocity ω about *y*-axis. The acceleration of the bead is

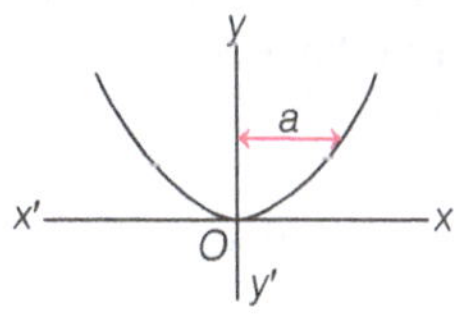

(a) $\frac{a}{4}\omega^2$ (b) $a\omega^2$ (c) zero (d) $\frac{\omega^2}{a}$

13. A particle starts travelling on a circle with constant tangential acceleration. The angle between velocity vector and acceleration vector, at the moment when particle completes half the circular track, is
(a) $\tan^{-1}(2\pi)$ (b) $\tan^{-1}(\pi)$
(c) $\tan^{-1}(3\pi)$ (d) zero

14. A rough horizontal plate rotates with angular velocity ω about a fixed vertical axis. A particle of mass *m* lies on the plate at a distance $\frac{5a}{4}$ from this axis. The coefficient of friction between the plate and the particle is $\frac{1}{3}$. The largest value of ω^2 for which the particle will continue to be at rest on the revolving plate is
(a) $\frac{g}{3a}$ (b) $\frac{4g}{5a}$ (c) $\frac{4g}{9a}$ (d) $\frac{4g}{15a}$

15. The radius of the curved road on a national highway is *R*. The width of the road is *b*. The outer edge of the road is raised by *h* with respect to the inner edge, so that a car with velocity *v* can pass safely over it. The value of *h* is
(a) $\frac{v^2b}{Rg}$ (b) $\frac{v}{Rgb}$ (c) $\frac{v^2R}{g}$ (d) $\frac{v^2b}{R}$

16. A tube of length *L* is filled completely with an incompressible liquid of mass *M* and closed at both the ends. The tube is then rotated in a horizontal plane about one of its ends with a uniform angular velocity ω. The force exerted by the liquid at the other end is
(a) $\frac{1}{2}M\omega^2L^2$ (b) $M\omega^2L$ (c) $\frac{1}{4}M\omega^2L$ (d) $\frac{1}{2}M\omega^2L$

17. A car is moving in a circular horizontal track of radius 10 m with a constant speed of 10 m/s. A plumb bob is suspended from the roof of the car by a light rigid rod. The angle made by the rod with the vertical is (Take $g = 10\,m/s^2$) **(1992 JEE)**

(a) zero (b) 30° (c) 45° (d) 60°

Multiple Correct Options

18. In circular motion of a particle,

(a) particle cannot have uniform motion
(b) particle cannot have uniformly accelerated motion
(c) particle cannot have net force equal to zero
(d) particle cannot have any force in tangential direction

19. A heavy particle is tied to the end *A* of a string of length 1.6 m. Its other end *O* is fixed. It revolves as a conical pendulum with the string making 60° with the horizontal. Then,

(a) its period of revolution is $\frac{4\pi}{7}$ s
(b) the tension in the string is $\frac{2}{\sqrt{3}}$ times the weight of the particle
(c) the speed of the particle is $2.8\sqrt{3}$ m/s
(d) the centripetal acceleration of the particle is $\frac{9.8}{\sqrt{3}}\,m/s^2$

20. A particle is acted upon by a force of constant magnitude which is always perpendicular to the velocity of the particle. The motion of the particle takes place in a plane. It follows that **(1987 JEE)**

(a) its velocity is constant
(b) its acceleration is constant
(c) its kinetic energy is constant
(d) it moves in a circular path

Match the Columns

21. A car is taking turn on a rough horizontal road without slipping as shown in figure. Let *F* is centripetal force, *f* the force of friction, N_1 and N_2 are two normal reactions. As the speed of car is increased, match the following two columns.

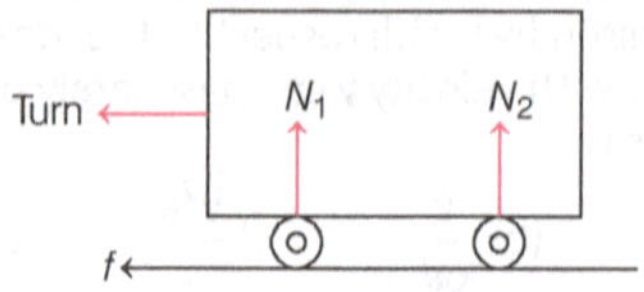

	Column I		Column II
(a)	N_1	(p)	will increase
(b)	N_2	(q)	will decrease
(c)	F/f	(r)	will remain unchanged
(d)	f	(s)	cannot say anything

Subjective Questions

22. A turn has a radius of 10 m. If a vehicle goes round it at an average speed of 18 km/h, what should be the proper angle of banking?

23. If the road of the previous problem is horizontal (no banking), what should be the minimum friction coefficient so that a scooter going at 18 km/h does not skid?

24. A circular road of radius 50 m has the angle of banking equal to 30°. At what speed should a vehicle go on this road so that the friction is not used?

25. A car driver going at speed *v* suddenly finds a wide wall at a distance *r*. Should he apply brakes or turn the car in a circle of radius *r* to avoid hitting the wall.

26. A 4 kg block is attached to a vertical rod by means of two strings of equal length. When the system rotates about the axis of the rod, the strings are extended as shown in figure.

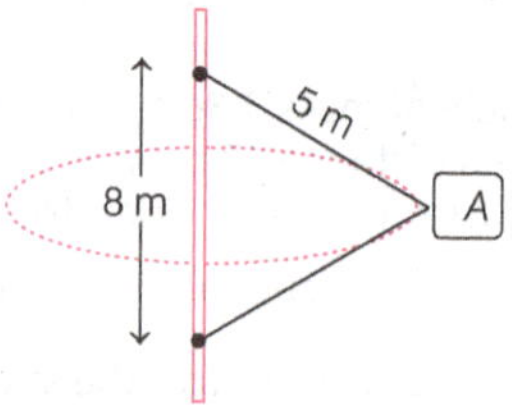

(a) How many revolutions per minute must the system make in order for the tension in the upper string to be 200 N?
(b) What is the tension in the lower string then?

27. A car moves at a constant speed on a straight but hilly road. One section has a crest and dip of the same 250 m radius.

(a) As the car passes over the crest the normal force on the car is one half the 16 kN weight of the car. What will be the normal force on the car as its passes through the bottom of the dip?
(b) What is the greatest speed at which the car can move without leaving the road at the top of the hill?
(c) Moving at a speed found in part (b) what will be the normal force on the car as it moves through the bottom of the dip? (Take, $g = 10\,m/s^2$)

28. Find the maximum speed at which a truck can safely travel without toppling over, on a curve of radius 250 m. The height of the centre of gravity of the truck above the ground is 1.5 m and the distance between the wheels is 1.5 m, the truck being horizontal.

29. A hemispherical bowl of radius *R* is rotating about its axis of symmetry which is kept vertical. A small ball kept in the bowl rotates with the bowl without slipping on its surface. If the surface of the bowl is smooth and the angle made by the radius through the ball with the vertical is α. Find the angular speed at which the bowl is rotating.

30. Show that the angle made by the string with the vertical in a conical pendulum is given by $\cos\theta = \frac{g}{L\omega^2}$, where L is the length of the string and ω is the angular speed.

31. A boy whirls a stone of small mass in a horizontal circle of radius 1.5 m and at height 2.9 m above level ground. The string breaks and the stone flies off horizontally and strikes the ground after travelling a horizontal distance of 10 m. What is the magnitude of the centripetal acceleration of the stone while in circular motion ?

32. A car moves with a constant tangential acceleration $a_t = 0.62\ \text{m/s}^2$ along a horizontal surface circumscribing a circle of radius $R = 40$ m. The coefficient of sliding friction between the wheels of the car and the surface is $\mu = 0.20$. What distance will the car ride without sliding if at the initial moment of time its velocity is equal to zero? **(Irodov 1.90)**

33. A particle of mass m moves along the internal smooth surface of a vertical cylinder of radius R. Find the force with which the particle acts on the cylinder wall if at the initial moment of time its velocity equals v_0 and forms an angle α with the horizontal. **(Irodov 1.94)**

34. A sleeve A can slide freely along a smooth rod bent in the shape of a half-circle of radius R. The system is set in rotation with a constant angular velocity ω about a vertical axis OO'. Find the angle θ corresponding to the steady position of the sleeve. **(Irodov 1.110)**

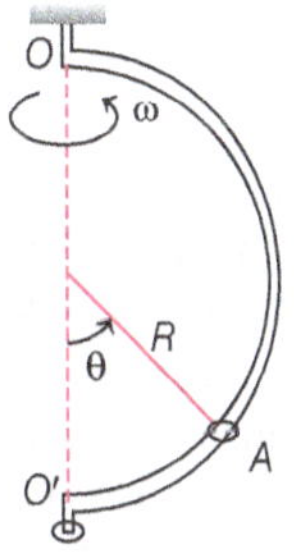

Answers

1. (a) **2.** (a) **3.** (c) **4.** (c) **5.** (d) **6.** (b) **7.** (c) **8.** (d) **9.** (d) **10.** (b)
11. (a) **12.** (b) **13.** (a) **14.** (d) **15.** (a) **16.** (d) **17.** (c) **18.** (a, b, c) **19.** (b, d) **20.** (c, d)
21. (a-q; b-p; c-r; d-p) **22.** $\tan^{-1}\left(\frac{1}{4}\right)$ **23.** 0.25 **24.** 17 ms^{-1} **25.** He should apply the brakes
26. (a) 39.6 rpm (b) 150 N **27.** (a) 24 kN (b) 50 ms^{-1} (c) 32 kN **28.** 35 m/s **29.** $\sqrt{\frac{g}{R\cos\alpha}}$
31. 113ms^{-2} **32.** $s = 60\text{m}$ **33.** $N = \frac{mv_0^2\cos^2\alpha}{R}$ **34.** $\theta = \cos^{-1}\left(\frac{g}{R\omega^2}\right)$

See hints on pages 584-588.

MODULE 3

Centrifugal Force

Newton's laws are valid only in inertial frames. In non-inertial frames a pseudo force $-m\mathbf{a}$ has to be applied. ($\mathbf{a}$ = acceleration of frame of reference). After applying the pseudo force one can apply Newton's laws in their usual form.

Now, suppose a frame of reference is rotating with constant angular velocity ω in a circle of radius 'r'. Then, it will become a non-inertial frame of acceleration $r\omega^2$ towards the centre.

Now, if we observe an object of mass 'm' from this frame then a pseudo force of magnitude $mr\omega^2$ will have to be applied to this object in a direction away from the centre. This pseudo force is called the centrifugal force.

After applying this force we can now apply Newton's laws in their usual form. Following examples will illustrate the concept more clearly:

Example 1 A particle of mass m is placed over a horizontal circular table rotating with an angular velocity ω about a vertical axis passing through its centre. The distance of the object from the axis is r. Find the force of friction f between the particle and the table.

Sol. Let us solve this problem from both frames. The one is a frame fixed on ground and the other is a frame fixed on particle itself.

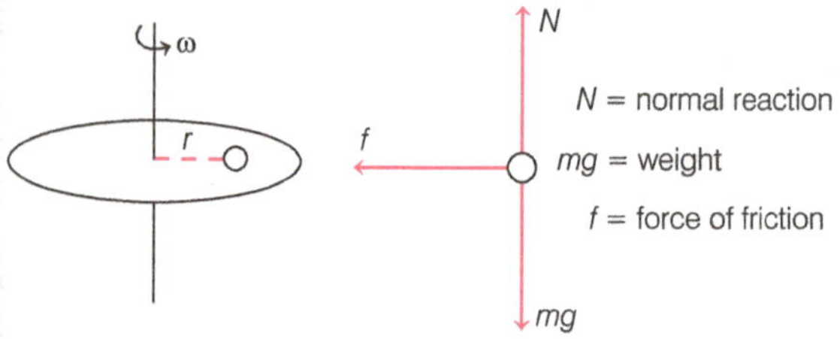

From Frame of Reference Fixed on Ground (Inertial)

Here, N will balance its weight and the force of friction f will provide the necessary centripetal force.

Thus, $$f = mr\omega^2$$ **Ans.**

From Frame of Reference Fixed on Particle Itself (Non-inertial)

In the free body diagram of particle with respect to particle itself, in addition to above three forces (N, mg and f) a pseudo force of magnitude $mr\omega^2$ will have to be applied in a direction away from the centre. But one thing should be clear that in this frame the particle is in equilibrium, i.e. N will balance its weight in vertical direction while f will balance the pseudo force in horizontal direction.

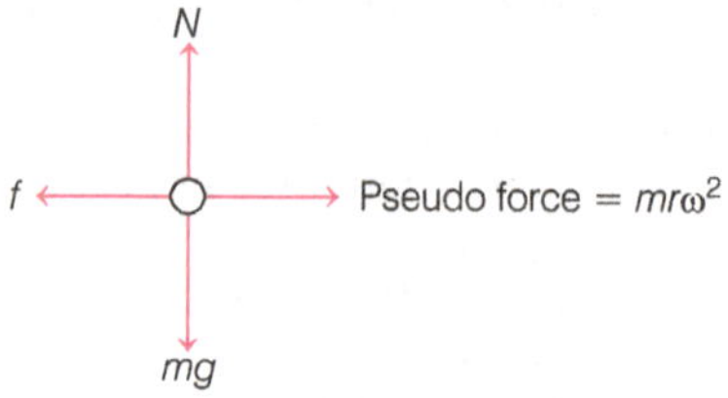

or $$f = mr\omega^2$$ **Ans.**

Thus, we can see that f comes out to be $mr\omega^2$ from both the frames.

Example 2 Two blocks A and B of masses 1 kg and 3 kg are attached with two massless strings as shown in figure. The system is kept over a smooth table and it is rotated about the axis shown in figure with constant angular speed $\omega = 2$ rad/s.

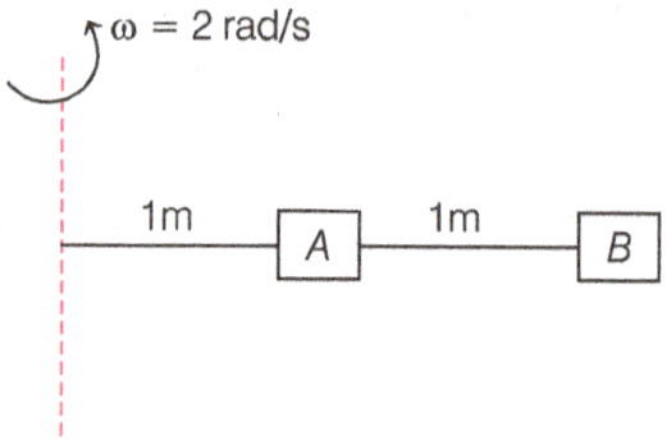

Find direction and magnitude of centrifugal force on

(a) A as observed by B

(b) B as observed by A

Sol. (a) Acceleration of B,

$$a_B = r_B\omega^2 = (2)(2)^2 = 8\ \text{m/s}^2 \quad \text{(towards centre)}$$

Mass of A $m_A = 1$ kg

$\therefore$ Centrifugal force (or pseudo force) on A,

$$F_A = m_A a_B = (1)(8) = 8\text{N}$$ **Ans.**

Direction of this force is in the opposite direction of $\mathbf{a}_B$. Therefore, direction of $\mathbf{F}_A$ is radially outwards.

(b) Acceleration of A,

$$a_A = r_A\omega^2 = (1)(2)^2 = 4\ \text{m/s}^2 \quad \text{(towards centre)}$$

Mass of B, $m_B = 3$ kg

$\therefore$ Centrifugal force on B,

$$F_B = m_B a_A = (3)(4) = 12\ \text{N}$$ **Ans.**

Direction of $\mathbf{F}_B$ is also radially outwards.

Note *From the above example we can see that pseudo force (or centrifugal force) does not make a pair of equal and opposite forces.*

Example 3 A table with smooth horizontal surface is turning at an angular speed ω about its axis. A groove is made on the surface along a radius and a particle is gently placed inside the groove at a distance a from the centre. Find the speed of the particle with respect to the table when its distance from the centre becomes L.

Sol. At distance x from centre,

Centrifugal force $= mx\,\omega^2$

$\therefore$ Acceleration $a = x\,\omega^2$

or $$v.\frac{dv}{dx} = x\,\omega^2$$

or $$\int_0^v v\,dv = \omega^2 \int_a^L x\,dx$$

or $$\frac{v^2}{2} = \frac{\omega^2}{2}(L^2 - a^2)$$

or $$v = \omega\sqrt{L^2 - a^2}$$ **Ans.**

Example 4 A small body of mass $m = 0.10$ kg moves in the reference frame rotating about a stationary axis with a constant angular velocity $\omega = 5.0$ rads^{-1}. What work does the centrifugal force of inertia perform during the transfer of this body along an arbitrary path from point 1 to 2 which are located at the distances $r_1 = 30$ cm and $r_2 = 50$ cm from the rotation axis? **(Irodov 1.128)**

Sol. At a distance r from the centre, centrifugal force,

$$F = mr\omega^2 \quad \text{(radially outwards)}$$

$$\therefore \quad W = \int_{r_1}^{r_2} F dr = \frac{m\omega(r_2^2 - r_1^2)}{2}$$

Substituting the values, we get

$$W = 0.20 \text{ J}$$

MODULE 4

Vertical Circular Motion

(I) In vertical circular motion a particle of mass m is attached to an inextensible light string of length R. The particle is given a velocity u at bottommost point.

(ii) As the particle moves up, its potential energy increases and kinetic energy (or speed) decreases.

In moving down, potential energy decreases and kinetic energy (or speed) increases.

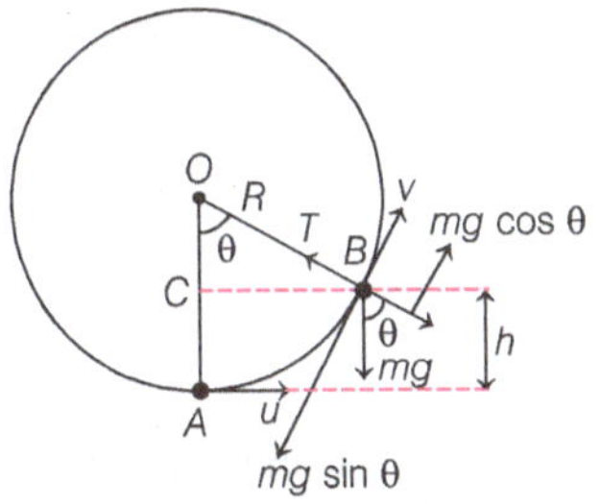

(iii) Thus vertical circular motion is a non-uniform circular motion in which speed of the particle continuously keeps on changing. Therefore, a_t and a_r both are non-zero. In moving upwards, speed decreases. So, a_t is in opposite direction of velocity. In moving downwards, speed increases. So, a_t is in the direction of velocity.

(iv) In circular motion, normally we resolve the forces in two directions, radial and tangential.

Here only two forces are acting on the particle, tension T and weight mg. Tension is always in the radial direction (towards centre). So, resolve mg along radial and tangential directions.

(v) Weight mg is a constant force, while tension T is variable. It is maximum at the bottommost point and minimum at the topmost point.

(vi) Following equations can be applied in vertical circular motion : (See the figure given above)

(a) $v < u$

(b) $h = OA - OC = R - R\cos\theta = R(1 - \cos\theta)$

(c) $\cos\theta = \dfrac{OC}{OB} = \dfrac{R - h}{R}$

(d) From energy conservation,

$$\boxed{v^2 = u^2 - 2gh} \quad \text{...(i)}$$

(e) $T - mg\cos\theta = ma_r = \dfrac{mv^2}{R}$

or $T = mg\cos\theta + \dfrac{mv^2}{R} = mg\left(\dfrac{R - h}{R}\right) + \dfrac{m}{R}(u^2 - 2gh)$

or $$T = mg + \frac{mu^2}{R} - \frac{3mgh}{R} \quad ...(ii)$$

(f) $mg\sin\theta = ma_t$

$\therefore \quad a_t = g\sin\theta$

Now depending upon the value of u following three cases are possible in vertical circular motion.

Case 1: If $u \ge \sqrt{5gR}$, then particle completes the circle.

Case 2: If $\sqrt{2gR} < u < \sqrt{5gR}$, then string slacks ($T = 0$) in upper half of the circle. So, only mg force acts on the particle and motion is projectile or parabolic after that (not circular) until string is taught once again.

Case 3: If $0 < u \le \sqrt{2gR}$, then particle oscillates in lower half of the circle. This is also a case of pendulum.

Now let us prove all three cases one by one.

Case 1. Condition of completing the circle

First Method: Tension is minimum at topmost point (where $h = 2R$). So, this is the most critical point in vertical circular motion because tension can't be negative. Otherwise string will slack. So, $T \ge 0$.

By putting $T = T_{min} = 0$ and $h = 2R$ in Eq. (ii), we get

$$u = \sqrt{5gR}$$

Therefore for completing the circle, velocity at bottommost point,

$$u \ge \sqrt{5gR}$$

From Eq. (ii) we can see that T decreases as h increases. So, T is minimum at topmost point where $h_{max} = 2R$.

Second method :

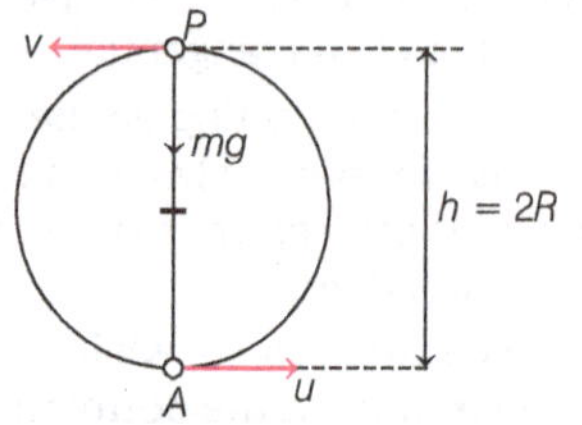

If we put $T = 0$ at topmost point then the only force acting on the particle is mg (downwards). If $v = 0$ at topmost point then particle will fall down due to this force mg. If this force mg is utilised as centripetal force, then only, the particle can remain there (without falling) by following a circular path.

So, minimum value of v at topmost point so that mg is 100% used in centripetal force can be obtained by using the equation,

$$\frac{mv^2}{R} = mg$$

$\therefore \quad v^2 = Rg$

or $\quad v = \sqrt{Rg}$

Now using Eq. (i) with $h = 2R$ we can find **minimum value of u at bottommost point**, which comes out to be $\sqrt{5gR}$.

Note *(a) If topmost velocity,*

(i) $v = \sqrt{gR}$, then mg is 100% used in centripetal force. This is the critical case where $T = 0$.

(ii) $v > \sqrt{gR}$, then centripetal force is greater than mg.

So, the remaining centripetal force is provided by tension.

(iii) $v < \sqrt{gR}$, then mg is not 100% used in centripetal force. So, the particle can't follow a circular path.

(b) Substituting $\theta = 0°$ and $u = \sqrt{5gR}$ in Eq. (ii), we get $T = 6$ mg or in the critical condition tension in the string at lowest position is 6 mg. This is shown in the figure given below.

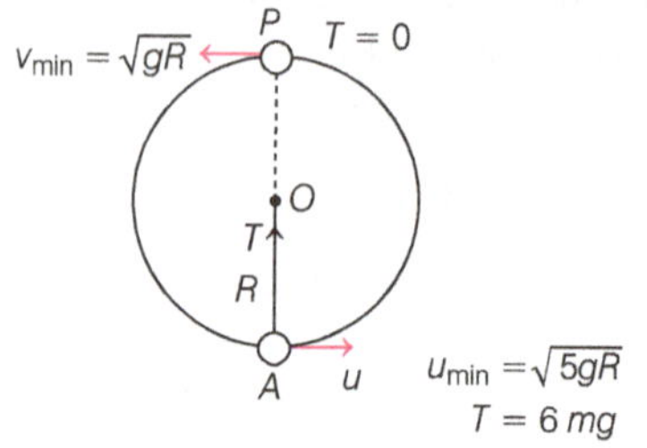

(c) In critical case, $u = \sqrt{5gR}$, $v = \sqrt{gR}$, $T_B = 6$ mg and $T_T = 0$.

If $u = \sqrt{8gR}$, means $5 \to 8$, increased by 3.

Then, $v = \sqrt{4gR}$, $T_B = 9mg$ and $T_T = 3mg$.

Here, u = bottommost velocity

v = topmost velocity

T_B = tension at bottommost point

and T_T = tension at topmost point.

Exercise : Prove the above result.

Case 2 and Case 3. Condition of leaving the circle and condition of oscillations.

(i) Eq. (i) discussed above tells us the velocity at height h if velocity at bottommost point u is given.

Now, let h_1 be the height where velocity becomes zero. So, putting $v = 0$ and $h = h_1$ in Eq. (i), we get

$$h_1 = \frac{u^2}{2g} \quad ...(iii)$$

(ii) Eq. (ii) discussed above gives us the tension at height h if velocity at bottommost point u is given.

Suppose h_2 be the height where tension becomes zero. Putting $T = 0$ and $h = h_2$ in Eq. (ii), we get

$$h_2 = \frac{u^2 + gR}{3g} \quad ...(iv)$$

Thus h_1 is the height were velocity becomes zero and h_2 is the height where tension becomes zero if velocity at bottommost point u is given.

Condition of Leaving the Circle ($\sqrt{2gR} < u < \sqrt{5gR}$)

The particle will leave the circle if tension in the string becomes zero but velocity is not zero or $T = 0$ but $v \ne 0$. Or, tension becomes zero before velocity becomes zero. This is possible only when

$$h_2 < h_1 \quad \text{or} \quad \frac{u^2 + gR}{3g} < \frac{u^2}{2g}$$

or $\quad 2u^2 + 2gR < 3u^2$

or $\quad u^2 > 2gR \quad \text{or} \quad u > \sqrt{2gR}$

Therefore, if $\sqrt{2gR} < u < \sqrt{5gR}$, the particle leaves the circle.

Further, $h > R$ if $u^2 > 2gR$. Thus, the particle will leave the circle when $h > R$ or $90° < \theta < 180°$. This situation is shown in the figure.

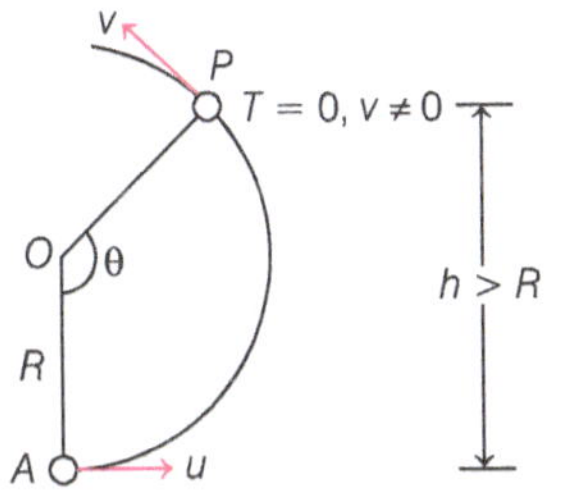

$$\sqrt{2gR} < u < \sqrt{5gR}$$

or $\quad 90° < \theta < 180°$

Note *(i) After leaving the circle, the particle will follow a parabolic path, as the particle comes under gravity.*

(ii) In this case h is actually $h_2 \left(= \frac{u^2 + gR}{3g} \right)$ *where* $T = 0$.

Condition of Oscillation ($0 < u \le \sqrt{2gR}$)

The particle will oscillate, if at extreme position velocity of the particle becomes zero but tension in the string is not zero. Or $v = 0$, but $T \ne 0$. Or, velocity becomes zero before tension becomes zero. This is possible when

$$h_1 < h_2 \quad \text{or} \quad \frac{u^2}{2g} < \frac{u^2 + gR}{3g}$$

or $\quad 3u^2 < 2u^2 + 2gR$

or $\quad u^2 < 2gR \quad \text{or} \quad u < \sqrt{2gR}$

Moreover, if $h_1 = h_2 = R$ then $u = \sqrt{2Rg}$. This implies that tension and velocity both becomes zero simultaneously at height R.

Further, $h \le R$ if $u \le \sqrt{2Rg}$. Thus, for $0 < u \le \sqrt{2gR}$, particle oscillates in lower half of the circle ($0° < \theta \le 90°$).

This situation is shown in the figure given below.

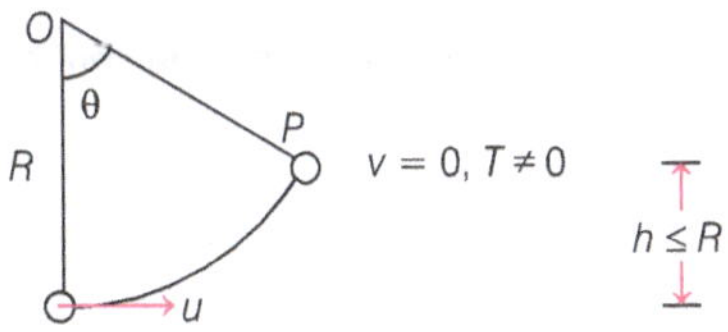

$$0 < u \le \sqrt{2gR} \quad \text{or} \quad 0° < \theta \le 90°$$

Note *Here h is actually* $h_1 \left(= \frac{u^2}{2g} \right)$ *where* $v = 0$.

Important POINTS

(i) The above three conditions have been derived for a particle moving in a vertical circle attached to a string. The same conditions apply, if a particle moves inside a smooth spherical shell of radius R. The only difference is that the tension is replaced by the normal reaction N.

Condition of Looping the Loop is $u \ge \sqrt{5\,gR}$

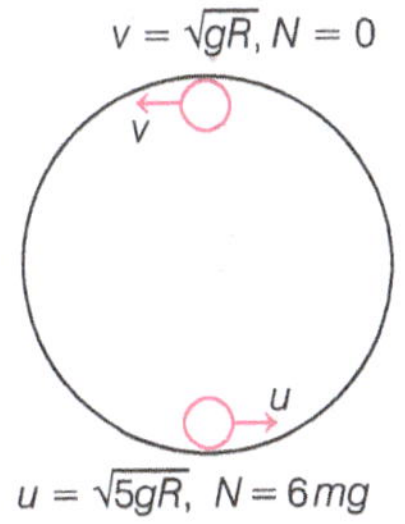

Condition of Leaving the Circle is $\sqrt{2gR} < u < \sqrt{5gR}$

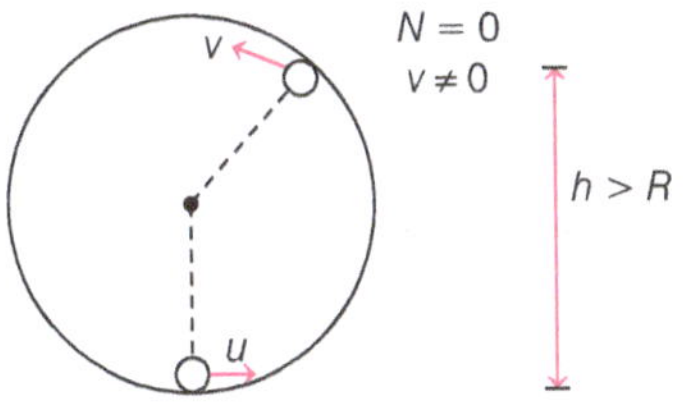

Condition of Oscillation is $0 < u \le \sqrt{2\,gR}$

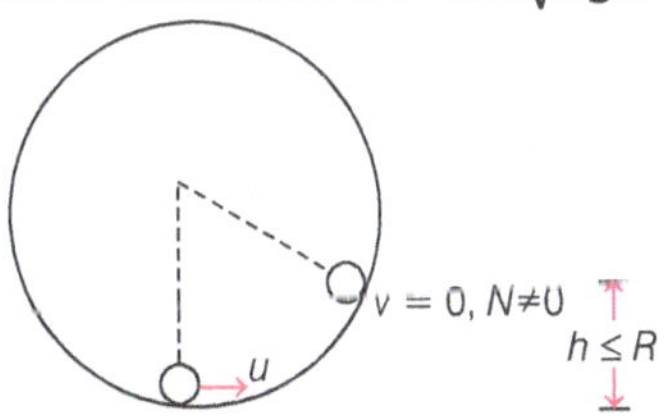

(ii) If a particle of mass m is connected to a light rod and whirled in a vertical circle of radius R, then to complete the circle, the minimum velocity of the particle at the bottommost point is not $\sqrt{5\,gR}$. Because in this case, velocity of the particle at the topmost point can be zero also. Using conservation of mechanical energy between points A and B as shown in figure (a), we get

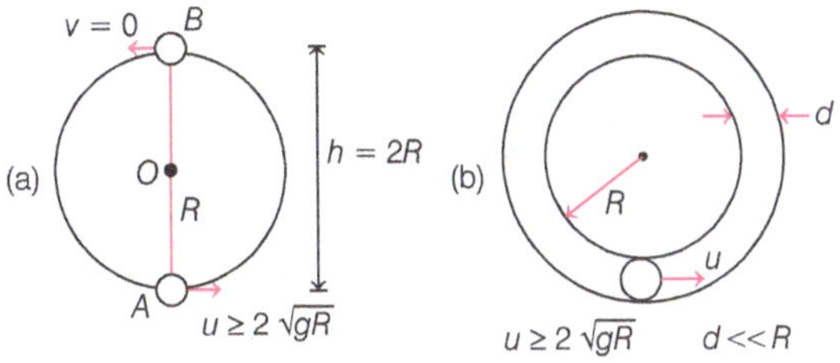

$$\frac{1}{2}m(u^2 - v^2) = mgh \quad \text{or} \quad \frac{1}{2}mu^2 = mg\,(2R) \quad (\text{as } v = 0)$$

$$\therefore \qquad u = 2\sqrt{gR} \quad \text{or} \quad \sqrt{4gR}$$

Therefore, the minimum value of u in this case is $2\sqrt{gR}$ or $\sqrt{4gR}$.

Same is the case when a particle is compelled to move inside a smooth vertical tube as shown in figure (b).

(iii) Oscillation of a pendulum is the part of a vertical circular motion. At point A and C since velocity is zero, net centripetal force will be zero. Only tangential force is present. From A to B or C to B speed of the bob increases. Therefore, tangential force is parallel to velocity. From B to A or B to C speed of the bob decreases. Hence, tangential force is antiparallel to velocity.

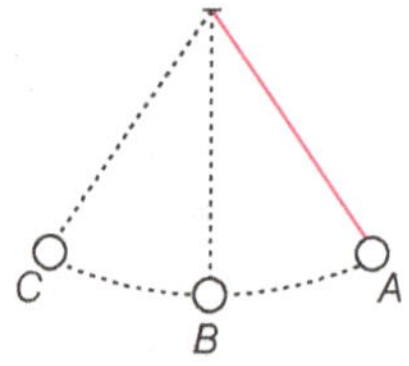

(iv) Motion of a ball over a smooth solid sphere

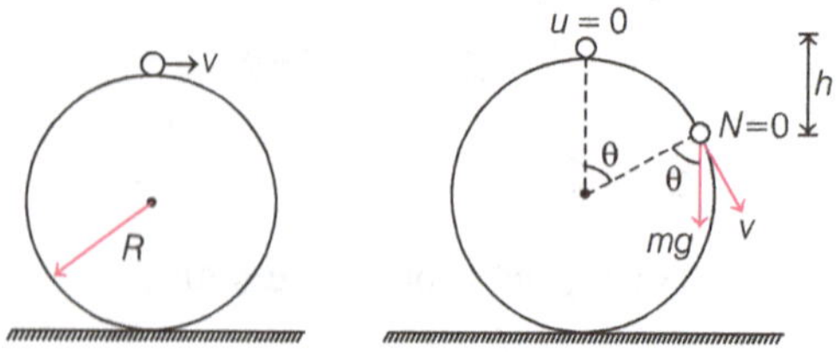

(a) (b)

Suppose a small ball of mass m is given a velocity v over the top of a smooth sphere of radius R. The equation of motion for the ball at the topmost point will be

$$mg - N = \frac{mv^2}{R}$$

or
$$N = mg - \frac{mv^2}{R}$$

From this equation, we see that the value of N decreases as v increases. Minimum value of N can be zero. Hence,

$$0 = mg - \frac{mv_{max}^2}{R}$$

or
$$v_{max} = \sqrt{Rg}$$

So, ball will lose contact with the sphere right from the beginning if velocity of the ball at topmost point $v > \sqrt{Rg}$. If $v < \sqrt{Rg}$ it will lose contact after moving certain distance over the sphere. Now, let us find the angle θ where the ball loses contact with the sphere if velocity at topmost point is tending to zero ($u = 0$). In figure (b)

$$h = R(1 - \cos\theta) \qquad \text{...(i)}$$

$$v^2 = 2gh \qquad \text{...(ii)}$$

Component of mg towards centre provides the centripetal force.

$$\therefore \qquad mg\cos\theta = \frac{mv^2}{R} \qquad (\text{as } N = 0) \qquad \text{...(iii)}$$

Solving Eqs. (i), (ii) and (iii), we get

$$\theta = \cos^{-1}\left(\frac{2}{3}\right) = 48.2°$$

Thus, the ball can move on the sphere maximum up to $\theta = \cos^{-1}\left(\frac{2}{3}\right)$.

Exercise : Find angle θ where the ball will lose contact with the sphere, if velocity at topmost point is $u = \frac{v_{max}}{2} = \frac{\sqrt{gR}}{2}$.

Ans. $\theta = \cos^{-1}\left(\frac{3}{4}\right) = 41.4°$

Hint: Only Eq. (ii) will change as,

$$v^2 = u^2 + 2gh \qquad (u \neq 0)$$

Example 1 A stone tied to a string of length L is whirled in a vertical circle with the other end of the string at the centre. At a certain instant of time the stone is at it lowest position and has a speed u. Find the magnitude of the change in its velocity as it reaches a position, where the string is horizontal. **(1998 JEE)**

Sol.

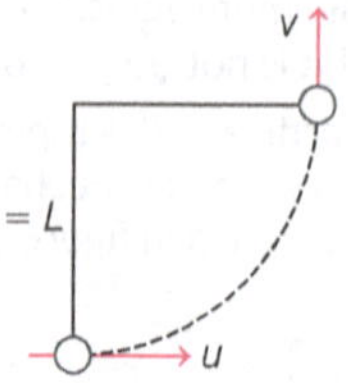

$$v = \sqrt{u^2 - 2gh} = \sqrt{u^2 - 2gL}$$

$$|\Delta \mathbf{v}| = |\mathbf{v}_f - \mathbf{v}_i| = \sqrt{v^2 + u^2 - 2v \cdot u\cos 90°}$$

$$= \sqrt{v^2 + u^2} = \sqrt{(u^2 - 2gL) + u^2}$$

$$= \sqrt{2\,(u^2 - gL)} \qquad \textbf{Ans.}$$

Note Here angle between $\mathbf{v}_f$ and $\mathbf{v}_i$ is 90°.

Example 2 With what minimum speed v must a small ball should be pushed inside a smooth vertical tube from a height h so that it may reach the top of the tube? Radius of the tube is R.

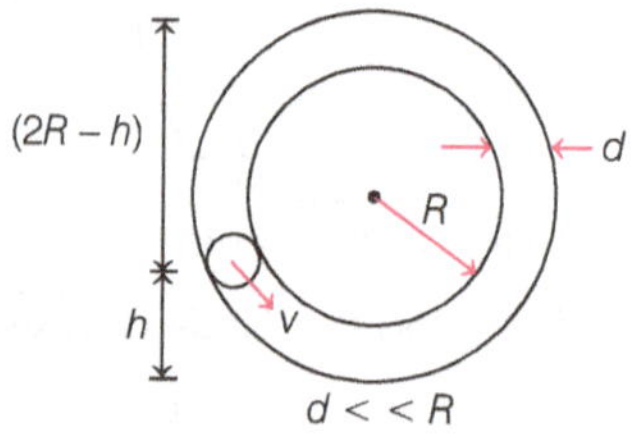

Sol. $v_{top} = \sqrt{v^2 - 2g\,(2R - h)}$

To just complete the vertical circle v_{top} may be zero.

$$\therefore \qquad 0 = \sqrt{v^2 - 2g\,(2R - h)}$$

or
$$v = \sqrt{2g\,(2R - h)} \qquad \textbf{Ans.}$$

Example 3 A particle is suspended from a fixed point by a string of length 5 m. It is projected from the equilibrium position with such a velocity that the string slackens after the particle has reached a height 8 m above the lowest point. Find the velocity of the particle, just before the string slackens. Find also, to what height the particle can rise further? Take $g = 9.8 \text{ m/s}^2$.

Sol. At P,

$$T = 0$$

$$\therefore \quad mg \cos\theta = \frac{mv^2}{R}$$

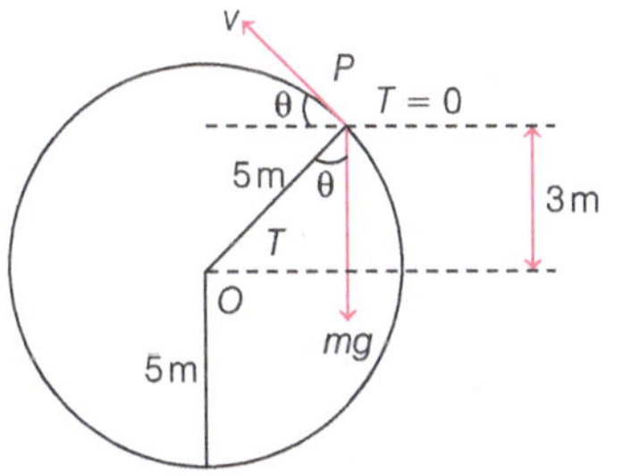

$$\text{or} \quad g\cos\theta = \frac{v^2}{R} \quad \text{or} \quad (9.8)\left(\frac{3}{5}\right) = \frac{v^2}{5}$$

$$\therefore \quad v = 5.42 \text{ m/s} \qquad \textbf{Ans.}$$

After point P motion is projectile

$$h = \frac{v^2 \sin^2\theta}{2g} = \frac{(5.42)^2 (4/5)^2}{2 \times 9.8}$$

$$= 0.96 \text{ m} \qquad \textbf{Ans.}$$

Example 4 A heavy particle hanging from a fixed point by a light inextensible string of length l is projected horizontally with speed $\sqrt{gl}$. Find the speed of the particle and the inclination of the string to the vertical when the tension in the string is equal to the weight of the particle.

Sol. Let $T = mg$ at angle θ as shown in figure.

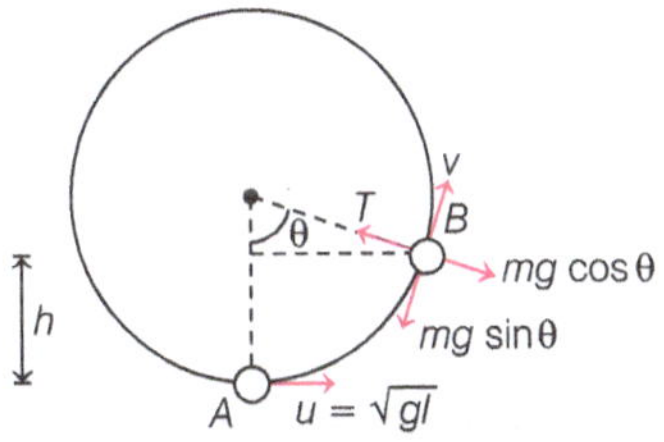

$$h = l(1 - \cos\theta) \qquad \text{...(i)}$$

Applying conservation of mechanical energy between points A and B, we get

$$\frac{1}{2} m(u^2 - v^2) = mgh$$

$$\therefore \quad v^2 = u^2 - 2gh \qquad \text{...(ii)}$$

$$\text{Here,} \quad u^2 = gl \qquad \text{...(iii)}$$

and v = speed of particle in position B

$$\text{Further,} \quad T - mg\cos\theta = \frac{mv^2}{l}$$

$$\text{or} \quad mg - mg\cos\theta = \frac{mv^2}{l} \qquad (\because T = mg)$$

$$\text{or} \quad v^2 = gl(1 - \cos\theta) \qquad \text{...(iv)}$$

Solving above four equations, we get

$$v = \sqrt{\frac{gl}{3}} \qquad \textbf{Ans.}$$

$$\text{Further,} \cos\theta = 1 - \frac{v^2}{gl} = 1 - \frac{1}{3} = \frac{2}{3}$$

$$\therefore \quad \theta = \cos^{-1}\left(\frac{2}{3}\right) \qquad \textbf{Ans.}$$

INTRODUCTORY Exercise

1. In the diagram shown in figure, a bob attached with a light string of radius R is given an initial velocity $u = \sqrt{4gR}$ at the bottommost point.

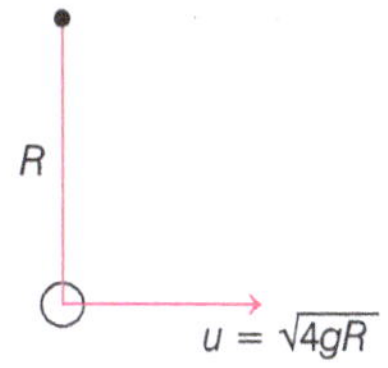

   (a) At what height will the string slack.
   (b) What is velocity of the bob just before slacking of the string.

2. In the above question, if $u = \sqrt{gR}$ then
   (a) after rotating an angle θ, velocity of the bob becomes zero. Find the value of θ.
   (b) If mass of the bob is 'm' then what is the tension in the string when velocity becomes zero?

3. In question number-1, if $u = \sqrt{7gR}$ then
   (a) What is the velocity at topmost point?
   (b) What is the tension at the topmost point?
   (c) What is the tension at the bottommost point?

4. A bob is suspended from a crane by a cable of length $l = 5$ m. The crane and load are moving at a constant speed v_0. The crane is stopped by a bumper and the bob on the cable swings out an angle of 60°. Find the initial speed v_0. ($g = 9.8 \text{ m/s}^2$)

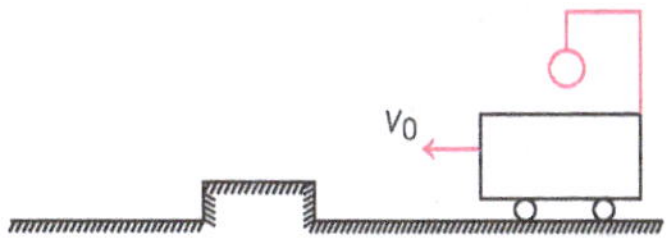

Answers

1. (a) $\frac{5}{3}R$ (b) $\sqrt{\frac{2}{3}gR}$
2. (a) 60° (b) $mg/2$
3. (a) $\sqrt{3gR}$ (b) $2\,mg$ (c) $8\,mg$
4. 7 m/s

See hints on page 588.

Typed Problems

Type 1. Based on different values in vertical circular motion.

Ex. 1 In the figure shown, $u=\sqrt{6gR}$ $(>\sqrt{5gR})$

Find h, v, a_r, a_t, T and F_{net} when

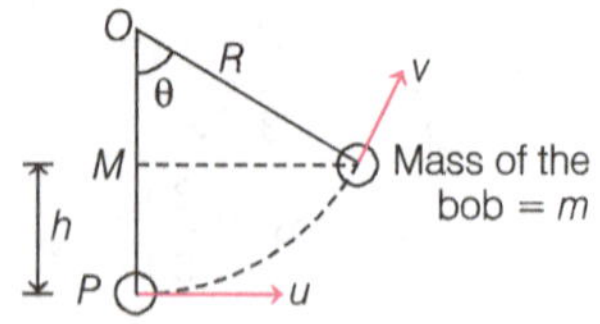

(a) $\theta=60°$ (b) $\theta=90°$

(c) $\theta=180°$

Sol. (a) **When $\theta=60°$**

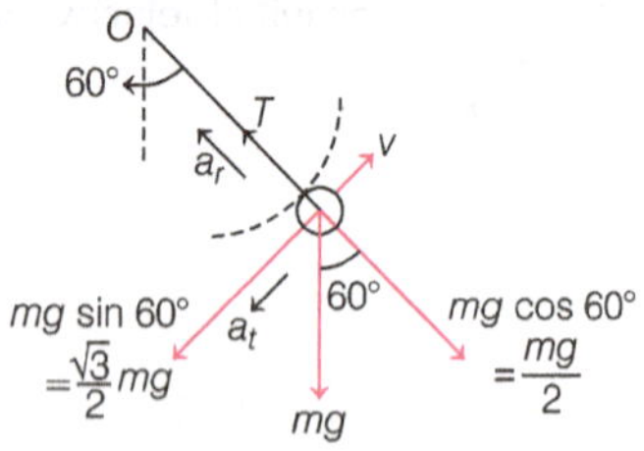

In the figure, we can see that,

$$h=PM=OP-OM=R-R\cos\theta$$

$$=R-R\cos 60°=R-\frac{R}{2}$$ **Ans.**

$$v=\sqrt{u^2-2gh}=\sqrt{6gR-2g\left(\frac{R}{2}\right)}=\sqrt{5gR}$$ **Ans.**

$$a_r=\frac{v^2}{R}=\frac{(\sqrt{5gR})^2}{R}=5g$$ **Ans.**

$$a_t=\frac{F_t}{m}=\frac{\sqrt{3}\,mg}{2m}=\frac{\sqrt{3}g}{2}$$

$$a=\sqrt{a_r^2+a_t^2}=\sqrt{(5g)^2+\left(\frac{\sqrt{3}g}{2}\right)^2}=\frac{\sqrt{103}}{2}g$$

$$T-\frac{mg}{2}=ma_r=\frac{mv^2}{R}=m(5g)$$

$\therefore$ $T=5.5mg$ **Ans.**

$$F_{net}=ma=\frac{\sqrt{103}}{2}mg$$ **Ans.**

(b) **When $\theta=90°$**

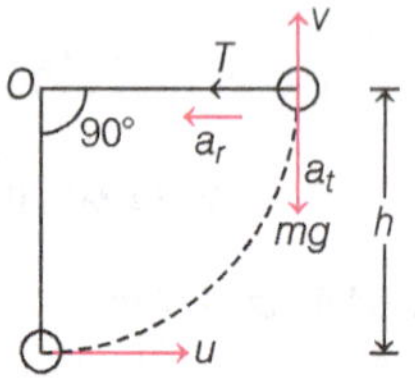

$h=R$ **Ans.**

$$v=\sqrt{u^2-2gh}=\sqrt{6gR-2gR}$$

$$=2\sqrt{gR}$$ **Ans.**

$$a_r=\frac{v^2}{R}=\frac{(2\sqrt{gR})^2}{R}=4g$$

$$a_t=\frac{F_t}{m}=\frac{mg}{m}=g$$

$$a=\sqrt{a_r^2+a_t^2}=\sqrt{(4g)^2+(g)^2}=\sqrt{17}g$$

$$T=ma_r=\frac{mv^2}{R}=m(4g)$$

$$=4mg$$ **Ans.**

$$F_{net}=ma=m(\sqrt{17}g)$$

$$=\sqrt{17}\,mg$$ **Ans**

(c) **When $\theta=180°$**

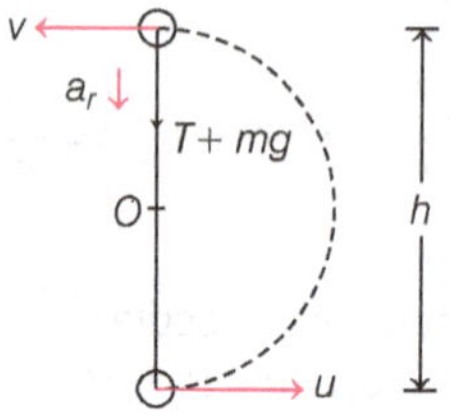

$h=2R$ **Ans.**

$$v=\sqrt{u^2-2gh}$$

$$=\sqrt{6gR-2g\times 2R}=\sqrt{2gR}$$ **Ans.**

$$a_r=\frac{v^2}{R}=\frac{(\sqrt{2gR})^2}{R}=2g$$ **Ans.**

$a_t=\frac{F_t}{m}=0$ (as $F_t=0$, both forces are radial)

$a=a_r$ (as $a_t=0$)

$=2g$ **Ans.**

$$T+mg=ma_r=\frac{mv^2}{R}=m(2g)$$

$\therefore$ $T=mg$

$F_{net}=ma=m(2g)=2mg$ **Ans.**

Note Points

(i) $F_{net}(=ma)$ is also the vector sum of two forces, T and mg acting on the body.

(ii)

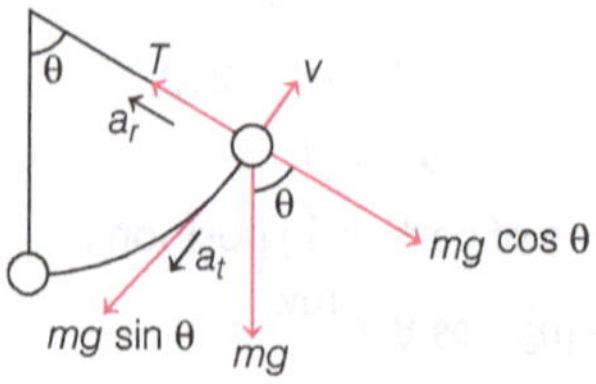

In general, $a_t=\frac{F_t}{m}=\frac{mg\sin\theta}{m}=g\sin\theta$

At $\theta=60°, 90°$ *and* $180°$*, this value is* $\frac{\sqrt{3}}{2}g$*, g and zero.*

Similarly, $T-mg\cos\theta=ma_r=\frac{mv^2}{R}$

$\Rightarrow \quad T=mg\cos\theta+\frac{mv^2}{R}$

(iii) At topmost and bottommost points, both forces act in radial direction.

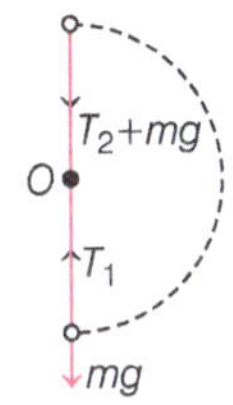

So, $\quad F_t=0$

$\Rightarrow \quad a_t=\frac{F_t}{m}=0$

Type 2. Based on motion of a pendulum.

Concept

Motion of a pendulum is the part of a vertical circular motion. It is the case of oscillation in vertical circular motion.Therefore velocity at bottommost point *C* should be less than or equal to $\sqrt{2gl}$.

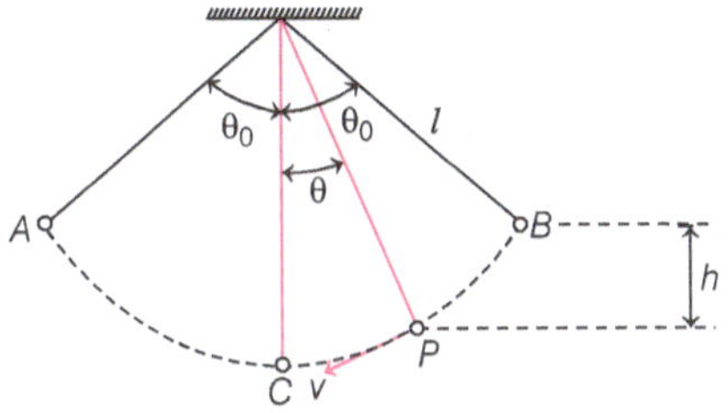

At extreme positions *A* and *B* where, $\theta=\pm\theta_0, v=0, T\neq 0,$

$a_r=\frac{v^2}{R}=0.$

$$T\neq 0$$

Therefore, $a_t=g\sin\theta_0$ and $T=mg\cos\theta_0$

At the bottommost point *C*, where $\theta=0°$

$$v=\text{maximum}$$

$$a_r=\text{maximum}$$

$$T=\text{maximum} \text{ and } a_t=0$$

At some intermediate point *P*, where $\theta=\theta$, neither of the terms discussed above is zero,

$$h=l\cos\theta-l\cos\theta_0$$

$$\Rightarrow \quad v=\sqrt{2gh}$$

$$a_r=\frac{v^2}{l}, a_t=g\sin\theta, a=\sqrt{a_r^2+a_t^2},$$

$$F_{net}=ma$$

and $\quad T-mg\cos\theta=\frac{mv^2}{l}=ma_r$

Ex. 2 A ball of mass '*m*' is released from point *A* where, $\theta_0=53°$. Length of pendulum is '*l*'. Find $v, a_r, a_t, a,$ *T* and F_{net} at

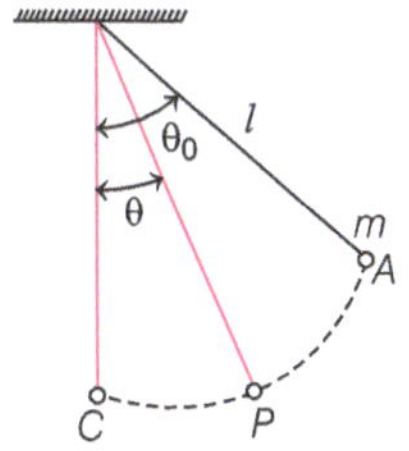

(a) point *A* (b) point *C*

(c) pont *P*, where $\theta=37°$

Sol. At point A

$$v=0 \Rightarrow a_r=\frac{v^2}{R}=0 \qquad (\because R-l)$$

$$a_t=g\sin\theta_0=g\sin 53°=\frac{4}{5}g$$

$$a=a_t=\frac{4}{5}g$$

$$T=mg\cos\theta_0=mg\cos 53°=\frac{3}{5}mg$$

$$F_{net}=ma=\frac{4}{5}mg$$

(b) **At point C**

$$h=OC-OM=l-l\cos 53°$$

$$=l-\frac{3}{5}l=\frac{2}{5}l$$

$$v=\sqrt{2gh}=\sqrt{2g\left(\frac{2}{5}l\right)}=\sqrt{\frac{4}{5}gl}$$

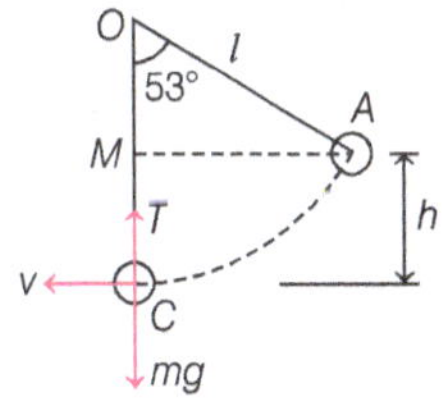

$$a_r=\frac{v^2}{R}=\frac{\left(\sqrt{\frac{4}{5}gl}\right)^2}{l}=\frac{4}{5}g$$

$$a_t=0$$

$$a=a_r=\frac{4}{5}g$$

$$T-mg=\frac{mv^2}{R}=\frac{4}{5}mg$$

$$\therefore \quad T=\frac{9}{5}mg$$

$$F_{net}=ma=\frac{4}{5}mg$$

(c) At point P

$$h = OM - ON$$
$$= l\cos 37° - l\cos 53°$$
$$= \frac{4}{5}l - \frac{3}{5}l = \frac{l}{5}$$

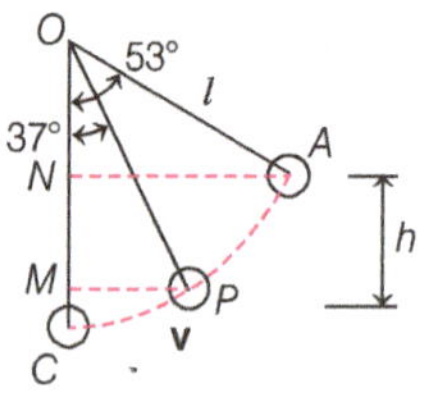

$$v = \sqrt{2gh} = \sqrt{2g\left(\frac{l}{5}\right)} = \sqrt{\frac{2}{5}gl}$$

$$a_r = \frac{v^2}{R} = \frac{\left(\sqrt{\frac{2}{5}gl}\right)^2}{l} = \frac{2}{5}g$$

$$a_t = g\sin\theta = g\sin 37° = \frac{3}{5}g$$

$$a = \sqrt{a_r^2 + a_t^2} = \sqrt{\left(\frac{2}{5}g\right)^2 + \left(\frac{3}{5}g\right)^2} = \frac{\sqrt{13}}{5}g$$

$$T - mg\cos\theta = ma_r$$

or $$T - mg\cos 37° = m\left(\frac{2}{5}g\right)$$

$\therefore$ $$T = \frac{6}{5}mg$$

$$F_{net} = ma = \frac{\sqrt{13}}{5}mg$$

Type 3. Based on condition of leaving the circle.

Ex. 3 A particle is suspended vertically from a point O by an inextensible massless string of length L. A vertical line AB is at a distance $L/8$ from O as shown in figure. The object is given a horizontal velocity u. At some point, its motion ceases to be circular and eventually the object passes through the line AB. At the instant of crossing AB, its velocity is horizontal. Find u. **(1999 JEE)**

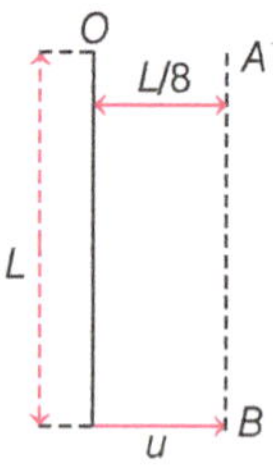

Sol. Let the string slacks at point Q as shown in figure. From P to Q path is circular and beyond Q path is parabolic. At point C, velocity of particle becomes horizontal, therefore

QD = half the range of the projectile.

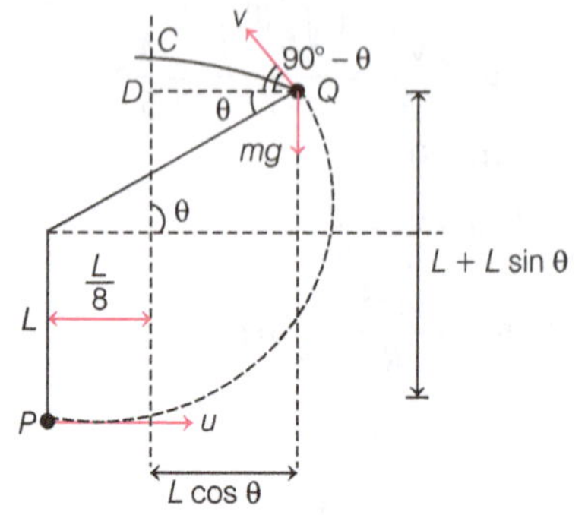

Now, we have following equations

(1) $T_Q = 0$. Therefore, $mg\sin\theta = \dfrac{mv^2}{L}$ ...(i)

(2) $v^2 = u^2 - 2gh = u^2 - 2gL(1 + \sin\theta)$ ...(ii)

(3) $QD = \dfrac{1}{2}$ (Range)

$$\Rightarrow \left(L\cos\theta - \frac{L}{8}\right) = \frac{v^2 \sin 2(90° - \theta)}{2g} = \frac{v^2 \sin 2\theta}{2g} \quad \text{...(iii)}$$

Eq. (iii) can be written as $\left(\cos\theta - \dfrac{1}{8}\right) = \left(\dfrac{v^2}{gL}\right)\sin\theta\cos\theta$

Substituting the value of $\left(\dfrac{v^2}{gL}\right) = \sin\theta$ from Eq. (i), we get

$$\left(\cos\theta - \frac{1}{8}\right) = \sin^2\theta \cdot \cos\theta = (1 - \cos^2\theta)\cos\theta$$

or $$\cos\theta - 1/8 = \cos\theta - \cos^3\theta$$

$\therefore$ $$\cos^3\theta = 1/8$$

or $$\cos\theta = 1/2 \quad \text{or} \quad \theta = 60°$$

From Eq. (i), $v^2 = gL\sin\theta = gL\sin 60°$

or $$v^2 = \frac{\sqrt{3}}{2}gL$$

$\therefore$ Substituting this value of v^2 in Eq. (ii), we get

$$u^2 = v^2 + 2gL(1 + \sin\theta)$$

$$u^2 = \frac{\sqrt{3}}{2}gL + 2gL\left(1 + \frac{\sqrt{3}}{2}\right) = \frac{3\sqrt{3}}{2}gL + 2gL$$

$$= gL\left(2 + \frac{3\sqrt{3}}{2}\right) \quad \therefore \quad u = \sqrt{gL\left(2 + \frac{3\sqrt{3}}{2}\right)}$$

Note *This is the case of leaving the circle.*

$\therefore$ $\sqrt{2gL} < u < \sqrt{5gL}$

Type 4. Based on motion of a ball over a smooth sphere.

Ex. 4 A small body of mass m slides without friction from the top of a hemisphere of radius r with initial velocity zero. At what height will the body be detached from the centre of the hemisphere ?

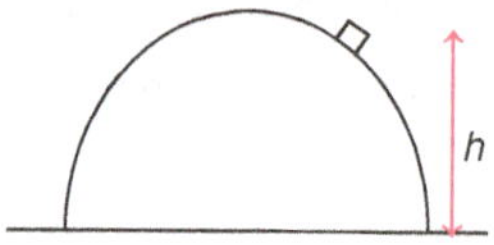

(a) $h = \dfrac{r}{2}$ (b) $h = \dfrac{r}{3}$ (c) $h = \dfrac{2r}{3}$ (d) $h = \dfrac{r}{4}$

Sol. When released from top with zero velocity block leaves contact at

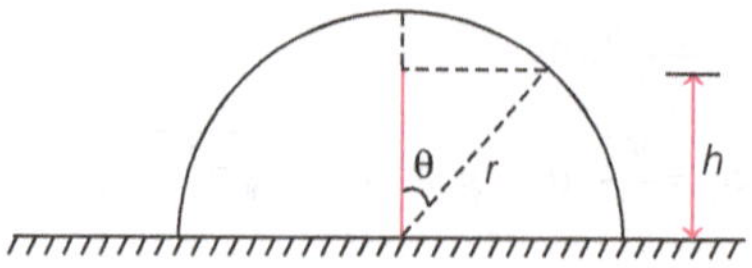

$$\cos\theta = \frac{2}{3} \Rightarrow h = r\cos\theta = \frac{2r}{3}$$

$\therefore$ The correct option is (c).

Ex. 5 A particle moves from rest at A on the surface of a smooth circular cylinder of radius r as shown. At B it leaves the cylinder. The equation relating α and β is

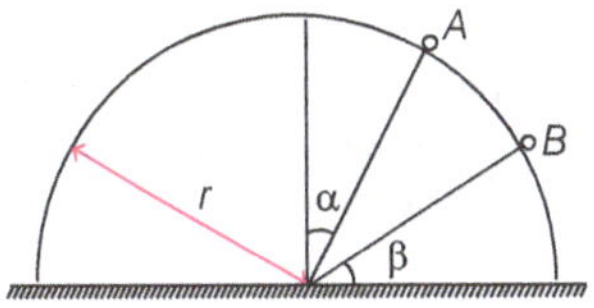

(a) $3\sin\alpha = 2\cos\beta$ (b) $2\sin\alpha = 3\cos\beta$
(c) $3\sin\beta = 2\cos\alpha$ (d) $2\sin\beta = 3\cos\alpha$

Sol. $h_{AB} = (r\cos\alpha - r\sin\beta)$

Velocity of particle at B,

$$v = \sqrt{2gh_{AB}} = \sqrt{2g(r\cos\alpha - r\sin\beta)}$$

Particle will leave contact at B, if component of weight is just equal to centripetal force (towards centre).

or $$mg\sin\beta = \frac{mv^2}{r}$$

or $$\sin\beta = \frac{v^2}{rg}$$

or $$\sin\beta = 2\cos\alpha - 2\sin\beta$$

$\therefore$ $$3\sin\beta = 2\cos\alpha$$

$\therefore$ The correct option is (c).

Ex. 6 A particle of mass m starts to slide down from the top of the fixed smooth sphere. What is the tangential acceleration when it breaks off the sphere ?

(a) $\frac{2g}{3}$ (b) $\frac{\sqrt{5}g}{3}$ (c) g (d) $\frac{g}{3}$

Sol. Particle breaks off the sphere at

$$\cos\theta = \frac{2}{3}$$

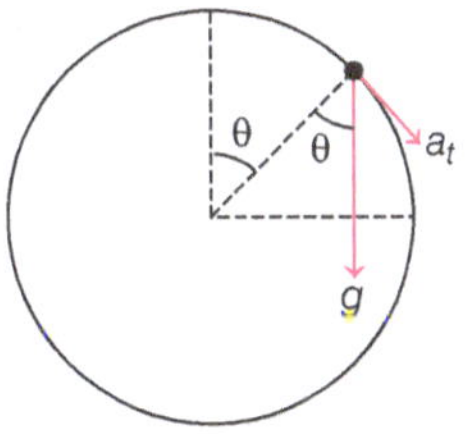

The tangential acceleration at this instant is

$$a_t = g\sin\theta = g\sqrt{1-\cos^2\theta}$$

$$= g\sqrt{1-\frac{4}{9}} = \frac{\sqrt{5}}{3}g \quad \textbf{Ans.}$$

Ex. 7 A spherical ball of mass m is kept at the highest point in the space between two fixed, concentric spheres A and B (see figure). The smaller sphere A has a radius R and the space between the two spheres has a width d. The ball has a diameter very slightly less than d. All surfaces are frictionless. The ball is given a gentle push (towards the right in the figure). The angle made by the radius vector of the ball with the upward vertical is denoted by θ **(2002 JEE)**

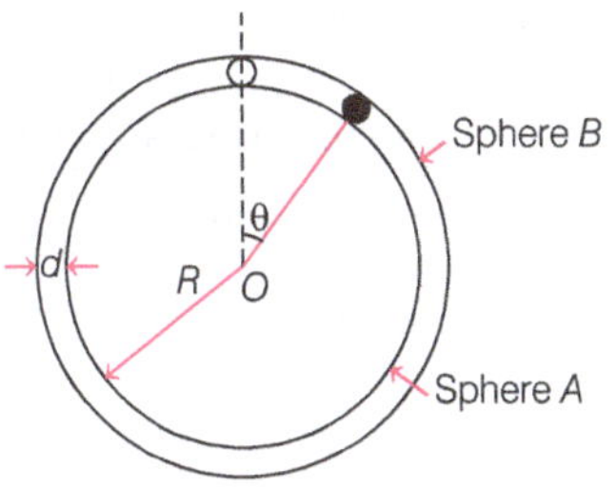

(a) Express the total normal reaction force exerted by the spheres on the ball as a function of angle θ.

(b) Let N_A and N_B denote the magnitudes of the normal reaction forces on the ball exerted by the spheres A and B, respectively. Sketch the variations of N_A and N_B as function of $\cos\theta$ in the range $0 \le \theta \le \pi$ by drawing two separate graphs in your answer book, taking $\cos\theta$ on the horizontal axis.

Sol. (a) $h = \left(R + \frac{d}{2}\right)(1-\cos\theta)$

Velocity of ball at angle θ is

$$v^2 = 2gh$$

$$= 2\left(R + \frac{d}{2}\right)(1-\cos\theta)\,g \quad \text{...(i)}$$

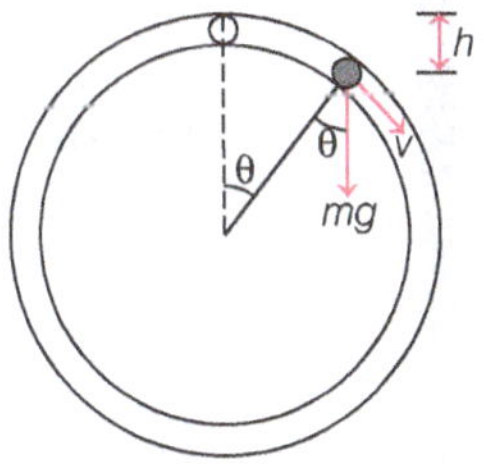

Let N be the normal reaction (away from centre) at angle θ. Then,

$$mg\cos\theta - N = \frac{mv^2}{\left(R + \frac{d}{2}\right)}$$

Substituting the value of v^2 from Eq. (i), we get

$$mg\cos\theta - N = 2\,mg\,(1-\cos\theta)$$

$\therefore$ $$N = mg\,(3\cos\theta - 2)$$

(b) The ball will lose contact with the inner sphere when

$$N = 0$$

or $$3\cos\theta - 2 = 0$$

or $$\theta = \cos^{-1}\left(\frac{2}{3}\right)$$

After this it makes contact with outer sphere and normal reaction starts acting towards the centre.

Thus for $\theta \le \cos^{-1}\left(\frac{2}{3}\right)$

$$N_B = 0$$

and $$N_A = mg\,(3\cos\theta - 2)$$

and for $\theta \ge \cos^{-1}\left(\frac{2}{3}\right)$

$$N_A = 0$$

and $$N_B = mg\,(2 - 3\cos\theta)$$

The corresponding graphs are as follows.

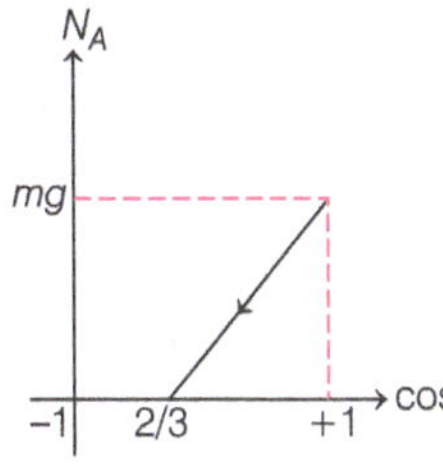

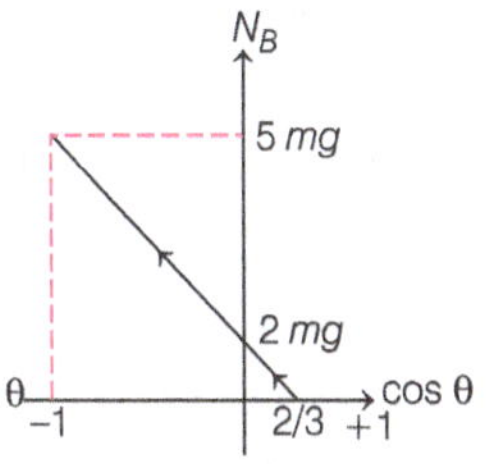

MODULE 4 Exercise

Assertion and Reason

See directions on page 547.

1. **Assertion :** In vertical circular motion, acceleration of bob at position A is greater than 'g'.

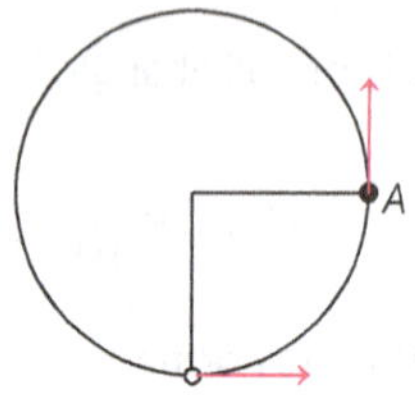

Reason : Net acceleration at A is resultant of tangential and radial components of acceleration.

2. **Assertion :** When water in a bucket is whirled fast overhead, the water does not fall out at the top of the circular path.

Reason : The centripetal force in this position on water is more than the weight of water.

Single Correct Option

3. A simple pendulum of length l has maximum angular displacement θ. Then, maximum kinetic energy of a bob of mass m is

(a) $\frac{1}{2}mgl$ (b) $\frac{1}{2}mgl\cos\theta$

(c) $mgl(1-\cos\theta)$ (d) $\frac{1}{2}mgl\sin\theta$

4. A particle is given an initial speed u inside a smooth spherical shell of radius R so that it is just able to complete the circle. Acceleration of the particle, when its velocity is vertical, is

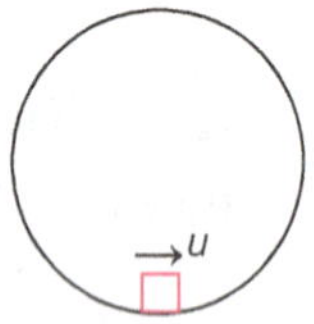

(a) $g\sqrt{10}$ (b) g (c) $g\sqrt{2}$ (d) $g\sqrt{6}$

5. A simple pendulum of length l and bob of mass m is displaced from its equilibrium position O to a position P, so that height of P above O is h. It is then released. What is the tension in the string when the bob passes through the equilibrium position O? Neglect friction, v is the velocity of the bob at O.

(a) $m\left(g-\frac{v^2}{l}\right)$ (b) $\frac{2\,mgh}{l}$

(c) $mg\left(1+\frac{h}{l}\right)$ (d) $mg\left(1+\frac{2h}{l}\right)$

6. A car moves along an uneven horizontal surface with a constant speed at all points. The normal reaction of the road on the car is

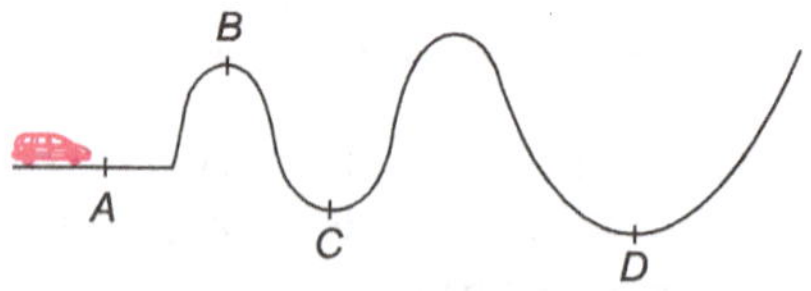

(a) $N_A = N_B = N_C = N_D$ (b) $N_C > N_D > N_A > N_B$
(c) $N_B > N_C > N_A > N_D$ (d) $N_C > N_D > N_B > N_A$

7. A mass is attached to the end of a string of length l which is tied to a fixed point O. The mass is released from the initial horizontal position of the string. Below the point O, at what minimum distance a peg P should be fixed, so that the mass turns about P and can describe a complete circle in the vertical plane?

(a) $\left(\frac{3}{5}\right)l$ (b) $\left(\frac{2}{5}\right)l$ (c) $\frac{l}{3}$ (d) $\frac{2l}{3}$

8. A frictionless track $ABCDE$ ends in a circular loop of radius R. A body slides down the track from point A which is at height $h = 5$ cm. Maximum value of R for a body to complete the loop successfully is

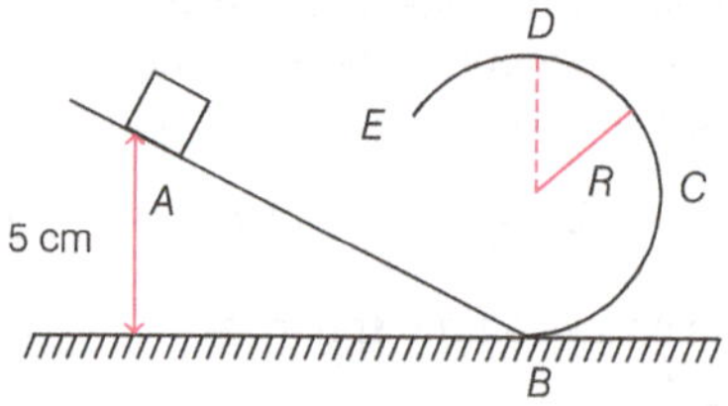

(a) 2 cm (b) $\frac{10}{3}$ cm (c) $\frac{15}{4}$ cm (d) $\frac{18}{3}$ cm

9. A particle suspended by a light inextensible thread of length l is projected horizontally from its lowest position with velocity $\sqrt{7gl/2}$. The string will slack after swinging through an angle equal to

(a) 30° (b) 90° (c) 120° (d) 150°

10. A skier plans to ski a smooth fixed hemisphere of radius R. He starts from rest from a curved smooth surface of height $\left(\frac{R}{4}\right)$. The angle θ at which he leaves the hemisphere is

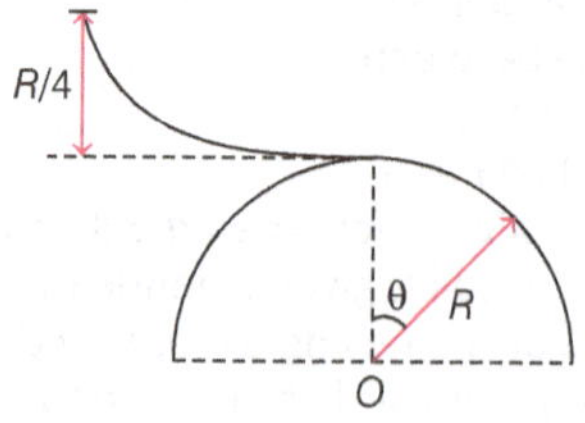

(a) $\cos^{-1}\left(\frac{2}{3}\right)$ (b) $\cos^{-1}\frac{5}{\sqrt{3}}$

(c) $\cos^{-1}\left(\frac{5}{6}\right)$ (d) $\cos^{-1}\left[\frac{5}{2\sqrt{3}}\right]$

11. A particle is projected with velocity u horizontally from the top of a smooth sphere of radius a, so that it slides down the outside of the sphere. If the particle leaves the sphere when it has fallen a height $\frac{a}{4}$, the value of u is

(a) $\sqrt{ag}$ (b) $\frac{\sqrt{ag}}{4}$ (c) $\frac{\sqrt{ag}}{2}$ (d) $\frac{\sqrt{ag}}{3}$

12. Two bodies of masses m and $4m$ are attached to a string as shown in the figure. The body of mass m hanging from string of length l is executing periodic motion with amplitude $\theta = 60°$ while other body is at rest on the surface.

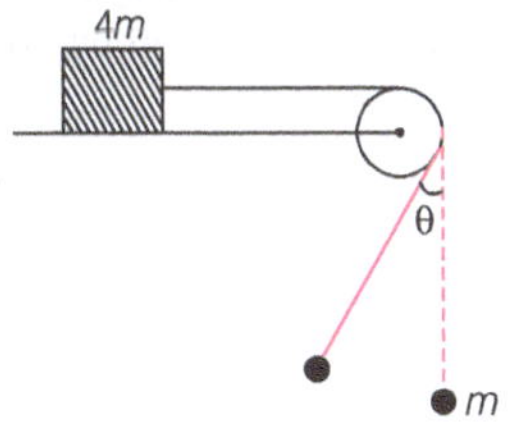

The minimum coefficient of friction between the mass $4m$ and the horizontal surface must be

(a) $\frac{1}{4}$ (b) $\frac{1}{3}$

(c) $\frac{1}{2}$ (d) $\frac{2}{3}$

13. The bob of a 0.2 m pendulum describes an arc of circle in a vertical plane. If the tension in the cord is $\sqrt{3}$ times the weight of the bob when the cord makes an angle 30° with the vertical, the acceleration of the bob in that position is

(a) g (b) $\frac{g}{2}$

(c) $\frac{\sqrt{3}g}{2}$ (d) $\frac{g}{4}$

14. A stone of mass 1 kg tied to a light inextensible string of length $L = \frac{10}{3}$ m, whirling in a circular path in a vertical plane. The ratio of maximum tension to the minimum tension in the string is 4. If g is taken to be 10 m/s^2, the speed of the stone at the highest point of the circle is

(a) 10 m/s (b) $5\sqrt{2}$ m/s

(c) $10\sqrt{3}$ m/s (d) 20 m/s

15. A 50 kg girl is swinging on a swing from rest. Then, the power delivered when moving with a velocity of 2 m/s upwards in a direction making an angle 60° (of string) with the vertical is

(a) 980 W (b) 490 W

(c) $490\sqrt{3}$ W (d) 245 W

16. A body of mass m hangs at one end of a string of length l, the other end of which is fixed. It is given a horizontal velocity, so that the string would just reach, where it makes an angle of 60° with the vertical. The tension in the string at bottommost point is

(a) $2\,mg$ (b) mg (c) $3\,mg$ (d) $\sqrt{3}\,mg$

17. In a simple pendulum, the breaking strength of the string is double the weight of the bob. The bob is released from rest when the string is horizontal. The string breaks when it makes an angle θ with the vertical. Then,

(a) $\theta = \cos^{-1}(1/3)$ (b) $\theta = 60°$

(c) $\theta = \cos^{-1}(2/3)$ (d) $\theta = 0$

18. The figure shows a light rod of length L rigidly attached to a small heavy block at one end and a hook at the other end. The system is released from rest with the rod in a horizontal position. There is a fixed smooth ring at a distance h below the initial position of the hook and the hook gets into the ring as it reaches there. What should be the minimum value of h, so that the block moves in a complete circle about the ring?

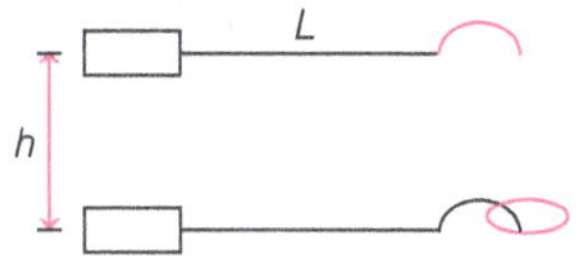

(a) $h = L$ (b) $h = 2L$

(c) $h = 3L$ (d) $h = 4L$

19. A boy holds a pendulum in his hand while standing at the edge of a circular platform of radius r rotating at an angular speed ω. The pendulum will hang at an angle θ with the vertical such that ($r >>$ length of pendulum)

(a) $\tan\theta = 0$ (b) $\tan\theta = \frac{\omega^2 r^2}{g}$

(c) $\tan\theta = \frac{r\omega^2}{g}$ (d) $\tan\theta = \frac{g}{\omega^2 r}$

20. A body tied to a string of length L is revolved in a vertical circle with minimum velocity. When the body reaches the uppermost point, the string breaks and the body moves under the influence of the gravitational field of earth along a parabolic path. The horizontal range AC of the body will be

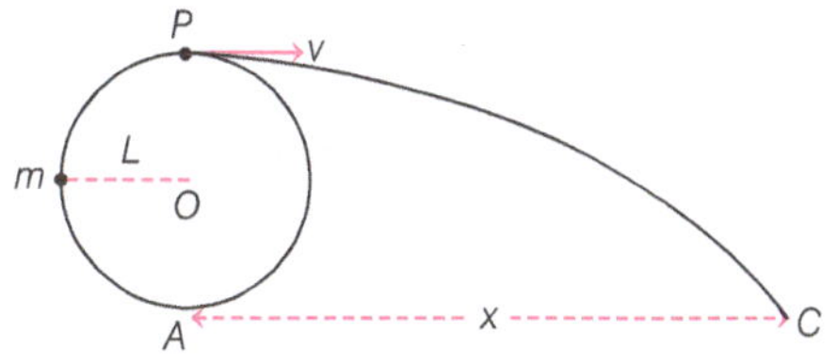

(a) $x = L$ (b) $x = 2L$

(c) $x = 2\sqrt{2}\,L$ (d) $x = \sqrt{2}\,L$

21. A simple pendulum is oscillating without damping. When the displacement of the bob is less than maximum, its acceleration vector **a** is correctly shown in **(2002 JEE)**

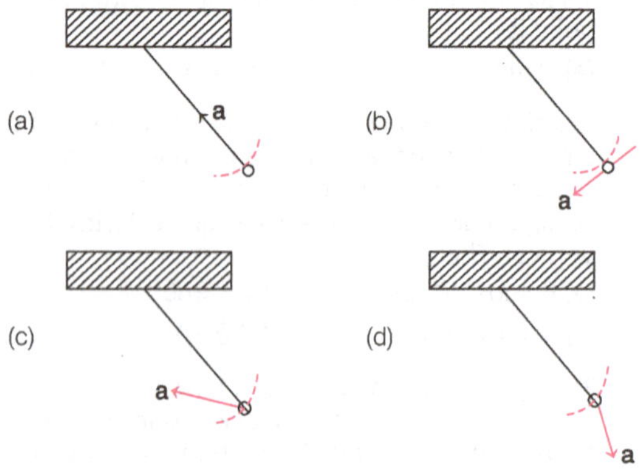

22. A wire, which passes through the hole in a small bead, is bent in the form of quarter of a circle. The wire is fixed vertically on ground as shown in the figure. The bead is released from near the top of the wire and it slides along the wire without friction. As the bead moves from A to B, the force it applies on the wire is **(2014 Main)**

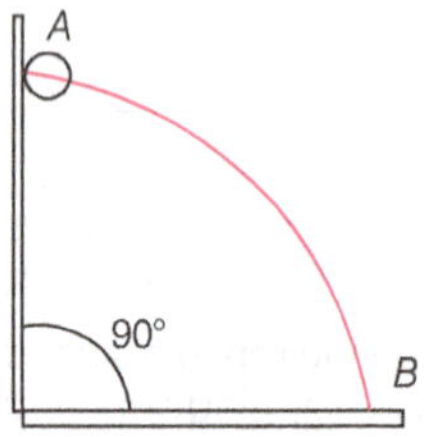

(a) always radially outwards
(b) always radially inwards
(c) radially outwards initially and radially inwards later
(d) radially inwards initially and radially outwards later

Multiple Correct Options

23. A ball tied to the end of a string swings in a vertical circle under the influence of gravity.
(a) When the string makes an angle 90° with the vertical, the tangential acceleration is zero and radial acceleration is somewhere between minimum and maximum
(b) When the string makes an angle 90° with the vertical, the tangential acceleration is maximum and radial acceleration is somewhere between maximum and minimum
(c) At no place in circular motion, tangential acceleration is equal to radial acceleration
(d) When radial acceleration has its maximum value, the tangential acceleration is zero

24. A small spherical ball is suspended through a string of length l. The whole arrangement is placed in a vehicle which is moving with velocity v. Now, suddenly the vehicle stops and ball starts moving along a circular path. If tension in the string at the highest point is twice the weight of the ball, then (assume that the ball completes the vertical circle)
(a) $v = \sqrt{5gl}$
(b) $v = \sqrt{7gl}$
(c) velocity of the ball at highest point is $\sqrt{gl}$
(d) velocity of the ball at the highest point is $\sqrt{3gl}$

Passage Based Questions

Passage (Q. 25 to Q.27)

A small particle of mass m attached with a light inextensible thread of length L is moving in a vertical circle. In the given case particle is moving in complete vertical circle and ratio of its maximum to minimum velocity is 2 : 1.

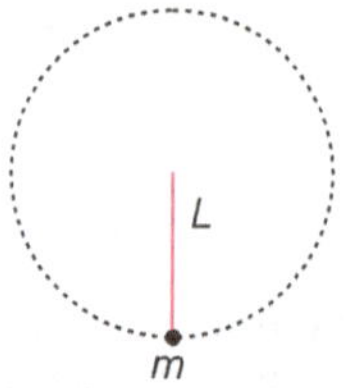

25. Minimum velocity of the particle is
(a) $4\sqrt{\frac{gL}{3}}$ (b) $2\sqrt{\frac{gL}{3}}$ (c) $\sqrt{\frac{gL}{3}}$ (d) $3\sqrt{\frac{gL}{3}}$

26. Kinetic energy of the particle at the lower most position is
(a) $\frac{4mgL}{3}$ (b) $2mgL$ (c) $\frac{8mgL}{3}$ (d) $\frac{2mgL}{3}$

27. Velocity of the particle when it is moving vertically downward is
(a) $\sqrt{\frac{10gL}{3}}$ (b) $2\sqrt{\frac{gL}{3}}$
(c) $\sqrt{\frac{8gL}{3}}$ (d) $\sqrt{\frac{13gL}{3}}$

Match the Columns

28. A bob of mass m is suspended from point O by a massless string of length l as shown. At the bottommost point, it is given a velocity $u = \sqrt{12gl}$ for $l = 1$ m and $m = 1$ kg, match the following two columns when string becomes horizontal (Take, $g = 10\ \text{ms}^{-2}$)

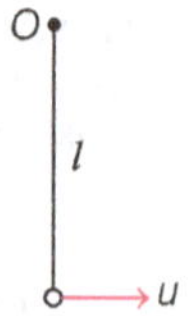

	Column I		Column II (SI units)
(a)	Speed of bob	(p)	10
(b)	Acceleration of bob	(q)	20
(c)	Tension in string	(r)	100
(d)	Tangential acceleration of bob	(s)	None of the above

Subjective Questions

29. A small block slides with velocity $0.5\sqrt{gr}$ on the horizontal frictionless surface as shown in the figure. The block leaves the surface at point C. Calculate angle θ in the figure.

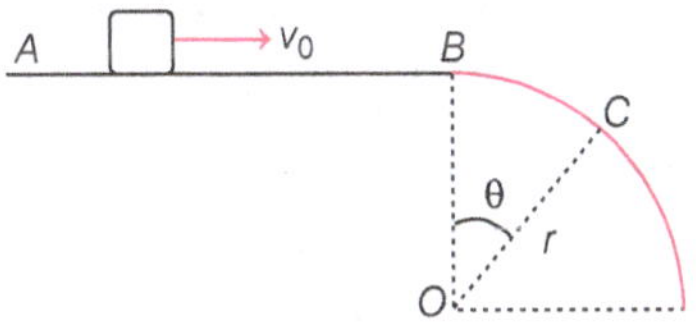

30. The bob of the pendulum shown in figure describes an arc of circle in a vertical plane. If the tension in the cord is 2.5 times the weight of the bob for the position shown. Find the velocity and the acceleration of the bob in that position.

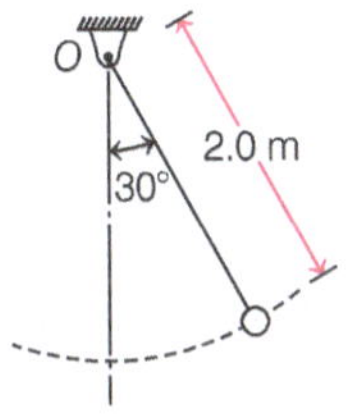

31. The sphere at A is given a downward velocity v_0 of magnitude 5 m/s and swings in a vertical plane at the end of a rope of length $l = 2$ m attached to a support at O. Determine the angle θ at which the rope will break, knowing that it can withstand a maximum tension equal to twice the weight of the sphere.

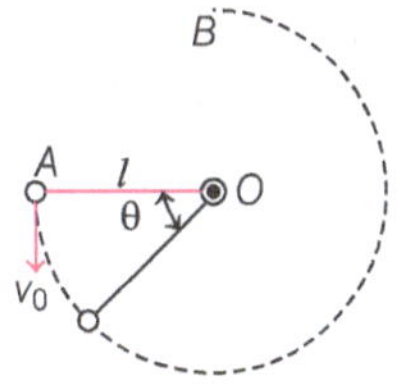

32. An aircraft loops the loop of radius $R = 500$ m with a constant velocity $v = 360$ km per hour. Find he weight of the flyer of mass $m = 70$ kg in the lower and upper points of the loop. **(Irodov 1.84)**

33. A ball suspended by a thread swings in a vertical plane so that its acceleration values in the extreme and the lowest position are equal. Find the thread deflection angle in the extreme position. **(Irodov 1.86)**

34. A small body A starts sliding off the top of a smooth sphere of radius R. Find the angle θ corresponding to the point at which the body breaks off the sphere, as well as the break off velocity of the body. **(Irodov 1.87)**

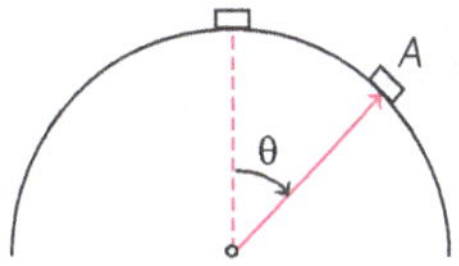

35. A ball of mass m is suspended by a thread of length l. With what minimum velocity has the point of suspension to be shifted in the horizontal direction for the ball to move along the circle about that point? What will be the tension of the thread at the moment it will be passing the horizontal position? **(Irodov 1.137)**

Answers

1. (a) **2.** (a) **3.** (c) **4.** (a) **5.** (d) **6.** (b) **7.** (a) **8.** (a) **9.** (c) **10.** (c)
11. (c) **12.** (c) **13.** (a) **14.** (a) **15.** (c) **16.** (a) **17.** (c) **18.** (a) **19.** (c) **20.** (b)
21. (c) **22.** (d) **23.** (b, d) **24.** (b, d) **25.** (b) **26.** (c) **27.** (a) **28.** (a-p; b-s; c-r; d-p)
29. $\theta = \cos^{-1}\left(\frac{3}{4}\right)$ **30.** 5.66 m/s, $16.75\ m/s^2$ **31.** $\theta = \sin^{-1}\left(\frac{1}{4}\right)$ **32.** 2.1 kN, 0.7 kN
33. $\theta \approx 53°$ **34.** $\theta \approx 48°$, $v = \sqrt{\frac{2gR}{3}}$ **35.** $v_{min} = \sqrt{5gl}, T = 3mg$

See hints on pages 588-593.

Unit Exercise (Cumulative)

Level 1

Single Correct Option

1. A disc is rotating in a room. A boy standing near the rim of the disc of radius R finds the water droplet falling from the ceiling is always falling on his head. As one drop hits his head, other one starts from the ceiling. If height of the roof above his head is H, then angular velocity of the disc is

(a) $\pi\sqrt{\frac{2gR}{H^2}}$ (b) $\pi\sqrt{\frac{2gH}{R^2}}$

(c) $\pi\sqrt{\frac{2g}{H}}$ (d) None of these

2. A particle is moving in a circle of radius R in such a way that at any instant the normal and tangential component of its acceleration are equal. If its speed at $t = 0$ is v_0, the time taken to complete the first revolution is

(a) $\frac{R}{v_0}$ (b) $\frac{R}{v_0}e^{-2\pi}$

(c) $\frac{R}{v_0}(1-e^{-2\pi})$ (d) $\frac{R}{v_0}(1+e^{-2\pi})$

3. A wet open umbrella is held vertical and whirled about the handle at a uniform rate of 21 rev in 44 s. If the rim of the umbrella is a circle of 1 m in diameter and the height of the rim above the floor is 4.9 m. The locus of the drop on floor is a circle of radius

(a) $\sqrt{2.5}$ m (b) 1 m (c) 3 m (d) 1.5 m

4. A ring kept on a smooth table of mass (2π) kg and of radius 0.25 m is making 300 rpm about an axis through its perpendicular to its plane. The tension (in newton) developed in ring is approximately

(a) 50 (b) 100 (c) 175 (d) 250

5. A smooth cone is rotated with an angular velocity ω as shown. A block A is placed at height h. Block has no motion relative to cone. Choose the correct options, when ω is increased

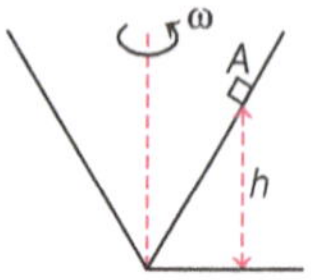

(a) net force acting on block will increase
(b) normal reaction acting on block will increase
(c) h will increase
(d) normal reaction will remain unchanged

6. A collar B of mass 2 kg is constrained to move along a horizontal smooth and fixed circular track of radius 5 m. The spring lying in the plane of the circular track and having spring constant 200 Nm^{-1} is undeformed when the collar is at A. If the collar starts from rest at B, the normal reaction exerted by the track on the collar when it passes through A is

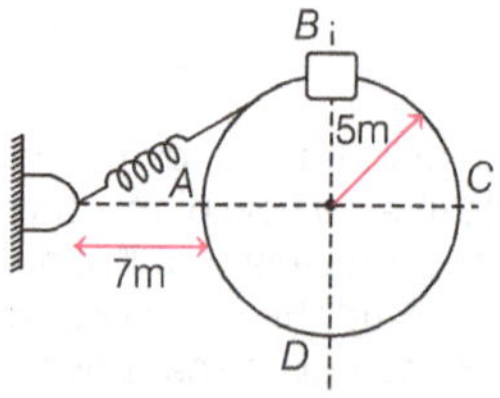

(a) 360 N (b) 720 N (c) 1440 N (d) 2880 N

7. A particle is at rest with respect to the wall of an inverted cone rotating with uniform angular velocity ω about its central axis. The surface between the particle and the wall is smooth. Regarding the displacement of particle along the surface up or down, the equilibrium of particle is

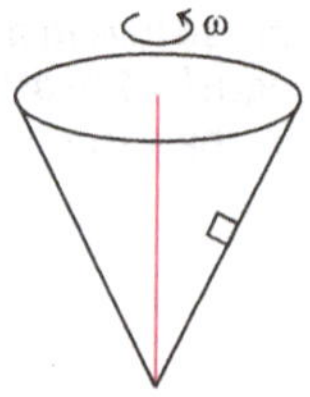

(a) stable (b) unstable
(c) neutral (d) None of these

8. A 10 kg ball attached at the end of a rigid massless rod of length 1 m rotates at constant speed in a horizontal circle of radius 0.5 m and period of 1.58 s, as shown in the figure. The force exerted by the rod on the ball is (Take, $g = 10$ ms^{-2})

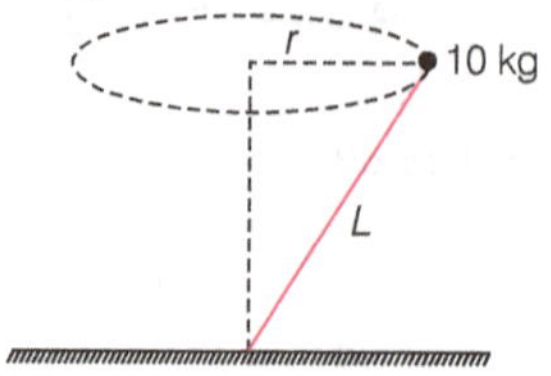

(a) 158 N (b) 128 N
(c) 110 N (d) 98 N

9. A particle A of mass m is attached to a vertical axis by two strings PA and QA of lengths $3L$ and $4L$, respectively. $PQ = 5L$. A rotates around the axis with an angular speed ω. The tensions in the two strings are T_1 and T_2. Then,

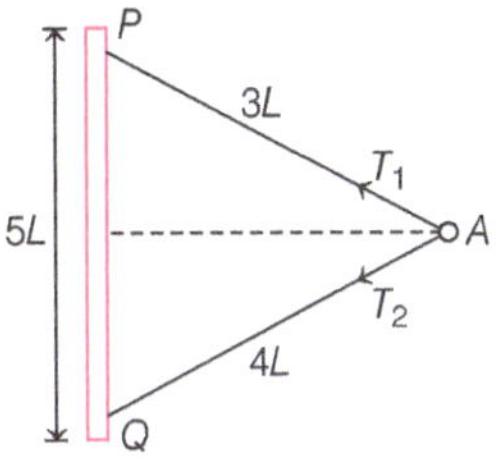

(i) $T_1 = T_2$
(ii) $3T_1 - 4T_2 = 5\ mg$
(iii) $4T_1 + 3T_2 = 12\ m\omega^2 L$

(a) (i) only (b) (i), (ii) (c) (i), (iii) (d) (ii), (iii)

10. A long horizontal rod has a bead which can slide along its length and is initially placed at a distance L from one end A of the rod. The rod is set in angular motion about A with a constant angular acceleration α. If the coefficient of friction between the rod and bead is μ, and gravity is neglected, then the time after which the bead starts slipping is **(2000 JEE)**

(a) $\sqrt{\frac{\mu}{\alpha}}$ (b) $\frac{\mu}{\sqrt{\alpha}}$
(c) $\frac{1}{\sqrt{\mu\alpha}}$ (d) infinitesimal

11. A smooth wire of length $2\pi r$ is bent into a circle and kept in a vertical plane. A bead can slide smoothly on the wire. When the circle is rotating with angular speed ω about the vertical diameter AB, as shown in figure, the bead is at rest with respect to the circular ring at position P as shown. Then, the value of ω^2 is equal to **(2019 Main)**

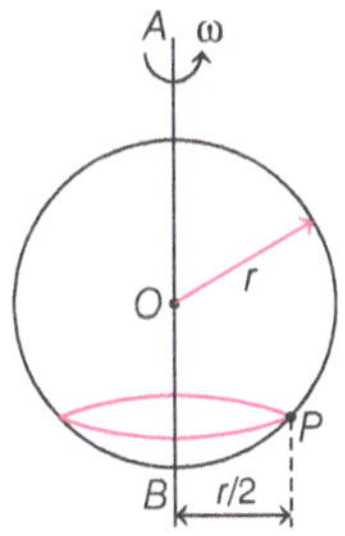

(a) $\frac{\sqrt{3}g}{2r}$ (b) $2g/(r\sqrt{3})$
(c) $(g\sqrt{3})/r$ (d) $2g/r$

12. Inside a smooth spherical cavity, particle A can slide freely. The block having this cavity is moving horizontally with constant acceleration $a_0 = 10\ m/s^2$.

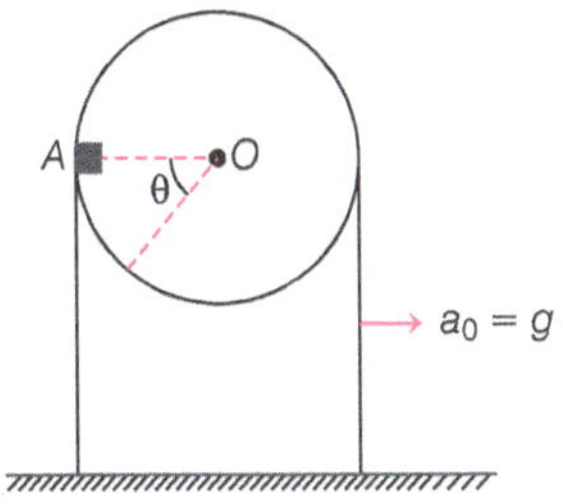

The particle is released from its initial position as shown in the figure. The angle θ with the vertical, when the particle will have maximum speed with respect to the block is

(a) $30°$ (b) $45°$
(c) $60°$ (d) $90°$

13. A section of fixed smooth circular track of radius R in vertical plane is shown in the figure. A block is released from position A and leaves the track at B. The radius of curvature of its trajectory just after it leaves the track at B is

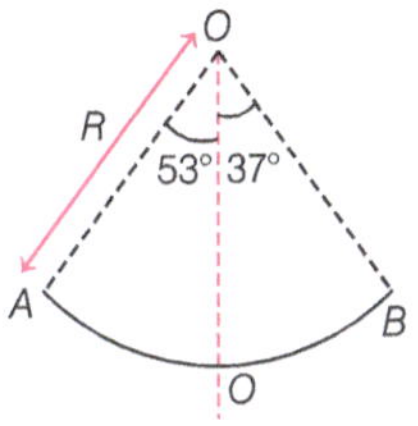

(a) R (b) $\frac{R}{4}$
(c) $\frac{R}{2}$ (d) $\frac{R}{3}$

14. A simple pendulum is released from rest with the string in horizontal position. The vertical component of the velocity of the bob becomes maximum, when the string makes an angle θ with the vertical. The angle θ is equal to

(a) $\frac{\pi}{4}$ (b) $\cos^{-1}\left(\frac{1}{\sqrt{3}}\right)$
(c) $\sin^{-1}\left(\frac{1}{\sqrt{3}}\right)$ (d) $\frac{\pi}{3}$

15. A particle is moving in a circular path in the vertical plane. It is attached at one end of a string of length l whose other end is fixed. The velocity at lowest point is u. The tension in the string is $\mathbf{T}$ and acceleration of the particle is $\mathbf{a}$ at any position. Then, $\mathbf{T} \cdot \mathbf{a}$ is zero at highest point, if

(a) $u > \sqrt{5gl}$
(b) $u = \sqrt{5gl}$
(c) Both (a) and (b) are correct
(d) Both (a) and (b) are wrong

16. In the above question, $\mathbf{T} \cdot \mathbf{a}$ is positive at the lowest point for

(a) $u \le \sqrt{2gl}$ (b) $u = \sqrt{2gl}$
(c) $u < \sqrt{2gl}$ (d) any value of u

17. A particle of mass 1 g executes an oscillatory motion on the concave surface of a spherical dish of radius 2 m placed on a horizontal plane. If the motion of the particle begins from a point on the dish at a height of 1 cm from the horizontal plane and the coefficient of friction is 0.01, the total distance covered by the particle before it comes to rest, is approximately

(a) 2.0 m (b) 10.0 m
(c) 1.0 m (d) 20.0 m

18. A simple pendulum is vibrating with an angular amplitude of 90° as shown in the figure. For what value of α, is the acceleration directed:

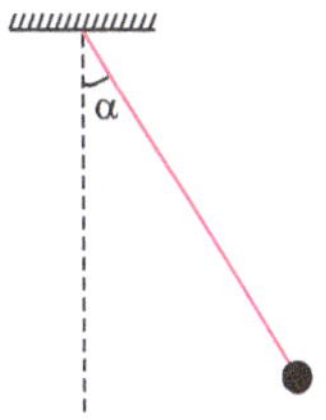

(i) Vertically upwards (ii) Horizontally
(iii) Vertically downwards

(a) $0°, \cos^{-1}\left(\frac{1}{\sqrt{3}}\right), 90°$ (b) $90°, \cos^{-1}\left(\frac{1}{\sqrt{3}}\right), 0°$

(c) $0°, \cos^{-1}\frac{1}{\sqrt{2}}, 90°$ (d) $\cos^{-1}\left(\frac{1}{\sqrt{3}}\right), 90°, 0°$

19. A bob of mass M is suspended by a massless string of length L. The horizontal velocity v at position A is just sufficient to make it reach the point B. The angle θ at which the speed of the bob is half of that at A, satisfies **(2008 JEE)**

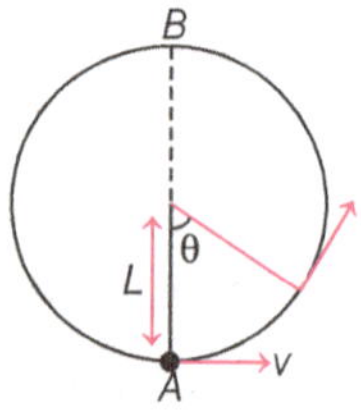

(a) $\theta = \frac{\pi}{4}$ (b) $\frac{\pi}{4} < \theta < \frac{\pi}{2}$

(c) $\frac{\pi}{2} < \theta < \frac{3\pi}{4}$ (d) $\frac{3\pi}{4} < \theta < \pi$

Multiple Correct Options

20. A particle is describing circular motion in a horizontal plane in contact with the smooth surface of a fixed right circular cone with its axis vertical and vertex down. The height of the plane of motion above the vertex is h and the semi-vertical angle of the cone is α. The period of revolution of the particle

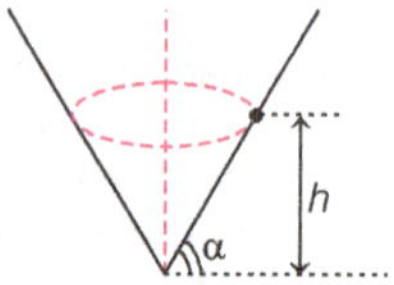

(a) increases as h increases
(b) decreases as h decreases
(c) increases as α increases
(d) decreases as α increases

21. A block of mass 10 kg is pulled along a smooth surface in the form of arc of circle of radius 10 m. The applied force F is 200 N as shown in figure. If the block starts from point A, then

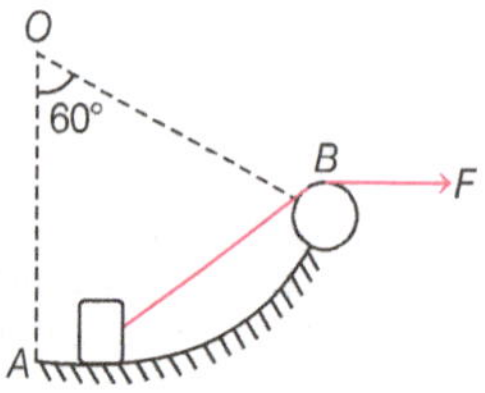

(a) the speed at point B is 17.32 m/s
(b) the speed at point B is 27 m/s
(c) the work done by gravity during motion from A to B is negative
(d) the work done by tension on block is zero

Passage Based Questions

Passage 1 (Q. 22 to Q.23)

A block of mass 1 kg is attached to a 1.25 m long light string tied to the apex of a triangular smooth block as shown in the figure. At $t = 0$, the triangular block is imparted a constant angular acceleration $\sqrt{\frac{40}{3}}$ rad/s^2 about an axis along its vertical side. Assume that the block is stationary with respect to triangular block.

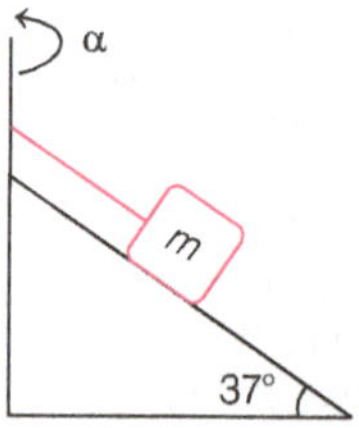

22. At what time will the block leave contact with the triangular block?
(a) 0.5 s (b) 1 s
(c) 2 s (d) 3 s

23. What would be the tension in the string at the instant when the block leaves the contact with the triangular block?
(a) 10 N (b) 16.67 N
(c) 30 N (d) 40 N

Passage 2 (Q. 24 to Q.25)

A small block of mass 1 kg is released from rest at the top of a rough track. The track is a circular arc of radius 40 m. The block slides along the track without toppling and a frictional force acts on it in the direction opposite to the instantaneous velocity. The work done in overcoming the friction up to the point Q, as shown in the figure, is 150 J. (Take the acceleration due to gravity, $g = 10\,\text{ms}^{-2}$) **(2013 JEE)**

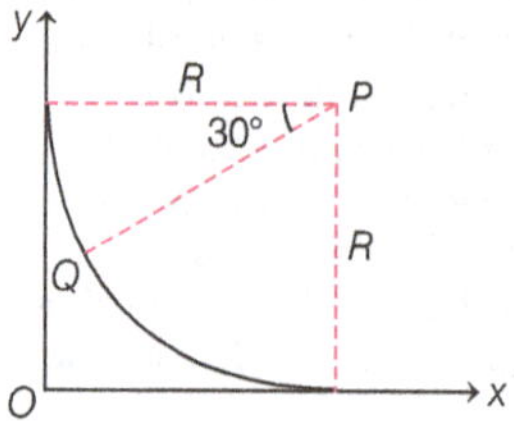

24. The speed of the block when it reaches the point Q is
(a) 5 ms^{-1} (b) 10 ms^{-1}
(c) $10\sqrt{3}$ ms^{-1} (d) 20 ms^{-1}

25. The magnitude of the normal reaction that acts on the block at the point Q is
(a) 7.5 N (b) 8.6 N
(c) 11.5 N (d) 22.5 N

Match the Columns

26. Speed of a particle moving in a circle of radius 2 m varies with time as $v = 2t$ (SI units). At $t = 1$ s match the following two columns.

	Column I		Column II (SI units)
(a)	$\mathbf{a} \cdot \mathbf{v}$	(p)	$2\sqrt{2}$
(b)	$\lvert\mathbf{a} \times \boldsymbol{\omega}\rvert$	(q)	2
(c)	$\mathbf{v} \cdot \boldsymbol{\omega}$	(r)	4
(d)	$\lvert\mathbf{v} \times \mathbf{a}\rvert$	(s)	None of the above

Here, symbols have their usual meanings.

27. Position vector (with respect to centre), velocity vector and acceleration vector of a particle in circular motion are $\mathbf{r} = (3\hat{i} - 4\hat{j})$ m, $\mathbf{v} = (4\hat{i} - a\hat{j})$ ms^{-1} and $\mathbf{a} = (-6\hat{i} + b\hat{j})$ ms^{-2}. Speed of particle is constant. Match the following two columns.

	Column I		Column II (SI units)
(a)	Value of a	(p)	8
(b)	Value of b	(q)	3
(c)	Radius of circle	(r)	5
(d)	$\mathbf{r}\cdot(\mathbf{v}\times\mathbf{a})$	(s)	None of the above

Subjective Questions

28. A point traversed half a circle of radius $R = 160$ cm during time interval $\tau = 10.0$ s. Calculate the following quantities averaged over that time:
   (a) the mean velocity $\langle \mathbf{v} \rangle$
   (b) the modulus of the mean velocity vector $|\langle \mathbf{v} \rangle|$
   (c) the modulus of the mean vector of the total acceleration $|\langle \mathbf{a} \rangle|$ if the point moved with constant tangential acceleration. **(Irodov 1.19)**

29. A point moves with deceleration along the circle of radius R so that at any moment of time its tangential and normal accelerations are equal in moduli. At the initial moment $t = 0$ the velocity of the point equals v_0. Find :
   (a) the velocity of the point as a function of time and as a function of the distance covered s
   (b) the total acceleration of the point as a function of velocity and the distance covered. **(Irodov 1.38)**

30. A point moves along an arc of a circle of radius R. Its velocity depends on the distance covered s as $v = \alpha\sqrt{s}$, where α is a constant. Find the angle θ between the vector of the total acceleration and the vector of velocity as a function of s. **(Irodov 1.39)**

31. A block of mass m is kept on a horizontal ruler. The friction coefficient between the ruler and the block is μ. The ruler is fixed at one end and the block is at a distance L from the fixed end. The ruler is rotated about the fixed end in the horizontal plane.
   (a) What can be the maximum constant angular speed for which the block does not slip ?
   (b) If the angular speed of the ruler is uniformly increased from zero at an angular acceleration α, at what angular speed will the block slip ?

32. A thin circular wire of radius R rotates about its vertical diameter with an angular frequency ω. Show that a small bead on the wire remains at its lowermost point for $\omega \le \sqrt{g/R}$. What is angle made by the radius vector joining the centre to the bead with the vertical downward direction for $\omega = \sqrt{2g/R}$? Neglect friction.

33. Two blocks tied with a massless string of length 3 m are placed on a rotating table as shown. The axis of rotation is 1 m from 1 kg mass and 2 m from 2 kg mass. The angular speed $\omega = 4$ rad/s. Ground below 2 kg block is smooth and below 1 kg block is rough. (Take, $g = 10$ m/s^2)

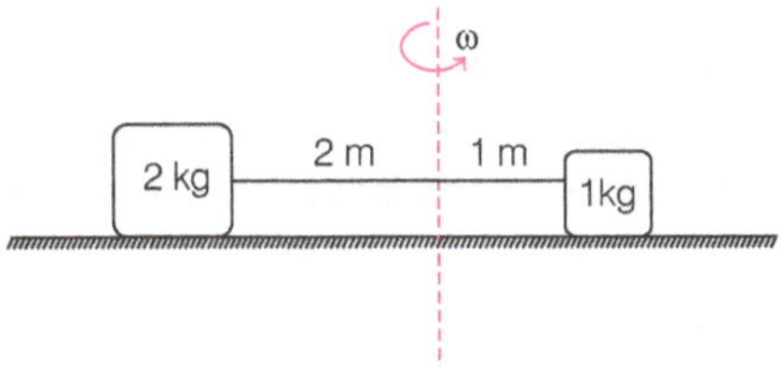

   (a) Find tension in the string, force of friction on 1 kg block and its direction.
   (b) If coefficient of friction between 1 kg block and ground is $\mu = 0.8$. Find maximum angular speed so that neither of the blocks slips.
   (c) If maximum tension in the string can be 100 N, then find maximum angular speed so that neither of the blocks slips.

Note *Assume that in part (b) tension can take any value and in parts (a) and (c) friction can take any value.*

34. A particle of mass m moves along a circle of radius R with a normal acceleration varying with time as $a_n = \alpha t^2$, where α is a constant. Find the time dependence of the power developed by all the forces acting on the particle, and the mean value of this power averaged over the first t seconds after the beginning of motion. **(Irodov 1.126)**

35. Two blocks of masses $m_1 = 10$ kg and $m_2 = 5$ kg connected to each other by a massless inextensible string of length 0.3 m are placed along a diameter of turn table. The coefficient of friction between the table and m_1 is 0.5 while there is no friction between m_2 and the table. The table is rotating with an angular velocity of 10 rad/s about a vertical axis passing through its centre O. The masses are placed along the diameter of table on either side of the centre O such that the mass m_1 is at a distance of 0.124 m from O. The masses are observed to be at rest with respect to an observer on the turn table. **(1997 JEE)**
   (a) Calculate the frictional force on m_1.
   (b) What should be the minimum angular speed of the turn table, so that the masses will slip from this position?
   (c) How should the masses be placed with the string remaining taut so that there is no frictional force acting on the mass m_1?

36. A toy car is moving on a circular path of radius 10 m on an inclined plane of inclination 30°. The coefficient of friction between the tyres and the inclined plane is $\mu = \frac{2}{\sqrt{3}}$. Find maximum constant speed of car such that car does not slip. Assume that the width of a car is negligible with respect to radius of circular path.

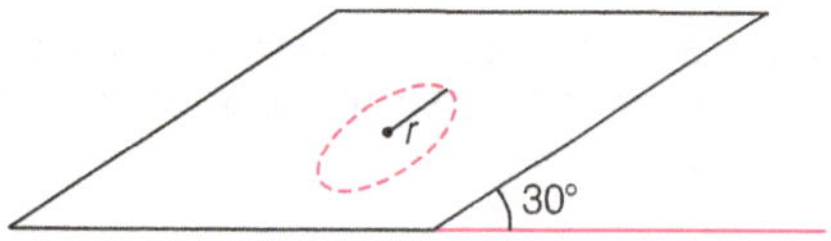

37. A smooth circular tube of radius R is fixed in a vertical plane. A particle is projected from its lowest point with a velocity just sufficient to carry it to the highest point. Show that the time taken by the particle to reach the end of the horizontal diameter is

$$\sqrt{\frac{R}{g}}\ln(1+\sqrt{2}).$$

Hint : $\int \sec\theta \cdot d\theta = \ln(\sec\theta + \tan\theta)$

38. A heavy particle slides under gravity inside a smooth vertical tube held in vertical plane. It starts from the highest point with velocity $\sqrt{2ag}$, where a is the radius of the circle. Find the angular position θ (as shown in figure) at which the vertical acceleration of the particle is maximum.

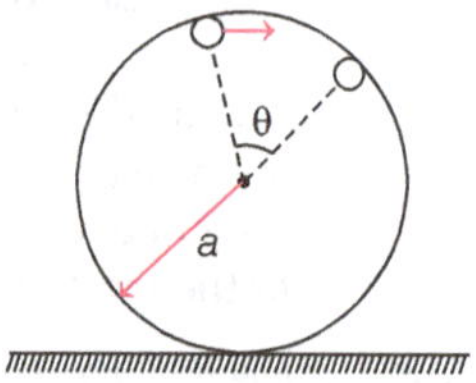

39. A particle of mass m is suspended by a string of length l from a fixed rigid support. A horizontal velocity $v_0 = \sqrt{3gl}$ is given to it. Calculate the angle made by the string with the vertical when the acceleration of the particle is inclined to the string by 45°.

40. The simple 2 kg pendulum is released from rest in the horizontal position. As it reaches the bottom position, the cord wraps around the smooth fixed pin at B and continues in the smaller arc in the vertical plane. Calculate the magnitude of the force R supported by the pin at B when the pendulum passes the position $\theta = 30°$. (Take, $g = 9.8\ \text{m/s}^2$)

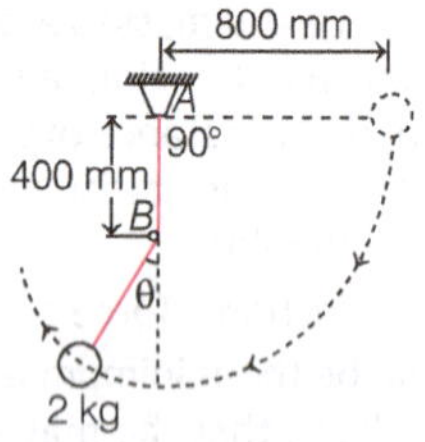

41. A small sphere of mass m suspended by a thread is first taken aside so that the thread forms the right angle with the vertical and then released. Find :
   (a) the total acceleration of the sphere and the thread tension as a function of θ, the angle of deflection of the thread from the vertical
   (b) the thread tension at the moment when the vertical component of the sphere's velocity is maximum
   (c) the angle θ between the thread and the vertical at the moment when the total acceleration vector of the sphere is directed horizontally. **(Irodov 1.85)**

42. A chain of length l is placed on a smooth spherical surface of radius R with one of its ends fixed at the top of the sphere. What will be the acceleration a of each element of the chain when its upper end is released? It is assumed that the length of the chain $l < \frac{1}{2}\pi R$. **(Irodov 1.107)**

43. A small body is placed on the top of a smooth sphere of radius R. Then the sphere is imparted a constant acceleration $\mathbf{a}_0$ in the horizontal direction and the body begins sliding down. Find: **(Irodov 1.108)**
   (a) the velocity of the body relative to the sphere at the moment of break-off
   (b) the angle θ_0 between the vertical and the radius vector drawn from the centre of the sphere to the break-off point; calculate θ_0 for $\mathbf{a}_0 = \mathbf{g}$.

44. A small body A starts sliding from the height h down an inclined groove passing into a half-circle of radius $h/2$.

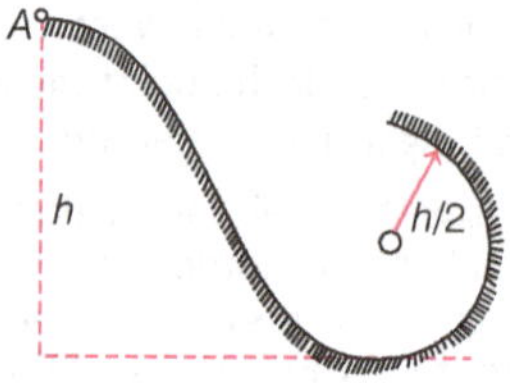

Assuming the friction to be negligible, find the velocity of the body at the highest point of its trajectory (after breaking off the groove). **(Irodov 1.136)**

45. A particle of mass m moves along a circle of radius R. Find the modulus of the average vector of the force acting on the particle over the distance equal to a quarter of the circle, if the particle moves
   (a) uniformly with velocity v
   (b) with constant tangential acceleration a, the initial velocity being equal to zero. **(Irodov 1.83)**

Level 2

Subjective Questions

46. A small disc A is placed on an inclined plane forming an angle α with the horizontal and is imparted an initial velocity v_0. Find how the velocity of the disc depends on the angle ϕ if the friction coefficient $\mu = \tan\alpha$ and at the initial moment $\phi_0 = \frac{\pi}{2}$. **(Irodov 1.106)**

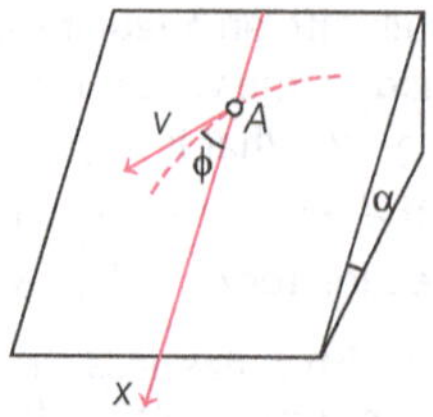

47. A block of mass m slides on a frictionless table. It is constrained to move inside a ring of radius R. At time $t = 0$, block is moving along the inside of the ring (i.e. in the tangential direction) with velocity v_0. The coefficient of friction between the block and the ring is μ. Find the speed of the block at time t.

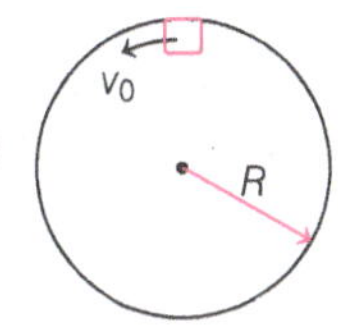

48. A device consists of a smooth L-shaped rod located in a horizontal plane and a sleeve A of mass m attached by a weightless spring to a point B. The spring stiffness is equal to k. The whole system rotates with a constant angular velocity ω about a vertical axis passing through the point O. Find the elongation of the spring. How is the result affected by the rotation direction? **(Irodov 1.88)**

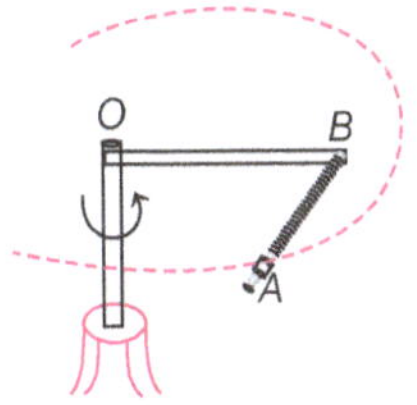

49. If the system shown in the figure is rotated in a horizontal circle with angular velocity ω. Find (Take, $g = 10 \text{m/s}^2$)

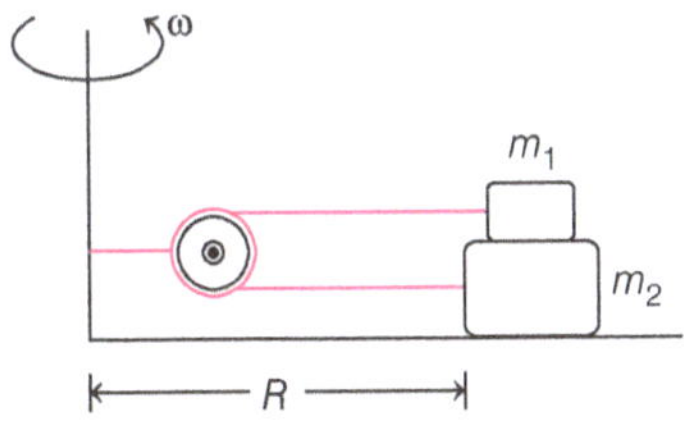

(a) the minimum value of ω to start relative motion between the two blocks.

(b) tension in the string connecting m_1 and m_2 when slipping just starts between the blocks.

The coefficient of friction between the two masses is 0.5 and there is no friction between m_2 and ground. The dimensions of the masses can be neglected. (Take, $R = 0.5$ m, $m_1 = 2$ kg, $m_2 = 1$ kg)

50. A cyclist rides along the circumference of a circular horizontal plane of radius R, the friction coefficient being dependent only on distance r from the centre O of the plane as $\mu = \mu_0 (1 - r/R)$, where μ_0 is a constant. Find the radius of the circle with the centre at the point along which the cyclist can ride with the maximum velocity. What is this velocity? **(Irodov 1.89)**

51. A chain of mass m forming a circle of radius R is slipped on a smooth round cone with half-angle θ. Find the tension of the chain if it rotates with a constant angular velocity ω about a vertical axis coinciding with the symmetry axis of the cone. **(Irodov 1.92)**

52. A horizontal plane supports a stationary vertical cylinder of radius R and a disc A attached to the cylinder by a horizontal thread AB of length l_0. An initial velocity v_0 is imparted to the disc as shown in the figure.

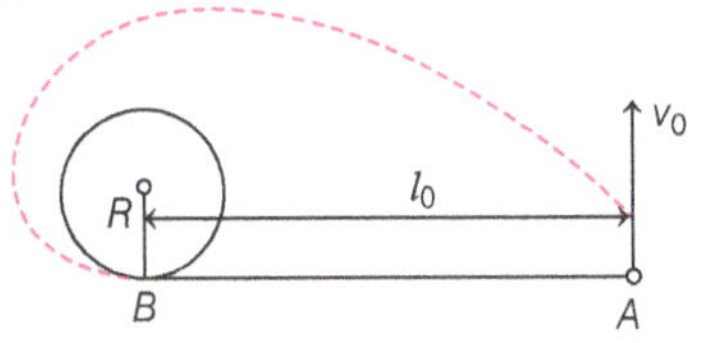

How long will it move along the plane until it strikes against the cylinder? The friction is assumed to be absent. **(Irodov 1.138)**

53. A smooth light horizontal rod AB can rotate about a vertical axis passing through its end A. The rod is fitted with a small sleeve of mass m attached to the end A by a weightless spring of length l_0 and stiffness k. What work must be performed to slowly get this system going and reaching the angular velocity ω? **(Irodov 1.142)**

54. A smooth semicircular wire track of radius R is fixed in a vertical plane (figure). One end of a massless spring of natural length $3R/4$ is attached to the lowest point O of the wire track. A small ring of mass m which can slide on the track is attached to the other end of the spring. The ring is held stationary at point P such that the spring makes an angle 60° with the vertical. The spring constant $k = mg/R$. Consider the instant when the ring is making an angle 60° with the vertical.

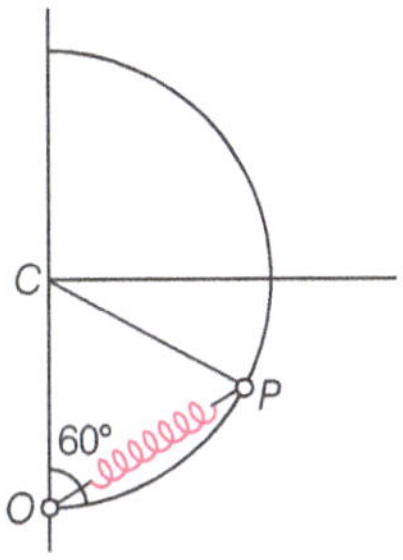

The spring is released, (a) Draw the free body diagram of the ring. (b) Determine the tangential acceleration of the ring and the normal reaction. **(1996 JEE)**

55. A circular tube of mass M is placed vertically on a horizontal surface as shown in the figure. Two small spheres, each of mass m, just fit in the tube, are released from the top. If θ gives the angle between radius vector of either ball with the vertical, obtain the value of the ratio M/m, if the tube breaks its contact with ground when $\theta = 60°$. Neglect any friction.

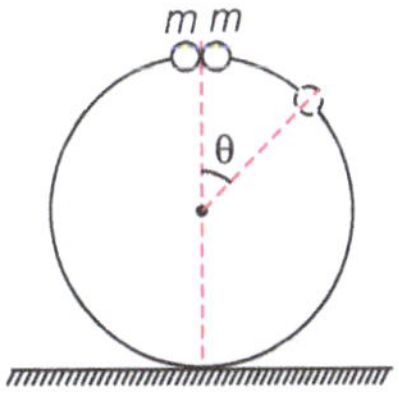

56. A ring of mass M hangs from a thread and two beads of mass m slides on it without friction. The beads are released simultaneously from the top of the ring and slides down in opposite sides. Show that the ring will start to rise, if $m > \dfrac{3M}{2}$.

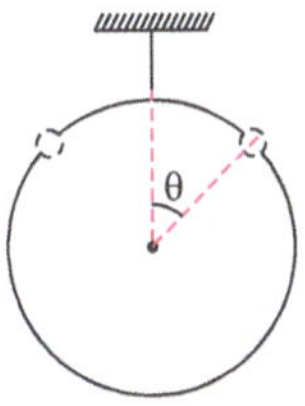

57. A vertical frictionless semicircular track of radius 1 m is fixed on the edge of a movable trolley (figure). Initially, the system is at rest and a mass m is kept at the top of the track. The trolley starts moving to the right with a uniform horizontal acceleration $a = 2g/9$. The mass slides down the track, eventually losing contact with it and dropping to the floor 1.3 m below the trolley. This 1.3 m is from the point, where mass loses contact. (Take, $g = 10\ \text{m/s}^2$)

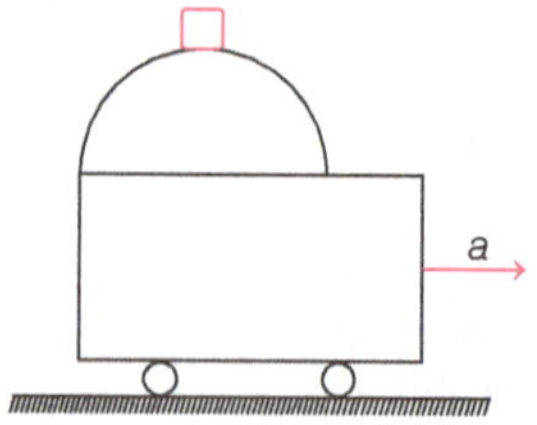

(a) Calculate the angle θ at which it loses contact with the trolley and
(b) the time taken by the mass to drop on the floor, after losing contact.

Answers

Level 1

1. (c) **2.** (c) **3.** (a) **4.** (d) **5.** (d) **6.** (c) **7.** (b) **8.** (b) **9.** (d) **10.** (a)
11. (b) **12.** (b) **13.** (c) **14.** (b) **15.** (b) **16.** (d) **17.** (c) **18.** (a) **19.** (d) **20.** (a,d)
21. (a,c) **22.** (b) **23.** (b) **24.** (b) **25.** (a) **26.** (a-r; b-p; c-s; d-r) **27.** (a-s; b-s; c-r; d-s)

28. (a) 50 cm/s (b) 32 cm/s (c) 10 cm/s^2 **29.** (a) $v = \dfrac{Rv_0}{v_0 t + R}$, $v = v_0 e^{-s/R}$ (b) $a_t = \dfrac{\sqrt{2}v_0^2}{R}$, $a = \dfrac{\sqrt{2}v_0^2}{Re^{2s/R}}$

30. $\tan\theta = 2s/R$ **31.** (a) $\sqrt{\dfrac{\mu g}{L}}$ (b) $\left[\left(\dfrac{\mu g}{L}\right)^2 - \alpha^2\right]^{1/4}$ **32.** 60°

33. (a) $T = 64$ N, $f = 48$ N (outwards) (b) 1.63 rad/s (c) 5 rad/s **34.** $P = m\alpha Rt$, $P_{av} = \dfrac{m\alpha RT}{2}$

35. (a) 36 N (b) $\omega_{min} = 11.67$ rad/s (c) $r_1 = 0.1$ m, $r_2 = 0.2$ m **36.** $5\sqrt{2}$ m/s **38.** $\theta = \cos^{-1}\left(\dfrac{2}{3}\right)$

39. $\dfrac{\pi}{2}$ **40.** $R = 45$ N **41.** (a) $T = 3\,mg\cos\theta$, $a = g\sqrt{1 + 3\cos^2\theta}$ (b) $T = \sqrt{3}\,mg$ (c) $\theta \approx 54.7°$

42. $a_t = \dfrac{Rg}{l}\left[1 - \dfrac{\cos l}{R}\right]$ **43.** (a) $v = \sqrt{\dfrac{2gR}{3}}$ (b) $\cos\theta_0 = \dfrac{2 + \eta\sqrt{5 + 9\eta^2}}{3(1 + \eta^2)}$, where $\eta = \dfrac{\omega_0}{g}$, $\theta \approx 1.7°$

44. $v = \dfrac{2}{3}\sqrt{\dfrac{gh}{3}}$

Level 2

45.(a) $|\mathbf{F}_{av}| = \dfrac{2\sqrt{2}mv^2}{\pi R}$ (b) $|\mathbf{F}_{av}| = ma$ **46.** $v = \dfrac{v_0}{1 + \cos\theta}$ **47.** $v = \dfrac{v_0}{1 + \dfrac{\mu v_0 t}{R}}$

48. $\Delta l = \dfrac{l}{\dfrac{x}{m\omega^2} - 1}$, independent of the rotation direction **49.** (a) $\omega_{min} = 6.32$ rad/s (b) $T = 30$ N

50. $v_{max} = \dfrac{1}{2}\sqrt{\mu_0 Rg}$ at $r = \dfrac{R}{2}$ **51.** $T = \dfrac{mg}{2\pi}\left(\cot\theta + \dfrac{\omega^2 R}{g}\right)$ **52.** $t = \dfrac{l_0^2}{2Rv_0}$

53. $W = \dfrac{kl_0^2(1 + \eta)}{2(1 + \eta)^2}$, where $\eta = \dfrac{m\omega^2}{k}$ **54.** (a) See the hints (b) $a_T = \dfrac{5\sqrt{3}}{8}g$, $N = \dfrac{3mg}{8}$ **55.** $\dfrac{1}{2}$

57. (a) 37° (b) 0.38 s

See hints on pages 594-607.

Hints & SOLUTIONS

Module 1.

Introductory Exercise

1. Direction of acceleration (acting towards centre) continuously keeps on changing. So, it is variable acceleration.

2. In uniform circular motion speed remains constant. In projectile motion (which is a curved path) acceleration remains constant.

3. (a) $a_r = \frac{v^2}{R} = \frac{(2t)^2}{1.0} = 4t^2$

At $t = 1\,s, a_r = 4\,cm/s^2$ **Ans.**

(b) $a_t = \frac{dv}{dt} = 2\,cm/s^2$ **Ans.**

(c) $a = \sqrt{a_r^2 + a_t^2} = \sqrt{(4)^2 + (2)^2}$

$= 2\sqrt{5}\,cm/s^2$ **Ans.**

4. $v_{av} = \frac{s}{t} = \frac{\sqrt{2}R}{(T/4)}$

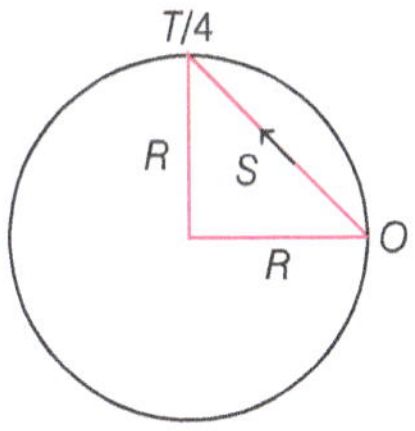

$= \frac{4\sqrt{2}R}{(2\pi R)/v}$

$\therefore \quad \frac{v_{av}}{v} = \frac{2\sqrt{2}}{\pi}$ **Ans.**

5. $(R\omega^2) = R\alpha$

$\therefore \quad \omega^2 = \alpha$

or $\quad (\alpha t)^2 = \alpha$

$\therefore \quad t = \frac{1}{\sqrt{\alpha}} = \frac{1}{\sqrt{4}}$

$= 0.5\,s$ **Ans.**

6. $a_t = \frac{dv}{dt} = 8t$

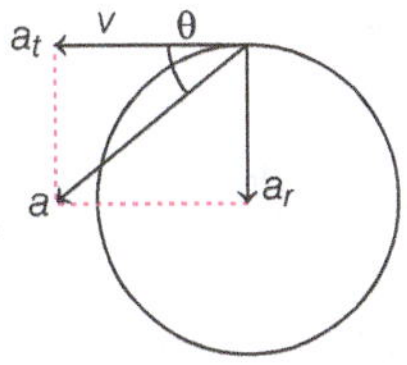

$$a_r = \frac{v^2}{R} = \frac{(4t^2)^2}{54} = \frac{8}{27}t^4$$

At $t = 3\,s$,

$a_t = 24\,m/s^2$

$a_r = 24\,m/s^2$

$\tan\theta = \frac{a_r}{a_t} = 1$

$\therefore \quad \theta = 45°$ **Ans.**

7. (a) $-a_r = a\cos 30° = (25)\frac{\sqrt{3}}{2}$

$= 21.65\,m/s^2$ **Ans.**

(b) $a_r = \frac{v^2}{R} \Rightarrow v = \sqrt{a_r R}$

or $\quad v = \sqrt{(21.65)(2.5)}$

$= 7.35\,m/s$ **Ans.**

(c) $a_t = a\sin 30° = (25)\left(\frac{1}{2}\right) = 12.5\,m/s^2$ **Ans.**

Exercise

1. **A to B** $\quad s = \sqrt{2}R$

and $\quad |\Delta \mathbf{v}| = \sqrt{v^2 + v^2 - 2v \cdot v\cos 90°}$

$= \sqrt{2}\,v$

Average acceleration $= \frac{|\Delta \mathbf{v}|}{t} = \frac{\sqrt{2}v}{t}$

and average velocity $= \frac{s}{t} = \frac{\sqrt{2}R}{t}$

The desired ratio is $\frac{v}{R} = \omega$

2. Component of acceleration perpendicular to velocity is centripetal acceleration.

$\therefore \quad a_c = \frac{v^2}{R}$ or $R = \frac{v^2}{a_c} = \frac{(2)^2}{2} = 2\,m$

3. $\omega_1 t - \omega_2 t = 2\pi$

$\therefore \quad t = \frac{2\pi}{\omega_1 - \omega_2}$

$= \frac{2\pi}{(2\pi/T_1) - (2\pi/T_2)}$

$= \frac{T_1 T_2}{T_2 - T_1} = \frac{(3600)(60)}{(3600) - (60)}$

$= \frac{3600}{59}\,s$ **Ans.**

4. At $\theta = 180°, |\Delta \mathbf{p}| = 2\,mv$ = maximum

At $\theta = 360°, |\Delta \mathbf{p}| = 0$ = minimum

5. $\frac{v^2}{R} = a_t = a$ (Here, $a_t = a$ say)

or $\frac{(at)^2}{R} = a$

$\therefore \quad t = \sqrt{\frac{R}{a}} = \sqrt{\frac{20}{5}} = 2\,\text{s}$ **Ans.**

6. $(\omega_1 - \omega_2)t = 2\pi$

or $t = \frac{2\pi}{\omega_1 - \omega_2} = \frac{2\pi}{(2\pi/T_1) - (2\pi/T_2)}$

$= \frac{T_1 T_2}{T_2 - T_1} = \frac{3 \times 1}{3 - 1}$

$= 1.5$ min **Ans.**

7. $\frac{d_B}{d_C} = \frac{v_B t}{v_C t} = \frac{2.5}{2} = \frac{5}{4}$ **Ans.**

8. $\omega_{\text{min}} = \frac{2\pi}{60}$ rad/min and $\omega_{\text{hr}} = \frac{2\pi}{12 \times 60}$ rad/min

$\therefore \quad \frac{\omega_{\text{min}}}{\omega_{\text{hr}}} = \frac{2\pi/60}{2\pi/12 \times 60} = 12$

9. $\alpha = \frac{d\omega}{dt} = -b$

$a_t = R\alpha = -Rb$ ...(i)

At $t = \frac{2a}{b}$, $\omega = -a$

$a_n = R\omega^2 = Ra^2$ ...(ii)

Now, $a = \sqrt{a_t^2 + a_n^2} = R\sqrt{a^4 + b^2}$

10. $|\mathbf{v}_{AB}| = |\mathbf{v}_A - \mathbf{v}_B|$

$= \sqrt{(R\omega)^2 + (R\omega)^2 - 2(R\omega)(R\omega)\cos\theta}$

$= 2R\,\omega \sin\left(\frac{\theta}{2}\right)$

11. $\frac{v^2}{r} = \frac{4}{r^2}$

$\Rightarrow \quad v = \frac{2}{\sqrt{r}}$

$\therefore \quad p = mv = \frac{2m}{\sqrt{r}}$

12. For a particle in a uniform circular motion

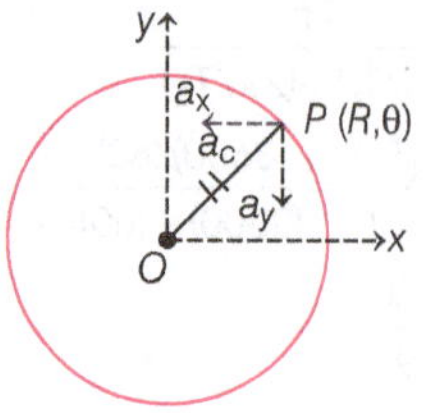

$\mathbf{a} = \frac{v^2}{R}$ towards centre of circle

[centripetal acceleration]

$\therefore \quad \mathbf{a} = \frac{v^2}{R}(-\cos\theta\,\hat{\mathbf{i}} - \sin\theta\,\hat{\mathbf{j}})$

or $\quad \mathbf{a} = -\frac{v^2}{R}\cos\theta\,\hat{\mathbf{i}} - \frac{v^2}{R}\sin\theta\,\hat{\mathbf{j}}$

13. Angle rotated by each particle in time duration from 0 to $\frac{\pi}{2\omega}$ is

$\theta = \omega \times t = \omega \times \frac{\pi}{2\omega} = \frac{\pi}{2}$ rad

So, positions of particles at $t = \frac{\pi}{2\omega}$ is as shown below

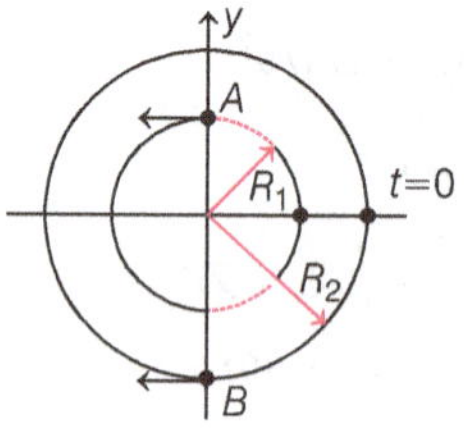

Velocities of particles at $t = \frac{\pi}{2\omega}$ are

$\mathbf{v}_A = -\omega R_1\hat{\mathbf{i}}$

and $\quad \mathbf{v}_B = -\omega R_2\hat{\mathbf{i}}$

The relative velocity of particles is

$\mathbf{v}_A - \mathbf{v}_B = -\omega R_1\hat{\mathbf{i}} - (-\omega R_2\hat{\mathbf{i}})$

$= \omega(R_2 - R_1)\hat{\mathbf{i}}$

14. $\Delta\mathbf{v} = \mathbf{v}_2 - \mathbf{v}_1$

$\therefore \quad |\Delta\mathbf{v}| = \sqrt{v^2 + v^2 - 2v\,v\cos\theta}$

$\Rightarrow \quad |\Delta\mathbf{v}| = 2v\sin\frac{\theta}{2} = 2v\sin 30°$

$= 2v \times \frac{1}{2} = v$

$\Rightarrow \quad |\Delta\mathbf{v}| = 10$ m/s (Given, $v = 10$ m/s)

Alternate method

$\because \quad \Delta\mathbf{v} = \mathbf{v}_2 - \mathbf{v}_1 = \mathbf{v}_2 + (-\mathbf{v}_1)$

$\therefore \quad |\Delta\mathbf{v}|^2 = v_1^2 + v_2^2 + 2v_1v_2\cos 120°$

$= v^2 + v^2 + 2v \times v \times \left(-\frac{1}{2}\right)$

$\Rightarrow \quad |\Delta\mathbf{v}| = v = 10$ m/s

15. Given, $s = t^3 + 5$

$\therefore \quad$ Speed, $v = \frac{ds}{dt} = 3t^2$

and rate of change of speed, $a_t = \frac{dv}{dt} = 6t$

$\therefore$ Tangential acceleration at $t = 2$ s,

$$a_t = 6 \times 2 = 12\,\text{ms}^{-2}$$

and at $t = 2$ s, $v = 3(2)^2 = 12\,\text{ms}^{-1}$

$\therefore$ Centripetal acceleration,

$$a_c = \frac{v^2}{R} = \frac{144}{20}\,\text{ms}^{-2}$$

$\therefore$ Net acceleration $= \sqrt{a_t^2 + a_c^2}$

$$= \sqrt{(12)^2 + \left(\frac{144}{20}\right)^2} = 14\,\text{ms}^{-2}$$

16. $\Delta t = \dfrac{\text{Distance travelled}}{\text{Speed}}$

$$= \frac{(2\pi R/6)}{v} = \frac{3.14 \times 300}{60 \times 3} = 5.23\text{ s}$$

(a) $|\Delta \mathbf{v}| = |\mathbf{v}_f - \mathbf{v}_i| = \sqrt{v^2 + v^2 - 2v \cdot v \cos 60^\circ}$

$$= 2v \sin 30^\circ = 60\text{ m/s}$$

(b) $a_i = \dfrac{v^2}{R} = 12\text{ m/s}^2$

(c) $|\mathbf{a}_{av}| = \dfrac{|\Delta \mathbf{v}|}{\Delta t} = \dfrac{60}{5.23}$

$$= 11.5\text{ m/s}^2$$

17. (a) Since, speed = constant

$\therefore$ Average speed = this constant value = 1 m/s.

(b) Average velocity $= \dfrac{s}{t}$

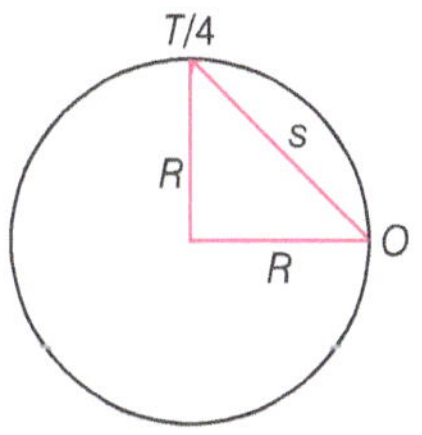

$$= \frac{\sqrt{2}R}{T/4} = \frac{4\sqrt{2}R}{(2\pi R/v)}$$

$$= \frac{4\sqrt{2}v}{2\pi} = \frac{2\sqrt{2}}{\pi}\text{ m/s}$$

(c) $|\mathbf{a}_{av}| = \dfrac{|\Delta \mathbf{v}|}{t}$

$$= \frac{\sqrt{v_f^2 + v_i^2 - 2v_f v_i \cos 90^\circ}}{(T/4)}$$

$$= \frac{4\sqrt{2}v}{T} \qquad (\text{as } v_i = v_f = v)$$

$$= \frac{(4\sqrt{2})v}{(2\pi R/v)} = \frac{2\sqrt{2}v^2}{\pi R}$$

$$= \sqrt{2}\text{ m/s}^2$$

(d) $s = \sqrt{2}R = \sqrt{2}\left(\dfrac{2}{\pi}\right) = \dfrac{2\sqrt{2}}{\pi}\text{ m}$

18. $a = \dfrac{v^2}{R}$

$$\Rightarrow \quad R = \frac{v^2}{a} = \frac{(u\cos\theta)^2}{g} \qquad \textbf{Ans.}$$

19. After 1 s $\quad \mathbf{v} = \mathbf{u} + \mathbf{a}t = 20\hat{\mathbf{i}} + 10\hat{\mathbf{j}}$,

$v = \sqrt{500}$ m/s

$= 10\sqrt{5}$ m/s

$\mathbf{a} = -10\hat{\mathbf{j}}$

$a_t = a\cos\theta$

$= \dfrac{\mathbf{a} \cdot \mathbf{v}}{v} = \dfrac{-100}{10\sqrt{5}}$

$= -2\sqrt{5}\text{ m/s}^2$ **Ans.**

$a_n = \sqrt{a^2 - a_t^2} = \sqrt{(10)^2 - (2\sqrt{5})^2}$

$= \sqrt{80}\text{ m/s}^2 = 4\sqrt{5}\text{ m/s}^2$ **Ans.**

$R = \dfrac{v^2}{a_n} = \dfrac{(10\sqrt{5})^2}{4\sqrt{5}}$

$= 25\sqrt{5}$ m **Ans.**

20. Suppose in time dt, point A travels from A to A' and $\angle AOA' = d\theta$.

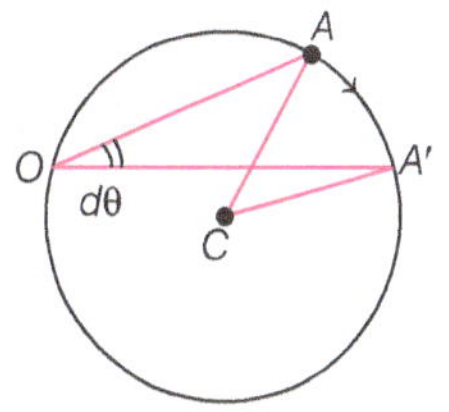

Given $\dfrac{d\theta}{dt} = \omega_0 = 0.4$ rad/s = constant

From the property of circle, $\angle ACA'$ will be $2(d\theta)$.

$\therefore \quad \omega_C = \dfrac{2(d\theta)}{dt} = 2\omega_0$

$= 0.8$ rad/s = constant

Now, $|\mathbf{v}|$ or $v = R\omega_C$

$= 0.5 \times 0.8$ m/s

$= 0.4$ m/s

$|\mathbf{a}|$ or $a = R\omega_C^2$

$= 0.5 \times (0.8)^2\text{ m/s}^2$

$= 0.32\text{ m/s}^2$

Note *In the formulae of circular motion, like $v = R\omega$ or $a = R\omega^2$, ω is the angular velocity about centre.*

Module 2.

2.

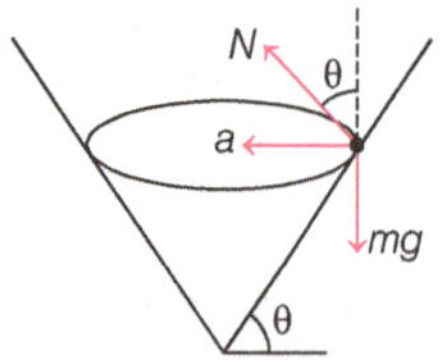

$$N \cos\theta = mg$$

$\Rightarrow$ $$N = mg \sec\theta$$

and $$N \sin\theta = \frac{mv^2}{R}$$

3. $N = mR\omega^2$

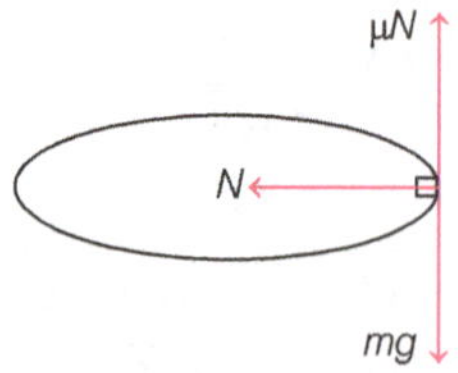

$$\mu N = mg$$

From these two equations. we get

$$\mu = \frac{g}{R\omega^2} = \frac{10}{2\times(5)^2}$$

$$= 0.2$$ **Ans.**

4. $\dfrac{mv^2}{R} = \mu mg$

$\therefore$ $$v = \sqrt{\mu Rg} = \sqrt{0.3\times300\times10}$$

$$= 30\ \text{m/s} = 108\ \text{km/h}$$ **Ans.**

5. $T\cos\theta = mg$ ...(i)

$T\sin\theta = mr\omega^2 = m(l\sin\theta)\,\omega^2$ ...(ii)

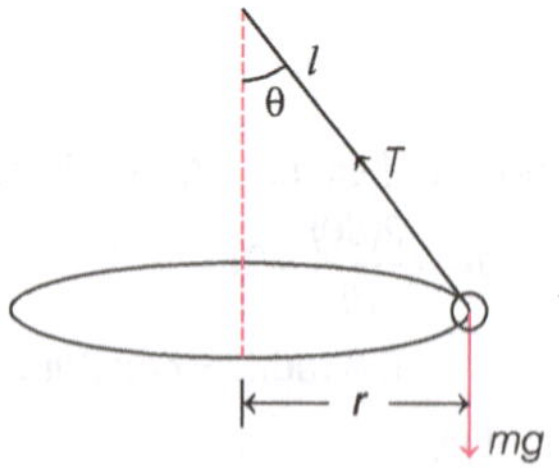

Solving these two equations, we get

$$\cos\theta = \frac{g}{l\omega^2} = \frac{g}{l(2\pi n)^2}$$

$$= \frac{10}{\left[2\pi\times\frac{2}{\pi}\right]^2} \qquad (\because l = 1\ \text{m})$$

$\therefore$ $$\theta = \cos^{-1}(5/8)$$ **Ans.**

6. $T = \dfrac{mg}{\cos\theta} = \dfrac{0.1\times10}{5/8} = \dfrac{8}{5}\text{N}$ **Ans.**

7. $f = ma = m\sqrt{a_t^2 + a_r^2}$

$$= m\sqrt{(R\alpha)^2 + (R\omega^2)^2}$$

$$= m\sqrt{(R\alpha)^2 + [R(\alpha t)^2]^2}$$

$$= 0.36\times10^{-3}\sqrt{\left(0.25\times\frac{1}{3}\right)^2 + \left[0.25\left(\frac{1}{3}\times2\right)^2\right]^2}$$

$$= 50\times10^{-6}\ \text{N}$$

$$= 50\ \mu\text{N}$$ **Ans.**

8. Necessary centripetal force to the coin is provided by friction. Thus,

$$mr\,\omega_{max}^2 = \mu mg$$

or $$r = \frac{\mu g}{\omega_{max}^2}$$

ω_{max} is made three times. Therefore, distance from centre r will remain $\frac{1}{9}$ times.

9. Maximum tension $= m\omega^2 r = m\times4\pi^2\times n^2\times r$

By substituting the values, we get $T_{max} = 87.64\ \text{N}$

10. $kx = \dfrac{mv^2}{r}$

$\Rightarrow$ $$v = \sqrt{\frac{kxr}{m}}$$

$$T = \frac{2\pi r}{v} = 2\pi\sqrt{\frac{mr}{kx}}$$

$\therefore$ $$T \propto \sqrt{\frac{r}{x}}$$

$$T' = T\sqrt{\frac{r'x}{rx'}}$$

$$= T\sqrt{\frac{3a\times a}{2a\times2a}} = \frac{\sqrt{3}}{2}T$$

11. $T\sin\theta = mr\omega^2$ or $T\sin\theta = m(l\sin\theta)\,\omega^2$

$\therefore$ $$T = ml\omega^2 = ml(2\pi f)^2$$

$$= ml\left(2\pi\times\frac{2}{\pi}\right)^2 = 16\ ml$$

12. The bead is moving on a circular path of radius a.

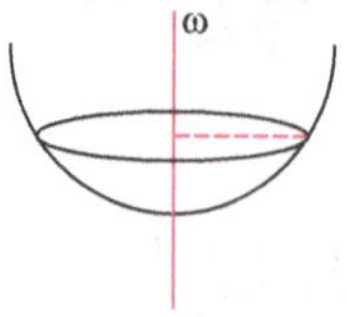

So, the acceleration is $a\omega^2$.

13. $v = \sqrt{2a_t s} = \sqrt{2a_t (\pi R)}$

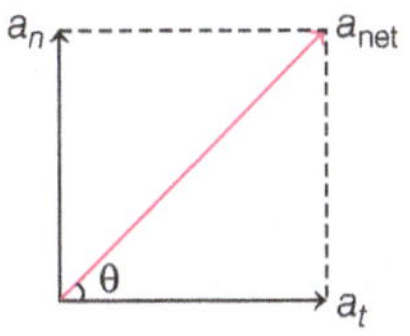

$$\therefore \quad a_n = \frac{v^2}{R} = 2\pi a_t \quad \text{or} \quad \frac{a_n}{a_t} = 2\pi$$

$$\tan\theta = \frac{a_n}{a_t} = 2\pi$$

$$\therefore \quad \theta = \tan^{-1}(2\pi)$$

14. $\mu mg = mR\omega^2$

$$\therefore \quad \omega^2 = \frac{\mu g}{R} = \frac{(1/3)g}{5a/4} = \frac{4g}{15a}$$

15. $\tan\theta = \frac{h}{b} = \frac{v^2}{Rg}$

$$\Rightarrow \quad h = \frac{v^2 b}{Rg}$$

16.

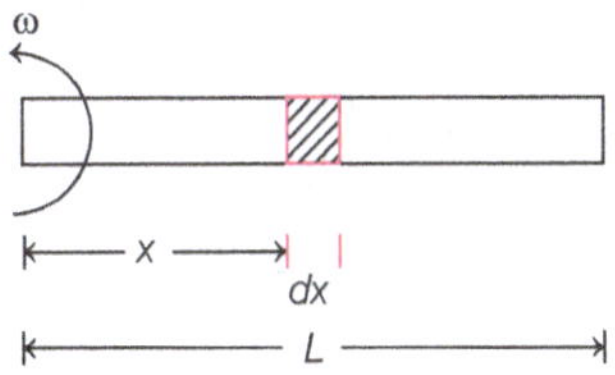

$$dm = \frac{M}{L} dx$$

$$dF = dm\,\omega^2 x$$

$$F = \frac{M}{L}\omega^2 \int_0^L x\,dx$$

$$= \frac{M}{L}\omega^2 \left|\frac{x^2}{2}\right|_0^L$$

$$= \frac{1}{2} M\omega^2 L$$

17. Force diagram of the bob is shown in figure.

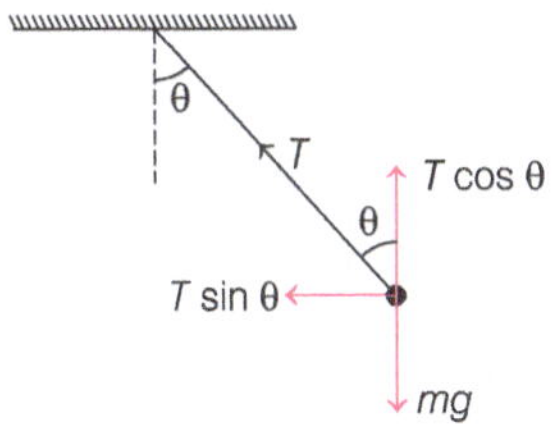

$$T \sin\theta = \frac{mv^2}{R}$$

$$\text{and} \quad T \cos\theta = mg$$

$$\therefore \quad \tan\theta = \frac{v^2}{Rg} = \frac{(10)^2}{(10)(10)}$$

$$\text{or} \quad \tan\theta = 1$$

$$\text{or} \quad \theta = 45°$$

Ans.

19. $R = 1.6 \cos 60° = 0.8$ m

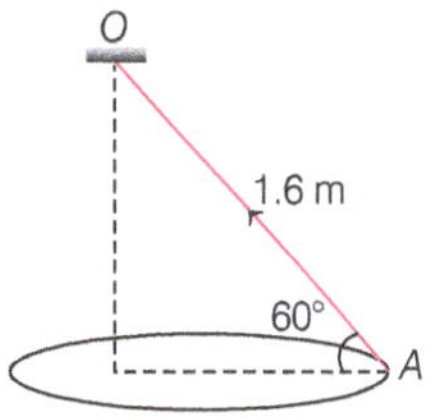

$$T \sin 60° = mg \quad \text{or} \quad T = \frac{2mg}{\sqrt{3}}$$

$$T \cos 60° = \frac{mv^2}{R}$$

$$\therefore \quad \frac{v^2}{R} = \frac{T}{2m} = \frac{g}{\sqrt{3}} = \frac{9.8}{\sqrt{3}}\,\text{m/s}^2$$

$$v = \sqrt{\frac{9.8}{\sqrt{3}} \times 0.8}\ \text{m/s}$$

$$\text{Time period} = 2\pi\frac{R}{v}$$

20. The given case is of uniform circular motion, in which speed or kinetic energy remains constant. Direction of velocity and acceleration keep on changing although its magnitude remains constant.

21. $f = \frac{mv^2}{R} = F$

$$\therefore \quad \frac{F}{f} = 1$$

With increase in speed f will increase but $\frac{F}{f}$ will remain same.

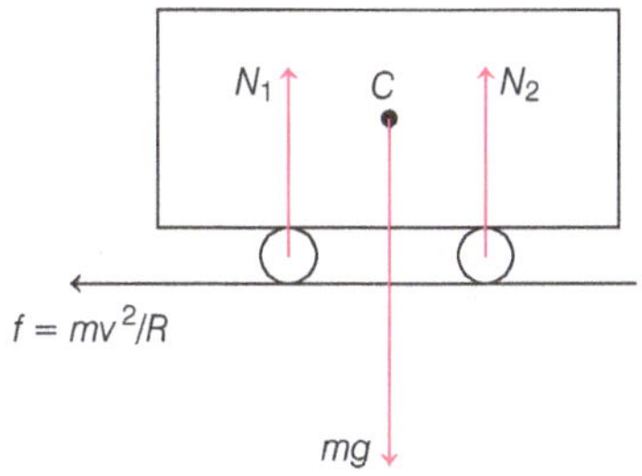

Further, with increase in the value of v, friction f will increase. Therefore, anti-clockwise moment of f about C will also increase. Hence, clockwise moment of N_2 should also increase. Thus, N_2 will increase. But $N_1 + N_2 = mg$. So, N_1 will decrease.

22. $v = \dfrac{18 \times 5}{18} = 5\,\text{m/s}$

$$\tan\theta = \frac{v^2}{Rg} = \frac{(5)^2}{10 \times 10} = \frac{1}{4}$$

$$\therefore \quad \theta = \tan^{-1}\left(\frac{1}{4}\right) \qquad \textbf{Ans.}$$

23. $v = \sqrt{\mu R g}$

$$\therefore \quad \mu = \frac{v^2}{gR} = \frac{(5)^2}{10 \times 10}$$

$$= 0.25 \qquad \textbf{Ans.}$$

24. $v = \sqrt{Rg \tan\theta} = \sqrt{50 \times 10 \times \tan 30^\circ}$

$$\approx 17\,\text{m/s} \qquad \textbf{Ans.}$$

25. If he applies the breaks, then to stop the car in a distance r

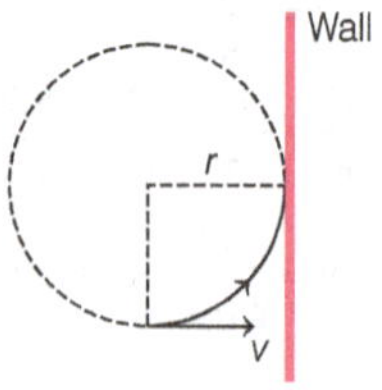

$$0 = v^2 - 2a_1 r$$

$\therefore \; a_1 = \dfrac{v^2}{2r}$ = minimum retardation required. (by friction)

If he takes a turn of radius r, the centripetal acceleration required is

$$a_2 = \frac{v^2}{r} \qquad \text{(provided again by friction)}$$

Since $a_1 < a_2$, it is better to apply brakes.

26.

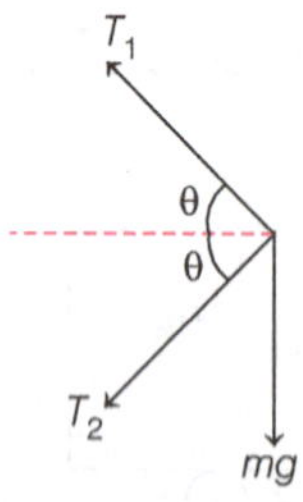

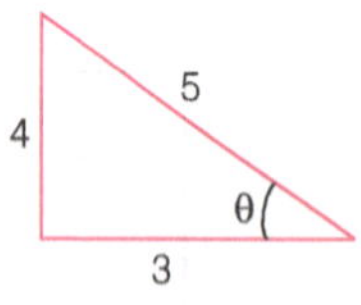

$$T_1 \cos\theta + T_2 \cos\theta = mr\omega^2 \qquad \text{...(i)}$$

$$\omega = (2n\pi)$$

Here, n = number of revolutions per second.

Substituting the proper values in Eq. (i),

$$200 \times \left(\frac{3}{5}\right) + T_2 \times \left(\frac{3}{5}\right) = (4)(3)(2n\pi)^2$$

or $\quad 600 + 3T_2 = 240\, n^2\pi^2 \qquad$ ...(ii)

Further, $\quad T_1 \sin\theta = T_2 \sin\theta + mg$

or $\quad 200 \times \dfrac{4}{5} = T_2 \times \dfrac{4}{5} + 4 \times 10$

or $\quad 800 = 4T_2 + 200 \qquad$ ...(iii)

Solving Eqs. (ii) and (iii), we get

$$T_2 = 150\,\text{N} \quad \text{and} \quad n = 0.66\,\text{rps}$$

$$= 39.6\,\text{rpm}$$

27. (a) $mg - N_1 = \dfrac{mv^2}{R}$

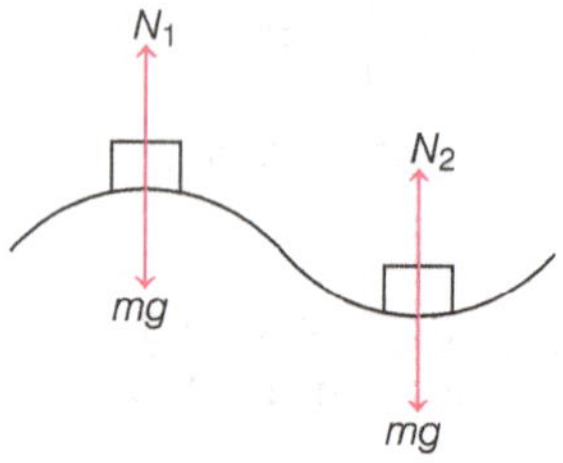

or $\quad mg - \dfrac{mg}{2} = \dfrac{mv^2}{R}$

$\therefore \quad \dfrac{mv^2}{R} = \dfrac{mg}{2} = \dfrac{16\text{ kN}}{2} = 8\text{ kN}$

Now, $\quad N_2 - mg = \dfrac{mv^2}{R}$

or $\quad N_2 = mg + \dfrac{mv^2}{R}$

$$= mg + \frac{mg}{2} = \frac{3}{2}mg$$

$$= \frac{3}{2}(16\,\text{kN}) = 24\,\text{kN} \qquad \textbf{Ans.}$$

(b) $mg - N = \dfrac{mv^2}{R}$

or $\quad N = mg - \dfrac{mv^2}{R}$

$$0 = mg - \frac{mv^2_{max}}{R}$$

$\therefore \quad v_{max} = \sqrt{gR} = \sqrt{10 \times 250}$

$$= 50\,\text{m/s} \qquad \textbf{Ans.}$$

(c) $N_2 - mg = \dfrac{mv^2}{R}$

$\therefore \quad N_2 = m\left(g + \dfrac{v^2}{R}\right)$

$$= \frac{16 \times 10^3}{10}\left[10 + \frac{2500}{250}\right]$$

$$= 32 \times 10^3\,\text{N}$$

$$= 32\,\text{kN} \qquad \textbf{Ans.}$$

28.

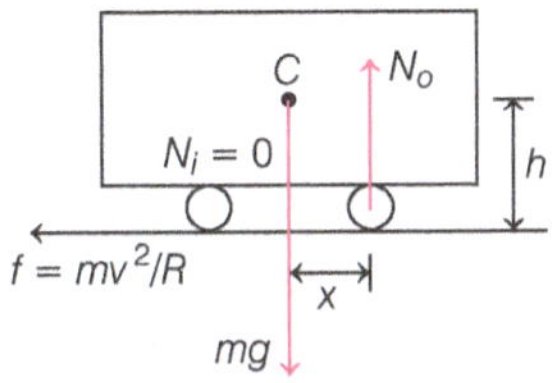

In critical case, normal reaction on inner wheel N_i will become zero. Normal reaction on outer wheel $N_o = mg$. Friction will provide the necessary centripetal force.

$$\therefore \qquad f = \frac{mv^2}{R}$$

Taking moment about C,

$$N_o(x) = f(h)$$

$$\text{or} \qquad (mg)x = \left(\frac{mv^2}{R}\right)h$$

$$\therefore \qquad v = \sqrt{\frac{gRx}{h}}$$

$$= \sqrt{\frac{9.8 \times 250 \times 0.75}{1.5}}$$

$$= 35\text{ m/s}$$

29. Let ω be the angular speed of rotation of the bowl.Two forces are acting on the ball.

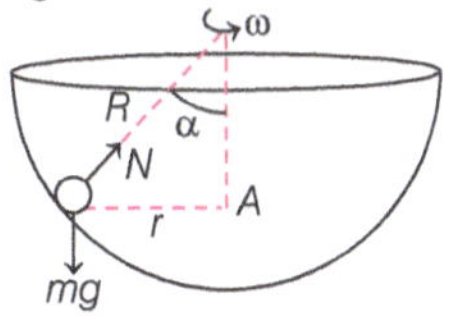

1. normal reaction N
2. weight mg

The ball is rotating in a circle of radius $r (= R \sin \alpha)$ with centre at A at an angular speed ω. Thus,

$$N \sin \alpha = mr\omega^2$$

$$= mR\omega^2 \sin \alpha \qquad \text{...(i)}$$

$$\text{and} \qquad N \cos \alpha = mg \qquad \text{...(ii)}$$

Dividing Eq. (i) by Eq. (ii), we get

$$\frac{1}{\cos \alpha} = \frac{\omega^2 R}{g}$$

$$\therefore \qquad \omega = \sqrt{\frac{g}{R \cos \alpha}}$$

30. $R = L \sin \theta$

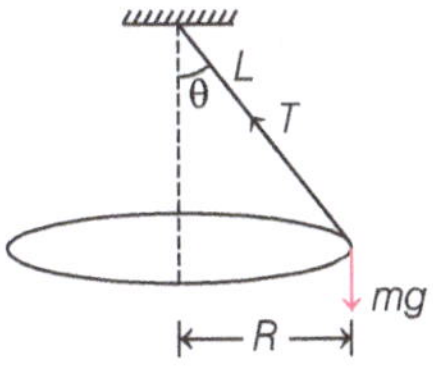

$$T \sin \theta = mR\omega^2 = m(L \sin \theta)\omega^2$$

$$\therefore \qquad T = mL\omega^2$$

Now, $\quad T \cos \theta = mg$

$$\therefore \qquad \cos \theta = \frac{mg}{T} = \frac{mg}{mL\omega^2} = \frac{g}{L\omega^2}$$

Hence proved.

31. $t = \sqrt{\dfrac{2h}{g}} = \sqrt{\dfrac{2 \times 2.9}{9.8}} = 0.77\text{ s}$

Horizontal distance $x = vt$

$$\therefore \qquad v = \frac{x}{t} = \frac{10}{0.77} \approx 13\text{ m/s}$$

$$a = \frac{v^2}{R} = \frac{(13)^2}{1.5}$$

$$\approx 113\text{ m/s}^2$$

Ans.

32. Total acceleration $a (= \sqrt{a_t^2 + a_r^2})$ is provided by friction.

Let car slips after travelling a distance s. Then,

$$v = \sqrt{2a_t s}$$

$$a_r = \frac{v^2}{R} = \frac{2a_t s}{R} = \frac{2 \times 0.62 \times s}{40}$$

$$= 0.031\text{ s}$$

Just before sliding,

$$ma = \text{maximum friction} = \mu N = \mu mg$$

$$\text{or} \qquad a = \mu g$$

$$\text{or} \qquad \sqrt{a_t^2 + a_r^2} = \mu g$$

$$\text{or} \qquad \sqrt{(0.62)^2 + (0.031\text{ s})^2} = 0.2 \times 9.8$$

Solving this equation, we get

$$s = 60\text{ m}$$

33. $v_0 \sin \alpha$ is vertical component of velocity,

$v_0 \cos \alpha$ is tangential to the horizontal circle shown in figure.

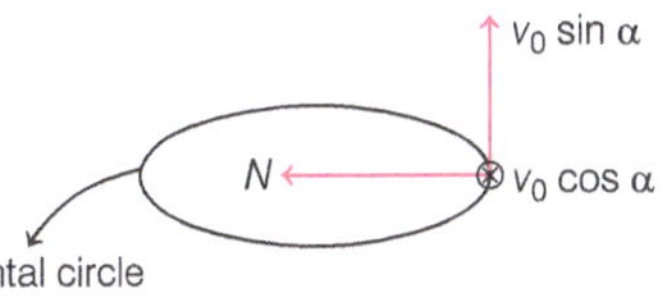

N is normal reaction on particle from the wall of cylinder (towards centre).

$\therefore N =$ centripetal force due to the tangential component $v_0 \cos \alpha$

$$\text{or} \qquad N = \frac{m(v_0 \cos \alpha)^2}{R}$$

$$\text{or} \qquad N = \frac{mv_0^2 \cos^2 \alpha}{R}$$

34.

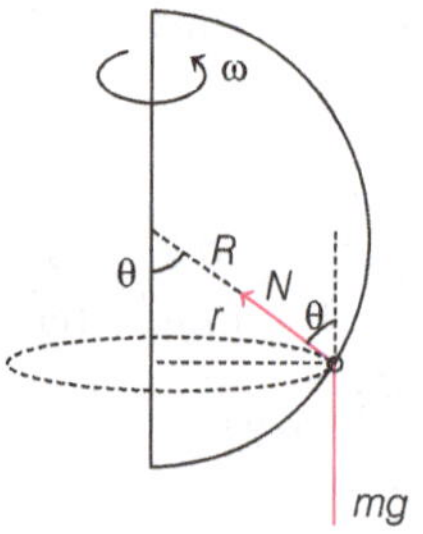

$$r = R\sin\theta$$

$$N\cos\theta = mg \quad \text{and} \quad N\sin\theta = mr\omega^2$$

Solving these three equations, we can find

$$\cos\theta = \frac{g}{R\omega^2}$$

Note *Since, $\cos\theta \not> 1$. Therefore if $R\omega^2 < g$, then there is only one steady state position, $\theta = 0°$. If $R\omega^2 < g$, then there are two steady state positions, one at $\theta = 0°$ and the other at $\cos\theta = \frac{g}{R\omega^2}$ or $\theta = \cos^{-1}\left(\frac{g}{R\omega^2}\right)$.*

Module 4.

Introductory Exercise

1. (a) String will slack at height h_2,

$$h_2 = \frac{u^2 + gR}{3g} = \frac{5}{3}R \qquad (\text{as } u^2 = 4gR)$$

(b) $v = \sqrt{u^2 - 2gh_2} = \sqrt{4gR - 2g \times \frac{5}{3}R}$

$$= \sqrt{\frac{2}{3}gR}$$

2. (a) Velocity becomes zero at height h_1,

$$h_1 = \frac{u^2}{2g} = \frac{gR}{2g} = \frac{R}{2}$$

Now, $\quad h_1 = R(1 - \cos\theta)$

$\therefore \quad \frac{R}{2} = R(1 - \cos\theta) \Rightarrow \theta = 60°$

(b) $T = mg\cos\theta = \frac{mg}{2}$ at $\theta = 60°$

3. (a) $v = \sqrt{u^2 - 2gh} = \sqrt{7gR - 2g(2R)} = \sqrt{3gR}$

(b) $T + mg = \frac{mv^2}{R} = \frac{m}{R}(mgR)$

$\therefore \quad T = 2mg$

(c) $T - mg = \frac{mu^2}{R} = \frac{m}{R}(7gR)$

$\therefore \quad T = 8mg$

4. $h = l - l\cos 60° = \frac{l}{2} = 2.5$ m

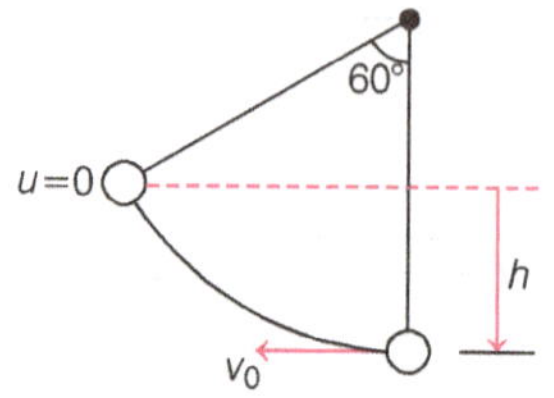

$$v_0 = \sqrt{2gh}$$

$$= \sqrt{2 \times 9.8 \times 2.5} = 7\text{ m/s} \qquad \textbf{Ans.}$$

Exercise

1. At *A*

Tangential acceleration is g. Radial acceleration is $\frac{v^2}{R}$.

$$\therefore \quad a_{net} = \sqrt{g^2 + \left(\frac{v^2}{R}\right)^2}$$

2. Weight (plus normal reaction) provides the necessary centripetal force or weight is used in providing the centripetal force.

3. $h = l - l\cos\theta = l(1 - \cos\theta)$

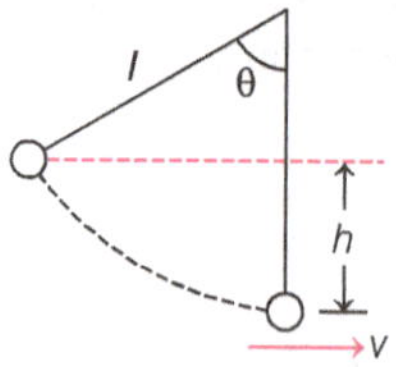

$$v^2 = 2gh = 2gl(1 - \cos\theta) = v^2_{max}$$

$$\therefore \quad K_{max} = \frac{1}{2}mv^2 = mgl(1 - \cos\theta) \qquad \textbf{Ans.}$$

4. $u = \sqrt{5gR}$

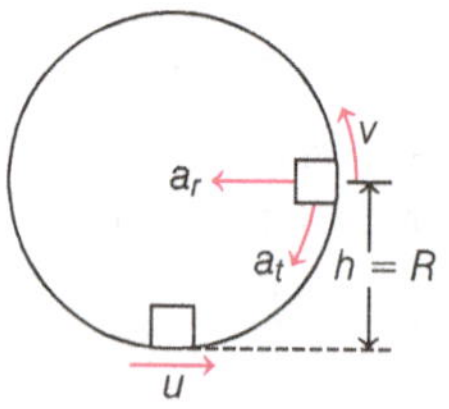

$$v^2 = u^2 - 2gh = (5gR) - 2gR$$

$$= 3gR$$

$$a_r = \frac{v^2}{R} = 3g$$

$$a_t = g$$

$$\therefore \quad a = \sqrt{a_r^2 + a_t^2}$$

$$= g\sqrt{10} \qquad \textbf{Ans.}$$

5. $v^2 = 2gh$

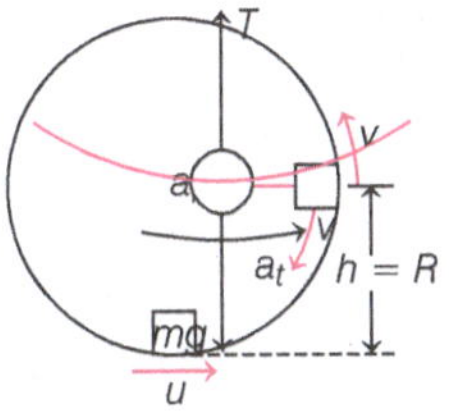

$$T - mg = \frac{mv^2}{l}$$

or $$T = mg + \frac{m(2gh)}{l}$$

$$= mg\left(1 + \frac{2h}{l}\right)$$ **Ans.**

6.

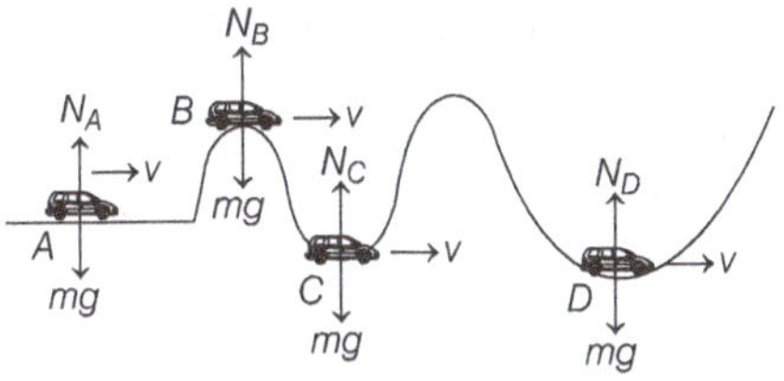

$A: N_A - mg = 0 \Rightarrow N_A = mg$

$B: mg - N_B = \frac{mv^2}{R_B} \Rightarrow N_B = mg - \frac{mv^2}{R_B}$

$C: N_C - mg = \frac{mv^2}{R_C} \Rightarrow N_C = mg + \frac{mv^2}{R_C}$

$D: N_D - mg = \frac{mv^2}{R_D} \Rightarrow N_D = mg + \frac{mv^2}{R_D}$

Since, $R_C < R_D, N_C > N_D$

$N_C > N_D > N_A > N_B$

7. $v^2 = 2gl = 5gR$

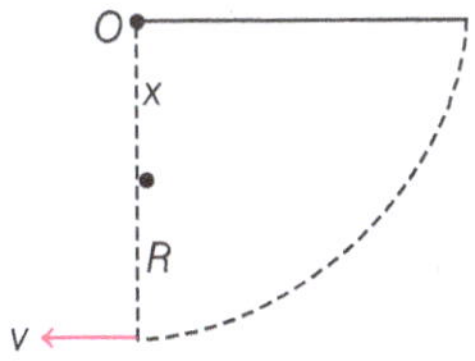

$\therefore$ $R = 0.4\,l$

or $x = l - R = 0.6\,l = \frac{3}{5}l$

8. $\sqrt{2gh} = \sqrt{5gR}$

$\therefore$ $R = \frac{2h}{5} = \frac{2 \times 5}{5} = 2\text{ cm}$

9. $h = l + l\sin\theta = l(1 + \sin\theta)$

$v^2 = u^2 - 2gh$

$= u^2 - 2gl(1 + \sin\theta)$

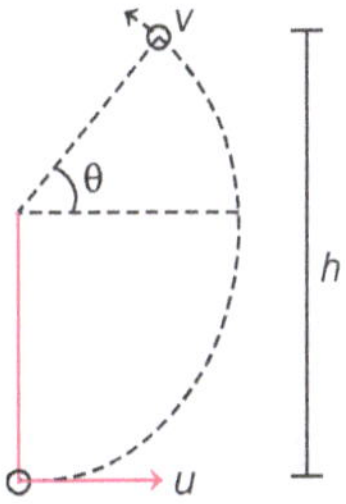

String will slack, where component of weight towards centre is just equal to centripetal force or,

$$mg\sin\theta = \frac{mv^2}{l} = \frac{m}{l}[u^2 - 2gl(1 + \sin\theta)]$$

Substituting $u^2 = \frac{7gl}{2}$, we get

$$\sin\theta = \frac{1}{2} \quad \text{or} \quad \theta = 30°$$

$\therefore$ The desired angle is $90° + 30°$ or $120°$.

10. At the time of leaving contact

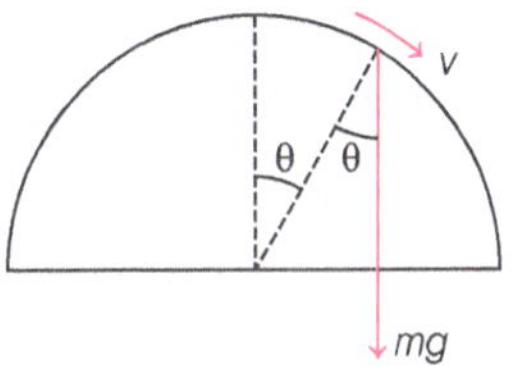

$N = 0$

$\therefore$ $mg\cos\theta = \frac{mv^2}{R} = \frac{m(2gh)}{R}$

$\therefore$ $\cos\theta = \frac{2h}{R} = \frac{2\left[\frac{R}{4} + R(1 - \cos\theta)\right]}{R}$

On solving this equation, we get

$\cos\theta = 5/6$

or $\theta = \cos^{-1}\left(\frac{5}{6}\right)$ **Ans.**

11. At the time of leaving contact at P,

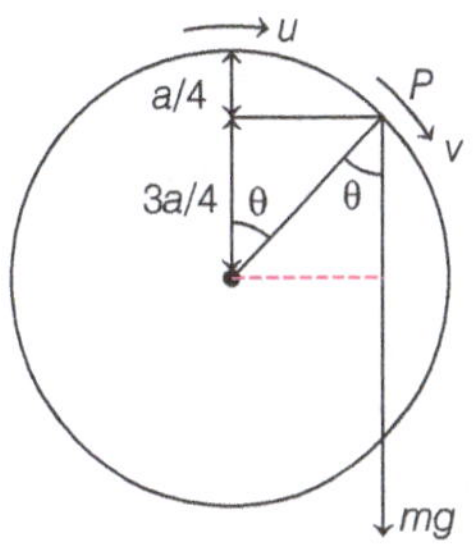

$N = 0$

$\therefore$ $mg\cos\theta = \frac{mv^2}{a} = \frac{m(u^2 + 2gh)}{a}$

$\therefore \quad g(3a/4) = \dfrac{u^2 + 2g(a/4)}{a}$

$\therefore \quad u = \dfrac{\sqrt{ag}}{2}$ **Ans.**

12. $h = l(1 - \cos 60°) = \dfrac{l}{2}$,

$v^2 = 2gh = gl$

Now, $T_{max} - mg = \dfrac{mv^2}{l}$ (at bottommost point)

$\therefore \quad T_{max} = 2mg \le \mu_s (4mg)$

$\therefore \quad \mu_s \ge 0.5$

13. $a = \sqrt{a_n^2 + a_t^2}$

$= \sqrt{\left(\dfrac{T - mg\cos 30°}{m}\right)^2 + (g\sin 30°)^2}$

$= g\sqrt{\left(\sqrt{3} - \dfrac{\sqrt{3}}{2}\right)^2 + \left(\dfrac{1}{4}\right)} = g$

14. Minimum tension is at topmost point (speed $= v$) and maximum tension at bottommost point (speed $= u$).

$$\frac{T_{max}}{T_{min}} = \frac{mg + \dfrac{mu^2}{L}}{-mg + \dfrac{mv^2}{L}} = 4$$

$$u^2 = v^2 + 2g(2L)$$

Solving, we get $v = 10$ m/s

15. Two forces are acting on girl, tension and weight. Power of tension will be zero and that of weight is,

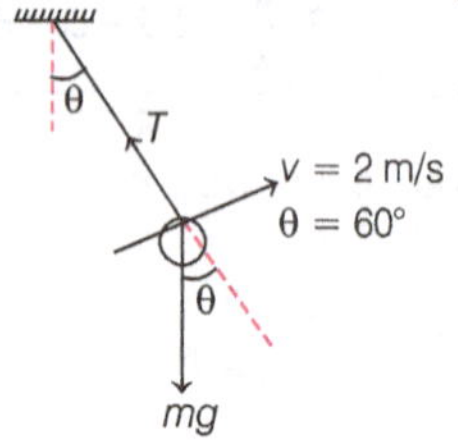

$P = mgv\cos(90° + \theta) = -mgv\sin 60°$

$= -50 \times 9.8 \times 2 \times \dfrac{\sqrt{3}}{2} = -490\sqrt{3}$ W

$\therefore$ Power delivered is $490\sqrt{3}$ W.

16. When body is released from the position P (inclined at angle θ from vertical), then velocity at mean position

$v = \sqrt{2gl(1 - \cos\theta)}$

$\therefore$ Tension at the lowest point

$= mg + \dfrac{mv^2}{l}$

$= mg + \dfrac{m}{l}[2gl(1 - \cos 60°)]$

$= mg + mg = 2mg$

17.

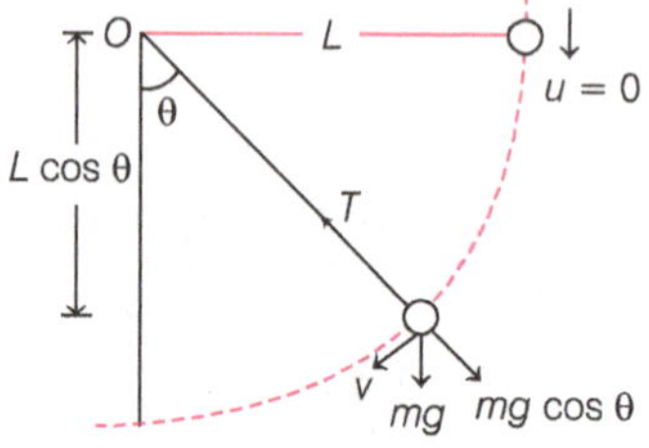

$v^2 = 0 + 2gL\cos\theta$

$T - mg\cos\theta = \dfrac{mv^2}{L} = 2mg\cos\theta$

$T = 3mg\cos\theta$

$= 3mg\cos\theta \le 2mg$

$\cos\theta \le \dfrac{2}{3}$

$\theta = \cos^{-1}(2/3)$

18. $mg(h + L) = \dfrac{1}{2}mv^2$

$\Rightarrow \quad v^2 = 2g(h + L)$

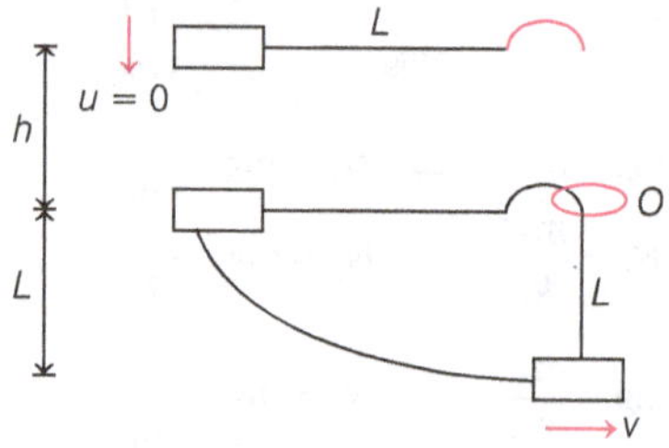

To complete the vertical circle about O,

$v \ge \sqrt{4gL}$

$v^2 \ge 4gL$

$2g(h + L) \ge 4gL$

$h \ge L$

$h_{min} = L$

19. $T\sin\theta = mr\omega^2$ ...(i)

$T\cos\theta = mg$ ...(ii)

$\therefore \quad \tan\theta = \dfrac{r\omega^2}{g}$

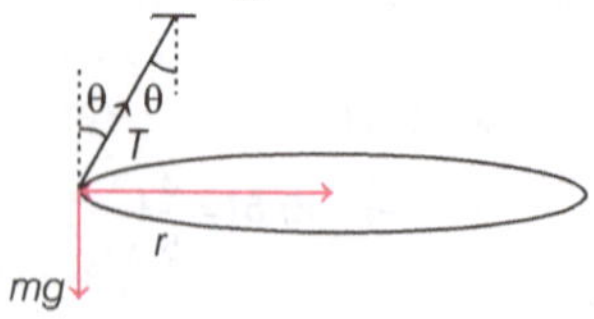

Note *Since, r >> length of pendulum. Hence, radius of bob of pendulum is also r.*

20. $t_{pc} = \sqrt{\dfrac{2h}{g}} = \sqrt{\dfrac{2(2L)}{g}} = 2\sqrt{\dfrac{L}{g}}$

$x = vt_{PC} = (\sqrt{gR})\left(2\sqrt{\dfrac{L}{g}}\right)$

But $\quad R = L$

$\therefore \quad x = 2L$

Note *At highest point, minimum velocity required is* $\sqrt{gR}$.

21. Net acceleration **a** of the bob in position B has two components.

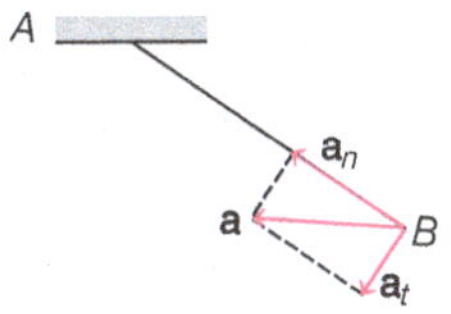

(a) $\mathbf{a}_n$ = radial acceleration (towards BA)

(b) $\mathbf{a}_t$ = tangential acceleration (perpendicular to BA)

Therefore, direction of **a** is correctly shown in option (c).

22. $h = R - R\cos\theta$

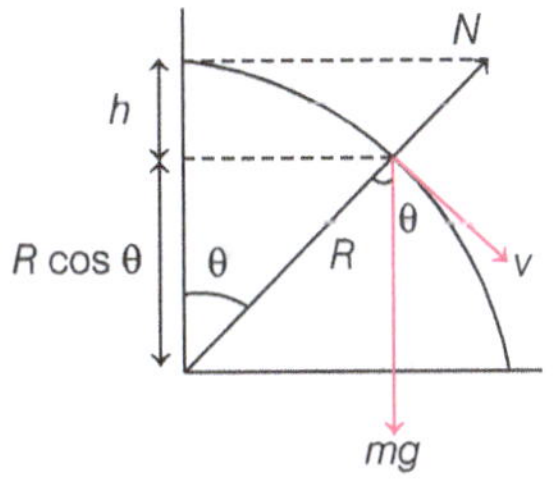

Using conservation of energy

$$mgR(1-\cos\theta) = \frac{1}{2}mv^2$$

Radial force equation is

$$mg\cos\theta - N = \frac{mv^2}{R}$$

Here, $\quad N$ = normal force on bead by wire

$$N = mg\cos\theta - \frac{mv^2}{R} = mg(3\cos\theta - 2)$$

$$N = 0 \text{ at } \cos\theta = \frac{2}{3}$$

So, normal force act radially outward on bead, if $\cos\theta > \frac{2}{3}$ and normal force act radially inward on bead, if $\cos\theta < 2/3$

$\therefore$ Force on ring is opposite to normal force on bead.

23. Radial acceleration is given by

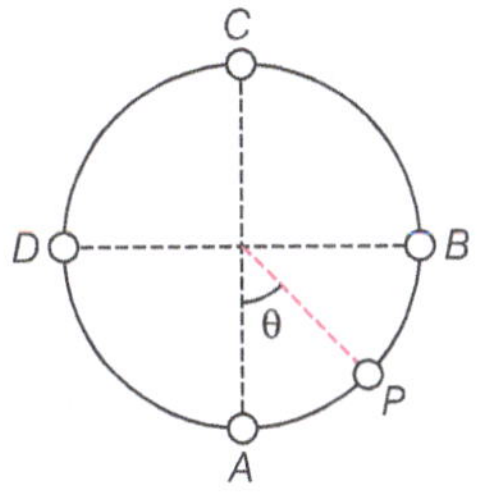

$$a_r = \frac{v^2}{R}$$

At A, speed is maximum.

Therefore, a_r is maximum.

At C, speed is minimum.

Therefore, a_r is minimum.

Tangential acceleration is $g\sin\theta$.

At point B, $\theta = 90°$.

Therefore, tangential acceleration is maximum $(= g)$.

24. $T + mg = \dfrac{mu^2}{l}$

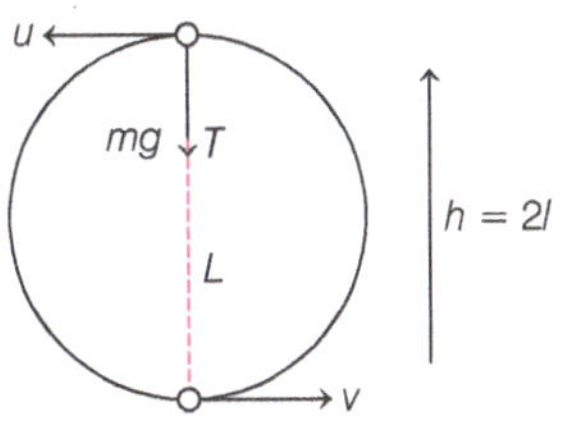

$$\therefore \quad 2mg + mg = \frac{mu^2}{l}$$

or $\quad u = \sqrt{3gl}$

$$v^2 = u^2 + 2gh$$

$$= 3gl + 2g(2l) = 7gl$$

$$\therefore \quad v = \sqrt{7gl}$$

25. Maximum velocity is at bottommost point and minimum velocity is at topmost point.

$$\frac{\sqrt{u_{min}^2 + 2g(2L)}}{u_{min}} = \frac{2}{1}$$

On solving, we get

$$u_{min} = 2\sqrt{\frac{gL}{3}} \qquad \textbf{Ans.}$$

26. $u_{max} = 2u_{min} = 4\sqrt{\dfrac{gL}{3}}$

$$\therefore \quad K_{max} = \frac{1}{2}mu_{max}^2 = \frac{8\,mgL}{3} \qquad \textbf{Ans.}$$

27. $v = \sqrt{u_{max}^2 - 2g(L)}$

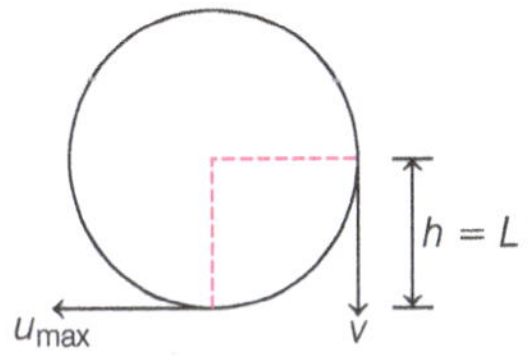

$$= \sqrt{\frac{16\,gL}{3} - 2gL} = \sqrt{\frac{10\,gL}{3}} \qquad \textbf{Ans.}$$

28. (a) $\quad v = \sqrt{u^2 - 2gh} = \sqrt{12gl - 2gl}$

$$= \sqrt{10gl} = \sqrt{10 \times 10 \times 1}$$

$$= 10\text{ m/s} \qquad \textbf{Ans.}$$

(b) $\quad a_r = \dfrac{v^2}{R}$ or $\dfrac{v^2}{l}$

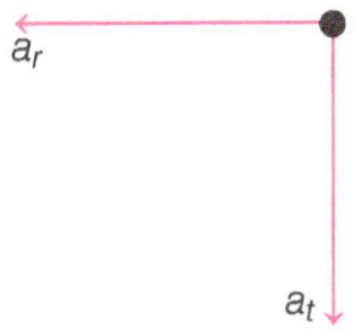

$$= \frac{(10)^2}{1} = 100\text{ m/s}^2$$

$$a_t = g = 10\text{ m/s}^2$$

$$\therefore \quad a = \sqrt{a_r^2 + a_t^2} = 100.49\text{ m/s}^2$$

(c) $\quad T = \dfrac{mv^2}{l} = \dfrac{(1)(10)^2}{1} = 100\text{ N}$

(d) $\quad a_t = g = 10\text{ m/s}^2$

29. $v^2 = v_0^2 + 2gh$

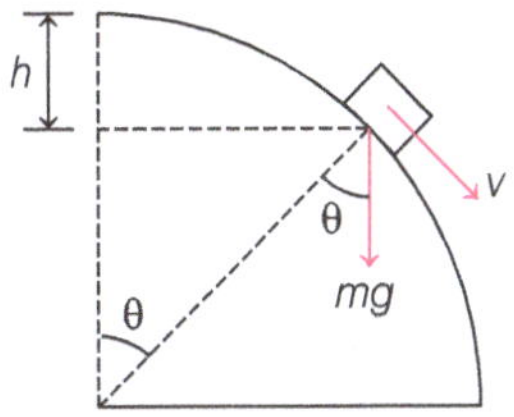

$$= (0.5\sqrt{gr})^2 + 2gr(1-\cos\theta)$$

$$= (2.25\,gr - 2gr\cos\theta)$$

At the time of leaving contact, $N = 0$

$$\therefore \quad mg\cos\theta = \frac{mv^2}{r}$$

$$= 2.25\,mg - 2\,mg\cos\theta$$

$$\therefore \quad \cos\theta = \frac{2.25}{3} = \frac{3}{4}$$

$$\therefore \quad \theta = \cos^{-1}(3/4) \qquad \textbf{Ans.}$$

30. $2.5\,mg - mg\cos 30^\circ = \dfrac{mv^2}{r}$

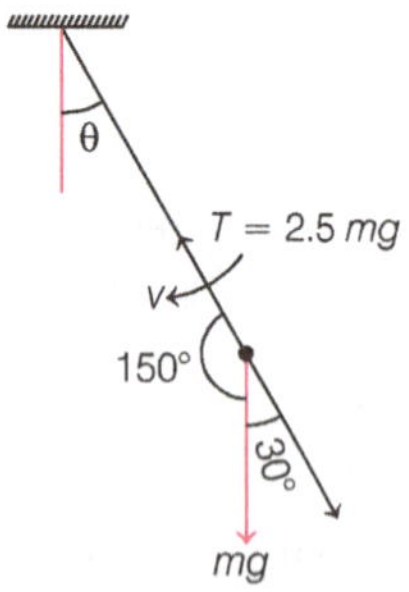

$\therefore\; 1.63\,g = \dfrac{v^2}{r} = \dfrac{v^2}{2}$

$\therefore\qquad v = 5.66\ \text{m/s}$

$F_{net} = \sqrt{(2.5\,mg)^2 + (mg)^2 + (2)(2.5\,mg)(mg)\cos 150^\circ}$

$= 1.7\,mg$

$a_{net} = \dfrac{F_{net}}{m} = 1.7\,g$

$\approx 16.75\ \text{m/s}^2$ **Ans.**

31. $v^2 = v_0^2 + 2\,gh = v_0^2 + 2\,gR\sin\theta$

$= (5)^2 + 2\times 10\times 2\sin\theta$

$= (25 + 40\sin\theta)$

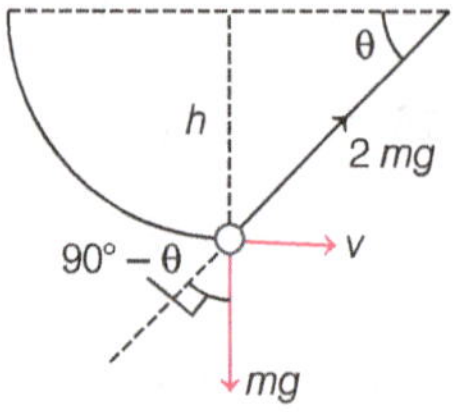

Now, $2\,mg - mg\sin\theta = \dfrac{mv^2}{R}$

or $\quad 2g - g\sin\theta = \dfrac{v^2}{R}$

or $2\times 10 - 10\sin\theta = \dfrac{25 + 40\sin\theta}{2}$

Solving this equation, we get

$$\theta = \sin^{-1}\left(\frac{1}{4}\right)$$ **Ans.**

32. $v = 360\ \text{km/h} = 360\times\dfrac{5}{18} = 100\ \text{m/s}$

$N_1 - mg = \dfrac{mv^2}{R}$

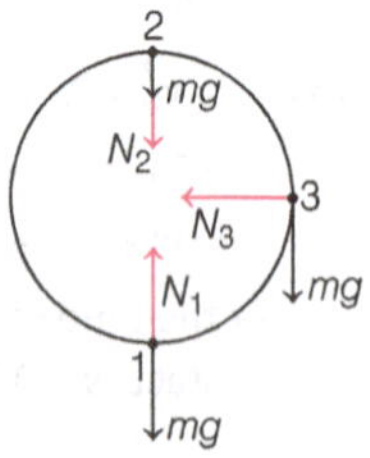

$\therefore\qquad N_1 = m\left(g + \dfrac{v^2}{R}\right)$

$= 70\left[9.8 + \dfrac{(100)^2}{500}\right]$

$= 2086\ \text{N}$ or $N_1 \approx 2.1\ \text{kN}$

$N_2 + mg = \dfrac{mv^2}{R}$

$\therefore\qquad N_2 = m\left(\dfrac{v^2}{R} - g\right)$

$= 70\left[\dfrac{(100)^2}{500} - 9.8\right] = 714\ \text{N}$

or $\qquad N_2 \approx 0.7\ \text{kN}$

33. $h = l - l\cos\theta$

$v^2 = 2gh = 2gl(1-\cos\theta)$

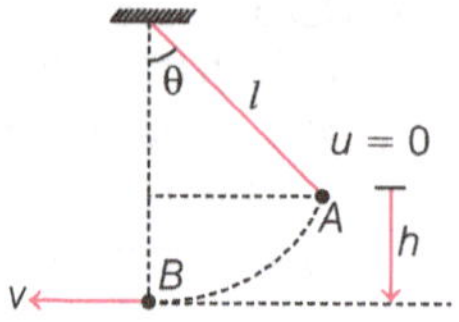

In position A, there is only tangential acceleration

$\left(a_r = 0, \text{as } a_r = \dfrac{u^2}{l} = 0\right)$.

$\therefore\quad a_A = a_t = \dfrac{mg\sin\theta}{m} = g\sin\theta$ ...(i)

In position B, there is only radial acceleration, because tangential force is zero. So, tangential acceleration is zero.

$\therefore\qquad a_B = a_r = \dfrac{v^2}{l} = 2g(1-\cos\theta)$ ...(ii)

Given, $\quad a_A = a_B$

Equating Eqs. (i) and (ii), we get

$$\theta \approx 53^\circ$$

34. $h = R - R\cos\theta$

$v^2 = 2gh = 2gR(1-\cos\theta)$ ...(i)

At the time of breaking off normal reaction will become zero. So component of mg towards centre C or $mg\cos\theta$ will be equal to centripetal force or $\dfrac{mv^2}{R}$. Thus,

$mg\cos\theta = \dfrac{mv^2}{R}$

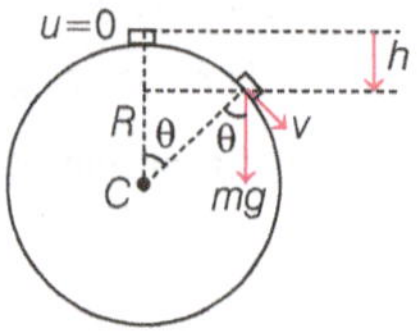

$= 2mg(1-\cos\theta)$

Solving this equation, we get

$\cos\theta = \dfrac{2}{3}$

or $\qquad \theta = \cos^{-1}(2/3) \approx 48^\circ$

Putting $\cos\theta = \dfrac{2}{3}$ in Eq. (i), we get

$$v = \sqrt{\frac{2gR}{3}}$$

35. $u_{min} = \sqrt{5gl}$, $T = 3mg$

In the frame of point of suspension,

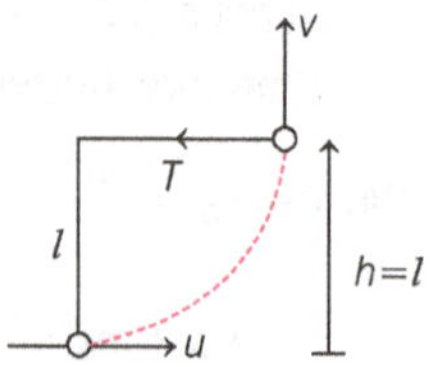

$u_{min} = \sqrt{5gl}$

$v^2 = u^2 - 2gh$

$= 5gl - 2gl = 3gl$

Now, $\qquad T = \text{centripetal force}$

$\therefore\qquad T = \dfrac{mv^2}{l} = \dfrac{m(3gl)}{l}$

or $\qquad T = 3mg$

Unit Exercise (Cumulative)

Level 1

1. $t = T$

$$\therefore \quad \sqrt{\frac{2H}{g}} = \frac{2\pi}{\omega}$$

$$\therefore \quad \omega = 2\pi\sqrt{\frac{g}{2H}} = \pi\sqrt{\frac{2g}{H}} \qquad \textbf{Ans.}$$

2. $a_t = a_r$

$$\therefore \quad \frac{dv}{dt} = \frac{v^2}{R} \qquad (v = \text{speed})$$

$$\therefore \quad \int_{v_0}^{v} v^{-2}\, dv = \frac{1}{R}\int_0^t dt$$

$$\left[\frac{1}{v_0} - \frac{1}{v}\right] = \frac{t}{R}$$

or $$\frac{1}{v} = \frac{1}{v_0} - \frac{t}{R} = \frac{R - v_0 t}{R v_0}$$

or $$v = \frac{R v_0}{R - v_0 t}$$

$$\therefore \quad \frac{dx}{dt} = \frac{R v_0}{R - v_0 t} \qquad (x = \text{distance travelled})$$

or $$\int_0^x dx = \int_0^t \frac{R v_0}{R - v_0 t}\, dt$$

$$x = R v_0 \left(-\frac{1}{v_0}\right)\left[\ln(R - v_0 t)\right]_0^t$$

$$= -R \ln\left(1 - \frac{v_0 t}{R}\right)$$

$$\therefore \quad 1 - \frac{v_0 t}{R} = e^{-x/R}$$

or $$t = t = \frac{R}{v_0}(1 - e^{x/R})$$

Putting $x = 2\pi R$, we get

$$t = \frac{R}{v_0}(1 - e^{-2\pi}) \qquad \textbf{Ans.}$$

3. $$v = \frac{d}{t_1} = \frac{21 \times 2\pi \times 0.5}{44} = 1.5 \text{ m/s}$$

$$t_2 = \sqrt{\frac{2h}{g}} = \sqrt{\frac{2 \times 4.9}{9.8}} = 1 \text{ s}$$

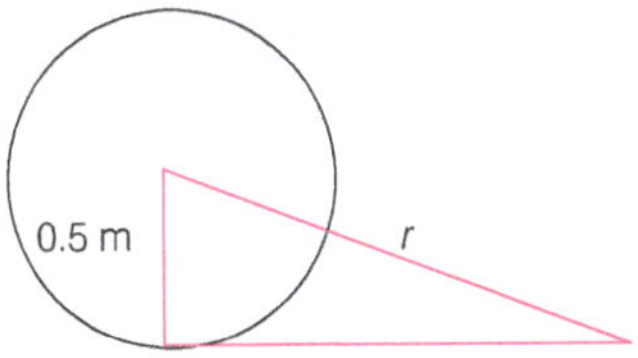

Horizontal distance travelled by drop $= vt_2 = 1.5$ m

$$\therefore \quad r = \sqrt{(1.5)^2 + (0.5)^2}$$

$$= \sqrt{2.5} \text{ m}$$

4. $\cos(d\theta)$ components of T are cancelled and $\sin(d\theta)$ components towards centre provide the necessary centripetal force to small portion PQ.

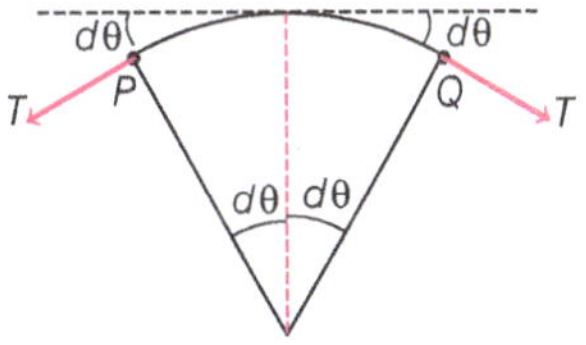

$$\therefore \quad 2T \sin(d\theta) = (m_{PQ})(R)\,\omega^2$$

For small angle, $\sin d\theta \approx d\theta$

$$\therefore \quad 2T\, d\theta = \left(\frac{m}{2\pi}\right)(2d\theta)(R)(2n\pi)^2$$

$$\therefore \quad T = 2\pi m n^2 R$$

Substituting the values, we get

$$T = (2\pi)(2\pi)(300/60)^2(0.25)$$

$$\approx 250 \text{ N} \qquad \textbf{Ans.}$$

5. $N\cos\theta = mg$ ...(i)

$N\sin\theta = mR\omega^2$ ...(ii)

From Eq. (i), we get

$$N = \frac{mg}{\cos\theta} = \text{constant}$$

as $\theta = \text{constant}$

Net force is the resultant of N and mg and both forces are constant.

Hence, net force is constant.

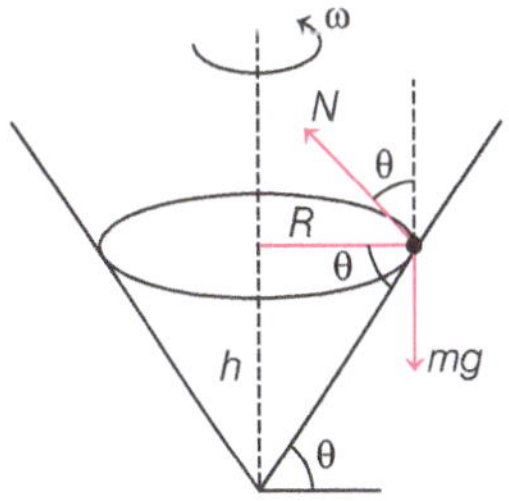

$$\omega = \sqrt{\frac{g}{R}\tan\theta}$$

But $$R = \frac{h}{\tan\theta}$$

$$\therefore \quad \omega = \sqrt{\frac{g}{h}}\tan\theta$$

or $$\omega \propto \frac{1}{\sqrt{h}} \qquad (\theta = \text{constant})$$

6. $E_A = E_B$

$\therefore \quad \frac{1}{2}mv_A^2 = \frac{1}{2} \times 200 \times (13-7)^2 \qquad (\because \ m = 2\text{ kg})$

$\therefore \quad v_A = 60\text{ m/s}$

At A, $\quad N = \frac{mv_A^2}{R} = \frac{(2)(60)^2}{5} = 1440\text{ N}$ **Ans.**

7. Let us see the FBD with respect to rotating cone (non-inertial)

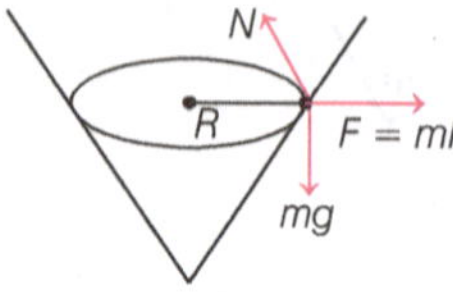

N, F and mg are balanced in the shown diagram. If displaced upwards, F will increase as R is increased. This will have a component up the plane. So, it will move upwards. Hence, the equilibrium is unstable.

8. $v = \frac{2\pi r}{T} = \frac{(2\pi)(0.5)}{1.58} \approx 2\text{ m/s}$

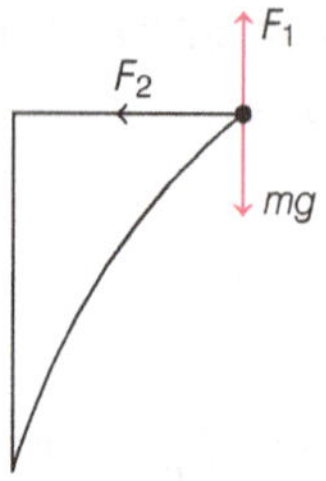

$F_1 = mg = 100\text{ N}$

$F_2 = \frac{mv^2}{r} = \frac{10 \times 4}{0.5}$

$= 80\text{ N}$

$\therefore$ Net force by rod on ball

$= \sqrt{F_1^2 + F_2^2} = 128\text{ N}$ **Ans.**

9.

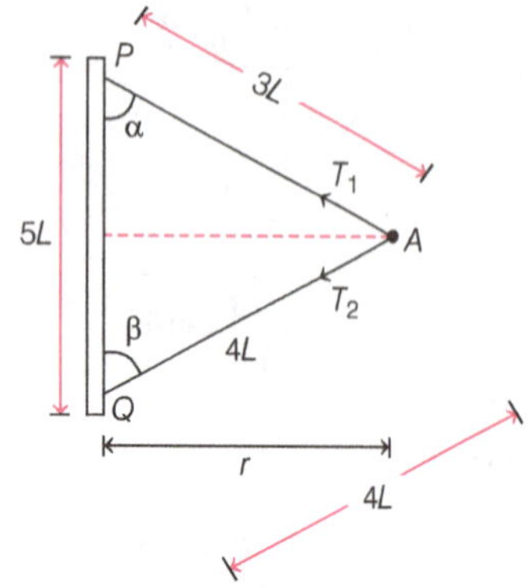

$\angle PAQ = 90°$

$\sin\alpha = \frac{4}{5}, \cos\alpha = \frac{3}{5}$

$\sin\beta = \frac{3}{5}, \cos\beta = \frac{4}{5}$

$r = 3L\sin\alpha = \frac{12L}{5}$

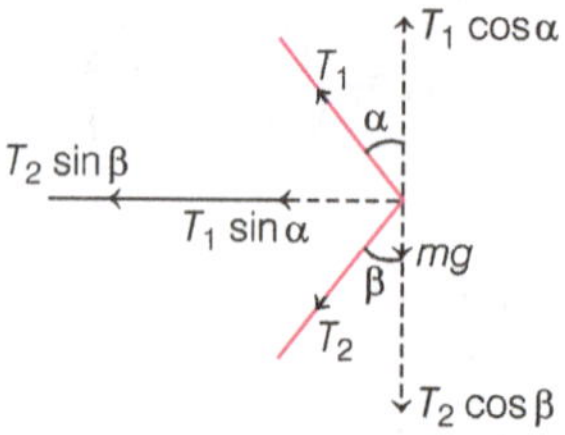

$T_1\cos\alpha = T_2\cos\beta + mg$

$T_1 \times \frac{3}{5} = T_2 \times \frac{4}{5} + mg$

$3T_1 - 4T_2 = 5mg \qquad \text{...(i)}$

$T_1\sin\alpha + T_2\sin\beta = m\omega^2 r$

$T_1 \times \frac{4}{5} + T_2 \times \frac{3}{5} = m\omega^2\left(\frac{12L}{5}\right)$

$4T_1 + 3T_2 = 12\,m\omega^2 L \qquad \text{...(ii)}$

10. Tangential force (F_t) of the bead will be given by the normal reaction (N), while centripetal force (F_c) is provided by friction (f_r). The bead starts sliding when the centripetal force is just equal to the limiting friction.

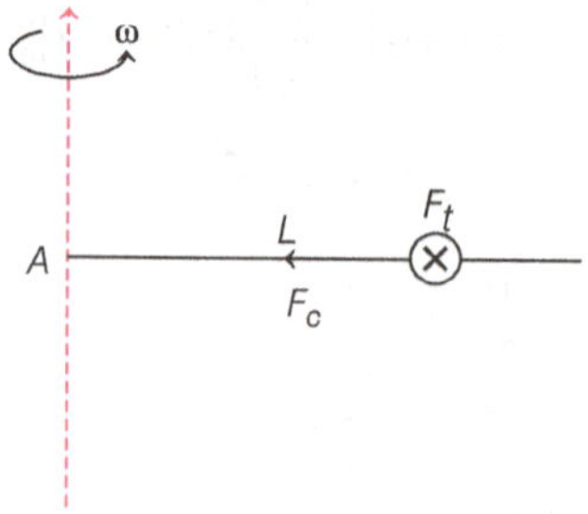

Therefore, $\quad F_t = ma_t = m\alpha L = N$

$\therefore$ Limiting value of friction

$(f_r)_{\max} = \mu N = \mu m\alpha L \qquad \text{...(i)}$

Angular velocity at time t is $\omega = \alpha t$

$\therefore$ Centripetal force at time t will be

$F_c = mL\omega^2 = mL\alpha^2 t^2 \qquad \text{...(ii)}$

Equating Eqs. (i) and (ii), we get

$t = \sqrt{\frac{\mu}{\alpha}}$

For $t > \sqrt{\frac{\mu}{\alpha}}$, $F_c > (f_r)_{\max}$ i.e. the bead starts sliding.

Note *In the figure, F_t is perpendicular to the paper inwards.*

11. Let N = normal reaction

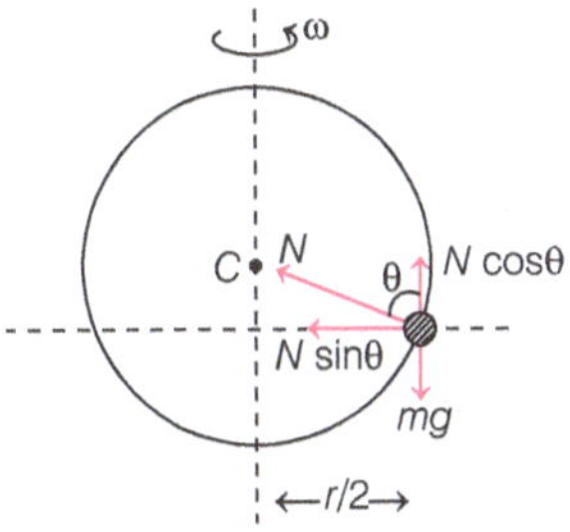

$$N\cos\theta = mg \qquad \text{...(i)}$$

and $$N\sin\theta = m\left(\frac{r}{2}\right)\omega^2 \qquad \text{...(ii)}$$

From Eqs. (i) and (ii), we have

$$\tan\theta = \frac{r\omega^2}{2g} \qquad \text{...(iii)}$$

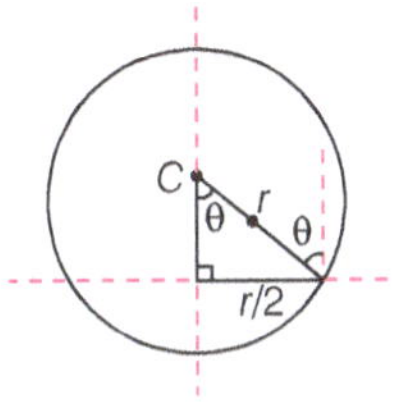

$$\tan\theta = \frac{\frac{r}{2}}{\sqrt{r^2-\left(\frac{r}{2}\right)^2}} = \frac{1}{\sqrt{3}} \qquad \text{...(iv)}$$

Putting this value in Eq. (iii), we get

$$\omega^2 = \frac{2g}{\sqrt{3}\,r}$$

12. At the maximum speed, the tangential acceleration of the particle with respect to the block should be zero.

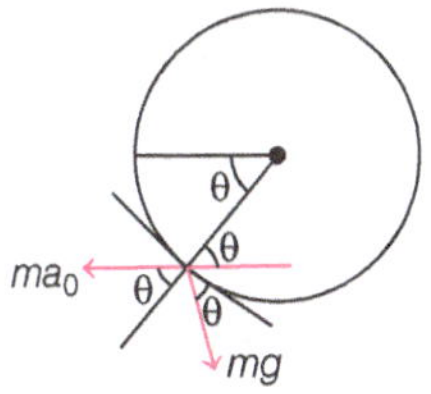

From figure, $\quad ma_0\sin\theta = mg\cos\theta$

$\therefore \quad \tan\theta = \frac{g}{a_0} = \frac{g}{g} = 1$

$\therefore \quad \theta = 45^\circ$

13. $h_{AB} = R\cos 37^\circ - R\cos 53^\circ$

$$= 0.8R - 0.6R = 0.2R$$

$\therefore \quad v_B = \sqrt{2gh_{AB}} = \sqrt{0.4gR}$

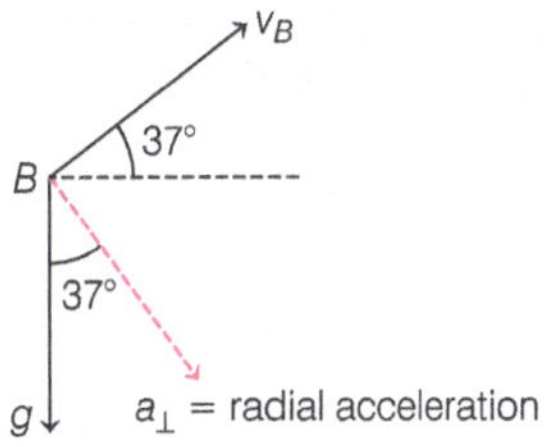

$$a_\perp = g\cos 37^\circ = (0.8g) = \frac{v_B^2}{r} = \frac{0.4\,gR}{r}$$

$\therefore \quad r$ = radius of curvature at B

$$= \frac{R}{2}$$

14. $h = R\cos\theta$

$$v_0 = \sqrt{2gh} = \sqrt{2gR\cos\theta}$$

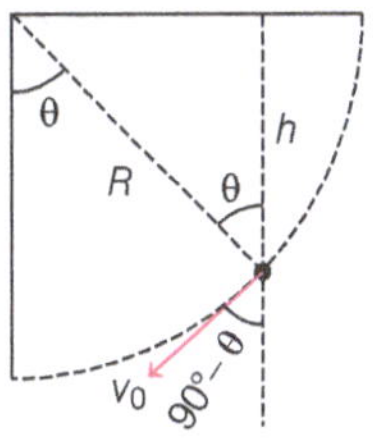

Vertical component of velocity is,

$$v = v_0\sin\theta$$
$$= \sin\theta\sqrt{2gR\cos\theta}$$

For v to be maximum

$$\frac{dv}{d\theta} = 0$$

$\therefore \quad [\cos\theta\sqrt{2gR\cos\theta}] + \frac{\sin\theta}{2\sqrt{2gR\cos\theta}}(-2gR\sin\theta) = 0$

$\therefore \quad 4gR\cos^2\theta - 2gR\sin^2\theta = 0$

or $\quad 2\cos^2\theta - (1-\cos^2\theta) = 0$

or $\quad \cos\theta = \frac{1}{\sqrt{3}}$

$\therefore \quad \theta = \cos^{-1}\left(\frac{1}{\sqrt{3}}\right)$ **Ans.**

15. If $u > \sqrt{5gl}$, then **T** and **a** are in same direction. Hence, $\mathbf{T}\cdot\mathbf{a}$ is positive.

If $u = \sqrt{5gl}$, then $T = 0$

$\therefore \quad \mathbf{T}\cdot\mathbf{a} = 0$

16. At lowest point **T** and **a** are always in same direction (towards centre).

$\therefore \quad \mathbf{T}\cdot\mathbf{a}$ is always positive.

17. Since, the particle is released from a small height, θ (angle of radius with vertical) will be very small.

Force of friction throughout the journey can be assumed to be μmg. Particle will finally come to rest when whole of its energy ($=mgh$) is lost in the work done against friction. Let particle stops after travelling a distance d. Then,

$$\mu mgd = mgh$$

or $$d = \frac{h}{\mu} = \frac{10^{-2}}{0.01} = 1.0 \text{ m}$$

18. When a is horizontal

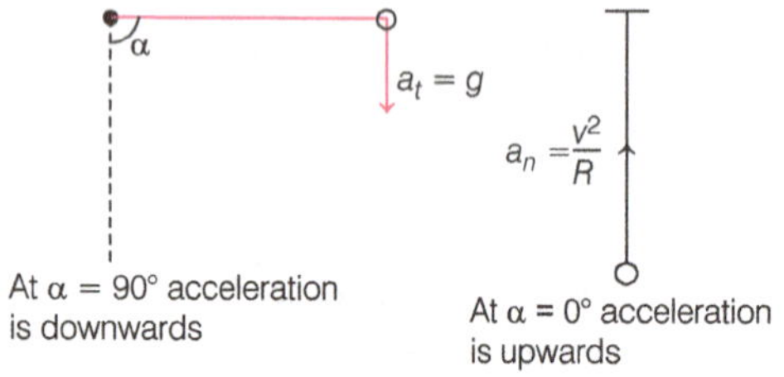

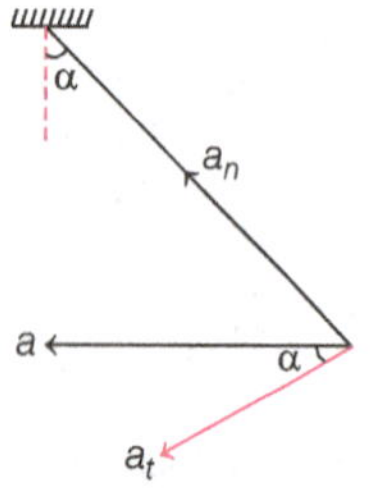

$$\tan\alpha = \frac{a_n}{a_t} = \frac{v^2/l}{g\sin\alpha}$$

$$= \frac{2gh/l}{g\sin\alpha} = \frac{2g\,l\cos\alpha/l}{g\sin\alpha}$$

$$= 2\cot\alpha$$

$$\tan\alpha = \sqrt{2}$$

or $$\cos\alpha = \frac{1}{\sqrt{3}}$$

$\therefore$ $$\alpha = \cos^{-1}\left(\frac{1}{\sqrt{3}}\right)$$

19. As, $$v = \sqrt{5gL} \quad \text{...(i)}$$

So, $$\left(\frac{v}{2}\right)^2 = v^2 - 2gh \quad \text{...(ii)}$$

$$h = L(1-\cos\theta) \quad \text{...(iii)}$$

Solving Eqs. (i), (ii) and (iii), we get

$$\cos\theta = -\frac{7}{8}$$

or $$\theta = \cos^{-1}\left(-\frac{7}{8}\right) = 151°$$

20. $R = h\cot\alpha$

$$N\cos\alpha = mg$$

$$N\sin\alpha = \frac{mv^2}{R}$$

Solving these two equations, we get

$$v = \sqrt{Rg\tan\alpha}$$

$$= \sqrt{(h\cot\alpha)(g\tan\alpha)}$$

$$= \sqrt{gh}$$

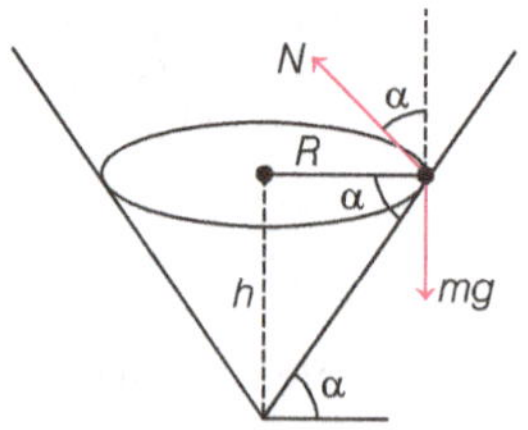

Now, $$T = \frac{2\pi R}{v} = \frac{2\pi h\cot\alpha}{\sqrt{gh}}$$

$$= 2\pi\sqrt{\frac{h}{g}}\cot\alpha$$

$\therefore$ $$T \propto \sqrt{h} \quad \text{and} \quad T \propto \cot\alpha$$

$\cot\alpha$ decreases as α increases.

21. According to work-energy theorem,

$$W_{\text{Total}} = \Delta \text{KE} = \frac{1}{2}mv^2 - \frac{1}{2}mu^2$$

or $$W_F + W_{mg} + W_N = \frac{1}{2}mv^2 \quad \text{...(i)}$$

$\boldsymbol{W_F}$**:** See miscellaneous example-15 of chapter work, energy and power,

$$W_F = 2000 \text{ J}$$

$\boldsymbol{W_{mg}}$**:** $W_{mg} = -mgh = -mgR(1-\cos 60°) = -500$ J

$\boldsymbol{W_N}$**:** $W_N = 0$

Substituting the values in Eq. (i), we get

$$2000 - 500 = \frac{1}{2} \times 10 \times v^2$$

$\therefore$ $$v = \sqrt{300}$$

$$= 10\sqrt{3} \text{ m/s}$$

22. In the frame of triangular block,

$$T\cos 37° = N\cos 53° + mr\omega^2 \text{ (in horizontal direction)}$$

or $$\frac{4}{5}T - \frac{3}{5}N = \omega^2$$

or $$4T - 3N = 5\omega^2 \quad \text{...(i)}$$

and $$T\sin 37° + N\sin 53° = mg$$

or $$\frac{3}{5}T + \frac{4}{5}N = 10$$

or $$3T + 4N = 50 \quad \text{...(ii)}$$

or $$3\left(\frac{5\omega^2 + 3N}{4}\right) + 4N = 50$$

or $$\frac{15}{4}\omega^2 + \frac{9}{4}N + 4N = 50$$

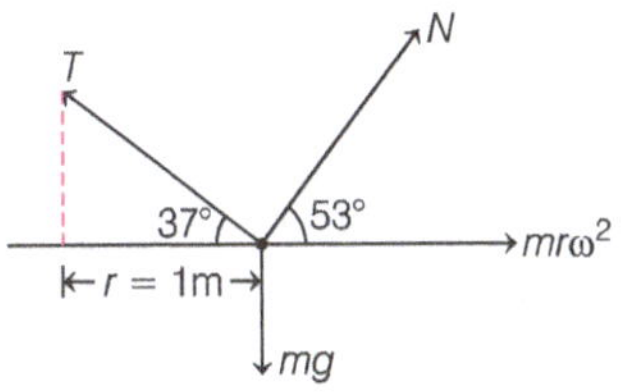

For leaving the surface, $N = 0$

$$\Rightarrow \quad \omega^2 = \frac{50 \times 4}{15} = \frac{40}{3}$$

$$\therefore \quad \omega = \sqrt{\frac{40}{3}}$$

$$\because \quad \omega = \alpha t$$

$$\therefore \quad t = \frac{\sqrt{40/3}}{\sqrt{40/3}} = 1\,s$$

23. $\because\ 3T + 4N = 50$

or $\quad 3T = 50 \quad (\because N = 0)$

$$\therefore \quad T = \frac{50}{3} = 16.67\,N$$

24. Height fallen up to point Q

$$h = R \sin 30^\circ$$
$$= 40 \times \frac{1}{2} = 20\,m$$

Work done against friction = Initial mechanical energy − Final mechanical energy

$$= mgh - \frac{1}{2}mv^2$$

Putting the values, we get

$$150 = 1 \times 10 \times 20 - \frac{1}{2} \times 1 \times v^2$$

$$\Rightarrow \quad v = 10\ m/s$$

25. At point Q, component of weight along PQ (radially outwards) is $mg\cos 60^\circ$ or $\frac{mg}{2}$.

Normal reaction is radially inwards

$$N - \frac{mg}{2} = \frac{mv^2}{R}$$

or

$$N = \frac{mg}{2} + \frac{mv^2}{R}$$
$$= \frac{1 \times 10}{2} + \frac{1 \times (10)^2}{40} = 7.5\,N$$

26. $a_r = \frac{v^2}{R} = \frac{(2t)^2}{2} = 2t^2$

$$a_t = \frac{dv}{dt} = 2\ m/s^2$$
$$v = 2t$$
$$\omega = \frac{v}{R} = \frac{2t}{2} = t$$

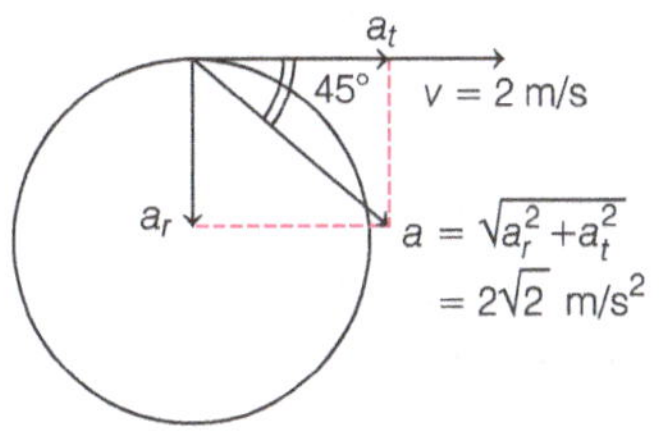

At 1 s,

$$a_r = 2\ m/s^2,$$
$$a_t = 2\ m/s^2,$$
$$v = 2\ m/s$$
and $\quad \omega = 1\ rad/s$

(a) $\mathbf{a} \cdot \mathbf{v} = av \cos 45^\circ$
$$= (2\sqrt{2})(2)(1/\sqrt{2}) = 4\ m^2/s^3$$

(b) $|\mathbf{a} \times \omega| = a\omega \sin 90^\circ$
$$= (2\sqrt{2})(1)(1)$$
$$= 2\sqrt{2}\ m/s^3$$

(c) $\mathbf{v} \cdot \omega = 0$ as $\theta = 90^\circ$

(d) $|\mathbf{v} \times \mathbf{a}| = va \sin 45^\circ = 4\ m^2/s^3$

27. $r = |\mathbf{r}| = \sqrt{(3)^2 + (-4)^2} = 5\,m$

$$v = |\mathbf{v}| = \sqrt{16 + a^2}$$
$$a_r = |\mathbf{a}| = \sqrt{36 + b^2}$$

In uniform circular motion, **v** is always perpendicular to **a**.

$$\therefore \quad \mathbf{v} \cdot \mathbf{a} = 0$$
or $\quad -24 - ab = 0$
$$\therefore \quad ab = 24 \quad \text{...(i)}$$
$$a_r = \frac{v^2}{r}$$
$$\therefore \quad \sqrt{36 + b^2} = \frac{16 + a^2}{5} \quad \text{...(ii)}$$

Solving these two equations, we can find the values of *a* and *b*.

(d) **r**, **v** and **a** lie in same plane. But $\mathbf{v} \times \mathbf{a}$ is perpendicular to this plane.

$$\therefore \quad \mathbf{r} \perp \mathbf{v} \times \mathbf{a}$$
or $\quad \mathbf{r} \cdot (\mathbf{v} \times \mathbf{a}) = 0$

28. In part (a) mean velocity is asked. It should be mean (or average) speed.

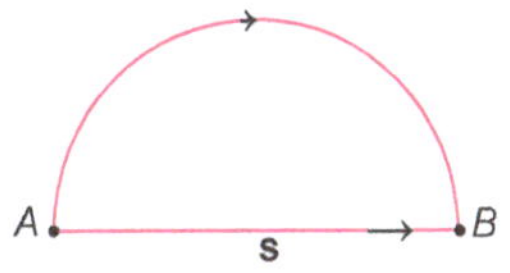

(a) Average speed $= \frac{\text{Total distance}}{\text{Total time}}$

$$= \frac{\pi R}{t} = \frac{\left(\frac{22}{7}\right)(160)}{10}$$

$$= 50.28 \text{ cm/s}$$

$$\approx 50 \text{ cm/s}$$

(b) Magnitude of mean (or average)

$$\text{velocity} = \frac{|\mathbf{s}|}{t} \qquad (\mathbf{s} = \text{displacement})$$

$$= \frac{2R}{t} = \frac{2 \times 160}{10}$$

$$= 32 \text{ cm/s}$$

(c) Let us assume that particle starts from rest. Tangential acceleration is a. Then,

$$d = \frac{1}{2} at^2 \qquad (d = \text{distance})$$

$$\Rightarrow \quad a = \frac{2d}{t^2}$$

$$v = at = \frac{2d}{t}$$

Magnitude of average acceleration,

$$|\mathbf{a}_{av}| = \frac{|\Delta \mathbf{v}|}{t} = \frac{v}{t} = \frac{2d}{t^2} = \frac{2\pi R}{t^2}$$

$$= \frac{(2)\left(\frac{22}{7}\right)(160)}{(10)^2} \approx 10 \text{ cm/s}^2$$

29. Given, $a_t = a_r$

or $$-\frac{dv}{dt} = -v \cdot \frac{dv}{ds} = \frac{v^2}{R} \qquad \text{...(i)}$$

$$\therefore \quad -\frac{dv}{v^2} = \frac{dt}{R}$$

Integrating this equation, we get

$$\frac{1}{v} = \frac{t}{R} + c_1 \qquad \text{...(ii)}$$

At $t = 0$, $v = v_0$

$$\Rightarrow \quad c_1 = \frac{1}{v_0}$$

Putting this value in Eq. (ii), we get

$$\frac{1}{v} = \frac{t}{R} + \frac{1}{v_0}$$

$$\therefore \quad v = \frac{Rv_0}{v_0 t + R}$$

Further, from Eq. (i), we get

$$-v\frac{dv}{ds} = \frac{v^2}{R}$$

or $$\frac{dv}{v} = -\frac{ds}{R}$$

Integrating, we get

$$\log_e v = -\frac{s}{R} + c_2 \qquad \text{...(iii)}$$

$$\therefore \quad v = e^{-\frac{s}{R} + c_2}$$

At $s = 0$, $v = v_0$

$$\therefore \quad e^{c_2} = v_0 \quad \text{or} \quad c_2 = \log_e v_0$$

Substituting in Eq. (iii), we get

$$\log_e v = -\frac{s}{R} + \log_e v_0$$

or $$\log_e v - \log_e v_0 = -\frac{s}{R}$$

or $$\log_e \left(\frac{v}{v_0}\right) = -\frac{s}{R}$$

$$\therefore \quad \frac{v}{v_0} = e^{-s/R}$$

or $$v = v_0 e^{-s/R}$$

Note *Velocity (or we can also call speed) will exponentially decrease with s from v_0 to zero. So, v - s graph is as under.*

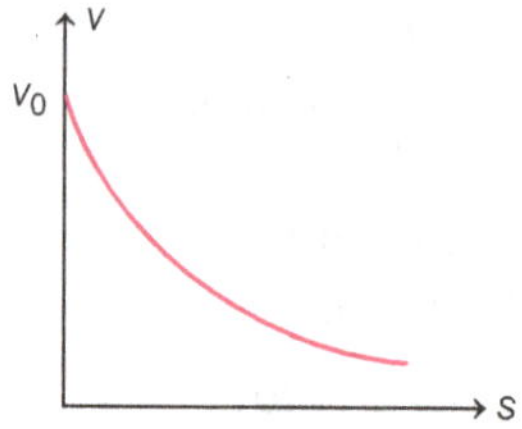

(b) $a_t = a_r = \frac{v^2}{R}$

$\therefore$ Total acceleration

$$a = \sqrt{a_r^2 + a_t^2} = \frac{\sqrt{2}\,v^2}{R}$$

Substituting $v = v_0 e^{-s/R}$, we get

$$a = \frac{\sqrt{2}\,v_0^2 e^{-\frac{2s}{R}}}{R}$$

or $$a = \frac{\sqrt{2}\,v_0^2}{R\,e^{2s/R}}$$

30. $v = \alpha\sqrt{s}$

$$\therefore \quad v^2 = \alpha^2 s$$

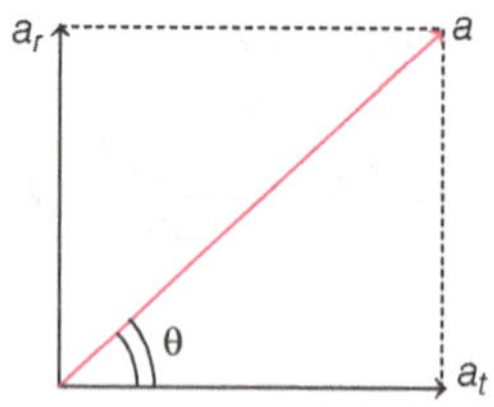

Comparing with $v^2 = 2a_t s$

where, a_t is tangential acceleration and s the distance, we have

$$2a_t = \alpha^2 \quad \text{or} \quad a_t = \frac{\alpha^2}{2}$$

Radial acceleration,

$$a_r = \frac{v^2}{R} = \frac{(\alpha\sqrt{s})^2}{R} = \frac{\alpha^2 s}{R} \qquad (a = \text{total acceleration})$$

$$\tan\theta = \frac{a_r}{a_t} = \frac{(\alpha^2 s/R)}{(\alpha^2/2)} = \frac{2s}{R}$$

$$\therefore \quad \tan\theta = \frac{2s}{R}$$

Note *Velocity vector is in the direction of a_t. Therefore the above angle is also the angle between total acceleration and velocity.*

31. (a) The frictional force provides the necessary centripetal force.

$$\therefore \quad mL\omega^2 = \mu mg$$

$$\text{or} \quad \omega = \sqrt{\frac{\mu g}{L}}$$

(b) Net force of circular motion will be provided by the friction

$$\omega = \alpha t \qquad \text{...(i)}$$

$$F_{\text{net}} = m\sqrt{a_t^2 + a_n^2}$$

$$\therefore \quad \mu mg = m\sqrt{(L\alpha)^2 + (L\omega^2)^2} \qquad \text{...(ii)}$$

Here, $a_t = L\alpha$ and $a_n = L\omega^2$

Substituting $\omega = \alpha t$ in Eq. (ii), we have

$$\mu g = \sqrt{L^2\alpha^2 + L^2\alpha^4 t^4}$$

$$\therefore \quad t = \left(\frac{\mu^2 g^2 - L^2\alpha^2}{L^2\alpha^4}\right)^{\frac{1}{4}}$$

Substituting the value of t in Eq. (i), we have

$$\omega = \left[\left(\frac{\mu g}{L}\right)^2 - \alpha^2\right]^{\frac{1}{4}}$$

32. $N\cos\theta = mg$

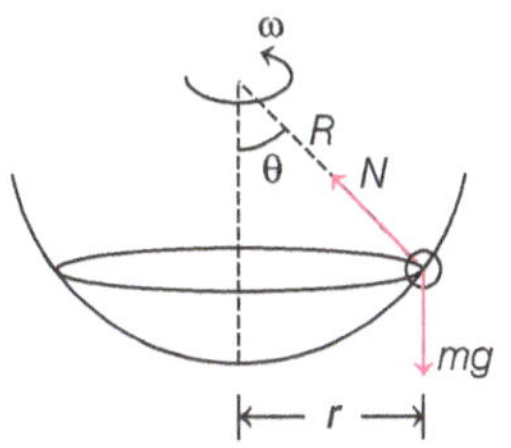

$$N\sin\theta = mr\omega^2 = m(R\sin\theta)\,\omega^2$$

Dividing these two equations, we get

$$\cos\theta = \frac{g}{R\omega^2} \qquad \text{...(i)}$$

$$\therefore \quad \omega = \sqrt{\frac{g}{R\cos\theta}}$$

At lowermost point, $\theta = 0°$

$$\therefore \quad \omega = \sqrt{\frac{g}{R}}$$

Substituting $\omega = \sqrt{2g/R}$ in Eq. (i), we have

$$\cos\theta = \frac{1}{2}$$

$$\therefore \quad \theta = 60° \qquad \textbf{Ans.}$$

33. (a)

$$T = m_2 r_2 \omega^2 = (2)(2)(4)^2 = 64\ \text{N}$$

Centripetal force required to 1 kg block is

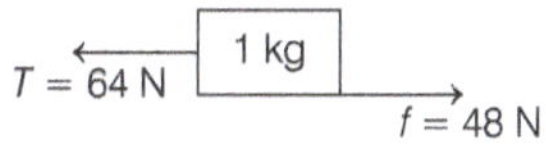

$$F_c = m_1 r_1 \omega^2 = (1)(1)(4)^2 = 16\ \text{N}$$

But available tension is 64 N. So, the extra force of 48 N is balanced by friction acting radially outward from the centre.

(b) $f_{max} = \mu m_1 g = 0.8 \times 1 \times 10 = 8$ N

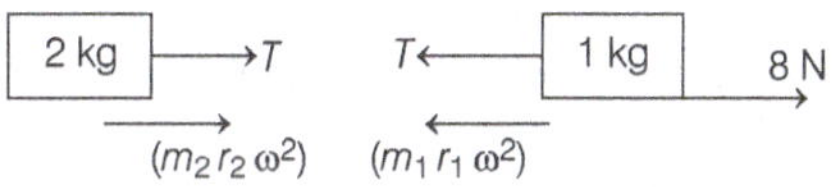

$$T = m_2 r_2 \omega^2 = (2)(2)\,\omega^2 = 4\,\omega^2 \qquad \text{...(i)}$$

$$T - 8 = m_1 r_1 \omega^2 = (1)(1)(\omega^2) = \omega^2 \qquad \text{...(ii)}$$

Solving Eqs. (i) and (ii), we get

$$\omega = 1.63\ \text{rad/s} \qquad \textbf{Ans.}$$

(c) $T = m_2 r_2 \omega^2$

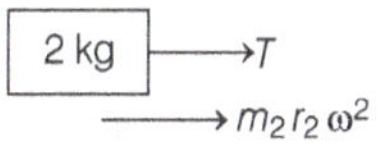

$$\therefore \quad 100 = (2)(2)\,\omega^2$$

$$\text{or} \quad \omega = 5\ \text{rad/s}$$

34. $a_n = \alpha t^2 = \dfrac{v^2}{R}$

$$\therefore \quad v = (\sqrt{\alpha R})\,t$$

Tangential acceleration,

$$a_t = \text{rate of change of speed}$$

$$= \frac{dv}{dt} = \sqrt{\alpha R}$$

Tangential force,

$$F_t = ma_t = m\sqrt{\alpha R}$$

Normal force is perpendicular to velocity. So, its power is always zero.

Time dependence of power

Only tangential force which is parallel to velocity will have power.

$$\therefore \quad P = F_t v \cos 0°$$
$$= (m\sqrt{\alpha R})(\sqrt{\alpha R})t$$
or $$P = m\alpha Rt$$

Mean/average power

$$P_{av} = \frac{\text{Total work done}}{\text{Total time}} = \frac{\frac{1}{2}mv^2}{t}$$
$$= \frac{\frac{1}{2}m[\sqrt{\alpha R}\,t]^2}{t} \quad \text{or} \quad P_{av} = \frac{m\alpha Rt}{2}$$

Note *Since, instantaneous power is a linear function of time. So, we can also find*

$$P_{av} = \frac{P_f + P_i}{2} = \frac{P_t + P_0}{2}$$
$$= \frac{(m\alpha Rt) + 0}{2} = \frac{m\alpha RT}{2}$$

35. Given, $m_1 = 10$ kg,
$m_2 = 5$kg, $\omega = 10$ rad/s
$r = 0.3$ m, $r_1 = 0.124$ m
$\therefore \quad r_2 = r - r_1 = 0.176$ m

(a) Masses m_1 and m_2 are at rest with respect to rotating table. Let f be the friction between mass m_1 and table.

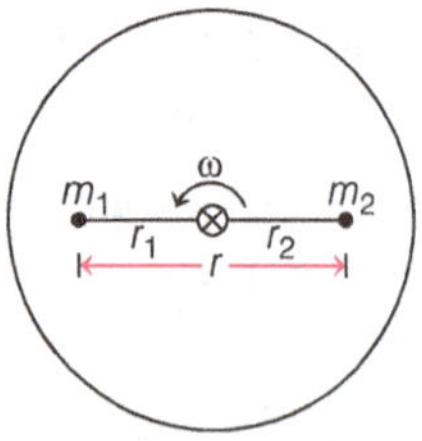

Free body diagram of m_1 and m_2 with respect to ground is shown below.

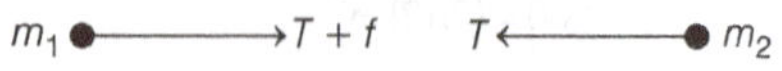

$$T = m_2 r_2 \omega^2 \quad \text{...(i)}$$
Since, $\quad m_2 r_2 \omega^2 < m_1 r_1 \omega^2$
Therefore, $\quad m_1 r_1 \omega^2 > T$
and friction on m_1 will be inward (toward centre)
$$f + T = m_1 r_1 \omega^2 \quad \text{...(ii)}$$
From Eqs. (i) and (ii), we get
$$f = m_1 r_1 \omega^2 - m_2 r_2 \omega^2 \quad \text{...(iii)}$$
$$= (m_1 r_1 - m_2 r_2)\,\omega^2$$
$$= (10 \times 0.124 - 5 \times 0.176)(10)^2 \text{ N}$$
$$= 36 \text{ N}$$
Therefore, frictional force on m_1 is 36 N (inwards)

(b) From Eq. (iii),
$$f = (m_1 r_1 - m_2 r_2)\,\omega^2$$

Masses will start slipping when this force is greater than f_{max} or $(m_1 r_1 - m_2 r_2)\,\omega^2 > f_{max} > \mu m_1 g$

$\therefore$ Minimum value of ω is
$$\omega_{min} = \sqrt{\frac{\mu m_1 g}{m_1 r_1 - m_2 r_2}}$$
$$= \sqrt{\frac{0.5 \times 10 \times 9.8}{10 \times 0.124 - 5 \times 0.176}}$$
or $\quad \omega_{min} = 11.67$ rad/s

(c) From Eq. (iii), frictional force $f = 0$
when $\quad m_1 r_1 = m_2 r_2$
or $$\frac{r_1}{r_2} = \frac{m_2}{m_1} = \frac{5}{10} = \frac{1}{2}$$
and $\quad r = r_1 + r_2 = 0.3$ m
$\therefore \quad r_1 = 0.1$ m and $r_2 = 0.2$ m
i.e. mass m_2 should be placed at 0.2 m and m_1 at 0.1m from the centre O.

36. From figure,

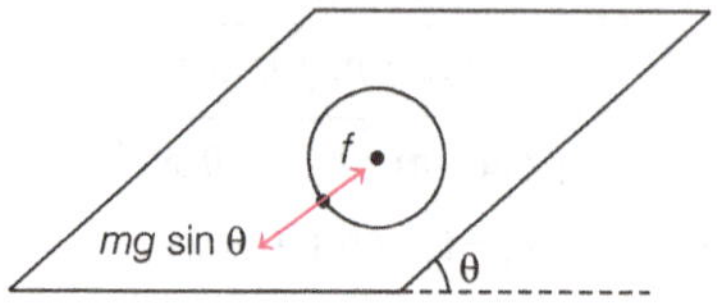

$$f - mg\sin\theta = \frac{mv^2}{r}$$
or $$\mu N - mg\sin\theta = \frac{mv_{max}^2}{r}$$
or $$\mu mg\cos\theta - mg\sin\theta = \frac{mv_{max}^2}{r}$$
or $$\mu g\cos\theta - g\sin\theta = \frac{v_{max}^2}{r} \quad (\text{here, } \theta = 30°)$$
$$\therefore \quad v_{max} = \sqrt{gr(\mu\cos\theta - \sin\theta)}$$
$$= \sqrt{gr(\mu\cos 30° - \sin 30°)}$$
$$= \sqrt{10 \times 10\left(\frac{2}{\sqrt{3}} \times \frac{\sqrt{3}}{2} - \frac{1}{2}\right)}$$
$$= \frac{10}{\sqrt{2}} = 5\sqrt{2} \text{ m/s}$$

37. Minimum velocity of particle at the lowest position to complete the circle should be $\sqrt{4gR}$ inside a tube.

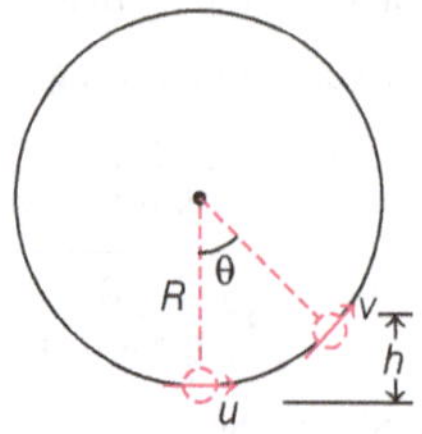

So, $u = \sqrt{4gR}$

$h = R(1 - \cos\theta)$

$\therefore \quad v^2 = u^2 - 2gh$

or $\quad v^2 = 4gR - 2gR(1 - \cos\theta)$

$= 2gR(1 + \cos\theta)$

or $\quad v^2 = 2gR\left(2\cos^2\frac{\theta}{2}\right)$

or $\quad v = 2\sqrt{gR}\cos\frac{\theta}{2}$

From $\quad ds = v \cdot dt$, we get

$R\,d\theta = 2\sqrt{gR}\cos\frac{\theta}{2} \cdot dt$

or $\quad \int_0^t dt - \frac{1}{2}\sqrt{\frac{R}{g}}\int_0^{\pi/2} \sec\left(\frac{\theta}{2}\right)d\theta$

or $\quad t = \sqrt{\frac{R}{g}}\left[\ln\left(\sec\frac{\theta}{2} + \tan\frac{\theta}{2}\right)\right]_0^{\pi/2}$

or $\quad t = \sqrt{\frac{R}{g}}\ln(1 + \sqrt{2})$ **Hence proved.**

38. At position θ,

$v^2 = v_0{}^2 + 2gh$

where, $\quad h = a(1 - \cos\theta)$

$\therefore \quad v^2 = (\sqrt{2ag})^2 + 2ag(1 - \cos\theta)$

or $\quad v^2 = 2ag(2 - \cos\theta)$ ...(i)

$N + mg\cos\theta = \frac{mv^2}{a}$

or $\quad N + mg\cos\theta = 2mg(2 - \cos\theta)$

or $\quad N = mg(4 - 3\cos\theta)$

Net vertical force,

$F = N\cos\theta + mg$

$= mg(4\cos\theta - 3\cos^2\theta + 1)$

This force (or acceleration) will be maximum when

$\frac{dF}{d\theta} = 0$

or $\quad -4\sin\theta + 6\sin\theta\cos\theta = 0$

So, either

$\sin\theta = 0, \ \theta = 0°$ or $\cos\theta = \frac{2}{3}$,

$\theta = \cos^{-1}\left(\frac{2}{3}\right)$

$\theta = 0°$ is unacceptable

Therefore, the desired position is at

$\theta = \cos^{-1}\left(\frac{2}{3}\right)$ **Ans.**

39. $h = l(1 - \cos\theta)$

$v^2 = v_0^2 - 2gh = 3gl - 2gl(1 - \cos\theta)$

$= gl(1 + 2\cos\theta)$

At 45° means radial and tangential components of acceleration are equal.

$\therefore \quad \frac{v^2}{l} = g\sin\theta$

or $\quad 1 + 2\cos\theta = \sin\theta$

Solving the equation, we get $\theta = 90°$ or $\frac{\pi}{2}$ **Ans.**

40. Speed of bob in the given position,

$v = \sqrt{2gh}$

Here, $\quad h = (400 + 400\cos 30°)$ mm

$= 746$ mm

$= 0.746$ m

$\therefore \quad v = \sqrt{2 \times 9.8 \times 0.746}$

$= 3.82$ m/s

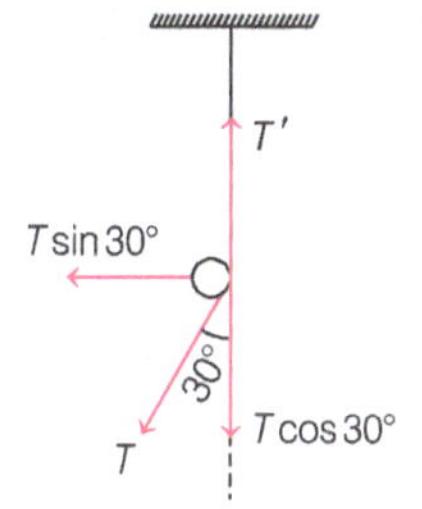

Now $T - mg\cos\theta = \frac{mv^2}{r}$

$T = 2 \times 9.8 \times \cos 30° + \frac{2 \times (3.82)^2}{(0.4)}$

or $\quad T = 90$ N

$\therefore \quad R = T\sin 30° = 45$ N **Ans.**

$T' = T\cos 30°$.

41. (a)

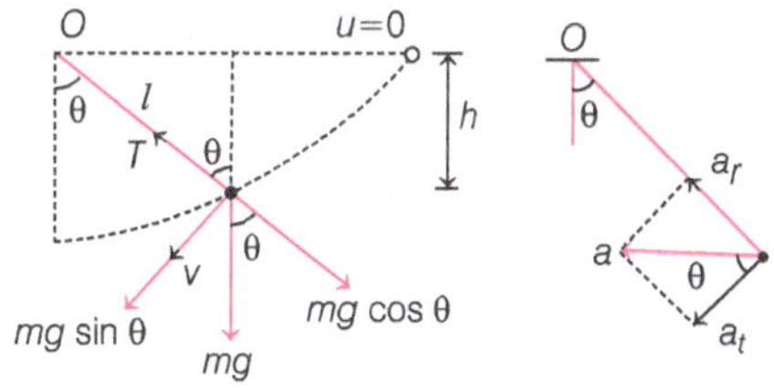

$h = l\cos\theta$

$v = \sqrt{2gh} = \sqrt{2gl\cos\theta}$

$T - mg\cos\theta = \frac{mv^2}{l} = 2mg\cos\theta$

$\therefore \quad T = 3mg\cos\theta$

Tangential acceleration, $a_t = \frac{mg\sin\theta}{m} = g\sin\theta$

Radial acceleration, $\quad a_r = \frac{v^2}{l} = 2g\cos\theta$

$\therefore$ Total acceleration, $a = \sqrt{a_t^2 + a_r^2}$

or $$a = g\sqrt{1 + 3\cos^2\theta}$$

(b) From the first figure, we can see that vertical component of velocity is,

$$V = v\sin\theta = [\sqrt{2gl\cos\theta}]\sin\theta$$

V will be maximum when

$$\frac{dV}{d\theta} = 0$$

Solving this equation, we get

$$\cos\theta = \frac{1}{\sqrt{3}} \Rightarrow T = 3mg\cos\theta$$

Putting $\cos\theta = \frac{1}{\sqrt{3}}$, we get

$$T = \sqrt{3}\,mg$$

(c) In the second figure total acceleration a is horizontal. Therefore angle between a and a_t is also θ.

$$\tan\theta = \frac{a_r}{a_t} = \frac{2g\cos\theta}{g\sin\theta}$$

or $$\tan\theta = \sqrt{2} \quad \text{or} \quad \theta = \tan^{-1}\sqrt{2}$$

$$\approx 54.7°$$

42.

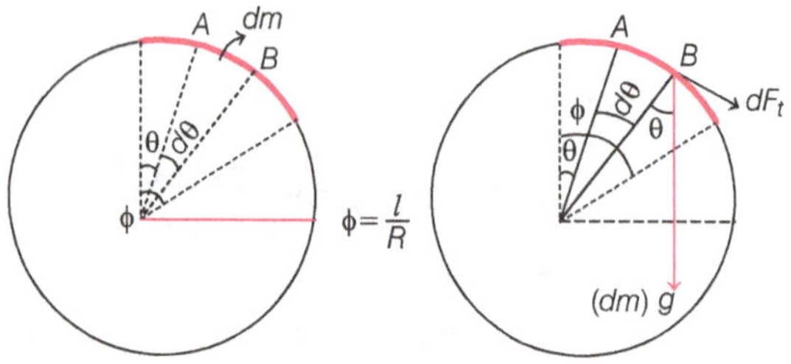

$$dm = \frac{m}{l}(Rd\theta)$$

Tangential force on dm,

$$dF_t = (dm)g\sin\theta = \left(\frac{mRg}{l}\right)\sin\theta \cdot d\theta$$

So, total tangential force,

$$F_t = \int_0^{\phi} dF_t$$

Tangential acceleration,

$$a_t = \frac{F_t}{m}$$

I leave mathematical part of this question for the students. Final answer comes out to be,

$$a_t = \frac{Rg}{l}\left[1 - \cos\frac{l}{R}\right]$$

43. Let us draw FBD of body with respect to sphere (non-inertial) with pseudo force ma_0 by assuming that sphere is moving towards right,

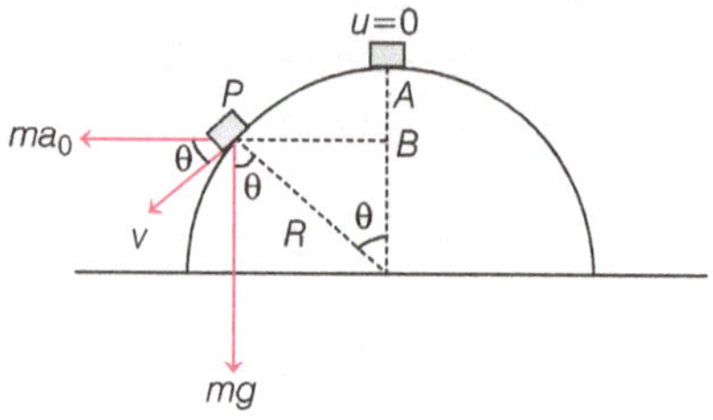

$$PB = R\sin\theta$$

$\Rightarrow$ $$AB = R(1 - \cos\theta)$$

From A to P, applying work-energy theorem (which can also be applied in a non-inertial frame)

$$W_{ma_0} + W_{mg} + W_N = \frac{1}{2}mv^2$$

N = normal reaction

$\therefore$ $$(ma_0)(PB) + (mg)AB + 0 = \frac{1}{2}mv^2$$

or $$(ma_0)(R\sin\theta) + (mg)(R)(1 - \cos\theta) = \frac{1}{2}mv^2 \quad \text{...(i)}$$

At the time of breaking off, $N = 0$ and

$$mg\cos\theta - ma_0\sin\theta = \frac{mv^2}{R} \quad \text{...(ii)}$$

Solving Eqs. (i) and (ii), we can find v and θ.

In the expression of θ, we can find θ_0 by putting $a_0 = g$.

Again mathematical part is left for students.

44. Velocity at bottommost point is,

$$u = \sqrt{2gh}$$

$\Rightarrow$ $$R = \frac{h}{2}$$

$\therefore$ $$u = \sqrt{4gR}$$

Since $\sqrt{2gR} < u < \sqrt{5gR}$

So, the body will leave contact with the groove at height given by,

$$H = \frac{u^2 + gR}{3g} \quad \text{...(i)}$$

(where $T = 0$ but $v \neq 0$)

$$= \frac{4gR + gR}{3g} = \frac{5}{3}R$$

Velocity at this point is,

$$v = \sqrt{u^2 - 2gH} = \sqrt{4gR - 2g \times \frac{5}{3}R}$$

$$= \sqrt{\frac{2gR}{3}}$$

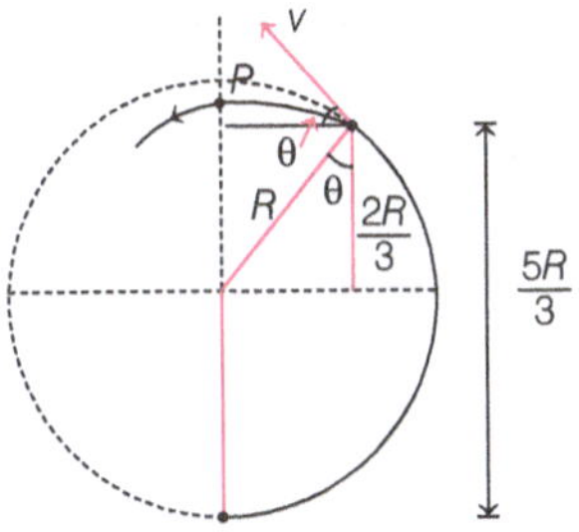

Velocity at highest point P,

$$v = v\cos\theta = \left[\sqrt{\frac{2gR}{3}}\right]\left(\frac{2R/3}{R}\right) \qquad \left(\because \cos\theta = \frac{2R/3}{R}\right)$$

$$= \frac{2}{3}\sqrt{\frac{g(2R)}{3}} \qquad (\because 2R = h)$$

$$= \frac{2}{3}\sqrt{\frac{gh}{3}}$$

Level 2

45. (a) $|\mathbf{F}_{av}| = m|\Delta \mathbf{a}| = m\dfrac{|\Delta \mathbf{v}|}{\Delta t}$ ...(i)

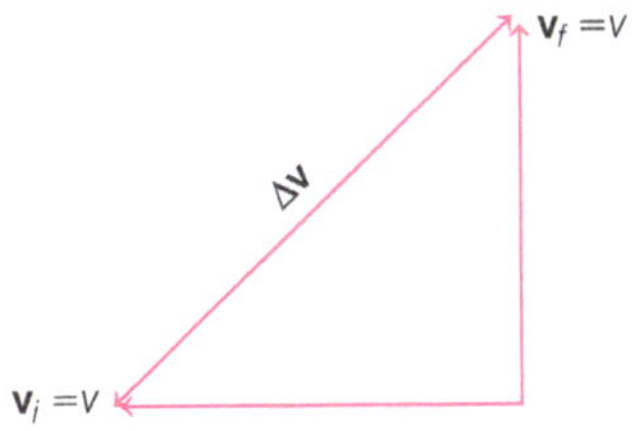

$$|\Delta \mathbf{v}| = |\mathbf{v}_f - \mathbf{v}_i| = \sqrt{2}\,v$$

$$\Delta t = \frac{T}{4} = \frac{2\pi R/v}{4} = \frac{\pi R}{2v}$$

Substituting in Eq. (i), we get

$$|\mathbf{F}_{av}| = \frac{2\sqrt{2}mv^2}{\pi R}$$

(b) In this case, $v_i = 0$

and $v_f = at$

$\therefore$ $|\Delta \mathbf{v}| = v_f = at$

and $\Delta t = t$

Substituting in Eq. (i), we get

$$|\mathbf{F}_{av}| = ma$$

46. f = maximum friction $= \mu mg\cos\alpha$

$$= mg\sin\alpha \qquad (\because \mu = \tan\alpha)$$

$$a_x = \frac{mg\sin\alpha - f\cos\phi}{m}$$

$$= \frac{mg\sin\alpha - mg\sin\alpha \cdot \cos\phi}{m}$$

$\therefore$ $a_x = g\sin\alpha(1 - \cos\phi)$ ...(i)

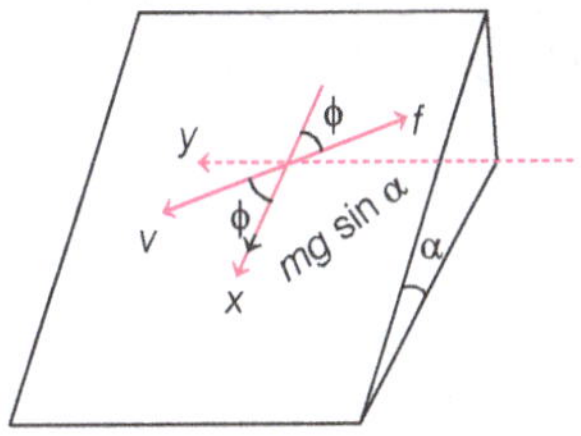

Tangential component of acceleration,

$$a_t = \frac{mg\sin\alpha\cos\phi - f}{m}$$

$$= \frac{mg\sin\alpha \cdot \cos\phi - mg\sin\alpha}{m}$$

$$= -g\sin\alpha(1 - \cos\phi) \qquad ..(ii)$$

$$= \text{Rate of change of speed or } \frac{dv}{dt}$$

From Eqs. (i) and (ii), we can see that

$$a_x = -a_t$$

or $\dfrac{dv_x}{dt} = -\dfrac{dv}{dt}$

or $\displaystyle\int_0^{v\cos\phi} dv_x = -\int_{v_0}^{v} dv$

$\therefore$ $v\cos\phi = v_0 - v$

or $v = \dfrac{v_0}{1 + \cos\phi}$

Note *Initially $v = v_0$ and $v_x = 0$. At the time of shown position, $v = v$ and $v_x = v\cos\phi$.*

47. $N = \dfrac{mv^2}{R}$

$$f_{max} = \mu N = \frac{\mu mv^2}{R}$$

$\therefore$ Retardation, $a = \dfrac{f_{max}}{m} = \dfrac{\mu v^2}{R}$

$\therefore$ $\left(-\dfrac{dv}{dt}\right) = \dfrac{\mu v^2}{R}$

or $\displaystyle\int_{v_0}^{v}\frac{dv}{v^2} = -\frac{\mu}{R}\int_0^t dt$

or $v = \dfrac{v_0}{1 + \dfrac{\mu v_0 t}{R}}$ **Ans.**

48. In this question original length of spring is also required, which is not given in the question. Suppose it is l.

Circular motion is horizontal. So, vertical mg force is balanced by the vertical normal reaction.

So, we can ignore the vertical forces and can make a simple two dimensional figure, showing only horizontal forces. Let elongation of the spring is Δl.

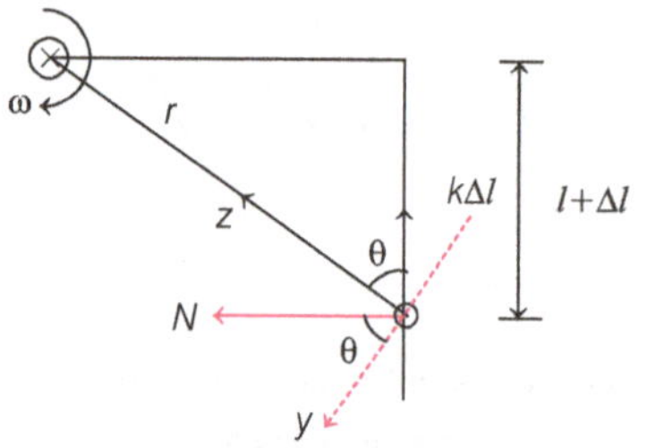

$$\therefore \quad r\cos\theta = (l+\Delta l)$$

$$\Sigma F_y = 0$$

$$\Rightarrow \quad N\cos\theta = (k\Delta l)\sin\theta \quad \text{...(i)}$$

$$\Sigma F_z = mr\omega^2$$

$$\text{or } N\sin\theta + (k\Delta l)\cos\theta = mr\omega^2 \quad \text{...(ii)}$$

Substituting value of N from Eq. (i) into Eq. (ii), we get

$$\frac{(k\Delta l)\sin^2\theta}{\cos\theta} + (k\Delta l)\cos\theta = mr\omega^2$$

Simplifying this equation, we get

$$k\Delta l = m\omega^2(r\cos\theta)$$

$$= m\omega^2(l+\Delta l)$$

$$\therefore \quad \Delta l = \frac{m\omega^2 l}{x - m\omega^2}$$

$$\text{or} \quad \Delta l = \frac{l}{\dfrac{x}{m\omega^2} - 1}$$

Note *This elongation Δl is independent of the rotation direction.*

49. (a) Force diagrams of m_1 and m_2 are as shown below

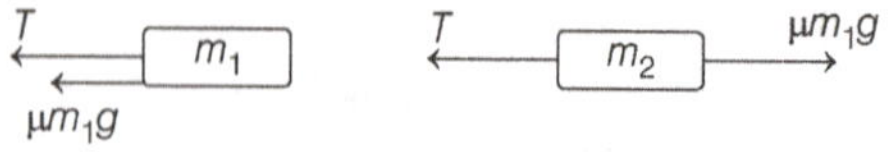

(Only horizontal forces have been shown)

Equations of motion are

$$T + \mu m_1 g = m_1 R\omega^2 \quad \text{...(i)}$$

$$T - \mu m_1 g = m_2 R\omega^2 \quad \text{...(ii)}$$

Solving Eqs. (i) and (ii), we have

$$\omega = \sqrt{\frac{2m_1\mu g}{(m_1 - m_2)R}}$$

Substituting the values, we have

$$\omega_{min} = 6.32 \text{ rad/s} \quad \textbf{Ans.}$$

(b) $T = m_2 R\omega^2 + \mu m_1 g$

$$= (1)(0.5)(6.32)^2 + (0.5)(2)(10)$$

$$\approx 30 \text{ N} \quad \textbf{Ans.}$$

50. Centripetal force will be provided by friction. In critical case,

$$\frac{mv^2}{r} = \text{maximum friction}$$

$$= \mu N \quad (\because N = mg)$$

$$= \mu_0\left(1 - \frac{r}{R}\right)mg$$

$$\therefore \quad v = \sqrt{\mu_0\left(r - \frac{r^2}{R}\right)g}$$

For v to be maximum, $\dfrac{dv}{dr} = 0$.

I am leaving this mathematical part for the students. Finally, we get

$$v_{max} = \frac{1}{2}\sqrt{\mu_0 Rg} \text{ at } r = \frac{R}{2}$$

51.

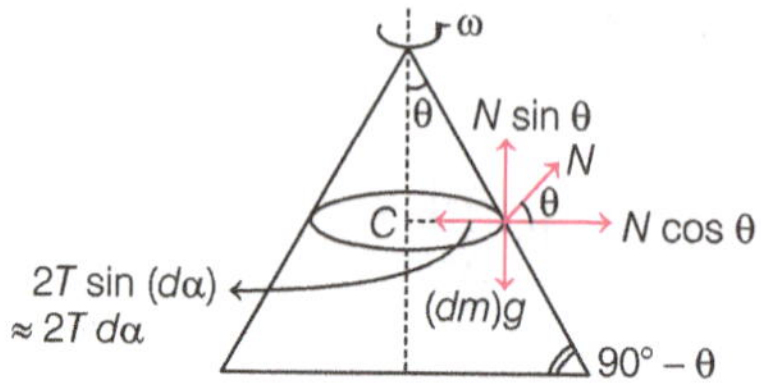

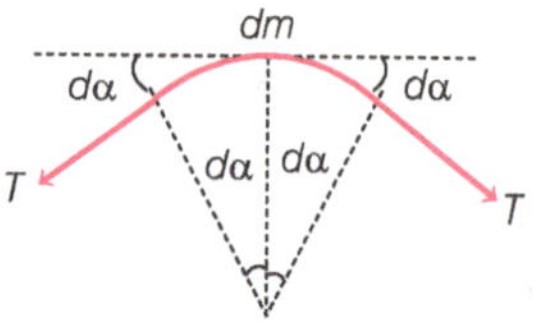

$$dm = \left(\frac{m}{2\pi}\right)(2d\alpha) = \left(\frac{m}{\pi}\right)d\alpha$$

$$N\sin\theta = (dm)g = \left(\frac{m}{\pi}d\alpha\right)g \quad \text{...(i)}$$

$$2Td\alpha - N\cos\theta = (dm)R\omega^2$$

$$\text{or} \quad N\cos\theta = 2Td\alpha - (dm)R\omega^2$$

$$\text{or} \quad N\cos\theta = 2Td\alpha - \left(\frac{m}{\pi}\right)(d\alpha)R\omega^2 \quad \text{...(ii)}$$

Dividing Eq. (i) by Eq. (ii), we have

$$\frac{\sin\theta}{\cos\theta} = \frac{\left(\frac{m}{\pi}\right)g}{2T - \left(\frac{m}{\pi}\right)R\omega^2} = \frac{g}{\left(\frac{2T\pi}{m}\right) - R\omega^2}$$

$$\Rightarrow \quad \left(\frac{2T\pi}{m}\right)\sin\theta - R\omega^2\sin\theta = g\cos\theta$$

$$\therefore \quad T = \frac{mg\cos\theta + mR\omega^2\sin\theta}{2\pi\sin\theta}$$

$$\text{or} \quad T = \frac{mg}{2\pi}\left(\cot\theta + \frac{\omega^2 R}{g}\right)$$

52. Suppose at time t, length x has been wrapped on the cylinder

$$x = R\theta = AB$$

The remaining length is $l = l_0 - x = l_0 - R\theta = BP$.

Net force on the disc is only tension and this is always perpendicular to velocity. Hence magnitude of velocity vector or speed does not change.

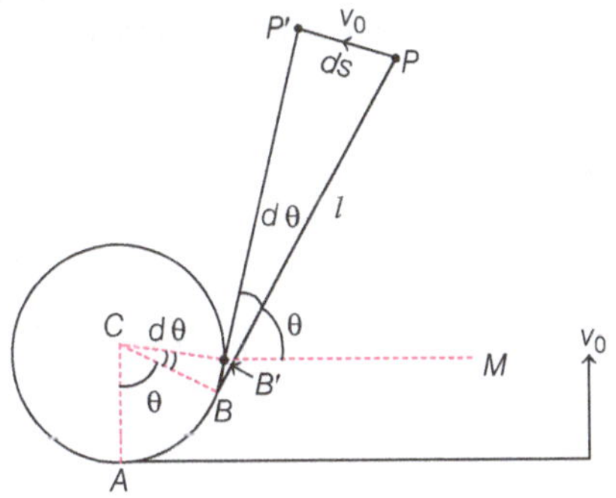

$$\angle ACB = \theta$$

$\therefore$ $\angle PBM$ is also θ.

Now, small distance travelled by disc,

$$ds = PP' = l d\theta = (l_0 - R\theta)\, d\theta$$

So, total distance travelled by the disc,

$$s = \int_{\theta=0}^{\theta_{max}} ds$$

where $\theta_{max} = \dfrac{l_0}{R}$

$$\therefore \quad s = \frac{l_0^2}{2R}$$

Now, speed v_0 is constant. Hence,

$$\text{time} = \frac{\text{total distance}}{\text{constant speed}}$$

$$= \frac{(l_0^2/2R)}{v_0}$$

or $$t = \frac{l_0^2}{2Rv_0}$$

Note *$BB' = Rd\theta$, $PP' = ld\theta$*

$BB' \neq PP'$, but distance travelled by disc is PP' not BB'.

53.

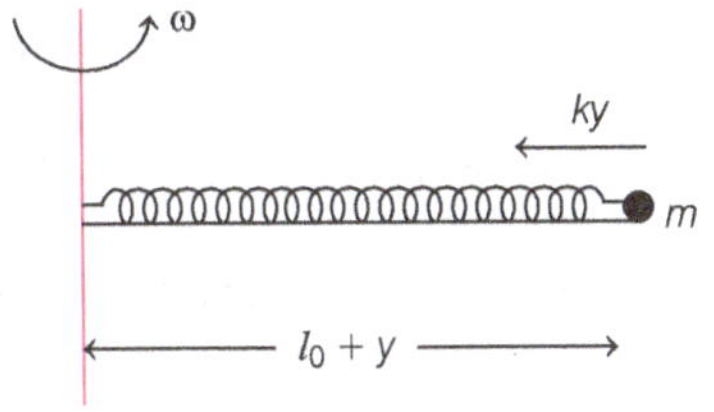

Let the spring is stretched by y. Then,

$$ky = \text{centripetal force}$$
$$= m(l_0 + y)\,\omega^2$$

$$\therefore \quad y = \frac{ml_0\omega^2}{k - m\omega^2} \quad \text{...(i)}$$

Now, W = change in mechanical energy

$$W = \frac{1}{2}ky^2 + \frac{1}{2}mv^2 \quad \text{...(ii)}$$

$$v = (l_0 + y)\,\omega \quad \text{...(iii)}$$

Solving these three equations, we get

$$W = \frac{kl_0^2\eta(1+\eta)}{2(1-\eta)^2}, \text{ where } \eta = \frac{m\omega^2}{k}$$

Note *Rod and spring are given massless. So in mechanical energy we have not taken their kinetic energy.*

54. (a) $CP = CO$

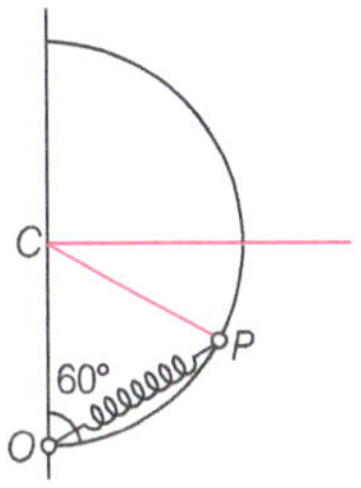

Radius of circle (R)

$\therefore$ $\angle CPO = \angle POC = 60°$

$\therefore$ $\angle OCP$ is also $60°$.

Therefore, ΔOCP is an equilateral triangle.

Hence, $OP = R$

Natural length of spring is $3R/4$.

$\therefore$ Extension in the spring,

$$x = R - \frac{3R}{4} = \frac{R}{4}$$

$\Rightarrow$ Spring force, $F = kx = \left(\dfrac{mg}{R}\right)\left(\dfrac{R}{4}\right) = \dfrac{mg}{4}$

The free body diagram of the ring will be as shown below.

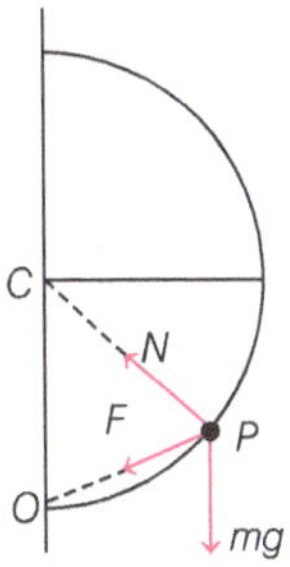

Here, $F = kx = \frac{mg}{4}$ and N = Normal reaction.

(b) **Tangential acceleration, a_T** The ring will move towards the x-axis just after the release. So, net force along x-axis

$$F_x = F \sin 60° + mg \sin 60°$$

$$= \left(\frac{mg}{4}\right)\frac{\sqrt{3}}{2} + mg\left(\frac{\sqrt{3}}{2}\right)$$

$$F_x = \frac{5\sqrt{3}}{8} mg$$

Therefore, tangential acceleration of the ring,

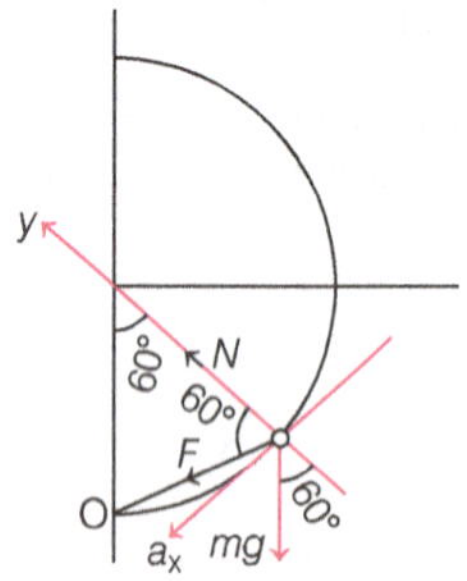

$$a_T = a_x = \frac{F_x}{m} = \frac{5\sqrt{3}}{8} g$$

$$\Rightarrow \quad a_T = \frac{5\sqrt{3}}{8} g$$

Normal reaction, N Net force along y-axis on the ring just after the release will be zero.

$$F_y = 0$$

$\therefore \quad N + F \cos 60° = mg \cos 60°$

$\therefore \quad N = mg \cos 60° - F \cos 60°$

$$= \frac{mg}{2} - \frac{mg}{4}\left(\frac{1}{2}\right) = \frac{mg}{2} - \frac{mg}{8}$$

$$N = \frac{3mg}{8}$$

55. Speed of each particle at angle θ is,

$$v = \sqrt{2gh} \quad \text{(from energy conservation)}$$

where, $h = R(1 - \cos \theta)$

$\therefore \quad v = \sqrt{2gR(1 - \cos \theta)}$

$$N + mg \cos \theta = \frac{mv^2}{R}$$

or $\quad N + mg \cos \theta = 2mg(1 - \cos \theta)$

or $\quad N = 2mg - 3mg \cos \theta \quad$ ...(i)

The tube breaks its contact with ground when $2N \cos \theta > Mg$

Substituting, $\quad 2N \cos \theta = Mg$

or $\quad 4mg \cos \theta - 6mg \cos^2 \theta = Mg$

Substituting, $\theta = 60°$

$$2mg - \frac{3mg}{2} = Mg$$

or $\quad \frac{M}{m} = \frac{1}{2}$ **Ans.**

Note *initially normal reaction on each ball will be radially outward and later it will be radially inward, so that normal reactions on tube is radially outward to break it off from the ground.*

56. Let R be the radius of the ring

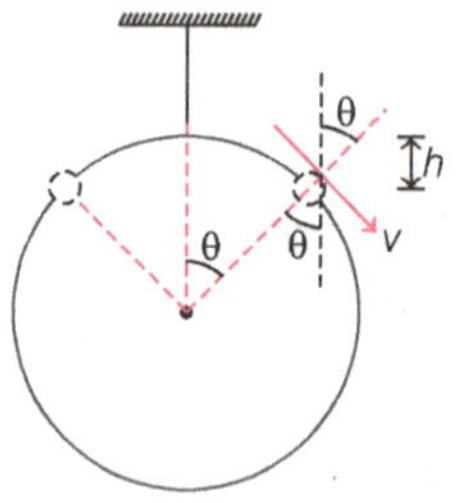

$$h = R(1 - \cos \theta)$$

$$v^2 = 2gh = 2gR(1 - \cos \theta)$$

$$\frac{mv^2}{R} = N + mg \cos \theta$$

or $\quad N = 2mg(1 - \cos \theta) - mg \cos \theta$

$$N = 2mg - 3mg \cos \theta$$

In the critical condition, tension in the string is zero and net upward force on the ring

$$F = 2N \cos \theta = 2mg(2 \cos \theta - 3 \cos^2 \theta) \quad \text{...(i)}$$

F is maximum when $\frac{dF}{d\theta} = 0$

or $\quad -2 \sin \theta + 6 \sin \theta \cos \theta = 0$

or $\quad \cos \theta = \frac{1}{3}$

Substituting in Eq. (i), we get

$$F_{max} = 2mg\left(2 \times \frac{1}{3} - 3 \times \frac{1}{9}\right) = \frac{2}{3} mg$$

$$F_{max} > Mg$$

or $\quad \frac{2}{3} mg > Mg$

or $\quad m > \frac{3}{2} M$ **Hence proved.**

57. (a) Let v_r be the velocity of mass relative to track at angular position θ.

From work-energy theorem, KE of particle relative to track

= Work done by force of gravity + work done by pseudo force

$\therefore \quad \frac{1}{2}mv_r^2 = mg(1-\cos\theta) + m\left(\frac{2g}{9}\right)\sin\theta$

or $\quad v_r^2 = 2g(1-\cos\theta) + \frac{4g}{9}\sin\theta \quad$ ...(i)

Particle leaves contact with the track, where $N = 0$

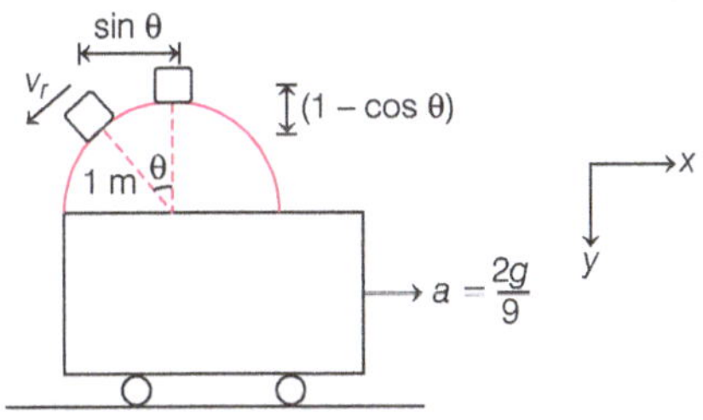

or $mg\cos\theta - m\left(\frac{2g}{9}\right)\sin\theta = mv_r^2$

or $\quad g\cos\theta - \frac{2g}{9}\sin\theta = 2g(1-\cos\theta) + \frac{4g}{9}\sin\theta$

or $\quad 3\cos\theta - \frac{6}{9}\sin\theta = 2$

Solving this, we get

$\theta \approx 37°$ **Ans.**

(b) From Eq. (i), we have

$$v_r = \sqrt{2g(1-\cos\theta) + \frac{4g}{9}\sin\theta}$$

or $\quad v_r = 2.58$ m/s at $\theta = 37°$

Vertical component of its velocity is

$$v_y = v_r \sin\theta$$

$$= 2.58 \times \frac{3}{5} = 1.55 \text{ m/s}$$

Now, $\quad 1.3 = 1.55t + 5t^2 \quad \left(\because s = ut + \frac{1}{2}gt^2\right)$

or $\quad 5t^2 + 1.55t - 1.3 = 0$

or $\quad t = 0.38$ s **Ans.**

Dimensional Formulae of Some of the Important Physical Quantities

Physical quantity	Relation with other quantities	Dimensional formula	SI unit
Area	length × breadth	$L \times L = L^2 = [M^0\ L^2\ T^0]$	m^2
Volume	length × breadth × height	$L \times L \times L = L^3 = [M^0\ L^3\ T^0]$	m^3
Density	$\frac{\text{mass}}{\text{volume}}$	$\frac{M}{L^3} = [M^1\ L^{-3}\ T^0]$	$kg\ m^{-3}$
Specific gravity	$\frac{\text{density of body}}{\text{density of water at } 4°C}$	$\frac{M/L^3}{M/L^3} = 1 = [M^0\ L^0\ T^0] \rightarrow$ no dimensions	No units
Speed or velocity	$\frac{\text{distance or displacement}}{\text{time}}$	$\frac{L}{T} = LT^{-1} = [M^0\ L^1\ T^{-1}]$	ms^{-1}
Linear momentum	mass × velocity	$M \times LT^{-1} = [M^1\ L^1\ T^{-1}]$	$kg\ ms^{-1}$
Acceleration	$\frac{\text{change in velocity}}{\text{time taken}}$	$\frac{L/T}{T} = LT^{-2} = [M^0\ L^1\ T^{-2}]$	ms^{-2}
Acceleration due to gravity (g)	$\frac{\text{change in velocity}}{\text{time taken}}$	$\frac{L/T}{T} = LT^{-2} = [M^0\ L^1\ T^{-2}]$	ms^{-2}
Force	mass × acceleration	$M \times LT^{-2} = [M^1 L^1 T^{-2}]$	N (newton)
Impulse	force × time	$M\ LT^{-2} \times T = [M^1\ L^1\ T^{-1}]$	Ns
Pressure	force/area	$\frac{MLT^{-2}}{L^2} = [M^1 L^{-1} T^{-2}]$	Nm^{-2}
Universal constant of gravitation (G)	From Newton's law of gravitation. $F = \frac{Gm_1m_2}{r^2}$ or $G = \frac{Fr^2}{m_1 m_2}$, where F is force between masses m_1, m_2 at a distance r	$G = \frac{MLT^{-2} \times L^2}{MM} = [M^{-1}\ L^3\ T^{-2}]$	$Nm^2\ kg^{-2}$
Work	force × distance	$M\ LT^{-2} \times L = [M^1\ L^2\ T^{-2}]$	J (joule)

Physical quantity	Relation with other quantities	Dimensional formula	SI unit
Moment of force	force × distance	$\text{M LT}^{-2} \times \text{L} = [\text{M}^1 \text{ L}^2 \text{ T}^{-2}]$	N-m
Power	$\frac{\text{work}}{\text{time}}$	$\frac{\text{M L}^2 \text{ T}^{-2}}{\text{T}} = [\text{M}^1 \text{ L}^2 \text{ T}^{-3}]$	W (watt)
Surface tension	$\frac{\text{force}}{\text{length}}$	$\frac{\text{M L T}^{-2}}{\text{L}} = [\text{M}^1 \text{ L}^0 \text{ T}^{-2}]$	Nm^{-1}
Surface energy	Energy of free surface	$[\text{M}^1 \text{ L}^2 \text{ T}^{-2}]$	J
Force constant	$\frac{\text{force}}{\text{displacement}}$	$\frac{\text{M L T}^{-2}}{\text{L}} = [\text{M}^1 \text{ L}^0 \text{ T}^{-2}]$	Nm^{-1}
Thrust	force	$[\text{M}^1 \text{ L}^1 \text{ T}^{-2}]$	N (newton)
Tension	force	$[\text{M}^1 \text{ L}^1 \text{ T}^{-2}]$	N (newton)
Stress, Pressure	$\frac{\text{force}}{\text{area}}$	$\frac{\text{M}^{-1} \text{ L}^{-1} \text{ T}^{-2}}{\text{L}^2} = [\text{M}^1 \text{ L}^{-1} \text{ T}^{-2}]$	Nm^{-2}
Strain	$\frac{\text{change in dimension}}{\text{original dimension}}$	$\frac{\text{L}}{\text{L}} = [\text{M}^0 \text{ L}^0 \text{ T}^0]$	No units
Coefficient of elasticity	$\frac{\text{stress}}{\text{strain}}$	$\frac{\text{M}^1 \text{ L}^{-1}\text{T}^{-2}}{1} = [\text{M}^1 \text{ L}^{-1} \text{ T}^{-2}]$	Nm^{-2}
Radius of gyration (K)	distance	$\text{L} = [\text{M}^0 \text{ L}^1 \text{ T}^0]$	m
Moment of inertia (I)	mass (radius of gyration)^2	$\text{ML}^2 = [\text{M}^1 \text{ L}^2 \text{ T}^0]$	kg m^2
Angle (θ) or Angular displacement (θ)	$\frac{\text{length } (l)}{\text{radius } (r)}$	$\frac{\text{L}}{\text{L}} = 1 = [\text{M}^0 \text{ L}^0 \text{ T}^0]$	radian
Angular velocity (ω)	$\frac{\text{angle}(\theta)}{\text{time } (t)}$	$\frac{1}{\text{T}} = \text{T}^{-1} = [\text{M}^0 \text{ L}^0 \text{ T}^{-1}]$	rad s^{-1}
Angular acceleration (α)	$\frac{\text{change in angular velocity}}{\text{time taken}}$	$\frac{1/\text{T}}{\text{T}} = \text{T}^{-2} = [\text{M}^0 \text{ L}^0 \text{ T}^{-2}]$	rad s^{-2}
Angular momentum	$I\omega$	$\text{ML}^2 \text{ T}^{-1} = [\text{M}^1 \text{ L}^2 \text{ T}^{-1}]$	$\text{kg m}^2 \text{ s}^{-1}$
Torque	$I\alpha$	$\text{ML}^2 \times \text{T}^{-2} = [\text{M}^1 \text{ L}^2 \text{ T}^{-2}]$	N-m

Physical quantity	Relation with other quantities	Dimensional formula	SI unit
Wavelength (λ)	length of one wave, i.e., distance	$L = [M^0\ L^1\ T^0]$	m
Frequency (ν)	number of vibrations/s	$\frac{1}{T} = T^{-1} = [M^0\ L^0\ T^{-1}]$	s^{-1} or Hz (hertz)
Angular frequency (ω)	$2\pi \times$frequency	$T^{-1} = [M^0\ L^0\ T^{-1}]$	radian/s
Velocity of light in vacuum (c)	$\frac{\text{distance travelled}}{\text{time taken}}$	$\frac{L}{T} = [M^0\ L^1\ T^{-1}]$	ms^{-1}
Velocity gradient	$\frac{\text{velocity}}{\text{distance}}$	$\frac{LT^{-1}}{L} = T^{-1} = [M^0\ L^0\ T^{-1}]$	s^{-1}
Rate of flow of fluid	$\frac{\text{volume}}{\text{time}}$	$\frac{L^3}{T} = L^3T^{-1} = [M^0\ L^3\ T^{-1}]$	m^3s^{-1}
Planck's constant (h)	$\frac{\text{energy } (E)}{\text{frequency } (\nu)}$	$\frac{ML^2T^{-2}}{T^{-1}} = [M^1\ L^2\ T^{-1}]$	J-s
Linear mass density (m)	$\frac{\text{mass}}{\text{length}}$	$\frac{M}{L} = [M^1\ L^{-1}\ T^0]$	kg m^{-1}
Distance travelled in nth second	$\frac{\text{distance}}{\text{time}}$	$\frac{L}{T} = [M^0\ L^1\ T^{-1}]$	ms^{-1}
Avogadro's number (N)	Number of atoms/ molecules in one gram atom/mole	$[M^0\ L^0\ T^0]$	$mole^{-1}$
Magnetic dipole moment (M)	$M = IA$	$AL^2 = [M^0\ L^2\ T^0\ A^1]$	Am^2
Pole strength (m)	$m = \frac{M}{2l}$	$\frac{AL^2}{L} = AL = [M^0\ L^1\ T^0\ A^1]$	Am
Magnetic permeability of free space (μ_0)	From Coulomb's law in magnetism, $F = \frac{\mu_0}{4\pi}\frac{m_1 m_2}{r^2}$, where m_1, m_2 are strengths of two poles; $[\mu_0] = \frac{4\pi\ [F]\ [r^2]}{m_1\ m_2}$	$\frac{MLT^{-2} \times L^2}{AL^2} = [M^1\ L^1\ T^{-2}\ A^{-2}]$	Hm^{-1}

Physical quantity	Relation with other quantities	Dimensional formula	SI unit
Resistance (R)	$\frac{\text{potential difference}}{\text{current}}$	$\frac{ML^2T^{-3}A^{-1}}{A} = [M^1\ L^2\ T^{-3}\ A^{-2}]$	Ω (ohm)
Capacitance (C)	$\frac{\text{charge}}{\text{potential difference}}$	$\frac{AT}{M\,L^2\ T^{-3}\ A^{-1}} = [M^{-1}\ L^{-2}\ T^4\ A^2]$	F (farad)
Surface density of charge	$\sigma = \frac{\text{charge}}{\text{area}}$	$\frac{AT}{L^2} = [M^0\ L^{-2}\ T^1\ A^1]$	Cm^{-2}
Electric dipole moment (p)	$q\,(2\,a)$	$AT\,[L] = [M^0\ L^1\ T^1\ A^1]$	Cm
Specific resistance or resistivity (ρ)	$\frac{Ra}{l}$	$\frac{ML^2T^{-3}A^{-2} \times L^2}{L} = [M^1\ L^3\ T^{-3}\ A^{-2}]$	ohm-m
Conductance (G)	$\frac{1}{R}$	$\frac{1}{ML^2T^{-3}A^{-2}} = [M^{-1}\ L^{-2}\ T^3\ A^2]$	ohm^{-1}
Conductivity (σ)	$\frac{1}{\rho}$	$\frac{1}{M^1L^3T^{-3}A^{-2}} = [M^{-1}\ L^{-3}\ T^3\ A^2]$	ohm^{-1}
Electric flux	electric field × area	$M^1L^1T^{-3}A^{-1} \times L^2 = [M^1\ L^3\ T^{-3}\ A^{-1}]$	$Nm^2\ C^{-1}$
Faraday constant	avogadro number × elementary charge	$\frac{1}{mol} \times AT = [M^0L^0T^1A^1mol^{-1}]$	C
Mass defect (Δm)	sum of masses of nucleons – mass of nucleus	$[M^1\ L^0\ T^0]$	kg
Binding energy of nucleus	(mass defect) × (speed of light)^2	$M \times L^2T^{-2} = [ML^2T^{-2}]$	J